Rhythms of Insect Evolution

Rhythms of Insect Evolution

Evidence from the Jurassic and Cretaceous in Northern China

Edited by
Dr. Dong Ren
Capital Normal University
BJ, China

Dr. Chungkun Shih
Capital Normal University
China
National Museum of Natural History
USA

Dr. Taiping Gao
Capital Normal University
BJ, China

Dr. Yongjie Wang
Capital Normal University
BJ, China

Dr. Yunzhi Yao
Capital Normal University
BJ, China

WILEY Blackwell

This edition first published 2019

Registered Offices
John Wiley & Sons, Inc., 111 River Street, Hoboken, NJ 07030, USA
John Wiley & Sons Ltd, The Atrium, Southern Gate, Chichester, West Sussex, PO19 8SQ, UK
Wiley-VCH Verlag GmbH & Co. KGaA, Boschstr. 12, 69469 Weinheim, Germany
John Wiley & Sons Singapore Pte. Ltd, 1 Fusionopolis Walk, #07-01 Solaris South Tower, Singapore 138628

Editorial Office
The Atrium, Southern Gate, Chichester, West Sussex, PO19 8SQ, UK

For details of our global editorial offices, customer services, and more information about Wiley products visit us at www.wiley.com.

Wiley also publishes its books in a variety of electronic formats and by print-on-demand. Some content that appears in standard print versions of this book may not be available in other formats.

Library of Congress Cataloging-in-Publication Data Applied for

9781119427988

Cover Design: Wiley
Cover Images: Front cover photos, clockwise from top left:
Oregramma illecebrosa Yang, Wang, Labandeira, Shih & Ren, 2014, pages 325, 633, 661
Lichnomesopsyche gloriae Ren, Labandeira & Shih, 2010, pages 5, 561, 632
Bellinympha filicifolia Wang, Ren, Liu & Engel, 2010, pages 5, 320, 656
Pseudopulex jurassicus Gao, Shih & Ren, 2012, pages 5, 600, 638
Back cover photo: *Cathayamodus fournieri* Gao, Shih, Labandeira & Ren, 2016, pages 611, 671

Set in 10/12pt WarnockPro by SPi Global, Chennai, India

10 9 8 7 6 5 4 3 2 1

Contents

Preface

Research and documentation of diverse and scientifically important fossil findings from the Middle Jurassic and the Early Cretaceous of Northern China, covering feathered dinosaurs, pterosaurs, birds, mammals, reptiles, amphibians, insects, and gymnosperm and angiosperm plants, have significantly enhanced and expanded our knowledge of paleontology, taxonomy, evolution, ecology and other related natural sciences. Highlighted by and based on the newly documented "silent stories" in numerous academic papers and media or on-line reports, many natural science textbooks on these subjects have been revised, and interest in these topics by the general public has greatly increased.

Yielding many contributions and serving important functions, insects performed interesting lifestyles of feeding, predation-avoidance and gene-propagation in their ecosystems during the Middle Jurassic and the Early Cretaceous. They encountered sudden deaths, most likely due to poisonous gases and volcanic ash, and were well-preserved as compression fossils. Hidden in layered sediments, with excellent preservation of amazing details and broad diversity, these insect fossils, together with plants and other animals, have been discovered in large numbers from Northern China. Many of these fossils, analyzed and studied with highly technical instrumentation and advanced software, have provided evidence and examples of the rhythms of insect evolution.

Started in 1923 and continued until 1935, Amadeus W. Grabau and Chi Ping carried out respectively the Early Foundational Study for fossil insects in China, then, from 1965 to 1985, Youchong Hong and Qibin Lin became professional paleoentomologists in China and accomplished the Early Taxonomic Studies. Over the period from 1985 to the present, Junfeng Zhang, Dong Ren (with his Team at the Capital Normal University [CNU]), Haichun Zhang (with his Team at the Nanjing Institute of Geology and Palaeontology, Chinese Academy of Sciences [NIGPAS]), and then Di-ying Huang (with his NIGPAS Team) have carried out the Major Taxonomic Studies of fossil insects and from 1991 to the present, these teams also conducted many Phylogenetic and Paleobiological Studies. For the stage of International Cooperative Studies from 2000 to the present, paleoentomology in China has become more open and international, while many extensive cooperative researches have been conducted actively with international scholars.

Integrating expertise and engaging experts from universities and institutions inside China and around the world in joint research projects has greatly improved the efficiency and effectiveness of uncovering rhythms of insect evolution from these fossils. New morphology and taxonomy of many taxa in 23 orders of insects have been described by paleoentomologists based on vast collections in CNU and NIGPAS in the past 18 years. Systematics, phylogeny, behavior, function, evolution, aberrational development of insects in distant time, and their interesting life styles of feeding, predation-avoidance and gene-propagation have been documented. Energized by these reported results and other relevant information in the literature, we present and summarize representative insect fossils from the Jurassic and Cretaceous of Northern China and the rhythms of insect evolution in this book.

Considering that readers of this book may be interested in areas other than just insect sciences, we highlighted this book with three focuses: natural science (presenting entomology, insect morphology, taxonomy, phylogeny, geology, ecosystem, pollination, mimicry, feeding, predation-avoidance, courtship and mating, etc.), popular science with human-interest (providing insect-related stories and cultures from China, USA and other countries) and artistic appreciation (demonstrating the beauty and elegance of fossil and extant insects in photos, line drawings and 3-D reconstruction artworks as visual arts). This book has a good balance of these three focuses.

Entomologists or paleoentomologists of China and foreign countries, as listed as co-authors of the literature and summarized in Chapter 4, have cooperated with us by providing their expertise and experience in these studies. Valuable contributions of taxonomy and morphology have been documented in the 23 insect orders (Chapters 5 to 27). Our Team also cover Jurassic–Cretaceous Non-marine Stratigraphy and Entomofaunas in Northern China (Chapter 1), Coexisting Animals and Plants in the Ecosystems (Chapter 2), Insects – In the Spotlight (Chapter 3), A History of Paleoentomology in China (Chapter 4), Insect Feeding (Chapter 28), Camouflage, Mimicry or Eyespot Warning (Chapter 29) and Gene Propagation – Courtship, Mating and Next Generation (Chapter 30). Large populations of insects with high diversity have certainly played critical roles in the evolution and ecosystems over eons, which in turn have made the Insecta a highly successful class. Utilization of high-quality and artistic photos of extant insects and plants, fossil and amber specimens, line drawings and 3-D computer-generated reconstruction artworks in this book will provide readers with clear and enjoyable impressions of our scientific findings. To provide global readers with broader information, we have included boxes to introduce insect-related stories from western and Chinese cultures.

Intended to appeal to a broad spectrum of readers (such as paleoentomologists, entomologists, evolutionists, botanists, biologists, geologists, paleoecologists, fossil collectors, naturalists, insect hobbyists and students), we believe that readers will have fun reading this book while gaining a new understanding and knowledge of insect-related natural science and enjoying the artistic presentations. One of our objectives for this book is to get young readers interested in the subjects, and hopefully, some of them will be inspired to become entomologists or paleontologists to create a bright, exciting future for natural sciences. New and important findings of more compression fossils from Northern China and innovative and in-depth future studies will be continued by us, so that we will further progress our journey of natural sciences.

Chungkun Shih and Dong Ren

List of Contributors

Yu Chang
Capital Normal University
Haidian District
Beijing
China

Sha Chen
Capital Normal University
Haidian District
Beijing
China

Zhenzhen Chen
Capital Normal University
Haidian District
Beijing
China

Yingying Cui
South China Normal University
Guangzhou
China

and

Capital Normal University
Haidian District
Beijing
China

Sile Du
Capital Normal University
Haidian District
Beijing
China

Hui Fang
Capital Normal University
Haidian District
Beijing
China

Cuiping Feng
Capital Normal University
Haidian District
Beijing
China

Taiping Gao
Capital Normal University
Haidian District
Beijing
China

Jun-Jie Gu
Sichuan Agricultural University
Chengdu
Sichuan
China

and

Capital Normal University
Haidian District
Beijing
China

Ye Han
Capital Normal University
Haidian District
Beijing
China

Shuo Huang
Capital Normal University
Haidian District
Beijing
China

Longfeng Li
Gansu Agricultural University
Lanzhou
Gansu
China

and

Capital Normal University
Haidian District
Beijing
China

Sheng Li
Capital Normal University
Haidian District
Beijing
China

Junhui Liang
Tianjin Natural History Museum
Hexi District
Tianjin
China

and

Capital Normal University
Haidian District
Beijing
China

Qingqing Lin
Capital Normal University
Haidian District
Beijing
China

Shan Lin
Capital Normal University
Haidian District
Beijing
China

Xiaodan Lin
Capital Normal University
Haidian District
Beijing
China

Xue Liu
Capital Normal University
Haidian District
Beijing
China

Zhenhua Liu
Sun Yat-sen University
Guangzhou
Guangdong
China

and

Capital Normal University
Haidian District
Beijing
China

Hong Pang
Sun Yat-sen University
Guangzhou
Guangdong
China

Dong Ren
Capital Normal University
Haidian District
Beijing
China

Mingyue Ren
Capital Normal University
Haidian District
Beijing
China

Chaofan Shi
Sun Yat-sen University
Guangzhou
Guangdong
China

and

Capital Normal University
Haidian District
Beijing
China

Chungkun Shih
Capital Normal University
Haidian District
Beijing
China

and

National Museum of Natural History
Smithsonian Institution
Washington, DC
USA

Mei Wang
Research Institute of Forest Ecology,
Environment and Protection
Chinese Academy of Forestry
Haidian District

Beijing
China

and

Capital Normal University
Haidian District
Beijing
China

Ruiqian Wang
Capital Normal University
Haidian District
Beijing
China

Yimo Wang
Capital Normal University
Haidian District
Beijing
China

Ying Wang
Beijing Museum of Natural History
Dongcheng District
Beijing
China

and

Capital Normal University
Haidian District
Beijing
China

Yongjie Wang
Capital Normal University
Haidian District
Beijing
China

Changyue Xing
Capital Normal University
Haidian District
Beijing
China

Hongru Yang
Capital Normal University
Haidian District
Beijing
China

Qiang Yang
Guangzhou University
Guangzhou
Guangdong
China

and

Capital Normal University
Haidian District
Beijing
China

Yunzhi Yao
Capital Normal University
Haidian District
Beijing
China

Xiuna Ye
Capital Normal University
Haidian District
Beijing
China

Yali Yu
Sun Yat-sen University
Guangzhou
Guangdong
China

and

Capital Normal University
Haidian District
Beijing
China

Kuiyan Zhang
Institute of Zoology
Chinese Academy of Sciences
Chaoyang District
Beijing
China

and

Capital Normal University
Haidian District
Beijing
China

Tingying Zhang
Capital Normal University
Haidian District
Beijing
China

Xiao Zhang
Capital Normal University

Haidian District
Beijing
China

Weiting Zhang
Hebei GEO University
Shijiazhuang
Hebei
China

and

Capital Normal University
Haidian District
Beijing
China

Zhipeng Zhao
Capital Normal University
Haidian District
Beijing
China

Acknowledgements

This book is supported by National Natural Science Foundation of China (No. 41688103, 31730087, 31872277), Beijing Natural Science Foundation (No. 5182004), Program for Changjiang Scholars and Innovative Research Team in University (IRT-17R75), Project of High-level Teachers in Beijing Municipal Universities (IDHT20180518, CIT&TCD201704090), Joint Fund of the Beijing Municipal Natural Science Foundation and Beijing Municipal Education Commission (KZ201810028046).

1

Jurassic-Cretaceous Non-Marine Stratigraphy and Entomofaunas in Northern China

Dong Ren

Capital Normal University, Haidian District, Beijing, China

1.1 Introduction

Northern China is an administrative and geographical region referring to the area located north of the Yellow River. Since the Late Triassic, the continental area of northern China was formed, and during the Jurassic and Cretaceous, almost all of northern China had become part of the Asian continent except for the southern Tibet, west of the Tarim basin of northwest China and Heilongjiang of northeast China. Thus, the Jurassic to Cretaceous strata of northern China are dominated by terrestrial sediments, volcanic rocks and volcanic sedimentary formations and coal-bearing beds. A great deal of fossils have been found and documented in numerous localities and horizons. The Jurassic and Cretaceous insects have been reported from several localities in Xinjiang, Gansu, Shaanxi, Inner Mongolia, Hebei, Beijing, Liaoning, Jilin and Shandong. Among them, the studies of the Jurassic and Cretaceous insect fossils from the Yumen-Jiuquan Basin, Gansu; Yanliao Area (Beijing-northern Hebei-western Liaoning-southeastern Inner Mongolia) and Laiyang Basin of Shandong have been more extensive, in-depth and detailed than the sporadic reports from other localities.

Using the non-marine insect-bearing stratigraphic occurrences, together with radiometric ages and accompanying fossils, the Jurassic and Cretaceous non-marine strata of northern China can be divided and correlated as shown in Table 1.1.

1.2 Non-marine Jurassic and Cretaceous Insect Fossil-Bearing Lithostratigraphic Division and Correlation in Northern China

Non-marine Jurassic and Cretaceous deposits are widely distributed in several basins in northern China, including variegated beds, red beds, coal-bearing horizons, evaporates and volcanics. These deposits, often of great thickness, contain a rich fossil biota and significant oil, coal and non-metallic mineral resources.

A relatively complete non-marine Jurassic and Cretaceous stratigraphic sequence and a stratigraphic correlation have been established in northern China (Table 1.1). There are two main representative stratigraphic sequences containing insect fossils; one occurs in the Yumen-Jiuquan Basin, Gansu, and the other in the Yanliao Area of Beijing-northern Hebei-western Liaoning-eastern Inner Mongolia. These deposits contain a diverse and abundant continental biota, including insects, conchostracans, ostracods, bivalves, gastropods, fish, dinosaurs, birds, reptiles, amphibians, mammals and plants. During the past decades, a number of studies have been published on the non-marine Jurassic and Cretaceous strata of this region [2–4]. Recent radiometric dating results have complemented biostratigraphic studies [5, 6]. During the last 10 years, we have undertaken a reassessment of the Jurassic and Cretaceous biostratigraphy of northern China, and have improved the assemblage sequence and zonation for insects and many accompanying fossil groups (Tables 1.2 and 1.3).

In northern China, Jurassic and Cretaceous non-marine insect fossil-bearing rocks can be recognized as two different depositional types: (i) large and stable inland depositional basins without volcanic material which is distributed in northwest China, represented by the Yumen-Jiuquan Basin and (ii) intermountain basins with abundant volcanic intercalations of lavas and tuffs separated and controlled by faults which are distributed in Northeastern China represented by the Beijing-northern Hebei-western Liaoning-southeastern Inner Mongolia basins (Yanliao Area).

1.2.1 Yumen-Jiuquan Basin in Gansu Province

During the mid-Mesozoic, the Yumen-Jiuquan Basin was a representative of large and stable inland depositional

Rhythms of Insect Evolution: Evidence from the Jurassic and Cretaceous in Northern China, First Edition.
Edited by Dong Ren, Chungkun Shih, Taiping Gao, Yongjie Wang, and Yunzhi Yao.

Table 1.1 Jurassic-Cretaceous non-marine lithostratigraphic division and correlation at selected insect-bearing localities in the northern China. Geological time scale is based on The ICS International Chronostratigraphic Chart (version 2013) [1].

System	Series	Stage	Time※ in Ma	Jiuquan, Gansu	Beijing	North Hebei	West Liaoning	Yanji, Jilin	Laiyang, Shandong
Cretaceous	Upper	Maastrichtian	66	Hiatus	Hiatus	Hiatus	Hiatus	Hiatus	Wangshi Group
		Campanian	72.1						
		Santonian	83.6					Longjing Fm	
		Coniacian	86.3						
		Turonian	89.8						
		Cenomanian	93.9		Sunjiawan Fm (Xiazhuang Fm)	Sunjiawan Fm (Tujinzhi Fm)	Sunjiawan Fm		
	Lower	Albian	100.5	Zhonggou Fm	Lushangfen Fm	Qinshila Fm	Fuxin Fm	Dalazi Fm Quanshuicun Fm	Dasheng Group
		Aptian	113.0	Xiagou Fm	Tuoli Fm	Jiufotang Fm	Jiufotang Fm	Changcai Fm	Qinshan Fm
		Barremian	125.0	Chijingpu Fm	Dahuichang Fm	Yixian Fm	Yixian Fm	Tuntianying Fm	Laiyang Fm
		Hauterivian	129.4	Chijingqiao Fm	Donglanggou Fm	Dabeigou Fm	Dabeigou Fm	Hiatus	Hiatus
		Valanginian	132.9	Hiatus	Zhangjiakou Fm	Zhangjiakou Fm	Zhangjiakou Fm ?		
		Berriasian	139.8						
Jurassic	Upper	Tithonian	145.0	Boluo Fm	Tuchengzi Fm (Houcheng Fm)	Tuchengzi Fm (Houcheng Fm)	Tuchengzi Fm		
		Kimmeridgian	152.1						
		Oxfordian	157.3	Xinhe Fm	Tiaojishan Fm	Tiaojishan Fm	Tiaojishan Fm (Lanqi Fm)		
	Middle	Callovian	163.5		Jiulongshan Fm	Jiulongshan Fm	Jiulongshan Fm (Haifanggou Fm)		
		Bathonian	166.1						Santai Fm
		Bajocian	168.3		Hiatus	Hiatus	Hiatus		
		Aalenian	170.3	Zhongjiangou Fm	Longmen Fm	Mentougou Fm (Xiahuayuan Fm)	Beipiao Fm		Fangzi Fm
	Lower	Toarcian	174.1		Yaopo Fm				Wennan Fm
		Pliensbachian	182.7	Dashankou Fm					
		Sinemurian	190.8		Xinglonggou Fm	Xinglonggou Fm	Xinglonggou Fm		Hiatus
		Hettangian	199.3 201.3	Hiatus					

Conformity   Disconformity   Unconformity   Hiatus

basins without volcaniclastic deposits in northwestern China. The Jurassic and Cretaceous non-marine rocks in the Yumen-Jiuquan basin comprise, in ascending order: the Dashankou, Zhongjiangou, Xinhe, Boluo, Chijingqiao, Chijingpu, Xiagou and Zhonggou Formations (Table 1.1).

The Dashankou Formation (maximum thickness, 510 m) consists of a basal gray-green conglomerate and sandstones intercalated with purple-red siltstones and mudstones. It unconformably overlies the Upper Triassic Nanyinger Formation.

The Zhongjiangou Formation (maximum thickness, 174 m) rests conformably on the Dashankou Formation and mainly comprises basal gray conglomerate, middle gray-green sandstones, mudstones and intercalated coals.

The Xinhe Formation (maximum thickness, 600 m) conformably overlies the Zhongjiangou Formation and is mainly composed of yellowish-green and dark gray sandstones intercalated siltstones and mudstones.

The Boluo Formation (maximum thickness, 700 m) conformably overlies the Xinhe Formation and includes purplish-red siltstones, sandstones and conglomerate.

The Chijinqiao Formation (maximum thickness, 200 m) unconformably overlies the Upper Jurassic Boluo Formation or other older strata. It is divided into a lower purplish-red conglomerate and sandstones and upper dark gray and gray-green siltstones, thin-bedded shales and mudstones. It yields abundant Jehol Fauna fossils [7].

The Chijinpu Formation (maximum thickness, 400 m), conformably overlying on the Chijinqiao Formation, is dominated by yellowish-green thick-bedded to massive sandstone and conglomerate beds in the lower part, and gray-black mudstone, sandstone, siltstone intercalated with shale and thin coal beds in the upper part. It contains abundant fossils of the conchostracan *Neodiestheria* sp. and bivalves *Sphaerium jeholensis*, *S. anderssoni*, *S. subphanum* and abundant Jehol Entomofauna fossils listed in Table 1.2 [7].

The Xiagou Formation (maximum thickness, 580 m) conformably overlies the Chijinpu Formation and consists of gray-green to purple-red siltstones, interbedded with thinly-bedded mudstone and shales, with abundant insect fossils.

The Zhonggou Formation (maximum thickness, 398 m), conformably overlying on the Xiagou Formation, is unconformably overlain by the Neogene Kuquan Formation and characterized by purple-red siltstones and siltstones.

1.2.2 Intermountain Volcanic Basins in Beijing-Northern Hebei-Western Liaoning-Southeastern Inner Mongolia

Volcanic activities in Northeastern China caused by tectonic movements have been intense throughout the mid Mesozoic period. The Jurassic-Cretaceous strata distributed in the Yanliao Area of Beijing-northern Hebei-western Liaoning-southeastern Inner Mongolia are representative of intermountain basins with abundant volcanic intercalations of lavas and tuffs separated and controlled by faults. Both the renowned Yanliao Biota and Jehol Biota have been found and named in this area originally. The Mesozoic non-marine rocks of Northeastern China have been studied since the 1920s [8], and hundreds of academic papers, including extensive monographs with a stratigraphic context, have been published [1–23].

The Jurassic and Cretaceous non-marine rocks in the Yanliao Area can been divided into 10 typical Formations, in ascending order: Xinglonggou, Mentougou, Jiulongshan, Tiaojishan, Tuchengzi, Zhangjiakou, Dabeigou, Yixian, Jiufotang, Fuxin and Sunjiawan Formations (Table 1.1).

The Xinglonggou Formation (maximum thickness, 905 m), unconformably overlies the Upper Triassic Kuntouboluo Formation, is mainly composed of basalts, andesites and tuffaceous conglomerate and is widely distributed in the Yanliao Area.

The Mentougou Formation (maximum thickness, 1330 m) rests disconformably on the Xinglonggou Formation and is characterized by coal-bearing beds, mainly consisting of lacustrine siltstones, mudstones, shales and sandstones, and rich in tuffaceous materials. These coal-bearing rocks have been called as Beipiao Formation in Liaoning Province or Xiahuayuan Formation in Hebei Province. In Beijing the Mentougou Formation can be divided into Upper Longmen Formation and Lower Yaopo Formation.

The Jiulongshan Formation (maximum thickness, 680 m) unconformably overlies the Mentougou Formation. It consists of varicolored fine-grained sandstones, fine conglomerates, siltstones, mudstones and shales, rich in fossils. To date, about 837 insect species in 476 genera, 166 families and 22 orders have been reported [2, 3, 24, 25]. The main contents of Yanliao Biota have been collected from this Formation. In Liaoning Province, the Jiulongshan Formation is also called the Haifanggou Formation. The fossil beds of Jiulongshan Formation in the Daohugou Village, Inner Mongolia, have been informally called the "Daohugou Formation", but this usage has received little acceptance because it is a junior synonym of the Jiulongshan Formation [5, 26–28]. There is a short hiatus between the Mentougou Formation and the Jiulongshan Formation in the Yanliao Area.

The Tiaojishan Formation (maximum thickness, 3500 m) rests disconformably or unconformably on the Jiulongshan Formation. It is dominated by thick-bedded to massive volcanic rocks, mainly basalts, andesites, trachyandensite and rhyolite. In Liaoning Province the Tiaojishan Formation is also called the Lanqi Formation.

The Tuchengzi Formation (maximum thickness, 2600 m), unconformably overlying the Tiaojishan Formation, is a suite of fluvial red and variegated coarse-grained, clastic sedimentary rocks and tuffaceous materials, mainly conglomerate, tuffs, coarse-grained sandstones, pebbly sandstones, alternating with mudstones and siltstones. In Hebei Province and Beijing, the Tuchengzi Formation is also called the Houcheng Formation.

The Zhangjiakou Formation (maximum thickness, 1640 m) unconformably overlies the Tuchengzi Formation. It is dominated by medium-acidic volcanic rocks including purplish-red andesites, rhyolite, acidic volcanic breccia and tuffs, occasionally with several sedimentary intercalations. It may be absent in western Liaoning.

The Dabeigou Formation (maximum thickness, 344 m) is composed of fluvial gray sandy conglomerates, coarse-, medium- and fine-grained sandstones, siltstones, mudstones, intercalating tuffaceous materials, unconformably overlying the Zhangjiakou Formation. It is also absent in western Liaoning. The foregoing data indicate that there may be a long hiatus between Lower Tuchengzi Formation and Upper Yixian Formation in western Liaoning [2, 3, 29]. In Beijing, the Dabeigou Formation is also called the Donglanggou Formation.

The Yixian Formation (maximum thickness, 2442 m) rests unconformably on the Tuchengzi Formation in western Liaoning, but conformably overlies the Dabeigou Formation in northern Hebei. It consists mainly of volcanic rocks with several fossil-bearing lacustrine sedimentary intercalations, mainly a basal tuffaceous conglomerate, gray to black and purplish-red andesites, basalts, grayish-green or grayish-yellow tuffs, tuffaceous sandstones, grits, sandy shales, mudstones, tuffaceous silty mudstones and siltstones and sandstones. It is rich in the well-known Jehol Biota such as insects, ostracods, conchostracan, bivalves, fish, reptiles, feathered dinosaurs, early birds, mammals and early angiosperms. To date, about 862 insect species in 573 genera, 204 families and 19 orders have been reported [2–4, 6, 29–34]. In Beijing, the Yixian Formation is also called Dahuichang Formation.

The Jiufotang Formation (maximum thickness, 2118 m) conformably or disconformably overlies the Yixian Formation. It is characterized by lacustrine medium gray to black, grayish-green or yellowish-gray shaley tuffaceous siltstones and shales, silty limestones, grained sandstones, thin-bedded tuffs, grained tuffaceous sandstones, tuffaceous grits and coarse-grained gravelly sandstones. Coal beds and oil shales sometimes occur in the upper part of the Formation. The Jiufotang Formation grades laterally toward southwest into clastic facies, with more coarse sedimentary rocks, which are called the Tuoli Formation in Beijing.

The Fuxin Formation (maximum thickness, 1550 m), conformably overlying the Jiufotang Formation, consists of coal-bearing siliciclastic rocks, with numerous cycles of gray and white sandstones with pebbles, grained sandstones, siltstones, mudstones, carbonaceous argillites and coal seams in the lower part and greenish-gray, yellowish-gray sandy conglomerates, coarse or fine-grained sandstones, siltstones, mudstones, thin-bedded coal beds and coal lenses in the upper part [9, 23, 35]. The Fuxin Formation is also called the Lushangfen Formation in Beijing and the Qinshila Formation in northern Hebei.

The Sunjiawan Formation (maximum thickness, 660 m), conformably or disconformably overlying the Fuxin Formation and overlain by Neogene rocks, is a suite of fluvial and flood accumulation facies, mainly with varicolored conglomerates and many intercalations of thin-bedded sandstones, siltstones and mudstones. This Formation is also called the Xiazhuang Formation in Beijing and the Tujinzhi Formation in northern Hebei.

Most of the Upper Cretaceous in the Yanliao Area is missing.

Non-marine Jurassic and Cretaceous rocks in the Jilin and Shandong are little different from those of the Yanliao Area. However, their stratigraphic sequence is incomplete; only the Laiyang Formation in Shandong has abundant insect fossils belonging to the Jehol Entomofauna.

1.3 Non-marine Jurassic and Cretaceous Entomofaunas in Northern China

Fossil insects are commonly abundant and widely distributed in the mid Mesozoic non-marine sediments that accumulated in freshwater fluvial or lacustrine environments. Over the past two decades, a large number of well-preserved insect fossils from the Jurassic and Cretaceous in northern China have been reported.

These fossil insects and their assemblage are a valuable group for providing zonation of non-marine rocks and assisting in their correlation. As a result, they can be useful for biostratigraphy subdivision and correlation of non-marine successions [2, 3, 36].

Insect fossils from the Jurassic and Cretaceous in northern China have been studied by palaeoentomologists since 1923 [8]. To date, three entomofaunas are known from the Jurassic and Cretaceous in northern China according to their stratigraphic and biogeographic occurrences. They are the Middle Jurassic Yanliao Entomofauna (Bathonian to Callovian in age), the Early Cretaceous Jehol Entomofauna (Hauterivian to Aptian in age), and late Early Cretaceous Fuxin Entomofauna (Albian in age) [2, 36] (Table 1.2).

Figure 1.1 Fossil excavation site at the Daohugou locality. Source: Photo by Dr. Chungkun Shih.

1.3.1 Yanliao Entomofauna

The Yanliao Entomofauna is an important component of the Yanliao Biota in the Middle Jurassic, which has become well-known in recent years because of the discovery of the best-preserved crown salamanders, the earliest known feathered dinosaurs, the earliest known gliding and aquatic mammaliaforms, the earliest known eutherian mammals and early angiosperms, etc. [2, 3, 5].

In 1983, abundant fossil insects collected from Haifanggou Formation have been named as Yanliao Entomofauna by Hong [24]. In 1995, the Yanliao Entomofauna has been extended into Yanliao Fauna by Ren *et al.* [2, 36], because a wide range of additional non-marine taxa including conchostracans, bivalves, fish, reptiles, and mammaliaforms coexisted with the insects. With abundant fossil plants found from the Jiulongshan Formation, the term "Yanliao Biota" has been used by multiple authors, but with different definitions [5, 34, 36]. The Yanliao Biota has also been referred to as the "Daohugou Biota", after many fossils found and documented from the Daohugou locality (Figure 1.1) in southeastern Inner Mongolia since 2002 [5, 26, 37, 38]. However, the term "Yanliao Biota" is gaining recognition among researchers since so-called "Daohugou Biota" is a junior synonym of the Yanliao Biota. Herein the Yanliao Entomofauna is preferred to include all insect fossils from the Middle Jurassic Haifanggou/Jiulongshan Formation and the Upper Jurassic Lanqi/Tiaojishan Formation because these Formations share strong similarities in terms of fossil content. A comprehensive review of the Yanliao Biota, building on previous reviews written from various perspectives, has been provided by Xu *et al.* [5, 25, 34, 36].

Hitherto, about 837 insect species in 476 genera, 166 families and 22 orders have been reported from the Yanliao Entomofauna [2, 3, 24, 25, 36, 39]. The most common taxa of the Yanliao insect fossil assemblages are listed in Table 1.2 and presented in Chapters 5–27.

Among abundant Yanliao herbivorous insects, the pollinating scorpionflies, including Mesopsychidae (e.g. *Lichnomesopsyche gloriae* Ren, Labandeira & Shih, 2010), Aneuretopsychidae (e.g. *Jeholopsyche liaoningensis* Ren, Shih & Labandeira, 2011) and Pseudopolycentropodidae (e.g. *Pseudopolycentropus janeannae* Ren, Shih & Labandeira, 2010) [40–43], have made significant contribution to the evolutionary development of pollination. These taxa had elongate, siphonate (tubular) proboscides and fed on ovular secretions of extinct gymnosperms, and likely engaged in pollination mutualisms with gymnosperms during the mid-Mesozoic, at least 70 million years before the similar and independent coevolution of nectar-feeding flies, moths, and beetles on angiosperms [43]; see also Chapters 24.3 and 28.2.

Amazingly, the herbivore katydid *Archaboilus musicus* Gu, Engel & Ren, 2012, with exceptionally well-preserved stridulatory structures in the forewings produced a low-frequency song to attract potential mates and constituted a normal part of the auditory background in the Yanliao Fauna ecosystem [44]; see also Chapter 30.2.

Yanliao Entomofauna is also characterized by some ectoparasitic insects. Some stem fleas of *Pseudopulex jurassicus* Gao, Shih & Ren, 2012, *P. wangi* Huang, Engel, Cai & Nel, 2013, *Hadropsylla sinica* Huang, Engel, Cai & Nel, 2013, etc., with long serrated stylets might have lived on and sucked blood of relatively large hosts, such as contemporaneous feathered dinosaurs or pterosaurs or medium-sized mammals. [28, 45–48]; see also Chapter 28.4.

Lacewings of *Bellinympha filicifolia* Wang, Ren, Liu & Engel, 2010 and *B. dancei* Wang Y., Ren, Shih & Engel, 2010 mimicked pinnate cycadophyte leaves [49] (see also Chapter 29.2.1), together with hangingfly *Juracimbrophlebia ginkgofolia* Wang Y., Labandeira,

Table 1.2 Biostratigraphic sequence of entomofaunas in Northern China during the Jurassic and Cretaceous.

System	Stage	Formation	Fauna	Ephemeroptera	Odonata	Plecoptera	Blattaria	Notoptera	Chresmodidae	Phasmatodea	Orthoptera	Dermaptera	Homoptera
Cretaceous	Maastrichtian												
	Campanian												
	Santonian												
	Coniacian												
	Turonian												
	Cenomanian	Sunjiawan (Xiazhuang)	Fuxin Entomofauna										
	Albian	Fuxin (Lushangfen)			*Hemeroscopus*; *Sinaeschnidia heishankoensis*								*Jiphara*; *Cretocercopis*; *Cretocixius stigmatosus*
	Aptian	Jiufotang	Jehol Entomofauna										
	Barremian	Yixian		*Ephemeropsis trisetalis*; *Epicharmeropsis*	*Sinaeschnidia cancellosa*; *Rudiaeschna limnobia*	*Sinosharaperla zhaoi*	*Perlucipecta*; *Habroblattula drepanoides*		*Sinochresmoda magnicornia*; *Chresmoda shihi*	*Hagiphasma paradoxa*; *Renphasma sinica*; *Cretophasmomima melanogramma*	*Parahagla sibirica*; *Vitimoilus ovatus*	*Sinoprotodiplatys zhangi*; *Barbderma oblongata*; *Cylindopygia falcate*	*Miracossus*; *Liaocossus*; *Lapicixius decorus*
	Hauterivian	Dabeigou											
	Valanginian	Zhangjiakou											
	Berriasian												
Jurassic	Tithonian	Tuchengzi											
	Kimmeridgian												
	Oxfordian	Tiaojishan	Yanliao Entomofauna										
	Callonian	Jiulongshan (Haifanggou)		*Jurassonurus amoenus*; *Mesobaetis*	*Sinokaratawia*; *Sinoeuthemis daohugouensis*	*Paranotonemoura pteroliriope*; *Karanemoura abrupta*	*Divocina noci, Fortiblatta cuspicolor*; *Fuzia dadao*	*Cseinitzia aristovi*; *Juramanto phasma*; *Juraperla grandis*; *Sinonele phasmoides*	*Jurachresmoda gaskelli*	*Adjacivena rasnitsyni*	*Liassophyllum caii*; *Archaboilus*; *Allaboilus*; *Locustopsis rhytofemoralis*	*Perissoderma triangulum*; *Palaeodermapteron*; *Belloderma arcuate*	*Daohugoucossus, Quadraticossus*; *Suljuktocossus, Synapocossus*; *Anthoscytina perpetua*; *Sunotettigarcia, Hirtaprosbole*
	Bathonian												
	Bajocian	Mentougou (Beipiao) (Xiahuayuan)											
	Aalenian												
	Toarcian												
	Plensbachian												
	Sinemurian	Xinglonggou (Nandaling)											
	Hettangian												

(Continued)

Table 1.2 (Continued)

Heteroptera	Neuroptera	Raphi-dioptera	Coleoptera	Hymenoptera	Diptera	Mecoptera	Siphona ptera	Trichoptera	Lepidoptera
			Longaevicupes macilentus *Zygadenia trachylenus* *Brochocoleus impressus* *Monticupe surrectus* *Cionocoleus magicus*						
Mesolygaeus laiyangensis *Byssoidecerus levigata* *Flexicorpus acutirostratus* *Primipentatoma peregrina* *Clavaticoris zhengi*	*Dipteromantispa brevisubcosta* *Choromyrmeleon aspoeckorum* *Aetheogramma speciosa* *Sophogramma lii*	*Chrysoraphidia relicta* *Baissoptera liaoningensis*	*Latocupes bella* *Furcicupes raucus* *Sinoporus lineatus* *Oxyporus yixianus* *Prophaenognatha robusta* *Paralithomerus exquisitus* *Sinopraecipuus bilobatus* *Abrocar brachyorhinos* *Coptoclava longipoda*	*Archoxyelyda mirabilis* *Rectilyda sticta* *Megapelecinus changi* *Acephialtitia colossa* *Shoushida regilla*	*Protapiocera megista* *Florinemestrius pulcherrimus* *Origoasilus pingquanensis* *Lichnoplecia kovalevi* *Dissup clausus*	*Typhothauma yixianensis* *Jeholopsyche liaoningensis* *Vitimopsyche kozlovi* *Megabittacus beipiaoensis*	*Tyrannopsylla beipiaoensis* *Saurophthirus exquisitus*	*Cathayamodus fournieri* *Sinomodus spatiosus* *Sinomodus peltatus*	
Miracorizus punctatus *Pumilanthocoris gracilis* *Peregrinpachymeridium comitcola* *Mirivena robusta*	*Bellinympha filicifolia* *Grammolingia boi* *Kallihemerobius aciedentatus* *Jurakempynus sinensis*	*Juroraphidia longicollum* *Ororaphidia bifurcata*	*Alloioscarabaeus cheni* *Jurotachinus breviantennatus* *Protagrypnus robustus* *Paradesmatus ponomarenkoi* *Loculitricoleus flatus*	*Mirolyda hirta* *Aethotoma aninomorpha* *Medilyda procera* *Mesoserphus venustus* *Synaphopterella patula* *Proephialtitia acantha*	*Eoptychopterina elenae* *Sinorhagio daohugouensis* *Archirhagio gracilentus* *Uranorhagio daohugouensis* *Praemacrochile dryasis*	*Tsuchingothauma shihi* *Lichnomesopsyche gloriae* *Pseudopolycentropus janeannae* *Juracimbrophlebia ginkgofolia*	*Pseudopulex jurassicus*	*Acisarcuatus variradius* *Pulchercylindratus punctatus* *Liadotaulius limus*	*Kladolepidopteron oviformis* *Ascololepidopterix multinerve* *Seresilepidopteron dualis*

Figure 1.2 A delegation of the Insect Fossil Conference visited the Daohugou Fossil Museum under construction in August, 2010. Source: Photo by Dr. Chungkun Shih.

Shih & Ren, 2012 mimicked a type of ginkgo leaf display several unusual mutualistic relationship and paleoecological features between insects and plants [50] (see also Chapter 29.2.2).

The majority of the fossils recovered from the Yanliao Entomofauna in Daohugou are complete, well-preserved and articulated with many fine-scale structures, including mouthparts, antennae and setae of insects. Some of the fossils of insects, plants and other animals are displayed in the Daohugou Fossil Museum (Figures 1.2 and 1.3) and the Ningcheng National Geological Park Museum (Figure 1.4). Many Yanliao insect fossils are even preserved in ways that provide behavioral information. For example, a male–female pair of the froghoppers *Anthoscytina perpetua* Li, Shih & Ren, 2013 preserved in a mating position is the earliest documented fossil record of copulating insects [51]; see also Chapter 30.4. Possible mechanisms of exceptional fossil preservation for the Yanliao Entomofauna could be the apparently sudden deaths and rapid burial caused by recurrent volcano eruptions and poisonous gases.

Studies of many insect fossils from the Yanliao Entomofauna indicated that the paleoenvironment and climate in the Daohugou paleolakes or swamp areas might have been a near-shore shallow lacustrine and wetland basin with a warm, humid climate, diverse and abundant vegetation, and highly aquiferous soil [3, 25, 52, 53]. Simultaneously, some high mountain insects, such as snakeflies *Mesoraphidia daohugouensis* Lyu, Ren & Liu, 2015 and *Ororaphidia bifurcata* Lyu, Ren & Liu, 2017 [54, 55], hairy-bodied tettigarctid *Hirtaprosbole erromera* Liu, Li, Yao & Ren, 2016 [56], support that there were high mountains above 800 m in paleoaltitude in the Daohugou area.

1.3.2 Jehol Entomofauna in the Yanliao Area

The Mesozoic fossils and strata in western Liaoning, northern Hebei and southeastern Inner Mongolia (Yanliao Area) have been studied for nearly 90 years. The first systematic palaeontological and stratigraphic studies on the Yanliao Area fossils were conducted by Grabau in 1923 [8]. During 1920–1940, the Yanliao Area was once called "Jehol" Province. The fossils from the Linyuan County, Liaoning Province was named "Jehol" by Grabau in 1923. In 1928, Grabau gave the name Jehol Fauna to the fossil community from the Jehol Series (= Yixian and Jiufotang Formations). In 1962, the Chinese paleontologist Gu Zhiwei first proposed the terms of "Jehol Group" and "Jehol Biota," representing an Early Cretaceous terrestrial ecosystem that is mainly distributed in Northeastern China [57]. The term "Jehol Biota" has been extensively used and is well-known today. The diversity of the Jehol Fauna has been summarized by some authors in a different period [2–4, 6, 17, 31–33, 58] (Table 1.3).

In 1995, insect fossils from the Yixian and Jiufotang Formations have been referred as Jehol Entomofauna by Ren [2, 3, 36]. The Early Cretaceous Jehol Entomofauna bears a close taphonomic resemblance to the Middle

Figure 1.3 The Daohugou Fossil Museum. Source: Photo by Dr. Chungkun Shih.

Figure 1.4 The Ningcheng National Geological Park Museum. Source: Photo by Dr. Chungkun Shih.

Figure 1.5 A delegation of the Insect Fossil Conference visited the Shihetun Fossil site in August, 2010. Source: Photo by Dr. Chungkun Shih.

Jurassic Yanliao Entomofauna [32]. The majority of the insect fossils recovered from the Yixian Formation are complete and articulated, with many fine-scale structures and details preserved.

Up to now, fossil insects, with their remarkable diversity, exceptional preservation and implications for evolution of many insect lineages, represent some of the most important discoveries among the Jehol Fauna, e.g. the Shihetun fossil site (Figure 1.5). To date, about 862 insect species in 573 genera, 204 families and 19 orders in the Jehol Fauna have been published [2–4, 6, 29–34, 39, 58] (Chapters 5–27). There are three distinct insect assemblages, approximately corresponding to the early, middle and late stages of the Jehol Entomofauna in Table 1.2 [32].

A hypothetical structure of ecosystem and their environmental settings of the Jehol Entomofauna have been preliminarily reconstructed. The insect assemblages in the Yixian Formation could been divided into four communities based on habitats, or five groups based on feeding habits. Of the four communities, the highest species diversity occurred in the forest community, followed by the aquatic, soil, and alpine communities. Of the five feeding groups, the highest species diversity appeared in the phytophagous group, followed by the carnivorous, parasitic, saprophagous, and heterophagous groups [3, 32, 58, 59].

Overall, the insect community in the Yixian Formation lacustrine ecosystem was relatively stable. The climate of this region was warm and humid, while there was seasonal arid and semi-arid climate in microenvironment. There were plenty of calm deep-water lakes, while other water bodies existed around the lakes, such as swamps and shallow water environment. The soil on the land was nutritious and moist enough for a variety of plants, insects and other animals to survive. There were mountains with alpines lakes and streams at least 800 m high in paleoaltitude [58].

More importantly, the insects played a key role in the entire Jehol ecosystem as they constituted prey for nearly all major groups of vertebrates: fishes, amphibians, choristoderans, turtles, lizards, birds, and pterosaurs. They might also have provided food for some dinosaurs and mammals [6, 58]. Some of the Jehol specimens of insects, plants and other animals are treasured in the Shihetun Fossil Museum (Figure 1.6) and the Chaoyang Paleontological Fossil Museum (Figure 1.7).

In addition, insects in the Yixian Formation also provide some important evidence for the co-evolution between insects and plants and animals. A large number of pollinating insect fossils, such as brachyceran flies of *Protonemestrius jurassicus* Ren, 1998, scorpionflies of *Vitimopsyche kozlovi* Ren, Labandeira & Shih, 2010, kalligrammatid lacewing *Sophogramma papilionacea* Ren & Guo, 1996, and flower bug *Vetanthocoris decorus*, Yao, Cai & Ren, 2006, etc., indicate that pollinating insects might have played an important role in the origin and early evolution of the angiosperms [43, 60–62] (see also Chapter 28.2). The discovery of the largest definitive

Figure 1.6 A delegation of the Insect Fossil Conference visited the Shihetun Fossil Museum in August, 2010. Source: Photo by Dr. Chungkun Shih.

Figure 1.7 The Chaoyang Paleontological Fossil Museum. Source: Photo by Dr. Chungkun Shih.

fleas of *Pseudopulex magnus* Gao, Shih & Ren, 2012 [45], *Saurophthirus exquisitus* Gao, Shih & Ren, 2013 [45] and *Tyrannopsylla beipiaoensis* Huang, Engel, Cai & Nel, 2013 [48] and blood-feeding true bugs of *Torirostratus pilosus* Yao, Shih & Engel, 2014 and *Flexicorpus acutirostratus* Yao, Cai & Engel, 2014 [61] from the Jehol Entomofauna highlight the diversity of Early Cretaceous ectoparasitic insects (Chapters 28.4 and 28.5).

1.3.3 Fuxin Entomofauna

Fossils from the Fuxin Formation, conformably overlying the Jiufotang Formation which overlies the Yixian Formation, have been given various names in different taxa by multiple authors. In 1981, plant fossils, found in the Sahai Formation and the Fuxin Formation of western Liaoning and other equivalent strata, were referred to as Fuxin Flora by Feng Chen [12, 13], which were characterized by the dominance of the Filicopsida, Ginkgopsida and Coniferopsida, with abundant Cycadopsida and Equisetales.

In 1981, fossil insects from the Lushangfen Formation in west hills, Beijing (equivalent to Fuxin Formation in western Liaoning) were named as Lushangfen Entomofauna by Hong [63]; later this was abandoned by Hong in 1993 [64]. In 1987, all animal and plant fossils from the Fuxin Formation in western Liaoning were called Fuxin Biota by Wuli Wang [14]. The most common elements of the Jehol Fauna fossil assemblages have never been found in the Fuxin Biota, including insects: *Ephemenopsis trisetalis, Coptoclava longipoda, Chironomaptera gregaria, Clyptostemma xyphidle, Sinaeschuidia heishankouensis* (Table 1.2); conchostracans: *Eosestheria-Diestheria-Liaoningestheria*; fish: *Lycoptera, Peipiaosteus* and dinosaur: *Psittacosaurus* (Table 1.3).

Considering the different contents of insect fossils from those in Jehol Entomofauna, the insect fossil assemblages from Lushangfen Formation have been termed the Fuxin Entomofauna by Ren [2, 36]. This Entomofauna contains about 20 reported fossil species. Although the number of species and fossil quantity are obviously less than those of the Jehol Entomofauna in the strata below, the Fuxin insect fossils are characterized by many fossil termites *Jitermes, Yanjingtermes* and *Yongdingia*; Odonata *Hemeroscopus* and Coleoptera *Cionocoleus* and *Monticupes, Diluticupes*.

1.4 Geological Ages of Non-marine Jurassic and Cretaceous Strata and Entomofaunas in Northern China

The Jurassic and Cretaceous was a period of violent tectonic movements, great paleogeographic, paleoclimatic and biotic changes, and the formation of vast quantities of endogenic and sedimentary (particularly coal and oil) deposits of economic value. The global Mesozoic has included both marine and non-marine strata and fossils. The standard global chronostratigraphic scale for the Mesozoic is based primarily on marine fossils; ammonoids and microfossils (foraminiferans and calcareous nannoplankton) integrated with a well-established global polarity timescale and a relative abundance of radioisotopic ages. Unfortunately, the practical isotopic data for the ages of the boundaries between stages and systems from Upper Jurassic to Lower Cretaceous in the International Stratigraphical Chart are deficient. These ages are defined through the averaging and interpolating method with the assumption that an ammonite subzone would last one million years and all the subzones have equal evolutionary rates [65, 66]. Consequently, there were different opinions on the "international chronostratigraphic framework chart

Table 1.3 Biostratigraphic sequence of entomofaunas with important accompanying fossils in northern China during the Jurassic and Cretaceous.

System	Stage	Formation	Biota	Conchostracans	Plants	Pisces	Mammalia	Aves	Pterosauria
Cretaceous	Maastrichtian								
	Campanian								
	Santonian								
	Coniacian								
	Turonian								
	Cenomanian	Sunjiawan (Xiazhuang)	Fuxin Biota *Hemeroscopus-Jiphara-Yanjiesheria-Xishanichtys-* etc.	*Yanjiestheria-Neodiestheria-Orthestheria-Eosestheria lushangfenensis*	*Ctenislyrata-Chilinia* ass. *Ruftordia-Dryopterites* ass. *Acanthopteris-Ginkgo coriacea* ass.				
	Albian	Fuxin (Lushanfen)				*Haizhoulepis changi* *Xishanichthys xiei*	*Endotherium niinomii*		
	Aptian	Jiufotang	Jehol Biota *Ephemeropsis-Sinaeschnidia-Habroblattula-Liaocossus-Baissoptera-Sophogramma-Vitimopsyche-Florinemestrius-Hagiphasma-* etc.	*Eosestheria-Diestheria* assm.	*Brachyphyllum-Izekanowskia-* *Solenites-Archaeofructus* etc.	*Lycoptera muroii*			*Sinosauropteryx-Sinopterus-Microraptor-Feilongus-Chaoyangopterus-Beipiaopterus-Caudipteryx* etc.
	Barremian	Yixian (Huajiying)				*Peipiaosteus pani-Lycoptera davidi-Protopsephurus liui*	*Jeholodens-Liaoconodon-Sinobaatar-Maotherium-Zhangheotherium-Sinodelphys-Eomaia*	*Confuciusornis-Changchengornis-Protopteryx-Jeholornis-Sapeornis-Longipteryx-Cathayornis-Liaoxiornis-Chaoyangia* etc.	
	Hauterivian	Dabeigou		*Nestoria-Keratestheria* *Nestoria-Jibeilimnadia*					
	Valanginian								
	Berriasian	Zhangjiakou							
Jurassic	Tithonian	Tuchengzi							
	Kimmeridgian								
	Oxfordian	Tiaojishan	Yanliao Biota *Jurassonurus-Archaboilus-Fuzia-Karanemoura-Anthoscytina-Suljuktocossus-Miracorizus-Pseudopulex-Juroraphidia-Grammolingia-Lichnomesopsyche-Eoptychopterina-Kladolepidopteron-Jurachresmoda-Belloderma-* etc.	*Triglypta pingquanensis-T. haifanggouensis-T. luanpingensis*	*Ningchengia jurassica-Anomozamites haifanggouensis-Weltricia daohugouensis-Yimaia captuliformis-Schizolepis daohugouensis-Juraherba bodae*				*Jeholopterus-Qinglongopterus-Jianchangopterus-Darwinopterus-Epidexipteryx-Aurornis-Eosinopteryx*
	Callonian	Jiulongshan (Haifanggou)				*Liaosteus hongi*	*Pseudotribos-Juramaia-Volaticotherium-Castrocauda-Arboroharamiya-Megaconus*		
	Bathonian								
	Bajocian								
	Aalenian	Mentougou (Beipiao) (Xiahuayuan)							
	Toarcian								
	Plensbachian								
	Sinemurian	Xinglonggou (Nandaling)							
	Hettangian								

for Jurassic/Cretaceous boundary" among geologists and the International Commission on Stratigraphy based on marine rocks and radiometric ages [1, 21–23].

Historically, different authors once placed the J/K boundary at 130, 135, 137, 142 or 145.5 Mya in the International Stratigraphic Charts [21, 65]. Until now, there is no reliable evidence to support whether to place the J/K boundary at 140 Mya, 142 Mya, 145.5 Mya, or at 135 Mya. This makes non-marine Jurassic and Cretaceous strata correlations challenging and very difficult to establish a non-marine Cretaceous stratigraphic framework chart in the global timescale [67–70]. To date, the position of the global Jurassic/Cretaceous boundary stratotype has not yet been formalized. A consensus on the age and correlation of the coal-, oil- and fossil bearing Jurassic-Cretaceous no-marine strata in northern China has not been achieved.

In this context, based on an updated time scale of The ICS (International Commission on Stratigraphy) International Chronostratigraphic Chart (version 2016) [1] and synthesized from the previous results, we correlate the non-marine Jurassic and Cretaceous rocks in northern China by the stratigraphic occurrences of insects, together with various radiometric ages of the intervening tuffs, tuffaceous rocks and lavas, other accompanying bivalves and dinoflagellate assemblage sequences found in the eastern Heilongjiang basins marine and non-marine interbeds [17–23].

In northern China, the ages of the Jiulongshan/Tiaojishan Formation bearing Yanliao Fauna and Yixian/Jiufotang Formations bearing Jehol Entomofauna are the focus of attention because major insect fossils are from these Formations.

The Jiulongshan/Haifanggou Formation has been widely accepted as the Middle Jurassic in age based on paleontological data [2, 3, 15, 24, 71]. In 2002, the strata section containing Yanliao biota at the Daohugou Village, Ningchen County, Inner Mongolia was measured, recognized and attributed to the Jiulongshan Formation and belonging to the Late Aalenian or Early Bajocian in age for the first time by Ren's Team [37]. In the past decade, a new batch of isotopic radiometric data have supported this age assessment and further indicated that the Jiulongshan is partially Bathonian but mainly Callovian [5]. For example, a $^{40}Ar/^{39}Ar$ age of 166.7 ± 1.0 Mya around the middle of the Haifanggou Formation and two recent $^{40}Ar/^{39}Ar$ dates of 159.5 ± 0.6 Mya for the Lanqi Formation were obtained in Beipiao, western Liaoning Province [72, 73]. At the Daohugou Locality a serial of $^{206}Pb/^{238}U$ SHRIMP ages (162 ± 2 Mya, 152 ± 2.3 Mya, 166 ± 1.5 Mya, 165 ± 2.4 Mya, 164 ± 1.2 Mya and 165 ± 1.2 Mya) and $^{40}Ar/^{39}Ar$ (159.8 ± 0.8 Mya, 164 ± 2.5 Mya) ages was obtained based on samples collected from the fossil-bearing layers respectively by different authors suggesting that the Daohugou beds are mostly Callovian [74–77].

The arguments over the age-range of the Jehol Group/Biota have arisen since 1920s. The age of the Yixian Formation, which yields the Jehol Biota, has been considered as late Tithonian, Early Cretaceous or spanning the Jurassic-Cretaceous boundary [2, 3, 6, 9–11, 15–21]. The point of view concerning Early Cretaceous age has been proven by new paleontological data and isotopic datings [22, 23].

A lot of age data using $^{40}Ar/^{39}Ar$ or U/Pb methods from tuff or lava flows of the lower Yixian Formation near Sihetun Village of Beipiao City, Liaoning, China have been obtained by many teams [78–82] ranging from the Barremian to the Aptian (from 130 to 120 Mya, interval about 10 Mya). Hence, the age data from the Yixian Formation may range from Hauterivian to Middle Aptian, but are mainly from around the Barremian-Aptian transition, and the age of the Jiufotang Formation is Middle Aptian.

Some SHRIMP U-Pb zircon age of 133.9 ± 2.5 Mya, 130.1 ± 2.5 from Dabeigou Formation in Luanping basin of northern Hebei and 135.8 ± 3.1 Mya, 136.3 ± 3.4 Mya, 135.4 ± 1.6 Mya from Zhangjiagou Formation have been obtained [83, 84]. This result not only supports a Hauterivian/Barremian-Early Albian age for the Jehol Biota, but also indicates that the non-marine Jurassic/Cretaceous boundary in the Yanliao Area is below the Yixian Formation, probably within the Tuchengzi Formation (Table 1.1).

References

1 Cohen, K.M., Finney, S.M., Gibbard, P.L., and Fan, J.X. (2013). The ICS international chronostratigraphic chart. *Episodes* **36** (3): 199–204.

2 Ren, D., Lu, L.W., Ji, S.A., and Guo, Z.G. (1995). *Faunae and Stratigraphy of Jurassic-Cretaceous in Beijing and the Adjacent Areas*, 1–222. Beijing: Seismic Publishing House.

3 Ren, D., Shih, C.K., Gao, T.P. *et al.* (2010). *SILENT STORIES–Insect Fossil Treasures from Dinosaur Era of the Northeastern China*, 1–332. Beijing: Science Press.

4 Chang, M.M., Chen, P.J., Wang, Y.Q., and Wang, Y. (2003). *The Jehol Biota*, 1–208. Shanghai: Shanghai Scientific & Technical Publishers.

5 Xu, X., Zhou, Z.H., Sullivan, C. *et al.* (2016). An updated review of the Middle-Late Jurassic Yanliao Biota: chronology, taphonomy, paleontology and paleoecology. *Acta Geologica Sinica (English Edition)* **90** (6): 2229–2243.

6 Zhou, Z.H., Wang, Y., Xu, X., and Ren, D. (2017). Jehol Biota: an exceptional window to the Early Cretaceous terrestrial ecosystem. In: *Terrestrial Conservation Lagerstatten – Windows into the Evolution of Life on Land* (ed. N.C. Fraser and H.D. Sues), 169–214. Scotland: Dunedin Academic Press Ltd.

7 Hong, Y.C. (1982). *Mesozoic Fossil Insects of Jiuquan Basin in Gansu Province*, 1–187. Beijing: Geological Publishing House.

8 Grabau, A.W. (1923). Cretaceous fossil from Shandong. *Bulletin of Geology Survey of China* **2** (5): 164–181.

9 Gu, Z.W. (1982). Correlation chart of the Jurassic in China with explanatory text. In: *Stratigraphical Correlation Chart in China with Explanatory Text* (ed. Nanjing Institute of Geology and Palaeontology, Academia Sinica), 223–240. Beijing: Science Press (in Chinese).

10 Gu, Z.W. (1983). On the boundary of non-marine Jurassic and Cretaceous, in China. In: *Stratigraphical Correlation Chart in China with Explanatory Text* (ed. Nanjing Institute of Geology and Palaeontology, Academia Sinica), 65–82. Beijing: Science Press (in Chinese).

11 Gu, Z.W. (2000). Non-marine Jurassic. In: *Stratigraphical Study in China (1979–1999)* (ed. Nanjing Institute of Geology and Palaeontolgy, Academia Sinica), 309–313. Hefei: University of Science and Technology of China Press (in Chinese).

12 Chen, F., Yang, G.X., and Zhou, H.Q. (1981). Lower Cretaceous flora in Fuxin Basin, Liaoning Province, China. *Earth Science: Journal of the Wuhan College of Geology* **2**: 39–51. (in Chinese with English abstract).

13 Chen, F., Meng, X.Y., Ren, S.Q. *et al.* (1988). *The Early Cretaceous Flora of Fuxin Basin and Tiefa Basin, Liaoning Province*, vol. **1–180**. Beijing: Geological Publishing House (in Chinese with English abstract).

14 Wang, W.L. (1987). On the Early Cretaceous Fuxin Biota of Northern China. *Bulletin of Shenyang Institute of Geology and Minerals, Chinese Academia of Geological Sciences* **16**: 53–58. (in Chinese).

15 Wang, W.L., Zheng, S.L., Zhang, L.J. *et al.* (1989). *Mesozoic Stratigraphy and Paleontology of Western Liaoning*, 168. Beijing: Geological Publishing House in Chinese.

16 Wang, S.E., Gao, L.Z., Pang, Q.Q. *et al.* (2015). Boundary between Jurassic and Cretaceous in Chinese non-marine basin and with International Stratigraphic Correlation: a Case Study of Biostratigraphic and Isotopic Dating in Luanping, Hebei Province. *Acta Geologica Sinica* **89** (8): 1331–1351. (in Chinese with English summary).

17 Chen, P.J. (1988). Distribution and migration of Jehol Fauna with reference to nonmarine Jurassic-Cretaceous boundary in China. *Acta Palaeontologica Sinica* **27**: 659–683. (in Chinese, English abstract).

18 Chen, P.J. and Chang, Z.L. (1994). Nonmarine Cretaceous stratigraphy of eastern China. *Cretaceous Research* **15** (3): 245–257.

19 Chen, P.J. (1994). Cretaceous conchostracan faunas of China. *Cretaceous Research* **15** (3): 259–269.

20 Chen, P.J., Li, J.J., Matsukawa, M. *et al.* (2006). Geological ages of dinosaur-track-bearing formations in China. *Cretaceous Research* **27** (1): 22–32.

21 Chen, P.J., Wang, Q.F., Zhang, H.C. *et al.* (2005). Jianshangou beds of the Yixian Formation in West Liaoning, China. *Science in China Series D: Earth Sciences* **48** (3): 298–312.

22 Sha, J.G., Chen, S.W., Cai, H.W. *et al.* (2006). Jurassic-Cretaceous boundary in Northeastern China: placement based on buchiid bivalves and dinoflagellate cysts. *Progress in Natural Sciences* **16** (suppl 1: 39–49).

23 Sha, J.G. (2007). Cretaceous stratigraphy of northeast China: non-marine and marine correlation. *Cretaceous Research* **28** (2): 146e170.

24 Hong, Y.C. (1983). *Middle Jurassic Fossil Insects in North China*, 1–223. Beijing: Geological Publishing House.

25 Liu, P.J., Huang, J.D., and Ren, D. (2010). Palaeocology of the Middle Jurassic Yanliao Entomofauna. *Acta Zootaxonomica Sinica* **35** (3): 568–584. (in Chinese with English abstract).

26 Zhang, J.F. (2002). Discovery of Daohugou Biota (Pre-Jehol Biota) with a discussion on its geological age. *Journal of Stratigraphy* **26** (3): 173–177.

27 Gao, K.Q. and Ren, D. (2006). Radiometric dating of ignimbrite from inner Mongolia provides no indication of a post-Middle Jurassic age for the Daohugou beds. *Acta Geologica Sinica (English Edition)* **80** (1): 42–45.

28 Huang, D.-Y. (ed.) (2016). *The Daohugou Biota*, 1–332. Shanghai: Shanghai Scientific & Technical Publishers.

29 Ren, D., Guo, Z.G., Lu, L.W. *et al.* (1997). A further contribution to the knowledge of the upper Jurassic Yixian Formation in western Liaoning. *Geological Review* **43** (5): 449–459.

30 Pan, Y.H., Sha, J.G., Fürsich, F.T. *et al.* (2012). Dynamics of the lacustrine fauna from the Early Cretaceous Yixian Formation, China: implications of volcanic and climatic factors. *Lethaia* **45** (3): 299–314.

31 Pan, Y.H., Sha, J.G., Zhou, Z.H., and Fürsich, F.T. (2013). The Jehol Biota: definition and distribution of exceptionally preserved relicts of a continental Early Cretaceous ecosystem. *Cretaceous Research* **44**: 30–38.
32 Zhang, H., Wang, B., and Fang, Y. (2010). Evolution of insect diversity in the Jehol Biota. *Science China, Earth Sciences* **53** (12): 1908–1917.
33 Zhou, Z. (2006). Evolutionary radiation of the Jehol Biota: chronological and ecological perspectives. *Geological Journal* **41** (3–4): 377–392.
34 Zhou, Z.H., Jin, F., and Wang, Y. (2010). Vertebrate assemblages from the Middle-Late Jurassic Yanliao Biota in northeast China. *Earth Science Frontiers* **17**: 252–254.
35 Liu, X.T., Liang, H.D., Sun, Z.C. *et al.* (1990). Fuxin basin. In: *Cretaceous in Oil and Gas-Bearing Areas of Northern China. Series on Stratigraphy and Palaeontology of Oil and Gas Bearing Areas in China* (ed. D.Q. Ye, X.C. Zhong, *et al.*), 99–122. Beijing: Petroleum Industry Press (in Chinese).
36 Ren, D., Lu, L.W., Ji, S.A. *et al.* (1996). Late Mesozoic fauna assemblages of Yanliao area, North China, and its paleoecological and paleogeographical significance. *Acta Geoscientia Sinica* **17**: 148–154. in Chinese with English Abstract.
37 Ren, D., Gao, K.Q., Guo, Z.G. *et al.* (2002). Stratigraphic division of the Jurassic in the Daohuguo area, Ningcheng, Inner Mongolia. *Geological Bulletin of China* **21** (8–9): 584–591.
38 Sullivan, C., Wang, Y., Hone, D.W.E. *et al.* (2014). The vertebrates of the Jurassic Daohugou Biota of Northeastern China. *Journal of Vertebrate Paleontology* **34** (2): 243–280.
39 Liu, P.J., Huang, J.D., Ren, D., and Zhao, Y.Y. (2009). Aquatic community succession and environmental changes of Late Mesozoic in Northern China. *Acta Zootaxonomica Sinica* **34** (4): 836–846. (in Chinese with English abstract).
40 Ren, D., Labandeira, C.C., and Shih, C.K. (2010). New Mesozoic Mesopsychidae (Mecoptera) from Northeastern China. *Acta Geologica Sinica (English Edition)* **84** (4): 720–731.
41 Ren, D., Shih, C.K., and Labandeira, C.C. (2010). New Jurassic pseudopolycentropodids from China (Insecta: Mecoptera). *Acta Geologica Sinica (English Edition)* **84** (1): 22–30.
42 Ren, D., Shih, C.K., and Labandeira, C.C. (2011). A well-preserved aneuretopsychid from the Jehol Biota of China (Insecta, Mecoptera, Aneuretopsychidae). *ZooKeys* **129**: 17–28.
43 Ren, D., Labandeira, C.C., Santiago-Blay, A.J. *et al.* (2009). A probable pollination mode before angiosperms: Eurasian, long-proboscid scorpionflies. *Science* **326** (5954): 840–847.
44 Gu, J.J., Montealegre-Zapata, F., Robert, D. *et al.* (2012). Wing stridulation in a Jurassic katydid (Insecta, Orthoptera) produced low-pitched musical calls to attract female. *Proceedings of the National Academy of Sciences USA (PNAS)* **109** (10): 3868–3873.
45 Gao, T.P., Shih, C.K., Xu, X. *et al.* (2012). Mid-Mesozoic flea-like ectoparasites of feathered or haired vertebrates. *Current Biology* **22**: 732–735.
46 Gao, T.P., Shih, C.K., Rasnitsyn, P.A. *et al.* (2013). New transitional fleas from China highlighting diversity of Early Cretaceous ectoparasitic insects. *Current Biology* **23** (13): 1261–1266.
47 Gao, T.P., Shih, C.K., Rasnitsyn, A.P. *et al.* (2014). The first flea with fully distended abdomen from the Early Cretaceous of China. *BMC Evolutionary Biology* **14**: 168.
48 Huang, D.-Y., Engel, M.S., Cai, C.Y. *et al.* (2012). Diverse transitional giant fleas from the Mesozoic era of China. *Nature* **483**: 201–204.
49 Wang, Y.J., Liu, Z.Q., Wang, X. *et al.* (2010). Ancient pinnate leaf mimesis among lacewings. *Proceedings of the National Academy of Sciences USA (PNAS)* **107** (37): 16212–16215.
50 Wang, Y.J., Labandeira, C.C., Shih, C.K. *et al.* (2012). Jurassic mimicry between a hangingfly and a ginkgo from China. *Proceedings of the National Academy of Sciences USA (PNAS)* **109** (50): 20514–20519.
51 Li, S., Shih, C.K., Wang, C. *et al.* (2013). Forever love: the hitherto earliest record of copulating insects from the Middle Jurassic of China. *PLoS ONE* **8** (11): e78188. https://doi.org/10.1371/journal.pone.0078188.
52 Tan, J.J. and Ren, D. (2006). Jurassic and Cretaceous Cupedomorpha (Insecta: Coleoptera) faunas of China. *Progress in Natural Science* **16** (suppl): 313–322.
53 Tan, J.J., Ren, D., and Shih, C.K. (2006). Palaeogeography, palaeoecology and taphonomy of Jurassic-Cretaceous Cupedomorpha faunas from China. In: *9th International Symposium on Mesozoic Terrestrial Ecosystems and Biota* (ed. S.E. Evans), 130–133. Manchester: Natural History Museum.
54 Lyu, Y.N., Ren, D., and Liu, X. (2017). Review of the fossil snakefly family Mesoraphidiidae (Insecta: Raphidioptera) in the Middle Jurassic of China, with description of a new species. *Alcheringa: An Australasian Journal of Palaeontology* **41** (3): 403–412.
55 Engel, M.S. and Ren, D. (2008). New snakeflies from the Jiulongshan Formation of Inner Mongolia, China (Raphidioptera). *Journal of the Kansas Entomological Society* **81** (3): 188–193.
56 Liu, X.H., Li, Y., Yao, Y.Z., and Ren, D. (2016). A hairy-bodied tettigarctid (Hemiptera: Cicadoidea)

from the latest Middle Jurassic of northeast China. *Alcheringa: An Australasian Journal of Palaeontology* **40**: 383–389.

57 Gu, Z.-W. (1962). *Jurassic and Cretaceous of China*, 84. Beijing: Science Press (in Chinese).

58 Liu, P.J., Huang, J.D., and Ren, D. (2009). Insect community palaeocology of the Yixian Formation in Northern Hebei and Western Liaoning. *Journal of Environmental Entomology* **31** (3): 254–274.

59 Zhang, H., Wang, B., and Fang, Y. (2015). *Mesozoic and Cenozoic Insects from Northern China*, 1–229. Shanghai: Shanghai Scientific & Technical Publishers.

60 Ren, D. (1998). Flower-associated Brachycera flies as fossil evidences for Jurassic angiosperm origins. *Science* **280**: 85–88.

61 Yao, Y.Z., Cai, W.Z., Xu, X. *et al.* (2014). Blood-feeding true bugs in the Early Cretaceous. *Current Biology* **24** (15): 1786–1792.

62 Labandeira, C.C., Yang, Q., Santiago-Blay, J.A. *et al.* (2016). The evolutionary convergence of mid-Mesozoic lacewings and Cenozoic butterflies. *Proceedings of the Royal Society B* **283**, 20152893.

63 Hong, Y.C. (1981). Discovery of New Early Cretaceous Insects from Xishan, Beijing. *Bulletin of Tianjing Institute of Geology and Mineral Resources, Chinese Academy of Geological Science* **1981** (4): 87–96.

64 Hong, Y.C. (1993). Origin, development, flourish and disappearance of Late Mesozoic Jehol biota in Eastern Asian Paleocontinent. *Geoscience* **7** (4): 373–383.

65 Wang, Q.F. and Chen, P.J. (2005). A brief introduction to the Cretaceous chronostratigraphic study. *Journal of Stratigraphy* **29** (2): 114–123.

66 Young, G.C. and Laurie, J. (1996). *Australian Phanerozoic Timescales*, 1–279. South Melbourne Britain: Oxford University Press.

67 Kennedy, W.J. and Odin, G.S. (1982). *The Jurassic and Cretaceous time scale, Numerical dating in stratigraphy* (ed. G.S. Odin), 557–592. Chichester: Wiley.

68 Gradstein, F.M., Ogg, J.G., and Smith, A.G. (2004). Chronostratigraphy: linking time and rock. In: *A Geological Time Scale* (ed. F.M. Gradstein, J.G. Ogg and A.G. Smith), 20–46. Cambridge: Cambridge University Press.

69 Gradstein, F.M., Ogg, K.G., Schmitz, M.D., and Ogg, G.M. (2012). *The Geologic Time Scale, Volume I*, 794–797. London: Elsevier.

70 Ogg, J.G. (2004). Status of divisions of the international geologic time scale. *Lethaia* **37** (2): 183–199.

71 Shen, Y.B., Chen, P.J., and Huang, D.-Y. (2003). Age of the fossil conchostracans from Daohugou of Ningcheng, Inner Mongolia. *Journal of Stratigraphy* **27** (4): 311–313.

72 Chang, S.C., Zhang, H.C., Renne, P.R., and Fang, Y. (2009). High-precision $^{40}Ar/^{39}Ar$ age constraints on the basal Lanqi Formation and its implications for the origin of angiosperm plants. *Earth and Planetary Science Letters* **279** (3–4): 212–221.

73 Chang, S.C., Zhang, H., Hemming, S.R. *et al.* (2014). $^{40}Ar/^{39}Ar$ age constraints on the Haifanggou and Lanqi Formations: when did the first flowers bloom. *Geological Society, London, Special Publications* **378** (1): 277–284.

74 Liu, Y.X., Liu, Y.Q., and Zhong, H. (2006). LA-ICPMS zircon U-Pb dating in the Jurassic Daohugou beds and correlative strata in Ningcheng of Inner Mongolia. *Acta Geologica Sinica (English Edition)* **80** (5): 733–742.

75 He, H.Y., Wang, X.L., Zhou, Z.H. *et al.* (2004). 40Ar/39Ar dating of ignimbrite from Inner Mongolia, Northeastern China, indicates a post-Middle Jurassic age for the overlying Daohugou bed. *Geophyical Research Letters* **31**: 1–4.

76 He, H.Y., Wang, X.L., Zhou, Z.H. *et al.* (2006). 40Ar/39Ar dating of Lujiatun bed (Jehol Group) in Liaoning, Northeastern China. *Geophysical Research Letters* **33**: L04303.

77 Yang, W. and Li, S.G. (2008). Geochronology and geochemistry of the Mesozoic volcanic rocks in Western Liaoning: implications for lithospheric thinning of the North China Craton. *Lithos* **102** (1): 88–117.

78 Chang, S.C., Zhang, H.C., Renne, P.R., and Fang, Y. (2009). High-precision $^{40}Ar/^{39}Ar$ age for the Jehol Biota. *Palaeogeography, Palaeoclimatology, Palaeoecology* **280** (1–2): 94–104.

79 Smith, P.E., Evensen, N.M., York, D. *et al.* (1995). Dates and rates in ancient lakes: 40Ar/39Ar evidence for an Early Cretaceous age for the Jehol Group, northeast China. *Canadian Journal of Earth Science* **32** (9): 1426–1431.

80 Swisher, C.C., Wang, Y., Wang, X. *et al.* (1999). Cretaceous age for the feathered dinosaurs of Liaoning, China. *Nature* **400**: 59–61.

81 Swisher, C.C., Wang, X., Zhong, Z. *et al.* (2002). Further support for a Cretaceous age for featured dinosaur beds of Liaoning Province, China: new 40Ar/39Ar dating of the Yixian and Tuchengzi Formations. *Chinese Science Bulletin* **47**: 135–138.

82 Zhu, R., Pan, Y., Shi, R. *et al.* (2007). Palaeomagnetic and 40Ar/39Ar dating constraints on the age of the Jehol Biota and the duration of deposition of the Sihetun fossil-bearing lake sediments, Northeast China. *Cretaceous Research* **28** (2): 171–176.

83 Niu, B.G., He, Z.J., and Song, B. (2003). Shrimp dating of the Zhangjiakou Formation volcanic rocks and implications. *Geological Bulletin of China* **22** (2): 140–141. (in Chinese with English abstract).

84 Liu, Y.Q., Li, P.X., and Tian, S.G. (2003). SHRIMP U-Pb zircon age of late Mesozoic tuff (lava) in Luanping basin, northern Hebei, and its implications. *Acta Petrologica et Mineralogica* **22** (3): 237–244.

2

Coexisting Animals and Plants in the Ecosystems

Chungkun Shih[1,2], Taiping Gao[1], and Dong Ren[1]

[1] *Capital Normal University, Haidian District, Beijing, China*
[2] *National Museum of Natural History, Smithsonian Institution, Washington, DC, USA*

2.1 Introduction

Based on numerous fossils reported from northern China, it is clear that there are many animals and plants coexisting with abundant insect fauna (Chapters 5–27) in the well-balanced ecosystems about 165–125 Mya. Reported fossils suggest that the climate was correspondingly humid and warm. There were various plants, such as mosses, ferns, seed ferns, cycads, ginkgos, conifers, and basal forms of angiosperm plants. Many species of arthropods (e.g. conchostracans, shrimps, crawfishes, insects, spiders, and harvestmen), fish (e.g. *Lycoptera* fish and sturgeon), amphibians (e.g. frogs, salamanders), reptiles (e.g. crocodiles, lizards, turtles), dinosaurs, pterosaurs, birds and mammals lived and coexisted in these ecosystems. As preserved in the fossils, some interesting and important evidence highlights the complex relationships and interactions such as mimesis and camouflage, parasitism vs. hosting, synergy vs. adversary, predators vs. prey, and competitors vs. cooperators. Living together, these animals and plants kept the ecosystems balanced, sustained and evolved.

Zhou and Wang, in 2017, reviewed and reported the documented vertebrate assemblages [1] of the Middle Jurassic Yanliao Biota and the Early Cretaceous Jehol Biota hitherto (Figure 2.1). The Jehol Biota has more fossil vertebrates with 171 species reported vs. 40 species from the Yanliao Biota. In both Biotas, mammals, dinosaurs and pterosaurs are well-presented. However, birds, lampreys, frogs, turtles and choristoderes (semi-aquatic diapsid reptiles) are not known from the Yanliao Biota. In the Jehol Biota, crocodylomorphs have not been reported. The Jehol Biota has 15 fish species vs. two species from the Yanliao Biota. To date, only one salamander genus of *Liaoxitriton* is believed to be shared by both Biotas. Furthermore, both Biotas share only a few insect genera, and none at the species level [2].

Carnivorous vertebrates might have been positioned at the top of the food chain; many vertebrates and insects fed on insects, others ate plant material, some even played the roles of scavengers, recyclers and cleaners. Some insects, with long siphonate mouthparts, sipped pollination drops from gymnosperm plants. On the other hand, some blood-sucking insects, with unique saw-teeth serrated mouthparts, drew blood from birds, dinosaurs, pterosaurs or mammals. Basal mammals, most fairly small in size, might have sought safety by hiding in abundant bushes or plants. But medium-sized mammals were reported to have preyed on small dinosaurs. For all creatures in the ecosystems, they strived to successfully carry out three main functions of life – to feed (Chapter 28), to avoid being eaten (Chapter 29) and to pass on their genes (Chapter 30). As Dickens wrote in *A Tale of Two Cities*: "It was the best of times, it was the worst of times."

2.2 Representative Fossils of Coexisting Animals

Dinosaur – *Sinosauropteryx prima* Ji & Ji, 1996 (Figure 2.2)

Sinosauropteryx prima, the first dinosaur covered with dense primitive plumage (protofeathers), was described by Ji and Ji in 1996 [3]. As indicated by the generic name, *S. prima* was first considered as a bird. However, in 1998, Chen *et al.* published a paper [4] and classified this as a theropod dinosaur in the order Saurischia and family Compsognathidae, which bears a close relationship with *Compsognathus*, the small dinosaurs found in the same Solenhofen site as the *Archaeopteryx*. After further studies, Currie and Chen [5] reported that *Sinosauropteryx* is important not only because of its integument, but also because it is

Rhythms of Insect Evolution: Evidence from the Jurassic and Cretaceous in Northern China, First Edition.
Edited by Dong Ren, Chungkun Shih, Taiping Gao, Yongjie Wang, and Yunzhi Yao.

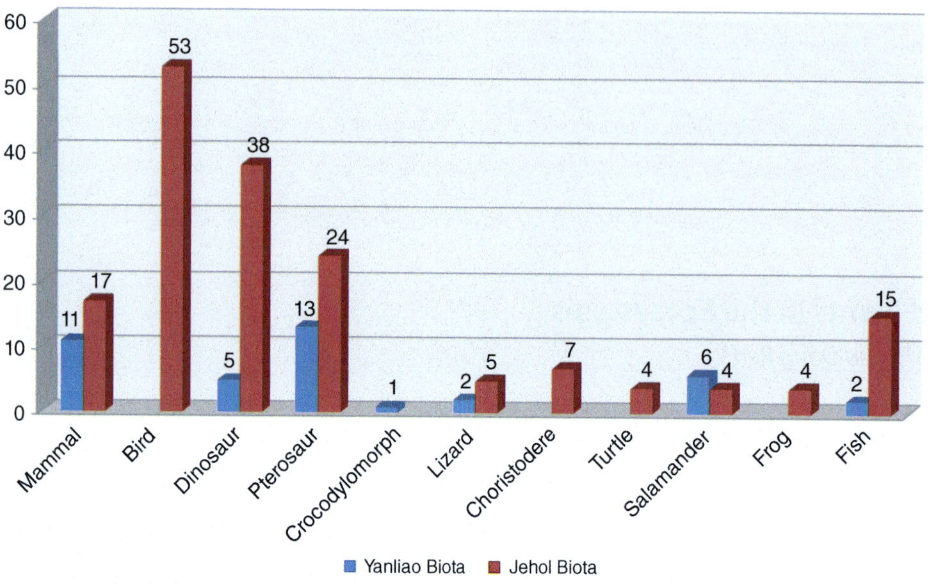

Figure 2.1 Comparison of vertebrate diversity of the Yanliao and Jehol Biotas, showing numbers of species for major vertebrate groups in both Biotas. Source: Modified from [1].

Figure 2.2 *Sinosauropteryx prima* Ji & Ji, 1996. Source: Photo provided by Dr. Shu'an Ji.

a basal coelurosaur and represents an important stage in theropod evolution that is poorly understood.

The largest discovered specimen has a body length up to 1.07 m with an estimated weight of 0.55 kg. The length of filament is about 13 mm on the head, 35 mm on the shoulder, and reach their maximum length midway down the tail at 40 mm. By examining melanosome structure and distribution, Zhang *et al.* [6] confirm the presence of light and dark bands of colors in the tail feathers of *Sinosauropteryx*. Furthermore, based on the presence of phaeomelanosomes, spherical melanosomes that make and store red pigment, they concluded that the darker feathers of *Sinosauropteryx* were chestnut or reddish brown in color. These protofeathers might have served for courtship display or to protect the skin and to keep warm – a suggestion that dinosaurs might have been warm-blooded. Findings later on show a specimen having unlaid eggs and some internal organs, another specimen containing the remains of a lizard in the gut region and another one with three mammal jawbones in the gut region. Hurum *et al.* [7] identified two of these jaws as belonging to *Zhangheotherium* and the third to *Sinobaatar*. Feathered dinosaur fossils provided the much needed evidence as a missing link to support the hypothesis that birds were evolved from dinosaurs. They suggested the view that feathers were first developed for warmth or display rather than flight.

Dinosaur – *Microraptor gui* Xu, Zhou & Wang, 2003 (Figure 2.3)

Xu *et al.* [8] described the first feathered *Microraptor*, *Microraptor zhaoianus* Xu, Zhou & Wang, 2000, as the first mature non-avian dinosaur to be found that is smaller than Archaeopteryx. The specific name is in honor of Prof. Xijin Zhao, a distinguished dinosaurologist who introduced Dr. Xu to the field of vertebrate paleontology. Three species have been named (*M. zhaoianus*, *M. gui*, and *M. hanqingi*), though further study has suggested that all of them represent variation

Figure 2.3 *Microraptor gui* Xu, Zhou & Wang, 2003. Source: Photo provided by Dr. Xing Xu.

in a single species, which is properly called *M. zhaoianus.* Alexander *et al.* [9] indicated that there were over 300 undescribed specimens attributable to *Microraptor* or its close relatives among the collections of several Chinese museums.

Microraptor gui, about 77 cm long, has feathers on its forelimbs, hind limbs and tail [10]. The specific name is in honor of Prof. Zhiwei Gu for his outstanding contribution to paleontology in China. This is the first dinosaur found to have feathers on its hind limbs. The feathers have asymmetrical vanes, a feature associated with bird flight. It is suggested that the forelimb and leg feathers made a perfect aerofoil to help this creature to glide between trees. The long rod-like tail with feathers might have provided the needed stability in its short glides and also served as a rudder to help steering the gliding direction. The structure of its feet is adapted for tree-climbing. This dinosaur with feathers on four limbs has been viewed as evidence to support the theory of evolution from dinosaur to bird and that gliding from tree down is a key step in the evolution toward flapping flights of fully-fledged birds. Feathers on hind limbs imply that the evolution of bird flights might have originated from gliding down from trees [10].

By analyzing the fossilized melanosomes, Li *et al.* [11] determined the plumage coloration of *Microraptor* is in a manner consistent with black, glossy coloration in modern birds.

Dinosaur – *Dilong paradoxus* Xu, Norell & Kuang, 2004 (Figure 2.4)

Dilong paradoxus, a small tyrannosauroid dinosaur, is one of the earliest and most primitive known tyrannosauroids and has a covering of "proto-feathers" in fossilized skin impressions from near the jaws and tail. The type specimen of *Dilong* (meaning emperor dragon in Chinese) is about 1.6 m in length, but it is thought to be a juvenile and might have been over 2 m long when fully grown. Xu *et al.* [12] speculated that the tyrannosauroids might have different skin coverings on different parts of their bodies – perhaps mixing scales and feathers. They also speculated that protofeathers might have been associated with juveniles for body warmth due to their small body size. When they grow larger, they might have shed the feathers and expressed only scales because they do not need feathers for insulation to stay warm, Turner *et al.*, in 2007, reanalyzed the relationships of coelurosaurian dinosaurs, including *Dilong*, and placed *Dilong* two steps above the tyrannosauroids in their phylogeny; more advanced than *Coelurus*, but more primitive than the Compsognathidae [13, 14]. However, other studies continued to find *Dilong* as a tyrannosauroid, and Carr and Williamson [15] found *Dilong* to fall within Tyrannosauroidea, not among the more advanced coelurosaurs.

Figure 2.4 *Dilong paradoxus* Xu, Norell & Kuang, 2004. Source: Photo provided by Dr. Xing Xu.

Dinosaur – *Mei long* Xu & Norell, 2004 (Figure 2.5)

Mei long, a duck-sized dinosaur in the Troodontidae and one of the most bird-like theropods, was described from the Early Cretaceous Jehol Biota. "*Mei*" sounds like a Chinese character meaning "soundly sleeping" and "*long*" is also a Chinese for "dragon." So far, *Mei* is the shortest generic name of any dinosaur [16].

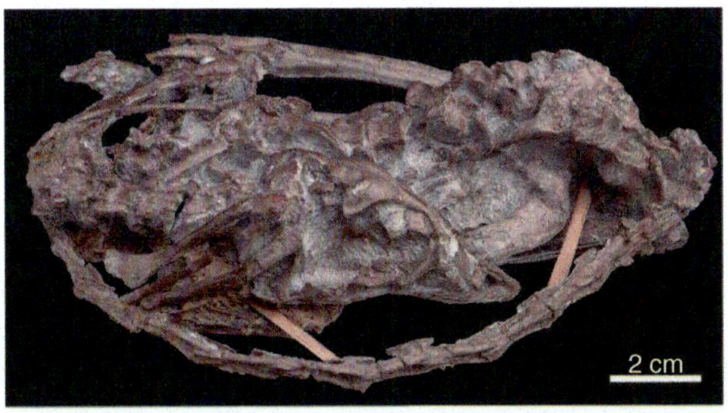

Figure 2.5 *Mei long* Xu & Norell, 2004 [16]. Source: Photo provided by Dr. Xing Xu.

The holotype specimen, well-preserved in three-dimensional detail, curls up with its head tucked under a forelimb, similar to the resting and sleeping position of modern birds. Besides morphological similarities, this posture provides a behavioral link between birds and dinosaurs. *Mei long* also has very large nostrils which represent a character of Troodontidae. The posture of holotype and chemical analysis of the fossil matrix suggest the living dinosaur was probably killed instantly by poisonous gases and then buried in volcanic ash. Some characters of *Mei long* support the theory that dinosaurs were warm-blooded and that small size was a prerequisite for flight. It might have been a carnivore with a body length of about 53 cm [16].

Dinosaur – *Anchiornis huxleyi* Xu, Zhao & Norell, 2009 (Figure 2.6)

Anchiornis huxleyi is a small, feathered, four-winged paravian dinosaur from the Tiaojishan Formation of the Upper Jurassic (Oxfordian age), 160 Mya. The first fossil was reported from the Yaolugou, Jianchang County, Liaoning, while the second (Figure 2.6), at the Daxishan of the same area. The generic name *Anchiornis* means "near bird," and the specific epithet of *huxleyi* is in honor of Thomas Henry Huxley who first proposed a close evolutionary relationship between birds and dinosaurs. This finding filled a gap in the transition between the body plans of avian birds and non-avian dinosaurs [17].

Anchiornis has a triangular skull, a feathered crest on the head, unusually long forelimbs with feathers, and very long hind legs with long flight/gliding feathers with symmetric vanes, similar to *Microraptor gui*, with exception of asymmetric vanes. Both the fore- and hind wings of *A. huxleyi* are shorter than those of *Microraptor*. The hind wing has 12–13 flight feathers anchored to the tibia (lower leg) and 10–11 to the metatarsus (upper foot). The feet (except for the claws) are completely covered with short feathers [19].

By comparing structures of melanosomes on well-preserved feathers of the second specimen with those that determine the color of feathers on living birds, Li *et al.* [20] reported that the colors of the dinosaur's feathers were mostly gray but the dinosaur was covered from head to toe with vivid plumage. The crest was a light brown color and both the fore- and hind wings were brilliant white while each white feather had a black tip. It is likely that *Anchiornis* exhibited such a vivid coloration for species identification, courtship display and attraction, or warning against potential predators. However, in 2015, Lindgren *et al.* [21] studied the third *Anchiornis* fossil at the Yizhou Fossil and Geology Park with similar procedures and found out that only gray-black type melanosomes were found even on the crest, which is different from the feather colors of the 2010 results.

Figure 2.6 *Anchiornis huxleyi* Xu, Zhao & Norell, 2009 [17]. Source: Photo provided by Dr. Xing Xu.

Pterosaur – *Jeholopterus ningchengensis* Wang, Zhou & Zhang, 2002 (Figure 2.7)

Jeholopterus ningchengensis, an adult or subadult rhamphorhynchoid pterosaur, is from Daohugou, Ningcheng, Inner Mongolia. With a wingspan of about 90 cm and a short neck and a short tail, it might have been a good flier. The skull is wider than long, having a similar shape like a frog skull. Hence, it is classified in the family of Anurognathidae (frog-jawed pterosaurs). This pterosaur has its wing membrane attached to the ankle of the hind limb. This specimen has well-preserved fibers on the wing membrane and hair-like structure on the neck, body and tail, a unique and rare finding on pterosaurs. The "hairs", covering the entire body from the neck to the tail, suggest that this pterosaur may be warm-blooded [22]. In 2009, Kellner *et al.* [23] reported the presence of three layers of fibers in the wing, allowing the animal to precisely adapt the wing profile. They also suggested Batrachognathinae for the clade comprising *Jeholopterus*, *Batrachognathus* and *Dendrorhynchoides*. With webbed pedal digits, *J. ningchengensis* might have lived near water and fed on fish or insects [22].

Pterosaur – Egg and Embryo (Figure 2.8)

Wang and Zhou in 2004 reported an embryo pterosaur fossil in an egg from the Yixian Formation of Jianshangou, Yixian, Liaoning [18]. This was the first record and proof

Figure 2.7 *Jeholopterus ningchengensis* Wang, Zhou & Zhang, 2002 [22]. Source: Photo provided by Dr. Shu'an Ji.

Figure 2.8 Pterosaur egg and embryo [18]. Source: Photo provided by Dr. Shu'an Ji.

that pterosaurs laid eggs. Ji *et al.* (2004) [24] and Chiappe *et al.* (2004) [25] also reported pterosaur eggs soon after. Lü *et al.* (2011) [26] and Wang *et al.* (2015) [27] reported adult pterosaurs having preserved eggs and embryos.

Bird – *Confuciusornis sanctus* Hou, Zhou & Gu, 1995 (Figure 2.9)

Fossil bird specimens were found in 1994 in Jianshangou and Huangbanjigou of Shangyuan near Beipiao City. In a 1995 *Nature* article, Hou *et al.* named one as *Confuciusornis sanctus*, classified it as Sauriurae – an ancient bird in the order of Confuciusornithiformes and the family of Confuciusornithidae. It is about 125 Mya, one of the most primitive birds except for the *Archaeopteryx* from the Upper Jurassic Solnhofen in Germany [28, 29].

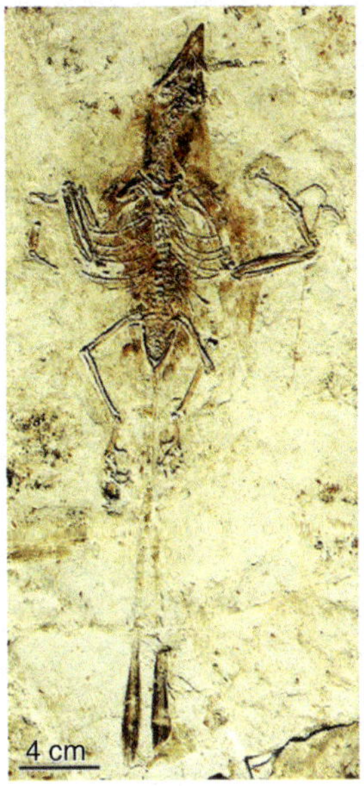

Figure 2.9 *Confuciusornis sanctus* Hou, Zhou & Gu, 1995 [28].

Besides *Confuciusornis sanctus*, three other species have been reported: *Confuciusornis chuonzhous* [30], *Confuciusornis suniae* [30] and *Confuciusornis dui* [31]. These birds have a fully developed horny beak without teeth; robust skull bones but not fused together and postorbital bone retained. Distinctive air pockets on the humerus highlight weight reduction similar to modern birds. Compared to modern birds the tail bones are slightly longer and the forelimb has three long fingers with strong claws, probably used for tree dwelling and climbing. Since then, many specimens were found in Sihetun, Shangyuan near Beipiao, some with males and females on the same fossil matrix. The male bird has a pair of long tail feathers about 20 cm.

Bird – *Dingavis longimaxilla* O'Connor, Wang & Hu, 2016 (Figure 2.10)

O'Connor *et al.* [32] described an ornithuromorph with an elongate rostrum from the Sihedang locality of the Lower Cretaceous Yixian Formation. Like the enantiornithines in the Longipterygidae, rostral elongation in *Dingavis longimaxilla* is achieved primarily through the maxilla, whereas neornithines elongate the premaxilla and rostralization is far more extreme than observed in early birds. Notably, in the

Figure 2.10 *Dingavis longimaxilla*, holotype, IVPP V20284, full slab photograph [32]. Source: Photo provided by Dr. Jingmai K. O'Connor.

rostrum of *Xinghaiornis lini* Wang, Chiappe, Teng & Ji, 2013 [33], the most "longirostrine" Early Cretaceous ornithuromorph, the premaxilla and maxilla contribute to the rostrum equally. These lineages together highlight the diversity of configurations in which early birds experimented with rostralization of the skull. Many Early Cretaceous birds have elongate maxillae, but this morphology was abandoned in more derived taxa and suggests that in Aves this skull configuration provided less structural stability [32].

Mammal – *Jeholodens jenkinsi* Ji, Luo & Ji, 1999 (Figure 2.11)

Jeholodens jenkinsi was described from the Lower Cretaceous Yixian Formation, in Sihetun, Shangyuan near Beipiao, Liaoning. Its specific epithet is in honor of Dr. Farish A. Jenkins Jr., a renowned paleomammologist. It is a primitive mammal having molars with three main cusps in a straight alignment, thus classified in the order of Triconodonta. Its dental formula is four spoon-shaped incisors, one canine tooth, two pre-molar teeth, three molar teeth on the upper jaw and four molar teeth on the lower jaw. *J. jenkinsi* has an advanced pectoral girdle – shoulder blade (scapula) and collarbone (clavicle), indicating that its forelimb was capable of an almost erect gait. It walked like those of derived therians.

Figure 2.11 *Jeholodens jenkinsi* Ji, Luo & Ji, 1999 [34]. Source: Photo provided by Dr. Shu'an Ji.

However, *J. jenkinsi* has a very primitive pelvis girdle, a sprawling hind limb, and a splayed hind foot, comparable to those of reptiles. Limb structures of *J. jenkinsi* suggest that it was probably a ground-dwelling animal in a terrestrial niche feeding on insects [34].

Mammal – *Juramaia sinensis* Luo, Yuan, Meng & Ji, 2011 (Figure 2.12)

Luo *et al.* reported *Juramaia sinensis* ("Jurassic mother from China"), a basal eutherian from the Daxigou

Figure 2.12 Holotype specimen of *Juramaia sinensis* Luo, Yuan, Meng & Ji, 2011, Beijing Museum of Natural History (BMNH) PM1143 [35]. Source: Photo provided by Dr. Ying Wang.

site of Jianchang County of Liaoning Province in the Upper Jurassic Tiaojishan Formation of China, 160 Mya. It is a shrew-like mammal with a body length of about 70–100 mm and body weight of about 15–17 g. This finding extends the first appearance of the eutherian–placental clade by about 35 Mya from the previous record, reducing and resolving a discrepancy between the previous fossil record and the molecular estimate for the placental–marsupial divergence [35].

This mammal fossil has a nearly complete skeleton (Figure 2.12), including an incomplete skull preserved with complete and detailed tooth structure of an insectivore, forelimb with climbing adaptation features, body covered by hairs and other anatomical features of eutherians [35].

Mammal – *Rugosodon eurasiaticus* Yuan, Ji, Meng & Tabrum, 2013 (Figure 2.13)

Yuan *et al.* [36] documented a rodent-like mammal from the Upper Jurassic Tiaojishan Formation (160 Mya) of Liaoning Province, China. The species is the oldest so far described in the Paulchoffatiidae, the basalmost family of the order Multituberculata.

The mammal is estimated to have a body weight of about 65–80 g, about that of an average chipmunk. This mammal has a wrinkled surface on its teeth (the generic name *Rugosodon* is Latin for "wrinkly tooth") which indicates that the animal was an omnivore, well-adapted to biting and feeding on both plants (fruits or seeds) and small animals (worms, insects or vertebrates). It has highly mobile ankle joints, highly mobile digits and a highly flexible spine, crucial for later the evolutionary success of multituberculates in the Cretaceous and Paleogene [36].

Figure 2.13 The Late Jurassic mammal *Rugosodon eurasiaticus* Yuan, Ji, Meng & Tabrum, 2013 [36]. Source: Photo provided by Dr. Ying Wang.

Reptile – *Monjurosuchus splendens* Endo, 1940 (Figure 2.14)

Monjurosuchus splendens is a medium-sized choristoderan reptile found in Niuyingzi of the Lingyuan City of Liaoning Province in the Yixian Formation. It is in the Subclass of Diapsida and the order Choristodera [37]. The length from snout to hip is 30 cm. The skull is flat and the orbit large. The frontal is very narrow. The teeth on the pterygoid are of various types – about 50 small conical teeth. Both fore and hind feet are webbed. Vertebral column consists of eight cervicals, 16 dorsals, three sacrals (hip bones), and 55 caudals. It has overlapping and imbricating scales covering its body. It might have lived in water, feeding on fish, amphibians, and invertebrates. The holotype specimen was lost during World War II, but Gao *et al.* described a newly discovered neotype with well-preserved soft tissue in 2000 [38]. Gao *et al.* in 2005 documented another Early Cretaceous Chinese choristodere in the genus *Philydrosaurus*, which was classified in the Monjurosuchidae with *Monjurosuchus* [39].

Spider – *Mongolarachne jurassica* Selden Shih & Ren, 2011 (Figure 2.15)

A pair of remarkable female and male spider fossils have been reported from the Middle Jurassic (~165 Mya) of Daohugou, Northeastern China. They are the largest known fossil spiders documented hitherto – the female with body length of 25 mm and first leg, 56.5 mm and the

Figure 2.14 *Monjurosuchus splendens* Endo, 1940. Source: Photo provided by Dr. Shu'an Ji.

Figure 2.15 *Mongolarachne jurassica* Selden, Shih & Ren, 2011 comb. nov., the largest known fossil spiders. Left: male, right: female [40]. Source: Female holotype donated and photo provided by Dr. Chungkun Shih.

male with body length of 16.5 mm and first leg, 58.2 mm. (Figure 2.15). The large female spider was originally named as *Nephila jurassica* in 2011, and placed in the modern family Nephilidae, on the basis of many morphological similarities but, as with many ancient fossils, the single specimen lacked synapomorphies of the family [41].

With the discovery and study of the giant male spider in 2013, the two sexes are considered conspecific based on their similar morphological features, size, and provenance. The male cannot be accommodated in the Nephilidae because of its pedipalp morphology. Therefore, the genus *Mongolarachne* and family Mongolarachnidae are introduced for the species. Comparison with possibly related families show that Mongolarachnidae are most likely on the orbicularian stem, close to other cribellate orbicularians (e.g. Deinopoidea), which suggests a greater diversity of cribellate orbicularians during the Middle Jurassic [40].

Harvestman – *Mesobunus dunlopi* Giribet, Tourinho. Shih & Ren, 2012 (Figure 2.16)

Opiliones ("harvestmen" or "daddy-long-legs") are the third largest arachnid order, after Acari (mites and ticks) and Araneae (spiders), with c. 6500 described

Figure 2.16 *Mesobunus dunlopi* Giribet, Tourinho. Shih & Ren, 2012 [42]. Source: Donated and photo provided by Dr. Chungkun Shih.

species and probably including up to 10 000 living species [43]. The oldest fossil record has been reported from the Early Devonian (c. 410 Mya) [44, 45].

Mesobunus dunlopi. From the Middle Jurassic (approx. 165 Mya) of Daohugou, Inner Mongolia, China, is assigned to the family of Sclerosomatidae which constitute the largest family of the arachnid order Opiliones, and one of the two families commonly found in the temperate regions of the northern Hemisphere. The specific epithet is in honor of Dr. Jason A. Dunlop, who has contributed enormously to Opiliones palaeontology, among many other arachnid groups [42].

Harvestmen have a sparse fossil record in the Mesozoic, with only three species known from the Jurassic. The other two are *Mesobunus martensi* Huang, Selden & Jason 2009 and *Daohugopilio sheari* Huang, Selden & Jason 2009 [46]. The *M. dunlopi* with the preservation quality and details of the penis and pedipalps allows us to place it in the extant sclerosomatid subfamilies Gagrellinae or Leiobuninae. The first recognizable fossil in this subfamily highlights morphological stasis over c. 165 Mya and the finding of this species along with lacustrine insects suggests a life mode similar to that of some modern sclerosomatids, and a possible connection between morphological and ecological stasis [46].

2.3. Representative Fossils of Coexisting Plants

Abundant and diverse plant fossils have been found from the Middle Jurassic to the Early Cretaceous in

Northeastern China. Examples are Bryophytes (mosses and liverworts), Lycopods, Equisetales, Filicales (ferns), Pteridospermae (seed ferns), Bennettitales (cycads and cycadeoids), Czekanowskiales, Ginkgoales (ginkgo), Coniferophytes (conifers, including pine family, cypress family, bald cypress family, and podocarp family), Gnetales, and Angiospermae, etc. Evidently, this area had lively and active eco-systems with lush plants of many different varieties. A book entitled *Early Angiosperms and Their Associated Plants from the Western Liaoning, China* by Sun *et al.*, gives a good account of plant fossils in the Yixian Formation [47]. We conducted a preliminary survey of diverse fossil plants and have provided some representative fossil plants of Northeastern China for illustration.

Angiospermae

Recent findings indicate that the first angiosperms (vascular flowering plants with seeds enclosed in an ovary) appeared in the eco-system during the Early Cretaceous (about 125 Mya). Two fossil plants from the Yixian Formation, called *Archaefructus liaoningensis* Sun, Dilcher, Zheng, & Zhou, 1998 [48] (Figure 2.17) and *Archaefructus sinensis* Sun, Ji, Dilcher & Nixon, 2002 [50] (Figure 2.18), are now widely accepted as two of the oldest angiosperm flowering plants. Leng and Friis [51] described an angiosperm with advanced features from the same locality, *Sinocarpus decussatesy.* Subsequently, Ji *et al.* [52] described a complete fossil plant specimen as *Archaefructus eoflora*, and claimed the reproductive organs at the tip of slender stalks represent bisexual flowers.

Figure 2.18 *Archaefructus sinensis* Sun, Ji, Dilcher & Nixon, 2002 [50]. Source: Photo provided by Dr. Chungkun Shih.

Figure 2.17 *Archaefructus liaoningensis* additional material [49]. Source: Donated and photo provided by Dr. Chungkun Shih.

Archaefructus liaoningensis Sun, Dilcher, Zheng & Zhou, 1998 (Figure 2.17)

Sun *et al.* described *Archaefructus liaoningensis* in 1998, on the cover of *Science*, which is recognized and accepted as one of the earliest angiosperm plants. *Archae* means ancient, *fructus* means fruit, and *liaoning* is named after the Liaoning Province. This finding pushed the date of the first angiosperm appearance earlier by about 15 million years [47, 48].

Wang and Zheng [49], in 2012, provided new observations that *Archaefructus* has ovules/seeds attached to the midrib on the abaxial side of the fruits and a whorled/opposite arrangement for the fruits on the axis and *Archaefructus sinensis* demonstrates that fruit pairs are inserted on the infructescence axis oppositely.

Archaefructus sinensis Sun, Ji, Dilcher & Nixon, 2002 (Figure 2.18)

Sun *et al.* described and published a second paper on the angiosperm plant, *Archaefructus sinensis*, in 2002 [50]. This well-preserved specimen, found in the Lingyuan City of Liaoning, consists of the entire plant with seed pods, seeds, stamen pairs, shoots, stems, leaves and roots. The "flower" of *Archaefructus* is a unique collection of female and male reproductive organs. The carpels, at the terminal ends of shoots, mature last, after the pollen has been dispersed and the anthers are lost on the same axis. The immature carpels are clustered close together and then become spaced out as the axis elongates and as they mature, most of which are arranged helically. Carpels mature into elongate follicles containing eight to 12 seeds. Petals, sepals or bracts are absent. The stamens are in pairs and remain attached to the stalks only while the carpels are young. As the carpels mature, the stamens abscise, leaving the short stalks that remain on the mature shoots. The *Archaefructaceae* probably are aquatic plants. The herbaceous nature of the plants is obvious by the thin stems that extend for some distance, which would require water for support [50].

Bennettitales

Bennettitales are in the order Cycadophyta, commonly called cycads and cycadeoids. These gymnosperm plants are usually short, evergreen, and woody. Large feather-like leathery compound leaves grow at the top of the stem. Young leaves are shaped like coiled tight fists. Male or female reproductive organs of cycads are on cone-like structures or on cones at the top of the stem. Usually, pollen is from one plant and the seed cone is grown on the other plant. Cycadeoids are related to cycads. They have flower-like reproductive organs. Cycadophyta started to appear in the Late Carboniferous and became more prolific in the Permian. They reached their peak in the Late Triassic to Early Cretaceous. Only 10 genera exist in today's world, with Cycads as a notable example. They are distributed in tropical and semi-tropical regions. Cycads and Cycadeoids are commonly found in the Northeastern Chinese fossils. A fossil of *Williamsonia* sp. is shown in Figure 2.19.

Czekanowskiales and Ginkgoales

Czekanowskiales (Figure 2.20) and Ginkgoales (Figure 2.21) are two orders of Ginkgophyta. There are separate male and female trees. Male reproductive organs are like "flower bundles"; those of the female are at the end of branches, having two ovules, but only one will grow into seed. Ginkgophyta started to appear in the Permian and reached their peak in the Jurassic to Early Cretaceous. Ginkgo plants occupied habitats all around the world and most lived in tropical or semi-tropical regions. They

Figure 2.19 *Williamsonia* sp. Source: Photo provided by Dr. Chungkun Shih.

Figure 2.20 Czekanowskiales, *Solenites* sp. Source: Photo provided by Dr. Chungkun Shih.

Figure 2.21 *Ginkgoites* sp. leaf and fruit. Source: Photo provided by Dr. Chungkun Shih.

were a key source for coal formation at that time. In the Early Cretaceous, they started to decline. Today, only one genus with one species (*Ginkgo biloba*) exists in a temperate region. When it was first found in a temple garden in China, it was called a living fossil plant. The leaves of today's plants are simple, fan-shaped and with veins that are also fan-shaped. However, ancient ginkgo leaves are glove-like with 4–7 notches. Young leaves of today's gingko plants still show notches which are the evolutionary remnant from a bygone era. Ginkgo leaves and seeds are used in Chinese herbal medicine. Czekanowskiales, the other order, have thin-thread or tongue-like leaves bundled at one end and attached to a short branch. They were found in the Triassic to the Early Cretaceous. Specimens of Ginkgoales and Czekanowskiales are discovered in small numbers in Northeastern Chinese fossils.

Figure 2.22 Conifer leaves and seeds. Source: Photo provided by Dr. Chungkun Shih.

Coniferophyta

Coniferophyta comprise Cordaitopsida and Coniferopsida. Cordaites were tall, slender trees with branches and leaves on the crown. The leaves were wide and strap-like. Cordaites plants were very abundant in the Late Carboniferous, and flourished during the Permian. However, after the Triassic, Cordaitopsida became extinct. Coniferopsida plants include conifers such as pine, cypress, bald cypress, and podocarp. Coniferopsida plants are considered advanced gymnosperm plants with many branches and small simple leaves. Most of them are tall, evergreen trees with woody stems; only a few are bushes. Male and female reproductive organs for most plants are on the same trees, with a few exceptions that have separate male and female trees. Seeds are grown on seed cones varying from species to species. Coniferopsida plants first appeared in the Early Carboniferous and spread all over the northern hemisphere in the Early Permian. In the Mesozoic and Cenozoic, they actually were the most abundant gymnosperm plants (Figures 2.22 and 2.23). Today, they are still distributed over a wide range of habitats around the world, forming evergreen needle-leaf forests in plains and mountains of different altitudes. Majestic redwood plants in forests of the Northwestern US are the oldest and tallest trees in the world. They are testimony that conifer trees have excellent durability, adaptation, and strength to live, growing and prospering over thousands of years. Specimens of conifer plants are very abundant in Northeastern Chinese fossils.

Figure 2.23 Conifer seeds. Source: Photo provided by Dr. Chungkun Shih.

Gnetales

Gnetales, an order in Gnetophyta, have stems and branches with nodes and segments. Branches and leaves are grown on the nodes which are slightly expanded. Branches are on the opposite side or in whorled shape from stem nodes at an angle between 30° and 60°. Leaves are at the base of nodes, usually two on the opposite sides. Female "flower-like" organs are located at the tip of branches. The seeds are long and oval-shaped

Figure 2.24 Seeds of *Ephedrites* sp. Source: Photo provided by Dr. Chungkun Shih.

(Figure 2.24). They grew on dry land in a tropical climate. Today's Ephedra are used in Chinese medicine. Specimens of Gnetales are seldom found in the Northeastern Chinese fossils.

References

1 Zhou, Z.H. and Wang, Y. (2017). Vertebrate assemblages of the Jurassic Yanliao Biota and the Early Cretaceous Jehol Biota: comparisons and implications. *Palaeoworld* **26**: 241–252. https://doi.org/10.1016/j.palwor.2017.01.002.

2 Huang, D.-Y. (2015). Yanliao Biota and the Yanshan movement. *Acta Palaeontologica Sinica* **54** (4): 501–546.

3 Ji, Q. and Ji, S.A. (1996). On discovery of the earliest bird fossil in China and the origin of birds. *Chinese Geology* **10** (233): 30–33.

4 Chen, P.J., Dong, Z.M., and Zhen, S.N. (1998). An exceptionally well-preserved theropod dinosaur from the Yixian Formation of China. *Nature* **391** (8): 147–152.

5 Currie, P.J. and Chen, P.J. (2001). Anatomy of *Sinosauropteryx prima* from Liaoning, Northeastern China. *Canadian Journal of Earth Sciences* **38** (1): 705–727.

6 Zhang, F., Kearns, S.L., Orr, P.J. *et al.* (2010). Fossilized melanosomes and the colour of Cretaceous dinosaurs and birds. *Nature* **463** (7284): 1075–1078. https://doi.org/10.1038/nature08740.

7 Hurum, J.H., Luo, Z.X., and Kielan-Jaworowska, Z. (2006). Were mammals originally venomous? *Acta Palaeontologica Polonica* **51** (1): 1–11.

8 Xu, X., Zhou, Z.H., and Wang, X.L. (2000). The smallest known non-avian theropod dinosaur. *Nature* **408**: 705–708.

9 Alexander, D.E., Gong, E., Martin, L.D. *et al.* (2010). Model tests of gliding with different hindwing configurations in the four-winged dromaeosaurid *Microraptor gui*. *Proceedings of the National Academy of Sciences USA* **107**: 2972–2976. https://doi.org/10.1073/pnas.0911852107.

10 Xu, X., Zhou, Z.H., Wang, X.L. *et al.* (2003). Four-winged dinosaurs from China. *Nature* **421** (6921): 335–340.

11 Li, Q.G., Gao, K.Q., Meng, Q.J. *et al.* (2012). Reconstruction of *Microraptor* and the evolution of iridescent plumage. *Science* **335** (6073): 1215–1219. https://doi.org/10.1126/science.1213780.

12 Xu, X., Norell, M.A., Kuang, X.W. *et al.* (2004). Basal tyrannosauroids from China and evidence for protofeathers in tyrannosauroids. *Nature* **431**: 680–684.

13 Turner, A.H., Pol, D., Clarke, J.A. *et al.* (2007). A basal dromaeosaurid and size evolution preceding avian flight. *Science* **317**: 1378–1381.

14 Turner, A.H., Pol, D., Clarke, J.A. *et al.* (2007). Supporting online material for: a basal dromaeosaurid and size evolution preceding avian flight. *Science* **317**: 1378–1381.

15 Carr, T.D. and Williamson, T.E. (2010). *Bistahieversor sealeyi*, gen. et sp. nov., a new tyrannosauroid from New Mexico and the origin of deep snouts in Tyrannosauroidea. *Journal of Vertebrate Paleontology* **30** (1): 1–16. https://doi.org/10.1080/02724630903413032.

16 Xu, X. and Norell, M. (2004). A new troodontid dinosaur from China with avian-like sleeping posture. *Nature* **431**: 838–841.

17 Xu, X., Zhao, Q., Norell, M.A. *et al.* (2009). A new feathered maniraptoran dinosaur fossil that fills a morphological gap in avian origin. *Chinese Science Bulletin* **54**: 430–435.

18 Wang, X.L. and Zhou, Z.H. (2004). Palaeontology: pterosaur embryo from the Early Cretaceous. *Nature* **429**: 621.

19 Hu, D.Y., Hou, L.H., Zhang, L.J., and Xu, X. (2009). A pre-Archaeopteryx troodontid theropod from China with long feathers on the metatarsus. *Nature* **461**: 640–643.

20 Li, Q.G., Gao, K.Q., Vinther, J. *et al.* (2010). Plumage color patterns of an extinct dinosaur. *Science* **327** (5971): 1369–1372. https://doi.org/10.1126/science.1186290.

21 Lindgren, J., Sjövall, P., Carney, R.M. *et al.* (2015). Molecular composition and ultrastructure of Jurassic paravian feathers. *Scientific Reports* **5**: 13520. https://doi.org/10.1038/srep13520.

22 Wang, X.L., Zhou, Z.H., Zhang, F.C., and Xu, X. (2002). A nearly completely articulated rhamphorhynchoid pterosaur with exceptionally well-preserved wing membranes and 'hairs' from Inner Mongolia, Northeast China. *Chinese Science Bulletin* **47** (3): 226–232.

23 Kellner, A.W.A., Wang, X.L., Tischlinger, H. *et al.* (2009). The soft tissue of *Jeholopterus* (Pterosauria, Anurognathidae, Batrachognathinae) and the structure of the pterosaur wing membrane. *Proceedings of Royal Society B* **277**: 321–329. https://doi.org/10.1098/rspb.2009.0846.

24 Ji, Q., Ji, S.A., Cheng, Y.N. *et al.* (2004). Palaeontology: pterosaur egg with a leathery shell. *Nature* **432**: 572.

25 Chiappe, L.M., Codorniú, L., Grellet-Tinner, G., and Rivarola, D. (2004). Argentinian unhatched pterosaur fossil. *Nature* **432**: 571.

26 Lü, J.C., Unwin, D.M., Deeming, D.C. *et al.* (2011). An egg-adult association, gender, and reproduction in pterosaurs. *Science* **331** (6015): 321–324. https://doi.org/10.1126/science.1197323.

27 Wang, X.L., Kellner, A.W.A., Cheng, X. *et al.* (2015). Eggshell and histology provide insight on the life history of a pterosaur with two functional ovaries. *Annals of the Brazilian Academy of Sciences* **87** (30): 1599–1609. https://doi.org/10.1590/0001-3765201520150364.

28 Hou, L.H., Zhou, Z.H., Martin, L.D., and Feduccia, A. (1995). A beaked bird from the Jurassic of China. *Nature* **377**: 616–618.

29 Hou, L.H., Zhou, Z.H., Gu, Y.C., and Zhang, H. (1995). Description of *Confuciusornis sanctus*. *Chinese Science Bulletin* **10**: 61–63.

30 Hou, L.H. (1997). *Mesozoic Birds of China*. Natou. Taiwan: Taiwan Feng Huang Gu Bird Garden.

31 Hou, L.H., Martin, L.D., Zhou, Z.H. *et al.* (1999). A diapsid skull in a new species of the primitive bird *Confuciusornis*. *Nature* **399**: 679–682.

32 O'Connor, J.K., Wang, M., and Hu, H. (2016). A new ornithuromorph (Aves) with an elongate rostrum from the Jehol Biota, and the early evolution of rostralization in birds. *Journal of Systematic Palaeontology* **14** (11): 939–948. https://doi.org/10.1080/14772019.2015.1129518.

33 Wang, X.R., Chiappe, L.M., Teng, F.F., and Ji, Q. (2013). *Xinghaiornis lini* (Aves: Ornithothoraces) from the Early Cretaceous of Liaoning: an example of evolutionary mosaic in early birds. *Acta Geologica Sinica (English Edition)* **87** (3): 686–689. https://doi.org/10.1111/1755-6724.12080.

34 Ji, Q., Luo, Z.X., and Ji, S.A. (1999). A Chinese triconodont mammal and mosaic evolution of the mammalian skeleton. *Nature* **398**: 326–330.

35 Luo, Z.X., Yuan, C.X., Meng, Q.J., and Ji, Q. (2011). A Jurassic eutherian mammal and divergence of marsupials and placentals. *Nature* **476**: https://doi.org/10.1038/nature10291.

36 Yuan, C.X., Ji, Q., Meng, Q.J. *et al.* (2013). Earliest evolution of multituberculate mammals revealed by a new Jurassic fossil. *Science* **341** (6147): 779–783. https://doi.org/10.1126/science.1237970.

37 Endo, R. (1940). A new genus of Thecodontia from the Lycoptera beds in Manchoukou. *Bulletin of the National Central Museum of Manchoukou* **2**: 1–14.

38 Gao, K.Q., Evans, S., Ji, Q. *et al.* (2000). Exceptional fossil material of a semi-aquatic reptile from China: the resolution of an enigma. *Journal of Vertebrate Paleontology* **20** (3): 417–421. https://doi.org/10.1671/0272-4634(2000)020[0417:efmoas]2.0.co;2.

39 Gao, K.Q. and Fox, R.C. (2005). A new choristodere (Reptilia: Diapsida) from the Lower Cretaceous of western Liaoning Province, China, and phylogenetic relationships of Monjurosuchidae. *Zoological Journal of the Linnean Society* **145** (3): 427–444. https://doi.org/10.1111/j.1096-3642.2005.00191.

40 Selden, P.A., Shih, C.K., and Ren, D. (2013). A giant spider from the Jurassic of China reveals greater diversity of the orbicularian stem group. *Naturwissenschaften* **100** (12): 1171–1181. https://doi.org/10.1007/s00114-013-1121-7.

41 Selden, P.A., Shih, C.K., and Ren, D. (2011). A golden orb-weaver spider (Araneae: Nephilidae: Nephila) from the Middle Jurassic of China. *Biology Letters* **7** (5): 775–778. https://doi.org/10.1098/rsbl.2011.0228.

42 Giribet, G., Tourinho, A.L., Shih, C.K., and Ren, D. (2012). An exquisitely preserved harvestman (Arthropoda, Arachnida, Opiliones) from the Middle Jurassic of China. *Organisms Diversity and Evolution* **12** (1): 51–56. https://doi.org/10.1007/s13127-011-0067-x.

43 Pinto-da-Rocha, R., Machado, G., and Giribet, G. (eds.) (2007). *Harvestmen: The Biology of Opiliones*. Cambridge, MA: Harvard University Press.

44 Dunlop, J.A., Anderson, L.I., Kerp, H., and Hass, H. (2003). Preserved organs of Devonian harvestmen. *Nature* **425**: 916.

45 Dunlop, J.A., Anderson, L.I., Kerp, H., and Hass, H. (2003). A harvestman (Arachnida: Opiliones) from the early Devonian Rhynie cherts, Aberdeenshire, Scotland. *Transactions of the Royal Society of Edinburgh. Earth Sciences* **94**: 341–354.

46 Huang, D.-Y., Selden, P.A., and Dunlop, J.A. (2009). Harvestmen (Arachnida: Opiliones) from the Middle Jurassic of China. *Naturwissenchaften* **96**: 955–962.

47 Sun, G., Zheng, S.L., Dilcher, D.L. *et al.* (2001). *Early Angiosperms and their Associated Plants from Western Liaoning, China*. Shanghai: Shanghai Scientific

and Technological Education Publishing House. 227 pp. (in Chinese and English).

48 Sun, G., Dilcher, D.L., Zheng, S.L., and Zhou, Z.K. (1998). In search of the first flower: a Jurassic angiosperm, *Archaefructus*, from Northeast China. *Science* **282**: 1692–1695.

49 Wang, X. and Zheng, X.T. (2012). Reconsiderations on two characters of early angiosperm *Archaefructus*. *Palaeoworld* **21** (3–4): 193–201.

50 Sun, G., Ji, Q., Dilcher, D.L. *et al.* (2002). Archaefructaceae, a new basal angiosperm family. *Science* **296**: 899–904.

51 Leng, Q. and Friis, E.M. (2003). *Sinocarpus decussatus* gen. et sp. nov., a new angiosperm with basally *synocarpous* fruits from the Yixian Formation of Northeast China. *Plant Systematics and Evolution* **241** (1–2): 77–88.

52 Ji, Q., Li, H.Q., Bowe, L.M. *et al.* (2004). Early Cretaceous *Archaefructus eoflora* sp. nov. with bisexual flowers from Beipiao, Western Liaoning, China. *Acta Geologica Sinica* **78** (4): 883–896.

3

Insects – In the Spotlight

Taiping Gao[1], *Chungkun Shih*[1,2], *and Dong Ren*[1]

[1] *Capital Normal University, Haidian District, Beijing, China*
[2] *National Museum of Natural History, Smithsonian Institution, Washington, DC, USA*

3.1 Introduction to Insects

Insects, the largest group of hexapod invertebrates within the Arthropoda, comprise winged groups of beetles, flies, butterflies, moths, wasps, dragonflies, mayflies, lacewings, etc., and some wingless groups of fleas, lice, silverfish, bristletails, diplurans, etc. [1]. Insects, as the most speciose and successful terrestrial animals, have adapted to broadly diverse environments [2]. In classification, Class Insecta Linnaeus, 1758 are assigned to Animalia (Kingdom), Arthropoda (Phylum), and Hexapoda (Subphylum). Insecta hitherto comprise 33 Orders of extant insects according to Gullan and Cranston in 2014 [1].

Rasnitsyn and Quicke summarized many extinct groups (Orders) of insects, described based on fossils [3]. In 2011, based on two adults and nearly 40 larvae, Staniczek *et al.* erected a new fossil insect Order, Coxoplectoptera, within the Palaeoptera from the Lower Cretaceous Crato Formation of Brazil [4]. Recently, many researchers have documented interesting and important insects in the mid-Cretaceous Burmese amber from Myanmar (*ca.* 99 Mya) [5, 6], which are summarized in an on-line checklist set up by Ross and nearly updated in real time [7]. Since 2016, there have been four new Orders of insects erected based on insects preserved in Burmese amber. Bai *et al.* erected a new insect Order of Alienoptera which have characters shared with cockroaches and mantids based on a single male insect, *Alienopterus brachyelytrus* Bai, Beutel, Klass, Wipfler & Zhang, 2016 in Burmese amber [8]. Huang *et al.* erected another new Order of Permopsocida to elucidate major radiation and evolution of suction feeding in hemimetabolous insects based on *Psocorrhyncha burmitica* [9]. Poinar and Brown erected the third new Order of Aethiocarenodea, including only one species of an exotic wingless female insect, *Aethiocarenus burmanicus* Poinar & Brown, 2017, in Burmese amber [10]. Mey *et al.* erected the fourth new Order of Tarachoptera in the Superorder Amphiesmenoptera and suggested its independent origin and evolution from an amphiesmenopteran ancestor which was not the ancestor of the Trichoptera-Lepidoptera clade [11].

Thousands of new species of insects, extant or fossil, have been described around the world every year to enhance our knowledge and understanding of the complex world of entomology now and then. In the area of paleoentomology, many well-preserved and interesting fossil insects have provided us a window of opportunity to study and understand the development and evolution of insects over hundred millions of years.

3.2 How to Identify an Insect

Insects are commonly called as "bugs", but in many cases those people confused by other crawling arthropods, may call a spider, a pseudoscorpion, a harvestman or a centipede an insect or a bug. The confusion is not surprising because the insects are so diverse, and many of them have different developmental stages as larvae, pupae and adults. In addition, many insects, such as orchid mantises, leaf butterflies, stick insects, etc., present exquisite mimicries so that they do not look like insects. Furthermore, other arthropods imitate insects, for example, ant spiders (Zodariidae) mimic the appearance of ants [12], which further increases the difficulty of insect identification. There are many informative books providing the morphology, classification and ecology of insects [1, 3, 13–19]. Herein, we briefly introduce morphological features and taxonomy for insects.

The life cycles of insects vary significantly. Nymphs, larvae or pupae of insects must break through the confinement of their exoskeleton as they grow or develop. They must slough off their old exoskeletons, and live on new and larger exoskeletons [1] (Figure 3.1). Some

Rhythms of Insect Evolution: Evidence from the Jurassic and Cretaceous in Northern China, First Edition.
Edited by Dong Ren, Chungkun Shih, Taiping Gao, Yongjie Wang, and Yunzhi Yao.

Figure 3.1 The conversion of pupa to adult of a lacewing mantispid. Source: Photos by Jason Shih.

insects, such as dragonflies, mayflies, stoneflies, true bugs, cockroaches, earwigs, stick insects, sucking lice, termites, grasshoppers, cicadas, etc., undergo three clear developmental stages: eggs, nymphs and adults. This developmental process is called "Incomplete Metamorphosis" or "Hemimetabolous". The hemimetabolous insects are those whose nymphs, called naiads, occupy aquatic habitats while the adults are terrestrial, for example, dragonflies, mayflies and stoneflies. However, paurometabolism applies to insects whose nymphs occupy the same environment as the adults, for example, true bugs, water striders, earwigs, etc. On the other hand, insects such as flies, lacewings, butterflies, moths, snakeflies, scorpionflies, caddisflies, wasps, etc., have a unique developmental stage called "pupation" between larvae and adults. During pupation, larvae usually reduce movement and enclose themselves within cocoons. These insects go through life cycles from eggs, larvae, pupae to adults, which are called "Complete Metamorphosis" or "Holometabolous". Pupation is commonly considered to be an effective mechanism for avoiding an unsuitable living environment.

In general, the two most important life purposes for insect nymphs, larvae and adults are feeding (Chapter 28), while avoiding being eaten (Chapter 29). Adult insects have various lengths of lifespan, for example, queen ants live for up to 30 years and worker ants 1–3 years, while queen termites, from 30 to 50 years, and workers and soldiers live approximately 1–2 years. On the other hand, mayflies and some caddisflies live a very short life without feeding. For all adult insects, their third life purpose is to mate and reproduce by passing on their genes to the next generation (Chapter 30).

The adult insects have segmented bodies, commonly divided as head, thorax and abdomen, all covered by exoskeletons. The heavily sclerotized head possesses a pair of antennae, a pair of compound eyes with ommatidia of hexagonal facets as sensory organs and complex mouthparts (Figure 3.2). Most insects have three ocelli, simple eyes, on the heads, but some insects of weak mobility, such as fleas, lice, etc., lost their ocelli, while their eyes may even show regression.

The thorax is considered to be the center of mobility, with three pairs of legs and two pairs of wings (only for

Figure 3.2 Head with antennae, eyes, and chewing mouthparts of a beetle. Source: Photo by Jason Shih.

winged insects). Fore, middle and hind legs are extended from the prothorax, mesothorax and metathorax, respectively. Having three pairs of legs is a very important diagnostic character that distinguishes insects from other arthropods; therefore insects have been known as the Subphylum Hexapoda. Each of the mesothorax and metathorax supports a pair of forewings and a pair of hind wings respectively for winged insects.

The soft abdomen is the place for the insect's digestive system, which generally includes 11–12 segments. Every abdominal segment is equipped for exchanging air via a pair of spiracles located at both sides. However, the numbers of abdominal segments or spiracles vary in different groups of insects. The abdomen is terminated by the exterior genitalia and sometime a pair of cerci in some stem insects. Some female insects have very specialized ovipositor adaptations to accommodate their reproductive life. For example, female parasitoid wasps possess very long ovipositors to lay eggs deeply into the host larvae in the tree trunk (Figure 3.3) [20].

Figure 3.3 A wasp of the species *Solenura ania* (Pteromalidae, Chalcidoidea) with an elongated ovipositor. Source: Photo by Jason Shih.

Compared to eyes, antennae are more important for insects to communicate with other insects and to detect chemical, mechanical or temperature signals in the environment so that insects can accomplish important tasks such as searching for food or suitable habitats, finding potential mates, and/or targeting a biological host [21, 22]. Typical antennae have many antennomeres, and the basal two are called scape and pedicel; the remaining is the flagellum, including many flagellomeres. There is broad diversity of antennal structures among different insects, for example, most cockroaches and crickets have filiform antennae, flies usually have aristate antennae, ants have geniculate antennae, and beetles display antennae with lamellate, serrate and flabellate forms. Various types of antennal sensilla, serving as chemoreceptors, mechanoreceptors, thermoreceptors, or hygroreceptors, are located on individual antennal flagellomeres, which collectively constitute the conspicuous part of the insect's antennae [23–25].

Insects are equipped with a wide range of mouthparts, adapted to diversified modes of feeding [26]. According to their structures and orientations, insects' mouthparts can be classified into three categories: hypognathous (mouthparts directed toward the ground, considered to be the primitive condition in insects such as cockroaches, mantids, silverfish, etc.); prognathous (mouthparts directed forward for carnivorous insects pursuing preys, e.g. earwigs, stick insects, larval neuropterans, etc.); and opisthognathous (mouthparts directed backward, e.g. cicadas, some true bugs and lice with sucking mouthparts). Generalized insect mouthparts consist of five basic structures: labrum, mandibles, maxillae, hypopharynx, and labium [14]. The detailed structures and type of mouthparts present on an insect may serve as diagnoses for identifying it. Some insects have chewing mouthparts, such as dragonflies, sawflies and some beetles (Figure 3.2), to chew prey or plant tissues for feeding or for depositing eggs [27]. Moths and butterflies possess siphonate mouthparts for sucking nectar or sugary fluids from flowers (Figure 3.4) [28]. Insects such as aphids and thrips, have piercing and sucking mouthparts with needle-like stylets to pierce the plant cells and suck fluid inside the cells [29].

As the second major tagma of the insect's body, the thorax is equipped to support the legs and wings. The typical insect leg consists of six main sections, the basal one is the coxa, which is connected with the sternites of thorax, followed by the trochanter, femur, tibia, tarsus and claws [1]. Diversification and specialization occur in legs due to their adaptation to living conditions. For example, lice, living an ectoparasitic lifestyle, have very robust legs with huge claws, but only 1–2 segments of tarsus [30, 31]. Mole crickets evolved with digging legs [32], honeybees have pollen-carrying hind legs (Figure 3.5) [33], insects living in water have swimming legs [34], and some

Figure 3.4 Butterfly with siphonate mouthparts. Source: Photo by Dr. Chungkun Shih.

Figure 3.5 Honeybees with pollen-carrying hind legs. Source: Photo by Dr. Chungkun Shih.

insects within the Polyneoptera, e.g. flies and katydids, developed a pad structure on their tarsi to increase friction or adhesion [35].

Insects, the earliest animals flying in the sky, evolved flight perhaps 170 million years before pterosaurs. The other two groups of animals having powered flight, i.e. birds and bats, appeared much later [18]. The ability of flight provides a great advantage for insects to adapt to different environments. Numerous papers studying the origin and early evolution of insect flight [36–43] and the functions of insect wings [44–46] have been published. The venation of the insect wing is an important diagnostic character for the pterygota insects. A pattern model, the Comstock-Needham system defining all the

Figure 3.6 A fishfly (Corydalidae) spreading its four wings with clear venation. Source: Photo by Jason Shih.

veins of insect wings, is commonly accepted by most entomologists [1, 14, 47]. Wing venation nomenclature is listed here for reference: C, Costa, Sc, Subcosta, ScP, posterior Subcosta; R, Radius, RA, anterior Radius; RP, posterior Radius; M, Media; MA, anterior Media; MP, posterior Media; Cu, Cubitus; CuA, anterior Cubitus; CuP, posterior Cubitus; AA: Anterior anal veins; AA1: first Anterior anal vein; A2, second Anterior anal vein; ra-rp, rp-ma indicate the two specific cross-veins connecting RA and RP, RP and MA, respectively. Right forewing and left forewing are indicated as RFW and LFW, respectively, and right hind wing and left hind wing as RHW and LHW, respectively (Figures 3.6 and 3.7). A number of cross-veins are present between the longitudinal veins. Venational pattern is greatly reduced in some insects, such as chalcidoids [27, 48]. The wings are even completely lost in fleas [49] and worker ants [50, 51].

The abdomen of an adult insect typically consists of 11–12 segments which incorporate the heart, mid-gut, other digestive organs and reproductive organs and appendages [52]. The abdomen is less sclerotized than the head or thorax, and is formed by the upper tergites and the lower sternites, and perhaps a pleurite. These structures are held together by a stretchable membrane [53]. The abdomen is usually simple in its anterior region, but quite complex in the last three segments where the external genitalia are used for mating (Figures 3.8 and 3.9) and/or laying eggs (Figure 3.10) [54]. Morphological features of external genitalia vary in different insects and are often used for taxonomic diagnoses [55]. Female cockroaches produce a special capsular structure, the ootheca, to enclose eggs for protection and maternal brood care [56, 57]. Female katydids have a large acinacifoliate ovipositor surround by a pair of short cerci, and many mayflies have short gonostyles and very long cerci. Most wasps have the "wasp waist" structures formed by the fusion of the first abdominal segment (propodeum)

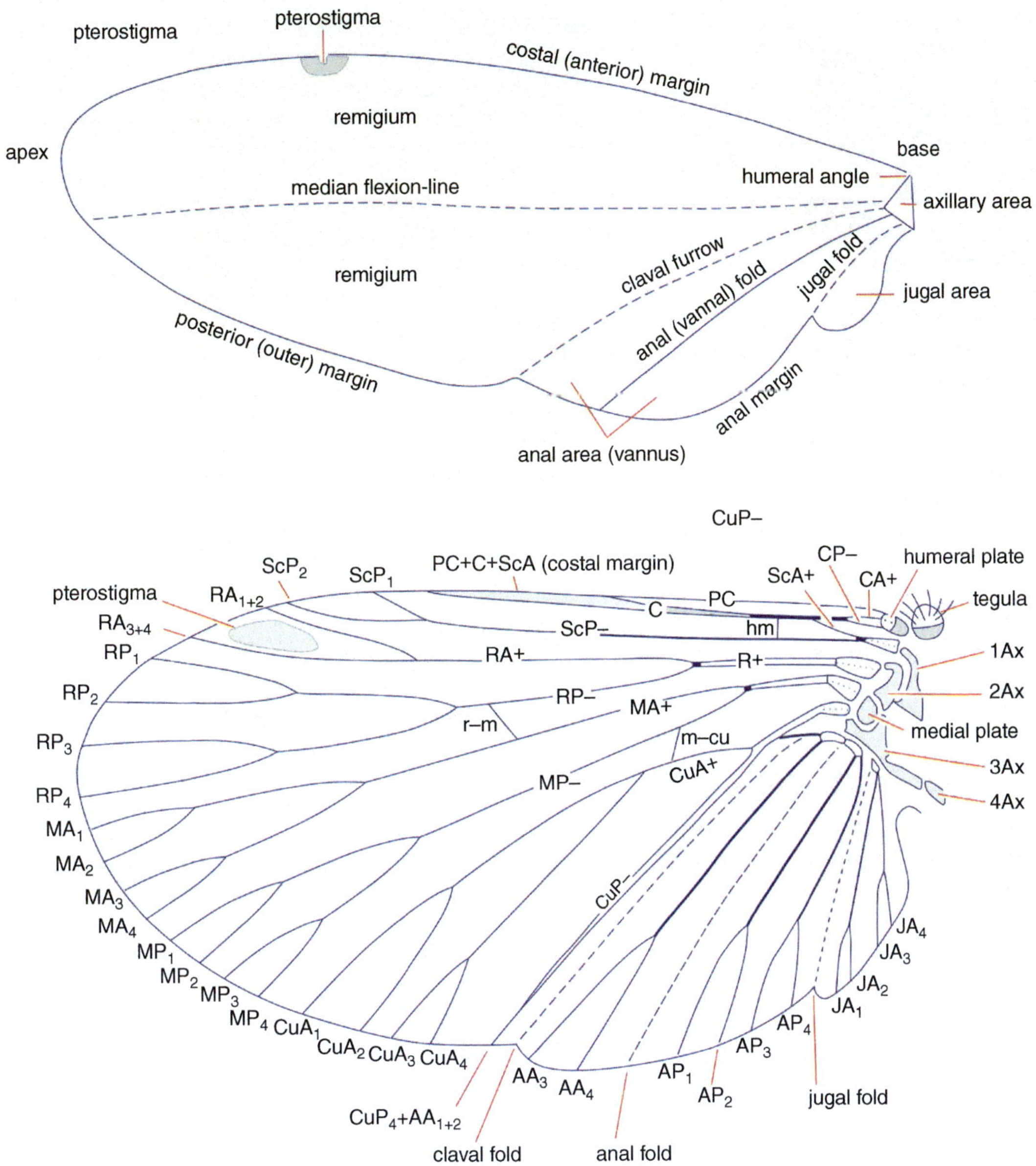

Figure 3.7 The terminologies of the insect wing. Source: Modified from [1].

and thorax, resulting in a constriction between the first abdominal segment and the second segment, which provides an adaptive mechanism to the reproductive style of wasps [18].

3.3 Origin and Evolution of Insects

There are over a million of species of insects formally documented so far, but the actual number of species should be much higher for these most remarkable lineages of insects in the long history of life on earth [18, 58, 59]. Many methods, such as bioinformatics-based molecular data or morphological data, anatomical research, especially paleontology, are used in studying the origin of insects [3, 18, 60]. The earliest known true insects have been reported from the Middle Devonian period based on the fossils from Gaspé in North America, corresponding to a time between 390 and 392 Mya [61, 62]. *Rhyniognatha hirsti* Tillyard, 1928 was re-described and assigned as an insect by Engel and Grimaldi in 2004 [62], based on its clear mandibular structure. In addition, Engel and Grimaldi suggested that the short and triangular morphology of the mouthpart of *R. hirsti* is typical of some Pterygota, indicating that it might have been a winged insect, although the definitive apomorphy, wings, are missing [63]. The origin of the winged insect is

Figure 3.8 Butterflies mating. Source: Photo by Dr. Chungkun Shih.

Figure 3.9 Flies mating. Source: Photo by Jason Shih.

hypothesized to have occurred during the Late Devonian or Early Carboniferous, since a diversity of wing morphologies and pterygote lineages have already been established [18, 64, 65]. *Stenodictya lobata* Brongniart, 1890, assigned to the order Paleodicotoptera and found in the Late Carboniferous in France, had two flaps on the first thoracic segment and four true wings on the second and third thoracic segments – a curiosity no longer found in insects [66, 67]. Protodonata with very diverse fossil records from the Carboniferous to the Permian had a wingspan of 70–75 cm and body length of 30–38 cm [68]. Flight capability gave significant advantages to flying insects by allowing them to avoid predation, to prey on others, and to move to new and more suitable environments. Several fossils insects, i.e. stem Hymenoptera, early Hemiptera and Psocodea, were discovered in the Bashkirian–Moscovian age (Pennsylvanian), but the

Figure 3.10 A lacewing laying eggs. Source: Photo by Jason Shih.

lineage of Eumetabola in the Neoptera radiated more successfully only after the mass extinctions at the end of the Permian [41]. Most modern insect families appeared from the Triassic to the Jurassic. Three families of Mecoptera and one family of Neuroptera equipped with siphonate mouthparts, with lengths ranging from 1 to 10 mm for Mecopterans and 5.4 to 18 mm for Kalligrammatidae, have been reported from the Mesozoic of China [69–71], and these insects provided pollination for gymnosperm plants from the Middle Jurassic to the Early

Figure 3.11 A bee visiting flowers. Source: Photo by Dr. Chungkun Shih.

Figure 3.12 A fly visiting flowers. Source: Photo by Jason Shih.

Figure 3.13 Two cantharids (soldier beetles) feeding on Texas thistle (*Cirsium texanum*) with siphonate mouthparts at the Savannah Oaks Ranch, Texas. Source: Photo by Dr. Chungkun Shih.

Cretaceous (165 to 125 Mya) before the appearance of robust and cupped angiosperm flowers (Chapter 20.3, Chapter 24.3 and Chapter 28.2.1 and 28.2.2). After the Early Cretaceous, some insects evolved and diversified as flower-visitors with the appearance of angiosperm flowering plants, which have resulted in the successful pollination mutualism by Lepidoptera (Figure 3.4), Hymenoptera (Figure 3.11), Diptera (Figure 3.12) and Coleoptera (Figure 3.13) [72–74].

Insects in Traditional Chinese Culture

Insects have been included and treated as an important part of Chinese traditional culture. *The Book of Songs*, the first anthology of poems and songs in ancient China from 1100 BCE to 600 CE, mentioned several groups of insects, such as mayflies, silkworms, crickets, and locusts, etc. In Chinese culture, mayflies are related to a short lifespan (see Chapter 5), while silkworms are considered to be a symbol of selflessness and dedication. Shangyin Li, a renowned poet in the late Tang Dynasty, wrote this poem: *"Spring silkworms keep spinning silk till the end, tears start to dry up after the candle burning to ashes."* This poem has touched many Chinese for more than a thousand years. Honeybees, as hard-working pollinators, are popular in literature and considered as the symbol of being diligent, productive and industrious. Another poet in the Tang Dynasty, Yin Luo, wrote a poem entitled "Honeybees": *"On the plain or on top of the hill, unlimited beautiful scenery is fully occupied. Collecting from hundreds of flowers to form honey, for whom you work so hard to provide the sweet."* Dragonflies and butterflies always represent being happy, energetic and beautiful, but flies and mosquitos denote being nasty and annoying. It is interesting that the filiform antennae of butterflies are used to describe the eyebrows of beautiful women.

The Last Emperor and Crickets

The movie of "The Last Emperor" presented the traumatic and tragic life of the last Emperor of the Qing Dynasty, Pu Yi, who, as a three-year-old boy, was crowned in 1909. During the ceremony, thousands of court officials, guests, monks, and guards, knelt down and kowtowed. Amid the loud shouting of "The first kowtow! Kowtow again! The third kowtow!," the young Pu Yi was attracted by the chirping sound of a cricket. "Cricket! Cricket! Where is the cricket?" the boy asked. A court official showed him a cricket in an exquisite cage, and told the Emperor, "See, the cricket is kowtowing to Your Majesty. Now, it is the Emperor's cricket."

After the Republic of China was formed in 1911, he stepped down as the last Emperor of the fading Qing Dynasty and became a dethroned Emperor confined in the Forbidden City. Before and during World War II, he was a traitor and took the role of a puppet figure head in Manchuria, Northeastern China under Japanese occupation. After the war, he was captured by Russian soldiers and moved to Russia. In 1950, he was returned to China as a war criminal in the Fushun Detention Center and went through reform and re-education. After 10 years of re-education, he was released to become a gardener in the Beijing Botanic Garden. Near the end of the movie, Pu Yi, as a fragile old man, went back alone to the palace as a commoner tourist in 1967. He met a young Red Guard boy and to prove that he was indeed the last Emperor, he climbed up to the throne and took out the elegant cricket cage under the throne cushion. The cricket with its cage became a symbol of a complete life cycle for the last Emperor.

References

1 Gullan, P.J. and Cranston, P.S. (2014). *The Insects: An Outline of Entomology*, 5e. Oxford: Wiley Blackwell.

2 Wilson, E.O. (2010). *The Diversity of Life*. Cambridge: Belknap Press.

3 Rasnitsyn, A.P. and Quicke, D.L.J. (2002). *History of Insects*. Dordrecht: Kluwer.

4 Staniczek, A., Bechly, G., and Godunko, R. (2011). Coxoplectoptera, a new fossil order of Palaeoptera (Arthropoda: Insecta), with comments on the phylogeny of the stem group of mayflies (Ephemeroptera). *Insect Systematics & Evolution* **42**: 101–138. https://doi.org/10.1163/187631211X578406.

5 Grimaldi, D.A., Engel, M.S., and Nascimbene, P.C. (2002). Fosiliferous Cretaceous amber from Myanmar (Burma): its rediscovery, biotic diversity, and paleontological significance. *American Museum Novitates* **3361**: 1–72.

6 Shi, G., Grimaldi, D.A., Harlow, G.E. *et al.* (2012). Age constraint on Burmese amber based on U–Pb dating of zircons. *Cretaceous Research* **37**: 155–163. https://doi.org/10.1016/j.cretres.2012.03.014.

7 Ross, A.J. (2018) Burmese (Myanmar) amber taxa, on-line checklist Version 2018.1.

8 Bai, M., Beutel, R.G., Klass, K.D. *et al.* (2016). Alienoptera - a new insect order in the roach–mantodean twilight zone. *Gondwana Research* **39**: 317–326. https://doi.org/10.1016/j.gr.2016.02.002.

9 Huang, D.-Y., Bechly, G., Nel, P. *et al.* (2016). New fossil insect order Permopsocida elucidates major radiation and evolution of suction feeding in hemimetabolous insects (Hexapoda: Acercaria). *Scientific Reports* **2016** (6): 23004. https://doi.org/10.1038/srep23004.

10 Poinar, G. and Brown, A.E. (2017). An exotic insect *Aethiocarenus burmanicus* gen. et sp. nov. (Aethiocarenodea ord. nov., Aethiocarenidae fam. nov.) from mid-Cretaceous Myanmar amber. *Cretaceous Research* **72**: 100–104. https://doi.org/10.1016/j.cretres.2016.12.011.

11 Mey, W., Wichard, W., Mueller, P., and Wang, B. (2017). The blueprint of the Amphiesmenoptera-Tarachoptera, a new order of insects from Burmese amber (Insecta, Amphiesmenoptera). *Fossil Record* **20**: 129–145.

12 Oliveira, P.S. (1988). Ant-mimicry in some Brazilian salticid and clubionid spiders (Araneae: Salticidae, Glubionidae). *Biological Journal of the Linnean Society* **33**: 1–15.

13 Imms, A.D. (1923). *A General Textbook of Entomology: Including the Anatomy, Physiology, Development and Classification of Insects*. Methuen, London.

14 Snodgrass, R.E. (1935). *Principles of Insect Morphology*. New York and London: McGraw-Hill Book Company, Inc.

15 Borror, D.J., Delong, D.M., and Triplehorn, C.A. (1976). *An Introduction to the Study of Insects*. New York: Holt, Rinehart and Winston.

16 Chapman, R.F. (1998). *The Insects Structure and Function*. Cambridge: Cambridge University Press.

17 Resh, V.H. and Cardé, R.T. (2003). *Encyclopedia of Insects*. Burlington: Academic Press.

18 Grimaldi, D.A. and Engel, M.S. (2005). *Evolution of the Insects*. New York: Cambridge University Press.

19 Price, P.W., Denno, R.F., Eubanks, M.D. *et al.* (2011). *Insect Ecology Behavior, Populations and Communities*. Cambridge: Cambridge University Press.

20 Gauld, I. and Bolton, B. (1988). *The Hymenoptera*. New York: Oxford University Press.

21 Cardé, R.T. and Minks, A.K. (1997). *Insect Pheromone Research: New Directions*. New York: Chapman & Hall.

22 Tegoni, M., Campanacci, V., and Cambillau, C. (2004). Structural aspects of sexual attraction and chemical communication in insects. *Trends in Biochemical Sciences* **29** (5): 257–264. https://doi.org/10.1016/j.tibs.2004.03.003.

23 Crook, D.J., Kerr, L.M., and Mastro, V.C. (2008). Distribution and fine structure of antennal sensilla in emerald ash borer (Coleoptera: Buprestidae). *Annals of the Entomological Society of America* **101**: 1103–1111. https://doi.org/10.1603/0013-8746-101.6.1103.

24 Barsagade, D.D., Tembhare, B., and Kadu, S.G. (2013). Microscopic stucture of antennal sensilla in the carpenter ant *Camponotus compressus* (Fabricius) (Formicidae: Hymenoptera). *Asian Myrmecology* **5**: 113–120. https://doi.org/10.20362/am.005012.

25 Gao, T.P., Shih, C., Labandeira, C.C. *et al.* (2016). Convergent evolution of ramified antennae in insect lineages from the Early Cretaceous of Northeastern China. *Proceedings of the Royal Society B: Biological Sciences* **283**: 20161448. https://doi.org/10.1098/rspb.2016.1448.

26 Labandeira, C.C. (1997). Insect mouthparts: ascertaining the paleobiology of insect feeding strategies. *Annual Review of Ecology and Systematics* **28**: 153–193. https://doi.org/10.1146/annurev.ecolsys.28.1.153.

27 Goulet, H. and Huber, J.T. (1993). *Hymenoptera of the World: An Identification Guide to Families*. Ottawa: Agriculture Canada Publication.

28 Krenn, H.W., Plant, J.D., and Szucsich, N.U. (2005). Mouthparts of flower-visiting insects. *Arthropod*

Structure & Development **34** (1): 1–40. https://doi.org/10.1016/j.asd.2004.10.002.

29 Peterson, A. (1915). Morphological studies on the head and mouth-parts of the Thysanoptera. *Annals of the Entomological Society of America* **8**: 20–59.

30 Price, R.D., Hellenthal, R.A., Palma, R.L. *et al.* (2003). The chewing lice world checklist and biological overview. In: *Illinois Natural History Survey*. Champaign.

31 Lehane, M.J. (2005). *The Biology of Blood-Sucking in Insects*. New York: Cambridge University Press.

32 Ulagaraj, S.M. (1975). Mole crickets: ecology, behavior, and dispersal flight (Orthoptera: Gryllotalpidae: Scapteriscus). *Environmental Entomology* **4**: 265–273.

33 Michener, C.D. (2000). *The Bees of the World*. Baltimore: Johns Hopkins University Press.

34 Voise, J. and Casas, J. (2010). The management of fluid and wave resistances by whirligig beetles. *Journal of The Royal Society Interface* **7** (43): 343–352. https://doi.org/10.1098/rsif.2009.0210.

35 Clemente, C.J. and Federle, W. (2008). Pushing versus pulling: division of labour between tarsal attachment pads in cockroaches. *Proceedings of the Royal Society B: Biological Sciences* **275**: 1329–1336. https://doi.org/10.1098/rspb.2007.1660.

36 Wigglesworth, V.B. (1963). Origin of wings in insects. *Nature* **197**: 97–98. https://doi.org/10.1038/385627a0.

37 Kukalova-Peck, J. (1978). Origin and evolution of insect wings and their relation to metamorphosis, as documented by the fossil record. *Journal of Morphology* **156**: 53–125.

38 Rasnitsyn, A.P. (1981). A modified paranotal theory of insect wing origin. *Journal of Morphology* **168**: 331–338.

39 Kaiser, J. (1994). A new theory of insect wing origins takes off. *Science* **266** (5184): 363–363. https://doi.org/10.1126/science.266.5184.363.

40 Averof, M. and Cohen, S.M. (1997). Evolutionary origin of insect wings from ancestral gills. *Nature* **385**: 627–630. https://doi.org/10.1038/385627a0.

41 Nel, A., Roques, P., Nel, P. *et al.* (2013). The earliest known holometabolous insects. *Nature* **503**: 257–261. https://doi.org/10.1038/nature12629.

42 Medved, V., Marden, J.H., Fescemyer, H.W. *et al.* (2015). Origin and diversification of wings: insights from a neopteran insect. *Proceedings of the National Academy of Sciences of the United States of America* **112** (52): 15946–15951. https://doi.org/10.1073/pnas.1509517112.

43 Prokop, J., Pecharová, M., Nel, A. *et al.* (2017). Paleozoic nymphal wing pads support dual model of insect wing origins. *Current Biology* **27** (2): 263–269. https://doi.org/10.1016/j.cub.2016.11.021.

44 Wootton, R.J. (1979). Function, homology and terminology in insect wings. *Systematic Entomology* **4** (1): 81–93. https://doi.org/10.1111/j.1365-3113.1979.tb00614.x.

45 Wootton, R.J. (1992). Functional morphology of insect wings. *Annual Review of Entomology* **37**: 113–140. https://doi.org/10.1146/annurev.en.37.010192.000553.

46 Wootton, R.J. and KukalovÁ-Peck, J. (2000). Flight adaptations in Palaeozoic Palaeoptera (Insecta). *Biological Reviews* **75** (1): 129–167.

47 Naumann, I.D., Came, P.B., Lawrence, J.F. *et al.* (1991). *Insects of Australia: A Textbook for Students and Research Workers*, 2e. Carlton: Melbourne University Publishing.

48 Snodgrass, R.E. (1911). The thorax of the Hymenoptera. *Proceedings of the United States National Museum* **39**: 37–91. +16 Pl.

49 Snodgrass, R.E. (1946). The skeletal anatomy of fleas (Siphonaptera). *Smithsonian Miscellaneous Collections* **104**: 1–89.

50 Wilson, E.O. (1963). The social biology of ants. *Annual Review of Entomology* **8**: 345–368. https://doi.org/10.1146/annurev.en.08.010163.002021.

51 Jongepier, E. and Foitzik, S. (2016). Fitness costs of worker specialization for ant societies. *Proceedings of the Royal Society B: Biological Sciences* **283** (1822): 20152572. https://doi.org/10.1098/rspb.2015.2572.

52 Matsuda, R. (1976). *Morphology and Evolution of the Insect Abdomen, with Special Reference to Developmental Patterns and their Bearings upon Systematics*. Oxford: Pergamon Press Ltd.

53 Snodgrass, R.E. (1933). Morphology of the insect abdomen. Part I. *Smithsonian Miscellaneous Collections* **85**: 31–128.

54 Matsuda, R. (1958). On the origin of the external genitalia of insects I. *Annals of the Entomological Society of America* **51**: 84–94.

55 Song, H. (2009). Species-specificity of male genitalia is characterized by shape, size, and complexity. *Insect Systematics & Evolution* **40**: 159–170. https://doi.org/10.1163/187631209X424571.

56 Mckittrick, F.A. (1964). *Evolutionary Studies of Cockroaches*. New York: Cornell University Agricultural Experiment Station.

57 Bell, W.J., Roth, L.M., and Nelepa, C.A. (2007). *Cockroaches: Ecology, Behavior, and Natural History*. Baltimore, Maryland: The Johns Hopkins University Press.

58 Glenner, H., Thomsen, P.F., Hebsgaard, M.B. *et al.* (2006). The origin of insects. *Science* **314**: 1883–1884. https://doi.org/10.1126/science.1129844.

59 Engel, M.S. (2015). Insect evolution. *Current Biology* **25**: 868–872. https://doi.org/10.1016/j.cub.2015.07.059.

60 Misof, B., Liu, S., Meusemann, K. *et al.* (2014). Phylogenomics resolves the timing and pattern of insect evolution. *Science* **346** (6210): 763–767. https://doi.org/10.1126/science.1257570.

61 Shear, W.A., Bonamo, P.M., Grierson, J.D. *et al.* (1984). Early land animals in North America: evidence from Devonian age arthropods from Gilboa, New York. *Science* **224** (4648): 492–494. https://doi.org/10.1126/science.224.4648.492.

62 Labandeira, C.C., Beall, B.S., and Hueber, F.M. (1988). Early insect diversification: evidence from a lower Devonian bristletail from Québec. *Science* **242** (4880): 913–916. https://doi.org/10.1126/science.242.4880.913.

63 Engel, M.S. and Grimaldi, D.A. (2004). New light shed on the oldest insect. *Nature* **427**: 627–630. https://doi.org/10.1038/nature02291.

64 Kukalová-Peck, J. (1983). Origin of the insect wing and wing articulation from the arthropodan leg. *Canadian Journal of Zoology* **61**: 1618–1669.

65 Ross, A.J. (2010). A review of the Carboniferous fossil insects from Scotland. *Scottish Journal of Geology* **46**: 157–168. https://doi.org/10.1144/0036-9276/01-413.

66 Handlirsch, A. (1906-1908). *Die Fossilen Insekten und die Phylogenie der rezenten Formen: Ein Handbuch für Paläontologen und Zoologen*, vol. **1–3**. Leipzig: Engelmann.

67 Hoell, H.V., Doyen, J.T., and Purcell, A.H. (1998). *Introduction to Insect Biology and Diversity*, 2e. Oxford: Oxford University Press.

68 Penney, D. and Jepson, J.E. (2014). *Fossil Insects: An Introduction to Palaeoentomology*. Manchester: Siri Scientific Press.

69 Ren, D., Labandeira, C., Santiago-Blay, J. *et al.* (2009). A probable pollination mode before angiosperms: Eurasian, long-proboscid scorpionflies. *Science* **326** (5954): 840–847. https://doi.org/10.1126/science.1178338.

70 Yang, Q., Wang, Y.J., Labandeira, C.C. *et al.* (2014). Mesozoic lacewings from China provide phylogenetic insight into evolution of the Kalligrammatidae (Neuroptera). *BMC Evolutionary Biology* **14**: 126. https://doi.org/10.1186/1471-2148-14-126.

71 Labandeira, C.C., Yang, Q., Santiago-Blay, J.A. *et al.* (2016). The evolutionary convergence of mid-Mesozoic lacewings and Cenozoic butterflies. *Proceedings of the Royal Society B: Biological Sciences* **283** (1824): 20152893. https://doi.org/10.1098/rspb.2015.2893.

72 Johnson, S.D. and Anderson, B. (2010). Coevolution between food-rewarding flowers and their pollinators. *Evolution: Education and Outreach* **3**: 32–39.

73 Labandeira, C.C. (2013). A paleobiologic perspective on plant–insect interactions. *Current Opinion in Plant Biology* **16** (4): 414–421. https://doi.org/10.1016/j.pbi.2013.06.003.

74 Labandeira, C.C. and Currano, E.D. (2013). The fossil record of plant-insect dynamics. *Annual Review of Earth and Planetary Sciences* **41**: 287–311. https://doi.org/10.1146/annurev-earth-050212-124139.

4

A History of Paleoentomology in China

Dong Ren[1], Chungkun Shih[1,2], and Taiping Gao[1]

[1] Capital Normal University, Haidian District, Beijing, China
[2] National Museum of Natural History, Smithsonian Institution, Washington, DC, USA

4.1 Introduction

Historically, Chinese Paleoentomology studies can be divided into five stages: Early foundational studies (1923–1935); Early taxonomic studies (1965–1984); Major taxonomic studies (1985–present); Phylogenetic and paleobiologic studies (1991–present); and International cooperative studies (2000–present). It is worth emphasizing that the four stages after 1965 are by no means distinctively separated; they have been gradually developed and evolved and sometimes overlapped. However, each stage has been distinguished by its central theme which has made a significant contribution to Paleoentomology in China.

4.2 Early Foundational Studies (1923–1935)

The study of Chinese paleoentomology started in 1923 and continued until 1935 with the publication of four papers on fossil insects by Amadeus W. Grabau and Chi Ping respectively.

Dr. Grabau (Figure 4.1) was born on 9 January 1870 in Cedarburg, Wisconsin, USA. Grabau took a boyhood delight in natural history. In 1896, he graduated with a BSc. in Geology and Mineralogy from Massachusetts Institute of Technology (MIT). He earned a DSc. in Geology in 1900 from Harvard University. Grabau taught at MIT (1892–1897), Rensselaer Polytechnic Institute (1899–1901), and Columbia University (1901–1919). Grabau's research covered a broad range. Besides paleontology and stratigraphy, he majored in petrology, mineralogy, ore deposits, petroleum geology, tectonic geology, geomorphology, glaciology, and paleoanthropology. Before working in China, Grabau was a distinguished geology and paleontology scholar in Europe and America.

In 1919, Grabau began working as a Professor in the Department of Geology, Peking University, the Chief Paleontologist for the Chinese Geological Survey and a part-time Professor at Tsinghua University. He taught the General Geology, Paleontology, Stratigraphy and History courses, cultivated and developed a lot of outstanding talent in the first-generation Chinese geologists and paleontologists. While staying in China, he published a series of important research results about Chinese Geology and Paleontology, including the famous books on the first volume of Chinese geological history (Paleozoic) and the second volume (Mesozoic) [1]. In these masterpieces, Grabau first named the Early Cretaceous fossils from northern Hebei and western Liaoning as "Jehol biota". In 1923, he published the paper "Cretaceous fossils from Shandong" [2]. This was the first Chinese paleoentomology paper in which four fossil insect genera and species were described.

Grabau was knowledgeable and attentive in his teaching, with emotion and thorough instruction. His kind manner, in-depth expertise and high quality standard won the respect and love of Chinese scholars and students. During World War II, he remained in Peking. Around 1941, he was interned by the Japanese Imperial Army. His health declined precipitously. On 22 March 1946, Grabau died in Beijing and was buried in the beautiful campus of Peking University according to his behest (Figure 4.2).

Dr. Ping was born on April 9, 1886 in Kaifeng City, Henan Province. In 1908, he graduated from the Pekin University. In 1909, he went to the College of Agriculture at Cornell University, USA, and learned entomology under the guidance of entomologist Dr. Geidam J. Needham. He received a BSc. in 1913 and a Ph.D. in 1918.

Rhythms of Insect Evolution: Evidence from the Jurassic and Cretaceous in Northern China, First Edition.
Edited by Dong Ren, Chungkun Shih, Taiping Gao, Yongjie Wang, and Yunzhi Yao.

Figure 4.1 Dr. Amadeus W. Grabau.

Figure 4.2 The memorial gravestone for Dr. Amadeus W. Grabau in the campus of Peking University in China. Source: Photo by Dr. Dong Ren.

He was the first Chinese scholar to obtain a Ph.D. in the USA.

Ping (Figure 4.3) conducted a wide range of studies, and achieved much pioneering work in the fields of entomology, neurophysiology, animal fauna, systematics, anatomy, vertebrate morphology, physiology and zoology. In 1955, he was elected as an academician of the National Academy of Sciences of China. Ping was the first Chinese academic to study insect fossils. From 1928 to 1935, he published three papers, which reported 32 genera and 35 species of insect fossils from the Lower Cretaceous Laiyang Formation and Yixian

Figure 4.3 Dr. Chi Ping.

Formation, the Upper Jurassic Maiyohkow, Turfan, Sinkiang (Xinjiang) and the Fushun Eocene amber of Liaoning, respectively [3–5].

4.3 Early Taxonomic Studies (1965–1985)

In the middle of the twentieth century, Youchong Hong and Qibin Lin became professional paleoentomologists in China. Youchong Hong (Figures 4.4 and 4.5) was born in November 1929 in Nanao County, Guangdong Province, and graduated from the Beijing Institute of Geology (China University of Geoscience) with a BSc. in Geology in 1953. From 1958 to 1960, he studied freshwater mollusk fossils at the USSR Academy of Sciences following his mentor, Dr. G.G. Martinson. Then, he studied insect fossils under the guidance of entomologist Dr. O. M. Martynova.

Hong worked hard in his career, published 160 papers, six individual monographs and 17 cooperative monographs [6–12]. In 1982, he proposed the concept of the Middle Jurassic Yanliao Entomofauna [10], which has been adopted by the Chinese paleontologists. In 1989, he retired from the Beijing Museum of Natural History. In 2009, he received the Li Siguang Award for Geological Sciences.

Qibin Lin was born in February, 1935, graduated from the Nanjing University and received his BSc. in Biology in 1959. After graduation, he worked in the

Figure 4.4 Dr. Youchong Hong. Source: Photo provided by Dr. Dong Ren.

Figure 4.5 Dr. Youchong Hong and Dr. Dong Ren in the Jiuquan basin of Gansu Province of China in 1989. Source: Photo by Dr. Dong Ren.

Nanjing Institute of Geology and Palaeontology, Chinese Academy of Sciences (NIGPAS) and engaged in the research of insect fossils. Lin (Figure 4.6) published 60 papers and one monograph [13–19]. His monograph "Early Mesozoic Fossil Insects from South China" [17] has had an important influence on Chinese paleoentomology.

4.4 Major Taxonomic Studies (1985–Present)

On the basis of Hong's and Lin's work, Junfeng Zhang, Dong Ren (with his Team), Haichun Zhang (with his

Figure 4.6 Dr. Qibin Lin. Source: Photo provided by Dr. Haichun Zhang.

Team), and then Di-ying Huang (with his Team) have carried out more detailed research work on fossil insects. Most of their research papers in English, especially on the Mesozoic fossil insects, have been published in the international peer-reviewed academic journals.

Junfeng Zhang was born in November 1945, and graduated from the Beijing Institute of Geology. From 1985, he carried out research work on the Cretaceous and Tertiary insect fossils in the Shandong and Yanliao areas. He worked in NIGPAS and has published 104 papers [20–24] and two monographs on Miocene fossil insects from Shanwang, Shandong [25, 26].

Haichun Zhang (Figure 4.7) was born in March 1965, graduated from Nanjing University and received his

Figure 4.7 Dr. Haichun Zhang. Source: Photo provided by Dr. Haichun Zhang.

BSc. in Paleontology in 1989. After getting his MSc. in Paleontology under the guidance of Qibin Lin from NIGPAS in 1992, he then worked in NIGPAS. In 1999, he received his Ph.D. in paleontology under the guidance of Peiji Chen and Junfeng Zhang. Zhang and his team have published more than 200 papers [27–32] and one monograph on Mesozoic fossil and Cenozoic insects from Northern China [33].

Di-ying Huang (Figure 4.8) was born in 1975, and graduated from the Earth Sciences Department, Nanjing University in 1997. After graduation, he worked in NIGPAS, and in 2005, he received his Ph.D. in paleontology from the Université Claude Bernard Lyon, France. Huang and his team have published more than 170 papers [34–37] and one monograph on Daohugou Biota [38].

In addition, many young researchers worked on the fossil insects and have obtained their Ph.D.s over the past 15 years. Some of them continued their studies of paleoentomology, while others pursued alternative careers. Their research and studies in graduate schools, research institutions and career assignments have established the backbone and promoted the rapid development of Chinese paleoentomology. Many of their published works will be documented in the following chapters of this book. They include Dr. Binglan Zhang, Dr. Yunzhi Yao, Dr. Ming Bai, Dr. Xinyue Liu, Dr. Ming Liu, Dr. Jingjing Tan, Dr. Yushuang Liu, Dr. Jiandong Huang, Dr. Kuiyan Zhang, Dr. Ying Wang, Dr. Bo Wang, Dr. Huali Chang, Dr. Yanli Yue, Dr. Junjie Gu, Dr. Weiting Zhang, Dr. Qiang Yang, Dr. Taiping Gao, Dr. Yingying Cui, Dr. Chaofang Shi, Dr. Qiaoling Ding, Dr. Yongjun Li, Dr. Chenyang Cai, Dr. Hao Wu, Dr. Yunan Wang, Dr. Yuanyuan Peng, Dr. Yali Yu, Dr. Chen Wang, Dr. Mei Wang, Dr. Longfeng Li, Dr. Congshuang Deng and Dr. Qi Zhang and Ms. Junhui Liang, and Ms. Yan Fang, *et al.*

Figure 4.8 Dr. Di-ying Huang. Source: Photo provided by Dr. Di-ying Huang.

4.5 Phylogenetic and Paleobiological Studies (1991–Present)

With the development of the research work from the first three stages, the Chinese paleoentomologists, with cooperation from many researchers from around the world, have made significant contributions not only to morphology, taxonomy and classification, but also to the phylogenetic relationships of fossil insects from different geological ages [39–44]. They have also contributed to knowledge of the interactions and co-evolution among insects with plants [45–48] and with vertebrates [34, 49–51] in their ecosystems. The 1994 paper "A cladistics study on the familial phylogeny of fossil and living Raphidioptera" by Ren and Hong marked the beginning of phylogenetic studies in Chinese paleoentomology [52]. The past 30 years witnessed significant accomplishments in Chinese studies of fossil insects.

Extant organisms have various and abundant morphological characters and genetic data. However, due to their long evolutionary history and the lack of a large number of important extinct members, most deduced phylogenies are a paraphyletic group or a polyphyletic group and cannot constitute a natural system. Only the fossil records can provide direct evidence and better spatio-temporal framework information for such processes. It is impossible to construct a natural system of the whole monophyletic group by using only macro- and micro-information of extant organisms.

From the viewpoint of natural history, both the concepts and characteristics of the monophyletic group, phylogenetic systematics and natural classification system have been clarified [53]. The natural classification system is a taxonomic system and natural evolution process of all members in the monophyletic group and their phylogenetic relationships. Only a monophyletic group including all extinct and extant members can be called a natural taxonomic system. There are two common misunderstandings in the process of phylogenetic reconstruction: (i) a monophyletic group is claimed as a natural system due to incomplete selection of in-group members; (ii) a parsimony or likely phylogenetic relationship through mathematics and program operations is claimed as a natural evolutionary system. With the accumulation of fossil material, more and more research work began to use the idea of natural classification to study the phylogenetic relationships of insects.

By combining the morphological character states and molecular data and integrating the whole ancient and modern members, a series of results on the inter-relationships of some important insects such as split-footed lacewings (Neuroptera, Nymphidae) [54], Pamphilioidea (Hymenoptera) [55], Archaeognatha [56], and Evanioidea (Hymenoptera) [57] etc., have been elucidated.

Due to the diversity and well-preservation of fossils found and documented in the Middle Jurassic of Jiulongshan Formation and the Lower Cretaceous of Yixian Formation in Northern and Northeastern China, many fossils of insects and co-existing animals and plants (Chapter 2) provided evidence and/or implications about their paleobiological associations. Many papers have been published to document interactions among insects with plants, vertebrates, and other insects. These paleobiological associations are highlighted and summarized in Chapter 28 Insect Feeding; Chapter 29 Camouflage, Mimicry or Eyespot Warning; and Chapter 30 Gene Propagation – Courtship, Mating and The Next Generation.

In addition, multiple methods and advanced technologies have been and/or are now being used in the study of fossil insects in China. Many findings and results from the studies of the Mesozoic fossils of Northern China will be reviewed and summarized in this book.

4.6 International Cooperative Studies (2000–Present)

Since the beginning of the twenty-first century, paleoentomology in China has become more open and international, while many extensive cooperative researches have been carried out with international scholars. Many international scholars, such as Dr. Dany Azar, Dr. Olivier Béthoux, Dr. Rolf G. Beutel, Dr. Steven R. Davis, Dr. David L. Dilcher, Dr. Michael S. Engel, Dr. David Grimaldi, Dr. Edmund A. Jarzembowski, Dr. Wiesław Krzemiński, Dr. Conrad C. Labandeira, Dr. Vladimir N. Makarkin, Dr. Maria E. McNamara, Dr. Fernando Montealegre-Z, Dr. André Nel, Dr. Julián F. Petrulevicius, Dr. Alexander G. Ponomarenko, Dr. Jakub Prokop, Dr. Alexandr P. Rasnitsyn, Dr. Jes Rust, Dr. Jörg W. Schneider, Dr. Paul A. Selden, Dr. Chungkun Shih, Dr. Nina D. Sinitshenkova, Dr. Adam Ślipiński, Dr. Jacek Szwedo, Dr. Peter Vršanský, Dr. Shaun L. Winterton, and Dr. Evgeny V. Yan, have maintained close cooperative research projects and activities with Chinese scholars and made important contributions to the development of the paleoentomology (Table 4.1).

In addition, other entomologists from all over the world have occasionally taken part in the research of fossil insects in China, and published many important articles. They include Dr. Ahrens Dirk, Dr. Daniil. S. Aristov, Dr. Alexei Bashkuev, Dr. Sven Bradler, Dr. Donald R. Davis, Dr. Suryendu Dutta, Dr. Martin Fikáček, Dr. Oliver Flint, Dr. Thorsten K. Frank, Dr. Gonzalo Giribet, Dr. Sam W. Heads, Dr. Carol L. Hotton, Dr. Katarzyna Kopec, Dr. Frédéric M. B. Jacques, Dr. James E. Jepson, Dr. Iwona Kania, Dr. Kopec Katarzyna, Dr. Alexander V. Khramov, Dr. Alexander G. Kirejtshuk, Dr. Boris C. Kondratieff, Dr. Richard A.B. Leschen, Dr. Antónia Monteiro, Dr. Alexey G. Moseyko, Dr. Alfred F. Newton, Dr. Georgiy V. Nikolajev, Dr. Michael Ohl, Dr. Jaime Ortega-Blanco, Dr. John D. Oswald, Dr. Martina Pecharová, Dr. Swagata Paul, Dr. Alexander A. Prokin, Dr. Chris Reid, Dr. David A. Rider, Dr. Ninon Robin, Dr. Jorge A. Santiago-Blay, Dr. Michael J. Sharkey, Mr. Matthew J. H. Shih, Dr. Jae-Cheon Sohn, Dr. Alexey Solodovnikov, Dr. Agnieszka Soszynska-Maj, Dr. Bill Stark, Dr. Frauke Stebner, Dr. Sergey Yu. Storozhenko, Dr. Ryszard Szadziewski, Dr. Sergey Tarasov, Dr. Margaret K. Thayer, Dr. Ana L. Tourinho, Dr. Torsten Wappler, Dr. Steffen Trümper, Dr. Jörg Wunderlich, *et al.*

During 20–28 August 2010, Capital Normal University was honored to organize and host the Fifth International Conference on Fossil Insects, Arthropods and Amber (FOSSIL X3) in Beijing, China. The objectives of the Conference were to discuss new research findings related to all the fields of Fossil Insects, Arthropods and Amber, to extend the goals of the International Palaeoentomological Society (IPS), and to provide local highlights and cultures to the Congress attendees. About 160 participants from 18 countries attended this Conference, including Argentina, Australia, Czech, France, Germany, India, Kazakhstan, Lebanon, Mexico, New Zealand, Poland, Russia, Slovak, Spain, Ukraine, United Kingdom, United States and China (Figure 4.9). A total of 53 papers and 113 abstracts were received for the abstract volume of the Conference. Ten oral sessions were presented and a three-day post-conference excursion was arranged to examine the geology and Jurassic-Cretaceous Biota of Northeastern China. In addition, a workshop on Advances in Fossil Diptera was conducted during this Conference. A special issue with 460 pages, comprising the preface and 33 papers and covering a broad spectrum of Paleoentomology, in the *Acta Geologica Sinica (English Edition)* (2010, Vol. 84, No. 4, edited by Dr. Dong Ren) was published before the opening of Conference.

Table 4.1 Contributions by researchers to Chinese paleoentomology in stages over time.

	Contributors	1923–1935	1936–1964	1965–1984	1985–1990	1991–1995	1996–2000	2001–2005	2006–2010	2011–2015	2016–2018
Early Foundational Studies	Grabau, Amadeus W.	–									
	Ping, Chi	–									
Early Taxonomic Studies	Hong, Youchong			–	–	–	–	–	–	–	–
	Lin, Qibin			–	–	–	–	–	–	–	–
Early Occasional Studies	Chen, Sicien			–							
	Tan, Chuan-Chieh			–							
	Zhou, Yao			–	–						
Major Taxonomic Studies	Zhang, Junfeng				–	–	–	–	–	–	
	Ren, Dong				–	–	–	–	–	–	–
	Zhang, Haichun				–	–	–	–	–	–	–
	Shih, Chungkun							–	–	–	–
	Huang, Di-ying						–	–	–	–	–
	Yang, Ding							–	–	–	–
	Cai, Wanzhi							–	–	–	–
	Yang, Xingke								–	–	–
	Pang, Hong							–	–	–	–
	Bai, Ming								–	–	–
	Yao, Yunzhi							–	–	–	–
	Liu, Xingyue								–	–	–
	Wang, Yongjie								–	–	–
	Wang, Bo								–	–	–
Major Phylogenetic and Palaeobiological Studies	Ren, Dong						–	–	–	–	–
	Zhang, Haichun								–	–	–
	Shih, Chungkun							–	–	–	–
	Huang, Di-ying									–	–
	Yao, Yunzhi								–	–	–
	Liu, Xingyue									–	–
	Wang, Yongjie								–	–	–
	Wang, Bo									–	–
Important International Cooperative Studies	Azar, Dany								–	–	–
	Béthoux, Olivier								–	–	–
	Beutel, Rolf G.									–	
	Davis, Steven R.								–	–	–
	Dilcher, David L.								–	–	
	Engel, Michael S.						–	–	–	–	–

(Continued)

Table 4.1 (Continued)

Contributors	1923–1935	1936–1964	1965–1984	1985–1990	1991–1995	1996–2000	2001–2005	2006–2010	2011–2015	2016–2018
Grimaldi, David						–		–		
Jarzembowski, Edmund A.								–	–	–
Krzemiński, Wiesław									–	–
Labandeira, Conrad C								–	–	–
Makarkin, Vladimir N.								–	–	–
McNamara, Maria E.									–	–
Montealegre-Z Fernando									–	–
Nel, André								–	–	–
Petrulevicius Julián F.,								–	–	
Ponomarenko, Alexander G.								–	–	
Prokop, Jakub								–	–	–
Rasnitsyn, Alexandr P.							–	–	–	–
Rust, Jes									–	–
Selden, Paul A.								–	–	–
Sinitshenkova, Nina D.							–	–		
Ślipiński, Adam									–	–
Szwedo, Jacek								–	–	–
Vršanský, Peter								–	–	–
Winterton, Shaun L.									–	–
Yan, Evgeny V.								–	–	–

Figure 4.9 About 160 participants from 18 countries attended the Fifth International Conference on Fossil Insects, Arthropods and Amber (FOSSIL X3) hosted by the Capital Normal University in Beijing, China. Source: Photo provided by Dr. Dong Ren.

References

1 Grabau, A.W. (1924–1928). *Stratigraphy of China (Part 1 and Part 2)*. Beijing: Geological Survey of China.

2 Grabau, A.W. (1923). Cretaceous fossils from Shantung. *Bulletin of the Geological Survey of China* **5**: 143–182.

3 Ping, C. (1928). Cretaceous fossil insects of China. *Palaeontoligia Sinica* **13**: 1–47.

4 Ping, C. (1932). On a Blattoid insect in the Fushun amber. *Bulletin of the Geological Society of China* **11**: 212–214.

5 Ping, C. (1935). On four fossil insects from Sinkiang. *The Chinese Journal of Zoology* **1**: 107–115.

6 Hong, Y.C. (1975). Eine neue fossile - Sinosiricidae (Hymenoptera: Siricoidea) in west-Weichangder Provinz Hebei. *Acta Entomologica Sinica* **18** (2): 235–241.

7 Hong, Y.C. (1980). *Fossil Insects. In Mesozoic Stratigraphy and Palaeontology of Shaan-Gan-Ning Basin*. Institute of Geology, Chinese Academy of Geological Sciences (eds.), 111–141. Beijing: Geological Publishing House.

8 Hong, Y.C. (1980). New fossil insects of late Paleozoic from Shanxi Province. *Geological Review* **26**: 89–95.

9 Hong, Y.C. (1982). *Mesozoic Fossil Insects of Jiuquan Basin in Gansu Province*. Beijing: Geological Publishing House.

10 Hong, Y.C. (1983). *Middle Jurassic Fossil Insects in North China*. Beijing: Geological Publishing House.

11 Hong, Y.C. (1990). *The Stratigraphy and Palaeontology of Laiyang Basin*. Beijing: Geological Publishing House.

12 Hong, Y.C. (2002). *Amber Insects of China*. Beijing: Science Technological Press.

13 Lin, Q.B. (1965). Two insects from the lower part of Jurassic, Inner Mongolia. *Acta Palaeontologica Sinica* **13** (2): 363–368.

14 Lin, Q.B. (1976). The Jurassic fossil insects from west Liaoning. *Acta Palaeontologica Sinica* **15**: 97–116.

15 Lin, Q.B. (1978). On the fossil Blattoidea of China. *Acta Entomologica Sinica* **21**: 335–342.

16 Lin, Q.B. (1980). Mesozoic insect fossils from Zhejiang and Anhui. In: *Division and Correlation of Mesozoic Vlovano-Sedimentary Formation in Zhejiang and Anhui Province*. NIOGA Palaeontologys (ed.), 211–238. Beijing: Science Press.

17 Lin, Q.B. (1986). *Early Mesozoic Fossil Insect from the South China*. Beijing: Science Press.

18 Lin, Q.B. (1992). Late Triassic insect fauna from Toksun, Xinjiang. *Acta Palaeontologica Sinica* **31**: 313–335.

19 Lin, Q.B. (1994). Cretaceous insects of China. *Cretaceous Research* **15**: 305–316.

20 Zhang, J.F. (1990). On *Chironomaptera* Ping 1928 (Diptera, Insecta), from late Mesozoic of East Asia. *Mesozoic Research* **2**: 237–247.

21 Zhang, J.F. (1992). Late Mesozoic entomofauna from Laiyang, Shandong Province, China, with discussion of its palaeoecological and stratigraphical significance. *Cretaceous Research* **13**: 133–145. https://doi.org/10.1016/0195-6671(92)90031-K.

22 Zhang, J.F. (2002). The most primitive earwigs (Archidermaptera, Dermaptera, Insecta) from the upper Jurassic of Nei Monggol Autonomous Region, Northeastern China. *Acta Micropalaeontologica Sinica* **19**: 348–362.

23 Zhang, J.F. and Zhang, H.C. (2003). *Kalligramma jurachegonium* sp. nov. (Neuroptera: Kalligrammatidae) from the Middle Jurassic of Northeastern China. *Oriental Insects* **37**: 301–308.

24 Zhang, J.F. (2013). Snipe flies (Diptera: Rhagionidae) from the Daohugou Formation (Jurassic), Inner Mongolia, and the systematic position of related records in China. *Palaeontology* **56**: 217–228. https://doi.org/10.1111/j.1475-4983.2012.01192.x.

25 Zhang, J.F. (1989). *Fossil Insects from Shanwang, Shandong, China*. Jinan: Shandong Science and Technology Publishing House.

26 Zhang, J.F., Sun, B., and Zhang, X.Y. (1994). *Miocene Insects and Spiders from Shanwang, Shandong*. Beijing: Science Press.

27 Zhang, H.C. (1997). Early Cretaceous insects from the Dalazi Formation of the Zhixin Basin, Jilin Province, China. *Palaeoworld* **7**: 75–103.

28 Zhang, H.C. and Zhang, J.F. (2000). Xyelid sawflies (Insecta, Hymenoptera) from the upper Jurassic Yixian Formation of Western Liaoning, China. *Acta Palaeontologica Sinica* **39**: 476–492.

29 Zhang, H.C. and Rasnitsyn, A.P. (2004). Pelecinid wasps (Insecta, Hymenoptera, Proctotrupoidea) from the Cretaceous of Russia and Mongolia. *Cretaceous Research* **25**: 807–825. https://doi.org/10.1016/j.cretres.2004.08.002.

30 Wang, B., Szwedo, J., and Zhang, H.C. (2012). New Jurassic Cercopoidea from China and their evolutionary significance (Insecta: Hemiptera). *Palaeontology* **55**: 1223–1243. https://doi.org/10.1111/j.1475-4983.2012.01185.x.

31 Wang, B., Zhang, H.C., Jarzembowski, E.A. *et al.* (2013). Taphonomic variability of fossil insects: a biostratinomic study of Palaeontinidae and Tettigarctidae (Insecta: Hemiptera) from the Jurassic Daohugou Lagerstätte. *Palaios* **28**: 233–242. https://doi.org/10.2110/palo.2012.p12-045r.

32 Wang, B., Rust, J., Engel, M.S. *et al.* (2014). A diverse Paleobiota in Early Eocene Fushun amber from China. *Current Biology* **24**: 1606–1610. https://doi.org/10.1016/j.cub.2014.05.048.

33 Zhang, H.C., Wang, B., and Fang, Y. (2015). *Mesozoic and Cenozoic Insects from Northern China*. Shanghai: Shanghai Scientific & Technical Publishers.

34 Huang, D.-Y., Engel, M.S., Cai, C. *et al.* (2012). Diverse transitional giant fleas from the Mesozoic era of China. *Nature* **483**: 201–204. https://doi.org/10.1038/nature10839.

35 Huang, D.-Y., Nel, A., Cai, C.Y. *et al.* (2013). Amphibious flies and paedomorphism in the Jurassic period. *Nature* **495**: 94–97. https://doi.org/10.1038/nature11898.

36 Cai, C.Y., Thayer, M.K., Engel, M.S. *et al.* (2014). Early origin of parental care in Mesozoic carrion beetles. *Proceedings of the National Academy of Science of the United States of America* **111** (39): 14170–14174. https://doi.org/10.1073/pnas.1412280111.

37 Cai, C.Y., Lawrence, J.F., Ślipiński, A., and Huang, D.-Y. (2015). Jurassic artematopodid beetles and their implications for the early evolution of Artematopodidae (Coleoptera). *Systematic Entomology* **40** (4): 779–788. https://doi.org/10.1111/syen.12131.

38 Huang, D.-Y. (2017). *The Daohugou Biota*. Shanghai: Shanghai Scientific & Technical Publishers.

39 Ren, D., Lu, L.W., Guo, Z.G., and Ji, S.A. (1995). *Faunae and Stratigraphy of Jurassic - Cretaceous in Beijing and the Adjacent Areas*. Beijing: Seismic Publishing House.

40 Ren, D., Shih, C.K., Gao, T.P. *et al.* (2010). *Silent Stories–Insect Fossil Treasure from Dinosaur Era of the Northeastern China*. Beijing, China: Science Press.

41 Shih, C.K., Feng, H., Liu, C.X. *et al.* (2010). Morphology, phylogeny, evolution, and dispersal of pelecinid wasps (Hymenoptera: Pelecinidae) over 165 Million Years. *Annals of the Entomological Society of America* **103** (6): 875–885.

42 Ren, D., Shih, C.K., Gao, T.P. *et al.* (2012). *Insect Fossil Treasures from the Mesozoic of the Northeastern China*. Beijing: Science Press (in Chinese).

43 Yang, Q., Wang, Y.J., Labandeira, C.C. *et al.* (2014). Mesozoic lacewings from China provide phylogenetic insight into evolution of the Kalligrammatidae (Neuroptera). *BMC Evolutionary Biology* **14** (126): 1–30.

44 Du, S.L., Yao, Y.Z., Ren, D., and Zhang, W. (2017). Dehiscensicoridae fam. nov. (Insecta: Heteroptera: Pentatomomorpha) from the upper Mesozoic of Northeast China. *Journal of Systematic Palaeontology* **15**: 991–1013. https://doi.org/10.1080/14772019.2016.1259665.

45 Ren, D. (1998). Flower-associated brachycera flies as fossil evidence for Jurassic angiosperm origins. *Science* **280**: 85–88. https://doi.org/10.1126/science.280.5360.85.

46 Ren, D., Labandeira, C.C., Santiago-Blay, J.A. *et al.* (2009). A probable pollination mode before angiosperms: Eurasian, long-proboscid scorpionflies. *Science* **326** (5954): 840–847. https://doi.org/10.1126/science.1178338.

47 Wang, Y.J., Liu, Z.Q., Wang, X. *et al.* (2010). Ancient pinnate leaf mimesis among lacewings. *Proceedings of the National Academy of Sciences of the United States of America* **107** (37): 16212–16215. https://doi.org/10.1073/pnas.1006460107.

48 Wang, Y.J., Labandeira, C.C., Shih, C.K. *et al.* (2012). Jurassic mimicry between a hangingfly and a ginkgo from China. *Proceedings of the National Academy of Sciences of the United States of America* **109** (50): 20514–20519. https://doi.org/10.1073/pnas.1205517109.

49 Gao, T.P., Shih, C.K., Xu, X. *et al.* (2012). Mid-Mesozoic flea-like ectoparasites of feathered or haired vertebrates. *Current Biology* **22** (8): 732–735. https://doi.org/10.1016/j.cub.2012.03.012.

50 Gao, T.P., Shih, C.K., Rasnitsyn, A.P. *et al.* (2013). New transitional fleas from China highlighting diversity of Early Cretaceous ectoparasitic insects. *Current Biology* **23** (13): 1261–1266. https://doi.org/10.1016/j.cub.2013.05.040.

51 Yao, Y.Z., Cai, W.Z., Xu, X. *et al.* (2014). Blood-feeding true bugs in the Early Cretaceous. *Current Biology* **24** (15): 1786–1792. https://doi.org/10.1016/j.cub.2014.06.045.

52 Ren, D. and Hong, Y.C. (1994). A cladistics study on the familial phylogeny of fossil and living Raphidioptera (Insecta). *Bulletin of the Chinese Academy of Geological Science* **29**: 103–117.

53 Ren, D. (2017). To re-construct an evolutionary tree conforming to the natural history process is the ultimate goal of systematic biology. *Acta Entomologica Sinica* **60** (6): 699–709. https://doi.org/10.16380/j.kcxb.2017.06.010.

54 Shi, C.F., Winterton, S.L., and Ren, D. (2015). Phylogeny of split-footed lacewings (Neuroptera, Nymphidae), with descriptions of new Cretaceous fossil species from China. *Cladistics* **31**: 455–490. https://doi.org/10.1111/cla.12104.

55 Wang, M., Rasnitsyn, A.P., Li, H. *et al.* (2016). Phylogenetic analyses elucidate the inter-relationships of Pamphilioidea (Hymenoptera, Symphyta). *Cladistics* **32**: 239–260. https://doi.org/10.1111/cla.12129.

56 Zhang, W.T., Li, H., Shih, C.K. *et al.* (2017). Phylogenetic analyses with four new Cretaceous bristletails reveal inter-relationships of Archaeognatha and Gondwana origin of Meinertellidae. *Cladistics* https://doi.org/10.1111/cla.12212.

57 Li, L.F., Rasnitsyn, A.P., Shih, C.K. *et al.* (2018). Phylogeny of Evanioidea (Hymenoptera, Apocrita), with descriptions of new Mesozoic species from China and Myanmar. *Systematic Entomology* **43**: 810–842. https://doi.org/10.1111/syen.12315.

5

Ephemeroptera – Mayflies

Mei Wang[1,2], *Qingqing Lin*[2], *Chungkun Shih*[2,3], *and Dong Ren*[2]

[1] *Research Institute of Forest Ecology, Environment and Protection, Chinese Academy of Forestry, Haidian District, Beijing, China*
[2] *Capital Normal University, Haidian District, Beijing, China*
[3] *National Museum of Natural History, Smithsonian Institution, Washington, DC, USA*

5.1 Introduction to Ephemeroptera

Ephemeroptera, commonly known as mayflies, are the most basal living linages of winged insects, comprising over 3100 known extant species in 42 families worldwide up to date. Mayflies, of ethereal beauty and with a slender and soft body, glossy abdomen and transparent wings, are often found in or near unpolluted and running freshwater (Figures 5.1 and 5.2). The mayfly adults (imagos) are short-lived, only from a few hours to a few days. Mayflies take no food during their adult life, but stay busy to fulfill one of their vital life missions to mate and to reproduce. Imagoes of most species have aerial mating flight either over the water or over nearby landmarks. Under the bright sunshine, these mayflies swarm in slow flights reflecting the sunlight with their majestic wings and glossy bodies while engaging their mating rituals.

The aquatic stage of mayfly nymphs takes much longer than the short lifespan of adults. The nymphs undergo a series of molts as they grow in the freshwater, and reach the stage of subimagoes (Figures 5.3 and 5.4), and soon after that, turning to imagoes. Male and female adults take to the air and form swarms, conduct aerial mating, and then females lay eggs in the water. Within a short time, all adults perish and become food for fish, birds and other predators.

The body sizes of Ephemeroptera range from 3 to 30 mm in length, with three body sections like all other insects, the head, thorax and abdomen, and two pairs of wings fluted with regular alternation of convex and concave veins. Heads have two big eyes, especially in males, and vestigial mouths as the adults don't feed at all. Antennae are short, filiform and with multi-antennomeres. The thorax is highly developed for flight, but mayflies are not good fliers. Fore legs of most males elongate, often equal to the body in length, and are used for seizing females during aerial mating. Wings are triangular, hind wings with reduced venations and much smaller than forewings. When they are at rest, both pairs of wings are held rigidly upright above the body. The abdomen has 10 segments, and the first segment is broadly joined to the metathorax.

Ephemeropterans are special for their prometabolous metamorphosis, and the nymphal stages of mayflies typically may be extended up to four years. Mayfly eggs are deposited in water, and then hatched into nymphs varying from a week to a year depending on the species and temperature. The nymphs are always aquatic, and undergo a series of molts as they grow [1, 2]. The morphology of nymphs differs from adults in the developed mouthparts, abdominal gills and the variable body form. Moreover, nymphs of various species are normally restricted to a particular type of habitat, ranging from lakes, large rivers to temporary ponds and streams, correlating with different morphological structures adapted for predations, scrapers or burrowers. The mayfly has a unique maturation stage in its life cycle, that is, the subimago between the nymph and imago. Subimagoes (Figures 5.3 and 5.4) look similar to the adults, but are immature in sex and have dull and smoky wings [3]. The subimago molts into the imago with clear and transparent wings within hours to two days. Most imagoes live no more than 24 hours, usually ranging from a few minutes to several days. Swarms of males fly above the water surface in a mating ritual, grabbing the females into the swarm. The copulation normally takes place in flight as they sink slowly toward the ground or the water surface, and mating only takes a few seconds and usually less than a minute.

They are sensitive to impurities and particularly dissolved oxygen levels and are reliable indicators of water quality [4]. They also help to purify freshwaters and are critical for filtering particulate matter from streams [5, 6], so they have considerable significance for aquatic ecosystems and human economies.

Rhythms of Insect Evolution: Evidence from the Jurassic and Cretaceous in Northern China, First Edition.
Edited by Dong Ren, Chungkun Shih, Taiping Gao, Yongjie Wang, and Yunzhi Yao.

Figure 5.1 A mayfly imago. Source: Photo by Jason Shih.

Figure 5.2 A mayfly imago. Source: Photo by Jason Shih.

Figure 5.3 A mayfly nymph molted to subimago. Source: Photo by Jason Shih.

In late August 2013, there was an amazing mayfly spectacle along the River Danube near the town of Tahitotfalu, about 29 km from Budapest, in Hungary. Millions of mayfly adults, *Ephoron virgo* Olivier, 1791, swarmed at night, gathering around street lights, dancing in the night sky, colliding with the faces of street walkers and

Figure 5.4 A mayfly subimago. Source: Photo by Jason Shih.

cyclists, or crawling and covering the parked cars and road surfaces. Next morning, mayfly corpses littered and piled up on the ground. However, many female adults, after mating, deposited their eggs in the river for the next generation. Efforts in reducing pollution and cleaning up the river have resulted in this mayfly reproduction swarm and natural wonder, enjoyed by local people and visiting tourists.

Mayflies in Ancient Chinese Literature

The beautiful wings and graceful dance movements of mayflies and their short life span as adults have been noticed and written about by Chinese literati and poets since ancient times. In *The Book of Songs*, the earliest collection of poems and songs in ancient China from 1100 BCE to the 600 CE, there is an ancient song on mayflies written in the Zhou Dynasty (1146 BCE to 256 BCE):

蜉蝣之羽,衣裳楚楚。心之忧矣,于我归处。

and in translation:

Mayfly's wings, like a fancy and bright dress. But it has such a short life span, I have great concerns about my destination for life voyage.

The other example is by Su Shi, also known as Su Tungpo, a prominent Chinese writer, poet, calligrapher, and a statesman of the Song dynasty. He wrote the Former Article on Red Cliff (Chibi) in the autumn of 1082 CE (Figure 5.5) with these sentences:

驾一叶之扁舟,举匏樽以相属。寄蜉蝣于天地,渺沧海之一粟。哀吾生之须臾,羡长江之无穷。

and in translation:

Riding in a narrow small boat on the river, we raise the glasses to toast each other. Like the mayflies in the vast space between the sky and earth, we are so small like a piece of millet in the ocean. We lament that our lives last only such a short instance, and envy the long Yangtze River without ending.

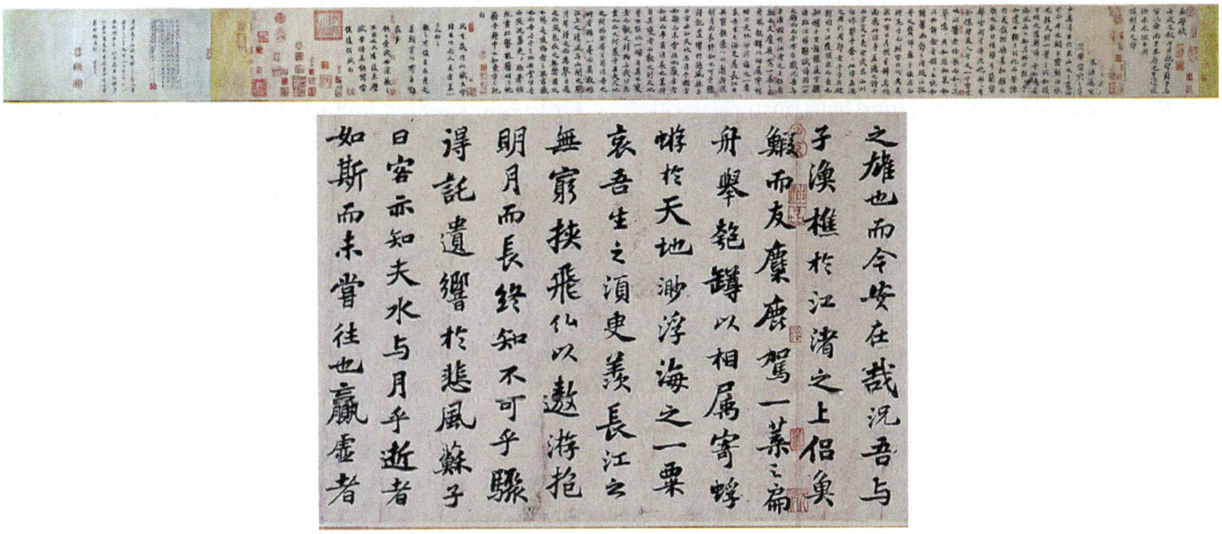

Figure 5.5 Calligraphy of Su Shi on his article of the Red Cliff in Song Dynasty (1082 CE) (A treasured collection of the Palace Museum of Taipei, Taiwan).

5.2 Progress in the Studies of Fossil Ephemeroptera

The study of ephemeropteran fossils started in the early nineteenth century, having more than 200 years of history up to now [7]. Mayfly fossils are believed to have existed since the Late Carboniferous and have survived and thrived ever since. Several extant families of Ephemeroptera were already in existence by the end of the Jurassic. Mayfly fossils are widely distributed in strata from the Late Carboniferous (305 Mya) to the Pliocene of Neogene (5.2 to 2.5 Mya) on six continents except for Antarctica. To date, 44 families, 142 genera and 268 species of fossil mayflies have been reported from all over the world.

Ephemeroptera, commonly known as mayflies, are the most basal living linages of winged insects, comprising over 3100 known extant species in 42 families worldwide to date. The oldest true mayfly fossils belong to the families of Jarmilidae and Oboriphlebiidae, dating back to the Early Permian, about 295–290 Mya. Recent phylogenetic research indicates that Ephemeroptera together with Odonata, considered as Paleoptera, form a sister group to all other modern winged insects of Neoptera [8, 9].

In China, the study of mayfly fossils started about 80 years ago, with only a few reports published. In 1928, Dr. Chi Ping re-described the first ephemeropteran fossil *Ephemeropsis trisetalis* from the Jehol Biota, Liaoning Province [10], and in 1935, three new genera and four species were reported from Turpan, Xinjiang [11]. Since then, more paleoentomologists have been involved in studies of new fossil findings, especially from Northeastern China. So far, there have been more than 23 species, 17 genera belonging to five families described in Northern China (Table 5.1). During this period, significant contributions have been made by Youchong Hong, Qibin Lin, Wuli Wang, Junfeng Zhang, Dong Ren, Chungkun Shih, Jiandong Huang, Nina. D. Sinitshenkova, and others.

5.3 Representative Fossils of Ephemeroptera from Northern China

Suborder Euplectoptera Tillyard, 1932

Infraorder Anteritorna Kluge, 1993

Family Fuyoidae Zhang & Kluge, 2007

Fuyoidae, an extinct family established with only one genus, are distinguished by forewing buds of larva widely separated from bases, hind wing buds half as long as forewing buds; inner margin of fore femur, tibia and tarsus covered with long setae; all leg claws bear a broad-based, dactyl-like appendage arising from claw base; cerci and paracercus subequal, each inner margin of cerci and lateral margin of paracercus with a regular row of long dense primary swimming setae. For imago, hind wing as long as half of forewing. Tarsus five-segmented, paracercus vestigial [12]. To date, the family contains only one described species.

Only one genus included from the Jurassic of Northern China: *Fuyous* Zhang & Kluge, 2007.

Fuyous Zhang & Kluge, 2007

Fuyous Zhang & Kluge, 2007, *Orient. Insects*, 41, 352–354 [12] (original designation).

Type species: *Fuyous gregarious* Zhang & Kluge, 2007.

Larva. Head nearly prognathous, maxilla seems to be strongly elongate and arched. Fore leg with femur thicker than those of middle and hind legs; posterolateral projections on abdominal segments prominent from the first to the ninth. All tergalii from the first to the seventh segments present; each tergalius unilamellate, with rounded tip; costal rib locates on fore (costal) margin, anal rib absent or invisible; on fore leg of female the first segment longest, about 1.3 times longer than the fifth, and twice longer than the second; the third slightly shorter than the second, and slightly longer than the fourth.

Distribution and age: Inner Mongolia; Middle Jurassic.

Only one species included from the Jurassic of Northern China (see Table 5.1).

Family Hexagenitidae Lameere, 1917

Hexagenitidae, an extinct family from the Early Jurassic to the mid-Cretaceous, are moderate to very large in body size. For imagos, CuA of forewing bifurcates to CuA_1 and CuA_2, from the bifurcation a vein "iCu" arises, which forms several (3–6) triads following one another: anterior branch of each triad forms next triad; each of these triads has anterior branch arched by its convexity anteriorly. $RSp+MA_1$, $iMA+MA_2$, MP_2+CuA, MP_1 and iMP are more or less approximated to $iMA+MA_2$ in forewings and hind wings. For larvae, tergalius VII is enlarged, distinctly longer than others, while tergalii I–VI have subequal length [13]. They have been recorded in Algeria, Brazil, Transbaikalia, China, German, Mongolia, and Ukraine. The earliest record of Hexagenitidae appeared in Transbaikalia and Mongolia from the Early to Middle Jurassic, the latest record of Hexagenitidae appeared in Myanmar (Burmese) amber from mid-Cretaceous hitherto [14]. To date, Hexagenitidae comprise 14 genera with 25 species.

Genera included from the Jurassic and Cretaceous of Northern China: *Ephemeropsis* Eichwald, 1864, *Siberiogenites* Sinitshenkova, 1985, *Epicharmeropsis* Huang, Ren & Shih, 2007 and *Shantous* Zhang & Kluge, 2007.

Siberiogenites Sinitshenkova, 1985

Siberiogenites Sinitshenkova, 1985, *The Jurassic Insects of Siberia and Mongolia*, 221, pp. 20–21 [15] (original designation).

Type species: *Siberiogenites angustatus* Sinitshenkova, 1985.

Larva mid-sized. Each tergalia has a rib on the costal margin and a strong anal rib at a distance from the anal margin. The paracercus and inner margins of the cerci are covered with long and dense swimming setae at a distance from their base.

Distribution and age: Liaoning; Early Cretaceous.

Only one species included from the Cretaceous of Northern China (see Table 5.1).

Epicharmeropsis Huang, Ren & Shih, 2007

Epicharmeropsis Huang, Ren & Shih, 2007, *Zootaxa*, 1629, 40–48 [13] (original designation).

Type species: *Epicharmeropsis hexavenulosus* Huang, Ren & Shih, 2007.

Imago. Mesonotal suture (MNs) in anterior part of mesonotum strongly stretched backwards medially, not transverse; paired medioparapsidal sutures (MPs) jointed at middle area of mesonotum, not parallel; membrane of forewing thickened at distal part of the field between C and RA; RSp unforked, the distinct intercalary veins existing between MP_2 and CuA_1 of forewing; cross-veins and intercalary veins numerous between longitudinal veins. Hind wing about half as long as forewing.

Distribution and age: Hebei, Liaoning; Early Cretaceous.

Two species included from the Cretaceous of Northern China (see Table 5.1).

Epicharmeropsis hexavenulosus Huang, Ren & Shih, 2007 (Figures 5.6 and 5.7)

Epicharmeropsis hexavenulosus Huang, Ren & Shih, 2007: *Zootaxa*, 1629, 40–42.

Locality and horizon: Shimen Village, Pingquan, Hebei, China; Lower Cretaceous, Yixian Formation.

The body length excluding caudalii is 31.0 mm, forewing length is 34.0 mm, hind wing length, 18.0 mm. Eyes relatively large, contiguous; metanotum long, with evident scutum; forewing broad and triangular with costal brace, MA_1 and MA_2 uniformly diverging,

Figure 5.6 *Epicharmeropsis hexavenulosus* Huang, Ren & Shih, 2007 (Holotype, CNU-E-YX-2007001-2). Source: Donated by Dr. Chungkun Shih.

Figure 5.7 A 3-D reconstruction of *Epicharmeropsis hexavenulosus* Huang, Ren & Shih, 2007. Source: Artwork by Dr. Chen Wang.

CuA_1 of forewing with six triads (loop-shaped veinlets) leading to wing margin. Hind wing broad with fairly obtuse tip; MA bifurcated near middle; MP bifurcated near the base of wing; more than four anal veins present. Three cerci, paracercus very short [13]. For Jehol Biota, three key representatives in the early studies are the conchostracan *Eosestheria middendorfii*, mayfly larva and imago *Ephemeropsis trisetalis*, and fish *Lycoptera jobolensis*. The mayfly *Ephemeropsis trisetalis* occurred in Transbaikalia of Russia and Mongolia. Based on the study of [13], the *Ephemeropsis*-like imago specimens reported before from China, which was classified as *Ephemeropsis trisetalis*, should be placed into *Epicharmeropsis* Huang, Ren & Shih, 2007.

Epicharmeropsis quadrivenulosus Huang, Sinitshenkova & Ren, 2007 (Figure 5.8)

Epicharmeropsis quadrivenulosus Huang, Sinitshenkova & Ren, 2007: *Zootaxa*, 1629, 42–48.

Locality and horizon: Jianshangou, Beipiao, Liaoning; Shimen Village, Pingquan, Hebei; Dakangpu Village, Yixian, Liaoning; Huangbanjigou, Beipiao, Liaoning. All in China, Lower Cretaceous, Yixian Formation.

Imago. The forewing length is 38.5 mm, hind wing length is 21.0 mm. CuA_1 of forewing with four triads (loop-shaped veinlets) leading to wing margin. RSa of left hind wing branched and iRSa absent, RSa of right hind wing unforked and iRSa present [13].

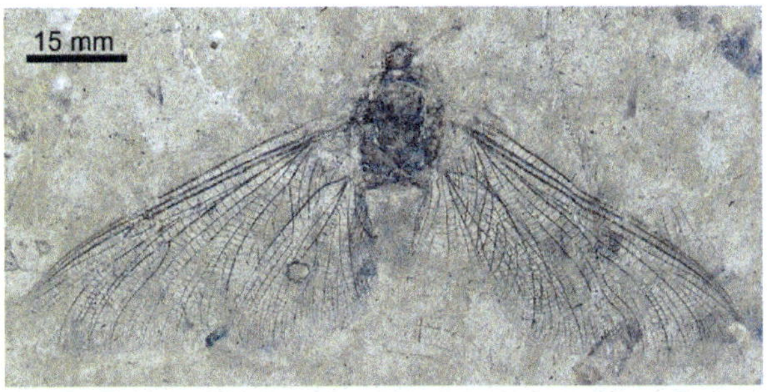

Figure 5.8 *Epicharmeropsis quadrivenulosus* Huang, Sinitshenkova & Ren, 2007 (Holotype, CNU-E-YX-2007002).

Shantous Zhang & Kluge, 2007

Shantous Zhang & Kluge, 2007, *Orient. Insects*, 41, 354–356 [12] (original designation).

Type species: *Shantous lacustris* Zhang & Kluge, 2007.

Larvae. The first to sixth tergalii very long (not shorter than the seventh tergalius) and lack anal ribs. The seventh tergalius widened, retains anal rib near middle of tergalius. Each tergalius from the first to the seventh has well-developed costal rib on costal margin.

Distribution and age: Inner Mongolia; Middle Jurassic.

Only one species included from the Jurassic of Northern China (see Table 5.1).

Family Mesonetidae Tshernova, 1969

Mesonetidae, an extinct family, have fossil larvae small to medium in body size, with abdominal segments broad but short, seven pairs of tergalii short and male forceps separated. To date, there are three genera and 25 species from the Late Triassic to Early Cretaceous. The earliest record of Mesonetidae is from the Middle Triassic in Russia [16]. The records of fossil Mesonetidae are from Russia, Mongolia, Ukraine and China.

Genera included from the Jurassic and Cretaceous of Northern China: *Furvoneta* Sinitshenkova, 1990 and *Clavineta* Sinitshenkova, 1991.

Furvoneta Sinitshenkova, 1990

Furvoneta Sinitshenkova, 1990, *The Late Mesozoic insects from Eastern Transbaikalia*, 239, 19 [17] (original designation).

Type species: *Mesoneta lata* Sinitshenkova, 1976.

Larva. The posterolateral projections on abdominal segments 1–8 are clearly extended backward; on the ninth, the posterolateral projections become very long and are nearly as long as the tenth abdominal tergum. Tergalii have a costal rib and anal rib; inner margin of fore femora, tibiae and tarsi are covered with long setae; the paracercus and inner margins of the cerci are covered with swimming setae.

Distribution and age: Liaoning; Middle Jurassic.

Only one species included from the Jurassic of Northern China (see Table 5.1).

Clavineta Sinitshenkova, 1991

Clavineta Sinitshenkova, 1991, *Paleont. Jour.*, 1, 119 [18] (original designation).

Type species: *Clavineta cantabilis* Sinitshenkova, 1991.

Larva. Abdominal segments short and wide, the width approximately four times as long as the length, posterolateral angles of all segments with short denticles, anterior angles slightly rounded, lateral margins almost straight. Anterior legs with long setae. Branchiae with thickened margins, at attachment sites of branchiae small outgrowths are sometimes discernible. Posterolateral denticles of the ninth tergite short, not longer than the preceding segment. Cercus long, paracercus shorter than cercus.

Distribution and age: Hebei, Middle Jurassic; Liaoning, Early Cretaceous.

Three species included from the Jurassic and Cretaceous of Northern China (see Table 5.1).

Superfamily Siphlonuroidea McCafferty, 1995

Family Siphlonuridae Bank, 1900

Siphlonuridae are not only a fairly large extant mayfly family, but also numerous in fossil records. The family is currently defined as monophyletic by the following apomorphies: penis lobes of male imagos cleft in distal half, consisting of a ventral and a dorsal part. In female imagoes, the seminal receptacle is sclerotized distally, forming a copulatory pouch in the eighth abdominal segment [19]. Siphlonuridae comprise two subfamilies: Siphlonurinae Banks, 1900 and Parameletinae Kluge, Studemann, Landolt & Gonser, 1995. There are four extant genera and 49 species described so far, which are only found in the Northern Hemisphere [20]. In addition, there are 24 extinct genera containing 35 species described hitherto from Siberia and Transbaikalia of Russia, Mongolia, China, Baltic region, Bavaria of Germany, France, Brazil, Victoria of Australia, and California, Colorado of USA [21]. The most ancient representative of the family is *Triassonurus doliiformis* Sinitshenkova, 2005 [22] which was collected from the Vosges in France (the early Middle Triassic).

Genera included from the Jurassic and Cretaceous of Northern China: *Olgisca* Demoulin, 1970, *Multiramificans* Huang, Liu, Sinitshenkova & Ren, 2007, *Cheirolgisca* Lin & Huang, 2008 and *Jurassonurus* Huang, Ren, Sinitshenkova & Shih, 2008.

Olgisca Demoulin, 1970

Olgisca Demoulin, 1970, *Bull. Inst. Roy. Sci. Nat. Belg.*, 46 (4), 4–7 [23] (original designation).

Type species: *Paedephemera schwertschlageri* Handlirsch, 1906.

Imago. Forewing triangular shape, almost eight veins going from CuA to basitornal margin; without intercalary veins between RSa_1 and RSa_2; MA stem is 1.5–2 times as long as MA fork; two cerci.

Distribution and age: Inner Mongolia; Middle Jurassic.

Only one species included from the Jurassic of Northern China (see Table 5.1).

Multiramificans Huang, Liu, Sinitshenkova & Ren, 2007

Multiramificans Huang, Liu, Sinitshenkova & Ren, 2007, *Ann. Zool.*, 57 (2), 222–223 [21] (original designation).

Type species: *Multiramificans ovalis* Huang, Liu, Sinitshenkova & Ren, 2007.

Imago. Metathorax unusually long, with evident scutum and scutellum; MA fork almost as long as MA stem; cubital area obviously large, CuA straight and connected to posterior margin by more than 15 veinlets; hind wing large with eight longitudinal veins behind medial area.

Distribution and age: Inner Mongolia; Middle Jurassic.

Only one species included from the Jurassic of Northern China (see Table 5.1).

Multiramificans ovalis Huang, Liu, Sinitshenkova & Ren, 2007 (Figure 5.9)

Multiramificans ovalis Huang, Liu, Sinitshenkova & Ren, 2007: *Ann. Zool.*, 57 (2), 222–223.

Locality and horizon: Daohugou, Ningcheng, Inner Mongolia, China; Middle Jurassic, Jiulongshan Formation.

Imago. Body length without the tenth abdominal segment is 18.5 mm; length of forewing fragment is 12.5 mm as preserved. RA and SC almost parallel to the costal margin; RS forming three subordinate triads; MA symmetrically forks before the middle area of wing; CuA straight, elongate and connected to wing margin by 18 simple veinlets. Hind wings: ovoid, relatively large; costal vein of hind wing smooth and without costal projection;

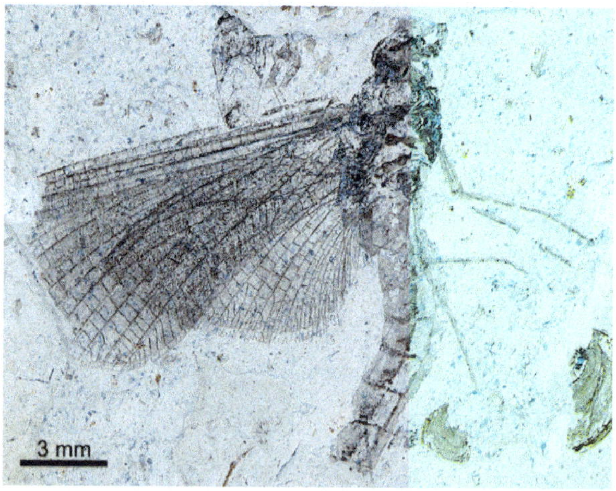

Figure 5.9 *Multiramificans ovalis* Huang, Liu, Sinitshenkova & Ren, 2007 (Holotype, CNU-E-DHG-2006001-1).

eight longitudinal veins parallel to wing tornoapical (outer) margin with numerous cross-veins among these longitudinal veins. Tibia of middle leg is evidently longer than hind one [21].

Cheirolgisca Lin & Huang, 2008

Cheirolgisca Lin & Huang, 2008, *Ann. Zool.*, 58(3), 522–523 [24] (original designation).

Type species: *Cheirolgisca ningchengensis* Lin and Huang, 2008.

Imago. The hind wing triangle-shaped and about one third as long as the forewing; three cerci, paracercus reduced; gonostyles 4-segmented, in forcipiform; penis not outwardly prominent.

Distribution and age: Inner Mongolia; Middle Jurassic.

Only one species included from the Jurassic of Northern China (see Table 5.1).

Jurassonurus Huang, Ren & Sinitshenkova, 2008

Jurassonurus Huang, Ren & Sinitshenkova, 2008, *Insect Sci.*, 15, 194–197 [25] (original designation).

Type species: *Jurassonurus amoenus* Huang, Ren and Sinitshenkova, 2008.

Imago. Prothorax relatively large, about one-third of pterothorax length; A1 unforked and an intercalary vein present between RSa1 and RSa2, few short intercalate veins at the wing margin. Hind wings are long oval with slightly developed costal projection. All legs with vestige of patella-tibial suture; femora of middle and hind legs evidently stronger than tibiae and tarsi; forceps with two apical segments and penis bifurcated in its apical part.

Distribution and age: Inner Mongolia; Middle Jurassic.

Only one species included from the Jurassic of Northern China (see Table 5.1).

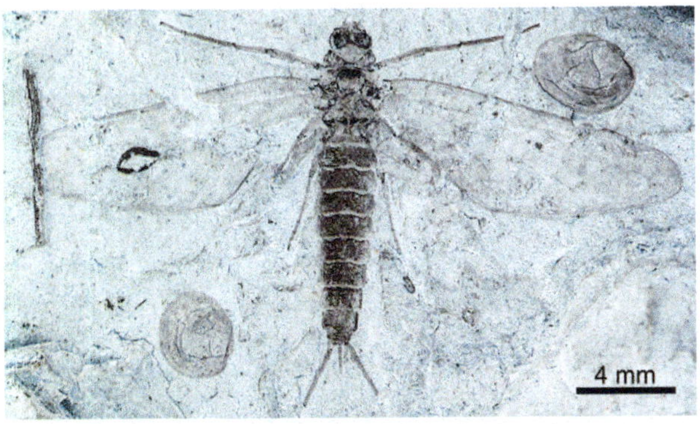

Figure 5.10 *Jurassonurus amoenus* Huang, Ren & Sinitshenkova, 2008 (Holotype, CNU-E-DHG-2006006-1). Source: Donated by Dr. Chungkun Shih.

Jurassonurus amoenus Huang, Ren & Sinitshenkova, 2008 (Figure 5.10)

Jurassonurus amoenus Huang, Ren & Sinitshenkova, 2008: *Insect Sci.*, 15, 194–197.

Locality and horizon: Daohugou, Ningcheng, Inner Mongolia, China; Middle Jurassic, Jiulongshan Formation.

Imago. The body length is 13.5 mm, forewing, 12.7 mm and hind wing, 6.0 mm. In male, eyes relatively large and contiguous, but small and separated in female; antenna only about two-thirds of head length. Thorax with clear basisternum (BS) and furcasternal protuberances (FSp) of mesothorax. A more or less wide furcasternal impression (FSi) appears between the furcasternal protubertances (FSp). MA branching after middle area of wing; cubital area strongly narrowing apically, with few simple veinlets arising from CuA. All tarsi five-segmented. Forceps are with two apical segments. Penis slightly V-shaped bifurcation in its apical part. Rudiment of paracercus is short. Cerci are very long, about 1.2 times as long as abdomen length [25].

Superfamily Siphlonuroidea McCafferty, 1995

Family Siphluriscidae Zhou & Peters, 2003

Siphluriscidae, a newly established family, are typified by imago's straight penis without accessory structures, the deeply divided styliger plate, hind wings more than half length of forewings, gill remnants on fore and middle coxae, and claws similar. For the larvae, gills labial-shaped, the third segment of the labial gills constrict, larval claw bears an unusually large flexible projection on inner side [26]. The family comprises two genera, the extant *Siphluriscus* Ulmer, 1920 and the Jurassic *Stackelbergisca* Tshernova, 1967. To date, the extant genus *Siphluriscus* only includes one Chinese species *Siphluriscus chinensis* Ulmer, 1920. The extinct genus *Stackelbergisca* contains three described species from Transbaikalia of Russia and China.

Only one genus included from the Jurassic of Northern China: *Stackelbergisca* Tshernova, 1967.

Stackelbergisca Tshernova, 1967

Stackelbergisca Tshernova, 1967, *Entomologicheskoe Obozrenie*, 46, 323–324 [27] (original designation).

Type species: *Stackelbergisca sibirica* Tshernova, 1967.

Imago, forewing triangular; anal margin long; CuA straight and connected to wing margin by a series of veinlets; CuP slightly curved. Larva, with seven pairs of foliate gills along both sides of abdomen.

Distribution and age: Inner Mongolia; Middle Jurassic.

Only one species included from the Jurassic of Northern China (see Table 5.1).

Family *Incertae sedis*

Caenoephemera Lin & Huang, 2001

Caenoephemera Lin & Huang, 2001, *Can. Entomol.*, 133, 748–751 [28] (original designation).

Type species: *Caenoephemera shangyuanensis* Lin & Huang, 2001.

Larva. Head rounded, triangular shape; mandibles well-developed with two apical teeth; pterothorax large, with two pairs of wing pads curved posteriorly; the hind pads not covered by the anterior pads and reduced in size, both articulated beside each other to the pterothorax; legs strong, each with five tarsal segments and one apical pretarsal claw; abdomen with seven pairs of gills, the first to the sixth pairs narrow and the seventh pair greatly enlarged, broad, and elliptical-elongate; three cerci shorter than body, the median filament bearing setae on both sides and the cerci with setae only on their inner surfaces.

Distribution and age: Liaoning; Early Cretaceous.

Only one species included from the Cretaceous of Northern China (see Table 5.1).

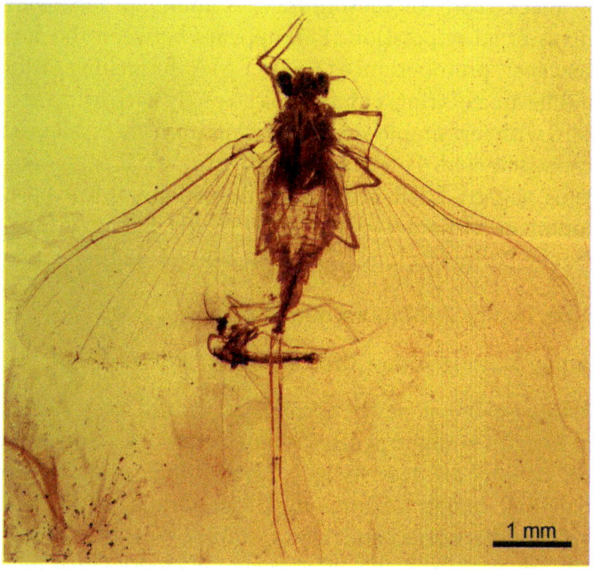

Figure 5.11 *Proximicorneus rectivenius* Lin, Shih & Ren, 2018 (Holotype, CNU-EPH-MA2017001).

A Mayfly in Myanmar Amber

A well-preserved female subimago mayfly *Proximicorneus rectivenius* Lin, Shih & Ren, 2018 of Prosopistomatidae (Figure 5.11) from the mid-Cretaceous Burmese amber has been described [29]. The amber species has the following characters: Moderate size, body length 4.0 mm excluding long cerci. AA vein of forewings situated almost at the tornus (vs. AA vein situated behind tornus in extant female *Prosopistoma*). Legs well-developed, with normal femur, tibia and 3-segmented tarsus; claws with one hooked and the other blunt (vs. legs reduced in great degree, tibia and tarsus fused to form an indefinite remnant but no more than 2-segmented tarsus in extant female *Prosopistoma*). Cerci long and fulfilled with setae while paracercus short, unsegmented and without setae. The new amber mayfly is the only and the earliest fossil record of Prosopistomatidae hitherto, dating back to the mid-Cretaceous and indicating the long-term presence of Prosopistomatidae in the Oriental region.

Table 5.1 A list of fossil Ephemeroptera from the Jurassic and Cretaceous of China.

Family	Species	Locality	Horizon/Age	Citation
Fuyoidae	*Fuyous gregarius* Zhang & Kluge, 2007	Ningcheng, Inner Mongolia	[b]Jiulongshan Fm., J_2	Zhang and Kluge [12]
Hexagenitidae	[a]*Ephemeropsis trisetalis* Eichwald, 1864	Luanping, Hebei	Dabeigou Fm., K_1	Wang [30]
	Shantous lacustris Zhang & Kluge, 2007	Ningcheng, Inner Mongolia	[b]Jiulongshan Fm., J_2	Zhang and Kluge [12]
	Epicharmeropsis hexavenulosus Huang, Ren & Shih, 2007	Pingquan, Hebei	Yixian Fm., K_1	Huang *et al.* [13]
	Epicharmeropsis quadrivenulosus Huang, Ren & Shih, 2007	Beipiao, Liaoning	Yixian Fm., K_1	Huang *et al.* [13]
		Pingquan, Hebei	Yixian Fm., K_1	Huang *et al.* [13]
	Siberiogenites branchicillus Huang, Sinitshenkova & Ren, 2011	Beipiao, Liaoning	Yixian Fm., K_1	Huang *et al.* [31]
Mesonetidae	[a]*Mesoneta antiqu* Brauer, Redtenbacher & Ganglbauer, 1889	Beipiao, Liaoning;	Haifanggou Fm., J_2	Hong [32]
		Luanping, Hebei	Jiulongshan Fm., J_2	Hong [32]
	[a]*Mesoneta beipiaoensis* Wang, 1980	Beipiao, Liaoning	Haifanggou Fm., J_2	Wang [30]

(*continued*)

Table 5.1 (Continued)

Family	Species	Locality	Horizon/Age	Citation
	[a]*Mesoneta* sp. Wang, 1980	Beipiao, Liaoning	[b]Tuchengzi Fm., J_3–K_1	Wang [30]
	Clavineta eximia Zhang, 2006	Luanping, Hebei	Jiulongshan Fm., J_2	Zhang [33]
	Clavineta excavata Huang, Sinitshenkova & Ren, 2011	Beipiao, Liaoning	Yixian Fm., K_1	Huang *et al.* [31]
	Clavineta brevinodia Huang, Sinitshenkova & Ren, 2011	Beipiao, Liaoning	Yixian Fm., K_1	Huang *et al.* [31]
	Furvoneta relicta Zhang, 2006	Beipiao, Liaoning	Haifanggou Fm., J_2	Zhang [33]
Siphlonuridae	[a]*Huizhougenia orbicularis* Lin, 1980	Huangshan, Anhui	Yantang Fm., J_3-K_1	Lin [35]
	[a]*Mesobaetis sibirica* Brauer, Redtenbacher & Ganglbauer, 1889	Tuha Basin, Xinjiang	Sanjianfang Fm., J_2	Hong *et al.* [34]
		Luanping, Hebei	Jiulongshan Fm., J_2	Wang [30]
		Chaoyang, Liaoning	[b]Jiulongshan Fm., J_2	Hong [32]
		Beipiao, Liaoning	Haifanggou Fm., J_2	Hong [32]
	[a]*Mesobaetis maculata* Hong, Liang & Hu, 1995	Tuha Basin, Xinjiang	Sanjianfang Fm., J_2	Hong *et al.* [34]
	[a]*Mesobaetis sanjianfangensis* Hong, Liang & Hu, 1995	Tuha Basin, Xinjiang	Sanjianfang Fm., J_2	Hong *et al.* [34]
	[a]*Mesobaetis latifilamentacea* Zhang, 2006	Chengde, Hebei	Jiulongshan Fm., J_2	Zhang [33]
	Multiramificans ovalis Huang, Liu & Sinitshenkova, 2007	Ningcheng, Inner Mongolia	Jiulongshan Fm., J_2	Huang *et al.* [21]
	Cheirolgisca ningchengensis Lin & Huang, 2008	Ningcheng, Inner Mongolia	Jiulongshan Fm., J_2	Lin and Huang [24]
	Jurassonurus amoenus Huang, Ren & Sinitshenkova, 2008	Ningcheng, Inner Mongolia	Jiulongshan Fm., J_2	Huang *et al.* [25]
	Olgisca angusticubitis Lin & Huang, 2008	Ningcheng, Inner Mongolia	Jiulongshan Fm., J_2	Lin and Huang [24]
Siphluriscidae	*Stackelbergisca cylindrata* Zhang, 2006	Ningcheng, Inner Mongolia	[b]Jiulongshan Fm., J_2	Zhang [33]
Family *Incertae sedis*	*Caenoephemera shangyuanensi* Lin & Huang, 2001	Beipiao, Liaoning	Yixian Fm., K_1	Lin and Huang [28]

a) The species is not presented in the main text because the original description, photo, and line-drawings are not precise and the holotype cannot be rechecked.

b) Horizon/Age revised from the original paper based on updated information and data.

References

1 Ruffieux, L., Sartori, M., and Eplattenier, G.L. (1996). Palmen body: a reliable structure to estimate the number of instars in *Siphlonurus aestivalis* (Eaton) (Ephemeroptera: Siphlonuridae). *International Journal of Insect Morphology and Embryology* **25**: 341–344. https://doi.org/10.1016/0020-7322(96)00004-9.

2 Brittain, J.E. and Sartori, M. (2003). Ephemeroptera. In: *Encyclopedia of Insects* (ed. V.H. Resh and R.T. Cardé), 373–380. Amsterdam: Academic Press.

3 Edmunds, G.F. and McCafferty, W.P. (1988). The mayfly subimago. *Annual Review of Entomology* **33**: 509–529. https://doi.org/10.1146/annurev.en.33.010188.002453.

4 Hubbard, M.D. and Peters, W.L. (1978). *Environmental Requirements and Pollution Tolerance of Ephemeroptera*, 461. Cincinnati: Environmental Protection Agency.

5 Merrit, R.W. and Cummins, K.W. (1978). *An Introduction to the Aquatic Insects of North America*, 441. Dubuque: Kendall Hunt Publishing Company.

6 Ward, J.V. (1992). *Aquatic Insect Ecology*, 438. New York: Wiley.

7 Hubbard, M.D. (1987). Ephemeroptera. In: *Fossilium Catalogus I: Animalia, Pars 129* (ed. F. Westphal), 99. Amsterdam: Kugler Publication.

8 Kukalova-Peck, J. (1991). Fossil history and the evolution of hexapod structures. In: *The Insects of Australia*, 2e, vol. **1** CSIRO (ed. I.D. Naumann), 141–179. Melbourne: Melbourne University Press.

9 Misof, B., Liu, S., Meusemann, K. *et al.* (2014). Phylogenomics resolves the timing and pattern of insect evolution. *Science* **346**: 763–767. https://doi.org/10.1126/science.1257570.

10 Ping, C. (1928). Cretaceous fossil insects of China. *Palaeontologia Sinica* **13**: 1–47.

11 Ping, C. (1935). On four fossil insects from Sinkiang. *The Chinese Journal of Zoology* **1**: 107–115.

12 Zhang, J.F. and Kluge, N.J. (2007). Jurassic larvae of mayflies (Ephemeroptera) from the Daohugou Formation in Inner Mongolia, China. *Oriental Insects* **41**: 351–366. https://doi.org/10.1080/00305316.2007.10417519.

13 Huang, J.D., Ren, D., Sinitshenkova, N.D., and Shih, C.K. (2007). New genus and species of Hexagenitidae (Insecta: Ephemeroptera) from Yixian Formation, China. *Zootaxa* **1629**: 39–50. https://doi.org/10.5281/zenodo.179369.

14 Lin, Q.Q., Kong, L.J., Shih, C.K. *et al.* (2018). The latest record of Hexagenitidae (Insecta: Ephemeroptera) with elongated abdominal sternum IX from mid-Cretaceous Myanmar amber. *Cretaceous Research* **91**: 140–146.

15 Sinitshenkova, N.D. (1985). The Jurassic mayflies (Ephemerida = Ephemeroptera) of South Siberia and West Mongolia. In: *The Jurassic Insects of Siberia and Mongolia*, vol. **211** (ed. A.P. Rasnitsyn), 11–23. Trudy Paleontologischeskogo Instituta Akademii Nauk, SSSR (in Russian).

16 Sinitshenkova, N.D. (2000). A review of Triassic mayflies, with a description of new species from Western Siberia and Ukraine (Ephemerida = Ephemeroptera). *Paleomological Journal* **34** (3): 275–283.

17 Sinitshenkova, N.D. (1990). The mayflies (Ephemerida). In: *The Late Mesozoic Insects from Eastern Transbaikalia*, vol. **239** (ed. A.P. Rasnitsyn), 13–20. Trudy Paleontologischeskogo Instituta Akademii Nauk, SSSR (in Russian).

18 Sinitshenkova, N.D. (1991). New Mesozoic mayflies from Transbaikal and Mongolia. *Paleont ologicheskii Zhurnal* **1**: 116–119. in Russian).

19 Kluge, N.J., Studemann, D., Landolt, P., and Gonser, T. (1995). A reclassification of Siphlonuroidea (Ephemeroptera). In: *Mitteilungen Der Schweizerischen Entomologischen Gesellschaft*, vol. **68**, 103–132.

20 Barber-James, M.H., Gattolliat, J.L., Sartori, M., and Hubbard, M.D. (2008). Global diversity of mayflies (Ephemeroptera, Insecta) in freshwater. *Hydrobiologia* **595**: 339–350. https://doi.org/10.1007/s10750-007-9028-y.

21 Huang, J.D., Liu, Y.S., Sinitshenkova, N.D., and Ren, D. (2007). A new fossil genus of Siphlonuridae (Insecta: Ephemeroptera) from the Daohugou, Inner Mongolia, China. *Annales Zoologici* **57** (2): 221–225.

22 Sinitshenkova, N.D., Marchal-Papier, F., Grauvogel-Stamm, L., and Gall, J.C. (2005). The Ephemeroptera (Insecta) from the Gres a Voltzia (early Middle Triassic) of the Vosges (NE France). *Palaontologische Zeitschrift* **79** (3): 377–397.

23 Demoulin, G. (1970). Contribution à l'étude morphologique systématique et phylogenique des Ephéméroptères jurassiques d'Europe centrale. V. Hexagenitidae = Paedephemeridae (syn. nov.). *Bulletin of the Royal Belgian Institute of Natural Sciences* **46** (4): 4–7.

24 Lin, Q.B. and Huang, D.Y. (2008). New Middle Jurassic mayflies (Insecta: Ephemeroptera: Siphlonuridae) from Inner Mongolia, China. *Annales Zoologici* **58** (3): 521–527. https://doi.org/10.3161/000345408X364346.

25 Huang, J.D., Ren, D., Sinitshenkova, N.D., and Shih, C.K. (2008). New fossil mayflies (Insecta: Ephemeroptera) from the Middle Jurassic of Daohugou, Inner Mongolia, China. *Insect Science* **15**: 193–198. https://doi.org/10.1111/j.1744-7917.2008.00200.x.

26 Zhou, C.F. and Peters, J.G. (2003). The nymph of *Siphluriscus chinensis* and additional imaginal description: a living mayfly with Jurassic origins (Siphluriscidae new family: Ephemeroptera). *Florida Entomologist* **86** (3): 345–352. https://doi.org/10.1653/0015-4040(2003)086[0345:TNOSCA]2.0.CO;2.

27 Tshernova, O.A. (1967). Mayflies of the recent family in Jurassic deposits of Transbaikalia (Ephemeroptera, Siphlonuridae). *Entomologicheskoe Obozrenie* **46**: 322–326. (in Russian).

28 Lin, Q.B. and Huang, D.Y. (2001). Description of *Caenoephemera shangyuanensis*, gen. nov., sp. nov. (Ephemeroptera) from the Yixian Formation. *The Canadian Entomologist* **133**: 747–754. https://doi.org/10.4039/Ent133747-6.

29 Lin, Q.Q., Shih, C.K., Zhao, Y.Y., and Ren, D. (2018). A new genus and species of Prosopistomatidae (Insecta: Ephemeroptera) from mid-Cretaceous Myanmar amber. *Cretaceous Research* **84**: 401–406. https://doi.org/10.1016/j.cretres.2017.11.020.

30 Wang, W.L. (1980). *Paleontological Atlas of Northeast China*, vol. **2** Mesozoic and Cnozoic, 131–132. Beijing: Geological Publishing House (in Chinese).

31 Huang, J.D., Sinitshenkova, N.D., and Ren, D. (2011). New mayfly nymphs (Insecta: Ephemeroptera) from Yixian Formation, China. *Paleontological Journal* **45** (2): 167–173. https://doi.org/10.1134/S0031030111020092.

32 Hong, Y.C. (1983). *Middle Jurassic Fossil Insects in North China*, 20–22. Beijing: Geological Publishing House (in Chinese).

33 Zhang, J.F. (2006). New mayfly nymphs from the Jurassic of northern and Northeastern China (Insecta: Ephemeroptera). *Paleontological Journal* (5): 80–86. https://doi.org/10.1134/S0031030106050091.

34 Hong, Y.C., Liang, S.J., and Hu, T. (1995). Study on geology and paleontological assemblage from Tuha Basin of Xinjiang, China. *Geosciences* **9**: 426–440. (in Chinese with English abstract).

35 Lin, Q.B. (1980). *Fossil Insects of Mesozoic from Zhejiang and Anhui Provinces*, 216–217. Beijing: Science Press (in Chinese).

6

Odonata – Dragonflies and Damselflies

Qiang Yang [1,2], Dong Ren [2], Hong Pang [3], and Chungkun Shih [2,4]

[1] Guangzhou University, Guangzhou, Guangdong, China
[2] Capital Normal University, Haidian District, Beijing, China
[3] Sun Yat-sen University, Guangzhou, Guangdong, China
[4] National Museum of Natural History, Smithsonian Institution, Washington, DC, USA

6.1 Introduction to Odonata

The Order Odonata, including dragonflies and damselflies, belong to Insecta: Pterygota: Odonatoptera. The superorder Odonatoptera comprise the three orders Geroptera, Protodonata and Odonata. The Geroptera with a single family, Eugeropteridae, lived in the early Late Carboniferous (Early Bashkirian). The Protodonata, noted for the giant "griffenflies", existed in the Late Carboniferous to the Permian. One of griffenflies, *Meganeuropsis permiana*, from the Early Permian of Elmo, Kansas and Midco, Oklahoma, with wingspan of about 710 mm, is the largest insect ever documented [1, 2]. The Odonata from the Permian to the present comprise three suborders: Anisoptera (dragonflies), Zygoptera (damselflies) and Anisozygoptera (damselflies, an ancient suborder including only three extant species, one in Japan, one in Nepal and the other in China) [3].

As one of the most basal groups of insects, Odonata are valuable for the study of evolution, especially the origination and development of wings. The Latin word of "Odon" means "tooth" highlighting the strong mandibles ("teeth") of these Odonata species. The extant odonatans vary widely in size with body length ranging from 15 to 135 mm and wingspan up to 190 mm. They have four elongated membranous wings with complicated veins. The wings have a notch, or node, at the center of the anterior margin; and a pterostigma near the tip of the anterior margin. The fore- and hind wings are almost equal in length and similar in shape for damselflies (Figure 6.1), but hind wings are broader at the base than forewings for dragonflies (Figure 6.2). Most damselflies, when at rest, hold the wings together above their body toward the rear, but dragonflies hold the wings horizontally outstretched. However, there are some exceptions. Dragonflies and damselflies are excellent fliers. Dragonflies, capable of flying at high speed, hovering at the same position and even flying backward, are good hunters in the sky. Odonata species can move their four wings independently for better maneuverability. Only lacewings (Neuroptera) and cicadas (Homoptera) can move their four wings independently, although they are not as good fliers as Odonata. Odonata heads are flexible and movable. Antennae are very short and bristle-like. There are about 6000 known species of Odonata worldwide, of which about 2700 are dragonflies.

Dragonflies and damselflies mate in the air. The mating rituals are displayed by pairs coupling together and flying in tandem. A male deposits a sperm packet in a secondary genitalia organ located below its second abdominal segment for a damselfly and below the third segment for a dragonfly. A male dragonfly grabs a female's head and neck (lower back part of head) with claspers (called cerci) at the end of the male's abdomen (Figure 6.3). But a male damselfly grabs a female's prothorax (Figure 6.4). The mating couple, in a distinctive heart-shape "mating wheel" position, fly together, while the male uses its secondary genitalia to pass the sperm to the tip of abdomen (tail) of the female.

The female has an ovipositor underneath the eighth abdominal segment. Some species use ovipositors to cut slits into aquatic plant stems for depositing eggs. Others just deposit eggs into water (Figure 6.5). Metamorphosis of Odonata is incomplete (hemimetabolous). The young, called naiads (or nymphs or larvae), are vicious aquatic predators. Naiads capture insects, tadpoles, small newts or even small fish. Naiads are equipped with a specially-designed and modified labium (lower lip) that can be extended forward quickly to catch its prey with a pair of spine or claw-like structure at its tip.

Rhythms of Insect Evolution: Evidence from the Jurassic and Cretaceous in Northern China, First Edition.
Edited by Dong Ren, Chungkun Shih, Taiping Gao, Yongjie Wang, and Yunzhi Yao.

Figure 6.1 Damselfly. Source: Photo by Dr. Chungkun Shih.

Figure 6.2 Dragonfly. Source: Photo by Dr. Chungkun Shih.

Figure 6.3 Mating dragonflies. Source: Photo by Dr. Chungkun Shih.

Figure 6.4 Mating damselflies. Source: Photo by Jason Shih.

Figure 6.5 Dragonflies laying eggs. Source: Photo by Jason Shih.

This unique device is hidden under the head when not in use.

A dragonfly naiad draws water into the gills in the rectum to get oxygen, and then expels water to give itself a jet propulsion movement. Some species of bottom dwelling dragonfly naiads, which burrow into the bottom sand or mud, have a siphon tube at the end of the abdomen and hold it above the sediment to siphon water into the rectum gills for breathing. A damselfly naiad has three leaf-like gills at the tail for breathing, in addition to their simple rectal gills. The naiads grow and undergo 10–20 molts before maturity. The time required for naiads to reach full development varies from one month to five to six years, depending on species, location, and climate conditions.

A fully developed naiad crawls out of water and climbs onto a stem of an aquatic plant or a solid surface at the edge of ponds, lakes or streams. Usually, this happens in the dark or in early morning to avoid predation. It splits open the back skin of the thorax and the head and the thorax comes out first. The legs are hardened for holding onto a support. Body fluid is pumped into the veins of wings, which then begin to expand. The abdomen is the last one to emerge. Wings and body parts hardened, the dragonfly or damselfly is ready to fly.

Adult dragonflies and damselflies are usually found near water – ponds, lakes, streams, rivers, marshes and surrounding areas. Adult damselflies live for about a week to two months, while adult dragonflies live a bit longer, about two weeks to three months. Adults and naiads of dragonflies and damselflies are considered to be beneficial insects because they prey on many small insects and help controlling mosquitoes, flies and other harmful small insects.

Odonata are thought of as "bird-watcher's" insects, because of their aerial displays and complex behaviors; they are almost entirely diurnal and have acute vision and an active, powerful, and maneuvered flight. They are numerous and diverse (approximately 6000 species), so dragonflies and damselflies have attracted a great deal of interest from professionals and amateurs. Many odonatologists have reviewed the biology of Odonata, and their taxonomy and systematics, and Grimaldi and Engel [4] suggested a cladogram of Odonatoptera.

The Largest Dragonfly or Damselfly

As mentioned before, one of the griffenflies, *Meganeuropsis permiana*, from the Early Permian of Elmo, Kansas and Midco, Oklahoma, with a wingspan of about 710 mm, is the largest insect ever documented [1, 2]. The "giant griffenflies" belong to the family of Meganeuridae in the suborder of Protodonata, not the true dragonflies in Odonata. In China, the late Carboniferous meganeurid *Shenzhousia qilianshanensis* is the largest insect found in China with a wingspan of 400–500 mm [5], but this species was emended in 2013 [6] with an estimated wing length of about 160 mm and a wingspan of about 330 mm (Figure 6.6).

The largest extant member of the Odonata is the damselfly *Megaloprepus caerulatus* (Drury) living in the Central and South America with a wingspan of up to 190 mm. In the fossil record of Odonata, the Middle

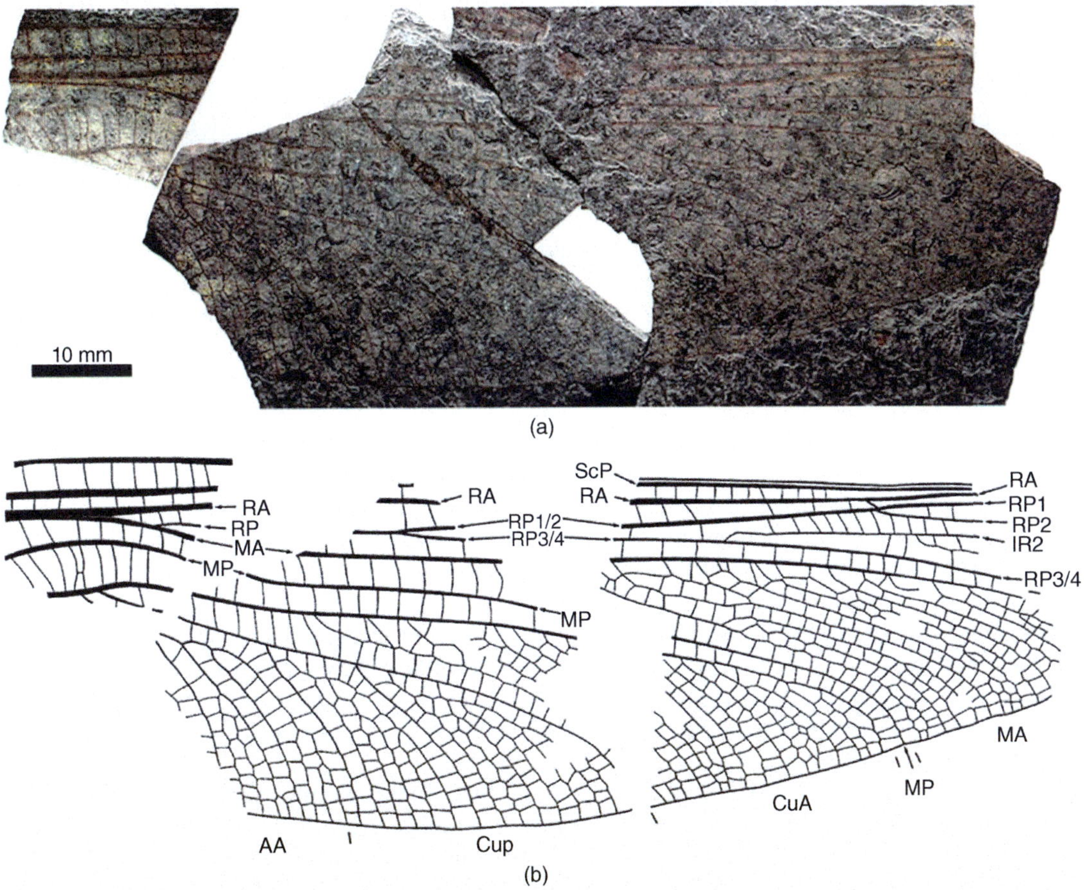

Figure 6.6 *Shenzhousia qilianshanensis* Zhang & Hong, 2006 [5]. (a) Photograph of specimen CNU-NX1-400. (b) Line drawing of specimen CNU-NX1-400 (right wing, positive imprint) [6].

Triassic *Triadotypus guillaumei* Grauvogel & Laurentiaux, 1952 from France is the largest known with a forewing of 136 mm in length and a wingspan of 280 mm (as estimated) [7]. The second largest one is the Middle Jurassic *Hemerobioides giganteus* Westwood, 1845 from the UK with a forewing of 120 mm in length (estimated) [8]. The third one is the Late Jurassic *Isophlebia aspasia* Hagen, 1866 from Germany with a forewing of 110 mm in length and a wingspan of 228 mm (estimated) [9]. The fourth one is the Middle Jurassic *Hsiufua chaoi* Zhang & Wang, 2006 from China with a forewing of 107.6 mm in length and a wingspan of 225 mm (estimated), which is the largest one known from China [10]. The fifth one is the Late Jurassic *Aeschnogomphus kuempeli* Bechly, 2000 from Germany with a forewing of 106 mm in length and a wingspan of 220 mm [11].

6.2 Progress in the Studies of Fossil Odonata

Study of fossil Odonata started in 1880s. In 1884, the famous French paleontologist Charles Brongniart reported *Meganura monyi* from France, which belongs to Protodonata – a basal group of "real" Odonata [12]. Since then, more than 95 families of fossil Odonata have been reported. Fossil Odonata specimens have been found from the Carboniferous to Neogene.

Fossil Odonata have been found from all over the world (including Antarctica) [13]. The earliest Odonata fossil recognized as a "modern" odonate is *Saxonagrion minutus* dated about 268 Mya from the Upper Permian [14]. It appears that odonatans, even in the Palaeozoic, had adapted to ecosystems and developed flying body structures similar to modern dragonflies and damselflies [15]. Since then, different groups rose and fell. Only Anisoptera, Zygoptera and two species of Anisozygoptera have continued until now.

Based on abundant studies by odonatologists, Grimaldi and Engel [4] set up six suborders: Protanisoptera, Archizygoptera, Triadophlebiomorpha, Tarsophlebioptera, Zygoptera and Epiprocta. Former suborders of Anisoptera and Anisozygoptera, which have been widely used by entomologists working on extant insects, are combined into Epiprocta.

In 1965, the Chinese paleontologist Youchong Hong described a new genus *Sinaeschnidia* from the Early Cretaceous of Chifeng, Inner Mongolia, which is the first report of Chinese fossil Odonata [16]. To date, 26 families, 71 genera and 90 species (including six indeterminate species) from China have been recorded. They are from the Late Triassic to the Early Cretaceous of the Mesozoic and the Miocene of the Cenozoic in Shaanxi, Xinjiang, Liaoning, Jilin, Hebei, Shandong, Inner Mongolia, Beijing, Shanxi, Gansu, Ningxia, and Zhejiang. Among them, 22 families, 61 genera and 74 species are from the Mesozoic in Northeastern China (including Liaoning, Jilin, Hebei, Shandong and Inner Mongolia) (Table 6.1).They are from Haifanggou Formation and Jiulongshan Formation of the Middle Jurassic; Chijinqiao Formation, Xiguayuan Formation, Dabeigou Formation and Hangjiahu Formation of the Late Jurassic; Yixian Formation of the Early Cretaceous; Lushangfen Formation, Jiufotang Formation, Shahai Formation, Dongshan Formation, Huliugou Formation, Zhidan Formation and Liupanshan Formation of the Late Cretaceous. Most of these specimens are well-preserved and valuable for morphological and evolutionary studies.

6.3 Representative Fossils of Odonata from Northern China

Family Aeschnidiidae Handlirsch, 1906

Aeschnidiidae are an extinct family which survived from the Late Jurassic to Early Cretaceous. They were usually found near seashores and rarely in deeply intra-continental environments [17]. As such, it has been regarded as a group associated with the seashore. Pritykina and Rasnitsyn (2002) thought it strange to find nymphs in great numbers in lakes thousands of kilometres away from the nearest sea, while aeschnidiid adults either could not be found in West Mongolia or could hardly be found in Transbaikalia [13]. By contrast, Aeschnidiidae are distributed widely in China. They are found from the Jehol Biota and Yixian Formation, both of which are far away from seashores. Furthermore, *Sinaeschnidia heishankowensis* is one of the typical representatives in the Jehol Biota. These findings in China indicate that the label "seashore group" may no longer be applicable to Aeschnidiidae.

Genera included from the Cretaceous of Northern China: *Dracontaeschnidium* Zhang & Zhang, 2001 and *Linaeschnidium* Huang, *Baudoin* & Nel, 2009.

Dracontaeschnidium Zhang & Zhang, 2001

Dracontaeschnidium Zhang & Zhang, 2001, Creta. Res. 22, 445 [18] (original designation).

Type species: *Dracontaeschnidium orientale* Zhang & Zhang, 2001

Hind wing wider basally, with intercalated veins between C and Sc, ending before nodus. Sc quite elongated, reaching basal portion of pterostigma, and then two rows of cells present from N to pterostigma between C and R_1. Primary antenodal cross-veins are far apart from each other. Three rows of cells between R_1 and R_2

before and beyond brace vein. Discoidal triangle nearly twice as high as wide, with its distal side clearly curved.

Distribution and age: Liaoning; Early Cretaceous.

Only one species included from the Cretaceous of Northern China (see Table 6.1).

Linaeschnidium Huang, Baudoin & Nel, 2009

Linaeschnidium Huang, Baudoin & Nel, 2009, Cretac. Res. 30, 805–806 [19] (original designation).

Type species: *Linaeschnidium sinensis* Huang, Baudoin & Nel, 2009.

Pterostigma colored, long, crossed by 13–15 concave veinlets, and distal side oblique and stronger; hind wing distal part of subdiscoidal space not very broad and rather transverse, presence of a vein AAspl and a well-angled AA1b; PsA vein reaching MP+ CuA; two (rarely three) rows of cells between Mspl and MA; three to four rows of cells between Rspl and IR2; two rows of cells between RA and RP1 distal of subnodus; fore- and hind wing antenodal supra-ScP beginning at wing base.

Distribution and age: Liaoning; Early Cretaceous.

Only one species included from the Cretaceous of Northern China (see Table 6.1).

Family Aeshnidae Rambur, 1842

They are the largest extant dragonflies found in North America and Europe and are among the largest dragonflies on the Earth. This family represent also the fastest flying dragonflies of the Odonata.

Genera included from the Cretaceous of Northern China: *Stylaeschnidium* Zhang & Zhang, 2001 and *Aeschna?* Chang & Sun, 2005.

Stylaeschnidium Zhang & Zhang, 2001

Stylaeschnidium Zhang & Zhang, 2001, Cretac. Res. 22, 444 [18] (original designation).

Type species: *Stylaeschnidium rarum* Zhang & Zhang, 2001.

Labial mask nearly flattened but only slightly concave medially. Wing rudiments parallel to one another. Female epiproct nearly triangular. Ovipositor of near-ultimate instar larva exceedingly elongated, only slightly shorter than abdomen. Paraproct almost uncurled distally, and not curved inward at tip.

Distribution and age: Liaoning; Early Cretaceous.

Only one species included from the Cretaceous of Northern China (see Table 6.1).

Aeschna? Chang & Sun, 2005

Aeschna? Chang & Sun, 2005, Glob. Geol. 24 (2), 108 [20] (original designation).

Type species: *Aeschna? acrodonta* Chang & Sun, 2005.

Larva characters only: legs short and thick, femur of foreleg shorter than that of hind leg; with two pairs of slender wing pads; the first and second abdomen narrow, then widen gradually from the third to the sixth; compound eyes big, not obviously, the distance between the eyes is shorter than the width of the back vision of the eyes.

Distribution and age: Liaoning; Early Cretaceous.

Only one species included from the Cretaceous of Northern China (see Table 6.1).

Family Aktassiidae Pritykina, 1968

Aktassiidae, a family as a member of the stem-group of the Petalurida Bechly, 1996, comprise two subfamilies: Pseudocymatophlebiinae Nel, Bechly & Jarzembowski, 1998 and Aktassiinae Pritykina 1968. This family have the characters that the forewing antenodal space is not shorter than the postnodal space and very dense wing venation with a distinctly increased number of cells. Thus, the family are restricted to Middle Jurassic to Lower Cretaceous strata, and at least two aktassiid genera had a very wide distribution in Eurasia. However, this family remain unknown in other parts of the world, unlike many other families of Mesozoic dragonflies.

Genera included from the Jurassic and Cretaceous of Northern China: *Pseudocymatophlebia* Nel, Bechly & Jarzembowski, 1998 and *Sinaktassia* Lin, Nel & Huang, 2010.

Pseudocymatophlebia Nel, Bechly & Jarzembowski, 1998

Pseudocymatophlebia Nel, Bechly & Jarzembowski, 1998, Paleontol. Lombarda, (N.S.), 10, 37 [21] (original designation).

Type species: *Pseudocymatophlebia hennigi* Nel, Bechly & Jarzembowski, 1998

The specific epithet is in honor of Prof. Willi Hennig, the founder of Phylogenetic Systematics. Wings very long with a dense meshwork of numerous cells; pterostigmal brace not very oblique and somewhat basally recessed; bridge cross-veins (Bqs) very numerous; no well-defined Rspl or Mspl but postdiscoidal area and area between IR2 and RP3/4 very wide with many cells; a secondarily very elongated and straight IR1 vanishing distally in area between RP1 and RP2 (autapomorphy), and not fused with the pseudo-IR1; cross-veins between MA and RP3/4 and between IR2 and RP2 oblique towards base of wing; many cross-veins in area between RA and RP basal of nodus.

Distribution and age: Inner Mongolia; Early Cretaceous.

Only one species included from the Cretaceous of Northern China (see Table 6.1).

Sinaktassia **Lin, Nel & Huang, 2010**

Sinaktassia Lin, Nel & Huang, 2010, *Zootaxa*, 2359, 61–62 [22] (original designation).

Sinaktassia: Zheng *et al.* 2016, *Cretac. Research*, 61, 86–90. [23]

Type species: *Sinaktassia tangi* Lin, Nel & Huang, 2010

The specific epithet is in honor of Mr. Yonggang Tang for his secular help to the authors in field work and research. Postnodal space very narrow, with many cells distal of pterostigma; IR1 basally strongly zigzagged, long; only one row of cells between RA and RP1 at level of pterostigma; space between RP1 and RP2 greatly expanded, with more than eight or nine rows of cells; PsA hypertrophied in forewing; subdiscoidal triangle widened in forewing, T transverse; both subdiscoidal triangle and T divided by cross-veins in forewing but free of cross-veins in hind wing; more than two rows of cells in basal part of postdiscoidal area between level of distal angle of T and level of midfork (first furcation of vein RP into RP1/2 and RP3/4); RP3/4 undulating and distally diverging from MA, with more than three rows of cells in-between; anal loop closed posteriorly and seven-celled [22, 23].

Distribution and age: Liaoning; Early Cretaceous.

Only one species included from the Cretaceous of Northern China (see Table 6.1).

Family Araripelibellulidae Bechly, 1996

Araripelibellulidae, an extinct family, comprise six genera and seven species to date, which have been reported from Spain, England, Brazil and China. They are small, elegant but robust, similar to some extant cordullids. The latter are usually active in pristine environment with flourished trees.

Only one genus included from the Cretaceous of Northern China: *Sopholibellula* Zhang, Ren & Zhou, 2006.

Sopholibellula **Zhang, Ren & Zhou, 2006**

Sopholibellula Zhang, Ren & Zhou, 2006, Acta Geol. Sin-Engl. 80 (3), 328 [24] (original designation).

Type species: *Sopholibellula eleganta* Zhang, Ren & Zhou, 2006.

Origins of RP and MA distinctly separated at the arculus in both pairs of wings; anal loop wider and shorter, with Y-shaped veins inside; MA and IR2 not zigzag; several small intercalary veins present in the postdiscoidal area of hind wing; cells smaller and much more dense, especially in the apex and hind margin; bigger in size.

Distribution and age: Liaoning; Early Cretaceous.

Two species included from the Cretaceous of Northern China (see Table 6.1).

Figure 6.7 *Sopholibellula eleganta* Zhang, Ren & Zhou, 2006, (Holotype, CNU-OD-LB2004002p) [24].

Sopholibellula eleganta **Zhang, Ren & Zhou, 2006 (Figure 6.7)**

Sopholibellula eleganta Zhang, Ren & Zhou, 2006: Acta Geol. Sin-Engl. 80 (3), 328.

Locality and horizon: Liaoning, China; Lower Cretaceous, Yixian Formation.

Male hind wings with anal triangle only divided into two cells by a longitudinal cross-vein; Ax2 of forewing at the station close to the distal side of discoidal triangle; long "cordulegastrid gap" and "libellulid gap" in both pairs of wings; anal loop four-celled, without midrib; costal side of both hypertriangle and discoidal triangle distinctly curved [24].

Family Campterophlebiidae Handlirsch, 1920

Campterophlebiidae are the largest family of "Anisozygoptera (damselflies)" which survived from the Jurassic to the Cretaceous. Since the first report in 1905, 27 genera and 46 species have been described. There have been no campterophlebiid records from China until 2002. Since then, five genera with nine species have been reported, all from Daohugou, Ningcheng, Inner Mongolia. They have a stout body, slender wings and strong legs with a sharp spine. According to their fierce appearance and the rather abundant fossil specimens collected, they might have been the dominant damsel-dragonfly group of Daohugou in the latest Middle Jurassic.

Genera included from the Jurassic and Cretaceous of Northern China: *Dorsettia* Whalley, 1985, *Bellabrunetia* Fleck & Nel, 2002, *Amnifleckia* Zhang, Ren & Cheng, 2006, *Parabrunetia* Huang, Fleck, Nel & Lin, 2006, *Sinokaratawia* Nel, Huang & Lin, 2007, *Zygokaratawia* Nel, Huang & Lin, 2008, *Karatawia* Martynov, 1925, *Parafleckium* Li, Nel, Ren & Pang,

2012, *Angustiphlebia* Li, Nel, Ren & Pang, 2013, *Hsiufua* Zhang & Wang, 2013 and *Junfengi* Zheng & Zhang, 2017

Dorsettia Whalley, 1985

Dorsettia Whalley, 1985, Bull. Br. Mus. Nat. Hist. Geol. Ser. 39, 107–187 [25] (original designation).

Type species: *Dorsettia laeta* Whalley, 1985.

The type species of the genus is from the Lower Jurassic of Dorset [25], and in 2016, Zheng *et al.* described a new species *Dorsettia sinica* from Tuziakeneigou [26] and emended the generic diagnosis with these characters: hind wing AA has a strong elbow near CuP, and is directed perpendicularly from its basal side to the distal edge of the wing; CuA strongly deviates from MP, and is divided into CuAa and CuAb far from MP; a very clear division of CuAa and CuAb is present, and CuAb moves basal to the distal edge of the wing; AA and CuAb are limited distally; the subdiscoidal cell reaches the edge of the wing; CuAa disappears distally, and the area between MP and CuAa is reduced; the subdiscoidal space is divided into five smaller cells; the areas between RP2 and IR1, RP3/4 and IR2, MA and MP all have one row of cells basally, and are distally constricted by an opposed curvature of these associated veins; pterostigma is elongate and not braced [25, 26].

Distribution and age: Xinjiang of China; Dorset, England; Lower Jurassic.

Only one species included from the Jurassic of Northern China (see Table 6.1).

Bellabrunetia Fleck and Nel, 2002

Bellabrunetia Fleck & Nel, 2002, Palaeontology, 45, 1123–1124 [27] (original designation).

Type species: *Bellabrunetia catherinae* Fleck & Nel, 2002.

A very long and straight gaff; primary antenodal brackets Ax1 and Ax2 parallel in hind wing, Ax2 oblique in forewing; AA reaching CuA in both pairs of wings; R and MA strongly approximate just distal of arculus in forewing; MAa and RP3/4 only slightly curved; CuAa very short; hind wing area between MP and CuA with two to three rows of cells; a single row of very long and oblique cross-veins in the postdiscoidal area and in the area between RP3/4 and IR2; IR2 not originating on RP3/4.

Distribution and age: Inner Mongolia; Middle Jurassic.

Only one species included from the Jurassic of Northern China (see Table 6.1).

Amnifleckia Zhang, Ren & Cheng, 2006

Amnifleckia Zhang, Ren & Cheng, 2006, Zootaxa, 1339, 53 [28] (original designation).

Type species: *Amnifleckia guttata* Zhang, Ren & Cheng, 2006.

Wings rather long and narrow; forewing distinctly longer than hind wing (by about 9 %); no cross-veins between Ax1 and Ax2; discoidal and subdiscoidal areas free in both pairs of wings; forewing discoidal space basally opened, unlike the closed discoidal space in hind wing; subdiscoidal space free, large, transverse and posteriorly limited by AA; pterostigma basally recessed, with at least two cross-veins below; forewing pterostigma distinctly shorter than that of hind wing; IR2 distinctly originating from RP.

Distribution and age: Inner Mongolia; Middle Jurassic.

Two species included from the Jurassicof Northern China (see Table 6.1).

Amnifleckia guttata Zhang, Ren & Cheng, 2006 (Figure 6.8)

Amnifleckia guttata Zhang, Ren & Cheng, 2006: Zootaxa, 1339, 53.

Locality and horizon: Daohugou, Ningcheng, Inner Mongolia, China; Middle Jurassic, Jiulongshan Formation.

MA and MP with two rows of cells between them, and IR2 and RP3/4 with two rows of cells between them, distinct guttula in the third and fourth abdominal segments [28].

Figure 6.8 *Amnifleckia guttata* Zhang, Ren & Cheng, 2006, (Holotype, D2834) [28].

Parabrunetia Huang, Fleck, Nel & Lin, 2006

Parabrunetia Huang, Fleck, Nel & Lin, 2006, Zootaxa, 1339, 63 [28] (original designation).

Type species: *Parabrunetia celinea* Huang, Fleck, Nel & Lin, 2006.

Ax1 distinctly basal of arculus. Presence of a dense net of irregular cells in anal area of hind wing; hind wing subdiscoidal space crossed by two veins; postdiscoidal area of fore- and hind wing strongly broadened near posterior wing margin; forewing subdiscoidal space distinctly narrower.

Distribution and age: Inner Mongolia; Middle Jurassic.

Only one species included from the Jurassic of Northern China (see Table 6.1).

Sinokaratawia Nel, Huang & Lin, 2007

Sinokaratawia Nel, Huang & Lin, 2007, Zootaxa, 1642, 14 [29] (original designation).

Type species: *Sinokaratawia prokopi* Nel, Huang & Lin, 2007

Forewing Ax2 strongly oblique; subdiscoidal cell posteriorly open in male but posteriorly closed or nearly so in female; MP straight, MAa with a smooth bend, distally zigzagged, a constricted area between MAa and MP; a very acute projecting anal angle in male hind wing; a distinct constriction of the area between RP3/4 and IR2; MAa distally zigzagged; two rows of cells in forewing postdiscoidal area and basal part of area between MP and CuA.

Distribution and age: Inner Mongolia; Middle Jurassic.

Only one species included from the Jurassic of Northern China (see Table 6.1).

Zygokaratawia Nel, Huang & Lin, 2008

Zygokaratawia Nel, Huang & Lin, 2008, European J. Entomol. 105, 784–786 [30] (original designation).

Type species: *Zygokaratawia reni* Nel, Huang & Lin, 2008.

The specific epithet is dedicated to Prof. Dong Ren. Wing characters only. Hind wing subdiscoidal space small and posteriorly closed; cubito-anal and anal areas very narrow in both fore and hind wings (no "anal triangle"); a long basal part to CuA before it branches; area between MP and CuA as broad as postdiscoidal area, which is distally constricted; MP straight; MAa zigzags and becomes much weakened distally; CuAa short; a distinct constriction in the area between RP3/4 and IR2 in the hind wing, but not in the forewing; pterostigma not basally recessed.

Distribution and age: Inner Mongolia; Middle Jurassic.

Only one species included from the Jurassic of Northern China (see Table 6.1).

Karatawia Martynov, 1925

Karatawia Martynov, 1925, Izvestiya Rossiiskoi Akademii Nauk, 17 (6), 569–598 [31] (original designation).

Type species: *Karatawia turanica* Martynov 1925.

The type species of the genus is from the Middle Jurassic of Turkestan [31], and in 2012, Li *et al.* described a new species *Karatawia sinica* from the Middle Jurassic Jiulongshan Formation [32] and amended the generic diagnosis with these characters: in both fore and hind wings, pterostigma elongate and relatively broad, not in a very proximal position, not braced, hind wing pterostigma longer than that of the forewing; IR1 weakly curved; RP2, IR2, RP3/4, MA and MP nearly straight; areas between RP2 / IR2, IR2 / RP3/4 and MA/MP barely expanded near posterior wing margin; MA weakly zigzagged in its distal half; basal area between MP and CuA broader than that of the postdiscoidal area; CuA short; MP + CuA straight [31, 32].

Distribution and age: Inner Mongolia of China, Turkestan, Mongolia, Tadjikistan; Early Jurassic, Middle Jurassic.

Only one species included from the Jurassic of Northern China (see Table 6.1).

Parafleckium Li, Nel, Ren & Pang, 2012

Parafleckium Li, Nel, Ren & Pang, 2012, Cretac. Res. 34, 341 [33] (original designation).

Type species: *Parafleckium senjituense* Li, Nel, Ren & Pang, 2012

The type species of the genus is from the Early Cretaceous of Xituyao [31], and in 2012, Li *et al.* described a new specimen of *Parafleckium senjituense* from Baitugou [34] and amended the generic diagnosis with these characters: In forewing: Distal area between RA and RP1 with two rows of cells and slightly widened; many secondary intercalary veins originating from MP. Gaff (basal part of CuA) straight and very long, correlated with proximal part of area between MP and CuA, about three times as wide as proximal part of area between MA and MP; Ax1 basal of arculus, very oblique; Ax2 almost perpendicular to ScP and RA; a secondary longitudinal vein in anal area parallel to AA and posterior wing margin; a long single cell covering the whole area between MP and CuA, just distal of gaff; MP and RP3/4 curved in mid part; numerous cells below narrow pterostigma; IR2 originating from RP3/4; oblique vein "O" very oblique and strong, only seven cells distal of base of RP2.

In male hind wing: proximal part of area between MP and CuA more than three times as wide as proximal part of postdiscoidal area between MA and MP; cubital area very broad, with seven rows of cells between CuAa and posterior wing margin; CuAa with three main posterior branches; CuAb long and straight, not distally reaching AA, correlated with a very long and posteriorly opened subdiscoidal cell subdivided into several small cells by transverse cross-veins; anal area triangular in shape, rather broad; male anal angle not very sharp [33, 34].

Distribution and age: Hebei, Inner Mongolia; Early Cretaceous.

Only one species included from the Cretaceous of Northern China (see Table 6.1).

Parafleckium senjituense Li, Nel, Ren & Pang, 2012 (Figure 6.9)

Parafleckium senjituense Li, Nel, Ren & Pang, 2012: Cretac. Res. 34, 341–342 and Zootaxa 3597, 54.

Locality and horizon: Baitugou, Nanyingpan, Sanbeigou, Duolun, Inner Mongolia, China; Yixian Formation, Lower Cretaceous.

Parafleckium senjituense is of great importance to clarify the phylogeny of the superfamily Isophlebioidea, as it has several significant structures considered as typical of either Campterophlebiidae or Isophlebiidae [34]. Since all the genera and species within these two families were established based on forewings only, so the new specimen, described by Li *et al.* in 2012, presented a precise description of some other important characters, especially those of the male hind wing. On the basis of this specimen, the diagnosis of *Parafleckium senjituense* has been emended [34].

Angustiphlebia Li, Nel, Ren & Pang, 2013

Angustiphlebia Li, Nel, Ren & Pang, 2013, J. Nat. Hist. 29–30 (47), 1954 [35] (original designation).

Type species: *Angustiphlebia mirabilis* Li, Nel, Ren & Pang, 2013

Hind wing very slender (length: width ratio about 6.0); Ax1 and Ax2 nearly perpendicular to ScP in both wings; arculus very oblique and curved; only one row of cells in area between IR2 and RP3/4 and in area between MA and MP; straight gaff (basal part of CuA) very long; CuAa short in both pairs of wings ending on posterior wing margin about midway between arculus and nodus; subdiscoidal cell free of cross-veins; AA separated into AA1 and AA2 after CuP-crossing; posterior margin of hind wing making a strong "hook" in anal area.

Distribution and age: Inner Mongolia; Middle Jurassic.

Only one species included from the Jurassic of Northern China (see Table 6.1).

Hsiufua Zhang & Wang, 2013

Hsiufua Zhang & Wang, 2013, Chin. Sci. Bull. 58, 1580 [10] (original designation).

Type species: *Hsiufua chaoi* Zhang & Wang, 2013

The specific epithet is dedicated to the famous Chinese entomologist, Prof. Hsiu-fu Chao. Forewing characters only. Forewing extremely narrow and elongate. Gaff (basal part of CuA) straight and very long; proximal part of area between MP and CuA about three times as wide as that between MA and MP; Ax1 basal of arculus, far from wing base and nearly perpendicular to ScP; Ax2 subparallel to Ax1; a secondary longitudinal vein in anal area parallel to AA and posterior wing margin; a long single cell covering basal part of area between MP and CuA, just distal of gaff; area between IR2 and RP3/4 clearly broadened in its mid part and greatly constricted near wing margin; CuAa short; pterostigma narrow and elongate, brown, far from wing apex, with a long, central unpigmented area; IR2 not originating from RP3/4; oblique vein "O" very oblique and strong, only four cells distal of base of RP2.

Distribution and age: Inner Mongolia; Middle Jurassic.

Only one species included from the Jurassic of Northern China (see Table 6.1).

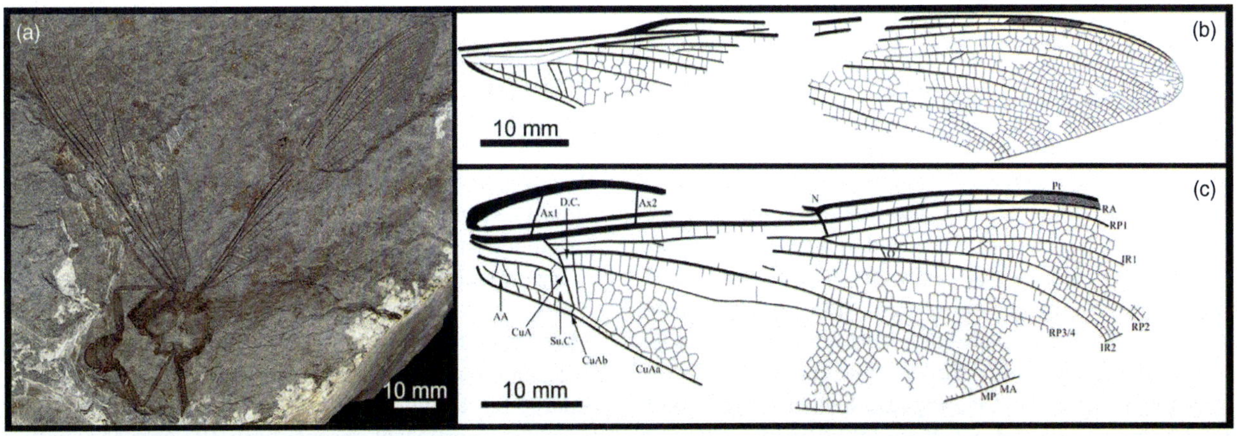

Figure 6.9 *Parafleckium senjituense* Li, Nel, Ren & Pang, 2012, (Holotype, CNU-ODO-LB2010002). (a) Habitus; (b) and (c) Line drawings of forewing and hind wing [33].

Figure 6.10 Photo of forewing of *Hsiufua chaoi* Zhang & Wang, 2013 (NIGP156221) [10]. Source: Photo provided by Dr. Haichun Zhang.

Hsiufua chaoi Zhang & Wang, 2013 (Figure 6.10)

Hsiufua chaoi Zhang & Wang, 2013: Chin. Sci. Bull. 58, 1580.

Locality and horizon: Daohugou, Ningcheng, Inner Mongolia, China; Middle Jurassic, Jiulongshan Formation.

Forewing hyaline, 107.6 mm long, 14.3 mm wide, a wingspan of about 225 mm (estimated). It is the largest odonate known in China.

Junfengi Zheng & Zhang, 2017

Junfengi Zheng & Zhang, 2017, Alcheringa, 41 (4), 509–513 [36] (original designation).

Type species: *Junfengi yulinensis* Zheng & Zhang, 2017

Dedicated to Dr. Junfeng Zhang for his contribution to paleoentomology. Male hind wing characters only: Ax1 and Ax2 parallel and of same obliquity; Arc closer to Ax1 than to Ax2; discoidal cell elongate, with distal side three times longer than basal side; subdiscoidal space wide, posteriorly closed, free and hexagonal; AA with four parallel posterior branches uncrossed in anal triangle; basal area between MA and MP of similar size to that between MP and CuAa; RP3/4 base closer to Arc than to N; IR1 originating closer to pterostigma than to RP2 base; three or four rows of cells between CuAa and posterior wing margin; absence of hook-like anal angle.

Distribution and age: Shaanxi; Middle Jurassic.

Only one species included from the Jurassic of Northern China (see Table 6.1).

Family Congqingiidae Zhang, 1992

Stout body, and of small size; head semicircular; eyes large, not as lateral swellings and separated by less than their width in dorsal view. Pronotum large, transversely elongate; synthorax stout, subcircular, mesothorax wider than long. Abdomen flat and wide, at least with basal three segments rather short and wide, the third one slightly longer than wide. Fore- and hind wings closely similar in shape and venation but quadri-lateral (q) differing in shape, base of wings gradually narrowed, not stalked; nodus (N) distinct, lying less than two-fifths of length of wing from base; only two antenodals (An); IR_3 and R_{4+5} arising somewhat nearer to arculus (Arc) than to subnodus (Sn); fork of R_{2+3} well beyond Sn; origin of Arc proximal to level of the second An and farther from N than from the base of the wing; q closed basally, trapezoid and with wide distal angle; subquadrilateral (sq) of fore- and hind wings similar in shape; anal cross-vein (Ac) near mid-way between position of An; anal vein (A) thick, well-developed as a separate vein at extreme base of wings; anterior branch of A (1A) long, reaching margin beyond middle position of wing; anal field of fore – and hind wings narrow but the former just slightly narrower than the latter, and with a single row of cells basally and two rows terminally.

Only one genus included from the Cretaceous of Northern China: *Congqingia* Zhang, 1992.

Congqingia Zhang, 1992

Congqingia Zhang, 1992, Odonatologica, 21 (3), 376 [37] (original designation).

Type species: *Congqingia rhora* Zhang, 1992

An Early Cretaceous genus of Congqingiidae with forewings longer than hind wings and quadrilateral shorter in forewings than in hind wings.

Distribution and age: Shandong; Early Cretaceous.

Only one species included from the Cretaceous of Northern China (see Table 6.1).

Family Corduliidae Selys-Longchamps, 1850

The Corduliidae comprise the emerald dragonflies or green-eyed skimmers. These dragonflies are usually black or dark brown with areas of metallic green or yellow, and most of them have large, emerald-green eyes. The larvae are black, hairy-looking, and usually semi-aquatic. Members of this family include the baskettails, emeralds, river cruisers, sundragons, shadowdragons, and boghaunters. They are quite common and found nearly worldwide, but some individual species are quite rare. Hine's emerald dragonfly (*Somatochlora hineana*), for example, is an endangered species in the United States.

Only one genus included from the Cretaceous of Northern China: *Mesocordulia* Ren & Guo, 1996.

Mesocordulia Ren & Guo, 1996

Mesocordulia Ren & Guo, 1996 Insect Sci. 3(2), 101 [38] (original designation).

Type species: *Mesocordulia boreala* Ren & Guo, 1996.

At least six antenodals in the forewing and four antenodals in the hind wing; basal side of triangle in the hind wing much closer to arculus, sectors of arculus arising separately; triangles, supratriangles and subtriangles free; Rspl and Mspl absent; both IR_3 and MA with a distinct branch; discoidal field parallel to the wing margin in the forewing and divergent from the beginning in the hind wing; anal loop compact and with eight to 11 cells; IR_2 well-developed; pterostigmal brace present and strongly slanted.

Distribution and age: Liaoning; Early Cretaceous.

Only one species included from the Cretaceous of Northern China (see Table 6.1).

Family Daohugoulibellulidae Nel & Huang, 2015

The Daohugoulibellulidae are the potential sister group of the Mesozoic Nannogomphidae. The distal part of the antesubnodal area is free of cross-veins; basal CuA before its branching distinctly prolonged but straight; no basal furcation of IR2 basal of lestine oblique vein; discoidal triangle transverse in forewing (apomorphy but convergent with the Paneurypalpidomorpha), but discoidal triangle of hind wing more elongate; pterostigma not exactly parallel-sided; hind wing CuA with four posterior branches; anal loop elongate, divided into four cells and without even a rudimentary zigzagged midrib; area between MP and CuA with only one row of cells in basal part; terminal branch of CuA not secondarily branched on CuA; RP3/4 and MA strictly parallel up to hind margin; area between RP2 and IR2 distinctly widened distally, with more than one cell row in distal half; Ax1 and Ax2 well-separated; arculus angled; posterior part (cross-vein) of arculus distinctly shorter than anterior part [RP and MA]; hind wing MP not shortened, ending well distal of level of nodus; first lestine oblique vein four cells distal of subnodus in both pairs of wings.

Only one genus included from the Jurassic of Northern China: *Daohugoulibellula* Nel & Huang, 2015.

Daohugoulibellula Nel & Huang, 2015

Daohugoulibellula Nel & Huang, 2015, Alcheringa 39, 526 [39] (original designation).

Type species: *Daohugoulibellula lini* Nel & Huang, 2015.

The specific epithet is dedicated to Prof. Qibin Lin for his contribution on fossil insects. Forewing discoidal triangle three-celled, whereas hind wing triangle is two-celled; only two double cells between MP and CuA in hind wing.

Distribution and age: Inner Mongolia; Middle Jurassic.

Only one species included from the Jurassic of Northern China (see Table 6.1).

Family Euthemistidae Pritykina, 1968

The Mesozoic Isophlebioptera Bechly, 1996 are a very large clade subdivided into four subgroups: Euthemistidae, Parazygoptera, Selenothemistidae, and Isophlebioidea. The family Euthemistidae Pritykina, 1968 comprise one genus and two species (both based on forewings). Our knowledge of the Euthemistidae remains indigent and its exact position is somewhat uncertain, even whether it is probably included in the Isophlebioptera. Several long intercalary veins between IR1 and RP1, and between IR1 and RP2, as well as between RP3/4 and IR2, and between IR2 and RP2 (intercalaries parallel to main longitudinal veins without apparent origin on them, but originating in cross-venation); extremely narrow postdiscoidal area (in forewings and probably also in hind wings); not petiolate; numerous secondary antenodal cross-veins between anterior wing margin and ScP distal of Ax2; discoidal cells opened in forewing and closed in hind wing; hind wing subdiscoidal area transverse, posteriorly closed, short and broad, with vein CuAb makes a strong angle with AA; hind wing gaff (basal CuA before its branching) not very long; RP2 aligned with subnodus; cross-veins in hind wing postdiscoidal space are not very long and not oblique.

Only one genus included from the Jurassic of Northern China: *Sinoeuthemis* Li, Nel, Shih, Ren & Pang, 2013.

Sinoeuthemis Li, Nel, Shih, Ren & Pang, 2013

Sinoeuthemis Li, Nel, Shih, Ren & Pang, 2013, ZooKeys 261, 43–44 [40] (original designation).

Type species: *Sinoeuthemis daohugouensis* Li, Nel, Shih, Ren & Pang, 2013.

Wings relatively short and very short CuA with weak posterior branches in both fore and hind wings.

Distribution and age: Inner Mongolia; Middle Jurassic.

Only one species included from the Jurassic of Northern China (see Table 6.1).

Sinoeuthemis daohugouensis Li, Nel, Shih, Ren & Pang, 2013 (Figure 6.11)

Sinoeuthemis daohugouensis Li, Nel, Shih, Ren & Pang, 2013: ZooKeys 261, 43–44.

Locality and horizon: Daohugou, Ningcheng, Inner Mongolia, China; Middle Jurassic, Jiulongshan Formation.

Sinoeuthemis daohugouensis is the first record of the isophlebiopteran family Euthemistidae from the Middle Jurassic of Northeastern China, while previously this family was restricted to the early Late Jurassic Kazakhstan. This finding provided evidence to amend the family diagnosis with hind wing characters. This species shows a mixture of characters alternatively present in different genera of the two families Euthemistidae and Sphenophlebiidae [39].

Family Gomphaeschnidae Tillyard and Fraser, 1940

Gomphaeschnidae are a relict group containing one extant genus, *Oligoaeschna* with 32 extant species living

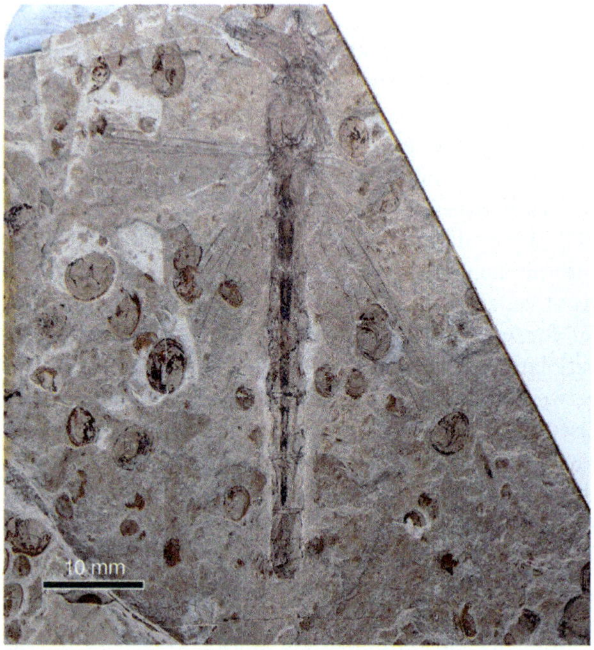

Figure 6.11 *Sinoeuthemis daohugouensis* Li, Nel, Shih, Ren & Pang, 2013, (Holotype, CNU-ODO-NN2012004 [40]). Source: Donated by Dr. Chungkun Shih.

in the Southeast Asia. *Oligoaeschna* is usually included in Aeshnidae by taxonomists working on extant dragonflies. Fossils of Gomphaeschnidae are represented by 25 species belonging to 10 genera. Up to now, they are all described from the Cretaceous or the Cenozoic of Europe, Asia and America.

Genera included from the Cretaceous of Northern China: *Sinojagoria* Bechly, 2001, *Falsisophoaeschna* Zhang, Ren & Pang, 2008 and *Sophoaeschna* Zhang, Ren & Pang, 2008.

Sinojagoria Bechly, 2001

Sinojagoria Bechly, 2001 N. paläont. Abh. 4, 159 [41] (original designation).

Type species: *Sinojagoria imperfecta* Bechly, 2001.

Forewing discoidal triangle longitudinally elongated and two-celled; hind wing discoidal triangle four-celled and rather stout; anal loop large, six or seven-celled, nearly as wide as long (possible autapomorphy); Ax2 on a level with the basal side of the discoidal triangle in the forewing; one secondary antenodal cross-vein between Ax1 and Ax2 in the forewing, and two in the hind wing; RP2 is weakly undulated; pterostigma covering two cells; only four antesubnodal cross-veins between RA and RP basal of the subnodus in the forewing; Rspl and Mspl more or less straight with a single row of cells between Rspl and IR2 and between Mspl and MA; IR2 straight; RP3/4 and MA more or less parallel and gently undulated veins; MA distally more undulated than RP3/4, resulting in a short widened area with two rows of cells in-between; MP diverging strongly from MA in the hind wing.

Distribution and age: Liaoning; Early Cretaceous.

Three species included from the Cretaceous of Northern China (see Table 6.1).

Falsisophoaeschna Zhang, Ren & Pang, 2008

Falsisophoaeschna Zhang, Ren & Pang, 2008, Prog. Nat. Sci. 18, 61–62 [42] (original designation).

Type species: *Falsisophoaeschna generalis* Zhang, Ren & Pang, 2008.

Anal loop much larger (seven celled vs. four celled); discoidal triangle in forewing longer (by 8%) than that of hind wing (vs. equal in length); RP2 more curved with four rows of cells between it and Pseudo-IR1 (vs. less curved, three rows of cells); CuAa basally strongly curved, thus the area between CuAa and MP with a distinct broader basal part than that of Sophoaeschna; wings, especially hind wings slightly broader, correlated with the broader area between IR2 and RP3/4.

Distribution and age: Inner Mongolia; Early Cretaceous.

Only one species included from the Cretaceous of Northern China (see Table 6.1).

Sophoaeschna Zhang, Ren & Pang, 2008

Sophoaeschna Zhang, Ren & Pang, 2008, Prog. Nat. Sci. 18, 60 [42] (original designation).

Type species: *Sophoaeschna frigida* Zhang, Ren & Pang, 2008.

Discoidal triangles of all wings equal in length, two celled, longitudinal elongated and with rather curving distal sides MAb; Ax2 situated on the level of the basal side of the discoidal triangle; Pseudo-IR1 long, originating beneath the distal side of the pterostigma; Rspl and Mspl developed well, more or less parallel to IR2 and MA, respectively, with only one row of cells in between; distinctly intercalary veins appeared between Rspl and RP3/4, as well as between Mspl and MP; pterostigma covers four cells, with basal side parallel to distal side; the pterostigmal brace vein straight, aligned with the basal side of the pterostigma; in hind wing, MP and CuAa converge to each other distally, with a single row of large cells in between; anal loop four celled; compound eyes large and confluent.

Distribution and age: Inner Mongolia; Early Cretaceous.

Only one species included from the Cretaceous of Northern China (see Table 6.1).

Sophoaeschna frigida Zhang, Ren & Pang, 2008 (Figure 6.12)

Sophoaeschna frigida Zhang, Ren & Pang, 2008: Prog. Nat. Sci. 18, 60.

Figure 6.12 *Sophoaeschna frigida* Zhang, Ren & Pang, 2008, (Holotype, CNU-OD-NN2004015) [42].

Locality and horizon: Liutiaogou, Ningcheng, Inner Mongolia, China; Lower Cretaceous, Yixian Formation.

Discoidal triangles of all wings equal in length, two celled, longitudinal elongated and with rather curving distal sides MAb; Ax2 situated on the level of the basal side of the discoidal triangle; Pseudo-IR1 long, originating beneath the distal side of the pterostigma; Rspl and Mspl developed well, more or less parallel to IR2 and MA, respectively, with only one row of cells in between; distinctly intercalary veins appeared between Rspl and RP3/4, as well as between Mspl and MP; pterostigma covers four cells, with basal side parallel to distal side; the pterostigmal brace vein straight, aligned with the basal side of the pterostigma; in hind wing, MP and CuAa converge to each other distally, with a single row of large cells in between; anal loop four celled; compound eyes large and confluent [42].

Family Gomphidae Rambur, 1842

The Gomphidae, commonly referred to as clubtail dragonflies, comprise about 90 genera and 900 species. The name refers to the club-like widening of the end of the abdomen (abdominal segments 7 through 9). However, this club is usually less pronounced in females and is entirely absent in some species.

Genera included from the Cretaceous of Northern China: *Liogomphus* Ren & Guo, 1996, *Gomphus?* Chang & Sun, 2005 and *Liaoninglanthus* Chang & Sun, 2006.

Liogomphus Ren & Guo, 1996

Liogomphus Ren & Guo, 1996, Insect Sci. 3 (2), 97–98 [38] (original designation).

Type species: *Liogomphus yixianensis* Ren & Guo, 1996.

No cross-veins in triangle, supertriangle and subtriangle; triangle of fore- and hind wings not very long, anterior sides subequal to inner side in length; three cross-veins under the pterostigma; the forewings with two cubito-anal cross-veins, hind wing with one; anal triangle of hind wing of male with three cells; pectinations of anal vein well-developed; Rsp1 and Msp1 absent; forewing without trigonal planate; R_3 smoothly curved (R_3 and IR_3), R_2 abruptly curved under pterostigma in left hind wing, because of a distinct distortion of distal part due to compression; IR_2 developed; IR_3 with two pectinate branches; two to three rows of cells between MA and Cup proximally, then cell-rows increasing to wing margin; six intermedian cross-veins in forewing and five in hind wing.

Distribution and age: Liaoning; Early Cretaceous.

Only one species included from the Cretaceous of Northern China (see Table 6.1).

Gomphus? Chang & Sun, 2005

Gomphus? Chang & Sun, 2005, Glob. Geol. 24 (2), 107 [20] (original designation).

Type species: *Gomphus? biconvexus* Chang & Sun, 2005.

Larva characters only: abdomen long, the widest part located at the posterior of abdomen, middle part of abdomen shrinking and dividing the abdomen into two parts, each part with middle part broad distinctly.

Distribution and age: Liaoning; Early Cretaceous.

Only one species included from the Cretaceous of Northern China (see Table 6.1).

Liaoninglanthus Chang & Sun, 2006

Liaoninglanthus Chang & Sun, 2006, Glob. Geol. 25(2), 107 [43] (original designation).

Type species: *Liaoninglanthus latus* Chang & Sun, 2006.

Forewing thin, short and broad; triangle cell nearly vertical, upper side slightly short, the other two equal in length; A_1 originated from the cu-a and basal, undulated, with a lot of cross-veins between Cu and A_1, forming dense comb-like cells, basal part cells bigger and fewer than posterior part; A_2 very short, originating near basal; thorax narrow; Rs_3 with many intercalary veins near the wing margin.

Distribution and age: Liaoning; Early Cretaceous.

Only one species included from the Cretaceous of Northern China (see Table 6.1).

Family Hemeroscopidae Pritykina, 1977

Hemeroscopidae, an extinct family from the Early Cretaceous, are known from fossils collected in the Transbaikal Region, Mongolia and China (from Lushangfen Formation of Xishan, Beijing and Jiufotang Formation of Yixian, Beipiao City respectively) [44].

Genera included from the Jurassic and Cretaceous of Northern China: *Hemeroscopus* Pritykina, 1977 and *Abrohemeroscopus* Ren, Liu & Cheng, 2003.

Hemeroscopus Pritykina, 1977

Hemeroscopus Pritykina, 1977. Pritykina & Rasnitsyn, 2002. Superorder Libellulidea Laicharting, 1781. Order Odonata Fabricius, 1792. The dragonflies. In: Rasnitsyn & Quicke (eds.) *History of Insects*. Dordrecht: Kluwer Academic Publishers. 97–103 [13] (original designation).

Type species: *Hemeroscopus baissicus* Pritykina, 1977.

Wings relatively short and very short CuA with weak posterior branches in both fore- and hind wings.

Distribution and age: Beijing of China, Siberia, Mongolia, Korea; Early Cretaceous.

Only one species included from the Cretaceous of Northern China (see Table 6.1).

Abrohemeroscopus Ren, Liu & Cheng, 2003

Abrohemeroscopus Ren, Liu & Cheng, 2003, Acta Entomol. Sinica, 46(5), 623 [44] (original designation).

Type species: *Abrohemeroscopus mengi* Ren, Liu & Cheng, 2003.

The specific epithet is dedicated to Mr. Qingjin Meng for loan of the material for study. Hind wing anal loop is smaller, only with 6–7 cells (plesiomorphy); Rspl is absent; the hind wing vein CuAa is curved and has five distinct posterior branches; the forewing MP is not shortened, reaching the posterior wing margin slightly beyond the level of the nodus; pterostigma more distinctly braced; the hind wing area between MP and CuAa is narrow, with only one row of cells near the discoidal triangle.

Distribution and age: Liaoning; Early Cretaceous.

Only one species included from the Cretaceous of Northern China (see Table 6.1).

Family Juralibellulidae Huang & Nel, 2007

Juralibellula is the sister group of the Neobrachystigmata Bechly, 1996. The character "hind wing MP distinctly curved towards the hind margin and thus somewhat shortened, ending basal of the level of the nodus" is the apomorphy of the Neobrachystigmata, but absent in Juralibellulidae.

Only one genus included from the Jurassic of Northern China: *Juralibellula* Huang & Nel, 2007.

Juralibellula Huang & Nel, 2007

Juralibellula Huang & Nel, 2007, N. Jb. Geol. Paläont. Abh. 246 (1), 64 [45] (original designation).

Type species: *Juralibellula ningchengensis* Huang & Nel, 2007.

Wing characters only. Distal part of antesubnodal area free of cross-veins; basal CuA before its branching distinctly prolonged but straight; no basal furcation of IR2 basal of lestine oblique vein; forewing discoidal triangle transverse (apomorphy but convergent with the Paneurypalpidomorpha), but hind wing one more elongate; both fore- and hind wing discoidal triangles three-celled; hind wing CuAa with five posterior branches, distinctly curved towards hind margin; anal loop elongate, divided into eight cells and with a rudimentary zigzagged midrib (apomorphy but convergent with the very advanced Mediocostida Bechly, 1996 = former Libellulidae and some corduliids); intercalary vein IR1 short; area between MP and CuA basally widened with two rows of cells; terminal branch of CuAa secondarily branched on CuA; RP3/4 and MA strictly parallel up to hind margin; area between RP2 and IR2 distally distinctly widened, with more than one cell row in distal half; Ax1 and Ax2 well apart; arculus rather straight; posterior part (cross-vein) of arculus distinctly shorter than anterior part [RP and MA]; hind wing MP not shortened, ending well distal of level of nodus; first lestine oblique vein four cells distal of subnodus in both pairs of wings.

Distribution and age: Inner Mongolia; Middle Jurassic.

Only one species included from the Jurassic of Northern China (see Table 6.1).

Family Liassogomphidae Tillyard, 1925

Pterostimata elongate; triangle cell with one cross-vein; subtriangle cell sometimes with forked cross-vein; 1A strongly curved distal to triangle cell, expand distal of notus; anal region well-developed, separated by a series of pectinate branches, with many rows of cells between branches.

Only one genus included from the Cretaceous of Northern China: *Chrysogomphus* Ren, 1994.

Chrysogomphus Ren, 1994

Chrysogomphus Ren, 1994, Geoscience, 8(3), 255 [46] (original designation).

Type species: *Chrysogomphus beipiaoensis* Ren, 1994.

Upper triangle cell with no cross-vein, three incomplete cross-veins between two strong straight antenodal cross-veins; forewing Cu-A region with one cross-vein, hind wing with two; hind wing R_{4+5} and MA not undulant, two rows of cells between R_2 and R_3 originate between nodus and pterostigma, and turned to four rows under pterostigma; IR_2 and Rspl not well-developed; two rows of cells formed under pterostigma between R_3 and IR_3, IR_3 with three pectinate branches; two rows of cells between R_{4+5} and MA formed near wing margin; basal part between MA and CuP developed four rows of cells, then turn dense to margin; MA with three pectinate branches.

Distribution and age: Liaoning; Early Cretaceous.

Only one species included from the Cretaceous of Northern China (see Table 6.1).

Family Liupanshaniidae Bechly, Nel & Martínez-Delclòs, 2001

Unique shape of the very elongated and narrow hind wing discoidal triangle (anterior side of discoidal triangle distally curved and ending on the anterior side MA of the hypertriangle; MAb is strongly sigmoidally curved with a very concave basal part and a strong angle in the distal part); hind wing discoidal triangle divided into at least three cells by parallel cross-veins; forewing discoidal triangle divided into three cells (but only known in two genera of *Araripeliupanshania* and *Paramesuropetala*); both pairs of wings (but especially the hind wing) with a strong convex secondary longitudinal vein (trigonal planate) in the postdiscoidal area, originating at the angle of MAb (convergent to several other groups of Aeshnoptera: Gomphides); the distal second oblique vein "O" between RP2 and IR2 is secondarily absent.

Only one genus included from the Cretaceous of Northern China: *Guyuanaeschnidia* Lin, 1982.

Guyuanaeschnidia Lin, 1982

Guyuanaeschnidia Lin, 1982. Mesozoic and Cenozoic insects. In: *Paleontological Atlas of Northwest China*. (III). Volume of Shaan-Gan-Ning Basin Insecta. Beijing: Geological Publishing House. 72 [47] (original designation).

Type species: *Guyuanaeschnidia eximia* Lin, 1982.

The type species of the genus is from the Early Cretaceous of Liupanshan [47], and in 2002, Lin *et al.* redescribed the holotype *Guyuanaeschnidia eximia* [48] and amended the generic diagnosis with these characters: pterostigmal brace aligned with basal side of pterostigma but not very oblique; RP2 undulate but not with a strong curve; pseudo-IR1 well-defined; primary IR1 long and well-defined; Rspl parallel to IR2, but rather zigzagged and weakly defined, with two rows of cells between it and IR2; hypertriangle divided into two cells; discoidal triangle elongated, narrow and divided into four cells; a very strong convex secondary longitudinal vein (trigonal planate) originating at angle of MAb in postdiscoidal area, trigonal planate distally forked, not vanishing in postdiscoidal area but ending in posterior wing margin; Mspl distinct and originating from two concave cross-veins just distal of discoidal triangle, with one row of cells between it and MA; anal loop well-defined, posteriorly closed and divided into five cells; MAb with a very strong angle and a very pronounced curve. [47, 48]

Distribution and age: Ningxia; Early Cretaceous.

Only one species included from the Cretaceous of Northern China (see Table 6.1).

Family Nodalulaidae Lin, Huang & Nel, 2007

The morphological disparity of the Cavilabiata (formerly Cordulegastroidea + Libelluloidea) was very important during the Late Jurassic/Early Cretaceous, even greater than nowadays, with about 10 extinct families, representing numerous stages of evolution for the wing venation of this clade. Nearly all these Mesozoic forms became extinct before the Late Cretaceous, probably during the Cenomanian–Turonian, probably in relation to the important changes in the lake water quality correlated to the rise and diversification of the Angiosperms (Fleck and Nel, 2003). The present discovery of a taxon representing a new family, somewhat "intermediate" between the Hemeroscopidae and the more modern Neobrachystigmata (a clade to which belong the recent families Corduliidae and Libellulidae, among others) is of great interest for our knowledge of the evolution of the whole group and its past diversity.

Only one genus included from the Cretaceous of Northern China: *Nodalula* Lin, Huang & Nel, 2007.

Nodalula Lin, Huang & Nel, 2007

Nodalula Lin, Huang & Nel, 2007, Zootaxa, 1469, 60 [49] (original designation).

Type species: *Nodalula dalinghensis* Lin, Huang & Nel, 2007.

Female ovipositor reduced; "cordulegastrid" gap present; relatively short pterostigma covering less than three complete cells; in the hind wing the basal CuA before its branching strongly prolonged and very straight; CuAa with four branches; terminal branch of CuAa apparently secondarily branched on CuA; area between MP and CuA slightly basally widened; RP3/4 and MA strictly parallel up to the hind margin; area between RP2 and IR2 distally distinctly widened, with more than one cell row in distal half; well-defined Rspl and Mspl, with one row of cells above them; a distinct pterostigmal brace; lestine oblique cross-vein shifted basally one cell distal of subnodus; hind wing anal cell rhombus-like, seven-celled; arculus not distinctly angled but more or less straight, with posterior part (cross-vein) distinctly shorter than anterior part; hind wing MP distinctly curved towards the hind margin and thus somewhat shortened, ending basal of level of nodus; forewing Ax1 shifted basal of level of distal angle of discoidal triangle; forewing discoidal triangle elongate; discoidal cells crossed; in both pairs of wings

Ax1 and Ax2 with two secondary antenodal cross-veins between them.

Distribution and age: Liaoning; Early Cretaceous.

Only one species included from the Cretaceous of Northern China (see Table 6.1).

Family Progobiaeshnidae Bechly, Nel & Martínez-Delclòs, 2001

Pterostigma relatively short, and pterostigmal brace vein perpendicular, not oblique like the basal side of the pterostigma; pseudo-IR I strongly reduced (very short and originating distinctly distal of the pterostigma); anal loop pentagonal, enlarged and divided into nine cells, correlated with a more pronounced elongation of the gaff (unknown in Gobiaeshna); several rows of cells between IR2 and Rspl which are more or less parallel; hind wing subdiscoidal triangle two-celled.

Genera included from the Jurassic and Cretaceous of Northern China: *Progobiaeshna* Bechly, Nel & Martínez-Delclòs, 2001, *Mongoliaeshna* Nel & Huang, 2010 and *Decoraeshna* Li, Nel, Ren & Pang, 2012.

Progobiaeshna Bechly, Nel & Martínez-Delclòs, 2001

Progobiaeshna Bechly, Nel & Martínez-Delclòs, 2001, N. Paläont. Abh., 4, 66–67 [41] (original designation).

Type species: *Progobiaeshna liaoningensis* Bechly, Nel & Martínez-Delclòs, 2001.

This genus is very similar to *Gobiaeshna occulta* Pritykina, 1977 (including the not oblique pterostigmal brace vein), but shows the following differences in the forewing: only a single row of cells between the parallel parts of RP1 and RP2 instead of two rows in *G. occulta*; about 13 secondary antenodal cross-veins distal of Ax2 instead of only nine in *G. occulta* (but the latter number is based on the "reconstruction" of Pritykina, since not all antenodals are preserved in the holotype); and about five secondary antenodal cross-veins between Axl and Ax2 instead of only three in *G. occulta*; PsA reduced to a simple cross-vein in submedian space, instead of being distinctly stronger than the other cubito-anal cross-veins.

Distribution and age: Liaoning; Early Cretaceous.

Only one species included from the Cretaceous of Northern China (see Table 6.1).

Mongoliaeshna Nel and Huang, 2010

Mongoliaeshna Nel and Huang, 2010, C. R. Palevol 9, 142 [50] (original designation).

Type species: *Mongoliaeshna sinica* Nel and Huang, 2010

Mongoliaeshna differs from *Progobiaeshna liaoningensis* Bechly, Nel & Martínez-Delclòs, 2001 in the anal loop divided into five cells instead of nine cells, forewing vein PsA angled with three cells in subdiscoidal space instead of two, basal part of area between MP and CuA with a longer zone with two rows of cells in hind wing, and basal part of area between RP1 and RP2 with two rows of cells just distal of second oblique vein, as in *Gobiaeshna occulta* Pritykina, 1977. *Mongoliaeshna* differs from this latter taxon in the subdiscoidal space three-celled. A further difference with both *Progobiaeshna* and *Gobiaeshna* is the free hypertriangle in *Mongoliaeshna*, instead of being crossed by two or three cross-veins.

Distribution and age: Inner Mongolia; Early Cretaceous.

Three species included from the Cretaceous of Northern China (see Table 6.1).

Decoraeshna Li, Nel, Ren & Pang, 2012

Decoraeshna Li, Nel, Ren & Pang, 2012, Geobios 45, 346 [51] (original designation).

Type species: *Decoraeshna preciosa* Li, Nel, Ren & Pang, 2012.

Proximal part of area between RP1 and RP2 with one row of cells up to close to level of base of pterostigma; free hypertriangles; four-celled discoidal triangles; rather oblique pterostigmal braces; short pterostigma; more than three antenodal cross-veins between Ax1 and Ax2.

Distribution and age: Liaoning; Early Cretaceous.

Only one species included from the Cretaceous of Northern China (see Table 6.1).

Decoraeshna preciosa Li, Nel, Ren & Pang, 2012 (Figure 6.13)

Decoraeshna preciosa Li, Nel, Ren & Pang, 2012: Geobios 45, 346.

Locality and horizon: Huangbanjigou, Liaoning, China; Lower Cretaceous, Yixian Formation.

Proximal part of area between RP1 and RP2 with one row of cells up to and close to level of base of pterostigma; free hypertriangles; four-celled discoidal triangles; rather oblique pterostigmal braces; short pterostigma;

Figure 6.13 *Decoraeshna preciosa* Li, Nel, Ren & Pang, 2012, (Holotype, CNU-ODO-LB2011003c) [51].

more than three antenodal cross veins between Ax1 and Ax2 [50].

Family Proterogomphidae Bechly, Nel, Martínez-Delclòs & Fleck, 1998

Triangles secondarily undivided; only two cells beneath the pterostigmata; vein pseudo-IR 1 very distinct and originating on RPl beneath the distal end-of the pterostigma; anal loop reduced to one or two cells; enlarged cell beneath the sub-basal space in the forewing; hind wing triangles more longitudinal elongate (convergent to Lindeniinae). All these characters seem to be autapomorphies. For further diagnostic characters see the emended diagnosis of the type genus below.

Only one genus included from the Cretaceous of Northern China: *Lingomphus* Vernoux, Huang, Jarzembowski & Nel., 2010.

Lingomphus Vernoux, Huang, Jarzembowski & Nel., 2010

Lingomphus Vernoux, Huang, Jarzembowski & Nel, 2010, Cretac. Res. 31, 94–95 [52] (original designation).

Type species: *Lingomphus magnificus* Vernoux, Huang, Jarzembowski & Nel, 2010

This genus with a "cordulegastrid gap"; hypertriangles, triangles and subtriangles free; pterostigma braced; 3–4 cells beneath pterostigma; pseudo-IR1 originating on RP1 beneath distal side of pterostigma; two antefurcal cross-veins in hind wing (but three in forewing); most basal postnodal cross-veins apparently oblique; hind wing discoidal triangle rather elongate but less than that in type genus of *Proterogomphus*; anal loop longer than wide and longitudinally divided into two cells; distal antefurcal cross-vein not oblique in the hind wing; hind wing anal and cubito-anal areas rather broad; CuAa with distinct posterior branches.

Distribution and age: Liaoning; Early Cretaceous.

Only one species included from the Cretaceous of Northern China (see Table 6.1).

Family Rudiaeschnidae Bechly, Nel & Martínez-Delclòs, 2001

Rspl distinctly curved and separated by at least three rows of cells from IR2; one to three convex oblique and undulated secondary veins anastomosing between IR2 and RP314 immediately basal of the origin of Rspl, at least in the hind wings (somewhat reduced in *Valdaeshna*; apparently present by convergence in *Aktassia*, but different); hind wing subdiscoidal triangle divided into two or three cells. A further putative autapomorphy could be the lateral expansions (genital lobes) along the third segment of the male abdomen that are present in Cymatophlebiinae and Rudiaeschnidae, but which are not visible (not present or not preserved) in the male holotype of *Valdaeshna surreyensis*, while the concerning character state is unknown for all other Valdaeshninae.

Genera included from the Jurassic and Cretaceous of Northern China: *Rudiaeschna* Ren & Guo, 1996 and *Fuxiaeschna* Lin, Zhang & Huang, 2004.

Rudiaeschna Ren & Guo, 1996

Rudiaeschna Ren & Guo, 1996, Insect Sci. 3(2), 96 [38] (original designation).

Type species: *Rudiaeschna limnobia* Ren & Guo, 1996.

The type species of the genus is from the Early Cretaceous of Huangbanjigou [38], in 2001, Bechly *et al.* erected the monotypic family Rudiaeschnidae [41], and in 2011, Li *et al.* redescribed a new complete specimen of *Rudiaeschna limnobia* from Huangbanjigou [53] and amended the generic diagnosis with these characters: PsA of the hind wing is more strongly zigzagged and distinctly weaker than that of the forewing; anal loop enlarged and gaff prolonged; RP1 and RP2 secondarily divergent; RP2 and IR2 distally distinctly diverging. A pair of symmetrically expanded and rounded lobes on abdominal segment III; with a row of small spines along distal two-thirds of exterior margin of these lobes; abdominal segment II distinctly constricted [38, 41, 53].

Distribution and age: Liaoning; Early Cretaceous.

Only one species included from the Cretaceous of Northern China (see Table 6.1).

Rudiaeschna limnobia Ren and Guo, 1996 (Figure 6.14)

Rudiaeschna limnobia Ren and Guo, 1996: Insect Sci. 3 (2), 96.

Locality and horizon: Huangbanjigou, Liaoning, China; Lower Cretaceous, Yixian Formation.

Pterostigma elongated and distinctly braced; secondary antenodal cross-veins not aligned; Ax2 on a level with distal angle of discoidal triangle; IR2 and RP2 smoothly undulated and distally divergent with about three rows of cells in-between; two oblique veins "O"; RP1 and RP2 basally slightly diverging with two or three rows of cells in-between basal of pterostigma; only a short pseudo IR1 present between RP I and RP2; anal loop much more distinct and broader; Rspl well-defined and slightly curved; two oblique secondary veins between IR2 and RP3/4 immediately basal of the origin of Rspl; MA and RP3/4 are smoothly undulated; a short and weakly developed Mspl with two or three rows of cells; distal side MAb of discoidal triangle straight; discoidal triangles elongated; hypertriangles divided by one to three cross-veins; PsA weaker in hind wing than in forewing and distinctly zigzagged. A pair of symmetrically expanded and rounded lobes on abdominal segment III; with a row of small spines along distal

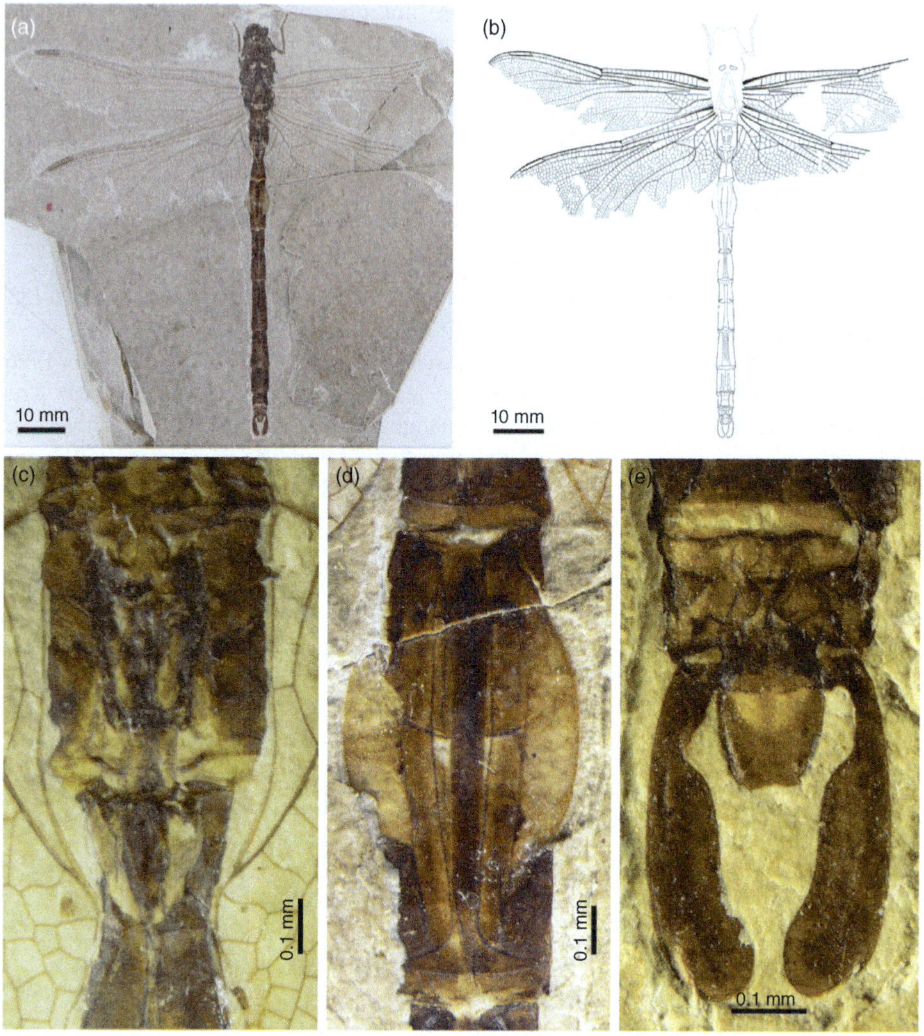

Figure 6.14 (a) Photograph of new material CNU-ODO-LB2010001; (b) Line drawing of specimen CNU-ODO-LB2010001; (c) Detail photograph of specimen CNU-ODO-LB2010001, segment II, to show the genitalia; (d) Detail photograph of specimen CNU-ODO-LB2010001, abdominal segment III; (e) Detail photograph of cerci for specimen CNU-ODO-LB2010001 [53].

two-thirds of exterior margin of these lobes; abdominal segment II distinctly constricted. Cerci foliate, apically rounded and without spine at their apex; inferior appendage (epiproct) broad and long, about rectangular in shape, and apically slightly bifid [38, 53].

***Fuxiaeschna* Lin, Zhang & Huang, 2004**

Fuxiaeschna Lin, Zhang & Huang, 2004, Odonatologica 33(4), 438 [54] (original designation).

Type species: *Fuxiaeschna hsiufunia* Lin, Zhang & Huang, 2004.

Compound eyes contiguous; anteclypeus with middle concave sutural groove, osteclypeus narrow; the venation: presence of a Mspl; the discoidal triangle with one cross-vein; the subdiscoidal triangle divided into three cells, PsA of the hind wing zigzagged; anal loop closed posteriorly, but weakly; a row of stout spines furnished along inner margin of tibia of fore and middle legs; abdomen with a middle carina on terga 3–8.

Distribution and age: Gansu; Early Cretaceous.

Only one species included from the Cretaceous of Northern China (see Table 6.1).

Family Stenophlebiidae Needham, 1903

The damsel-dragonfly clade Stenophlebioptera is exclusively distributed in the Mesozoic, flourishing between the Late Jurassic and the Early Cretaceous. Its youngest known representative was described from the Cenomanian (99.6 to 93.6 Mya) of France [55]. This clade is currently considered as the putative sister group of the Anisoptera and has been the subject of a relatively recent revision [56]. Stenophlebiidae is an extinct family comprising seven genera, widespread in Germany, Spain, England, France, China, Mongolia and Kazakhstan.

Genera included from the Jurassic and Cretaceous of Northern China: *Stenophlebia* Hagen, 1866, *Yixianstenophlebia* Nel & Huang, 2015 and *Liaostenophlebia* Zheng, Wang & Jarzembowski, 2016.

Stenophlebia Hagen, 1866

Stenophlebia Hagen, 1866, Paleontographica 15, 57–96 [57] (original designation).

Type species: *Stenophlebia amphitrite* Hagen, 1862 [58].

Wing narrow and short, 42 mm long; Cr long, with two cells below it, and well-aligned with distal part of ScP; Sn short, one cell long; supplementary veinlet below Sn aligned with RP2 and three cells long; Arc midway between Ax1 and Ax2; base of IR1 two cells distal of base of RP2.

Distribution and age: Liaoning of China, Germany, England; Late Jurassic, Early Cretaceous.

Only one species included from the Cretaceous of Northern China (see Table 6.1).

Yixianstenophlebia Nel & Huang, 2015

Yixianstenophlebia Nel & Huang, 2015, Creta. Res. 56, 422 [59] (original designation).

Type species: *Yixianstenophlebia magnifica* Nel & Huang, 2015.

Wing characters only. Discoidal triangles well transverse with a nearly right angle between MAa and MAb; hind wing with a broad anal area with three rows of cells between AA and posterior wing margin; wings very shortly petiolate; subnodus Sn not aligned with nodal Cr; hind wing vein CuAa short, ending on posterior wing margin well basal of nodus level; presence of four rows of cells in postdiscoidal area just distal of discoidal triangle (autapomorphy, as all Stenophlebiomorpha have three row of cells or less).

Distribution and age: Inner Mongolia; Early Cretaceous.

Only one species included from the Cretaceous of Northern China (see Table 6.1).

Liaostenophlebia Zheng, Wang & Jarzembowski, 2016

Liaostenophlebia Zheng, Wang & Jarzembowski, 2016, Creta. Res. 67, 60 [60] (original designation).

Type species: *Liaostenophlebia yixianensis* Zheng, Wang & Jarzembowski, 2016.

Hind wing base characters only. Ax2 opposite MAb; Arc slightly distal of Ax1; T elongate, transverse and three-celled; hypertriangle broad and three-celled, with anterior side quite curved; subdiscoidal space foot-shaped, narrow and four-celled; Hal well-defined, transverse, closed and two-celled; AA with strong elbow distally, directly perpendicular to base and towards ventral edge of wing; four rows of cells present in basal postdiscoidal area.

Distribution and age: Liaoning; Early Cretaceous.

Only one species included from the Cretaceous of Northern China (see Table 6.1).

Family Tarsophlebiidae Handlirsch, 1906

The Tarsophlebiidae are an extinct family of medium-sized fossil odonates from the Upper Jurassic and Lower Cretaceous of Eurasia. They are either the most basal member of the damsel dragonfly grade (Anisozygoptera) within the stem group of Anisoptera (dragonflies), or the sister group of all Recent odonates. They are characterized by the basally open discoidal cell in both pairs of wings, very long legs, paddle-shaped male cerci, and a hypertrophied ovipositor in females. Hind wings with hypertrophied subdiscoidal cell that is developed as "pseudo discoidal cell"; fusion of veins MAb + MP + CuA for a considerable distance before separation of MP and CuA in hind wing; vein AA strongly bent at insertion of CuP-crossing; extremely acute distal angles of forewing discoidal and subdiscoidal cell. Distinctly prolonged legs, three-segmented tarsi, with basal tarsomere very long; and absence of subapical tooth on tarsal claw; male cerci with paddle-like distal expansions. The extremely prolonged female ovipositor could be a further synapomorphy for the family.

Only one genus included from the Cretaceous of Northern China: *Turanophlebia* Pritykina, 1968.

Turanophlebia Pritykina, 1968

Turanophlebia Pritykina, 1968, "Strekozy Karatau (Odonata)" [Dragonflies of Karatau (Odonata)]. In Rohdendorf *Yurskoy Nasekomiye Karatau* [*Jurassic Insects of Karatau*] (in Russian). Moscow: Academy of Sciences of the USSR, Section of General Biology, Publishing House "Nauka." 26–54 [61] (original designation).

Type species: *Turanophlebia martynovi* Pritykina, 1968.

Wings with more dense venation; presence of more than 25 postnodal cross-veins (11–16 in *Tarsophlebia*); six or more rows of cells between CuA and posterior hind wing margin (less than five rows in *Tarsophlebia*); more than 10 secondary antenodal cross-veins in hind wing (less than 10 in *Tarsophlebia*); IR1 longer than in *Tarsophlebia*; presence of long secondary longitudinal not zigzagged veins in area between IR2 and RP2.

Distribution and age: Liaoning of China, Kazakhstan, Mongolia, England, Russia; Late Jurassic, Early Cretaceous.

Only one species included from the Cretaceous of Northern China (see Table 6.1).

Family *Incertae sedis*

Genera included from the Jurassic and Cretaceous of Northern China: *Samarura* Bauer, Redtenbacher & Ganglbauer, 1889, *Parapetala* Huang, Nel & Lin, 2003, *Telmaeshna* Zhang, Ren & Pang, 2008 and *Sinocymatophlebiella* Li, Nel, Ren & Pang, 2011.

Samarura Brauer, Redtenbacher & Ganglbauer, 1889

Samarura Brauer, Redtenbacher & Ganglbauer, 1889, Pritykina (1968). "Strekozy Karatau (Odonata)" [Dragonflies of Karatau (Odonata)]. In Rohdendorf BB. *Yurskoy Nasekomiye Karatau [Jurassic Insects of Karatau]* (in Russian). Moscow: Academy of Sciences of the USSR, Section of General Biology, Publishing House "Nauka." 26–54 [61] (original designation).

Type species: *Samarura gigantea* Brauer, Redtenbacher & Ganglbauer, 1889.

Larva characters only: body soft, wing pad reaching the third segment of abdomen, terminal part with three caudal gill.

Distribution and age: Hebei of China, Russia; Early Cretaceous.

Two species included from the Cretaceous of Northern China (see Table 6.1).

Parapetala Huang, Nel & Lin, 2003

Parapetala Huang, Nel & Lin, 2003, Cretac. Res. 24, 141–142 [62] (original designation).

Type species: *Parapetala liaoningensis* Huang, Nel & Lin, 2003.

Eyes distinctly separated; wing veins RP1 and RP2 long, parallel, basal of pterostigma; veins Rspl and Mspl absent in both fore- and hind wings; anal loop small, three-celled, closed, and not elongate; discoidal triangles unicellular, that of forewing being distinctly broader than that of hind wing; arculus close to Ax1; pterostigmal braces well-defined, aligned with basal sides of pterostigma; RP2 and IR2 not convergent close to posterior wing margin; RP3/4 and MA undulate and parallel from base to apex; trigonal planates present in both forewing and hind wing, more developed in hind wing.

Distribution and age: Liaoning; Early Cretaceous.

Only one species included from the Cretaceous of Northern China (see Table 6.1).

Telmaeshna Zhang, Ren & Pang, 2008

Telmaeshna Zhang, Ren & Pang, 2008, Zootaxa, 1681, 63 [63] (original designation).

Type species: *Telmaeshna paradoxica* Zhang, Ren & Pang, 2008.

All known characters from a hind wing. No Anx present between Ax0 and Ax1; Ax1 and Ax2 relatively close together with only one intermediary antenodal cross-vein; two oblique veins ("O") present; Pterostigma very long and strong; Pseudo-IR1 not well-defined (zigzag, very short and originating distinctly distal of the pterostigma); base of RP2 not aligned with subnodus; area between RP1 and RP2 broad with three rows of cells up to pterostigma; Mspl absent; median space free of cross-veins; hind wing hypertriangle and discoidal triangle divided into five or more cells respectively, subdiscoidal triangle three-celled; approximately pentagonal anal loop large and nine-celled, with prolonged gaff; hind wing distinctly broad and rather round in shape.

Distribution and age: Liaoning; Early Cretaceous.

Only one species included from the Cretaceous of Northern China (see Table 6.1).

Sinocymatophlebiella Li, Nel, Ren & Pang, 2011

Sinocymatophlebiella Li, Nel, Ren & Pang, 2011, *Zootaxa* 2927:58 [64] (original designation).

Type species: *Sinocymatophlebiella hasticercus* Li, Nel, Ren & Pang, 2011.

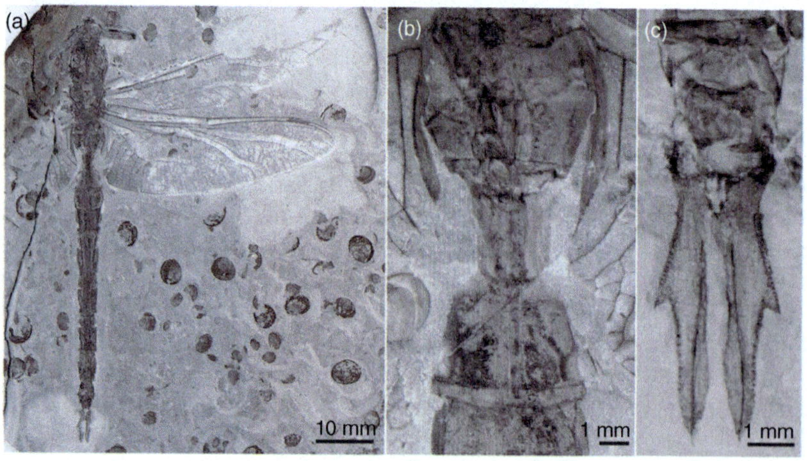

Figure 6.15 *Sinocymatophlebiella hasticercus* Li, Nel, Ren & Pang, 2012, (Holotype, CNU-ODO-NN2010004p). (a) Habitus; (b) Abdomen segment II; (c) Cerci [64].

Only one cross-vein beneath pterostigma; pseudo-IR1 short, originating on RP1 distal of pterostigma; two oblique veins; RP1 and RP2 basally parallel with one row of cells between them basal of the pterostigma; RP2 smoothly undulate at its mid part and IR2 weakly undulate; Rspl poorly developed and zigzagged, with one row of cells between it and IR2; MAb straight; hypertriangle unicellular; discoidal and subdiscoidal triangle divided by cross-veins; subdiscoidal area only crossed by CuP-crossing, and no other cross-veins between CuP-crossing and PsA; fore- and hind wing discoidal triangles of same shape and slightly longitudinally elongated; RP3/4 and MA strongly undulate; Mspl reduced; CuAa with numerous branches; anal loop suppressed; guff very short; male hind wing with two posterior branches of AA between CuAb and anal triangle; distal side of anal triangle has a broad and long anal membranule; abdomen segment II strongly constricted; cerci hastate.

Distribution and age: Inner Mongolia; Middle Jurassic.

Only one species included from the Jurassic of Northern China (see Table 6.1).

***Sinocymatophlebiella hasticercus* Li, Nel, Ren & Pang, 2011 (Figure 6.15)**

Sinocymatophlebiella hasticercus Li, Nel, Ren & Pang, 2011: *Zootaxa* 2927, 58.

Locality and horizon: Daohugou, Ningcheng, Inner Mongolia, China; Middle Jurassic, Jiulongshan Formation.

Sinocymatophlebiella hasticercus shows important similarities with the Jurassic genus *Cymatophlebiella* from Karatau, suggesting they could belong to the same family, but the latter genus is too poorly known to accurately establish its affinities. The discovery of *Sinocymatophlebiella hasticercus* supports the evolutionary scenario of a Jurassic rapid and massive diversification

Table 6.1 List of fossil Odonata from the Jurassic and Cretaceous of China.

Family	Species	Locality	Horizon/Age	Citation
Aeschinidiidae	[a]*Brunneaeschnidia jiuquanensis* Hong, 1982	Yumen, Gansu	[b]Chijinqiao Fm., K_1	Hong [65]
	Dracontaeschnidium orientale Zhang & Zhang, 2001	Beipiao, Liaoning	Yixian Fm., K_1	Zhang and Zhang [18]
	[a]*Hebeiaeschnidia fengningensis* Hong, 1982	Fengning, Hebei	[b]Yixian Fm., K_1	Hong [65]
	Linaeschnidium sinensis Huang, Baudoin & Nel, 2009	Beipiao, Liaoning	Yixian Fm., K_1	Huang *et al.* [19]
Aeschnidiidae	*Aeschna? acrodonta* Chang & Sun, 2005	Beipiao, Liaoning	Yixian Fm., K_1	Chang and Sun [20]
	Sinaeschnidia cancellosa Ren, 1995	Beipiao, Liaoning	Yixian Fm., K_1	Ren [66]
	[a]*Sinaeschnidia heishankowensis* Hong, 1965	Chifeng, Inner Mongolia	Shahai Fm., K_1	Hong [16]
	Sinaeschnidia huzhouensis Zhou & Wei, 1980	Huzhou, Zhejiang	Hangjiahu Fm., J_3	Zhou and Wei [67]
	Stylaeschnidium rarum Zhang & Zhang, 2001	Yixian, Liaoning	Yixian Fm., K_1	Zhang and Zhang [18]
Aktassiidae	*Pseudocymatophlebia boda* Li, Nel & Ren, 2012	Chifeng, Inner Mongolia	Yixian Fm., K_1	Li *et al.* [68]
	Sinaktassia tangi Lin, Nel & Huang, 2010	Chifeng, Inner Mongolia	Yixian Fm., K_1	Lin *et al.* [22], Zheng *et al.*, [23]
Araripelibellulidae	*Sopholibellula amoena* Zhang, Ren & Zhou, 2006	Beipiao, Liaoning	Yixian Fm., K_1	Zhang *et al.* [24]
	Sopholibellula eleganta Zhang, Ren & Zhou, 2006	Beipiao, Liaoning	Yixian Fm., K_1	Zhang *et al.* [24]

(Continued)

Table 6.1 (Continued)

Family	Species	Locality	Horizon/Age	Citation
Campterophlebiidae	*Angustiphlebia mirabilis* Li, Nel, Ren & Pang, 2013	Ningcheng, Inner Mongolia	Jiulongshan Fm., J_2	Li *et al.* [35]
	Amnifleckia guttata Zhang, Ren & Cheng, 2006	Ningcheng, Inner Mongolia	Jiulongshan Fm., J_2	Zhang *et al.* [28]
	Amnifleckia splendida Huang, Fleck, Nel & Lin, 2006	Ningcheng, Inner Mongolia	Jiulongshan Fm., J_2	Zhang *et al.* [28]
	Bellabrunetia catherinae Fleck & Nel, 2002	Ningcheng, Inner Mongolia	Jiulongshan Fm., J_2	Fleck and Nel [27]
	Dorsettia sinica Zheng, Nel & Wang, 2016	Karamay, Xinjiang	Badaowan Fm., J_1	Zheng *et al.* [26]
	Hsiufua chaoi Zhang & Wang, 2013	Ningcheng,Inner Mongolia	Haifanggou Fm., J_2	Zhang *et al.* [10]
	Junfengi yulinensis Zheng & Zhang, 2017	Yulin,Shaanxi	Yanan Fm.; J_2	Zheng *et al.* [36]
	Karatawia sinica Li, Nel & Ren, 2012	Ningcheng, Inner Mongolia	Jiulongshan Fm., J_2	Li *et al.* [32]
	Parabrunetia celinea Huang, Fleck, Nel & Lin, 2006	Ningcheng, Inner Mongolia	Jiulongshan Fm., J_2	Huang *et al.* [28]
	Parafleckium senjituense Li, Nel, Ren & Pang, 2012	Fengning, Hebei	Yixian Fm., K_1	Li *et al.* [33, 34]
	Sinokaratawia daohugouica Zhang, Ren & Pang, 2008	Ningcheng, Inner Mongolia	Jiulongshan Fm., J_2	Zhang *et al.* [69]
	Sinokaratawia gloriosa Zhang, Ren & Pang, 2008	Ningcheng, Inner Mongolia	Jiulongshan Fm., J_2	Zhang *et al.* [69]
	Sinokaratawia magica Zhang, Ren & Pang, 2008	Ningcheng, Inner Mongolia	Jiulongshan Fm., J_2	Zhang *et al.* [69]
	Sinokaratawia prokopi Nel, Huang & Lin, 2007	Ningcheng,Inner Mongolia	Jiulongshan Fm., J_2	Nel *et al.* [29]
	Zygokaratawia reni Nel, Huang & Lin, 2008	Ningcheng,Inner Mongolia	Jiulongshan Fm., J_2	Nel *et al.* [30]
Congqingiidae	*Congqingia rhora* Zhang, 1992	Laiyang, Shandong	[b]Laiyang Fm., K_1	Zhang [37]
Corduliidae	*Mesocordulia boreala* Ren & Guo, 1996	Beipiao, Liaoning	Yixian Fm., K_1	Ren and Guo [38]
Daohugoulibellulidae	*Daohugoulibellula lini* Nel & Huang, 2015	Ningcheng, Inner Mongolia	Jiulongshan Fm., J_2	Nel and Huang [39]
Euthemistidae	*Sinoeuthemis daohugouensis* Li, Nel, Shih, Ren & Pang, 2013	Ningcheng, Inner Mongolia	Jiulongshan Fm., J_2	Li *et al.* [40]
Gomphaeschnidae	*Falsisophoaeschna generalis* Zhang, Ren & Pang, 2008	Chifeng, Inner Mongolia	Yixian Fm., K_1	Zhang *et al.* [42]
	Sinojagoria cancellosa Li, Nel, Ren & Pang, 2012	Beipiao, Liaoning	Yixian Fm., K_1	Li *et al.* [51]
	Sinojagoria imperfecta Bechly, 2001	Beipiao, Liaoning	Yixian Fm., K_1	Bechly *et al.* [41]
	Sinojagoria magna Li, Nel, Ren & Pang, 2012	Beipiao, Liaoning	Yixian Fm., K_1	Li *et al.* [51]
	Sophoaeschna frigida Zhang, Ren & Pang, 2008	Chifeng, Inner Mongolia	Yixian Fm., K_1	Zhang *et al.* [42]
Gomphidae	[a]*Archaeogomphus labius* Lin, 1976	Yixian, Liaoning	Yixian Fm., K_1	Lin [70]
	[a]*Cercus clavas* Hong, 1982	Yumen, Gansu	[b]Chijinqiao Fm., K_1	Hong [71]

(Continued)

Table 6.1 (Continued)

Family	Species	Locality	Horizon/Age	Citation
	[a]*Dissurus liaoyuanensis* Hong, 1982	Lingyuan, Liaoning	[b]Yixian Fm., K_1	Hong [71]
	[a]*Dissurus qinquanensis* Hong, 1982	Yumen, Gansu	[b]Chijinqiao Fm., K_1	Hong [71]
	Gomphus? biconvexus Chang & Sun, 2005	Beipiao, Liaoning	Yixian Fm., K_1	Chang and Sun [20]
	[a]*Jibeigomphus xinboensis* Hong, 1984	Weichang, Hebei	[b]Dabeigou Fm., K_1	Hong [72]
	Liaoninglanthus latus Chang & Sun, 2006	Beipiao, Liaoning	Yixian Fm., K_1	Chang and Sun [43]
	Liogomphus yixianensis Ren & Guo, 1996	Beipiao, Liaoning	Yixian Fm., K_1	Ren andGuo [38]
	Neimenggologomphus dongwugaiensis Hong, 1982	Wulat, Inner Mongolia	Huliugou Fm., K_1	Hong [71]
	[a]*Sinogomphus taushanensis* Hong, 1982	Naiman Banner, Jilin	[b]Chijinqiao Fm., K_1	Hong [71]
	[a]*Pseudosamarura largina* Lin, 1976	Yixian, Liaoning	Yixian Fm., K_1	Lin [70]
Hemeroscopidae	*Abrohemeroscopus mengi* Ren, Liu & Cheng, 2003	Yixian, Liaoning	Jiufotang Fm., K_1	Ren *et al.* [44]
	Hemeroscopus baissicus Pritykina, 1977	Xishan, Beijing	Lushangfen Fm., K_1	Huang and Lin [73]
Juralibellulidae	*Juralibellula ningchengensis* Huang & Nel, 2007	Ningcheng, Inner Mongolia	Jiulongshan Fm., J_2	Huang and Nel [45]
Liassogomphidae	*Chrysogomphus beipiaoensis* Ren, 1994	Beipiao, Liaoning	Yixian Fm., K_1	Ren [46]
Liassophlebiidae	*Paraliassophlebia chengdeensis* Hong, 1983	Chengde, Hebei	Jiulongshan Fm., J_2	Hong [74]
Liupanshaniidae	*Guyuanaeschnidia eximia* Lin, 1982	Guyuan, Ningxia	Liupanshan Fm., K_1	Lin [49]
	[a]*Liupanshania sijiensis* Hong, 1982	Jiuquan, Gansu	Dongshan Fm., K_1	Hong [71]
Nodalulaidae	*Nodalula dalinghensis* Lin, Huang & Nel, 2007	Beipiao, Liaoning	Yixian Fm., K_1	Lin *et al.* [49]
Progobiaeshnidae	*Decoraeshna preciosa* Li, Nel, Ren & Pang, 2012	Beipiao, Liaoning	Yixian Fm., K_1	Li *et al.* [51]
	Mongoliaeshna exiguusens Li, Nel, Ren & Pang, 2012	Ningcheng, Inner Monglia.	Yixian Fm., K_1	Li *et al.* [51]
	Mongoliaeshna hadrens Li, Nel, Ren & Pang, 2012	Ningcheng, Inner Monglia.	Yixian Fm., K_1	Li *et al.* [51]
	Mongoliaeshna sinica Nel & Huang, 2010	Ningcheng, Inner Mongolia	Yixian Fm., K_1	Nel and Huang [50]
	Progobiaeshna liaoningensis Bechly, Nel & Martinez-Delclos, 2001	Liaoning	Yixian Fm., K_1	Bechly *et al.* [41]
Proterogomphidae	*Lingomphus magnificus* Vernoux, Huang, Jarzembowski & Nel., 2010	Beipiao, Liaoning	Yixian Fm., K_1	Vernoux *et al.* [52]

(Continued)

Table 6.1 (Continued)

Family	Species	Locality	Horizon/Age	Citation
Rudiaeschnidae	*Fuxiaeschna hsiufunia* Lin, Zhang & Huang, 2004	Huating, Gansu	Luohandong Fm., K_1	Lin *et al.* [54]
	Rudiaeschna limnobia Ren & Guo, 1996	Beipiao, Liaoning	Yixian Fm., K_1	Ren and Guo [38]
Stenophlebiidae	*Liaostenophlebia yixianensis* Zheng, Wang & Jarzembowski, 2016	Beipiao, Liaoning	Yixian Fm., K_1	Zheng *et al.* [60]
	Stenophlebia liaoningensis Zheng, Nel & Wang, 2016	Beipiao, Liaoning	Yixian Fm., K_1	Zheng *et al.* [75]
	[a)]*Sinostenophlebia zhanjiakouensis* Hong, 1984	Shidongzi, Hebei	Yixian Fm., K_1	Hong [71]
	Yixianstenophlebia magnifica Nel & Huang, 2015	Ningcheng, Inner Monglia.	Yixian Fm., K_1	Nel and Huang [59]
Tarsophlebiidae	*Turanophlebia sinica* Huang & Nel, 2009	Beipiao, Liaoning	Yixian Fm., K_1	Huang and Nel [76]
Family *incertae sedis* [of uncertain placement]	[a)]*Huabeia liugouensis* Hong, 1983	Chengde, Hebei	Jiulongshan Fm., J_2	Hong [74]
	Parapetala liaoningensis Huang, Nel & Lin, 2003	Yixian, Liaoning	Yixian Fm., K_1	Huang *et al.* [62]
	Samarura gigantea Brauer, Redtenbacher & Ganglbauer, 1889	Beipiao, Liaoning	Haifanggou Fm., J_2	Pritykina [61]
	[a)]*Samarura punctaticaudata* Hong, 1983	Chengde, Hebei	Jiulongshan Fm., J_2	Hong [74]
	Sinocymatophlebiella hasticercus Li, Nel, Ren & Pang, 2011	Ningcheng, Inner Mongolia	Jiulongshan Fm., J_2	Li *et al.* [64]
	Telmaeshna paradoxica Zhang, Ren & Pang, 2008	Beipiao, Liaoning	Yixian Fm., K_1	Zhang *et al.* [63]

a) The species is not present in the main text because the original description, photo and line-drawings are not precise and the holotype cannot be rechecked.

b) Horizon/Age revised from the original paper based on updated information and data.

References

1 Carpenter, F.M. (1939). The Lower Permian insects of Kansas. Part 8. Additional Megasecoptera, Protodonata, Odonata, Homoptera, Psocoptera, Protelytroptera, Plectoptera, and Protoperlaria. *Proceedings of the American Academy of Arts and Sciences* **73**: 29–70.

2 Carpenter, F.M. (1947). Lower Permian insects from Oklahoma. Part 1. Introduction and the orders Megasecoptera, Protodonata, and Odonata. *Proceedings of the American Academy of Arts and Sciences* **76**: 25–54.

3 Li, J.K., Nel, A., Zhang, X.P. *et al.* (2012). A third species of the relict family Epiophlebiidae discovered in China (Odonata: Epiproctophora). Systematic Entomology **37** (2): 408–412. https://doi.org/10.1111/j.1365-3113.2011.00610.x.

4 Grimaldi, D. and Engel, M.S. (2005). Evolution of the Insect, 1–755. Cambridge: Cambridge University Press.

5 Zhang, Z.J., Hong, Y.C., Lu, L.W. *et al.* (2006). *Shenzhousia qilianshanensis* gen. et sp. nov. (Protodonata, Meganeuridae), a giant dragonfly from the Upper Carboniferous of China. *Progress in Natural Science* **16** (3): 328–330.

6 Li, Y.J., Béthoux, O., Pang, H., and Ren, D. (2013). Early Pennsylvanian Odonatoptera from the Xiaheyan locality (Ningxia, China): new material, taxa, and perspectives. *Fossil Record* **16** (1): 117–139. https://doi.org/10.1002/mmng.201300006.

7 Nel, A., Béthoux, O., Bechly, G. *et al.* (2001). The Permo-Triassic Odonatoptera of the "Protodonate" grade (Insecta: Odonatoptera). *Annales de la Société Entomologique de France* **37**: 501–525.

8 Handlirsch, A. (1907). *Die fossilen Insekten und die Phylogenie der rezenten Formen, Parts V-II. Ein Handbuch fur Palaontologen und Zoologen*, 641–1120. Leipzig: Wilhelm Engelmann.

9 Riek, E.F. (1976). A new collection of insects from the Upper Triassic of South Africa. *Annals of the Natal Museum* **22**: 791–820.

10 Zhang, H.C., Zheng, D.R., Wang, B. *et al.* (2013). The largest known odonate in China: Hsiufua chaoi Zhang et Wang, gen. et sp. nov. from the Middle Jurassic of Inner Mongolia. *Chinese Science Bulletin* **58**: 1579–1584. https://doi.org/10.1007/s11434-012-5567-3.

11 Bechly, G. (2000). Two new fossil dragonfly species (Insecta: Odonata: Pananisoptera: Aeschnidiidae and Atassiidae) from the Solnhofen lithographic limestones (Upper Jurassic, Germany). *Stuttgarter Beiträge zur Naturkunde, Serie B* **288**: 1–9.

12 Brongniart, C. (1884). Sur un gigantesque néurorthoptère, provenant des terrains houillers de Commentry (Allier). *Comptes Rendus Hebdomadaires des Seances de l'Academie des Sciences* **98**: 832–833.

13 Pritykina, L.N. and Rasnitsyn, A.P. (2002). Superorder Libellulidea Laicharting, 1781. Order Odonata Fabricius, 1792. The dragonflies. In: *History of Insects* (ed. A.P. Rasnitsyn and D.L.J. Quicke), 97–103. Dordrecht: Kluwer Academic Publishers.

14 Nel, A., Fleck, G., Béthoux, O. *et al.* (1999). *Saxonagrion minutes* nov. gen. et sp., the oldest damselfly from the Upper Permian of France (Odonatoptera, Panodonata, Saxonagrionidae nov. fam.). *Geobios* **32**: 883–888. https://doi.org/10.1016/S0016-6995(99)80870-5.

15 Wootton, R.J. and Kukalová-Peck, J. (2000). Flight adaptations in Palaeozoic Palaeoptera (Insecta). *Biological Reviews* **75**: 129–167.

16 Hong, Y.C. (1965). A new fossil dragonfly, *Sinaeschnidia* Hong gen. nov. (Odonata, Insecta). *Acta Entomologica Sinica* **14** (2): 171–176. (in English with Russian abstract).

17 Pritykina, L.N. (1993). First dragonflies (Odonata; Aeschnidiidae) from Cenomanian of Crimea. *Paleontological Journal* **27** (1A): 179–181.

18 Zhang, J.F. and Zhang, H.C. (2001). New findings of larval and adult aeschnidiids (Insecta: Odonata) in the Yixian Formation, Liaoning Province, China. *Cretaceous Research* **22**: 443–450. https://doi.org/10.1006/cres.2001.0273.

19 Huang, D.-Y., Baudoin, A., and Nel, A. (2009). A new aeschnidiid genus from the Early Cretaceous of China (Odonata: Anisoptera). *Cretaceous Research* **30**: 805–809. https://doi.org/10.1016/j.cretres.2008.12.014.

20 Chang, J.P. and Sun, Z.S. (2005). New discovery of fossil dobsons in Jehol Biota from western Liaoning, China. *Global Geology* **24** (2): 105–111. (in Chinese with English abstract).

21 Nel, A., Bechly, G., Jarzembowski, E.A., and Martínez-Delclòs, X. (1998). A revision of the fossil petalurid dragonflies (Insecta: Odonata: Anisoptera: Petalurida). *Paleontologia Lombarda, (Nuova Serie)* **10**: 1–68.

22 Lin, Q.B., Nel, A., and Huang, D.-Y. (2010). *Sinaktassia tangi*, a new Chinese Mesozoic genus and species of Aktassiidae (Odonata: Petaluroidea). *Zootaxa* **2359**: 61–64. https://doi.org/10.5281/zenodo.193619.

23 Zheng, D.R., Nel, A., Wang, B. *et al.* (2016). The discovery of the hind wing of the Early Cretaceous dragonfly *Sinaktassia tangi* Lin, Nel & Huang, 2010 (Odonata, Aktassiidae) in Northeastern China. *Cretaceous Research* **61**: 86–90. https://doi.org/10.1016/j.cretres.2015.12.015.

24 Zhang, B.L., Ren, D., Zhou, C.Q., and Pang, H. (2006). New Genus and Species of Fossil Dragonflies (Insecta: Odonata) from the Yixian Formation of Northeastern China. *Acta Geologica Sinica (English Edition)* **80** (3): 327–335. https://doi.org/10.1111/j.1755-6724.2006.tb00252.x.

25 Whalley, P.E.S. (1985). The systematics and palaeogeography of the Lower Jurassic insects of Dorset, England: Bulletin of the British Museum (Natural History). *Geology* **39**: 107–187.

26 Zheng, D.R., Nel, A., Wang, B. *et al.* (2016). A new damsel-dragonfly from the Lower Jurassic of northwestern China and its palaeobiogeographic significance. *Journal of Paleontology* **90** (3): 485–490. https://doi.org/10.1017/jpa.2016.59.

27 Fleck, G. and Nel, A. (2002). The first Isophlebioid dragonfly (Odonata: Isophlebioptera: Campterophlebiidae) from the Mesozoic of China. *Palaeontology* **45** (6): 1123–1136. https://doi.org/10.1111/1475-4983.00278.

28 Zhang, B.L., Fleck, G., Huang, D.-Y. *et al.* (2006). New isophlebioid dragonflies (Odonata:Isophlebioptera: Campterophlebiidae) from the Middle Jurassic of China. *Zootaxa* **1339**: 51–68. https://doi.org/10.5281/zenodo.174364.

29 Nel, A., Huang, D.-Y., and Lin, Q.B. (2007). A new genus of isophlebioid damsel-dragonflies (Odonata: Isophlebioptera: Campterophlebiidae) from the Middle Jurassic of China. *Zootaxa* **1642**: 13–22. https://doi.org/10.5281/zenodo.179672.

30 Nel, A., Huang, D.-Y., and Lin, Q.B. (2008). A new genus of isophlebioid damsel-dragonflies with "calopterygid" – like wing shape from the Middle Jurassic of China (Odonata: Isophlebioidea: Campterophlebiidae). *European Journal of Entomology* **105**: 783–787. https://doi.org/10.14411/eje.2008.103.

31 Martynov, A.V. (1925). On the knowledge of fossil insects from Jurassic beds in Turkestan. *Izvestiya Rossiiskoi Akademii Nauk [Bulletin de l'Académie des Sciences de l'URSS]* **6** (17): 569–598.

32 Li, Y.J., Nel, A., Ren, D. *et al.* (2012). Reassessment ofthe Jurassic damsel-dragonfly genus Karatawia (Odonata: Campterophlebiidae). *Zootaxa* **3417**: 64–68.

33 Li, Y.J., Nel, A., Ren, D., and Pang, H. (2012). A new damsel-dragonfly from the Lower Cretaceous of China enlightens the systematics of the Isophlebioidea (Odonata: Isophlebioptera: Campterophlebiidae). *Cretaceous Research* **34**: 340–343. https://doi.org/10.1016/j.cretres.2011.11.019.

34 Li, Y.J., Nel, A., Ren, D., and Pang, H. (2012). Redescription ofthe damsel-dragonfly *Parafleckium senjituense* on the basis of a more complete specimen (Odonata: Isophlebioptera: Campterophlebiidae). *Zootaxa* **3597**: 53–56.

35 Li, Y.J., Nel, A., Ren, D., and Pang, H. (2013). A new damsel-dragonfly from the Mesozoic of China with a hook-like male anal angle (Odonata: Isophlebioptera: Campterophlebiidae). *Journal of Natural History* **47** (29-30): 1953–1958. https://doi.org/10.1080/00222933.2012.759287.

36 Zheng, D.R., Dong, C., Wang, H. *et al.* (2017). The first damsel-dragonfly (Odonata: Isophlebioidea: Campterophlebiidae) from the Middle Jurassic of Shaanxi Province, northwestern China. *Alcheringa: An Australasian Journal of Palaeontology* https://doi.org/10.1080/03115518.2017.1318170.

37 Zhang, J.F. (1992). *Congqingia rhora* gen. nov., sp. nov., a new dragonfly from the Upper Jurassic of eastern China (Anisozygoptera: Congqingiidae fam. nov.). *Odonatologica* **21** (3): 375–383.

38 Ren, D. and Guo, Z.G. (1996). Three new genera and three new species of dragonflies from the Late Jurassic of Northeast China (Anisoptera: Aeshnidae, Gomphidae, Corduliidae). *Insect Science* **3** (2): 95–105. https://doi.org/10.1111/j.1744-7917.1996.tb00212.x.

39 Nel, A. and Huang, D.-Y. (2015). A new family of 'libelluloid' dragonflies from the Middle Jurassic of Daohugou, Northeastern China (Odonata: Anisoptera: Cavilabiata). *Alcheringa: An Australasian Journal of Palaeontology* **39** (4): 525–529. https://doi.org/10.1080/03115518.2015.1050316.

40 Li, Y.J., Nel, A., Shih, C.K. *et al.* (2013). The first euthemistid damsel-dragonfly from the Middle Jurassic of China (Odonata, Epiproctophora, Isophlebioptera). *ZooKeys* **261**: 41–50. https://doi.org/10.3897/zookeys.261.4371.

41 Bechly, G., Nel, A., Martínez-Delclòs, X. *et al.* (2001). A revision and phylogenetic study of Mesozoic Aeshnoptera, with description of several new families, genera and species (Insecta: Odonata: Anisoptera). *Neue Paläontogische Abhandlungen* **4**: 1–219.

42 Zhang, B.L., Ren, D., and Pang, H. (2008). New dragonflies (Insecta: Odonata: Gomphaeschnidae) from the Yixian Formation of Inner Mongolia, China. *Progress in Natural Science* **18**: 59–64. https://doi.org/10.1016/j.pnsc.2007.07.005.

43 Chang, J.P. and Sun, Z.S. (2006). New taxa of Gomphidae (Insecta: Odonata) in Jehol Biota from western Liaoning. *China. Global Geology* **25** (2): 105–112. (in Chinese with English abstract).

44 Ren, D., Liu, J.Y., and Cheng, X.D. (2003). A new hemeroscopid dragonfly from the Lower Cretaceous of Northeast China (Odonata: Hemeroscopidae). *Acta Entomologica Sinica* **46** (5): 622–628.

45 Huang, D.-Y. and Nel, A. (2007). Oldest "libelluloid" dragonfly from the Middle Jurassic of China (Odonata: Anisoptera: Cavilabiata). *Neues Jahrbuch für Geologie Paläontologie Abhandlungen* **246** (1): 63–68. https://doi.org/10.1127/0077-7749/2007/0246-0063.

46 Ren, D. (1994). A new genera and species of Liassogomphidae (Insecta, Odonata) from late Jurassic of Liaoning Province, China. *Geoscience* **8** (3): 254–259. (in Chinese with English abstract).

47 Lin, Q.B. (1982). Mesozoic and Cenozoic insects. In: *Paleontological Atlas of Northwest China*. (III). Volume of Shaan-Gan-Ning Basin Insecta. (ed. Xi'an Institute of Geology and Mineral Resources), 70–83. Beijing: Geological Publishing House (in Chinese).

48 Lin, Q.B., Nel, A., and Huang, D.-Y. (2002). Phylogenetic analysis of the Mesozoic dragonfly family Liupanshaniidae (Insecta: Aeshnoptera: Odonata). *Cretaceous Research* **23** (4): 439–444. https://doi.org/10.1006/cres.2002.1013.

49 Lin, Q.B., Huang, D.-Y., and Nel, A. (2007). A new family of Cavilabiata from the Lower Cretaceous Yixian Formation, China (Odonata: Anisoptera). *Zootaxa* **1469**: 59–64.

50 Nel, A. and Huang, D.-Y. (2010). A new Mesozoic Chinese genus of aeshnopteran dragonflies. *Comptes Rendus Palevol* **9**: 141–145. https://doi.org/10.1016/j.crpv.2010.05.005.

51 Li, Y.J., Nel, A., Ren, D., and Pang, H. (2012). New gomphaeschnids and progobiaeshnids from the Yixian Fonnation in Liaoning Province (China) illustrate the tremendous Upper Mesozoic diversity ofthe aeshnopteran dragonflies. *Geobios* **45**: 339–350. https://doi.org/10.1016/j.geobios.2011.11.016.

52 Vernoux, J., Huang, D.-Y., Jarzembowski, E.A., and Nel, A. (2010). The Proterogomphidae: a worldwide Mesozoic family of gomphid dragonflies(Odonata: Anisoptera: Gomphides). *Cretaceous Research* **31**: 94–100. https://doi.org/10.1016/j.cretres.2009.09.010.

53 Li, Y.J., Nel, A., Ren, D. *et al.* (2011). A new Chinese Mesozoic dragontly clarifies the relationships between

Rudiaeschnidae and Cymatophlebiidae (Odonata: Aeshnoptera). *Zootaxa* **2802**: 51–57.

54 Lin, Q.B., Zhang, S., and Huang, D.-Y. (2004). Fuxiaeschna Hsiufunia gen. nov., spec. nov., A new Lower Cretaceous dragonfly from Northwestern China (Anisoptera: Rudiaeschnidae). *Odonatologica* **33** (4): 437–442.

55 Nel, A., Fleck, G., Garcia, G. *et al.* (2015). New dragonflies from the lower Cenomanian of France enlighten the timing of the odonatan turnover at the Early-Late Cretaceous boundary. *Cretaceous Research* **52**: 108–117. https://doi.org/10.1016/j.cretres.2015.06.008.

56 Fleck, G., Bechly, G., Martínez-Delclòs, X. *et al.* (2003). Phylogeny and classification of the Stenophlebioptera (Odonata, Epiproctophora). *Annales de la Société entomologique de France. (N.S.)* **39** (1): 55–93. https://doi.org/10.1080/0037927.2003.10697363.

57 Hagen, H.A. (1866). Die Neuropteren aus dem Lithographischen Schiefer im Bayern. *Paleontographica* **15**: 57–96.

58 Hagen, H.A. (1862). *Über die Neuropteren aus dem Lithographischen Schiefer im Bayern. Paleontographica*, vol. **10**, 96–145.

59 Nel, A. and Huang, D.-Y. (2015). A new genus and species of damsel-dragonfly (Odonata: Stenophlebiidae) from the Lower Cretaceous of Inner Mongolia. *Cretaceous Research* **56**: 421–425. https://doi.org/10.1016/j.cretres.2015.06.008.

60 Zheng, D.R., Wang, H., Jarzembowski, E.A. *et al.* (2016). New data on Early Cretaceous odonatans (Stenophlebiidae, Aeschnidiidae) from northern China. *Cretaceous Research* **67**: 59–65. https://doi.org/10.1016/j.cretres.2016.07.005.

61 Pritykina, L.N. (1968). "Strekozy Karatau (Odonata)" [Dragonflies of Karatau (Odonata)]. In: *Yurskoy Nasekomiye Karatau [Jurassic Insects of Karatau] (in Russian)* (ed. B.B. Rohdendorf), 26–54. Moscow: Academy of Sciences of the USSR, Section of General Biology, Publishing House "Nauka".

62 Huang, D.-Y., Nel, A., and Lin, Q.B. (2003). A new genus and species of aeshnopteran dragonfly from the Lower Cretaceous of China. *Cretaceous Research* **24**: 141–147. https://doi.org/10.1016/S0195-6671(03)00027-2.

63 Zhang, B.L., Ren, D., and Pang, H. (2008). Telmaeshna paradoxica gen. et sp. nov., a new fossil dragonfly (Insecta: Odonata: Anisoptera) from the Yixian Formation, Liaoning, China. *Zootaxa* **1681**: 62–68. https://doi.org/10.5281/zenodo.180415.

64 Li, Y.J., Nel, A., Ren, D., and Pang, H. (2011). A new genus and species ofhawker dragonfly of uncertain affinities from the Middle Jurassic of China (Odonata: Aeshnoptera). *Zootaxa* **2927**: 57–62.

65 Hong, Y.C. (1982). *Mesozoic Fossil Insects of Jiuquan Basin in Gansu Province*, 187. Beijing: Geological Publishing House (in Chinese).

66 Ren, D., Lu, L.W., Guo, Z.G., and Ji, S.A. (1995). *Faunae and stratigraphy of Jurassic-Cretaceous in Beijing and the Adjacent Areas*, 47–49. Beijing: Seismic Publishing House (in Chinese and English).

67 Zhou, W.B. and Wei, J.L. (1980). A new species of *Sinaeschnidia* Hong. *Entomotaxonomia* **2** (1): 77–79. (in Chinese with English abstract).

68 Li, Y.J., Han, G., Nel, A. *et al.* (2012). A new fossil petalurid dragonfly (Odonata: Petaluroidea: Aktassiidae) from the Cretaceous of China. *Alcheringa: An Australasian Journal of Palaeontology* **36**: 319–322. https://doi.org/10.1080/03115518.2012.644995.

69 Zhang, B.L., Ren, D., and Pang, H. (2008). New isophlebioid dragonflies from the Middle Jurassic *of Inner* Mongolia, China (Insecta: Odonata: Isophlebioptera: Campterophlebiidae). *Acta Geologica Sinica (English Edition)* **82** (6): 1104–1114. https://doi.org/10.1111/j.1755-6724.2008.tb00710.x.

70 Lin, Q.B. (1976). The Jurassic fossil insect from western Liaoning. *Acta Palaeontologica Sinica* **15** (1): 97–116. (in Chinese).

71 Hong, Y.C. (1982). Class Insecta. In: *The Mesozoic Stratigraphy and Paleontology of the Guyang Coal-Bearing Basin, Neimenggol Autonomous Region* (ed. Geological Bureau of the Inner Mongolia Autonomous Region), 86–87. China, Beijing: Geological Publishing House (in Chinese).

72 Hong, Y.C. (1984). Mesozoic Volume. In: *Palaeontological Atlas of North China* (ed. Tianjin Institute of Geology and Mineral Resources), 132–139. Beijing: Geological Publishing House (in Chinese).

73 Huang, D.-Y. and Lin, Q.B. (2001). The Early Cretaceous Hemeroscopid larva fossils from Beijing, China. *Chinese Science Bulletin* **46** (17): 1477–1481.

74 Hong, Y.C. (1983). Order Odonata. In: *Middle Jurassic Fossil Insects in North China* (ed. Y. Hong), 22–25. Beijing: Geological Publishing House (in Chinese).

75 Zheng, D.R., Nel, A., Wang, B. *et al.* (2016). The first Early Cretaceous damsel-dragonfly (Stenophlebiidae: Stenophlebia) from western Liaoning, China. *Cretaceous Research* **61**: 124–128. https://doi.org/10.1016/j.cretres.2016.01.003.

76 Huang, D.-Y. and Nel, A. (2009). The first Chinese Tarsophlebiidae from the Lower Cretaceous Yixian Formation, with morphological and phylogenetic implications (Odonatoptera: Panodonata). *Cretaceous Research* **30**: 429–433. https://doi.org/10.1016/j.cretres.2008.08.005.

7

Blattaria – Cockroaches

Junhui Liang[1,2], Chungkun Shih[2,3], and Dong Ren[2]

[1] *Tianjin Natural History Museum, Hexi District, Tianjin, China*
[2] *Capital Normal University, Haidian District, Beijing, China*
[3] *National Museum of Natural History, Smithsonian Institution, Washington, DC, USA*

7.1 Introduction to Blattaria

Blattaria, commonly referred to as "true" cockroaches, prefer moist surroundings and dwell in forests, grasslands, farmlands, marshes, deserts, frozen tundra, underground, caves, and most annoyingly, in residential areas. About 5000 extant and 1000 extinct [1] cockroach species have been described in the world, however, only about 16 species [2] are deemed as household pests. Among them, *Blattella germanica* and *Periplaneta americana* cause harm and damage to human lives, thus making all cockroaches suffer from the stigma of being ugly, dirty and disgusting. However, in reality, flies, mosquitoes and fleas are much worse than cockroaches in carrying and passing diseases, damaging food sources and harming humans.

Due to cockroaches' surprisingly high diversity, most cockroaches are beneficial insects to their ecosystems. Some cockroaches act as scavengers to consume and help recycling dead plants, fallen leaves, and other organic matter in forests and other environments (see Figure 7.1). For example, Australian wood cockroaches (*Panesthia cribrata* Saussure, 1864) feed on wood in rotten logs in the forest of Australia. In some zoos, the large flightless Madagascar hissing cockroach (*Gromphadorhina portentosa* Schaum, 1853) is a popular pet, delighting children when it hisses by forcing air to expel through its spiracles on the fourth abdominal segment.

In spite of the long history and broad diversity of cockroaches, the family-level classification, unfortunately, is unsettled with various views on the high-level classification. Some even consider Isoptera (termites) as an infraorder within the cockroach order Blattodea (see Chapter 8, Termitoidae). Roth presented a traditional classification, based largely on the work of McKittrick [3], which includes six families: Blaberidae, Blattidae, Blattellidae, Cryptocercidae, Nocticolidae, and Polyphagidae [4].

Cockroaches are highly adaptive to whatever ecosystem they live in. The flattened oval body and the ability to fold hind wings below the forewings allow cockroach to pass through small cracks or hide in a narrow space. The nocturnal feeding activities help them to avoid many predators under daylight. Female cockroaches carry their oothecae for a while before depositing them in concealed places, which improve the survivability of the next generation. *P. americana* has a lifespan of about 4 years and a female is capable of laying more than 250 eggs in a year. Typically, it takes only one year for their nymphs to reach adulthood and reproduce.

Cockroaches even exhibit a rare insect behavior – maternal care. Female Australian giant burrowing cockroaches, *Macropanesthia rhinoceros* Saussure, 1895 live with the nymphs in burrows and feed them with dried leaves until they reach adulthood. In South America, the females of *Paratropes* carry their nymphs between their wings and abdomen to protect them. In Malaysia, the females of *Perisphaeria* have small openings on the abdomen to allow nymphs to attach their mouthparts for nutrients and protection.

Due to cockroaches' tolerance, endurance, and adaptability, they have been very successful in developing varieties and diversity of body structures and functions to fit the changes of environment and climate. Since they first appeared in the Late Carboniferous, about 315 million years ago (Mya), cockroaches are a shining example of survivors of the fittest (see Figure 7.2).

Rhythms of Insect Evolution: Evidence from the Jurassic and Cretaceous in Northern China, First Edition.
Edited by Dong Ren, Chungkun Shih, Taiping Gao, Yongjie Wang, and Yunzhi Yao.

Figure 7.1 Cockroach. Source: Photo by Jason Shih.

Figure 7.2 Cockroaches mating. Source: Photo by Jason Shih.

Cockroaches as a Traditional Chinese Medicine

In traditional Chinese medicine, the dried whole bodies of female *Eupolyphaga sinensis* and *Steleophaga plancyi*, both in the Corydiidae and commonly known as "sand cockroaches" or "wingless cockroaches", have been used clinically for more than 2000 years in China. Its medicinal uses have been clearly documented by almost all popular pharmacopeia from the Han to Qing Dynasties. The wild specimens can be captured in the summer and then raised all year round. The preparations are to kill the female insects in a pot of boiling water and then dry them. As a traditional Chinese medicine for regulating blood condition and trauma, ground-up wingless cockroaches are widely used for bruises, swelling, pain, injury, abdominal masses and amenorrhea. It is also used for curing injured tendons, joints, sinews, contusions, fractures and lacerations [5]. A cream made from powdered cockroaches is in use in some Chinese hospitals as a treatment for burns. A syrup invented by a pharmaceutical company promises to cure gastroenteritis, duodenal ulcers and pulmonary tuberculosis. Recent study shows that *P. americana* contains mannitol small peptides and sugar amino acids which are claimed to be effective for hepatitis, cardiovascular disease, lung cancer and malignant fibroids [6].

7.2 Progress in the Studies of Fossil Blattodea

During the Late Carboniferous, cockroaches already thrived on the Earth with a worldwide distribution. In terms of quantity, cockroach fossils are only second to beetle fossils [7, 8]. Cockroaches have adapted to multiple types of ecosystem and reached a high level of ecological and morphological diversity over the course of their evolution, including cryptic cockroaches, pollinating cockroaches [9], aquatic cockroaches [10, 11], jumping cockroaches [12, 13], carnivorous cockroaches [14, 15], beetle-like cockroaches [16], translucent cockroaches [17], and light-mimicking cockroaches [18, 19].

Fossil Blattaria research in China was initiated by Grabau in 1923 (Chapter 4.2). During the past 90 years, there have been many significant studies of Chinese fossil cockroaches by Qibin Lin, Youchong Hong, Dong Ren, Junfeng Zhang, Wuli Wang, Haichun Zhang, Chungkun Shih, O. Béthoux, P. Vršanský, and the CNU Team (Junhui Liang, Taiping Gao, Tiantian Wang, Yinxia Guo, Dandan Wei, Chongda Wang and Xin Liu). Hitherto, nine families (Blattidae, Blattulidae, Caloblattinidae, Ectobiidae, Fuziidae, Liberiblattinidae, Mesoblattinidae, Raphidiomimidae and Umenocoleidae) in fossils and amber (Fushun, Liaoning), comprising 33 genera and 88 species, have been described from Northern China.

The earliest known fossil "roachoid" [cockroach-like insect], from China is *Qilianiblatta namurensis* Zhang, Schneider & Hong, 2012 (Archimylacrididae), found in the Tupo Formation in Zhongwei City, Ningxia Hui Autonomous Region. In age, this is early Bashkirian, Early Pennsylvanian (315 Mya), the earliest Late Carboniferous (see box) [20–23]. Complex veins and long ovipositor are the main characters of the cockroach body in the Paleozoic; even the wing venation of some species show similarity to pinnules of the widespread Carboniferous and Permian tree ferns. Until the Late Triassic, morphological characters of the cockroaches became more similar to those of the modern groups, such as more simplified wings and shorter ovipositors. From the Jurassic to the Cretaceous, cockroaches developed more diversity, while jumping, carnivorous, beetle-like, earwig-like cockroaches appeared.

The Earliest "Roachoid" in China

"Roachoids" became increasingly common during the Late Carboniferous (Pennsylvanian), which was the main period in the development and differentiation of the winged insects. Blattodea have been among the most abundant and diverse groups at most insect fossil sites. Important sites are distributed in China, France, North America, England, Siberia, Morocco and Germany. The well-preserved "roachoid" specimens from the Pennsylvanian (315 Mya), Tupo Formation (Ningxia Hui Autonomous Region, China) are the earliest fossil record from China hitherto (Figure 7.3). They provide preserved evidence to address the questions of homology in the insect wing venation character system, which have been debated for more than 100 years. Observed differences between left and right forewings allow conjectures of primary homology in forewing venation of Blattodea and Pennsylvanian "cockroachoids" to be clarified: (i) the radial system is organized with distal branches of RA translocated to RP in *Q. namurensis* and in crown-Blattodea; and (ii) branches of the median system are translocated to/fused with CuA in Pennsylvanian "cockroachoids" (Figure 7.3) [21, 22].

Figure 7.3 New material (CNU-NX1–303) of *Qilianiblatta namurensis* Zhang, Schneider & Hong, 2012. Source: modified from [22].

7.3 Representative Fossils of Blattaria from Northern China

Family Blattidae Handlirsch, 1932

Blattidae have existed from the Cretaceous till now. The oldest Blattidae is *Sinoblatta laiyangensis* Grabau, 1923 from the Early Cretaceous of Shandong, China [24]. Up to date, only one species has been reported from the Jurassic and Cretaceous of Northern China.

Only one genus included from the Cretaceous of Northern China: *Sinoblatta* Grabau, 1923.

Sinoblatta Grabau, 1923

Sinoblatta Grabau, 1923, *Bull. Geol. Surv. China*, 5, 167–173 [24] (original designation).

Type species: *Sinoblatta laiyangensis* Grabau, 1923.

Large oval head with large eyes, the heart-shaped form of the prothorax, and the relatively long femora of the legs. The scapular vein of the tegmina very strong.

Distribution and age: Shandong; Early Cretaceous.

Only one species included from the Cretaceous of Northern China (see Table 7.1).

Family Blattulidae Vishniakova, 1982

Representatives of the Blattulidae were placed in the Mesoblattinidae, until Vishniakova [25] established the family and commented on their relationship with Polyphagoidea. The Blattulidae, a family of cosmopolitan Mesozoic cockroaches, have these typical characters: size small, Sc with a few branches or simple, R proximal branches regular and simple, M forking into two branches or reduced to a simple straight vein and the anal region smallish, with four to six veins. However, their diversity at the generic level is low with only 13 genera reported so far in their more than 100 million years of existence [26, 27]. To date, only four genera with 13 species in this family have been described from the Jurassic and Cretaceous of Northern China.

Genera included from the Jurassic and Cretaceous of Northern China: *Elisama* Giebel, 1856, *Blattula* Handlirsch, 1908, *Habroblattula* Wang, Liang & Ren, 2007 and *Macaroblattula* Wang, Ren & Liang, 2007.

Elisama Giebel, 1856

Elisama Giebel, 1856, Leipzig. XVIII [28] (original designation).

Ctenoblattina Scudder, 1886, *Memoirs of the Boston Society of Natural History*. 3 (13), 439–485 [29]; Syn. by Vršanský, 2003, *Entomological Problems*, 33 (1–2), 119–151 [16].

Type species: *Elisama minor* Giebel, 1856.

Forewing has regular and simple venation with distinct maculae; Sc simple; R arcuate; Clavus without cross-veins; M and Cu reduced to several veins.

Distribution and age: Liaoning of China, Middle Jurassic; Hebei and Liaoning of China, Early Cretaceous; Russia, Cretaceous; Brazil, England and Lebanon,

Early Cretaceous; Mongolia, Early Cretaceous and Late Jurassic.

Three species included from the Jurassic and Cretaceous of Northern China (see Table 7.1).

Blattula Handlirsch, 1906

Blattula Handlirsch, 1906, Leipzig. 1433 [30] (original designation).

Mesoblattula Handlirsch 1906–1908, Die Fossilen Insekten und die Phylogenie der rezenten Formen: Ein Handbuch für Paläntologen und Zoologen. 430–431 [30]; Syn. by Vršanský & Ansorge, 2007, *African Invertebrates*, 48 (1), 103–126 [31].

Parablattula Handlirsch 1920, *Annalen des Naturhistorischen Museum Wien* 49, 1–240; Syn. by Vršanský & Ansorge, 2007, *African Invertebrates*, 48 (1), 103–126 [31].

Type species: *Blattina langfeldti* Geigitz, 1880.

Forewing ovoid, anterior margin and posterior margin parallel; Sc simple; humeral area wide; R straight, and branches of R parallel; M not developed; CuA with pectinate branches. Hind wing Sc straight; M branched; CuA straight and with pectinate branches.

Distribution and age: Hebei and Liaoning of China, Early Cretaceous and Middle Jurassic; Beijing of China, Early Cretaceous; Russia, Early Jurassic and Middle Jurassic; Korea and England, Early Cretaceous; Kazakhstan and Mongolia, Late Jurassic; Germany, Kyrgyzstan, Tajikistan and Australia, Early Jurassic.

Eight species included from the Jurassic and Cretaceous of Northern China (see Table 7.1).

Habroblattula Wang, Liang & Ren, 2007

Habroblattula Wang, Ren & Liang, 2007, *Zootaxa*, 1443, 17–27 [32] (original designation).

Type species: *Habroblattula drepanoides* Wang, Liang & Ren, 2007.

Large species (forewing length/width: 11.0–12.0/3.2–4.0 mm), forewing elongate and elliptical. Wings, pronotum and legs with bright and dark coloration. Forewing Sc terminally branched, CuP strongly curved, all branches strong and dark, intercalaries and cross-veins distinct. Hind wing with pterostigma (dark macula in the R1), R1 and RS differentiated, CuA with abundant secondary branches.

Distribution and age: Liaoning; Early Cretaceous.

Only one species included from the Cretaceous of Northern China (see Table 7.1).

Habroblattula drepanoides Wang, Liang & Ren, 2007 (Figure 7.4)

Habroblattula drepanoides Wang, Liang & Ren, 2007: *Zootaxa*, 1443, 17–27.

Locality and horizon: Huangbanjigou, Beipiao, Liaoning, China; Lower Cretaceous, Yixian Formation.

Forewing with characteristic coloration like "falcate"; costal space narrow; Sc straight and somewhat thickened; R 12–17; M 3–8; CuA 3–6; A 6–8. Hind wing length 10.0–11.0 mm. Sc simple and weak; R1 5–6 and RS 9–11; M 2–5; CuA 6–9; A1 forked. The species of *H. drepanoides*, a "colored" cockroach species, possibly indicates warm and moist climate. For *H. drepanoides*, the total number of veins (excluding SC and A) reaching the wing margin is found to be the least variable character, suggesting active flight, which is also supported by the asymmetrical difference between left and right forewings. Higher variability of the hind wing might suggest *Habroblattula* was an advanced genus, with vein strength control mechanisms [32].

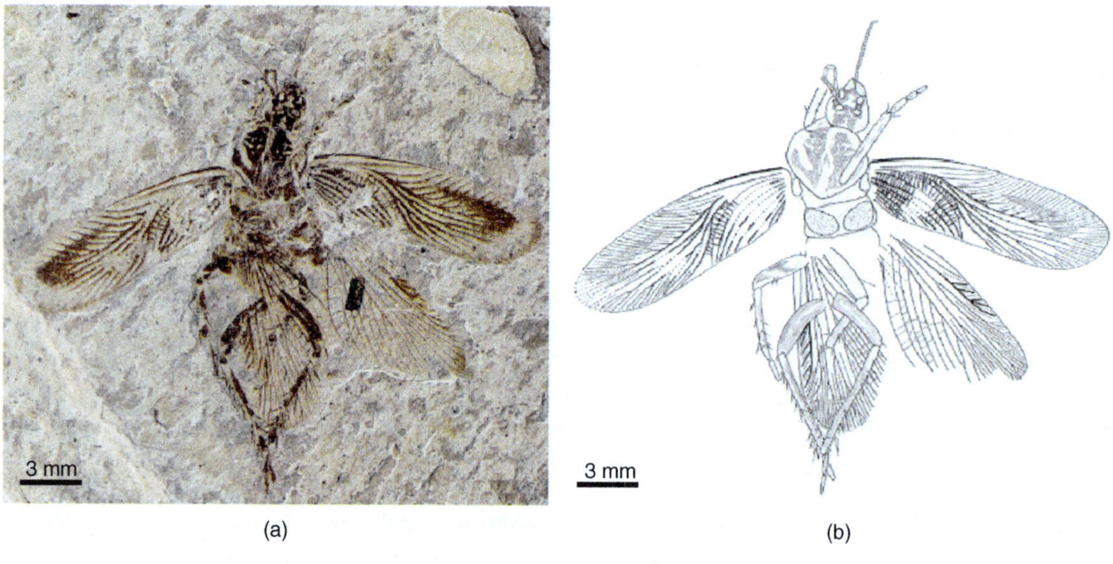

Figure 7.4 *Habroblattula drepanoides* Wang, Liang & Ren, 2007 (Holotype, CNU-B-LB-2006369. (a), Photograph; (b), Line drawing [32].

Macaroblattula Wang, Ren & Liang, 2007

Macaroblattula Wang, Ren & Liang, 2007, *Ann. Zool.*, 57 (3), 483–495 [33] (original designation).

Type species: *Macaroblattula ellipsoides* Wang, Ren & Liang, 2007.

Forewing Sc area with a dark macula and Sc terminally branched; base of R strong, slightly curved, with simple veins; M branched more basal than CuA; CuP strongly curved, all branches strong and dark. Hind wing with pterostigma, R1 and RS differentiated, CuA strong, with several veins. Dark coloration at the tip of hind wing.

Distribution and age: Liaoning; Early Cretaceous.

Only one species included from the Cretaceous of Northern China (see Table 7.1).

Family Caloblattinidae Vršanský & Ansorge, 2000

Caloblattinidae, a large extinct family, have been reported from the Middle Triassic to the Late Cretaceous deposits throughout Europe, Asia, South America, South Africa, Greenland and Australia [16]. The Caloblattinidae are distinguished by large body, fairly long ovipositors in females, and both wings with dark coloration and veins multiple-branched [34]. Most species in this family have advanced morphology and conspicuous coloration, which indicate diurnal and/or crepuscular activities. Amber specimens of caloblattinids are extraordinarily rare, probably due to their large size. Up to now, 40 species within nine genera have been reported from the Jurassic and Cretaceous of Northern China.

Genera included from the Jurassic and Cretaceous of Northern China: *Rhipidoblattina* Handlirsch, 1906, *Samaroblatta* Tillyard, 1919, *Euryblattula* Martynov, 1937, *Samaroblattula* Martynov, 1937, *Sogdoblatta* Martynov, 1937, *Taublatta* Martynov, 1937, *Nipponoblatta* Fujiyama, 1974, *Fusiblatta* Hong, 1980 and *Nuurcala* Vršanský, 2003.

Rhipidoblattina Handlirsch, 1906

Rhipidoblattina Handlirsch, 1906, Leipzig, 1433 [30]. Trans. from Mesoblattinidae by Vršanský, 2008, *Paleontological J.*, 42 (1), 36–42 [35].

Type species: *Rhipidoblattina geikiei* Scudder, 1886.

Forewing 8.5–28 mm, long and narrow, with distinct intercalary veins and cross-veins; humeral area narrow, as long as anal area; Sc simple, with one to two veins; R slightly curved at the middle of forewing, with many veins; M simple; CuA forking more basal than M; anal area long.

Distribution and age: Liaoning and Hebei of China, Early Jurassic and Middle Jurassic; Beijing and Hebei of China, Late Jurassic; Jilin, Liaoning, Shandong and Gansu of China, Early Cretaceous; Mongolia, Russia, Korea, and England, Early Cretaceous; Tajikistan, Australia, Kyrgyzstan, Early Jurassic; Kazakhstan, Late Jurassic.

Eighteen species included from the Jurassic and Cretaceous of Northern China (see Table 7.1).

Samaroblatta Tillyard, 1919

Samaroblatta Tillyard, 1919, *Proc. Linn. Soc. N. S. W.* 44, 358–382 [36]. Trans. from Mesoblattinidae by Vršanský, 2008, *Paleontological J.*, 42 (1), 36–42 [35].

Type species: *Samaroblatta reticulata* Tillyard, 1919.

Medium-sized tegmina. Humeral area wide, as the anal area. Sc with a single vein; Main stem of R with fairly strong double curves; anal area somewhat cultriform.

Distribution and age: Hebei and Inner Mongolia of China, Middle Jurassic; Gansu of China, Early Cretaceous; Korea, Early Cretaceous; Mongolia, Middle Jurassic; Russia, South Africa and Tajikistan, Early Jurassic.

Four species included from the Jurassic and Cretaceous of Northern China (see Table 7.1).

Euryblattula Martynov, 1937

Euryblattula Martynov, 1937, *Trud. Pal. Inst. Akad. Nauk SSSR.*, 7 (1), 1–231 [37]. Trans. from Mesoblattinidae by Vršanský, 2008, *Paleontological J.*, 42 (1), 36–42 [35].

Type species: *Euryblattula sparsa* (Martynov, 1937).

Forewing wide; Sc simple, with one branch, humeral area as long as anal area; R short and curved, with some branches; M forking distal to CuA forking; CuA distinctly curved, CuP strongly curved; anal area short.

Distribution and age: Liaoning of China, Early Jurassic; Hebei of China, Middle Jurassic; Beijing of China, Late Jurassic; Kyrgyzstan, Early Jurassic.

Five species included from the Jurassic of Northern China (see Table 7.1).

Samaroblattula Martynov, 1937

Samaroblattula Martynov, 1937, *Trudy paleontologicheskogo instituta akademii nauk SSSR.*, 7 (1), 1–231 [37]. Trans. from Mesoblattinidae by Vršanský, 2008, *Paleontological J.*, 42 (1), 36–42 [35].

Type species: *Samaroblattula subacuta* Martynov, 1937.

Forewing wide, with cross-veins and intercalary veins. Humeral area wide; Sc as long as anal area, with two to three branches; M forking to two main branches, media area narrow; CuA forking more basal than M forking; anal area long, with cross-vein. Hind wing Sc with a few branches; R, M, CuA almost forking at the same level; R with many branches; M not developed; CuA with many veins.

Distribution and age: Qinghai of China, Middle Jurassic; Beijing of China, Late Jurassic; Mongolia, Middle Jurassic; Tajikistan, Early Jurassic.

Five species included from the Jurassic of Northern China (see Table 7.1).

Sogdoblatta Martynov, 1937

Sogdoblatta Martynov, 1937, *Trud. Pal. Inst. Acad. Nunk SSSR.*, 7 (1), 1–231 [37]. Trans. from Mesoblattinidae by Vršanský, 2008, *Paleontological J.*, 42 (1), 36–42 [35].

Type species: *Sogdoblatta robusta* Martynov, 1937.

Forewing very long, with developed intercalary veins and cross-veins; Sc, with more than three branches, as long as anal area; R slightly curved; CuA branched earlier than M, and veins more than M; anal area big. Hind wing, Sc very long; R with double forked branches, R and M almost forking at the same level; CuA forking more basal, with many branches.

Distribution and age: Hebei and Liaoning of China, Middle Jurassic; Beijing of China, Late Jurassic; Kyrgyzstan and Tajikistan, Early Jurassic; Mongolia and Russia, Middle Jurassic.

Three species included from the Jurassic of Northern China (see Table 7.1).

Taublatta Martynov, 1937

Taublatta Martynov, 1937, *Trud. Pal. Inst. Acad. Nunk SSSR*, 7 (1), 1–231 [37]. Trans. from Mesoblattinidae by Vršanský, 2008, *Paleontological J.*, 42 (1), 36–42 [35].

Type species: *Taublatta curvata* Martynov, 1937.

Forewing wide and short, with developed intercalary veins and cross-veins, the length of forewing 15–20 mm; Sc longer than anal area, with two to three branches; humeral area as long as anal area or slightly longer; R strongly curved; CuA with many branches.

Distribution and age: Liaoning of China, Early Jurassic; Kyrgyzstan and Tajikistan, Early Jurassic; Mongolia, Middle Jurassic.

Only one species included from the Jurassic of Northern China (see Table 7.1).

Nipponoblatta Fujiyama, 1974

Nipponoblatta Fujiyama, 1974, *Bull. Natn. Sci. Mus.*, 17 (4), 311–314 [38]. Trans. from Mesoblattinidae by Vršanský, 2008, *Paleontological J.*, 42 (1), 36–42 [35].

Type species: *Nipponoblatta suzugaminae* Fujiyama, 1974.

The specific epithet is dedicated to Dr. Sotoji Imamura and to Miss Hiroko Ishine of the Suzugamine Women's College for collecting this material. Large in size. Costal margin moderately curved, posterior margin nearly straight; humeral area large, nearly half the length of forewing; anal area large, slightly shorter than half the length of humeral area; Sc with a long branch; R with 10 branches, stem of R moderately curved sigmoidally; M moderately branching off; intercalary veins developed, without cross-veins.

Distribution and age: Liaoning of China, Early Cretaceous; Japan, Early Jurassic.

Only one species included from the Cretaceous of Northern China (see Table 7.1).

Fusiblatta Hong, 1980

Fusiblatta Hong, 1980, *Bull. Chin. Acad. Geol. Sci.*, 4, 49–60 [39]. Trans. from Mesoblattinidae by Vršanský, 2008, *Paleontological J.*, 42(1), 36–42 [35].

Type species: *Fusiblatta arcuata* Hong, 1980.

Forewing fusiformis; intercalary and cross-veins not very developed; humeral area narrow, but very long, almost as long as anal area; Sc with one to two branches; R with simple branches; M with two branches, anterior and posterior branches with double forks; CuA branched more basal than M; Anal area extremely wide.

Distribution and age: Liaoning and Hebei; Middle Jurassic.

Two species included from the Jurassic of Northern China (see Table 7.1).

Nuurcala Vršanský, 2003

Nuurcala Vršanský, 2003, *Ento. Prob.*, 33 (1–2), 119–151 [16] (original designation).

Type species. *Nuurcala popovi* Vršanský, 2003

The specific epithet is dedicated to Dr. Popov for his excellent work on paleoentomology. Medium-sized with compact body. Females with long and straight ovipositors. Pronotum transverse ovoid, with simple coloration. Forewing with distinct cubital space, rather wide, but with more or less parallel margins and characteristic forewing coloration (dark coloration with pale area in R). Sc branched, R rich, M branched, Cu veins ending prior to the apex of the wing. Vein A branched.

Distribution in China and age: Liaoning of China, Early Cretaceous; Mongolia, Early Cretaceous.

Only one species included from the Cretaceous of Northern China (see Table 7.1).

Nuurcala obesa Wang & Ren, 2013 (Figure 7.5)

Nuurcala obesa Wang & Ren, 2013: *ZooKeys*, 318, 35–46.

Locality and horizon: Huangbanjigou, Beipiao, Liaoning, China; Lower Cretaceous; Yixian Formation.

Forewing length/width: 25.2/9.9 mm. Pronotum, shield-like, simple symmetrical zonal coloration at the margin. Forewing with one dark maculae at the edge of Sc area, dark coloration with pale area in R; intercalaries thick; Sc 3, shorter than clavus; R 15; M slightly curved with nine branches, most branches at wing apex; CuA 10;

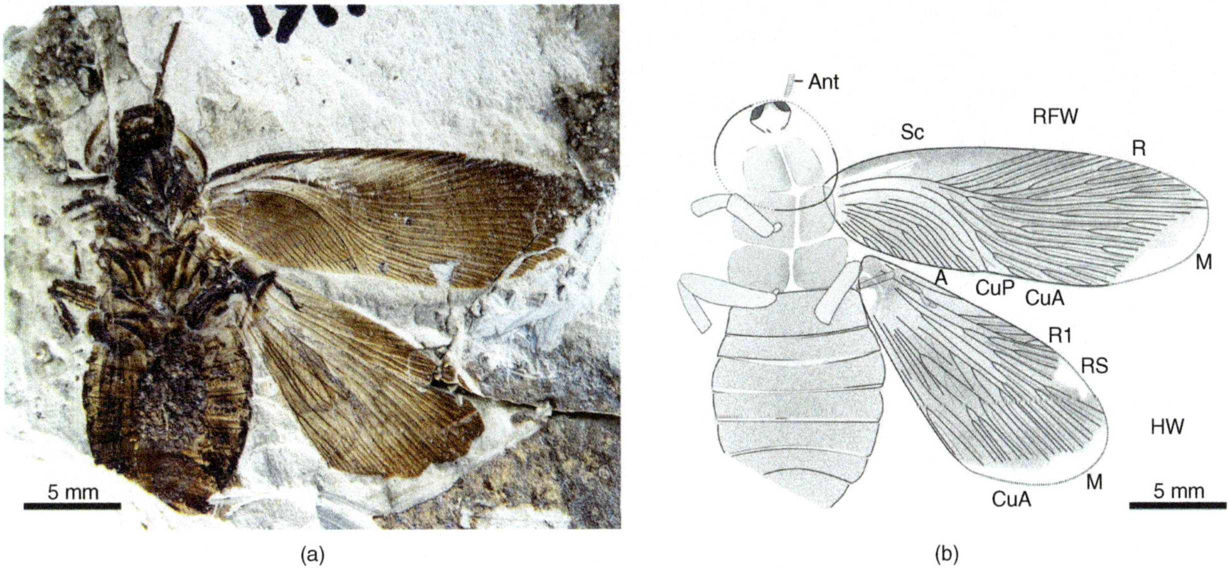

Figure 7.5 *Nuurcala obesa* Wang & Ren, 2013 (Holotype, CNU-BLA-LB-2012055). (a), Photograph; (b), Line drawing. (Note: the specimen number is corrected from "NN" in [40] to "LB").

clavus long, more than a third of the wing length; A with tertiary branches; fan-like pleating present and visible on forewing. Hind wing with branched Sc; R1 and RS with about nine veins; M four branches; Cu reticulate [40].

In the Yixian Formation, fossils of Blattulidae were abundant, but Caloblattinidae, quite rare. This is in contrast to the rich presence of caloblattinids in the assemblages in Karatau, South Kazakhstan [14, 41]. Yixian Formation is characterized by a high proportion of cockroach fossils with sophisticatedly colored wings [32, 33, 42, 43], e.g. Blattulidae. Dry habitats are rather characterized by monochromatic and pale cockroach individuals [43, 44]. The colored *Nuurcala obesa* supports the notion that the Yixian Formation was humid [45].

Family Ectobiidae Brunner von Wattenwyl, 1865

Ectobiidae, the largest family of extant cockroaches, have a very long evolutionary history from the Early Cretaceous. Many species have been described from the Zaza Formation of Transvaikalian Russia [46] and the Crato Formation of Brazil [47], but only one species, *Piniblattella yixianensis* Gao, Shih & Ren, 2018, reported from the Early Cretaceous of China.

Only one genus included from the Cretaceous of Northern China: *Piniblattella* Gao, Shih & Ren, 2018.

Piniblattella Vršanský, 1997

Piniblattella Vršanský, 1997, *Entomological Problems*, 28 (1), 67–79 [46] (original designation).

Type species. *Piniblattella vitimica* Vršanský, 1997.

Forewing extending beyond apex of abdomen, elliptical; Sc simple or terminally forked; CuA branches meeting posterior wing margin. M rich, branched with 12 to 23 veinlets; R without R1. Hind wing with CuA secondarily branched, A1 terminally forked, R1 distinct and Rs with apical forked rami.

Distribution and age: Liaoning of China, Early Cretaceous; Brazil, Mongolia and Russia, Early Cretaceous.

Only one species included from the Cretaceous of Northern China (see Table 7.1).

Piniblattella yixianensis Gao, Shih & Ren, 2018 (Figures 7.6 and 7.7)

Piniblattella yixianensis Gao, Shih & Ren, 2018: *J. Syst. Palaeontol.*, doi: 10.1080/14772019.2018.1426059, 2.

Locality and horizon: Huangbanjigou, Beipiao, Liaoning, China; Lower Cretaceous; Yixian Formation.

To provide maternal protection and care, most extant female cockroaches produce special, relatively hardened capsules, named oothecae, which extend posteriorly from abdominal terminalia encasing two rows of elongate, ovate eggs [3, 49]. Oothecae of cockroaches provide protection for eggs and offspring either to avoid predators' attack or to survive under inclement environmental conditions until first-instar nymphs establish independent life habits [49–52] *Piniblattella yixianensis* Gao, Shih & Ren, 2018 in the family of Ectobiidae Brunner von Wattenwyl, 1865 discovered from the Lower Cretaceous of Yixian Formation is the earliest known fossil cockroach group possessing an internally partitioned ootheca [48]. This species is erected based on six specimens with clear ootheca structures. The

(a) (b)

Figure 7.6 *Piniblattella yixianensis* Gao, Shih & Ren, 2018, (Holotype, CNU-BLA-LB2013800p). (a), Photograph; (b), Line drawing [48].

Figure 7.7 *Piniblattella yixianensis* Gao, Shih & Ren, 2018, Source: Artwork by Dr. Chen Wang.

ootheca of this species measures 46–62% of its body length, contains 60–70 eggs, and exhibits the oviparity B reproduction mode, similar to that of the extant ectobiid and blattid cockroaches. It is suggested that cockroach reproductive mode using the complete oothecate structure occurred during or before the mid Early Cretaceous. Maternal protection and care associated with the oothecate condition by these Early Cretaceous cockroaches added a unique and major life-history trait, which might have been a key factor in the subsequent evolutionary diversification and ecological expansion of modern cockroach lineages [48].

Family Fuziidae Vršanský, Liang & Ren, 2009

Fuziidae, an indigenous Chinese family, are characterized by a very small head covered by a very large pronotum; forewing widest in the apical third and costal area very long and wide. Forewing Sc strong, simple or with terminal branch and RS slightly differentiated in most individuals. This family is further distinguished by its unique structure of the male's elongated body and forceps of earwig-like cerci (used for attachment to the external ovipositor during courtship). So far, six species within four genera have been described from the Jurassic and Cretaceous of Northern China.

Genera included from the Jurassic and Cretaceous of Northern China: *Fuzia* Vršanský, Liang & Ren, 2009, *Parvifuzia* Guo and Ren, 2011, *Arcofuzia* Wei, Shih & Ren, 2012 and *Colorifuzia* Wei, Liang & Ren, 2013.

Fuzia Vršanský, Liang & Ren, 2009

Fuzia Vršanský, Liang & Ren, 2009, *Geol. Carpath.*, 60 (6), 449–462 [44] (original designation).

Type species: *Fuzia dadao* Vršanský, Liang & Ren, 2009.

Medium-sized male specimens with length/width: 10–15.5/3.6–6.5 mm (in females 12.5–13/4.2–4.5 mm). Pronotum large, with wide paranotalia and two colored central stripes. Forewing with very wide costal area, simple and strong partially colored Sc; RS is differentiated in most individuals; anal area very short, with simple branches of A. Male body narrow and longer than the forewings, with cerci forming forceps (with notches). The external ovipositor very rigidly curved.

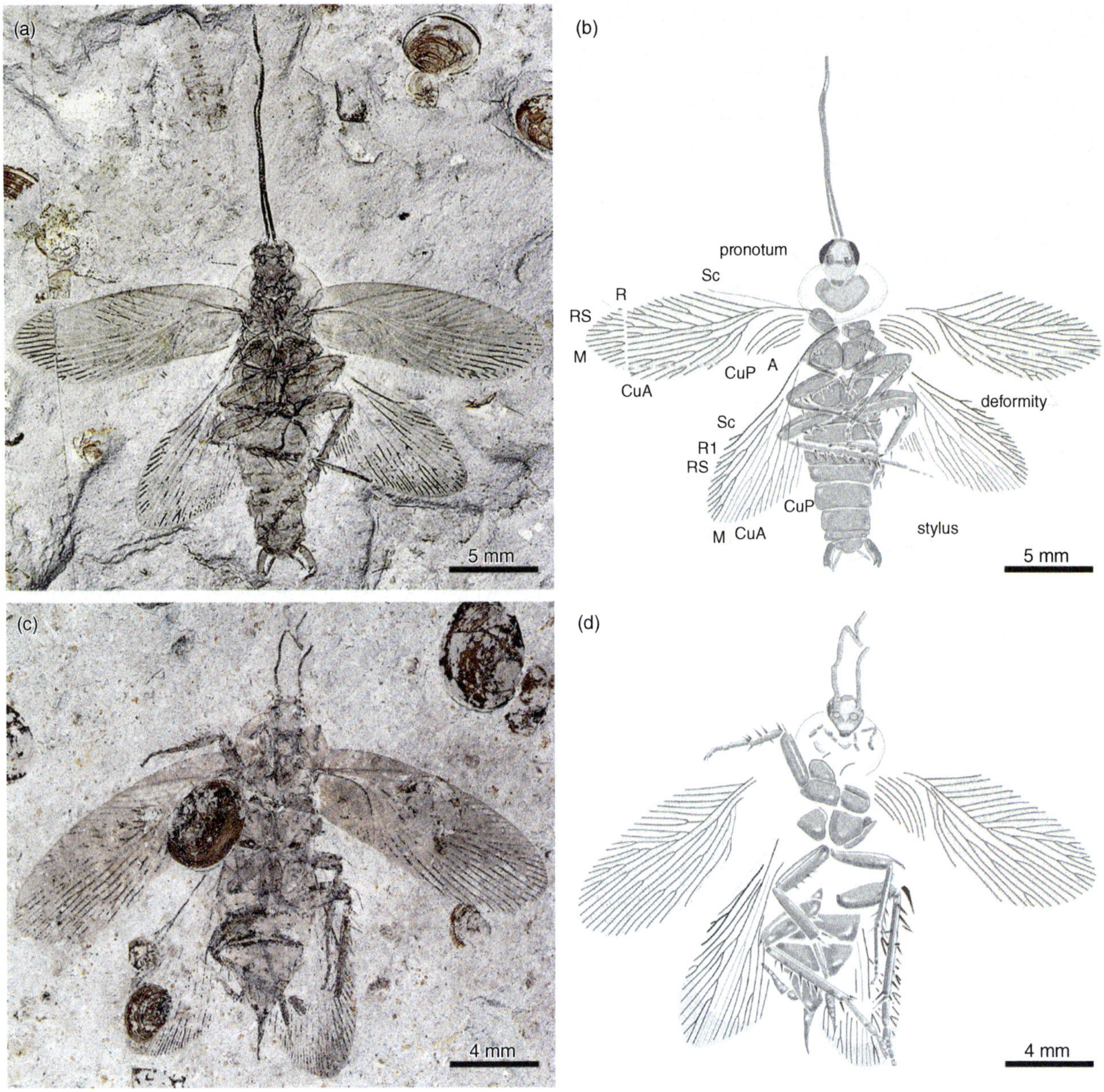

Figure 7.8 *Fuzia dadao* Vršanský, Liang & Ren, 2009 (Holotype, female CNU-B-NN-2006666; Paratype, male CNU-B-NN-2006301). (a, c), Photographs; (b, d), Line drawings [44]. (Both donated by Dr. Chungkun Shih).

Distribution and age: Inner Mongolia; Middle Jurassic.

Only one species included from the Jurassic of Northern China (see Table 7.1).

***Fuzia dadao* Vršanský, Liang & Ren, 2009 (Figure 7.8)**

Fuzia dadao Vršanský, Liang & Ren, 2009: *Geol. Carpath.*, 60 (6), 449–462.

Locality and horizon: Daohugou, Ningcheng, Inner Mongolia, China; Middle Jurassic, Jiulongshan Formation.

The pronotum in female (2.5–3.1/4.2–4.5 mm) (2.2–2.6/3.2–3.8 mm in male). RS differentiated in most individuals; R 15–25, apical R1 with multiple branches; M 3–10; CuA 4–13; Clavus very short, A 4–7. Hind wing with simple Sc; R1 3–4 and RS 9–13; M 2–6; CuA 6–11, secondarily or tertiarily branched, CuP simple.

Fuzia dadao has no coloration in both sexes and enlargement of the pronotum combined with the large forewing costal area, especially in females, which suggest a protective adaptation to a cryptic way of life. Forewings of *F. dadao* are not symmetrical, but radial veins had a very low coefficient of variation (CV) (14.44), which support comparatively good flight. Deformities (vein fusions) in the wing appear relatively common and distinct with seven vein fusions in 54 wings of 36 specimens, indicating that the species was undergoing rapid evolution. The male notches fit to the ridge formed by the base and the strong tubercle of narrowed lateral margins of the female ovipositor outer valves, suggesting that the forceps might have been used primarily for attaching to females during courtship. The morphology of the female ovipositor suggests that it served for oviposition into hard substrates, such as roots, straws or, into decaying wood [44].

Parvifuzia Guo & Ren, 2011

Parvifuzia Guo & Ren, 2011, *Acta Geol. Sin.*, 85 (2), 501–506 [53] (original designation).

Type species: *Parvifuzia marsa* Guo & Ren, 2011

Small species (forewing length/width: 6.3–6.4/2.1–2.1 mm). Apex of cerci strongly curved inward and rounded in shape, with a narrow gap at the center; forewing narrow; venation simple and intercalaries unclear; R strongly curved; anal area wide; fluently curved CuA.

Distribution and age: Inner Mongolia; Middle Jurassic.

Three species included from the Jurassic of Northern China (see Table 7.1).

Arcofuzia Wei, Shih & Ren, 2012

Arcofuzia Wei, Shih & Ren, 2012, *Zootaxa*, 3597, 25–32 [54] (original designation).

Type species: *Arcofuzia cana* Wei, Shih & Ren, 2012

Forewing extremely wide, Sc thick and without branches, with many small cross-veins; wing venation rich (43–54 veins); intercalary veins thick; cambered markings located near the apex of forewing from anterior of R to the anterior of CuA.

Distribution and age: Inner Mongolia; Middle Jurassic.

Only one species included from the Jurassic of Northern China (see Table 7.1).

Colorifuzia Wei, Liang & Ren, 2013

Colorifuzia Wei, Liang & Ren, 2013, *Geodiversitas*, 35 (2), 335–343 [42] (original designation).

Type species: *Colorifuzia agenora* Wei, Liang & Ren, 2013.

Relatively large body size (length/width: 14.9–16.1/5.5–6.7 mm), coloration markings located in the area of R and M of forewing; forewing venation rich and comparatively wide.

Distribution and age: Inner Mongolia; Middle Jurassic.

Only one species included from the Jurassic of Northern China (see Table 7.1).

Family Liberiblattinidae Vršanský, 2002

Liberiblattinidae, a small extinct family, distinguished from other families by branched Sc, slightly curved R with indicated Rs, comparatively weak M, and simplified A. Simplified M, and CuA without tertiary branching in hind wing. It retains plesiomorphies such as branched Sc, R not reaching apex of wing, branched anal veins, representing the ancestry for eusocial cockroaches, Mantodea, Skokidae, Umenocoleidae and a lineage leading to all living cockroaches [55]. The family is distributed at the Eurasian, from Upper Jurassic to Lower Cretaceous sediments and amber.

Only one genus included from the Jurassic of Northern China: *Entropia* Vršanský, 2012.

Entropia Vršanský, 2012

Entropia Vršanský, 2012, *Orient. Insects*, 46(1), 12–18 [55] (original designation).

Type species: *Entropia initialis* Vršanský, 2012.

Robust. Pronotum with well-developed paranotalia. Forewing wide, with long branched Sc; R slightly curved, without tertiary branching; M and Cu rich; CuP sharply curved. Hind wing with simple Sc, RS differentiated, M expanded and CuA tertiary branched.

Distribution and age: Inner Mongolia; Middle Jurassic.

Only one species included from the Jurassic of Northern China (see Table 7.1).

Family Mesoblattinidae Handlirsch, 1906

Mesoblattinidae, representing more derived cockroach groups, originated during the Early Jurassic from Caloblattinidae, and then evolved into three families, Blattidae, Blattellidae and Blaberidae [16, 31, 34, 43]. The Mesoblattinidae have irregular branches and no branching at the apex of M, which are the plesiomorphies. On the other hand, Mesoblattinidae express homology with Blattellidae and Blattidae, i.e. regular veins, parallel forewing margins, RS differentiated and A and CuP simple. However, Mesoblattinidae are considered as a "waste basket," because a large number of genera and species have been classified into this family. For example, many species of Caloblattinidae, Blattulidae, Raphidiomimidae and other families are described in the Mesoblattinidae [34, 43, 56]. Hitherto, 16 species within seven genera have been reported from the Jurassic and Cretaceous of Northern China.

Genera included from the Jurassic and Cretaceous of Northern China: *Mesoblattina* Genitz, 1880, *Triassoblatta* Tillyard, 1919, *Laiyangia* Grabau, 1923, *Karatavoblatta* Vishniakova, 1968, *Jingyuanoblatta* Lin, 1982, *Basiblattina* Zhang, 1997 and *Perlucipecta* Wei & Ren, 2013.

Mesoblattina Geinitz, 1880

Mesoblattina Geinitz, 1880, *Zeitschrift der deutschen geologischen Gesellschaft*, 32, 510–530 [57] (original designation).

Type species: *Mesoblattina protypa* Genitz, 1880.

Forewing narrow, anterior and posterior wing margins nearly parallel; Sc simple, not branched; humeral area narrow, as long as or slightly longer than anal area; R straight or slightly curved; MA simple, MP not branched or simple; anal area wide and round.

Distribution and age: Liaoning and Gansu of China, Early Jurassic; Liaoning of China, Middle Jurassic; Gansu of China, Early Jurassic; Korea, Spain and England, Early Cretaceous; Germany, Kyrgyzstan, Mongolia and Switzerland, Early Jurassic.

Five species included from the Jurassic and Cretaceous of Northern China (see Table 7.1).

Triassoblatta Tillyard, 1919

Triassoblatta Tillyard, 1919, *Proc. Linn. Soc. N. S. W*. 44, 358–382 [36] (original designation).

Type species: *Triassoblatta typica* Tillyard, 1919.

The tip of forewing sharp, leaf-shaped; Sc with a few branches; CuA and CuP joining at the base of wing; no intercalary veins.

Distribution and age: Hebei; Early Jurassic and Middle Jurassic.

Four species included from the Jurassic of Northern China (see Table 7.1).

Laiyangia Grabau, 1923

Laiyangia Grabau, 1923, *Bull. Geol. Surv. Chin.*, 5, 167–173 [24] (original designation).

Type species: *Laiyangia paradoxiformis* Grabau, 1923.

Veins simple, the principal veins beginning far basal. Sc meeting the wing anterior margin near the mid-length; Radial sector branching off near the base of the radial; M much branched; CuA with four veins.

Distribution and age: Shandong; Early Cretaceous.

Two species included from the Cretaceous of Northern China (see Table 7.1).

Karatavoblatta Vishniakova, 1968

Karatavoblatta Vishniakova, 1968, In: Rohdendorf (ed.). *Yurskie nasekomye Karatau* (Jurassic Insects of Karatau), Nauka, Moscow. 55–86 [41] (original designation).

Type species: *Karatavoblatta longicaudata* Vishniakova, 1968.

Head out of pronotum; body with long ovipositor; R slightly curved, not reaching the apex of forewing; M forking distal to CuA; intercalary veins developed and reticulations not developed.

Distribution and age: Liaoning of China; Early Cretaceous; Kazakhstan, Late Jurassic.

Only one species included from the Cretaceous of Northern China (see Table 7.1).

Jingyuanoblatta Lin, 1982

Jingyuanoblatta Lin, 1982, *Atlas of Palaeontology from Northwest Region*, Shaanganning division. 77–79 [58] (original designation).

Hind wing Sc long, simply; R branched at the middle of wing; M forked later than R, R and M with fewer veins; CuA with abundant branches, reticulations present in CuA-CuP space.

Distribution and age: Gansu; Early Cretaceous.

Only one species included from the Cretaceous of Northern China (see Table 7.1).

Basiblattina Zhang, 1997

Basiblattina Zhang, 1997, *Palaeoworld*, 7, 75–103 [59] (original designation).

Type species: *Basiblattina conformis* Zhang, 1997

Tegmen elongated with humeral area broadened. Sc with two branches; R strongly arcuate at the length of anal area near base and then becoming smooth toward apex, with many comb-like branches; anal area high, nearly as long as humeral area, A with eight branches.

Distribution and age: Jilin; Early Cretaceous.

Only one species included from the Cretaceous of Northern China (see Table 7.1).

Perlucipecta Wei & Ren, 2013

Perlucipecta Wei & Ren, 2013, *Geol. Carpath.*, 64 (4), 291–304 [43] (original designation).

Type species: *Perlucipecta aurea* Wei & Ren, 2013

Pronotum elliptical, transversal, transparent near edges and colored at the center. Abdomen broad, with dark-colored longitudinal stripes. Forewing long, with rich venation (43–54 veins). Sc bifurcated; R rich and stem straight; M with abundant branches; CuA sparse, CuP radian concave; A rich. Hind wing Sc simple; R and Rs rich and branched; M simple or branched; CuA with abundant dichotomized branches and blind rami.

Distribution and age: Liaoning of China, Early Cretaceous; Brazil, Early Cretaceous.

Two species included from the Cretaceous of Northern China (see Table 7.1).

Perlucipecta aurea Wei & Ren, 2013 (Figure 7.9)

Perlucipecta aurea Wei & Ren, 2013: *Geol. Carpath.*, 64 (4), 291–304.

Locality and horizon: Huangbanjigou, Beipiao, Liaoning, China; Lower Cretaceous, Yixian Formation.

Forewing with length/width: 12.0–17.0/4.0–5.6 mm, with intercaleries and numerous cross-veins; Sc with one to four veins (simplified expression "Sc 1–4" for this chapter), subcostal area shorter than clavus; R 18–24,

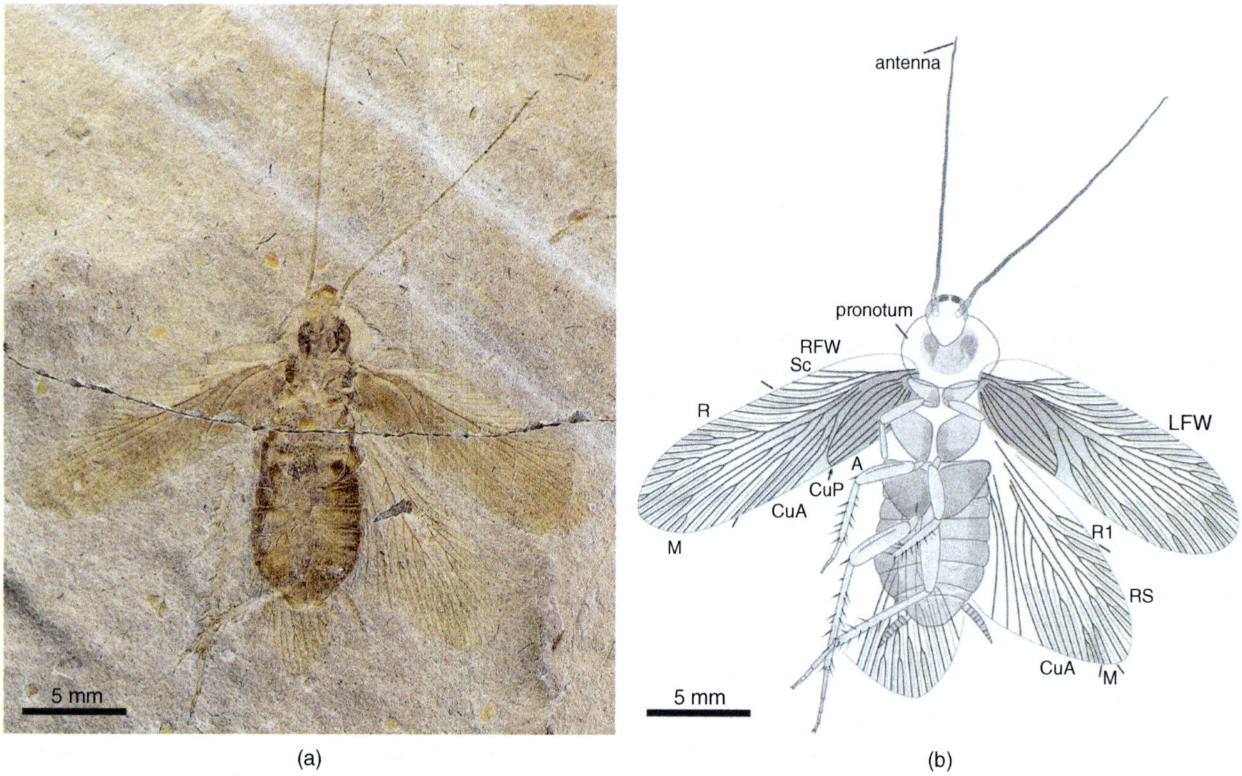

Figure 7.9 *Perlucipecta aurea* Wei & Ren, 2013 (Holotype, CNU-B-NN-2011610p). (a), Photograph; (b), Line drawing [34].

posteriorly pectinate with secondary forks; medial 7–15, media area wide; cubital 3–10, area narrow and gently curved; clavus broad; A 6–10, some reaching CuP. Hind wing with length 11.0–13.9 mm; R1 2–4 and Rs 8–14; CuA 9–11 [43].

The male tergal of *Pelucipecta aurea* is different from that of *P. vršanský* – another species of this genus, suggesting that they occupied different source habitats. Forewing of *P. aurea* had sophisticatedly colored parts, implying that the habitats of the Yixian Formation inhabited by these cockroaches were humid [45]. The coefficient of variation (CV) in the total number of veins (per wing) of *P. aurea* is relatively low (6.23%). This can be associated with the more advanced age of the group at that time (since this taxon was originated in the earliest Jurassic), and fits well with the trend of the decreasing variability of insect species over time [34].

Family Raphidiomimidae Vishniakova, 1973

Raphidiomimidae, a highly specialized family, appear to have been predatory. The family have many autapomorphies: long and prognathous head entirely extending out of the pronotum; long maxillary palps and labial palps; and elongate forewing with broad anal area and narrowed costal space. The forelegs appear to have been held forward, and probably convergent with those of mantises. Many species in this family have colorful wing patterns and are known from the Late Jurassic of Karatau and England, the Middle Jurassic of Inner Mongolia of China, and the Late Cretaceous Myanmar (Burmese) amber [14, 15, 31, 60–62]. To date, seven species within five genera have been reported from the Jurassic and Cretaceous of Northern China.

Genera included from the Jurassic and Cretaceous of Northern China: *Liadoblattina* Handlirsch, 1908, *Fortiblatta* Liang, Vršanský & Ren, 2009, *Divocina* Liang, Vršanský & Ren, 2012, *Graciliblatta* Liang, Huang & Ren, 2012 and *Falcatusiblatta* Liang, Shih & Ren, 2017.

Liadoblattina Handlirsch, 1908

Liadoblattina Handlirsch, 1908, Leipzig, 1433 [30]. Trans. from Mesoblattinidae by Vršanský & Ansorge, 2007, *Afr. Invertebr.*, 48 (1), 103–126 [31].

Type species: *Liadoblattina blakei* (Scudder, 1886).

The specific epithet is dedicated to Dr. Blake for his research on the Lias insects. The shape of wing elongated with a slightly stronger curved anterior edge; R strongly curved, and reaching the anterior fore margin of forewing, area of R small, and the main veins of R not forking; the anterior branch of M richly branched, and the posterior branch with three branches.

Distribution and age: Hebei and Liaoning of China, Early Cretaceous; Mongolia, Early Cretaceous; Germany, Russia and England, Early Jurassic.

Two species included from the Cretaceous of Northern China (see Table 7.1).

***Fortiblatta* Liang, Vršanský & Ren, 2009**

Fortiblatta Liang, Vršanský & Ren, 2009, *Zootaxa*, 1974, 17–30 [15] (original designation).

Type species: *Fortiblatta cuspicolor* Liang, Vršanský & Ren, 2009.

Large specimens (forewing length/width: 21.1–26.0/6.5–8.0 mm). Pronotum vaulted, slightly elongate with dark coloration at center and margins. Humeral area long and narrow, Sc branched; R not reaching the apex; the base of anal region colored. Hind wing with simple Sc; M usually with two branches; CuA basal most branches strong, CuP simple; reticulations present in CuA-CuP space, joined with intercalaries. Ovipositor outer valves internalized.

Distribution and age: Inner Mongolia; Middle Jurassic.

Only one species included from the Jurassic of Northern China (see Table 7.1).

***Divocina* Liang, Vršanský & Ren, 2012**

Divocina Liang, Vršanský & Ren, 2012, *Revista Mexicana de Ciencias Geológicas*, 29 (2), 411–421 [62] (original designation).

Type species: *Divocina noci* Liang, Vršanský & Ren, 2012.

Small species (forewing, length/width: 12.3–16/3.2–4.8 mm; hind wing, 10.3–14 mm long). Wings dark, with a narrow pale line at anterior margin. Sc branched; R slightly curved, not reaching apex; M and CuA rich. Hind wing with simple Sc; RS differentiated; M rich; CuA with quarterly branching veins; Stem and posteriormost CuA branch meeting CuP with reticulations, CuP and A1 simple, A2 branched. Ovipositor apparently external.

Distribution and age: Inner Mongolia; Middle Jurassic.

Only one species included from the Jurassic of Northern China (see Table 7.1).

***Divocina noci* Liang, Vršanský & Ren, 2012 (Figure 7.10)**

Divocina noci Liang, Vršanský & Ren, 2012: *Rev. Mex. Cienc. Geol.*, 29 (2), 411–421.

Locality and horizon: Daohugou, Ningcheng, Inner Mongolia, China; Middle Jurassic, Jiulongshan Formation.

Pronotum longitudinal length/width: 2.5–3.5/2.5–3.0 mm. Sc 2–6; R 8–17; M 5–16, CuA 4–11 and A 5–10. Hind wing length: 10.3–14 mm. R1 3–7, RS 5–9; M 2–8; CuA 5–9.

Divocina noci has been suggested as a crepuscular and/or nocturnal predatory cockroach based on the following characters: head with a typical supporting ridge on ventral side, strong mandibles, long mechanoreceptors approximating eyes and dark coloration with pale anterior margin of both wings, which resemble some crepuscular and nocturnal ice-crawlers.

The overall coefficients of variation (CV) for the number of veins of *D. noci* in the distal forewing margin and total number of veins at the margin (CV total = 7.65; hind wing CV total = 7.54) are low when compared with the wing center. The low variations, together with wing symmetry, suggest they had comparatively good flight capability. Among 44 complete fossil wings studied, only two wing deformities, i.e. a fusion joining R and M in a forewing and fusion of two M veins, are present, suggesting a stabilized morphotype. Internalized inner valvae of *D. noci* suggests they might have laid eggs in conglomerates or even within ootheca [62].

***Graciliblatta* Liang, Huang & Ren, 2012**

Graciliblatta Liang, Huang & Ren, 2012, *Zootaxa*, 3449, 62–68 [61] (original designation).

Type species: *Graciliblatta bella* Liang, Huang & Ren, 2012.

(a) (b)

Figure 7.10 *Divocina noci* Liang, Vršanský & Ren, 2012 (Holotype, CNU-B-NN-2006067). (a), Photograph; (b), Line drawing [62].

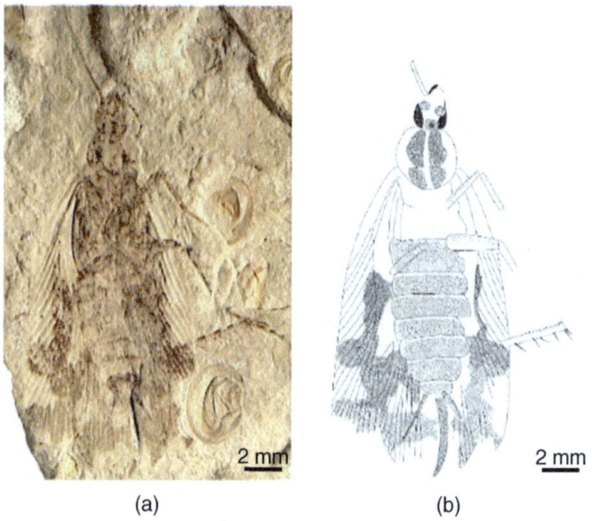

Figure 7.11 *Falcatusiblatta gracilis* Liang, Shih & Ren, 2017 (Holotype, TNP 42411p). (a), Photograph; (b), Line drawing [56]. Source: Photo provided by Junhui Liang.

Mid-sized forewing with length/width: 17.5–19.0/5.2–5.5 mm. Head and pronotum elongated. Sc richly branched, with seven to eight branches; R not reaching apical margin, slightly curved and basally with dark coloration. CuA and CuP slightly curved. Anal veins with tertiary branches. Hind wing with simple Sc; R1 and Rs differentiated, and R1 secondarily branched.

Distribution and age: Inner Mongolia; Middle Jurassic.

Only one species included from the Jurassic of Northern China (see Table 7.1).

Falcatusiblatta Liang, Shih & Ren, 2017

Falcatusiblatta Liang, Shih & Ren, 2017, *Alcheringa* 42 (1), 101–109 [56] (original designation).

Type species: *Falcatusiblatta gracilis* Liang, Shih & Ren, 2017.

Small-sized with forewing length/width: 11.1–13.4/3.1–4.9 mm. Two dark stripes at the center of pronotum expanding medially and parallel. Forewing with an irregular pattern of light and dark patches; ovipositor very long, knife-shaped; cerci heteronomous.

Distribution and age: Inner Mongolia; Middle Jurassic.

Two species included from the Jurassic of Northern China (see Table 7.1).

Falcatusiblatta gracilis Liang, Shih & Ren, 2017 (Figure 7.11)

Falcatusiblatta gracilis Liang, Shih & Ren, 2017: *Alcheringa*, 42 (1), 101–109.

Locality and horizon: Daohugou, Ningcheng, Inner Mongolia, China; Middle Jurassic, Jiulongshan Formation.

Pronotum slightly elongate (length/width: 2.4–3.4/2.4–3.3 mm). Sc 3–4; R 13–17, without reaching the tip of forewing; M 3–11; CuA 8–12; A 8–13. Long cerci with 9–10 heterogeneous articles.

The forewing markings provided disruptive camouflage to avoid predation by abundant predators in the Middle to Late Jurassic ecosystem, such as feathered dinosaurs, pterosaurs, reptiles, mammals or other predatory insects. Heteronomous cerci (i.e. those having basal and terminal segments of different morphology) are very rare in cockroaches and unknown in fossils except for members of Fuziidae. The reasons for heterogeneity of the segments are ambiguous. *Falcatusiblatta* possess a long external ovipositor, a primitive feature inherited from its Caloblattinidae ancestors, which is longer than in other Raphidiomimidae species from the same locality. Generally, cockroaches with such long ovipositors have been reported from the Mississippian/Pennsylvanian to Permian, but they also persisted to the Late Jurassic, even to the Cretaceous. Long ovipositor suggested they inserted eggs into soil and crevices within rotting wood and humus. Venational variability and instability have been observed in four wings of three specimens. This symmetry present in both forewings indicates a stable heritable character, correlated with the ecological stress [56].

Family Umenocoleidae Chen & Tan, 1973

Umenocoleidae, documented entirely from the Cretaceous, were once considered as the most primitive beetles in Coleoptera [63], and then assigned to Blattaria [64]. This family is known based on fossils from the Early Cretaceous of Siberia, China, Brazil, and amber from Lebanon and New Jersey. Umenocoleidae are characterized by the heavily elytrized forewing, highly reduced venation, narrow pronotum exposing most of the head, the head quite broad and the eyes distantly separated and short external ovipositor. To date, only one species has been reported from the Jurassic and Cretaceous of Northern China.

Only one genus included from the Cretaceous of Northern China: *Umenocoleus* Chen & Tan, 1973.

Umenocoleus Chen & Tan, 1973

Umenocoleus Chen & Tan, 1973, *Acta Ento. Sin.*, 16 (2), 169–178 [63] (original designation). Trans. from Coleoptera by Vršanský, 2003, *AMBA Projekty*, 7(1), 3–30 [64].

Type species: *Umenocoleus sinuatus* Chen & Tan, 1973.

Forewing not colored, radius poorly branched, fused with simple M; Cu strongly curved, S-shaped; anal veins multiple. All veins distinct. Pronotum roughly punctuate.

Distribution and age: Jilin, Gansu; Early Cretaceous.

Two species included from the Cretaceous of Northern China (see Table 7.1).

Table 7.1 A List of Fossil Blattaria from the Jurassic and Cretaceous of China.

Family	Species	Locality	Horizon/Age	Citation
Blattidae	[a] *Zhujiblatta anofissilis* Lin, 1980	Zhuji, Zhejiang	Shouchang Fm., K_1	Lin [65]
	Sinoblatta laiyangensis Grabau, 1923	Laiyang, Shandong	Laiyang Fm., K_1	Grabau [24]
	[a] *Strictiblatta longanusis* Lin, 1980	Hefei, Anhui	Yantang Fm., K_2	Lin [65]
Blattulidae	*Blattula apicifurca* (Lin, 1986)	Liuyang, Hunan	Guanyintan Fm., J_1	Lin [66]
	Blattula chengdeensis (Hong, 1980)	Chengde, Hebei	Jiulongshan Fm., J_2	Hong [39]
	Blattula ctinoida Lin, 1986	Qiyang, Hunan	Guanyintan Fm., J_1	Lin [66]
	Blattula curvula (Ren, 1995)	Luanping, Hebei	Jiulongshan Fm., J_2	Ren *et al.* [67]
	Blattula delicatula Ren, 1995	Chengde, Hebei	Yixian Fm., K_1	Ren *et al.* [67]
	Blattula hymena Lin, 1986	Zhongshan, Guangxi	Shiti Fm., J_1	Lin [66]
	Blattula kellos Zhang, 1986	Chengde, Hebei	Jiulongshan Fm., J_2	Zhang [68]
	Blattula kiensis (Martynov, 1937)	Hanshan, Anhui	Hanshan Fm., J_2	Lin [69]
	Blattula liaoningensis Hong, 1986	Huludao, Liaoning	Haifanggou Fm., J_2	Hong [70]
	Blattula mirta (Zhang, 1986)	Chengde, Hebei	Jiulongshan Fm., J_2	Zhang [68]
	Blattula pachohymena Lin, 1985	Hanshan, Anhui	Hanshan Fm., J_2	Lin [69]
	Blattula platypa Ren, 1995	Chengde, Hebei	Yixian Fm., K_1	Ren *et al.* [67]
	Blattula rudis Ren, 1995	Xishan, Beijing	Lushangfen Fm., K_1	Ren *et al.* [67]
	Blattula sincera (Lin, 1985)	Hanshan, Anhui	Hanshan Fm., J_2	Lin [69]
	Blattula tuodianensis (Zhang & Hong, 2003)	Shuangbai, Yunnan	Tuodian Fm., J_3	Zhang *et al.* [71]
	Blattula zaoshangensis Lin, 1986	Liuyang, Hunan	Zaoshang Fm., J_1	Lin [66]
	Elisama cuboides Wang, Liang & Ren, 2007	Beipiao, Liaoning	Yixian Fm., K_1	Wang *et al.* [33]
	Elisama extenuata (Ren, 1995)	Chengde, Hebei	Yixian Fm., K_1	Wang *et al.* [33]
	Elisama dignata (Wang, 1987)	Beipiao, Liaoning	Haifanggou Fm., J_2	Wang [72]
	Habroblattula drepaniodes Wang, Liang & Ren, 2007	Beipiao, Liaoning	Yixian Fm., K_1	Wang *et al.* [32]
	Macaroblattula ellipsoids Wang, Liang & Ren, 2007	Beipiao, Liaoning	Yixian Fm., K_1	Wang *et al.* [33]
Caloblattinidae	*Euryblattula beipiaoensis* Wang, 1987	Beipiao, Liaoning	Beipiao Fm., J_1	Wang [72]
	Euryblattula chaoyangensis Wang, 1987	Chaoyang, Liaoning	Beipiao Fm., J_1	Wang [72]
	Euryblattula flabelliformis Wang, 1980	Benxi, Liaoning	Changliangzi Fm., J_1	Wang [73]
	Euryblattula huapenensis Hong, 1997	Yanqing, Beijing	[b] Tuchengzi Fm., J_3-K_1	Hong and Xiao [74]
	Euryblattula lepta Lin, 1986	Lanshan, Hunan	Guanyintan Fm., J_1	Lin [66]
	Euryblattula lingulata Hong, 1980	Chengde, Hebei	Jiulongshan Fm., J_2	Hong [39]
	Euryblattula monchis Lin, 1986	Qiyang, Hunan	Guanyintan Fm., J_1	Lin [66]

Table 7.1 (Continued)

Family	Species	Locality	Horizon/Age	Citation
	Euryblattula opima Lin, 1986	Lanshan, Hunan	Guanyintan Fm., J_1	Lin [66]
	Euryblattula obliqua Lin, 1986	Lanshan, Hunan	Guanyintan Fm., J_1	Lin [66]
	Euryblattula pura Lin, 1986	Lanshan, Hunan	Guanyintan Fm., J_1	Lin [66]
	Fusiblatta arcuata Hong, 1980	Luanping, Hebei	Jiulongshan Fm., J_2	Hong [39]
	Fusiblatta dongchangtaiensis Hong, 1980	Xinbin, Liaoning	Houjiatun Fm., J_2	Hong [39]
	Nipponoblatta acerba Ren, 1995	Beipiao, Liaoning	Yixian Fm.,K_1	Ren *et al.* [67]
	Nipponoblatta deformis Lin, 1986	Lanshan, Hunan	Guanyintan Fm., J_1	Lin [66]
	Nuurcala obesa Wang & Ren, 2013	Beipiao, Liaoning	Yixian Fm., K_1	Wang *et al.* [40]
	Rhipidoblattina chichengensis Hong, 1997	Chicheng, Hebei	[b]Tuchengzi Fm., J_3-K_1	Hong [75]
	Rhipidoblattina decoris Lin, 1978	Yixian, Liaoning	Dalazi Fm., K_1	Lin [76]
	Rhipidoblattina emacerata Zhang, 1986	Luanping, Hebei	Huayuan Fm., J_1	Zhang [68]
	Rhipidoblattina fuxinensis Lin, 1976	Fuxin, Liaoning	Fuxin Fm., K_1	Lin [77]
	Rhipidoblattina forticrusa Lin, 1986	Lanshan, Hunan	Guanyintan Fm., J_1	Lin [66]
	Rhipidoblattina jilinensis Hong, 1992	Jiutai, Jilin	Yingcheng Fm., K_1	Hong [78]
	Rhipidoblattina lanceolata Hong, 1980	Chengde, Hebei	Jiulongshan Fm., J_2	Hong [39]
	Rhipidoblattina mayingziensis Wang, 1987	Chaoyang, Liaoning	Beipiao Fm., J_1	Wang [72]
	Rhipidoblattina magna Zhang, 1997	Longjing, Jilin	Dalazi Fm., K_1	Zhang [59]
	Rhipidoblattina nanligezhuangensis Hong & Wang, 1990	Laiyang, Shandong	Laiyang Fm., K_1	Hong *et al.* [79]
	Rhipidoblattina radipinguis Lin, 1986	Jiangyong, Hunan	Shiti Fm., J_1	Lin [76]
	Rhipidoblattina spathulata Hong, 1982	Yumen, Gansu	Chijinpu Fm., K_1	Hong [80]
	Rhipidoblattina shulanensis Hong, 1992	Jiutai, Jilin	Yingcheng Fm., K_1	Hong [78]
	Rhipidoblattina jidongensis Chang & Wang, 1993	Funing, Hebei	Beipiao Fm., J_1	Chang and Wang [81]
	Rhipidoblattina (*Rhipidoblattina*) *beipiaoensis* Hong, 1983	Beipiao, Liaoning	Haifanggou Fm., J_2	Hong [82]
	Rhipidoblattina (*Rhipidoblattina*) *liugouensis* Hong, 1983	Xiaofanzhangzi, Hebei	Jiulongshan Fm., J_2	Hong [82]
	Rhipidoblattina (*Rhipidoblattina*) *liaoningensis* Hong, 1980	Chaoyang, Liaoning	Yixian Fm., K_1	Hong [39]
	Rhipidoblattina (*Canaliblatta*) *tenuis* Hong, 1983	Xiaofanzhangzi, Hebei	Jiulongshan Fm., J_2	Hong [82]
	Rhipidoblattina (*Canaliblatta*) *yanqingensis* Hong, 1997	Yanqing, Beijing	[b]Tuchengzi Fm., J_3-K_1	Hong and Xiao [74]

(Continued)

Table 7.1 (Continued)

Family	Species	Locality	Horizon/Age	Citation
	Rhipidoblattina (*Canaliblatta*) *hebeiensis* Hong, 1983	Zhouyingzi, Hebei	Jiulongshan Fm., J_2	Hong [82]
	Samaroblatta frondoidis Lin, 1978	Dongsheng, Inner Mongolia	Yanan Fm., J_2	Lin [76]
	Samaroblatta gausis Lin, 1982	Jingyuan, Gansu	Wangjiashan Fm., K_1	Lin [58]
	Samaroblatta nitida Lin, 1986	Lanshan, Hunan	Guanyintan Fm., J_1	Lin [66]
	Samaroblatta rhypha Lin, 1986	Lanshan, Hunan	Guanyintan Fm., J_1	Lin [66]
	Samaroblatta turanica Martynov, 1937	Hanshan, Anhui	Hanshan Fm., J_2	Lin [66]
	Samaroblatta wangyingziensis Hong, 1980	Luanping, Hebei	Jiulongshan Fm., J_2	Hong [39]
	Samaroblatta zhouyingziensis Hong, 1980	Luanping, Hebei	Jiulongshan Fm., J_2	Hong [39]
	Samaroblattula houchengensis Hong & Xiao, 1997	Yanqing, Beijing	[b]Tuchengzi Fm., J_3-K_1	Hong and Xiao [74]
	Samaroblattula lata Hong & Xiao, 1997	Yanqing, Beijing	[b]Tuchengzi Fm., J_3 -K_1	Hong and Xiao [74]
	Samaroblattula lineata Hong & Xiao, 1997	Yanqing, Beijing	[b]Tuchengzi Fm., J_3-K_1	Hong and Xiao [83]
	Samaroblattula lingulata Hong & Xiao, 1997	Yanqing, Beijing	[b]Tuchengzi Fm., J_3-K_1	Hong and Xiao [83]
	Samaroblattula reticulate Hong, 1982	Dachaidan, Qinghai	Dameigou Fm., J_2	Hong [80]
	Samaroblattula subacuta Martynov, 1937	Jiangyong, Hunan	Shiti Fm., J_1	Lin [66]
	Samaroblattula scabra Lin, 1986	Jiangyong, Hunan	Shiti Fm., J_1	Lin [66]
	Sogdoblatta compressa Martynov, 1937	Liuyang, Hunan	Zaoshang Fm., J_1	Lin [66]
	Sogdoblatta haifanggouensis Hong, 1983	Beipiao, Liaoning	Haifanggou Fm., J_2	Hong [82]
	Sogdoblatta heiheensis Hong & Xiao, 1997	Yanqing, Beijing	[b]Tuchengzi Fm., J_3-K_1	Hong and Xiao [83]
	Sogdoblatta luanpingensis Hong, 1980	Luanping, Hebei	Jiulongshan Fm., J_2	Hong [39]
	Taublatta hesta Lin, 1986	Qiyang, Hunan	Guanyintan Fm., J_1	Lin [66]
	Taublatta niujiaoshiensis Lin, 1986	Jiangyong, Hunan	Shiti Fm., J_1	Lin [66]
	Taublatta ninghuaensis Lin, 1978	Ninghua, Fujian	Lishan Fm., J_1	Lin [76]
	Taublatta semifoliosa Lin, 1986	Jiangyong, Hunan	Shiti Fm., J_1	Lin [66]
	Taublatta siccitifoliosa Lin, 1986	Jiangyong, Hunan	Shiti Fm., J_1	Lin [66]
	Taublatta strenis Lin, 1986	Qiyang, Hunan	Guanyintan Fm., J_1	Lin [66]
	Taublatta yangshugouensis Wang, 1987	Kazuo, Liaoning	Beipiao Fm., J_1	Wang [72]
Ectobiidae	*Piniblattella yixianensis* Gao, Shih & Ren, 2018	Beipiao, Liaoning	Yixian Fm., K_1	Gao *et al.* [48]

Table 7.1 (Continued)

Family	Species	Locality	Horizon/Age	Citation
Fuziidae	*Arcofuzia cana* Wei, Shih & Ren, 2012	Ningcheng, Inner Mongolia	Jiulongshan Fm., J_2	Wei *et al.* [54]
	Colorifuzia agenora Wei, Liang & Ren, 2013	Ningcheng, Inner Mongolia	Jiulongshan Fm., J_2	Wei *et al.* [42]
	Fuzia dadao Vršanský, Liang & Ren, 2010	Ningcheng, Inner Mongolia	Jiulongshan Fm., J_2	Vršanský *et al.* [44]
	Parvifuzia brava Guo & Ren, 2011	Ningcheng, Inner Mongolia	Jiulongshan Fm., J_2	Guo *et al.* [53]
	Parvifuzia marsa Guo & Ren, 2011	Ningcheng, Inner Mongolia	Jiulongshan Fm., J_2	Guo *et al.* [53]
	Parvifuzia peregrina Wei, Liang & Ren, 2012	Ningcheng, Inner Mongolia	Jiulongshan Fm., J_2	Wei *et al.* [84]
Liberiblattinidae	*Entropia initialis* Vršanský, Liang & Ren, 2012	Ningcheng, Inner Mongolia	Jiulongshan Fm., J_2	Vršanský *et al.* [55]
Mesoblattinidae	*Basiblattina conformis* Zhang, 1997	Longjing, Jilin	Dalazi Fm., K_1	Zhang [59]
	Jingyuanoblatta pluma Lin, 1982	Jingyuan, Gansu	Wangjiashan Fm., K_1	Lin [58]
	Karatavoblatta formosa Ren, 1995	Lingyuan, Liaoning	Yixian Fm., K_1	Ren *et al.* [67]
	Laiyangia delicatula Zhang, 1985	Laiyang, Shandong	Laiyang Fm., K_1	Zhang [85]
	Laiyangia paradoxiformis Grabau, 1923	Laiyang, Shandong	Laiyang Fm., K_1	Grabau [24]
	Mesoblattina cretacea Hong, 1982	Yumen, Gansu	Zhonggou Fm., K_1	Hong [80]
	Mesoblattina multivenosa Martynov, 1937	Zigui, Hunan	Guanyintan Fm., J_1	Lin [66]
	Mesoblattina protypa Geinitz, 1880	Qiyang, Hunan	Guanyintan Fm., J_1	Lin [66]
	Mesoblattina paucivenose Hong, 1982	Yumen, Gansu	Dashankou Fm., J_1	Hong [80]
	Mesoblattina simplicis Hong, 1980	Hengren, Liaoning	Changliangzi Fm., J_1	Hong [39]
	Mesoblattina wuweiensis Lin, 1978	Wuwei, Gansu	Yanan Fm., K_1	Lin [76]
	Mesoblattina wanbeiensis Lin, 1985	Hanshan, Anhui	Hanshan Fm., J_2	Lin [69]
	Mesoblattina xiangnanensis Lin, 1986	Lanshan, Hunan	Guanyintan Fm., J_1	Lin [66]
	Mesoblattina sinica Ping, 1928	Beipiao, Liaoning	Haifanggou Fm., J_2	Ping [86]
	Perlucipecta aurea Wei & Ren, 2013	Beipiao, Liaoning	Yixian Fm., K_1	Wei *et al.* [43]
	Perlucipecta vrsanskyi Wei & Ren, 2013	Beipiao, Liaoning	Yixian Fm., K_1	Wei *et al.* [43]
	[a)] *Soliblatta lampra* Lin, 1986	Lanshan, Hunan	Guanyintan Fm., J_1	Lin [66]
	[a)] *Summatiblatta colorata* Lin, 1986	Lanshan, Hunan	Guanyintan Fm., J_1	Lin [66]

(Continued)

Table 7.1 (Continued)

Family	Species	Locality	Horizon/Age	Citation
	Triassoblatta damiaoliangensis Hong, 1983	Chengde, Hebei	Jiulongshan Fm., J_2	Hong [82]
	Triassoblatta fusiformis Hong, 1983	Xiaofanzhangzi, Hebei	Jiulongshan Fm., J_2	Hong [82]
	Triassoblatta longitriangulata Chang & Wang, 1993	Funing, Hebei	Beipiao Fm., J_1	Chang *et al.* [81]
	Triassoblatta shimenzhaiensis Chang & Wang, 1993	Funing, Hebei	Beipiao Fm., J_1	Chang *et al.* [81]
Raphidiomimidae	*Divocina noci* Liang, Vršanský & Ren, 2012	Ningcheng, Inner Mongolia	Jiulongshan Fm., J_2	Liang *et al.* [62]
	Fortiblatta cuspicolor Liang, Vršanský & Ren, 2009	Ningcheng, Inner Mongolia	Jiulongshan Fm., J_2	Liang *et al.* [15]
	Falcatusiblatta gracilis Liang, Shih & Ren, 2017	Ningcheng, Inner Mongolia	Jiulongshan Fm., J_2	Liang *et al.* [56]
	Falcatusiblatta qiandaohuiensis Liang, Shih & Ren, 2017	Ningcheng, Inner Mongolia	Jiulongshan Fm., J_2	Liang *et al.* [56]
	Graciliblatta bella Liang, Huang & Ren, 2012	Ningcheng, Inner Mongolia	Jiulongshan Fm., J_2	Liang *et al.* [61]
	Liadoblattina heishanyaoensis Chang & Wang, 1993	Funing, Hebei	Beipiao Fm., J_1	Chang *et al.* [81]
	Liadoblattina laternoforma Lin, 1978	Yixian, Liaoning	Yixian Fm., K_1	Lin [76]
Umenocoleidae	*Umenocoleus nervosus* Zhang, 1997	Yanbian, Jilin	Dalazi Fm., K_1	Zhang [59]
	Umenocoleus sinuatus Chen & Tan, 1973	Yumen, Gansu	Chijinpu Fm., K_1	Chen and Tan [63]
Family *Incertae sedis*	[a] *Prolaxta haianensis* Lin, 1989	Jiangsu	Taizhou Fm., K_2	Lin [87]

a) The species is not presented in the main text because the original description, photo, and line-drawings are not precise and the holotype cannot be rechecked.

b) Horizon/Age revised from the original paper based on updated information and data.

References

1 Vršanský, P., Vishnikova, V.N., and Rasnitsyn, A.P. (2002). Order Blattida Latreille, 1810. The cockroaches. In: *History of insect*, 263–270. American: Kluwer Academic Publishers.

2 Liu, X.W. (1999). Class Insecta: order Blattodea. In: *Insect classification*, 169–181. Nanjing Normal University Press.

3 McKittrick, F.A. (1964). *Evolutionary Studies of Cockroaches*, 197. New York: Cornell University Agricultural Experiment Station.

4 Roth, L.M. (1991). Blattodea. Blattaria (cockroaches). In: *The insects of Australia. A textguide for students and researchers*, vol. **1**, 320–329.

5 Hu, Y.F., Lv, X.M., Wang, Y.M., and Peng, F. (2008). Research advance in medicinal value of *Periplaneta Americana*. *Medical Recapitulate* **14** (18): 108–110. (In Chinese with English summary).

6 Jiang, Y.X., Wang, X.C., Jin, C.G. *et al.* (2006). Inhibitory effect of *Periplaneta Americana* extract on 3LL lung cancer in mice. *Chinese Journal Lung Cancer* **9** (6): 488–491. (In Chinese with English summary).

7 Carpenter, F.M. (1953). The geological history and evolution of insects. *American Scientist* **41**: 256–270.

8 Carpenter, F.M. (1992). Order Blattaria Latreille, 1810. In: *Treatise on Invertebrate Palaeontology. Part R. Arthropoda 4 (3). Superclass Hexapoda*, 134–137. The Geological Society of America, Inc. and the

University of Kansas, Boulder, Colorado and Lawrence.

9 Nagamitsu, T. and Inoue, T. (1997). Cockroach pollination and breeding system of *Uvaria elmeri* (Annonaceae) in a lowland mixed–dipterocarp forest in Sarawak. *American Journal of Botany* **84**: 208–213.

10 Shelford, R. (1909). Note on some amphibious cockroaches. *Records of the Australian Museum* **3**: 125–127.

11 Takahashi, R. (1926). Observations on the aquatic cockroach, *Opistoplatia maculata*. *Dobutsugaku Zassi* **38**: 89.

12 Vršanský, P. (2007). Jumping cockroaches (Blattaria, Skokidae fam. N.) from the Late Jurassic of Karatau in Kazakhstan. *Biologia, Bratislava* **62** (5): 588–592.

13 Picker, M., Colville, J.F., and Burrows, M. (2012). A cockroach that jumps. *Biology Letters* **8** (3): 390–392. https://doi.org/10.1098/rsbl.2011.1022.

14 Vishniakova, V.N. (1973). *Problems of the Insect Palaeontology. Lectures on the XXIV Annual Readings in Memory of N.A. Kholodkovsky (1–2 April, 1971)* (ed. E.P. Narchuk), 64–77. Leningrad: Nauka.

15 Liang, J.H., Vršanský, P., Ren, D., and Shih, C.K. (2009). A new Jurassic carnivorous cockroach (Insecta, Blattaria, Raphidiomimidae) from the Inner Mongolia in China. *Zootaxa* **1974**: 17–30.

16 Vršanský, P. (2003). Unique assemblage of Dictyoptera (Insecta – Blattaria, Mantodea, Isoptera) from the Lower Cretaceous of bon Tsagaan Nuur in Mongolia. *Entomological Problems* **33** (1–2): 119–151.

17 Vidlička, L., Vršanský, P., and Shcherbakov, D.E. (2003). Two new troglobitic cockroach species of the genus *Speleoblatta* (Blattaria: Nocticollidae) from North Thailand. *Journal of Natural History* **37**: 107–114.

18 Zompro, O. and Fritzsche, I. (1999). *Lucihormetica* n. gen. n. sp., the first record of luminescence in an orthopteroid insect (Dictyoptera: Blaberidae: Blaberinae: Brachycolini). *Amazoniana* **15**: 211–219.

19 Vršanský, P., Chorvát, D., Fritzsche, I. *et al.* (2012). Light-mimicking cockroaches indicate tertiary origin of recent terrestrial luminescence. *Naturwissenschaften* http://dx.doi.org/10.1007/s00114-012-0956-7.

20 Lu, L., Fang, X., Ji, S., and Pang, Q. (2002). A contribution to the knowledge of the Namurian in Ningxia. *Acta Geoscientia Sinica* **23**: 165–168.

21 Zhang, Z.J., Schneider, J.W., and Hong, Y.C. (2012). The most ancient roach (Blattida): a new genus and species from the earliest late Carboniferous (Namurian) of China, with discussion on the phylomorphogeny of early blattids. *Journal of Systematic Palaeontology* **11**: 27–40. https://doi.org/10.1080/14772019.2011.634443.

22 Guo, Y.X., Béthoux, O., Gu, J.J., and Ren, D. (2013). Wing venation homologies in Pennsylvanian 'cockroachoids' (Insecta) clarified thanks to a remarkable specimen from the Pennsylvanian of Ningxia (China). *Journal of Systematic Palaeontology* **11**: 41–46. https://doi.org/10.1080/14772019.2011.637519.

23 Wei, D.D., Béthoux, O., Guo, Y.X. *et al.* (2013). New data on the singularly rare 'cockroachoids' from Xiaheyan (Pennsylvanian; Ningxia, China). *Alcheringa* **17**: 1–11. https://doi.org/10.1080/03115518.2013.808863.

24 Grabau, A.W. (1923). Cretaceous fossils from Shandong. *Bulletin of the Geological Survey of China* (5): 167–173.

25 Vishniakova, V.N. (1982). Jurassic cockroaches of the new family Blattulidae from Siberia. *Palaeontologica Journal* **2**: 67–77.

26 Vršanský, P. (2005). A fossil insect in a drilling core sample-cockroach *Kridla stastia* gen. et sp. nov. (Blattulidae) from the Cretaceous of the Verkhne-Bureinskaya depression in eastern Russia. *Entomological Problems* **35** (2): 115–116.

27 Cifuentes-Ruiz, P., Vršanský, P., Vega, F.J. *et al.* (2006). Campanian terrestrial arthropods from the Cerro del Pueblo Formation, Difunta group in Northeastern Mexico. *Geologica Carpathica* **57** (5): 347–354.

28 Giebel, C.G.A. (1856). Fauna der Vorwelt mit steter Berücksichtigung der lebenden Thiere. In: *Monographisch dargestellt. Zweiter Band. Glieder–thiere Erste Abtheilung. Insecten und Spinnen*. 411 pp. Leipzig: F.A. Brockhaus.

29 Scudder, S. (1886). A review of Mesozoic cockroaches. *Memoirs of the Boston Society of Natural History* **3** (13): 439–485.

30 Handlirsch, A. (1906–1908). *Die fossilen Insekten und die Phylogenie der rezenten Formen: ein Handbuch für Paläontologen und Zoologen*, 1–1430. Leipzig, Germany: Wilhelm Engelmann.

31 Vršanský, P. and Ansorge, J. (2007). Lower Jurassic cockroaches (Insecta: Blattaria) from Germany and England. *African Invertebrates* **48** (1): 103–126.

32 Wang, T.T., Liang, J.H., and Ren, D. (2007). Variability of *Habroblattula drepanoides* gen. Et. Sp. nov. (Insecta: Blattaria: Blattulidae) from the Yixian Formation in Liaoning, China. *Zootaxa* **1443**: 17–27.

33 Wang, T.T., Ren, D., Liang, J.H., and Shih, C.K. (2007). New Mesozoic cockroaches (Blattaria: Blattulidae) from Jehol Biota of western Liaoning in China. *Annal Zoologici (Warszawa)* **57** (3): 483–495.

34 Vršanský, P. (2000). Decreasing variability-from the Carboniferous to the present! (validated on independent lineages of Blattaria). *Paleontological Journal* **34** (3): 374–379.

35 Vršanský, P. (2008). Late Jurassic cockroaches (Insecta, Blattaria) from the Houtiyn-Hotgor locality in Mongolia. *Paleontological Journal* **42** (1): 36–42.

36 Tillyard, R.J. (1919). Mesozoic insects of Queensland. No. 6. Blattoidea. *Proceedings of the Linnean Society of New South Wales* **44**: 358–382.

37 Martynov, A.V. (1937). Liassic insects of Shurab and Kyzyl-Kiya. *Transactions of the Paleontological Instutute of the USSR Academy of Sciences* [Trudy Paleontologicheskogo instituta Akademii nauk SSSR] 7 (1): 1–231. (in Russian).

38 Fujiyama, I. (1974). A Liassic cockroach from Toyora, Japan. *Bulletin of the National Science Museum* **17** (4): 311–314.

39 Hong, Y.C. (1980). New genus and species of Mesoblattinidae in China. *Bulletin Chinese Academic Geological Science* **1** (4): 49–60. (In Chinese with English summary).

40 Wang, C.D. and Ren, D. (2013). *Nuurcala obesa* sp. n. (Blattida, Caloblattinidae) from the Lower Cretaceous Yixian Formation in Liaoning Province, China. *ZooKeys* **318**: 35–46.

41 Vishniakova, V.N. (1968). New cockroaches (Insecta: Blattodea) from the Upper Jurassic of Karatau mountains. In: *Jurassic insects from Karatau* (ed. B.B. Rohdendorf), 55–86. Moscow: Nauka.

42 Wei, D.D., Liang, J.H., and Ren, D. (2013). A new fossil genus of the Fuziidae (Insecta, Blattida) from the Middle Jurassic of Jiulongshan Formation, China. *Geodiversitas* **35** (2): 335–343.

43 Wei, D.D. and Ren, D. (2013). Completely preserved cockroaches of the family Mesoblattinidae from the Upper Jurassic-Lower Cretaceous Yixian Formation (Liaoning Province, NE China). *Geologica Carpathica* **64** (4): 291–304.

44 Vršanský, P. (2009). Advanced morphology and behaviour of extinct earwig-like cockroaches (Blattaria: Fuziidae fam. nov.). *Geologica Carpathica* **60** (6): 449–462.

45 Ding, Q.H., Zhang, L.D., Guo, S.Z. *et al.* (2003). Paleoclimatic and palaeoenvironmental proxies of the Yixian Formation in the Beipiao area, western Liaoning. *Geological Bulletin of China* **22** (3): 186–191.

46 Vršanský, P. (1997). *Piniblattella* gen. nov. – the most ancient genus of the family Blattellidae (Blattodea) from the Lower Cretaceous of Siberia. *Entomological Problems* **28** (1): 67–79.

47 Lee, S.W. (2016). Taxonomic diversity of cockroaches assemblages (Blattaria, Insecta) of the Aptian Crato Formation (Cretaceous, NE Brazil). *Geologica Carpathica* **67**: 433–450.

48 Gao, T.P., Shih, C.K., Labandeira, C.C. *et al.* (2018). Maternal care by Early Cretaceous cockroaches. *Journal of Systematic Palaeontology* https://doi.org/10.1080/14772019.2018.1426059.

49 Roth, L.M. (1970). Evolution and taxonomic significance of reproduction in Blattaria. *Annual Review of Entomology* **15** (1): 75–96.

50 Roth, L.M. and Willis, E.R. (1957). An analysis of oviparity and viviparity in the Blattaria. *Transactions of the American Entomological Society* **83** (4): 221–238.

51 Roth, L.M. (2003). Systematics and phylogeny of cockroaches (Dictyoptera: Blttaria). *Oriental Insects* **37** (1): 1–186.

52 Bell, W.J., Roth, L.M., and Nelepa, C.A. (2007). *Cockroaches: Ecology, Behavior, and Natural History*. Baltimore, Maryland: The Johns Hopkins University Press, 230 pp.

53 Guo, Y.X. and Ren, D. (2011). A new cockroach genus of the family Fuziidae from Northeastern China (Insecta: Blattida). *Acta Geologica Sinica* (English Edition) **85** (2): 501–506.

54 Wei, D.D., Shih, C.K., and Ren, D. (2012). *Arcofuzia cana* gen. Et sp. n. (Insecta, Blattaria, Fuziidae) from the Middle Jurassic sediments of Inner Mongolia, China. *Zootaxa* **3597**: 25–32.

55 Vršanský, P., Liang, J.H., and Ren, D. (2012). Malformed cockroach (Blattida: Liberiblattinidae) in the Middle Jurassic sediments from China. *Oriental Insects* **46** (1): 12–18.

56 Liang, J.H., Shih, C.K., and Ren, D. (2017). New Jurassic predatory cockroaches (Blattaria: Raphidiomimidae) from Daohugou, China and Karatau, Kazakhstan. *Alcheringa* **42** (1): 101–109. https://doi.org/10.1080/03115518.2017.1374460.

57 Geinitz, F.E. (1880). Der Jura in Mecklenburg und seine Versteinerungen. *Zeitschrift der deutschen geologischen Gesellschaft* **22**: 510–535.

58 Lin, Q.B. (1982). Mesozoic and Cenozoic insects. In: *Atlas of Palaeontology from Northwest Region, Shaanganning division*, 77–79. Beijing: Geological Publishing House (In Chinese with English summary).

59 Zhang, H.C. (1997). Early Cretaceous insects from the Dalazi Formation of the Zhixin Basin, Jilin Province, China. *Palaeoworld* **7**: 75–103.

60 Grimaldi, D. and Ross, A. (2004). *Raphidiomimula*, an enigmatic new cockroach in Cretaceous amber from Myanmar (Burma) (Insecta: Blattodea: Raphidiomimidae). *Journal of Systematic Palaeontology* **2**: 101–104.

61 Liang, J.H., Huang, W.L., and Ren, D. (2012). *Graciliblatta bella* gen. et sp. n. – a rare carnivorous cockroach (Insecta, Blattida, Raphidiomimidae) from

the Middle Jurassic sediments of Daohugou in Inner Mongolia, China. *Zootaxa* **3449**: 62–68.

62 Liang, J.H., Vršanský, P., and Ren, D. (2012). Variability and symmetry of a Jurassic nocturnal predatory cockroach (Blattida: Raphidiomimidae). *Revista Mexicana de Ciencias Geológicas* **29** (2): 411–421.

63 Chen, S.X. and Tan, J.J. (1973). A new family of Coleoptera from the Lower Cretaceous of Kansu. *Acta Entomologica Sinica* **16** (2): 169–178.

64 Vršanský, P. (2003). Umenocoleidae – an amazing lineage of aberrant insects (Insecta, Blattaria). *AMBA projekty* 7 (1): 3–30.

65 Lin, Q.B. (1980). Mesozoic insect fossils from Zhejiang and Anhui. In: *Division and Correlation of Mesozoic Volcano-Sedimentary Formation in Zhejiang and Anhui Province*, 211–238. Beijing: Science Press (In Chinese with English summary).

66 Lin, Q.B. (1986). Early Mesozoic fossil insect from the South China. In: *Palaeontologica Sinica*, Series B, No. 21, 28–53. Beijing: Science Press (In Chinese with English summary).

67 Ren, D., Lu, L.W., Guo, Z.G., and Ji, S.A. (1995). *Faunae and stratigraphy of Jurassic-Cretaceous in Beijing and the adjacent areas*, 50–56. Beijing: Seismie Publishing House (In Chinese with English summary).

68 Zhang, J.F. (1986). Some fossil insects from the Jurassic of Northern Hebei, China. *Paleontology and Stratigraphy of Shandong, Shandong* 74–84. (In Chinese with English summary).

69 Lin, Q.B. (1985). Insect fossils from the Hanshan Formation at Hanshan County, Anhui Province. *Acta Palaeontologica Sinica* **24** (3): 300–304. (In Chinese with English summary).

70 Hong, Y.C. (1986). New fossil insects of Haifanggou Formation, Liaoning Province. *Journal Changchun College Geological* **4**: 10–16. (In Chinese with English summary).

71 Zhang, Z.J., Lu, L.W., Jin, Y.G. *et al.* (2003). The first found insect fossil in Tuodian Formation, Dianzhong. *Geological Bulletin China* **22** (6): 452–455. (In Chinese with English summary).

72 Wang, W.L. (1987). The Early Mesozoic fossil insects in Western Liaoning. In: *Mesozoic stratigraphy and palaeontology of Western Liaoning III* (ed. X.H. Yu, W.L. Wang, X.T. Liu, *et al.*), 206–210. Beijing: Geological Publishing House (In Chinese with English summary).

73 Wang, W.L. (1980). *Palaeontological Atlas of North China II*, 134–136. Beijing: Geological Publishing House (In Chinese with English summary).

74 Hong, Y.C. and Xiao, Z.Z. (1997). Blattoid fossil of Yanqing County, Houcheng Formation in Beijing 2: 1–6. *Beijing Geological* (In Chinese with English summary).

75 Hong, Y.C. (1997). Discovery of fossil insects from Houcheng Formation of Hebei Province and establish of Houcheng Entomofauna. *Beijing Geological* **1**: 1–10. (In Chinese with English summary).

76 Lin, Q.B. (1978). On the fossil Blattoidea of China. *Acta Entomologica Sinica* **21** (3): 335–342. (In Chinese with English summary).

77 Lin, Q.B. (1976). The Jurassic fossil from Western Liaoning. *Acta Palaeontologica Sinica* **15** (1): 97–118. (In Chinese with English summary).

78 Hong, Y.C. (1992). Middle and Late Jurassic insects. In: *Palaeontological Atlas of Jilin Province*, 410–417. In Chinese with English summary).

79 Hong, Y.C. and Wang, W.L. (1990). Fossil insects from the Laiyang Basin, Shandong Province. In: *Stratigraphy and Palaeontology of Laiyang Basin, Shandong Province*, 44–189. Beijing: Geological Publishing House (In Chinese with English summary).

80 Hong, Y.C. (1982). *Mesozoic fossil insects of Jiuquan Basin in Gansu Province*, 63–70. Beijing: Geological Publishing House (In Chinese with English summary).

81 Chang, J.P. and Wang, W.D. (1993). A few new species of insect in Lower Jurassic near the Fushuizhai District of the Funing, Hebei. *Journal of Changchun University of Earth Science* **23** (1): 10–15.

82 Hong, Y.C. (1983). *Middle Jurassic fossil insects in North China*, 26–37. Beijing: Geological Publishing House (In Chinese with English summary).

83 Hong, Y.C. and Xiao, Z.Z. (1997). New fossil Blattodea, Coleoptera and Mecoptera (Insecta) from Houcheng Formation of Yanqing County, Beijing. *Beijing Geological* **3**: 1–10. (In Chinese with English summary).

84 Wei, D.D., Liang, J.H., and Ren, D. (2012). A new species of Fuziidae (Insecta, Blattida) from the Inner Mongolia, China. *ZooKeys* **217**: 53–61. https://doi.org/10.3897/zookeys.217.3508.

85 Zhang, J.F. (1985). New data of Mesozoic insect fossils from Laiyang in Shandong. *Shandong Geological* **1** (2): 23–39. (In Chinese with English summary).

86 Ping, C. (1928). Cretaceous fossil insects of China. *Palaeontologica Sinica, Series B* **13** (1): 1–35.

87 Lin, Q.B. and Wang, Q.S. (1989). Late Cretaceous insects from North Jiangsu Basin. In: *Stratigraphy and Palaeontology of the Taizhou Formation and the first member of the Funing Formation North Jiangsu Basin*, 193–197. Nanjing: Nanjing University Press.

8

Termitoidae – Termites

Zhipeng Zhao[1], Dong Ren[1], and Chungkun Shih[1,2]

[1] *Capital Normal University, Haidian District, Beijing, China*
[2] *National Museum of Natural History, Smithsonian Institution, Washington, DC, USA*

8.1 Introduction to Termitoidae

Termites are wood-feeding eusocial insects commonly found and observed by many people. Hitherto, about 3000 fossil and extant termite species have been described [1]. The distribution of termites tends to be near the warm regions, mainly in tropical and subtropical zones [2]. They have been classified as the order Isoptera ("Iso-" meaning "equal" and "ptera" meaning "winged"), separated from the cockroach order of Blattodea. However, recent phylogenetic studies indicate that they evolved from close ancestors of cockroaches during the Jurassic, thus, they are treated as Termitoidae, which are an epifamily of cockroaches [3, 4]. Termites have nearly equal-sized fore- and hind wings. However, there are exceptions in that some termites may have intra-individual variations of wings and the hind wings of Mastotermitidae present anal lobes.

Termites show clear morphological polymorphism based on the castes, which basically include imagoes (reproductive alates), workers and soldiers. All three castes of termites, generally developed from larvae, go through incomplete metamorphosis while being cared for by workers. All of them share similar morphologies such as prognathous-chewing mouthparts, moniliform antenna, well-developed pronotum, 10-segmented abdomen, and sternal glands for communication. The reproductive winged imagoes are produced at certain times of the year, and huge swarms emerge from the colony when nuptial flights begin. These imagoes have four reticulate wings with basal sutures which will be shed after nuptial flights, but Mastotermitidae and some basal termites don't have distinct basal suture of hind wings. Termite workers generally have similar morphology to imagoes except for the absence of wings, weakly sclerotized body, and higher length proportion of head to body than those of imagoes.

The worker termites, as the majority and valuable contributors of a colony, are with females and males developed from larvae (Figure 8.1). Depending on species, male and female workers may have different roles in a termite colony [5]. The role of workers is typical and common, while nymphs of Archotermopsidae, Kalotermitidae, Stolotermitidae play the role of workers as the pseudergates (blind wingless nymphs, performing some of the functions of a worker). [6, 7].

Termite soldiers are stronger with more specialized body parts, especially the head, than other castes. The soldiers' length proportion of head to body is much higher than those of imagoes and workers mainly due to their extension of head capsule or mandibles. The heads of termite soldiers are generally equipped with asymmetric mandibles, roughly sorted into three basic types: extended head with elongate mandibles (Figure 8.2), plug-shaped head and rhino-shaped head (Figure 8.3) with vestigial mandibles. Furthermore, some soldiers of higher termites are able to spray chemicals through the fontanelle or a horn-like frontal nozzle on the head capsule to dispel the enemies. These specialized structures of soldiers are intended for defense, while most soldiers lack eyes or developed genitalia, but a few soldiers present vestigial eyes. Due to defensive jaws, termite soldiers must be fed by workers.

Cryptocercidae (wood cockroaches in the order Blattodea) show evidence of being the sister group with Termitoidae. They both share some homologously functional microorganisms. For example, *Trichonympha*, a genus of parabasalid excavates that live in the intestines of many termite species, collected in the Cryptocercidae can be transplanted into the hind gut of some lower termites [8, 9].

In recent years, many molecular phylogenetic studies proved that termites can no longer be an order of Dictyoptera but should be downgraded [3]. The formal name and classification of the termites have been debated

Rhythms of Insect Evolution: Evidence from the Jurassic and Cretaceous in Northern China, First Edition.
Edited by Dong Ren, Chungkun Shih, Taiping Gao, Yongjie Wang, and Yunzhi Yao.

Figure 8.1 Worker Formosan subterranean termites (*Coptotermes formosanus*). Source: Photo by Jason Shih.

Figure 8.2 A soldier of Formosan subterranean termites (*Coptotermes formosanus*). Source: Photo by Jason Shih.

[4, 10]. Among termite researchers and published literature, "Isoptera" (Infraorder) and Termitoidae (epifamily) have been used [1, 11–13]. In this chapter, we use the terminology of epifamily of Termitoidae to be consistent with the phylogenetic position of termites.

Twelve families of Termitoidae have been documented so far: Cratomastotermitidae, Mastotermitidae, Termopsidae, Hodotermitidae, Archotermopsidae, Stolotermitidae, Kalotermitidae, Archeorhinotermitidae, Stylotermitidae, Rhinotermitidae, Serritermitidae, and Termitidae. Many researchers have tried to elucidate the relationship of these families through morphological and molecular phylogenetic studies [14–17]. Mastotermitidae is certainly the most basal family while Termitidae is the most derived family of living termites.

Termites are believed to be the first and earliest insect group to have eusociality, which might have been potentially induced by wood-feeding [8, 18, 19]. The microbes located in the hind gut of termites help them

Figure 8.3 A soldier termite of *Nasutitermes takasagoensis*; inhabits from underground to trees. Source: Photo by Mei-Ling Lo.

to digest the cellulose [20]. Such a diet allows them to have superfluous food, and achieve ecological dominance [16]. Different from non-social insects, termites show many special social behaviors, e.g. building and taking care of the nest, complicated communication, undertaking, trophallaxis, foraging, dealation, defense, and tending to the queen.

Termite nests are stable ecosystems in which termites construct and maintain the structure and environment by maintaining temperature and humidity, and cultivating fungi. Termite nests are generally in wood or underground. In the nests, the communication among termites can be divided into chemical and physical. Most termites are blind, so communication primarily occurs through chemical, mechanical and pheromonal cues [21, 22]. The chemical secretion of pheromone can regulate the caste differentiation [23] and act on the odor trails in foraging [24]. In addition, physical communication includes tactile stimulation in proctodeal feeding [6] and vibration signals for issuing warnings.

Not unexpectedly, such sanctuaries are also favorable for termitophiles which have been found within several insect orders: Diptera, Hemiptera, Hymenoptera, Lepidoptera, Thysanura, and Coleoptera [6, 25]. They utilize the materials and energy from the termite nests to develop their own populations, while some have mutual benefit relationships with the host termites. Some ants, which are termites' main enemies, prey on termites (Figure 8.4) or even raid termite nests for inhabitants. To fight these defensive wars, termite soldiers carry "weapons" to guard and ensure the safety of the community. They have several strategies to defend against enemies, which basically include: snapping the invaders with the mandibles, spraying chemicals from the fontanelle, or blocking the entrances by a plug-shaped head. Sternal glands play an important role in setting up odor trails for blind termites and the secretion of odor

Figure 8.4 A Formosan subterranean termite under attack by an ant. Source: Photo by Jason Shih.

helps foraging termites communicate with each other outside the nest and to be well-organized.

Termite Eusociality

Termites' eusocial behaviors, including the caste system, communication, sharing symbiotic protozoa or other microbes in the guts and roles and functions of individuals, significantly enhance the survival and/or development of their colonies (see box, Insect Eusociality in Chapter 22). Termites are one of the most successful insect taxa on Earth, widely distributed around the world except for Antarctica. Their colonies range in size from several hundred to several million individuals. Termite queens have lifespan up to 30–50 years, which is the longest for any insects.

Depending on their feeding habits, termites are designated into two groups: the lower termites and higher termites. The lower termites, with guts containing protozoa and many species of bacteria, mostly feed on wood or fungus-infected wood. The higher termites, with guts containing a few species of bacteria but no protozoa [26], consume various materials, e.g. feces, humus, grass, leaves and roots [27].

Termite nests are used as a living space for the colony while providing protection against environmental factors and potential predators. There are three types of termite nests: subterranean (completely underground), epigeal (protruding above the soil surface), and arboreal (building on trees) [28]. Many termites use underground nests. Some termites build their nests inside wooden structures such as logs, stumps or the dead parts of trees, thus, in some cases, causing wood structure damage inside human buildings, Termite mounds, built above the subterranean nests by using soil and mud, are found in the dry and well-drained areas in Africa, Australia and South America. The structure of the mounds is practical and efficient, including an extensive system of tunnels and conduits that serves as a ventilation and temperature control system for the underground nest.

Inside a complex nest, enclosed with complex tunnels and ventilation systems, it is important to maintain cleanliness, avoid decomposition and prevent infection. If the dead individuals are not properly undertaken, the clean operation of the colony may be negatively impacted. Interestingly, worker termites have undertaking behaviors for the dead, including circumvention [29], cannibalism [30, 31], corpse removal or burying them in fecal material, debris or soil [6, 29, 32 36]. The signal for corpse identification comes from odor (chemical), tactile (physical) and visual (biological) [37].

Wood feeding termites must rely on protozoans or other microbes in their hind gut to accomplish digestion of cellulose. The "digesters" are not innate but acquired, even after molting. Therefore, the individuals in need of new protozoans or microbes have to exchange food, saliva or feces with other termites, and the processes are called "trophallaxis". Trophallaxis include two types: stomodeal and proctodeal. The stomodeal means the recipients use their antennae to caress the head of the donors or tap the mandibles of the donors, thus, receiving fluid droplets from mouth to mouth [6]. Such an "intimate motion" allows the recipient to ingest saliva or regurgitated fluid containing wood fragments and "digester". Proctodeal, on the other hand, involves the recipients using their antennae, palps or forelegs to touch the perianal region of the donors. As a consequence, the recipients ingest the excreta from the donor's rectum. Both stomodeal and proctodeal methods are generally indispensable, e.g. termite soldiers, unable to feed due to specialized mandibles, must be fed through the stomodeal method, and newly-hatched larvae need the proctodeal method to gain protozoans or other microbes from other termites.

8.2 Progress in the Studies of Fossil Termites

Cretatermes carpenteri is the first described Cretaceous termite fossil from Labrador, Canada. [38], and the earliest termite fossil is *Baissatermes lapideus* from the Early Cretaceous (ca. 140 Mya) of the Berriasian Zaza Formation in Baissa, Siberia [39]. Fossil termites show the broadest diversity in Dominican amber, with well-preserved specimens [40–42]. Hitherto, 34 genera and 40 species of fossil termites have been described from the Cretaceous from around the world. Most of the described species are imagoes, while the soldiers or workers were rarely preserved. The earliest soldiers and workers have been reported in Myanmar (Burmese) amber (98.79 ± 0.62 Mya) [1, 43]. To date, only seven species of termites in six genera have been described from the Lower Cretaceous Lushangfen Formation, at the Lushangfen Village, southwest Beijing, China [44]. However, no termite fossils have been documented from the Jurassic or any earlier geological ages.

8.3 Representative Fossil Termites from Northern China

Family Hodotermitidae Desneux, 1904

Hodotermitids are called "harvester" termites, because the cellulose of their feed mostly comes from foraging dry grasses. Hodotermitids can be diagnosed by the following characters: no fontanelle and ocelli; antenna with 23–32 antennomeres; beside the apical teeth of mandibles, the left mandible with three marginal teeth and the right mandible with two; pronotum saddle-shaped and narrower than head; tarsus with four tarsomeres and tibial spur formula as 3 : 2–5 : 3–5; only one sternal gland presents on the fourth abdominal sternite; cerci with one to five segments and the styli present generally [7].

Genera included from the Cretaceous of Northern China: *Jitermes* Ren, 1995, *Yanjingtermes* Ren, 1995, *Yongdingia* Ren, 1995, *Huaxiatermes* Ren, 1995, *Asiatermes* Ren, 1995 and *Mesotermes* Ren, 1995.

Jitermes Ren, 1995

Jitermes Ren, 1995, *Faunae and Stratigraphy of Jurassic-Cretaceous in Beijing and the Adjacent Areas.* Geological Publishing House, Beijing. 56–57 [44]. (original designation)

Type species: *Jitermes tsaii* Ren, 1995.

The specific epithet is in honor of Prof. Cai Banghua for his dedication to entomology. Diagnosis: Humeral suture straight; the first branch of Rs close to wing base, superior branches which head to costal margin without secondary branch, inferior branches which head to posterior margin close to wing apex; M between Rs and CuA with same distance; the first branch of M at about distal third of wing, weakly posterior to the first branch of Rs; Medial field broad, with six branches; CuA with few branches, the longest vein exceeding half of posterior margin.

Distribution and age: Beijing; Early Cretaceous.

Only one species included from the Cretaceous of Northern China (see Table 8.1).

Yanjingtermes Ren, 1995

Yanjingtermes Ren, 1995, *Faunae and Stratigraphy of Jurassic-Cretaceous in Beijing and the Adjacent Areas.* Geological Publishing House, Beijing. 57–58 [44] (original designation).

Type species: *Yanjingtermes giganteus* Ren, 1995.

R_1 with two branches; radial field broad; the first branch of Rs close to the wing base, superior branches with secondary branches, inferior branches well-developed; M close to CuA, the first branch is more anterior than the first branch of Rs, medial field broad, with many branches, cubital field narrow.

Distribution and age: Beijing; Early Cretaceous.

Only one species included from the Cretaceous of Northern China (see Table 8.1).

Yongdingia Ren, 1995

Yongdingia Ren, 1995, *Faunae and Stratigraphy of Jurassic-Cretaceous in Beijing and the Adjacent Areas.* Geological Publishing House, Beijing. 58–59 [44] (original designation).

Type species: *Yongdingia opipara* Ren, 1995.

Humeral suture straight; wing posterior margin longer than costal margin; the terminal of Rs broad but less obvious than that of *Yanjingtermes*, anterior branches of Rs short, have large angle with costal margin, secondary branches present; inferior branches developed without secondary branches. M independently starting from humeral suture, with similar distance to Rs and Cu despite the middle part closer to Cu than Rs. Cu terminal after half of the posterior margin.

Distribution and age: Beijing; Early Cretaceous.

Only one species included from the Cretaceous of Northern China (see Table 8.1).

Huaxiatermes Ren, 1995

Huaxiatermes Ren, 1995, *Faunae and Stratigraphy of Jurassic-Cretaceous in Beijing and the Adjacent Areas.* Geological Publishing House, Beijing. 59–60 [44] (original designation).

Type species: *Huaxiatermes huangi* Ren, 1995.

The specific epithet is in honor of Prof. Huang Fusheng for his dedication to entomology. Diagnosis: Humeral suture curved; wing posterior margin longer than costal margin; Rs without inferior branch, Rs first branch posterior; M starts from the stem of Rs, the first branch more anterior than the first branch of Rs, with many branches; Cu starts on the humeral suture, close to Rs, with many branches, terminating weakly proximal to apex to posterior margin.

Distribution and age: Beijing; Early Cretaceous.

Only one species included from the Cretaceous of Northern China (see Table 8.1.)

Asiatermes Ren, 1995

Huaxiatermes Ren, 1995, *Faunae and Stratigraphy of Jurassic-Cretaceous in Beijing and the Adjacent Areas.*

Geological Publishing House, Beijing. 60 [44] (original designation).

Type species: *Asiatermes reticulatus* Ren, 1995.

Similar to *Cretatermes* Emerson, 1967; Costal field quite narrow, Rs inferior branches undeveloped; anterior branch of M and Cu combined.

Distribution and age: Beijing; Early Cretaceous.

Only one species included from the Cretaceous of Northern China (see Table 8.1).

Mesotermes Ren, 1995

Huaxiatermes Ren, 1995, *Faunae and Stratigraphy of Jurassic-Cretaceous in Beijing and the Adjacent Areas.* Geological Publishing House, Beijing. 60–61 [44] (original designation).

Type species: *Mesotermes incompletes* Ren, 1995.

Rs with many long superior branches, secondary branches absent, inferior branches undeveloped; the first branch of M posterior to Rs; medial field relatively narrow; M with four to five branches; cubital field broad; Cu with many branches terminating distal to the middle of posterior margin; reticulate veins undeveloped.

Distribution and age: Beijing; Early Cretaceous.

Two species included from the Cretaceous of Northern China (see Table 8.1).

Table 8.1 A list of fossil termites from the Cretaceous of China.

Family	Species	Locality	Horizon/Age	Citation
Hodotermitidae	*Jitermes tsaii* Ren, 1995	Lushangfen Village, Beijing	Lushangfen Fm., K_1	Ren *et al.* [44]
	Yanjingtermes giganteus Ren, 1995	Lushangfen Village, Beijing	Lushangfen Fm., K_1	Ren *et al.* [44]
	Yongdingia opipara Ren, 1995	Lushangfen Village, Beijing	Lushangfen Fm., K_1	Ren *et al.* [44]
	Huaxiatermes huangi Ren, 1995	Lushangfen Village, Beijing	Lushangfen Fm., K_1	Ren *et al.* [44]
	Asiatermes reticulatus Ren, 1995	Lushangfen Village, Beijing	Lushangfen Fm., K_1	Ren *et al.* [44]
	Mesotermes incompletes Ren, 1995	Lushangfen Village, Beijing	Lushangfen Fm., K_1	Ren *et al.* [44]
	Mesotermes latus Ren, 1995	Lushangfen Village, Beijing	Lushangfen Fm., K_1	Ren *et al.* [44]

References

1 Engel, M.S., Barden, P., Riccio, M.L., and Grimaldi, D.A. (2016). Morphologically specialized termite castes and advanced sociality in the Early Cretaceous. *Current Biology* **26** (4): 522–530. https://doi.org/10.1016/j.cub.2015.12.061.

2 Eggleton, P. (2000). Global patterns of termite diversity. In: *Termites: Evolution, Sociality, Symbioses, Ecology* (ed. T. Abe, D.E. Bignell and M. Higashi), 25–51. Dordrecht: Springer.

3 Inward, D., Beccaloni, G., and Eggleton, P. (2007). Death of an order: a comprehensive molecular phylogenetic study confirms that termites are eusocial cockroaches. *Biology Letters* **3** (3): 331–335. https://doi.org/10.1098/rsbl.2007.0102.

4 Eggleton, P., Beccaloni, G., and Inward, D. (2007). Response to Lo *et al. Biology Letters* **3** (5): 564–565. https://doi.org/10.1098/rsbl.2007.0367.

5 Korb, J. (2008). Termites, hemimetabolous diploid white ants? *Frontiers in Zoology* **5** (1): 15. https://doi.org/10.1186/1742-9994-5-15.

6 Krishna, K. and Weesner, T.M. (1969). *Biology of Termites.* New York and London: Academic Press.

7 Krishna, K., Grimaldi, D.A., Krishna, V., and Engel, M.S. (2013). *Treatise on the Isoptera of the World,* 200. New York: Bulletin of the Museum of Natural History.

8 Wilson, E.O. (1971). *The Insect Societies.* Cambridge: Belknap Press of Harvard University Press.

9 Lo, N., Tokuda, G., Watanabe, H. *et al.* (2000). Evidence from multiple gene sequences indicates that termites evolved from wood-feeding cockroaches. *Current Biology* **10** (13): 801–804. https://doi.org/10.1016/S0960-9822(00)00561-3.

10 Lo, N., Engel, M.S., Cameron, S. *et al.* (2007). Save Isoptera: a comment on Inward *et al. Biology Letters* **3** (5): 562–563. https://doi.org/10.1098/rsbl.2007.0264.

11 Cameron, S.L., Lo, N., Bourguignon, T. *et al.* (2012). A mitochondrial genome phylogeny of termites (Blattodea: Termitoidae): robust support for interfamilial relationships and molecular synapomorphies define major clades. *Molecular Phylogenetics and Evolution* **65** (1): 163–173. https://doi.org/10.1016/j.ympev.2012.05.034.

12 Engel, M.S. (2011). Family-group names for termites (Isoptera), redux. *Zookeys* **148**: 171–184. https://doi.org/10.3897/zookeys.148.1682.

13 Xiao, B., Chen, A., Jiang, G. *et al.* (2012). Complete mitochondrial genomes of two cockroaches, *Blattella*

germanica and *Periplaneta americana*, and the phylogenetic position of termites. *Current Genetics* **58**: 65–77. https://doi.org/10.1007/s00294-012-0365-7.

14 Inward, D.J., Vogler, A.P., and Eggleton, P. (2007). A comprehensive phylogenetic analysis of termites (Isoptera) illuminates key aspects of their evolutionary biology. *Molecular Phylogenetics and Evolution* **44** (3): 953–967. https://doi.org/10.1016/j.ympev.2007.05.014.

15 Legendre, F., Whiting, M.F., Bordereau, C. *et al.* (2008). The phylogeny of termites (Dictyoptera: Isoptera) based on mitochondrial and nuclear markers: implications for the evolution of the worker and pseudergate castes, and foraging behaviors. *Molecular Phylogenetics and Evolution* **48** (2): 615–627. https://doi.org/10.1016/j.ympev.2008.04.017.

16 Engel, M.S., Grimaldi, D.A., and Krishna, K. (2009). Termites (Isoptera): their phylogeny, classification, and rise to ecological dominance. *American Museum Novitates* **3650**: 1–27.

17 Bourguignon, T., Lo, N., Cameron, S.L. *et al.* (2015). The evolutionary history of termites as inferred from 66 mitochondrial genomes. *Molecular Biology and Evolution* **32** (2): 406–421. https://doi.org/10.1093/molbev/msu308.

18 Grimaldi, D.A. and Engel, M.S. (2005). *Evolution of the Insects*. 238–252. New York: Cambridge University Press.

19 Nalepa, C.A. (2010). Altricial development in wood-feeding cockroaches: the key antecedent of termite eusociality. In: *Biology of Termites: A Modern Synthesis* (ed. D.E. Bignell, Y. Roisin and N. Lo), 69–96. The Netherlands: Springer.

20 Wier, A., Dolan, M., Grimaldi, D.A. *et al.* (2002). Spirochete and protist symbionts of a termite (*Mastotermes electrodominicus*) in Miocene amber. *Proceedings of the National Academy of Sciences of the United States of America* **99** (3): 1410–1413.

21 Costa-Leonardo, A.M. and Haifig, I. (2010). Pheromones and exocrine glands in Isoptera. *Vitamins and Hormones* **83**: 521–549.

22 Costa-Leonardo, A.M. and Haifig, I. (2013). Termite communication during different behavioral activities. In: *Biocommunication of Animals* (ed. W. Guenther). The Netherlands: Springer.

23 Bordereau, C. (1985). The role of pheromones in termite caste differentiation. In: *Caste Differentiation in Social Insects* (ed. J. Watson, B. Okot-Kober and C. Noirot), 221–226. Oxford: Pergamon Press.

24 Bordereau, C. and Pasteels, J.M. (2010). Pheromones and chemical ecology of dispersal and foraging in termites. In: *Biology of Termites: A Modern Synthesis* (ed. D.E. Bignell, Y. Roisin and N. Lo), 1–26. The Netherlands: Springer.

25 Cai, C., Huang, D., Newton, A.F. *et al.* (2017). Early evolution of specialized termitophily in Cretaceous rove beetles. *Current Biology* **27** (8): 1229–1235. https://doi.org/10.1016/j.cub.2017.03.009.

26 Breznak, J.A. and Brune, A. (1993). Role of microorganisms in the digestion of lignocellulose by termites. *Annual Review of Entomology* **39** (1): 453–487. https://doi.org/10.1146/annurev.en.39.010194.002321.

27 Radek, R. (1999). Flagellates, bacteria, and fungi associated with termites: diversity and function in nutrition – a review. *Ecotropica* **5**: 183–196.

28 Noirot, C. and Darlington, J.P.E.C. (2000). Termite nests: architecture, regulation and defence. In: *Termites: Evolution, Sociality, Symbioses, Ecology* (ed. T. Abe, D.E. Bignell and M. Higashi), 121–139. Dordrecht: Springer.

29 Su, N.-Y. (2005). Response of the *Formosan subterranean* termites (Isoptera: Rhinotermitidae) to baits or nonrepellent termiticides in extended foraging arenas. *Journal of Economic Entomology* **98**: 2143–2152.

30 Kramm, K.R., West, D.F., and Rockenbach, P.G. (1982). Termite pathogens: transfer of the entomopathogen *Metarhizium anisopliae* between *Reticulitermes* sp. termites. *Journal of Invertebrate Pathology* **40**: 1–6.

31 Rosengaus, R. and Traniello, J. (2001). Disease susceptibility and the adaptive nature of colony demography in the dampwood termite *Zootermopsis angusticollis*. *Behavioral Ecology and Sociobiology* **50** (6): 546–556.

32 Zoberi, M.H. (1995). Metarhizium anisopliae, a fungal pathogen of *Reticulitermes flavipes* (Isoptera: Rhinotermitidae). *Mycologia* **87**: 354–359.

33 Myles, T.G. (2002). Alarm, aggregation, and defense by *Reticulitermes flavipes* in response to a naturally occurring isolate of *Metarhizium anisopliae*. *Soiobiology* **40**: 13.

34 Chouvenc, T., Robert, A., Semon, E., and Bordereau, C. (2012). Burial behaviour by dealates of the termite *Pseudacanthotermes spiniger* (Termitidae, Macrotermitinae) induced by chemical signals from termite corpses. *Insectes Sociaux* **59**: 119–125.

35 Ulyshen, M.D. and Shelton, T.G. (2012). Evidence of cue synergism in termite corpse response behavior. *Naturwissenschaften* **99**: 89–93.

36 Sun, Q., Haynes, K.F., and Zhou, X. (2013). Differential undertaking response of a lower termite to congeneric and conspecific corpses. *Scientific Reports* **3**: 1650.

37 Sun, Q. and Zhou, X. (2013). Corpse management in social insects. *International Journal of Biological Sciences* **9** (3): 313–321.

38 Emerson, A.E. (1967). Cretaceous insects from Labrador a new genus and species of Termite (Isoptera Hodotermitidae). *Psyche* **74**: 276–289.
39 Engel, M.S., Grimaldi, D.A., and Krishna, K. (2007). Primitive termites from the Early Cretaceous of Asia (Isoptera). *Stuttgarter Beiträge zur Naturkunde Serie B* **371**: 1–32.
40 Krishna, K. and Grimaldi, D.A. (1991). A new fossil species from Dominican amber of the living Australian termite genus *Mastotermes* (Isoptera: Mastotermitidae). *American Museum Novitates* **3021**: 1–10.
41 Krishna, K. (1996). New fossil species of termites of the subfamily Nasutitermitinae from Dominican and Mexican amber (Isoptera,Termitidae). *American Museum Novitates* **3176**: 1–8.
42 Engel, M.S. and Krishna, K. (2007). New *Dolichorhinotermes* from Ecuador and in Mexican amber (Isoptera: Rhinotermitidae). *American Museum Novitates* **3592**: 1–8.
43 Shi, G., Grimaldi, D.A., Harlow, G.E. *et al.* (2012). Age constraint on Burmese amber based on U–Pb dating of zircons. *Cretaceous Research* **37**: 155–163. https://doi.org/10.1016/j.cretres.2012.03.014.
44 Ren, D., Lu, L.W., Guo, Z.G., and Ji, S.A. (1995). *Faunae and Stratigraphy of Jurassic-Cretaceous in Beijing and the Adjacent Areas.* (In Chinese with English summary). ed, 222. Beijing: Geological Publishing House.

9

Orthoptera – Grasshoppers and Katydids

Jun-Jie Gu[1,2], Chungkun Shih[2,3], and Dong Ren[2]

[1] *Sichuan Agricultural University, Chengdu, Sichuan, China*
[2] *Capital Normal University, Haidian District, Beijing, China*
[3] *National Museum of Natural History, Smithsonian Institution, Washington, DC, USA*

9.1 Introduction to Orthoptera

Orthoptera are one of the oldest orders of insects, known from the Late Carboniferous to recent, encompassing grasshoppers, locusts, katydids (bush-crickets), crickets, weta, and their relatives. To date, over 25 700 known extant species worldwide have been described. Orthopterans have incomplete metamorphosis. The body sizes of orthopterans are usually large, only very few species smaller than 5.0 mm, the largest ones longer than 120.0 mm in length. They are widely distributed in all faunas, with broad species diversity in tropical areas, but less diversity in high elevation or high latitude areas.

As important terrestrial insects, most orthopterans are characterized by their well-developed saltatorial hind legs for jumping. They have a hypognathous head, filiform (thread-like) antennae, shielded thorax and usually two pairs of wings with relatively straight veins. The forewings are leather-like, usually modified for courtship; hind wings are membranes with a broad anal area. Chewing mouthparts are equipped with various mandibles correlated to diverse feeding habits.

The Orthoptera are classified into two suborders, Ensifera and Caelifera. Ensifera are the oldest group of Orthoptera, which comprise 11 families in seven superfamilies, i.e. Gryllotalpoidea, Grylloidea, Hagloidea, Rhaphidophoroidea, Schizodactyloidea, Stenopelmatoidea and Tettigonioidea (Figures 9.1–9.4). The ensiferan insects, with over 12 000 species in almost 2000 genera, are characterized by long and filiform antennae usually longer than their body lengths, three or four segmented tarsi, modified and stridulated forewings and sword- or sickle-shaped ovipositors. Most males, even some females of these insects can produce sounds by rubbing their stridulate apparatus on the forewings. It is a kind of acoustic communication which is normally correlated with mating or fighting. Their feeding habits are very broad, including herbivorous, omnivorous or predaceous feeding. Some even have the highly specialized habits of feeding on nectar and pollen or seeds and fruits.

Caelifera comprise 28 families in nine superfamilies, i.e. Acridoidea, Eumastacoidea, Pneumoroidea, Proscopioidea, Pyrgomorphoidea, Tanaoceroidea, Tetrigoidea, Tridactyloidea and Trigonopterygoidea (Figure 9.4). They have antennae shorter than half body length, short ovipositors, enlarged hind femora and three or fewer tarsal segments. Most are diurnal and herbivorous. Some can migrate in swarms and cause serious plant or crop damage. For example, locusts are the most notorious grasshoppers that have been among the most destructive agricultural pests throughout the course of human history. Although most grasshoppers are harmful pests, they serve as an option of high protein and popular food. In Southeast Asia, China and Mexico, some people treat grasshoppers and some other insects as a delicacy, thus, the dishes of fried grasshoppers are very popular in many restaurants and food markets.

Serenading with Love Songs

Most katydids and crickets can produce songs, which are related to their mating behavior. Normally, the males produce the songs by rubbing files on their wings against a plectrum on the other wing. Some exceptional katydids and most crickets can produce pure-tone calls, which are more musical than broadband calls, for attracting females at night over a long distance.

The history of cricket culture in China can be traced back to 2000 years ago. During the Tang Dynasty (618–906 CE), people started to keep crickets in cages and enjoyed their songs while in captivity [2]. According to an ancient Chinese book *Kaiyuan Tianbao YiShi* about the Tang Dynasty, the court ladies would obtain crickets

Rhythms of Insect Evolution: Evidence from the Jurassic and Cretaceous in Northern China, First Edition.
Edited by Dong Ren, Chungkun Shih, Taiping Gao, Yongjie Wang, and Yunzhi Yao.

Figure 9.1 Leaf feeding by an "Oblong-Winged" katydid (*Amblycorypha oblongifolia*), Source: Photo by Peter Jian Ming Shih.

Figure 9.2 Mating katydids. Source: Photo by Jason Shih.

Figure 9.3 Molting of a katydid. Source: Photo by Jason Shih.

in the early autumn and kept them in small golden cages. They placed the cricket cages near their pillows so that they could hear the cricket songs during the long night. Most of the court ladies were concubines to the Tang Emperor, who typically had many concubines confined inside the luxurious palace. A similarity can be drawn between the concubines and their captive crickets in their golden cages. Rather than enjoying the sweet love songs of the crickets, many concubines heard music reflecting their own sadness and loneliness in the cricket's chirp [2]. Later on, keeping crickets as pets was widespread and became popular for common people in China.

9.2 Progress in the Studies of Fossil Orthoptera

The first attempt to summarize the fossil orthopterans and their relatives was performed by Handlirsch [3]. Handlirsch [4] interpreted them as a superorder Orthoptera which included the Paleozoic orders of Protorthoptera, Saltatoria, Demaptera, Phasmida, Diploglossata and Thysanoptera. The order Protorthoptera were divided into Oedischoidea, Caloneuroidea, Geraroidea, Cacurgoidea and Cnemidolestodea.

Zeuner [5] and Carpenter [6] followed the systems proposed by Handlirsch. Sharov [7] divided Protorthoptera into three orders: Orthoptera, Protoblattodea and Paraplecoptera. Then the orders of Caloneuroidea and Glosselytrodea were removed from the superorder Orthopteroidea. Sharov [8] defined the Orthoptera to include two suborders of Ensifera and Caelifera. Ensifera comprise three superfamilies: Oedischiidea, Tettigonioidea and Gryllidea and Caelifera, four superfamilies: Locustopseidea, Tridactylidea, Pneumoridea and Acridoidea. Gorochov [9] contributed a comprehensive summary on ensiferans including structure, classification, ecology, phylogeny and evolution. The classification of Ensifera was recognized as four infraorders: Oedischiidea, Elcanidea, Tettigoniidea and Gryllidea. Oedischiidea have two superfamilies: Oedischioidea and Xenopteroidea; Elcanidea with two

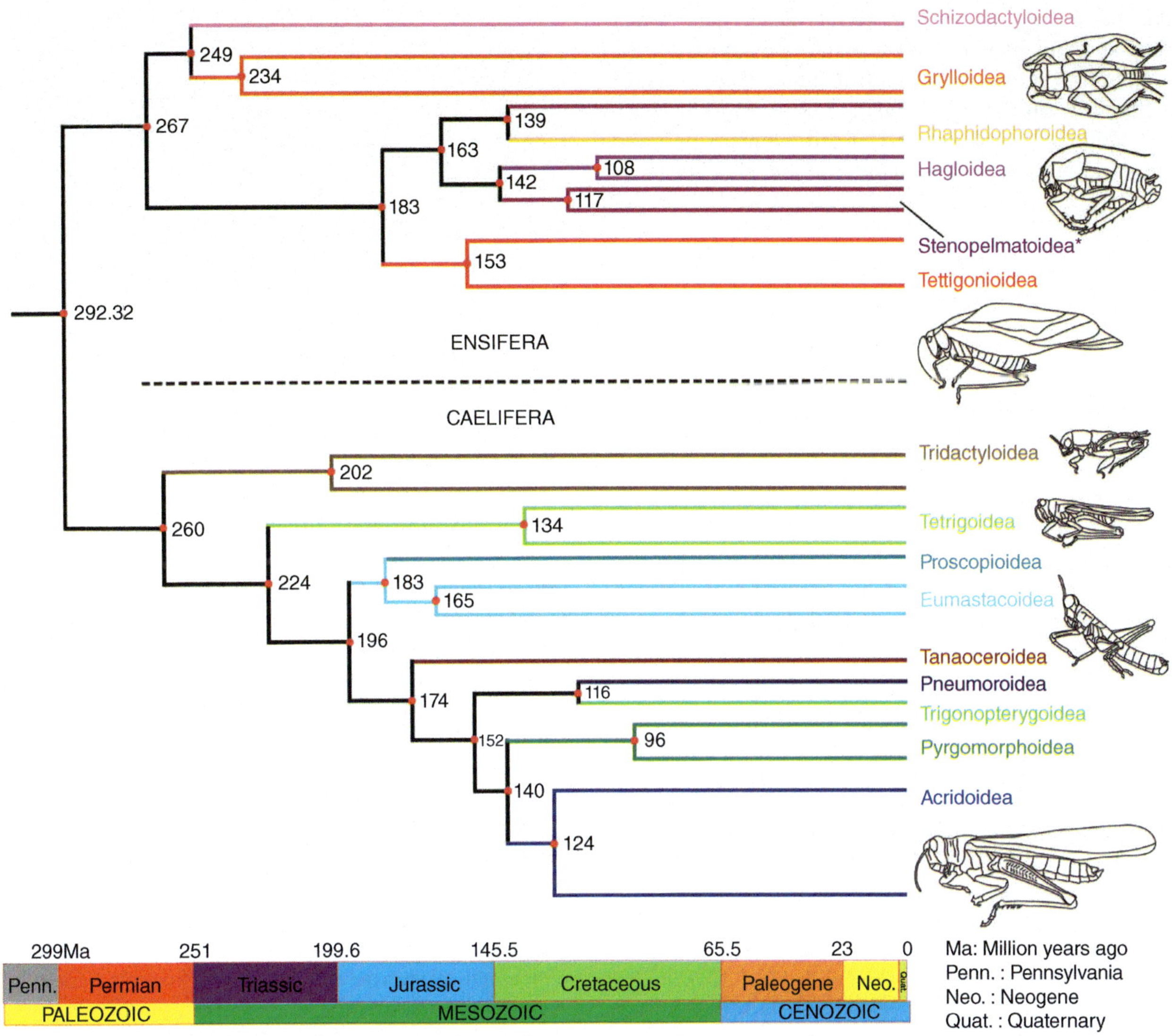

Figure 9.4 A possible summary phylogeny of Orthoptera with divergence time-estimation. Source: Modified from [1].

superfamilies: Elcanoidea and Permoraphidioidea; and Tettigoniidea, three superfamilies: Hagloidea, Tettigonioidea and Stenopelmatoidea.

Despite a long history of studies and numerous findings of fossils from the Carboniferous to the Cenozoic, a comprehensive cladistic analysis of this group including the extinct and extant lineages has still not been carried out. Sharov and Gorochov made abundant contributions on the classification and phylogeny [7–13]. However, they did not follow the principles of phylogenetic systematics [14], instead, they employed traditional systematics based on a priori weighting of selected characters and without any outgroup comparisons. Then Béthoux and Nel [15], for the first time, made a cladistic analysis based on new venation nomenclature, including taxa ranging mainly from the Carboniferous to the Cretaceous and one modern taxon. In this study, some novel phylogenetic hypotheses concerning the relationships within "orthopteroid" lineage were proposed. The group Hagloidea sensu Gorochov [9], which include extinct and extant representatives, was deemed to be paraphyletic.

In China, the study of orthopteran fossils started over 50 years ago. In 1965, Lin Qibin reported the first fossil Orthoptera species, *Sinohagla anthoides*, from the Lower Jurassic of Inner Mongolia, China [16]. From then on, more paleoentomologists have been involved in studies with new fossil discoveries, ranging from the Early Jurassic to the Paleocene of China. So far, there have been more than 63 species, 40 genera belonging to five families described in the Jurassic to the Cretaceous China. During this period, significant contributions have been made by Youchong Hong, Qibin Lin, Junfeng Zhang,

Haichun Zhang, Dong Ren, Wenli Wang, Wuli Wang, Yan Fang, Jun-Jie Gu, and He Wang. We appreciate the cooperation and contribution from our collaborators, O. Béthoux, F. Montealegre-Zapata, J. Prokop, M.S. Engel, A.V. Gorochov; and our CNU Team members past and present, Xiangmin Meng, Lianmei Li and He Tian.

9.3 Representative Fossils of Orthoptera from Northern China

Suborder "Ensifera" Chopard, 1920

Superfamily Hagloidea Handlirsch, 1906

Family Haglidae Handlirsch, 1906

Haglidae, a major constituent of the Hagloidea, have existed from the Triassic to the Cretaceous but flourished from the Triassic to the Middle Jurassic [10, 17]. This group is characterized by their strongly curved cross-veins between CuPb and CuPaβ, the presence of "handle" vein and "fan" and a broad area between RP (posterior branch of radius) and MA. The oldest fossil representatives are *Notopamphagopsis bolivari* Cabrera, 1928 and *Prohagla superba* Riek 1954 documented from the Middle Triassic in Argentina and Australia. Species of this extinct family are plentiful and diverse, classified into eight subfamilies: Haglopterinae Gorochov 1986, Haglinae Handlirsch 1906, Isfaropterinae Martynov 1937, Triassaginae Gorochov & Maehr 2008, Voliopinae Gorochov 1986, Bachariinae Gorochov 1988, Angarohaglinae Gorochov 1995 [9, 18] and Cyrtophyllitinae Zeuner 1935 [19]. However, the family Haglidae seem to be a paraphyletic group in a cladistic analysis by Bèthoux and Nel [15]. To date, there are 69 genera containing 92 species described from the Mesozoic of Africa, Central Asia, China, Europe, Oceania, Russia, and South America [20].

Genera included from the Jurassic and Cretaceous of Northern China: *Liassophylum* Zeuner, 1935, *Archaboilus* Martynov, 1937, *Isfaroptera* Martynov, 1937, *Archaeohagla* Lin, 1965, *Sinohagla* Lin, 1965, *Yenshania* Hong, 1982, *Laiyangohagla* Wang & Liu, 1996 and *Vitimoilus* Gorochov, 1996.

Liassophylum Zeuner, 1935

Liassophylum Zeuner, 1935, *Stylops* 4, 106 [19] (original designation).

Type species: *Liassophyllum abbreviatum* Zeuner, 1935.

Vein R is simple for a long distance, strongly arched toward anterior margin basal of its divergence; base of RP is strongly curved; MA is very close to base of RP, nearly touching it; area between R and MA is distinctly broad.

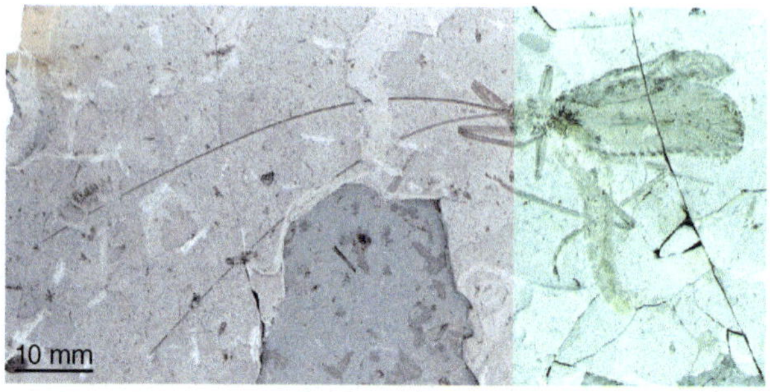

Figure 9.5 *Liassophylum caii* Gu, Qiao & Ren, 2012 (Paratype, CNU-ORT-NN-2009013). Source: Donated by Hongtao Cai.

Distribution and age: Inner Mongolia; Middle Jurassic.

Only one species included from the Jurassic of Northern China (see Table 9.1).

Liassophylum caii Gu, Qiao & Ren, 2012 (Figure 9.5)

Liassophylum caii Gu, Qiao & Ren, 2012: *Alcheringa*, 36 (1), 27–34.

Locality and horizon: Daohugou, Ningcheng, Inner Mongolia, China; Middle Jurassic, Jiulongshan Formation.

The specific epithet is dedicated to Hongtao Cai, Director of Hongtao Palaeontological Museum, for his donation of fossil specimens. Large sized with the body length around 60 mm. Antennae are located between the eyes, and much longer than the body length; the dorsal margins of the sockets are not below the ventral margins of the compound eyes; scape is cylindrical with a pedicel shorter than the scape; fore-tibia and hind femora have two rows of ventral spines; cerci are highly elongate and slender. Wings are short, similar to abdomen in length; male wing apex not distinctly pointed; R is simple for a long distance, strongly arched toward anterior margin distal to the redirection of ScP; base of RP nearly forming a right-angle; MA is undulating and nearly touching base of RP [21].

Archaboilus Martynov, 1937

Archaboilus Martynov, 1937, *Trud. Inst. Paleont. Acad. Sci. URSS.* 7 (1), 51 [22] (original designation).

Type species: *Archaboilus kisylkiensis* Martynov, 1937.

Vein ScA is long, the branches of ScP with a secondary vein among them, formed by two rows of cells; MP undulating, its base is strongly curved toward posterior wing margin; a vein separating the area between MA and RP into two parts.

Distribution and age: Inner Mongolia; Middle Jurassic.

Only one species included from the Jurassic of Northern China (see Table 9.1).

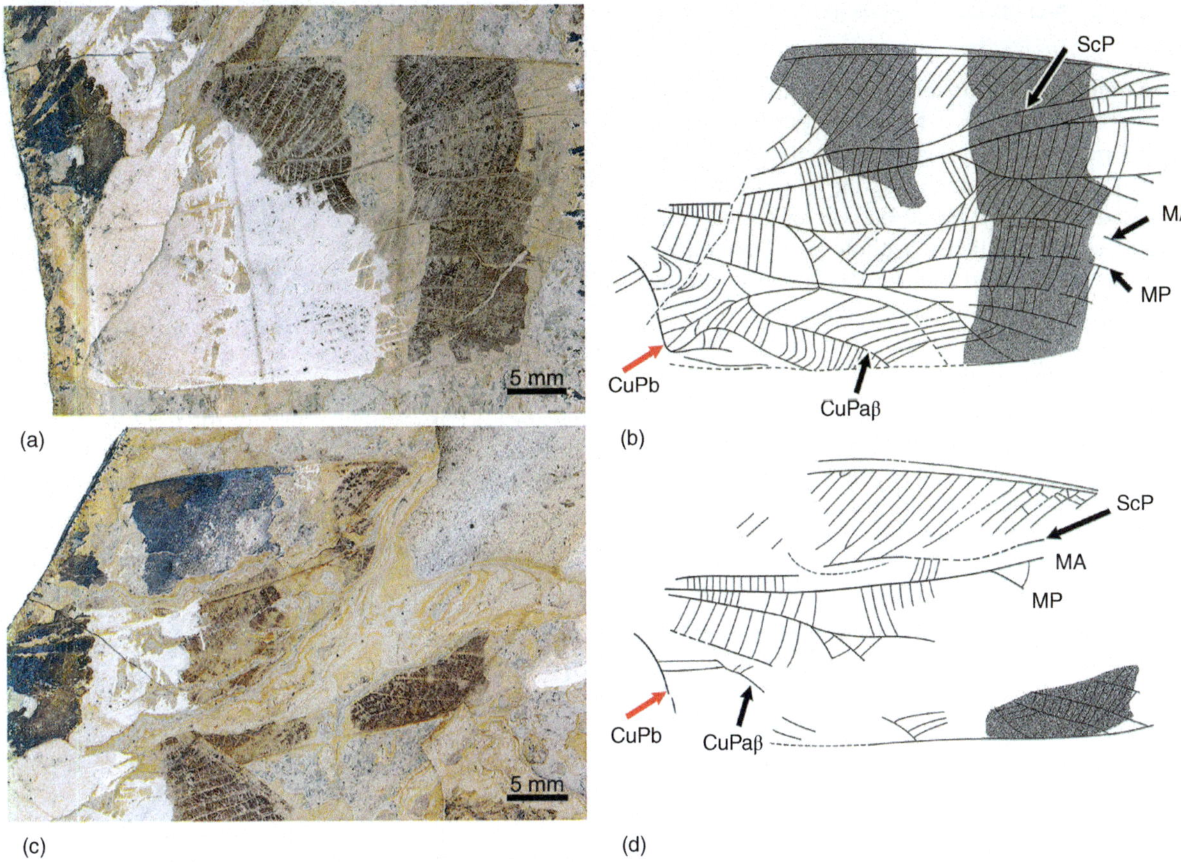

Figure 9.6 *Archaboilus musicus* Gu, Engel & Ren 2012, (a–d) photographs and line drawings of right and left wing of the holotype. Red arrows show the location of the stridulatory file (CuPb). Source: Modified from [23].

Archaboilus musicus **Gu, Engel & Ren, 2012 (Figure 9.6)**

Archaboilus musicus Gu, Engel & Ren, 2012: *P. Natl. Acad. Sci. USA*. 109 (10), 3868.

Locality and horizon: Daohugou, Ningcheng, Inner Mongolia, China; Middle Jurassic, Jiulongshan Formation.

Forewing is large (>7 cm in length), with broad-striped coloration; RP is ramifying earlier than RA (anterior branch of radius); MA is not distinctly undulated; CuPaβ has broken section at level of the "handle" and directed toward CuA and CuPaα. The stridulatory file is located on CuPb of both forewings, the LW (left forewing) file of the fossil shows 107 teeth, probably missing only a few in the median area; the RW (right wing) file exhibits 96 well-preserved teeth, missing approximately 11 toward the basal end (Figures 9.6 and 9.7). Teeth are asymmetrical, with lateral flaps projecting anteriorly. Males could produce low-pitched musical calls [23].

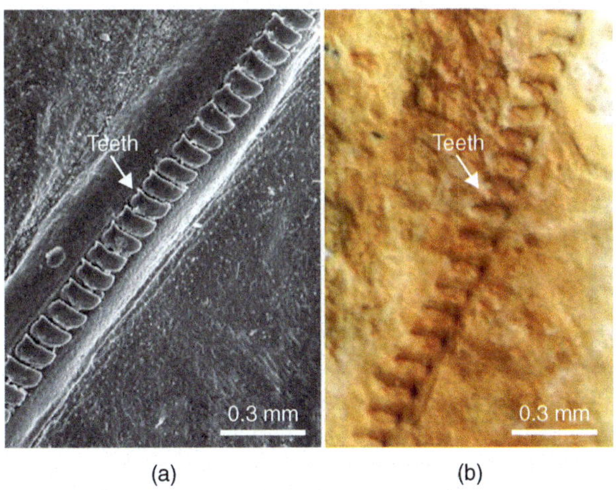

Figure 9.7 File morphology of *Cyphoderris buckelli* and *A. musicus*. (a) SEM picture of the left file of *Cyphoderris buckelli*. (b) Left file morphology of *A. musicus* (dorsal view). Source: Modified from [23].

Musical Calls by *Archaboilus musicus* in the Middle Jurassic

It is rare to find fine structure of insects' acoustic apparatus on compression fossils. A fossil species of Haglidae, *Archaboilus musicus* Gu, Engel and Ren, 2012, with exceptionally preserved stridulate file and teeth, has been reported from the Middle Jurassic of China [23]. It presents a set of detailed morphological data which has never been described before. Firstly, most extant katydids exhibit conspicuous wing asymmetry: the LW is acoustically damped, but has a functional file; the RW exhibits sound-radiating cells, a plectrum, and an atrophied or absent file. In contrast, *A. musicus* exhibits highly symmetric wings with identical files which are similar to the extant species of *Cyphoderris* spp. which is in the Prophalangopsidae Kirby, 1906 of Hagloidea. Secondly, the files of both forewings bear asymmetrical teeth, with lateral flaps projecting anteriorly, which are more similar to the files of *Cyphoderris* spp. than to the typical katydid's files. The tooth distribution pattern of *A. musicus* indicates that its tooth spacing increases linearly toward the basal ends of both files. Such tooth spacing is just as well-organized as that of extant species producing pure-tone calls [23] (Figure 9.7). Together, this tooth distribution and wing symmetry suggest that the stridulatory files of *A. musicus* have already been adapted to producing pure-tone sounds, during the closing of the wings, in the same way as extant species [23].

Isfaroptera Martynov, 1937

Isfaroptera Martynov, 1937, *Trau. Inst. Paleont. Acad. Sci. URSS*, 7 (1), 50 [22] (original designation).

Type species: *Isfaroptera grylliformis* Martynov, 1937.

Vein RP is simple, MA is simple and without fusion with RP; MP fuses with CuA, then separates from CuA + CuPaα.

Distribution and age: Liaoning; Middle Jurassic.

Only one species included from the Jurassic of Northern China (see Table 9.1).

Archaeohagla Lin, 1965

Archaeohagla Lin, 1965, *Acta Palaeontol. Sin.* 13, 365 [16] (original designation).

Type species: *Archaeohagla sinensis* Lin, 1965.

Stem of R is undulate, area between RP and MA with two kinds of cross-veins.

Distribution and age: Inner Mongolia; Early Jurassic.

Only one species included from the Jurassic of Northern China (see Table 9.1).

Sinohagla Lin, 1965

Sinohagla Lin, 1965, *Acta Palaeontol. Sin.* 13, 364 [16] (original designation).

Type species: *Sinohagla anthoides* Lin, 1965

Vein ScP is straight, RA has five branches reaching wing apex, the first branch of CuA + CuPaα is directed to anterior wing margin at its beginning.

Distribution and age: Inner Mongolia; Early Jurassic.

Only one species included from the Jurassic of Northern China (see Table 9.1).

Laiyangohagla Wang & Liu, 1996

Laiyangohagla Wang & Liu, 1996, *Mem. Beijing Nat. Hist. Mus.* 55, 70 [24] (original designation).

Type species: *Laiyangohagla beipoziensis* Wang & Liu, 1996.

Vein ScP is oblique, R has many branches reaching wing apex, RP is ramifying distally, the inner tympanum is smaller than the outer one.

Distribution and age: Shandong; Early Cretaceous.

Only one species included from the Cretaceous of Northern China (see Table 9.1).

Vitimoilus Gorochov, 1996

Vitimoilus Gorochov, 1996, *Paleontol. Zhur.* 3, 73–82 [25] (original designation).

Type species: *Vitimoilus captiosus* Gorochov, 1996.

Forewings are broad oval and large sized; R is forking distally; M forks into MA and MP distally, at the level of the mid-length of wing; CuA fuses with CuPaα distal of the middle of the wing length; cross-veins in basal part of CuPb-CuPaβ area are not strongly curved.

Distribution and age: Hebei and Liaoning; Early Cretaceous.

Only one species included from the Cretaceous of Northern China (see Table 9.1).

Vitimoilus ovatus Gu, Tian, Yin, Shi & Ren 2017 (Figure 9.8)

Vitimoilus ovatus Gu, Tian, Yin, Shi & Ren 2017: *Cretac. Res.*74, 151–154.

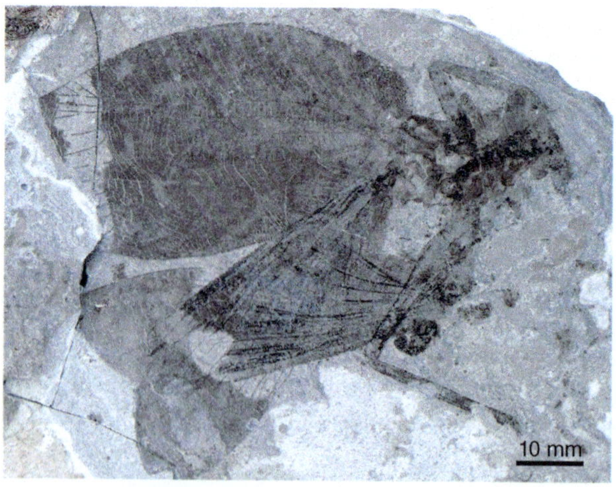

Figure 9.8 *Vitimoilus ovatus* Gu, Tian, Yin, Shi & Ren 2017 (Holotype, CNU-ORT-HF-2010004p).

Locality and horizon: Fengning, Hebei, China; Lower Cretaceous, Dabeigou Formation. Jianchang, Liaoning, China; Lower Cretaceous, Yixian Formation.

Large sized, wing length from 55.0–73.7 mm; forewing broad and oval shaped; R is forking relatively basal and at the level of the divergence of M; RP obviously curved toward anterior wing margin basally; free M extremely short; basal part of MP strongly directed toward posterior margin, then bent to wing apex; CuA + CuPaα has five to six branches, area between branches is covered by irregular reticulated cross-veins [26].

Family Prophalangopsidae Kirby, 1906

Prophalangopsidae, a rather peculiar family of Ensifera, were prosperous from the Middle Jurassic to the Early Cretaceous. They are characterized by ScA reaching costal wing margin at level or distal of origin of CuA from M + CuA, CuA + CuPaα with the third branch anteriorly ramified and branches of ScA aligned with branches of the area between ScA and ScP. It is the only family of ensiferans which can be traced from the Early Jurassic to present. This lineage is now represented by the relict subfamily Prophalangopsinae and Cyphoderrinae [1, 9, 12, 13, 27–29]. The extant prophalangopsids are disjunctively distributed and recorded from Northwestern America, Eastern Siberia, Southwestern China and Northern India. They usually inhabit high-altitude mountains and prefer cool and humid environment. The extinct representatives are classified into five extinct subfamilies, i.e. Protaboilinae, Aboilinae, Chifengiinae, Termitidiinae and Tettohaglinae. The earliest fossil record was collected from the Lower Jurassic Upkurgan coal deposit, Kyrgyzstan, and assigned as *Protaboilus praedictus* Gorochov, 1988 in Protaboilinae.

Genera included from the Jurassic and Cretaceous of Northern China: *Pycnophlebia* Deichmüller, 1886, *Aboilus* Martynov, 1925, *Pseudohagla* Sharov, 1962, *Parahagla* Sharov, 1968, *Ashanga* Zherikhin, 1985, *Bacharaboilus* Gorochov, 1988, *Shangxiania* Zhang, 1993, *Mesohagla* Zhang, 1996, *Allaboilus* Ren & Meng, 2006, *Angustaboilus* Li, Ren & Meng, 2007, *Circulaboilus* Li, Ren & Wang, 2007, *Novaboilus* Li, Ren & Meng, 2007, *Sigmaboilus* Fang, Zhang, Wang & Zhang, 2007, *Ashangopsis* Lin, Huang & Nel, 2008 and *Scalpellaboilus* Gu, Qiao & Ren, 2010.

Pycnophlebia Deichmüller, 1886

Pycnophlebia Deichmüller, 1886, *Mitt. K. miner.-geol. Prähist. Mus. Dresden*. 7, 1–88 [30] (original designation).

Type species: *Pycnophlebia speciosa* (Germar, 1839).

Body large-sized, stoutly built, with long antennae and with open auditory tympana in the fore tibia. Pre-anal portion of forewing is sub-triangular, RP has about 10 branches which are separated by intercalated zig-zag veins.

Distribution and age: Liaoning; Middle Jurassic.

Only one species included from the Jurassic of Northern China (see Table 9.1).

Aboilus Martynov, 1925

Aboilus Martynov, 1925, *Bull. Acad. Sci. USSR* 19, 581 [31] (original designation).

Type species: *Aboilus fasciatus* Martynov 1925.

Male forewing is oblong or broadly oval; ScA is slightly arched, ending at anterior margin before the middle length of tegmen; RP is diverging from R slightly before the middle length of tegmen; both RA and RP give more than four branches; origin of MP is much closer to the fusion point of CuA + CuPaα than to the origin of RP.

Distribution and age: Inner Mongolia; Middle Jurassic.

Seven species included from the Jurassic of Northern China (see Table 9.1).

Pseudohagla Sharov, 1962

Pseudohagla Sharov, 1962, *In Rodendorf. Chlenistonogie. Trakheinyeikhelitserovye. Osnovypaleontologii [Fundamentals of Paleontology]* 9, 145 [32] (original designation).

Type species: *Pseudohagla pospelovi* Sharov, 1968.

Vein ScA is cutting branches of ScP and reaching anterior margin before the middle of the wing, RA, RP and CuA + CuPaα have numerous branches [33].

Distribution and age: Inner Mongolia; Middle Jurassic.

Only one species included from the Jurassic of Northern China (see Table 9.1).

Parahagla Sharov, 1968

Parahagla Sharov, 1968, *Trudy Paleontologicheskogo Instituta*. 118, 178 [8] (original designation).

Hebeihagla Hong, 1982, *Scientia Sin. series B*, 25 (10), 1118–1129 [34]; *Habrohagla* Ren, Lu, Guo & Ji, 1995, *Seismic Publishing House, Beijing* [35]; *Grammohagla* Meng & Ren, 2006, *Acta Zootaxonomica. Sin.* 31 (2), 282–288 [36]; *Trachohagla* Meng, Ren & Li, 2006, *Acta Zootaxonomica. Sin.* 31 (4), 752–757 [37]. Syn. by Gu, Qiao & Ren, 2010, *Jour. Orth. Res.* 19 (1), 45–56 [38].

Type species: *Parahagla sibirica* Sharov, 1968.

Forewing: ScA is reduced and parallel with branches of ScP, area between MA and MP is distinctly narrower in basal part and about two times as broad in middle part, CuPaβ broken by a straight "handle" in male forewing. Hind wing: ScP is sigmoidal, posterior branch of CuA + CuPaα is not reaching CuPaβ; hind tibia has four spurs; ovipositor is well-developed and apparently longer than pronotum.

Distribution and age: Liaoning, Hebei, Inner Mongolia; Early Cretaceous.

Only one species included from the Cretaceous of Northern China (see Table 9.1).

Ashanga Zherikhin, 1985

Ashanga Zherikhin, 1985, *Trudy Paleontologicheskogo Instituta.* 211, 175 [39] (original designation).

Type species: *Ashanga clara* Zherikhin, 1985.

Vein ScA is oblique and straight, fusing with the branch of ScP, CuA + CuPaα is running for a moderate distance before ramification.

Distribution and age: Gansu and Liaoning; Early Cretaceous.

Three species included from the Cretaceous of Northern China (see Table 9.1).

Bacharaboilus Gorochov, 1988

Bacharaboilus Gorochov, 1988, *Paleont. Zhur.* 2, 65 [11] (original designation).

Type species: *Bacharaboilus mongolicus* Gorochov, 1988.

Posterior wing margin is posteriorly bowed in its basal half. ScA is slightly curved, reaching anterior margin before the middle length of forewing; area between base of RP and MA is markedly broad; RP is basally curved; CuA + CuPaα has numerous branches [40].

Distribution and age: Inner Mongolia; Middle Jurassic.

Two species included from the Jurassic of Northern China (see Table 9.1).

Bacharaboilus lii Gu, Qiao & Ren, 2011 (Figure 9.9)

Bacharaboilus lii Gu, Qiao & Ren, 2011: *Zootaxa*, 2909, 65.

Locality and horizon: Daohugou, Ningcheng, Inner Mongolia, China; Middle Jurassic, Jiulongshan Formation.

The specific epithet is dedicated to Mr. Li Yuezhuo, Director of Sanya Palaeontological Museum, for providing the holotype for this study. Head is typically hypognathous; antenna is about 80 mm long (much longer than the body) and inserted between the eyes; scape cylindrical; mandibles are stout and thick with shortened apical teeth and well-defined molar dentes, which reflects an herbivorous feeding habit. RP is branching basal of the first branch of RA; the section of CuPaβ between "handle" and posterior wing margin is undulated; AA1 is fusing with CuPb for a moderate distance [40].

Shangxiania Zhang, 1993

Shangxiania Zhang, 1993, *Palaeoworld*, 2, 49 [41] (original designation).

Figure 9.9 *Bacharaboilus lii* Gu, Qiao & Ren, 2011 (Holotype, CNU-ORT-NN2011001p).

Type species: *Shangxiania fengjiashanensis* Zhang, 1993.

Female forewing is narrow, costal and subcostal area is short, ScA is reaching anterior margin basal of the middle of the wing.

Distribution and age: Shaanxi; Early Cretaceous.

Only one species included from the Cretaceous of Northern China (see Table 9.1).

Mesohagla Zhang, 1996

Mesohagla Zhang, 1996, *Entomotaxonomia*, 18 (4), 250 [42] (original designation).

Type species: *Mesohagla xinjiangensis* Zhang, 1996.

Female forewing is three times as long as wide, Costal margin nearly straight and strongly arched near apex, ScP reaches anterior margin near apex, RA has two branches.

Distribution and age: Xinjiang; Early Jurassic.

Only one species included from the Jurassic of Northern China (see Table 9.1).

Allaboilus Ren & Meng, 2006

Allaboilus Ren & Meng, 2006, *Acta Zootaxonomica. Sin.* 31 (3), 513 [43] (original designation).

Flexaboilus Li, Ren & Meng, 2007, *Acta Zootaxonomica. Sin.* 32 (1), 174–181 [44]; *Furcaboilus* Li, Ren &

Meng, 2007, *Acta Zootaxonomica. Sin.* 32 (1), 174–181 [44]. Syn. by Gu, Qiao & Ren, 2010, *Jour. Orth. Res.* 19 (1), 45–56 [38].

Type species: *Allaboilus dicrus* Ren & Meng, 2006.

Forewing large-sized, ScA is long and reaching anterior margin close to midlength of forewing, ScP is slightly sigmoidal, RA, RP and CuA + CuPaα have numerous branches, ovipositor is long and stout.

Distribution and age: Inner Mongolia; Middle Jurassic.

Four species included from the Jurassic and Cretaceous of Northern China (see Table 9.1).

Angustaboilus Li, Ren & Meng, 2007

Angustaboilus Li, Ren & Meng, 2007, *Acta Zootaxonomica. Sin.* 32 (1), 175 [44] (original designation).

Type species: *Angustaboilus fangianus* Li, Ren & Meng, 2007.

Vein ScA is short and arched, costal area is sigmoidal shape, ScP is straight and reaching anterior margin at ¾ of its length.

Distribution and age: Inner Mongolia; Middle Jurassic.

Only one species included from the Jurassic of Northern China (see Table 9.1).

Circulaboilus Li, Ren & Wang, 2007

Circulaboilus Li, Ren & Wang, 2007, *Acta Zootaxonomica. Sin.* 32 (2), 412 [33] (original designation).

Type species: *Circulaboilus aureus* Li, Ren & Wang, 2007.

Forewing is oval, apex pointed, ScA is long, distinctly beyond the mid-length of the wing, branches of ScP are spaced out, R forks distally, area between divergence of R and M is markedly widening and broad, the "handle" is long and straight in male.

Distribution and age: Inner Mongolia; Middle Jurassic.

Two species included from the Jurassic of Northern China (see Table 9.1).

Novaboilus Li, Ren & Meng, 2007

Novaboilus Li, Ren & Meng, 2007, *Acta Zootaxonomica. Sin.* 32 (1), 177 [44] (original designation).

Type species: *Novaboilus multifurcatus* Li, Ren & Meng, 2007.

Vein ScA is slightly sigmoidal, area between anterior margin and ScP is broad, branches of ScP has secondary veinlets, distal branches of them are more oblique.

Distribution and age: Inner Mongolia; Middle Jurassic.

Only one species included from the Jurassic of Northern China (see Table 9.1).

Sigmaboilus Fang, Zhang, Wang & Zhang, 2007

Sigmaboilus Fang, Zhang, Wang & Zhang, 2007, *Zootaxa*, 1637, 56 [45] (original designation).

Type species: *Sigmaboilus gorochovi* Fang, Zhang, Wang & Zhang, 2007.

The specific epithet is in honor of A.V. Gorochov, an eminent paleoentomologist. Forewing: ScA is long, mildly undulating, reaching anterior wing margin beyond midlength; ScP long with numerous branches ending in stem ScA; cross-veins at the base of area between M + CuA and CuP sigmoidal. Hind wing: ScP and R with basal common stem, basal free part of R slightly curved; R diverging nearer wing base than that in forewing.

Distribution and age: Inner Mongolia; Middle Jurassic.

Five species included from the Jurassic of Northern China (see Table 9.1).

Ashangopsis Lin, Huang & Nel, 2008

Ashangopsis Lin, Huang & Nel, 2008, *C. R. Palevol.* 7 (4), 206 [46] (original designation).

Type species: *Ashangopsis daohugouensis* Lin, Huang & Nel, 2008.

Pronotum is narrow and long, forewing is broad and shortened. ScA is straight and nearly parallel with ScP, main longitudinal veins are stout and strong [38].

Distribution and age: Inner Mongolia; Middle Jurassic.

Only one species included from the Jurassic of Northern China (see Table 9.1).

Scalpellaboilus Gu, Qiao & Ren, 2010

Scalpellaboilus Gu, Qiao & Ren, 2010, *Jour. Orth. Res.* 19 (1), 42 [38] (original designation).

Type species: *Scalpellaboilus angustus* Gu, Qiao & Ren, 2010.

Forewing is long and narrow, the ratio of length to width is >5; C is developed, not fused with anterior margin and ScA, R is undulated, M + CuA curved.

Distribution and age: Inner Mongolia; Middle Jurassic.

Only one species included from the Jurassic of Northern China (see Table 9.1).

Superfamily Elcanoidea Handlirsch, 1906

Family Elcanidae Handlirsch, 1906

Elcanidae, a specialized family of Orthoptera, existed from the Triassic to the Late Cretaceous. These insects have long, filiform antennae as others in Ensifera, but particular wing venation resembling neither tettigoniids nor gryllids. The phylogenetic position of the Elcanidae

(Elcanoidea) is still unsettled. This clade is considered as a sister group of Caelifera by Sharov [8], and this relationship is supported by a cladistic analysis [15]. In another opinion, Elcanoidea and Permoraphidoidea are hypothesized to be the sister group to all other orthopteran superfamilies [17]. As a revision by Gorochov *et al.* [47], the Elcanidae comprise two subfamilies: Elcaninae and Archelcaninae. Up to date, with some new taxa discovered from amber, this family include 28 species in 12 genera distributed from the Jurassic and Cretaceous localities in Brazil, China, Eurasia and Myanmar [8, 47–49].

Only one genus included from the Cretaceous of Northern China: *Panorpidium* Westwood, 1854.

Panorpidium Westwood, 1854

Panorpidium Westwood, 1854, *Q. J. geol. Soc. Lond.* 10, 394 [50] (original designation).

Type species: *Panorpidium tessellatum* Westwood, 1854.

Forewing is usually medium or large in size with relatively few branches of ScA; presence of three or four longitudinal branches between basal part of RP and CuA + CuPaα; fusion of distal parts of CuPaβ, CuPb and 1A with each other.

Distribution and age: Liaoning; Early Cretaceous.

Only one species included from the Cretaceous of Northern China (see Table 9.1).

Panorpidium yixianensis Fang, Wang, Zhang, Wang, Jarzembowski, Zheng, Zhang, Li & Liu, 2015 (Figure 9.10)

Panorpidium yixianensis Fang, Wang, Zhang, Wang, Jarzembowski, Zheng, Zhang, Li & Liu, 2015: *Cretac. Res.* 52, 323–328.

Figure 9.10 *Panorpidium yixianensis* Fang, Wang, Zhang, Wang, Jarzembowski, Zheng, Zhang, Li & Liu, 2015 (Holotype, NIGP 159068). Source: Photo provided by Dr. Haichun Zhang.

Locality and horizon: Yangshuwanzi, Ningcheng, Chifeng, Inner Mongolia, China; Lower Cretaceous, Yixian Formation.

Forewings are very large with relatively few branches of ScA; there are four longitudinal branches between basal part of RP and CuA + CuPaα; area between MP and anal edge of forewing are narrow and relatively short, and divided by short transverse or partly oblique cross-veins [49].

Suborder "Caelifera" Ander, 1936

Superfamily Locustopsoidea Handlirsch, 1906

Family Locustopsidae Handlirsch, 1906

Locustopsidae, an extinct family of Caelifera, existed from the Triassic to the Late Cretaceous. The Locustopsidae are characterized by the base of CuPaβ is displaced from the basal part of the tegmen and CuA + CuPaα has relatively few (one to three) branches. This family comprise two subfamilies, Locustopsinae and Araripelocustinae [47]. They have been recorded from Germany, France, England, Russia, Central Asia, China, Egypt, North America, Brazil and Australia. Among them, Locustopsinae are known from the Late Triassic to the Late Cretaceous, while the oldest representatives of *Locustopsis* spp. have been documented from the Late Triassic of UK. Araripelocustinae have been only discovered from the Early Cretaceous. To date, 17 genera and 70 species of Locustopsidae have been described [20].

Only one genus included from the Jurassic of Northern China: *Locustopsis* Handlirsch, 1906.

Locustopsis Handlirsch, 1906

Locustopsis Handlirsch, 1906, *Ein Handbuch für Paläontologen und Zoologen.*1 (3–4), 421 [3] (original designation).

Type species: *Locustopsis elegans* Handlirsch, 1906.

Small to medium in size, MA of forewing has three branches; CuA + CuPaα has two branches.

Distribution and age: Inner Mongolia; Middle Jurassic.

Only one species included from the Jurassic of Northern China (see Table 9.1).

Locustopsis rhytofemoralis Gu, Yue, Shi, Tian & Ren, 2016 (Figure 9.11)

Locustopsis rhytofemoralis Gu, Yue, Shi, Tian & Ren, 2016: *Zootaxa*, 4169 (2), 377–380.

Locality and horizon: Daohugou, Ningcheng, Inner Mongolia, China; Middle Jurassic, Jiulongshan Formation.

Male is small sized, with body length 15.0 mm long, measured from the head to the abdominal apex; Head is

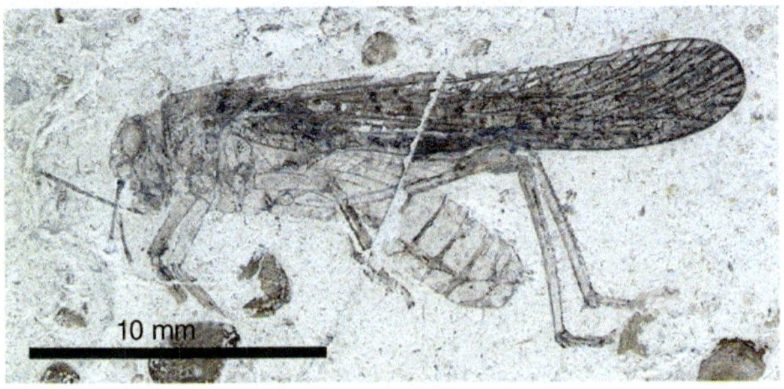

Figure 9.11 *Locustopsis rhytofemoralis* Gu, Yue, Shi, Tian & Ren, 2016 (Holotype, CNU-ORT-NN-2011010p).

hypognathous; antennal sockets are strongly rimmed; scape are slightly longer than pedicel, both of them are distinctly wider than flagella; mandibles are stout, subocular furrow is S-shaped; pronotum is saddle-shaped, with large hind lobe of disc which is extending posteriorly and covering the basal part of the tegmina; lateral lobe is well-developed; ScA is long, reaching anterior margin close to mid-length of wing, distal of the divergence of M; MA forks into MA1 and MA2 basally, at the level of the origin of CuPaβ [51].

Table 9.1 A list of fossil Orthoptera from the Jurassic and Cretaceous of China.

Family	Species	Locality	Horizon/age	Citation
	Suborder "Ensifera" Chopard, 1920			
Elcanidae	*Panorpidium yixianensis* Fang, Wang, Zhang, Wang, Jarzembowski, Zheng, Zhang, Li & Liu, 2015	Ningcheng, Inner Mongolia	Yixian Fm., K_1	Fang *et al.* [49]
Haglidae	*Archaboilus musicus* Gu, Engel & Ren, 2012	Ningcheng, Inner Mongolia	Jiulongshan Fm., J_2	Gu *et al.* [23]
	Archaeohagla sinensis Lin, 1965	Ordos, Inner Mongolia	Fm. Unknown, J_1	Lin [16]
	[a]*Isfaroptera? Yujiagouensis* Hong, 1983	Beipiao, Liaoning	Haifanggou Fm., J_2	Hong [52]
	Laiyangohagla beipoziensis Wang & Liu, 1996	Laiyang, Shandong	Laiyang Fm., K_1	Wang and Liu [24]
	Liassophylum caii Gu, Qiao & Ren, 2012	Ningcheng, Inner Mongolia	Jiulongshan Fm., J_2	Gu *et al.* [21]
	Sinohagla anthoides Lin, 1965	Ordos, Inner Mongolia	Fm. Unknown, J_1	Lin [16]
	Vitimoilus ovatus Gu, Tian, Yin, Shi & Ren, 2017	Fengning, Hebei	Dabeigou Fm., K_1	Gu *et al.* [26]
	[a]*Yenshania hebeiensis* Hong, 1982	Longhua, Hebei	[b]Yixian Fm., K_1	Hong [34]
Prophalangopsidae	*Aboilus lamina* (Lin, 1982)	Kangxian, Gansu	Fm. Unknown, J_1	Lin and Huang [53]
	Aboilus chinensis Fang, Zhang & Wang, 2009	Ningcheng, Inner Mongolia	Jiulongshan Fm. J_2	Fang *et al.* [54]
	Aboilus cornutus Li, Ren & Wang, 2007	Ningcheng, Inner Mongolia	Jiulongshan Fm. J_2	Li *et al.* [33]
	Aboilus jiyuanensis Lin & Huang, 2006	Jiyuan, Henan	Ma'ao Group, J_2	Lin and Huang [53]
	Aboilus perbellus Wang, Li, Zhang, Fang, Wang & Zhang, 2015	Ningcheng, Inner Mongolia	Jiulongshan Fm., J_2	Wang *et al.* [55]
	Aboilus stratosus Li, Ren & Wang, 2007	Ningcheng, Inner Mongolia	Jiulongshan Fm., J_2	Li *et al.* [33]
	Aboilus tuzigouensis Lin & Huang, 2006	Kelamayi, Xinjiang	Badaowan Fm., J_1	Lin and Huang [53]
	Allaboilus dicrus Ren & Meng, 2006	Ningcheng, Inner Mongolia	Jiulongshan Fm., J_2	Ren and Meng, [43]

(Continued)

Table 9.1 (Continued)

Family	Species	Locality	Horizon/Age	Citation
	Allaboilus gigantus Ren & Meng, 2006	Ningcheng, Inner Mongolia	Jiulongshan Fm., J_2	Ren and Meng [43]
	Allaboilus hani Gu, Qiao & Ren, 2010	Beipiao, Liaoning	Yixian Fm., K_1	Gu *et al.* [38]
	Allaboilus robustus Gu, Qiao & Ren, 2010	Ningcheng, Inner Mongolia	Jiulongshan Fm., J_2	Gu *et al.* [38]
	[a]*Alloma faciata* Hong, 1982	Jianchang, Liaoning	Yixian Fm., K_1	Hong [56]
	[a]*Alloma huanghuachunensis* Hong, 1982	Kazuo, Liaoning	Jiufotang Fm., K_1	Hong [34]
	Angustaboilus fangianus Li, Ren & Meng, 2007	Ningcheng, Inner Mongolia	Jiulongshan Fm., J_2	Li *et al.* [44]
	Ashanga borealis Fang, Zhang, Wang & Zheng, 2013	Linyuan, Liaoning	Yixian Fm., K_1	Fang *et al.* [57]
	Ashanga hongi Meng & Ren, 2006	Beipiao, Liaoning	Yixian Fm., K_1	Gu *et al.* [38]
	Ashanga jiuquanensis Wang & Zhang, 2017	Jiuquan, Gansu	Zhonggou Fm., K_1	Wang *et al.* [58]
	Ashangopsis daohugouensis Lin, Huang & Nel, 2008	Ningcheng, Inner Mongolia	Jiulongshan Fm., J_2	Lin *et al.* [46]
	Bacharaboilus lii Gu, Qiao & Ren, 2011	Ningcheng, Inner Mongolia	Jiulongshan Fm., J_2	Gu *et al.* [40]
	Bacharaboilus jurassicus Li, Ren & Wang, 2007	Beipiao, Liaoning	Yixian Fm., K_1	Li *et al.* [33]
	[a]*Brunneus haifanggouensis* Hong, 1983	Ningcheng, Inner Mongolia	Haifanggou Fm., J_2	Hong [52]
	[a]*Chifengia batuyingziensis* Wang, 1987	Beipiao, Liaoning	Haifanggou Fm., J_2	Wang 1987 [59]
	[a]*Chifengia mosaica* Hong, 1982	Chifeng, Liaoning	Yixian Fm., K_1	Hong [56]
	Circulaboilus aureus Li, Ren & Wang, 2007	Ningcheng, Inner Mongolia	Jiulongshan Fm., J_2	Li *et al.* [33]
	Circulaboilus priscus Gu, Qiao & Ren, 2010	Ningcheng, Inner Mongolia	Jiulongshan Fm., J_2	Gu *et al.* [38]
	Mesohagla xinjiangensis Zhang, 1996	Kelamayi, Xinjiang	Badaowan Fm. J_1	Zhang [42]
	[a]*Mesoprophalangopsis liaoxiensis* Hong, 1986	Beipiao, Liaoning	Haifanggou Fm., J_2	Hong [60]
	Novaboilus multifurcatus Li, Ren & Meng, 2007	Ningcheng, Inner Mongolia	Jiulongshan Fm., J_2	Li *et al.* [44]
	Parahagla sibirica Sharov, 1968	Beipiao, Liaoning; Luanping, Hebei; Chengde, Hebei; Jiuquan, Gansu	Yixian Fm., Chijinpu Fm., K_1	Gu *et al.* [38]
	[a]*Parahaglopsis posteria* Hong, 1992	Jiutai County, Jilin,	Yincheng Fm., K_1	Hong [61]
	Protaboilus amblus Ren & Meng, 2006	Ningcheng, Inner Mongolia	Jiulongshan Fm., J_2	Ren and Meng [43]
	Protaboilus rudis Ren & Meng, 2006	Ningcheng, Inner Mongolia	Jiulongshan Fm., J_2	Ren and Meng [43]
	Pseudohagla shihi Li, Ren & Wang, 2007	Ningcheng, Inner Mongolia	Jiulongshan Fm., J_2	Li *et al.* [33]

(Continued)

Table 9.1 (Continued)

Family	Species	Locality	Horizon/Age	Citation
	Pycnophlebia obesa Wang, 1987	Beipiao, Liaoning	Haifanggou Fm., J_2	Wang [59]
	Scalpellaboilus angustus Gu, Qiao & Ren, 2010	Ningcheng, Inner Mongolia	Jiulongshan Fm., J_2	Gu *et al.* [38]
	Shangxiania fengjiashanensis Zhang, 1993	Shang county, Shaanxi	Fengjiashan Fm. K_1	Zhang [41]
	[a]*Shanxius reticulates* Hong, 1984	Shanxi	Fm. Unknown, J_1	Hong [62]
	[a]*Shanxius meileyingziensis* Hong, 1988	Kazuo, Liaoning	Jiufuotang Fm. K_1	Hong [63]
	Sigmaboilus calophlebius Wang, Fang & Zhang, 2018	Ningcheng, Inner Mongolia	Jiulongshan Fm., J_2	Wang *et al.* [64]
	Sigmaboilus fuscus Gu, Zhao & Ren, 2009	Ningcheng, Inner Mongolia	Jiulongshan Fm., J_2	Gu *et al.* [65]
	Sigmaboilus gorochovi Fang, Zhang, Wang & Zhang, 2007	Ningcheng, Inner Mongolia	Jiulongshan Fm., J_2	Fang *et al.* [45]
	Sigmaboilus longus Fang, Zhang, Wang & Zhang, 2007	Ningcheng, Inner Mongolia	Jiulongshan Fm., J_2	Fang *et al.* [45]
	Sigmaboilus peregrinus Gu, Zhao & Ren, 2009	Ningcheng, Inner Mongolia	Jiulongshan Fm., J_2	Gu *et al.* [65]
	Sigmaboilus sinensis Fang, Zhang, Wang & Zhang, 2007	Ningcheng, Inner Mongolia	Jiulongshan Fm., J_2	Fang *et al.* [45]
	[a]*Sinoprophalangopsis reticulata* Hong, 1983	Beipiao, Liaoning	Haifanggou Fm., J_2	Hong [52]
	[a]*Sunoprophalangopsis clathrata* Hong, 1982	Luanping, Hebei	Jiulongshan Fm., J_2	Hong [34]
	[a]*Sunoprophalangopsis elegantis* Hong, 1982	Luanping, Hebei	Jiulongshan Fm., J_2	Hong [34]
	[a]*Sunoprophalangopsis scupta* Hong, 1982	Luanping, Hebei	Jiulongshan Fm., J_2	Hong [34]
	[a]*Zhemengia sinica* Hong, 1982	Tongliao, Inner Mongolia	Fm. Unknown, J_1	Hong [34]
		Suborder "Caelifera" Ander, 1936		
Eumastacridae	*Taphacris turgis* Lin, 1980	Zhuji, Zhejiang	Hengshan Fm. K_1	Lin [66]
Locustopsidae	*Locustopsis rhytofemoralis* Gu, Yue, Shi, Tian & Ren 2016	Ningcheng, Inner Mongolia	Jiulongshan Fm., J_2	Gu *et al.* [51]
	[a]*Mesolocustopsis sinica* Hong, 1990	Laiyang, Shandong	Laiyang Fm., K_1	Hong [67]
	Pseudoacrida costata Lin, 1982	Guyuan, Ningxia; Laiyang, Shandong	Liupanshan Group, K_1; Laiyang Fm., K_1	Lin [68]
Family *Incertae sedis*	[a]*Falsirameus ravus* Zhang, 1985	Laiyang, Shandong	Laiyang Fm., K_1	Zhang [69]
	[a]*Yuxiania jurassica* Hong, 1997	Yuxian, Hebei	Houcheng Fm. J_3	Hong [70]

a) The species is not presented in the main text because the original description, photo, and line-drawings are not precise and the holotype cannot be rechecked.

b) Horizon/Age revised from the original paper based on updated information and data.

References

1 Song, H.J., Amédégnato, C., Cigliano, M.M. *et al.* (2015). 300 million years of diversification: elucidating the patterns of orthopteran evolution based on comprehensive taxon and gene sampling. *Cladistics* **6**: 621–651. https://doi.org/10.1111/cla.12116.

2 Jin, X.B. (1994). Chinese cricket culture. *Cultural Entomology Digest* **3**: 21–29.

3 Handlirsch, A. (1906-1908). *Die fossilen Insekten und die Phylogenie der rezenten Formen: Ein Handbuch für Paläontologen und Zoologen*. W. Engelmann.

4 Handlirsch, A. (1925). Palaeontologie. In: *Handbuch der Entomologie Band 3* (ed. C. Schröder), 117–306. Jena: Fischer.

5 Zeuner, F.E. (1939). *Fossil Orthoptera Ensifera*. London: British Museum (Natural History).

6 Carpenter, F.M. (1950). The lower Permian insects of Kansas: part 10. The order Protorthoptera: the family Liomopteridae and its relatives. *Proceedings of the American Academy of Arts and Sciences* **78** (4): 187–219.

7 Sharov, A.G. (1962). A new Permian family of Orthoptera. *Paleontologicheskij Zhurnal* **1962**: 112–116.

8 Sharov, A.G. (1968). Filogeniya orthopteroidnykh nasekomykh. *Trudy Paleontologicheskogo instituta, Akademiya Nauk SSSR* **118**: 1–216.

9 Gorochov, A.V. (1995). System and Evolution of the Suborder Ensifera (Orthoptera) Part I. In: *Proceedings of the Zoological Institute, Russian Academy of Sciences*. S. Petersburg: Zoological Institute, Russian Academy of Sciences.

10 Gorochov, A.V. (1986). Triassic insects of the superfamily Hagloidea (Orthoptera). *USSR Academy of Sciences, Proceedings of the Zoological Institute, Leningrad* **143**: 65–100.

11 Gorochov, A.V. (1988). The Lower and Middle Jurassic superfamily hagloidea (orthoptera). *Paleontologicheskii Zhurnal* **2**: 50–61.

12 Gorochov, A.V. (2001). The most interesting finds of orthopteroid insects at the end of the 20th century and a new recent genus and species. *Journal of Orthoptera Research* **10**: 353–367.

13 Gorochov, A.V. (2003). New data on taxonomy and evolution of fossil and recent Prophalangopsidae (Orthoptera: Hagloidea). *Acta Zoological Cracoviensia* **46**: 117–127.

14 Hennig, W. (1981). *Inscet Phylogeny*. New York: Wiley.

15 Béthoux, O. and Nel, A. (2002). Venational pattern and revision of Orthoptera *sensu n.* and sister group phylogeny of Palaeozoic and Mesozoic Orthoptera *sensu n. Zootaxa* **96**: 1–88.

16 Lin, Q.B. (1965). Two insects from the lower part of Jurassic, Inner Mongolia. *Acta Palaeontologica Sinica* **13** (2): 363–368.

17 Gorochov, A.V. and Rasnitsyn, A.P. (2002). Superorder Gryllidea Laicharting, 1781 (Orthopteroidea Handlirsch, 1903). In: *History of Insects* (ed. A.P. Rasnitsyn and D.L.J. Quicke), 293–303. Dordrecht: Kluwer Academic Publishers.

18 Gorochov, A.V. and Maehr, M.D. (2008). New names for some fossil taxa of the infraclass Polyneoptera (Insecta). *Zoosystematica Rossica* **17** (1): 60.

19 Zeuner, F.E. (1935). The recent and fossil Prophalangopsidae. *Stylops: A Journal of Taxonomic Entomology* **4** (5): 102–108.

20 Cigliano, M.M., Braun H., Eades D.C. and Otte D. (2017) Orthoptera Species File. Version 5.0/5.0. http://Orthoptera.SpeciesFile.org (accessed 10.9.18).

21 Gu, J.J., Qiao, G.X., and Ren, D. (2012). The first discovery of Cyrtophyllitinae (Orthoptera, Haglidae) from the Middle Jurassic and its morphological implications. *Alcheringa* **36** (1): 27–34. https://doi.org/10.1080/03115518.2011.576535.

22 Martynov, A.V. (1937). Liassic insects from Shurab and Kyzyl-Kiya. *Trudy Paleontologičeskogo Instituta, Akademiya Nauk SSR* **7** (1): 80–160.

23 Gu, J.J., MontealegreZ, F., Robert, D. *et al.* (2012). Wing stridulation in a Jurassic katydid (Insecta, Orthoptera) produced low-pitched musical calls to attract females. *Proceedings of the National Academy of Sciences of the United States of America* **109** (10): 3868–3873. https://doi.org/10.1073/pnas.1118372109.

24 Wang, W.L. and Liu, M.W. (1996). A new genus and species of Haglidae from late Mesozoic of China, with description of its auditory organs. *Memoris of Beijing Natural History Museum* **55**: 69–77.

25 Gorochov, A.V. (1996). New Mesozoic insects of the superfamily Hagloidea (Orthoptera). *Paleontologicheskii Zhurnal* **3**: 73–82. (In Russian).

26 Gu, J.J., Tian, H., Yin, X.C. *et al.* (2017). A new species of Cyrtophyllitinae (Insecta: Ensifera) from the Cretaceous China. *Cretaceous Research* **74**: 151–154. https://doi.org/10.1016/j.cretres.2016.12.023.

27 Morris, G.K. and Gwynne, D.T. (1978). Geographical distribution and biological observations of Cyphoderris (Orthoptera: Haglidae) with a description of a new species. *Psyche* **85**: 147–167.

28 Liu, X.W., Zhou, M., Bi, W.X., and Tang, L. (2009). New data on taxonomy of recent Prophalangopsidae (Orthoptera: Hagloidea). *Zootaxa* **2026**: 53–62.

29 Zhou, Z.J., Zhao, L., Liu, N. *et al.* (2017). Towards a higher-level Ensifera phylogeny inferred from

mitogenome sequences. *Molecular Phylogenetics and Evolution* **108**: 22–33. https://doi.org/10.1016/j.ympev.2017.01.014.
30 Deichmüller, J.V. (1886). Die Insecten aus dem lithographischen Schiefer mit dresdener Museum. *Mittheilungen aus dem Kœniglichen Mineralogisch Geologischen und Præhistorischen Museum in Dresden* **37**: 1–84.
31 Martynov, A.V. (1925). To the knowledge of fossil insects from Jurassic beds in Turkestan. *Bulletin de lí Académie des Sciences de líURSS* **19**: 569–598.
32 Sharov, A.V. (1962). Podtip Mandibulata zhvalonosnye chlenistonogie, in Osnovy paleontologii. Chlenistonogie. Trakheinye I khelitserovye [Fundamentals of Palaeontology. Arthropoda, Tracheata, Chelicerata.] (ed. B.B. Rodendorf.). *Izdatel'stvo Akademii Nauk SSSR, Moscow* **9**: 145–146.
33 Li, L.M., Ren, D., and Wang, Z.H. (2007). New prophalangopsids from late Mesozoic of China (Orthoptera, Prophalangopsidae, Aboilinae). *Acta Zootaxonomica Sinica* **32** (2): 412–422.
34 Hong, Y.C. (1982). Fossil Haglidae (Orthoptera) in China. *Scientia Sinica (series B)* **25** (10): 444–456.
35 Ren, D., Lu, L.W., Guo, Z.G., and Ji, S.A. (1995). *Faunae and Stratigraphy of Jurassic-Cretaceous in Beijing and the Adjacent Areas*. Beijing: Seismic Publishing House.
36 Meng, X.M. and Ren, D. (2006). Late Jurassic prophalangopsids from Northeast China (Orthoptera, Prophalangopsidae, Chifengiinae). *Acta Zootaxonomica Sinica* **31** (2): 282–288. (in Chinese with English abstract).
37 Meng, X.M., Ren, D., and Li, L.M. (2006). New prophalangopsid fossils from Liaoning, China (Orthoptera, Prophalangopisdae, Chifengiinae). *Acta Zootaxonomica. Sinica* **31** (4): 752–757. (in Chinese with English abstract).
38 Gu, J.J., Qiao, G.X., and Ren, D. (2010). Revision and new taxa of fossil Prophalangopsidae (Orthoptera: Ensifera). *Journal of Orthoptera Research* **19** (1): 41–56.
39 Zherikhin, V.V. (1985). The Jurassic Orthoptera in South Siberia and West Mongolia. In: *Jurassic Insects of Siberia and Mongolia* (ed. A.P. Rasnitsyn) Trudy Paleontol. Inst. Akademii Nauk SSSR 211, 171–184. Moscow: Nauka.
40 Gu, J.J., Qiao, G.X., and Ren, D. (2011). A exceptionally-preserved new species of *Bacharaboilus* (Orthoptera: Prophalangopsidae) from the Middle Jurassic of Daohugou, China. *Zootaxa* **2909**: 64–68.
41 Zhang, J.F. (1993). The late Mesozoic insect fossils in south Shaanxi and South Henan. *Palaeoworld* **2**: 49–56.
42 Zhang, H.C. (1996). First discovery of fossil Haglidae (Orthoptera) in Northwest China. *Entomotaxonomia* **18** (4): 249–251.
43 Ren, D. and Meng, X.M. (2006). New Jurassic Protaboilins from China (Orthoptera, Prophalangopsidae, Protaboilinae). *Acta Zootaxonomica Sinica* **31**: 513–519.
44 Li, L.M., Ren, D., and Meng, X.M. (2007). New fossil prophalangopsids from China (Orthoptera, Prophalangpsidae, Aboiliane). *Acta Zootaxonomica Sinica* **32** (1): 174–181.
45 Fang, Y., Zhang, H.C., Wang, B., and Zhang, Y.T. (2007). New taxa of Aboilinae (Insecta, Orthoptera, Prophalangopsidae) from the Middle Jurassic of Daohugou, Inner Mongolia, China. *Zootaxa* **1637**: 55–62.
46 Lin, Q.B., Huang, D.-Y., and Nel, A. (2008). A new genus of Chifengiine (Orthoptera: Ensifera: Prophalangopsidae) from the Middle Jurassic (Jiulongshan Formation) of Inner Mongolia, China. *Comptes Rendus Palevol* **7**: 205–209.
47 Gorochov, A.V., Jarzembowski, E.A., and Coram, R.A. (2006). Grasshoppers and crickets (Insecta: Orthoptera) from the Lower Cretaceous of Southern England. *Cretaceous Research* **27**: 641–662. https://doi.org/10.1016/j.cretres.2006.03.007.
48 Peñalver, E. and Grimaldi, D.A. (2010). Latest occurrences of the Mesozoic family Elcanidae (Insecta: Orthoptera), in Cretaceous amber from Myanmar and Spain. *Annales De La societe entomologique De France* **46** (1–2): 88–99.
49 Fang, Y., Wang, B., Zhang, H. *et al.* (2015). New Cretaceous Elcanidae from China and Myanmar (Insecta, Orthoptera). *Cretaceous Research* **52**: 323–328. https://doi.org/10.1016/j.cretres.2014.05.004.
50 Westwood, J.O. (1854). Contribution to fossil entomology. *Quarterly Journal of the Geological Society of London* **10**: 378–396.
51 Gu, J.J., Yue, Y.L., Shi, F.M. *et al.* (2016). First Jurassic grasshopper (Insecta, Caelifera) from China. *Zootaxa* **4169** (2): 377–380.
52 Hong, Y.C. (1983). *Middle Jurassic Fossil Insects in North China*. Beijing: Geological Publishing House.
53 Lin, Q.B. and Huang, D.-Y. (2006). Revision of "*Parahagla lamina*" Lin, 1982 and two new species of *Aboilus* (Orthoptera: Prophalangopsidae) from the Early-Middle Jurassic of Northwest China. *Progress in Natural Science* **16**: 303–307.
54 Fang, Y., Zhang, H.C., and Wang, B. (2009). A new species of Aboilus (Insecta, Orthoptera, Prophalangopsidae) from the Middle Jurassic of Daohugou, Inner Mongolia, China. *Zootaxa* **2249**: 63–68.

55 Wang, H., Li, S., Zhang, Q. *et al.* (2015). A new species of *Aboilus* (Insecta, Orthoptera) from the Jurassic Daohugou beds of China, and discussion of forewing coloration in *Aboilus*. *Alcheringa* **39**: 250–258. https://doi.org/10.1080/03115518.2015.993297.
56 Hong, Y.C. (1982). *Mesozoic Fossil Insects of Jiuquan Basin in Gansu Province*. Beijing: Geological Publishing House.
57 Fang, Y., Zhang, H.C., Wang, B. *et al.* (2013). A new Chifengiinae species (Orthoptera: Prophalangopsidae) from the Lower Cretaceous Yixian Formation (Liaoning, P.R. China). *Insect Systematics & Evolution* **44** (2): 141–147. https://doi.org/10.1163/1876312X-44022099.
58 Wang, H., Zheng, D.R., Lei, X.J. *et al.* (2017). A new species of Chifengiinae (Orthoptera: Prophalangopsidae) from the Lower Cretaceous Zhonggou Formation of the Jiuquan Basin, Northwest China. *Cretaceous Research* **73**: 60–64. https://doi.org/10.1016/j.cretres.2017.01.006.
59 Wang, W.L. (1987). The Early Mesozoic insect fossils in Western Liaoning. In: *Mesozoic Stratigraphy and Palaeontology of Western Liaoning III* (ed. X.H. Yu, W.L. Wang, X.T. Liu and W. Zhang), 212–214. Beijing: Geological Publishing House.
60 Hong, Y.C. (1986). New fossil insects of Haifanggou Formation, Liaoning Province. *Journal of Changchun College of Geology* **4**: 10–16.
61 Hong, Y.C. (1992). Middle and Late Jurassic fossil insects. In: *Palaeontological Atlas of Jilin Province* (ed. Bureau of Geology and Mineral of Jilin Province), 420–421. Changchun: Jilin Science and Technology Publishing House.
62 Hong, Y.C. (1984). Mesozoic Volume. In: *Palaeontological Atlas of North China* (ed. Institute of Geology and Mineral Resources of Tianjin), 148–152. Beijing: Geological Publishing House.
63 Hong, Y.C. (1988). Early Cretaceous Orthoptera, Neuroptera, Hymenoptera (Insecta) of Kezuo in West Liaoning Province. *Entomotaxonomia* **10** (1–2): 120–124.
64 Wang, H., Fang, Y., Zhang, Q. *et al.* (2018). New material of *Sigmaboilus* (Insecta, Orthoptera, Prophalangopsidae) from the Jurassic Daohugou Beds, Inner Mongolia, China. *Earth and Environmental Science Transactions of the Royal Society of Edinburgh* https://doi.org/10.1017/S1755691017000172.
65 Gu, J.J., Zhao, Y.Y., and Ren, D. (2009). New fossil Prophalangopsidae (Orthoptera, Hagloidea) from the Middle Jurassic of Inner Mongolia, China. *Zootaxa* **2004**: 16–24.
66 Lin, Q.B. (1980). Fossil insects from the Mesozoic of Zhejiang and Anhui. In: *Division and Correlation of Stratigraphy of Mesozoic Volcanic Sediments from Zhejiang and Anhui* (ed. Nanjing Institute of Geology and Palaeontology), 217–218. Beijing: Science Press.
67 Hong, Y.C. (1990). *The Stratigraphy and Palaeontology of Laiyang Basin, Shandong Province*. Beijing: Geological Publishing House.
68 Lin, Q.B. (1982). Mesozoic and Cenozoic insects. In: *Atlas of Palaeontology from Northwest Region, Shaanganning Division* (ed. Institute of Geology and Mineral Resources of Xi'an), 76–77. Beijing: Geological Publishing House.
69 Zhang, J.F. (1985). New data of Mesozoic insect fossil insects from Laiyang in Shandong. *Geology of Shandong* **1** (2): 23–39.
70 Hong, Y.C. (1997). Discovery of fossil insects from Houcheng Formation of Hebei Province and establish of Houcheng Entomofauna. *Beijing Geology* **9** (1): 1–10.

10

Notoptera – Rock Crawlers and Ice Crawlers

Yingying Cui[1,2], *Chungkun Shih*[2,3], *and Dong Ren*[2]

[1] South China Normal University, Guangzhou, China
[2] Capital Normal University, Haidian District, Beijing, China
[3] National Museum of Natural History, Smithsonian Institution, Washington, DC, USA

10.1 Introduction to Notoptera (Mantophasmatodea and Grylloblattodea)

The Mantophasmatodea, commonly called "rock crawlers", "heelwalkers" or "gladiators", have been named as the most recently discovered order of extant insects by Klass *et al.* [1]. Extant Mantophasmatodea, known from Namibia, Tanzania, and South Africa [2–5], are wingless insects looking like a mix between praying mantids and stick insects (Figure 10.1). They are predatory carnivores, using their well-equipped spiny fore- and mid-legs to catch their prey of small insects or other arthropods at night. Each of their legs has an enlarged pretarsal arolium with long setae. Molecular evidence indicates that they are most closely related to the equally enigmatic group Grylloblattodea [6, 7].

Grylloblattodea, commonly called "ice crawlers" and referred to as "Grylloblattida" or "Grylloblattidae," comprise a group of rare insects characterized by lack of wings, a comparatively uniform morphology, a very limited diversity (32 species), and constrained ecological preferences and geographical distribution in a very cold environment [8–11]. Presently, 32 extant species have been described in five genera: *Galloisiana, Grylloblatta, Grylloblattella, Grylloblattina* and *Namkungia*. Among them, only *Grylloblatta* occurs in northwestern North America, the other four genera are distributed in Japan, the Korean Peninsula, Northeastern China, the Russian Pacific coast, and the Altai and Sayan mountain ranges [11]. Grylloblattodea species occur mostly in mountainous and high alpine regions, often at the edges of glaciers. As temperature is the main barrier for migration of modern Grylloblattodea, the range of each species is extremely limited. Species of *Galloisiana, Namkungia*, and *Grylloblattina* live in dense forests and some are specialized cave dwellers.

Following the recommendation of Engel and Grimaldi [12], Arillo and Engel [13] combined the Mantophasmatodea and Grylloblattodea into a single order, employing the name Notoptera as proposed by Crampton [14] for the composite lineage including both the living and fossil taxa documented.

In China, only two Grylloblattodea species have been found and described so far. The first one, *Galloisiana sinensis* Wang 1987, was found by Wang Shuyong in Changbai Mountain, Northeastern China [15]. Subsequently, in 1998, it was declared by China as one of the National First Class Protective Animals, one of the only two insects with this status. In 2009, the second species, *Grylloblattella cheni* Bai, Wang & Yang, 2010 [8], a female, was found in Kanas of Xingjiang, China (Figure 10.2). The specific epithet is in honor of Prof. Sicien Chen (Shixiang Chen), Fellow of Chinese Academy of Sciences, the founder and former director of the Institute of Zoology, Chinese Academy of Sciences, who has made great contributions to entomological research of China.

Adult extant Grylloblattodeas range from 12 to 35 mm in body lengths. They are characterized by an elongate, slightly flattened body without wings. The head is short and prognathous with chewing mouthparts and robust mandibles. Antenna is filiform, with 28–50 segments. Compound eyes are small or absent, ocelli absent. Three thoracic segments are similar in shape and mobile. Three pairs of thin and long legs are similar too, with 5-segmented tarsi. The abdomen typically has 10 segments with long cerci, having 8–10 segments. The asymmetrical external genitalia of the male are located

Rhythms of Insect Evolution: Evidence from the Jurassic and Cretaceous in Northern China, First Edition.
Edited by Dong Ren, Chungkun Shih, Taiping Gao, Yongjie Wang, and Yunzhi Yao.

Figure 10.1 *Karoophasma biedouwense* Klass, Picker, Damgaard, van Noort & Tojo, 2003 [2], female, from South Africa. Source: Photo by Monika Eberhard.

Figure 10.2 *Grylloblattella cheni* Bai, Wang & Yang, 2010. Source: Photo provided by Dr. Ming Bai.

in the ninth abdominal segment, with styli at the terminal. The female has a long sword-shaped ovipositor to deposit eggs into the soil.

Ice-crawlers live at elevations between 200 and 3200 m. They are adapted to cold-temperate habitats with a preferred temperature range of about 3–16°C. Mortality rate increases if higher than 20°C. It is reported that Grylloblattodea may die even if held only briefly in a human hand. Intolerance to mildly high temperature and lack of wings are the limiting factors for their migration, spreading and genetic exchange.

Fossil species of Grylloblattodea have wings which are in contrast to the wingless extant ones. The group with all the fossil species, Grylloblattida sensu Storozhenko, 2002 [16], are based on two forewing character states, namely (i) a desclerotized basal portion of M/MP, and (ii) a particular branching pattern of CuA, namely with a posterior stem (CuA2) simple, and an anterior stem (CuA1) branched only well distal to the CuA1-CuA2 fork. The identification of the stem-representatives of Grylloblattodea remains a matter of debate, because of (i) the lack of wings in the extant representatives, and (ii) a lack of suitably preserved body remains other than wings in the fossil material. Therefore, the data on body parts of fossils other than wings would be of particular relevance to test the assumed phylogenetic link between the fossil Grylloblattida sensu Storozhenko, 2002 [16] and Grylloblattodea, and perhaps their closest extant relatives, Mantophasmatodea.

Overall, extant species have a limited capability for migration resulting in their confined distribution. Based on the distribution of extinct and extant families, the evolutionary trends for Grylloblattodea are from winged insects to wingless, from warm climate to cold climate, from broad distribution to narrow distribution. They are precious insect groups which may face potential extinction.

10.2 Progress in the Studies of Fossil Notoptera (Mantophasmatodea and Grylloblattodea)

Fossil rock crawlers in the Mantophasmatodea have been described from the Eocene Baltic amber [1, 13, 17, 18] and the Middle Jurassic of Daohugou, Inner Mongolia, China [19].

Fossils assigned to Grylloblattodea [20] were rare in the late Carboniferous [21]. Less than 10 species are known, recovered from the localities of Mazon Creek (USA) and Commentry (France), among other localities in Europe, and also in Tunguska Basin (Russia) and Ningxia (China). The Grylloblattodea reached peak species-richness and abundance in the Permian [22]. A remarkable diversity and wide distribution of these taxa have been documented, with more than 30 families and nearly 200 species from the Permian sites of Lodève (France), Obora (Czech Republic), Elmo and Midco (USA), and multiple localities in Germany and Russia. However, the actual diversity at the species and genus level in particular localities might have been over-estimated.

During the Mesozoic, the diversity of Grylloblattodea groups declines. Less than 60 species are known from the Triassic. One of them was found in the Kemerovo region (Russia) and one at Cerro Cacheuta (Argentina). The bulk of the species are from a rich assemblage in Central Asia, the Madygen Formation. The diversity at the species and genus levels might have also been over-estimated. Other records are from the South Africa and the northeast of Australia.

The diversity in the Jurassic is presently intensively investigated. Up to now, about 40 species placed in six families have been described. The detailed description of specimens with well-preserved body parts is crucial in a phylogenetic context. Very important and well-preserved specimens have been documented recently at the Chinese Daohugou locality [23–28]. No fossils are known later than the Early Cretaceous. In this period only three species have been recorded [29].

The study of this order in China is still going on. Up to now, the researchers in China have described a total of 12 species. Those species were attributed to five extinct families (Table 10.1). All of these specimens are from the Middle Jurassic Jiulongshan Formation of Daohugou, Ningcheng, Inner Mongolia. The Middle Jurassic is a period during which the diversity of Grylloblattodea started to decline, thus, it is of great significance to study the fossil specimens from this period.

10.3 Representative Fossils of Notoptera (Mantophasmatodea and Grylloblattodea) from Northern China

Suborder Mantophasmatodea Klass, Zompro, Kristensen & Adis, 2002

Family Mantophasmatidae Klass, Zompro, Kristensen & Adis, 2002

Subfamily Raptophasmatinae Zompro, 2005

This subfamily was erected for species found in the Eocene Baltic amber [18]. The fossil specimens reported from the Middle Jurassic of Northeastern China indicate that their geological range is wider and the age is much older.

Only one genus included from the Jurassic of Northern China: *Juramantophasma* Huang, Nel, Zompro & Walker, 2008.

***Juramantophasma* Huang, Nel, Zompro & Walker, 2008**

Juramantophasma Huang, Nel, Zompro & Walker, 2008, Naturwissenschaften. 95, 948–950 [19] (original designation).

Type species: *Juramantophasma sinica* Huang, Nel, Zompro & Walker, 2008.

Juramantophasma exhibits several characters currently considered as apomorphies of the Mantophasmatodea, i.e. a third tarsomere with a sclerotized elongated dorsal process; enlarged and fan-like pretarsal arolia with a clearly visible row of dorsal setae, identical to those of the modern *Mantophasma zephyra* [1, 4, 30]; the last tarsomere making a right angle with the others, keeping it up in the air; female gonoplacs (valves 3) short and claw-shaped; and egg with a circular ridge [19].

Distribution and age: Inner Mongolia; Middle Jurassic.

Only one species included from the Jurassic of Northern China (see Table 10.1).

***Juramantophasma sinica* Huang, Nel, Zompro & Walker, 2008 (Figure 10.3)** *Juramantophasma sinica* Huang, Nel, Zompro & Walker, 2008: *Naturwissenschaften*, 95, 948–950.

Locality and horizon: Daohugou, Ningcheng, Inner Mongolia, China; Middle Jurassic, Jiulongshan Formation.

The female holotype is 34 mm long. The hypognathous head is covered with fine fur-like setae, 5.1 mm long and 4.4 mm wide; large eyes, 1.7 mm long and 1.3 mm wide, but no visible ocelli; antennae filiform, with basal five antennomeres preserved; large scape, 0.8 mm long; pedicel, 0.25 mm long; the first flagellomere, 0.95 mm long; the second flagellomere, 0.8 mm long.

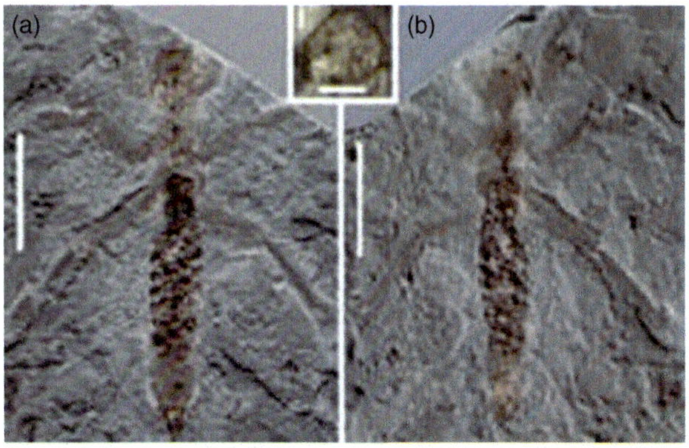

Figure 10.3 *Juramantophasma sinica* Huang, Nel, Zompro & Walker, 2008 (Holotype, NIGPAS 142171a-b) [19]. Source: Photo provided by Dr. Di-ying Huang.

The following characters of *J. sinica* are further shared with the Mantophasmatodea: hypognathous head, wings absent, antennae probably long, and filiform, ocelli not visible, probably absent, pronotum longer than meso- and metanotum; hind legs very thin but longer than the middle and forelegs; fore femora are the widest; four basal tarsomeres very short, and bearing very large euplantulae (very small in modern Grylloblattodea); abdominal segments transverse; ovipositor elongate, nearly reaching the apex of abdomen; egg large, elongate, and a chorion with a pattern of small spots and a central gibbosity. The mandibles of *J. sinica* look similar to those of the recent crickets and katydids. *Juramantophasma* was probably a predator based on its mandible with strong teeth and broad fore legs with strong hook-like claws, but less specialized than the extant Mantophasmatids [19].

Suborder Grylloblattodea Brues & Melander, 1932

Family Juraperlidae Huang & Nel, 2007

The extinct family Juraperlidae, with a single monotypic genus of *Juraperla*, are characterized by having a secondary vein between anterior wing margin and ScP. This character is unique among Grylloblattoidea insects [31].

Only one genus included from the Jurassic of Northern China: *Juraperla* Huang & Nel, 2007.

Juraperla Huang & Nel, 2007

Juraperla Huang & Nel, 2007, *Eur. J. Entomol.*, 104, 837–840 [31] (original designation).

Type species: *Juraperla daohugouensis* Huang & Nel, 2007.

The monotypic genus, *Juraperla*, has two species, *J. daohugouensis* Huang & Nel, 2007 and *J. grandis* Cui, Béthoux, Shih & Ren, 2010. These two species shared the same characters, i.e. simple RP and the first forks of M and CuA located near or opposite the arculus [26, 31].

Distribution and age: Inner Mongolia; Middle Jurassic.

Two species included from the Jurassic of Northern China (see Table 10.1).

Juraperla grandis Cui, Béthoux, Shih & Ren, 2010 (Figure 10.4)

Juraperla grandis Cui, Béthoux, Shih & Ren, 2010: *Acta. Geol. Sin.-Engl.*, 84, 710–713.

Locality and horizon: Daohugou, Ningcheng, Inner Mongolia, China; Middle Jurassic, Jiulongshan Formation.

The species *Juraperla grandis* was erected based on the well-preserved specimen with four wings and partial structure of body. The differences between *J. grandis* and *J. daohugouensis* are: *J. grandis* has body size much larger than that of *J. daohugouensis*; CuA1 of *J. grandis* has three branches vs. four braches for *J. daohugouensis* and two rows of cells occurring between MP and the anterior branch of CuA1 which is unique to *J. grandis* [26, 31].

Family Plesioblattogryllidae Huang, Nel & Petrulevičius, 2008

The extinct family Plesioblattogryllidae were erected based on a single complete female specimen with preserved details of the head, legs and eggs. The wing venation of family Plesioblattogryllidae is nearly identical to that of *Blattogryllus karatavicus* Rasnitsyn, 1976, which belongs to the family Blattogryllidae Rasnitsyn, 1976, known from many localities [19]. Both of these families are of particular importance, as they have been considered as the closest relatives of extant Grylloblattodea [16, 23, 25, 32].

Only one genus included from the Jurassic of Northern China: *Plesioblattogryllus* Huang, Nel & Petrulevičius, 2008.

Plesioblattogryllus Huang, Nel & Petrulevičius, 2008

Plesioblattogryllus Huang, Nel & Petrulevičius, 2008, *Zoo. J. Linnean. Soc*, 152: 17–24 [23] (original designation).

Type species: *Plesioblattogryllus magnificus* Huang, Nel & Petrulevičius, 2008.

The genus has two species, *P. magnificus* [23], and *P. minor* [33]. The diagnoses for this family and genus are: mandibles very strong with a sharp pointed apical tooth, and a few marginal teeth with a broad base; compound eyes and ocelli present; antenna slightly longer than head, with antennomeres 4–6 (7?) shorter than others; tarsomeres 1–4 with a pair of rather large euplantulae (not well-documented in *P. minor*); pretarsi with strong claws but no arolia; eggs olive-shaped; cerci segmented. In forewing, ScP ending on the anterior wing margin; area between anterior margin and ScP narrow; RA simple, nearly parallel to ScP; RP posteriorly pectinate, with several branches; M fused with CuA in basal part; MA usually simple; MP with 1–3 branches; CuA1 with two branches. In hind wing: ScP ending on the anterior wing margin; RA parallel to ScP, simple; RP posteriorly pectinate, with four branches; MA simple; MP forked; CuA with three branches.

Distribution and age: Inner Mongolia; Middle Jurassic.

Two species included from the Jurassic of Northern China (see Table 10.1).

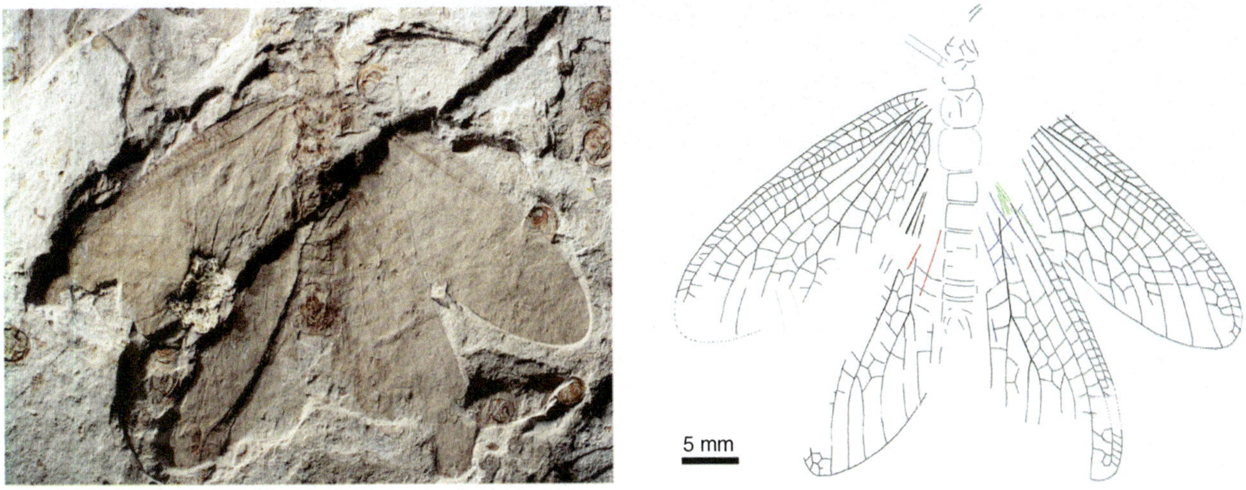

(a) (b)

Figure 10.4 *Juraperla grandis* Cui, Béthoux, Shih & Ren, 2010 (Holotype, CNU-GRY-NN2009006). (a) Photograph and (b) Line drawing [26]. Source: Donated by Chungkun Shih.

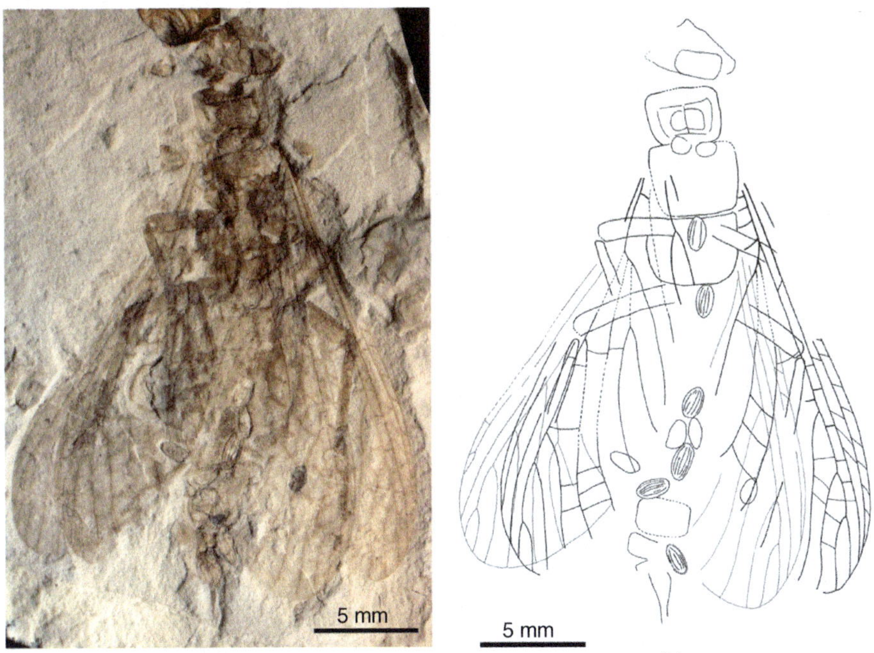

(a) (b)

Figure 10.5 *Plesioblattogryllus minor* Ren & Aristov, 2011 (CNU-GRY-NN2011001), female. (a) Photograph and (b) Line drawing [25].

***Plesioblattogryllus minor* Ren & Aristov, 2011 (Figure 10.5)**
Plesioblattogryllus minor Ren & Aristov, 2011: *Paleontol. J.*, 45, 273–274.

Locality and horizon: Daohugou, Ningcheng, Inner Mongolia, China; Middle Jurassic, Jiulongshan Formation.

The species *P. minor* Ren & Aristov, 2011 is smaller in size (forewing about 20 mm long, hind wing 15 mm long) than the species *P. magnificus*. Other diagnostic characters are: pronotum rectangular, nearly same width as head, or narrower; MA fused with RP slightly distal to the origin of RP; MP simple or two-branched, rarely three-branched [33].

Four complete specimens of this species have been reported by Cui in 2012 [25] and one is shown in Figure 10.5. Due to the good preservation of these specimens, evidence and implications on the phylogenetic relationships between extant Grylloblattodea and the fossil Blattogryllidae and Plesioblattogryllidae were provided. First, the apomorphic asymmetry of gonocoxites is confirmed for genus *Blattogryllus*, while this character is confirmed with Grylloblattodea. Then the features of the female genital region in *P. minor* show a remarkable similarity with those of Grylloblattodea, though it is not possible to extract shared apomorphies from this character system. Also, the head of *P. minor* also bears much resemblance with that in Grylloblattodea. Therefore, these results on the relationships between Grylloblattodea, Blattogryllidae and Plesioblattogryllidae are highly important as such a link enables the evolution of the wingless extant representatives to be traced back to the early radiation of Neoptera by use of wing venation characters [25].

Family *Incertae sedis*

Duoduo Cui, 2012

Duoduo Cui, 2012, *Arthropod Syst. Phylo.*, 70: 167–180 [25] (original designation).

Type species: *Duoduo qianae* Cui, 2012.

The specific epithet is in honor of Dr. Y. Qian (Institute of Applied Entomology, Yangzhou University, China), for her help to Y. Cui. The specimen exhibits a combination of character states unknown in other genera of both Blattogryllidae Rasnitsyn, 1976 and Plesioblattogryllidae Huang, Nel & Petrulevičius, 2008.

Distribution and age: Inner Mongolia; Middle Jurassic.

Only one species included from the Jurassic of Northern China (see Table 10.1).

Family Geinitziidae Handlirsch, 1906

The extinct family Geinitziidae Handlirsch, 1906 [34] comprise eight genera: *Fletchizia* Riek, 1976 [35], *Geinitzia* Handlirsch, 1906 [34], *Megasepididontus* Huang & Nel, 2008 [24], *Permoshurabia* Aristov, 2009 [36], *Prosepididontus* Handlirsch, 1920 [37], *Shurabia* Martynov, 1937 [38], *Sinosepididontus* Huang & Nel, 2008 [24] and *Stegopterum* Sharov, 1961 [39].

Genera included from the Jurassic of Northern China: *Geinitzia* Handlirsch, 1906; *Shurabia* Martynov, 1937 and *Sinosepididontus* Huang & Nel, 2008.

Geinitzia Handlirsch, 1906

Geinitzia Handlirsch, 1906, *Ein Handbuch für paläontologen und Zoologen*: 640 pp. [34] (original designation).

Type species: *Gryllacris schlieffeni* Geinitz, 1884, by subsequent designation [40].

One species of the genus from the Middle Jurassic of China, exhibiting a character that forewing CuA with five branches.

Distribution and age: Inner Mongolia; Middle Jurassic.

Only one species included from the Jurassic of China (see Table 10.1).

Figure 10.6 *Geinitzia aristovi* Cui, Storozhenko & Ren, 2012 (Holotype, CNU-GRY-NN2009005) [28].

Geinitzia aristovi Cui, Storozhenko & Ren, 2012 (Figure 10.6)

Geinitzia aristovi Cui, Storozhenko & Ren, 2012: *Alcheringa*, 36, 251–261.

Locality and horizon: Daohugou, Ningcheng, Inner Mongolia, China; Middle Jurassic, Jiulongshan Formation.

The specific epithet is dedicated to Dr. Aristov (Moscow) for his contribution to studies of Grylloblattida and other insect fossils. This species has CuA in forewing with five branches vs four branches of CuA in forewing of other species of *Geinitzia* [28].

Shurabia Martynov, 1937

Shurabia Martynov, 1937, *Trudy Paleontologicheskogo Instituta Akademii Nauk SSSR*: 232 pp. [38] (original designation).

Type species: *Shurabia ovata* Martynov, 1937.

One species, *Shurabia grandis,* is reported from the Middle Jurassic of China [24].

Distribution and age: Inner Mongolia; Middle Jurassic.

Only one species included from the Jurassic of Northern China (see Table 10.1).

Figure 10.7 *Shurabia grandis* (Huang & Nel, 2008) (new material, CNU-GRY-NN2009001) [24].

***Shurabia grandis* (Huang & Nel, 2008) (Figure 10.7)** *Shurabia grandis* Cui, Storozhenko & Ren, 2012: *Alcheringa*, 36, 251–261.

Megasepididontus grandis Huang & Nel, 2008, *Alcheringa*, 32: 395–403 [24].

Syn. by Cui, Storozhenko & Ren, 2012: *Alcheringa*, 36, 251–261 [28].

Locality and horizon: Daohugou, Ningcheng, Inner Mongolia, China; Middle Jurassic, Jiulongshan Formation.

The genus *Magasepididontus* Huang & Nel, 2008 was synonymized under *Shuabia*, therefore, the species *Magasepididontus grandis* was transferred to *Shurabia*. Due to the well-preservation of the specimens, the body and appendage characters described for *Shurabia grandis* helped to clarify the morphological structure of the genus [24].

Sinosepididontus Huang & Nel, 2008

Sinosepididontus Huang & Nel, 2008, *Alcheringa*, 32: 395–403 [24] (original designation).

Type species: *Sinosepididontus chifengensis* Huang & Nel, 2008.

According to Huang and Nel [24], the diagnosis for this genus are the simple RA, relatively broad costal area and anal veins forming a single veinlet prior to reaching the hind wing margin.

Distribution and age: Inner Mongolia; Middle Jurassic.

Only one species included from the Jurassic of Northern China (see Table 10.1).

Family Bajanzhargalanidae Storozhenko, 1992

Bajanzhargalanidae, one of the most poorly known families of Grylloblattida sensu Storozhenko [16], were established to accommodate a single species, *Bajanzhargalana magna* Storozhenko, 1988, known from the Khoutiyn-Khotgor locality, (Mongolia; Late Jurassic; Rasnitsyn and Zherikhin [41]). To date, this family comprise four genera, i.e. *Bajanzhargalana* Storozhenko, 1988 (the type genus) [42], *Nele* Ansorge, 1996 [43], *?Sylvafossor* Atristov, 2004 [44], and *Sinonele* Cui, Béthoux, Klass & Ren, 2015 [27].

Only one genus included from the Jurassic of Northern China: *Sinonele* **Cui, Béthoux, Klass and Ren, 2015.**

Sinonele Cui, Béthoux, Klass & Ren, 2015

Sinonele Cui, Béthoux, Klass & Ren, 2015, *Arthropod Struct. Dev.*, 44, 688–716 [27] (original designation).

Type species: *Sinonele fangi* Cui, Béthoux, Klass & Ren, 2015.

The diagnosis of the genus are: *Antenna*: the second flagellomere of antenna much shorter than others (character not documented in other genera of Bajanzhargalanidae). *Forewing*: ScP reaching anterior wing margin near 3/4 of wing length; area occupied by RP branches located at a level anterior of the apex; occurrence of a long secondary intercalary vein in the area between CuA2 and CuP (documented only in the large sample of *S. fangi*); occurrence of a long secondary intercalary vein in the area between CuP and AA1. *Male postabdomen* (known for *S. fangi* only): ventral side of segment 9 with all sclerites fused into a subgenital plate that bears several fields of tubercles, with coxal lobes fused into an unpaired subgenital lobe (with median notch) that bears a pair of styli, and with anterior apodemes (va9); cerci quite short, curved, division in cercomeres indistinct [27].

Distribution and age: Inner Mongolia; Middle Jurassic.

Four species included from the Jurassic of Northern China (see Table 10.1).

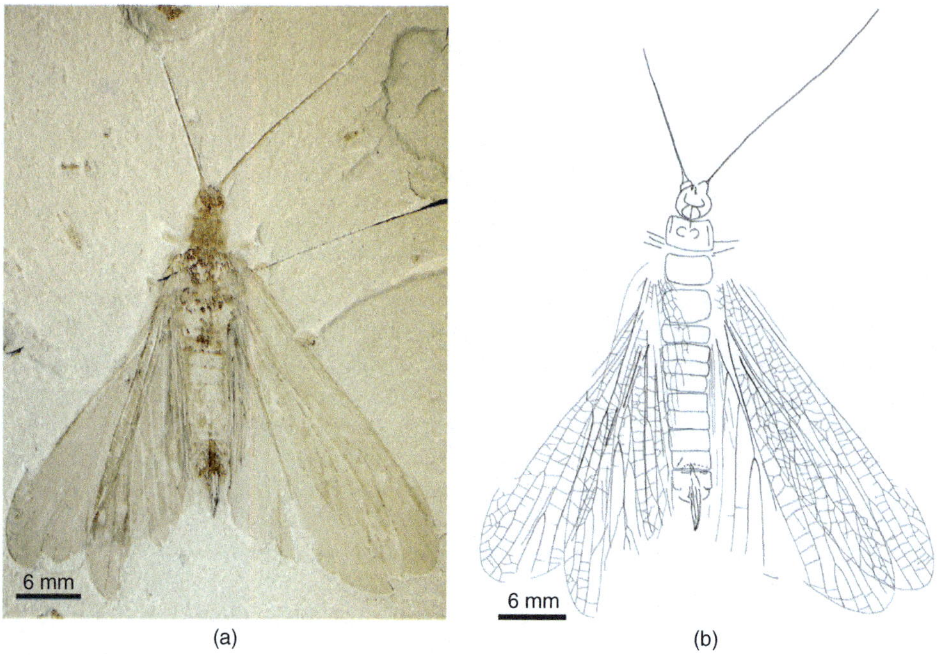

Figure 10.8 *Sinonele fangi* Cui, Béthoux, Klass & Ren, 2015 (Holotype, CNU-GRY-NN2010004). (a) Photograph and (b) Line drawing [27]. Source: Donated by Liang Fang.

***Sinonele fangi* Cui, Béthoux, Klass & Ren, 2015 (Figure 10.8)**
Sinonele fangi Cui, Béthoux, Klass & Ren, 2015: *Arthropod Struct. Dev.*, 44, 688–716.

Locality and horizon: Daohugou, Ningcheng, Inner Mongolia, China; Middle Jurassic, Jiulongshan Formation.

The specific epithet is dedicated to Mr. Liang Fang for his donation of this specimen. The diagnosis characters for this species: Medium size; main veins and cross-veins dark, membrane more lightly pigmented, darker along the anterior wing margin. Compared to other fossil species of Bajanzhargalanidae from the Middle Jurassic of Daohugou Locality, *S. fangi* has the highest number of specimens. Thirteen selected specimens, including four specimens with post-abdomen well-preserved, have been described in Cui *et al.* [27].

***Sinonele hei* Cui, Béthoux, Klass & Ren, 2015 (Figure 10.9)**
Sinonele hei Cui, Béthoux, Klass & Ren, 2015: *Arthropod Struct. Dev.*, 44, 688–716.

Locality and horizon: Daohugou, Ningcheng, Inner Mongolia, China; Middle Jurassic, Jiulongshan Formation.

Only one single specimen, the holotype, has been documented for *S. hei*. The diagnosis characters for this species are: dark coloration all over forewing and hind wing remigium (anterior wing margin darker), except for the most apical area, where the membrane is translucent [27].

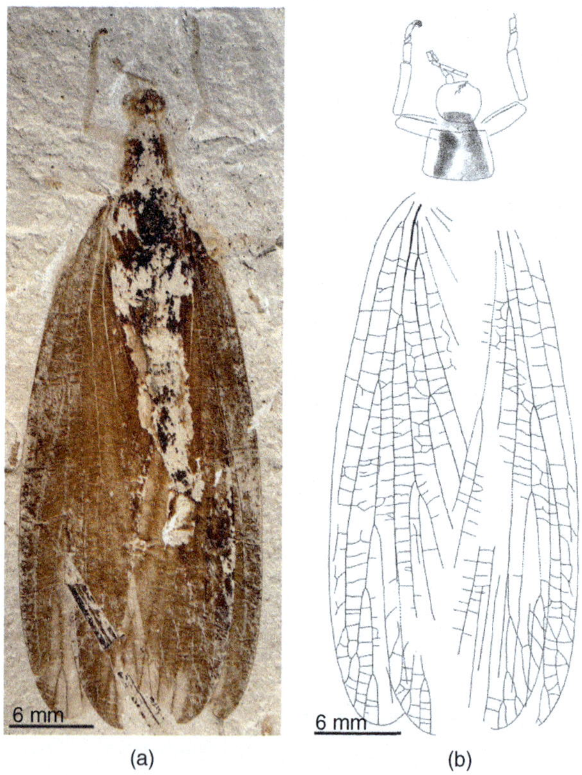

Figure 10.9 *Sinonele hei* Cui, Béthoux, Klass & Ren, 2015 (Holotype, CNU-GRY-NN2010030). (a) Photograph and (b) Line drawing [27].

Table 10.1 A list of fossil Notoptera (Mantophasmatodea and Grylloblattodea) from the Jurassic and Cretaceous of China.

Family	Species	Location	Horizon/Age	Citation
Suborder Mantophasmatodea Klass, Zompro, Kristensen & Adis, 2002				
Mantophasmatidae	*Juramantophasma sinica* Huang, Nel, Zompro & Walker, 2008	Ningcheng, Inner Mongolia	Jiulongshan Fm., J_2	Huang *et al.* [19]
Suborder Grylloblattodea Brues & Melander, 1932				
Bajanzhargalanidae	*Sinonele fangi* Cui, Béthoux, Klass & Ren, 2015	Ningcheng, Inner Mongolia	Jiulongshan Fm., J_2	Cui *et al.* [27]
	Sinonele hei Cui, Béthoux, Klass & Ren, 2015	Ningcheng, Inner Mongolia	Jiulongshan Fm., J_2	Cui *et al.* [27]
	Sinonele phasmoides Cui, Béthoux, Klass & Ren, 2015	Ningcheng, Inner Mongolia	Jiulongshan Fm., J_2	Cui *et al.* [27]
	Sinonele mini Cui, Béthoux, Klass & Ren, 2015	Ningcheng, Inner Mongolia	Jiulongshan Fm., J_2	Cui *et al.* [27]
Geinitziidae	*Geinitzia aristovi* Cui, Storozhenko & Ren, 2012	Ningcheng, Inner Mongolia	Jiulongshan Fm., J_2	Cui *et al.* [28]
	Shurabia grandis (Huang & Nel, 2008)	Ningcheng, Inner Mongolia	Jiulongshan Fm., J_2	Huang & Nel [24]
	Sinosepididontus chifengensis Huang & Nel, 2008	Ningcheng, Inner Mongolia	Jiulongshan Fm., J_2	Huang & Nel [24]
Juraperlidae	*Juraperla daohugouensis* Huang & Nel, 2007	Ningcheng, Inner Mongolia	Jiulongshan Fm., J_2	Huang & Nel [31]
	Juraperla grandis Cui, Béthoux, Shih & Ren, 2010	Ningcheng, Inner Mongolia	Jiulongshan Fm., J_2	Cui *et al.* [26]
Plesioblattogryllidae	*Plesioblattogryllus magnificus* Huang, Nel & Petrulevicius, 2008	Ningcheng, Inner Mongolia	Jiulongshan Fm., J_2	Huang *et al.* [23]
	Plesioblattogryllus minor Ren & Aristov, 2009	Ningcheng, Inner Mongolia	Jiulongshan Fm., J_2	Ren & Aristov [33]
Family *Incertae sedis*	*Duoduo qianae* Cui, 2012	Ningcheng, Inner Mongolia	Jiulongshan Fm., J_2	Cui [25]

References

1 Klass, K.-D., Zompro, O., Kristensen, N.P., and Adis, J. (2002). Mantophasmatodea: a new insect order with extant members in the afrotropics. *Science* **296** (5572): 1456–1459. https://doi.org/10.1126/science.1069397.

2 Klass, K.-D., Picker, M.D., Damgaard, J. *et al.* (2003). The taxonomy, genitalic morphology, and phylogenetic relationships of southern African Mantophasmatodea (Insecta). *Entomologische Abhandlungen (Dresden)* **61**: 3–67.

3 Picker, M., Griffiths, C., and Weaving, A. (2002). *Field Guide to Insects of South Africa*. New Holland Books.

4 Zompro, O., Adis, J., and Weitschat, W. (2002). A review of the order Mantophasmatodea (Insecta). *Zoologischer Anzeiger-A Journal of Comparative Zoology* **241** (3): 269–279. https://doi.org/10.1078/0044-5231-00080.

5 Zompro, O., Adis, J., Bragg, P.E. *et al.* (2003). A new genus and species of Mantophasmatidae (Insecta: Mantophasmatodea) from the Brandberg Massif, Namibia, with notes on behaviour. *Cimbebasia* **19**: 13–24.

6 Terry, M.D. and Whiting, M.F. (2005). Mantophasmatodea and phylogeny of the lower neopterous insects. *Cladistics* **21** (3): 240–257. https://doi.org/10.1111/j.1096-0031.2005.00062.x.

7 Cameron, S.L., Barker, S.C., and Whiting, M.F. (2006). Mitochondrial genomics and the new insect order Mantophasmatodea. *Molecular Phylogenetics and Evolution* **38** (1): 274–279. https://doi.org/10.1016/j.ympev.2005.09.020.

8 Bai, M., Jarvis, K., Wang, S.Y. *et al.* (2010). A second new species of ice crawlers from China (Insecta:

Grylloblattodea), with thorax evolution and the prediction of potential distribution. *PLoS One* **5**: e12850. https://doi.org/10.1371/journal.pone.0012850.

9 Schoville, S.D. and Roderick, G.K. (2010). Evolutionary diversification of cryophilic Grylloblatta species (Grylloblattodea: Grylloblattidae) in alpine habitats of California. *BMC Evolutionary Biology* **10** (1): 163. https://doi.org/10.1186/1471-2148-10-163.

10 Schoville, S.D., Uchifune, T., and Machida, R. (2013). Colliding fragment islands transport independent lineages of endemic rock-crawlers (Grylloblattodea: Grylloblattidae) in the Japanese archipelago. *Molecular Phylogenetics and Evolution* **66** (3): 915–927. https://doi.org/10.1016/j.ympev.2012.11.022.

11 Wipfler, B., Bai, M., Schoville, S. *et al.* (2014). Ice Crawlers (Grylloblattodea) – the history of the investigation of a highly unusual group of insects. *Journal of Insect Biodiversity* **2** (2): 1–25. https://doi.org/10.12976/jib/2014.2.2.

12 Engel, M.S. and Grimaldi, D.A. (2004). A new rock crawler in Baltic amber, with comments on the order (Mantophasmatodea: Mantophasmatidae). *American Museum Novitates* **3431**: 1–11. https://doi.org/10.1206/0003-0082(2004)431<0001:ANRCIB>2.0.CO;2.

13 Arillo, A. and Engel, M.S. (2006). Rock crawlers in Baltic amber (Notoptera: Mantophasmatodea). *American Museum Novitates* **3539**: 1–10. https://doi.org/10.1206/0003-0082(2006)3539[1:RCIBAN]2.0.CO;2.

14 Crampton, G.C. (1915). The thoracic sclerites and the systematic position of *Grylloblatta campodeiformis* Walker, a remarkable annectant, "orthopteroid" insect. *Entomological News* **26** (10): 337–350.

15 Wang, S.Y. (1987). The discovery of Grylloblattodea in china and the description of a new species. *Acta Entomologica Sinica* **30** (4): 423–429.

16 Storozhenko, S.Y. (2002). Order Grylloblattida Walker, 1914 (= Notoptera, Crampton, 1915, = Grylloblattodea Brues et Melander, 1932, + Protorthoptera Handlirsch, 1906, = Paraplecoptera Martynov, 1925, + Protoperlaria Tillyard, 1928). In: *History of Insects* (ed. A.P. Rasnitsyn and D.L.J. Quicke), 278–281. Dordrecht: Kluwer Academic Publishers.

17 Zompro, O. (2001). The Phasmatodea and *Raptophasma* n. gen., Orthoptera *incertae sedis*, in Baltic amber (Insecta: Orthoptera). *Mitteilungen des Geologisch-Paläontologischen Institutes der Universität Hamburg* **85**: 229–261.

18 Zompro, O. (2005). Inter- and intra-ordinal relationships of the Mantophasmatodea, with comments on the phylogeny of polyneopteran orders (Insecta: Polyneoptera). *Mitteilungen des Geologisch-Paläontologischen Institutes der Universität Hamburg* **89**: 85–116.

19 Huang, D.-Y., Nel, A., Zompro, O., and Walker, A. (2008). Mantophasmatodea now in the Jurassic. *Naturwissenschaften* **95** (10): 947–952. https://doi.org/10.1007/s00114-008-0412-x.

20 Storozhenko, S.Y. (1998). *Sistematika, filogeniya i evolyutsiya grilloblattidovykh nasekomykh (Insecta: Grylloblattida)*. Vladivostok: Dal'nauka.

21 Béthoux, O. and Nel, A. (2010). Description of a new grylloblattidan insect from Montceau-les-Mines (Pennsylvanian; France) and definition of *Phenopterum* Carpenter, 1950. *Systematic Entomology* **35** (3): 546–553. https://doi.org/10.1111/j.1365-3113.2010.00527.x.

22 Beckemeyer, R.J. and Hall, J.D. (2007). The entomofauna of the Lower Permian fossil insect beds of Kansas and Oklahoma, USA. *African Invertebrates* **48** (1): 23–39.

23 Huang, D.-Y., Nel, A., and Petrulevičius, J.F. (2008). New evolutionary evidence of Grylloblattida from the Middle Jurassic of Inner Mongolia, north-east China (Insecta, Polyneoptera). *Zoological Journal of the Linnean Society* **152** (1): 17–24. https://doi.org/10.1111/j.1096-3642.2007.00351.x.

24 Huang, D.-Y. and Nel, A. (2008). New 'Grylloblattida' related to the genus *Prosepididontus* Handlirsch, 1920 in the Middle Jurassic of China (Insecta: Geinitziidae). *Alcheringa* **32** (4): 395–403. https://doi.org/10.1080/03115510802417893.

25 Cui, Y. (2012). New data on the Blattogryllidae-Plesioblattogryllidae-Grylloblattidae complex (Insecta: Grylloblattida: Blattogryllopterida tax. n.). *Arthropod Systematics & Phylogeny* **70** (3): 167–180.

26 Cui, Y., Béthoux, O., Shih, C.K., and Ren, D. (2010). A new species of the family Juraperlidae (Insecta: Grylloblattida) from the Middle Jurassic of China. *Acta Geologica Sinica (English edition)* **84** (4): 710–713. https://doi.org/10.1111/j.1755-6724.2010.00274.x.

27 Cui, Y., Béthoux, O., Klass, K.-D., and Ren, D. (2015). The Jurassic Bajanzhargalanidae (Insecta: Grylloblattida?): new genera and species, and data on postabdominal morphology. *Arthropod Structure & Development* **44** (6): 688–716. https://doi.org/10.1016/j.asd.2015.04.008.

28 Cui, Y., Storozhenko, S.Y., and Ren, D. (2012). New and little-known species of Geinitziidae (Insecta: Grylloblattida) from the Middle Jurassic of China, with notes on taxonomy, habitus and habitat of these insects. *Alcheringa* **36** (2): 251–261. https://doi.org/10.1080/03115518.2012.628806.

29 Vršanský, P., Storozhenko, S.Y., Labandeira, C.C., and Ihringova, P. (2001). *Galloisiana olgae* sp. nov. (Grylloblattodea: Grylloblattidae) and the paleobiology of a relict order of insects. *Annals of the Entomological*

Society of America **94** (2): 179–184. https://doi.org/10.1603/0013-8746(2001)094[0179:GOSNGG]2.0.CO;2.

30 Beutel, R.G. and Gorb, S.N. (2006). A revised interpretation of the evolution of attachment structures in Hexapoda with special emphasis on Mantophasmatodea. *Arthropod Systematics & Phylogeny* **64** (1): 3–25.

31 Huang, D.-Y. and Nel, A. (2007). A new Middle Jurassic "grylloblattodean" family from China (Insecta: Juraperlidae fam. n.). *European Journal of Entomology* **104** (4): 837–840. https://doi.org/10.14411/eje.2007.104.

32 Rasnitsyn, A.P. (1976). Grylloblattidy – sovremennye predstaviteli otryada Protoblattod (Insecta, Protoblattodea) [Grylloblattidae: modern representatives of order Protoblattodea (Insecta)]. *Doklady Akademii nauk SSSR* **228** (2): 502–504.

33 Ren, D. and Aristov, D.S. (2011). A new species of *Plesioblattogryllus* Huang, Nel et Petrulevičius (Grylloblattida: Plesioblattogryllidae) from the Middle Jurassic of China. *Paleontological Journal* **45** (3): 273–274. https://doi.org/10.1134/S0031030111030178.

34 Handlirsch, A. (1906). *Die Fossilen Insekten und die Phylogenie der rezenten Formen: Ein Handbuch für paläontologen und Zoologen*. Leipzig: Engelmann.

35 Riek, E.F. (1976). A new collection of insects from the Upper Triassic of South Africa. *Annals of the Natal Museum* **22** (3): 791–820.

36 Aristov, D.S. (2009). Review of the stratigraphic distribution of Permian Grylloblattida (Insecta) with description of new taxa. *Paleontological Journal* **43** (6): 37–45. https://doi.org/10.1134/S0031030109060070.

37 Handlirsch, A. (1920). Kapitel 7. In: *Handbuch der Entomologie. III* (ed. C. Schröder), 117–304. Jena: G. Fischer.

38 Martynov, A.V. (1937). Liassic insects of Shurab and Kizil-Kija. *Trudy Paleontologicheskogo Instituta Akademii Nauk SSSR* **7**: 1–232.

39 Sharov, A.V. (1961). Paleozoic insects of the Kuznetsk Basin: the orders Protoblattodea and Paraplecoptera. *Trudy Paleontologicheskogo Instituta Akademii Nauk SSSR* **85**: 157–234.

40 Zeuner, F.T. (1939). *Fossil Orthoptera Ensifera*. London: British Museum of Natural History.

41 Rasnitsyn, A.P. and Zherikhin, V.V. (2002). Alphabetic list of selected insect fossil localities – impression fossils. In: *History of Insects* (ed. A.P. Rasnitsyn and D.L.J. Quicke), 437–444. Dordrecht: Kluwer Academic Publishers https://doi.org/10.1007/0-306-47577-4_4.

42 Storozhenko, S.Y. (1988). A review of the family Grylloblattidae. *Articulata* **3** (5): 167–181.

43 Ansorge, J. (1996). Insekten aus dem oberen Lias von Grimmen (Vorpommern, Norddeutschland). *Neue Paläontologische Abhandlungen* **2**: 1–132.

44 Aristov, D.S. (2004). The fauna of grylloblattid insects (Grylloblattida) of the Lower Permian locality of Tshekarda. *Paleontological Journal* **38** (Suppl 2): 80–145.

11

Dermaptera – Earwigs

Mingyue Ren[1], Chungkun Shih[1,2], Changyue Xing[1], and Dong Ren[1]

[1] Capital Normal University, Haidian District, Beijing, China
[2] National Museum of Natural History, Smithsonian Institution, Washington, DC, USA

11.1 Introduction to Dermaptera

Dermaptera, meaning "skin wings", highlight earwigs having short leather-like, but rarely used, forewings. The name "earwig" is derived from old western folklore that claimed that earwigs at night would crawl through people's ears and lay eggs in the brain, which is totally unfounded. Earwigs are easy to recognize by their forceps-like pincers, the cerci, at the terminal part of their abdomens. When disturbed, they usually heft the abdomen and open the forceps to disturb and scare foes. Most earwigs are active at night, feeding or scavenging on a wide variety of food, including living and dead plants, insects and small arthropods, while often hiding during the day in small, moist crevices or under stones or plants. Due to their nocturnal habit, feeding on decaying matter and/or emitting a foul smell, earwigs are often deemed as mysterious and unwelcome.

However, female earwigs do present a positive side, namely, maternal care, by protecting and taking care of their eggs and nymphs until their second molt. After laying the eggs, the mother usually sits above or beside the eggs while cleaning them to prevent fungal infection (Figure 11.1). When the eggs or nymphs are threatened, the mother reacts with an attack using her cerci. Some female earwigs provide the 1st instar nymphs with food or guide them to food sources. The females' special maternal care for the eggs and nymphs might play an important role in enhancing the survival of their offspring.

Dermaptera species usually have a slender and flattened body with brownish or black coloring, and body lengths ranging from 4 to 50 mm. The head is prognathous and compound eyes well-developed. Ocelli are vestigial or absent, but they are present in some basal earwig fossils. The antennae are thread-like, typically with 10–30 antennomeres, but some species have more than 50 antennomeres, which are used to sense the surroundings. Earwigs have wings or are wingless. The forewings, also known as tegmina, are leathery and short, covering the dorsal sides of the thorax at rest and protecting the hind wings, but not used for flight. The hind wings are large, membranous, fan-shaped or semi-circular with expanded anal lobe and radiated venation. In general, winged species are not good at flying and cannot fly for a long distance. The abdomen is long, with 8–10 exposed segments which are freely bendable. The 3rd and the 4th segments of some species have scent glands to secrete a foul-smelling fluid for protection. It has two pincer-like cerci (also called forceps) at the rear of the abdomen. The cerci are used for defending, mating, feeding and folding the hind wings. There are some differences between the cerci of males and females, i.e. the cerci are more curved and larger for males, but straight, less curved and shorter for females.

Most earwigs are omnivorous, but phytophagous or predacious species are also known. Some species live on decaying material. They are active and are easily attracted by lights at night. Female earwigs lay their egg clusters, about 30–50 eggs, not long after mating, on leaves, under a rock or a log, in a nest cavity or in a tunnel underground. It takes six weeks to one year for the nymphs to mature into adults. The nymphs share with adults in their living environment. Young nymphs are usually agile, but the adults become increasingly less mobile due to aging.

The Dermaptera are classified into three suborders according to Engel [1]: Archidermaptera, Eodermaptera, and Neodermaptera. Archidermaptera and Eodermaptera are only reported from fossil records. All extant species, about 2000 described, belong to the Neodermaptera. They are found around the world except for extremely cold regions.

Rhythms of Insect Evolution: Evidence from the Jurassic and Cretaceous in Northern China, First Edition.
Edited by Dong Ren, Chungkun Shih, Taiping Gao, Yongjie Wang, and Yunzhi Yao.

Figure 11.1 Maternal care for cleaning and protecting eggs. Source: Photos by Jason Shih.

11.2 Progress in the Studies of Fossil Dermaptera

The study of fossil earwigs was initiated in the middle of the nineteenth century. In 1856, Massalongo reported the first fossil earwig, which was a Forficulidae earwig from the Eocene of Italy [2]. Due to the significant differences between the extant and extinct groups and the poorly preserved fossil earwigs, they are easily confused with some groups of Coleoptera. The taxonomic position of Dermaptera changed many times. Until 1910, Burr put the Dermaptera as an independent order [3]. In recent years, the fossil records of Dermaptera have been steadily growing. To date, 50 genera and 86 species in nine families have been described. The earliest known earwig fossil was found from the Middle Jurassic of Northeastern China [4, 5].

In China, the study of fossil Dermaptera began in 1935. Ping described *Mesoforficula sinkianensis* Ping, 1935 from the Late Jurassic of Xinjiang [6]. Since then, many species have been reported. Hitherto, about 23 fossil species have been described in China. In 2010, Zhao et al. reported *Belloderma* Zhao, Shih & Ren 2010 in Bellodermatidae as a transition from Archidermaptera to Eodermaptera [5]. Their phylogenetic analyses suggested that the oldest Dermaptera might be traced back to the Late Triassic to the Early Jurassic. Up to now, 17 species of 14 genera belonging to four families have been described from the Middle Jurassic and Early Cretaceous of Northern China (Table 11.1). They are mainly from Northeastern and Northwestern China. Most fossils were from the Middle Jurassic Jiulongshan Formation and the Lower Cretaceous Yixian Formation.

11.3 Representative Fossils of Dermaptera from Northern China

Suborder Archidermaptera Bey-Bienko, 1936

Superfamily Protodiplatyoidea Martynov, 1925

Family Dermapteridae Vishniakova, 1980

Dermapteridae, one of the extinct families, are different from the extant earwigs. Antenna filiform with more than 19 antennomeres. Two ocelli present. Pronotum wider than long, shaped diversiform. Tegmina strongly convex and heavily sclerotized, relatively long. Claws well-developed and bearing broom-shaped arolia. Abdomen slender and elongated. Male paramere crookedly or directly back. Cerci sharp and short. Hitherto, three genera and three species of Northern China fossils have been described [4, 7].

Genera included from the Jurassic of Northern China: *Sinopalaeodermata* Zhang, 2002, *Jurassimedeola* Zhang, 2002 and *Palaeodermapteron* Zhao, Ren & Shih, 2011.

Sinopalaeodermata Zhang, 2002

Sinopalaeodermata Zhang, 2002, *Acta Micropalaeontol. Sin.*, 19 (4), 351 [4] (original designation).

Type species: *Sinopalaeodermata neimonggolensis* Zhang, 2002.

Antenna with more than 19 antennomeres, Prosternum quite narrow but elongated longitudinally, and slightly constricted near coxal cavities of prothorax. Claws well-developed and bearing broom-shaped arolia. Cerci soft, elongated, filiform and multi-articulated.

Distribution and age: Inner Mongolia; Middle Jurassic.

Only one species included from the Jurassic of Northern China (see Table 11.1).

***Jurassimedeola* Zhang, 2002**

Jurassimedeola Zhang, 2002, *Acta Micropalaeontol. Sin.*, 19 (4), 353 [4] (original designation).

Type species: *Jurassimedeola orientalis* Zhang, 2002.

Antenna moderately robust, with the 1st flagellomere clearly thicker and longer than the 3rd. Tegmina with longitudinal veins weakly developed, and hardly visible. In hind wing, Sc and R elongated, ending at costal margin. Exposed vaginulae short, nearly as long as the last abdominal segment.

Distribution and age: Inner Mongolia; Middle Jurassic.

Only one species included from the Jurassic of Northern China (see Table 11.1).

***Palaeodermapteron* Zhao, Ren & Shih, 2011**

Palaeodermapteron Zhao, Ren & Shih, 2011, *Acta Geol. Sin.*, 85 (1), 75–80 [7] (original designation).

Type species: *Palaeodermapteron dicranum* Zhao, Ren & Shih, 2011.

Head relatively small, antenna thin; pronotum transverse, anterior margin almost as wide as the posterior corners, with two sharp and frontad anterior corners. Tegmina with longitudinal veins strongly developed, costal margin convex. Pygidium transverse. The 1st segment of cercus shorter than the succeeding ones.

Distribution and age: Inner Mongolia; Middle Jurassic.

Only one species included from the Jurassic of Northern China (see Table 11.1).

Family Protodiplatyidae Martynov, 1925

Fossil records of Protodiplatyidae (=Longicerciatidae) are quite rare. To date, only 12 genera and 18 species of this family have been documented from four countries: Kazakhstan, UK, Mongolia, and China [7–14]. Diagnostic characters of Protodiplatyidae include retention of venation (albeit quite reduced) in their tegmina, long and segmented cerci, antenna with at least 26 antennomeres and pentamerous tarsi.

Genera included from the Jurassic and Cretaceous of Northern China: *Longicerciata* Zhang, 1994, *Sinoprotodiplatys* Nel, Aria, Garrouste & Waller, 2012, *Barbderma* Xing, Shih & Ren, 2016, *Perissoderma* Xing, Shih & Ren, 2016 and *Abrderma* Xing, Shih & Ren, 2016.

***Longicerciata* Zhang, 1994**

Longicerciata Zhang, 1994, *Acta Palaeontol. Sin.*, 33, 231 [12] (original designation).

Type species: *Longicerciata mesozoica* Zhang, 1994.

Head large; the 3rd flagellomere longer than the 4th flagellomere. Hind wing is less than half the length of the abdomen. The cercus is segmented with at least 36 articles, as long as the body.

Distribution and age: Shandong; Early Cretaceous.

Two species included from the Cretaceous of Northern China (see Table 11.1).

***Sinoprotodiplatys* Nel, Aria, Garrouste & Waller, 2012**

Sinoprotodiplatys Nel, Aria, Garrouste & Waller, 2012, *Cretac. Res.*, 33, 190 [13] (original designation).

Type species: *Sinoprotodiplatys zhangi* Nel, Aria, Garrouste & Waller, 2012.

The specific epithet is dedicated to Dr. J.F. Zhang, a Chinese paleoentomologist. Body rather large, antenna with 18 antennomeres; mandibles with two apical teeth; pronotum with anterior and posterior margins subequal in width; tegmina without visible veins; hind wing extending beyond the fourth abdominal segment; cercus filiform, very long and with about 20 articles.

Distribution and age: Liaoning; Early Cretaceous.

Two species included from the Jurassic and Cretaceous of Northern China (see Table 11.1).

***Barbderma* Xing, Shih & Ren, 2016**

Barbderma Xing, Shih & Ren, 2016, *Cretac. Res*, 64, 60 [14] (original designation).

Type species: *Barbderma oblonguata* Xing, Shih & Ren, 2016.

Antenna with 19 antennomeres, the scape broader than others. Pronotum approximately oblong or trapezoidal and anterior margin almost as wide as or wider than the posterior. Tegmina with longitudinal veins; pygidium small; cercus long and about half as long as body.

Distribution and age: Liaoning; Early Cretaceous.

Only one species included from the Cretaceous of Northern China (see Table 11.1).

***Perissoderma* Xing, Shih & Ren, 2016**

Perissoderma Xing, Shih & Ren, 2016, *Zootaxa*, 4205 (2), 182 [15].

Type species: *Perissoderma triangulum* Xing, Shih & Ren, 2016.

Body densely setae. Antenna with 17 antennomeres. Abdominal segments posterior to the sixth segment compacted into a short structure. Tegmina with longitudinal veins, costal margin arched. Pygidium small.

Cercus long, about half as long as body.

Distribution and age: Inner Mongolia; Middle Jurassic.

Only one species included from the Jurassic of Northern China (see Table 11.1).

Abrderma Xing, Shih & Ren, 2016

Abrderma Xing, Shih & Ren, 2016, *Zootaxa*, 4205 (2), 185 [15] (original designation).

Type species: *Abrderma gracilentum* Xing, Shih & Ren, 2016.

Head triangular and small. Antenna with 17–19 antennomeres. Mandibles with two apical teeth. Two ocelli present. Tegmen large and long with longitudinal veins strongly developed, sutural margin straight, costal margin arched. Cerci short. Tarsomeres 1–4 shortened, horseshoe-like.

Distribution and age: Inner Mongolia; Middle Jurassic.

Only one species included from the Jurassic of Northern China (see Table 11.1).

Abrderma gracilentum Xing, Shih & Ren, 2016 (Figure 11.2)

Abrderma gracilentum Xing, Shih & Ren, 2016: *Zootaxa*, 4205 (2), 185.

Locality and horizon: Daohugou, Ningcheng, Inner Mongolia, China; Middle Jurassic, Jiulongshan Formation.

The body length excluding antenna and cercus is 17.4 mm. Head subtriangular, small. Ocelli present; compound eye large, prominent and located near posterior margin of head. Pronotum approximately elliptical, 1.3 mm wide and 2.5 mm long. Tegmen with longitudinal veins strongly developed, 5.3 mm long and 2.5 mm wide. Costal margin slightly arched, sutural margin straight, posterior margin apical. Abdomen distally with external ovipositor, 2.4 mm long, slot present at the middle. Pygidium small [15].

Figure 11.2 *Abrderma gracilentum* Xing, Shih & Ren, 2016, (Holotype, CNU-DER-NN2016002p).

Suborder Eodermaptera Engel, 2003

Family Bellodermatidae Zhao, Shih & Ren, 2010

Bellodermatidae, bridging the missing link between suborders of Archidermaptera and Eodermaptera, have greatly enhanced the understanding of early evolution of Dermaptera [5]. The family can be diagnosed by: tegmina elongate along the sutural margin, with costal margin and outer margin strongly arched, while retaining venation; tarsi three-segmented and forficulid-type; female with exposed ovipositor; cerci multi-segmented but short, especially asymmetrical.

Only one genus included from the Jurassic of Northern China: *Belloderma* Zhao, Shih & Ren, 2010.

Belloderma Zhao, Shih & Ren, 2010

Belloderma Zhao, Shih & Ren, 2010, *BMC Evol. Biol*, 10, 344 [5] (original designation).

Type species: *Belloderma arcuata* Zhao, Shih & Ren, 2010.

Head relatively large, and posterior margin strongly notched. Antenna with the 1st flagellomere broad but shorter than or equal to the 3rd. Eyes large. Pronotum transverse, anterior margin almost as wide as posterior one. Mesoscutellum exposed or not. Tarsi shorter than tibiae. Pygidium small. The 1st article of cercus longer than succeeding ones.

Distribution and age: Inner Mongolia; Middle Jurassic.

Two species included from the Jurassic of Northern China (see Table 11.1).

Belloderma arcuata Zhao, Shih & Ren, 2010 (Figures 11.3 and 11.4).

Belloderma arcuata Zhao, Shih & Ren, 2010: *BMC Evol. Biol*, 10, 344.

Locality and horizon: Daohugou, Ningcheng, Inner Mongolia, China; Middle Jurassic, Jiulongshan Formation.

Body (excluding antenna and cercus) 15.5 mm long, covered with pubescence. Head 2.3 mm long, 2.6 mm wide and subtriangular. Pronotum about 1.8 times as wide as long; anterior margin straight, other margins round. Tegmina about three times as long as wide,

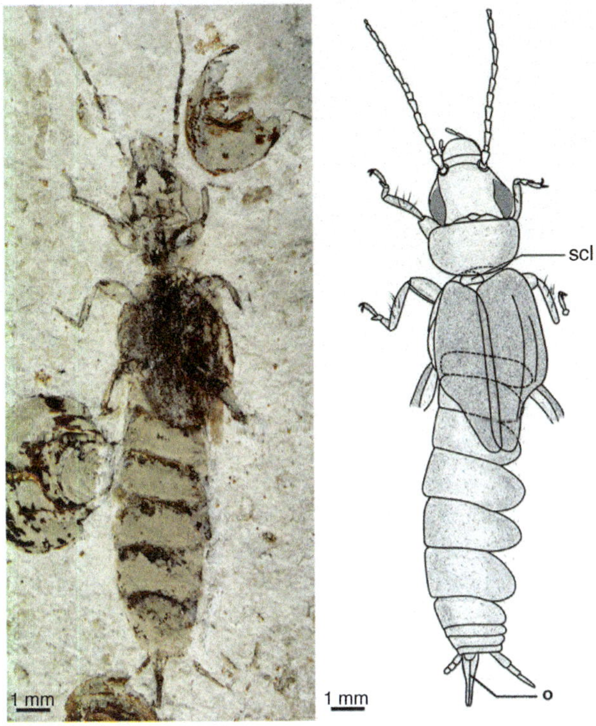

(a) (b)

Figure 11.3 *Belloderma arcuata* Zhao, Shih & Ren, 2010, (Holotype, CNU-DER-NN2008002), (a) Photograph and (b) Line drawing. Source: Donated by Dr. Chungkun Shih.

Figure 11.4 Reconstruction of *Belloderma arcuata* Zhao, Shih & Ren, 2010. Source: Artwork by Dr. Chen Wang.

extending backward to the 2nd abdominal segment. Length ratio of foreleg femur: tibia: tarsus is 1.2 : 1.2 : 0.9, and 1.3 : 1.6 : 1.1 for midleg. Abdomen longer than head and thorax combined; abdomen distally with external ovipositor, 2 mm long, slot present in middle. Pygidium small [5].

Suborder Neodermaptera Engel, 2003

Superfamily Pygidicranoidea Verhoeff, 1902

Family Pygidicranidae Verhoeff, 1902

The extant Pygidicranidae [1, 16], which is a widely distributed family, include 10 extant subfamilies and an extinct subfamily. Pygidicranidae are typified by the antenna, the 4th–6th antennomeres longer than width, and the broad and flattened abdomen. Up to now, 14 genera and 15 species of fossil Pygidicranidae have been reported ranging from the Jurassic to the Cenozoic. Among them, three genera and three species are from Northern China [11, 17, 18].

Genera included from the Cretaceous of Northern China: *Archaeosoma* Zhang, 1994, *Geosoma* Zhang, 1997 and *Cylindopygia* Yang, Ren & Shih, 2015.

Archaeosoma Zhang, 1994

Archaeosoma Zhang, 1994, *Acta Palaeontol. Sin.*, 33, 234 [12] (original designation).

Type species: *Archaeosoma serratum* Zhang, 1994.

Antenna with at least 24 flagellomeres, ocelli present, compound eyes large, located near posterior margin of head. Hind wings narrow and small. Cercus short, without dentition.

Distribution and age: Shandong; Early Cretaceous.

Only one species included from the Cretaceous of Northern China (see Table 11.1).

Geosoma Zhang, 1997

Geosoma Zhang, 1997, *Palaeoworld*, 7 (8), 83 [17] (original designation).

Type species: *Geosoma prodromum* Zhang, 1997.

Antenna elongate, over 26 flagellomeres with scape massive. Ocelli absent, compound eyes large, located at both sides of the head base. Pronotum nearly reversely trapezoidal and slightly narrower than head. Elytron short, nearly trapezoid-shaped.

Distribution and age: Jilin; Early Cretaceous.

Only one species included from the Cretaceous of Northern China (see Table 11.1).

Cylindopygia Yang, Ren & Shih, 2015

Cylindopygia Yang, Ren & Shih, 2015, *Cretac. Res.*, 52 (1), 330 [18] (original designation).

Type species: *Cylindopygia falcate* Yang, Ren & Shih, 2015.

Large body size 19.5–21 mm long, (excluding antennae and forceps). Antenna with >22 antennomeres, scape long and robust with apex part wider than basal. Pronotum approximately trapezoidal. Abdomen subcylindrical and densely setose; forceps evenly curved, falciform, with longitudinal ridges dorsoventrally.

Figure 11.5 *Cylindopygia falcate* Yang, Ren & Shih, 2015, (Holotype, CNU-DER-LB2013001p).

Distribution and age: Liaoning; Early Cretaceous.

Only one species included from the Cretaceous of Northern China (see Table 11.1).

***Cylindopygia falcate* Yang, Ren & Shih, 2015 (Figure 11.5)**

Cylindopygia falcate Yang, Ren & Shih, 2015: *Cretac. Res.*, 52 (1), 330.

Locality and horizon: Huangbanjigou, Beipiao, Liaoning, China; Lower Cretaceous, Yixian Formation.

Body large, 19.5 mm long, excluding antennae and forceps, 5.0 mm wide. Antenna 8.0 mm long, at least 22 antennomeres; scape long and robust (0.7 mm long) with apex wider than basal (the widest part 0.4 mm). Pronotum approximately trapezoidal. Tegmina without venation, symmetrical, 5.9 mm long, anterior margin truncated, outer margins convex, inner margins straight. Forceps sickle shaped, possessing longitudinal ridges dorsoventrally. Pygidium relatively large and broad connecting with each of forceps. Valvula present but not protruding from abdominal apex [18].

An Earwig in Myanmar Amber

A well-preserved adult female earwig *Gracilipygia canaliculata* Ren, Zhang, Shih & Ren, 2017 of Pygidicranidae from the mid-Cretaceous Myanmar amber has been described (Figure 11.6). Body small with a total length of 5.89 mm, excluding antennae and cercal forceps (vs. the extant species of Pygidicranidae usually have longer body of over 10 mm). Compound eyes extremely large and exophthalmic (vs. the extant pygidicranids are slightly exophthalmic). Pronotum subquadrate; femora crassate, arolium present. Pygidium prominent. This is the first amber species assigned to the subfamily Pyragrinae. The big and simple arolium, the prominent pygidium and the elongate cercal forceps suggest that *G. canaliculata* represents a stem taxa of Pyragrinae [19].

Figure 11.6 *Gracilipygia canaliculata* Ren, Zhang, Shih & Ren, 2017. (Holotype, CNU-DER-MA2016001).

Family *Incertae sedis*

***Mesoforficula* Ping, 1935**

Mesoforficula Ping, 1935, *Chin. Jour. Zool.*, 1, 108 [6] (original designation).

Type species: *Mesoforficula sinkianensis* Ping, 1935.

The fore and hind wings are narrow. The antenna is comparatively short, the thorax and abdomen slender, the cerci simple and short.

Distribution and age: Xinjiang; Late Jurassic.

Only one species included from the Jurassic of Northern China (see Table 11.1).

***Atopderma* Zhao, Ren & Shih, 2010**

Atopderma Zhao, Ren & Shih, 2010, *Insect Sci.*, 17, 460 [20] (original designation).

Type species: *Atopderma ellipta* Zhao, Ren & Shih, 2010.

Head relatively large. Antenna with the 1st antennomere broad and longer than or equal to the 3rd. Eyes medium sized. Pronotum transverse. Mesoscutellum not exposed. Tegmina costal margin convex. Tarsi longer than or equal to tibiae [20].

Distribution and age: Inner Mongolia; Middle Jurassic.

Only one species included from the Jurassic of Northern China (see Table 11.1).

Table 11.1 A list of fossil Dermaptera from the Jurassic and Cretaceous of China.

Family	Species	Locality	Horizon/Age	Citation
	Suborder Archidermaptera Bey-Bienko, 1936			
Dermapteridae	*Jurassimedeola orientalis* Zhang, 2002	Ningcheng, Inner Mongolia	Jiulongshan Fm., J_2	Zhang [4]
	Palaeodermapteron dicranum Zhao, Ren & Shih, 2011	Ningcheng, Inner Mongolia	Jiulongshan Fm., J_2	Zhao *et al.* [7]
	Sinopalaeodermata neimonggolensis Zhang, 2002	Ningcheng, Inner Mongolia	Jiulongshan Fm., J_2	Zhang [4]
Protodiplatyidae (=Longicerciatidae)	*Abrderma gracilentum* Xing, Shih & Ren, 2016	Ningcheng, Inner Mongolia	Jiulongshan Fm., J_2	Xing *et al.* [15]
	Barbderma oblonguata Xing, Shih & Ren, 2016	Beipiao, Liaoning	Yixian Fm., K_1	Xing *et al.* [14]
	Longicerciata mesozoica Zhang, 1994	Laiyang, Shandong	Laiyang Fm., K_1	Zhang [12]
	Longicerciata rumpens Zhang, 1994	Laiyang, Shandong	Laiyang Fm., K_1	Zhang [12]
	Perissoderma triangulum Xing, Shih & Ren, 2016	Ningcheng, Inner Mongolia	Jiulongshan Fm., J_2	Xing *et al.* [15]
	Sinoprotodiplatys zhangi Nel, Aria, Garrouste & Waller, 2012	Beipiao, Liaoning	Yixian Fm., K_1	Nel *et al.* [13]
	Sinoprotodiplatys ellipsoideuata Xing, Shih & Ren, 2016	Beipiao, Liaoning	Yixian Fm., K_1	Xing *et al.* [14]
	Suborder Eodermaptera Engel, 2003			
Bellodermatidae	*Belloderma arcuate* Zhao, Ren & Shih, 2010	Ningcheng, Inner Mongolia	Jiulongshan Fm., J_2	Zhao *et al.* [5]
	Belloderma ovata Zhao, Ren & Shih, 2010	Ningcheng, Inner Mongolia	Jiulongshan Fm., J_2	Zhao *et al.* [5]
	Suborder Neodermaptera Engel, 2003			
Pygidicranidae	*Archaeosoma serratum* Zhang, 1994	Laiyang, Shandong	Laiyang Fm., K_1	Zhang [12]
	Cylindopygia falcate Yang, Ren & Shih, 2015	Beipiao, Liaoning	Yixian Fm., K_1	Yang *et al.* [18]
	Geosoma prodromum Zhang, 1997	Zhixin Basin, Jilin	Dalazi Fm., K_1	Zhang [17]
Family *Incertae sedis*	*Atopderma ellipta* Zhao, Ren & Shih, 2010	Ningcheng, Inner Mongolia	Jiulongshan Fm., J_2	Zhao *et al.* [20]
	Mesoforficula sinkianensis Ping, 1935	Turfan, Sinkiang	Maiyohkow Fm., J_3	Ping [6]

References

1 Engel, M.S. (2003). The earwigs of Kansas, with a key to genera north of Mexico (Insecta: Dermaptera). *Transactions of the Kansas Academy of Science* **106**: 115–123.

2 Massalongo, A.B.P. (1856). *Prodromo di un' entomologia fossile del M. Bolca*, 11–21. Verona, Antonelli: Studii Paleontologia [School Program].

3 Burr, M. (1910). *Fauna of British India, including Ceylon and Burma. Dermaptera*, 1–217. Published under the authority of the Secretary of State for India in Council, London.

4 Zhang, J.F. (2002). The most primitive earwigs (Archidermaptera, Dermaptera insect) from the Upper Jurassic of Nei Monggol Autonomous Region, Northeastern China. *Acta Micropalaeontologica Sinica* **17**: 459–464.

5 Zhao, J.X., Zhao, Y.Y., Shih, C.K. *et al.* (2010). Transitional fossil earwigs – a missing link in Dermaptera

evolution. *BMC Evolutionary Biology* **10** (344): 1–10. https://doi.org/10.1186/1471-2148-10-344.

6 Ping, C. (1935). On four fossil insects from Sinkiang. *Chinese Journal of Zoology* **1**: 107–115.

7 Zhao, J.X., Shih, C.K., Ren, D., and Zhao, Y.Y. (2011). New primitive fossil earwig from Daohugou, Inner Mongolia, China (Insecta: Dermaptera: Archidermaptera). *Acta Geologica Sinica – English Edition* **85** (1): 75–80.

8 Martynov, A.V. (1925). To the knowledge of fossil insects from Jurassic beds in Turkestan. 2. Raphidioptera (continued), Orthoptera (s.l.), Odonata. *Neuroptera Bulletin de l'Academie des Sciences de l'Union des Republiques Sovietiques Socialistes* **19**: 569–598.

9 Vishniakova, V.H. (1980). Earwig from the Upper Jurassic the Karatau range (insect, Forficulida). *Paleontological Journal* 78–94. (in Russian).

10 Vishniakova, V.H. (1986). Earwigs. Forficulida (= Dermaptera). In Nasekomye v rannemelovykh ekosistemakh zapadnoy Mongolii. In: *The Joint Soviet-Mongolian Palaeontological Expedition*, vol. **28** (ed. A.P. Rasnitsyn), 1–171. Moscow: Nauka (in Russian).

11 Whalley, P.E.S. (1985). The systematics and palaeogeography of the Lower Jurassic insects of Dorset, England. *Bulletin of the British Museum (Natural History). Geology Series* **39** (3): 107–189.

12 Zhang, J.F. (1994). Discovery of primitive fossil earwigs (Insecta) from the Late Jurassic of Laiyang, Shandong and its significance. *Acta Palaeontologica Sinica* **33**: 229–245.

13 Nel, A., Aria, C., Garrouste, R., and Waller, A. (2012). Evolution and palaeosynecology of the Mesozoic earwigs (Insecta: Dermaptera). *Cretaceous Research* **33**: 189–195. https://doi.org/10.1016/j.cretres.2011.10.002.

14 Xing, C.Y., Shih, C.K., Zhao, Y.Y., and Ren, D. (2016). New protodiplatyids (Insecta: Dermaptera) from the Lower Cretaceous Yixian Formation of Northeastern China. *Cretaceous Research* **64**: 59–66. https://doi.org/10.1016/j.cretres.2016.03.016.

15 Xing, C.Y., Shih, C.K., Zhao, Y.Y., and Ren, D. (2016). New earwigs in Protodiplatyidae (Insecta: Dermaptera) from the Middle Jurassic Jiulongshan Formation of Northeastern China. *Zootaxa* **4205** (2): 180–188.

16 Verhoeff, K.W. (1902). Über Dermapteren. 1. Aufsatz: Versuch eines neuen, natürlichen Systems auf vergleichend-morphologischer Grundlage und über den Mikrothorax der Insecten. *Zoologischer Anzeiger* **25** (665): 181–208.

17 Zhang, H.C. (1997). Early Cretaceous insects from the Dalazi Formation of the Zhixin Basin, Jilin Province, China. *Palaeoworld* **7**: 75–103.

18 Yang, D., Shih, C.K., and Ren, D. (2015). The earliest pygidicranid (Insecta: Dermaptera) from the Lower Cretaceous of China. *Cretaceous Research* **52** (1): 329–335. https://doi.org/10.1016/j.cretres.2014.03.008.

19 Ren, M.Y., Zhang, W.T., Shih, C.K., and Ren, D. (2017). A new earwig (Dermaptera: Pygidicranidae) from the Upper Cretaceous Myanmar amber. *Cretaceous Research* **74**: 137–141. https://doi.org/10.1016/j.cretres.2017.02.012.

20 Zhao, J.X., Ren, D., and Shih, C.K. (2010). Enigmatic earwig-like fossils from Inner Mongolia, China. *Insect Science* **17**: 459–464. https://doi.org/10.1111/j.1744-7917.2010.01315.x.

12

Chresmodidae – Water-Walking Insects

Chaofan Shi[1,2], *Chungkun Shih*[2,3], *and Dong Ren*[2]

[1] *Sun Yat-sen University, Guangzhou, Guangdong, China*
[2] *Capital Normal University, Haidian District, Beijing, China*
[3] *National Museum of Natural History, Smithsonian Institution, Washington, DC, USA*

12.1 Introduction to Chresmodidae

Chresmodidae, an enigmatic and extinct insect family, superficially resembled water striders (Hemiptera: Gerridae) with large body and extremely long legs. Like water striders, chresmodids were aquatic. The long legs covered with dense and short setae are suitable for them to walk and ski across water surface [1–3]. However, considering their noteworthy large size and heavy weight, chresmodids might have needed floating vegetation to support their bodies on the water surface [4]. They may have been carnivorous, and fed on small insects or other small creatures which lived in water or fell onto the water surface, trapped by surface tension [3, 5–9].

The phylogenetic position of the family has been debated since the type genus *Chresmoda* was set up by Germar in 1839 [10–12]. To date, three genera have been confidently assigned to Chresmodidae. All the chresmodids have been found from the Middle Jurassic to the Late Cretaceous, distributed in the Eurasian Region and Brazil [3, 4, 7, 8, 13–16].

Chresmodids are medium to large insects. The head is small, covered with short and fine setae. The antenna is mostly filiform, covered with short or long setae. Sexual dimorphism is present in the antenna of this family. For *Sinochresmoda magnicornia* Zhang, Ren & Pang, 2008, the scape of the male is expanded more than that of the female and the first flagellomere of the male is incurvated, but straight for that of the female [14]. Compound eyes are rounded. Mouthparts are prognathous, probably for chewing [4, 7, 8, 13–16].

All legs are extremely long and narrow, with a series of dense setae alongside. Three pairs of legs are almost of the same length. Leg lengths vary slightly in different species. However, the length ratio of femur and tibia, especially in forelegs, shows interspecific differentiation, which has been used to distinguish species [14–16]. Tarsi are long and highly specialized, with up to more than 40 tarsomeres, which is unique among Insecta [4]. The "ultra-articulated tarsi", named by Nel *et al.* [7, 8], have been found in both adults and nymphs [15]. The high articulation of tarsi was probably adapted for water surface skiing.

It was known that nymph and male chresmodids were apterous, until the discovery of *Sinochresmoda*. To date, all the female chresmodids and male *Sinochresmoda*, as well as two male *Chresmoda*, have been found to have four wings, probably with flight capability. The wing lengths vary among genera and species. The pterygote chresmodids folded their wings roof-like over the abdomen when at rest. Very few fossils were found with unfolded wings, which illuminated the wing venation of this family [4, 14, 16]. Cerci are mostly short, with the exception of *Jurachresmoda*, which bears long and segmented cerci. Ovipositors are long, strong and sword-like, with a row of inner indentations along the posterior end. The serrated ovipositor was probably adapted to endophytic egg laying in floating or aquatic plants [7, 14, 15].

Walking on Water

One of the most impressive features of Chresmodidae is their extremely long and thin legs, which suggests their capability of walking on water surface. In fact, walking on water is not rare for creatures around the world, ranging from insects, spiders to lizards [17].

Several Heteroptera families have been observed to be able to walk or ski on water, including Gerridae (water striders), Mesoveliidae (water treaders), Hydrometridae

Rhythms of Insect Evolution: Evidence from the Jurassic and Cretaceous in Northern China, First Edition.
Edited by Dong Ren, Chungkun Shih, Taiping Gao, Yongjie Wang, and Yunzhi Yao.

(marsh treaders) and Hermatobadidae (doral treaders). Gerridae, water striders, are the most common and popular ones. On warm days, we can find gerrids walking or skiing on the water surface of a nearby pond, lake or calm stream [18]. Gerridae have a similar appearance to Chresmodidae, except that they have shorter forelegs than middle and hind legs, and four antennomeres vs. many more antennomeres.

Water striders, also known as "Jesus bugs," are able to walk on top of water due to a combination of the high surface tension of water and their long, hydrophobic legs (Figure 12.1). The long, strong and flexible legs allow the body weight to be evenly distributed over a large surface area, and flow with the water movement. The body and legs are covered with several thousand fine setae per square millimeter, to prevent wetting from waves, rain, or spray. The fine setae would trap air if the body accidentally became submerged. The air bubbles throughout the body would act as buoyancy to bring the water strider to the water surface again, as well as help breathing from underwater [18]. The momentum transfer of skiing on water is primarily in the form of subsurface vortices. The legs with setae are used like oars to create vortices or spirals in the water that generate a forward speed of up to 150 cm s^{-1} [19].

Six-spotted fishing spiders, *Dolomedes triton,* of the family Pisauridae are representatives of water-skiing spiders in Arachnida. They live near ponds or streams, and hunt during the day. These spiders are often seen with their legs sprawled out on the water while waiting for prey. They wait patiently for hours, and hunt by running over the surface of water or diving under up to 18 cm. They feed on insects, tadpoles and occasionally small fish underwater as well as preying on those falling on the water surface or traveling on water, such as water striders [20].

Basiliscus are large corytophanid lizards, famous for running across water for a significant distance. They are commonly known as the "Jesus Christ lizard," or simply the "Jesus lizard", in reference to the biblical story of Jesus walking on water. When fleeing from predators, the *Basiliscus* can run across the water for a short distance while holding most of their body out of the water. The hind feet of *Basiliscus* are large, with scaly fringes on the sides of the third, fourth, and fifth toes. These are compressed between the toes when the lizard walks on land. When it senses danger, the lizard jumps into the water, opening up the fringes against the water surface. This increases the surface area and pockets of air, thus giving it the lift needed to run across water. On water, a *Basiliscus* can run at a speed of 1.5 m s^{-1} for nearly 4.5 m before sinking. Juveniles can run up to 10–20 m [21].

Figure 12.1 Water striders (Gerridae) mating and walking on water. Source: Photo by Jason Shih.

12.2 Progress in the Studies of Fossil Chresmodidae

As mentioned before, phylogenetic placement of Chresmodidae as a family has been a long-lasting debate, ranging from Hemiptera to Paraplecoptera, Grylloneans = Polyneoptera and Orthopterida. Chresmodids were first described by Germar in 1839 [10], with the type species *Chresmoda obscura* assigned to Mantodea from the Upper Jurassic of Solnhofen-Eichstätt, Germany. The taxonomic placement of *Chresmoda* has been reinterpreted over the last two centuries, being considered related to or in Gerridae (Hemiptera), Mantodea, Paraplecoptera, and Phasmatodea. Handlirsch (1906–1908) [22] set up the family Chresmodidae in the order of Phasmatodea. Later, Martynov (1928) [23] established a suborder Chresmododea as stem-group of Phasmatodea, including four Mesozoic families: Chresmodidae (Jurassic-Cretaceous), Aeroplanidae (Triassic), Necrophasmatidae and Aerophasmidae (both Jurassic). The assignment of chresmodids to Phasmatodea was also held by Sharov in 1968 [24]. Popov, in 1980, placed the family in the Gerromorpha (Hemiptera) based on the aquatic characters of chresmodids and resemblance of gerrids [25]. Ponomarenko, in 1985, excluded the family from Paraneoptera [2] because of the well-developed cerci and 5-articulated tarsi, and referred it to the extinct Paraplecoptera, a stem-group of Polyneoptera, which was agreed by Martínez-Delclòs in 1991 [6]. Rasnitsyn and Quicke, in 2002, considered the family as unplaced within Polyneoptera [12], while Grimaldi & Engel, in 2005, tentatively placed it into Orthopterida [3]. Nel *et al.*, in 2005 [8], assigned Chresmodidae to the clade Archaeorthoptera Béthoux & Nel, 2002 [26], which does not correspond exactly to the Orthopterida sensu

Grimaldi & Engel, 2005 [3], by excluding the modern Phasmatodea. Delclòs *et al.* in 2008, also classified Chresmodidae as polyneopteran insects of the Archaeorthoptera [4], based on study of the venation of Spanish and Brazilian fossil specimens possessing unfolded forewings and hind wings.

Study of Chresmodidae in China started since 1949, when Esaki described *Chresmoda orientalis* from the Early Cretaceous of Northeastern China [27]. Zhang *et al.*, in 2008 and 2010, described five species of three genera based on several well-preserved fossil specimens from the Middle Jurassic of Inner Mongolia and the Early Cretaceous of Inner Mongolia, Liaoning and Hebei in China [14–16], as listed in Table 12.1. These taxa demonstrated sexual dimorphism in antenna and wing venation of the family, and extended our knowledge of their geological history back to the Middle Jurassic.

12.3 Representative Fossils of Chresmodidae from Northern China

Family Chresmodidae Handlirsch, 1908

Genera included from the Jurassic and Cretaceous of Northern China: *Chresmoda* Germar, 1839, *Sinochresmoda* Zhang, Ren & Pang, 2008, *Jurachresmoda* Zhang, Ren & Shih, 2008

Jurachresmoda Zhang, Ren & Shih, 2008

Jurachresmoda Zhang, Ren & Shih, 2008, *Zooxata*. 1762, 54 [15] (Original designation).

Type species: *Jurachresmoda gaskelli* Zhang, Ren & Shih, 2008.

Antenna is filiform, with more than 18 antennomeres. Mesothorax is significantly longer than prothorax and metathorax. Foretibia is longer than half of the forefemur length. Mid femur is the longest. Cercus is long. Female wings are long, exceeding the body length (including the ovipositor), almost twice as long as the abdomen. Nymph cercus is articulated.

Distribution and age: Inner Mongolia; Middle Jurassic.

Two species included from the Jurassic of Northern China (see Table 12.1).

Jurachresmoda gaskelli Zhang, Ren & Shih, 2008 (Figure 12.2)

Jurachresmoda gaskelli Zhang, Ren & Shih, 2008: *Zootaxa*, 1762, 54.

Locality and horizon: Daohugou, Ningcheng, Inner Mongolia, China; Middle Jurassic, Jiulongshan Formation.

The specific name is dedicated to Mr. Tony Gaskell for being an excellent business leader and role-model, providing guidance, motivation and encouragement to Chungkun Shih. A female adult and nymphs of the species were described from the Middle Jurassic of China, which represent the earliest fossil record of the family, about 165 million years ago. Legs are extremely long, especially the mid legs. Fringing setae are present on mid tibia and tarsi of *J. gaskelli*, which are considered to assist water-skiing. Also, the articulation of cerci was first found in Chresmodidae nymphs [15].

Chresmoda Germar, 1839

Chresmoda Germar, 1839, *Nova Acta Academia Leopoldiana Carola*. XIX, 201 [10] (Original designation).

Type species: *Chresmoda obscura* Germar, 1839.

Antennae of both male and female are filiform, covered with short setae. Legs are long and slender. Tarsi are extraordinary long, with more than 40 tarsomeres. Fringing setae are absent along the wing margins. Forewing costal area is broad near wing base and narrow distally. ScP and R are long and straight. MA has two to three branches. MP forked around the mid-length of the wing. CuA fused with MP basally. Cercus is short and unarticulated. The ovipositor is long, with two serrated valvulae.

Distribution and age: Liaoning in China, Myanmar, Lebanon, Germany, Spain and Brazil; Cretaceous.

Three species included from the Cretaceous of Northern China (see Table 12.1).

Chresmoda multinervis Zhang, Ren & Shih, 2010 (Figure 12.3)

Chresmoda multinervis Zhang, Ren & Shih, 2010: *Acta Geol. Sin.-Engl.* 84 (1), 39–41.

Locality and horizon: Fengning, Hebei, China; Lower Cretaceous, Yixian Formation.

The antenna is short, with 20 antennomeres. Scape is expanded and distinctly larger than the remaining antennomeres. Foretibia is long, slightly longer than half of forefemur length. Wings are long, exceeding the length of abdomen. Body and wings are covered with dense and short setae. Cercus is short, about 13–14% of abdomen length. Pterygote males of this species and *C. shihi* below were described, which represent the first reported and rare pterygote males in the genus *Chresmoda*, in contrast to all the apterous male *Chresmoda* from other places previously published [16].

Delclòs *et al.* have assigned Chresmodidiae to the Archaeorthoptera mainly based on the structure of median and cubital veins [4], which were not preserved on the wing of *C. multinervis*. However,

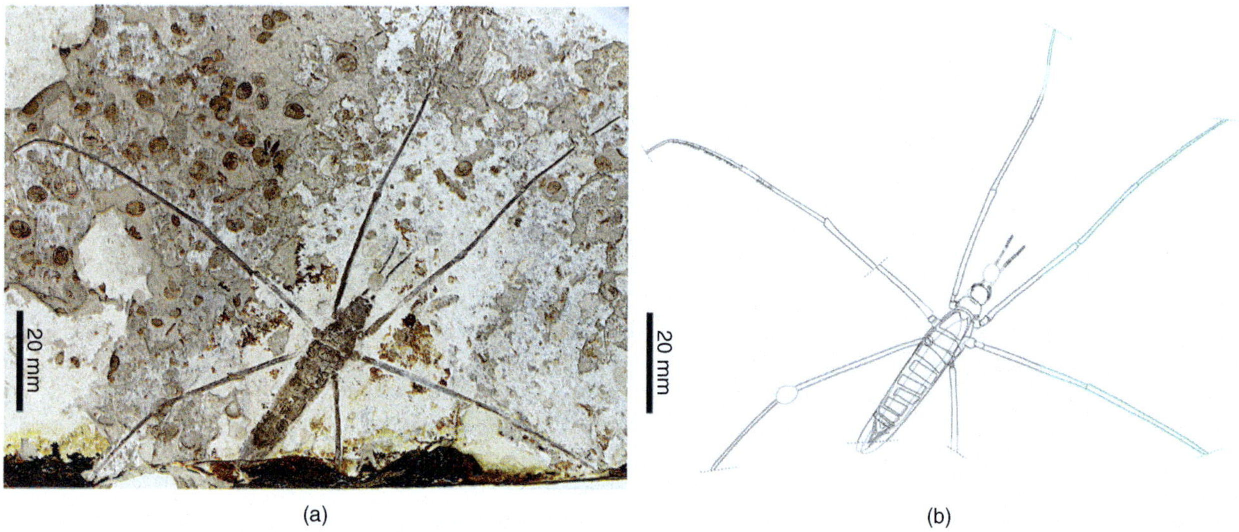

Figure 12.2 *Jurachresmoda gaskelli* Zhang, Ren & Shih, 2008 (Holotype, CNU-CH-DHG2007010). (a) Photograph; (b) Line drawing. Source: Modified from [15]. Donated by Dr. Chungkun Shih.

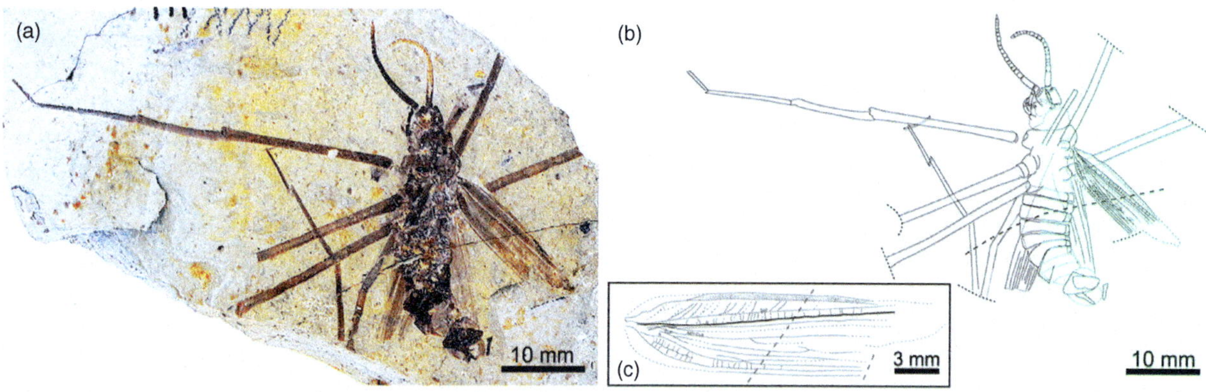

Figure 12.3 *Chresmoda multinervis* Zhang, Ren & Shih, 2010 (Holotype, CNU-CH-HF2007007). (a) Photograph; (b) Line drawing; and (c) Line drawing of wing. Source: modified from [16].

the general appearance of the forewing of this specimen, especially mostly parallel longitudinal veins, show more similarity with Phasmatodea [16].

Chresmoda shihi Zhang, Ren & Shih, 2010 (Figures 12.4 and 12.5)

Chresmoda shihi Zhang, Ren & Shih, 2010: *Acta Geol. Sin.-Engl.* 84 (1), 41–42.

Locality and horizon: Fengning, Hebei, China; Lower Cretaceous, Yixian Formation.

The specific name is dedicated to Mr. Jason Zhong Wen Shih for providing inspiration and support to Dr. Shih in his paleontology studies. Jason also provided exquisite and high quality photographs of extant insects for this book and two previous insect books published by our Capital Normal University (CNU) Team.

The antenna is short. Scape is expanded and larger than the remaining antennomeres. Foretibia is about one-third of forefemur length. Wings are long, exceeding the length of abdomen. Fringing setae are present on the mid legs in this species, as well as both *Jurachresmoda* species, which are also present in modern gerrids. Fringing setae are considered to assist water-skiing, functioning similar as gerrids' fringing setae. The presence of fringing setae only on the mid legs suggests that chresmodids might have used the mid legs for locomotional strokes, while using the fore- and hind legs for navigating, balancing or assisting its water-skiing [16].

Sinochresmoda Zhang, Ren & Pang, 2008

Sinochresmoda Zhang, Ren & Pang, 2008, *Zootaxa*. 1702, 27 [14] (Original designation).

Figure 12.4 *Chresmoda shihi* Zhang, Ren & Shih, 2010 (Holotype, CNU-CH-HF20070010). Source: Donated by Dr. Chungkun Shih.

Figure 12.5 Photo of Jason Shih who took and provided many insect photos in this book. Source: Photo by Jason Shih.

Type species: *Sinochresmoda magnicornia* Zhang, Ren & Pang, 2008.

The head is small. The antenna is normally filiform, but exhibits sexual dimorphism. The scape of the male

Figure 12.6 *Sinochresmoda magnicornia* Zhang, Ren & Pang, 2008 (Holotype, CNU-CH-NN2007004).

is expanded more than that of the female and the first flagellomere of the male is incurvated, but straight for that of the female. Both male and female have wings. Dense fringing setae present along the wing margins.

Distribution and age: Inner Mongolia; Early Cretaceous.

Only one species included from the Cretaceous of Northern China (see Table 12.1).

Sinochresmoda magnicornia **Zhang, Ren & Pang, 2008 (Figure 12.6)**

Sinochresmoda magnicornia Zhang, Ren & Pang, 2008: *Zootaxa*, 1702, 27.

Locality and horizon: Liutiaogou, Ningcheng, Inner Mongolia, China; Lower Cretaceous, Yixian Formation.

Foretibia is short, shorter than half of forefemur length. Wings are short, not exceeding the abdomen, approximately 60% of the body length. Before both male and female of *S. magnicornia* with wings were reported from the Early Cretaceous of Inner Mongolia, China, it was known that male and nymph chresmodids were apterous. Also, *S. magnicornia* possessed sexual dimorphism in the antenna, which was first reported for the Chresmodidae [14].

Table 12.1 A list of fossil Chresmodidae from the Jurassic and Cretaceous of Northern China.

Family	Species	Locality	Horizon/Age	Citation
Chresmodidae	*Jurachresmoda gaskelli* Zhang, Ren & Shih, 2008	Ningcheng, Inner Mongolia	Jiulongshan Fm., J_2	Zhang *et al.* [15]
	Jurachresmoda sanyica Zhang, Ren & Pang, 2010	Ningcheng, Inner Mongolia	Jiulongshan Fm., J_2	Zhang *et al.* [16]
	Chresmoda multinervis Zhang, Ren & Shih, 2010	Fengning, Hebei	Yixian Fm., K_1	Zhang *et al.* [16]
	Chresmoda orientalis Esaki, 1949	Lingyuan, Liaoning	Yixian Fm., K_1	Esaki [27]
	Chresmoda shihi Zhang, Ren & Shih, 2010	Fengning, Hebei	Yixian Fm., K_1	Zhang *et al.* [16]
	Sinochresmoda magnicornia Zhang, Ren & Pang, 2008	Ningcheng, Inner Mongolia	Yixian Fm., K_1	Zhang *et al.* [14]

References

1 Baudoin, R. (1980). Sur les Gerris des miroirs d'eau actuels et les *Chresmoda* des lagunes post-recifales portlandiennes de Solnhofen. *Annales des Sciences Naturelles* **2**: 111–116.

2 Ponomarenko, A.G. (1985). Fossil insects from the Tithonian "Solnhofener Plattenkalke" in the Museum of Natural History, Vienna. *Annalen des Naturhistorischen Museums in Wien* **87A**: 135–144.

3 Grimaldi, D.A. and Engel, M.S. (2005). *Evolution of the Insects*. New York: Cambridge University Press.

4 Delclòs, X., Nel, A., Azar, D. *et al.* (2008). The enigmatic, Mesozoic family Chresmodidae (Polyneoptera: Archaeorthoptera): new palaeobiological and phylogenetic data, with the description of a new species from the Lower Cretaceous of Brazil. *Neues Jahrbuch für Geologie und Paläontologie, Abhandlungen* **247** (3): 353–381. https://doi.org/10.1127/0077-7749/2008/0247-0353.

5 Martínez-Delclòs, X. (1989). *Chresmoda aquatica* n. sp. insecto Chresmodidae del Cretácico inferior de la Sierra del Montsec (Lleida, España). *Revista Española de Paleontología* **4**: 67–74.

6 Martínez-Delclòs, X. (1991) Insectes hemimetàbols del Cretaci Inferior d'Espanya. Tafonomia i Paleoautoecologia. Doctor thesis. Departament de Geologia Dinamica, Geofisica i Paleontologia, Universitat de Barcelona.

7 Nel, A., Azar, D., and Martínez-Delclòs, X. (2004). A new Upper Cretaceous species of *Chresmoda* from Lebanon – a latest representative of Chresmodidae (Insecta: Polyneoptera inc. sed.): first record of homeotic mutations in the fossil record of insects. *European Journal of Entomology* **101** (1): 145–151.

8 Nel, A., Martínez-Delclòs, X., Béthoux, O. and Azar, D. (2005). *Chresmoda*, an enigmatic Mesozoic insect that is finally placed. 3rd International Congress of Palaeoentomology (FossilsX3) Abstracts, Pretoria, South Africa: 48.

9 Ren, D., Shih, C.K., Gao, T.P. *et al.* (2010). *Silent Stories – Insect Fossil Treasures from Dinosaur Era of the Northeastern China*. Beijing: Science Press.

10 Germar, E.F. (1839). *Die Versteinerten Insecten Solnhofen. Nova Acta Academia Leopoldiana Carola* **XIX**: 187–222.

11 Carpenter, F.M. (1992). Superclass Hexapoda. In: *Treatise on Invertebrate Paleontology, R, Arthropoda 4* (ed. R.C. Moore and R.L. Kaesler). Boulder, Clorado, and Lawrence, Kansas: Geological Society of America & University of Kansas pp. xxi+617.

12 Rasnitsyn, A.P. and Quicke, D.L.J. (2002). *History of Insects*. Hingham: Kluwer Academic Publishers.

13 Zhang, W.W., Cai, W.Z., Li, W.Z. *et al.* (2017). A new species of Chresmodidae from mid-Cretaceous amber discovered in Myanmar. *Zoological Systematics* **42** (2): 243–247. https://doi.org/10.11865/zs.201714.

14 Zhang, X.W., Ren, D., Pang, H., and Shih, C.K. (2008). A new genus and species of Chresmodidae (Insecta: Gryllones) from Upper Jurassic–Lower Cretaceous of Yixian Formation, Inner Mongolia, China. *Zootaxa* **1702**: 26–40.

15 Zhang, X.W., Ren, D., Pang, H., and Shih, C.K. (2008). A water-skiing chresmodid from the Middle Jurassic in Daohugou, Inner Mongolia, China (Polyneoptera: Orthopterida). *Zootaxa* **1762**: 53–62.

16 Zhang, X.W., Ren, D., Pang, H., and Shih, C.K. (2010). Late Mesozoic chresmodids with forewing from Inner

Mongolia, China (Polyneoptera: Archaeorthoptera). *Acta Geologica Sinica – English Edition* **84** (1): 38–46. https://doi.org/10.1111/j.1755-6724.2010.00168.x.

17 Bush, J.W.M. and Hu, D.L. (2006). Walking on water: biolocomotion at the Interface. *The Annual Review of Fluid Mechanics* **38**: 339–369.

18 Ward, J.V. (1992). *Aquatic Insect Ecology: 1. Biology and Habitat*. New York: Wiley.

19 Hu, D.L., Chan, B., and Bush, J.W.M. (2003). The hydrodynamics of water strider locomotion. *Nature* **424**: 663–666. https://doi.org/10.1038/nature01793.

20 Wise, D. (1993). *Spiders in Ecological Webs*. New York: Cambridge University Press.

21 Roach, J. (2004) How "Jesus Lizards" Walk on Water. National Geographic News, Nov 16.

22 Handlirsch, A. (1906–1908). *Die Fossilen Insekten und die Phylogenie der rezenten Formen: Ein Handbuch für Paläontologen und Zoologen*. Leipzig: Verlagvon Wilhelm Engelmann.

23 Martynov, A.V. (1928). A new fossil form of Phasmatodea from Galkino (Turkestan), and on Mesozoic Phasmids in general. *The Annals and Magazine of Natural History; Zoology, Botany, and Geology* **10** (1): 319–328. https://doi.org/10.1080/00222932808672788.

24 Sharov, A.G. (1968). *Filogeniya ortopteroidnykh nasekomykh*, 118. Moskva: Trudy Paleontologicheskogo Instituta, Akademiya Nauk S.S.S.R.. [Cin Russian, Translated in English in 1971: *Phylogeny of the Orthopteroidea*. Israel program for scientific translations, Keter Press, Jerusalem.].

25 Popov, Y.A. (1980). Superorder Cimicidea Laicharting, 1781. Order Cimicina Laicharting, 1781. In: *Historical Development of the Class of Insects*, vol. **175** (ed. B.B. Rohdendorf and A.P. Rasnitsyn), 58–69. Trudy Paleontologicheskogo Instituta, Akademiya Nauk S.S.S.R.

26 Béthoux, O. and Nel, A. (2002). Venation pattern and revision of Orthoptera *sensu nov.* and sister groups. Phylogeny of Palaeozoic and Mesozoic Orthoptera *sensu nov. Zootaxa* **96**: 1–88.

27 Esaki, T. (1949). The occurrence of the Mesozoic insect Chresmoda in the Far East. *Insecta Matsumurana* **17**: 4–5.

13

Phasmatodea – Stick Insects and Leaf Insects

Chaofan Shi[1,2], Chungkun Shih[2,3], Sha Chen[2], and Dong Ren[2]

[1] *Sun Yat-sen University, Guangzhou, Guangdong, China*
[2] *Capital Normal University, Haidian District, Beijing, China*
[3] *National Museum of Natural History, Smithsonian Institution, Washington, DC, USA*

13.1 Introduction to Phasmatodea

Phasmatodea, commonly known as "stick or leaf insects", are a rather small insect order with more than 3000 known species in the world and most of them live in tropics and subtropics [1, 2]. The Greek word of "*phasma*" means "phantom", indicating their remarkable camouflage to imitate the stems, twigs, and leaves. The order is a monophyly mainly based on the anterior dorsolateral glands on the prothorax, which provide effective defense. Also, the male has a vomer on sternite X, and the female has a praeopercular organ on sternite VII.

Phasmatodea are divided into two suborders of Timematodea and Euphasmatodea. The previously widely used two suborders of Areolatae and Anareolatae have been proven to be paraphyletic and polyphyletic, respectively, by Bradler in 2003 [3] and Grimaldi and Engel in 2005 [4]. Timematodea consist of only one family of Timematidae and a single genus of *Timema* with 21 species. They were restricted in western North America, principally in California [3–13]. Timematodea are monophyletic, supported by three-subdivided tarsi, development of a mesal lobe on the right cercus and egg-laying behavior, in which females ingest soil and then coat the eggs with this material [14]. The Euphasmatodea comprise five extant families of Phasmatidae, Heteronemiidae, Bacillidae, Pseudophasmatidae and Phyllidae (leaf insects).

Stick insects usually have an elongated, stick-like body (Figure 13.1) and the longest insect in the world is female *Phryganistria chinensis* Zhao, 2016 (informal name) with a body length (excluding antennae and legs) of 370 mm (see box below). Leaf insects (Phyllidae), instead, bear a flat, leaf-like body. Antennae are filiform. Compound eyes are small and ocelli can only be found in some winged males. Mouthparts are prognathous and chewing. Pronotum is short, with apertures of the defensive glands. Mesonotum and metanotum are elongated and metanotum usually fused with abdominal tergite I. Legs are gressorial, usually long and slender. Tarsus comprises five tarsomeres, rarely three tarsomeres. Cercus has one article. Most, but not all, phasmids are apterous. For the winged species, the forewing, called the tegmen, is reduced and short. However, the hind wing is usually developed and membranous, either folded tightly against the long and slender body, or modified to mimic leaves, like the forewing in such species. Most males are smaller and more slender than females.

Phasmatodea have a hemimetabolous life cycle with three stages: eggs, nymphs and adults. Many phasmids are parthenogenic and females lay eggs without the need of mating with males [15]. The eggs are usually oval or barrel-shaped with hard shells. They look like seeds so that ants would mistakenly take them back to their nests for burial. The buried eggs are better protected by the ants. A female may produce 100–1200 eggs after mating, depending on the species [16]. Female phasmids have various ways to lay eggs. For example, some females drop the eggs from a high place on a tree. When dropping to the ground, eggs roll into dry leaves or debris. This egg-dropping behavior is adopted by most stick insects in China. Alternatively, *Macellina digitate* glues the eggs to plants. Some other females insert their abdomen tips into loose sand and lay eggs [15]. Afterwards, the females bury the eggs with soil using their legs. This way of laying eggs has high crypticity. The eggs usually take 13–114 days to hatch, depending on the species [16]. Stick insects in temperate regions often undergo diapause, especially in winter, e.g. eggs of *Diapheromera femorata* take two years to hatch due to diapause [17].

The nymphs of stick insects have the basic morphological characters of adults except in having fewer antennomeres, rudimentary reproductive organ and wings [18]. Nymphs live in the same surroundings as

Rhythms of Insect Evolution: Evidence from the Jurassic and Cretaceous in Northern China, First Edition.
Edited by Dong Ren, Chungkun Shih, Taiping Gao, Yongjie Wang, and Yunzhi Yao.

Figure 13.1 A stick insect with elongated stick-like body. Source: Photo by Jason Shih.

adults. Nymphal growth and development can last for several months and nymphs need to undergo a series of molts (Figure 13.2). Many phasmids have six molts for females to reach adulthood and five molts for males. The male adults generally live for four to five weeks and females live for eight to nine weeks [15].

"Tree Lobster" – The Revived "Extinct" Insect

Dryococelus australis Montrouzier, 1855, commonly known as the "Lord Howe Island stick insect" or "tree lobster" [19], is a species of stick insect living on the Lord Howe Island Group, which are small volcanic remnant islands in the Tasman Sea between Australia and New Zealand. In the 1920s, the species was thought to be extinct in its largest habitat, Lord Howe Island, because of the invasion of black rats, *Rattus rattus* Linnaeus, 1758, from an aground supply ship in 1918.

However, dead remains were rediscovered by climbers on Ball's Pyramid in the 1960s [19]. In 2001, a small population of 24 individuals of *D. australis* living on the small islet of Ball's Pyramid was discovered [19, 20]. Because of this small population, they were called "one of the rarest insects in the world" [20]. The adults of *D. australis* are about 130 mm in length. Sexes are easily distinguishable [19]. Females are about 25% smaller than males, oblong in shape, and with sturdy legs. Males have very strong and spiny metafemora and form a bond with females. It was observed that males always follow females no matter where the females go, and males always hold females tightly using their legs at night. This behavior is exceptional for insects. "Tree lobsters" are wingless, but they can run quickly. The females lay eggs while hanging upside down on the branch and incubation lasts for nine months. The nymphs are bright green and active during the day, but the adults are black and active at night. A breeding pair captured in 2003 were used for captive breeding successfully in the Melbourne Zoo. Eggs have been sent to three zoos in England, USA, and Canada to establish insurance populations.

Record-Breaking Stick Insects

Phryganistria is a genus of stick insects belonging to the family of giant stick insects (Phasmatidae), distributed in Southeastern Asia. A specimen of this genus, dubbed *P. chinensis* Zhao (informal name), was collected in Liuzhou City, Guangxi, China in 2014, representing the world's longest known extant insect with a body length (excluding antennae and legs) of 370 mm. It displaced the previous length-champion insect of *Phobaeticus chani* Bragg, 2008 with a body length (excluding antennae and legs) of 357 mm. *Phobaeticus chani* was selected as one of "The IISE 2008 Top 10 New Species" by the International Institute for Species Exploration (IISE) at the Arizona State University and an international committee of taxonomists.

Another species *Phryganistria tamdaoensis* Bresseel and Constant, 2014 was also listed as one of "The IISE 2014 Top 10 New Species" by the IISE in 2015, because of their body shape and color, which makes them "masters of camouflage" [21]. The specimens were firstly collected from Tam Đảo National Park in northwestern Vietnam in August 2010 by J. Constant and P. Limbourg, both from the Royal Belgian Institute of Natural Sciences. The males have a unique color pattern, with golden-brown anterior and posterior regions of head, black margin of mesonotum, blue carinae of femora, black spines and pinkish-brown tibiae and tarsomeres. Females are light brown to dull green with bluish-green neck. Being 230 mm in length and 93 mm in width, the species is one of the smallest species of giant stick insects [22].

13.2 Progress in the Studies of Fossil Phasmatodea

The study of fossil Phasmatodea was initiated in 1839 when Germar described the first fossil stick insect from the Jurassic of Germany. However, the assignment of the fossil taxa related to Phasmatodea have been contentious for a long time. According to Sharov [23], Carpenter [24], Rasnitsyn and Quicke [25], and Willmann [26], the earliest stick insects can be traced back to the Late Permian from Central Mongolia, assigned to Permophamatidae, followed by the Late Triassic representatives of

Figure 13.2 Molting by a nymph of stick insects. Source: Photos by Jason Shih.

Xiphopteridae, Aeroplanidae, Prochresmodidae, from the Madygen Formation, Kirghizia and Ipswich of Queensland, Australia (only Aeroplanidae). On the other hand, Tilgner [27] concluded that the phasmatodean affinities of these fossils are mostly based on characters which are not diagnostic for Phasmatodea. However, Tilgner [27] admitted that the Early Cretaceous fossils of Hagiphasmatidae from China [28] have actual affinities with Phasmatodea, based on the female operculum and the enlarged abdominal sternum VIII. Hagiphasmatidae, along with Susumaniidae, formed the Susumanioidea within Phamatodea, which have been found from the Middle Jurassic to the Eocene [25, 28–31]. So far, 24 genera and 47 species of fossil and amber phasmids have been described from the Mesozoic of the world, mainly distributed in Russia, Kazakhstan, Kirghizia, Myanmar and China [26, 32–35], but most species only have isolated wings or incompletely preserved bodies [33]. Fossil stick and leaf insects are also known from the Late Eocene Baltic amber, the Early Oligocene Florissant fossils in Colorado, USA, and the Early Miocene Dominican amber. Among the extant families, only Phasmatidae and Phyllidae have fossil records from the Cenozoic [25].

Phasmatodea are renowned for their camouflage to imitate the stems, twigs, and leaves. The stick mimicry in fossils were firstly reported by Zompro in 2001, represented by an unnamed phasmid nymph of Sucinophasmatinae (Archipseudophasmatidae), from Eocene Baltic amber [36–38]. Wedmann et al. [39] described the oldest fossil record of leaf insects (Phyllidae) from the Eocene of Messel. Wang *et al.* [31], in 2014, described an Early Cretaceous fossil of Susumanioidea from China, which possessed leaf (or leaf-shaped organ) mimicking capabilities.

Fossil Phasmatodea were studied in China from 1997 when Ren erected a family Hagiphasmatidae with three genera and three species, based on well-preserved specimens from the Early Cretaceous of Liaoning and Hebei [28]. Later on, Nel and Delfosse [40], Shang *et al.* [30] and Wang *et al.* [31] described fossil stick insects from the Late Mesozoic of Northeastern China, which were inferred as genuine stem-Phasmatodea. To date, six genera with six species have been reported, collected from Liaoning, Hebei and Inner Mongolia.

Stick Insects in Amber Reveal Diverse Tarsal Pads in the Mid-Cretaceous

Many extant insects have developed pad structure, euplantulae or arolia on their tarsi to increase friction or enhance adhesion for better mobility. Fossil polyneopteran insects, such as Grylloblattodea, Mantophasmatodea and Orthoptera, have been described with euplantulae from the Mesozoic. However, the origin and early evolution of euplantulae in stick insect are poorly understood due to rare fossil records. *Tumefactipes prolongates* Chen, Shih, Ren & Gao, 2018, from the mid-Cretaceous Myanmar (Burmese) amber, possessed extremely specialized and expanded euplantulae on their tarsomere II [35]. These findings represent the first known and the earliest fossil records of euplantula structure within the Phasmatodea, demonstrating the diversity of euplantulae in Polyneoptera during the Mesozoic. Such tarsal pads might have increased friction and helped these mid-Cretaceous stick insects to climb more firmly on various surfaces, such as broad leaves, wetted tree branches or ground. *Tumefactipes prolongates* Chen, Shih, Ren & Gao, 2018, along with *Granosicorpes lirates* Chen, Shih, Ren & Gao, 2018 from the same amber deposit, are the earliest fossil records of Timematodea hitherto. These taxa provide more morphological data for us to understand the relationships of Timematodea, Euphasmatodea, Orthoptera and Embioptera (Figure 13.3).

Figure 13.3 *Tumefactipes prolongates* Chen, Shih, Ren & Gao, 2018, (Holotype, BU-001232). (a) Habitus. (b) Right metatarsus in ventral view. (c) 3D reconstructions based on micro-CT data, showing the ventral view of right metatarsus. (d) The euplantula of tarsomere II of left mesotarsus, arrows indicating the long setae. (e) Enlargement of the white rectangle in (d), showing the small square lines that may be setae. Source: Modified from [35].

13.3 Representative Fossils of Phasmatodea from Northern China

Superfamily Susumanioidea Gorochov, 1988

Family Hagiphasmatidae Ren, 1997

Hagiphasmatidae were erected by Ren in 1997 [28] consisting of the first described fossil stick insects in China. Hagiphasmatidae are typified by large body size, not leaf-like, long and filiform antenna, and well-developed forewing. In the forewing, Rs arises from R near wing base, with two terminal branches; MA with two branches. MP forked distally. In the hind wing, Rs is single, fused with MA for a long distance. Females possess elongated operculum, a preopercular organ and a fold of membrane on the posterior margin of sternum VII, indicating close affinities of the family and Euphasmatodea.

Genera included from the Cretaceous of Northern China: *Aethephasma* Ren, 1997, *Hagiphasma* Ren, 1997 and *Orephasma* Ren, 1997.

Aethephasma Ren, 1997

Aethephasma Ren, 1997, *Acta Zootaxon Sin.*, 22 (3), 273–274 [28] (original designation).

Type species: *Aethephasma megista* Ren, 1997.

In forewing, Sc terminated in C approximately at the middle of wing length. CuA_1 fused with MP for a long distance. 2A is present. In hind wing, Sc terminated in C

Figure 13.4 *Aethephasma megista* Ren, 1997 (Holotype, CNU-PHA-LB1997003).

beyond the middle of wing length. MA separated from Rs before the terminus of Sc.

Distribution and age: Hebei; Early Cretaceous.

Only one species included from the Cretaceous of Northern China (see Table 13.1).

***Aethephasma megista* Ren, 1997 (Figure 13.4)**

Aethephasma megista Ren, 1997: *Acta Zootaxon Sin.*, 22 (3), 273.

Locality and horizon: Pingquan, Hebei, China; Lower Cretaceous, Yixian Formation.

Mandible is robust. Both forewings and hind wings are large and well-developed. Female abdomen is 11-segmented. Tergite IX distinctly descended laterally. The posterior part of the Sternite VII possesses an obvious oval trace of praeopercular organ. The operculum distinctly exceeded the end of the abdomen. Cerci are slightly shorter than the operculum, covered with thin setae. The species probably had a phytophagous habit, like the other members of this family. The strong mandibles suggest feeding on hard tissue of plant leaves or bark [28].

***Hagiphasma* Ren, 1997**

Hagiphasma Ren, 1997, *Acta Zootaxon Sin.*, 22 (3), 271 [28] (original designation).

Type species: *Hagiphasma paradoxa* Ren, 1997.

Figure 13.5 *Hagiphasma paradoxa* Ren, 1997 (CNU-PHA-LJ2010001c).

In the forewing, Sc terminated in C slightly beyond the middle of wing length. MA forked basal to the level of Rs forking. MP forked distally. CuA_1 fused with MP for a short distance. 2A is absent. In the hind wing, Sc terminated in C at the middle of wing length. MA separated from Rs far distal to the level of terminus of Sc. CuA fused with CuP distally.

Distribution and age: Liaoning; Early Cretaceous.

Only one species included from the Cretaceous of Northern China (see Table 13.1).

***Hagiphasma paradoxa* Ren, 1997 (Figure 13.5)**

Hagiphasma paradoxa Ren, 1997: *Acta Zootaxon Sin.*, 22 (3), 271.

Locality and horizon: Jianchang, Liaoning, China; Lower Cretaceous, Yixian Formation.

Femora, tibiae and tarsi are straight, carinated but without spines. Arolia are strongly developed. Both forewings and hind wings are large and well-developed. Female abdomen is 11-segmented. Tergite I is not fused with metanotum. The posterior area of the sternite VII possesses a pear-like trace of praeopercular organ. The subgenital plate or operculum is keel-shaped, extending far beyond the apex of the abdomen, with deep notch at the apical part and covered with granules. Its lateral margins bear thin and dense teeth-like ornament. The trace of alimentary canal is obvious. It comprises esophagus, crop, midgut, and hindgut, but with no convolution.

The character combination of the species indicates its adaption to living on plants and to phytophagy. The strong arolia might be related to the ability to climb up plants. More importantly, the well-preserved trace of alimentary canal clearly shows the normal crop, the atrophied gizzard, with no convolutions and gastric caeca, suggesting the species might have fed on hard tissue of plant leaves or bark. Also, the black trace of gut was formed by the carbonaceous food remains from complex plant tissue [28].

Orephasma Ren, 1997

Orephasma Ren, 1997, *Acta Zootaxon Sin.*, 22 (3), 275 [28] (original designation).

Type species: *Orephasma eumorpha* Ren, 1997.

Sc terminated near the wing apex. MA forked approximately at the level of forking of Rs. MP forked in the basal half of wing length. CuA_1 fused with MP basally for a short way, then separated from it, and fused with CuA_2 distally till the termination, forming a single and combined $CuA_1 + CuA_2$. 2A closely paralleled to 1A. In hind wing, Sc long, ended near the wing apex. MA separated from Rs before the terminus of Sc. MP and CuA respectively fused with CuP distally.

Distribution and age: Hebei; Early Cretaceous.

Only one species included from the Cretaceous of Northern China (see Table 13.1).

Family Susumaniidae Gorochov, 1988

Susumaniidae, along with Hagiphasmatidae, formed Susumanioidea, which have been considered as stem-Phasmatodea. As pointed out above, female hagiphasmatids possess an operculum as their extant counterparts [28]. In Susumaniidae, a vomer in male *Renphasma sinica* [40], an operculum in female and forceps-like processes in male of *Adjacivena rasnitsyni* [30], an operculum in females and a "shoulder pad" in all specimens of *Cretophasmomima melanogramma* [31], show high similarity to the extant taxa. Also, the broad cerci in the male *C. melanogramma* is unusual among Susumaniidae, but common in extant phasmid insects, which indicates the diversity of external genitalic morphology in stem-Phasmatodea [31, 41]. To date, susumaniids have been found from the Middle Jurassic to the Late Cretaceous of China, Denmark, Kazakhstan and Russia [32]. Susumaniidae are typified by reduction of secondary C, proximal origin of RP in the basal half of wing, MA and MP with few branches, CuA simple in forewing, and RP and MA_1 fused over some distance in hind wing [30, 40].

Genera included from the Jurassic and Cretaceous of Northern China: *Cretophasmomima* Kuzmina, 1985, *Adjacivena* Shang, Béthoux & Ren, 2011 and *Renphasma* Nel & Delfosse, 2011.

Cretophasmomima Kuzmina, 1985

Cretophasmomima Kuzmina, 1985, *Pal. J.*, 19, 61 [42] (original designation).

Type species: *Cretophasmomima vitimica* Kuzmina, 1985.

Forewing Rs separated from R near basal one-third of wing length. CuA + CuPaα is simple. CuA_2 is absent.

Distribution and age: Inner Mongolia in China, and Russia; Cretaceous.

Only one species included from the Cretaceous of China (see Table 13.1).

Cretophasmomima melanogramma Wang, Béthoux and Ren, 2014 (Figure 13.6)

Cretophasmomima melanogramma Wang, Béthoux and Ren, 2014: *PLoS ONE* 9 (3), e91290.

Locality and horizon: Liutiaogou, Ningcheng, Inner Mongolia, China; Lower Cretaceous, Yixian Formation.

Head is globular. Mandible has incisivi and lacinia has two apical teeth. Prothorax and mesothorax are trapeze-shaped. Metathorax is quadrangular. Forefemur is straight. Forewing is long and well-developed. Main veins and intercalary veins are surrounded by dark coloration in most area of the forewing and in distal parts of the hind wing. A shoulder pad is present on the forewing. Female is slightly larger than the male, both with equal wing size. The male has two dark brown cerci, with no visible articulation. Female gonapophysis 8 and 9 are elongated. Operculum is slender, as long as the gonapophyses. The remarkable wing coloration in this species suggests leaf mimicry in the Early Cretaceous, when the diverse small-sized arboreal insectivore birds

Figure 13.6 *Cretophasmomima melanogramma* Wang, Béthoux and Ren, 2014 (Holotype, CNU-PHA-NN2012002).

and mammals have probably triggered the acquisition of such primary defense mechanisms. Based on these Cretaceous phasmids, it is proposed that leaf mimicry preceded the appearance of twig and bark mimicry in Phasmatodea [31].

Adjacivena Shang, Béthoux & Ren, 2011

Adjacivena Shang, Béthoux & Ren, 2011, *Eur. J. Entomol.*, 108, 678 [30] (original designation).

Type species: *Adjacivena rasnitsyni* Shang, Béthoux & Ren, 2011.

Forewing MA_2 approaches $MP + CuA_1$ a short distance before the middle of the wing (main diagnostic character). $MP + CuA_1$ forked (known only for type species). Hind wing (known only for type species) MA_1 fused over a moderate distance with RP. MA_2 fused with MA_1 distal to its divergence from $RP + MA_1$. Legs are slender, moderately long and apparently not armed.

Distribution and age: Inner Mongolia; Middle Jurassic.

Only one species included from the Jurassic of China (see Table 13.1).

Adjacivena rasnitsyni Shang, Béthoux & Ren, 2011 (Figure 13.7)

Adjacivena rasnitsyni Shang, Béthoux & Ren, 2011: *Eur. J. Entomol.*, 108, 679.

Locality and horizon: Daohugou, Ningcheng, Inner Mongolia, China; Middle Jurassic, Jiulongshan Formation.

The specific name is dedicated to Dr. Alexandr Rasnitsyn (Paleontological Institute, Russian Academy of Sciences, Moscow). Legs are slender and moderately long, not armed and with a cover of fine setae. Female forewing is darkly colored, with pale spots forming four transverse stripes. Forewing of the presumed male bears alternating pale and dark areas. Female ovipositor is concealed by an operculum. In the male, the 10th tergite extended, with a thorn pad. The species also shows intra-specific variability in wing venation, i.e.

Figure 13.7 *Adjacivena rasnitsyni* Shang, Béthoux & Ren, 2011 (Holotype, CNU-PHA-NN2009002p).

number of forewing RP branches and the location of the forking of MA_1. Sexual dimorphism is present in this species, exemplified by difference in forewing coloration and males slender than females. This is the earliest stick insect fossils documented up to date [30].

Renphasma Nel & Delfosse, 2011

Renphasma Nel & Delfosse, 2011, *Acta pal. Pol.*, 56 (2), 429 [40] (original designation).

Type species: *Renphasma sinica* Nel & Delfosse, 2011.

The generic name is after Dr. Ren Dong and *Phasma*. Tegmen RP originated near middle of wing length, with short branches. MA forked opposite the base of RP, with rather long branches. MP separated from CuA rather far from MP + CuA base. MP forked, with rather short branches.

Distribution and age: Liaoning; Early Cretaceous.

Only one species included from the Cretaceous of China (see Table 13.1).

Table 13.1 A list of fossil Phasmatodea from the Jurassic and Cretaceous of China.

Family	Species	Locality	Horizon/Age	Citation
Hagiphasmatidae	*Hagiphasma paradoxa* Ren, 1997	Beipiao, Liaoning	Yixian Fm., K_1	Ren [28]
	Aethephasma megista Ren, 1997	Pingquan, Hebei	Yixian Fm., K_1	Ren [28]
	Orephasma eumorpha Ren, 1997	Pingquan, Hebei	Yixian Fm., K_1	Ren [28]
Susumaniidae	*Renphasma sinica* Nel & Delfosse, 2011	Beipiao, Liaoning	Yixian Fm., K_1	Nel & Delfosse [40]
	Adjacivena rasnitsyni Shang, Béthoux & Ren, 2011	Ningcheng, Inner Mongolia	Jiulongshan Fm., J_2	Shang *et al.* [30]
	Cretophasmomima melanogramma Wang, Béthoux & Ren, 2014	Ningcheng, Inner Mongolia	Yixian Fm., K_1	Wang *et al.* [31]

References

1 Bragg, P.E. (1995). The phasmid database version 1.5. *Phasmid Studies* **3**: 41–42.

2 Tilgner, E.H. (2002) Systematics of Phasmida. Doctor dissertation. University of Georgia.

3 Bradler, S. (2003). Phasmatodea, Gespenstschrecken. In: *Lehrbuch der Speziellen Zoologie, Band I: Wirbellose Tiere. 5. Teil: Insecta* (ed. H.H. Dathe), 251–261. Heidelberg, Germany: Spektrum Akademischer Verlag.

4 Grimaldi, D.A. and Engel, M.S. (2005). *Evolution of the Insects*. New York, NY: Cambridge University Press.

5 Kristensen, N.P. (1975). The phylogeny of hexapod "orders". A critical review of recent accounts. *Journal of Zoological Systematics and Evolutionary Research* **13**: 1–44. https://doi.org/10.1111/j.1439-0469.1975.tb00226.x.

6 Vickery, V.R. (1993). Revision of *Timema* Scudder (Phasmatoptera: Timematodea) including three new species. *The Canadian Entomologist* **125**: 657–692. https://doi.org/10.4039/Ent125657-4.

7 Sandoval, C.P. and Vickery, V.R. (1996). *Timema douglasi* (Phasmatoptera: Timematodea), a new parthenogenetic species from southwestern Oregon and northern California, with notes on other species. *Canadian Entomologist* **128**: 79–84.

8 Sandoval, C.P. and Vickery, V.R. (1998). *Tinema coffmani* (Phasmatoptera: Timematodea) a new species from Arizona and description of the female of *Timema ritensis*. *Journal of Orthoptera Research* (**7**): 103–106. https://doi.org/10.2307/3503503.

9 Vickery, V.R. and Sandoval, C.P. (1997). *Timema bartmani* (Phasmatoptera: Timematodea: Timematidae), a new species from southern California. *The Canadian Entomologist* **129**: 933–936.

10 Vickery, V.R. and Sandoval, C.P. (1998). *Timema monikensis* sp. nov. (Phasmatoptera: Timematodea: Timematidae), a new parthenogenetic species in California. *Lyman Entomological Museum and Research Laboratory* **22**: 1–3.

11 Vickery, V.R. and Sandoval, C.P. (1999). Two new species of *Timema* (Phasmatoptera: Timematodea: Timematidae), one parthenogenetic, in California. *Journal of Orthoptera Research* **8**: 45–47. https://doi.org/10.2307/3503424.

12 Vickery, V.R. and Sandoval, C.P. (2001). Descriptions of three new species of *Timema* (Phasmatoptera: Timematodea: Timematidae) and notes on three other species. *Journal of Orthoptera Research* **10**: 53–61. https://doi.org/10.1665/1082-6467.

13 Bradler, S. (1999). The vomer of *Timema* Scudder, 1895 (Insecta: Phasmatodea) and its significance for phasmatodean phylogeny. *CFS Courier Forschungsinstitut Senckenberg* **215**: 43–47.

14 Tilgner, E.H., Kiselyova, T.G., and McHugh, J.V. (1999). A morphological study of *Timema cristinae* Vickery with implications for the phylogenetics of Phasmida. *Deutsche Entomologische Zeitschrift* **46**: 149–162.

15 Zheng, L.Y. and Gui, H. (1999). *Insect Classification*. Jiangsu: Nan Jing Normal University Press.

16 Bedford, G.O. (1978). Biology and ecology of the Phasmatodea. *Annual Review of Entomology* **23** (1): 125–149. https://doi.org/10.1146/annurev.en.23.010178.001013.

17 Resh, V.H. and Cardé, R.T. (2009). *Encyclopedia of Insects*, 1168. Academic Press.

18 Poinar, G. Jr. (2011). A walking stick, *Clonistria dominicana* n. sp. (Phasmatodea: Diapheromeridae) in Dominican amber. *Historical Biology* **23** (02–03): 223–226.

19 Priddel, D., Carlile, N., Humphrey, M. *et al.* (2003). Rediscovery of the 'extinct' Lord Howe Island stick-insect (*Dryococelus australis* (Montrouzier)) (Phasmatodea) and recommendations for its conservation. *Biodiversity and Conservation* **12** (7): 1391–1403. https://doi.org/10.1023/A:1023625710011.

20 Buckley, T.R., Attanayake, D., and Bradler, S. (2009). Extreme convergence in stick insect evolution: phylogenetic placement of the Lord Howe Island tree lobster. *Proceedings of the Royal Society B: Biological Sciences* **276** (1659): 1055–1062. https://doi.org/10.1098/rspb.2008.1552.

21 Gray, R. (2015) "Top 10 new species of 2015: Cartwheeling spiders, 'chicken from hell' and a giant stick insect celebrated in the year's first official list." MailOnline (21 May 2015).

22 Bresseel, J. and Constant, J. (2014). Giant sticks from Vietnam and China, with three new taxa including the second longest insect known to date (Phasmatodea, Phasmatidae, Clitumninae, Pharnaciini). *European Journal of Taxonomy* (104): 1–38. https://doi.org/10.5852/ejt.2014.104.

23 Sharov, A.G. (1968). Phylogeny of the orthopteroid insects. *Trudy Paleontologicheskogo Instituta Akademia Nauk SSSR* **118**: 1–216. [in Russian].

24 Carpenter, F.M. (1992). Superclass Hexapoda. In: *Treatise on Invertebrate Paleontology, Part R, Arthropoda 3–4* (ed. R.L. Kaesler). Boulder, CO: Geological Society of America, p. xxii 655.

25 Rasnitsyn, A.P. and Quicke, D.L.J. (2002). *History of Insects*. Dordrecht: Kluwer Academic Publishers, 684.

26 Willmann, R. (2003). Die phylogenetischen Beziehungen der Insecta: Offene Fragen und Probleme. *Verhandlungen Westdeutscher Entomologentag* **2001**: 1–64.

27 Tilgner, E.H. (2000). The fossil record of Phasmida (Insecta: Neoptera). *Insect Systematics & Evolution* **31**: 473–480. https://doi.org/10.1163/187631200X00507.

28 Ren, D. (1997). First record of fossil stick-insects from China with analyses of some palaeobiological features (Phasmatodea: Hagiphasmatidae fam. nov.). *Acta Zootaxonomica Sinica* **22** (3): 268–282.

29 Gorochov, A.V. (2000). Phasmomimidae: are they Orthoptera or Phasmatoptera? *Paleontological Journal* **34**: 295–300.

30 Shang, L.J., Béthoux, O., and Ren, D. (2011). New stem-Phasmatodea from the Middle Jurassic of China. *European Journal of Entomology* **108** (4): 677–685. https://doi.org/10.14411/eje.2011.086.

31 Wang, M.M., Béthoux, O., Bradler, S. *et al.* (2014). Under cover at pre-angiosperm times: a cloaked phasmatodean insect from the Early Cretaceous Jehol Biota. *PLoS One* **9** (3): e91290. https://doi.org/10.1371/journal.pone.0091290.

32 Wang, M.M. and Ren, D. (2013). Progress in the research of fossil Phasmatodea. *Acta Zootaxonomica Sinica* **38** (3): 626–633. (in Chinese).

33 Engel, M.S., Wang, B., and Alqarni, A.S. (2016). A thorny, 'anareolate' stick-insect (Phasmatidae s.l.) in Upper Cretaceous amber from Myanmar, with remarks on diversification times among Phasmatodea. *Cretaceous Research* **63**: 45–53. https://doi.org/10.1016/j.cretres.2017.01.007.

34 Chen, S., Zhang, W.W., Shih, C.K., and Ren, D. (2017). Two new species of Archipseudophasmatidae (Insecta: Phasmatodea) from Upper Cretaceous Myanmar amber. *Cretaceous Research* **73**: 65–70. https://doi.org/10.1016/j.cretres.2017.01.007.

35 Chen, S., Deng, S.W., Shih, C.K. *et al.* (2018). The earliest timematids in Burmese amber reveal diverse tarsal pads of stick insects in the mid-Cretaceous. *Insect Science* https://doi.org/10.1111/1744-7917.12601.

36 Zompro, O. (2001). The Phasmatodea and *Raptophasma* n. gen., Orthoptera *incertae sedis*. In: *Baltic Amber (Insecta: Orthoptera)*. Mitteilungen des Geologisch-Paläontologischen Institutes der Universität Hamburg, vol. **85**, 229–261.

37 Zompro, O. (2004). Revision of the genera of the Areolatae, including the status of *Timena* and *Agathemera* (Insecta, Phasmatodea). *Abhandlungen des Naturwissenschaftlichen Vereins in Hamburg* **37**: 1–327.

38 Wedmann, S. (2010). A brief review of the fossil history of plant masquerade by insects. *Palaeontographica Abteilung B* **283**: 175–182. https://doi.org/10.1127/palb/283/2010/175.

39 Wedmann, S., Bradler, S., and Rust, J. (2007). The first fossil leaf insect: 47 million years of specialized cryptic morphology and behavior. *Proceedings of the National Academy of Sciences of the United States of America* **104** (2): 565–569. https://doi.org/10.1073/pnas.0606937104.

40 Nel, A. and Delfosse, E. (2011). A new Chinese Mesozoic stick insect. *Acta Palaeontologica Polonica* **56**: 429–432. https://doi.org/10.4202/app.2009.1108.

41 Bradler, S. (2009). Die Phylogenie der Stab- und Gespentschrecken (Insecta: Phasmatodea). *Species, Phylogeny and Evolution* **2**: 3–139.

42 Kuzmina, S.A. (1985). New orthopterans of the family Phasmomimidae from the Lower Cretaceous of Transbaikalia. *Paleontological Journal* **19** (3): 56–63.

14

Plecoptera – Stoneflies

Yingying Cui[1,2], *Chungkun Shih*[2,3], *and Dong Ren*[2]

[1] South China Normal University, Guangzhou, China
[2] Capital Normal University, Haidian District, Beijing, China
[3] National Museum of Natural History, Smithsonian Institution, Washington, DC, USA

14.1 Introduction to Plecoptera

Plecoptera, meaning "folded wings," are named for a small order of hemimetabolous insects with the hind wings folding under the front wings while the insects are at rest (Figure 14.1). They are commonly known as "stoneflies." Up to now, there are about 3500 extant species described in the world.

The sizes of stoneflies vary greatly, with body lengths ranging from 4 to 70 mm. Stoneflies can be diagnosed by the following morphological characters: body soft, tarsi three-segmented, antennae elongate filiform, mouthparts mandibulated, two compound eyes, two or three ocelli, two usually long cerci, abdomen 10-segmented with the eleventh segment vestige, and colorful wings in some species [1].

Nymphs of stoneflies are entirely aquatic. Stonefly nymphs mainly live in cool, clean and well-oxygenated running water, with some exceptional species found in lakes. Therefore, together with mayflies (Ephemeroptera, Chapter 5) and caddisflies (Trichoptera, Chapter 26), these insects are commonly accepted as international EPT bioindicators of water quality (E – Ephemeroptera, P – Plecoptera, T – Trichoptera). They feed on insect larvae, algae, plant fragments or decomposed organic matter.

Adult stoneflies typically are terrestrial or sub-aerial, with the exception of *Capnia lacustra* Jewett, 1965, whose adults have been collected only at a depth of 60–80 m in Lake Tahoe (USA) [2], and the adult female of *Zapada cinctipes* (Banks, 1897) that can remain under water for a long time (20–60 minutes) for ovipositing [3]. The adults have two pairs of membranous wings, some species with wings reduced or absent. Hind wings, with anal fan large or small, can be folded horizontally over and around the abdomen when at rest, and covered by the forewings on top, and the costal and subcostal areas folding down (in some families they roll slightly). The adults prefer to hover near rivers or lakes; or rest on the plants and rocks near the water, but, only feed seldom.

The wing venation varies significantly among stoneflies, mostly on the number of branches and cross-veins [4]. The diagnostic characters of wing venation for Plecoptera are: presence of an arculus in both fore- and hind wings, M (Media) two-branched, and in hind wings, presence of fusion of RP with M. Although wing venation characters are presumably insufficient to provide a reliable phylogenetic signal by themselves [5], they do have the significant advantage of being applicable on fossil taxa, mostly recorded as wing imprints.

In this chapter, we follow the serial insect wing venation ground plan [6, 7]. Wing venation nomenclature is listed here for reference: ScP, posterior Subcosta; RA, anterior Radius; RP, posterior Radius; M, Media; MA, anterior Media; MP, posterior Media; Cu, Cubitus; CuA, anterior Cubitus; CuP, posterior Cubitus; AA: anterior Analis; AA1: first anterior Analis; AA2, second anterior Analis; ra-rp, rp-ma, mp-cua indicate the three specific cross-veins connecting RA and RP, RP and MA, MP and CuA, respectively. Right forewing and left forewing are indicated as RFW and LFW, respectively, and right hind wing and left hind wing as RHW and LHW, respectively. Specific terms, such as "antero-apical area," and conjectures of topographic homology follow Béthoux [4] and Béthoux *et al.* [8].

The Plecoptera include 16 extant families, distributed over all continents except Antarctica. The most widely accepted classification is by Zwick [9], which recognizes two suborders of Antarctoperlaria and Arctoperlaria. As indicated by their names, they are distributed mostly in Southern and Northern Hemispheres respectively. Colonization of the Southern Hemisphere has occurred

Rhythms of Insect Evolution: Evidence from the Jurassic and Cretaceous in Northern China, First Edition.
Edited by Dong Ren, Chungkun Shih, Taiping Gao, Yongjie Wang, and Yunzhi Yao.

Figure 14.1 Stoneflies. Source: Photos by Jason Shih.

in some derived genera of Arctoperlaria – the family Perlidae colonized to South America, and the Notonemouridae to Australia, South America and Africa [9, 10]. The definition of these two suborders are based on characters that are unknown or very unlikely to be preserved as fossils, such as muscles. Therefore, to deeply understand the evolution, the characters of wing venation are obviously important, considering wings are the most commonly preserved structure in fossils.

14.2 Progress in the Studies of Fossil Plecoptera

Since the studies of German Baltic ambers in the middle of nineteenth century [11], fossil stoneflies have been studied for 150 years. Later, many paleontologists found and described numerous new genera and species from all over the world. Among them, the Russian paleontologist Dr. Sinitshenkova has made great contributions in this area. The fossil stoneflies are mostly distributed in Eurasia from the Jurassic to Cretaceous.

There are numerous and diverse Plecoptera fossils in China. Abundant adults and nymphs are well-preserved in northern China from the Middle Jurassic (165 Mya) and the Early Cretaceous (125 Mya). However, stonefly fossil research in China did not start until 1928. Ping [12, 13], Lin [14–17], and Hong [18, 19] have described 11 genera with 14 species. Later on, Liu *et al.* with the assistance of Dr. Sinitshenkova, published numerous genera and species [20–24]. However, with the continuous research recently based on newly discovered fossils and improved methods and techniques, we have enhanced the knowledge of fossil stoneflies and documented some revisions [8, 10, 25, 26]. Many studies of stonefly fossils are still ongoing.

The earliest fossil record of Plecoptera is *Gulou carpenteri* Béthoux, Cui, Kondratieff, Stark & Ren, 2011 (see box below and Figure 14.2), which has been described from the Late Carboniferous of China, about 310 Mya [27].

The Oldest Stonefly

Stoneflies have been expected to occur very early during the evolution of insects, owing to their presumed "basal" position in insect phylogeny. The oldest stonefly, *G. carpenteri* Béthoux *et al.* [27] (Figure 14.2), has been documented from the Pennsylvanian, 310 Mya, which is considered as the only genuine Plecopteran from this period. The specific epithet is in honor of Prof. F.M. Carpenter, for his important contribution to the study of fossil insects, in particular Paleozoic fauna. The specimens were collected from a fossil site near Xiaheyan Village, Zhongwei City, Ningxia Hui Autonomous Region, China. Apart from a few Permian genuine Plecoptera [28, 29], *G. carpenteri* is the only compelling Paleozoic stem-Plecopteran, and is the earliest one. Although it provides no conclusive evidence on relationships of this group with other recognized major insect lineages, the wing morphology of this species does allow some plesiomorphic character states to be outlined for crown-Plecoptera.

14.3 Representative Fossils of Plecoptera from Northern China

Family Leuctridae Klapálek, 1905

Leuctridae, commonly known as Rolled-wing Stoneflies and Needle Stoneflies, are an extant family with about 330 species. These stoneflies can reach a body length of 5–13 mm, but most of the species are less than 10 mm. To date, three fossil species have been assigned as

(a)

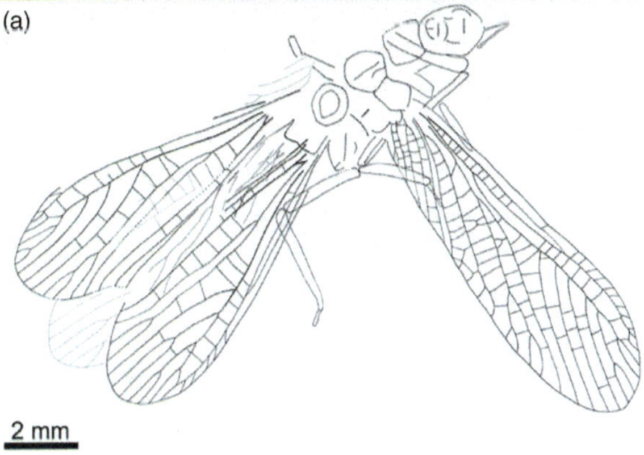

(b)

Figure 14.2 The oldest stonefly fossil, *Gulou carpenteri* Béthoux, Cui, Kondratieff, Stark & Ren, 2011 (Holotype, CNU-NX1-143). (a) Photograph; (b) Line drawing [27].

crown-leuctrids [8], including one species, *Aristoleuctra yehae* Liu, Ren & Sinitshenkova, 2006 from the Middle Jurassic of Northern China [20].

Only one genus included from the Jurassic of Northern China: *Aristoleuctra* Liu, Ren & Sinitshenkova, 2006.

Aristoleuctra Liu, Ren & Sinitshenkova, 2006

Aristoleuctra Liu, Ren & Sinitshenkova, 2006, *Ann. Zool.*, 56, 549–554 [20] (original designation).

Béthoux, Kondratieff, Grímsson, Ólafsson & Wappler, 2015, *Syst. Entomol.*, 40: 322–341 [8] (revised designation).

Type species: *Aristoleuctra yehae* Liu, Ren & Sinitshenkova, 2006.

The specific epithet is dedicated to Madam Ruiqin Yeh, mother of Chungkun Shih, for her love, care and guidance. In the forewings, base of M is distinct from base of R; arculus connected to M shortly after its divergence from R; 3/5 cross-veins in the area between M/MP and CuA (RFW/LFW; in addition to the arculus); MP originating very obliquely from M; six cross-veins in the area between CuA and CuP. CuP strong at mid-course; in hind wings, MP originating very obliquely from M; mp-cua cross-vein short.

Distribution and age: Inner Mongolia; Middle Jurassic.

Only one species included from the Jurassic of Northern China (see Table 14.1).

Aristoleuctra yehae Liu, Ren & Sinitshenkova, 2006 (Figure 14.3)

Aristoleuctra yehae Liu, Ren & Sinitshenkova, 2006: *Syst. Entomol.*, 40, 322–341.

Locality and horizon: Daohugou, Ningcheng, Inner Mongolia, China; Middle Jurassic, Jiulongshan Formation.

Based on a reinvestigation of the holotype, more data on wing venation have been reported, i.e. aspects of forewing bases, the orientation and position of the base of MP, and the location of the mp-cua cross-vein in hind wing [8]. Furthermore, instead of the statement "hind wings with enlarged anal area" in the original publication [20], a tentative reconstruction of hind wing vannus is proposed, which indicates that this species possesses a reduced vannus.

The systematic placement of *A. yehae* is proposed as a member of family Leuctridae, by sharing characters with other extant species of Leuctridae, namely, "in fore- and hind wing, area delimited by the anterior wing margin and RA, and beyond the end of (the basal free part of) ScP, without cross-veins, veinlet or vein." Therefore, instead of being a member of the extinct family Baleyopterygidae Sinitshenkova, 1985, i.e. stem to both Capniidae and Leuctiedae according to Sinitshenkova [30], *A. yehae* is a crown-Leuctridae.

Family Notonemouridae Ricker, 1950

Notonemouridae, an extant family with more than 120 species, are commonly found and documented in the Southern Hemisphere: South America, South Africa, Australia, and New Zealand. This family are closely related to Nemouridae [9], and therefore, might have been derived from a Northern Hemisphere lineage. Only one definite fossil species of the family has been recorded in China.

Only one genus included from the Jurassic of Northern China: *Paranotonemoura* Cui & Béthoux, 2018.

Subfamily Paranotonemourinae Cui & Béthoux, 2018

Paranotonemoura Cui & Béthoux, 2018

Paranotonemoura Cui & Béthoux, 2018, *J. Sys. Palaeontol.* [10] (original designation).

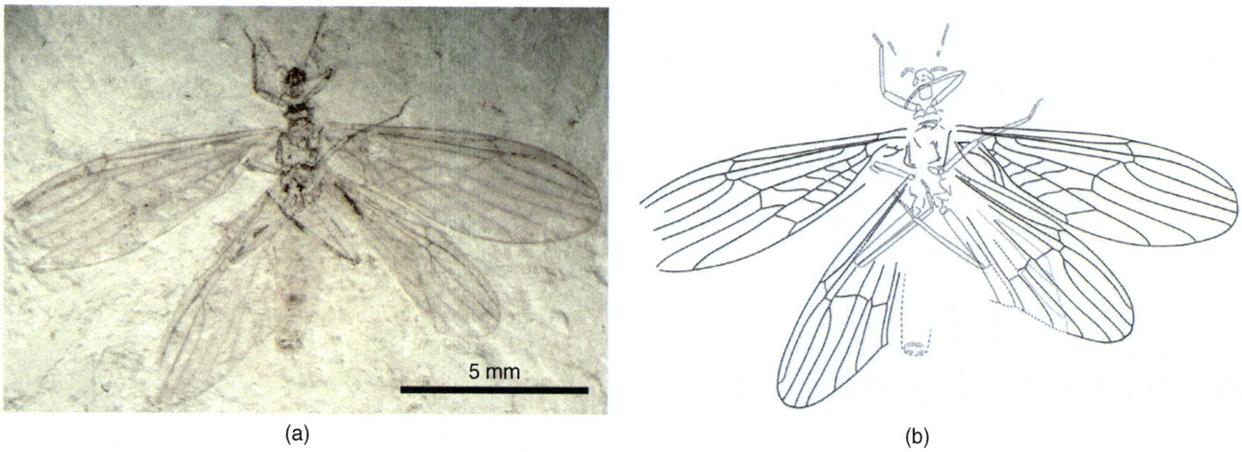

Figure 14.3 *Aristoleuctra yehae* Liu, Ren & Sinitshenkova, 2006 (Holotype, CNU-PLE-NM2005002). (a) Photograph; (b) Line drawing [8, 20]. Source: Donated by Dr. Chungkun Shih.

Type species: *Paranotonemoura zwicki* Cui & Béthoux, 2018.

The specific epithet is in honor of Dr. Zwick for his important contribution on the study of extant stoneflies. In the forewing, near the two thirds of wing length, ScP fused with RA for some distance, then diverging from it and reaching the anterior wing margin; area between anterior wing margin and ScP usually with two to three cross-veins; the basal-most one very near the wing base ("humeral cross-vein"), and the distal one (or two) in the narrowest area between anterior wing margin and ScP; in the area delimited by the anterior wing margin and RA, and distal to the distal free part of ScP, occurrence of dark coloration; R forked into RA and RP in the basal fifth of wing length; RA simple; RP forked near the ending point of ScP on RA; in the area between RA and RP, only one cross-vein (ra-rp) exists exactly/slightly before the fork of RP; M forked into MA and MP in the middle of the wing length; only one cross-vein (rp-ma) in area between RP and M/MA; area between M/MP and CuA with three to four cross-veins, including the very curved, oblique and long mp-cua; the mp-cua cross-vein more than twice as long as the longest cross-vein in the area between CuA and CuP; CuP reaching the posterior wing margin distal to the fork of M in the type species.

Distribution and age: Inner Mongolia; Middle Jurassic.

Only one species included from the Jurassic of Northern China (see Table 14.1).

Paranotonemoura zwicki **Cui & Béthoux, 2018 (Figure 14.4)**

Paranotonemoura zwicki Cui & Béthoux, 2018: *J. Syst. Palaeontol.*

Locality and horizon: Daohugou, Ningcheng, Inner Mongolia, China; Middle Jurassic, Jiulongshan Formation.

It is the first time that a fossil species of Notonemouridae has been unambiguously identified. Our discovery demonstrates that divergence date of the corresponding lineage is estimated to be before 165 Mya. The identification and discovery of this fossil species provide a new stonefly paleobiogeography, namely the origin and

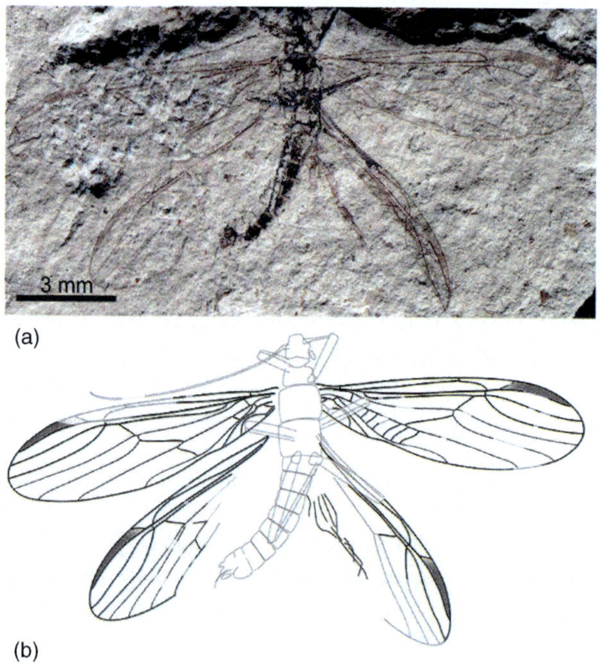

Figure 14.4 *Paranotonemoura zwicki* Cui & Béthoux, 2018 (Holotype, CNU-PLE-NN2016103). (a) Photograph; (b) Line drawing [10].

distribution of a novel taxon from the Northern Hemisphere to the Southern Hemisphere. The fossil record demonstrates that the family Notonemouridae occurred in the Northern Hemisphere 165 Mya, but faced subsequent extinction in this area. A north-to-south dispersion probably took place between 220 and 160 Mya, when Laurussia and Gondwana elements were not fully differentiated [10].

Superfamily Perloidea Latreille, 1802

Family Perlidae Latreille, 1802

Perlidae, commonly known as Summer Stoneflies, are an extant family with about 1000 species occurring nearly all over the world. Among them, more than 500 species are distributed in Asia.

Only one genus included from the Cretaceous of Northern China: *Sinosharaperla* Liu, Sinitshenkova & Ren, 2007.

Sinosharaperla Liu, Sinitshenkova & Ren, 2007

Sinosharaperla Liu, Sinitshenkova & Ren, 2007, *Cretac. Res.*, 28, 322–326 [22] (original designation).

Archaeoperla Liu, Ren & Sinitshenkova, 2008, (type species: *Archaeoperla rarissimus* Liu, Ren & Sinitshenkova, 2008) *Acta. Geol. Sin.-Engl.*, 82, 249–256 [21]. Syn. by Cui, Béthoux, Kondratieff, Liu & Ren, 2015, *J. Syst. Palaeontol.*, 13, 884 [25].

Type species: *Sinosharaperla zhaoi* Liu, Sinitshenkova & Ren, 2007.

The specific epithet is in honor of Chengxiang Zhao for his donation of this fossil to CNU. In the forewing, CuA with seven to eight distal branches (in female?; possibly fewer branches in male?); in the distal half, numerous sigmoid cross-veins in the areas ranging from the anterior branch of RP to the anterior branches of CuA; dark pigmentation along the ra-rp cross-vein, along the first rp-ma cross-vein, along cross-veins in the area between the anterior wing margin and ScP, and along M at its fork.

Distribution and age: Liaoning; Early Cretaceous.

Only one species included from the Cretaceous of Northern China (see Table 14.1).

Sinosharaperla zhaoi Liu, Sinitshenkova & Ren, 2007 (Figure 14.5)

Sinosharaperla zhaoi Liu, Sinitshenkova & Ren, 2007: *Cretac. Res.*, 28, 322–326.

Cui, Béthoux, Kondratieff, Liu & Ren, 2015, *J. Syst. Palaeontol.*, 13, 884 [25] (Figure 14.5).

Archaeoperla rarissimus Liu, Ren & Sinitshenkova, 2008, *Archaeoperla ralus* Liu, Ren & Sinitshenkova, 2008, *Acta. Geol. Sin.-Engl.*, 82, 249–256 [21]. Syn. by Cui, Béthoux, Kondratieff, Liu & Ren, 2015, *J. Syst. Palaeontol.*, 13, 884 [25].

Locality and horizon: Chaomidian, Beipiao, Liaoning, China; Lower Cretaceous, Yixian Formation.

Figure 14.5 *Sinosharaperla zhaoi* Liu, Sinitshenkova & Ren, 2007 (Holotype, CNU, BPYX1). (a) Photograph; (b) Line drawing [22, 25]. Source: Donated by Chengxiang Zhao.

The species *Sinosharaperla zhaoi* Liu, Sinitshenkova & Ren, 2007 was assigned to the fossil family Siberioperlidae by Liu *et al.* [22], which is classified as a stem-Antarctoperlaria. Since all extant Antarctoperlarian families (Gripopterygidae, Diamphipnoidae, Eustheniidae, and Austroperlidae) are distributed in the Southern Hemisphere, the assignment by Liu *et al.* would imply that the group once occurred in the Northern Hemisphere and that an extensive extinction affected the species in the Northern Hemisphere.

Cui *et al.* [25] re-investigated the specimen and assigned this species to the superfamily Perloidea, probably in Perlidae. This transfer supports the less dramatic scenario predicting that disjunct distribution is the result of a "divergent evolution of a common stock following the breakup of the Paleozoic super-continent Pangaea into Laurasia and Gondwana" (p. 734 in [9]). The occurrence of a probable Perlidae (at least infraorder Systellognatha) in the Northern Hemisphere is consistent with the current distribution of the group [9].

Superfamily Pteronarcyoidea Newman, 1853

Family Pteronarcyidae Newman, 1853

Pteronarcyidae, commonly known as Giant Stoneflies or Salmonflies, are an extant family with only 13 species in two genera. Among them, 11 species occurred in North/Central America (10 in North) and two species in Asia. To date, only one single fossil species, *Pteroliriope sinitshenkovae* Cui, Shih & Ren, 2016, has been recorded as a crown Pteronarcyidae.

Only one genus included from the Jurassic of Northern China: *Pteroliriope* Cui, Béthoux, Kondratieff, Shih & Ren, 2016.

Pteroliriope Cui, Béthoux, Kondratieff, Shih & Ren, 2016

Pteroliriope Cui, Béthoux, Kondratieff, Shih & Ren, 2016, *BMC Evol. Biol.*, 16, 217 [26] (original designation).

Type species: *Pteroliriope sinitshenkovae* Cui, Shih & Ren, 2016.

The specific epithet honors Dr. N. Sinitshenkova for her important contributions to the study of fossil stoneflies. This species shows a combination of several wing venation characters among genera of Pteronarcyidae. In the forewing, area between R/RP and M basal to the fork of M of regular width, without cross-veins (shared with *Pteronarcella* spp.; opposite condition in *Pteronarcys* spp.); MA distinct from RP (shared with *Pteronarcella* spp., fusion occasionally present in *Pteronarcys* spp.); AA2 with more than three branches (shared with *Pteronarcys* spp.; less than three branches in *Pteronarcella* spp.); except for the aa1-aa2 cross-vein, no cross-vein between AA2 branches (cross-veins present in both *Pteronarcella* spp. and *Pteronarcys* spp.). Hind wing: CuA branched (simple in *Pteronarcella* spp.; branched in *Pteronarcys* spp.).

Distribution and age: Inner Mongolia; Middle Jurassic.

Only one species included from the Jurassic of Northern China (see Table 14.1).

Pteroliriope sinitshenkovae Cui, Shih & Ren, 2016 (Figure 14.6)

Pteroliriope sinitshenkovae Cui, Shih & Ren, 2016: *BMC Evol. Biol.*, 16, 217.

Locality and horizon: Daohugou, Ningcheng, Inner Mongolia, China; Middle Jurassic, Jiulongshan Formation.

The first discovery of a fossil member of the Pteronarcyidae demonstrates that the Pteronarcyoidea and the Perloidea (sensu Zwick, 2000) had already diverged c. 165 million years ago, and that divergence events within the Pteronarcyoidea had already occurred. This discovery concurs with the view that divergence of most stonefly families took place very early, probably in the Triassic, or even in the Permian [26].

Family Pronemouridae Liu, Sinitshenkova & Ren, 2011

According to Liu *et al.* [31], the extinct family Pronemouridae was erected by having plesiomorphic characters that differ from nemourids, which are: short multi-segmented cerci and a CuA vein with forks. Liu *et al.* [31] also stated that Pronemouridae can be related to the five extant families in the superfamily Nemouroidea, but the Pronemouridae differs from the Taeniopterygidae by having the short second tarsal segment, and it differs from the other four families by having CuA with two branches. The two previously known genera, *Dimoula* Sinitshenkova, 2005 (an imago) and *Nemourisca* Sinitshenkova, 1987 (a nymph) from the Lower Cretaceous of Eastern Transbaikalia were transferred to the Pronemouridae in [31].

Only one genus included from the Jurassic of Northern China: *Pronemoura* Liu, Sinitshenkova & Ren, 2011.

Pronemoura Liu, Sinitshenkova & Ren, 2011

Pronemoura Liu, Sinitshenkova & Ren, 2011, *Palaeontol*, 54, 4 [31] (original designation).

Type species: *Pronemoura shii* Liu, Sinitshenkova & Ren, 2011.

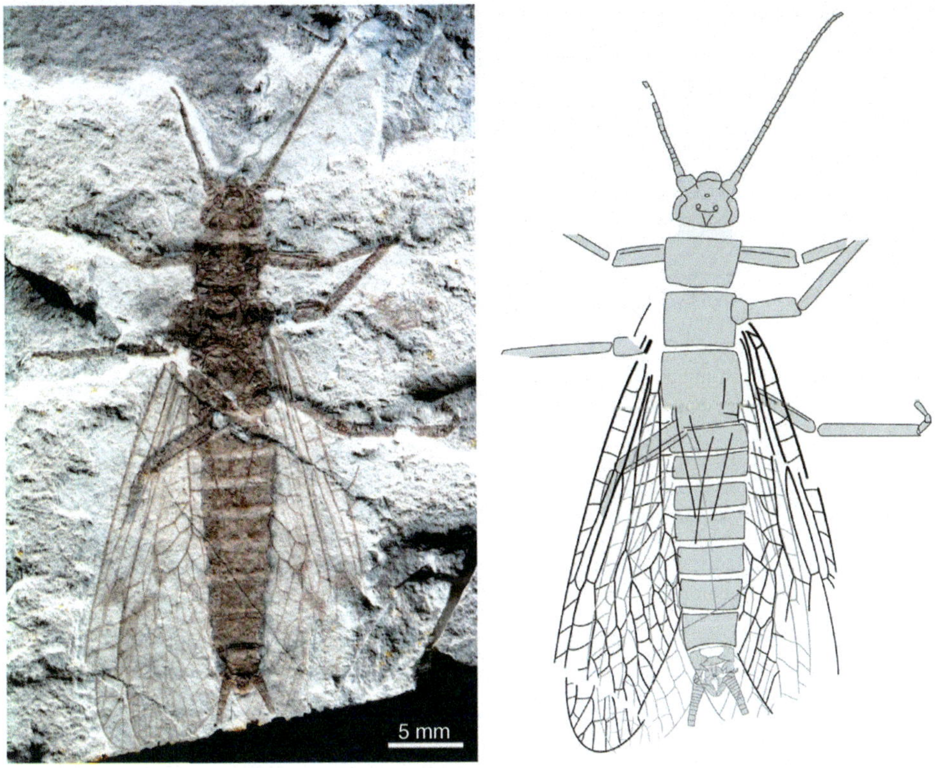

(a) (b)

Figure 14.6 *Pteroliriope sinitshenkovae* Cui, Shih & Ren, 2016 (Holotype, CNU-PLE-NN2015001). (a) Photograph; (b) Line drawing [26].

Figure 14.7 *Pronemoura shii* Liu, Sinitshenkova & Ren, 2011 (Holotype, CNU, NMDHG3) [31].

The genus *Pronemoura* can be distinguished from the previous known genus *Dimoula* Sinitshenkova, 2005 by having the characteristics of "lacking the multiple costal cross-veins on all wings, and anal region of hind wing broadened." There are five species in this genus. Some species have consistent venation in their left and right wings, for example, *P. shii*, *P. angustithorax* and *P. peculiaris*; but other species are slightly different, such as *P. longialata* and *P. minuta*. On the other hand, there are also slight wing variations in different individuals (including male vs. female) or the same species [31].

Distribution and age: Inner Mongolia; Middle Jurassic.

Five species included from the Jurassic of Northern China (see Table 14.1).

Pronemoura shii **Liu, Sinitshenkova & Ren, 2011 (Figure 14.7)**

Pronemoura shii Liu, Sinitshenkova & Ren, 2011: *Palaeontol*, 54, 4.

Locality and horizon: Daohugou, Ningcheng, Inner Mongolia, China; Middle Jurassic, Jiulongshan Formation.

The specific epithet is dedicated to Dr. Fuming Shi, for his affection, care and guidance to Dr. Y.S. Liu. Head large, antenna short, shorter than combined length of head and thorax, wings large, "X" well expressed, posterior branch of Rs deeply curved, not S-shaped, and the broadened anal area of hind wing with at least five longitudinal veins. The left and right wings have consistent venation [31].

Table 14.1 A list of fossil Plecoptera from the Jurassic and Cretaceous of China.

Family	Species	Locality	Age/Horizon	Citation
Perlidae?	*Sinosharaperla zhaoi* Liu, Sinitshenkova & Ren, 2007	Beipiao, Liaoning	Yixian Fm, K_1	Liu *et al.* [22]
Notonemouridae	*Paranotonemoura zwicki* Cui & Béthoux, 2018	Ningcheng, Inner Mongolia	Jiulongshan Fm, J_2	Cui *et al.* [10]
Pteronarcyidae	*Pteroliriope sinitshenkovae* Cui, Shih & Ren, 2016	Ningcheng, Inner Mongolia	Jiulongshan Fm, J_2	Cui *et al.* [26]
Platyperlidae	*Platyperla kingi* Ping, 1935	Tulufan, Xingjiang	Meiyaogou Fm, J	Ping [13]
	Platyperla platypoda Brauer, Redtenbacher & Ganglbauer, 1889	Beipiao, Liaoning	Haifanggou Fm, J_2	Hong [18]
Perlomorpha *Incertae sedis*	*Triassoperla yongrenensis* Lin,1977	Yongren, Yunnan	Nalajing Fm, T_3	Lin [14]
Mesoleuctridae	*Mesoleuctra peipiaoensis* Ping, 1928	Beipiao, Liaoning	Beipiao Fm, K_1	Ping [12]
	Capitiperla tonicopoda Lin, 1992	Tuokexun, Xinjiang	Huangshanjie Fm, T_3	Lin [17]
Perlariopseidae	*Sinoperla abdominalis* Ping, 1928	Beipiao, Liaoning	Beipiao Fm, K_1	Ping [12]
	Perlariopsis peipiaoensis Ping, 1928	Beipiao, Liaoning	Beipiao Fm, K_1	Ping [12]
	Rectonemoura yujiagouensis Hong, 1983	Beipiao, Liaoning	Haifanggou Fm, J_2	Hong [18]
	Sinotaeniopteryx chendeensis Hong, 1983	Chengde, Hebei	Jiulongshan Fm, J_2	Hong [18]
	Sinotaeniopteryx luanpingensis Hong, 1983	Luanping, Hebei	Jiulongshan Fm, J_2	Hong [18]
	Karanemoura mancus Liu, Sinitshenkova & Ren, 2009	Ningcheng, Inner Mongolia	Jiulongshan Fm, J_2	Liu *et al.* [24]
	Karanemoura abrupta Sinitshenkova, 1987	Ningcheng, Inner Mongolia	Jiulongshan Fm, J_2	Sinitshenkova [32]
Leuctridae	*Aristoleuctra yehae* Liu, Ren & Sinitshenkova, 2007	Ningcheng, Inner Mongolia	Jiulongshan Fm, J_2	Liu *et al.* [22]
Taeniopterygidae	*Jurataenionema stigmaeus* Liu & Ren, 2007	Ningcheng, Inner Mongolia	Jiulongshan Fm, J_2	Liu *et al.* [23]
	Jurataenionema inornatus Liu & Ren, 2007	Ningcheng, Inner Mongolia	Jiulongshan Fm, J_2	Liu *et al.* [23]
	Protaenionema fuscalatus Liu & Shih, 2007	Ningcheng, Inner Mongolia	Jiulongshan Fm, J_2	Liu *et al.* [23]
	Mengitaenioptera multiramis Liu & Ren, 2008	Ningcheng, Inner Mongolia	Jiulongshan Fm, J_2	Liu and Ren [33]
	Noviramonemoura trinervis Liu & Ren, 2008	Ningcheng, Inner Mongolia	Jiulongshan Fm, J_2	Liu and Ren [33]
	Liaotaenionema tenuitibia Liu, Ren & Sinitshenkova, 2008	Lingyuan, Liaoning	Yixian Fm, K_1	Liu *et al.* [21]

(Continued)

Table 14.1 (Continued)

Family	Species	Locality	Age/Horizon	Citation
Pronemouridae	*Pronemoura shii* Liu, Sinitshenkova & Ren, 2011	Ningcheng, Inner Mongolia	Jiulongshan Fm, J_2	Liu *et al.* [31]
	Pronemoura angustithorax Liu, Ren & Shih, 2011	Ningcheng, Inner Mongolia	Jiulongshan Fm, J_2	Liu *et al.* [31]
	Pronemoura longialata Liu, Sinitshenkova & Ren, 2011	Ningcheng, Inner Mongolia	Jiulongshan Fm, J_2	Liu *et al.* [31]
	Pronemoura minuta Liu, Sinitshenkova & Ren, 2011	Ningcheng, Inner Mongolia	Jiulongshan Fm, J_2	Liu *et al.* [31]
	Pronemoura peculiaris Liu, Sinitshenkova & Ren, 2011	Ningcheng, Inner Mongolia	Jiulongshan Fm, J_2	Liu *et al.* [31]
Nemouridae	*Parvinemoura parvus* Liu, Ren & Shih, 2008	Beipiao, Liaoning	Yixian Fm, K_1	Liu *et al.* [21]
Capniidae	*Dobbertiniopteryx Juracapnia* Liu, Sinitshenkova & Ren, 2009	Ningcheng, Inner Mongolia	Jiulongshan Fm, J_2	Liu *et al.* [24]
Perlida *Incertae sedis*	*Fluminiperla hastis* Lin,1980	Zhuji, Zhejiang	Shouchang Fm, K_1	Lin [15]
Questionable species	*Marciperla curta* Lin,1986	Hexian, Guangxi	Shiti Fm, J_1	Lin [16]
	Sinonemoura grabaul Ping, 1928	Chifeng, Hebei	Chifeng Fm, K_1	Ping [12]
	Sinoperla(?) liaoningensis Hong, 1992	Huairen, Liaoning	Xiahuapidianzi Fm, J_3	Hong [19]

References

1 Fochetti, R. and De Figueroa, J.M.T. (2008). Global diversity of stoneflies (Plecoptera; Insecta) in freshwater. *Hydrobiologia* **595** (1): 365–377. https://doi.org/10.1007/s10750-007-9031-3.

2 Jewett, S.G. (1963). A stonefly aquatic in the adult stage. *Science* **139** (3554): 484–485. https://doi.org/10.1126/science.139.3554.484.

3 Tozer, W. (1979). Underwater behavioural thermoregulation in the adult stonefly, *Zapada cinctipes*. *Nature* **281**: 566–567. https://doi.org/10.1038/281566a0.

4 Béthoux, O. (2005). Wing venation pattern of Plecoptera (Insecta: Neoptera). *Illiesia* **1** (9): 52–81.

5 Grimaldi, D. (2001). Insect evolutionary history from Handlirsch to Hennig, and beyond. *Journal of Paleontology* **75** (6): 1152–1160. https://doi.org/10.1017/S0022336000017200.

6 Lameere, A. (1922). Sur la nervation alaire des Insectes. *Bulletin de la Classe des Sciences de l'Académie Royale de Belgique* **8**: 138–149.

7 Lameere, A. (1923). On the wing-venation of insects. *Psyche* **30** (3–4): 123–132. https://doi.org/10.1155/1923/16920.

8 Béthoux, O., Kondratieff, B., Grímsson, F. *et al.* (2015). Character state-based taxa erected to accommodate fossil and extant needle stoneflies (Leuctridae–*Leuctrida tax. n.*) and close relatives. *Systematic Entomology* **40** (2): 322–341. https://doi.org/10.1111/syen.12102.

9 Zwick, P. (2000). Phylogenetic system and zoogeography of the Plecoptera. *Annual Review of Entomology* **45** (1): 709–746. https://doi.org/10.1146/annurev.ento.45.1.709.

10 Cui, Y., Ren, D., and Béthoux, O. (2018). The Pangean journey of 'south forestflies' (Insecta: Plecoptera) revealed by their first fossil. *Journal of Systematic Palaeontology* https://doi.org/10.1080/14772019.2017.1407370.

11 Pictet, F.J. and Hagen, H. (1856). Die im Bernstein befindlichen Neuropteren der Vorwelt. In: *Die im*

Bernstein befindlichen organischen Reste der Vorwelt, Band 2 (ed. G.C. Berendt), 40–122. Berlin: Nicolai.
12 Ping, C. (1928). Study of the Cretaceous fossil insects of China. *Palaeontologia Sinica* **13** (1): 1–47.
13 Ping, C. (1935). On four fossil insects from Sinkiang. *Chinese Journal of Zoology* **1** (107): e115.
14 Lin, Q.B. (1977). Fossil insects from Yunnan. In: *Mesozoic Fossils from Yunnan*, 373–381. Beijing: Science Press.
15 Lin, Q.B. (1980). Fossil insects from the Mesozoic of Zhejiang and Anhui. In: *Division and Correlation of Stratigraphy of Mesozoic Volcanic Sediments from Zhejiang and Anhui*, 223–224. Beijing: Science Press.
16 Lin, Q.B. (1986). Early Mesozoic fossil insects from South China. *Palaeontologica Sinica, Series B* **170** (21): 69–82.
17 Lin, Q.B. (1992). Late Triassic insect fauna from Toksun, Xinjiang. *Acta Palaeontologica Sinica* **31** (3): 313–335.
18 Hong, Y.C. (1983). *Middle Jurassic Fossil Insects in North China*. Beijing: Geological Publishing House.
19 Hong, Y.C. (1992). *Palaeontological Atlas of Jilin Province*. Jilin: Jilin Science and Technology Press.
20 Liu, Y.S., Ren, D., Sinitshenkova, N.D., and Shih, C.K. (2006). A new Middle Jurassic stonefly from Daohugou, Inner Mongolia, China (Insecta: Plecoptera). *Annales Zoologici (Warszawa)* **56** (3): 549–554.
21 Liu, Y.S., Ren, D., Sinitshenkova, N.D., and Shih, C.K. (2008). Three new stoneflies (Insecta: Plecoptera) from Yixian Formation of Liaoning, China. *Acta Geologica Sinica (English Edition)* **82** (2): 249–256.
22 Liu, Y.S., Sinitshenkova, N.D., and Ren, D. (2007). A new genus and species of stonefly (Insecta: Plecoptera) from the Yixian Formation of China. *Cretaceous Research* **28** (2): 322–326. https://doi.org/10.1016/j.cretres.2006.08.003.
23 Liu, Y.S., Sinitshenkova, N.D., Ren, D., and Shih, C.K. (2007). The oldest known record of Taeniopterygidae in the Middle Jurassic of Daohugou, Inner Mongolia, China (Insecta: Plecoptera). *Zootaxa* **1521** (1): 1–8. https://doi.org/10.11646/zootaxa.1439.1.1.
24 Liu, Y.S., Sinitshenkova, N.D., and Ren, D. (2009). A revision of *Dobbertiniopteryx* and *Karanemoura* (Insecta: Plecoptera) from Daohugou, China, with the description of a new species. *Paleontological Journal* **43** (2): 183–190. https://doi.org/10.1134/S0031030109020099.
25 Cui, Y., Béthoux, O., Kondratieff, B. *et al.* (2015). *Sinosharaperla zhaoi* (Insecta: Plecoptera; Early Cretaceous), a Gondwanian element in the northern hemisphere, or just a misplaced species? *Journal of Systematic Palaeontology* **13** (10): 883–889. https://doi.org/10.1080/14772019.2014.960903.
26 Cui, Y., Béthoux, O., Kondratieff, B. *et al.* (2016). The first fossil salmonfly (Insecta: Plecoptera: Pteronarcyidae), back to the Middle Jurassic. *BMC Evolutionary Biology* **16** (1): 217–230. https://doi.org/10.1186/s12862-016-0787-9.
27 Béthoux, O., Cui, Y., Kondratieff, B. *et al.* (2011). At last, a Pennsylvanian stem-stonefly (Plecoptera) discovered. *BMC Evolutionary Biology* **11** (1): 248–259. https://doi.org/10.1186/1471-2148-11-248.
28 Sharov, A.G. (1991). Order Plecoptera. In: *Fundamentals of Paleontology Arthropoda, Tracheata, Chelicerata* (ed. B.B. Rohdendorf), 173–179. Washington, DC: Smithsonian Institution Libraries.
29 Carpenter, F.M. (1992). Superclass Hexapoda. In: *Treatise on Invertebrate Paleontology* (ed. R.L. Kaesler). Boulder, CO: The Geological Society of America and the University of Kansas, pp. xxii+655.
30 Sinitshenkova, N. (2002). Order Perlida Latreille, 1810. The stoneflies (=Plecoptera Burmeister, 1839). In: *History of Insects* (ed. A.P. Rasnitsyn and D.L.J. Quicke), 281–287. Dordrecht: Kluwer Academic Publishers.
31 Liu, Y.S., Sinitshenkova, N.D., Ren, D., and Shih, C.K. (2011). Pronemouridae fam. nov. (Insecta: Plecoptera), the stem group of Nemouridae and Notonemouridae, from the Middle Jurassic of Inner Mongolia, China. *Palaeontology* **54** (4): 923–933. https://doi.org/10.1111/j.1475-4983.2011.01063.x.
32 Sinitshenkova, N.D. (1987). Istoritcheskoe rasvitie vesnjanok. *Trudy Paleontologicheskova Instituta Akademii nauk SSSR* **221**: 1–143.
33 Liu, Y.S. and Ren, D. (2008). Two new Jurassic stoneflies (Insecta: Plecoptera) from Daohugou, Inner Mongolia, China. *Progress in Natural Science* **18** (8): 1039–1042. https://doi.org/10.1016/j.pnsc.2008.03.014.

15

Psocoptera – Barklice and Booklice

Ruiqian Wang[1], Yunzhi Yao[1], Dong Ren[1], and Chungkun Shih[1,2]

[1] Capital Normal University, Haidian District, Beijing, China
[2] National Museum of Natural History, Smithsonian Institution, Washington, DC, USA

15.1 Introduction to Psocoptera

Psocoptera, commonly known as "barklice" or "booklice", have about 10 000 described species from all over the world. Although the Psocoptera are called "lice", they are not parasites, instead they are free living insects. The scientific name comes from the Greek psocus (to gnaw or grind) and pteron (wing), referring to the psocopteran's mandibles, which are shaped like a pestle and mortar to grind food [1]. Typically, barklice tend to live outdoors, feeding mostly on organic matter such as lichens, algae, plant spores, dead plant and insect material; while booklice live with humans and may become pests, feeding on stored products and paper materials. Barklice have body lengths ranging from 1 to 10 mm, but most are less than 6 mm [2]. Booklice are smaller ranging from 1 to 2 mm.

Psocoptera have chewing mandibles, and the central lobe of the maxilla is modified into a slender rod. They also have a relatively large head, large compound eyes, three ocelli and slender, long antennae with 12–50 antennomeres. The wings are usually present in barklice, but absent in booklice. When present, the wings are membranous, with reduced venation, and held roof-like over the abdomen when at rest [3]. The forewings are larger than the hind wings. The legs of Psocoptera are slender and adapted for walking.

Psocoptera have incomplete metamorphosis. Most male psocopterans perform a mating dance, and after fertilization, the female lays eggs under bark, leaves or a silk mat. Some barklice are gregarious, living in small colonies beneath a silk blanket spun with glands in their mouths. Interestingly, both adults and larvae can spin silk from these glands [4].

Recently, morphological and molecular evidence has shown that the parasitic lice (Phthiraptera) evolved from within the psocopteran suborder Troctomorpha [5, 6]. The Psocoptera are classified into three suborders, Trogiomorpha (booklice), Troctomorpha (booklice), and Psocomorpha (barklice). Trogiomorpha have antennae with 20–50 antennomeres, and always three-segmented tarsi [7]. Trogiomorpha comprise three infraorders of Atropetae, Prionoglaridetae and Psyllipsocetae [8]. Troctomorpha ordinarily have antennae with 15–17 antennomeres and two-segmented tarsi. Troctomorpha comprise two infraorders of Amphientometae, Nanopsocetae, and the order Phthiraptera (parasitic lice) [8]. The suborder Psocomorpha have antennae with 13 antennomeres, and 2 or 3-segmented tarsi. Psocomorpha have the most species (c. 3600 species in 24 families) in four infraorders of Epipsocetae, Caeciliusetae, Homilopsocidea, and Psocetae [9].

15.2 Progress in the Studies of Fossil Psocoptera

The research of the fossil Psocoptera started in the middle of the nineteenth century. Pictet was the first person studying the fossil Psocoptera and reported *Amphientomum paradoxum* from Baltic amber [10].

Handlirsch erected the family Archipsyllidae Handlirsch, 1906 [11]. Tillyard made a major contribution to the study of fossil Psocoptera when studying the insects from the Early Permian deposits in Kansas and other fossils from the Late Permian deposits in Australia [12]; Martynov subsequently described fossil taxa from the Late Permian deposits in Russia [13]. Carpenter studied fossils from the Kansas beds [14]. More recently, Becker-Migdisova [15, 16] and Vishnyakova [17] made a contribution to the description of fossils in different ages from Russia and Germany. Nowadays, many scien-

Rhythms of Insect Evolution: Evidence from the Jurassic and Cretaceous in Northern China, First Edition.
Edited by Dong Ren, Chungkun Shih, Taiping Gao, Yongjie Wang, and Yunzhi Yao.

tists from different countries are studying the fossil and amber Psocoptera [18–22].

The research into the fossil Psocoptera started late in China. Qibin Lin and Youchong Hong are the specialists of this field in the early studies [23, 24]. In 2008, Huang *et al.* have reported a new species from the Mesozoic [25]. In addition, new species in Myanmar (Burmese) amber have been reported by different scientists [18–22].

15.3 Representative Fossils of Psocoptera from Northern China

Family Archipsyllidae Handlirsch, 1906

Archipsyllids are distinguished by having elongated, oval forewings, about thrice as long as broad. The Sc reduced to a small basal section joining costa and a distal section at base of pterostigma. Pterostigma is short and broad, angled distally (or basally). There is a cross-vein from the pterostigma to Rs. Rs arising near the wing base, and forked distally. The cross-vein between Rs to M present in midwing. M fused basally with R, and four branched. Cu1 forking. The areola postica (AP) is long and low. Cu1b is short. The anal area is reduced [26].

The mouthparts of archipsyllids have some different characters from those of other barklice or booklice. Their mandibles are elongate, with long and narrow labra. In 2016, Huang *et al.* placed *Psocorrhyncha burmitica* from mid-Cretaceous Myanmar (Burmese) amber into a new insect order Permopsocida [27], based on their phylogenetic analyses. Yoshizawa and Lienhard, in 2016, also indicated that Archipsyllidae is outside Psocodea [28], based on morphological characters and phylogenetic analysis of *Mydiognathus eviohlhoffae*, also in Myanmar amber. However, due to un-clear or un-identifiable characters in fossil specimens, we still follow the original taxonomic designation of Archipsyllidae here, pending more fossil specimens with more evidence for future studies.

Only one genus included from the Jurassic of Northern China: *Archipsylla* Handlirsch, 1906.

Archipsylla Handlirsch, 1906

Archipsylla Handlirsch, 1906, *Die fossilen Insekten und die Phylogeny der rezenten Formen*, 502 [11] (original designation).

Type species: *Archipsylla primitiva* Handlirsch, 1906. Diagnosis same as for the family.

Distribution and age: Inner Mongolia of China, Middle Jurassic; Russia, Late Jurassic.

Only one species included from the Jurassic of Northern China (see Table 15.1).

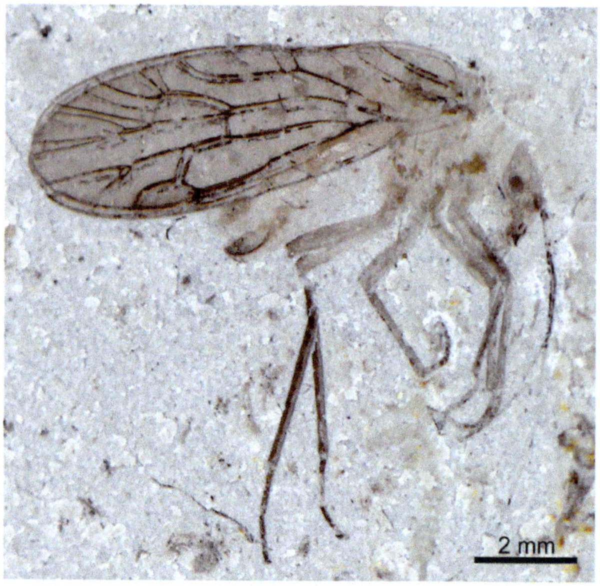

Figure 15.1 *Archipsylla sinica* Huang, Nel, Azar, and Nel, 2008. Holotype, NIGP 142194 [25]. Source: Photo provided by Dr. Di-Ying Huang.

Archipsylla sinica Huang, Nel, Azar & Nel, 2008 (Figure 15.1)

Archipsylla sinica Huang, Nel, Azar & Nel, 2008: *Geobios*, 461–464.

Locality and horizon: Daohugou, Ningcheng, Inner Mongolia, China; Middle Jurassic, Jiulongshan Formation.

Antenna filiform with 11 annulated flagellomeres. Forewing with no nodulus. Pterostigma sclerotized, relatively short but distinctly longer than wide and not posteriorly angled and related to Rs by a cross-vein, rs-m, ending on anterior branch of M. M_4 present. AP elongate and free. 1A closer to Cu_2 than to 2A. Hind wing with nearly the same size and venation structure as forewing. Tarsus four-segmented; claws symmetrical, with a secondary tooth and a pulvillus at its base, an arolium. The presence of four-segmented tarsi in Archipsyllidae suggests that the reduction in number of tarsomeres occurred independently at least twice in the modern paraneopteran lineages Psocodea ("Psocoptera" + Phthiraptera) and Condylognatha (Thysanoptera + Hemiptera) [25].

Table 15.1 A species of fossil Psocoptera from the Jurassic of China.

Family	Species	Locality	Horizon/age	Citation
Archipsyllidae	*Archipsylla sinica* Huang, Nel, Azar & Nel, 2008	Ningcheng, Inner Mongolia	Jiulongshan Fm, J_2	Huang *et al.* [25]

References

1 Amateur Entomologists' Society (2017). Booklice and Barklice (Order: Psocoptera), www.amentsoc.org/insects/fact-files/orders/psocoptera.html.

2 Dellinger, T.A. and Day, E. (2015). Psocids: Barklice and Booklice, Psocodea: various families. *Virginia Cooperative Extension, Virginia Tech.* 2015, ENTO-143NP.

3 Gullan, P.J. and Cranston, P.S. (2014). Taxobox 17: Psocodea: "Psocoptera" (bark lice and book lice). In: *The Insects: An Outline of Entomology*, 5e, 509. Oxford: Wiley Blackwell.

4 Daly, H.V., Doyen, J.T. and Purcell, A.H. (1998). Psocoptera, in *Introduction to Insect Biology and Diversity* (2nd edition), Oxford University Press, New York, NY, pp. 404–406.

5 Johnson, K.P., Yoshizawa, K., and Smith, V.S. (2004). Multiple origins of parasitism in lice. *Proceedings of the Royal Society of London* **271** (1550): 1771–1776. https://doi.org/10.1098/rspb.2004.2798.

6 Yoshizawa, K. and Johnson, K.P. (2006). Morphology of male genitalia in lice and their relatives and phylogenetic implications. *Systematic Entomology* **31** (2): 350–361. https://doi.org/10.1111/j.1365-3113.2005.00323.x.

7 Li, F.S. (2002). Keys to infraorders of Psocoptera. In: *Psocoptera of China*, 46. Beijing: National Natural Science Foundation of China, Science Press.

8 Yoshizawa, K., Lienhard, C. and Johnson, K.P. (2006). Molecular systematics of the suborder Trogiomorpha (Insecta: Psocodea: 'Psocoptera'). *Zoological Journal of the Linnean Society* **146** (2): 287–299. https://doi.org/10.1111/j.1096-3642.2006.00207.x.

9 Lienhard, C. and Smithers, C.N. (2002). *Psocoptera (Insecta): World Catalogue and Bibliography.* Instrumenta Biodiversitatis., 5. Geneva: Muséum d'histoire naturelle.

10 Pictet, F.J. (1854). *Traité de paléontologie ou histoire naturelle des animaux fossiles considérés dans leur rapports zoologiques et géologiques*, 2e, 376. Paris: Bailliere.

11 Handlirsch, A. (1906). Die mesozoische insektenfauna. In: *Die fossilen Insekten und die Phylogenie der rezenten Formen*, 1161–1173. Leipzig: Verlag von Wilhelm Engelmann.

12 Tillyard, R.J. (1926). Kansas Permian insects. 8. The order Copeognatha. *American Journal of Science* **11**: 315–349.

13 Martynov, A. (1928). Permian fossil insects of Northeast Europe. *Travaux du Musée géologique de l'Académie des Sciences de l'U.S.S.R* **4**: 1–118.

14 Carpenter, F.M. (1926). Fossil insects from the Lower Permian of Kansas. *Bulletin of the Museum of Comparative Zoology, Harvard College* **67**: 437–444.

15 Becker-Migdisova, E.F. (1953). Obzar faun ravnokrylykh i senoidou erunakovska I kuznetskai svit kuzbassa. (Review of the Homoptera and Copeognatha of the Erunakov and Kuznetsk Formations of the Kuznetsk Basin). *Doklady Akademii nauk Soyuza Sovetskihk Sotsialistichechesckihk Respublik* **90** (1).

16 Becker-Migdisova, E.F. (1962). Some new Heteroptera and Psocoptera. *Paleontologicheskii Zhurnal* **1962**: 89–104.

17 Vishnyakova, V.N. (1967). Order Psocoptera. In: *Keys to the Insects of the European USSR* (ed. G.I.A. Bei-Bienko, B.E. Bykhovskii and G.S. Medvedev), 362–384. Zoologicheskii institut (Akademiia nauk SSSR)/Israel Program for Scientific Translations.

18 Nel, A. and Waller, A. (2007). The first fossil Compsocidae from Cretaceous Burmese amber (Insecta, Psocoptera, Troctomorpha). *Cretaceous Research* **28** (6): 1039–1041. https://doi.org/10.1016/j.cretres.2007.02.002.

19 Azar, D., Huang, D.-Y., Cai, C.Y. and Nel, A. (2014). The earliest records of pachytroctid booklice from Lebanese and Burmese Cretaceous ambers (Psocodea, Troctomorpha, Nanopsocetae, Pachytroctidae). *Cretaceous Research* **52**: 336–347. https://doi.org/10.1016/j.cretres.2014.04.005.

20 Azar, D., Hakim, M., and Huang, D.-Y. (2016). A new compsocid booklouse from the Cretaceous amber of Myanmar (Psocodea: Troctomorpha: Amphientometae: Compsocidae). *Cretaceous Research* **68**: 28–33. https://doi.org/10.1016/j.cretres.2016.08.003.

21 Azar, D., Huang, D.-Y., El-Hajj, L. *et al.* (2017). New Prionoglarididae from Burmese amber (Psocodea: Trogiomorpha: Prionoglaridetae). *Cretaceous Research* **75**: 146–156. https://doi.org/10.1016/j.cretres.2017.03.028.

22 Grimaldi, D. and Engel, M.S. (2005). The paraneopteran orders. In: *Evolution of the Insects*, 261–271. New York, NY: Cambridge University Press.

23 Lin, Q. (1976). The Jurassic fossil insects from Western Liaoning. *Acta Palaeontologica Sinica* **15** (1).

24 Hong, Y.C. (2002). Psocoptera. In: *Atlas of Amber insects of China*, 88–101. Beijing: Beijing Science & Technology Press.

25 Huang, D.-Y., Nel, A., Azar, D. and Nel, P. (2008). Phylogenetic relationships of the Mesozoic paraneopteran family Archipsyllidae (Insecta: Psocodea). *Geobios* **41** (4): 461–464. https://doi.org/10.1016/j.geobios.2007.11.00.

26 Smithers, C.N. (1972). The classification and phylogeny of the Psocoptera. The Australian Museum Trustees, Sydney, Australia. (Ph.D. dissertation).
27 Huang, D.-Y., Bechly, G., Nel, P. *et al.* (2016). New fossil insect order Permopsocida elucidates major radiation and evolution of suction feeding in hemimetabolous insects (Hexapoda: Acercaria). *Scientific Reports* **6**: 23004. https://doi.org/10.1038/srep23004.
28 Yoshizawa, K. and Lienhard, C. (2016). Bridging the gap between chewing and sucking in the hemipteroid insects: new insights from Cretaceous amber. *Zootaxa* **4079** (2): 229–245. https://doi.org/10.11646/zootaxa.4079.2.5.

16

Homoptera – Cicadas and Hoppers

Ying Wang[1,2]*, Xiao Zhang*[2]*, Tingying Zhang*[2]*, Xue Liu*[2]*, Chungkun Shih*[2,3]*, Yunzhi Yao*[2]*, and Dong Ren*[2]

[1] *Beijing Museum of Natural History, Dongcheng District, Beijing, China*
[2] *Capital Normal University, Haidian District, Beijing, China*
[3] *National Museum of Natural History, Smithsonian Institution, Washington, DC, USA*

16.1 Introduction to Homoptera

Homoptera, meaning "similar wings," are insects with four uniformly membranous wings. Metamorphosis of Homoptera is usually simple, and incomplete. Aphids, scale insects, whiteflies, planthoppers, treehoppers, leafhoppers, spittlebugs and cicadas comprise a group of insects having the unique structures of wings and the rostra. Recent studies based on morphological and molecular cladistics show that the phylogenetic relationship of Homoptera is not monophyletic. Many support the combination of Homoptera and Heteroptera to form the largest non-holometabolan insect Order of the Hemiptera. However, this book still treats Homoptera and Heteroptera in two separate Chapters (16 and 17).

Homopterans, as a paraphyletic group, are now divided into three monophyletic Suborders, Sternorrhyncha, Auchenorrhyncha, and the relict group Coleorrhyncha [1]. Sternorrhyncha are typically classified into four superfamilies, Aleyrodoidea (whiteflies), Aphidoidea (aphids), Coccoidea (scale insects) and Psylloidea (jumping plant lice). Auchenorrhyncha, morphologically defined by their possession of a tymbal acoustic system, an aristate antennal flagellum and reduction of the proximal median plate in the wing base, comprise the Infraorders of Fulgoromorpha and Cicadomorpha [2]. Fulgoromorpha comprise three superfamilies, Surijokocixioidea, Coleoscytoidea and Fulgoroidea [3]. They are commonly called "planthoppers" (Figure 16.1) with nymphs coated with hydrophobic wax. Cicadomorpha comprise four superfamilies, Cicadoidea (cicadas) (Figure 16.2), Cercopoidea ("spittlebugs/froghoppers"), Membracoidea (also called Cicadelloidea, "leafhoppers" and "treehoppers") (Figure 16.3) and Palaeontinoidea (extinct cicada-like insects) [2]. Most cicadomorphans have a mid-gut filter chamber to remove excess water from their intake of xylem or phloem sap [4]. For fossil Cicadomorpha described from the mid Mesozoic of northern China, we have documented taxa assigned to three additional superfamilies, Prosboloidea Handlirsch, 1906, Scytinopteroidea, Handlirsch 1906 and Pereborioidea, Zalessky 1930. Coleorrhyncha, commonly called "moss bugs" or "beetle bugs," have a single extant family, the Peloridiidae, and three fossil families. Details of the relict Coleorrhyncha will be presented in Chapter 17, Heteroptera.

Homoptera are distributed worldwide. It is estimated that more than 40 000 species have been described in the world. All species of Auchenorrhyncha and Sternorrhyncha have long, stylet mouthparts, feeding on plants sap. Some of them, by transmitting viral and mycoplasmatic diseases, are pests of agricultural crops and trees. However, some species have beneficial values, such as the production of the cochineal dyestuff by scale insects *Dactylopius coccus* and shellac resin secreted by the female lac insects *Kerria lacca* [1].

Homoptera species vary greatly in size with wingspans ranging from a few millimeters (mm) for aphids and scale insects to 200 mm for cicadas. In some families (e.g. leafhoppers), the forewings and hind wings are almost equal in length, but in other families (e.g. cicadas), forewings are longer than the hind wings. They are good fliers, and when at rest, usually hold wings roof-like over their bodies. The mouthparts of cicadas and their kin form a long stylet (called beak or rostrum) from the lower side of the rear of their heads. The mouthparts are used to suck plant fluid or juice. Antennae of cicadas and leafhoppers are short and bristle-like, barely visible. As all Auchenorrhyncha and Sternorrhyncha feed on plants, the great majority of auchenorrhynchans feeding on angiosperms, their diversity and speciose might have been related to angiosperm radiations.

Rhythms of Insect Evolution: Evidence from the Jurassic and Cretaceous in Northern China, First Edition.
Edited by Dong Ren, Chungkun Shih, Taiping Gao, Yongjie Wang, and Yunzhi Yao.

Figure 16.1 Mating of planthopper nogodinids. Source: Photo by Jason Shih.

Honeydew Makers

All homopterans are terrestrial plant feeders and many share a common biology of producing honeydew and being ant-attended, especially those in the Sternorrhyncha. Aphids are iconic examples of this behavior (Figures 16.4 and 16.5). They typically produce honeydew several times their body weight per day. Honeydew production is widespread in different homopteran lineages and probably originated as early as the Permian. In fact, its appearance could have stimulated the development of nectar production in various plants, initially as a mechanism for attracting predacious insects so as to reduce the level of herbivory [5].

Honeydew is eagerly collected by ants, which in many cases are solicited from the homopterans using rapid strokes of their antennae [6]. Ants seem preadapted for "milking" homopterans since trophallaxis, or liquid food exchange, is an intricate part of their social life, made possible by the structure of their proventriculus. Many studies on ant–hemipteran symbiosis have focused on aphids. An extreme example of this symbiosis involves *Lasius neoniger* ants tending *Aphis maidiradicis*: the ants overwinter the aphids in their colonies and then deliver them back to the host plants in the spring [1].

Poinar and Poinar illustrated an ant, in Dominican amber, carrying a scale insect (coccid) in its mandibles, demonstrating a relationship in which the ants farm the scale insects for their sugary exudate; and the relationship is still extant (p. 180 in [7] and figure 108 in [8]). Although not an aphid, this example illustrates a similarly coevolved relationship. Perkovsky, in 2006, described Eocene–Oligocene Saxonian amber with an association of ants and aphids that is suggestive of the typical ant–aphid relationship [9].

Musicians in the Summer

Cicadas are best known for their ability to produce a loud sound. Different cicada species sing at different times of the day. They use the sounds for communication, especially for courtship. Each species has its own distinctive song and attracts only females of its own kind. The organs of sound production are the tymbals, a pair of ribbed cuticular membranes located on either side of the first abdominal tergite. In many species, the tymbals are partly or entirely concealed by tymbal covers, platelike anterior projections of the second abdominal tergite. Contraction of internal tymbal muscles causes the tymbals to buckle inward, and relaxation of these muscles allows the tymbals to pop back to their original position. The sound produced is amplified by the substantially hollow abdomen, which acts as a resonator [4].

The Suborder Auchenorrhyncha have complex tymbal acoustic system. On each side of the tergites, one or two are a tymbal and tympanum; the former is composed of thin plates that vibrate [10, 11]. Most auchenorrhynchans produce vibrations inaudible to humans, but cicadas are very loud because the tymbals are large and elaborate. The sounds produced by both groups are used for communication among individuals and even for defense [12]. Cicadas produce sound by alternate muscular distortion and relaxation of a specialized area of elastic cuticle, and tymbal, to give individual clicks or variably modulated pulses of sound.

16.2 Progress in the Studies of Fossil Homoptera

Compression fossils of hoppers were first described by Brodie, in 1845, from the Mesozoic of England [13]. The first Cenozoic hoppers from the Baltic amber were named by Germar seven years later [14].

The basal Homoptera, Archescytinidae Tillyard, 1926, were first recorded from the Early Artinskian (Early Permian) of Moravia (Obora) with a limited number of species. The Archescytinidae, occurring from the Early Permian to the Late Triassic, are believed to be the most basal Homoptera as well-preserved rostra in some of the fossils [15]. A Permian Auchenorrhyncha, *Scytinoptera kokeni* Handlirsch, 1904 (in Scytinopteridae), was first reported by Handlirsch from Russia in 1904 [14]. The earliest modern taxa of the Fulgoromorpha appeared in the late Middle Permian (about 253 Mya). During the latter part of the Late Cretaceous, homopteran fauna became quite similar to current forms, containing the modern Auchenorrhyncha and most of the Sternorrhyncha families [15].

Figure 16.2 A cicada emerging from nymph exoskeleton. Source: Photos by Jason Shih.

The aphids (Aphidoidea) are uncommon before the Early Cretaceous around the world, but became abundant and highly diverse during the Early Cretaceous in the northern temperate zone only. The Protopsyllidiidae appear to be an extinct sister group to the

Figure 16.3 A froghopper, *Machaerota* sp. (Machaerotidae in Superfamily of Cercopoidea) near the tubular structure of mainly calcium carbonate made by the nymphs. Source: Photo by Jason Shih.

Sternorrhyncha [16], even though the family have been traditionally believed to be closely related to or even within the Psyllomorpha [15, 17, 18]. The family have an extensive record, with at least 30 genera and 50 species of compression fossils from the Late Permian to the mid-Cretaceous. The oldest whiteflies (Aleyrodidae in Aleyrodoidea) have been described from the Late Jurassic and the oldest scale insects from the Early Cretaceous [5]. Hitherto, it is estimated that more than 63 families, 855 genera, and 1425 species of fossil Homoptera have been described.

A homopteran fossil in China was first reported by Lin, in 1977, from the Cenozoic of Mengla County in Yunnan Province. He described *Nisocercopis validis* Lin, 1977, which triggered studies of Chinese fossil Homoptera [19]. During the past 40 years, there have been many significant studies of Chinese fossil homopterans by Qibin Lin, Juanjie Tan, Youchong Hong, Dong Ren, Junfeng Zhang, Wenli Wang, Haichun Zhang, Bo Wang, Ying Wang, Chungkun Shih, Shu Li, Jacek Szwedo, Haijing Hu, Guang Yang, Yi Li, Xiaohui Liu, Dong Chen, among others. The Chinese Mesozoic fossils are not only abundant and diverse, but also well-preserved [20]. Collectively, the Chinese researchers, with corroborations

Figure 16.4 An aphid delivering a baby. Source: Photo by Jason Shih.

Figure 16.5 Ants feeding honeydew from aphids which are attacked by a parasitoid flower fly (Syrphidae). Source: Photo by Jason Shih.

of foreign experts, have described Chinese homopteran fossils comprising 24 families, 82 genera, and 140 species hitherto (Table 16.1).

Suborder Sternorrhyncha Amyot and Serville, 1843:

Superfamily Aphidoidea, 4 families, 10 genera, 16 species

Superfamily Naibioidea, 1 family, 1 genus, 1 species

Superfamily Palaeoaphidoidea, 1 family, 1 genus, 1 species

Family *Incertae sedis*, 2 genera, 2 species

Superfamily Protopsyllidioidea, 1 family, 2 genera, 2 species

Suborder Auchenorrhyncha Dumeril, 1806:

Superfamily Fulgoroidea, 4 families, 7 genera, 10 species

Superfamily Cercopoidea, 3 families, 15 genera, 27 species

Superfamily Cicadoidea, 1 family, 11 genera, 16 species

Superfamily Membracoidea, 3 families, 4 genera, 5 species

Superfamily Prosboloidea, 2 families, 3 genera, 3 species

Superfamily Scytinopteroidea, 1 family, 1 genus, 1 species

Superfamily Palaeontinoidea, 1 family, 23 genera, 53 species

Superfamily Pereborioidea, 1 family, 1 genus, 2 species

Suborder Paleorrhyncha Carpenter, 1931:

16.3 Representative Fossils of Homoptera from Northern China

Suborder Sternorrhyncha Amyot and Serville, 1843

Superfamily Aphidoidea

Family Aphididae Latreille, 1802

Aphididae, characterized by large compound eyes with triommatidia, are an extant family of superfamily Aphidoidea, comprising two subfamilies of Aphidinae Latreille, 1802 and Macrosiphinae Wilson, 1910. Up to now, only two genera and three species of fossil Aphididae have been reported from the Early Jurassic to the Cenozoic. Among them, two representative fossils are *Sunaphis shandongensis* Hong & Wang, 1990 and *Sunaphis laiyangensis* Hong & Wang, 1990, from Shandong, China [21]. They appear on various plants and have host alternations during their life history mainly in autumn.

Only one genus included from the Cretaceous of Northern China: *Sunaphis* Hong & Wang, 1990.

Sunaphis Hong & Wang, 1990

Sunaphis Hong & Wang, 1990, *The Stratigraphy and Palaeontology of Laiyang Basin, Shandong Province*, 81 [21] (original designation).

Type species: *Sunaphis shandongensis* Hong & Wang, 1990.

Antenna with five or six antennomeres; Rs arising from the front or middle of pterostigma; M arising from the base of pterostigma, with three branches; CuA_1 far from

CuA_2, separately arising from Sc + R + M; the vestige of siphunculi rounded.

Distribution and age: Shandong; Early Cretaceous.

Two species included from the Cretaceous of Northern China (see Table 16.1).

Family Hormaphididae Mordvilko, 1908

Hormaphididae, comprising three aphid subfamilies of Cerataphidinae Baker, 1920, Nipponaphidinae Ghosh, 1988 and Hormaphidinae Mordvilko, 1908, are the only extant aphid family from the Mesozoic of China with more than 200 species worldwide. The Hormaphididae are characterized by the possession of several intrinsically fascinating biological characteristics. All known soldier aphids are members of the Hormaphididae and the Pemphigidae [22]. In the Hormaphididae, strong host specificity is well-defined and represented with different patterns of host association among tribes. The Cerataphidinae are primarily associated with Styracaceae, and the primary hosts of Hormaphidinae and Nipponaphidinae are Hamamelidaceae. The secondary host association is more relaxed, with Cerataphidinae on Compositae, Gramineae, Loranthaceae, Palmaceae, and Zingiberaceae; Hormaphidinae on Betulaceae and Pinaceae and Nipponaphidinae on Fagaceae, Lauraceae, and Moraceae [23]. Host alternation of most species by alate sexuparae in the autumn, two-year cycles, producing galls on the primary host. Sexuales apterous with well-developed rostra.

Genera included from the Cretaceous of Northern China: *Petiolaphioides* Hong & Wang, 1990 and *Petiolaphis* Hong & Wang, 1990.

Petiolaphioides Hong & Wang, 1990

Petiolaphioides Hong & Wang, 1990, *The Stratigraphy and Palaeontology of Laiyang Basin, Shandong Province*, 88 [21] (original designation).

Type species: *Petiolaphioides shandongensis* Hong & Wang, 1990.

Antenna with six antennomeres; Antennomere III long; secondary rhinaria rounded and arranged transversely, intensively and regularly; bases of M completed and clear, arising from the Sc + R + M; the common bases of CuA_1 and CuA_2 long; head big and wide.

Distribution and age: Shandong; Early Cretaceous.

One species included from the Cretaceous of Northern China (see Table 16.1).

Petiolaphis Hong & Wang, 1990

Petiolaphis Hong & Wang, 1990, *The Stratigraphy and Palaeontology of Laiyang Basin, Shandong Province*, 86 [21] (original designation).

Type species: *Petiolaphis laiyangensis* Hong & Wang, 1990.

Antenna with six antennomeres; secondary rhinaria rounded; bases of M complete and clear, arising from the Sc + R + M; the common bases of CuA_1 and CuA_2 long; body short and wide.

Distribution and age: Shandong; Early Cretaceous.

One species included from the Cretaceous of Northern China (see Table 16.1).

Family Oviparosiphidae Shaposhnikov, 1979

Oviparosiphidae, an extinct aphid family, are diagnosed by having both the ovipositor and siphunculi; radial sector arising from the middle of the elongate pterostigma; vein M with three branches; cubitus branches leaving the main vein at points rather close to each other; antenna short and three antennomeres apparently with transverse secondary rhinaria. Records of fossil Oviparosiphidae are scarce and only 10 species from the Mesozoic of China have been reported [24]. Among them, more Oviparosiphidae are known from the Early Cretaceous but absent from the Late Cretaceous [24]. The oldest species is *Daoaphis magnalata* Huang, Wegierek, Żyła & Nel, 2015 from the late Middle Jurassic of Inner Mongolia, China [25].

Genera included from the Jurassic and Cretaceous of Northern China: *Oviparosiphum* Shaposhnikov, 1979, *Expansaphis* Hong & Wang, 1990, *Sinoviparosiphum* Ren, 1995, *Archeoviparosiphum* Żyła, Homan, Franielczyk & Wegierek, 2015 and *Daoaphis* Huang, Wegierek, Żyła & Nel, 2015.

Oviparosiphum Shaposhnikov, 1979

Oviparosiphum Shaposhnikov, 1979, *Paleontol. J.*, 13, 458 [26] (original designation).

Type species: *Oviparosiphum jakovlevi* Shaposhnikov, 1979.

Antenna with seven antennomeres; secondary rhinaria slightly ellipsoidal, large; siphunculi in the form of short truncate cones; ovipositor large with valvae I and III well-developed [27].

Distribution and age: Shandong of China, Mongolia; Early Cretaceous.

Only one species included from the Cretaceous of Northern China (see Table 16.1).

Oviparosiphum stictum Fu, Yao & Qiao, 2017 (Figure 16.6)

Oviparosiphum stictum Fu, Yao & Qiao, 2017: *Cretac. Res.*, 75, 157–161.

Locality and horizon: Huangbanjigou, Beipiao, Liaoning, China; Yixian Formation, Lower Cretaceous.

The appearance of developed ovipositors in many extinct families further demonstrates that oviparity

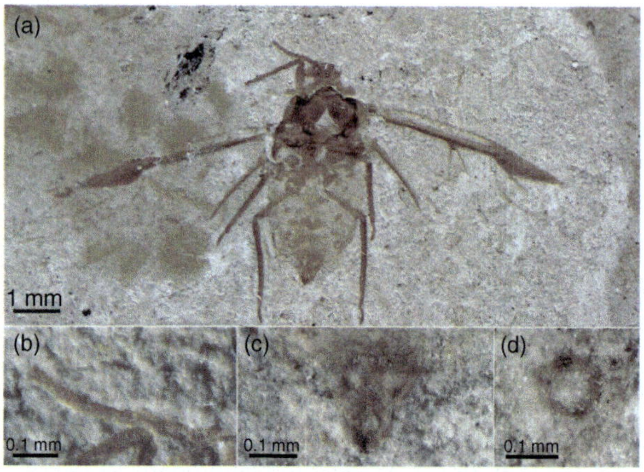

Figure 16.6 *Oviparosiphum stictum* Fu, Yao & Qiao, 2017. (a), Habitus of Holotype, CNU-HET-LB2016001; (b), Antennomeres; (c), Ovipositor; (d), Siphunculus.

is a plesiomorphic state [28]. In addition, this fossil provides new evidence for the Cretaceous as a rapid radiation period of aphids [29]. Alate specimen. Body length 3.89 mm. Length of antennomere I 0.03 mm (Figure 16.6a). Antennomere III, 0.35 mm elongated, thicker than other antennomeres. Antennomere IV (0.18 mm) nearly half as long as antennomere III, almost three times as long as wide. Antennomere III, IV, V with abundant rows of elliptical rhinaria (Figure 16.6b). Mesothorax highly sclerotized. Pterostigma long and broad, its length about 3.5 times as long as its greatest width. The bases of CuA_1 and CuA_2 separately extend from a main vein (Sc + R + M). Vein Rs slightly curved, starting from proximal part of pterostigma. Main stem of veins on hind wing thick. Basal diameter of siphunculus 0.19 mm (Figure 16.6d). Ovipositor well-developed, two valves visible (Figure 16.6c) [27].

Expansaphis Hong & Wang, 1990

Expansaphis Hong & Wang, 1990, *The Stratigraphy and Palaeontology of Laiyang Basin, Shandong Province*, 77 [21] (original designation).

Type species: *Expansaphis ovata* Hong & Wang, 1990.

Antenna with five antennomeres, shorter than or equal to head and thorax; secondary rhinaria ellipsoidal, arranged regularly; bases of CuA_1 and CuA_2 swollen; ovipositor developed.

Distribution and age: Shandong; Early Cretaceous.

Two species included from the Cretaceous of Northern China (see Table 16.1).

Sinoviparosiphum Ren, 1995

Sinoviparosiphum Ren, 1995, *Faunae and Stratigraphy of Jurassic-Cretaceous in Beijing and the Adjacent Areas*, 71 [30] (original designation).

Type species: *Sinoviparosiphum lini* Ren, 1995.

The specific epithet is dedicated to Prof. Qibin Lin for his contribution to the research of fossil aphids. Antenna with four antennomeres, with processus terminalis; two monocular eyes located inside the compound eyes; prothorax longer than head; CuA_2 short, almost perpendicular; M near to Rs.

Distribution and age: Hebei; Early Cretaceous.

Only one species included from the Cretaceous of Northern China (see Table 16.1).

Archeoviparosiphum Żyła, Homan, Franielczyk & Wegierek, 2015

Archeoviparosiphum Żyła, Homan, Franielczyk & Wegierek, 2015, *Zookeys*, 483, 9–22 [31] (original designation).

Mesoviparosiphum Zhang, Zhang, Hou & Ma, 1989, *Geology of Shandong*, 5, 28–46. Syn. by Heie & Węgierek, 2011, A list of fossil aphids (Hemiptera, Sternorrhyncha, Aphidomorpha), 45 [24].

Paroviparosiphum Zhang, Zhang, Hou & Ma, 1989, *Geology of Shandong*, 5, 28–46. Syn. by Heie & Węgierek, 2011, A list of fossil aphids (Hemiptera, Sternorrhyncha, Aphidomorpha), 51 [24].

Type species: *Oviparosiphum baissense* Shaposhnikov & Wegierek, 1989.

In 2011, Heie and Węgierek transferred all four species in *Paroviparosiphum* Zhang, Zhang, Hou & Ma 1989 and *Mesoviparosiphum* Zhang, Zhang, Hou & Ma 1989 to *Oviparosiphum* Shaposhnikov, 1979 [24]. Then, in 2015, Żyła *et al.* revised the *Oviparosiphum* and set up a new combination in the genus of *Archeoviparosiphum*, including four species originally belonging to *Paroviparosiphum* and *Mesoviparosiphum* and two species originally belonging to *Oviparosiphum*. Among them, *Oviparosiphum baissensis* Shaposhnikov & Wegierek was designated as the type species and renamed as *Archeoviparosiphum baissense* (Shaposhnikov & Wegierek, 1989) [31].

Antenna with seven antennomeres; secondary rhinaria slightly ellipsoidal, small; pterostigma at most five times as long as wide; abdomen without setae; siphunculi in the form of pores; ovipositor small and rudimentary.

Distribution and age: Shandong; Early Cretaceous.

Five species included from the Cretaceous of Northern China (see Table 16.1).

Daoaphis Huang, Wegierek, Żyła & Nel, 2015

Daoaphis Huang, Wegierek, Żyła & Nel, 2015, *Euro. J. of Entomol.*, 112 (1), 187 [25] (original designation).

Type species: *Daoaphis magnalata* Huang, Wegierek, Żyła & Nel, 2015.

Antenna with seven antennomeres, long and bead-like, reaching the anterior part of the abdomen; secondary rhinaria ellipsoidal, arranged in numerous transverse rows;

forewings much longer than the body; bases of CuA_1 and CuA_2 close to each other, but CuA_2 not connected with the Sc + R + M.

Distribution and age: Inner Mongolia; Middle Jurassic.

Only one species included from the Jurassic of Northern China (see Table 16.1).

Daoaphis magnalata Huang, Wegierek, Żyła & Nel, 2015 (Figure 16.7)

Daoaphis magnalata Huang, Wegierek, Żyła & Nel, 2015: *Euro. J. Entomol.*, 112 (1), 187.

Locality and horizon: Daohugou, Ningcheng, Inner Mongolia, China; Middle Jurassic, Jiulongshan Formation.

Antenna equal to half of the body length. The apical part of the rostrum between middle and hind coxae. The pronotum wider than the head. Metascuta wide apart, with a poorly sclerotized area between them. The mesopostnotum forms a short U-shaped slat, widely overlapping the sides of the body. Legs short. Hind coxae very close to each other. Width of the forewing, 1.80–1.92 mm. Apical part well-developed. Cubital veins slightly thickened. Hind wing longer than half of the body length, with two cubital veins [25].

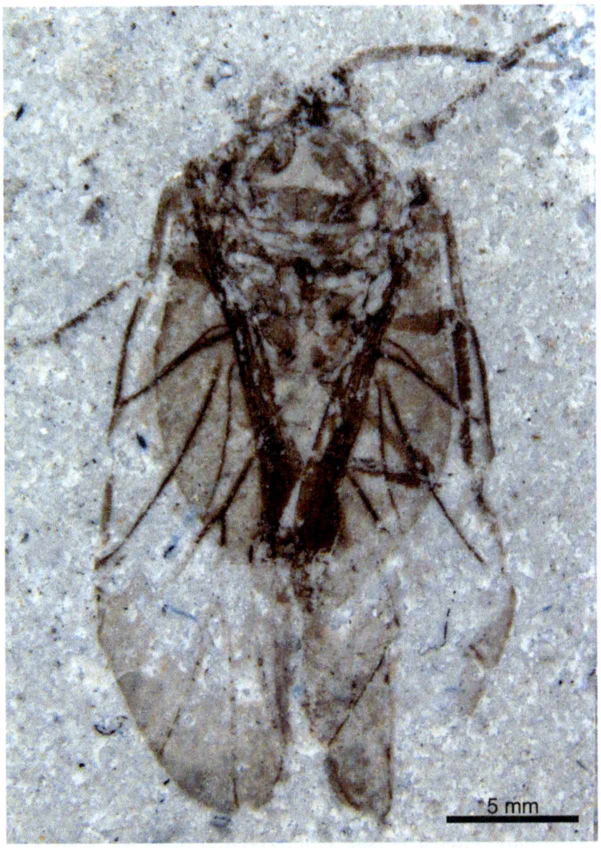

Figure 16.7 *Daoaphis magnalata* Huang, Wegierek, Żyła & Nel, 2015, (Holotype, NIGP 140801). Source: Photo provided by Dr. Di-ying Huang.

Family Sinaphididae Zhang, Zhang, Hou & Ma, 1989

The Sinaphididae are similar to Oviparosiphidae which are characterized by having both the ovipositor and siphunculi. However, the Sinaphididae are distinguished from the Oviparosiphidae by its circular, disorderly arranged secondary rhinaria, CuA_1 and CuA_2 separated at bases and undeveloped basal parts of M and CuA_1 [32]. Sinaphidids are scarce and only nine species from the Mesozoic of China have been described [24].

Genera included from the Cretaceous of Northern China: *Sinaphidium* Zhang, Zhang, Hou & Ma, 1989 and *Tartaraphis* Zhang, Zhang, Hou & Ma, 1989.

Sinaphidium Zhang, Zhang, Hou & Ma, 1989

Sinaphidium Zhang, Zhang, Hou & Ma, 1989, *Geology of Shandong*, 5, 34 [32] (original designation).

Type species: *Sinaphidium epichare* Zhang, Zhang, Hou & Ma, 1989.

Head nearly semicircular; Antenna about four times as long as head, with the antennomere III 5.5 times as long as wide; pterostigma large; M with considerably long basal part lost, and all branches curved; CuA_2 curved toward wing base, CuA_1 arising reversely.

Distribution and age: Shandong; Early Cretaceous.

Only one species included from the Cretaceous of Northern China (see Table 16.1).

Tartaraphis Zhang, Zhang, Hou & Ma, 1989

Tartaraphis Zhang, Zhang, Hou & Ma, 1989, *Geology of Shandong*, 5, 35 [32] (original designation).

Type species: *Tartaraphis peregrina* Zhang, Zhang, Hou & Ma, 1989.

Antenna with seven antennomeres, the antennomeres III and VII with rhinaria sparse; tarsi of leg slender and long; forewing with CuA_2 less than a half as long as CuA_1. Ovipositor spherical, moderate-sized.

Distribution and age: Shandong; Early Cretaceous.

Only one species included from the Cretaceous of Northern China (see Table 16.1).

Superfamily Naibioidea

Family Sinojuraphididae Huang & Nel, 2008

Pre-Cretaceous aphid taxa and specimens are much less numerous than those in the Cretaceous. The extinct Jurassic Sinojuraphididae, which have been added to the Naibioidea, confirm the persistence of ancient lineages among the Jurassic aphids together with juraphidids [33]. However, the phylogenetic relationships of Naibioidea are still debated. Originally Shcherbakov, in 2007, believed Naibioidea to be the "basal" group of Coccidomorpha [34]. However, Heie & Wegierek, in 2011, placed Naibioidea (as a superfamily comprising Naibiidae Shcherbakov, 2007, Sinojuraphididae Huang & Nel, 2008 and Dracaphididae Hong, Zhang, Guo & Heie, 2009) in Aphidomorpha [24]. Since the Daohugou flora contained lower plants, pteridophytes, seed-ferns and gymnosperms, but no angiosperms, sinojuraphids were likely to have lived on gymnosperms [33].

Only one genus included from the Jurassic of Northern China: *Sinojuraphis* Huang & Nel, 2008.

Sinojuraphis Huang & Nel, 2008

Sinojuraphis Huang & Nel, 2008, *Palaeontology*, 51 (3), 715 [33] (original designation).

Type species: *Sinojuraphis ningchengensis* Huang & Nel, 2008.

Forewing common stem of CuA very long, distinctly longer than the two distal free branches; stem of CuA without any posterior groove along it; M forked; very narrow pterostigma; R_1 reaching wing apex; rostrum shorter than body; 12 antennomeres, with basal antennomeres short.

Distribution and age: Inner Mongolia; Middle Jurassic.

Only one species included from the Jurassic of Northern China (see Table 16.1).

Superfamily Palaeoaphidoidea

Family Ellinaphididae Kania & Wegierek, 2008

The representatives of extinct Cretaceous aphid family Ellinaphididae are numerous, comprising nine genera and 32 species [35]. Most of them are described from Baissa (Lower Cretaceous, Russia), while only one species of the genus *Ellinaphis* from China (Shandong, Laiyang Basin, Tuanwang) [32]. An analysis of the number of representatives of selected genera (*Annulaphis, Caudaphis* and *Ellinaphis*) indicates that they occurred more frequently in cool strata [35]. Antennae in Ellinaphididae follow a model which can be found in many Early Cretaceous aphids (with small rhinaria in many transverse rows on all antennomeres of flagellum), particularly in Oviparosiphidae and Shaposhnikoviidae [26]. According to Kania & Wegierek (2008), Ellinaphididae and Palaeoaphididae are sister families and share the similar structure of the wings, including CuA_1 and CuA_2 distinctly thickened in the basal part and forming a common stem [35].

Only one genus included from the Cretaceous of Northern China: *Ellinaphis* Shaposhnikov, 1979.

Ellinaphis Shaposhnikov, 1979

Ellinaphis Shaposhnikov, 1979, *Paleontol. J.*, 13, 449–461 [26] (original designation).

Type species: *Ellinaphis sensoriata* Shaposhnikov, 1979.

Antennomere VII not narrowed as in *Rallotopaphis*; secondary rhinaria ellipsoidal, large, and antennomere VII up to three and half times as long as wide; two cubital veins on hind wings [24].

Distribution and age: Shandong of China, Mongolia, Russia; Early Cretaceous.

Only one species included from the Cretaceous of Northern China (see Table 16.1).

Family *Incertae sedis*

Dataiphis Lin, 1995

Dataiphis Lin, 1995, *Acta Palaeontol. Sin.*, 34 (2), 195 [36] (original designation).

Type species: *Dataiphis conderis* Lin, 1995.

Thorax greater than abdomen; Rs arising from the base of pterostigma; M straight, with three branch, arising from half of Sc + R + M; M_1 close to Rs, M_2 almost equal to M_3; CuA_1 and CuA_2 very close, separately arising from Sc + R + M; CuA_1 longer than CuA_2.

Distribution and age: Gansu; Early Cretaceous.

Only one species included from the Cretaceous of Northern China (see Table 16.1).

Superfamily Protopsyllidioidea

Family Protopsyllidiidae Carpenter, 1931

Protopsyllidiidae are an extinct family of psyllina in the Homoptera Suborder Sternorrhyncha. This family, known from the Late Permian (ca. 260 Mya) to the Late Cretaceous (90 Mya), comprises 31 genera and 57 species described so far [16]. There are 28 species known from the Permian [37–45], two species from the Triassic [17, 46], 20 species from the Jurassic [9, 17, 41, 43, 45, 47–51], and seven species from the Cretaceous [16, 45]. These species have been recorded from China, Mongolia, Burma, Kyrgyzstan, Kazakhstan, Tajikistan, Australia, South Africa, America (New Jersey), England, Germany, and Russia [1, 15, 52].

Genera included from the Jurassic and Cretaceous of Northern China: *Sinopsocus* Lin, 1976 and *Poljanka* Klimaszewski, 1995.

Sinopsocus Lin, 1976

Sinopsocus Lin, 1976, *Acta Palaeonto. Sin.*, 1976, 1 [50] (original designation).

Type species: *Sinopsocus oligovenus* Lin, 1976.

Forewing nearly ovate, the length is two times as long as the width. Sc disappeared. R straight and long, Rs with no branches and separated from the stem of R, M and Cu merging into the same stem, the stem of M shorter, branching after the branching of R and curtly close to costal margin. CuA branched from basal of the wing and short branches, vein with spines.

Distribution and age: Liaoning; Early Cretaceous.

Only one species included from the Cretaceous of Northern China (see Table 16.1).

Poljanka Klimaszewski, 1995

Poljanka Klimaszewski, 1995, *Acta Biologica Silesiana*, 27, 33–43 [45] (original designation).

Type species: *Cicadellopsis shurabensis* Bekker-Migdisova, 1985.

Forewing broadened in subapical part; stem of R slightly shorter than M + CuA; M with two long branches; cell m_{1+2} variously elongated, always longer than cell cua_1; CuA forked into long CuA_1 and short CuA_2; anal field long, almost half length of forewing. Hind tibiae longer than fore and mid tibiae; tarsi two-segmented, the first segment longer than the second.

Distribution and age: Kyrgyzstan, Early to Middle Jurassic; Inner Mongolia of China, Middle Jurassic; Kazakhstan, Middle to Late Jurassic.

Only one species included from the Jurassic of Northern China (see Table 16.1).

Poljanka hirsuta Yang, Yao & Ren, 2012 (Figure 16.8)

Poljanka hirsuta Yang, Yao & Ren, 2012: *Zootaxa*, 3274, 36–42.

Locality and horizon: Daohugou, Ningcheng, Inner Mongolia, China; Middle Jurassic, Jiulongshan Formation.

Scape 1.50 times thicker than pedicel, pedicel two times thicker than flagellomeres, apical flagellomere swollen; femora about two times thicker than corresponding tibiae, hind tibiae 1.46–1.57 times as long as fore and mid tibiae, the first tarsomere of fore legs 1.33 times as long as the second, the first tarsomere of mid legs 1.50 times as long as the second, the first tarsomere of hind legs 2.31 times as long as the second; R1 reduced, vein M + CuA about 1.57 times as long as vein R, branches of M about 2.22–2.39 times as long as M [53].

Suborder Auchenorrhyncha Dumeril, 1806

Infraorder Fulgoromorpha Evans, 1946

Superfamily Fulgoroidea Latreille, 1807

Family Cixiidae Spinola, 1839

Cixiidae, the most basal family of Fulgoroidea with a cosmopolitan distribution [54], comprise three subfamilies: Bothriocerinae Muir, 1923, Borystheninae Emeljanov, 1989 and Cixiinae Spinola, 1839 [55]. Cixiidae are a group of planthoppers with over 1500 described extant species in over 170 genera, which may be about 40% of the actual worldwide fauna [56]. Fossil records of Cixiidae are not very rich, and numerous fossil taxa previously ascribed to Cixiidae need to be re-examined and revised. The oldest fossils are known from the Lower Cretaceous deposits of England, Brazil and China [30, 57], and Lebanese amber [58]. Cixiidae are diagnosed by plesiomorphic features: median ocellus located just above the frontoclypeal suture, a row of spines on the second hind tarsomere and females have a well-developed "orthopteroid-type" ovipositor [59].

Genera included from the Cretaceous of Northern China: *Yanducixius* Ren, Lu & Guo, 1995 and *Lapicixius* Ren, Yin & Dou, 1998.

Yanducixius Ren, Lu & Guo, 1995

Yanducixius Ren, Lu & Guo, 1995, *Faunae and Stratigraphy of Jurassic-Cretaceous in Beijing and the Adjacent Areas*, 64–73 [30] (original designation).

Type species: *Yanducixius yihi* Ren, Lu & Guo, 1995.

The specific epithet is dedicated to a renowned Geologist Prof. Liangfu Yih. Forewing costal margin nearly parallel to posterior margin; Sc parallel to costal margin, disappear before pterostigma; R_1 with three branches; Rs with five branches; M with four branches; CuA divides from CuP or clavus, with two branches; vein A Y-shaped; cross-veins developed.

Distribution and age: Beijing; Early Cretaceous.

Two species included from the Cretaceous of Northern China (see Table 16.1).

Lapicixius Ren, Yin & Dou, 1998

Lapicixius Ren, Yin & Dou, 1998, *Acta Zootaxo. Sin.*, 23 (3), 281–288 [57] (original designation).

Type species: *Lapicixius decorus* Ren, Yin & Dou, 1998.

Head in dorsal view narrower than pronotum. Vertex broad, divided by a transverse curved carina medially into fore and hind part, both parts with an obsolete longitudinal midline. Posterior margin of vertex straight or slightly convex. Eyes large and prominent. Frons slightly longer than broad, with a longitudinal carina. Median ocellus present. Epistemic suture slightly curved. Postclypeus about as long as frons. Rostrum surpassing hind coxae. Pronotum longest at middle, anterior margin weakly concave or broadly incised, posterior margin convex. Mesoscutum about as long as broad, with five carinae, the median one continuous with that of pronotum. Hind tibia without spines laterally. Forewing exceeding apex of abdomen, apical margin evenly rounding. Pterostigma well-developed. All veins without granule. Rs arising slightly basad of middle of tegmen, with three to four veinlets. M forking at middle

Figure 16.8 *Poljanka hirsuta* Yang, Yao & Ren, 2012. Photographs: (a). Holotype, CNU-PSY-NN2011008p; (b). Right forewing; (c). Left foreleg and antennae; (d). Rostrum.

wing, and giving rise to five to seven terminal branches. CuA forking before the level of origin of Rs, with four distal branches.

Distribution and age: Liaoning; Early Cretaceous.

Only one species included from the Cretaceous of Northern China (see Table 16.1).

Planthoppers in Myanmar (Burmese) Amber

Planthopper fossils from the Jurassic and Cretaceous of China are not abundant. To date, only four families have been documented, among which most are only based on single wings preserved in compression fossils without or with insufficient body characters. However, the diversity of planthoppers in Myanmar amber is broader with at least seven families reported so far. These amber planthoppers usually have a well-preserved and detailed body and wing characters for in-depth taxonomic studies.

Perforissidae, an extinct planthopper family only known from the Cretaceous, can be diagnosed by a pronotum deeply cleft posteriorly and rounded sensory pits persisting in adults. A species of Perforissidae, *Foveopsis heteroidea* Zhang, Ren & Yao, 2017 recently reported from Myanmar (Burmese) amber, has well-preserved sensory pits and wing-coupling apparatus (Figure 16.9). However, the abundance and distribution of sensory pits is variable based on various specimens of this species, suggesting that sensory pits cannot be considered as a diagnostic character. Well-preserved wing-coupling apparatus and nodal line indicate this planthopper had a capability of moderately versatile flights. Its broad geographical distribution in its fossil records of 40 million years in the Cretaceous might have been associated with or caused by their flight capability assisted by favorable and helpful wind flows. The sexual differences between male and female *F. heteroidea* in Myanmar amber are consistent with fossil planthoppers of *Perforissus muiri* Shcherbakov, 2007 from the Late Cretaceous of New Jersey [60].

Family Fulgoridiidae Handlirsch, 1939

Fulgoridiidae, an extinct "cixiid-like" family, are known so far from the Jurassic and the Cretaceous deposits of Europe and Asia. Up to date, over 150 species have been ascribed to Fulgoridiidae, but the vast majority of them may need revisions. Fulgoridiidae, as a paraphyletic assemblage, comprise at least two lineages [58]. The "true" Fulgoridiidae share a long basal vein RP and a short stem of CuA due to its early double forking, and body structure resembling the recent representatives of Cixiidae. This family are treated as roots of Fulgoroidea [61].

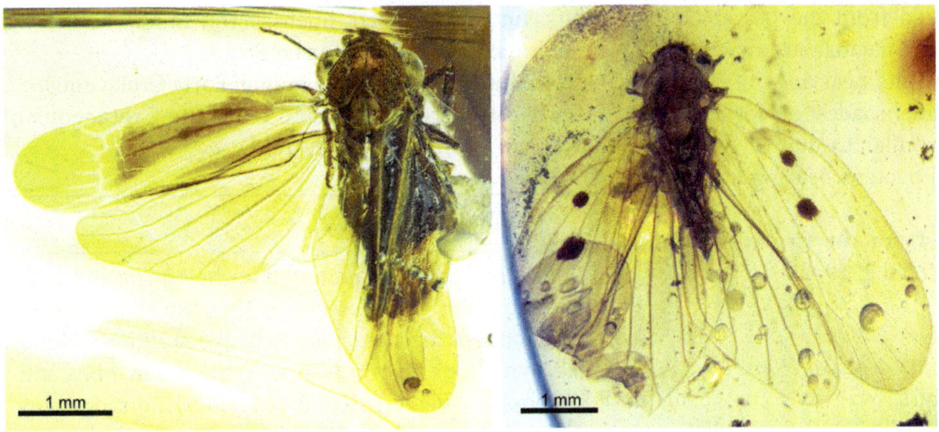

(a) (b)

Figure 16.9 *Foveopsis heteroidea* Zhang, Ren & Yao, 2017. (a). Holotype, CNU-HOM-MA2017003, female. (b). Paratype, CNU-HOM-MA2017004, male.

Genera included from the Jurassic and Cretaceous of Northern China: *Eofulgoridium* Martynov, 1937 and *Fenghuangor* Li & Szwedo, 2011.

Eofulgoridium Martynov, 1937

Eofulgoridium Martynov, 1937, *Akad. Nauk SSSR, Trudy Paleontol. Inst.*, 7, 95–97 [47] (original designation).

Type species: *Eofulgoridium kisylkiense* Martynov, 1937.

Forewing Sc at about middle of R and C, dividing point of M at about wing midlength, M with three branches. Hind wing costal margin concave, Rs arising beyond midwing, dividing point of M and CuA beyond midwing.

Distribution and age: Xinjiang of China and Kyrgyzstan, Early Jurassic; Gansu of China, Early Cretaceous.

Two species included from the Jurassic and Cretaceous of Northern China (see Table 16.1).

Fenghuangor Li & Szwedo, 2011

Fenghuangor Li & Szwedo, 2011, *Zootaxa*, 3094, 52–62 [62] (original designation).

Type species: *Fenghuangor imperator* Li & Szwedo, 2011.

This genus is clearly differentiated from all other known genera of Fulgoridiidae from Europe and Asia by its size and venation pattern, with short common portion of branches M_{3+4} and CuA_{1a}. It differs from the genus *Eofulgoridium* Martynov, 1939 by the length/width ratio of tegmen of 2.1; costal area distinctly narrower than costal cell; costal area slightly widened in apical portion; branch $ScRA_1$ connected to Pc + CP basad of costal area apex; cell c5 with a single transverse veinlet icua; cell c5a widely cuneiform and branch CuA_1 forked twice before the margin.

Distribution and age: Inner Mongolia; Middle Jurassic.

Figure 16.10 *Fenghuangor imperator* Li & Szwedo, 2011. Holotype. CUN-HEM-2006266p.

Only one species included from the Jurassic of Northern China (see Table 16.1).

Fenghuangor imperator Li & Szwedo, 2011 (Figure 16.10)

Fenghuangor imperator Li & Szwedo, 2011: *Zootaxa*, 3094, 54.

Locality and horizon: Daohugou, Ningcheng, Inner Mongolia, China; Middle Jurassic, Jiulongshan Formation.

Venation without distinct nodal and apical lines of transverse veinlets; cell c2 about 4.8 times as long as wide; cell c3 more than three times as long as wide [62].

Family Lalacidae Hamilton, 1990

Lalacidae, an extinct family placed in "cixiid-like" group, comprise three subfamilies, Lalacinae Hamilton, 1990, Ancoralinae Hamilton, 1990 and Protodelphacinae Hamilton, 1990, including nine genera with 24 species. Fossils reported mainly from the Lower Cretaceous of

Brazil, only one genus reported from the Lower Cretaceous of Beijing, China. The Lalacidae are diagnosed by head with a pair of shallow pits near apex of crown, ambient vein (appendix) narrow, broadest on costa, with sclerous striations, venation similar to that of modern Kinnaridae, and hind leg with pectens bearing movable setae [63].

Only one genus included from the Cretaceous of Northern China: *Cretocixius* Zhang, 2002.

Cretocixius Zhang, 2002

Cretocixius Zhang, 2002, *Acta Zootaxo. Sin.*, 27 (1), 20–23 [64] (original designation).

Type species: *Cretocixius stigmatosus* Zhang, 2002.

Costal margin of tegmen straight and incrassated obviously; appendix narrow, only apical margin with sclerous striations; pterostigma well-developed; Sc 2-branched in the pterostigma area; vein R_1 simple and enclosing the posterior margin of pterostigma; CuA with six terminal branches; all cross-veins thick obviously, r-m long; with six subapical and 14 apical cells.

Distribution and age: Beijing; Early Cretaceous.

Only one species included from the Cretaceous of Northern China (see Table 16.1).

Infraorder Cicadomorpha Evans, 1946

Superfamily Cercopoidea Leach, 1815

Family Cercopidae Leach, 1815

Cercopidae are popularly known as "spittlebugs", because their nymphs usually produce masses of white foams for thermal and moisture control and for protection against predators [65]. Their adults are also called "froghoppers" for jumping around plants. Cercopidae are diagnosed by their bright color patterns, the presence of one or two lateral spines and an apical crown of spines on the hind tibia [66]. Fossil records of Cercopidae are abundant, from the Jurassic to the Cenozoic, widely distributed around the world, comprising 13 genera with 46 species. During the Cenozoic, taxa of Cercopidae have increased rapidly [67].

Genera included from the Jurassic and Cretaceous of Northern China: *Sinocercopis* Hong, 1982 and *Hebeicercopis* Hong, 1983.

Sinocercopis Hong, 1982

Sinocercopis Hong, 1982, *Mesozoic Fossil Insects of Jiuquan Basin in Gansu Province*, 86 [68] (original designation).

Type species: *Sinocercopis liaoyuanensis* Hong, 1982.

Forewing with length two times as long as the width; the stem among Rs and M and CuA linear; the divided points between Rs and R, and M and CuA also linear, separately near the point of apex 1/4 and 1/3; all branches long. Hind wing broad, open sector, m-cua before the branches of CuA. Rough surface.

Distribution and age: Liaoning; Early Cretaceous.

Only one species included from the Cretaceous of Northern China (see Table 16.1).

The High Jump Champion

"Spittlebug" nymphs are noted for being covered with foaming froth for protection. The adults are called "froghoppers" because they look like and hop around like small frogs among plants. Spittlebug nymphs and froghoppers feed on plant juices and are considered as serious pests of plants.

A commonly acknowledged high jumping champion has been the flea (Order Siphonaptera) which, with a body length of 3 mm, can jump 200 mm high and 330 mm wide. A study by researchers at the Cambridge University in 2003 showed that a 6 mm long spittlebug, *Philaenus spumarius,* can jump about 700 mm high in the air. It has a take-off velocity of up to 3.1 m s^{-1} [69]. The ratio of jumping height to body length is about 116. Imagine this is equivalent to a human being with a height of 1.7 m jumping 197 m, higher than a 70-story skyscraper. Froghoppers can store tremendous amount of energy in their robust hind legs, and by releasing the energy in an instant, froghoppers are capable of jumping from harm's way.

Family Procercopidae Handlirsch, 1906

Procercopidae, an extinct family, have been recorded as the most basal family of Cercopoidea [70]. They had a slender forewing, Rs arising in basal 1/3 of wing, and M and CuA branching in distal 1/3 of wing. They have been described from the Early Jurassic to the Early Cretaceous in Germany, Russia, Central Asia, Southeast Asia and China [71], comprising 15 genera with 37 species. Procercopidae were dominant taxa of the Jurassic Cercopoidea. Although being a common group in the Yanliao Biota at Daohugou of China, the Procercopidae had a comparatively low diversity [72].

Genera included from the Jurassic and Cretaceous of Northern China: *Procercopis* Handlirsch, 1906, *Procercopina* Martynov, 1937, *Anthoscytina* Hong, 1983, *Chengdecercopis* Hong, 1983, *Cretocercopis* Ren, Lu & Guo, 1995, *Anomoscytina* Ren, Yin & Dou, 1998, *Jurocercopis* Wang & Zhang, 2009, *Stellularis* Chen, Yao & Ren, 2015 and *Titanocercopis* Chen, Zhang & Wang, 2015.

Procercopis Handlirsch, 1906

Procercopis Handlirsch, 1906, *Ein Handbuch fur Palaontologen und Zoologen*, 500–501 [73] (original designation).

Type species: *Procercopis alutacea* Handlirsch, 1906.

Forewing elongate, about four times as long as wide, M at least with three branches, apical area with several cross-veins.

Distribution and age: Xinjiang of China, Germany and Kyrgyzstan; Early Jurassic.

Only one species included from the Jurassic of Northern China (see Table 16.1).

Procercopina Martynov, 1937

Procercopina Martynov, 1937, *Akad. Nauk SSSR, Trudy Paleontol. Inst.*, 7, 99–101 [47] (original designation).

Type species: *Procercopina asiatica* Martynov, 1937.

R convex with a short branch just before its termination. Rs convex almost as strongly as R, branching into two slightly beyond level of branching of R, with anterior branch connected to R by a short cross-vein. M branching into two greatly distal of terminal middle length with a cross-vein present between anterior and posterior branches, and just distal of cross-vein (between Rs and anterior branch of M). CuA oblique, with two branches; CuP simple, with two branches, m-cu (between posterior branch of M and anterior branch of CuA) slightly distad of branching of M. Tegmental surface finely tuberculate. Tegmen, as preserved, length 12 mm and width 4 mm.

Distribution and age: Xinjiang of China, Germany and Kyrgyzstan, Early Jurassic; Thailand, Middle Jurassic.

Only one species included from the Jurassic of Northern China (see Table 16.1).

Anthoscytina Hong, 1983

Anthoscytina Hong, 1983, *Middle Jurassic Fossil Insects in North China*, 61–62 [74] (original designation).

Paracicadella Hong, 1983, *Middle Jurassic Fossil Insects in North China*, 62–63 [74]. Syn. by Shcherbakov, 1988. *Paleontol. J.*, 22 (4), 55–66 [75].

Type species: *Anthoscytina longa* Hong, 1983.

Tegmen longer than broad at widest part, costal margin not sinuate, apical margin evenly rounding. Hind wing with a marginal vein distinctly. Abdomen with nine visible segments. The female genitalia with three pairs of oblong valvulae extending far beyond the abdominal apex.

Distribution and age: Liaoning, Hebei, Inner Mongolia, Middle Jurassic; Hebei, Beijing, Early Cretaceous.

Nine species included from the Jurassic and Cretaceous of Northern China (see Table 16.1).

Anthoscytina perpetua Li, Shih & Ren, 2013 (Figure 16.11)

Anthoscytina perpetua Li, Shih & Ren, 2013: *PloS ONE*, 8 (11), e78188.

Locality and horizon: Daohugou, Ningcheng, Inner Mongolia, China; Middle Jurassic, Jiulongshan Formation.

Forewing slender, RA simple; M branching into MA and MP at distal fifth of wing; CuA_1 twice as long as CuA_2; cross-vein ir at level of cross-vein r-m, apical of the cross-vein m-cua. Hind wing with cross-vein r-m between MA and RP, slightly distal to bifurcation of M. They exhibit belly-to-belly mating position as preserved, with male's aedeagus inserting into the female's bursa copulatrix. Abdominal segments eight to nine of male are disarticulated suggesting these segments were twisted and flexed during mating. Due to potential taphonomic effect, we cannot rule out that they might have taken side-by-side position, as in extant froghoppers. Genitalia

(a)

(b)

Figure 16.11 *Anthoscytina perpetua* Li, Shih & Ren, 2013. (a). Holotype, male on the right, CNU-HEM-NN2012002, Allotype, female on the left, CNU-HEM-NN2012003; (b). Extant froghoppers in copulation. Source: Photo by Jason Shih.

of male and female, based on paratypes, show symmetric structures. Please see Section 30.4 and [76].

Chengdecercopis Hong, 1983

Chengdecercopis Hong, 1983, *Middle Jurassic Fossil Insects in North China*, 59–60 [73] (original designation).

Type species: *Chengdecercopis xiaofanzhangziensis* Hong, 1983.

Body with total length about 6.3 mm, head nearly rounded, width more than length; the heart-shaped pronotum, meso-praescutum small triangulate; abdomen wide, nine segments; forewing about 2.5 to 3 times as long as wide, with reticulate pattern at costal margin. ScA reaching the mid-wing; ScP curved and merged with R + M and CuA at the same point; vein R_1 no branch, Rs leaving R_2 about 1/3 vein length to margin. M with four branches; the central cell complete; branch CuA after M; anal lobe developed with A_2.

Distribution and age: Hebei; Middle Jurassic.

Only one species included from the Jurassic of Northern China (see Table 16.1).

Cretocercopis Ren, Lu & Guo, 1995

Cretocercopis Ren, Lu & Guo, 1995, *Faunae and Stratigraphy of Jurassic-Cretaceous in Beijing and the Adjacent Areas*, 70 [30] (original designation).

Type species: *Cretocercopis yii* Ren, Lu & Guo, 1995.

The specific epithet is in honor of a renowned paleontologist Prof. Yongen Yi for his guidance. Forewing about 3.5 times as long as wide, R_1 with three branches, Rs with four branches, the bifurcation of CuA basad of the bifurcation of M. Hind wing, M, CuA and CuP all have two branches respectively.

Distribution and age: Beijing; Early Cretaceous.

Only one species included from the Cretaceous of Northern China (see Table 16.1).

Anomoscytina Ren, Yin & Dou, 1998

Anomoscytina Ren, Yin & Dou, 1998, *Acta Zootaxo. Sin.*, 23 (3), 281–288 [57] (original designation).

Type species: *Anomoscytina anomala* Ren, Yin & Dou, 1998.

Hind tibia with seven stout spines laterally. Rostrum reaching between median coxae. In hind wing, M forking before level of forking of CuA. The two terminal branches of CuA convergent apically.

Distribution and age: Liaoning; Early Cretaceous.

Only one species included from the Cretaceous of Northern China (see Table 16.1).

Jurocercopis Wang & Zhang, 2009

Jurocercopis Wang & Zhang, 2009, *Geobios*, 42 (2), 243–253 [77] (original designation).

Type species: *Jurocercopis grandis* Wang & Zhang, 2009.

Tegmen large and slender, length about 20 mm, length/width ratio 2.8–3.0; costal margin convex at about basal 0.4 wing length (basal 0.3 wing length in *Cretocercopis* Ren, 1995); Sc + RA arising from Sc + R at about basal third of wing; RA with some branches; M with more than four branches; cross-veins m absent; CuA branching distinctly basal of M branching; CuA_1 slightly longer than CuA_2. Antenna half the length of fore tibia. Rostrum moderate, extending to mid coxae. Pronotum hexagon, widest at middle; its posterior margin straight. Fore femur shorter than hind femur. Hind tibia slender, about 1.8 times as long as hind femur. Hind tibia with a short lateral spine at 3/5 its length. Hind tibia with two rows of apical denticles, and each row consisting of about six teeth. Both basitarsomere and mid tarsomere of hind tarsi with a row of small teeth apically. Ovipositor slender, curved upwards.

Distribution and age: Inner Mongolia; Middle Jurassic.

Only one species included from the Jurassic of Northern China (see Table 16.1).

Stellularis Chen, Yao & Ren, 2015

Stellularis Chen, Yao & Ren, 2015, *Cretac. Res.*, 52, 402–406 [78] (original designation).

Type species: *Stellularis longirostris* Chen, Yao & Ren, 2015.

Forewing slender, RA and M single, cross-vein ir at level of r-m, apical of m-cua, hind wing cross-vein r-m distad of bifurcation of M. Rostrum apex reaching hind coxae.

Distribution and age: Liaoning; Early Cretaceous.

Three species included from the Cretaceous of Northern China (see Table 16.1).

Stellularis longirostris Chen, Yao & Ren, 2015 (Figure 16.12)

Stellularis longirostris Chen, Yao & Ren, 2015: *Cretac. Res.*, 52, 402–406.

Locality and horizon: Huangbanjigou, Beipiao, Liaoning, China; Lower Cretaceous, Yixian Formation.

Head not flattened, narrower than pronotum, eyes nearly globose. Postclypeus swollen with distinct transverse grooves and large, rostrum very long, extending to hind coxae. Pronotum greatly expanded, Hind tibia with one lateral spine, tarsus with a little distinctly visible palmula. Forewing elongate, length 10–12 mm, width 2.5–3.5 mm, length/width ratio 3–4. Hind wing M single, CuA branching into CuA_1 and CuA_2. Anal tube elongate. Ovipositor long, distinctly exceeding tip of tegmina [78].

Titanocercopis Chen, Zhang & Wang, 2015

Titanocercopis Chen, Zhang & Wang, 2015, *Entomol. Sci.*, 18 (2), 147–152 [79] (original designation).

Figure 16.12 *Stellularis longirostris* Chen, Yao & Ren, 2015 (Holotype, CNU-HEM-LB2013001).

Type species: *Titanocercopis borealis* Chen, Zhang & Wang, 2015.

Body large, with a total length of about 30 mm. Tegmen slender, about 25 mm long; clavus and costal area long; stem ScP + R short, nearly straight, and branching into ScP + RA and RP near basal 1/3 of wing length; RA multi-branched; RP single or two-branched; M forking basal of or almost at the same level of CuA; cross-veins im present; CuP long, almost straight, terminating just basal of CuA_2 ending.

Distribution and age: Inner Mongolia; Middle Jurassic.

Only one species included from the Jurassic of Northern China (see Table 16.1).

Family Sinoalidae Wang & Szwedo, 2012

Sinoalidae are an extinct family erected by Wang & Szwedo in 2012. Up to now, only four genera with six species have been reported from the Jurassic and Cretaceous of China. This family are closely related to Procercopidae based on pronotum with a median incision at the hind margin; tegmen slender, partly punctuate; basal portion of Sc not exceeding apex of basal cell, and hind legs with lateral spines. This family with some ancestral characters and some derived characters, are tentatively placed in the Cercopoidea [70].

Genera included from the Jurassic and Cretaceous of Northern China: *Huabeicercopis* Hong, 1983, *Luanpingia* Hong, 1983, *Jiania* Wang & Szwedo, 2012 and *Sinoala* Wang & Szwedo, 2012.

Huabeicercopis Hong, 1983

Huabeicercopis Hong, 1983, *Middle Jurassic Fossil Insects in North China*, 57–58 [74] (original designation).

Type species: *Huabeicercopis yangi* Hong, 1983.

The specific epithet is dedicated to Prof. Zunyi Yang. Tegmen about 4.1 times as long as wide, with apex rounded; basal cell about 0.25 times as long as tegmen length, closed with anastomosis; ScP, R + M and CuA leaving basal cell at common point; ScRA about 4.4 times as long as Sc + R; CuA at base mildly convex, forked at level of claval apex; stigmal cell very narrow, about half as wide as radial cell; cross-vein m-cua basad of CuA forking.

Distribution and age: Hebei; Middle Jurassic.

Only one species included from the Jurassic of Northern China (see Table 16.1).

Luanpingia Hong 1983

Luanpingia Hong, 1983, *Middle Jurassic Fossil Insects in North China*, 54 [74] (original designation).

Type species: *Luanpingia longa* Hong, 1983.

Tegmen about 2.9 times as long as wide, with apex truncate; basal cell about 0.23 times as long as tegmen length, closed with anastomosis; ScR, M and CuA leaving basal cell at common point; ScRA about 1.3 times as long as Sc + R; terminal $ScRA_1$ short, about half as long as ScRA; CuA at base mildly convex, forked distad of claval apex; stigmal cell as wide as radial cell; cross-vein m-cua absent.

Distribution and age: Hebei, Middle Jurassic; Hebei, Early Cretaceous.

Two species included from the Jurassic and Cretaceous of Northern China (see Table 16.1).

Jiania Wang & Szwedo, 2012

Jiania Wang & Szwedo, 2012, *Palaeontology*, 55 (6), 1223–1243 [70] (original designation).

Type species: *Jiania crebra* Wang & Szwedo, 2012.

Tegmen about 3.5 times as long as wide, with apex widely rounded; basal cell about 0.17 times as long as tegmen length, closed with anastomosis; ScR + M very short; ScRA about 1.5 times as long as ScR; CuA at base distinctly convex, forking basad of claval apex; stigmal cell narrow, about half as wide as radial cell.

Distribution and age: Inner Mongolia; Middle Jurassic

Two species included from the Jurassic of Northern China (see Table 16.1).

Jiania crebra Wang & Szwedo, 2012 (Figure 16.13)

Jiania crebra Wang & Szwedo, 2012: *Palaeontology*, 55 (6), 1223–1243.

Locality and horizon: Daohugou, Ningcheng, Inner Mongolia, China; Middle Jurassic, Jiulongshan Formation.

Tegmen about 3.4 times as long as wide, with costal margin thickened and curved at base; costal cell punctate; cell c1 over twice as long as adjoining apical cell; cell

Figure 16.13 *Jiania crebra* Wang & Szwedo, 2012 (Holotype, NIGP154598a, part). Source: Photo provided by Dr. Haichun Zhang.

c3 about as long as adjoining apical cell; ovipositor distinctly exceeding tip of tegmina, about 1.2 times as long as hind tibia [70].

Sinoala Wang & Szwedo, 2012

Sinoala Wang & Szwedo, 2012, *Palaeontology*, 55 (6), 1223–1243 [70] (original designation).

Type species: *Sinoala parallelivena* Wang & Szwedo, 2012.

Tegmen about 3.4 times as long as wide, with apex truncate; basal cell about 0.12 times as long as tegmen length, closed with anastomosis; ScR, M and CuA leaving basal cell at common point; ScRA about 10 times as long as Sc + R; CuA at base mildly convex, forking at level of claval apex; stigmal cell very narrow, 0.4–0.5 width of radial cell; cross-vein m-cua distad of stem CuA forking [70].

Distribution and age: Inner Mongolia; Middle Jurassic.

Only one species included from the Jurassic of Northern China (see Table 16.1).

Superfamily Cicadoidea Latreille, 1802

Family Tettigarctidae Distant, 1905

Tettigarctidae, the most ancient and relict group of Cicadoidea, are the sister group to Cicadidae [80]. Their most distinct characters are expanded pronotum, concealing much of the mesonotum. Tettigarctidae comprise two subfamilies, Cicadoprosbolinae Becker-Migdisova, 1947 and Tettigarctinae Distant, 1905, including 29 genera with 42 extinct species and two extant species [81]. Fossil records of Tettigarctidae, ranging from the Late Triassic to the Miocene, are distributed mainly in the Northern Hemisphere, but extant tettigarctids live in the Southern Hemisphere, e.g. Australia [81–83]. The paleobiogeography of Tettigarctidae needs further systematic study [84]. Unlike other cicadas, the male tettigarctids, not producing loud airborne acoustic signals, can only produce substrate-transmitted acoustic signals [85].

Genera included from the Jurassic and Cretaceous of Northern China: *Shuraboprosbole* Becker-Migdisova, 1949, *Protabanus* Hong, 1982, *Sunotettigarcta* Hong, 1983, *Quadrisbole* Lin, 1986, *Sinocicadia* Hong & Wang, 1990, *Shaanxiarcta* Shcherbakov, 2008, *Tianyuprosbole* Chen, Wang & Zhang, 2014, *Hirtaprosbole* Liu, Yao & Ren, 2016, *Macrotettigarcta* Chen & Wang, 2016, *Maculaprosbole* Zheng, Chen & Wang, 2016 and *Sanmai* Chen, Zhang & Wang, 2016.

Shuraboprosbole Becker-Migdisova, 1949

Shuraboprosbole Becker-Migdisova, 1949, *Akad. Nauk SSSR, Trudy Paleontol. Inst.*, 22, 23–24 [86] (original designation).

Type species: *Shuraboprosbole plachutai* Becker-Migdisova, 1949.

Head with inflated, declivous crown. Rostrum long, extending to or beyond hind coxae. Pronotum greatly expanded, with posterior half transversely rugose. Median length of pronotum three times that of vertex. Mesonotum partly exposed. Legs with dense setae. Forewing elongate, with nodes at about basal 0.6 of wing length; costal area wide; stem R + M branching at about basal 0.2 of wing length; M four branched; R half or 1/3 as long as stem R + M; RA two-branched; M fused with CuA for a distance; CuP ending at about midpoint of wing. Hind wing and ovipositor moderate, depressed to pygofer.

Distribution and age: Kyrgyzstan and England, Early Jurassic; Inner Mongolia of China, Middle Jurassic.

Three species included from the Jurassic of Northern China (see Table 16.1).

Protabanus Hong, 1982

Protabanus Hong, 1982, *Mesozoic Fossil Insects of Jiuquan Basin in Gansu Province*, 171 [68] (original designation).

Type species: *Protabanus chaoyangensis* Hong, 1982.

Nodal line beyond mid-wing; MP forked just before the nodal line; M cross-joined with CuA beyond the basal cell. Costal area wider than intercubital one, two anterior branches of RA, M forked at about 2/3 of the distance from basal cell to nodal line, and prenodal part coriaceous.

Distribution and age: Liaoning; Middle Jurassic.

Only one species included from the Jurassic of Northern China (see Table 16.1).

Sunotettigarcta Hong, 1983

Sunotettigarcta Hong, 1983, *Middle Jurassic Fossil Insects in North China*, 55 [74] (original designation).

Type species: *Sunotettigarcta hebeiensis* Hong, 1983.

Tegmen elongate. Costal margin strongly arched. Antecubital area widest at nodal line. Costal area wider than antecubital one. Nodes at about basal half of wing length. RP separated from stem R before mid-part of wing, straight, reaching margin. M forked before nodal line. CuA curved and fused with M at a point. Transverse reclined and nearer to wing tip than the nodal line.

Distribution and age: Inner Mongolia and Hebei; Middle Jurassic.

Two species included from the Jurassic of Northern China (see Table 16.1).

***Sinocicadia* Hong & Wang, 1990**

Sinocicadia Hong & Wang, 1990, *Fossil insects from the Laiyang Basin, Shandong Province*, 44–189 [21] (original designation).

Type species: *Sinocicadia shandongensis* Hong & Wang, 1990.

Hind wing 17 mm long and 13 mm wide, elongate, costal area wide, with transversely wrinkled margin, coupling lobe not reaching the mid-length. Hind wing with a marginal vein distinctly. Several micro-spines near the vein and on the anal areas.

Distribution and age: Shandong; Early Cretaceous.

Only one species included from the Cretaceous of Northern China (see Table 16.1).

***Shaanxiarcta* Shcherbakov, 2008**

Shaanxiarcta Shcherbakov, 2008, *Russian Entomol. J.*, 16, 343–348 [81] (original designation).

Involuta Zhang, 1993, *Palaeoworld*, 3, 49–56 [87]. Syn. by Shcherbakov, 2008, *Russian Entomol. J.*, 17, 343–348 [81].

Type species: *Involuta perrara* Zhang, 1993.

Female forewing long and narrow; costal area wide and short, having slightly branched, shortcut Sc; Rs four branched to cover the apex of wing area. CuP simple and short.

Distribution and age: Shaanxi; Early Cretaceous.

Only one species included from the Cretaceous of Northern China (see Table 16.1).

***Tianyuprosbole* Chen, Wang & Zhang, 2014**

Tianyuprosbole Chen, Wang & Zhang, 2014, *Zootaxa*, 3764 (5), 581–586 [88] (original designation).

Type species: *Tianyuprosbole zhengi* Chen, Wang & Zhang, 2014.

The specific epithet is in honor of Prof. Xiaoting Zheng, who is the curator of the Shandong Tianyu Museum of Nature. Body small and stout. Antennae well-developed; scape thick; pedicel slightly thinner and longer than scape; flagellum aristiform. Eyes large, semi-circular in lateral view. Postclypeus greatly inflated. Pronotum exceedingly expanded and concealing mesonotum, with posterior 2/3 transversely rugose. Fore femur slightly thick; hind tibia slender and long. Tegmen short, nearly trapezoid. Clavus and costal area long and broad. Postnodal area ornamented with ruffles, short, almost half as long as prenodal area. Nodal line wide but not well-defined. Nodus distinct. ScP fused with R + M at wing base, dividing from ScP + RA at nodal line, and ending at nodus. Stem R + M branching at basal 0.21 of wing length. R about 0.10 wing length and two-branched; RA two-branched; RP single, strongly curved at nodal line. M fused with CuA for a distance about 0.10 of wing length, branching into M_{1+2} and M_{3+4} at basal 0.48 of wing length; M_{1+2} and M_{3+4} branching into M_1, M_2, M_3 nearly at the same level. CuA nearly straight, but curved just beyond nodal line and then branching into CuA_1 long and sinuous. CuP straight, ending at nodal line. Pcu ending at about midpoint of wing. A_1 longitudinal [88].

Distribution and age: Inner Mongolia; Middle Jurassic.

Only one species included from the Jurassic of Northern China (see Table 16.1)

***Hirtaprosbole* Liu, Yao & Ren, 2016**

Hirtaprosbole Liu, Yao & Ren, 2016, *Alcheringa*, 40 (3), 383–389 [89] (original designation).

Type species: *Hirtaprosbole erromera* Liu, Yao & Ren 2016.

Body densely pilose. Rostrum extending beyond hind coxae. Postclypeus bulge with distinct transverse grooves. Number of antennomeres (including scape and pedicel) five or higher. Pronotum with one fissure positioned more or less centrally and covered by setae. Forewing, ScP ending at middle of anterior margin; stem ScP + R longer than stem ScP + R + M; RA divided into three branches; M fused with CuA at one point; nodal line at middle of wing; cell a6 nearly quadrate; cell a8 subequal to cell a10 in length.

Distribution and age: Inner Mongolia; Middle Jurassic.

Only one species included from the Jurassic of Northern China (see Table 16.1).

***Hirtaprosbole erromera* Liu, Yao & Ren, 2016 (Figure 16.14)**

Hirtaprosbole erromera Liu, Yao & Ren, 2016: *Alcheringa*, 40 (3), 383–389.

Locality and horizon: Daohugou, Ningcheng, Inner Mongolia, China; Middle Jurassic, Jiulongshan Formation.

The body length 17.5–24.0 mm, covered by dense and long setae, especially postclypeus and pronotum; eyes large, antenna slender with setae, rostrum extending beyond hind coxae, pronotum not transversely rugose, forewing membranous, nodal line at midwing. Hind wing shorter than forewing, ovipositor long, hind tibia slender, unarmed, distitarsus longest. The long and dense setae of *H. erromera* might have contributed

Figure 16.14 *Hirtaprosbole erromera* Liu, Yao & Ren, 2016 (Holotype, CNU-HEM-NN2012399).

to heat preservation; this species probably adapted to life in high mountains and cold places [89].

Macrotettigarcta Chen & Wang, 2016

Macrotettigarcta Chen & Wang, 2016, *Spixiana*, 39 (1), 119–124 [90] (original designation).

Type species: *Macrotettigarcta obesa* Chen & Wang, 2016.

Body thick. Ovipositor relatively short. Mesonotum with longitudinal carinae. Forewing large, length about 50 mm; basal cell short and broad; nodal incision at about basal 0.6 of wing length; stem ScP + R + M + CuA branching at about basal 0.18 wing length; stem ScP + R very short; RA three-branched; M + CuA branching into M and CuA at basal 0.2 of wing length, just beyond its connection with cross-veins cua-cup; stem M_{1+2} connected with M_3 by cross-veins im; CuA_2 long, S-shaped; CuP ending at about midpoint of wing.

Distribution and age: Inner Mongolia; Middle Jurassic.

Only one species included from the Jurassic of Northern China (see Table 16.1).

Maculaprosbole Zheng, Chen & Wang, 2016

Maculaprosbole Zheng, Chen & Wang, 2016, *ZooKeys*, 632, 47–55 [91] (original designation).

Type species: *Maculaprosbole zhengi* Zheng, Chen & Wang, 2016.

The specific epithet is dedicated to Prof. Xiaoting Zheng, who is the founder of Shandong Tianyu Museum of Nature and donated the type material. Forewing large, with color patterns, nodal line about middle of wing length, nodus distinct, RA with three branches, RP single, M with four branches, CuA with two branches, CuP almost straight, ending at about 2/5 of wing length.

Distribution and age: Inner Mongolia; Middle Jurassic.

Only one species included from the Jurassic of Northern China (see Table 16.1).

Sanmai Chen, Zhang & Wang, 2016

Sanmai Chen, Zhang & Wang, 2016, *Acta Palaeontol. Polonica*, 61 (4), 853–862 [92] (original designation).

Type species: *Sanmai kongi* Chen, Zhang & Wang, 2016.

The specific epithet is in reference to the family name, Kong, of Confucius (Kung Fu-Tsy), the founder of Confucianism. Postclypeus swollen. Compound eye large, oval or semicircular in lateral view. Antenna with scape slightly thicker than pedicel; flagellum aristiform, with five antenomeres. Pronotum expanded, anterior half with tiny granular protuberances and posterior area transversely rugose. Mesonotum partly exposed, with posterior 1/3 area transversely rugose. Femora with tiny granules; tibiae densely setose, with distinct ridges, hind tibia with two lateral spines; tarsidensely setose, with three tarsomeres; claws well-developed. Ovipositor ensiform, upcurved, adpressed to pygofer extended just below anal tube. Forewing with dark membrane colored with light and irregular speckles and longitudinal stripes; apical cells eight; R bifurcating at about basal 1/3 of wing length; RP fused with nodal line for a distance or sinuous near nodal line; M three-branched; M_{3+4} unbranched; CuA bifurcating just beyond nodal line; distal section of CuA_2 running along wing margin. Hind wing with M three-branched and M_{1+2} simple [92].

Distribution and age: Inner Mongolia; Middle Jurassic.

Three species included from the Jurassic of Northern China (see Table 16.1).

Superfamily Membracoidea Rafinesque, 1815

Family Archijassidae Becker-Migdisva, 1962

Archijassidae, an extinct family, comprise three subfamilies: Archijassinae Becker-Migdisova, 1962, Karajassinae Shcherbakov, 1992 and Dellasharinae Shcherbakov, 2012. Up to date, 16 genera with 25 species from the Triassic to the Cretaceous of Australia and Eurasia have been described. This family are diagnosed by tegmen with six to eight full-sized apical cells, hind wing with five to six full-sized apical cells, RA long and MP connected or completely fused to CuA_1 [93].

Genera included from the Cretaceous of Northern China: *Archijassus* Handlirsch, 1906 and *Mesoccus* Zhang, 1985.

Archijassus Handlirsch, 1906

Archijassus Handlirsch, 1906, *Ein Handbuch fur Palaontologen und Zoologen*, 501–502 [73] (original designation).

Type species: *Archijassus heeri* Geinitz, 1880.

Forewing costal margin strongly angular, RS arising beyond midwing; M with four branches.

Distribution and age: Germany and Switzerland, Early Jurassic; Kazakhstan, Middle Jurassic; Shandong of China, Early Cretaceous.

Only one species included from the Cretaceous of Northern China (see Table 16.1).

Mesoccus Zhang, 1985

Mesoccus Zhang, 1985, *Shandong Geology*, 1 (2), 25–26 [94] (original designation).

Type species: *Mesoccus lutarius* Zhang, 1985.

Body long, head narrower than pronotum, eye very small, antenna near frontal edge of eye, frons not obvious, and postclypeus developed. Hind tibia with two rows of long spines, veins of forewing apical area clear, and ovipositor large and triangular.

Distribution and age: Shandong; Early Cretaceous.

Two species included from the Cretaceous of Northern China (see Table 16.1).

Family Cicadellidae Latreille, 1802

Cicadellidae, with over 30 000 extant species, are one of the most speciose and most diverse families of Hemiptera [95]. To date, there are over 40 subfamilies [96]. Fossil representatives of the Cicadellidae are known since the Early Cretaceous of Laiyang, China [94]. Paleogene records seem to be quite rich [97]. Cicadellids are diagnosed by the presence of strong and sometimes long setae on the hind tibia, hind femur with a few strong spines placed distally, hind coxae parallel and connected by the uniting pegs [98].

Only one genus included from the Cretaceous of Northern China: *Sinojassus* Zhang, 1985.

Sinojassus Zhang, 1985

Sinojassus Zhang, 1985, *Shandong Geology*, 1 (2), 26 [94] (original designation).

Type species: *Sinojassus brevispinatus* Zhang, 1985.

Postclypeus long and triangular; rostrum short and reaching coxae of foreleg and eye near postclypeus. Hind wing Rs, M and CuA all with two branches; cross-veins rs-m and m-cu visible.

Distribution and age: Shandong; Early Cretaceous.

Only one species included from the Cretaceous of Northern China (see Table 16.1).

Superfamily Palaeontinoidea Handlirsch, 1906

Family Palaeontinidae Handlirsch, 1906

Palaeontinidae Handlirsch, 1906 are an extinct group of Homoptera known from the Triassic to the Early Cretaceous in Eurasia, South America, Africa and Australia [17, 47, 73, 99, 100]. Palaeontinids are diagnosed by a small head, narrower than pronotum; well-developed mesonotum; forewing with distal part larger than basal part; the stems of R and M separated before or close to mid-wing and M four-branched. Hind wing with a prominent indentation on the costal margin [52]. To date, 53 species within 23 genera have been reported from the Jurassic and Cretaceous of Northern China, widely distributed in Xinjiang, Hebei, Liaoning, Gansu and Inner Mongolia [30, 68, 74, 101–121].

Genera included from the Jurassic and Cretaceous of Northern China: *Martynovocossus* (Martynov, 1931), *Palaeontinodes* Martynov, 1937, *Palaeontinopsis* Martynov, 1937, *Plachutella* Becker-Migdisova, 1949, *Suljuktocossus* Becker-Migdisova, 1949, *Gansucossus* (Hong, 1982), *Sinopalaeocossus* Hong, 1983, *Ilerdocossus* Gomez, 1984, *Jibeicossus* Hong, 1984, *Yanocossus* Ren, Lu & Guo, 1995, *Liaocossus* Ren, Yin & Dou, 1998, *Miracossus* Ren, Yin & Dou, 1998, *Daohugoucossus* Wang, Zhang & Fang, 2006, *Eoiocossus* Wang, Zhang & Fang, 2006, *Aborocossus* Wang, Zhang & Fang, 2007, *Hamicossus* Wang & Ren, 2007, *Quadraticossus* Wang & Ren, 2007, *Neimenggucossus* Wang, Zhang & Fang, 2007, *Cladocossus* Wang & Ren, 2009, *Cricocossus* Wang & Ren, 2009, *Ningchengia* Wang, Zhang & Fang, 2009, *Synapocossus* Wang, Shih & Ren, 2013 and *Kallicossus* Chen, Zhang & Wang, 2014.

Martynovocossus (Martynov, 1931)

Pseudocossus Martynov, 1931, *Ann. Soc. Palaeontol. Russia*, 9, 93–122 [122]; Homonymum. With Kenrick, 1914, *Trans. of the Entomol. Soc. of London*, 1914, 587–602 [123]; Gaede, 1933, Cossidae, 807–824 [124]. Syn. by Wang & Zhang, 2008, *Geol. J.*, 43, 1–18 [111].

Type species: *Pseudocossus zemcuznicovi* Martynov, 1931.

This generic name is in honor of the late Professor Andrey V. Martynov, a Russian paleontomologist. Forewing subtriangular, length 31–78 mm; nodal indentation at about basal 0.4 of wing length; ScP with four to eight veinlets; RA free from vein R for a short distance; cross-vein r + m-cua transverse; nodal line dividing discal cell into two parts; no cross-vein in discal cell. Hind wing length 20–35 mm, length/width ratio about 1.4; costal area wide, maximal width in basal 1/3; nodal indentation at basal 0.4 of wing length; Sc terminating in costal margin slightly basal of nodal indentation; RP fused with branch M_1 for a distance; M branching into veins M_1 and M_{2+3+4} at wing base; M_{3+4} connected with CuA by cross-vein m-cua, and bifurcating into M_3 and M_4 at the same level of nodal indentation; CuA bifurcating into CuA_1 and CuA_2 distal of the level of nodal indentation; CuP and Pcu simple [111, 113].

Distribution and age: Xinjiang of China and Russia, Early Jurassic; Inner Mongolia of China, Middle Jurassic; Kazakhstan, Early and Middle Jurassic.

Six species included from the Jurassic of Northern China (see Table 16.1).

Palaeontinodes Martynov, 1937

Palaeontinodes Martynov, 1937, *Trud. Pal. Inst. Acad. Nauk SSSR.*, 7 (1), 101–103 [47] (original designation).

Type species: *Palaeontinodes shabarovi* Martynov, 1937.

The specific epithet is dedicated to H.V. Shabarovi for his first finding and collection of fossil insects from the Sulyukta Formation in Tajikistan. Forewing triangular, length 22–58 mm; anterior margin fringed with a thick border; nodal indentation weak; apex pointed; ScA distinguishable and ending at the bottom of indentation; ScP unbranched and fused with RA at about basal 0.3 of wing length; cross-vein r + m-cua long and subvertical; discal cell divided by a long cross-vein m-cua; Pcu simple; A_2 short and fused with A_1 basally; clavus about 0.3 times as long as wing length; nodal line distinct [104, 109].

Distribution and age: Tajikistan, Russia and Kyrgyzstan, Early Jurassic; Inner Mongolia and Hebei of China and Russia, Middle Jurassic.

Five species included from the Jurassic of Northern China (see Table 16.1).

Plachutella Becker-Migdisova, 1949

Plachutella Becker-Migdisova, 1949, *Akad. Nauk USSR., Trudy Paleontol. Inst.*, 22, 11–15 [86] (original designation).

Type species: *Plachutella rotundata* Becker-Migdisova, 1949.

All species within this genus were erected on the basis of hind wings. Hind wing anterior margin straight for distal part; RP fused with M_1; M_{1+2} and M_{3+4} independent; M_2 close to and at most connected with M_{3+4} at one point but not coalesced with it [104].

Distribution and age: Xinjiang of China, Kyrgyzstan and Russia, Early Jurassic; Hebei and Inner Mongolia of China and Kazakhstan, Middle Jurassic.

Three species included from the Jurassic of Northern China (see Table 16.1).

Suljuktocossus Becker-Migdisova, 1949

Suljuktocossus Becker-Migdisova, 1949, *Akad. Nauk USSR., Trudy Paleontol. Inst.*, 22, 8 [86] (original designation).

Type species: *Suljuktocossus prosboloides* Becker-Migdisova, 1949.

Forewing with pointed apex; costal area and clavus reduced; ScP arising basally, unbranched and coalesced with RA; cross-vein r + m-cua short and oblique; antenodal and postnodal regions trapezoid; clavus about 1/3 of wing length. Hind wing small with indentation at about basal 0.4 of wing length; ScP coalesced with R for a short distance; RP connected with M_1 by a short cross-vein r-m or fused with it for a distance; M_1 arising just before stem M_{3+4} departing from M_2; M_{3+4} connected with CuA by a cross-vein at about basal 0.25 of wing length [105, 109, 112].

Distribution and age: Tajikistan, Early Jurassic; Inner Mongolia of China, Middle Jurassic.

Three species included from the Jurassic of Northern China (see Table 16.1).

Suljuktocossus yinae Wang & Ren, 2007 (Figure 16.15)

Suljuktocossus yinae Wang & Ren, 2007: *Zootaxa*, 1576, 58.

Locality and horizon: Daohugou, Ningcheng, Inner Mongolia, China; Middle Jurassic, Jiulongshan Formation.

This species is named after Ms. Ruiyue Yin who collected and donated this fossil to CNU. Body compressed with both forewings and hind wings. Forewing length 56 mm, width 27 mm. CP present and ending at nodal indentation. Sc without branches. R bifurcating at the same level of M separation into M_{1+2} and M_{3+4}. M_{1+2} separated distad of M_{3+4}. Clavus small with two anal veins. A_2 with two branches, shorter branch curved downward

Figure 16.15 *Suljuktocossus yinae* Wang & Ren, 2007 (Holotype, CNU-H-NN2007001p). Source: Donated by Ms. Ruiyue Yin.

to margin at base; longer branch fused with A_1 slightly before distal end of clavus. Nodal line traceable as indentation fused with R_1 for a short distance and cutting Rs little distad of point of Sc fusing with R_1, then reaching M just distad of M forking, finally fused with m4-cua to connection with CuA_2, following latter for a short distance and then reaching CuP at distal end of clavus. Hind wing length 35 mm, width 25 mm with obvious indentation. Sc arising basally. R parallel Sc before branching into R_{1+2} and Rs. Rs curved and fused with M_1 for a long interval after dividing from R. M four-branched. M_1 separated from M basally; M_2 straight and arising from M distad of the point of M_1 separated from M. Stem of M_{3+4} long, connected with CuA by a short cross-vein m3 + 4-cua and slightly curved when cross-vein meeting M_{3+4}. M_{3+4} branched into M_3 and M_4 at the same level of indentation. CuA bifurcated slightly basad than M_{3+4}. CuP simple. Vein A two branched. A_2 short and curved downward to anal margin near base. Both forewings and hind wings with distinct color pattern consisting of regular dark patches on a pale background [115].

Gansucossus (Hong, 1982)

Yumenia Hong, 1982, *Mesozoic fossil insects of Jiuquan basin in Gansu Province*, 81 [68]. Homonymum. With Hou, 1958, *Memoirs of the Institute of Palaeontology, Academia Sinica*, 1, 93 [125]. Syn. by Wang, Zhang & Fang, 2006, *Zootaxa*, 1268, 59–68 [105].

Type species: *Yumenia pectinata* Hong, 1982.

Hind wing length 22–32 mm, length/width ratio 1.5–1.9; costal area wide (length/width ratio 3.6–5.5); Sc terminating at indentation; RP fused with M_1 for a considerable interval; Stem M with four branches; M_{3+4} connected to CuA by cross-vein m-cua; media area (between M_{2+3+4} and CuA) wide; CuA arising basally; A_1 terminating at anal margin [105].

Distribution and age: Gansu, Early-Middle Jurassic; Hebei and Inner Mongolia, Middle Jurassic.

Three species included from the Jurassic of Northern China (see Table 16.1).

Sinopalaeocossus Hong, 1983

Sinopalaeocossus Hong, 1983, *Middle Jurassic Fossil Insects in North China*, 52 [73] (original designation).

Type species: *Sinopalaeocossus scabratus* Hong, 1983.

Forewing length/width ratio about 1.8; nodal indentation at about middle portion; vein ScP with several obscure branches, and fused with RA at basal 1/4 of wing length; M bifurcating at the same level of R forking; M_4 geniculate basally at junction of cross-vein m_4-cua; discal cell about 1/5 of wing length, antenodal region square, postnodal region slightly longer than antenodal region; clavus about 1/5 of wing length; CuP curved anteriorly, terminating in the distal end of clavus; Pcu simple; A_1 curved posteriorly; nodal line distinct; membrane covered with color pattern. Hind wing oval, length/width ratio 1.46–1.57, nodal indentation at basal 0.25–0.33 of wing length; R arising from stem R + M basally, fused with Sc and branching into Sc + RA and RP; RP fused with M_1 for a short distance; M bifurcating into M_{1+2} and M_{3+4} near its base; M_{3+4} unbranched and close to CuA at one point; membrane with color pattern [106].

Distribution and age: Hebei and Inner Mongolia; Middle Jurassic.

Two species included from the Jurassic of Northern China (see Table 16.1).

Ilerdocossus Gomez, 1984

Ilerdocossus Gomez-Pallerola, 1984, *Boletín Geológico y Minero*, 304, photo 4 [99] (original designation).

Wonnacottella Whalley & Jarzembowski, 1985, *Bull. Br. Mus. Nat. Hist., London (Grol.)*, 38 (5), 381–412 [126]. Syn. by Menon *et al.*, 2005, *Cretac. Res.*, 26, 837–844 [127].

Type species: *Ilerdocossus villaltai* Gomez, 1984.

Body stout; mesonotum with about 15 longitudinal carinae. Forewing Sc unbranched; cross-vein r + m-cua long and oblique; RA, RP and M separating at one point; cross-vein m_4-cua oblique; discal cell about 1/4 of wing length; antenodal region trapezoid; postnodal region reduced; CuA bifurcating basal of the level of nodal indentation; CuA_1 and CuA_2 long; clavus narrow, about 1/3 of wing length. Hind wing with small costal area; M_1 fused with RA for a long distance [110].

Distribution and age: Inner Mongolia of China, Middle Jurassic; Spain, Early Cretaceous.

Only one species included from the Jurassic of Northern China (see Table 16.1).

Yanocossus Ren, Lu & Guo, 1995

Yanocossus Ren, Lu & Guo, 1995, *Faunae and Stratigraphy of Jurassic-Cretaceous in Beijing and the Adjacent Areas*, 64–65 [30] (original designation).

Type species: *Yanocossus guoi* Ren, Lu & Guo, 1995.

The specific epithet is dedicated to Mr. Ziguang Guo who collected and donated this specimen. Forewing length 33–40 mm, the maximum width (at middle of wing) 15–17 mm; nodal indentation obscure, at basal 0.35 of wing length; Sc unbranched; cross-vein r + m-cua long and oblique; RA, RP and M separating at one point; M_4 vertical basally, and then geniculate at junction with cross-vein m_4-cua; cross-vein m_4-cua longitudinal; discal cell about 1/3 of wing length; antenodal region trapezoid; postnodal region reduced, nearly hemicycle; clavus narrow, less than 1/3 of wing length [30, 110].

Distribution and age: Hebei; Early Cretaceous.

Only one species included from the Cretaceous of Northern China (see Table 16.1).

***Liaocossus* Ren, Yin & Dou, 1998**

Liaocossus Ren, Yin & Dou, 1998, *Entomologia Sin.*, 5 (3), 222–232 [102] (original designation).

Type species: *Liaocossus hui* Ren, Yin & Dou, 1998.

The specific epithet is dedicated to Mr. Guozhong Hu, Mayor of Beipiao City, for supporting this project. Forewing triangular. The anterior margin indented where the nodal line reaching the margin. The membrane not only pitted distad of the nodal line but also with color markings. Sc without serial branches basally, extending beyond the nodal line and terminating on RA. RP arising from RA basally. Hind wing small, without a prominent indentation on costal margin. M_1 coalesced with RP for a long distance. M with four branches. CuA forked deeply [102].

Distribution and age: Liaoning, Hebei; Early Cretaceous.

Five species included from the Cretaceous of Northern China (see Table 16.1).

***Miracossus* Ren, Yin & Dou, 1998**

Miracossus Ren, Yin & Dou, 1998, *Entomologia Sin.*, 5 (3), 222–232 [102] (original designation).

Type species: *Miracossus ingentius* Ren, Yin & Dou, 1998.

Body large-sized, with a very long rostrum extending to the middle of abdomen. Forewing, length 37–46 mm, the maximum width 14–21 mm; Sc curved and unbranched; cross-vein r + m-cua long and oblique; RA, RP and M separating at one point; cross-vein r-r present near nodal line; M_4 vertical basally, and then geniculate at junction with cross-vein m_4-cua; cross-vein m_4-cua longitudinal; discal cell about 1/3 of wing length; nodal line thick and S-shaped in discal cell; CuA strongly curved between cross-veins r + m-cua and m_4-cua. Hind wing with a small costal area. RP connected to M_1 by a short cross-vein r-m at about basal 0.3 of wing length; M branching into M_1 and M_{2+3+4} basally; M_{3+4} branching into M_3 and M_4 at the same level of cross-vein r-m; M_4 connected to CuA just distal of cross-vein r-m [102, 110].

Distribution and age: Liaoning; Early Cretaceous.

Only one species included from the Cretaceous of Northern China (see Table 16.1).

***Daohugoucossus* Wang, Zhang & Fang, 2006**

Daohugoucossus Wang, Zhang & Fang, 2006, *Zootaxa*, 1268, 59–68 [105] (original designation).

Type species: *Daohugoucossus solutus* Wang, Zhang & Fang, 2006.

Forewings Sc reduced and united anteriorly with R, departing from R at the point of basal cross-vein m-cua connected with M, and then fused with R again, at last extending beyond the nodal line and terminating on R_1; M_{1+2} separating distad of M_{3+4}; CuA connected with M by two transverse veins; A bifurcated and A_2 with branches. Tegmina elongated, narrow in the basal part. Hind wing elongate and sharp apically, Sc coalesced with R just distad of R ramification; Rs coalesced with M_1 for a considerable interval before separating; Rs, M_1 and M_2 straight for distal half; M_{3+4} long, in contact but not coalesced with CuA; all wings with distinct color patterns [105, 118].

Distribution and age: Inner Mongolia; Middle Jurassic.

Four species included from the Jurassic of Northern China (see Table 16.1).

***Daohugoucossus shii* Wang, Ren & Shih, 2007 (Figure 16.16)**

Daohugoucossus shii Wang, Ren & Shih, 2007: *Science in China Series D, Earth Sciences*, 16 (1), 483.

Locality and horizon: Daohugou, Ningcheng, Inner Mongolia, China; Middle Jurassic, Jiulongshan Formation.

This species, dedicated to Mr. Yan Shi for his assistance and contribution in collecting fossils, is established based on a well-preserved specimen consisting of a body with a pair of stretched-out forewings and hind wings. Head with a pair of large eyes; a long rostrum with two visible segments, extended to the thorax. Postclypeus narrower than frons, with blurry boundary between them. On the ventral view, legs partly preserved; femora, tibia and part of tarsi and pretarsi of all legs visible, femora stout and subequal in length, tarsi three-segmented, the third segment of tarsi longer than the other two segments. Abdomen with six visible segments and clothed with dense setae. Forewing tegmina. CP and nodal line present. Sc reduced and united anteriorly with R, departing from R at the point of basal cross-vein m-cua connected with M, and then coalesced with R again nearly at the same level of M branching into M_{1+2} and M_{3+4}, at last extending to the nodal line and terminating on R_1. R_1 little upwards before ending in the distal, Rs simple and deriving from R slightly before Sc fused with R; there is a short cross-vein r-m between

Figure 16.16 *Daohugoucossus shii* Wang, Ren & Shih, 2007 (Holotype, CNU-H-NN2006005p). Source: Donated by Dr. Chungkun Shih.

Rs and M. M four-branched, and bifurcating into M_{1+2} and M_{3+4} before the nodal line; M_{1+2} separating distad of M_{3+4}. CuA connected with the stem of M by two transverse veins and also connected with M_4 by a long cross-vein; CuA originating at wing base, curving sharply before it bifurcating at nodal line; CuA_2 nearly paralleling to CuA_1. CuP simple and straight. A two-branched. Tegmina elongated, narrow in the basal part. Hind wing smaller than forewing, costal margin curved posteriorly, formed a prominent indentation, Rs coalesced with M_1 for a considerable interval before separating. M_{3+4} fused with CuA and soon departed from the latter. All wings with distinct color pattern. Body length 36 mm, width 18 mm; forewing length 48 mm, width 23 mm; hind wing length 30 mm, width 16 mm [118].

Eoiocossus Wang, Zhang & Fang, 2006

Eoiocossus Wang, Zhang & Fang, 2006, *Annales Zoologici (Warszawa)*, 56 (4), 757–762 [107] (original designation).

Papilioncossus Wang, Ren & Shih, 2007, *Prog. Nat. Sc.*, 17 (1), 112–116 [117]. Syn. by Wang, Zhang & Szwedo, 2009, *Palaeontology*, 52 (1), 53–64 [112].

Type species: *Eoiocossus validus* Wang, Zhang & Fang, 2006.

Forewing length 61–78 mm; posterior margin fringed with a narrow border; Sc with several branches; Sc coalesced with RA at basal 0.25–0.29 wing length; transverse vein m_4-cua coalesced with nodal line for a considerable interval, and then curved posteriorly and subparallel to M_3 for the remaining part; discal cell about 1/3 of wing length, postnodal region long; CuA_2 with two branches; clavus small, less than 1/6 of wing length; CuP and Pcu curved; A_1 curved, fused with A_2 basally [107, 117].

Distribution and age: Inner Mongolia; Middle Jurassic.

Four species included from the Jurassic of Northern China (see Table 16.1).

Abrocossus Wang, Zhang & Fang, 2007

Abrocossus Wang, Zhang & Fang, 2007, *Alavesia*, 1, 89–104 [109] (original designation).

Type species: *Abrocossus longus* Wang, Zhang & Fang, 2007.

Forewing nearly triangular, with pointed apex; ScP unbranched, coalesced with R for a very short distance; cross-vein r + m-cua vertical and very short; m_4-cua long, coalesced with nodal line for a considerable interval; discal cell about 1/3 of wing length; antenodal region hemicycle; postnodal region nearly trapezoid; clavus slightly less than 1/3 of wing length; Pcu curved anteriorly; A_1 and A_2 short. Membrane infuscate with some pale speckles [109].

Distribution and age: Inner Mongolia; Middle Jurassic.

Only one species included from the Jurassic of Northern China (see Table 16.1).

Hamicossus Wang & Ren, 2007

Hamicossus Wang & Ren, 2007, *Zootaxa*, 1390, 41–49 [114] (original designation).

Type species: *Hamicossus laevis* Wang & Ren, 2007.

Forewing triangular. Sc with several oblique veinlets, fused with R + M at base, separating from R + M for a short distance, and fusing with R again. M_4 sharply flexed. CuA connected with M_4 by horizontal cross-vein. Hind wing with distinct costal indentation [114].

Distribution and age: Inner Mongolia; Middle Jurassic.

Only one species included from the Jurassic of Northern China (see Table 16.1).

Quadraticossus Wang & Ren, 2007

Quadraticossus Wang & Ren, 2007, *Zootaxa*, 1390, 41–49 [114] (original designation).

Type species: *Quadraticossus fangi* Wang & Ren, 2007.

The specific epithet is dedicated to Mr. Liang Fang for his contribution in collecting Daohugou fossils, including this specimen. Forewing triangular. Sc separated from R at base, and then fused with R, extending to near the indentation and terminating on R_1. Rs simple. M four-branched, M_4 sharply flexed after it arising from M_{3+4}. CuA connected with M_4 by a horizontal cross-vein. Vein A two branched, and A_2 with branches. Nodal line divided discal cells into two parts, and the shape of 2nd discal cell nearly rectangular. Hind wing small, with distinct indentation. The stem of M_{1+2} long, and M_4 not divided from M_{3+4}. Anal vein single [114].

Distribution and age: Inner Mongolia; Middle Jurassic.

Three species included from the Jurassic of Northern China (see Table 16.1).

Neimenggucossus Wang, Zhang & Fang, 2007

Neimenggucossus Wang, Zhang & Fang, 2007, *Alavesia*, 1, 89–104 [109] (original designation).

Type species: *Neimenggucossus normalis* Wang, Zhang & Fang, 2007.

Hind wing oval, length/width ratio about 1.40; costal area length/width ratio 3.8, maximal width at the middle of wing; nodal indentation at basal 0.38 of wing length; RP fused with M_1 for a considerable interval, and terminating near apex; M four branched; M_{3+4} divided from stem M basally, close to CuA at one point, and then bifurcating at the same level of RP fused with M_1 [109].

Distribution and age: Inner Mongolia; Middle Jurassic.

Only one species included from the Jurassic of Northern China (see Table 16.1).

Cladocossus Wang & Ren, 2009

Cladocossus Wang & Ren, 2009, *Acta Geol. Sin.-Engl.*, 83 (1), 33–38 [116] (original designation).

Type species: *Cladocossus undulatus* Wang & Ren, 2009.

Forewing Sc with several oblique veinlets, separated from R + M at base, and fused with R_1 distal to R branched dichotomously; M five-branched and M_3 with two branches; CuA_2 curved, S-shaped. CuA connected with R + M by cross-vein r + m-cua, with M by m-cua, with M_4 by m_4-cua; forewing with prominent color pattern consisting of hyaline patches on a dark background [116].

Distribution and age: Inner Mongolia; Middle Jurassic.

Only one species included from the Jurassic of Northern China (see Table 16.1).

Cricocossus Wang & Ren, 2009

Cricocossus Wang & Ren, 2009, *Acta Geol. Sin.-Engl.*, 83 (1), 33–38 [116] (original designation).

Type species: *Cricocossus paradoxus* Wang & Ren, 2009.

Forewing triangular, Sc with oblique veinlets, separated from R + M at base, and fused with R_1 distal to R branched dichotomously; M five-branched and M_3 with branches; CuA connected with M by a short cross-vein m-cua at basal, with M_{3+4} by m_{3+4}-cua; forewing with color pattern consisting of hyaline patches on a dark background [116].

Distribution and age: Inner Mongolia; Middle Jurassic.

Only one species included from the Jurassic of Northern China (see Table 16.1).

Ningchengia Wang, Zhang & Szwedo, 2009

Ningchengia Wang, Zhang & Szwedo, 2009, *Palaeontology*, 52 (1), 53–64 [112] (original designation).

Type species: *Ningchengia aspera* Wang, Zhang & Szwedo, 2009.

Forewing medium-sized, length about 33 mm; indentation at about basal 0.42 of wing length; ScP without branches; M distinctly angled at junction with cross-vein r + m-cua; r + m-cua vertical and very short; Cu curved basally; marginal membrane much thicker than vein width. Hind wing RP strongly curved at connection with M_1 by cross-vein r-m; M_{3+4} departing from M_{2+3+4} basally; M_{3+4} bifurcating distal of the level of nodal indentation; M_{3+4} in contact with CuA or connected with it by an oblique cross-vein m-cua basally [112].

Distribution and age: Inner Mongolia; Middle Jurassic.

Two species included from the Jurassic of Northern China (see Table 16.1).

Synapocossus Wang, Shih & Ren, 2013

Synapocossus Wang, Shih & Ren, 2013, *Alcheringa*, 37, 19–30 [120] (original designation).

Type species: *Synapocossus sciacchitanoae* Wang, Shih & Ren, 2013.

Relatively small body size, length 16–19 mm and width 9–11 mm; RP and M_1 coalescing for an interval on forewing; and M_{3+4} without bifurcation on hind wing [120].

Distribution and age: Inner Mongolia; Middle Jurassic.

Only one species included from the Jurassic of Northern China (see Table 16.1).

Synapocossus sciacchitanoae Wang, Shih & Ren, 2013 (Figure 16.17)

Synapocossus sciacchitanoae Wang, Shih & Ren, 2013: *Alcheringa*, 37 (1), 20.

Locality and horizon: Daohugou, Ningcheng, Inner Mongolia, China; Middle Jurassic, Jiulongshan Formation.

The specific epithet is dedicated to Ms. Fran Sciacchitano for being a colleague and friend providing assistance, motivation and inspiration to Dr. Shih. Head (including compound eyes) about 3/5 of the width of the pronotum, about 2/5 of the width of the mesonotum; postclypeus swollen, about 1/3 of the width of head (including compound eyes). Fore femora slightly thickened. Forewing triangular, M_{1+2} forked slightly apicad of M_{3+4} forking. Branch RP coalesces with branch M_1 for an interval on both left and right wings (some intra-specific variation with variable length in the coalescence of RP with branch M_1 from very short to long). Clavus reaching 1/3 of forewing length, claval veins A_1 and A_2 fused in basal half. Discal cell about three times as long as wide. Hind wing small. RP coalesced

Figure 16.17 *Synapocossus sciacchitanoae* Wang, Shih & Ren, 2013 (Paratype, CNU-HEM-NN2007008p). Source: Donated by Dr. Chungkun Shih.

with M_1 for a long interval at about middle 1/3 of hind wing length. Basal portions of M_{3+4} and CuA connected, veinlet m-cua absent. Abdomen covered with setae [120].

Kallicossus Chen, Zhang & Wang, 2014

Kallicossus Chan, Zhang & Wang, 2014, *Acta Palaeontol. Sin.*, 53 (3), 345–351 [121] (original designation).

Type species: *Kallicossus ningchengensis* Chen, Zhang & Wang, 2014.

Hind wing oval, length/width ratio about 1.65. Nodal indentation strong, at basal 0.37 wing length. Costal area triangular, length/width ratio about 3.28, maximal width in basal 1/3. Stem R fused with Sc, wide, slightly curved. Stem R branching into RA and RP at 0.28 wing length. Vein RA two-branched. Branch RA_1 short, terminating at costal margin distal of indentation. Branch RA_2 terminating at costal margin. Branch RP fused with M_1 at the level of nodal indentation, and then parallel to RA_2. Stem M bifurcating into M_{1+2} and M_{3+4}. Vein M_{1+2} long. Crossvein m-cua wide and long. Vein CuA two-branched. Vein CuP simple. Vein A_1 long. Marginal membrane distinct and wide at outer margin [121].

Distribution and age: Inner Mongolia; Middle Jurassic.

Only one species included from the Jurassic of Northern China (see Table 16.1).

Superfamily Pereborioidea Zalessky, 1930

Family Pereboriidae Zalessky, 1930

Pereboriidae (pro Pereboridae Zalessky, 1930), an extinct family belonging to the basal Cicadomorpha [128], are fossils recorded from the Permian to the Early Cretaceous of Brazil, Russia, South Africa and China. Pereboriidae, including seven genera with nine species, are diagnosed by membranous forewing, extensive branches of R, RS and CuA [129].

Only one genus included from the Cretaceous of Northern China: *Jiphara* Ren, Lu & Guo, 1995.

Jiphara Ren, Lu & Guo, 1995

Jiphara Ren, Lu & Guo, 1995, *Faunae and Stratigraphy of Jurassic-Cretaceous in Beijing and the Adjacent Areas*, 64–73 [30] (original designation).

Type species: *Jiphara wangi* Ren, Lu & Guo, 1995.

The specific epithet is in honor of a renowned geologist, Prof. Zhuquan Wang. Sc_1 apparently longer than Sc_2, stem R_1 curved toward the apex, having numerous branches and branching again. Rs from the middle of R_1; R forking basad to that of M and CuA; M_1 shorter than M_2; A_2 developed from the posterior margin.

Distribution and age: Beijing; Early Cretaceous.

Two species included from the Cretaceous of Northern China (see Table 16.1).

Table 16.1 A list of fossil Homoptera from the Jurassic and Cretaceous of China.

Family	Species	Locality	Horizon/Age	Citation
	Suborder Sternorrhyncha Amyot and Serville, 1843			
Aphididae	*Sunaphis laiyangensis* Hong & Wang, 1990	Laiyang, Shandong	Laiyang Fm., K_1	Hong *et al.* [21]
	Sunaphis shandongensis Hong & Wang, 1990	Laiyang, Shandong	Laiyang Fm., K_1	Hong *et al.* [21]
Ellinaphididae	*Ellinaphis leptoneura* (Zhang, Zhang, Hou & Ma, 1989)	Laiyang, Shandong	Laiyang Fm., K_1	Zhang *et al.* [32]
Hormaphididae	*Petiolaphioides shandongensis* Hong & Wang, 1990	Laiyang, Shandong	Laiyang Fm., K_1	Hong *et al.* [21]
	Petiolaphis laiyangensis Hong & Wang, 1990	Laiyang, Shandong	Laiyang Fm., K_1	Hong *et al.* [21]
Oviparosiphidae	*Archeoviparosiphum camptotropum* (Zhang, Zhang, Hou & Ma, 1989)	Laiyang, Shandong	Laiyang Fm., K_1	Zhang *et al.* [32]; Żyła *et al.* [31]
	Archeoviparosiphum latum (Hong & Wang, 1990)	Laiyang, Shandong	Laiyang Fm., K_1	Hong *et al.* [21]; Żyła *et al.* [31]
	Archeoviparosiphum malacum (Zhang, Zhang, Hou & Ma, 1989)	Laiyang, Shandong	Laiyang Fm., K_1	Zhang *et al.* [32]; Żyła *et al.* [31]
	Archeoviparosiphum opimum (Zhang, Zhang, Hou & Ma, 1989)	Laiyang, Shandong	Laiyang Fm., K_1	Zhang *et al.* [32]; Żyła *et al.* [31]
	Archeoviparosiphum tuanwangense (Zhang, Zhang, Hou & Ma, 1989)	Laiyang, Shandong	Laiyang Fm., K_1	Zhang *et al.* [32]; Żyła *et al.* [31]

(Continued)

Table 16.1 (Continued)

Family	Species	Locality	Horizon/Age	Citation
	Daoaphis magnalata Huang, Wegierek & Żyła, 2015	Ningcheng, Inner Mongolia	Jiulongshan Fm., J_2	Huang *et al.* [25]; Żyła *et al.* [31]
	Expansaphis laticosta Hong & Wang, 1990	Laiyang, Shandong	Laiyang Fm., K_1	Hong *et al.* [21]; Żyła *et al.* [31]
	Expansaphis ovata Hong & Wang, 1990	Laiyang, Shandong	Laiyang Fm., K_1	Hong *et al.* [21]; Żyła *et al.* [31]
	Oviparosiphum stictum Fu, Yao & Qiao, 2016	Beipiao, Liaoning	Yixian Fm., K_1	Fu *et al.* [27]
	Sinoviparosiphum lini Ren, 1995	Chengde, Hebei	Yixian Fm., K_1	Ren *et al.* [30]; Żyła *et al.* [31]
Protopsyllidiidae	*Poljanka hirsuta* Yang, Yao & Ren, 2012	Ningcheng, Inner Mongolia	Jiulongshan Fm., J_2	Yang *et al.* [53]
	Sinopsocus oligovenus Lin, 1976	Beipiao, Liaoning	Yixian Fm., K_1	Lin [50]
Sinaphididae	*Sinaphidium epichare* Zhang, Zhang, Hou & Ma, 1989	Laiyang, Shandong	Laiyang Fm., K_1	Zhang *et al.* [32]
	Tartaraphis peregrina Zhang, Zhang, Hou & Ma, 1989	Laiyang, Shandong	Laiyang Fm., K_1	Zhang *et al.* [32]
Sinojuraphididae	*Sinojuraphis ningchengensis* Huang & Nel, 2008	Ningcheng, Inner Mongolia	Jiulongshan Fm., J_2	Huang *et al.* [33]
Family *Incertae sedis*	*Dataiphis coniferis* Lin, 1995	Huating, Gansu	Luohandong Fm., K_1	Lin *et al.* [36]
	[a]*Penaphis* circa Lin, 1980	Jiande, Zhejiang	Baishuiling Fm., K_1	Lin [130]
		Suborder Auchenorrhyncha Dumeril, 1806		
Archijassidae	*Archijassus plurinervis* Zhang, 1985	Laiyang, Shandong	Laiyang Fm., K_1	Zhang [94]
	Mesoccus advenus Zhang, 1985	Laiyang, Shandong	Laiyang Fm., K_1	Zhang [94]
	Mesoccus lutarius Zhang, 1985	Laiyang, Shandong	Laiyang Fm., K_1	Zhang [94]
Cercopidae	*Sinocercopis liaoyuanensis* Hong, 1982	Liaoyuan, Liaoning	[b]Yixian Fm., K_1	Hong [68]
	[a]*Hebeicercopis triangulata* Hong, 1983	Luanping, Hebei	Jiulongshan Fm., J_2	Hong [74]
Cicadellidae	*Sinojassus brevispinatus* Zhang, 1985	Laiyang, Shandong	Laiyang Fm., K_1	Zhang [94]
Cixiidae	*Lapicixius decorus* Ren, Yin & Dou, 1998	Beipiao, Liaoning	Yixian Fm., K_1	Ren *et al.* [57]
	Yanducixius pardalinus Ren, Lu & Guo, 1995	Xishan, Beijing	Lushangfen Fm., K_1	Ren *et al.* [30]
	Yanducixius yihi Ren, Lu & Guo, 1995	Xishan, Beijing	Lushangfen Fm., K_1	Ren *et al.* [30]
Dysmorphoptilidae	[a]*Stigmocercopis parvis* Lin, 1986	Jiangyong, Hunan	Shiti Fm., J_1	Lin [132]
Fulgoridiidae	*Eofulgoridium chanmaense* Hong, 1982	Yumen, Gansu	Chijinpu Fm., K_1	Hong [68]
	Eofulgoridium tenellum Zhang, Wang & Zhang, 2003	Jimsar, Xinjiang	Sangonghe Fm., J_1	Zhang *et al.* [133]
	Fenghuangor imperator Li & Szwedo, 2011	Ningcheng, Inner Mongolia	Jiulongshan Fm., J_2	Li *et al.* [62]
	[a]*Valvifulgoria pingkuiensis* Lin, 1986	Zhongshan, Guangxi	Shiti Fm., J_1	Lin [132]
	[a]*Valvifulgoria tiantungensis* Lin, 1986	Zhongshan, Guangxi	Shiti Fm., J_1	Lin [132]
Lalacidae	*Cretocixius stigmatosus* Zhang, 2002	Jingxi, Beijing	Lushangfen Fm., K_1	Zhang [64]

(Continued)

Table 16.1 (Continued)

Family	Species	Locality	Horizon/Age	Citation
Membracidae	[a]*Tegulicicada plana* Lin, 1986	Zhongshan, Guangxi	Shiti Fm., J_1	Lin [132]
Palaeontinidae	*Abrocossus longus* Wang, Zhang & Fang, 2007	Ningcheng, Inner Mongolia	Jiulongshan Fm., J_2	Wang *et al.* [109]
	Cladocossus undulatus Wang & Ren, 2009	Ningcheng, Inner Mongolia	Jiulongshan Fm., J_2	Wang and Ren [116]
	Cricocossus paradoxus Wang & Ren, 2009	Ningcheng, Inner Mongolia	Jiulongshan Fm., J_2	Wang and Ren [116]
	Daohugoucossus solutus Wang, Zhang & Fang, 2006	Ningcheng, Inner Mongolia	Jiulongshan Fm., J_2	Wang *et al.* [105]
	Daohugoucossus shii Wang, Ren & Shih, 2007	Ningcheng, Inner Mongolia	Jiulongshan Fm., J_2	Wang *et al.* [118]
	Daohugoucossus parallelivenius Wang, Ren & Shih, 2007	Ningcheng, Inner Mongolia	Jiulongshan Fm., J_2	Wang *et al.* [118]
	Daohugoucossus lii Wang, Ren & Shih, 2007	Ningcheng, Inner Mongolia	Jiulongshan Fm., J_2	Wang *et al.* [118]
	Eoiocossus validus Wang, Zhang & Fang, 2006	Ningcheng, Inner Mongolia	Jiulongshan Fm., J_2	Wang *et al.* [107]
	Eoiocossus conchatus (Wang, Ren & Shih, 2007)	Ningcheng, Inner Mongolia	Jiulongshan Fm., J_2	Wang *et al.* [117]
	Eoiocossus giganteus (Wang, Ren & Shih, 2007)	Ningcheng, Inner Mongolia	Jiulongshan Fm., J_2	Wang *et al.* [117]
	Eoiocossus pteroideus (Wang, Ren & Shih, 2007)	Ningcheng, Inner Mongolia	Jiulongshan Fm., J_2	Wang *et al.* [117]
	Gansucossus pectinata (Hong, 1982)	Subei, Gansu	Dashankou Fm., J_1-J_2	Hong [68]
	Gansucossus luanpingensis (Hong, 1983)	Luanping, Hebei	Jiulongshan Fm., J_2	Hong [74]
	Gansucossus typicus Wang, Zhang & Fang, 2006	Ningcheng, Inner Mongolia	Jiulongshan Fm., J_2	Wang *et al.* [105]
	Hamicossus laevis Wang & Ren, 2007	Ningcheng, Inner Mongolia	Jiulongshan Fm., J_2	Wang and Ren [114]
	Ilerdocossus ningchengensis Wang, Zhang & Fang, 2008	Ningcheng, Inner Mongolia	Jiulongshan Fm., J_2	Wang *et al.* [110]
	[a]*Jibeicossus qingshilaense* Hong, 1984	Shidongzi, Hebei	Jiulongshan Fm., J_2	Hong [134]
	Kallicossus ningchengensis Chen, Zhang & Wang, 2014	Ningcheng, Inner Mongolia	Jiulongshan Fm., J_2	Chen *et al.* [121]
	Liaocossus beipiaoensis Ren, Yin & Dou, 1998	Beipiao, Liaoning	Yixian Fm., K_1	Ren *et al.* [102]
	Liaocossus exiguus Ren, Yin & Dou, 1998	Beipiao, Liaoning	Yixian Fm., K_1	Ren *et al.* [102]
	Liaocossus hui Ren, Yin & Dou, 1998	Beipiao, Liaoning	Yixian Fm., K_1	Ren *et al.* [102]
	Liaocossus fengningensis Ren, Yin & Dou, 1998	Fengning, Hebei	Yixian Fm., K_1	Ren *et al.* [102]
	Liaocossus pingquanensis Ren, Yin & Dou, 1998	Pingquan, Hebei	Yixian Fm., K_1	Ren *et al.* [102]
	Martynovocossus punctulosus (Wang & Ren, 2006)	Ningcheng, Inner Mongolia	Jiulongshan Fm., J_2	Wang and Ren [113]; Wang *et al.* [111]
	Martynovocossus bellus (Wang & Ren, 2006)	Ningcheng, Inner Mongolia	Jiulongshan Fm., J_2	Wang and Ren [113]; Wang *et al.* [111]
	Martynovocossus ancylivenius (Wang & Ren, 2006)	Ningcheng, Inner Mongolia	Jiulongshan Fm., J_2	Wang and Ren [113]; Wang *et al.* [111]
	Martynovocossus cheni Wang, Zhang & Fang, 2008	Ningcheng, Inner Mongolia	Jiulongshan Fm., J_2	Wang *et al.* [111]

(Continued)

Table 16.1 (Continued)

Family	Species	Locality	Horizon/Age	Citation
	Martynovocossus decorus Wang, Zhang & Fang, 2009	Ningcheng, Inner Mongolia	Jiulongshan Fm., J_2	Wang *et al.* [111]
	Martynovocossus strenus (Zhang, 1997)	Karamai, Xinjiang	Badaowan Fm., J_1	Zhang [103]; Wang *et al.* [111]
	Miracossus ingentius Ren, Yin & Dou, 1998	Beipiao, Liaoning	Yixian Fm., K_1	Ren *et al.* [102]
	Neimenggucossus normalis Wang, Zhang & Fang, 2007	Ningcheng, Inner Mongolia	Jiulongshan Fm., J_2	Wang *et al.* [109]
	Ningchengia minuta (Wang, Zhang & Fang, 2006)	Ningcheng, Inner Mongolia	Jiulongshan Fm., J_2	Wang *et al.* [104]; Wang *et al.* [112]
	Ningchengia aspera Wang, Zhang & Szwedo, 2009	Ningcheng, Inner Mongolia	Jiulongshan Fm., J_2	Wang *et al.* [112]
	Palaeontinodes haifanggouensis Hong, 1983	Beipiao, Liaoning	Haifanggou Fm., J_2	Hong [74]
	Palaeontinodes reshuitangensis Wang & Zhang, 2007	Lingyuan, Liaoning; Ningcheng, Inner Mongolia	Jiulongshan Fm., J_2	Wang *et al.* [108]
	Palaeontinodes daohugouensis Wang, Zhang & Fang, 2007	Ningcheng, Inner Mongolia	Jiulongshan Fm., J_2	Wang *et al.* [109]
	Palaeontinodes locellus Wang, Zhang & Fang, 2007	Ningcheng, Inner Mongolia	Jiulongshan Fm., J_2	Wang *et al.* [109]
	Palaeontinodes separatus Wang, Zhang & Fang, 2007	Ningcheng, Inner Mongolia	Jiulongshan Fm., J_2	Wang *et al.* [109]
	[a)] *Palaeontinopsis liaoxiensis* Hong, 1986	Huludao, Liaoning	Haifanggou Fm., J_2	Hong [135]
	[a)] *Palaeontinopsis sinensis* Hong, 1986	Huludao, Liaoning	Haifanggou Fm., J_2	Hong [135]
	Plachutella zhouyingziensis (Hong, 1983)	Luanping, Hebei	Jiulongshan Fm., J_2	Hong [74]
	Plachutella exculpta Zhang, 1997	Karamai, Xinjiang	Badaowan Fm., J_1	Zhang [103]
	Plachutella magica Wang, Zhang & Fang, 2006	Ningcheng, Inner Mongolia	Jiulongshan Fm., J_2	Wang *et al.* [104]
	Quadraticossus fangi Wang & Ren, 2007	Ningcheng, Inner Mongolia	Jiulongshan Fm., J_2	Wang and Ren [114]
	Quadraticossus longicaulis Wang & Ren, 2007	Ningcheng, Inner Mongolia	Jiulongshan Fm., J_2	Wang and Ren [114]
	Quadraticossus eumorphus Wang & Ren, 2007	Ningcheng, Inner Mongolia	Jiulongshan Fm., J_2	Wang and Ren [114]
	Sinopalaeocossus scabratus Hong, 1983	Luanping, Hebei	Jiulongshan Fm., J_2	Hong [74]
	Sinopalaeocossus trinervus Wang, Zhang & Fang, 2006	Ningcheng, Inner Mongolia	Jiulongshan Fm., J_2	Wang *et al.* [106]
	Suljuktocossus coloratus Wang, Zhang & Fang, 2006	Ningcheng, Inner Mongolia	Jiulongshan Fm., J_2	Wang *et al.* [104]
	Suljuktocossus chifengensis Wang, Zhang & Fang, 2007	Ningcheng, Inner Mongolia	Jiulongshan Fm., J_2	Wang *et al.* [109]
	Suljuktocossus yinae Wang & Ren, 2007	Ningcheng, Inner Mongolia	Jiulongshan Fm., J_2	Wang and Ren [115]
	Synapocossus sciacchitanoae Wang, Shih & Ren, 2013	Ningcheng, Inner Mongolia	Jiulongshan Fm., J_2	Wang *et al.* [120]
	Yanocossus guoi Ren, Lu & Guo, 1995	Chengde, Hebei	Yixian Fm., K_1	Ren *et al.* [30]
Pereboriidae	*Jiphara wangi* Ren, Lu & Guo, 1995	Xishan, Beijing	Lushangfen Fm., K_1	Ren *et al.* [30]
	Jiphara reticulata Ren, Lu & Guo, 1995	Xishan, Beijing	Lushangfen Fm., K_1	Ren *et al.* [30]
Procercopidae	*Anomoscytina anomala* Ren, Yin & Dou, 1998	Beipiao, Liaoning	Yixian Fm., K_1	Ren *et al.* [57]

(Continued)

Table 16.1 (Continued)

Family	Species	Locality	Horizon/Age	Citation
	Anthoscytina brevineura Chen,Wang & Zhang, 2015	Ningcheng, Inner Mongolia	Jiulongshan Fm., J_2	Chen *et al.* [71]
	Anthoscytina elegans Chen, Wang & Zhang, 2015	Ningcheng, Inner Mongolia	Jiulongshan Fm., J_2	Chen *et al.* [71]
	Anthoscytina hongi Chen, Wang & Zhang, 2015	Huludao, Liaoning	Haifanggou Fm., J_2	Hong [135]; Chen *et al.* [71]
	Anthoscytina liugouensis (Hong, 1983)	Luanping, Hebei	Jiulongshan Fm., J_2	Hong [74]
	Anthoscytina longa Hong, 1983	Beipiao, Liaoning	Haifanggou Fm., J_2	Hong [74]
	Anthoscytina parallelica Ren, Lu & Guo, 1995	Luanping, Hebei	Yixian Fm., K_1	Ren *et al.* [30]
	Anthoscytina perpetua Li, Shih & Ren, 2013	Ningcheng, Inner Mongolia	Jiulongshan Fm., J_2	Li *et al.* [76]
	*Anthoscytina pustulosus (*Ren, Lu & Guo, 1995)	Xishan, Beijing	Lushangfen Fm., K_1	Ren *et al.* [30]
	Anthoscytina trinervus (Ren, Lu & Guo, 1995)	Xishan, Beijing	Lushangfen Fm., K_1	Ren *et al.* [30]
	Chengdecercopis xiaofanzhangziensis Hong,1983	Chengde, Hebei	Jiulongshan Fm., J_2	Hong [74]
	Cretocercopis yii Ren, Lu & Guo, 1995	Xishan, Beijing	Lushangfen Fm., K_1	Ren *et al.* [30]
	Jurocercopis grandis Wang & Zhang, 2009	Ningcheng, Inner Mongolia	Jiulongshan Fm., J_2	Wang and Zhang [77]
	Procercopina delicata Zhang, Wang & Zhang, 2003	Karamay, Xinjiang	Badaowan Fm., J_1	Zhang *et al.* [133]
	Procercopis shawanensis Zhang, Wang & Zhang, 2003	Shawan, Xinjiang	Badaowan Fm., J_1	Zhang *et al.* [133]
	Stellularis aphthosa (Ren, Yin & Dou, 1998)	Beipiao, Liaoning	Yixian Fm., K_1	Ren *et al.* [57]
	Stellularis longirostris Chen, Yao & Ren, 2015	Beipiao, Liaoning	Yixian Fm., K_1	Chen *et al.* [78]
	Stellularis macula (Hu, Yao & Ren, 2014)	Beipiao, Liaoning	Yixian Fm., K_1	Hu *et al.* [136]
	Titanocercopis borealis Chen, Zhang & Wang, 2015	Ningcheng, Inner Mongolia	Jiulongshan Fm., J_2	Chen *et al.* [79]
Prosbolidae	[a]*Longimaxilla sinnica* Hong, 1982	Yumen, Gansu	Xiagou Fm., K_1	Hong [68]
	[a]*Permocicada beipiaoensis* Wang, 1987	Beipiao, Liaoning	Jiulongshan Fm., J_2	Wang [137]
Qiyangiricaniidae	[a]*Qiyangiricania cesta* Lin, 1986	Yongzhou, Hunan	Guanyintan Fm., J_1	Lin [132]; Szwedo *et al.* [138]
Scytinopteridae	[a]*Sunoscytinopteris lushangfenensis* Hong, 1984	Fangshan, Beijing	Lushangfen Fm., K_1	Hong [101]
Sinoalidae	*Jiania crebra* Wang & Szwedo, 2012	Ningcheng, Inner Mongolia	Jiulongshan Fm., J_2	Wang *et al.* [70]
	Jiania gracila Wang & Szwedo, 2012	Ningcheng, Inner Mongolia	Jiulongshan Fm., J_2	Wang *et al.* [70]
	Huabeicercopis yangi Hong, 1983	Luanping, Hebei	Jiulongshan Fm., J_2	Hong [74]
	Luanpingia longa Hong 1983	Luanping, Hebei	Jiulongshan Fm., J_2	Hong [74]
	Luanpingia senjituensis Hong 1984	Luanping, Hebei	[b]Yixian Fm., K_1	Hong [101]
	Sinoala parallelivena Wang & Szwedo, 2012	Ningcheng, Inner Mongolia	Jiulongshan Fm., J_2	Wang *et al.* [70]
Tettigarctidae	*Hirtaprosbole erromera* Liu, Yao & Ren, 2016	Ningcheng, Inner Mongolia	Jiulongshan Fm., J_2	Liu *et al.* [89]
	Macrotettigarcta obesa Chen & Wang, 2016	Ningcheng, Inner Mongolia	Jiulongshan Fm., J_2	Chen and Wang [90]

(Continued)

Table 16.1 (Continued)

Family	Species	Locality	Horizon/Age	Citation
	Maculaprosbole zhengi Zheng, Chen & Wang, 2016	Ningcheng, Inner Mongolia	Jiulongshan Fm., J_2	Zheng *et al.* [91]
	Protabanus chaoyangensis Hong, 1982	Chaoyang, Liaoning	[b]Jiulongshan Fm., J_2	Hong [68]
	[a]*Quadrisbole stenis* Lin, 1986	Liuyang, Hunan	Zaoshang Fm., J_1	Lin [132]
	Sanmai kongi Chen, Zhang & Wang, 2016	Ningcheng, Inner Mongolia	Jiulongshan Fm., J_2	Chen *et al.* [92]
	Sanmai mengi Chen, Zhang & Wang, 2016	Ningcheng, Inner Mongolia	Jiulongshan Fm., J_2	Chen *et al.* [92]
	Sanmai xuni Chen, Zhang & Wang, 2016	Ningcheng, Inner Mongolia	Jiulongshan Fm., J_2	Chen *et al.* [92]
	Shaanxiarcta perrara (Zhang, 1993)	Shangxian, Shaanxi	Fengjiashan Fm., K_1	Zhang [87]; Shcherbakov [81]
	Shuraboprosbole daohugouensis Wang & Zhang, 2009	Ningcheng, Inner Mongolia	Jiulongshan Fm., J_2	Wang and Zhang [77]
	Shuraboprosbole minuta Wang & Zhang, 2009	Ningcheng, Inner Mongolia	Jiulongshan Fm., J_2	Wang and Zhang [77]
	Shuraboprosbole media Wang & Zhang, 2009	Ningcheng, Inner Mongolia	Jiulongshan Fm., J_2	Wang and Zhang [77]
	Sinocicadia shandongensis Hong and Wang, 1990	Laiyang, Shandong	Laiyang Fm., K_1	Hong and Wang [21]
	Sunotettigarcta hebeiensis Hong, 1983	Luanping, Hebei	Jiulongshan Fm., J_2	Hong [74]
	Sunotettigarcta hirsuta Li, Wang & Ren, 2012	Ningcheng, Inner Mongolia	Jiulongshan Fm., J_2	Li *et al.* [139]
	Tianyuprosbole zhengi Chen, Wang & Zhang, 2014	Ningcheng, Inner Mongolia	Jiulongshan Fm., J_2	Chen *et al.* [88]
		Suborder Paleorrhyncha Carpenter, 1931		
Archescytinidae	[a]*Lepidoscytina miaobaoensis* Wang, 1980	Benxi, Liaoning	Changliangzi Fm., J_1	Wang [131]

a) The species is not present in the main text because the original description, photos and line-drawings are not precise and the holotype cannot be rechecked.

b) Horizon/Age revised from the original paper based on updated information and data.

References

1 Grimaldi, D. and Engel, M.S. (2005). *Evolution of the Insects*. New York: Cambridge University Press.

2 Gullan, P.J. and Cranston, P.S. (2005). *The Insects: An Outline of Entomology*, 3e. Blackwell Publishing, Ltd.

3 Bourgoin, T. and Szwedo, J. (2008). The 'cixiid-like' fossil planthopper families. *Bulletin of Insectology* **61** (1): 107–108.

4 Dietrich, C.H., Resh, V.H., and Carde, R.T. (2003). *Encyclopedia of Insects*. San Diego, CA: Academic Press.

5 Rasnitsyn, A.P. and Quicke, D.L.J. (2002). *History of Insects*. Dordrecht: Kluwer Academic Publishers.

6 Buckley, R. (1987). Ant-plant-homopteran interactions. *Advances in Ecological Research* **16**: 53–85.

7 Poinar, G.O. and Poinar, R. (1994). *The Quest for Life in Amber*. New York: Addison-Wesley.

8 Poinar, G.O. and Poinar, R. (1999). *The Amber Forest*. Princeton, NJ: Princeton University Press.

9 Perkovsky, E.E. (2006). Vstrechaemost sininklozov muravev (Hymentoptera, Formicidae) I trei (Homoptera, Aphidinea) v Saksonskom I Rovenskom yantaryakh. *Paleontologicheskii Zhurnal* **2**: 72–74.

10 Claridge, M.F. (1985). Acoustic signals in the Homoptera: behavior, taxonomy, and evolution. *Annual Review of Entomology* **30**: 297–317.

11 Ossianilsson, F. (1949). Insect drummers, a study on the morphology and function of the sound-producing organ of Swedish Homoptera

Auchenorrhyncha with notes on their sound production. *Opuscula Entomologia, Supplementum* **10**: 139.
12 Cocroft, R. (1999). Thornbug to thornbug: the inside story of insect song. *Natural History* **10**: 53–57.
13 Brodie, P.B. (1845). *A History of the Fossil Insects in the Secondary Rocks of England*. London: John van Voorst.
14 Shcherbakov, D.E. (2002). The 270 million year history of Auchenorrhyncha (Homoptera). *Denisia* **4**: 29–35.
15 Shcherbakov, D.E. and Popov, Y.A. (2002). Superorder Cimicidea Laicharting, 1781 Order Hemiptera Linné, 1758. The bugs, cicadas, plantlice, scale insects, etc. (=Cimicida Laicharting, 1781, = Homoptera Leach, 1815 + Heteroptera Latreille, 1810). In: *History of Insects* (ed. A.P. Rasnitsyn and D.L.J. Quicke), 143–157. Dordrecht: Kluwer Academic Publishers.
16 Grimaldi, D.A. (2003). First amber fossils of the extinct family Protopsyllidiidae, and their phylogenetic significance among Hemiptera. *Insect Systematics and Evolution* **34**: 329–344.
17 Evans, J.W. (1956). Palaeozoic and Mesozoic Hemiptera (insect). *Australian Journal of Zoology* **4**: 165–258.
18 Hennig, W. (1981). *Insect Phylogeny*. Chichester: Wiley.
19 Lin, Q.B. (1977). Fossil insects from Yunnan. In: *Mesozoic Fossils from Yunnan, 2* (ed. Nanjing Institute of Geology and Palaeontology), 373–381. Beijing: Science Press.
20 Wang, Y., Ren, D., Liang, J.H. *et al.* (2006). The fossil Homoptera of China: a review of present knowledge. *Acta Zootaxonomica Sinica* **31** (2): 294–303. (in Chinese with English abstract).
21 Hong, Y.C. and Wang, W.L. (1990). Fossil insects from the Laiyang basin, Shandong Province. In: *The stratigraphy and palaeontology of Laiyang Basin, Shandong Province* (ed. Bureau of Geology and Mineral of Shandong Province), 44–189. Beijing: Geological Publishing House [in Chinese with English summary].
22 Stern, D.L. (1994). A phylogenetic analysis of soldier evolution in the aphid family Hormaphididae. *Proceedings of the Royal Society of London B* **256**: 203–209.
23 Chen, J., Jiang, L.Y., and Qiao, G.X. (2013). A total-evidence phylogenetic analysis of hormaphidinae (hemiptera: aphididae), with comments on the evolution of galls. *Cladistics-the International Journal of the Willi Hennig Society* **30** (1): 26–66.
24 Heie, O.E. and Węgierek, P. (2011). *A List of Fossil Aphids (Hemiptera, Sternorrhyncha, Aphidomorpha)*. Upper Silesian Museum: Department of Natural History.
25 Huang, D., Wegierek, P., Zyla, D., and Nel, A. (2015). The oldest aphid of the family Oviparosiphidae (Hemiptera: Aphidoidea) from the Middle Jurassic of China. *European Journal of Entomology* **112** (1): 187–192.
26 Shaposhnikov, G.K. (1979). Late Jurassic and Early Cretaceous aphids. *Paleontological Journal* **13**: 449–461.
27 Fu, Y., Yao, Y.Z., Qiao, G.X. *et al.* (2017). A new species of Oviparosiphidae (Hemiptera: Aphidomorpha) from the Lower Cretaceous of China. *Cretaceous Research* **75**: 157–161.
28 Heie, O.E. (1987). Paleontology and phylogeny. In: *Aphids: Their Biology, Natural Enemies and Control*, vol. **2** (ed. A.K. Minks and P. Harrewijn), 367–391. Elsevier Science Publishers B.V.
29 Von Dohlen, C.D. and Moran, N.A. (2000). Molecular data support a rapid radiation of aphids in the Cretaceous and multiple origins of host alternation. *Biological Journal of the Linnean Society* **71**: 689–717.
30 Ren, D., Lu, L.W., Guo, Z.G., and Ji, S.A. (1995). *Faunae and Stratigraphy of Jurassic-Cretaceous in Beijing and the Adjacent Areas*. Beijing: Seismic Publishing House.
31 Żyła, D., Homan, A., Franielczyk, B., and Wegierek, P. (2015). Revised concept of the fossil genus *oviparosiphum* shaposhnikov, 1979 with the description of a new genus (hemiptera, sternorrhyncha, aphidomorpha). *Zookeys* **483**: 9–22.
32 Zhang, J.F., Zhang, S., Hou, F., and Ma, G. (1989). Late Jurassic aphids (Homoptera, Insecta) from Shandong Province, China. *Geology of Shandong* **5**: 28–46. [In Chinese with English summary].
33 Huang, D. and Nel, A. (2008). A new Middle Jurassic aphid family (Insecta: Hemiptera: Sternorrhyncha: Sinojuraphididae fam. nov.) from Inner Mongolia, China. *Palaeontology* **51** (3): 715–719.
34 Shcherbakov, D.E. (2007). Extinct four-winged precoccids and the ancestry of scale insects and aphids (Hemiptera). *Russian Entomological Journal* **16**: 47–62.
35 Kania, I. and Wegierek, P. (eds.) (2008). *Palaeoaphididae (Hemiptera, Sternorrhyncha) from Lower Cretaceous Baissa Deposits. Morphology and Classification*. Kraków: Wydawnictwa instytut Systematyki i Ewolucji Zwierząt Polskiej Akademii Nauk, Ltd.
36 Lin, Q.B. (1995). On *penaphis* lin, 1980 of Cretaceous (Callaphididae, Homoptera) and its coevolutionary relationship. *Acta Palaeontologica Sinica* **34** (2): 194–204.

37 Tillyard, R.J. (1926). Upper Permian insects of New South Wales. Part 1. Introduction and the order Hemiptera. *Proceedings of the Linnaean Society of New South Wales* **51**: 1–30.

38 Davis, C. (1942). Hemiptera and Copeognatha from the Upper Permian of New South Wales. *Proceedings of the Linnaean Society of New South Wales* **67**: 111–122.

39 Evans (1943). Upper Permian homoptera from new South Wales. *Records of the Australian Museum* **21**: 180–198.

40 Evans, J.W. (1943). Two interesting upper Permian homoptera from new South Wales. *Transactions of the Royal Society of South Australia* **67**: 7–9.

41 Bekker-Migdisova, E.E. (1959). Some new representatives of a group of Sternorrhyncha from the Permian and Mesozoic of the USSR. *Materialy kosnovam paleontologii* **3**: 104–116.

42 Bekker-Migdisova, E.E. (1960). New Permian Homoptera from European USSR. *Trudy Paleontologicheskogo Instituta Akademii Nauk SSSR* **76**: 1–112.

43 Bekker-Migdisova, E.E. (1985). Fossil Psyllomorpha insects. *Trudy Paleontologicheskogo Instituta Akademii Nauk SSSR* **206**: 1–93.

44 Riek, E.F. (1976). New Upper Permian insects from Natal, South Africa. *Annals of the Natal Museum* **22** (3): 755–789.

45 Klimaszewski, S.M. (1995). Supplement to the knowledge of Protopsyllidiidae (Homoptera, Psyllomorpha). *Acta Biologica Silesiana* **27**: 33–43.

46 Tillyard, R.J. (1917). Permian and Triasic insects from New South Wales in the collection of Mr. John Mitchell. *Proceedings of the Linnaean Society of New South Wales* **42**: 720–756.

47 Martynov, A.V. (1937). Liassic insects of Shurab and Kizilkya Mongolia. *Trudy Paleontologicheskogo Instituta Akademii Nauk SSSR* **7** (1): 1–232.

48 Handlirsch, A. (1939). Neue Untersuchungen über die fossilen Insekten mit Ergänzungen und Nachträgen sowie Ausblicken auf phylogenetische, palaeogeographische und allgemein biologische Probleme.II Teil. *Annalen des Naturhistorischen Museums in Wien* **49**: 1–240.

49 Bekker-Migdisova, E.E. (1968). Protopsyllidiids and their morphology (Homoptera, Protopsyllidiidae). *Jurassic Insects of Karatau* **25**: 87–98.

50 Lin, Q.B. (1976). The Jurassic fossil insects from Western Liaoning. *Acta Palaeontologica Sinica* **15**: 97–116.

51 Ansorge, J. (1996). Insekten aus dem Oberen Lias von Grimmen (Vorpommern, Norddeutschland). *Neue Palaontologische Abhandlungen* **2**: 1–132.

52 Carpenter, F.M. (1992). *Treatise on Invertebrate Paleontology, Part R, Arthropoda 4*, vol. **3**. Lawrence/Kansas: Geological Society of America and University of Kansas.

53 Yang, G., Yao, Y.Z., and Ren, D. (2012). A new species of Protopsyllidiidae (Hemiptera, Sternorrhyncha) from the Middle Jurassic of China. *Zootaxa* **3274**: 36–42.

54 Fennah, R.G. (1987). A new genus and species of Cixiidae (Homoptera, Fulgoroidea) from Lower Cretaceous amber. *Journal of Natural History, London* **21**: 1237–1240.

55 Emeljanov, A.F. (2002) Contribution to classification and phylogeny of the family Cixiidae (Hemiptera, Fulgoromorpha). Zikaden – Leafhoppers, Planthoppers and Cicadas (Insecta: Hemiptera: Auchenorrhyncha), Denisia 04, zugleich Kataloge des OÖ, 176, 103–112.

56 Larivière, M.C. (1999). Cixiidae (Insecta: Hemiptera: Auchenorrhyncha). *Fauna of New Zealand* **40**: 1–93.

57 Ren, D., Yin, J.C., and Dou, W.X. (1998). New planthoppers and froghoppers from the late Jurassic of Northeast China (Homoptera: Auchenorrhyncha). *Acta Zootaxonomica Sinica* **23** (3): 281–288.

58 Szwedo, J., Bourgoin, T., and Lefebvre, F. (eds.) (2004). *Fossil Planthoppers (Hemiptera: Fulgoromorpha) of the World: An Annotated Catalogue with Notes on Hemiptera Classification.* Polish Academy of Sciences: Museum and Institute of Zoology.

59 Asche, M. 1987 Preliminary thoughts on the phylogeny of Fulgoromorpha (Homoptera, Auchenorrhyncha). In: Proceedings of the 6th Auchenorrhyncha Meeting, Turin, Italy, **7**, 47–53.

60 Zhang, X., Ren, D., and Yao, Y.Z. (2017). A new species of *Foveopsis* Shcherbakov (Hemiptera: Fulgoromorpha: Fulgoroidea: Perforissidae) from mid-Cretaceous Myanmar amber. *Cretaceous Research* **79**: 35–42.

61 Szwedo, J. (2010). Jurassic Fulgoridiidae and roots of Fulgoroidea (Insecta: Hemiptera: Fulgoromorpha). *Earth Science Frontiers* **17**: 222–223.

62 Li, S., Szwedo, J., Ren, D., and Pang, H. (2011). *Fenghuangor imperator* gen. et sp. nov. of Fulgorididae from the Middle Jurassic of Daohugou Biota (Hemiptera: Fulgoromorpha). *Zootaxa* **3094**: 52–62.

63 Hamilton, K.G.A. (1990). Homoptera. Insects from the Santana Formation, Lower Cretaceous, of Brazil. *Bulletin of the American Museum of Natural History* **195**: 82–122.

64 Zhang, Z.J. (2002). New Early Cretaceous lalacid from Jingxi Basin of Beijing, China (Homoptera: Fulgoroidea). *Acta Zootaxonomica Sinica* **27** (1): 20–23.

65 Keskinen, E. and Meyer-Rochow, V.B. (2004). Post-embryonic photoreceptor development and dark/light adaptation in the spittle bug Philaenus spumarius (L.)(Homoptera, Cercopidae). *Arthropod Structure and Development* **33** (4): 405–417.
66 Paladini, A. and Carvalho, G.S. (2013). Descriptions of two new species of *Sphenorhina* (Hemiptera, Cercopidae, Tomaspidinae) from the Neotropical region. *Revista Brasileira de Entomologia* **57** (2): 165–168.
67 Hu, H.J., Yao, Y.Z., and Ren, D. (2012). The review of Cercopoidea (Hemiptera) fossils. *Journal of Environmental Entomology* **34** (4): 483–496.
68 Hong, Y.C. (1982). *Mesozoic Fossil Insects of Jiuquan Basin in Gansu Province*, 80–173. Beijing: Geological Publishing House.
69 Burrows, M. (2003). Biomechanics: froghopper insects leap to new heights. *Nature* **424**: 509. https://doi.org/10.1038/424509a.
70 Wang, B., Szwedo, J., and Zhang, H.C. (2012). New Jurassic Cercopoidea from China and their evolutionary significance (Insecta: Hemiptera). *Palaeontology* **55** (6): 1223–1243.
71 Chen, J., Wang, B., Zhang, H.C. *et al.* (2015). New fossil Procercopidae (Hemiptera: Cicadomorpha) from the Middle Jurassic of Daohugou, Inner Mongolia, China. *European Journal of Entomology* **112** (2): 373–380.
72 Wang, B. and Zhang, H.C. (2009). A remarkable new genus of Procercopidae (Hemiptera: Cercopoidea) from the Middle Jurassic of China. *Comptes Rendus Palevol* **8** (4): 389–394.
73 Handlirsch, A. (1906–1908). *Die fossilen Insekten und die Phylogenie der rezenten Formen*. Leipzig: Verlagvon Wilhelm Engelmann.
74 Hong, Y.C. (1983). *Middle Jurassic Fossil Insects in North China*, 48–64. Beijing: Geological Publishing House.
75 Shcherbakov, D.E. (1988). New cicadas (Cicadina) from the Late Mesozoic of Transbaikalia. *Paleontological Journal* **22** (4): 55–66.
76 Li, S., Shih, C.K., Wang, C. *et al.* (2013). Forever love: the hitherto earliest record of copulating insects from the Middle Jurassic of China. *PloS ONE* **8** (11): e78188.
77 Wang, B. and Zhang, H.C. (2009). Middle Jurassic Tettigarctidae from Daohugou, China (Insecta: Hemiptera: Cicadoidea). *Geobios* **42** (2): 243–253.
78 Chen, D., Yao, Y.Z., and Ren, D. (2015). A new species of fossil Procercopidae (Hemiptera, Cicadomorpha) from the Lower Cretaceous of Northeastern China. *Cretaceous Research* **52**: 402–406.
79 Chen, J., Zhang, H.C., Wang, B. *et al.* (2015). High variability in tegminal venation of primitive cercopoids (Hemiptera: Cicadomorpha), as implied by the new discovery of fossils from the Middle Jurassic of China. *Entomological Science* **18** (2): 147–152.
80 Moulds, M.S. (2005). An appraisal of the higher classification of cicadas (Hemiptera: Cicadoidea) with special reference to the Australian fauna. *Records of the Australian Museum* **57**: 375–446.
81 Shcherbakov, D.E. (2008). Review of the fossil and extant genera of the cicada family Tettigarctidae (Hemiptera: Cicadoidea). *Russian Entomological Journal* **17**: 343–348.
82 Wappler, T. (2003). Systematik, Phylogenie, Taphonomie und Palaookologie der Insekten aus dem Mittle Eowan des Eckfelder marares. *Vulkaneifel, Clausthaler Geozissenchafen* **2**: 1–241.
83 Kaulfuss, U. and Moulds, M. (2015). A new genus and species of tettigarctid cicada from the early Miocene of New Zealand: Paratettigarcta zealandica (Hemiptera, Auchenorrhyncha, Tettigarctidae). *Zookeys* **484**: 83–94.
84 Nel, A., Zarbout, M., Barale, G., and Philippe, M. (1998). *Liassotettigarcta africana* sp. n. (Auchenorrhyncha: Cicadoidea: Tettigarctidae), the first Mesozoic insect from Tunisia. *European Journal of Entomology* **95**: 593–598.
85 Claridge, M.F., Morgan, J.C., and Moulds, M.S. (1999). Substrate-transmitted acoustic signals of the primitive cicada, Tettigarcta crinita Distant (Hemiptera Cicadoidea, Tettigarctidae). *Journal of Natural History* **33** (12): 1831–1834.
86 Becker-Migdisova, E.E. (1949). Mesozoic Homoptera of Middle Asia. *Trudy Paleontologicheskogo Instituta Akademii Nauk USSR* **22**: 1–68.
87 Zhang, J.F. (1993). A contribution to the knowledge of insects from the late Mesozoic in southern Shaanxi and Henan provinces, China. *Palaeoworld* **3**: 49–56.
88 Chen, J., Wang, B., Zhang, H.C., and Wang, X.L. (2014). A remarkable new genus of Tettigarctidae (Insecta, Hemiptera, Cicadoidea) from the Middle Jurassic of Northeastern China. *Zootaxa* **3764** (5): 581–586.
89 Liu, X.H., Li, Y., Yao, Y.Z., and Ren, D. (2016). A hairy-bodied tettigarctid (Hemiptera: Cicadoidea) from the latest Middle Jurassic of Northeast China. *Alcheringa: An Australasian Journal of Palaeontology* **40** (3): 383–389.
90 Chen, J. and Wang, B. (2016). A giant tettigarctid cicada from the Mesozoic of Northeastern China. *Spixiana* **39** (1): 119–124.

91 Zheng, Y., Chen, J., and Wang, X.L. (2016). A new genus and species of Tettigarctidae from the Mesozoic of Northeastern China (Insecta, Hemiptera, Cicadoidea). *ZooKeys* **632**: 47–55.

92 Chen, J., Zhang, H.C., Wang, B. *et al.* (2016). New Jurassic tettigarctid cicadas from China with a novel example of disruptive coloration. *Acta Palaeontologica Polonica* **61** (4): 853–862.

93 Shcherbakov, D.E. (2012). More on Mesozoic Membracoidea (Homoptera). *Russian Entomological Journal* **21**: 15–22.

94 Zhang, J.F. (1985). New data of Mesozoic fossil insects from Laiyang in Shandong. *Shandong Geology* **1** (2): 25–26.

95 Szwedo, J., Gebicki, C., and Kowalewska, M. (2010). *Microelectrona cladara* gen. et sp. nov.: a new Protodikraneurini from the Eocene Baltic amber (Hemiptera: Cicadomorpha: Cicadellidae: Typhlocybinae). *Acta Geologica Sinica (English Edition)* **84** (4): 696–704.

96 Dietrich, C.H. (2005). Keys to the families of Cicadomorpha and subfamilies and tribes of Cicadellidae (Hemiptera: Auchenorrhyncha). *Florida Entomologist* **88** (4): 502–517.

97 Szwedo, J. (2005). *Jantarivacanthus kotejai* gen. et sp. n. from Eocene Baltic amber, with notes on the Bathysmatophorini and related taxa (Hemiptera: Cicadomorpha: Cicadellidae). *Polskie Pismo Entomologiczne* **74**: 251–276.

98 Emeljanov, A.F. (1987). Filogenia tsikadovykh (Homoptera, Cicadina) po sravnitel'no-morfologicheskim dannym. [Phylogeny of Cicadina (Homoptera) on comparatively morphological data.]. *Trudy Vesoyuznogo Entomologicheskogo Obshchestva* **69**: 19–109. (In Russian).

99 Gomez-Pallerola, J.E. (1984). Nuevos Paleontínidos del yacimiento Infracretácico de la "Pedrera de Meiá" (Lérida). *Boletín Geológico y Minero* **95** (4): 301–309. (In French).

100 Martins-Neto, R.G. (1998). Novos registros de paleontinídeos (Insecta, Hemiptera) na Formação Santana (Cretáceo Inferior), Bacia do Araripe, nordeste do Brasil. *Acta Geologica Leopoldensia* **21** (46/47): 69–74.

101 Hong, Y.C. (1984). *Palaeontological Atlas of North China*. Beijing: Geological Publishing House.

102 Ren, D., Yin, J.C., and Dou, W.X. (1998). Late Jurassic Palaeontinids (Homoptera: Auchenorrhyncha) from Hebei and Liaoning Provinces in China. *Entomologica Sinica* **5** (3): 222–232.

103 Zhang, H.C. (1997). Jurassic palaeontinids from Karamai, Xinjiang, with a discussion of Palaeontinidae (Homoptera: Palaeontinidae) in China. *Entomologia Sinica* **4** (4): 312–323.

104 Wang, B., Zhang, H.C., and Fang, Y. (2006). Some Jurassic Palaeontinidae (Insecta, Hemiptera) from Daohugou, Inner Mongolia, China. *Palaeoworld* **15**: 115–125.

105 Wang, B., Zhang, H.C., and Fang, Y. (2006). *Gansucossus*, a replacement name for *Yumenia* Hong, 1982 (Insecta, Hemiptera, Palaeontinidae), with description of a new genus. *Zootaxa* **1268**: 59–68.

106 Wang, B., Zhang, H.C., Fang, Y., and Duan, Y. (2006). Revision of the genus *Sinopalaeocossus* Hong (Hemiptera: Palaeontinidae), with description of a new species from the Middle Jurassic of China. *Zootaxa* **1349**: 37–45.

107 Wang, B., Zhang, H.C., Fang, Y., and Zhang, Z.L. (2006). A new genus and species of Palaeontinidae (Insecta: Hemiptera) from the Middle Jurassic of Daohugou, China. *Annales Zoologici (Warszawa)* **56** (4): 757–762.

108 Wang, B., Zhang, H.C., and Fang, Y. (2007). *Palaeontinodes reshuitangensis*, a new species of Palaeontinidae (Hemiptera, Cicadomorpha) from the Middle Jurassic of Reshuitang and Daohugou of China. *Zootaxa* **1500**: 61–68.

109 Wang, B., Zhang, H.C., and Fang, Y. (2007). Middle Jurassic Palaeontinidae (Insecta, Hemiptera) from Daohugou of China. *Alavesia* **1**: 89–104.

110 Wang, B., Zhang, H.C., Fang, Y., and Wang, D.J. (2008). New data on Cretaceous Palaeontinidae (Insecta: Hemiptera) from China. *Cretaceous Research* **29**: 551–560.

111 Wang, B., Zhang, H.C., Fang, Y., and Zhang, Y.T. (2008). A revision of Palaeontinidae (Insecta: Hemiptera, Cicadomorpha) from the Jurassic of China with descriptions of new taxa and new combinations. *Geological Journal* **43**: 1–18.

112 Wang, B., Zhang, H.C., and Szwedo, J. (2009). Jurassic Palaeontinidae from China and the higher systematics of Palaeontinoidea (Insecta: Hemiptera: Cicadomorpha). *Palaeontology* **52** (1): 53–64.

113 Wang, Y. and Ren, D. (2006). Middle Jurassic *Pseudocossus* fossils from Daohugou, Inner Mongolia in China (Homoptera, Palaeontinidae). *Acta Zootaxonomica Sinica* **31** (2): 289–293.

114 Wang, Y. and Ren, D. (2007). Two new genera of fossil palaeontinids from the Middle Jurassic in Daohugou, Inner Mongolia, China (Hemiptera, Palaeontinidae). *Zootaxa* **1390**: 41–49.

115 Wang, Y. and Ren, D. (2007). Revision of the genus *Suljuktocossus* Becker-Migdisova, 1949 (Hemiptera, Palaeontinidae), with description of a new species from Daohugou, Inner Mongolia, China. *Zootaxa* **1576**: 57–62.

116 Wang, Y. and Ren, D. (2009). New fossil palaeontinids from the Middle Jurassic of Daohugou, Inner Mongolia, China (Insecta, Hemiptera). *Acta Geologica Sinica-English Edition* **83** (1): 33–38.

117 Wang, Y., Ren, D., and Shih, C.K. (2007). Discovery of Middle Jurassic palaeontinids from Inner Mongolia, China (Homptera: Palaeontinidae). *Progress in Natural Science* **17** (1): 112–116.

118 Wang, Y., Ren, D., and Shih, C.K. (2007). New discovery of palaeontinid fossils from the Middle Jurassic in Daohugou, Inner Mongolia (Homoptera, Palaeontinidae). *Science in China Series D, Earth Sciences* **50** (4): 481–486.

119 Wang, Y., Wang, L., and Ren, D. (2008). Revision of genera *Quadraticossus*, *Martynovocossus* and *Fletcheriana* (Insecta, Hemiptera) from the Middle Jurassic of China with description of a new species. *Zootaxa* **1855**: 56–64.

120 Wang, Y., Shih, C.K., Szwedo, J., and Ren, D. (2013). New fossil palaeontinids (Hemiptera, Cicadomorpha, Palaeontinidae) from the Middle Jurassic of Daohugou, China. *Alcheringa* **37** (1): 19–30. https://doi.org/10.1080/03115518.2012.690972.

121 Chen, J., Zhang, H.C., Wang, B. *et al.* (2014). A new genus and species of Palaeontinidae (Insecta, Hemiptera) from the Middle Jurassic of Daohugou, Inner Mongolia. *Acta Palaeontologica Sinica* **53** (3): 345–351.

122 Martynov, A.V. (1931). To the morphology and systematical position of the fam. Palaeontinidae Handl., with a description of a new form from Ust-baley, Siberia. *Ezhegodnik Ruskogo Paleontologicheskogo Obshchetstva* **9**: 93–122.

123 Kenrick, G.H. (1914). New or little known Heterocera from Madagascar. *Transactions of the Entomological Society of London* **61** (4): 587–602.

124 Gaede, M. (1933). Cossidae. In: *Die Gross-Schmetterlinge der Erde, II. Abteilung: Exotische Fauna 10 (die Indo-Australischen Spinner und Schwarmer)* (ed. A. Seitz), 807–824. Stuttgart: Alfred Kernen Publisher.

125 Hou, Y.C. (1958). Cretaceous nonmarine ostracods of the subfamily Cyprideinae from north-western and north-eastern regions of China. *Memoirs of the Institute of Palaeontology, Academia Sinica* **1**: 33–104. (in Chinese with English summary).

126 Whalley, P.E.S. and Jarzembowski, E.A. (1985). Fossil insects from the lithographic limestone of Montsech (Late Jurassic-Early Cretaceous), Lerida Province, Spain. *Bulletin of the British Museum (Natural History). London: British Museum (Natural History)* **38** (5): 381–412.

127 Menon, F., Heads, S.W., and Martil, D.M. (2005). New Palaeontinidae (Insecta: Cicadomorpha) from the Lower Cretaceous Crato Formation of Brazil. *Cretaceous Research* **26**: 837–844.

128 Vassilenko, D.V., Shcherbakov, D.E., and Karasev, E.V. (2014). Biodamage on Phylladoderma leaves from the Upper Permian of the Pechora Basin. *Paleontological Journal* **48** (4): 447–450.

129 Becker-Migdisova, E.E. (1962). Otryad Homoptera, Otryad Heteroptera [Orders Homoptera and Heteroptera]. (ed. Orlov, Y.A.). *Osnovy Paleontologii [Fundamentals of Paleontology]* **9**: 162–224.

130 Lin, Q.B. (1980). *Mesozoic Insects from Zhejiang and Anhui Provinces*. Beijing: Science Press.

131 Wang, W.L. (1980). *Paleontological Atlas of Northeast China Part 2*. Beijing: Geological Publishing House. (in Chinese).

132 Lin, Q.B. (1986). *Early Mesozoic Fossil Insects from South China*, 56–68. Beijing: Science Press.

133 Zhang, H.C., Wang, Q.F., and Zhang, J.F. (2003). Some Jurassic homopteran insects from the Junggar basin, Xinjiang, China. *Acta Palaeontologica Sinica* **42** (4): 548–551.

134 Hong, Y.C. (1984). Mesozoic Volume. In: *Palaeontological Atlas of North China* (ed. Institute of Geology and Mineral Resources), 152–156. Beijing: Geological Publishing House (in Chinese).

135 Hong, Y.C. (1986). New fossil insects of Haifanggou Formation, Liaoning Province. *Journal of Changchun Collage of Geological* **4**: 10–16. (in Chinese).

136 Hu, H.J., Yao, Y.Z., and Ren, D. (2014). New Fossil Procercopidae (Hemiptera, Cicadomorpha) from the Early Cretaceous of Northeastern China. *Acta Geologica Sinica (English Edition)* **88** (3): 725–729.

137 Wang, W.L. (1987). *Mesozoic Stratigraphy and Palaeontology of Western Liaoning III*. Beijing: Geological Publishing House.

138 Szwedo, J., Wang, B., and Zhang, H.C. (2011). An extraordinary Early Jurassic planthopper from Hunan (China) representing a new family Qiyangiricaniidae fam. nov. (Hemiptera, Fulgoromorpha, Fulgoroidea). *Acta Geologica Sinica-English Edition* **85** (4): 739–748.

139 Li, S., Wang, Y., Ren, D., and Pang, H. (2012). Revision of the genus *Sunotettigarcta* Hong, 1983 (Hemiptera, Tettigarctidae), with a new species from Daohugou, Inner Mongolia, China. *Alcheringa* **36** (4): 501–507.

17

Heteroptera – True Bugs

Sile Du[1], Shan Lin[1], Chungkun Shih[1,2], Dong Ren[1], and Yunzhi Yao[1]

[1] Capital Normal University, Haidian District, Beijing, China
[2] National Museum of Natural History, Smithsonian Institution, Washington, DC, USA

17.1 Introduction to Heteroptera

Heteroptera, commonly called "true bugs", are part of the successful radiation of non-holometabolous insects, comprising more than 40 000 known species worldwide today. They are generally treated as a suborder of the Hemiptera, and more than 80 extant families have been reported in all the Continents (except Antarctica) and in many islands. Their adaptabilities over long evolutionary time have resulted in extreme structural and biological diversity. According to the morphology and phylogeny, researchers divide Heteroptera into seven infraorders: Enicocephalomorpha, Dipsocoromorpha, Gerromorpha, Nepomorpha, Leptopodomorpha, Cimicomorpha, and Pentatomomorpha. However, the relationships among the seven infraorders of Heteroptera are not well-established [1, 2]. In addition, Coleorrhyncha, commonly known as "moss bugs" and treated as a suborder of Hemiptera, have some affinities with the Heteroptera, but they are distinct. We will include Coleorrhyncha in this chapter.

The diagnostic characters of Heteroptera are antenna with four to five segments; piercing-sucking mouthparts; pronotum and scutellum robust; forewings of most species with both membranous and corium portions (called hemelytra); with unique scent glands for defending natural enemies and attracting mates.

The feeding habits in Heteroptera are divided into four types, i.e. phytophagous, fungivorous, predaceous, and hematophagia. Most heteropterans are phytophagous, feeding on vegetative and reproductive organs, e.g. leaves, fruits, and seeds, etc. (Figures 17.1 and 17.2). They are often considered as pests of agriculture and forestry. Fungivorous taxa, such as flat bugs (Aradidae), feed on fungal hyphae. Predaceous species, often with a stout beak and grasping legs for preying, feed on other insects and/or other small arthropods (Figure 17.3). Hematophagia groups, such as Cimicidae, Triatominae, and Polyctenidae, suck blood from bats, birds and/or human.

Coleorrhyncha, usually known as "moss bugs", are often called "living fossils". The extant moss bugs are easily recognized by their strange appearance. Their bodies are flattened, generally ranging from 2 to 4 mm in length with broad shape. They have a strange head with widely separated eyes. The anterior or top surface of the wing is hardened and holds a network of veins and areolae or closed cells with different shapes. Most moss bugs have indurative forewings and do not possess hind wings, so they cannot fly. In addition, the geographical distribution of this group is also very specialized. Fossils are widely distributed in the Eurasian Continent and Australia, but the 36 species of the exclusive extant family are present in New Caledonia, New Zealand, southeastern of Australia and south of South America, mainly in the swamp area [3, 4].

Scent-Emitting

In the autumn with temperature dropping steadily, the leaves start to showcase their beautiful red and yellow colors before falling off. Under the influence of cold weather, the "brown marmorated stink bugs", *Halyomorpha halys* Stål, 1855 [5] in the family of Pentatomidae, living on plants (Figure 17.4), start their migration into houses in New Jersey, via any openings or cracks, which provide them suitable hiding places and warmth away from the freezing winter. This stink bug was accidentally introduced to the US from Asia and first officially reported from the western hemisphere in Allentown, Pennsylvania in 2001 [6]. After hibernation for the winter, they emerge from hiding in late spring, crawling on the walls or flying clumsily into light fixtures. When startled

Rhythms of Insect Evolution: Evidence from the Jurassic and Cretaceous in Northern China, First Edition.
Edited by Dong Ren, Chungkun Shih, Taiping Gao, Yongjie Wang, and Yunzhi Yao.

Figure 17.1 A "lychee giant shield bug", *Tessaratoma papillosa* (Tessaratomidae), providing maternal care of eggs and offspring. Source: Photos by Jason Shih.

Figure 17.2 A brilliantly colored nymph of the "lychee giant shield bug", *Tessaratoma papillosa* (Tessaratomidae). Source: Photo by Jason Shih.

or disturbed, they secrete dark-colored fluid and emit pungent odor from their scent glands. Some of them, escaping from the house, return to nature and resume their feeding and breeding activities.

As indicated by the common name of "stink bugs", these insects are well-equipped with many scent glands on the dorsal side of the abdomen and the ventral side of the thorax to emit the pungent odor. Both adults and nymphs produce strong smelling and irritating defensive chemicals for defense and repelling predators. Hexenyl, octenal, and hexenal acetate were found by scientists in secretions of bugs [7]. A 2006 study done at Clemson University has determined at least two of the compounds present in the stink bug odor: the aldehydes trans-2-decenal and trans-2-octenal. It was reported that coreid and pentatomid bugs defend themselves against small vertebrate and invertebrate predators by using their scent glands [8–14]. Some bugs even have the ability to spray pungent chemicals into the eyes of birds, though sometimes this is ineffective [15, 16].

Apart from defense against predators, the secretions from the scent glands in some aquatic and soil-inhabiting bugs are able to inhibit the growth of certain microorganisms. For instance, the secretions from metathoracic scent glands of *Scaptocoris divergens* can inhibit the growth of spores and mycelium of *Fusarium*, a pathogen of bananas that causes banana wilt [11]. In addition, the scent also acts as a sex pheromone for attracting mates. Sexual dimorphisms are demonstrated in gland structures and their secretions.

Blood-Feeding Bugs

The sanguinivorous behavior, i.e. blood-feeding, is a feeding strategy among modern insects [17]. There are four primary orders, i.e. Phthiraptera (lice), Siphonaptera (fleas, Chapter 25), Diptera (true flies, Chapter 23), and Heteroptera (true bugs), from which critical hematophages have arisen [18]. Key blood-feeding representatives of these four orders, e.g. sucking lice, fleas, mosquitoes, horseflies, bedbugs, etc., are well-known to many people. In addition, some blood-feeding heteropterans can feed on the blood of warm-blooded animals, while transmitting fatal viruses and other pathogens to humans.

The species of Cimicidae and Triatominae are sources of terror. The Cimicidae, commonly called "bed bugs", are a family of parasitic insects frequently discovered in beds. This family, including about 90 species, are mainly distributed in temperate and tropical areas. All cimicids are small, oval-shaped, and flat in appearance, but their

Figure 17.3 An "assassin bug", *Epidaus famulus* (Stål, 1863) (Reduviidae). Source: Photo by Jason Shih.

Figure 17.4 A "brown marmorated stink bug", *Halyomorpha halys*, in the garden. Source: Photo by Dr. Chungkun Shih.

bodies become bulged after feeding. They have a pair of small non-functional wing pads, but they cannot fly. With the beak-like mouthparts, they can pierce the skin and suck the blood of their hosts.

The Triatominae are a subfamily of Reduviidae, commonly known as "conenose bugs", "kissing bugs", or "vampire bugs". More than 130 species of this subfamily are hematophagous, and most of them feed on vertebrate blood; only a very few species feed on invertebrates [19, 20]. They are mainly widespread in the Americas, with a few species present in Asia, Africa, and Australia through migration. The species of Triatominae have a cone-shaped body about 25 mm long. Chagas disease, a serious tropical parasitic disease caused by the protozoan *Trypanosoma cruzi*, may be transmitted by every species of the Triatominae. Once acute infection goes into remission, patients may develop chronic disease, which in up to 30–40% of cases are characterized by cardiomyopathy, arrhythmias, megacolon, megaoesophagus, and, more rarely, polyneuropathy and stroke [21]. It is estimated that 6.6 million people, mostly in Mexico, Central America, and South America, have Chagas disease as of 2015. In 2015, Chagas was estimated to result in 8000 deaths.

Blood-feeding habits have evolved independently numerous times among arthropods [22, 23] and at least three times within the true bugs [24, 25]. To date, the earliest evidence of such feeding behavior among true bugs is the Torirostratidae Yao, Cai, Shih & Engel, 2014 from the Lower Cretaceous Yixian Formation in Northeastern China with two species: *Torirostratus pilosus* Yao, Shih & Engel, 2014 and *Flexicorpus acutirostratus* Yao, Cai & Engel, 2014 [26] (see Chapter 28.5). This discovery extends the geological record of hematophagous true bugs by approximately 30 million years and provides a new method for studying the fossil insects of blood-feeding.

17.2 Progress in the Studies of Fossil Heteroptera and Fossil Coleorrhyncha

The fossils of Heteroptera were first recorded by Fabricius, who was the first person to study the terrestrial Heteroptera and reported the first Reduviidae fossil species in 1775 and a Lygaeidae fossil species in 1794. In the early twentieth century, Handlirsch (1906–1908, 1939) [27, 28] made a lot of progress. Meanwhile, Tillyard [29] and Martynov [30] made a systematic study of extinct insects. These led to a qualitative leap in the research on fossil species of Heteroptera. New and productive localities of fossils have been found in Europe, Asia, North America, and Australia. Research scopes have been expanded from taxonomy and morphology to the systematics, phylogenetics, geographic history, paleoecology, etc.

In the later part of the twentieth century, the studies of Heteroptera fossils have made further progress. Yuri A. Popov, a renowned paleoentomologist from Russia, made great contributions to the taxonomy, morphology, phylogeny and evolution of fossil and extant hemipteran insects (for both Heteroptera and Coleorrhyncha). Since he published his first paper on fossil Heteroptera in 1961 until his death in November 2016, he studied fossils from the Mesozoic and Cenozoic around the world, and authored more than 170 papers. These covered all the superfamilies of extant and extinct species, and described more than 20 new families and subfamilies and over 300 new genera and species.

In China, the studies of Heteroptera fossils started more than 80 years ago, but with only a few reports published before 1980. In 1928, Chi Ping described the first Heteroptera fossil *Mesolygaeus laiyangensis* from Laiyang of Shandong [31]. Qibin Lin reported two species, *Karataviella pontoforma* and *Karataviella chinensiss*, from Liaoning, China [32]. Since then, more paleoentomologists have been involved in the research of Heteroptera fossils, especially from the Northeastern China. So far, more than 100 species, 86 genera in 23 families have been described in China. Fossil specimens with well-preserved morphological characters from China are essential to the study of early evolution of Heteroptera. During this period, significant contributions have been made by Chi Ping, Youchong Hong, Wenli Wang, Qibin Lin, Junfeng Zhang, Haichun Zhang, Dong Ren, Wanzhi Cai, Yunzhi Yao, Chungkun Shih, and Weiting Zhang.

The insects of Coleorrhyncha have existed and thrived since the Late Permian and they have been a special group of evolutionary significance [33]. The research of Coleorrhyncha fossils around the world has a history of about 100 years. In 1906, Handlirsch, for the first time, studied the insect fossils of the Early Jurassic in the Wainlode area of England and established two genera and two species (*Eocimex liasinus* and *Progonocimex jurassicus* Handlirsch, 1906), both of which were classified into the Heteroptera [27]. In 1926, Tillyard named a fossil *Actinoscytina belmontensis* Tillyard, 1926, which is the oldest fossil record of the Coleorrhyncha from the Late Permian of Australia [34]. Myers and China, in 1929, classified this group into Coleorrhyncha [35]. After 10 years, Handlirsch established two genera and four species from the fossils from Dobbertin of Germany in 1939 [28]. From the 1940s to the mid-1950s, there was no relevant literature reported. However, from the mid-late 1950s to the present, the research on fossil Coleorrhyncha ushered in a rapid development period. Evans [36], Becker-Migdisova [37], Wootton [38], Popov [39–46], Hong [47], Popov and Shcherbakov [33], Wang *et al.* [48] have carried out a lot of great work for the Coleorrhyncha fossils and described many genera and species. Popov has made the most significant contribution, describing 65 fossil species of 11 genera in detail, creating a solid foundation for the further study of Coleorrhyncha fossils. In 2009, Burckhardt used the morphological data to study the phylogenetic systematics of Peloridiidae which is the only extant family of Coleorrhyncha and described the characteristics of every extant species in detail [3]. At the same time, he also carried out some biogeographical research. To date, scholars from around the world have published nearly 30 papers on the classification of Coleorrhyncha fossils, including two families, 29 genera, and 100 species (including Chinese fossil species), which are distributed in eight countries.

Popov, a Renowned Paleoentomologist

Dr. Yuri A. Popov, a renowned paleoentomologist from Russia, made great contributions to the taxonomy, morphology, phylogeny and evolution of fossil and extant hemipteran insects (Heteroptera and Coleorrhyncha). From 1961 when he published his first paper on fossil Heteroptera to his death in November 2016, he studied fossils from the Mesozoic and Cenozoic around the world. He co-authored more than 170 papers covering all superfamilies of extant and extinct species and described more than 20 new families and subfamilies and over 300 new genera and species. Besides being an accomplished scientist, Yuri was a wonderful friend – always kind and considerate to us and generous and cooperative in helping our joint projects. We are very appreciative that he provided us with many literature papers and lots of guidance for our research of Heteroptera. We always enjoyed every opportunity to see him and work with him. We still remember his visit to our Capital Northern University (CNU) Laboratory in 2010 (Figure 17.5).

17.3 Representative Fossils of Heteroptera from Northern China

Infraorder "Gerromorpha" Popov, 1971

Superfamily Mesovelioidea Douglas & Scott, 1867

Family Mesoveliidae Douglas & Scott, 1867

The members of the Mesoveliidae are commonly known as "water treaders" or "pondweed bugs" [49]. All species are predators, feeding mainly on arthropods. They live in moist surroundings, either humid terrestrial settings, marginal aquatic habitats (litter, moss), or on

Figure 17.5 Prof. Ren (left), Prof. Popov (middle) and Prof. Yao (right) in the CNU Laboratory in August 2010. Source: Photo by Dr. Chungkun Shih.

water surface extensively covered with floating leaves of aquatic plants [49]. The family contain two subfamilies of Madeoveliinae Poisson, 1959 and Mesoveliinae Douglas & Scott, 1867 with 11 extant genera and 45 extant species [49–53].

However, only two unquestionable fossil species of mesoveliids have been reported: *Mesovelia dominicana* Garrouste & Nel, 2010 from the middle Miocene Dominican amber and *Sinovelia mega* Yao, Zhang & Ren, 2012 from the Early Cretaceous of China. Three other genera and species with equivocal relationships have been assigned to this family: *Duncanovelia extensa* Jell & Duncan, 1986, *Karanabis kiritschenkoi* Bekker-Migdisova, 1962 and *Engynabis tenuis* Bode, 1953 [54–58]. In addition, one undescribed specimen has been reported from the Lower Cretaceous (late Aptian or early Albian [59]), Crato Formation in Northeastern Brazil [60].

Only one genus included from the Cretaceous of Northern China: *Sinovelia* Yao, Zhang & Ren, 2012.

Sinovelia Yao, Zhang & Ren, 2012

Sinovelia Yao, Zhang & Ren, 2012, *Alcheringa*, 36, 108 [55] (original designation).

Type species: *Sinovelia mega* Yao, Zhang & Ren, 2012.

Head rounded triangular, with distinct median groove (or ridge) on dorsal head. Rostrum second segment longer than head. Antenna with four segments, length almost half of body. Pronotum without collar, with labial groove on thoracic sternum and the first segment of abdomen. Fore tarsus almost one-third of fore tibiae.

Distribution and age: Liaoning; Early Cretaceous.

Two species included from the Cretaceous of Northern China (see Table 17.1).

Sinovelia mega Yao, Zhang & Ren, 2012 (Figure 17.6)

Sinovelia mega Yao, Zhang & Ren, 2012: *Alcheringa*, 36, 108.

Locality and horizon: Huangbanjigou, Beipiao, Liaoning, China; Lower Cretaceous, Yixian Formation.

Rostrum second segment longer than head. Eyes far from anterior margin of pronotum. The fourth segment of antenna subequal to the third in length. The second segment of tarsus longest [55].

Infraorder "Leptopodomorpha" Popov, 1971

Family Saldidae Amyot & Serville, 1843

Saldidae, with about 335 extant species described [61], are the largest family in the Leptopodomorpha. Saldids have large and reniform compound eyes, costal fracture in corium and usually four or five closed cells in the membrane. Wing polymorphism is a frequent phenomenon in Saldidae [62]. Because saldids are usually found on rocky shores of streams and lakes [63, 64], they are commonly called "shore bugs". They are distributed

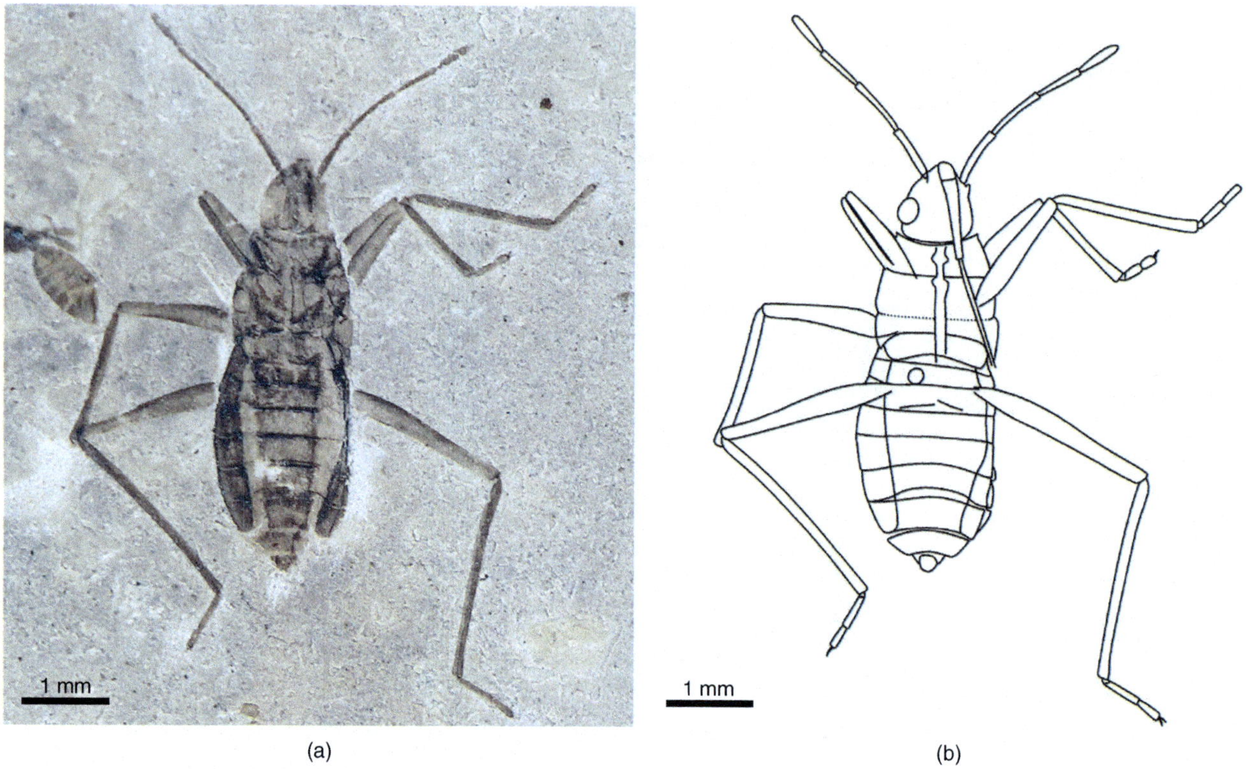

Figure 17.6 *Sinovelia mega* Yao, Zhang & Ren, 2012 (Holotype, CNU-HE-LB2010224p). (a) Photograph; (b) Line drawing [55].

worldwide, especially in the Northern Hemisphere [61, 65].

To date, 15 fossil species within nine genera have been reported. In the past decade, three fossil species are from the Lower Cretaceous of Yixian Formation in China: *Brevrimatus pulchalifer* Zhang, Yao & Ren, 2011; *Venustsalda locella* Zhang, Song, Yao & Ren, 2012 and *Luculentsalda maculosa* Zhang, Yao & Ren, 2013 [66–68]. With the locality moving to the north, Ryzhkova described two species from the Lower Cretaceous in Mongolia, *Ulanocoris femoralis* Ryzhkova, 2012 and *Ulanocoris grandis* Ryzhkova, 2012 [69]. After three years, he recorded three species, *Baissotea infanta* Ryzhkova, 2015, *B. peregrine* Ryzhkova, 2015, *B. popovi* Ryzhkova, 2015 from the Lower Cretaceous of Siberia and Mongolia [70]. Most of other fossil species are described from the Cenozoic, such as the Eocene species, *Salda exigua* Germar & Berendt, 1856 from Baltic amber; the Upper Oligocene species, *Oligosaldina rottensis* Statz & Wagner, 1950, *O. rhenana* Statz & Wagner, 1950, and *O. aquatilis* Statz & Wagner, 1950 from Germany; the Miocene species, *Propentacora froeschneri* Lewis, 1969 (= *Oreokora froeschneri*) from USA, and *Salda littoralis* Jessen, 1923 found in the recent Glacial clay [71–73].

Genera included from the Cretaceous of Northern China: *Brevrimatus* Zhang, Yao & Ren, 2011, *Venustsalda* Zhang, Song, Yao & Ren, 2012 and *Luculentsalda* Zhang, Yao & Ren, 2013.

Brevrimatus Zhang, Yao & Ren, 2011

Brevrimatus Zhang, Yao & Ren, 2011, *ZooKeys*, 130, 187 [66] (original designation).

Type species: *Brevrimatus pulchalifer* Zhang, Yao & Ren, 2011.

Head relatively short. Corium with large pale spots, medial fracture short, costal fracture of hemelytra very long, membrane with five closed cells. Posterior margin of female sternum VII concave along the midline. Base of ovipositor exposed.

Distribution and age: Duolun County, Inner Mongolia; Early Cretaceous.

Only one species included from the Cretaceous of Northern China (see Table 17.1).

Brevrimatus pulchalifer Zhang, Yao & Ren, 2011 (Figure 17.7)

Brevrimatus pulchalifer Zhang, Yao & Ren, 2011: *ZooKeys*, 130, 189.

Locality and horizon: Baitugou, Nanyingpan, Duolun, Inner Mongolia, China; Lower Cretaceous, Yixian Formation.

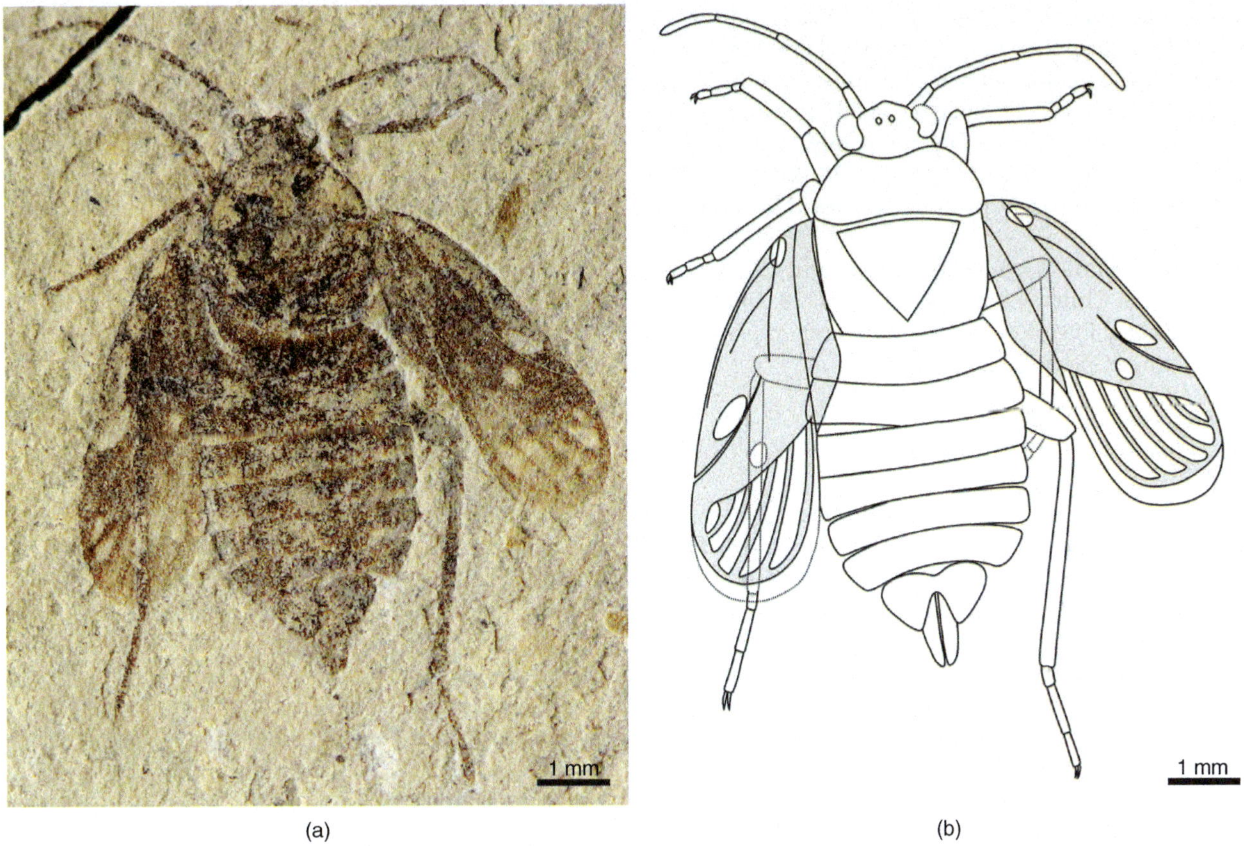

Figure 17.7 *Brevrimatus pulchalifer* Zhang, Yao & Ren, 2011 (Holotype, CNU-HE-ND2010334p). (a) Photograph; (b) Line drawing [66].

Body length 8.00 mm and width 3.18 mm. Head relatively short, length 0.84 mm and width 1.24 mm. The last segment of antenna slightly swollen. Corium with three large pale spots, medial fracture short, costal fracture of hemelytra very long; membrane with five cells, apex of the innermost cell of membrane extending past apex of the outermost cell. Posterior margin of female sternum VII extremely concave along the midline. Hemelytra with only slight modification for mating, the embolar region slightly thickened [66].

***Venustsalda* Zhang, Song, Yao & Ren, 2012**

Venustsalda Zhang, Song, Yao & Ren, 2012, *Zootaxa.* 3273, 64 [67] (original designation).

Type species: *Venustsalda locella* Zhang, Song, Yao & Ren, 2012.

Compound eyes large, covering most of head in lateral view. Posterior pronotal margin indented distinctly. Forewing medial fracture and costal fracture long; vein on corium reaching the outermost cell; membrane with six closed cells. In male, parandrium tapered.

Distribution and age: Liaoning; Early Cretaceous.

Only one species included from the Cretaceous of Northern China (see Table 17.1).

***Luculentsalda* Zhang, Yao & Ren, 2013**

Luculentsalda Zhang, Yao & Ren, 2013, *Zootaxa,* 3647 (4), 586 [68] (original designation).

Type species: *Luculentsalda maculosa* Zhang, Yao & Ren, 2013.

Ocelli present and widely separated. Pronotum with callus. Forewing costal fracture long, not connected with medial fracture, medial fracture short; membrane with five closed cells, the outermost cell almost as long as the adjacent cell. In male, abdomen VIII short.

Distribution and age: Liaoning; Early Cretaceous.

Only one species included from the Cretaceous of Northern China (see Table 17.1).

Family Archegocimicidae Handlirsch, 1906

The sizes of the Mesozoic Archegocimicidae range from 3 to 10 mm; body elongate ovoid; compound eyes protrusive, contiguous with pronotum; pronotum large, transverse, divided into anterior and posterior lobes; scutellum more or less broad, nearly equilateral; corium and membrane not distinctly separated; costal fracture present; R, Rs, M, and CuA generally with parallel branches running distally, forming several closed cells on membrane. Legs slender, tarsi relatively long, with two pretarsal claws. The Archegocimicidae comprise 26 genera and 36 species dating from the Early Jurassic to the Early Cretaceous, but some are exceptionally poorly preserved and their placements should be considered as dubious. Fossils in this family are mainly found in Germany, Russia, the UK, and China [27, 28, 43, 46, 58, 74], and they were apparently quite widespread by the Early Jurassic [45].

Genera included from the Jurassic and Cretaceous of Northern China: *Mesolygaeus* Ping, 1928, *Longianteclypea* Zhang, Engel, Yao & Ren, 2014 and *Propritergum* Zhang, Engel, Yao & Ren, 2014.

Mesolygaeus Ping, 1928

Mesolygaeus Ping, 1928, *Paleontol. Sin. Series B*, 13 (1), 5–51 [31] (original designation).

Enicocoris Popov, 1980, *Trans. joint Soviet-Mongolian Paleontol. Exped.*, 13, 48–51 [75]; Popov, 1986, *Proc. Joint Sov. Mongol. Paleont. Exped.*, 28, 47–84 [42]. Syn. by Hong & Wang, 1990, Fossil insects from the Laiyang Formation, 89–105 [76].

Sinolygaeus Hong, 1980, 137 (type species: *Sinolygaeus naevius* Hong, 1980), in Wang, 1980, *Paleontological Atlas of North East China Part 2*. 130–153 [77]. Syn. by Zhang, Engel, Yao, Ren & Shih., 2014. *J. Syst. Palaeontol.*, 12 (1), 93–111 [78].

Xishania Hong, 1981, *Tianjin Inst. Geol. & Miner. Resour. Bull.*, 4, 87–94 [79]; Hong, 1984, Insecta, 128–185 [80]. Syn. by Hong & Wang, 1990, Fossil insects from the Laiyang Formation, 89–105 [76].

Jiaodongia Hong, 1984, *Prof. Pap. Stratigr. Palaeontol.*, 11, 31–34 [81]. Syn. by Zhang, 1991, *Acta Palaeontol. Sin.*, 30, 679–704 [82].

Type species: *Mesolygaeus laiyangensis* Ping, 1928.

A deep transverse depression separated anterior and posterior lobes of pronotum, lateral margins of anterior lobe straight, parallel or slightly divergent posteriorly, anterior angles feebly rounded, posterior margin of pronotum convex. Membrane on forewing with five cells.

Distribution and age: Shandong, Beijing, Gansu, Liaoning, Inner Mongolia; Early Cretaceous.

Two species included from the Cretaceous of Northern China (see Table 17.1).

Longianteclypea Zhang, Engel, Yao & Ren, 2014

Longianteclypea Zhang, Engel, Yao & Ren, 2014, *J. Syst. Palaeontol.*, 12(1), 103 [78] (original designation).

Type species: *Enicocoris tibialis* Popov, 1986.

Anteclypeus and paraclypeus elongate. Anterior lob of pronotum much narrower than posterior lobe, with distinctly concave; membrane on forewing with four cells. Connexiva present.

Distribution and age: Liaoning; Early Cretaceous.

Only one species included from the Cretaceous of Northern China (see Table 17.1).

Longianteclypea tibialis (Popov, 1986) (Figure 17.8)

Longianteclypea tibialis Zhang, Engel, Yao & Ren, 2014: *J. Syst. Palaeontol.*, 12 (1), 104.

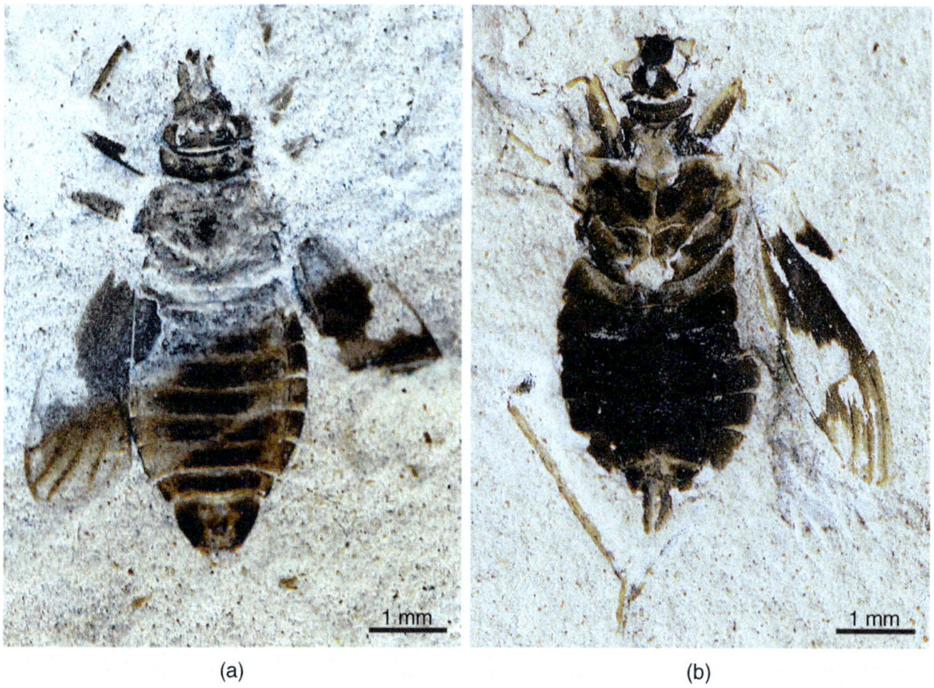

Figure 17.8 *Longianteclypea tibialis* (Popov, 1986) (a) Photograph of CNU-HET-LB2010331, male. (b) Photograph of CNU-HET-LB2010332, female [78].

Enicocoris tibialis Popov, 1986, *Trans. joint Soviet-Mongolian Paleontol. Exped,* 28, 47–84 [42]. Syn. by Zhang, Engel, Yao, Ren & Shih, 2014. *J. Syst. Palaeontol.*, 12 (1), 93–111 [78].

Mesolygaeus laiyangensis Ping, 1928, *Paleontol. Sin. Series B,* 13 (1), 5–51 [31]; Zhang, 1991, *Acta Palaeontol. Sin.*, 30, 679–704 [82] (erroneous synonymy).

Mesolygaeus laiyangensis Ping, 1928, *Paleontol. Sin. Series B,* 13 (1), 5–51 [31]; Hong, 1995, *Acta Geol. Gansu,* 4 (1), 1–13 [83] (erroneous synonymy).

Locality and horizon: Huangbanjigou, Beipiao, Liaoning, China; Lower Cretaceous, Yixian Formation.

This species was previously placed in *Enicocoris*, a synonym of *Mesolygaeus*. The species differs greatly from *Mesolygaeus* as diagnosed below, thus, it was transferred in 2014 to *Longianteclypea* [78].

Body elongate oval, length 6.34 mm and width 2.37 mm. Head triangular, anteclypeus and paraclypeus extremely long, anteclypeus length 0.40 mm, paraclypeus length 0.37 mm. Compound eyes relatively occupying nearly one-half sides of head, moderately protrusive, contiguous with anterior margin of pronotum. Pronotum divided into distinct anterior and posterior lobes; anterior margin of anterior lobe distinctly concave, anterior angles acute, lateral margins slightly converging posteriorly; posterior lobe trapezoidal, posterior margin distinctly concave. Scutellum large, nearly triangular, longer than pronotum on midline. Male genital segment semicircular; female with ovipositor elongate [78].

Propritergum Zhang, Engel, Yao & Ren, 2014

Propritergum Zhang, Engel, Yao & Ren, 2014, *ZooKeys*, 130, 102 [78] (original designation).

Type species: *Propritergum opimum* Zhang, Engel, Yao & Ren, 2014.

Pronotum divided into anterior and posterior lobes, margin of pronotum distinctly explanate. Forewing costal fracture very long; membrane with cells. Connexiva absent.

Distribution and age: Liaoning; Early Cretaceous.

Only one species included from the Cretaceous of Northern China (see Table 17.1).

Propritergum opimum Zhang, Engel, Yao & Ren, 2014 (Figure 17.9)

Propritergum opimum Zhang, Engel, Yao & Ren, 2014: *J. Syst. Palaeontol.*, 12 (1), 10.

Locality and horizon: Huangbanjigou, Beipiao, Liaoning, China; Lower Cretaceous, Yixian Formation.

Body ovoid, length 5.87 mm and width 2.79 mm. Head broad, length 0.75 mm and width 1.23 mm. Antennal segment I obviously thickened, segment II longest. Compound eyes occupying nearly entire sides of head. Pronotum trapezoidal, margins explanate, anterior lobe occupying most part of pronotum, anterior margin of pronotum slightly concave, posterior margin convex. Scutellum large. Forewing reaching end of abdomen; costal fracture long, reaching middle of corium; membrane with cells. Metatibia with a row of stout setae on lateral margin, metatarsomere I very short, metatarsomere II almost subequal to metatarsomere III in length. Female genitalia triangular, ovipositor long [78].

Phylogenetic Study of Archegocimicidae

Although previous researches have demonstrated the monophyly of Leptopodomorpha [1, 84–86, 144], the systematic position of Archegocimicidae is still controversial. In order to clarify the relationships among the families in Leptopodomorpha, we used 45 morphological characters and expanded previous treatment of the clade to include living and fossil representatives of all of the principal groups within the Infraorder. Based on the result shown in Figure 17.10, the monophyly of Leptopodomorpha was reconfirmed, Saldoidea and Letopodoidea were recognized as monophyletic sister clades, the former containing Saldidae, Aepophilidae, and Archegocimicidae and the latter including Palaeoleptidae, Omaniidae, and Leptopodidae. However, the sister relationship between Aepophilidae and Archegocimicidae is weakly supported in our result, therefore, the relationships of Aepophilidae, Archegocimicidae, and Saldidae are still contentious [78].

Infraorder "Nepomorpha" Popov, 1968

Superfamily Ochteroidea Kirkaldy, 1906

Family Ochteridae Kirkaldy, 1906

Ochteridae are a small family with three extant genera and 55 species, distributed in the tropical and warm-temperate regions of the world [88, 89]. These bugs are usually found along the shores of ponds or streams and feed on soil microfauna of the littoral zone [90]. Looking like Saldidae, they have long and filiform antennae, head moderately transverse and eyes sessile. There are only five described fossil ochterid species: *Propreocoris maculatus* Popov, Dolling & Whalley, 1994, from the Lower Jurassic (Lower Lias) Charmouth fauna in England, and four other species from the Lower Cretaceous Yixian Formation in China [91, 92].

Genera included from the Cretaceous of Northern China: *Pristinochterus* Yao, Cai & Ren, 2007, *Floricaudus* Yao, Ren & Shih, 2011 and *Angulochterus* Yao, Zhang & Ren, 2011.

Pristinochterus Yao, Cai & Ren, 2007

Pristinochterus Yao, Cai & Ren, 2007, *Eur. J. Entomol.*, 104, 827 [91] (original designation).

Type species: *Pristinochterus zhangi* Yao, Cai & Ren, 2007.

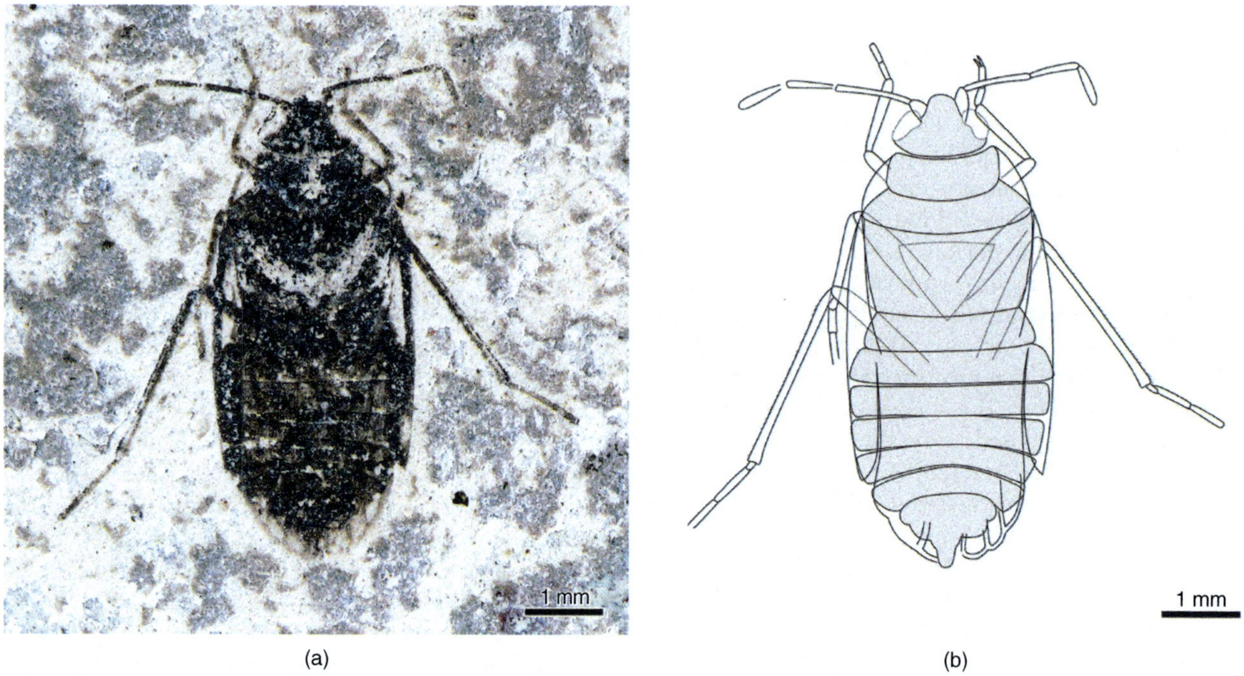

Figure 17.9 *Propritergum opimus* Zhang, Engel, Yao & Ren, 2012 (Holotype, CNU-HET-LB2006653). (a) Photograph; (b) Line drawing [78].

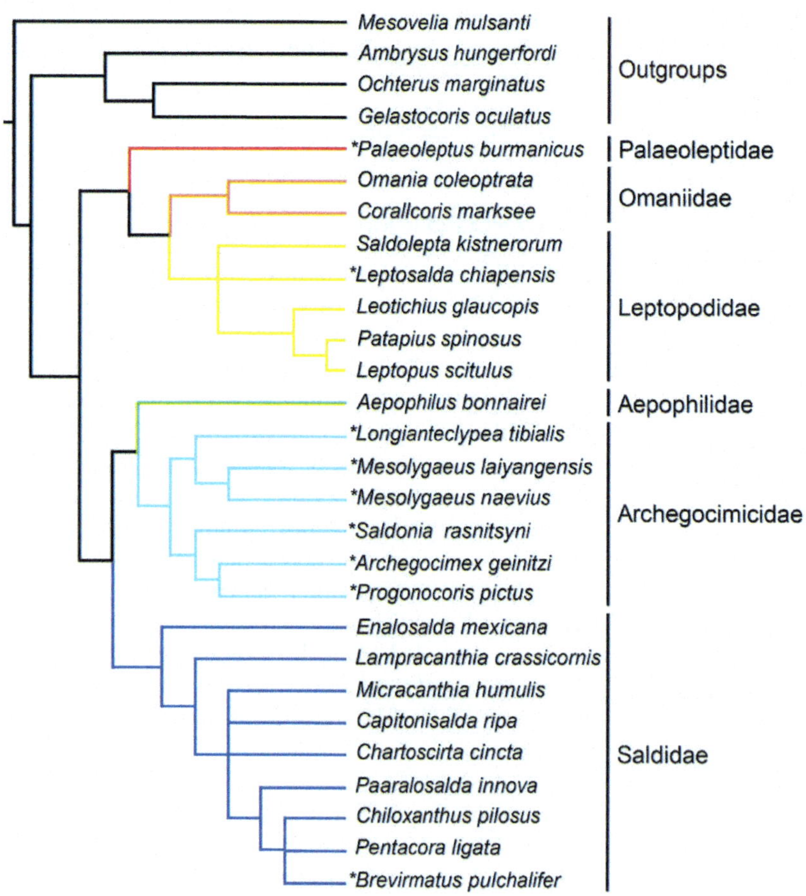

Figure 17.10 Phylogeny of Infraorder Leptopodomorpha (modified from [78]). Strict consensus tree yielded by NONA.

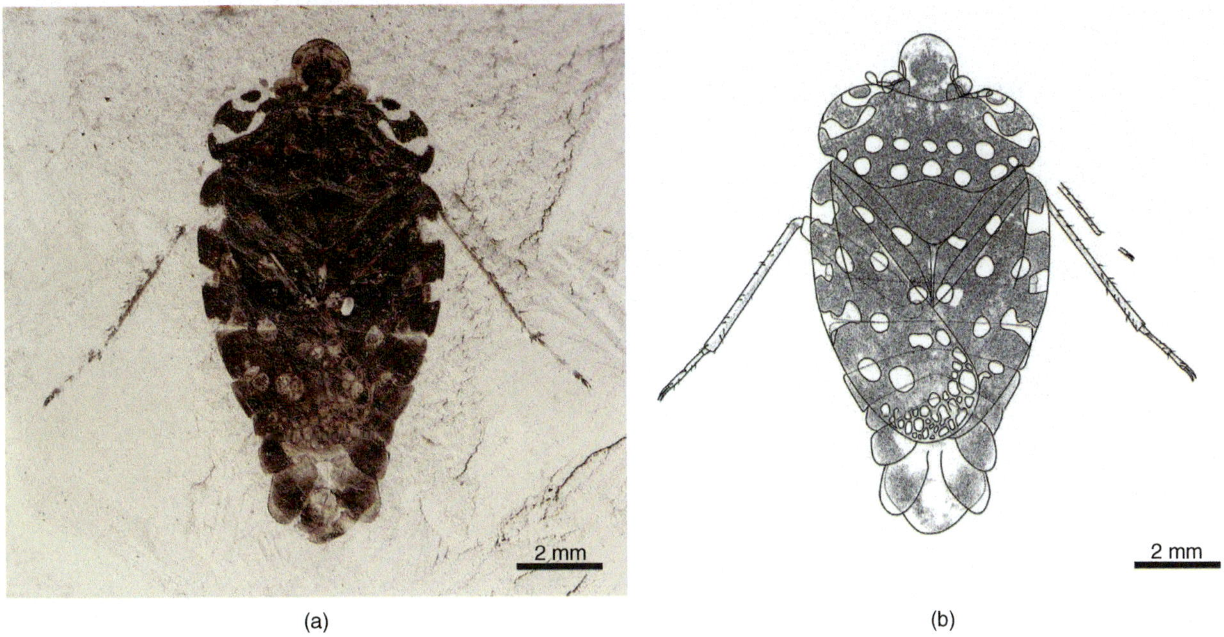

Figure 17.11 *Floricaudus multilocellus* Yao, Ren & Shih, 2011 (Holotype, male, CNU-HE-LB2008006). (a) Photograph; (b) Line drawing. Source: Donated by Dr. Chungkun Shih [92].

The specific epithet is in honor of Lin Zhang for assistance and contribution in collecting Liaoning fossils. Body large, lengths ranging 10–14.3 mm. Antenna four-segmented, basal two segments slender, apical two segments stout, all segments subequal in length. Eye relatively small, diameter shorter than half of head width. Lateral margins of pronotum strongly expanded. Costal fracture absent, membrane with about 30 cells. Tibiae with setae and stout spines, spines not placed in rows, tarsal segments 2-2-3.

Distribution and age: Liaoning, Inner Mongolia; Early Cretaceous.

Two species included from the Jurassic and Cretaceous of Northern China (see Table 17.1).

Floricaudus Yao, Ren & Shih, 2011

Floricaudus Yao, Zhang, Ren & Shih, 2011, *Syst. Entomol.*, 36, 593 [92] (original designation).

Type species: *Floricaudus multilocellus* Yao, Ren & Shih, 2011.

Two hook-like posterior angles on pronotum. Apex of abdomen exposed, not covered by hemelytra, lateral margins of sterna VII, VIII, and IX in males flat and strongly expanded.

Distribution and age: Liaoning; Early Cretaceous.

Only one species included from the Cretaceous of Northern China (see Table 17.1).

Floricaudus multilocellus Yao, Ren & Shih, 2011 (Figure 17.11)

Floricaudus multilocellus Yao, Ren & Shih, 2011: *Syst. Entomol.*, 36, 593.

Locality and horizon: Huangbanjigou, Beipiao, Liaoning, China; Lower Cretaceous, Yixian Formation.

Body length 12.6 mm, twice of width. Pronotum with two hook-like posterior angles. Hemelytron not reaching tip of abdomen, with a deep median fracture, membrane with over 30 cells [92].

Angulochterus Yao, Zhang & Ren, 2011

Angulochterus Yao, Zhang & Ren, 2011, *Syst. Entomol.*, 36, 595 [92] (original designation).

Type species: *Angulochterus quatrimaculatus* Yao, Zhang & Ren, 2011.

Anterior margin of head foliate-like. Eyes large, contiguous to anterior margin of pronotum; rostrum extending beyond mid coxae. Pronotum and corium punctate, pronotum with two hook-like posterior angles; tibiae spines not placed in rows. Corium with deep median fracture and costal fracture, veins on membrane far from apex of forewing. Male abdominal segments distal of segments VI and VII symmetrical.

Distribution and age: Liaoning; Early Cretaceous.

Only one species included from the Cretaceous of Northern China (see Table 17.1).

Angulochterus quatrimaculatus Yao, Zhang & Ren, 2011 (Figure 17.12)

Angulochterus quatrimaculatus Yao, Zhang & Ren, 2011: *Syst. Entomol.*, 36, 595.

Locality and horizon: Huangbanjigou, Beipiao, Liaoning, China; Lower Cretaceous, Yixian Formation.

Body length 8.2 mm. Two markings near anterior angles of pronotum, four light markings arranged in a

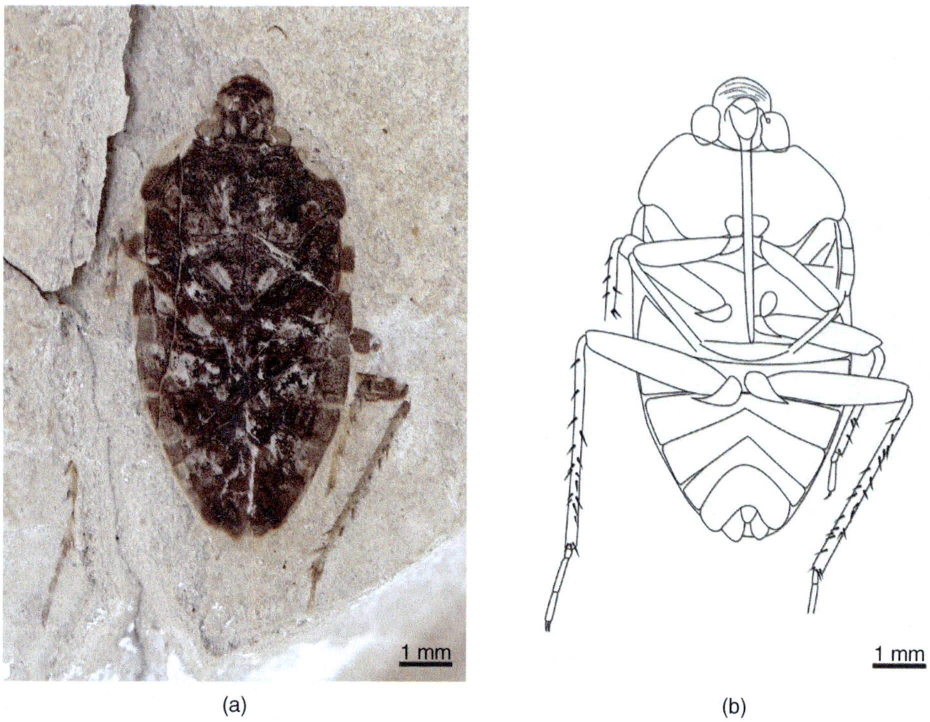

Figure 17.12 *Angulochterus quatrimaculatus* Yao, Zhang & Ren, 2011 (Holotype, male, CNU-HE-LB2006644p). (a) Photograph; (b) Line drawing [92].

row on posterior part of pronotum. Hemelytron with six markings. Each side of scutellum with two markings. Tibiae with dense spines, spines shorter than diameter of tibia. Hemelytron reaching tip of abdomen [92].

Family Naucoridae Fallen, 1814

Naucoridae, commonly called "creeping water bugs", are a cosmopolitan family, especially in Neotropical and Oriental Regions [97]. The extant naucorids comprise 40 genera and 395 species [89], most prefer to live in rapidly moving fresh water and only some live in ponds or other still waters. Naucoridae can be recognized by the following characters: antenna short, not extending beyond head; rostrum short and thick, not surpassing prosterum, and the anteocular portion of head only slightly produced in front of eyes; the tarsus of fore leg with one or two segments [89]. To date, 19 genera and 22 species of fossil Naucoridae [27, 41, 42, 93–96] have been reported, ranging from the Early Cretaceous to the Cenozoic. Only two species have been described from the Yixian Formation in China [93].

Genera included from the Cretaceous of Northern China: *Exilcrus* Zhang, Yao & Ren, 2011 and *Miroculus* Zhang, Yao & Ren, 2011.

Exilcrus Zhang, Yao & Ren, 2011

Exilcrus Zhang, Yao & Ren, 2011, *Acta Geol. Sin-Engl.*, 85 (2), 492 [93] (original designation).

Type species: *Exilcrus cameriferus* Zhang, Yao & Ren, 2011.

Rostrum not extending to the posterior margin of prosternum. Pronotum anterior and lateral margins nearly straight, posterior margin distinctly convex. Corium spotted and with distinct large cells.

Distribution and age: Liaoning; Early Cretaceous.

Only one species included from the Cretaceous of Northern China (see Table 17.1).

Miroculus Zhang, Yao & Ren, 2011

Miroculus Zhang, Yao & Ren, 2011, *Acta Geol. Sin.-Engl.*, 85 (2), 495 [93] (original designation).

Type species: *Miroculus laticephlus* Zhang, Yao & Ren, 2011.

Rostrum short and stout. Anterior of pronotum concave; lateral margins slightly convex, posterior margin straight. Scutellum transverse triangular.

Distribution and age: Liaoning; Early Cretaceous.

Only one species included from the Cretaceous of Northern China (see Table 17.1).

Miroculus laticephlus Zhang, Yao & Ren, 2011 (Figure 17.13)

Miroculus laticephlus Zhang, Yao & Ren, 2011: *Acta Geol. Sin.-Engl.*, 85 (2), 496.

Locality and horizon: Baitugou, Nanyingpan, Duolun, Inner Mongolia, China; Lower Cretaceous, Yixian Formation.

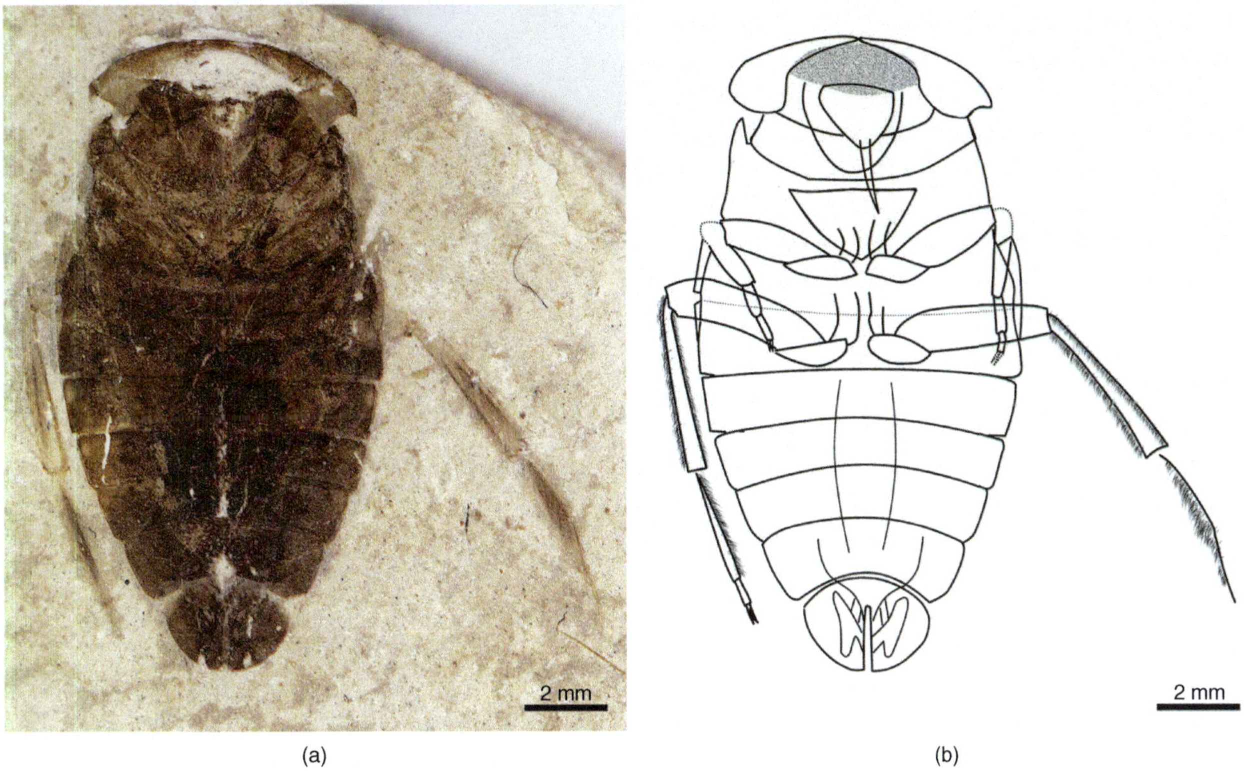

Figure 17.13 *Miroculus laticephlus* Zhang, Yao, Ren & Zhao, 2011 (Holotype, CNU-HE-LB2006692). (a) Photograph; (b) Line drawing [93].

Body ovate, length 14.2 mm, maximum width 7.7 mm. Head short and broad, width 6.3 mm and length 1.9 mm, eyes extended, extremely convergent anteriorly. Rostrum reaching posterior margin of prothorax. Pronotum distinctly transverse, shorter than head. Scutellum triangular, length 1.5 mm and width 3.0 mm. Corium spotted. Mid leg relatively short, femur 2.3 mm long, tibia 1.8 mm long, and tarsus, 1.4 mm [93].

Family Notonectidae Leach, 1815

Notonectidae are considered as the second most diverse family in Nepomorpha, just inferior to Corixidae [97]. There are more than 400 extant species in two subfamilies, Notonectinae and Anisopinae. Notonectidae, a worldwide group, are well-represented in temperate and tropical regions [89]. Notonectidae species, commonly called "backswimmers," swim on their "backs". Atmospheric air is taken up posteroventrally by the insects, making them lighter than water [98]. They float passively on the water surface unless they swim underwater or anchor themselves to submerged vegetation [63]. Backswimmers are predaceous, feeding generally on small aquatic arthropods such as chironomids, mosquito larvae or anything they can overpower [99, 100].

Up to now, there are five genera with seven species described from the Mesozoic of Kazakhstan, Russia, Argentina, Germany, and China. In China, the reported specimens have been respectively collected from the Fangyan Formation in Zhejiang [101], Chijinqiao Formation in Gansu [102], Laiyang Formation in Shandong [76], and Yixian Formation in Liaoning [103]. Three fossil species *Notonecta xyphiale* [104, 105], *Tarsonecta mecopoda* Zhang, 1989, and *Notonectopsis sinica* Hong & Wang, 1990 have been described from China. Among them, *Notonecta xyphiale* was recorded from the Upper Jurassic to Lower Cretaceous Chijinqiao Formation [106], Yingcheng Formation [107], Yixian Formation [80], and Laiyang Formation [76, 152]. *Notonectopsis sinica* was recovered from the Lower Cretaceous Laiyang Formation of Shandong Province [76].

Only one genus included from the Cretaceous of Northern China: *Notonecta* Linnaeus, 1758.

Notonecta Linnaeus, 1758

Notonecta Linnaeus, 1758, *Entomotaxonomia*, 3: 198 [109] (original designation).

Type species: *Notonecta glauca* Linnaeus, 1758.

All tarsi two-segmented; with two claws on fore and middle tarsi; claws of the hind legs greatly reduced and inconspicuous. Venter of abdomen concave with median keel; keel and ridges fringed with setae forming air chambers; abdominal apex symmetrical.

Distribution and age: Gansu, Shandong, Hebei of China, Early Cretaceous; Russia, Late Jurassic.

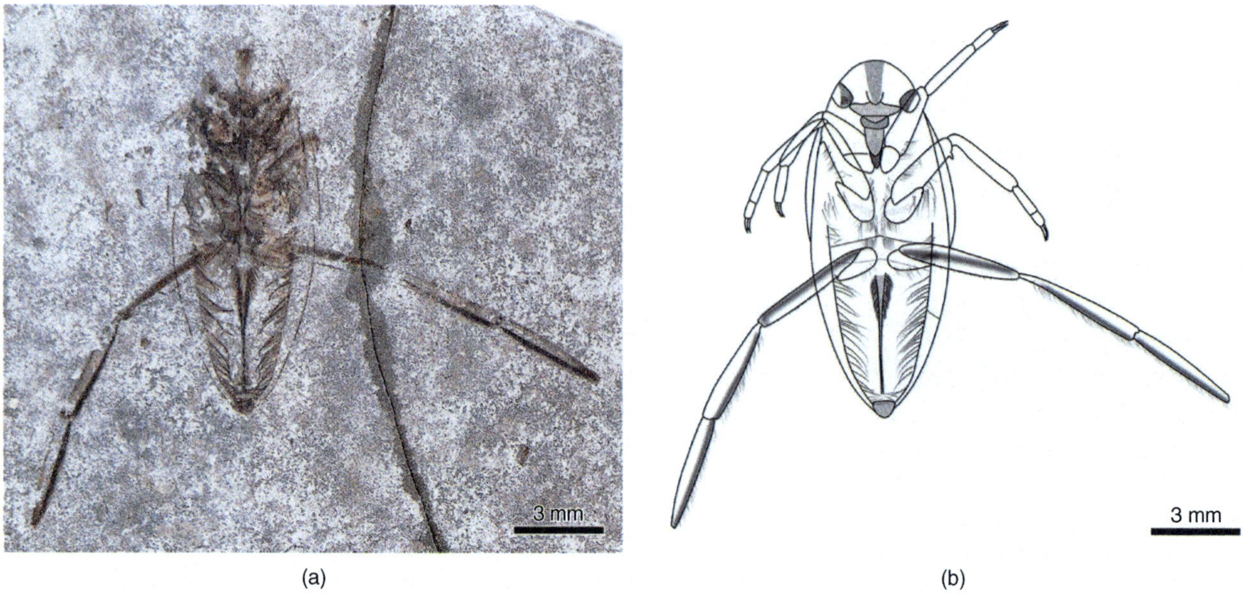

Figure 17.14 *Notonecta vetula* Zhang, Yao & Ren, 2012 (Holotype, CNU-HET-GY2010001). (a) Photograph; (b) Line drawing [103].

Two species included from the Jurassic and Cretaceous of Northern China (see Table 17.1).

***Notonecta vetula* Zhang, Yao & Ren, 2012 (Figure 17.14)**
Notonecta vetula Zhang, Yao & Ren, 2012: *Alcheringa*, 36 (2), 240.

Locality and horizon: Yumen City, Gansu, China; Upper Jurassic, Chijinqiao Formation.

Body concave ventrally. Head with vertex protruding slightly, Median carina on the frons of the fifth-instar broad, eyes relatively small, widely separated. Labrum apex rounded. Abdomen venter concave, with median keel extending to visceral segment. Keel and ridges fringed with setae, forming air chambers. Terminal abdomen relatively narrow and round. Abdominal apex symmetrical [103].

Family Corixidae Leach, 1815

Corixidae, commonly called "water boatmen" because of their oar-like hind legs, are the most ancient group and the most speciose of Nepomorpha. Their body sizes range from 2.5 to 15 mm. They have general appearance of flattened backswimmers, but they swim with their dorsal sides up [89]. Most species are known to live in stagnant fresh waters, such as ponds, lakes, and water reservoirs, but a few on saline waters [110–112]. Popov made a great contribution to the classification of fossil Corixidae and erected five extinct subfamilies: Archaecorixinae, Ijannectinae, Velocorixinae, Corixonectinae, and Diapherininae [42, 43, 94, 95]. He also erected the related Jurassic extinct family Shurabellidae [95]. Sixty-seven fossil species of Corixidae have been documented from the Late Triassic to the Oligocene, among which 18 genera and 25 species from China.

Genera included from the Jurassic and Cretaceous of Northern China: *Karataviella* Becker-Migdisove, 1949, *Sigarella* Popov, 1971, *Daohugocorixa* Zhang, 2010 and *Jiulongshancorixa* Zhang, 2010.

***Karataviella* Becker-Migdisove, 1949**
Karataviella Becker-Migdisove, 1949, *Tr. Paleontol. Inst. Akad. Nauk*. SSSR 22, 1–68 [113] (original designation).

Type species: *Karataviella brachyptera* Becker-Migdisove, 1949.

Compound eyes large and divided, occupying the main part of the head. Pronotum large, apical angles blunt rounded. Scutellum broad, triangular. Membrane of forewing small without pattern. Hind leg with basitarsus thin and relatively short, about equal to the length of hind tibia, fringed with long setae; the second tarsomere about half as long as the tarsus or shorter.

Distribution and age: Liaoning, Hebei, and Shandong, Early Cretaceous; Inner Mongolia; Middle Jurassic.

Five species included from the Jurassic and Cretaceous of Northern China (see Table 17.1).

***Sigarella* Popov, 1971**
Sigarella Popov, 1971, *Trudy Paleontologicheskogo Instituta Akademii Nauk SSSR, Moscow*. 139, 89–105 [95] (in Russian) (original designation).

Type species: *Corixe florissantella* Cockerell, 1906.

Body brown. Pronotum large with a median carina. Vein R fused with M forming R + M, with Cu forming

R + M + Cu. Scutellum small; claval with an anal vein. Corial and scutellum distinctly boundary without membrane. Abdomen with eight segments, asymmetrical.

Distribution and age: Colorado; Oligocene; Shandong of China; Early Cretaceous.

Only one species included from the Cretaceous of Northern China (see Table 17.1).

Daohugocorixa Zhang, 2010

Daohugocorixa Zhang, 2010, *Paleontol. J.*, 44 (5), 520 [114] (original designation).

Type species: *Daohugocorixa vulcanica* Zhang, 2010.

Head equal to or slightly shorter than pronotum. Interocular length much wider than eye; posterior margin of pronotum clearly convex, with lateral angles rounded, long transversed and arched wrinkles near the middle. Costal fracture present, overlapping in repose, very narrow of hemelytral membrane areas.

Distribution and age: Inner Mongolia; Middle Jurassic.

Only one species included from the Jurassic of Northern China (see Table 17.1).

Jiulongshanocorixa Zhang, 2010

Jiulongshanocorixa Zhang, 2010, *Paleontol. J.*, 44 (5), 519 [114] (original designation).

Type species: *Jiulongshanocorixa vulcanica* Zhang, 2010.

Head longer than pronotum; interocular length much wider than eye. Pronotum obtusely triangular, with lateral angles, without wrinkles, with long carina closing to anterior and lateral margins. Costal fracture present. Parts of hemelytral membrane overlapped opposite wing in repose, very narrow. Claval commissure more than twice as long as scutellum.

Distribution and age: Hebei; Early Cretaceous.

Only one species included from the Cretaceous of Northern China (see Table 17.1).

Infraorder "Cimicomorpha" Leston, Pendergrast & Southwood, 1954

Family Torirostratidae Yao, Cai, Shih & Engel, 2014

Torirostratidae are an extinct family erected based on fossils from the Lower Cretaceous Yixian Formation in Northeastern China. The fossil specimens of this family often have these diagnostic characters: mouthparts are preserved with the rostra laterally oriented or extended anteriorly, rostrum four-segmented and the second segment longest with base swollen. By applying geochemical methods for determining their feeding diets and combining taphonomic and morphological data, Yao *et al.*, in 2014, demonstrated that these species of Torirostratidae represent the earliest evidence of blood feeding among true bugs, and extend the geological record of such lineages by about 30 million years [26]. In addition, these records enhance the ecological and phylogenetic diversity of blood-feeding insects during the Early Cretaceous, expanding our knowledge of paleoecological associations in these ancient and unknown ecosystems (see Section 28.5).

Genera included from the Jurassic and Cretaceous of Northern China: *Torirostratus* Yao, Shih & Engel, 2014 and *Flexicorpus* Yao, Cai & Engel, 2014.

Torirostratus Yao, Shih & Engel, 2014

Torirostratus Yao, Shih & Engel, 2014, *Curr Biol.*, 24, 1786 [26] (original designation).

Type species: *Torirostratus pilosus* Yao, Shih & Engel, 2014.

Body length over 12 mm. Rostrum extending beyond procoxae. Apex of corium is finger-like, cubital ridge situated along claval suture, contiguous to apex of clavaus. Width of abdomen as long as pronotum, the fourth segment widest.

Distribution and age: Liaoning, Inner Mongolia; Hebei; Early Cretaceous.

Only one species included from the Cretaceous of Northern China (see Table 17.1).

Torirostratus pilosus Yao, Shih & Engel, 2014 (Figure 17.15)

Torirostratus pilosus Yao, Shih & Engel, 2014: *Curr. Biol.*, 24, 1786.

Locality and horizon: Huangbanjigou, Beipiao, Liaoning; Dawangzhangzi, Lingyuan, Liaoning; Liutiaogou, Dashuangmiao, Ningcheng, Chifeng, Inner Mongolia; Shimen, Yangshuling, Pingquan, Hebei; All in China; Lower Cretaceous, Yixian Formation.

Body about three times as long as wide. Head length longer than width; anteocular section longer than postocular; interocular space longer than diameter of compound eye in dorsal view, interocellar space as long as diameter of ocellus. Connexivum broad, with black square symbols at posterior angles of segments III–V and a cuneiform symbol between segments VI and VII [26].

Flexicorpus Yao, Cai & Engel, 2014

Flexicorpus Yao, Cai & Engel, 2014, *Curr. Biol.*, 24, 1786 [26] (original designation).

Type species: *Flexicorpus acutirostratus* Yao, Cai & Engel, 2014.

Body length less than 10 mm. Rostrum extending to procoxae. Ocelli large, diameter of ocellus slightly wider than interocellar space. Ovipositor extending through the last two abdominal segments.

Distribution and age: Liaoning; Early Cretaceous.

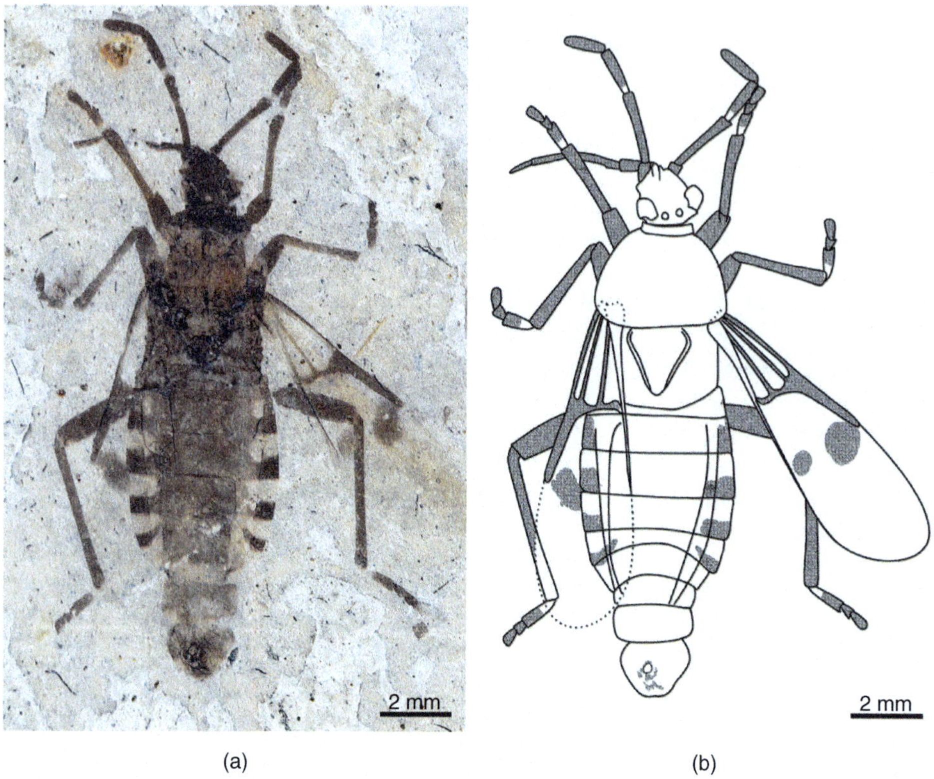

Figure 17.15 *Torirostratus pilosus* Yao, Shih & Engel, 2014 (Holotype, CNU-HE-LB2006506p). (a) Photograph; (b) Line drawing. Source: Donated by Dr. Chungkun Shih [26].

Only one species included from the Cretaceous of Northern China (see Table 17.1).

Superfamily Cimicoidea Stephen, 1829

Family Vetanthocoridae Yao, Cai & Ren, 2006

Vetanthocoridae, an extinct family, are important supplement to Anthocoridae ("flower bugs") records and provide new impetus for studying their origin and evolution. They have these characters: body lengths range from 4.3 to 13.7 mm; rostrum four-segmented, not reaching base of abdomen; antenna four-segmented; pronotum trapezoidal, hind margin concave; abdomen with connexivum; and ovipositor long, extending through the last two abdominal segments. Vetanthocoridae contain two tribes: Vetanthocorini and Crassicerini, with 12 genera and 16 species, among which 13 species from the Lower Cretaceous Yixian Formation in China; two species from the Middle Jurassic of Jiulongshan Formation in China, which are the oldest fossil records of Vetanthocoridae. Also, *Liaoxia longa* Hong, 1987 from Shahai Formation in Lower Cretaceous was transferred to this family in 2006 [115–120].

Genera included from the Jurassic and Cretaceous of Northern China: *Vetanthocoris* Yao, Cai & Ren, 2006, *Collivetanthocoris* Yao, Cai & Ren, 2006, *Byssoidecerus* Yao, Cai & Ren, 2006, *Mecopodus* Yao, Cai & Ren, 2006, *Curticerus* Yao, Cai & Ren, 2006, *Pustulithoracalis* Yao, Cai & Ren, 2006, *Curvicaudus* Yao, Cai & Ren, 2006, *Crassicerus* Yao, Cai & Ren, 2006, *Pumilanthocoris* Hou, Yao & Ren, 2012, *Longilanceolatus* Tang, Yao & Ren, 2015 and *Punctivetanthocoris* Tang, Yao & Ren, 2017.

Vetanthocoris Yao, Cai & Ren, 2006

Vetanthocoris Yao, Cai & Ren, 2006, *Zootaxa*, 1360, 8 [116] (original designation).

Type species: *Vetanthocoris decorus* Yao, Cai & Ren, 2006.

Pronotum lacking collar, about three times as wide as long. Hemelytra at most only slightly extending beyond apex of abdomen. Pronotum shorter than abdomen.

Distribution and age: Liaoning; Early Cretaceous.

Two species included from the Cretaceous of Northern China (see Table 17.1).

Collivetanthocoris Yao, Cai & Ren, 2006

Collivetanthocoris Yao, Cai & Ren, 2006, *Zootaxa*, 1360, 16 [116] (original designation).

Type species: *Collivetanthocoris rapax* Yao, Cai & Ren, 2006.

Pronotum with collar, about 0.5 times as long as wide. Hemelytra distinctly longer than abdominal tip. Pronotum wider than abdomen.

Distribution and age: Liaoning; Early Cretaceous.

Only one species included from the Cretaceous of Northern China (see Table 17.1).

Byssoidecerus **Yao, Cai & Ren, 2006**

Byssoidecerus Yao, Cai & Ren, 2006, *Zootaxa*, 1360, 19 [116] (original designation).

Type species: *Byssoidecerus levigata* Yao, Cai & Ren, 2006.

The third and fourth segments quite slender, subequal in length, thinner than half diameter of the second segment; embolium wide, almost 0.2 times as broad as long.

Distribution and age: Liaoning; Early Cretaceous.

Only one species included from the Cretaceous of Northern China (see Table 17.1).

Mecopodus **Yao, Cai & Ren, 2006**

Mecopodus Yao, Cai & Ren, 2006, *Zootaxa*, 1360, 22 [116] (original designation).

Type species: *Mecopodus xanthos* Yao, Cai & Ren, 2006.

The third and fourth segments thicker than half diameter of the second segment, embolium narrow, almost 0.1 times as broad as long. Pronotum and male abdomen are normal.

Distribution and age: Liaoning; Early Cretaceous.

Only one species included from and Cretaceous of Northern China (see Table 17.1).

Curticerus **Yao, Cai & Ren, 2006**

Curticerus Yao, Cai & Ren, 2006, *Zootaxa*, 1360, 31 [116] (original designation).

Type species: *Curticerus venustus* Yao, Cai & Ren, 2006.

Head, pronotum, and scutellum combined is longer than antenna, the third and fourth segments together longer than the second segment. Pronotum about 0.5 times as long as wide, the third to seventh sterna of abdomen are almost equal in width.

Distribution and age: Liaoning; Early Cretaceous.

Only one species included from the Cretaceous of Northern China (see Table 17.1).

Pustulithoracalis **Yao, Cai & Ren, 2006**

Pustulithoracalis Yao, Cai & Ren, 2006, *Zootaxa*, 1360, 35 [116] (original designation).

Type species: *Pustulithoracalis gloriosus* Yao, Cai & Ren, 2006.

Body small with length less than 6 mm, the third and fourth segments about same thickness as the second segment. Pronotum, scutellum, clavus and corium with punctures.

Distribution and age: Liaoning; Early Cretaceous.

Only one species included from the Cretaceous of Northern China (see Table 17.1).

Curvicaudus **Yao, Cai & Ren, 2006**

Curvicaudus Yao, Cai & Ren, 2006, *Zootaxa*, 1360, 23 [116] (original designation).

Type species: *Curvicaudus ciliatus* Yao, Cai & Ren, 2006.

Pronotum with two longitudinal carinae or sulci medially. Corium with deep embolar fracture. Apex of male abdomen bending to left.

Distribution and age: Liaoning; Early Cretaceous.

Two species included from the Cretaceous of Northern China (see Table 17.1).

Crassicerus **Yao, Cai & Ren, 2006**

Crassicerus Yao, Cai & Ren, 2006, *Zootaxa*, 1360, 29 [116] (original designation).

Type species: *Crassicerus furtivus* Yao, Cai & Ren, 2006.

Head, pronotum, and scutellum combined subequal to antenna in length, the second segment subequal to the third and fourth segments together. Pronotum approximately three times as wide as long, the third to sixth sterna of abdomen are almost equal in width [119].

Distribution and age: Liaoning; Early Cretaceous.

Two species included from the Cretaceous of Northern China (see Table 17.1).

Pumilanthocoris **Hou, Yao & Zhang, 2012**

Pumilanthocoris Hou, Yao, & Zhang, 2012, *Eur. J. Entomol.*, 109, 282 [117] (original designation).

Type species: *Pumilanthocoris gracilis* Hou, Yao & Zhang, 2012.

The first segment of rostrum slightly shorter than the third, the second segment of rostrum shorter than the fourth. Pronotum with collar.

Distribution and age: Inner Mongolia; Middle Jurassic.

Two species included from the Jurassic of Northern China (see Table 17.1).

Longilanceolatus **Tang, Yao & Ren, 2015**

Longilanceolatus Tang, Yao & Ren, 2015, *Cretac. Res.*, 56, 505 [118] (original designation).

Type species: *Longilanceolatus tenellus* Tang, Yao & Ren, 2015.

Head comparatively round, posterior margin distinct convex at midline. Eyes round, distant from anterior margin of pronotum. Ovipositor very long, slightly projecting beyond the apex of the abdomen.

Distribution and age: Liaoning; Early Cretaceous.

Only one species included from the Cretaceous of Northern China (see Table 17.1).

Punctivetanthocoris Tang, Yao & Ren, 2017

Punctivetanthocoris Tang, Yao & Ren, 2017, *J. Syst. Palaeontol.*, 15 (9), 699 [120] (original designation).

Type species: *Punctivetanthocoris pubens* Tang, Yao & Ren, 2017.

Dorsal surface of pronotum, clavus and corium glabrous; rostrum four-segmented, length of the second segment equal to the fourth. Pronotum without collar. Corium without costal fracture.

Distribution and age: Liaoning; Early Cretaceous.

Only one species included from the Cretaceous of Northern China (see Table 17.1).

Punctivetanthocoris pubens Tang, Yao & Ren, 2017 (Figure 17.16)

Punctivetanthocoris pubens Tang, Yao & Ren, 2017: *J. Syst. Palaeontol.*, 15 (9), 700.

Locality and horizon: Huangbanjigou, Beipiao, Liaoning, China; Lower Cretaceous, Yixian Formation.

Dorsal surface of pronotum, clavus and corium covered with bushy setae and dense punctation. Rostrum four-segmented, the third segment longest, the fourth segment slightly shorter than the second; antenna four-segmented, the second longest, as long as the third and fourth segments combined. Pronotum with collar. Hemelytra extending beyond apex of abdomen, corium with distinct median fracture and costal fracture, scutellum longer than claval commissure. Abdomen wider than pronotum, with wide dorsal laterotergites, ovipositor long [120].

Phylogenetic Study of Vetanthocoridae

In order to confirm the position of Vetanthocoridae within Cimiciformes and clarify the phylogenetic relationships with the extant flower bugs, two morphological data sets were analyzed. The first phylogenetic analysis was conducted for the Cimiciformes. By analyzing 17 terminal taxa and 53 morphological characters, the phylogenetic result confirmed that the monophyly of Cimiciformes and Cimicoidea, and also demonstrated that the extinct family Vetanthocoridae is treated as a member of Cimicoidea (Figure 17.17). The second phylogenetic analysis was conducted for the Cimicoidea. By analyzing nine species and 21 morphological characters, the result confirmed that the three extant families of

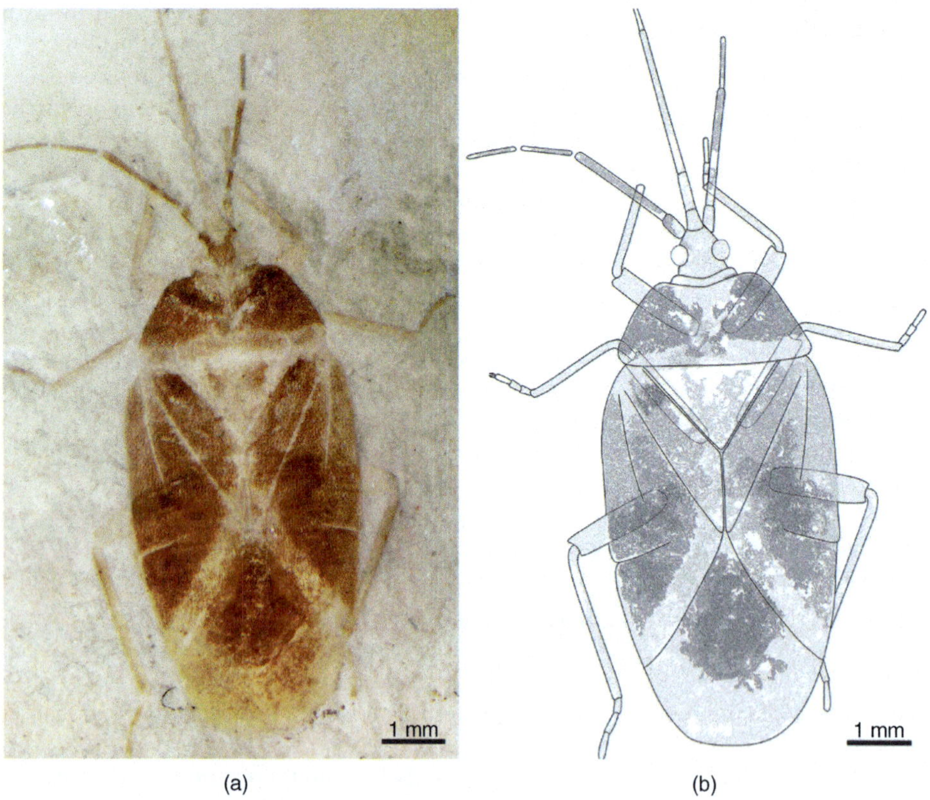

Figure 17.16 *Punctivetanthocoris pubens* Tang, Yao & Ren, 2017 (Holotype, CNU-HET-LB2012089p). (a) Photograph; (b) Line drawing [120].

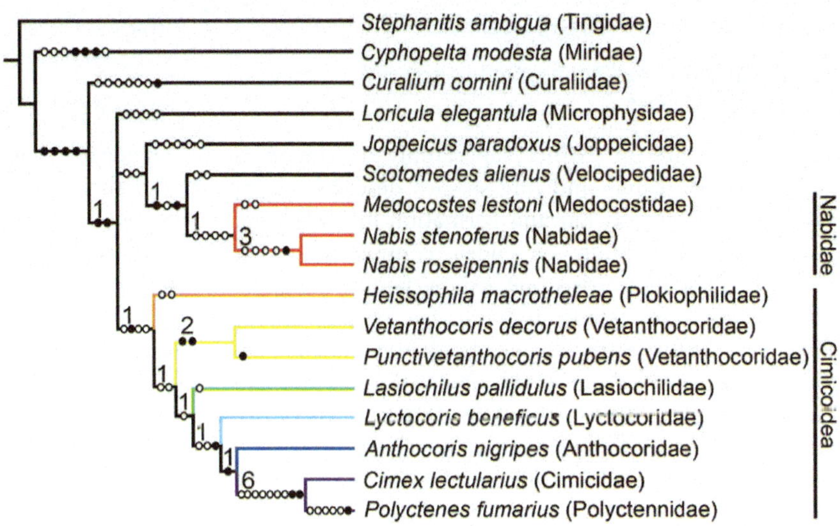

Figure 17.17 The strict consensus tree for Cimiciformes (modified from [120]): tree length = 126 steps; consistency index = 0.53; retention index = 0.64. •, non-homoplastic changes; o, homoplastic changes. The larger red (bold) numbers above the branches are Bremer support values.

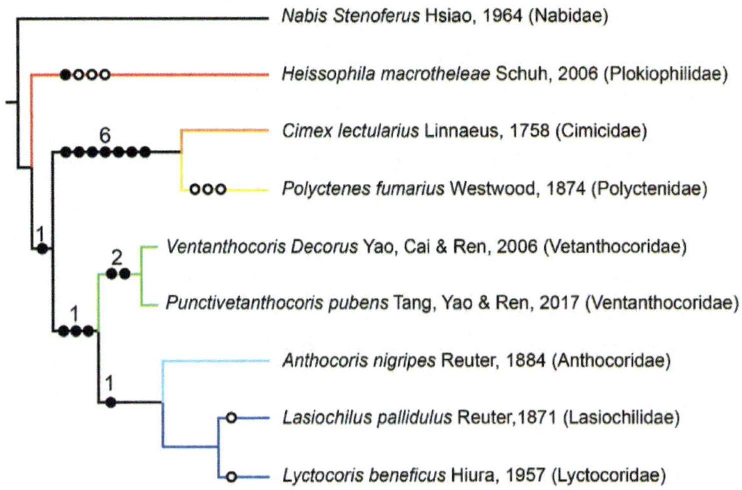

Figure 17.18 The strict consensus tree for Cimicoidea (modified from [120]): tree length = 32 steps; consistency index = 0.75; retention index = 0.73. •, non-homoplastic changes; o, homoplastic changes. The larger red (bold) numbers above the branches are Bremer support values.

flower bugs form a monophyletic clade and Vetanthocoridae thought to be closely related to Anthocoridae (*sensu lato*), which is considered as a representative of the Cimicoidea (Figure 17.18) [120].

Superfamily Miroidea Hahn, 1833

Family Miridae Hahn, 1833

Miridae are the largest family of Heteroptera, with about 10 000 living species in 1400 genera [89]. The sizes of Miridae range from less than 2 mm to about 15 mm. They do not have any ocelli. Most species of Miridae are notorious agricultural pests that pierce plant tissues, feed on the sap, and sometimes transmit viral plant diseases. In addition, some species are predatory. Miridae have a lot of fossil records. Among them, mirids are quite abundant in Eocene Baltic amber, with over 40 recorded amber species. One amber species of Miridae was described from the Eocene of Northeastern China [121], but its subfamily is uncertain. The oldest fossil record of Miridae, *Mirivena robusta* Yao, Cai & Ren, 2007, has been described from the Middle Jurassic Jiulongshan Formation in China.

Only one genus included from the Jurassic of Northern China: *Mirivena* Yao, Cai & Ren, 2007.

Mirivena Yao, Cai & Ren, 2007

Mirivena Yao, Cai & Ren, 2007, *Zootaxa*, 1442, 38 [122] (original designation).

Type species: *Mirivena robusta* Yao, Cai & Ren, 2007.

Body length about 11.0 mm, head 1.4 times as wide as long. Eyes widely separated from anterior margin of pronotum. Corium without CuP, R reaching anterior margin of the forewing.

Distribution and age: Inner Mongolia; Middle Jurassic.

Only one species included from the Jurassic of Northern China (see Table 17.1).

Superfamily Tingoidea Laporte, 1833

Family Ignotingidae Zhang, Golub, Popov & Shcherbakov, 2005

Ignotingidae are an extinct superfamily of Tingoidea with only one genus and one species described from the Early Cretaceous Laiyang Formation in Shandong, China. The superfamily Tingoidea, commonly known as "lace bugs", comprise one extant family, Tingidae and three extinct family, Ignotingidae Zhang, Golub, Popov & Shcherbakov, 2005, Ebboidae Perrichot, Nel, Guilbert & Néraudeau, 2006, and Hispanocaderidae Golub, Popov & Arillo, 2012 [123–125]. Because of the appearance of their pronotum and hemelytra, Ignotingidae are closely related to Tingidae. But, Ignotingidae are differentiated from Tingidae by these characters: head and thorax with coarsely punctate, just like Vianaididae; head without protruding mandibular plates or bucculae; antennae very long, extending beyond length of body; rostrum thick and moderately long, directed caudally but not appressed to body venter, movable at base. Pronotum with areolate dorsal projection and paranota; Scutellum exposed; corium extend to apex of hemelytron, with several cells; without costal fracture; R + M strongest vein; clavus large, triangular, commissura clavi long; membrane without areolate; rotatory hind coxae more widely separated than mid and fore ones; tarsi of leg three-segmented; the second and third abdominal sternites fused; laciniate ovipositor not concealed by paratergites [123].

Only one genus included from the Cretaceous of Northern China: *Ignotingis* Zhang, Golub, Popov & Shcherbakov, 2005 [123].

Ignotingis Zhang, Golub, Popov & Shcherbakov, 2005

Ignotingis Zhang, Golub, Popov & Shcherbakov, 2005, *Cretac. Res.*, 26, 784 [123] (original designation).

Type species: *Ignotingis mirifica* Zhang, Golub, Popov & Shcherbakov, 2005.

Head without protruding mandibular plates or prominent bucculae. Antenna very long, extending beyond length of body; the third segment longest, the second much longer than the first. Pronotum with moderately narrow collar and rather narrow paranota, expanded onto mesonotum and scutellum exposed. Without costal fracture, membrane (zone of hemelytral overlap) not areolate, long and narrow.

Distribution and age: Shandong; Early Cretaceous.

Only one species included from the Cretaceous of Northern China (see Table 17.1).

Infraorder "Pentatomomorpha" Leston, Pendergrast & Southwood, 1954

Family Venicoridae Yao Ren & Cai, 2012

Venicoridae, with four species, have only been reported from the Early Cretaceous of China [126, 127]. Their diagnostic characters are: antennal segment 2 very long, distinctly longer than segment 3, subequal to segments 3 and 4 combined; pronotum without distinct callus; claval apices close together but not contiguous, not concealed by scutellum; connexivum completely exposed; R, M, and Cu veins inosculated at basal of corium; Sc absent, R and M veins diverging at basal of corium, only 1A veins on clavus; and spiracles on abdominal segment VIII in males absent.

Genera included from the Cretaceous of Northern China: *Venicoris* Yao, Ren & Rider, 2012, *Clavaticoris* Yao, Ren & Cai, 2012 and *Halonatusivena* Du, Yao & Ren, 2016.

Venicoris Yao, Ren & Rider, 2012

Venicoris Yao, Ren & Rider, 2012, *PLoS ONE*, 7 (5), 13 [126] (original designation).

Type species: *Venicoris solaris* Yao, Ren & Rider, 2012.

Rostrum extending to mid coxae of thorax. Pronotum trapezoidal, without collar, prosternum carinae present; Scutellum triangular; R, M, and Cu inosculated at basal of corium, membrane with reticulate venation, 1A on clavus, clavus tapering, hind wings with a branching distal sector of R + M.

Distribution and age: Liaoning, Early Cretaceous.

Only one species included from the Cretaceous of Northern China (see Table 17.1).

Venicoris solaris Yao, Ren & Rider, 2012 (Figure 17.19)

Venicoris solaris Yao, Ren & Rider, 2012: *PLoS ONE*, 7 (5), 13.

Locality and horizon: Huangbanjigou, Beipiao, Liaoning, China; Lower Cretaceous, Yixian Formation.

Antenna longer than half of body. Scutellum marginally longer than pronotum at median line; black round markings at basal angles of scutellum. Femora with pale annulus, tibiae with two pale annuli. Base of membrane with three black markings, outer ones largest, middle ones smallest; costal margin of hemelytron weakly convex. Abdomen broad, connexivum with black square markings at posterior angles of each segment [126].

Clavaticoris Yao, Ren & Cai, 2012

Clavaticoris Yao, Ren & Cai, 2012, *PLoS ONE*, 7 (5), 15 [126] (original designation).

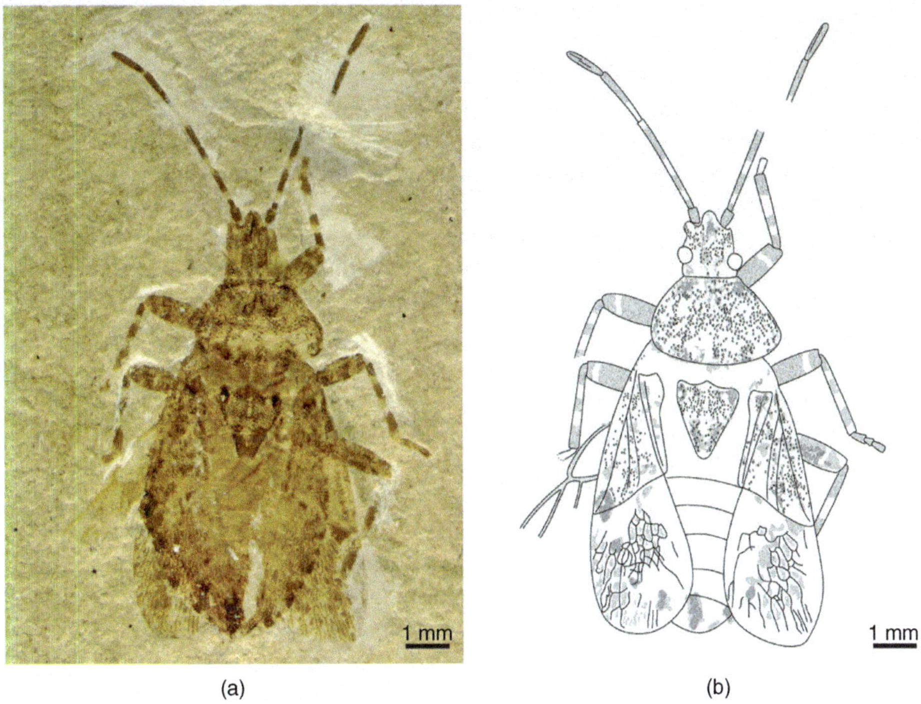

Figure 17.19 *Venicoris solaris* Yao, Ren & Rider, 2012 (Holotype, male, CNU-HE-LB2006526). (a) Photograph; (b) Line drawing. [126].

Type species: *Clavaticoris zhengi* Yao, Ren & Cai, 2012.

The specific epithet is in honor of Dr. Leyi Zheng (Institute of Entomology, College of Life Science, Nankai University, Tianjin, China) for his outstanding contribution to the study of Chinese Heteroptera. Body length 17 mm, scutellum triangular, extending to the second abdominal segment, apex rounded. Rostrum reaching the third sternite of abdomen. R rising from basal part of corium, ending within apical 1/6 of costal margin of corium, M and Cu rising from basal part of R, parallel, basal clavus extended.

Distribution and age: Liaoning; Early Cretaceous.

Only one species included from the Cretaceous of Northern China (see Table 17.1).

Halonatusivena Du, Yao, Ren, 2016

Halonatusivena Du, Yao, Ren, 2016, *Cretac. Res.*, 68, 23 [127] (original designation).

Type species: *Halonatusivena shii* Du, Yao & Ren, 2016.

The specific epithet is in honor of Yan Shi for his assistance and contribution to collect Liaoning fossils. Antennal segment 2 shorter than segments 3 and 4 combined in length. Side of head glossy, head relatively round, eyes outer convex. Apical apical half of scutellum narrowed. Clavus tapering; membrane with reticulate venation, edges of veins irregular, similar to irregular stain, vasiform structure in veins reduced. Ovipositor long, extending through the last four abdominal segments.

Distribution and age: Liaoning; Early Cretaceous.

Two species included from the Cretaceous of Northern China (see Table 17.1).

Phylogenetic Study on Venicoridae

To study the relationships among the extinct family of Venicoridae and other families of Pentatomomorpha, Yao *et al.* carried out phylogenetic analyses by selecting 130 morphological characters for 34 extant taxa and 5 fossil taxa to analyze their relationships. The major conclusions, shown in Figure 17.20, from the phylogenetic analyses include: Pentatomomorpha is a monophyletic group; Aradoidea and the Trichophora are sister groups; Venicoridae is a family, may be ancestral to the Trichophora and as the sister group of the remainder. Origin time of Pentatomomorpha should be earlier than the existing fossil records, which may be tracked back to the Middle or Early Triassic [126].

Superfamily Coreoidea Reuter, 1910

Family Pachymeridiiae Handlirsch, 1906–1908

Popov *et al.*, in 1977, considered the extinct family of Pachymeridiidae ancestral to the Coreoidea (embracing both Pyrrhocoroidea and Lygaeoidea) [39]. Pachymeridiidae flourished during the Jurassic, but their extinction during the mid-Cretaceous remains unexplained [90]. A large number of well-preserved fossil pachymeridiids were collected from the Yixian Formation and the Jiulongshan Formation [128, 129].

Genera included from the Jurassic and Cretaceous of Northern China: *Sinopachymeridium* Yao, Cai & Ren, 2006, *Beipiaocoris* Yao, Cai & Ren, 2008, *Bellicoris* Yao, Cai & Ren, 2008, *Nitoculus* Yao, Cai & Ren, 2008 and *Viriosinervis* Yao, Cai & Ren, 2008.

Sinopachymeridium Yao, Cai & Ren, 2006

Sinopachymeridium Yao, Cai & Ren, 2006, *Annales Zoologici.*, 56 (4), 754 [128] (original designation).

Type species: *Sinopachymeridium popovi* Yao, Cai & Ren, 2006.

The specific epithet is in honor of Dr. Yuri A. Popov for his outstanding contribution to the study of fossil Heteroptera. Rostrum extending to metacoxae; tarsi three-segmented. Forewing divided into a distinct corium and membrane; C present, separating at basal 1/4 of corium from Sc + R + M; Sc, R, and M diverging at a single point, membrane with seven simple longitudinal veins, the first vein situated remote from anterior margin of membrane, the fourth and fifth veins merging into one before middle of membrane.

Distribution and age: Inner Mongolia; Middle Jurassic.

Only one species included from the Jurassic of Northern China (see Table 17.1).

Sinopachymeridium popovi Yao, Cai & Ren, 2006 (Figure 17.21)

Sinopachymeridium popovi Yao, Cai & Ren, 2006: *Annales Zoologici.*, 56 (4), 754.

Locality and horizon: Daohugou, Ningcheng, Inner Mongolia, China; Middle Jurassic, Jiulongshan Formation.

Fore legs simple, with femora slightly shorter than tibiae, two claws at apex of tarsi. The hind legs are distinctly longer than the fore and mid legs, femur 1.6 times as long as tibia. Procoxae situated remotely from mesocoxae. Apical 1/5 of corium with a costal fracture. Abdominal sterna III–VI subequal in width, the fourth widest [128].

Beipiaocoris Yao, Cai & Ren, 2008

Beipiaocoris Yao, Cai & Ren, 2008, *Acta Geol. Sin.-Engl.*, 82 (1), 37 [129] (original designation).

Type species: *Beipiaocoris multifurcus* Yao, Cai & Ren, 2008.

Clypeus beyond mandibular plates; Interocellar space narrower than interocular space. Pronotum with collar. Four similar large cells on corium, membrane with five bifurcating emanating veins, arising from cross-veins at corium-membrane boundary joined.

Distribution and age: Liaoning; Early Cretaceous.

Only one species included from the Cretaceous of Northern China (see Table 17.1).

Bellicoris Yao, Cai & Ren, 2008

Bellicoris Yao, Cai & Ren, 2008, *Acta Geol. Sin.-Engl.*, 82 (1), 39 [129] (original designation).

Type species: *Bellicoris mirabilis* Yao, Cai & Ren, 2008.

Eyes situated at middle of head; rostrum extending to near mid coxae. Pronotum without collar. Hemelytron reaching beyond tip of abdomen, clavus large, claval commissure subequal in length to scutellum, membrane with four free wide veins, pale, similar to the plaque.

Distribution and age: Liaoning; Early Cretaceous.

Only one species included from the Cretaceous of Northern China (see Table 17.1).

Nitoculus Yao, Cai & Ren, 2008

Nitoculus Yao, Cai & Ren, 2008, *Acta Geol. Sin.-Engl.*, 82 (1), 40 [129] (original designation).

Type species: *Nitoculus regillus* Yao, Cai & Ren, 2008.

Eyes large and round, anteocular portion distinctly longer than postocular. Rostrum extending to hind coxae. Pronotum without collar. C present, Sc, R, and M diverging at a single point, membrane with a few free veins.

Distribution and age: Liaoning; Middle Jurassic.

Only one species included from the Cretaceous of Northern China (see Table 17.1).

Viriosinervis Yao, Cai & Ren, 2008

Viriosinervis Yao, Cai & Ren, 2008, *Acta Geol. Sin.-Engl.*, 82 (1), 42 [129] (original designation).

Type species: *Viriosinervis stolidus* Yao, Cai & Ren, 2008.

Pronotum without collar. Scutellum triangular, longer than pronotum at median line, wider than length, dorsal surface with dimples. Hemelytron long and narrow, C present, with Sc + R inosculated at basal of forewing, connected with Sc, R, and M diverging at a single point, R connected with C + Sc at apex of corium. Claval commissure shorter than lateral margin of scutellum.

Distribution and age: Liaoning; Middle Jurassic.

Only one species included from the Cretaceous of Northern China (see Table 17.1).

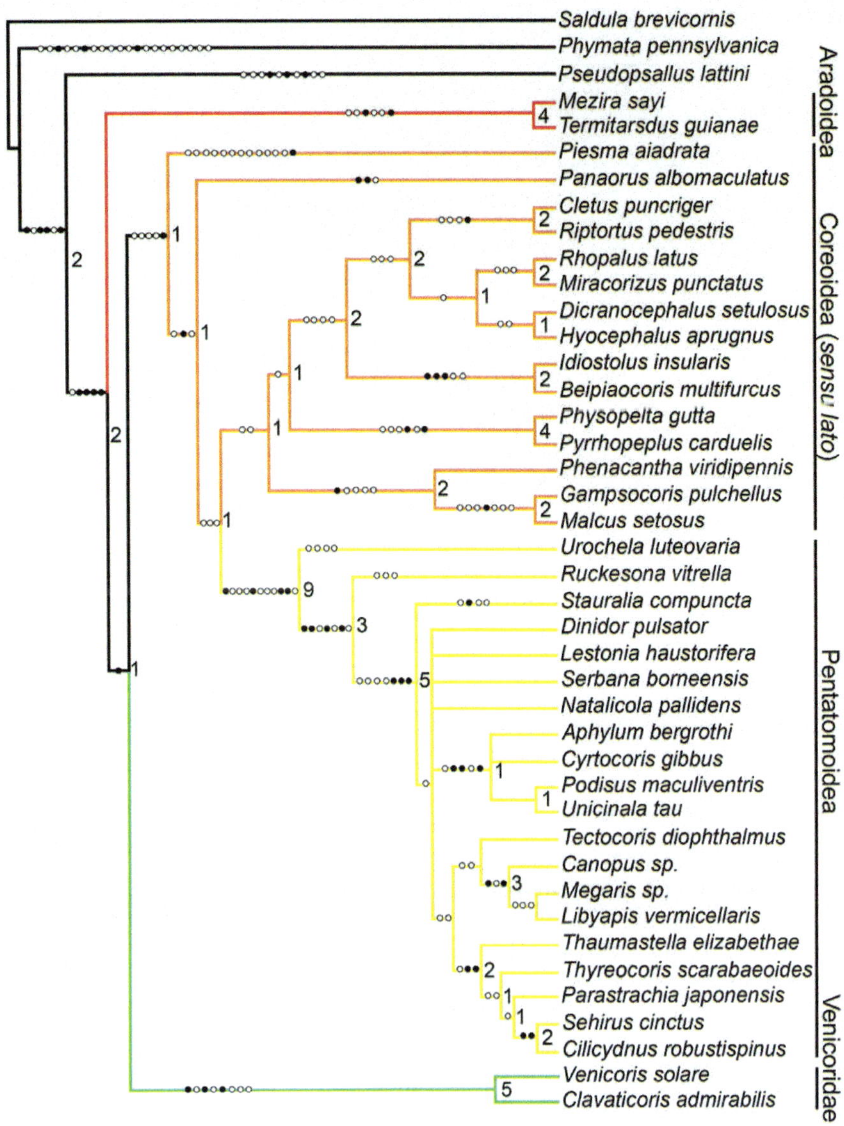

Figure 17.20 Phylogeny of extant and extinct Pentatomomorpha (modified from [126]). The strict consensus tree of 15 most parsimonious trees with Bremer support values (near branch nodes). • Nonhomoplasious; o homoplasious.

***Peregrinpachymeridium* Lu, Yao & Ren, 2011**

Peregrinpachymeridium Lu, Yao & Ren, 2011, *Zootaxa*, 2835, 42 [130] (original designation).

Type species: *Peregrinpachymeridium comitcola* Lu, Yao & Ren, 2011.

Collar present. Corium with clear marking. Sc, R, and M diverging at a single point, Cu arising claval suture, connected with 1A at apex, a cross-vein at corium-membrane boundary joined with C + Sc, R, M, and Cu; 1A near clavus suture, posterior claval vein 2A along hind margin of clavus. Claval commissure longer than scutellum; membrane with six free wide veins, arising from cross-veins at corium-membrane boundary joined.

Distribution and age: Inner Mongolia; Middle Jurassic.

Only one species included from the Jurassic of Northern China (see Table 17.1).

***Corollpachymeridium* Lu, Yao & Ren, 2011**

Corollpachymeridium Lu, Yao & Ren, 2011, *Zootaxa*, 2835, 43 [130] (original designation).

Type species: *Corollpachymeridium heteroneurus* Lu, Yao & Ren, 2011.

Rostrum extending to the middle of mid coxae and hind coxae. Membrane with five free wide veins, C connected with Sc in front of costal fracture, R and C + Sc connected with membrane at two different points.

Distribution and age: Inner Mongolia; Middle Jurassic.

Only one species included from the Jurassic of Northern China (see Table 17.1).

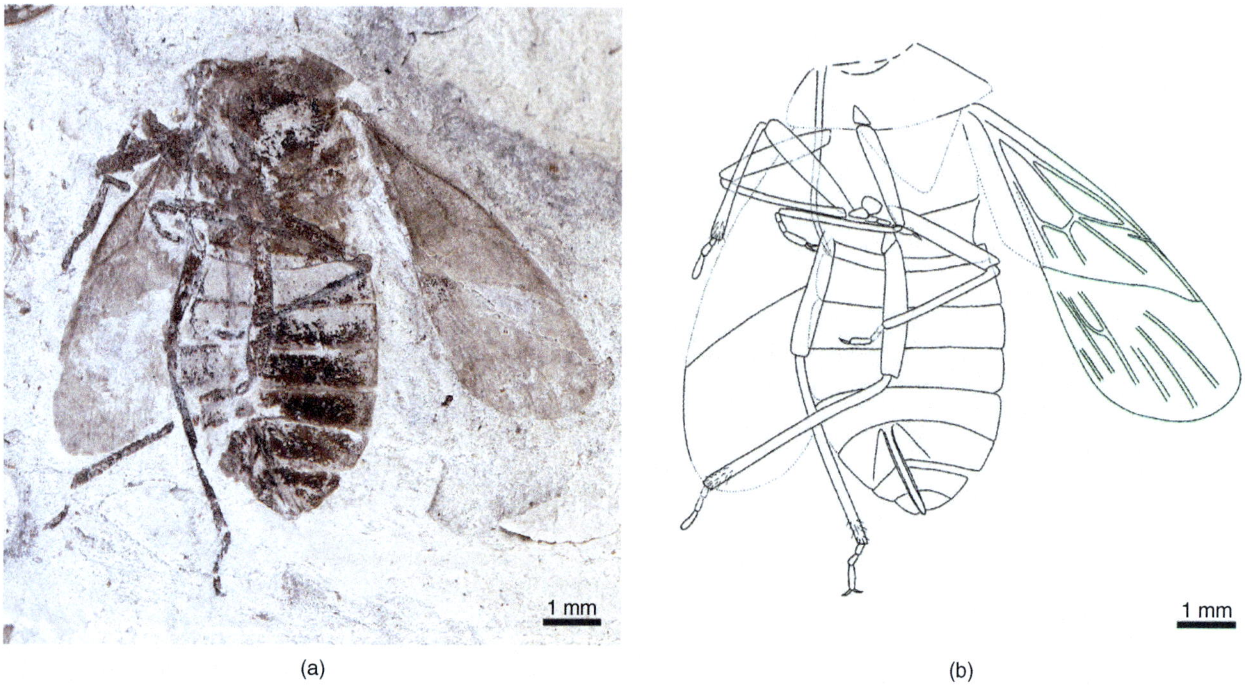

Figure 17.21 *Sinopachymeridium popovi* Yao, Cai & Ren, 2006 (Holotype, CNU-HE-NN2005001). (a) Photograph; (b) Line drawing [128].

Family Rhopalidae Amyot & Serville, 1843

Rhopalidae, known as "scentless plant bugs," are usually light-colored and tiny, compared to coreids. So far, eight extant genera with over 200 species of rhopalids are known around the world. Fossil Rhopalidae have been found only in China. These fossils obviously differ from the extant group in their lanceolate ovipositor and abdominal sternum VII separated by ovipositor into two parts [131, 132].

Genera included from the Jurassic and Cretaceous of Northern China: *Miracorizus* Yao, Cai & Ren, 2006, *Originicorizus* Yao, Cai & Ren, 2006, *Quatlocellus* Yao, Cai & Ren, 2006, *Longiclavula* Yao, Cai & Ren, 2006, *Grandicaputus* Yao, Cai & Ren, 2006 and *Vescisalignus* Chen, Yao & Ren, 2015.

Miracorizus Yao, Cai & Ren, 2006

Miracorizus Yao, Cai & Ren, 2006, *Zootaxa*, 1269, 58 [131] (original designation).

Type species: *Miracorizus punctatus* Yao, Cai & Ren, 2006.

Posterior region of pronotum with two longitudinal carinae. Clavus with a vein-like carina and a vein arising at basal point, both crossing claval suture, carina reaching middle of corium, veins through posterior region of corium into membrane; membrane with a black spot at the front.

Distribution and age: Inner Mongolia; Middle Jurassic.

Only one species included from the Jurassic of Northern China (see Table 17.1).

Longiclavula Yao, Cai & Ren, 2006

Longiclavula Yao, Cai & Ren, 2006, *Zootaxa*, 1269, 63 [131] (original designation).

Type species: *Longiclavula calvata* Yao, Cai & Ren, 2006.

Corium with only one wavy longitudinal vein; clavus tapering, without carina and vein; membrane with numerous veins. Abdomen oval, all sutures of abdominal straight, the third to the fifth sterna subequal in width, distinctly wider than other sterna.

Distribution and age: Inner Mongolia; Middle Jurassic.

Only one species included from the Jurassic of Northern China (see Table 17.1).

Origincorizus Yao, Cai & Ren, 2006

Origincorizus Yao, Cai & Ren, 2006, *Zootaxa*, 1384, 42 [132] (original designation).

Type species: *Originicorizus pyriformis* Yao, Cai & Ren, 2006.

Body length less than 7 mm; rostrum long, extending to the third abdominal sternite. Vein 1A on middle of clavus, posterior claval vein 2A along hind margin of clavus and ending at apex; membrane with some longitudinal veins.

Distribution and age: Inner Mongolia; Middle Jurassic.

Only one species included from the Jurassic of Northern China (see Table 17.1).

Quatlocellus Yao, Cai & Ren, 2006

Quatlocellus Yao, Cai & Ren, 2006, *Zootaxa*, 1384, 45 [132] (original designation).

Type species: *Quatlocellus liae* Yao, Cai & Ren, 2006.

Body longer than 8 mm. Anterior portion longer than postocular. Rostrum short, not extending to mesocoxae. Vein 2A along hind margin of clavus, ending at apex.

Distribution and age: Inner Mongolia; Middle Jurassic.

Only one species included from the Jurassic of Northern China (see Table 17.1).

Grandicaputus Yao, Cai & Ren, 2006

Grandicaputus Yao, Cai & Ren, 2006, *Zootaxa*, 1384, 51 [132] (original designation).

Type species: *Grandicaputus binpunctatus* Yao, Cai & Ren, 2006.

Clypeus distinctly surpassing mandibular plates. Tibia with densely setae. Corium with claval commissure; membrane with nearly 20 longitudinal veins distally.

Distribution and age: Inner Mongolia; Middle Jurassic.

Only one species included from the Jurassic of Northern China (see Table 17.1).

Vescisalignus Chen, Yao & Ren, 2015

Vescisalignus Chen, Yao, Ren, 2015, *Zootaxa*, 4058 (1), 135 [133] (original designation).

Type species: *Vescisalignus indecorus* Chen, Yao, Ren, 2015.

Head triangular. Clypeus surpassing mandibular plates. Scutellum triangular, apex angle blunt. Corium with large hyaline areas, membrane with more than 10 free longitudinal veins, base with a comma-like marking. Ovipositor laciniate, splitting sternite VII, and extending through the last four abdominal segments.

Distribution and age: Liaoning; Early Cretaceous.

Only one species included from the Cretaceous of Northern China (see Table 17.1).

Vescisalignus indecorus Chen, Yao & Ren, 2015 (Figure 17.22)

Vescisalignus indecorus Chen, Yao & Ren, 2015: *Zootaxa*, 4058 (1), 136.

Locality and horizon: Huangbanjigou, Beipiao City, Liaoning Province, China. Lower Cretaceous, Yixian Formation.

Apex of head surpassing the first antennal segment, anterior margin of pronotum distant from eyes, subequal to pronotum in length and about half as wide as pronotum. Scutellum triangular, apex angle blunt, base margin wider than length; femora distinctly thicker and shorter than tibiae, tarsus three-segmented, second longest, third ovoid. Sutures on the fifth abdominal segment straight; ovipositor laciniate [133].

Family Dehiscensicoridae Du, Yao & Ren, 2017

Du *et al.*, in 2017, erected an extinct family of Dehiscensicoridae with five genera and five species from the Lower Cretaceous Yixian Formation in Liaoning, China [134]. Based on a combination of fossil and extant morphological characters, a cladistic analysis was conducted to clarify the phylogenetic status of the Dehiscensicoridae. Dehiscensicoridae are distinguished by the following characters: head wider than half of pronotum; visible ocelli; rostrum four-segmented; antenna four-segmented with short scape which slightly or not extending beyond head apex; scutellum small triangular; clavus are clubbed with close but not contiguous apices and without being concealed by scutellum which are not claval commissure. In the wings, vein C is on the anterior margin of corium forming a narrow and thick edge; the end of corium margin with a wave-like appearance; veins on the membrane appear weakly ribbon-like; a cross-vein on corium-membrane boundary is without joined claval suture. Tarsal segments 3-3-3, claws are with visible pulvilli; abdomen broad connexivum with laciniate ovipositor.

Genera included from the Cretaceous of Northern China: *Dehiscensicoris* Du, Yao & Ren, 2017, *Pingquanicoris* Du, Yao & Ren, 2017, *Changirostrus* Du, Yao & Ren, 2017, *Crassiantenninus* Du, Yao & Ren, 2017 and *Minuticoris* Du, Yao & Ren, 2017.

Dehiscensicoris Du, Yao & Ren, 2017

Dehiscensicoris Du, Yao & Ren, 2017, *J. Syst. Palaeontol.*, 15 (12), 995 [134] (original designation).

Type species: *Dehiscensicoris sanctus* Du, Yao & Ren, 2017.

Base of head formed a neck. The second antennal segment subequal to the third. Ocelli visible, situated near level of posterior margins of compound eyes, interocellar space wider than diameter of ocellus. Sc split from R at base of corium.

Distribution and age: Liaoning; Early Cretaceous.

Only one species included from the Cretaceous of Northern China (see Table 17.1).

Dehiscensicoris sanctus Du, Yao & Ren, 2017 (Figure 17.23)

Dehiscensicoris sanctus Du, Yao & Ren, 2017: *J. Syst. Palaeontol.*, 15 (12), 995.

Locality and horizon: Huangbanjigou, Beipiao, Liaoning Province, China. Lower Cretaceous, Yixian Formation.

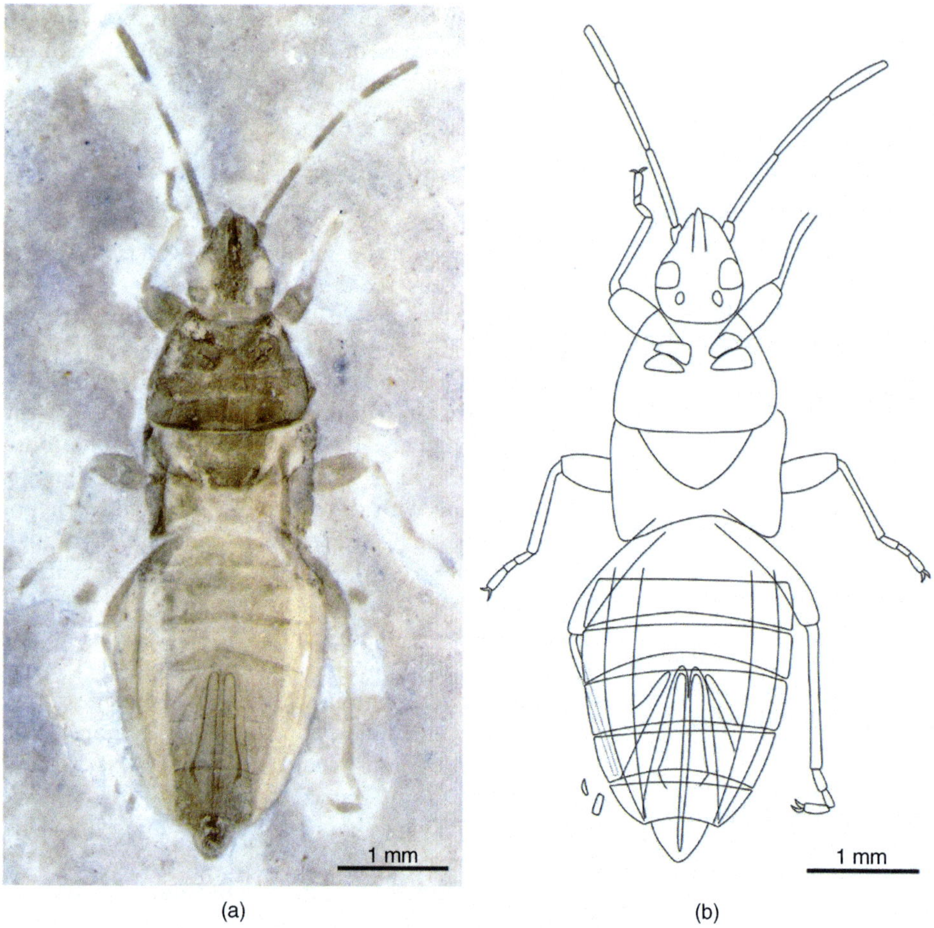

Figure 17.22 *Vescisalignus indecorus* Chen, Yao & Ren, 2015 (Holotype, CNU-HET-LB2010249). (a) Photograph; (b) Line drawing [133].

Body of moderate size; diameter of ocellus narrower than interocellar space; rostrum reaching hind coxae; antenna longer than head, pronotum and scutellum combined, length of the second segment subequal to the third, dorsal surface on pronotum and scutellun has densely punctate; tibiae have dense setae, corium small, C vein present, a cross-vein between Sc to R and R to M, R and M fused at base of membrane. Abdomen subequal to pronotum in width, connexivum absent, ovipositor long, extending through the last three abdominal segments [134].

Pingquanicoris Du, Yao & Ren, 2017

Pingquanicoris Du, Yao & Ren, 2017, *J. Syst. Palaeontol.*, 15 (12), 1000 [134] (original designation).

Type species: *Pingquanicoris punctatus* Du, Yao & Ren, 2017.

Base of head without a neck. Diameter of ocellus narrower than interocellar space. Sc fusion with R at 2/3 of corium, a closed cell formed in distal part of corium.

Distribution and age: Hebei; Early Cretaceous.

Only one species included from the Cretaceous of Northern China (see Table 17.1).

Changirostrus Du, Yao & Ren, 2017

Changirostrus Du, Yao & Ren, 2017, *J. Syst. Palaeontol.*, 15 (12), 1001 [134] (original designation).

Type species: *Changirostrus maculatus* Du, Yao & Ren, 2017.

Rostrum extending beyond hind coxae. Ocelli closer to another than to compound eyes, diameter of ocellus narrower than interocellar space. The third antennal segment shorter than the second, the second segment longest. Head without conspicuous postocular constriction. Ovipositor long, extending through the last three abdominal segments.

Distribution and age: Liaoning; Early Cretaceous.

Only one species included from the Cretaceous of Northern China (see Table 17.1).

Crassiantenninus Du, Yao & Ren, 2017

Crassiantenninus Du, Yao & Ren, 2017, *J. Syst. Palaeontol.*, 15 (12), 1004 [134] (original designation).

Type species: *Crassiantenninus minutus* Du, Yao & Ren, 2017.

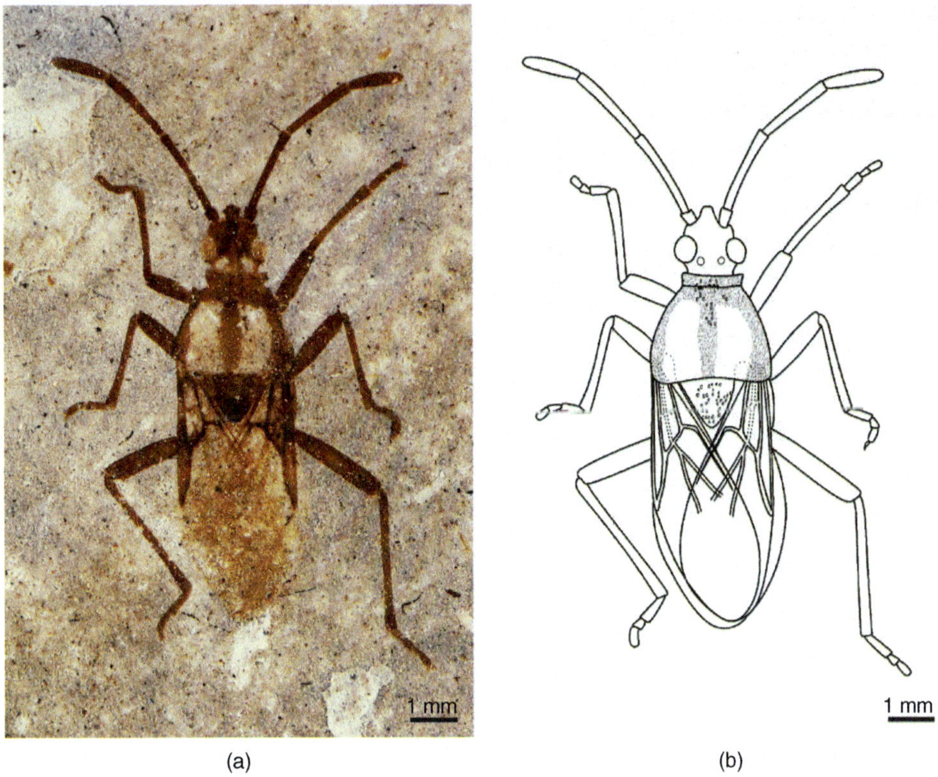

Figure 17.23 *Dehiscensicoris sanctus* Du, Yao & Ren, 2017 (Holotype, CNU-HE-LB2006152p). (a) Photograph; (b) Line drawing [134].

Rostrum reaching mid coxae. Eyes reniform, curved inner margin. Diameter of ocellus subequal to interocellar space. Antenna large, slightly shorter than body length or nearly as long as body length, the third segment longer than the second.

Distribution and age: Liaoning; Early Cretaceous.

Only one species included from the Cretaceous of Northern China (see Table 17.1).

***Minuticoris* Du, Yao & Ren, 2017**

Minuticoris Du, Yao & Ren, 2017, *J. Syst. Palaeontol.*, 15 (12), 1009 [134] (original designation).

Type species: *Minuticoris brunneus* Du, Yao & Ren, 2017.

Eyes reniform, widely separated; ocelli present, nearly at the level of posterior margins of compound eyes, diameter of ocellus subequal to interocellar space. Antenna only slightly longer than half of body length. Pronotum wider than long.

Distribution and age: Liaoning; Early Cretaceous.

Only one species included from the Cretaceous of Northern China (see Table 17.1).

Phylogenetic Study of Dehiscensicoridae

Coreoidea (*sensu lato*) are one of the most important groups of extant Pentatomomorpha, which are usually considered to comprise four superfamilies: Coreoidea (*sensu stricto*), Pyrrhocoroidea, Idiostoloidea and Lygaeoidea [89, 90, 135–137]. During the last two decades, many researchers have proposed different schemes of phylogenetic relationships among Coreoidea (*sensu lato*) from the perspective of molecular and morphological evidence [87, 89, 138–144].

In our recent study, we collected some well-preserved fossils which provide new information for studying the origin and evolution of the Coreoidea (*sensu lato*). We have selected two fossil taxa and 23 extant taxa, and chosen 72 morphological characters to conduct phylogenetic analyses. The results, in Figure 17.24, confirm that the Pyrrhocoroidea, Coreoidea, Idiostoloidea, Lygaeoidea, and Dehiscensicoridae are monophyletic groups. The Idiostoloidea are sister groups with Lygaeoidea. The Piesmatidae belong to Lygaeoidea. The results can be simplified and presented in the following way: Pyrrhocoroidea + (Dehiscensicoridae + (Coreoidea + (Idiostoloidea + Lygaeoidea))) [134].

Superfamily Pentatomoidea Reuter, 1910

Family Cydnidae Billberg, 1820

Cydnidae, known as "burrower bugs", can be easily recognized by their spiny tibiae for burrowing and digging

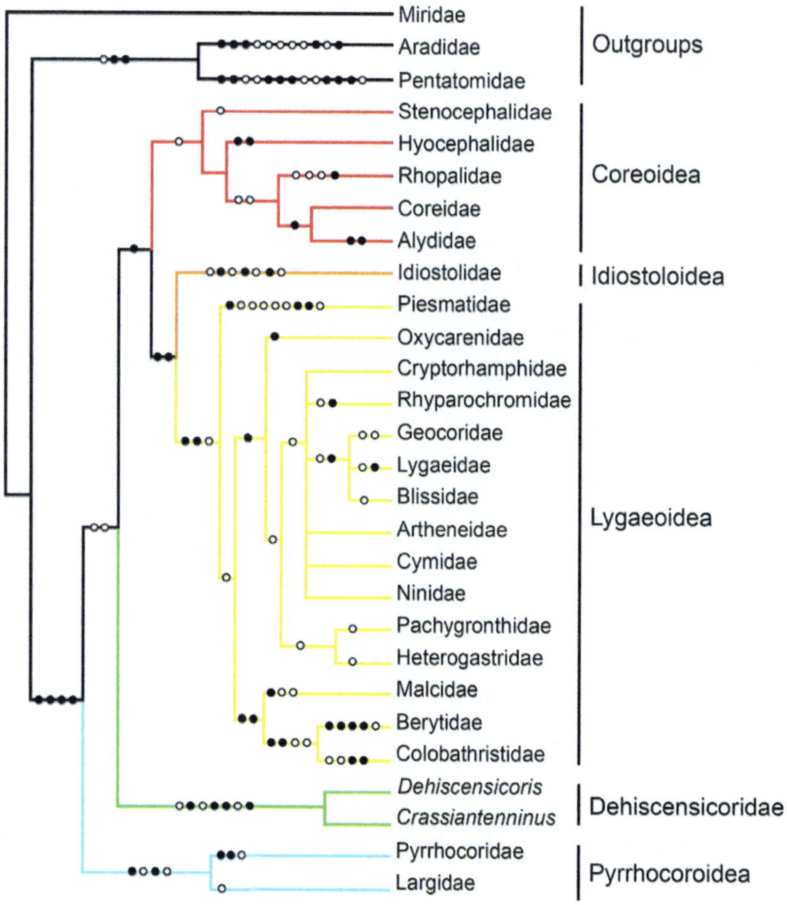

Figure 17.24 Phylogeny of extant and extinct Dehiscensicoridae (modified from [134]). Strict consensus tree of the three most parsimonious trees, • non-homoplastic changes; o homoplastic changes.

in the soil or litters. During the Late Mesozoic, members of Cydnidae were widespread throughout all zoogeographical regions. So far, only ten species of the Mesozoic Cydnidae have been reported. Two species, *Cilicydnus robustispinus* Yao, Cai & Ren, 2007, and *Orienicydnus hongi* Yao, Cai & Ren, 2007, have been described from the Lower Cretaceous Yixian Formation in Liaoning, China. These are the first discovery of fossil Cydnidae in China [145].

Genera included from the Cretaceous of Northern China: *Cilicydnus* Yao, Cai & Ren, 2007 and *Orienicydnus* Yao, Cai & Ren, 2007.

***Cilicydnus* Yao, Cai & Ren, 2007**

Cilicydnus Yao, Cai & Ren, 2007, *Zootaxa*, 1388, 60 [145] (original designation).

Type species: *Cilicydnus robustispinus* Yao, Cai & Ren, 2007.

Scutellum transverse. Femora stout, tibiae with dense stout spines, tarsomeres subequal in thickness; mid and hind coxae narrowly separated, coxae and trochanter rounded triangular. Corium with deep median fracture, clavus large, claval commissure shorter than scutellum.

Distribution and age: Liaoning; Early Cretaceous.

Only one species included from the Cretaceous of Northern China (see Table 17.1).

***Orienicydnus* Yao, Cai & Ren, 2007**

Orienicydnus Yao, Cai & Ren, 2007, *Zootaxa*, 1388, 62 [145] (original designation).

Type species: *Orienicydnus hongi* Yao, Cai & Ren, 2007.

The specific epithet is dedicated to Dr. Youchong Hong for his great contribution to insect fossil studies in China. Dorsum of head, pronotum, scutellum, and corium with long dense setae. Head half round, wider than long, subequal to pronotum in length, anterior margin with nine primary setae. Rostrum extending to posterior margin of pronotum. Eyes subtriangular, contiguous to anterior margin of pronotum. Femora very stout, evidently thicker than tibiae, tibiae with dense and stout spines, tarsi three-segmented.

Distribution and age: Liaoning; Early Cretaceous.

Only one species included from the Cretaceous of Northern China (see Table 17.1).

***Orienicydnus hongi* Yao, Cai & Ren, 2007 (Figure 17.25)**

Orienicydnus hongi Yao, Cai & Ren, 2007: *Zootaxa*, 1388, 62.

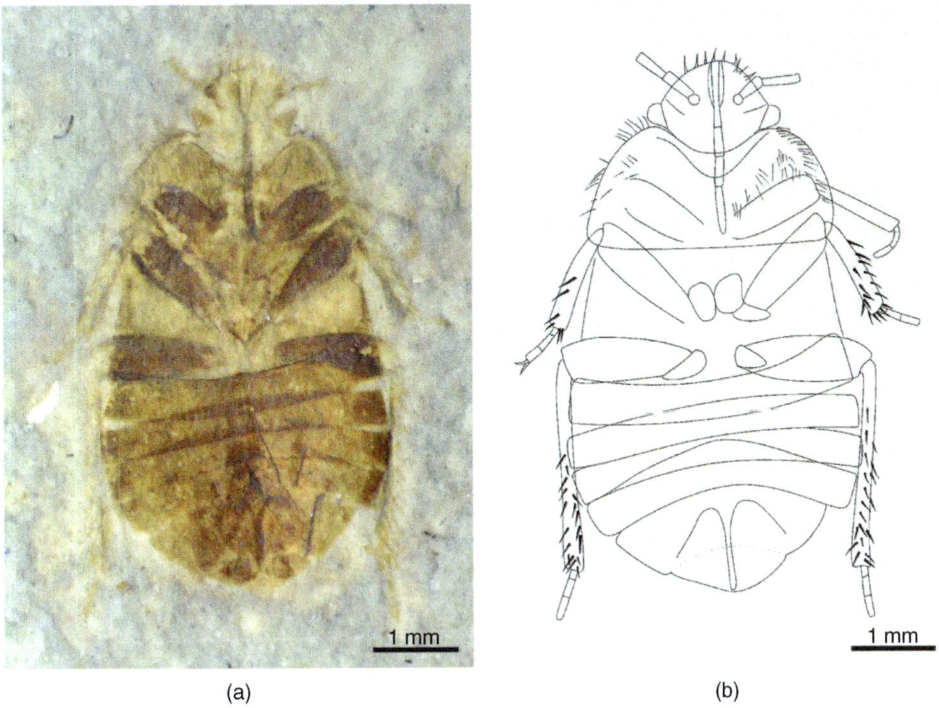

Figure 17.25 *Orienicydnus hongi* Yao, Cai & Ren, 2007 (Holotype, CNU-HE-LB2006300). (a) Photograph; (b) Line drawing [145].

Locality and horizon: Huangbanjigou, Beipiao, Liaoning, China. Lower Cretaceous, Yixian Formation.

Rostral segments subequal in thickness, fourth segment distinctly longer than third. Eyes subtriangular, contiguous to anterior margin of pronotum. Second tarsomere subequal to third. Anterior margin of pronotum very concave, lateral margins convex, posterior margin nearly straight, anterior angles rounded, posterior angles rounded, about 2.7 times as wide as long [145].

Family Primipentatomidae Yao, Cai, Rider & Ren, 2013

Primipentatomidae, an extinct family of Pentatomoidea, were erected by Yao, Cai, Rider & Ren in 2013, including four genera and five species from the Lower Cretaceous deposits of Northeastern China. Based on a combination of fossil and extant morphological characters, a cladistic analysis clarifies the phylogenetic status of the family. Primipentatomidae are distinguished by the following characters: Medium-sized body. Head triangular. Antenna with four segments, the first segment is very short. Rostrum elongate and extending beyond fore coxae, the first segment slightly thickened and hidden between bucculae. Pronotum trapezoidal with tongue-shaped scutellum, with an area that is elevated and luniform-shaped basally, less than half of the length of abdomen. Frena is very long and reaching the apex of scutellum. Legs are common and lacking spines. Tarsi with three segments. Sternum of thorax with distinctly median carina. Forewing is macropterous with small corium which is shorter than the scutellum and with a big cell. One vein radiates from the cell into the membrane. Clavus tapering and set under scutellum in repose, claval suture not formed. The membrane is with a large basal cell and two veins radiate from the cell. Ovipositor plate-like, abdominal sternum is eight-segmented without fusion. The eighth paratergites is relatively large, covering the ninth paratergites [146].

Genera included from the Cretaceous of Northern China: *Primipentatoma* Yao, Cai, Rider & Ren, 2013, *Breviscutum* Yao, Cai, Rider & Ren, 2013, *Oropentatoma* Yao, Cai, Rider & Ren, 2013 and *Quadrocoris* Yao, Cai, Rider & Ren, 2013.

Primipentatoma Yao, Cai, Rider & Ren, 2013

Primipentatoma Yao, Cai, Rider & Ren, 2013, *J. Syst. Palaeontol.*, 11, 1, 67 [146] (original designation).

Type species: *Primipentatoma peregrina* Yao, Cai, Rider & Ren, 2013.

Antenna remote from eyes, the third antennal segment subequal to the fourth in length. Rostrum extended to near mid coxae. Eyes with interocular space longer than diameter of eye. Pronotum extended to the fourth abdominal segment. Middle coxae situated closer to hind coxae than to fore coxae. Abdomen oval, connexivum broad.

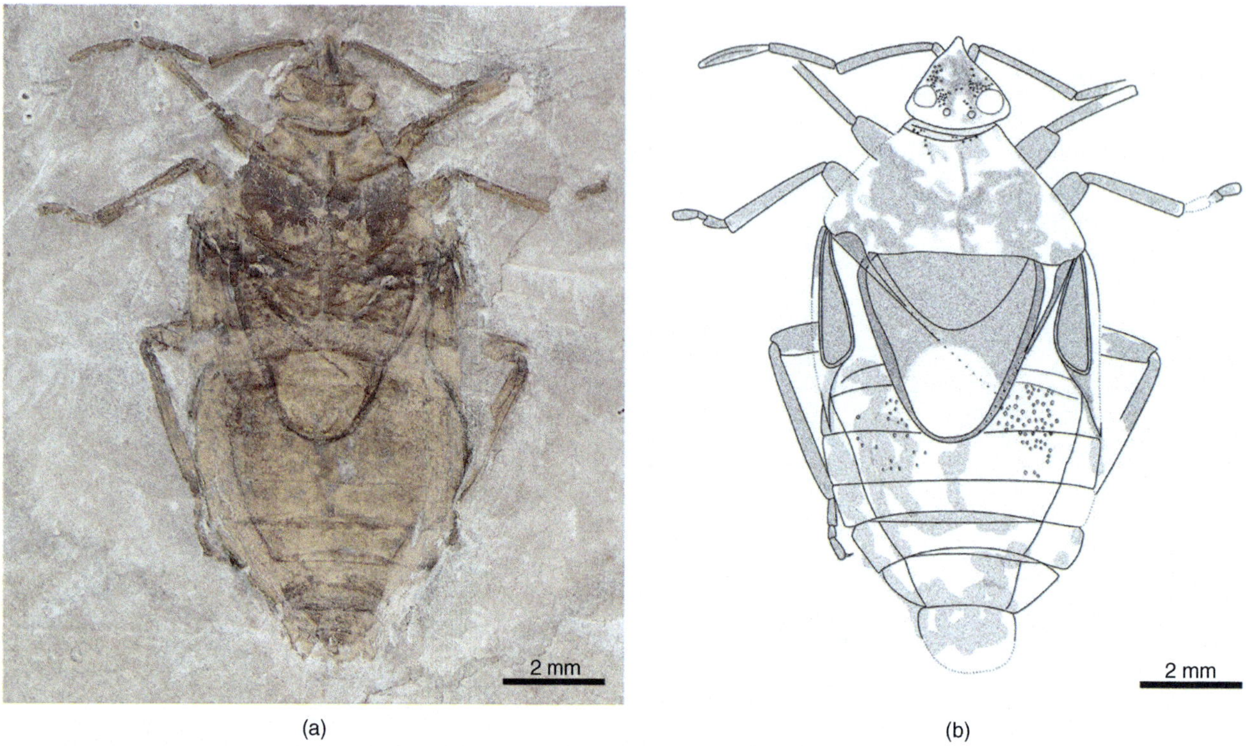

Figure 17.26 *Primipentatoma peregrina* Yao, Cai, Rider & Ren, 2013 (Holotype, CNU-HE-LB2006001). (a) Photograph; (b) Line drawing [146].

Distribution and age: Liaoning; Early Cretaceous.

Only one species included from the Cretaceous of Northern China (see Table 17.1).

Primipentatoma peregrina **Yao, Cai, Rider & Ren, 2013 (Figure 17.26)**

Primipentatoma peregrina Yao, Cai, Rider & Ren, 2013: *J. Syst. Palaeontol.*, 11, 1, 71.

Locality and horizon: Huangbanjigou, Beipiao and Dawangzhangzi, Lingyuan, Liaoning, China. Lower Cretaceous, Yixian Formation.

The body length is 12.2 mm; maximal width of body 6.6 mm; the second antennal segment longest. Pronotum occupying one third of body length, anterior disk with two oval spots and two cuneate spots. Scutellum with two oval spots on raised luniform area. Hemelytron not reaching apex of abdomen, clavus and corium shorter than lateral margin of scutellum. Connexivum broad, with black square markings at the posterior angles of each segments [146].

Breviscutum **Yao, Cai, Rider & Ren, 2013**

Breviscutum Yao, Cai, Rider & Ren, 2013, *J. Syst. Palaeontol.*, 11, 1, 74 [146] (original designation).

Type species: *Breviscutum lunatum* Yao, Cai, Rider & Ren, 2013.

Body length more than 10 mm; the third antennal segment longer than the fourth; rostrum exceeding fore coxae. Pronotum with a distinct collar, lacking callus. Scutellum extending to the third abdominal segment, distinctly wider than long, apex very broadly rounded or blunt.

Distribution and age: Liaoning; Early Cretaceous.

Only one species included from the Cretaceous of Northern China (see Table 17.1).

Oropentatoma **Yao, Cai, Rider & Ren, 2013**

Oropentatoma Yao, Cai, Rider & Ren, 2013, *J. Syst. Palaeontol.*, 11, 1, 76 [146] (original designation).

Type species: *Oropentatoma epichara* Yao, Cai, Rider & Ren, 2013.

Body elongated. The third antennal segment longer than the fourth. Rostrum almost extending to mid coxae. Pronotum with a distinct collar, lacking callus. Scutellum extending to the third abdominal segment, width subequal to length, apex rounded.

Distribution and age: Liaoning; Early Cretaceous.

Only one species included from the Cretaceous of Northern China (see Table 17.1).

Quadrocoris **Yao, Cai, Rider & Ren, 2013**

Quadrocoris Yao, Cai, Rider & Ren, 2013, *J. Syst. Palaeontol.*, 11, 1, 77 [146] (original designation).

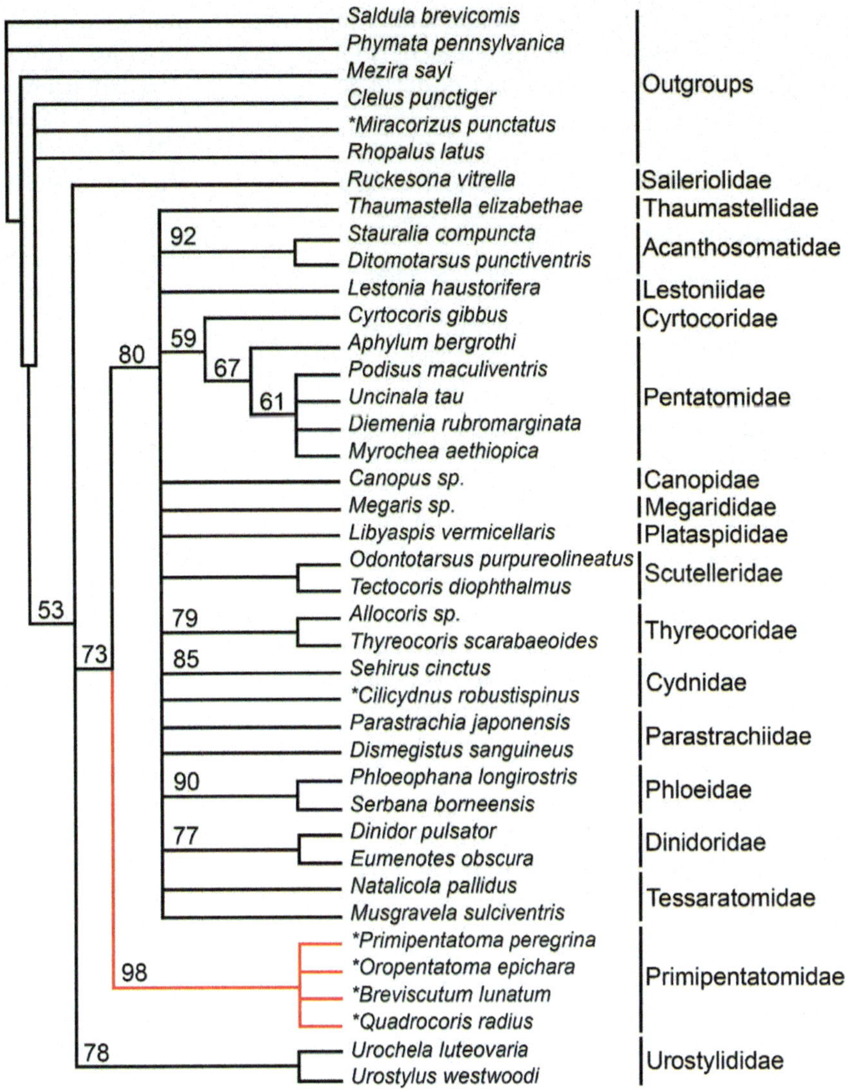

Figure 17.27 Phylogeny of extinct Primipentatomidae and other families of Pentatomomorpha Source: modified from [146]. The strict consensus tree of all most parsimonious trees with bootstrap values above branches; *: fossil species.

Type species: *Quadrocoris radius* Yao, Cai, Rider & Ren, 2013.

Body length less than 10 mm. Rostrum arising at apex of head, extending to near hind coxae. Pronotum with distinct callus. Scutellum extending to the third abdominal segment. Abdomen oval, with seven visible segments.

Distribution and age: Liaoning; Early Cretaceous.

Only one species included from the Cretaceous of Northern China (see Table 17.1).

Phylogenetic Study of Primipentatomidae

We used phylogenetic analyses to evaluate the relationships among Primipentatomidae and other families of Pentatomomorpha [146]. We chose a total of 72 morphological characters, among which 56 were taken from Grazia *et al.* [147]; but three characters were changed based on [147]. For these 72 characters, 48 are binary and 24 multistate. Nine characters were analyzed as non-additive. The results of our phylogenetic analyses show that the Primipentatomidae represent a new family, the monophyly of which is supported by a robust bootstrap value. Also, the monophyly of Acanthosomatidae, Pentatomidae, Thyreocoridae, Dinidoridae, Phloeidae, and Scutelleridae are strongly supported by high bootstrap values in the analyses (Figure 17.27). Meanwhile, the family Primipentatomidae is interpreted as the sister group to all Pentatomoidea, exclusive of Saileriolidae + Urostylididae. However, the monophyly of the extremely variable family Cydnidae is not supported. Urostylididae (*sensu lato*) is a paraphyletic group and the relationships among the remaining Pentatomoidea, exclusive of Urostylididae *sensu lato* and Primipentatomidae, is less resolved, but the results of [146] are very similar to the most parsimonious tree of [147].

Suborder Coleorrhyncha Myers & China, 1929

Superfamily Progonocimicoidea Handlirsch, 1906

Family Progonocimicidae Handlirsch, 1906

As mentioned before, Coleorrhyncha are usually known as "moss bugs" and often called "living fossils". The Progonocimicidae are an extinct fossil family in Coleorrhyncha, established by Handlirsch [27]. This family comprise approximately 60 species in 20 genera. Only four fossil species in Progonocimicidae of Coleorrhyncha are from China. These groups with the simple vein on forewing, dSc has long distance from basal cell and vein A-R is short, generally horizontal, M_{3+4} alone or converged with CuA1. Hind wing with four large cells. Width of head is shorter than 0.6 times of pronotum. Antenna with six or seven segments, setaceous, or four to six segments, base segment thick. Hind tarsi with three segments, the first segment is the longest. Hind tibia with spurs. The ninth promerou is lacking a gonostylus.

Genera included from the Jurassic of Northern China: *Cicadocoris* Becker-Migdisova, 1958 and *Karabasia* Wang, Szwedo, Zhang & Lin, 2011.

Cicadocoris Becker-Migdisova, 1958

Cicadocoris Becker-Migdisova, 1958, *Mater Fundam Paleontol.*, 2, 57–67 [37].

Type species: *Cicadocoris kuliki* Becker-Migdisova, 1958.

Tegmen apically rounded, meeting point of M_{3+4} with margin; vein costal margin arcuate; medial cell shorter than 1/2 length of tegmen; basal cell longer than 1/3 length of radial cell; R_1 closer to Rs than dSc; abdomen oval, longer than wide, with the fourth sternite broadest, ovipositor slightly beyond IX sternite.

Distribution and age: Russia, Early Jurassic; Inner Mongolia of China, Middle Jurassic.

Three species included from the Jurassic of Northern China (see Table 17.1).

Cicadocoris anisomeridis Dong, Yao & Ren, 2014 (Figure 17.28)

Cicadocoris anisomeridis Dong, Yao & Ren, 2014: *Syst. Entomol.*, 39 (4), 775.

Locality and horizon: Daohugou, Ningcheng, Inner Mongolia, China; Middle Jurassic, Jiulongshan Formation.

Head no more than half as wide as pronotum; basitarsomere twice as long as mid-tarsomere. Abdomen with laterosternites approximately about 4.10–5.40 times narrower than sternites. Tegmen narrowing in apex of calvus, usually narrower than 1/3 of full length at claval apex. C vein slightly arcuate; costal area narrow. R_1 about 1.11–1.50 times as long as dSc, Rs about 1.3 –1.9 times as long as R_1 [148].

Phylogenetic Study of Coleorrhyncha

In order to evaluate the phylogenetic relationships and to construct a robust hypothesis of the relationships among the genera of Coleorrhyncha, we chose representatives of the suborder Cicadomorpha (*Oncotympana maculaticollis*) as the outgroups [33, 48]. Ten ingroup taxa include eight fossils and two extant taxa. Twenty characters from the adult morphology were identified and scored for all taxa. Results of the phylogenetic analyses, as shown in Figure 17.29, show that the monophyly of Progonocimicidae is supported based on R three-branched and head length less than 0.6 times as long as pronotum. However, the monophyly of Progonocimicinae and Cicadocorinae are not supported by any characters in the analyses,

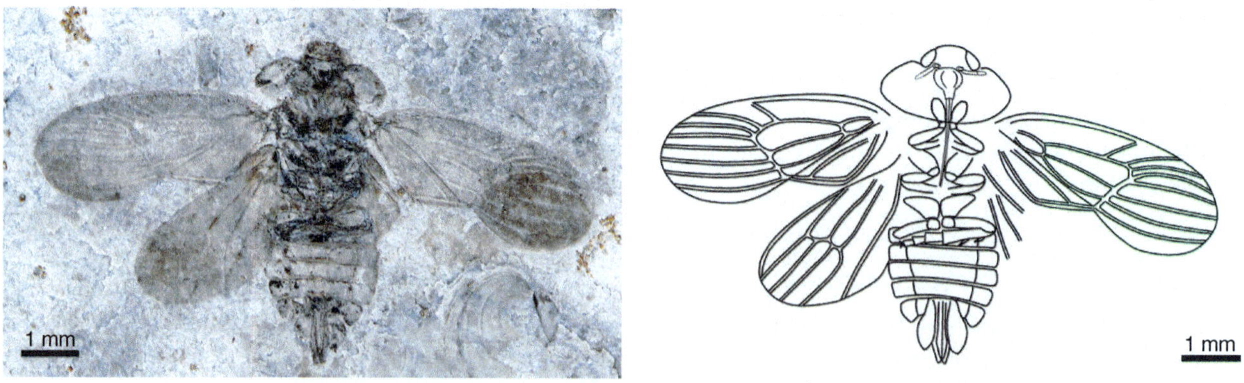

(a) (b)

Figure 17.28 *Cicadocoris anisomeridis* Dong, Yao & Ren, 2014 (Holotype, CNU-PRO-NN2012240p). (a) Photograph; (b) Line drawing [148].

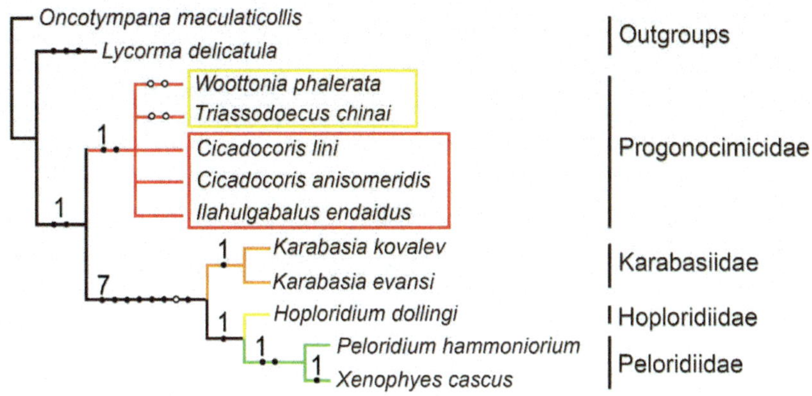

Figure 17.29 The strict consensus tree (modified from [147]): tree length = 26 steps; consistency index = 0.84; retention index = 0.92. • non-homoplasious; o homoplasious. The blue number above the branches are Bremer support values. The species in yellow rectangles belong to Progonocimicinae; the species in red rectangles belong to Cicadocorinae.

and the genera form an unresolved polytomy. Therefore, the two subfamilies should be abolished and we keep just the Progonocimicidae. The Karabasiidae have also traditionally been separated into two subfamilies: Karabasiinae and Hoploridiinae. The results show that the family Karabasiidae as currently understood is paraphyletic with respect to Peloridiidae, with the latter being placed as a sister group to the Karabasiid subfamily Hoplorodiidae [148].

Table 17.1 A list of fossil Heteroptera from the Jurassic and Cretaceous of China.

Family	Species	Locality	Horizon/age	Citation
		Infraorder Gerromorpha Popov, 1971		
Mesoveliidae	*Sinovelia mega* Yao, Zhang & Ren, 2012	Beipiao, Liaoning	Yixian Fm., K_1	Yao *et al.* [55]
	Sinovelia popovi Yao, Zhang & Ren, 2012	Beipiao, Liaoning	Yixian Fm., K_1	Yao *et al.* [55]
		Infraorder Leptopodomorpha Popov, 1971		
Archegocimicidae	*Longianteclypea tibialis* Zhang, Engel, Yao & Ren, 2014	Beipiao, Liaoning	Yixian Fm., K_1	Zhang *et al.* [78]
	Mesolygaeus laiyangensis Ping, 1928	Laiyang, Shandong	Laiyang Fm., K_1	Ping [31]
	Mesolygaeus naevius (Hong in Wang, 1980) Zhang, Engel, Yao & Ren, 2014	Beipiao, Liaoning	Yixian Fm., K_1	Wang [77]
	Propritergum opimum Zhang, Engel, Yao & Ren, 2014	Beipiao, Liaoning	Yixian Fm., K_1	Zhang *et al.* [78]
Saldidae	*Brevrimatus pulchalifer* Zhang, Yao & Ren, 2011	Duolun, Inner Mongolia	Yixian Fm., K_1	Zhang *et al.* [66]
	Venustsalda locella Zhang, Song, Yao & Ren, 2012	Beipiao, Liaoning	Yixian Fm., K_1	Zhang *et al.* [67]
	Luculentsalda maculosa Zhang, Yao & Ren, 2013	Beipiao, Liaoning	Yixian Fm., K_1	Zhang *et al.* [68]
		Infraorder Nepomorpha Popov, 1968		
Belostomatidae	[a]*Sinobelostoma liui* Chou & Hong, 1989	Pingliang, Gansu	Huachi Fm., K_1	Chou and Hong [149]
Corixidae	[a]*Corixopsis tuanwangensis* Hong & Wang 1990	Laiyang, Shandong	Laiyang Fm., K_1	Hong and Wang [76]
	[a]*Crypsacorixa tachis* Lin, 1992	Tuokexun, Xinjiang	Huangshanjie, Fm., T_3	Lin [150]

(Continued)

Table 17.1 (Continued)

Family	Species	Locality	Horizon/Age	Citation
	[a]*Cutitegmena oonis* Lin, 1992	Tuokexun, Xinjiang	Huangshanjie, Fm., T_3	Lin [150]
	Daohugocorixa vulcanica Zhang, 2010	Ningcheng, Inner Mongolia	[b]Jiulongshan Fm., J_2	Zhang [114]
	[a]*Huatingicorixa laozhuanensis* Hong, 1995	Pingliang, Gansu	Huachi Fm., K_1	Hong [83]
	Jiulongshancorixa genuina Zhang, 2010	Luanping, Hebei	Jiulongshan Fm., J_2	Zhang [114]
	Karataviella macra Zhang, 1986	Luanping, Hebei	Xiahuayuan Fm., J_2	Zhang [151]
	Karataviella pontoforma Lin, 1976	Beipiao, Liaoning	Yixian Fm., K_1	Lin [32]
	Karataviella shandongensis Zhang, 1985	Laiyang, Shandong	[b]Laiyang Fm., K_1	Zhang [152]
	Karataviella stolida Zhang, 1986	Chengde, Hebei	Jiulongshan Fm., J_2	Zhang [151]
	Karataviella popovi Zhang, 2010	Ningcheng, Inner Mongolia	[b]Jiulongshan Fm., J_2	Zhang [114]
	[a]*Lacocorixa divena* Lin, 1992	Tuokexun, Xinjiang	Huangshanjie, Fm., T_3	Lin [150]
	[a]*Lufengnecta corragis* Lin, 1977	Lufeng, Yunnan	[b]Pinglang Fm., K_2	Lin [153]
	[a]*Linicorixa odota* Lin, 1980	Shouchang, Zhejiang	Laocun Fm., J_3	Lin [101]
	[a]*Mesocorixa nanligezhuangensis* Hong & Wang, 1990	Laiyang, Shandong	Laiyang Fm., K_1	Hong & Wang [76]
	[a]*Ratiticorixa stenorhinchis* Lin, 1980	Huangshan, Anhui	Yantang Fm., J_3-K_1	Lin [101]
	[a]*Siculicorixa estria* Lin, 1980	Jinhua, Zhejiang	Fangyan Fm., K_1	Lin [101]
	Sigarella tennuis Hong & Wang 1990	Laiyang, Shandong	Laiyang Fm., K_1	Hong and Wang [76]
	[a]*Venacorixa xiangzhongensis* Lin, 1986	Yongzhou, Hunan	[b]Guanyintan Fm., J_1	Lin [154]
	[a]*Vulcanicorixa dorylis* Lin, 1980	Shouchang, Zhejiang	Shouchang Fm., J_3	Lin [101]
	[a]*Yanliaocorixa (Karataviella) chinensis* Lin, 1976	Beipiao, Liaoning	Haifanggou Fm., J_2	Lin [32]
Naucoridae	*Exilcrus cameriferus* Zhang, Yao & Ren, 2011	Beipiao, Liaoning	[b]Yixian Fm., K_1	Zhang *et al.* [93]
	Miroculus laticephlus Zhang, Yao & Ren, 2011	Beipiao, Liaoning	[b]Yixian Fm., K_1	Zhang *et al.* [93]
Notonectidae	*Notonecta (Clypostemma) xyphiale* Popov, 1964	Luanping, Hebei;	Yixian Fm., K_1	Popov [104]
	Notonecta vetula Zhang, Yao & Ren, 2012	Yumen, Gansu	[b]Chijinqiao Fm., K_1	Zhang *et al.* [103]
	[a]*Notonectopsis sinica* Hong & Wang, 1990	Laiyang, Shandong	Laiyang Fm., K_1	Hong and Wang [76]

(Continued)

Table 17.1 (Continued)

Family	Species	Locality	Horizon/Age	Citation
	[a]*Clypostemma limna* Lin, 1980	Lanxi, Zhejiang	Fangyan Fm., K_1	Lin [101]
	Clypostemma petila Zhang, 1985	Laiyang, Shandong	[b]Laiyang Fm., K_1	Zhang [152]
	[a]*Clemmys hydatidosa* Hong, 1982	Jiuquan, Gansu	[b]Chijinqiao Fm., K_1	Hong [106]
Ochteridae	*Angulochterus quatrimaculatus* Yao, Zhang & Ren, 2011	Beipiao, Liaoning	Yixian Fm., K_1	Yao *et al.* [92]
	Floricaudus multilocellus Yao, Ren & Shih, 2011	Beipiao, Liaoning	Yixian Fm., K_1	Yao *et al.* [92]
	Pristinochterus zhangi Yao, Cai & Ren, 2007	Beipiao, Liaoning;	[b]Yixian Fm., K_1	Yao *et al.* [91]
	Pristinochterus ovatus Yao, Zhang & Ren, 2011	Beipiao, Liaoning	Yixian Fm., K_1	Yao *et al.* [92]
	Infraorder Cimicomorpha Leston, Pendergrast & Southwood, 1954			
Anthocoridae	[a]*Mesanthocoris brunneus* Hong & Wang, 1990	Laiyang, Shandong	Laiyang Fm., K_1	Hong and Wang [76]
Ignotingidae	*Ignotingis mirifica* Zhang, Golub & Popov, 2005	Laiyang, Shandong	[b]Laiyang Fm., K_1	Zhang *et al.* [123]
Miridae	*Mirivena robusta* Yao, Cai & Ren, 2007	Ningcheng, Inner Mongolia	Jiulongshan Fm., J_2	Yao *et al.* [122]
Torirostratidae	*Torirostratus pilosus* Yao, Shih & Engel, 2014	Beipiao, Liaoning	Yixian Fm., K_1	Yao *et al.* [26]
	Flexicorpus acutirostratus Yao, Cai & Engel, 2014	Beipiao, Liaoning	Yixian Fm., K_1	Yao *et al.* [26]
Vetanthocoridae	*Byssoidecerus levigatus* Yao, Cai & Ren, 2006	Beipiao, Liaoning	[b]Yixian Fm., K_1	Yao *et al.* [116]
	Collivetanthocoris rapax Yao, Cai & Ren, 2006	Beipiao, Liaoning	[b]Yixian Fm., K_1	Yao *et al.* [116]
	Crassicerus furtivus Yao, Cai & Ren, 2006	Beipiao, Liaoning	[b]Yixian Fm., K_1	Yao *et al.* [116]
	Crassicerus limpiduspterus Tang, Yao & Ren, 2016	Beipiao, Liaoning	Yixian Fm., K_1	Tang *et al.* [119]
	Curticerus venustus Yao, Cai & Ren, 2006	Beipiao, Liaoning	[b]Yixian Fm., K_1	Yao *et al.* [116]
	Curvicaudus ciliatus Yao, Cai & Ren, 2006	Beipiao, Liaoning	[b]Yixian Fm., K_1	Yao *et al.* [116]
	Curvicaudus spinosus Tang, Yao & Ren, 2015	Beipiao, Liaoning	Yixian Fm., K_1	Tang *et al.* [118]
	[a]*Liaoxia longa* Hong, 1987	Kazuo, Liaoning	Shahai Fm., K_1	Hong [115]
	Longilanceolatus tenellus Tang, Yao & Ren, 2015	Beipiao, Liaoning	Yixian Fm., K_1	Tang *et al.* [118]
	Mecopodus xanthos Yao, Cai & Ren, 2006	Beipiao, Liaoning	[b]Yixian Fm., K_1	Yao *et al.* [116]
	Pumilanthocoris gracilis Hou, Yao & Zhang, 2012	Ningcheng, Inner Mongolia	Jiulongshan Fm., J_2	Hou *et al.* [117]
	Pumilanthocoris obesus Hou, Yao & Zhang, 2012	Ningcheng, Inner Mongolia	Jiulongshan Fm., J_2	Hou *et al.* [117]

(Continued)

Table 17.1 (Continued)

Family	Species	Locality	Horizon/Age	Citation
	Punctivetanthocoris pubens Tang, Yao & Ren, 2016	Beipiao, Liaoning	Yixian Fm., K_1	Tang *et al.* [120]
	Pustulithoracalis gloriosus Yao, Cai & Ren, 2006	Beipiao, Liaoning	[b]Yixian Fm., K_1	Yao *et al.* [116]
	Vetanthocoris decorus Yao, Cai & Ren, 2006	Beipiao, Liaoning	[b]Yixian Fm., K_1	Yao *et al.* [116]
	Vetanthocoris longispicus Yao, Cai & Ren, 2006	Beipiao, Liaoning	[b]Yixian Fm., K_1	Yao *et al.* [116]
	Infraorder Pentatomomorpha Leston, Pendergrast & Southwood, 1954			
Coreidae	[a]*Bibiticen hebeiensis* Hong, 1984	Chengde, Hebei	[b]Yixian Fm., K_1	Hong [80]
	[a]*Coriodes longus* Hong, 1987	Kazuo, Liaoning	Shahai Fm., K_1	Hong [115]
	[a]*Kazuocoris liaonngensis* Hong, 1987	Kazuo, Liaoning	Shahai Fm., K_1	Hong [115]
	[a]*Hebeicoris longa* Hong, 1983	Luanping, Hebei	Jiulongshan Fm., J_2	Hong [47]
	[a]*Hebeicoris luanpingensis* Hong, 1983	Luanping, Hebei	Jiulongshan Fm., J_2	Hong [47]
	[a]*Hebeicoris xinboensis* Hong, 1984	Weichang, Hebei	[b]Dabeigou Fm., K_1	Hong [80]
	[a]*Sinocoris oblonga* Hong, 1983	Beipiao, Liaoning	Haifanggou Fm., J_2	Hong [47]
	[a]*Sinocoris ovata* Hong,1983	Beipiao, Liaoning	Haifanggou Fm., J_2	Hong [47]
	[a]*Weichangicoris daobaliangengsis* Hong, 1984	Weichang, Hebei	[b]Dabeigou Fm., K_1	Hong [80]
Cydnidae	*Cilicydnus robustispinus* Yao, Cai & Ren, 2007	Beipiao, Liaoning	Yixian Fm., K_1	Yao *et al.* [145]
	Orienicydnus hongi Yao, Cai & Ren, 2007	Beipiao, Liaoning	Yixian Fm., K_1	Yao *et al.* [145]
Dehiscensicoridae	*Changirostrus maculatus* Du, Yao & Ren, 2017	Beipiao, Liaoning	Yixian Fm., K_1	Du *et al.* [134]
	Crassiantenninus minutus Du, Yao & Ren, 2017	Beipiao, Liaoning	Yixian Fm., K_1	Du *et al.* [134]
	Dehiscensicoris sanctus Du, Yao & Ren, 2017	Beipiao, Liaoning	Yixian Fm., K_1	Du *et al.* [134]
	Minuticoris brunneus Du, Yao & Ren, 2017	Beipiao, Liaoning	Yixian Fm., K_1	Du *et al.* [134]
	Pingquanicoris punctatus Du, Yao & Ren, 2017	Pingquan, Hebei	Yixian Fm., K_1	Du *et al.* [134]
Lygaeidae	[a]*Oligacanchus damiaoense* Hong, 1980	Chifeng, Inner Mongolia	[b]Jiufotang Fm., K_1	Wang [77]
	[a]*Sinolygaeus naevius* Hong, 1980	Chifeng, Inner Mongolia	[b]Jiufotang Fm., K_1	Wang [77]
Pachymeridiidae	*Beipiaocoris multifurcus* Yao, Cai & Ren, 2008	Beipiao, Liaoning	Yixian Fm., K_1	Yao *et al.* [129]
	Bellicoris mirabilis Yao, Cai & Ren, 2008	Beipiao, Liaoning	[b]Yixian Fm., K_1	Yao *et al.* [129]

(Continued)

Table 17.1 (Continued)

Family	Species	Locality	Horizon/Age	Citation
	Corollpachymeridium heteroneurus Lu, Yao & Ren, 2011	Ningcheng, Inner Mongolia	Jiulongshan Fm., J_2	Lu *et al.* [130]
	Nitoculus regilus Yao, Cai & Ren, 2008	Ningcheng, Inner Mongolia	Jiulongshan Fm., J_2	Yao *et al.* [129]
	Sinopachymericium popovi Yao, Cai & Ren, 2006	Ningcheng, Inner Mongolia	Jiulongshan Fm., J_2	Yao *et al.* [128]
	Peregrinpachymeridium comitcola Lu, Yao & Ren, 2011	Ningcheng, Inner Mongolia	Jiulongshan Fm., J_2	Lu *et al.* [130]
	Viriosinervis stolidus Yao, Cai & Ren, 2008	Ningcheng, Inner Mongolia	Jiulongshan Fm., J_2	Yao *et al.* [129]
Primipentatomidae	*Breviscutum lunatum* Yao, Cai, Rider & Ren, 2013	Beipiao, Liaoning	Yixian Fm., K_1	Yao *et al.* [146]
	Oropentatoma epichara Yao, Cai, Rider & Ren, 2013	Beipiao, Liaoning	Yixian Fm., K_1	Yao *et al.* [146]
	Quadrocoris radius Yao, Cai, Rider & Ren, 2013	Beipiao, Liaoning	Yixian Fm., K_1	Yao *et al.* [146]
	Primipentatoma fangi Yao, Cai, Rider & Ren, 2013	Huludao, Liaoning	Jiufotang Fm., K_1	Yao *et al.* [146]
Primipentatomidae	*Primipentatoma peregrina* Yao, Cai, Rider & Ren, 2013	Beipiao, Liaoning	Yixian Fm., K_1	Yao *et al.* [146]
Pyrrhocoridae	[a]*Mesopyrrhocorix fasciata* Hong & Wang, 1990	Laiyang, Shandong	Laiyang, Fm., K_1	Hong & Wang [76]
Rhopalidae	*Longiclavula calvata* Yao, Cai & Ren, 2006	Ningcheng, Inner Mongolia	Jiulongshan Fm., J_2	Yao *et al.* [131]
	Miracorizus punctatus Yao, Cai & Ren, 2006	Ningcheng, Inner Mongolia	Jiulongshan Fm., J_2	Yao *et al.* [131]
	Grandicaputus binpunctatus Yao, Cai & Ren, 2006	Ningcheng, Inner Mongolia	Jiulongshan Fm., J_2	Yao *et al.* [132]
	Originicorizus pyriformis Yao, Cai & Ren, 2006	Ningcheng, Inner Mongolia	Jiulongshan Fm., J_2	Yao *et al.* [132]
	Quatlocellus liae Yao, Cai & Ren, 2006	Ningcheng, Inner Mongolia	Jiulongshan Fm., J_2	Yao *et al.* [132]
	Vescisalignus indecorus Chen, Yao & Ren, 2015	Beipiao, Liaoning	Yixian Fm., K_1	Chen *et al.* [133]
Venicoridae	*Clavaticoris zhengi* Yao, Ren & Cai, 2012	Beipiao, Liaoning	[b]Yixian Fm., K_1	Yao *et al.* [126]
	Halonatusivena shii Du, Yao & Ren, 2016	Beipiao, Liaoning	Yixian Fm., K_1	Du *et al.* [127]
	Halonatusivena nervosus Du, Yao & Ren, 2016	Beipiao, Liaoning	Yixian Fm., K_1	Du *et al.* [127]
	Venicoris solaris Yao, Ren & Rider, 2012	Beipiao, Liaoning	[b]Yixian Fm., K_1	Yao *et al.* [126]
	Suborder Coleorrhyncha Myers & China, 1929			
Karabasiidae	[a]*Karabasia plana* Lin, 1986	Xiwan, Guangxi Zhuang Autonomous Region	Shiti Fm, J_2	Lin [155]

(Continued)

Table 17.1 (Continued)

Family	Species	Locality	Horizon/Age	Citation
Progonocimicidae	*Cicadocoris anisomeridis* Dong, Yao & Ren, 2012	Ningcheng, Inner Mongolia	Jiulongshan Fm, J_2	Dong, *et al.* [148]
	Cicadocoris assimilis Dong, Yao & Ren, 2012	Ningcheng, Inner Mongolia	Jiulongshan Fm, J_2	Dong, *et al.* [154]
	Cicadocoris varians Dong, Yao & Ren, 2012	Ningcheng, Inner Mongolia	Jiulongshan Fm, J_2	Dong, *et al.* [108]
	[a]*Mesocimex(Mesoscytina) brunnea* Hong, 1983	Beipiao, Liaoning	Haifanggou Fm, J_2	Hong [47]
	[a]*Mesocimex lini* Wang, Szwedo & Zhang, 2009	Ningcheng, Inner Mongolia	Jiulongshan Fm, J_2	Wang, *et al.* [48]
	[a]*Mesocimex sinensis* Hong, 1983	Beipiao, Liaoning	Haifanggou Fm, J_2	Hong [47]

a) The species is not presented in the main text because the original description, photos, and line-drawings are not precise and the holotype cannot be rechecked.
b) Horizon/Age revised from the original paper based on updated information and data.

References

1 Weirauch, C. and Schuh, R.T. (2011). Systematics and evolution of Heteroptera: 25 years of progress. *Annual Review of Enotomology* **56**: 487–510.

2 Li, H., Leavengood, J.M.J., Chapman, E.G. *et al.* (2017). Mitochondrial phylogenomics of Hemiptera reveals adaptive innovations driving the diversification of true bugs. *Proceedings of the Royal Society B: Biological Sciences* **284**: 20171223. https://doi.org/10.1098/rspb.2017.1223.

3 Burckhardt, D. (2009). Taxonomy and phylogeny of the Gondwanan moss bugs or Peloridiidae (Hemiptera, Coleorrhyncha). *Deutsche Entomologische Zeitschrift* **56** (2): 173–235.

4 Burckhardt, D., Bochud, E., Damgaard, J. *et al.* (2011). A review of the moss bug genus Xenophyes (Hemiptera: Coleorrhyncha: Peloridiidae) from New Zealand: systematics and biogeography. *Zootaxa* **2** (923): 1–26.

5 Stål, C. (1855). Pentatoma halys. *Ofversigt af Kongliga Svenska Vetenskaps-Akademiens Förhandlingar* **12**: 182.

6 Hamilton, G.C. and Shearer, P.W. (2003). Brown marmorated stink bug – a new exotic insect in New Jersey. FS002. Rutgers Cooperative Extension, New Jersey Agricultural Experiment Station, Rutgers, NJAES.

7 Aldrich, J.R. (1988). Chemical ecology of the Heteroptera. *Annual Review of Entomology* **33**: 211–238.

8 Conradl, A.F. (1904). Variations in the protective value of odoriferous secretions of some Heteroptera. *Science* **19**: 393–394.

9 Blum, M.S. and Traynham, J.G. (1960). The chemistry of the pentatomid scent gland. *XI International Congress of Entomology, Vienna* **3**: 48–52.

10 Blum, M.S. (1961). The presence of 2–hexenal in the scent gland of the pentatomid (*Brochymena quadripustulata*). *Annals of the Entomological Society of America* **54**: 410–412.

11 Roth, L.M. (1961). A study of the odoriferous glands of *Scaptocoris divergens* (Hemiptera: Cydnidae). *Annals of the Entomological Society of American* **54**: 900–911.

12 Waterhouse, D.F., Forss, D.A., and Hackman, R.H. (1961). Characteristic odour components of the scent of stink bugs. *Journal of Insect Physiology* **6**: 113–121.

13 Remold, H. (1962). Über die biologische Bedeutung der Duftdrüsen bei den Landwanzen (Geocorisae). *Journal of Comparative Physiology* **45**: 636–694.

14 Staddon, B.W. (1979). The scent glands of Heteroptera. *Advances in Insect Physiology* **14**: 351–418.

15 Alcock, J. (1973). The feeding response of hand-reared red-winged blackbirds (*Agelaius phoeniceus*) to a stinkbug (*Euschistlls consperslls*). *American Midland Naturalist* **89**: 307–313.

16 Schlee, M.A. (1986). Avian predation on Heteroptera: experiments on the European blackbird *Turdus m. merula* L. *Ethology* **73**: 1–18.

17 Lehane, M.J. (2005). *The Biology of Blood-Sucking in Insects*. New York: Cambridge University Press.

18 Adams, T.S. (1999). Hematophagy and hormone release. *Annals of the Entomological Society of America* **92**: 1–13.

19 Sandoval, C.M., Joya, M., Gutiérrez, R., and Angulo, V.M. (2000). Cleptohaemathophagia of the Triatomine bug *Belminus herreri*. *Medical and Veterinary Entomology* **14**: 100–101.

20 Sandoval, C.M., Duarte, R., Gutiérrez, R. *et al.* (2004). Feeding sources and natural infection of *Belminus herreri* (Hemiptera, Reduviidae, Triatominae) from dwellings in Cesar, Colombia. *Memórias Do Instituto Oswaldo Cruz* **99** (2): 137–140.

21 Pérez-Molina, J.A. and Molina, I. (2018). Chagas disease. *Lancet* **391**: 82–94. https://doi.org/10.1016/S0140-6736(17)31612-4.

22 Balashov, Y.S. (1984). Interaction between blood-sucking arthropods and their hosts, and its influence on vector potential. *Annual Review of Entomology* **29**: 137–156.

23 Ribeiro, J.M.C. (1995). Blood-feeding arthropods: live syringes or invertebrate pharmacologists? *Infectious Agents and Disease* **4**: 143–152.

24 Grimaldi, D.A. and Engel, M.S. (eds.) (2005). *Evolution of the Insects*. New York: Cambridge University Press.

25 Ribeiro, J.M.C., Assumpção, T.C., and Francischetti, I.M.B. (2012). An insight into the sialomes of blood-sucking Heteroptera. *Psyche Stuttgart* e470436. https://doi.org/10.1155/2012/470436.

26 Yao, Y.Z., Cai, W.Z., Xu, X. *et al.* (2014). Blood-feeding true bugs in the Early Cretaceous. *Current Biology* **24**: 1786–1792.

27 Handlirsch, A. (1906–1908). *Die fossilen insekten und die phylogenie der rezentlen formen. Ein handbuch für paläontologen und zoologlen*, 1–1430. Leipzig: Engelmann (in German).

28 Handlirsch, A. (1939). Neue Untersuchungen über die fossilen Insekten, part 2. *Annalen des Naturhistorischen Museums in Wien* **49**: 1–240. (in German).

29 Tillyard, R.J. (1918). Mesozoic insects of Queensland. 3. Odonata and Protodonata. *Proceedings of the Linnean Society of New South Wales* **43**: 417–435.

30 Martynov, A.V. (1938). Permian fossil insects from the Arkhangelsk district. Part V. The family Euthygrammatidae and its relationships (with the description of a new genus and family from Chekarda). *Trudy Paleontologicheskogo institut, Akademiya Nauk SSSR* **7**: 69–80.

31 Ping, C. (1928). Cretaceous fossil insects of China. *Palaeontologia Sinica, Series B* **13** (1): 5–51.

32 Lin, Q.B. (1976). The Jurassic fossil insects from Western Liaoning. *Acta Palaeontologica Sinica* **15** (1): 97–116.

33 Popov, Y.A. and Shcherbakov, D.E. (1991). Mesozoic Peloridioidea and their ancestors (Insecta: Hemiptera, Coleorrhyncha). *Geology Palaeontology* **25**: 215–235.

34 Tillyard, R.J. (1926). Kansas Permian insects. Part 6. Additions to the orders Protohymenoptera and Odonata. *American Journal of Science, Series 5* **11** (2): 58–73.

35 Myers, J.G. and China, W.E. (1929). The systematic position of the Peloridiidae as elucidated by a further study of the external anatomy of *Hemiodoecus leai* China. *Annals and Magazine of Natural History* **3**: 282–294.

36 Evans, J.W. (1956). Palaeozoic and Mesozoic Hemiptera (Insecta). *Australian Journal of Zoology* **4**: 165–258.

37 Becker-Migdisova, E.E. (1958). New fossil Homoptera. Pt. 1. *Material for the Fundamentals of Paleontology* **2**: 57–67. (In Russian).

38 Wootton, R.J. (1963). Actinoscytinidae (Hemiptera: Heteroptera) from the Upper Triassic of Queensland. *Annals and Magazine of Nature History* **6**: 249–255.

39 Popov, Y.A. and Wootton, R.J. (1977). The upper Liassic Heteroptera of Mecklenburg and Saxony. *Systematic Entomology* **2**: 333–351.

40 Popov, Y.A. (1982). Upper Jurassic hemipterans, genus *Olgamartynovia* (Hemiptera, Progonocimicidae) from Central Asia. *Paleontology Journal* **2**: 78–94.

41 Popov, Y.A. (1985). Jurassic bugs and Coleotthyncha of southern Siberia and western Mongolia. In: *Jurassic Insects of Siberia and Mongolia* (ed. I.D. Sukacheva and A.P. Rasnitsyn), 28–47. Moscow: Trudy Paleontology Instuitut, Akademie Nauk (in Russian).

42 Popov, Y.A. (1986). Peloridiinas and bugs: suborders Peloridiina (= Coleorrhyncha) et Cimicina (= Heteroptera). In: *Insects in Early Cretaceous Ecosystems of Western Mongolia*, vol. **28** (ed. A.P. Rasnitsyn), 50–83. Moscow: Tr. Sovm. Sovet. – Mongol. Paleontol. Eksped, Publish House "Nauka" (in Russian).

43 Popov, Y.A. (1988). New Mesozoic water boatmen (Corixidae, Shurabellidae). In: *New Species of Fossil Invertebrates of Mongolia*, vol. **33** (ed. A.Y. Rozanov), 63–71. Moscow: Tr. Sovm. Sovet. – Mongol. Paleontol. Eksped, Publish House "Nauka" (in Russian).

44 Popov, Y.A. (1988). New Mesozoic Coleorrhyncha and Heteroptera from eastern Transbaikalia. *Paleontologicheski Zhurnal* **4**: 67–77. (in Russian).

45 Popov, Y.A. (1989). Some aspects of systematic of Leptopodoidea. *Acta Biologica Silesiana* **13** (30): 63–68.

46 Popov, Y.A., Dolling, W.R., and Whalley, P.E.S. (1994). British Upper Triassic and Lower Jurassic Heteroptera and Coleorrhyncha (Insecta: Hemiptera). *Genus* **5** (4): 307–347.

47 Hong, Y.C. (1983). *Middle Jurassic Fossil Insects in North China*. Beijing: Geological Publishing House (in Chinese).

48 Wang, B., Szwedo, J., and Zhang, H.C. (2009). Jurassic Progonocimicidae (Hemiptera) from China and phylogenetic evolution of Coleorrhyncha. *Science in China, Series D, Earth Sciences* **52**: 1953–1961.

49 Andersen, N.M. and Weir, T.A. (2004). Mesoveliidae, Hebridae, and Hydrometridae of Australia (Hemiptera: Heteroptera: Gerromorpha), with a reanalysis of the phylogeny of semiaquatic bugs. *Invertebrate Systematics* **18**: 467–522.

50 Andersen, N.M. and Polhemus, J.T. (1980). Four new genera of Mesoveliidae (Hemiptera, Gerromorpha) and the phylogeny and classification of the family. *Insect Systematics & Evolution* **11** (4): 369–392.

51 Andersen, N.M. (1982). The semiaquatic bugs (Hemiptera, Gerromorpha). Phylogeny, adaptations, biogeography and classification. *Systematic Zoology* **32** (4): 69–71.

52 Aukema, B. and Rieger, C. (1995). *Catalogue of the Heteroptera of the Palaearctic Region. Vol. 1. Enicocephalomorpha, Dipsocoromorpha, Nepomorpha, Gerromorpha and Leptopodomorpha*, 77–79. Amsterdam: Netherlands Entomological Society.

53 Chen, P.P., Nieser, N., and Zettel, H. (2005). *The Aquatic and Semiaquatic Bugs (Heteroptera: Nepomorpha & Gerromorpha) of Malesia*. Leiden-Boston: Brill Academic Publishers.

54 Garrouste, R. and Nel, A. (2010). First semi-aquatic bugs Mesoveliidae and Hebridae (Hemiptera: Heteroptera: Gerromorpha) in Miocene Dominican amber. *Insect Systematics & Evolution* **41**: 93–102.

55 Yao, Y.Z., Zhang, W.T., and Ren, D. (2012). The first report of Mesoveliidae (Heteroptera: Gerromorpha) from the Yixian Formation of China and its taxonomic significance. *Alcheringa: An Australasian Journal of Palaeontology* **36** (1): 107–116.

56 Jell, P.A. and Duncan, P.M. (1986). Invertebrates, mainly insects, from the freshwater, Lower Cretaceous, Koonwarra fossil bed (Korumburra group), South Gippsland, Victoria. *Memoirs of the Association of Australasian Palaentologists* **3**: 111–205.

57 Bekker-Migdisova, E.E. (1962). Order Heteroptera: Heteropterans, or True Bugs. In: *Osnovy Paleontologii. Tom 9. Chlenistonogie. Trakheinye i Khelitserovye (Fundamentals of Palaentology: Vol. 9. Arthropoda: Tracheata and Chelicerata)* (ed. B.B. Rohdendorf), 208–225. Moscow: Aademy of Sciences (in Russian).

58 Bode, A. (1953). Die Insektenfauna des Ostniedersächsischen Oberen Lias. *Palaeontographica* **103A**: 1–517.

59 Martill, D.M. (2007). The age of the Cretaceous Santana Formation fossil Konservat Lagerstätte of north-east Brazil: a historical review and an appraisal of the biochronostratigraphic utility of its palaeobiota. *Cretaceous Research* **28**: 895–920.

60 Popov, Y.A. and Bechly, G. (2007). Heteroptera: bugs. In: *The Crato Fossil Beds of Brazil: Window into an Ancient World* (ed. D.M. Martill, G. Bechly and R.F. Loverige), 317–328. Cambridge: Cambridge University Press.

61 Schuh, R.T. and Polhemus, J.T. (2009). Revision and analysis of Pseudosaldula Cobben (Insecta: Hemiptera: Saldidae): a group with a classic Andean distribution. *Bulletin of the American Museum of Natural History* **323**: 1–102.

62 Polhemus, J.T. and Chapman, H.C. (1979). Family Saldidae/shore bugs in the semiaquatic and aquatic Hemiptera of California (Heteroptera: Hemiptera). In: *Bulletin of the California Insect Survey*, vol. **21** (ed. A.S. Menke), 16–33. Berkeley: University of California Press.

63 Brooks, A.R. and Kelton, L.A. (1967). Aquatic and semiaquatic Heteroptera of Alberta, Saskatchewan, and Manitoba (Hemiptera). *Memoirs of the Entomological Society of Canada* **51**: 1–92.

64 Poinar, G. and Buckley, R. (2009). *Palaeoleptus burmanicus* n. gen., n. sp., an Early Cretaceous shore bug (Hemiptera: Palaeoleptidae n. fam.) in Burmese amber. *Cretaceous Research* **30**: 1000–1004.

65 Schuh, R.T., Galil, B., and Polhemus, J.T. (1987). Catalog and bibliography of Leptopodomorpha (Heteroptera). *American Museum of Natural History* **185**: 243–406.

66 Zhang, W.T., Yao, Y.Z., and Ren, D. (2011). New shore bug (Hemiptera, Heteroptera, Saldidae) from the Early Cretaceous of China with phylogenetic analyses. *ZooKeys* **130**: 185–198.

67 Zhang, W.T., Song, J.J., Yao, Y.Z., and Ren, D. (2012). A new fossil Saldidae (Hemiptera: Heteroptera: Leptopodomorpha) from the Early Cretaceous in China. *Zootaxa* **3273**: 63–68.

68 Zhang, W.T., Yao, Y.Z., and Ren, D. (2013). A new Early Cretaceous shore bug (Hemiptera: Heteroptera: Saldidae) from China. *Zootaxa* **3647** (4): 585–592.

69 Ryzhkova, O.V. (2012). New Saldoid bugs of the family Enicocoridae (Hemiptera: Heteroptera: Leptopodomorpha) from the Lower Cretaceous of Mongolia. *Paleontological Journal* **46** (5): 485–494.

70 Ryzhkova, O.V. (2015). New Saldidae-Enicocorinae (Heteroptera: Leptopodomorpha) from the Lower Cretaceous of Siberia and Mongolia. *Paleontological Journal* **49** (2): 153–161.

71 Germar, E.F. and Berendt, G.C. (1856). Die im Bernstein befindlichen Hemipteren und Orthopteren der Vorwelt. In: *Die im Bernstein befindlichen organischen Reste der Vorwelt gesammelt in Verbindung mit Mehrenen* (ed. G.C. Berendt), 1–140. Berlin (Germany), 2 (1): Bearbeitet und herausgegeben.

72 Statz, G. and Wagner, E. (1950). Geocorisae (Landwanzen) aus den Oberoligocäner Ablagerungen von Rott. *Palaeontographica*, (A) **48**: 97–136.

73 Lewis, S.E. (1969). Fossil insects of the Latah Formation (Miocene) of Eastern Washington and Northern Idaho. *Northwest Science* **41** (3): 99–115.

74 Handlirsch, A. (1925). Palaeontologie. In: *Handbuch der Entomologie Band 3: Geschichte, Literatur, Technik, Paläontologie, Phylogenie, Systematik* (ed. C. Schröder), 117–306. Jena: Verlag von Gustav Fischer.

75 Popov, Y.A. (1980). Heteroptera from the Lower Cretaceous deposits of locality Manlay. *Transactions of the joint Soviet-Mongolian Paleontological Expedition* **13**: 48–51. (in Russian).

76 Hong, Y.C. and Wang, W.L. (1990). Fossil insects of the Laiyang Formation. In: *The Stratigraphy and Palaeontogy of Laiyang Basin, Shandong Province* (ed. Regional Geological Surveying Team and Shandong Bureau of Geology and Mineral Resources), 44–189. Beijing: Geological Publishing House (in Chinese).

77 Wang, W.L. (1980). Class Insecta. In: *Paleontological Atlas of North East China Part 2* (ed. Shenyang Institute of Geology and Mineral), 137–142. Beijing: Geological Publishing House (in Chinese).

78 Zhang, W.T., Engel, M.S., Yao, Y.Z. *et al.* (2014). The Mesozoic family Archegocimicidae and phylogeny of the infraorder Leptopodomorpha (Hemiptera). *Journal of Systematic Palaeontology* **12** (1): 93–111.

79 Hong, Y.C. (1981). Discovery of new Early Cretaceous insects from Xishan Beijing. *Tianjin Institute of Geology and Mineral Resources Bulletin* **4**: 87–94. (in Chinese).

80 Hong, Y.C. (1984). Insecta. In: *Paleontological Atlas of North China. Mesozoic Volume* (ed. Tianjin Institute of Geology and Mineral Resources), 128–185. Beijing: Geological Publishing House (in Chinese).

81 Hong, Y.C. (1984). New fossil insects of Laiyang group from Laiyang Basin, Shandong Province. *Professional Papers of Stratigraphy and Paleontology* **18**: 31–34. (in Chinese).

82 Zhang, J.F. (1991). Going further into late Mesozoic Mesolygaeids (Heteroptera, Insecta). *Acta Palaeontologica Sinica* **30**: 679–704. (in Chinese with English abstract).

83 Hong, Y.C. (1995). Fossil insects of the southern Ordos Basin. *Acta Geologica Gansu* **4** (1): 1–13. (in Chinese).

84 Schuh, R.T. and Polhemus, J.T. (1980). Analysis of taxonomic congruence among morphological, ecological, and biogeographic data sets for the Leptopodomorpha (Hemiptera). *Systematic Zoology* **29** (1): 1–26.

85 Wheeler, W.C., Schuh, R.T., and Bang, R. (1993). Cladistic relationships among higher groups of Heteroptera: congruence between morphological and molecular data sets. *Entomologica Scandinavica* **24**: 121–137.

86 Forero, D. (2008). The systematics of the Hemiptera. *Revista Colombiana de Entomología* **34** (1): 1–21.

87 Li, H.M., Deng, R.Q., Wang, J.W. *et al.* (2005). A preliminary phylogeny of the Pentatomomorpha (Hemiptera: Heteroptera) based on nuclear 18S rRNA and mitochondrial DNA sequences. *Molecular Phylogenetic and Evolution* **37**: 313–326.

88 Baehr, M. (1989). Review of the Australian Ochteridae (Insecta, Heteroptera). *Spixiana (München)* **11** (2): 111–126.

89 Schuh, R.T. and Slater, J.A. (1995). *True Bugs of the World (Hemiptera: Heteroptera) Classification and Natural History*. Ithaca: Cornell University Press, New York.

90 Shcherbakov, D.E. and Popov, Y.A. (2002). Superorder Cimicidea Laicharting, 1781. Order Hemiptera Linné, 1758. In: *The Bugs, Cicadas, Plantlice, Scale Insects, etc. History of Insects* (ed. A.P. Rasnitsyn and D.L.J. Quicke), 143–157. Dordrecht: Kluwer Academic Publishers.

91 Yao, Y.Z., Cai, W.Z., and Ren, D. (2007). *Pristinochterus* gen. n. (Hemiptera: Ochteridae) from the upper Mesozoic of Northeastern China. *European Journal of Entomology* **104**: 827–835.

92 Yao, Y.Z., Zhang, W.T., Ren, D., and Shih, C.K. (2011). New fossil Ochteridae (Hemiptera: Heteroptera: Ochteroidea) from the Upper Mesozoic of north-eastern China, with phylogeny of the family. *European Systematic Entomology* **36**: 589–600.

93 Zhang, W.T., Yao, Y.Z., and Ren, D. (2011). First description of fossil Naucoridae (Heteroptera: Nepomorpha) from late Mesozoic of China. *Acta Geologica Sinica* **85** (2): 490–500.

94 Popov, Y.A. (1968). True Hemipterous insects of the Jurassic Karatau Fauna (Heteroptera). In: *Jurassic Insects of the Karatau* (ed. B.B. Rodendorf), 99–113. Moscow: Nauka (in Russian).

95 Popov, Y.A. (1971). Historical development of Hemiptera infraorder Nepomorpha (Heteroptera). *Trudy Paleontologicheskogo Instituta Akademii Nauk SSSR* **139**: 1–230. (in Russian).

96 Ruf, M.L., Goodwyn, P.P., and Martins-Neto, R.G. (2005). New Heteroptera (Insecta) from the Santana Formation, Lower Cretaceous (Northeastern Brazil), with description of a new family and new taxa of Naucoridae and Gelastocoridae. *Gaea* **1** (2): 68–74.

97 Polhemus, J.T. and Polhemus, D.A. (2008). Global diversity of true bugs (Heteroptera; Insecta) in freshwater. *Hydrobiologia* **595**: 379–391.

98 Parsons, M.C. (1970). Respiratory significance of the thoracic and abdominal morphology of the three aquatic bugs Ambrysus, Notonecta and Hesperocorixa (Insecta, Heteroptera). *Zoomorphology* **66**: 242–298.

99 Usinger, R.L. (1956). Aquatic Hemiptera. In: *Aquatic Insects of California* (ed. R.L. Usinger), 182–229. Berkeley: University of California Press.

100 Truxal, F.S. (1979). Family Notonectidae: backswimmers. In: *The Semiaquatic and Aquatic Hemiptera of California (Heteroptera: Hemiptera)*, vol. **21** (ed. A.S. Menke), 139–148. Bulletin of the California Insect Survey.

101 Lin, Q.B. (1980). *Fossil Insects from the Mesozoic of Zhejiang and Anhui in Division and Correlation of the Mesozoic Volcano Sedimentary Beds in Zhejiang and Anhui, China*. Beijing: Nanjing Institute of Geology and Paleontology, Academia Sinica Publish, Science Press (in Chinese).

102 Lin, Q.B. (1982). Mesozoic and Cenozoic insects. In: *Atlas of Palaeontology from Northwest Region, Shaanganning Division* (ed. Xian Institute of Geology and Mineral Resources), 70–83. Beijing: Geological Publishing House (in Chinese).

103 Zhang, W.T., Yao, Y.Z., and Ren, D. (2012). Phylogenetic analysis of a new fossil Notonectidae (Heteroptera: Nepomorpha) from the Late Jurassic of China. *Alcheringa: An Australasian Journal of Palaeontology* **36** (2): 239–250.

104 Popov, Y.A. (1964). A new subfamily of the Mesozoic water bugs (Heteroptera) from Transbaikalia. *Palaeontologicheskii Zhurnal* **2**: 63–71. (in Russian).

105 Zhang, W.T., Yao, Y.Z., and Ren, D. (2012). A revision of the species *Notonecta xyphiale* (Popov, 1964) (Heteroptera: Notonectidae). *Cretaceous Research* **33**: 159–164.

106 Hong, Y.C. (1982). *Mesozoic Fossil Insects of Jiuquan Basin in Gansu Province*. Beijing: Geological Publishing House (in Chinese).

107 Hong, Y.C. (1992). Middle and Late Jurassic insects. In: *Palaeontological Atlas of Jilin Province* (ed. Bureau of Geology and Mineral Resources Jilin Province), 410–425. Jilin: Jilin Science and Technology Press (in Chinese).

108 Dong, Q.P., Yao, Y.Z., and Ren, D. (2012). A new species of Progonocimicidae (Hemiptera, Coleorrhyncha) from Northeastern China. *Zootaxa* **3495**: 73–78.

109 Linnaeus, C.V. (1758). *Systema Naturae per Regna Tria Naturae: Secundum Classes, Ordines, Genera, Species, Cum Characteribus Differentiis Synonymis, Locis*, 10e, Revised Edition. Holmiae: Laurentius Salvius.

110 Scudder, G.G.E. (1976). Water-boatmen of saline waters (Hemiptera: Corixidae). In: *Marine Insects* (ed. L. Cheng), 263–289. Amsterdam: North Holland Publishers.

111 Scudder, G.G.E. (1983). A review of factors governing the distribution of two closely related Corixids in the Saline Lakes of British Columbia. *Hydrobiologia* **105**: 143–154.

112 Lauck, D.R. (1979). "Family Corixidae", the semiaquatic and aquatic Hemiptera of California (Heteroptera: Hemiptera). *Bulletin California Insect Survey* 87–123.

113 Becker-Migdisove, E.E. (1949). Mesozoic homoptera of middle Asia. *Trudy Paleontologicheskogo Instituta Akadamii Nauk. SSSR* **22**: 1–68.

114 Zhang, J.F. (2010). Revision and description of water boatmen from the Middle–Upper Jurassic of Northern and Northeastern China (Insecta: Hemiptera: Heteroptera: Corixidae). *Paleontological Journal* **44** (5): 515–525.

115 Hong, Y.C. (1987). The study of Early Cretaceous fossil insects of Kazuo, West Liaoning. In: *Professional Papers of Stratigraphy and Palaeontology*, vol. **18**, 76–85. (in Chinese).

116 Yao, Y.Z., Cai, W.Z., and Ren, D. (2006). Fossil flower bugs (Heteroptera: Cimicomorpha: Cimicoidea) from the Late Jurassic of Northeast China, including a new family, Vetanthocoridae. *Zootaxa* **1360**: 1–40.

117 Hou, W.J., Yao, Y.Z., and Ren, D. (2012). The earliest fossil flower bugs (Heteroptera: Cimicomorpha: Cimicoidea: Vetanthocoridae) from the Middle Jurassic of Inner Mongolia, China. *European Journal of Entomology* **109**: 281–288.

118 Tang, D., Yao, Y.Z., and Ren, D. (2015). New fossil flower bugs (Heteroptera: Cimicomorpha: Cimicoidea: Vetanthocoridae) with uniquely long ovipositor from the Yixian Formation (Lower Cretaceous), China. *Cretaceous Research* **56**: 504–509.

119 Tang, D., Yao, Y.Z., and Ren, D. (2016). A new species of Vetanthocoridae (Heteroptera: Cimicomorpha) from the Lower Cretaceous of China. *Cretaceous Research* **64**: 30–35.

120 Tang, D., Yao, Y.Z., and Ren, D. (2017). Phylogenetic position of the extinct insect family Vetanthocoridae (Heteroptera) in Cimiciformes. *Journal of Systematic Palaeontology* **15** (9): 697–708.

121 Hong, Y.C. (2002). *Amber Insects of China*. Beijing: Beijing Science and Technology Press (in Chinese).

122 Yao, Y.Z., Cai, W.Z., and Ren, D. (2007). The oldest fossil plant bug (Heteroptera: Miridae) from Middle Jurassic of Inner Mongolia, China. *Zootaxa* **1442**: 37–41.

123 Zhang, J.F., Golub, V.B., Popov, Y.A., and Shcherbakov, D.E. (2005). Ignotingidae fam. nov. (Insecta: Heteroptera: Tingoidea), the earliest lace bugs from the Upper Mesozoic of eastern China. *Cretaceous Research* **26**: 783–792.

124 Perrichot, V., Nel, A., Guilbert, E., and Néraudeau, D. (2006). Fossil Tingoidea (Heteroptera: Cimicomorpha) from French Cretaceous amber, including Tingidae and a new family, Ebboidae. *Zootaxa* **1203**: 57–68.

125 Golub, V.B., Popov, Y.A., and Arillo, A. (2012). Hispanocaderidae n. fam. (Hemiptera: Heteroptera: Tingoidea), one of the oldest lace bugs from the Lower Cretaceous Álava amber (Spain). *Zootaxa* **3270**: 41–50.

126 Yao, Y.Z., Ren, D., and Rider, D.A. (2012). Phylogeny of the Infraorder Pentatomomorpha based on fossil and extant morphology, with description of a new fossil family from China. *PLoS One* **7** (5): e37289. https://doi.org/10.1371/journal.pone.0037289.

127 Du, S.L., Yao, Y.Z., and Ren, D. (2016). New fossil species of the Venicoridae (Heteroptera: Pentatomomorpha) from the Lower Cretaceous of Northeast China. *Cretaceous Research* **68**: 21–27.

128 Yao, Y.Z., Cai, W.Z., and Ren, D. (2006). *Sinopachymeridium popovi* gen. And sp. nov. – a new fossil true bug (Heteroptera: Pachymeridiiae) from the Middle Jurassic of Inner Mongolia, China. *Annales Zoologici (Warszawa)* **56** (4): 753–756.

129 Yao, Y.Z., Cai, W.Z., and Ren, D. (2008). Jurassic fossil true bugs of the Pachymeridiidae (Hemiptera: Pentatomomorpha) from Northeast China. *Acta Geologica Sinica* **82** (1): 35–47.

130 Lu, Y., Yap, Y.Z., and Ren, D. (2011). Two new genera and species of fossil true bugs (Hemiptera: Heteroptera: Pachymeridiidae) from Northeastern China. *Zootaxa* **2835**: 41–52.

131 Yao, Y.Z., Cai, W.Z., and Ren, D. (2006). The first discovery of fossil rhopalids (Heteroptera: Coreoidea) from Middle Jurassic of Inner Mongolia, China. *Zootaxa* **1269**: 57–68.

132 Yao, Y.Z., Cai, W.Z., Ren, D., and Shih, C.K. (2006). New fossil rhopalids (Heteroptera: Coreoidea) from the Middle Jurassic of Inner Mongolia, China. *Zootaxa* **1384**: 41–58.

133 Chen, X.T., Yao, Y.Z., and Ren, D. (2015). A new genus and species of Rhopalidae (Hemiptera: Heteroptera) from the Early Cretaceous of Liaoning Province, China. *Zootaxa* **4058** (1): 135.

134 Du, S.L., Yao, Y.Z., Ren, D., and Zhang, W.T. (2017). Dehiscensicoridae fam. nov. (Insecta: Heteroptera: Pentatomomorpha) from the Upper Mesozoic of Northeast China. *Journal of Systematic Palaeontology* **15** (12): 991–1013. https://doi.org/10.1080/14772019.2016.1259665.

135 Carver, M., Gross, G.F., and Woodward, T.E. (1991). Hemiptera (bugs, leafhoppers, cicadas, aphids, scale insects etc.). In: *The Insects of Australia*, 2e (ed. I.D. Naumann), 429–509. Melbourne: Melbourne University Press.

136 Zheng, L.Y. (1999). Class Insecta: order Hemiptera: suborder Heteroptera (= order Hemiptera s. str.). In: *Insect Classification* (ed. L.Y. Zheng and H. Gui), 442–520. Nanjing: Nanjing Normal University Press (in Chinese).

137 Henry, T.J. (2009). Biodiversity of the Heteroptera. In: *Insect Biodiversity: Science and Society* (ed. R.G. Foottit and P.H. Adler), 223–263. Oxford: Wiley-Blackwell.

138 Henry, T.J. and Froeschner, R.C. (1988). *Catalog of the Heteroptera, or True Bugs, of Canada and the Continental United States*, 12–28. Boca Raton: St. Lucie Press.

139 Li, X.Z. (1994). A preliminary study on the phylogeny of Rhopalidae (Hemiptera: Coreoidea). *Acta Zootaxonomica Sinica* **3947** (1): 527–542.

140 Schaefer, C.W. (1993). The Pentatomomorpha (Hemiptera: Heteroptera): an annotated outline of its systematic history. *Europe Journal of Entomology* **90**: 105–122.

141 Henry, T.J. (1997). Phylogenetic analysis of family groups within the infraoder Pentatomomorpha (Hemiptera: Heteroptera), with emphasis on the Lygaeoidea. *Annals of the Entomological Society of America* **90**: 275–301.

142 Xie, Q., Bu, W.J., and Zheng, L.Y. (2005). The Bayesian phylogenetic analysis of the 18S rRNA sequences from the main lineages of Trichophora (Insecta: Heteroptera: Pentatomomorpha). *Molecular Phylogenetic and Evolution* **37** (2): 313–326.

143 Hua, J.M., Li, M., Dong, P.Z. *et al.* (2008). Comparative and phylogenomic studies on the mitochondrial

genomes of Pentatomomorpha (Insecta: Hemiptera: Heteroptera). *BMC Genomics* **9**: 610.

144 Xie, Q., Tian, Y., Zheng, L.Y., and Bu, W.J. (2008). 18S rRNA hyper-elongation and the phylogeny of Euhemiptera (Insecta: Hemiptera). *Molecular Phylogenetics and Evolution* **47**: 463–471.

145 Yao, Y.Z., Cai, W.Z., and Ren, D. (2007). The first fossil Cydnidae (Hemiptera: Pentatomoidea) from the Late Mesozoic of China. *Zootaxa* **1388**: 59–68.

146 Yao, Y.Z., Cai, W.Z., Rider, D.A., and Ren, D. (2013). Primipentatomidae fam. nov. (Hemiptera: Heteroptera: Pentatomomorpha), an extinct insect family from the Cretaceous of north-eastern China. *Journal of Systematic Palaeontology* **11** (1): 63–82.

147 Grazia, J., Schuh, R.T., and Wheeler, W.C. (2008). Phylogenetic relationships of family groups in Pentatomoidea based on morphology and DNA sequences (Insecta: Heteroptera). *Cladistics* **24**: 1–45.

148 Dong, Q.P., Yao, Y.Z., and Ren, D. (2014). New fossil Progonocimicidae (Hemiptera: Coleorrhyncha: Progonocimicoidea) from the Upper Mesozoic of Northeastern China, with a phylogeny of Coleorrhyncha. *Systematic Entomology* **39**: 773–782.

149 Chou, Y. and Hong, Y.C. (1989). An Early Cretaceous new genus and species in Shaanganning basin (Insecta: Heteroptera). *Entomotaxonomia* **3**: 197–205. (in Chinese with English abstract).

150 Lin, Q.B. (1992). Late Triassic insect Fauna from Toksun, Xinjiang. *Acta Palaeontologica Sinica* **31** (3): 313–335.

151 Zhang, J.F. (1986). Some fossil insects from the Jurassic of northern Hebei, China. In: *The Paleontology and Stratigraphy of Shandong* (ed. Paleontology Society of Shandong, China), 74–84. Beijing: Ocean Press (in Chinese).

152 Zhang, J.F. (1985). New data of the Mesozoic fossil insects from Laiyang in Shandong. *Geology of Shandong* **1** (2): 23–39.

153 Lin, Q.B. (1977). Insect fossils f rom South China. In: *Insect Fossils from Yunnan Province*, vol. **2**, 373–381. Beijing: Science Press.

154 Dong, Q.P., Yao, Y.Z., and Ren, D. (2012). A new species of Progonocimicidae (Hemiptera, Coleorrhyncha) from the Middle Jurassic of China. *Alcheringa* **37**: 31–37.

155 Lin, Q.B. (1986). Early Mesozoic fossil insects from South China. In: *Palaeontologla Sinica, New Series B*, 54–56. Beijing: Science Press (in Chinese).

18

Megaloptera – Dobsonflies, Fishflies, and Alderflies

Yongjie Wang[1], *Chungkun Shih*[1,2], *and Dong Ren*[1]

[1] Capital Normal University, Haidian District, Beijing, China
[2] National Museum of Natural History, Smithsonian Institution, Washington, DC, USA

18.1 Introduction to Megaloptera

Megaloptera, as one of the three orders of Neuropterida, are a small order of holometabolous insects, comprising ca. 373 extant species within two families of Corydalidae (dobsonflies and fishflies) and Sialidae (alderflies). The megalopterans are characterized by large body size, flat and prognathous adult head, broad anal area in hind wing and aquatic larval stages [1]. The adults of megalopterans are often found near the streams or swamps owing to their exclusively aquatic larvae. The order is commonly considered as the most basal lineage within Neuropterida, and recent phylogeny confirms its sister relationship with Neuroptera [2, 3].

Corydalidae, including two subfamilies of Corydalinae (dobsonflies) and Chauliodinae (fishflies), is the largest family within Megaloptera, comprising 295 extant species in 27 genera [4]. Corydalinae (dobsonflies), with 160 extant species, are restricted in North and South America, South Africa, and Asia. Some species of dobsonflies with wingspans beyond 200 mm are among the largest insects. The aquatic larvae of the Corydalinae have eight pairs of lateral gills in adapting the life under the water. The adults of male dobsonflies generally possess extremely protruded mandibles which are used during the courtship to attract the females, but these formidable insects are harmless to humans. However, the reduced mandibles are also present in some dobsonflies such as *Neoneuromus*, *Protohermes*, etc. Liu *et al.* considered that the nuptial gifts (spermatophores) may possibly be an alternative explanation on the reduced mandibles [5].

Chauliodinae (fishflies), the other subfamily of Corydalidae, have ca. 135 extant species and distributed in North and South America, South Africa, Asia, New Zealand, and Australia. The fishflies live mainly in the subtropical or warm temperate regions, however, some species can adapt to certain harsh habitats and range in a broad area. The fishflies have dimorphic antennae, i.e. pinniform in females (Figure 18.1) and filiform in males, which are distinctly different from those of dobsonflies.

The family Sialidae (alderflies), as a group of "living fossils", comprise 12 extant genera and 87 species [6]. Compared with Corydalidae, adult alderflies are diagnosed by relatively small body size, the absence of ocelli, shortened pronotum and dilated and bilobed fourth tarsomeres.

The larvae of Megaloptera, including dobsonflies, fishflies and alderflies, provide food to trout, bass and other fresh-water fish. Their life cycle, in comparison with larvae of mayflies, stoneflies and caddisflies, does not create as good fly fishing opportunities. However, these larvae provide good food and good fishing for the fresh-water fishermen.

A paraphyletic Megaloptera was proposed by some researchers who supported the idea of Sialidae branching off from Megaloptera and being sister to Raphidioptera [2, 7, 8]. However, the recent phylogenetic results, using both morphological and molecular data, have confirmed the monophyly of Megaloptera [9–14]. Although the interrelationships within Sialidae are not fully resolved, the basalmost clade of Megaloptera is corroborated (Figure 18.2). According to divergence time estimated by Wang *et al.*, the origin time of Megaloptera was ca. 297 Mya and the divergence time of Sialidae and Corydalidae was ca. 251 Mya at the beginning of Triassic [3].

18.2 Progress in the Studies of Fossil Megaloptera

Similar to other extant species-poor orders, the Megaloptera fossils are also scarce including about 47 fossil species [15]. The oldest megalopterans in an extinct

Rhythms of Insect Evolution: Evidence from the Jurassic and Cretaceous in Northern China, First Edition.
Edited by Dong Ren, Chungkun Shih, Taiping Gao, Yongjie Wang, and Yunzhi Yao.

Figure 18.1 A female fishfly (Corydalidae: Chauliodinae). Source: Photo by Jason Shih.

family Parasialidae have been documented from the Late Permian of Russia and Mongolia [16, 17]. In fact, many megalopteran-like fossils have been transferred to other families, e.g. Euchauliodidae, Permosialidae, and Tychtodelopteridae [18]. Notably, numerous earlier fossil larvae were putatively assigned to the Megaloptera, due to the gill-like appendages on the abdomen resembling the extant Corydalidae, but the absence of the adults implied the systematic position of these larvae are still questionable. In China, only two genera of fishflies, i.e. *Jurochauliodes* Wang & Zhang, 2010 [19] and *Eochauliodes* Liu, Wang, Shih, Ren & Yang, 2012 [20], have been described from the Middle Jurassic of Inner Mongolia, which are also the oldest convincing evidence of Corydalidae [19, 20].

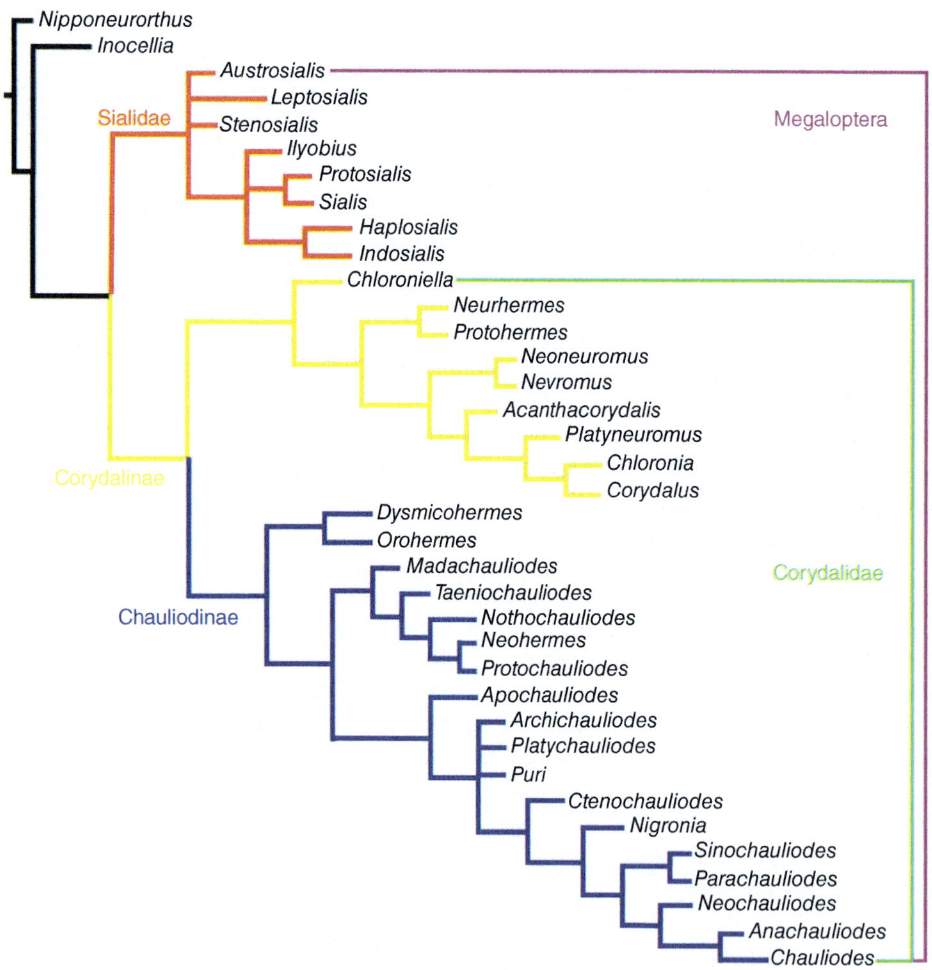

Figure 18.2 Phylogeny of Megaloptera. Source: modified from [14]).

18.3 Representative Fossils of Megaloptera from Northern China

Family Corydalidae Leach, 1815

Most fossil corydalids belong to the subfamily Corydalinae (dobsonflies) that are known from the Paleogene, with putative stem-group dobsonflies in the Early Cretaceous [4]. The earliest chauliodines (fishflies) are known from the Middle Jurassic of China.

Genera included from the Jurassic of Northern China: *Jurochauliodes* Wang & Zhang, 2010 and *Eochauliodes* Liu, Wang, Shih, Ren & Yang, 2012.

Phylogeny of Fishflies in Chauliodinae

The Middle Jurassic fishflies from China as the earliest fishfly adults promote the understanding of the evolution of the Chauliodinae [20]. In the phylogenetic results (Figure 18.3), the earliest Jurassic genus *Jurochauliodes* as the basalmost clade is sister to other genera which is clearly distributed in distinct clades with the other Jurassic genus *Eochauliodes* and the recent Asian genera. It indicates that these insects with close current distributions should have no direct ancestor-descendant relationships. The *Archichauliodes* clade as the second divergence is distributed in western Northern America. Interestingly, the other Jurassic genus *Eochauliodes* and an Early Cretaceous *Cretochaulus* are grouped to be sisters to the extant *Protochauliodes* clade that are restricted in southern South America, eastern Australia, Madagascar, and South Africa. It implies the origin and global dispersion of fishflies should have completed before the breakup of Pangaea which was much earlier than the current fossil records.

Jurochauliodes Wang & Zhang, 2010

Jurochauliodes Wang & Zhang, 2010, *J. Paleontol.* 84 (4), 777. [19] (original designation).

Type species: *Jurochauliodes ponomarenkoi* Wang & Zhang, 2010.

The specific epithet is in honor of Alexandr G. Ponomarenko, an outstanding Russian paleontologist and entomologist. The genus was firstly established by Wang and Zhang in 2010, based on a larva from the Middle Jurassic of China [19]. Liu *et al.*, in 2012, supplemented the adult features and re-described the genus [20]. The genus is characterized by medium body size (forewing length, 35.0 mm); broadly elliptical wings; Rs 4-branched, each branch bifurcated; MA bifurcated

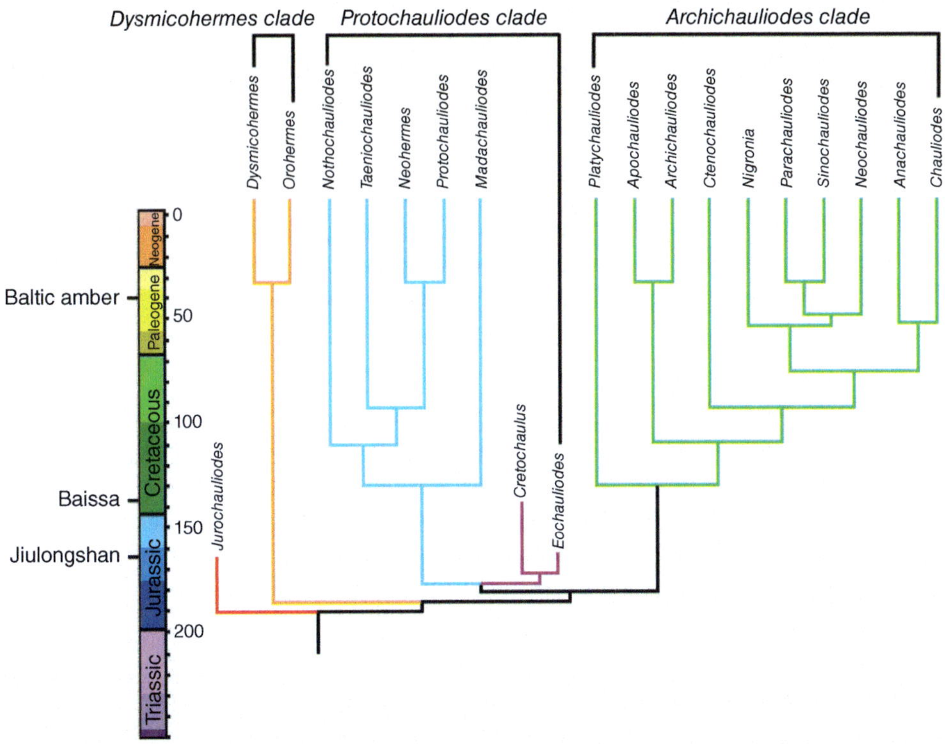

Figure 18.3 Phylogeny of fishflies, both fossil and extant. Source: modified from [20]).

Figure 18.4 *Eochauliodes striolatus* Liu, Wang, Shih, Ren & Yang, 2012. (a), Habitus of holotype CNU-MEG-NN2011004; (b), Larva, CNU-MEG-NN2011008. Source: Larva fossil donated by Dr. Chungkun Shih.

distad; MP 2-branched, with anterior branch distally 3 or 4-branched, and with posterior branch usually bifurcated; 1A, 2A, and 3A all present in forewing, each with two branches and straightly extending; 1A and 2A connected by a short cross-vein in forewing; larva: medium-sized (body length of final instar larva about 40.0 mm by estimation). Pronotum subquadrate, distinctly wider than long, and also wider than head; lateral abdominal gills on segments nearly as long as width of respective segments, and slightly shorter than hind legs; Spiracles on segment 8 not protruding as a pair of tubes [20].

Distribution and age: Inner Mongolia; Middle Jurassic.

Only one species included from the Jurassic of Northern China (see Table 18.1).

***Eochauliodes* Liu, Wang, Shih, Ren & Yang, 2012**

Eochauliodes Liu, Wang, Shih, Ren & Yang, 2012, *PLoS ONE*, 7 (7), e40345 [20] (original designation).

Type species: *Eochauliodes striolatus* Liu, Wang, Shih, Ren & Yang, 2012.

Adult medium-sized (forewing length, 30.0 mm). Antenna filiform. Ocelli close to each other. Wings narrowly elongated, nearly 3.0 times as long as wide; membrane hyaline, with a few dark stripes along longitudinal veins. Nygmata present between MA and MP. Rs 2-branched, each branch with a distal fork; MA bifurcated distad; MP 2-branched, with anterior branch distally forked; 1A, 2A, and 3A all present in forewing, each with two branches and curved near wing margin; 1A and 2A connected by a short cross-vein in forewing. Larva medium-sized (body length of final instar larvae around 40.0 mm by estimation). Antenna with terminal two segments nearly as long as second segment. Pronotum subquadrate, slightly wider than long. Lateral abdominal gills on segments 1–5 distinctly longer than width of respective segments, and also longer than hind legs. Spiracles on segment 8 not protruding as a pair of tubes [20].

Distribution and age: Inner Mongolia; Middle Jurassic.

Only one species included from the Jurassic of Northern China (see Table 18.1).

***Eochauliodes striolatus* Liu, Wang, Shih, Ren & Yang, 2012 (Figure 18.4)**

Eochauliodes striolatus Liu, Wang, Shih, Ren & Yang, 2012: *PLoS ONE*, 7 (7), e40345.

Locality and horizon: Daohugou, Ningcheng, Inner Mongolia, China; Middle Jurassic, Jiulongshan Formation.

The species is distinguished in appearance from all other species of Chauliodinae by the longitudinal dark stripes on the wings, bifurcated MA, bifurcated anterior branch of MP in the forewing, the 2A connected to 1A by a short cross-vein; larva, head and thorax rather dark, pronotum slightly wider than long, and lateral abdominal gills distinctly longer than hind legs [20].

Table 18.1 A list of fossil Megaloptera from the Jurassic of China.

Family	Species	Locality	Horizon/Age	Citation
Corydalidae	*Eochauliodes striolatus* Liu, Wang, Shih, Ren & Yang, 2010	Ningcheng, Inner Mongolia	Jiulongshan Fm., J_2	Liu *et al.* [20]
	Jurochauliodes ponomarenkoi Wang & Zhang 2010	Ningcheng, Inner Mongolia	Jiulongshan Fm., J_2	Wang and Zhang [19]

References

1 New, T.R. and Theischinger, G. (1993). Megaloptera (Alderflies, Dobsonflies). *Handbuch der Zoologie (Berlin)* **4**: 1–97.

2 Winterton, S.L., Hardy, N.B., and Wiegmann, B.M. (2010). On wings of lace: phylogeny and Bayesian divergence time estimates of Neuropterida (Insecta) based on morphological and molecular data. *Systematic Entomology* **35**: 349–378.

3 Wang, Y.Y., Liu, X.Y., Garzón-Orduña, I.J. *et al.* (2016). Mitochondrial phylogenomics illuminates the evolutionary history of Neuropterida. *Cladistics* **33**: 617–636.

4 Engel, M.S., Winterton, S.L., and Breitkreuz, L.C. (2018). Phylogeny and evolution of Neuropterida: where have wings of lace taken us. *Annual Review of Entomology* **63**: 531–551.

5 Liu, X.Y., Hayashi, F., Lavine, L.C., and Yang, D. (2015). Is diversification in male reproductive traits driven by evolutionary trade-offs between weapons and nuptial gifts? *Proceedings of the Royal Society B: Biological Sciences* **282**: 20150247. https://doi.org/10.1098/rspb.2015.0247.

6 Liu, X.Y., Hayashi, F., and Yang, D. (2014). Phylogeny of the family Sialidae (Insecta: Megaloptera) inferred from morphological data, with implications for generic classification and historical biogeography. *Cladistics* **31** (1): 1–32. https://doi.org/10.1111/cla.12071.

7 Hennig, W. (1953). Kritische Bemerkungen zum phylogenetischen system der Insekten. *Beiträgezur zur Entomologie* **3**: 1–85.

8 Štys, P. and Bilinski, S. (1990). Ovariole types and the phylogeny of hexapods. *Biological Reviews* **65**: 401–429.

9 Aspöck, U., Plant, J.D., and Nemeschkal, H.L. (2001). Cladistic analysis of Neuroptera and their systematic position within Neuropterida (Insecta: Holometabola: Neuroptera). *Systematic Entomology* **26**: 73–86.

10 Aspöck, U., Haring, E., and Aspöck, H. (2012). The phylogeny of the Neuropterida: long lasting and current controversies and challenges (Insecta: Endopterygota). *Arthropod Systematics and Phylogeny* **70**: 119–129.

11 Haring, E. and Aspöck, U. (2004). Phylogeny of the Neuropterida: a first molecular approach. *Systematic Entomology* **29**: 415–430.

12 Wang, Y.Y., Liu, X.Y., Winterton, S.L., and Yang, D. (2012). The first mitochondrial genome for the fishfly subfamily Chauliodinae and implications for the higher phylogeny of Megaloptera. *PLoS One* **7**: e47302.

13 Zhao, C.J., Liu, X.Y., and Yang, D. (2014). Wing base structural data support the sister relationship of Megaloptera and Neuroptera (Insecta: Neuropterida). *PLoS One* **9**: e114695.

14 Liu, X.Y., Lü, Y.N., Aspöck, H. *et al.* (2016). Homology of the genital sclerites of Megaloptera (Insecta: Neuropterida) and their phylogenetic relevance. *Systematic Entomology* **41** (1): 256–286.

15 Oswald, J.D. (2017) Neuropterida species of the world. Version 5.0. http://lacewing.tamu.edu. Last accessed 27 September 2018.

16 Ponomarenko, A.G. (1977). Paleozoic alderflies (Insecta, Megaloptera). *Paleontological Journal* **11**: 73–81.

17 Ponomarenko, A.G. (2000). New alderflies (Megaloptera: Parasialidae) and glosselytrodeans (Glosselytrodea: Glosselytridae) from the Permian of Mongolia. *Paleontological Journal* **34**: S309–S311.

18 Ansorge, J. (2001). *Dobbertinia recticulata* Handlirsch 1920 from the Lower Jurassic of Dobbertin (Mecklenburg/Germany) – the oldest representative of Sialidae (Megaloptera). *Neues Jahrbuch Fur Geologie Und Palaontologie Monatshefte* **9**: 553–564.

19 Wang, B. and Zhang, H.C. (2010). Earliest evidence of fishflies (Megaloptera: Corydalidae): an exquisitely preserved larva from the Middle Jurassic of China. *Paleontological Journal* **84**: 774–780.

20 Liu, X.Y., Wang, Y.J., Shih, C.K. *et al.* (2012). Early evolution and historical biogeography of fishflies (Megaloptera: Chauliodinae): implications from a phylogeny combining fossil and extant taxa. *PloS One* **7**: e40345.

19

Raphidioptera – Snakeflies

Hui Fang [1], Yongjie Wang [1], Dong Ren [1], and Chungkun Shih [1,2]

[1] Capital Normal University, Haidian District, Beijing, China
[2] National Museum of Natural History, Smithsonian Institution, Washington, DC, USA

19.1 Introduction to Raphidioptera

Raphidioptera, commonly called "snakeflies", constitute a distinctive, small holometabolous order, belonging to the superorder Neuropterida. The adult snakeflies are characterized by a prognathous head, narrowly elongate prothorax extended like a neck, transparent wings with intricate venation (Figure 19.1), and a long ovipositor in females (Figure 19.2). The long head can be raised high above the body, wings are folded roof-like when at rest [1].

The extant Raphidioptera comprise *ca.* 33 genera and 240 species, placed in only two families of Raphidiidae and Inocelliidae [1]. They are generally entomophagous at both larval and adult stages, but the adults of Inocelliidae have not been observed to feed, and some adult snakeflies have been reported to feed on pollen [2]. Female snakeflies use their needle-like ovipositors to deposit eggs under bark crevices of coniferous and deciduous trees [3]. Snakefly larvae, similar to certain beetle larvae, are predatory on eggs, larvae of various insects, Auchenorrhyncha (cicadas, leafhoppers, treehoppers, planthoppers, and spittlebugs), and adults of minute arthropods such as mites, spiders, springtails, barklice, Sternorrhyncha (aphids, whiteflies, and scale insects) [4]. The larvae usually take about two years (from one to three years) to reach maturity [3]. Pupae are exarate with free appendages and active. Adults are also generalist predators, consuming prey similar to those of the larvae. As we know, snakeflies serve as beneficial insects for controlling some pests. However, snakefly larvae are hosts for parasitic wasps, such as Ichneumonidae, Braconidae, and Perilampidae [3].

Two prerequisites for the occurrence of extant snakeflies are arboreal biotopes and a climate characterized by markedly low temperatures [1, 2, 5]. Low temperature triggers pupation and is a precondition for wing development of adult snakeflies. Nowadays, snakeflies are mainly distributed in the Holarctic region where typically cold winters occur, while there are a few species in the Oriental region and Central America in some high-altitude mountainous areas [1]. The western Palearctic has the richest species diversity of extant Raphidioptera, while East and Southeast Asia are becoming to be known as another diversity center of snakeflies due to the rapidly increasing number of species described there since 2013 [6].

19.2 Progress in the Studies of Fossil Raphidioptera

The first Mesozoic snakefly was described by Martynov in 1925 [7]. It is remarkable that snakeflies were prosperous in the Mesozoic, with high species diversity and global distribution. Fossil Raphidioptera comprise *ca.* 100 species in six extinct families: Chrysoraphidiidae, Baissopteridae, Mesoraphidiidae, Priscaenigmatidae, Metaraphidiidae and Juroraphidiidae from Eurasia, North America and South America in compression fossils and amber [6, 8].

Many extinct Mesozoic snakeflies had a relatively broad and short prothorax, although extant snakeflies have a narrowly elongate prothorax. Jurassic snakeflies are relatively scarce, with only seven genera and 15 species described within the Priscaenigmatidae, Mesoraphidiidae, and Metaraphidiidae. However, Cretaceous snakeflies seem to have been extraordinarily diverse based on their fossil records, representing more than half of all fossil snakefly species [1]. Due to significant extinction at the end of the Cretaceous, extant snakefly species have become a relict group restricted to the northern hemisphere [1, 5, 9].

Fossil Raphidioptera from China comprise 16 genera and 33 species in four families ranged from the Middle Jurassic to the Early Cretaceous (Table 19.1),

Rhythms of Insect Evolution: Evidence from the Jurassic and Cretaceous in Northern China, First Edition.
Edited by Dong Ren, Chungkun Shih, Taiping Gao, Yongjie Wang, and Yunzhi Yao.

Figure 19.1 *Agulla* sp. (in Raphidiidae) with elongate prothorax and intricate venation. Source: Photo by Dr. Xingyue Liu.

Figure 19.2 *Inocellia japonica* (in Inocelliidae) with a long ovipositor. Source: Photo by Dr. Xingyue Liu.

representing nearly 1/3 of the described fossil snakeflies in the world. The first fossil species was reported by Hong in 1982 [10].

In 2014, Liu, Ren and Yang provided a phylogenetic tree of Raphidioptera including fossil and extant snakeflies, which shed light on the evolution of Raphidioptera (Figure 19.3) [1]. In their work, Raphidioptera is confirmed to be monophyletic, and the monophyly of Priscaenigmatomorpha and Raphidiomorpha is also confirmed. Juroraphidiidae is assigned to be the sister-group of Raphidiomorpha. Priscaenigmatomorpha is the sister-group of the clade consisting of Juroraphidiidae and Raphidiomorpha. Within Priscaenigmatomorpha, Chrysoraphidiidae is the sister group of Priscaenigmatidae. *Hondelagia* and *Priscaenigma* form a monophyletic group, supporting the monophyly of Priscaenigmatidae. Within Raphidiomorpha, Baissopteridae is the sister-group of the remaining four families, which are split into two lineages, Mesoraphidiidae + Metaraphidiidae and Raphidiidae + Inocelliidae [1].

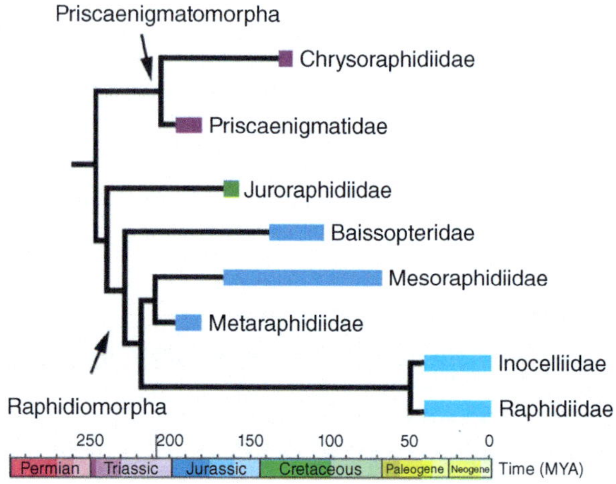

Figure 19.3 Evolutionary chronogram of Raphidioptera based on present result. Source: modified from [1].

As the sister-group of Megaloptera + Neuroptera, Raphidioptera should also have diverged from Neuropterida during the Late Permian (*ca.* 250 Mya), based on molecular clock approaches [11, 12]. However, the earliest currently known fossil snakefly is of the Early Jurassic age [1]. The Early Jurassic appears to be a crucial period for the diversification of snakefly suborders, which possibly indicates a much earlier origination of Raphidioptera than the Early Jurassic (Figure 19.3). During the Middle and Late Jurassic, the subtropical desert regions reduced in size and snakeflies could have dispersed north-south during this period [1]. Raphidioptera were widely distributed in Laurasia and Gondwana during the Early Cretaceous, which was probably caused by north-south dispersal of some snakefly lineages before the breakup of Pangaea in the Late Jurassic (*ca.* 155 Mya). The Cretaceous snakefly fossils from the Southern Hemisphere show close affinity with the Mesozoic snakeflies from the Northern Hemisphere [1].

19.3 Representative Fossils of Raphidioptera from Northern China

Family Baissopteridae Martynova, 1961

Pterostigma usually with cross-vein closing it proximally, with at least one incorporated branch of RA. Forewing: rich venation with 3–5 branches of RP, many cross-veins: 3–5 ra-pr, 5–15 ir, 4–7 r-m, 3–5 im, 2–3 icu. Hind wing: many cross-veins: 4–5 ra-pr, 5–14 ir, 4–6 r-m, 2–4 m-cu [13].

Genera included from the Cretaceous of Northern China: *Baissoptera* Martynova, 1961 and *Microbaissoptera* Lyu, Ren & Liu, 2017.

Figure 19.4 *Baissoptera bicolor* Lyu, Ren & Liu, 2017, (Holotype, CNU-RAP-LB-2017028).

***Baissoptera* Martynova, 1961**

Baissoptera Martynova, 1961, *Paleontol. Zh.*, 3, 80 [13] (original designation).

Type species: *Baissoptera martinsoni* Martynova, 1961.

The specific epithet is dedicated to Dr. G.G. Martinson for the studies of deposits of Transbaikalia and Mongolia. Ocelli present; pronotum shorter than head length. Wings broad, less than 4.0 times as long as wide; ScP terminating into costal margin posteriad wing midpoint; cross-veins in radial and medial areas forming two or more gradate series; numbers of cross-veins in radial and medial areas forming three or more closed radial cells [13–15].

Distribution and age: Liaoning; Early Cretaceous.

Five species included from the Cretaceous of Northern China (see Table 19.1).

***Baissoptera bicolor* Lyu, Ren & Liu, 2017 (Figure 19.4)**

Baissoptera bicolor Lyu, Ren & Liu, 2017: *Cretac. Res.*, 80, 17.

Locality and horizon: Huangbanjigou, Beipiao, Liaoning, China; Lower Cretaceous, Yixian Formation.

Body medium-sized; pronotum slightly longer than wide. Wings broad, approximately three to four times as long as wide; pterostigma pale-colored proximally and dark-colored distally, ca one-fifth of wing length, and a pterostigmal cross-vein present; cross-veins in radial and medial areas forming two gradate series in posterior part of fore- and hind wings [15].

***Microbaissoptera* Lyu, Ren & Liu, 2017**

Microbaissoptera Lyu, Ren & Liu, 2017, *Cretac. Res.*, 80, 22 [15] (original designation).

Type species: *Microbaissoptera monosticha* Lyu, Ren & Liu, 2017.

Body small-sized (body length probably less than 14.3 mm, forewing length 8.7–10.8 mm); ocelli present; prothorax elongate. Fore- and hind wings broad; cross-veins in radial and medial areas forming only one gradate series in posterior portion and only a single cross-vein present between RP branches [15].

Distribution and age: Liaoning; Early Cretaceous.

Only one species included from the Cretaceous of Northern China (see Table 19.1).

***Microbaissoptera monosticha* Lyu, Ren & Liu, 2017 (Figure 19.5)**

Microbaissoptera monosticha Lyu, Ren & Liu, 2017: *Cretac. Res.*, 80, 24.

Locality and horizon: Huangbanjigou, Beipiao, Liaoning, China; Lower Cretaceous, Yixian Formation.

Body 14.3 mm long, head ovoid. Flagellum with more than 40 flagellomeres. Both fore- and hind wings bearing uniformly colored pterostigma with an oblique pterostigmal cross-vein. Costal space preserved with seven cross-veins; ScP terminating into posterior costal margin at wing midpoint; two simple RA veinlets present; a single series of gradate cross-veins in radial and medial areas present in both wings [15].

Family Chrysoraphidiidae Liu, Ren & Yang, 2014

Wings relatively broad; pterostigma distinct in both wings; Sc closely to C; Rs + MA separating from R near wing base; forewing MA simple; two series of gradate cross-veins in radial to mediocubital areas in forewing; 1A pectinate in forewing [1, 6].

Figure 19.5 *Microbaissoptera monosticha* Lyu, Ren & Liu, 2017, (Holotype, CNU-RAP-LB-2017061).

Only one genus included from the Cretaceous of Northern China: *Chrysoraphidia* Liu, Makarkin, Yang & Ren, 2013.

Chrysoraphidia Liu, Makarkin, Yang & Ren, 2013

Chrysoraphidia Liu, Makarkin, Yang & Ren, 2013, *Cretac. Res.*, 45, 307 [6] (original designation).

Type species: *Chrysoraphidia relicta* Liu, Makarkin, Yang & Ren, 2013.

Medium-sized; others same as family characters.

Distribution and age: Liaoning; Early Cretaceous.

Only one species included from the Cretaceous of Northern China (see Table 19.1).

Family Juroraphidiidae Liu, Ren & Yang, 2014

Prothorax much longer than combined lengths of meso- and metathorax; pterostigma long; costal region rather narrow; forewing MP with two simple main branches; only one discoidal cell present between main branches of MP; CuA and CuP having a distinct common stem; 1A short, simple [1].

Only one genus included from the Jurassic of Northern China: *Juroraphidia* Liu, Ren & Yang, 2014.

Juroraphidia Liu, Ren & Yang, 2014

Juroraphidia Liu, Ren & Yang, 2014, *BMC Evol. Biol.*, 14: 84, 2 [1] (original designation).

Type species: *Juroraphidia longicollum* Liu, Ren & Yang, 2014.

Small-sized raphidiopterans (forewing 6.1–8.6 mm long). Others as family characters.

Distribution and age: Inner Mongolia; Middle Jurassic.

Only one species included from the Jurassic of Northern China (see Table 19.1).

Juroraphidia longicollum Liu, Ren & Yang, 2014 (Figure 19.6)

Juroraphidia longicollum Liu, Ren & Yang, 2014: *BMC Evol. Biol.*, 14: 84, 3.

Locality and horizon: Daohugou, Ningcheng, Inner Mongolia, China; Middle Jurassic, Jiulongshan Formation.

Body well-preserved except for metathorax and abdomen; head length 2.1 mm; ocelli not observed; mouthparts with labrum and mandibles preserved; pronotum 3.1 mm long; mesothorax 1.3 mm long. Forewing 8.9 mm long, 2.7 mm wide; costal area nearly as wide as subcostal area; Sc terminate near distal ending of pterostigma by weak fusion with C; pterostigma elongate, nearly half of forewing length; Rs with four simple branches; one gradate cross-vein series present; MP deeply branched into two long simple branches. Hind wing 7.2 mm long and 2.3 mm wide, shorter than forewing, with anal area distinctly narrowed [1].

Figure 19.6 *Juroraphidia longicollum* Liu, Ren & Yang, 2014, (Paratype, CNU-RAP-NN-2013001p).

Family Mesoraphidiidae Martynov, 1925

Body medium to large. Head short or elongate. Three ocelli positioned between posterior half of compound eyes. Wings narrowly elongate, with relatively few cross-veins in radial and medial fields. ScP terminating into C near wing mid-length; a single scp-ra cross-vein; pterostigma short or elongate, proximally closed by a cross-vein scp-ra, with or without pterostigmal cross-vein, distally closed by a veinlet of RA. Forewing: 2ra-rp cross-vein posteriorly connecting to anterior branch of RP; MA originating close to first branching point of MP and coalescent with stem of RP for a short distance; two cross-veins present between MA and MP, forming two long medial cells; CuA connecting to M distad of separation of M from R; three discoidal cells posterior to MP, forming a triangle; CuA distally with three branches; 1A and 2A simple. Hind wing: 3ra-rp cross-vein posteriorly connecting to anterior branch of RP; MA originating from stem of R, and coalescent with RP for a short distance; CuA distally bifurcate; one discal cell and one or two discoidal cells present [7, 16–18].

Genera included from the Jurassic and Cretaceous of Northern China: *Mesoraphidia* Martynov, 1925; *Ororaphidia* Engel & Ren, 2008; *Stenoraphidia* Lyu, Ren & Liu, 2018 and *Styporaphidia* Engel & Ren, 2008.

Mesoraphidia Martynov, 1925

Mesoraphidia Martynov, 1925, *Izv. Ross. Akad. Nauk*, 19, 235 [7] (original designation).

Type species: *Mesoraphidia grandis* Martynov, 1925.

Body length 6.5–21.0 mm; three ocelli present. Forewing length 5.5–19.0 mm; few cross-veins in radial and medial fields; ScP terminating into C near wing mid-length; pterostigma short or elongate, with or without pterostigmal cross-vein; two long medial cells between MA and MP; three discoidal cells posterior to MP [7, 18].

Distribution and age: Inner Mongolia and Liaoning, Middle Jurassic; Liaoning, Early Cretaceous.

Eleven species included from the Jurassic and Cretaceous of Northern China (see Table 19.1).

***Mesoraphidia daohugouensis* Lyu, Ren & Liu, 2015 (Figure 19.7)**
Mesoraphidia daohugouensis Lyu, Ren & Liu, 2015: *Zootaxa*, 3999, 561.

Locality and horizon: Daohugou, Ningcheng, Inner Mongolia, China; Middle Jurassic, Jiulongshan Formation.

Body of medium size; pronotum less than half of combined lengths of meso- and metanotum. Wings narrowly elongate; ScP terminating into C around wing mid-length; pterostigma about one-third of wing length, without pterostigmal cross-vein; forewing RP trifurcate; hind wing RP with anterior branch simple or distally bifurcate with simple posterior branch; MA with two simple branches in both wings [8, 18].

***Ororaphidia* Engel & Ren, 2008**
Ororaphidia Engel & Ren, 2008, *J. Kans. Entomol. Soc.*, 81 (3), 188 [19] (original designation).

Type species: *Ororaphidia megalocephala* Engel & Ren, 2008.

Pronotum with lateral margins, ovipositor not long. A cross-vein in pterostigmal area; forewing Sc terminating into C beyond wing mid-length; forewing MA originating before MP fork; MP with three medial cells; 1A simple; hind wing MP with a single medial cell [18, 19].

Figure 19.7 *Mesoraphidia daohugouensis* Lyu, Ren & Liu, 2015, (Holotype, CNU-RAP-NN-2015001).

Figure 19.8 *Ororaphidia bifurcata* Lyu, Ren & Liu, 2017, (Holotype, CNU-RAP-NN-2016016p).

Distribution and age: Inner Mongolia; Middle Jurassic.

Two species included from the Jurassic of Northern China (see Table 19.1).

***Ororaphidia bifurcata* Lyu, Ren & Liu, 2017 (Figure 19.8)**
Ororaphidia bifurcata Lyu, Ren & Liu, 2017: *Alcheringa*, 41, 407.

Locality and horizon: Daohugou, Ningcheng, Inner Mongolia, China; Middle Jurassic, Jiulongshan Formation.

Body length about 8.4 mm, forewing length 8.2–8.3 mm; pterostigma proximally closed by a scp-ra,

Figure 19.9 *Stenoraphidia obliquivenatica* (Ren, 1994), (additional material, CNU-RAP-LB-2017079).

with one inclined pterostigmal cross-vein; RP and MA with two simple branches individually in both wings; marginal ra-rp cross-vein posteriorly connecting to the stem of RP in both wings (2ra-rp in forewing and 3ra-rp in hind wing, respectively); rp-ma cross-vein posteriorly connecting to the stem of MA in both wings; in forewing, MA originating close to first branching point of MP and coalescent with the stem of RP for a short distance; two radial cells present [18].

Stenoraphidia Lyu, Ren & Liu, 2018

Stenoraphidia Lyu, Ren & Liu, 2018, *Cretac. Res.*, 89, 120 [20] (original designation).

Type species: *Alloraphidia obliquivenatica* Ren, 1994.

Head elongate, with an extremely prolonged occiput nearly half the length of the head, which is a remarkable feature previously unknown in other snakeflies. Prothorax elongate, nearly twice as long as wide. Pterostigma long; a cross-vein usually present at subdistal portion of pterostigma [20].

Distribution and age: Liaoning; Early Cretaceous.

Two species included from the Cretaceous of Northern China (see Table 19.1).

Stenoraphidia obliquivenatica (Ren, 1994) (Figure 19.9)

Stenoraphidia obliquivenatica Lyu, Ren & Liu, 2018: *Cretac. Res.*, 89: 120.

Alloraphidia obliquivenatica Ren, 1994, *Prof. Pap. of Stratigr. and Palaeontol.*, 25, 133 [21]. Syn. by Lyu, Ren & Liu, 2018, *Cretac. Res.*, 89, 119–125 [20].

Locality and horizon: Huangbanjigou, Beipiao, Liaoning, China; Lower Cretaceous, Yixian Formation.

Large-sized snakeflies (body length 14.1–23.0 mm, forewing length 13.8–16.9 mm). Prothorax with anterior half pale-colored, but posterior half dark-colored. Pterostigma long, uniformly colored, width of the oblique pterostigmal cross-vein clearly broadening at distal portion; RP with three to four pectinate branches; MA with three branches [20].

Styporaphidia Engel & Ren, 2008

Styporaphidia Engel & Ren, 2008, *J. Kans. Entomol. Soc.*, 81 (3), 189 [19] (original designation).

Type species: *Styporaphidia magia* Engel & Ren, 2008.

Pronotum quadrate, shorter than combined lengths of meso- and metanotum. Two cross-veins present in the apical half of the pterostigma; forewing MA originating before tangent of MP forking; MP with three medial cells; few number of medial cells; 1A simple [19].

Distribution and age: Inner Mongolia; Middle Jurassic.

Only one species included from the Jurassic of Northern China (see Table 19.1).

Figure 19.10 *Styporaphidia magia* Engel & Ren, 2008, (additional material, CNU-RAP-NN-2016029).

Styporaphidia magia Engel & Ren, 2008 (Figure 19.10)

Styporaphidia magia Engel & Ren, 2008: *J. Kans. Entomol. Soc.*, 81 (3), 189.

Locality and horizon: Daohugou, Ningcheng, Inner Mongolia, China; Middle Jurassic, Jiulongshan Formation.

Sex indeterminate. Total body length 13 mm; abdomen shorter than wings, parallel-sided, with eight preserved segments, abdominal apex not preserved. Forewing length 10 mm, width 4 mm; a single sc-r cross-vein present basal separation of Rs from the stem of R; CuA connecting to M distal to separation of M from R, with a single terminal branch; 1A simple [19].

Table 19.1 A list of fossil Raphidioptera from the Jurassic and Cretaceous of China.

Family	Species	Locality	Horizon/Age	Citation
Baissopteridae	*Baissoptera bicolor* Lyu, Ren & Liu, 2017	Beipiao, Liaoning	Yixian Fm., K_1	Lyu *et al.* [15]
	Baissoptera euneura Ren, 1997	Beipiao, Liaoning	Yixian Fm., K_1	Ren [22]
	Baissoptera grandis Ren, 1995	Beipiao, Liaoning	Yixian Fm., K_1	Ren [23]

(Continued)

Table 19.1 (Continued)

Family	Species	Locality	Horizon/Age	Citation
	Baissoptera liaoningensis Ren, 1994	Beipiao, Liaoning	Yixian Fm., K_1	Ren [21]
	Baissoptera sinica Lyu, Ren & Liu, 2017	Beipiao, Liaoning	Yixian Fm., K_1	Lyu *et al.* [15]
	Microbaissoptera monosticha Lyu, Ren & Liu, 2017	Beipiao, Liaoning	Yixian Fm., K_1	Lyu *et al.* [15]
Chrysoraphidiidae	*Chrysoraphidia relicta* Liu, Makarkin, Yang & Ren, 2013	Beipiao, Liaoning	Yixian Fm., K_1	Liu *et al.* [6]
Juroraphidiidae	*Juroraphidia longicollum* Liu, Ren & Yang, 2014	Ningcheng, Inner Mongolia	Jiulongshan Fm., J_2	Liu *et al.* [1]
Mesoraphidiidae	[a)]*Alloraphidia anomala* Ren, 1997	Beipiao, Liaoning	Yixian Fm., K_1	Ren [22]
	[a)]*Huaxiaraphidia shandongensis* Hong, 1992	Laiyang, Shandong	Laiyang Fm., K_1	Hong [23]
	[a)]*Huaxiaraphidia sinensis* Hong, 1992	Laiyang, Shandong	Laiyang Fm., K_1	Hong [23]
	[a)]*Jilinoraphidia dalaziensis* Hong & Chang, 1989	Yanji, Jilin	Dalazi Fm, K_1	Hong and Chang [24]
	[a)]*Kezuoraphidia kezuoensis* (Hong, 1992)	Kazuo, Liaoning	Shahai Fm., K_1	Hong [25]; Willmann [26]
	Mesoraphidia amoena Ren, 1997	Beipiao, Liaoning	Yixian Fm., K_1	Ren [22]
	Mesoraphidia daohugouensis Lyu, Liu & Ren, 2015	Ningcheng, Inner Mongolia	Jiulongshan Fm., J_2	Lyu *et al.*[8]
	Mesoraphidia furcivenata Ren in Ren *et al.*, 1995	Beijing	Lushangfen Fm., K_1	Ren *et al.*[14]
	Mesoraphidia glossophylla (Ren, 1997)	Beipiao, Liaoning	Yixian Fm., K_1	Engel [9]
	Mesoraphidia heteroneura Ren, 1997	Beipiao, Liaoning	[b)]Yixian Fm., K_1	Ren[22]; Pérez-de la Fuente *et al.*[27]
	Mesoraphidia longistigmosa (Ren, 1994)	Beipiao, Liaoning	Yixian Fm., K_1	Engel [9]
	Mesoraphidia myrioneura (Ren, 1997)	Beipiao, Liaoning	Yixian Fm., K_1	Engel [9]
	Mesoraphidia obliquivenatica (Ren, 1994)	Beipiao, Liaoning	Yixian Fm., K_1	Engel [9]
	Mesoraphidia polyphlebia (Ren, 1994)	Beipiao, Liaoning	Yixian Fm., K_1	Engel [9]
	Mesoraphidia shangyuanensis (Ren, 1994)	Beipiao, Liaoning	Yixian Fm., K_1	Engel [9]
	Mesoraphidia sinica Ren, 1997	Beipiao, Liaoning	Yixian Fm., K_1	Ren [22]
	Ororaphidia bifurcata Lyu, Ren & Liu, 2017	Ningcheng, Inner Mongolia	Jiulongshan Fm., J_2	Lyu *et al.*[18]
	Ororaphidia megalocephala Engel & Ren, 2008	Ningcheng, Inner Mongolia	Jiulongshan Fm., J_2	Engel and Ren [19]
	[a)]*Siboptera fornicata* (Ren, 1994)	Beipiao, Liaoning	Yixian Fm., K_1	Ren [21]
	[a)]*Sinoraphidia viridis* Hong, 1982	Kazuo, Liaoning	Shahai Fm., K_1	Hong [10]

(Continued)

Table 19.1 (Continued)

Family	Species	Locality	Horizon/Age	Citation
	Stenoraphidia longioccipitalis Lyu, Ren & Liu, 2018	Beipiao, Liaoning	Yixian Fm., K_1	Lyu *et al.* [20]
	Stenoraphidia obliquivenatica (Ren, 1994)	Beipiao, Liaoning	Yixian Fm., K_1	Lyu *et al.* [20]
	Styporaphidia magia Engel & Ren, 2008	Ningcheng, Inner Mongolia	Jiulongshan Fm., J_2	Engel and Ren [19]
	[a]*Xuraphidia liaoxiensis* Hong, 1992	Kazuo, Liaoning	Shahai Fm., K_1	Hong [25]
	[a]*Yanoraphidia gaoi* Ren, 1995	Chengde, Hebei	[b]Yixian Fm., K_1	Ren *et al.* [14]; Ren [22]; Pérez-de la Fuente *et al.* [27]

a) The species is not present in the main text because the original description, photos and line-drawing are not precise and the holotype cannot be rechecked.

b) Horizon/Age revised from the original paper based on updated information and data.

References

1 Liu, X.Y., Ren, D., and Yang, D. (2014). New transitional fossil snakeflies from China illuminate the early evolution of Raphidioptera. *BMC Evolutionary Biology* **14**: 84. https://doi.org/10.1186/1471-2148-14-84.

2 Aspöck, H. (2002). The biology of Raphidioptera: a review of present knowledge. *Acta Zoologica Academiae Scientiarum Hungaricae* **48** (Suppl 2): 35–50.

3 Ren, D., Shih, C.K., Gao, T.P. *et al.* (2010). Chapter 16: Raphidioptera – looking out and looking forward. In: *Silent Stories - Insect Fossil Treasures from Dinosaur Era of the Northeastern China*, 158–164. Beijing: Science Press.

4 Grimaldi, D.A. and Engel, M.S. (2005). *The Evolution of the Insects*, 338. New York: Cambridge University Press.

5 Aspöck, H. (1998). Distribution and biogeography of the order Raphidioptera: updated facts and a new hypothesis. *Acta Zoologica Fennica* **209**: 33–44.

6 Liu, X.Y., Makarkin, V.N., Yang, Q., and Ren, D. (2013). A remarkable new genus of basal snakeflies (Insecta: Raphidioptera: Priscaenigmatomorpha) from the Early Cretaceous of China. *Cretaceous Research* **45**: 306–313. https://doi.org/10.1016/j.cretres.2013.06.001.

7 Martynov, A.V. (1925). To the knowledge of fossil insects from Jurassic beds in Turkestan. 1. Raphidioptera. *Izvestiya Rossiiskoi Akademii Nauk* **19** (6): 233–246.

8 Lyu, Y.N., Liu, X.Y., and Ren, D. (2015). First record of the fossil snakefly genus *Mesoraphidia* (Insecta: Raphidioptera: Mesoraphidiidae) from the Middle Jurassic of China, with description of a new species. *Zootaxa* **3999**: 560–570. https://doi.org/10.11646/zootaxa.3999.4.6.

9 Engel, M.S. (2002). The smallest snakefly (Raphidioptera: Mesoraphidiidae) A new species in Cretaceous amber from Myanmar, with a catalog of fossil snakeflies. *American Museum Novitates* **3363**: 1–22. https://doi.org/10.1206/0003-0082(2002)363<0001:TSSRMA>2.0.CO;2.

10 Hong, Y.C. (1982). *Mesozoic Fossil Insects of Jiuquan Basin in Gansu Province*, 187. Beijing: Geological Publishing House (in Chinese).

11 Wang, Y.Y., Liu, X.Y., Winterton, S.L. *et al.* (2012). The first mitochondrial genome for the fishfly subfamily Chauliodinae and implications for the higher phylogeny of Megaloptera. *PLoS One* **7** (10): e47302. https://doi.org/10.1371/journal.pone.0047302.

12 Winterton, S.L., Hardy, N.B., and Wiegmann, B.M. (2010). On wings of lace: phylogeny and Bayesian divergence time estimates of Neuropterida (Insecta) based on morphological and molecular data. *Systematic Entomology* **35**: 349–378. https://doi.org/10.1111/j.1365-3113.2010.00521.x.

13 Martynova, O.M. (1961). Modern and extinct snakefly. *Paleontological Journal* **3**: 73–83.

14 Ren, D., Lu, L.W., Guo, Z.G. *et al.* (1995). *Faunae and Stratigraphy of Jurassic-Cretaceous in Beijing and the Adjacent Areas*, 98–99. Beijing: Seismic Publishing House (in Chinese).

15 Lyu, Y.N., Ren, D., and Liu, X.Y. (2017). Systematic revision of the fossil snakefly family Baissopteridae (Insecta: Raphidioptera) from the Lower Cretaceous of China, with description of a new genus and three new species. *Cretaceous Research* **80**: 13–26. https://doi.org/10.1016/j.cretres.2017.08.004.

16 Martynov, A.V. (1925). To the knowledge of fossil insects from Jurassic beds in Turkestan. 2. Raphidioptera (continued), Orthoptera (s.l.), Odonata, Neuroptera. *Izvestiya Rossiiskoi Akademii Nauk* **19** (6): 569–598.

17 Engel, M.S., Lim, J.D., and Baek, K.S. (2006). Fossil snakeflies from the Early Cretaceous of southern Korea (Raphidioptera: Mesoraphidiidae). *Neues Jahrbuch für Geologie und Paläontologie* **4**: 249–256.

18 Lyu, Y.N., Ren, D., and Liu, X.Y. (2017). Review of the fossil snakefly family Mesoraphidiidae (Insecta: Raphidioptera) in the Middle Jurassic of China, with description of a new species. *Alcheringa: An Australasian Journal of Palaeontology* **41** (3): 403–412. https://doi.org/10.1080/03115518.2017.1298840.

19 Engel, M.S. and Ren, D. (2008). New snakeflies from the Jiulongshan Formation of Inner Mongolia, China (Raphidioptera). *Journal of the Kansas Entomological Society* **81** (3): 188–193. https://doi.org/10.2317/JKES-802.19.1.

20 Lyu, Y.N., Ren, D., and Liu, X.Y. (2018). A remarkable new genus of the snakefly family Mesoraphidiidae (Insecta: Raphidioptera) from the Lower Cretaceous of China, with description of a new species. *Cretaceous Research* **89**: 119–125. https://doi.org/10.1016/j.cretres.2018.02.018.

21 Ren, D. (1994). A new discovery of snake-flies (Insecta) from the Late Jurassic of Beipiao, Liaoning. *Professional Papers of Stratigraphy and Palaeontology* **25**: 131–140. (in Chinese with English abstract).

22 Ren, D. (1997). Studies on the late Mesozoic snake-flies of China (Raphidioptera: Baissopteridae, Mesoraphidiidae, Alloraphidiidae). *Acta Zootaxonomica Sinica* **22**: 172–188. (in Chinese with English summary).

23 Hong, Y.C. (1992). A new family of Mesozoic snake-flies (Insecta, Raphidioptera) from the Laiyang Basin, China. *Paleontologicheskii Zhurnal* **3**: 101–105. (in Russian with English summary).

24 Hong, Y.C. and Chang, J.P. (1989). A new family "Jilinoraphidiidae" of Raphidioptera (Insecta). *Geoscience* **3** (3): 290–295. (in Chinese with English abstract).

25 Hong, Y.C. (1992). The study of Early Cretaceous Coleoptera, Raphidioptera, Diptera of Kezuo, West Liaoning Province. *Acta Geologica Gansu* **1**: 1–13. (in Chinese with English summary).

26 Willmann, R. (1994). Raphidiodea aus dem Lias und die Phylogenie der Kamelhalsfliegen (Insecta: Holometabola). *Paleaontologische Zeitschrift* **68**: 167–197.

27 Pérez-de La Fuente, R., Penalver, E., Delclos, X., and Engel, M.S. (2012). Snakefly diversity in Early Cretaceous amber from Spain (Neuropterida, Raphidioptera). *ZooKeys* **204**: 1–40. https://doi.org/10.3897/zookeys.204.2740.

20

Neuroptera – Lacewings

Zhenzhen Chen[1], Shuo Huang[1], Yu Chang[1], Yongjie Wang[1], Chungkun Shih[1,2], and Dong Ren[1]

[1] Capital Normal University, Haidian District, Beijing, China
[2] National Museum of Natural History, Smithsonian Institution, Washington, DC, USA

20.1 Introduction to Neuroptera

The Neuroptera, as a relic order of the Holometabola, comprise about 6000 extant species across 16 families: i.e. Ascalaphidae, Berothidae, Chrysopidae, Coniopterygidae, Dilaridae, Hemerobiidae, Ithonidae s.l., Mantispidae, Myrmeleontidae, Nemopteridae, Nevrorthidae, Nymphidae, Osmylidae, Psychopsidae, Rhachiberothidae, and Sisyridae [1]. These insects, characterized by their equal or subequal membranous wings laced with reticulated venation, are commonly called as lacewings, such as green lacewings, brown lacewings, lance lacewings, pleasing lacewings, beaded lacewings, silk lacewings, dusty lacewings, etc.

In the field, neuropterans are easily recognized based on their characteristic features, which exhibit diverse morphological plasticity. The wingspans of adult neuropterans show a very broad range from the minute to very large. The smallest "dusty lacewings" (Coniopterygidae) have the wingspans between 1.8 and 5 mm, while the largest "antlions" (Myrmeleontidae) with the wingspans exceeding 150 mm. Most lacewings have the moderate wingspans from 20 to 30 mm (Figure 20.1). Adults generally have chewing mouthparts, but the mouthparts in some taxa of Nemopteridae, especially in Crocinae [2], can prolong into a long rostrum [3]. They typically have filiform antennae with tapering apex, but exceptions are present in some taxa, i.e. clubbed in some antlions and owlflies (Figures 20.2 and 20.3) and pectinated in some male dilarids. Compound eyes are prominent, usually bare, but ocelli commonly reduced, except that Osmylidae retain three distinct ocelli. Pro-, meso- and metathorax are well-defined. Prothorax is commonly short, but remarkably elongated in mantispids (Figure 20.4) and some of berothids and rhachiberothids. Legs are usually cursorial and fit for walking, while Mantispidae and Rhachiberothidae possess strong raptorial forelegs to help them catch preys like a praying mantis. Two pairs of membranous and equal-sized wings are often held rooflike above the bodies of lacewings when at rest. However, the hind wings may be remarkably modified into unique and different forms in Nemopteridae, while some taxa of Hemerobiidae and Coniopterygidae possess the reduced hind wings. Generally, most neuropterans have the intricate venation with an exception of Coniopterygidae that possess only the major longitudinal veins without most cross-veins, while the genus *Conwentzia* with only vestigial hind wings and females of *Helicoconis* are wingless. Occasionally, some neuropterans, especially the large lacewings (e.g. antlions or owlflies), are mistaken as the extraordinary "dragonflies". In fact, the dragonflies (in Division of Palaeoptera, Chapter 6) and lacewings (in Superorder of Endopterygota) can be conveniently distinguished by the form of antennae, i.e. long and diversiform in Neuroptera vs. short and setaceous in dragonfly.

Neuropterans, both adults and larvae, have a very wide range of feeding modes. Some are well-known as generalist predators, e.g. Chrysopidae, Hemerobiidae and Coniopterygidae are considered as the effective natural enemies in controlling agricultural pests. Because most lacewings are slow fliers, they adopt various effective tactics of predation. Green lacewings (Figure 20.5) and brown lacewings are apparently walking hunters, often actively moving in their search for prey. On the other hand, owlflies are aerial hunters with excellent flight capability to catch prey in the air. The predation by larvae appears to be even more diverse. Besides the typical walking hunters of the larvae of Chrysopidae (Figure 20.6), Hemerobiidae, Coniopterygidae and Osmylidae, the predation of pit-dwelling myrmeleontid (antlion) larvae is particularly specialized. The antlion larvae, after building various pits on the sand surface, hide beneath the pit to wait for the falling victims. Moreover, larvae of some neuropterans have parasitic behavior, for example, some berothid larvae are

Rhythms of Insect Evolution: Evidence from the Jurassic and Cretaceous in Northern China, First Edition.
Edited by Dong Ren, Chungkun Shih, Taiping Gao, Yongjie Wang, and Yunzhi Yao.

Figure 20.1 Lance lacewing (*Osmylus* sp.). Source: Photo by Han Xu.

Figure 20.2 Antlion lacewing. Source: Photo by Jason Shih.

Figure 20.3 Owlfly lacewing. Source: Photo by Jason Shih.

Figure 20.4 Mantispid lacewing. Source: Photo by Jason Shih.

Figure 20.5 Green lacewing. Source: Photo by Dr. Chungkun Shih.

able to kill and feed on the termites by injecting neurotoxin, while the larvae of mantispids are obligate parasites of spider eggs (Figure 3.1) [4]. Nevertheless, the non-predatory lacewings are also recorded for both larvae and adults. The larvae of Australian *Ithone* with typically reduced mouthparts prefer feeding on decaying subterranean material [5]. Since adult nemopterids are often observed to visit flowers in the field, Tjeder in 1967 considered that the South African nemopterines possibly live on pollen due to the virtual absence of animal remains in their guts, only fragments of vegetation [3]. Interestingly, some chrysopids, such as *Chrysoperla carnea*, are typical honeydew feeders [4, 6].

Unequivocally, Neuroptera are cosmopolitan insects, existing in all major zoogeographical regions. However, the distribution patterns across the families are uneven. Mostly, a widely distributed family just represents the combination of some disjunct and restricted subfamilies or genera. For example, the archaic family Osmylidae have been recorded in most zoogeographical regions except the Nearctic [7]. However, most subfamilies of

Figure 20.6 A larva of lacewing carrying debris for camouflage. Source: Photo by Jason Shih.

Osmylidae are narrowly distributed and restricted in one or two regions with the exception of Spilosmylinae occurring in four realms, i.e. Palearctic, Indian/Indomalayan, Afrotropic and Australasian realms. The absence of recent osmylids in the Nearctic seems to segregate the continuous distributions. The report that some Eocene osmylids from Northern America fill the biogeographical gaps [8] indicates the absence of the recent Nearctic osmylids was caused by a historical extinction. Similar conditions are common among the extant Neuroptera. Therefore, the answer to the global distribution of Neuroptera is far from resolved.

The monophyly of Neuroptera is widely accepted, however, the interrelationships among the families are far from resolved, especially the involvements of molecular evidence in the phylogenetic reconstruction [11–13]. The earlier pioneers inferred the phylogeny of Neuroptera primarily based on the morphological evidence presented by Withycombe [9] and MacLeod [14] (Figure 20.7). Until 2001, Aspöck *et al.* first reconstructed the phylogeny of Neuroptera according to both adult and larval characters in the computerized cladistic analysis, and divided Neuroptera into three suborders, i.e. Nevrorthiformia, Myrmeleontiformia, and Hemerobiiformia [10]. In this result, the family Nevrorthidae represented the earliest divergence of Neuroptera, which was also supported by the subsequent works from morphological analysis [15, 16]. Haring and Aspöck in 2004 conducted the first molecular phylogeny of Neuroptera that was moderately different from the morphological result, i.e. Nevrorthidae, Sisyridae and Osmylidae representing the earliest divergence and the splitting of Hemerobiiformia [17]. This result was supported by the evidence from the genital sclerites and the anatomic features of adult head [18, 19]. However the phylogenies from the larger molecular data sets are distinctively different from the morphological works [11, 12] of which the particular family Coniopterygidae was sister to the other neuropteran families indicating an obviously different evolution of Neuroptera (Figure 20.8). As an archaic group, it was unfortunate that only minor fossil taxa have been considered in the phylogenetic reconstruction of Neuroptera. It is expected that the described fossil neuropterans will be indispensable in the future to study and resolve the phylogeny of Neuroptera.

20.2 Progress in the Studies of Fossil Neuroptera

The Neuroptera can be traced back to the earliest Permian with a low family-level diversification in the earlier stage [20, 21]. Until the Late Permian, Neuroptera underwent the first radiation and completed the global dispersion recorded from Europe, Asia, Australia and South America. Nevertheless, most Permian neuropteran-like insects merely show some superficial resemblances of wing features to the extant lineages, but their interrelationships are still not well-illustrated.

After the mass extinction at the end of the Permian, Neuroptera had the most significant diversification resulting in the origination of almost all major extant lineages during the Mesozoic. The Triassic fauna of Neuroptera became more diverse compared to the Permian in that about 10 extinct families were described from around the world. However, the Triassic specimens commonly possess the plesiomorphic characters of Neuroptera that result in the uncertain systematic positions of most families. Two extant families of Osmylidae and Psychopsidae were first recorded in the Triassic [22, 23]. The Jurassic lacewings started the rapid radiation mainly found from Eurasia. Some Triassic species also survived to the Early Jurassic [24, 25]. Many particularly endemic lacewings have been described from the Middle and Late Jurassic that are generally deduced to be closely related to the contemporaneous environments. The most remarkable lacewings should be the Kallgrammatidae which are also called "Mesozoic butterflies". The marvelous large insects with the long-proboscid siphonate mouthparts were inferred to be the pollinators of the Jurassic Bennettitalean plants representing an extinct gymnosperm-insect relationship [26, 27]. The other

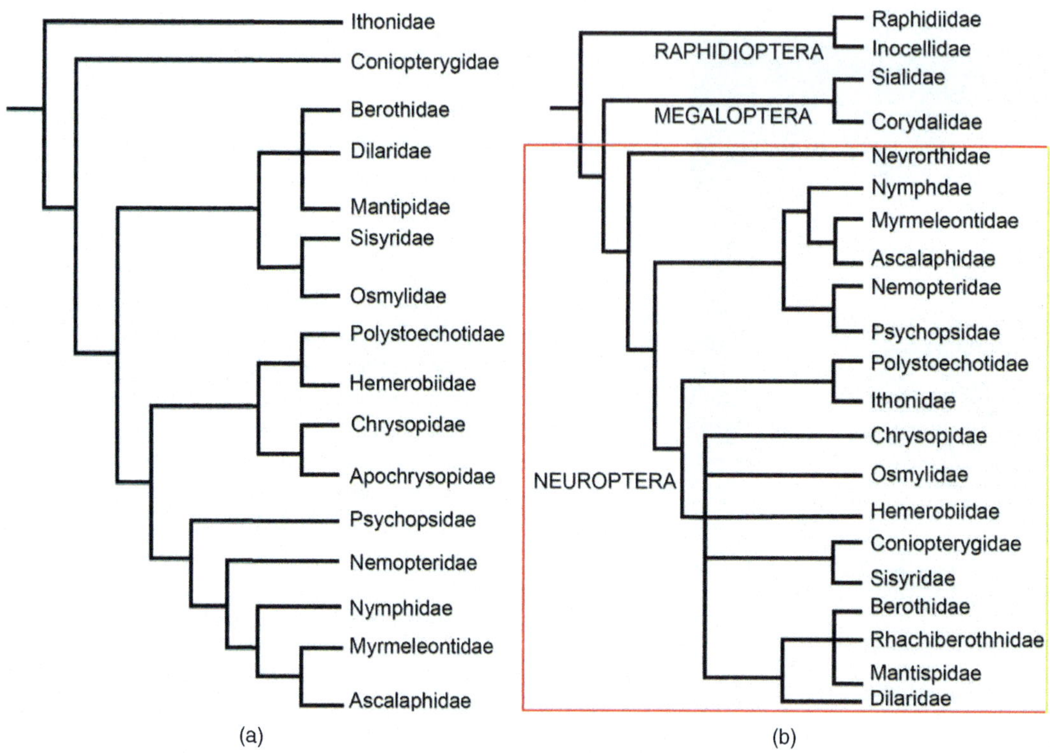

Figure 20.7 Phylogeny of Neuroptera. (a). Source: Modified from [9]; (b). Modified from [10].

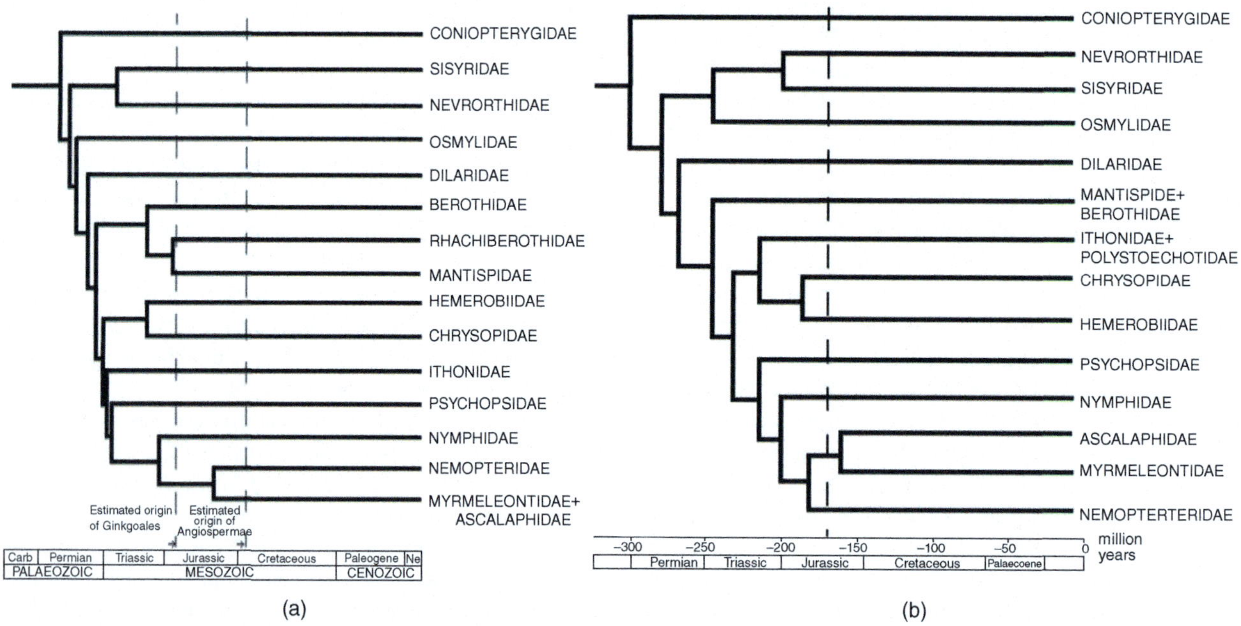

Figure 20.8 Phylogeny of Neuroptera combining morphological and molecular data. (a). Modified from [12]; (b). Modified from [11].

interesting Jurassic family is the Saucrosmylidae which developed wings to mimick the pinnate leaves of the Cycadales or Bennettitales [28].

The Cretaceous lacewings followed the trends as displayed in the Jurassic lacewings, but the distributions of fossil localities are much broader. The lacewings have been reported from some new Cretaceous localities, e.g. Santana assemblage of Brazil, Lebanese amber, New Jersey amber, and Myanmar (Burmese) amber, etc. In the Cretaceous, some particular lacewings have been described, e.g. the two-winged enigmatic family Dipteromantispidae Makarkin, Yang & Ren, 2013 [29, 30] and

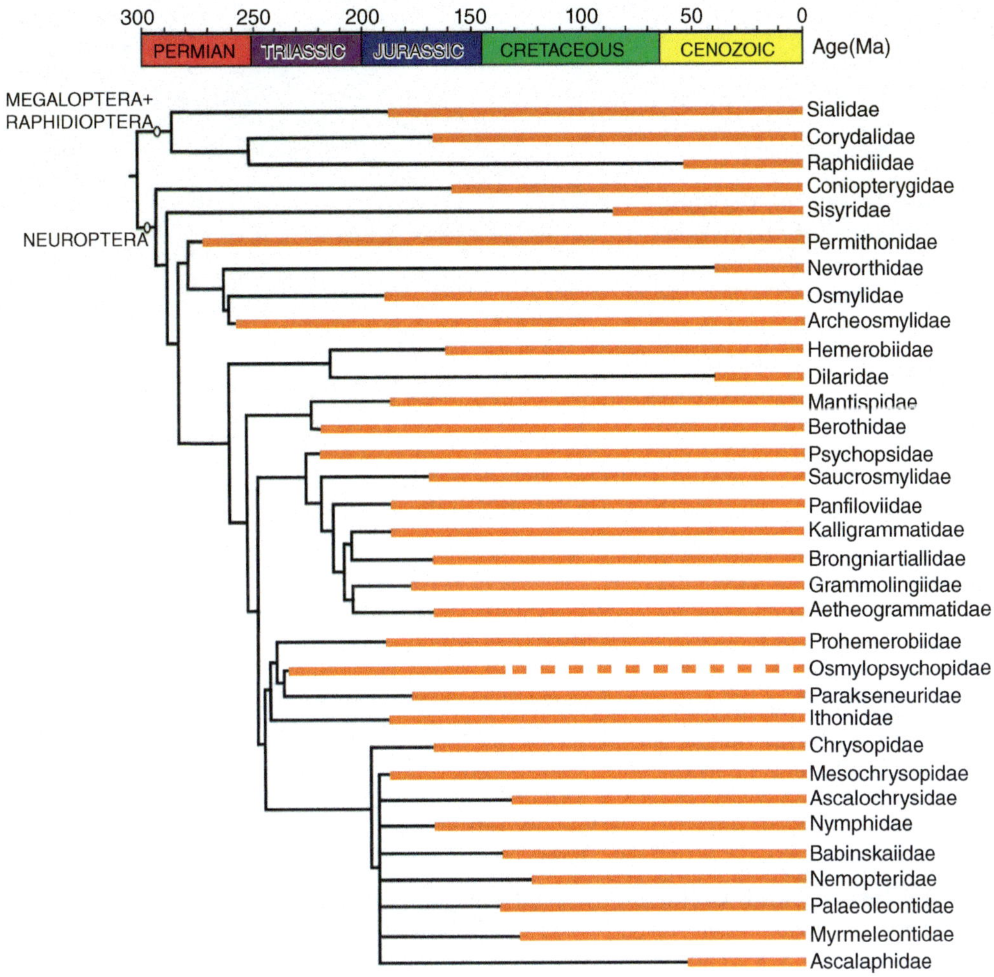

Figure 20.9 Phylogeny of Neuroptera. Source: Modified from [32].

the spider-associated green lacewing [31]. At the end of the Mesozoic, all living families of Neuroptera had originated. Yang *et al.* [32] conducted a phylogenetic analysis to reconstruct the phylogeny of Neuroptera covering 30 extant and extinct families based on both morphological and molecular data (Figure 20.9). Because molecular data are mainly referred to Winterton *et al.* [11], the primary topology of the recent Neuroptera is not modified. Nevertheless, the inclusion of extinct groups is partly helpful to understand the systematic positions of these insects. Two families of Coniopterygidae and Sisyridae represent the first divergence of the Neuroptera, and the most basal family, Permithonidae, branched after the first clade implying the origin of Neuroptera should be much earlier than the current fossil records. Interestingly, the monophyletic Myrmeleontiformia is also well-reconstructed, but most Mesozoic extinct groups are nested in this clade, implying the high diversification of this lineage in the history. Considering the weak support of some clades within this tree and the absence of some extinct lineages, more work is still required to explore the evolution and phylogeny of Neuroptera.

The Neuroptera fossils are very diverse from the Jurassic and Cretaceous of Northern China, and hitherto about 21 families, 89 genera and 153 species have been described (Table 20.1). In 1928, Dr. Ping first described the species *Mesohemerobius jeholensis* from the Yixian Formation of Liaoning [33]. Dr. Hong, in 1980, described the oldest neuropteran *Minonymphites orthophlebes* from the Triassic of Tongchuan, Shaanxi [34]. Dr. Lin, in 1986, reported another early species *Idiastogyia fatisca* from the Early Jurassic of Jiexi, Guangdong [35]. Dr. Ren conducted pioneer work (1996 up to now) on the Jiulongshan Formation and Yixian Formation to enhance our understanding of the diversity the Mesozoic Neuroptera, especially for various extinct families, e.g. Kalligrammatidae, Saucrosmylidae, Grammolingiidae, Parakseneuridae, Aetheogrammatidae, etc. In addition, other Chinese paleoentomologists, i.e. Junfeng Zhang, Di-ying Huang, Bo Wang, Yongjie Wang, Chungkun Shih, Qiang Yang, Chaofan Shi, Yuanyuan Peng, etc. have also made significant contributions to the Mesozoic Neuroptera.

20.3 Representative Fossils of Neuroptera from the Jurassic and Cretaceous of Northern China

Family Aetheogrammatidae Ren & Engel, 2008

Aetheogrammatidae, considered as one of the most specialized families within the Suborder of Myrmeleontiformia in Neuroptera, have been erected for *Aetheogramma speciosa* Ren & Engel, 2008 [36]. They are characterized by oval wings, with rounded apices, subequal in length, with broad fuscous bands running from costal margin to posterior margin; pterostigma absent; trichosors and microtrichia absent; wings lacking conspicuous "eyespots"; cross-veins numerous and regular over surface of wing, not forming gradate series and principal longitudinal veins (Sc, R_1, Rs, MA, and MP) without twigging along wing margins. They are closely related to another highly specialized Mesozoic family, the Kalligrammatidae. The oldest fossil representatives are known from the Middle Jurassic Jiulongshan Formation of Inner Mongolia, China. Aetheogrammatidae comprise two subfamilies: Ectopogrammatinae and Aetheogrammatinae. To date, there are four extinct genera containing five species described from the Middle Jurassic to the Lower Cretaceous of China and *Kalligrammina areolata* Panfilov, 1980 [36, 37] from the Middle Jurassic Karabastau Formation at Karatau in Kazakhstan.

Genera included from the Jurassic and Cretaceous of Northern China: *Aetheogramma* Ren & Engel, 2008, *Ectopogramma* Engel, Huang & Lin, 2011, *Cyclicogramma* Yang, Makarkin, Shih & Ren, 2015 and *Curtogramma* Yang, Makarkin, Shih & Ren, 2015.

Aetheogramma Ren & Engel, 2008

Aetheogramma Ren & Engel, 2008, *J. Kans. Entomol. Soc.*, 81, 163 [36] (original designation).

Type species: *Aetheogramma speciosa* Ren & Engel, 2008.

In forewing, cross-veins between subcostal veinlets absent; branches of RP and MP strongly curved; M forked relatively close to wing base. In hind wing, hind margin straight; branches of ORB_1 irregularly spaced, strongly curved [38].

Distribution and age: Liaoning; Early Cretaceous.

Two species included from the Cretaceous of Northern China (see Table 20.1)

Aetheogramma bistriatum Yang, Makarkin, Shih & Ren, 2015 (Figure 20.10)

Aetheogramma bistriatum Yang, Makarkin, Shih & Ren, 2015: *Cretac. Res.*, 55, 27.

Locality and horizon: Huangbanjigou, Beipiao, Liaoning, China; Lower Cretaceous, Yixian Formation.

Wing fuscous, with two white broad transverse interrupted fasciae (distal one along outer margin; proximal one slightly proximad mid-point of wing), and white patch between these fasciae anteriorly. Cross-veins very dense over entire wing. In forewing, ORB_2 originating in distal part of wing, preserved costal space moderately broad. Humeral veinlet not recurrent. Subcostal veinlets simple (one forked) [38].

Figure 20.10 *Aetheogramma bistriatum* Yang, Makarkin, Shih & Ren, 2015 (Holotype, CNU-NEU-LB2014008) [38].

Ectopogramma Engel, Huang & Lin, 2011

Ectopogramma Engel, Huang & Lin, 2011, *J. Kans. Entomol. Soc.*, 84 (4), 316 [37] (original designation).

Type species: *Ectopogramma kalligrammoides* Engel, Huang & Lin, 2011.

Forewing without recurrent humeral vein; c-sc cross-veins (costal veinlets) angled toward apex, with unique basal forks, some apical forks near wing apex, with interconnecting intercalary veinlets between individual cross-veins [37].

Distribution and age: Inner Mongolia; Middle Jurassic.

Only one species included from the Jurassic of Northern China (see Table 20.1).

Curtogramma Yang, Makarkin, Shih & Ren, 2015

Curtogramma Yang, Makarkin, Shih & Ren, 2015, *Cretac. Res.*, 55, 29 [38] (original designation).

Type species: *Curtogramma ovatum* Yang, Makarkin, Shih & Ren, 2015.

Large insect, forewing about 60 mm long. Trichosors not detected. Pterostigma absent. Costal cross-veins simple. ScP long, entering margin close to wing apex. Presence of three ORBs. M forking at about the middle of wing, MA and MP strongly arched. CuA and CuP similarly archedly branched. Anal region well-developed [38].

Distribution and age: Liaoning; Early Cretaceous.

Only one species from the Cretaceous of Northern China (see Table 20.1).

Cyclicogramma Yang, Makarkin, Shih & Ren, 2015

Cyclicogramma Yang, Makarkin, Shih & Ren, 2015, *Cretac. Res.*, 55, 28 [38] (original designation).

Figure 20.11 *Cyclicogramma rotundum* Yang, Makarkin, Shih & Ren, 2015 (Holotype, CNU-NEU-LB2014009) [38]. Source: Donated by Dr. Chungkun Shih.

Type species: *Cyclicogramma rotundum* Yang, Makarkin, Shih & Ren, 2015.

Only the hind wing preserved, less than 30 mm long. Costal region narrow. Presence of two ORBs. M proximally branched, MA single branched and slightly curved at middle, MP deeply dichotomized. CuA with two to three secondary sub-branches, CuP simply biforked [38].

Distribution and age: Liaoning; Early Cretaceous.

Only one species included from the Cretaceous of Northern China (see Table 20.1).

***Cyclicogramma rotundum* Yang, Makarkin, Shih & Ren, 2015 (Figure 20.11)**

Cyclicogramma rotundum Yang, Makarkin, Shih & Ren, 2015: *Cretac. Res.*, 55, 29.

Locality and horizon: Huangbanjigou, Beipiao, Liaoning, China; Lower Cretaceous, Yixian Formation.

Hind wing broadly oval, trichosors not detected. Costal space very narrow basally, gradually slightly dilated toward apex. Humeral veinlet cross-vein like. All subcostal veinlets simple, nearly straight, not connected by cross-veins. Pterostigma absent. ScP relatively short. Subcostal space much broader than costal space; with dense cross-veins. ScP distally not fused with RA. RA simple. RP consists of two ORBs. MA concave, simple, slightly arched. MP deeply forked somewhat proximad wing mid-point. Cu dividing into CuA and CuP near wing base, proximad fork of M. Cross-veins very dense over entire wing. Wing fuscous, with two white transverse interrupted fasciae (one near middle, the other near wing margin) [38].

Family Ascalochrysidae Ren & Makarkin, 2009

Ascalochrysidae are an enigmatic family within Neuroptera, which is close to Mesochrysopidae but showing the superficial myrmeleontoid appearances, e.g. the typically large body size, posteriad of termination of Sc and R_1 and complicated venation. However, Ren and Makarkin [39] proposed the chrysopoid affinity according to the analysis of venation, i.e. the absence of trichosors, nygmata and the basal sinuous cross-vein r-m, the distal fusion of Sc and R_1, the reduction of CuP to a part of apparent basal cross-vein between CuA and 1A, the entire loss (or strong reduction) of 2A and 3A, the loss or poor development of the humeral lobe and presence of frenulum of coupling apparatus sharing with Chrysopidae. Notably, the venation of the family is particular for the presence of the putative "M_5" in hind wing interpreted by Ren and Makarkin, which is rare within the recent Neuroptera except for the single species *Nipponeurorthus pallidinervis* Nakahara, 1958 in Nevrorthidae [40]. Meanwhile, Ren and Makarkin declared the condition of presence of "M_5" occurred in more ancient neuropterans (Permithonidae), thus, considered it is likely a plesiomorphy of Neuroptera. In fact, the homology of "M_5" is not well-illustrated across the Neuroptera due to the lack of the specimens. In any case, the occurrence of Ascalochrysidae indicates the diversification of the chrysopoid clade in the Mesozoic.

Only one genus included from the Cretaceous of Northern China: *Ascalochrysa* Ren & Makarkin, 2009.

***Ascalochrysa* Ren & Makarkin, 2009**

Ascalochrysa Ren & Makarkin, 2009, *Cretac. Res.*, 30 (5), 1218 [39] (original designation).

Type species: *Ascalochrysa megaptera* Ren & Makarkin, 2009.

Large lacewing (hind wing approximately 60 mm long). Hind wing with these characters: trichosors and nygmata absent; Sc and R_1 fused distally, $Sc + R_1$ entering wing margin well beyond apex; M branched very close to wing base; CuA relatively short, concave; 1A well-developed; not forming gradate series or regular reticulation [39].

Distribution and age: Liaoning; Early Cretaceous.

Only one species included from the Cretaceous of Northern China (see Table 20.1).

Family Berothidae Handlirsch, 1906

Berothidae, a small neuropteran family, comprise about 100 species discontinuously distributed mainly across tropical and warm-temperate regions of the world [41–43]. This family is generally related to the other homologous family Rhachiberothidae, which was originally treated as a subfamily of Berothidae for sharing similar venational characters [44]. Although both families are easily identified by the specialization of the raptorial forelegs, the involvement of Mantispidae with the similar raptorial forelegs raised the question about the relationships of these three families [45]. In fact, the relationships among these three families are still not well-documented in the recent phylogenetic work [11, 12]. The berothid fossils are quite scarce, and so far about 40 fossil species have been described. The oldest convincing berothid fossil is found from the Middle Jurassic of Daohugou, Inner Mongolia, China [46]. The Jurassic berothids are scarce and found from five localities hitherto, i.e. Daohugou, Novospasskoe, Khoutiyn-Khotgor, Bakhar, and Karatau [47]. The Cretaceous berothids underwent a rapid specific diversification that are mainly recorded from Archingeay, Myanmar (Burmese) amber, Canadian amber, Crato Formation, Lebanese amber, Obeshchayushchiy, Raritan (New Jersey) amber, Spanish amber and Yixian Formation.

Genera included from the Jurassic and Cretaceous of Northern China: *Sinosmylites* Hong, 1983 and *Oloberotha* Ren & Guo, 1996.

Sinosmylites Hong, 1983

Sinosmylites Hong, 1983, *Middle Jurassic Fossil Insects in North China*, 94 [48] (original designation).

Type species: *Sinosmylites pectinatus* Hong, 1983.

Forewing with costal space strongly narrowed basally; humeral veinlet not recurrent and branched; costal cross-veins mostly simple; Sc and R_1 fused distally; all subcostal veinlets simple; MP forked far distal to origin of Rs; CuA pectinate, with seven branches; few cross-veins in radial space arranged mainly in 1–2 inner gradate series [46].

Distribution and age: Inner Mongolia and Liaoning; Middle Jurassic.

Three species included from the Jurassic of Northern China (see Table 20.1).

Sinosmylites rasnitsyni Makarkin, Yang & Ren, 2011 (Figure 20.12)

Sinosmylites rasnitsyni Makarkin, Yang & Ren, 2011: *ZooKeys*, 130, 204.

Locality and horizon: Daohugou, Ningcheng, Inner Mongolia, China; Middle Jurassic, Jiulongshan Formation.

The specific epithet is in honor of the distinguished Russian paleoentomologist Dr. A. P. Rasnitsyn. Antenna moniliform, incomplete; preserved segments transverse (wider than long). Prothorax short. Mesonotum of usual neuropteran morphology. Legs covered with short setae; fore and mid-legs relatively short; tibia of hind-leg long; basitarsus of fore and hind-leg is the longest segment of tarsus. Abdomen very poorly preserved. Forewing with broad and rounded apex. Costal space moderately broad, strongly dilated at proximal 1/5 of wing length, narrowed basally. Subcostal veinlets simple, regularly arranged, closely spaced. Sc distally fused with R_1 far from wing apex; $Sc+R_1$ with 9–11 simple veinlets. Subcostal space broad, with one basal cross-vein located

Figure 20.12 *Sinosmylites rasnitsyni* Makarkin, Yang & Ren, 2011 (Holotype, CNU-NEU-NN2011002p) [46].

immediately after origin of Rs. M appears fused basally for a short distance; forked much distal to origin of Rs_1. MA and MP almost parallel, distally with one or two long forked branches respectively. Cu divided into CuA and CuP proximal to origin of Rs. CuA pectinate, with seven branches, some forked once. CuP once deeply forked. Hind wing poorly preserved, costal space narrow, distal part only slightly dilated. Subcostal veinlets simple, rather closely spaced [46].

Oloberotha Ren & Guo, 1996

Oloberotha Ren & Guo, 1996, *Acta Zootaxon. Sin.*, 21 (4), 468 [49] (original designation).

Type species: *Oloberotha sinica* Ren & Guo, 1996.

Antenna short, scape elongate, Costal area widest in basal third. Humeral vein curved and terminated in costa basally and with some basal fork in forewing. Subcostal sinuate and terminating in R_1, connected basally to R_1 by a single cross-vein. R_1 and Rs connected by three cross-veins. Rs with seven or eight branches. Numerous outer graduate cross-veins present between branches of Rs and M [49].

Distribution and age: Liaoning; Early Cretaceous.

Only one species included from the Cretaceous of Northern China (see Table 20.1).

Family Chrysopidae Hagen, 1866

Chrysopidae, the second largest family within Neuroptera, comprise *ca.* 85 extant genera and 1300–2000 species worldwide. Chrysopids, called "green lacewings", usually have bright green to greenish-brown body and wing veins. Compared to the remarkable diversity of extant chrysopids, the fossil green lacewings are much less diverse, including *ca.* 28 genera and 70 species so far. The earliest chrysopid-like lacewing is from the

Lower Jurassic of Dobbertin [50], but it was transferred to the Mantispidae [51]. A key question is the conflict interrelationships among the chrysopid-like insects. In 2005, Nel *et al.* [52] proposed the classification system of the superfamily Chrysopoidea including five families (Allopteridae, Mesochrysopidae, Tachinymphidae, Limaliidae, and Chrysopidae). In the same year, Makarkin and Menon reported to use the Mesochrysopidae *s.l.* to assemble these chrysopid-like fossil taxa [53]. Herein, we adopt their opinions and retain two valid families, i.e. Mesochrysopidae (including Allopteridae, Tachinymphidae, and Limaliidae) and Chrysopidae. Therefore the earliest true green lacewings have been documented from the Middle Jurassic of Daohugou, China [54]. The known Jurassic green lacewings are quite sparse, only about five other species reported from the Upper Jurassic of Karatau, Kazakhstan [52, 54–56].

Genera included from the Jurassic and Cretaceous of Northern China: *Mesypochrysa* Martynov, 1927, *Drakochrysa* Yang & Hong, 1990 and *Lembochrysa* Ren & Guo, 1996.

Mesypochrysa Martynov, 1927

Mesypochrysa Martynov, 1927, *Izv. Akad. Nauk SSSR*, 21 (5), 764–768 [55] (original designation).

Type species: *Mesypochrysa latipennis* Martynov, 1927.

The type species of the genus is from the Upper Jurassic of Karatau, and in 2015, Khramov *et al.* described a new species *Mesypochrysa sinica* from Daohugou and emended the generic diagnosis with these characters: presence of two gradate series of cross-veins, proximal connection of 1m cell and stem of Rs by the cross-vein, MP connected with CuA distally to a point of bifurcation by one cross-vein, CuP bifurcated terminally forming quadrangle (sometimes a trapezoid) with wing margin, Cu + MP with 3–8 unforked branches, Rs with 5–20 branches once or few times forked terminally, A_1 and A_2 long and simple and sometimes forked terminally, A_1 close to CuP connected with it by short cross-vein or even fused, A_3 short [54, 55].

Distribution and age: Inner Mongolia; Middle Jurassic.

Only one species included from the Jurassic of Northern China (see Table 20.1).

Drakochrysa Yang & Hong, 1990

Drakochrysa Yang & Hong, 1990, *Geoscience*, 4 (4), 24 [57] (original designation).

Type species: *Drakochrysa sinica* Yang & Hong, 1990.

Sc long and curved; subcostal space with 1–2 cross-veins, located after the basalmost of Rs; MP_1 and MP_2 circuitous between the two gradate series, forming the part of psm and psc; m-cu is close to the apex of the wing; CuA forked distally with four branches [57].

Distribution and age: Shandong; Early Cretaceous (Barremian).

Only one species included from the Cretaceous of Northern China (see Table 20.1).

Lembochrysa Ren & Guo, 1996

Lembochrysa Ren & Guo, 1996, *Acta Zootaxon. Sin.*, 21 (4), 469 [49] (original designation).

Type species: *Lembochrysa miniscula* Ren & Guo, 1996.

The genus was first established based on the specimens from the Yixian Formation. They provided the following apomorphic characters: Sc and R_1 unfused distally, two cross-veins between Sc and R_1 in forewing, MA connected to MP_1 by a cross-vein in forewing. In hind wing, MP forking deeply, not coalesced with Rs + MA, CuP short and with two branches [49].

Distribution and age: Liaoning; Early Cretaceous.

Two species included from the Cretaceous of Northern China. (see Table 20.1).

Family Dipteromantispidae Makarkin, Yang and Ren, 2013

The neuropterans usually bear four well-developed wings. However, there are a few brachypterous, micropterous, or apterous lacewing species, found in several extant families. Grimaldi described an enigmatic fossil neuropteran genus *Mantispidiptera* Grimaldi, 2000 and its two minute species *Mantispidiptera enigmatica* Grimaldi and *Mantispidiptera henryi* Grimaldi, with forewings 2.63–3.12 mm long, from the Late Cretaceous (Turonian) amber of New Jersey (USA) [29]. This genus is noteworthy for its special forewing venation and hind wing reduced to small structures resembling the halteres of Diptera. The type genus of Dipteromantispidae resembles *Mantispidiptera* but is much larger, yet its forewing venation is even further reduced. The reduced venation in this larger specimen supports the proposal of Wedmann and Makarkin [51] that these character states are not explained by minute size and that its raptorial foreleg characters that resemble those of the Mantispidae do not constitute sufficient reasons to place it in Mantispidae.

Only one genus included from the Cretaceous of Northern China: *Dipteromantispa* Makarkin, Yang & Ren, 2013.

Dipteromantispa Makarkin, Yang & Ren, 2013

Dipteromantispa Makarkin, Yang & Ren, 2013, *Foss. Rec.*, 16 (1), 68 [30] (original designation).

Type species: *Dipteromantispa brevisubcosta* Makarkin, Yang & Ren, 2013.

Larger than *Mantispidiptera*. Hind wing reduced to small structures resembling the halteres of Diptera. Forewing about 8 mm long (vs. 2.6–3.1 mm in *Mantispidiptera*); Sc very short, joining margin at approximately mid-wing (vs. Sc longer, joining margin at approximately distal one-third wing length in *Mantispidiptera*); MP forked once; CuA simple [30].

Distribution and age: Liaoning; Early Cretaceous.

Only one species included from the Cretaceous of Northern China (see Table 20.1).

***Dipteromantispa brevisubcosta* Makarkin, Yang & Ren, 2013 (Figure 20.13)**

Dipteromantispa brevisubcosta Makarkin: Yang & Ren, 2013: *Foss. Rec.*, 16 (1), 68.

Locality and horizon: Huangbanjigou, Beipiao, Liaoning, China; Lower Cretaceous, Yixian Formation.

Head nearly as wide as prothorax; posterocular lobe expanded. Antennae widely spaced, long; scape short, length approximately equal to width; distinctly longer than other antennomeres; pedicel short, only slightly longer than first flagellomere; flagellum with about 40 flagellomeres. Prothorax rather short, narrow and rounded anteriorly, with long, quite dense, laterally directed setae. Basisternum small, narrow, with short median longitudinal ridge; Furcasternum indistinct, weakly sclerotized. Forelegs raptorial; procoxa elongate, stout, covered with dense, rather short setae. Mesothorax largest segment of thorax, weakly sclerotized ventrally. Metathorax smaller than mesothorax, weakly sclerotized ventrally.

Abdomen: Gonocoxites of 7th segment paired (with finger-like process). Gonocoxites of 9th segment (= gonopophyses laterales) ovoid (in ventral view), with small anterior projection (hypocaudae). Gonostyli not detected, probably absent. Forewing ovate in shape, without maculation. Costal space strongly dilated at its middle length, narrowed basally, distally. Sc very short, entering margin approximately at wing mid-point. R_1 entering margin before wing apex. Subcostal space narrow proximally, broadens toward termination of Sc. Pterostigma absent. Rs with three simple, widely spaced branches. Two cross-veins between R and M systems: basal 1r-m, distal 2r-m. M basally fused with R for long distance. Two intramedian cross-veins (1im and 2im) located proximad cross-veins 1r-m and 2r-m respectively. MA simple, MP rather deeply forked. Two intracubital cross-veins. CuA simple, rather short. CuP parallel to CuA, probably with one branch. Possible 1A fragmentarily preserved [30].

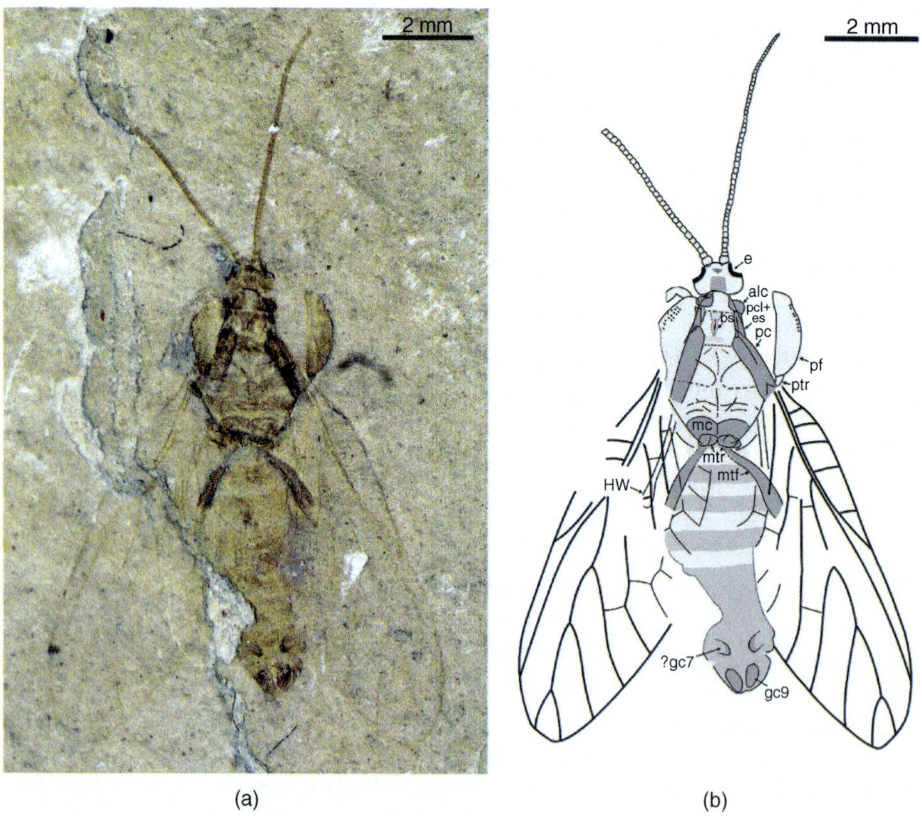

Figure 20.13 *Dipteromantispa brevisubcosta* Makarkin, Yang & Ren, 2013 (Holotype, CNU-NEU-LB2011013). (a), Photograph; (b), Line drawing [30].

Family Grammolingiidae Ren, 2002

Grammolingiidae are a small family found in the Jurassic of Central and East Asia. Three genera and nine species have been described previously in this family: eight species from the Middle Jurassic Daohugou, Inner Mongolia, China [58, 59]; while only one species *Leptolingia shartegica* Khramov, 2010 from the Late Jurassic Sharteg, Mongolia [60]. *Leptolingia* was erected by Ren [58] with two species and Shi *et al.* [59] emended this genus, identifying definitive generic characters and added a species, *L. calonervis*. The fossil species have been recorded from Daohugou, China; Houtiyn-Hotgor, Mongolia; Shar-Teg, Mongolia; Sai-Sagul, Kyrgyzstan; and Karatau, Kazakhstan. To date, five genera with 17 species have been reported [58–65].

The Grammolingiidae are similar to the Late Mesozoic lacewing families of Panfiloviidae, Kalligrammatidae, Aetheogrammatidae and Saucrosmylidae in these characters: large wings; closely-spaced cross-veins in radial area, not arranged in gradate series; MP dichotomously forked near wing base; CuA and CuP separated at wing base, with CuA multi-branched in the middle part of the wings and closely spaced cross-veins in the cubital area. However, Grammolingiidae are distinguished by the costal area with one series of veinlets between cross-veins, resulting in two rows of cells in the costal area along entire wing length, or at least at the basal part. Grammolingiidae are distinguished by the conspicuous stripes on wings, which could be classified into four patterns. The coloration and patterns on wings of grammolingiids could be their predation avoidance strategy, by mimicking the surroundings or providing camouflage to avoid detection or recognition of their large wings by predators (see Section 29.1.5).

Genera included from the Jurassic of Northern China: *Grammolingia* Ren, 2002, *Leptolingia* Ren, 2002, *Litholingia* Ren, 2002 and *Chorilingia* Shi, Wang, Yang & Ren, 2012.

Grammolingia Ren, 2002

Grammolingia Ren, 2002, *Entomol. Sin.*, 9 (12), 54 [58] (original designation).

Type species: *Grammolingia boi* Ren, 2002.

The specific epithet is dedicated to Dr. Bo for the loan of fossil material for study. Forewing 1A terminated on the posterior margin beyond the separation of the first branch of Rs from Rs. CuA forks before the fork of CuP. Both CuA and CuP forked beyond the fork of the first branch of Rs from Rs. 1A terminated beyond the separation of the first branch of Rs [58].

Distribution and age: Inner Mongolia; Middle Jurassic.

Four species included from the Jurassic of Northern China (see Table 20.1).

Figure 20.14 *Grammolingia uniserialis* Shi, Wang & Ren, 2013 (Holotype, CNU-NEU-NN-2010512p) [64].

Grammolingia uniserialis Shi, Wang & Ren, 2013 (Figure 20.14)

Grammolingia uniserialis Shi, Wang & Ren, 2013: *Foss. Rec.*, 16, 172.

Locality and horizon: Daohugou, Ningcheng, Inner Mongolia, China; Middle Jurassic, Jiulongshan Formation.

A relatively small grammolingid. Compound eyes protruding at both sides of head. Prothorax small, almost the same size of head. Meso- and metathorax dilated, and mesothorax a little bigger than metathorax. Distinct punctures present at mesonotum. Mesoscutellum large, triangular. Wings nearly elliptical, somewhat falcate. All wings hyaline, with parallel fuscous stripes across entire wings. Pale speckles sparsely present inside stripes. Trichosors present along distal part of costal and outer margins. Forewing costal area dilated at some distance from the base, strongly narrowed basally and smoothly narrowed to the middle; costal cells double at basal part, single from the middle to distal part; MA forked at near distal one fifth of forewing length. Hind wing narrower than forewing. Costal area narrow, with one row of cells along entire wing length. Sc, R_1, Rs stem parallel with C, but subcostal area and area between R_1 and Rs broader than costal area. Rs pectinate, with more than eight branches parallel and curved toward outer margin. MA and the anterior branch of MP dichotomously forked at distal part of hind wing. Posterior branch of MP pectinate at middle part, with four main branches. CuA curved, forked proximal to the fork of posterior branch of MP, with rich branches. CuP and anal veins indistinct [64].

Leptolingia Ren, 2002

Leptolingia Ren, 2002, *Entomol. Sin.*, 9 (4), 62 [58] (original designation).

Type species: *Leptolingia jurassica* Ren, 2002.

Forewing with vein Rs arising close to the base of the wing, with six to seven branches. MP dichotomously branched basally, MP_2 forked at the level of the separation of the second branch of Rs from Rs. CuA forked after the fork of CuP. On the hind wing, MP forked much before the separation of MA from Rs [58].

Distribution and age: Inner Mongolia; Middle Jurassic.

Four species included from the Jurassic of Northern China (see Table 20.1).

***Litholingia* Ren, 2002**

Litholingia Ren, 2002, *Entomol. Sin.*, 9 (4), 57 [58] (original designation).

Type species: *Litholingia rhora* Ren, 2002.

Forewing relatively broad oval. MP forked beyond the separation of MA from Rs. CuA separated after the fork of CuP. CuP dichotomously branched. 1A terminated at the posterior margin beyond the separation of the first branch of Rs from Rs. At pterostigmatic area, costal cross-veins pectinate, and interlinked by several veinlets between adjacent cross-veins [59].

Distribution and age: Inner Mongolia; Middle Jurassic.

Four species included from the Jurassic of Northern China (see Table 20.1).

***Chorilingia* Shi, Wang, Yang & Ren, 2012**

Chorilingia Shi, Wang, Yang & Ren, 2012, *Alcheringa*, 36 (3), 310 [65] (original designation).

Type species: *Chorilingia euryptera* Shi, Wang, Yang & Ren, 2012.

Wings narrowly elliptical. The new genus is distinguished from the three other genera of Grammolingiidae by the first branch of Rs separating from the Rs stem distal to the forks of CuA and CuP, about at one-third of the wing length from base [65].

Distribution and age: Inner Mongolia; Middle Jurassic.

Four species included from the Jurassic of Northern China (see Table 20.1).

***Chorilingia euryptera* Shi, Wang, Yang & Ren, 2012 (Figure 20.15)**

Chorilingia euryptera Shi, Wang, Yang & Ren, 2012: *Alcheringa*, 36 (3), 311.

Locality and horizon: Daohugou, Ningcheng, Inner Mongolia, China; Middle Jurassic, Jiulongshan Formation.

Moderate to large lacewing. Compound eyes large. Antenna filiform with only three segments visible. Prothorax shorter than wide, possibly due to distortion when preserved. Meso- and metathorax dilated and isometric. Mesonotum with fully developed sclerites, scutum large, scutellum triangular. Metanotum consisting of complete sclerites, scutellum slightly smaller and triangular. Abdomen with eight segments visible. Forewing: costal

Figure 20.15 *Chorilingia euryptera* Shi, Wang, Yang & Ren, 2012 (Holotype, CNU-NEU-NN2010513) [65].

cells doubled along entire, or most of wing length; MA separated from Rs stem near wing base, forked in the distal part of forewing; MP dichotomously branched shortly after MA separated from Rs stem; CuP forked before CuA. Wings elongated, somewhat falcate. Hind wings longer than forewings. Posterior margin convex and outer margin concave in hind wing. All wings hyaline, with fuscous stripes parallel across the entire wings. The fuscous stripes are represented from the wing base on the forewings, and are also present at the apex. Hind wing hyaline at base, stripes ranging across the entire wings and present at the apex. Small pale spots are sparsely present within the stripes. Trichosors present along distal part of costal and outer margin [65].

Family Ithonidae Newman, 1853 *sensu* Winterton and Makarkin, 2010

Based on their phylogenetic analysis incorporating both molecular and morphological data, Winterton and Makarkin, in 2010, formally synonymized Polystoechotidae and Rapismatidae with Ithonidae [66]. Therefore, Ithonidae *sensu lato* currently have four distinct lineages namely ithonid genus-group (former Ithonidae), polystoechotid genus-group (former Polystoechotidae), rapismatid genus-group (former Rapismatidae) and extinct *Principiala* genus-group. The Ithonidae *sensu stricto*, as a relict family, are characterized by larva grub-like with "C" shape, adult body stout and with a lot of setae, head broad and short, retracted under pronotum and costal space tapered-like. The recent Ithonidae *sensu lato* comprise 10 extant genera and 39 species, which have disjunctive distribution in Asia, Oceania, North and South America. To date, there are 13 genera and 28 species of fossil records described. Noteworthy, only the polystoechotid genus-group have been documented in the early strata from the Early to Late Jurassic [21, 56, 58, 67], while major diversification of the rapismatid genus-group seems to have started in the Early Cretaceous [68, 69].

Genera included from the Jurassic and Cretaceous of Northern China: *Mesopolystoechus* Martynov, 1937, *Lasiosmylus* Ren & Guo, 1996, *Jurapolystoechotes* Ren, Engel & Lü, 2002 and *Guithone* Zheng, Ren & Wang, 2016.

Mesopolystoechus Martynov, 1937

Mesopolystoechus Martynov, 1937, *Tr. Paleontol. Inst. Akad. Nauk SSSR*, 7, 38 [21] (original designation).

Type species: *Mesopolystoechus apicalis* Martynov, 1937.

Costal cross-veins densely arranged, with dichotomous forks distally; the width of costal space present twice as subcostal space; Sc and R_1 fused distally; the terminal of Sc and R_1 distinctly bent posteriad, ending close to the wing apex; two gradate series of cross-veins arranged in the radial area; brown stripes irregularly dispersed [48].

Distribution and age: Hebei; Early Cretaceous.

Only one species included from the Cretaceous of Northern China (see Table 20.1).

Lasiosmylus Ren & Guo, 1996

Lasiosmylus Ren & Guo, 1996, *Acta Zootaxon. Sin.*, 21 (4), 466 [49] (original designation).

Type species: *Lasiosmylus newi* Ren & Guo, 1996.

The specific epithet is dedicated to Dr. Tim R. New for his contribution to the research of fossil Neuroptera. Wings with many fuscous spots; humeral veinlet recurrent, with several simple branches; Sc and R_1 separated distally, entering the margin before the wing apex; one or two sc-r_1 cross-veins; relatively few cross-veins present in radial area. The genus was first assigned to Osmylidae [49], but Makarkin *et al.* [71], in 2012, transferred it to Ithonidae. Zheng *et al.* [70], in 2016, re-examined the type specimen of the genus and emended the diagnosis of the genus: stout body covered with dense setae; head hypognathous, protruding from pronotum partly; forewing membranous area with many fuscous spots; humeral plate distinct; costal space dilated basally and narrowed distally; humeral veinlet recurrent, with several simple branches; costal cross-veins simple, moderately curved distally in the apical half of the costal space; Sc and R_1 separated distally, entering the margin before the wing apex [70].

Distribution and age: Liaoning; Early Cretaceous.

Two species included from the Cretaceous of Northern China (see Table 20.1).

Lasiosmylus longus Zheng, Ren & Wang, 2016 (Figure 20.16)

Lasiosmylus longus Zheng, Ren & Wang, 2016: *ZooKeys*, 636, 46.

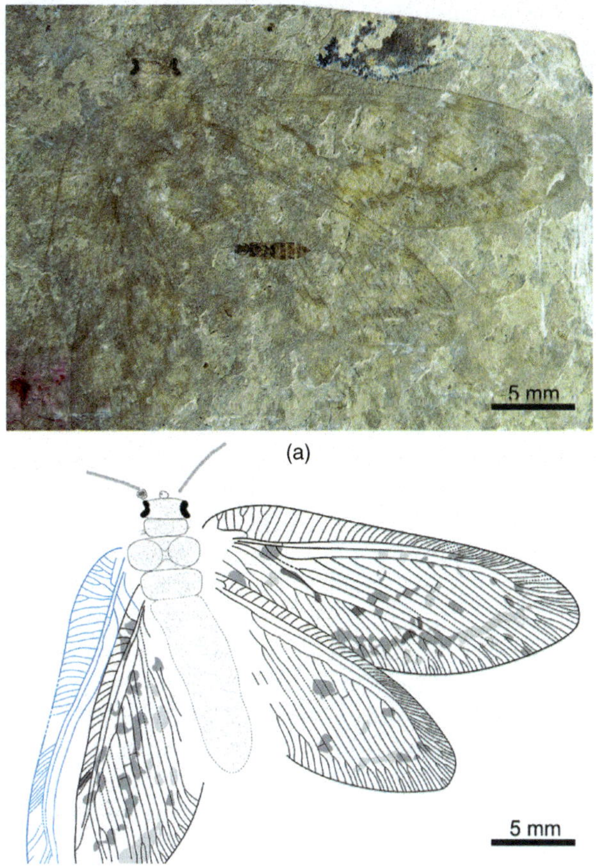

Figure 20.16 *Lasiosmylus longus* Zheng, Ren & Wang, 2016 (Holotype, CNU-NEU-LB2015003). (a), Photograph; (b), Line drawing [70].

Locality and horizon: Huangbanjigou, Beipiao, Liaoning, China; Lower Cretaceous, Yixian Formation.

Body *ca.* 16.3 mm long; head hypognathous, retracted into pronotum partly; antenna filiform (*ca.* 4.0 mm) and incompletely preserved; compound eye large, ocelli absent; pronotum quadrate, numerous long setae concentrated laterally; mesonotum and metanotum stout; abdomen and legs indiscernible. Forewing *ca.* 22.7 mm long and 7.9 mm wide; slender and membranous with numerous fuscous spots; humeral plate discernible; veins covered by dense setae, particular along wing margin; trichosors and nygmata undetectable; costal space broad basally (max. width = 2.1 mm), narrowed distally; recurrent humeral veinlet with several branches; costal cross-veins simple and with the occasional distal dichotomous forks, densely arranged distally; Sc and R_1 separated distally; one subcostal cross-vein close to the origin of Rs; R_1 with many pectinately branches distally, entering the anterior margin; Rs branches regularly arranged with about 13 branches; few cross-veins present between branches of Rs; MA simple; MP first fork distant from wing base, close to the MA divergence from Rs; one mp_1-mp_2 cross-vein detected; CuA branched near the middle of wing, with 10 pectinate branches; CuP with three simple branches; anal veins partly preserved, 1A with three branches and forked proximally, 2A proximally forked. Hind wing *ca.* 18.0 mm long and 7.3 mm wide, venation similar to forewing except costal space narrow; cubitus veins and anal veins not well-preserved [70].

Jurapolystoechotes Ren, Engel & Lü, 2002

Jurapolystoechotes melanolomus Ren, Engel & Lü, 2002, *J. Kans. Entomol. Soc.*, 75 (3), 191 [72] (original designation).

Type species: *Jurapolystoechotes melanolomus* Ren, Engel & Lü, 2002.

Wings medium-sized; one subcostal cross-vein present near the wing base; numerous cross-veins present between R_1 and Rs; Sc and R_1 fused distally; the terminal of Sc and R_1 distinctly bent posteriad, ending close to the wing apex; two gradate series of cross-veins arranged in the radial area; twig of marginal veins prominent, mainly beyond the outer gradate series of cross-veins [72].

Distribution and age: Inner Mongolia; Middle Jurassic.

Only one species included from the Jurassic of Northern China (see Table 20.1).

Guithone Zheng, Ren & Wang, 2016

Guithone Zheng, Ren & Wang, 2016, *Acta Palaeontol. Pol.*, 61 (4), 850 [73] (original designation).

Type species: *Guithone bethouxi* Zheng, Ren & Wang, 2016.

The specific epithet is in honor of Olivier Béthoux to acknowledge his sincere help to B.Y. Zheng. *Guithone* is the first true fossil "moth lacewing" for sharing many characters with the extant ithonids, e.g. disjunct Sc and R_1 present "S" shape distally, and entering the margin before the apex of wings; the MP fork distant to the proximal wing, with simply dichotomous branching. The genus can be identified by the following characters: trichosors present along the anterior margin; humeral veinlet simple, not recurrent; Sc and R_1 separated distally, reaching the anterior margin straightly upon the wing apex; few cross-veins present in the radial area; MP forked beyond the divergence of MA from Rs [73].

Distribution and age: Inner Mongolia; Middle Jurassic.

Only one species included from the Jurassic of Northern China (see Table 20.1).

Guithone bethouxi Zheng, Ren & Wang, 2016 (Figure 20.17)

Guithone bethouxi Zheng, Ren & Wang, 2016: *Acta Palaeontol. Pol.*, 61 (4), 850.

Figure 20.17 *Guithone bethouxi* Zheng, Ren & Wang, 2016 (Holotype, CNU-NEU-NN2015003p) [73].

Locality and horizon: Daohugou, Ningcheng, Inner Mongolia, China. Middle Jurassic, Jiulongshan Formation.

Body robust, with few setae on the surface; head hypognathous. Antennae not preserved; compound eyes large, as wide as the half of frons; ocelli absent; mandible stout, apical portion sharp; thorax stout, less hairy; scutums in mesoscutum nearly round on both sides; mesoscutum of similar size as metascutum; legs medium-sized, covered by abundant setae; tarsal claws short, sharp, with arolium at base of each; abdomen stout. Forewing slender, dense setae along margin and veins. Trichosors distributed along the proximal half of anterior margin; nygmata not detected; costal area shrinking gradually from the base to apex; humeral veinlet not recurrent; costal cross-veins dichotomously forked distally (rarely trichotomous); Sc and R_1 separated distally, reaching the anterior margin before wing apex; Rs slightly zigzagged in distal portion, with 17 regularly pectinate branches; MA simple, and the divergence of MA close to the origin of Rs; MP forked beyond the divergence point of MA from Rs; MP_2 only partly preserved, first fork near mid-length of wing; CuA simple; CuP pectinately branched; anal region well-developed, A_1–A_3 each with 2–4 distal branches. Hind wing incompletely preserved, main veins hardly identifiable; trichosores not detected; frenulum well-preserved, visible on both hind wings. Wing membrane tinged with irregular pale brown shades [73].

Family Kalligrammatidae Handlirsch, 1906

Kalligrammatidae, possibly the most spectacular lacewings among the Mesozoic insects, are occasionally referred to as "butterflies of the Jurassic" [74], because of their long maxillary palps, some with long siphonate mouthparts, large body sizes and long wingspans, and wings with patterned markings, including eyespots (Section 29.3). These features provide an appearance similar to some large, extant lepidopterans. Kalligrammatids have a mosaic of structural features displayed in other neuropterans. The typical large body size (longest wingspan *ca.* 160 mm) and complicated venation imply their affinity with the Suborder Myrmeleontiformia. The typical long ovipositor in *Oregramma* (see Figure 29.21) superficially resembles the extant "pleasing lacewings" in Dilaridae. Their siphonate, long-proboscid mouthparts do not occur in extant Neuroptera, implying these fossil kalligrammatids had different feeding habits which became extinct later. Labandeira *et al.*, in 2016, documented that Kalligrammatidae had evolved the specialized gymnosperm-insect association in the Mesozoic [26]. Hitherto, 12 genera and 29 species are known from the Jurassic to the Early Cretaceous [24, 49, 56, 75–89], clearly indicating the remarkable diversity of Kalligrammatidae in the Mesozoic.

Phylogenetic Relationships for the Genera of Kalligrammatidae

In order to evaluate the phylogenetic relationships among the genera of Kalligrammatidae, Yang *et al.* conducted a phylogenetic analysis based on a set of morphological data containing 28 species and 30 morphological characters [27]. The results of the phylogenetic analysis indicate that the monophyly of the Kalligrammatidae are firmly recovered, sharing the synapomorphic characters of a complex MA bifurcation and a broad triangular forewing and hind wing. The Kalligrammatidae are divided into five primary clades, allocated to the five subfamilies of Sophogrammatinae, Meioneurinae, Oregrammatinae, Kalligrammatinae, and Kallihemerobiinae. Based on the parsimony results (Figure 20.18), the Sophogrammatinae represents the earliest divergence of the Kalligrammatidae, and is the sister-group to all other kalligrammatids. The Sophogrammatinae can be differentiated from the other four subfamilies by the trajectory of the MP, in which both branches are parallel and bent posteriorly and distally, versus the alternate condition in which the MP_1 diverges from the MP_2 close to the wing base, and forms a large triangular region, present in the other four subfamilies. The other four subfamilies share the two synapomorphic characters of eyespot presence and a distinctively triangular MP region. The sister-group relationships of the Meioneurinae and Oregrammatinae are supported by the two homoplasious characters of less interlinked veinlets along the costal region and simple distal Cu_1 forks. In the phylogenetic results, the monophyly of Kallihemerobiinae is partially recovered, principally based on sharing the single plausible character of a proximal bifurcation of the 1A vein, retrieved from the parsimony analysis. Based on the phylogeny of Kalligrammatidae, it is clear that the conspicuous insects underwent a rapid evolutionary diversification during the mid-Mesozoic [26].

Genera included from the Jurassic and Cretaceous of Northern China: *Kalligramma* Walther, 1904, *Kalligrammula* Handlirsch, 1919, *Sophogramma* Ren & Guo, 1996, *Kallihemerobius* Ren & Oswald, 2002, *Oregramma* Ren, 2003, *Limnogramma* Ren, 2003, *Sinokalligramma* Zhang, 2003, *Apochrysogramma* Yang, Makarkin & Ren, 2011, *Protokalligramma* Yang, Makarkin & Ren, 2011, *Affinigramma* Yang, Wang, Labandeira, Shih & Ren, 2014, *Stelligramma* Yang, Wang, Labandeira, Shih & Ren, 2014, *Abrigramma* Yang, Wang, Labandeira, Shih & Ren, 2014, *Ithigramma* Yang, Wang, Labandeira, Shih & Ren, 2014 and *Huiyingogramma* Liu, Zheng, Wang, Fang & Zhang, 2014.

Kalligramma Walther, 1904

Kalligramma Walther, 1904, *Denk. Med-Nat Ges. Jena*, 11, 184 [75] (original designation).

Type species: *Kalligramma haeckeli* Walther, 1904.

One eyespot is present respectively in both fore and hind wings. In forewing, humeral veinlet not recurrent, simple; subcostal veinlets mainly forked; two ORBs; branches MP pectinate anteriorly directed. In hind wing, ScP and RA fused apically; CuP forked distally [90].

Distribution and age: Inner Mongolia; Middle Jurassic.

Twelve species included from the Jurassic of Northern China (see Table 20.1).

Kalligramma elegans Yang, Makarkin & Ren, 2014 (Figure 20.19)

Kalligramma elegans Yang, Makarkin & Ren, 2014: *Ann. Entomol. Soc. Am.*, 107 (5), 919.

Locality and horizon: Daohugou, Ningcheng, Inner Mongolia, China; Middle Jurassic, Jiulongshan Formation.

Most closely related to *Kalligramma paradoxum*, but may be separated from it (and other species of *Kalligramma*) by structure of the eyespot: distance between inner and outer gray rings proximad eyespot much less than diameter of central dark spot, and distance between three gray rings distad eyespot approximately equal to their widths. Forewing costal space moderately broad; slightly narrowed basally. ScA

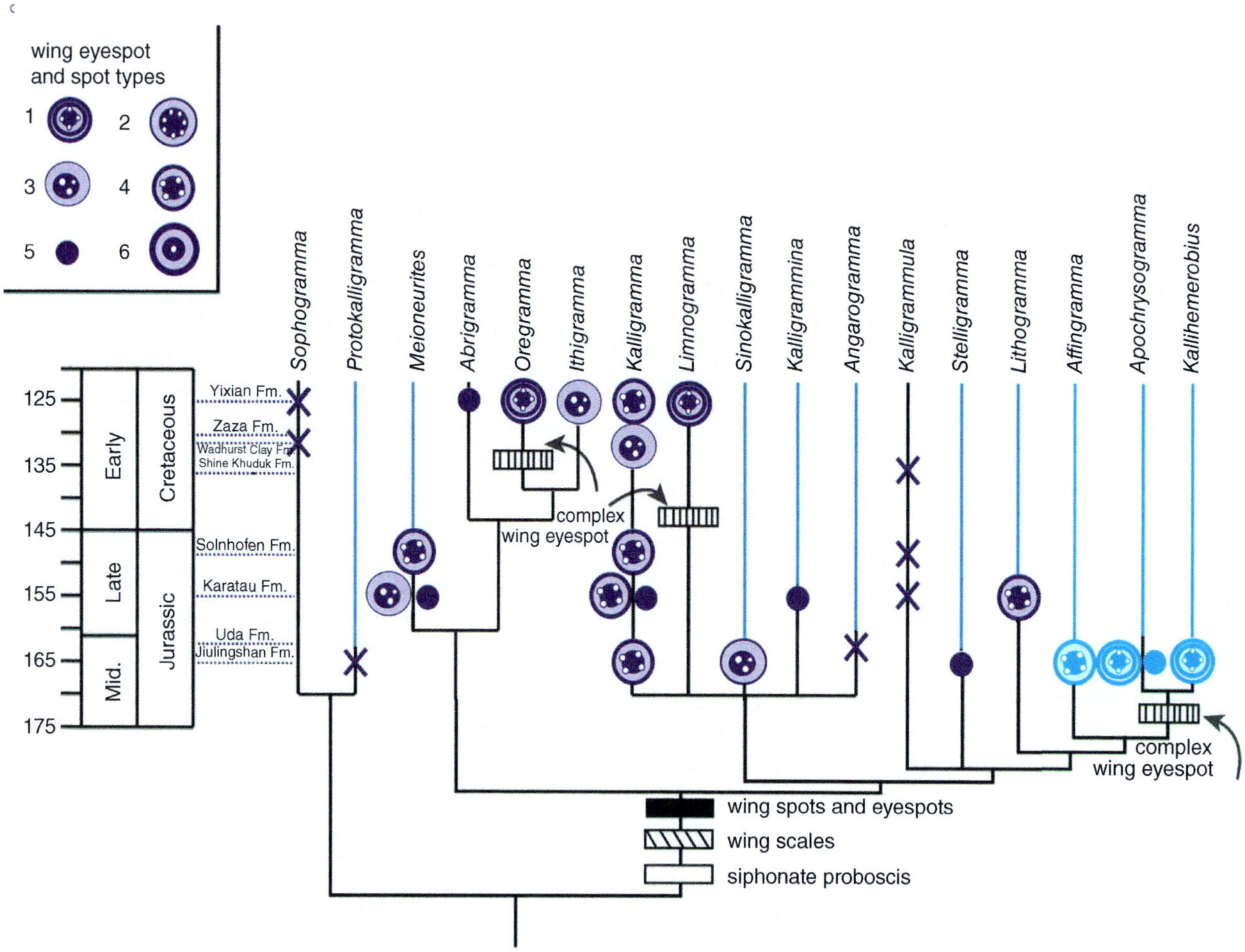

Figure 20.18 Phylogenetic relationships within the Kalligrammatidae. Source: Modified from [26].

Figure 20.19 *Kalligramma elegans* Yang, Makarkin & Ren, 2014 (Holotype, CNU-NEU-NN-2013005p) [90].

well-developed, strongly curved terminating at ScP. Humeral veinlet simple, not recurrent. Cross-veins very dense over entire wing; many cross-veins abnormally forked in radial, intracubital, intra anal spaces; two cross-veins abnormally fused medially or connected by additional cross-veins. Costa anteriorly covered with very short, dense setae; thinner veins, cross-veins covered with rather rare, longer setae arranged in one line. Membrane setae present, very dense in costal space, sparse in other wing areas. Eyespot well-developed, consisting of blackish central structure, several smaller brown rounded structures, grayish rings [75, 90].

Kalligrammula Handlirsch, 1919

Kalligrammula Handlirsch, 1919, *Senckenbergiana*, 1, 62 [76] (original designation).

Type species: *Kalligrammula senckenbergiana* Handlirsch, 1919.

Deeply bifurcated CuP in hind wings (CuP shallowly forked in other genera, except for *Abrigramma*), stem of Rs originating from R near the wing base [91].

Distribution and age: Inner Mongolia; Middle Jurassic.

Only one species included from the Jurassic of Northern China (see Table 20.1).

Sophogramma Ren & Guo, 1996

Sophogramma Ren & Guo, 1996, *Acta Zootaxon. Sin.*, 21 (4), 462 [49] (original designation).

Type species: *Sophogramma papilionacea* Ren & Guo, 1996.

Forewing eyespot absent. Humeral recurrent vein present. All costal veinlets sinuate and forked. Rs with about 12–15 primary branches. MA with five parallel pectinate branches. CuA pectinately forked and terminated at hind margin of wing. CuP single. 1A running parallel to CuP for a long distance and with numerous branches [49].

Distribution and age: Liaoning; Late Jurassic.

Three species included from the Jurassic of Northern China (see Table 20.1).

Kallihemerobius Ren & Oswald, 2002

Kallihemerobius Ren & Oswald, 2002, *Stutt. Beit. Naturk.*, 317, 4 [82] (original designation).

Type species: *Kallihemerobius pleioneurus* Ren & Oswald, 2002.

Mouthparts siphonate. Forewing with eight ORBs between Rs and MA; MP with nine pectinate branches, the second pectinately forked. Hind wing with eight ORBs, all dichotomously forked; MA with three pectinate branches [27].

Distribution and age: Inner Mongolia; Middle Jurassic.

Five species included from the Jurassic of Northern China (see Table 20.1)

Oregramma Ren, 2003

Oregramma Ren, 2003, *Acta Zootaxon. Sin.*, 28 (1), 105 [83] (original designation).

Type species: *Oregramma gloriosa* Ren, 2003.

An eyespot present on forewing. Humeral recurrent vein absent. All costal veinlets straight, unforked. Sc terminated on R_1. Rs with 12 branches. MA with four pectinate branches near wing margin. MP at base forked, with four pectinate branches. CuA forked pectinately. CuP single. Both 1A and 2A parallel with CuP for a long distance. 3A simple [83].

Distribution and age: Liaoning; Early Cretaceous.

Three species included from the Cretaceous of Northern China (see Table 20.1).

Oregramma aureolusa Yang, Wang, Labandeira, Shih & Ren, 2014 (Figure 20.20)

Oregramma aureolusa Yang, Wang, Labandeira, Shih & Ren, 2014: *BMC Evol. Biol.*, 14 (126), 18.

Locality and horizon: Liutiaogou, Ningcheng, Inner Mongolia, China; Lower Cretaceous, Yixian Formation.

Compound eyes large; palpi elongate. Forewing lacking humeral recurrent veins (Vr). MA pectinately forked; MP sinuate, coursing in an arc, with eight pectinate branches. CuA and CuP branched near wing margin. 1A dichotomously forked near terminus; 2A pectinately forked.

Forewing with eyespots; pterostigma indistinct. Hind wing pyriform in gross shape; all costal veinlets arched with some cross-veins forking near posterior wing margin. R_1 simple, with three upturned pectinate branches near terminus. MA dichotomously forked along middle section of wing [27].

Limnogramma Ren, 2003

Limnogramma Ren, 2003, *Acta Zootaxon. Sin.*, 28 (1), 107 [83] (original designation).

Figure 20.20 *Oregramma aureolusa* Yang, Wang, Labandeira, Shih & Ren, 2014 (Holotype, CNU-NEU- NN2009-032p). (a), Photograph; (b), Line drawing [27].

Type species: *Limnogramma mira* Ren, 2003.

An eyespot present in hind wing. Humeral recurrent vein present. All costal veinlets straight, most of them unforked, cross-veins present between them. Rs with five branches at least, most of them forked before marginal twiggings. MA single. MP at base forked pectinately, with three branches. CuA single and simple. CuP at base forked at wing base [83].

Distribution and age: Liaoning; Early Cretaceous.

Three species included from the Cretaceous of Northern China (see Table 20.1).

Sinokalligramma Zhang, 2003

Sinokalligramma Zhang, 2003, *Acta Geol. Sin.*, 77, 143 [84] (original designation).

Type species: *Sinokalligramma jurassicum* Zhang, 2003.

Hind wing short and broad, obtusely triangular. Eyespot well-developed. Humeral recurrent vein absent. C clearly arched, with costal area moderately broadened, costal veinlets sinuate and forked. Sc and R_1 almost joining near wing apex. R_1 pectinately forked anteriorly near wing apex. Rs with seven primary branches [84].

Distribution and age: Inner Mongolia; Late Jurassic.

Only one species included from the Jurassic of Northern China (see Table 20.1).

Apochrysogramma Yang, Makarkin & Ren, 2011

Apochrysogramma Yang, Makarkin & Ren, 2011, *Zootaxa*, 2873, 65 [92] (original designation).

Type species: *Apochrysogramma rotundum* Yang, Makarkin & Ren, 2011.

Apochrysogramma is the second genus in the subfamily Kallihemerobiinae. Its forewing easily differs from that of the type genus by the more rounded shape, much more numerous and closely spaced subcostal veinlets, branches of Rs and MP, less dense cross-veins, and eyespot constructed differently [92].

Distribution and age: Inner Mongolia; Middle Jurassic.

Only one species included from the Jurassic of Northern China (see Table 20.1).

Protokalligramma Yang, Makarkin & Ren, 2011

Protokalligramma Yang, Makarkin & Ren, 2011, *Zootaxa*, 2873, 61 [92] (original designation).

Type species: *Protokalligramma bifasciatum* Yang, Makarkin & Ren, 2011.

Costal space strongly narrowed toward apex; simply constructed MP, not pectinately branched. Cross venation relatively scarce. Eyespot absent, CuP and 1A relatively short, not running parallel to hind margin [92].

Distribution and age: Inner Mongolia; Middle Jurassic.

Figure 20.21 *Protokalligramma bifaciatum* Yang, Makarkin & Ren, 2011 (Holotype, CNU-NEU- NN2009026) [92].

Olny one species included from the Jurassic of Northern China (see Table 20.1).

Protokalligramma bifasciatum Yang, Makarkin & Ren, 2011 (Figure 20.21)

Protokalligramma bifasciatum Yang, Makarkin & Ren, 2011: *Zootaxa*, 2873, 62.

Locality and horizon: Daohugou, Ningcheng, Inner Mongolia, China; Middle Jurassic, Jiulongshan Formation.

Forewing broad, round-ovate. Entire wing membrane densely covered with setae, longest in basal part of costal space (in region of humeral veinlet), long in anterior portion and short to very short in other regions. Trichiation on veins relatively short. Trichosors, not visible due to dense, short trichiation on and near margins, probably absent. Costa stout. Costal space most dilated at 1/4 proximal length; slightly narrowed toward wing base, narrowed toward apex. All subcostal veinlets in proximal half dichotomously branched, connected by two to four cross-veins forming three to four irregular costal series. Subcostal space relatively narrow, with rather widely spaced cross-veins. CuA relatively short, CuP few-branched, forked twice in distal half. Wing membrane fuscous, blackish in costal space, paler in distal and posterior portion of wing, with color pattern consisting of two transverse blackish bands, and several small dark brown or blackish patches basally and near hind margin [92].

Affinigramma Yang, Wang, Labandeira, Shih & Ren, 2014

Affinigramma Yang, Wang, Labandeira, Shih & Ren, 2014, *BMC Evol. Biol.*, 14 (126), 5 [27] (original designation).

Type species: *Affinigramma myrioneura* Yang, Wang, Labandeira, Shih & Ren, 2014.

Siphonate proboscis present. Humeral recurrent veins absent; all costal veinlets sinuate and forked. Rs with

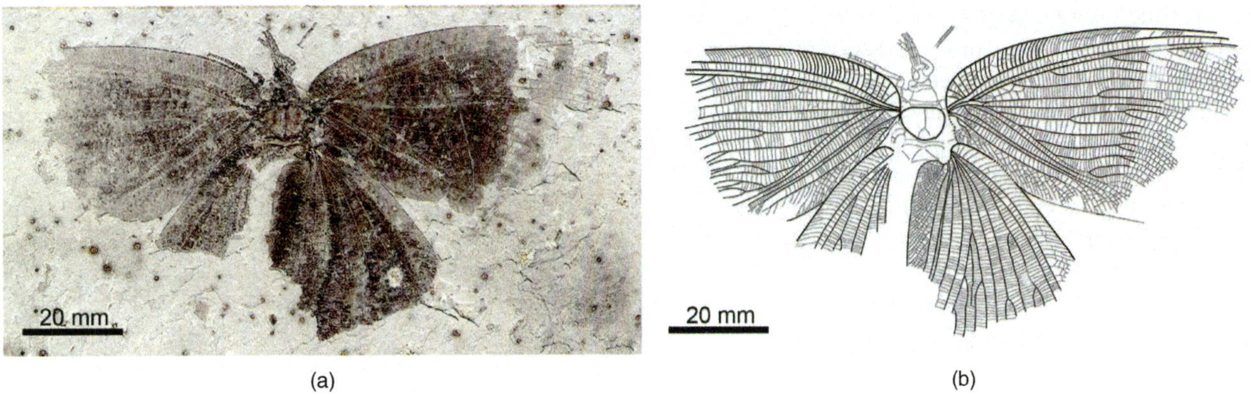

Figure 20.22 *Abrigramma calophleba* Yang, Wang, Labandeira, Shih & Ren, 2014 (Holotype, CNU-NEU- HP2009-001p). (a), Photograph; (b), Line drawing [27].

at least six pectinately forked branches. Only one ORB, between Rs and MA. MA originating from R_1, initially dichotomously forked, then pectinately forked and continuing to the end of vein. CuA single; CuP pectinately forked [27].

Distribution and age: Inner Mongolia; Middle Jurassic.

Only one species included from the Jurassic of Northern China (see Table 20.1).

***Stelligramma* Yang, Wang, Labandeira, Shih & Ren, 2014**

Stelligramma Yang, Wang, Labandeira, Shih & Ren, 2014, *BMC Evol. Biol.*, 14 (126), 9 [27] (original designation).

Type species: *Stelligramma allochroma* Yang, Wang, Labandeira, Shih & Ren, 2014.

Forewing without eyespot markings. Humeral recurrent veins evident. Most costal veinlets curved, deeply forked near wing margin, bearing one to three interlinked veinlets. Rs with pectinate branches, all bifurcate. MA forked at vein midsection. CuA pectinately forked terminally and CuP dichotomously forked at midvein [27].

Distribution and age: Inner Mongolia; Middle Jurassic.

Only one species included from the Jurassic of Northern China (see Table 20.1).

***Abrigramma* Yang, Wang, Labandeira, Shih & Ren, 2014**

Abrigramma Yang, Wang, Labandeira, Shih & Ren, 2014, *BMC Evol. Biol.*, 14 (126), 16 [27] (original designation).

Type species: *Abrigramma calophleba* Yang, Wang, Labandeira, Shih & Ren, 2014.

Humeral recurrent veins absent. Forewing Rs with at least nine primary branches, all single. MA pectinately forked, having a false origin from MP. Hind wing Rs almost originating from R_1 at 1/4 of the distance from its base [27].

Distribution and age: Hebei; Early Cretaceous.

Only one species included from the Cretaceous of Northern China (see Table 20.1).

***Abrigramma calophleba* Yang, Wang, Labandeira, Shih & Ren, 2014 (Figure 20.22)**

Abrigramma calophleba Yang, Wang, Labandeira, Shih & Ren, 2014: *BMC Evol. Biol.*, 14 (126), 16.

Locality and horizon: Pingquan, Chengde, Hebei, China; Lower Cretaceous, Yixian Formation.

Humeral recurrent veins (Vr) absent. Forewing Rs with at least nine primary branches, all single. MA pectinately forked, having a false origin from MP. 2A with *ca.* 13 pectinately forked branches. Hind wing Rs almost originating from R_1 at 1/4 of the distance from its base, branching at *ca.* apical third of wing length, forming at least two primary branches. MP first branch forked nearly at base of wing, the third branch terminates at the wing's posterior border, before wing apex. 2A with at least nine pectinate branches; 3A short, with three pectinate, dichotomous branches [27].

***Ithigramma* Yang, Wang, Labandeira, Shih & Ren, 2014**

Ithigramma Yang, Wang, Labandeira, Shih & Ren, 2014, *BMC Evol. Biol.*, 14 (126), 16 [27] (original designation).

Type species: *Ithigramma multinervium* Yang, Wang, Labandeira, Shih & Ren, 2014.

Large insects; densely setose or scaled throughout the body and wings. Antenna filiform and short. Mouthparts with elongate palpi, exceeding the head length. Wings with indistinct pterostigma, wing eyespot present [27].

Distribution and age: Inner Mongolia; Early Cretaceous.

Only one species included from the Cretaceous of Northern China (see Table 20.1).

***Huiyingogramma* Liu, Zheng, Wang, Fang & Zhang, 2014**

Huiyingogramma Liu, Zheng, Zhang, Wang, Fang & Zhang, 2014, *Alcheringa*, 38, 68 [93] (original designation).

Type species: *Huiyingogramma formosum* Liu, Zheng, Wang, Fang & Zhang, 2014.

Forewing broad. Eyespot well-developed. Cross-veins dense over entire wing. Humeral recurrent vein present. Costal space broad; cross-veins present between costal veinlets; Sc and R_1 fused distally; Rs with several primary branches, most of them deeply forked; MA simple; MP dichotomously forked several times; both CuA and CuP deeply forked; 1A and 2A well-developed; 3A short [93].

Distribution and age: Inner Mongolia; Middle Jurassic.

Only one species included from the Jurassic of Northern China (see Table 20.1).

Family Mantispidae Leach, 1815

Mantispidae, also known as "mantidflies" or "mantisflies", are a highly specialized family of hemerobiiform Neuroptera characterized by their raptorial forelegs, large eyes and elongation of the pronotum behind the forelegs. Their specialization extends beyond morphology to their behavior and life history, with some species being parasites of Araneae or aculeate Hymenoptera. Their larvae develop within the egg cases of spiders (Figure 3.1) or inside the nests of wasps. Indeed, some tropical species are remarkable mimics of their hymenopteran hosts, a feature that affords them some protection from predators. The family, most diverse and abundant in tropical regions, comprise approximately 400 extant species classified into four extant subfamilies: Symphrasinae, Drepanicinae, Calomantispinae, and Mantispinae [51, 74, 94, 95] and the extinct subfamily Mesomantispinae [51, 96].

Mantidflies are rare in the fossil records [51], but have a geological range dating back to the Lower Jurassic. The oldest mantidflies are *Liassochrysa stigmatica* Ansorge and Schluter, 1990 [50] from the Lower Jurassic of Dobbertin, Germany, and *Promantispa similis* Panfilov, 1980 [56] from the Upper Jurassic of Karatau. In the Cretaceous, the family is represented by *Mesomantispa sibirica* Makarkin, 1997 [96] from Baissa, East Siberia, *Gerstaeckerella asiatica* Makarkin, 1990 [97] from Kzyl-Zhar, Kazakhstan, and *Doratomantispa burmanica* Poinar and Buckley, 2011 [98] from Myanmar (Burmese) amber and three genera from the Lower Cretaceous of Yixian Formation in Northeastern China [99] (see below). Mantispidae of the Cenozoic are known from the Eocene oil shales of Grube Messel, Germany, with *Symphrasites eocenicus* Wedmann and Makarkin, 2007 [51], an unnamed larva from the Eocene Baltic amber [100], the Eocene-Oligocene Bembridge Marls of the Isle of Wight, England, with *Vectispa relicta* Cockerell, 1921 [101, 102], *Prosagittalata oligocenica* Nel, 1989 [103] from the Late Oligocene of Cereste and *Climaciella henrotayi* Nel, 1989 [103] from Dauphin, France, and the youngest from Dominican and Mexican ambers (*Feroseta priscus* Poinar, 2006 [104], *Dicromantispa moronei* Engel and Grimaldi, 2007 and *D. electromexicana* Engel and Grimaldi, 2007 [105]).

Genera included from the Jurassic of Northern China: *Archaeodrepanicus* Jepson, Heads, Makarkin & Ren, 2013, *Clavifemora* Jepson, Heads, Makarkin & Ren, 2013 and *Sinomesomantispa* Jepson, Heads, Makarkin & Ren, 2013.

Archaeodrepanicus Jepson, Heads, Makarkin & Ren, 2013

Archaeodrepanicus Jepson, Heads, Makarkin & Ren, 2013, *Palaeontology*, 56, 606 [99] (original designation).

Type species: *Archaeodrepanicus nuddsi* Jepson, Heads, Makarkin & Ren, 2013.

The specific epithet is in honor of Dr. John R. Nudds, paleontologist at the University of Manchester. Proximal segments of antenna elongate, forewing elongate; Rs with only six branches; between branches of Rs one gradate series of cross-veins; color pattern, consisting of two to three dark bands [99].

Distribution and age: Liaoning; Early Cretaceous.

Two species included from the Cretaceous of Northern China (see Table 20.1).

Archaeodrepanicus nuddsi Jepson, Heads, Makarkin & Ren, 2013 (Figure 20.23)

Archaeodrepanicus nuddsi Jepson, Heads, Makarkin & Ren, 2013: *Palaeontology*, 56, 606.

Locality and horizon: Huangbanjigou, Beipiao, Liaoning, China; Lower Cretaceous, Yixian Formation.

Head elongate; postocular lobe strongly expanded; coronal suture distinct. Antenna incomplete; 13 flagellomeres preserved of fairly uniform size. Prothorax wide, comparatively short. Forelegs are incomplete, raptorial, setae preserved. Coxa elongate, stout. Femur with two rows of ventral spines: approximately 22 minor spines, many disarticulated. Tibia incomplete; with prostrate setae. Tarsus incomplete; four segments preserved. Both midlegs preserved, however very incomplete, covered with setae. In the hind legs, femur is incomplete. Tibia

Figure 20.23 *Archaeodrepanicus nuddsi* Jepson, Heads, Makarkin & Ren, 2013 (Holotype, CNU-NEU-LB2011001p) [99].

about 3 mm long. Tarsus incomplete. Forewing: trichosors present on anterior and posterior wing margins. Costal space narrowed toward wing apex. Subcostal veinlets closely spaced; all forked, some dichotomously. Humeral vein recurrent, few branched. Sc sharply bends toward R_1 distally, however, probably terminates on anterior wing margin. Between Sc and R four cross-veins. Sc and R parallel for most of length. R_1 long, entering wing margin before wing apex. Rs originates near wing base. Rs with six primary branches; some forking before mid-point of vein. Between branches of Rs, one gradate series of cross-veins visible; fourth series not detected (although one cross-vein visible). Cross-vein 2r-m preserved. M originates near wing base; M deeply forked. MA and MP simple for most of their length; forked distally; MP dichotomously forked, proximal to fork of MA. Cross-veins 1m-cu, 2m-cu, and 3m-cu preserved. Cu primary fork close to origin. CuA pectinately branched with six primary branches, proximal-most branch deeply forked; distal-most branches of CuA terminated on posterior wing margin after midpoint of wing. CuP pectinately branched, with three primary branches preserved (probably five when complete); some deeply forked. All branches of Rs, M and Cu with end twigging. 1A incomplete, at least once forked. 2A and 3A not preserved. Abdomen is incomplete, four segments partially preserved [99].

Clavifemora Jepson, Heads, Makarkin & Ren, 2013

Clavifemora Jepson, Heads, Makarkin & Ren, 2013, *Palaeontology*, 56, 610 [99] (original designation).

Type species: *Clavifemora rotundata* Jepson, Heads, Makarkin & Ren, 2013.

The genus has the prominent stout club-like forelegs that are easily distinguished from other genera of Mesomantispinae [99].

Distribution and age: Inner Mongolia; Middle Jurassic.

Only one species included from the Jurassic of Northern China (see Table 20.1).

Sinomesomantispa Jepson, Heads, Makarkin & Ren, 2013

Sinomesomantispa Jepson, Heads, Makarkin & Ren, 2013, *Palaeontology*, 56, 609 [99] (original designation).

Type species: *Sinomesomantispa microdentata* Jepson, Heads, Makarkin & Ren, 2013.

Wide profemur with small spines with regard to femur; proximal segments of antenna transverse; costal margin of the forewing concave before wing mid-point; distinct color patterning of forewing [99].

Distribution and age: Liaoning; Early Cretaceous.

Only one species included from the Cretaceous of Northern China (see Table 20.1).

Family Mesochrysopidae Handlirsch, 1906

Mesochrysopidae are a Mesozoic extinct chrysopid-like family, whose systematic position and composition are poorly documented. The taxonomy of this family remained obscure and unresolved for a long time. Some authors believed that this is a fossil subfamily (variously composed) of Chrysopidae [106–109], while others treated this as a separate family. Again, its generic composition varied among authors [20, 56, 110]. Herein, we accept the opinion of Makarkin and Menon [53] that Mesochrysopidae comprise 13 genera, i.e. *Protoaristenymphes* from Lower Jurassic of Luxembourg; *Macronympha* and *Aristenymphes* from Upper Jurassic of Kazakhstan; *Mesochrysopa* and *Mesotermes* from Upper Jurassic of Germany; *Allopterus, Kareninoides, Longicellochrysa, Mesascalaphus, Siniphes* from Lower Cretaceous of China; *Tachinymphes* from Lower Cretaceous of Siberia; and *Armandochrysopa, Triangulochrysopa, Karenina* from Lower Cretaceous of Brazil [52, 53, 111, 112].

Genera included from the Jurassic and Cretaceous of Northern China: *Allopterus* Zhang, 1991, *Tachinymphes* Ponomarenko, 1992, *Protoaristenymphes* Nel & Henrotay, 1994, *Mesascalaphus* Ren, 1995, *Longicellochrysa* Ren, Makarkin & Yang, 2010 and *Kareninoides* Yang, Makarkin & Ren, 2012.

Allopterus Zhang, 1991

Allopterus Zhang, 1991, *Sci. China (Ser. B)*, 34 (9), 1106 [113] (original designation).

Type species: *Allopterus luianus* Zhang, 1991.

The genus belongs to a specialized group within Mesochrysopidae, characterized by the distinctive differences between fore and hind wings in shape, size and venation [113].

Distribution and age: Shandong; Early Cretaceous.

Only one species included from the Cretaceous of Northern China. (see Table 20.1).

Tachinymphes Ponomarenko, 1992

Tachinymphes Ponomarenko, 1992, *Paleontol. J.*, 26 (3), 45 [114] (original designation).

Type species: *Tachinymphes ascalaphoides* Ponomarenko, 1992.

Hind wing anal area very reduced, with AP and A_4 absent or rudimentary and very short; hind wing CuP very short, there is a simple cross-vein between CuA and AA; hind wing cells c_1 and c_2 posteriorly open; forewing MA strongly approximating MP_{1+2} at its base but not fused with it; CuA, MP_{3+4}, MP_{1+2}, and MA not distally fused in all wings; fore and hind wing costal areas not widened; ScP and RA distally fused. Antennae very short [114].

Distribution and age: Liaoning; Early Cretaceous.

Three species included from the Cretaceous of Northern China (see Table 20.1).

***Protoaristenymphes* Nel & Henrotay, 1994**

Protoaristenymphes Nel & Henrotay, 1994, *Ann. Soc. Entomol. Fr.*, 30, 293–318 [115] (original designation).

Type species: *Protoaristenymphes bascharagensis* Nel and Henrotay, 1994.

Medium-sized, forewing length 16–25 mm; in forewing, M forked somewhat distal to or at level of origin of RP; 1m long, five to six times as long as wide; MP with four branches; AA_{3+4} deeply forked [116].

Distribution and age: Inner Mongolia; Middle Jurassic.

Only one species included from the Jurassic of Northern China (see Table 20.1).

***Mesascalaphus* Ren, 1995**

Mesascalaphus Ren, 1995, *Faunae and stratigraphy of Jurassic-Cretaceous in Beijing and the adjacent areas*, 99 [117] (original designation).

Type species: *Mesascalaphus yangi* Ren, 1995.

The specific epithet is dedicated to the entomologist Professor Jikun Yang. Eyes small. Pronotum subquadrate. Wings very long and slender at base, only single row of costal cells throughout. No cross-veins between Sc and R_1. Rs with 12 branches in forewing, 11 in hind wing. One MA in forewing and two MP in hind wing. Both MA and MP single. CuA area of forewing distinctly broader than hind wing. 2A in forewing [117].

Distribution and age: Liaoning; Early Cretaceous.

Only one species included from the Cretaceous of Northern China (see Table 20.1).

***Siniphes* Ren & Yin, 2002**

Siniphes Ren & Yin, 2002, *Acta Zootaxon. Sin.*, 27 (2), 269 [118] (original designation).

Type species: *Siniphes delicatus* Ren & Yin, 2002.

Moderate-sized. Wings without trichosors or covering of microtrichia. Pterostigma well-defined. Costal area of both fore and hind-wing narrower. In forewing, $Sc + R_1$ entering margin almost at wing apex. Subcostal cross-vein absent. MP forked almost at the separation of Rs from R. CuA longer than CuP and 1A forked. In the hind wing, anterior Banksian line not conspicuous, both CuA and CuP shorter than those in the forewing. The genus was considered as a synonym of *Tachinymphes* by Nel *et al.* in 2005 [52, 118].

Distribution and age: Liaoning; Late Jurassic.

Only one species included from the Jurassic of Northern China (see Table 20.1).

***Longicellochrysa* Ren, Makarkin & Yang, 2010**

Longicellochrysa Ren, Makarkin & Yang, 2010, *Zootaxa*, 2523, 51 [119] (original designation).

Figure 20.24 *Kareninoides lii* Yang, Makarkin & Ren, 2012 (Holotype, CNU-NEU-LY2011001p) (Donated by Yanjun Li) [116].

Type species: *Longicellochrysa yixiana* Ren, Makarkin & Yang, 2010.

Large lacewing. Antenna long, more than the lengths of head and prothorax combined. Pterostigma well-defined. Long hypostigmal cell absent. Rs smooth, slightly zigzagged distally, with 16 regularly pectinate zigzagged branches. Cell 1m long, about five times as long as the maximal width. Basal cross-vein 1m-cu long, at origin of M; 2m-cu connecting im and CuA long [119].

Distribution and age: Liaoning; Early Cretaceous.

Only one species included from the Cretaceous of Northern China (see Table 20.1).

***Kareninoides* Yang, Makarkin & Ren, 2012**

Kareninoides Yang, Makarkin & Ren, 2012, *Zootaxa*, 3597, 6 [116] (original designation).

Type species: *Kareninoides lii* Yang, Makarkin & Ren, 2012.

The prothorax of *Kareninoides* is elongate as most other genera. Forewing clearly elongate, with hind margin smoothly slightly curved; hind wing with similar length to forewing (cf. distinctly shorter in other genera); both Banksian lines well-developed [116].

Distribution and age: Liaoning; Early Cretaceous.

Only one species included from the Cretaceous of Northern China (see Table 20.1).

***Kareninoides lii* Yang, Makarkin & Ren, 2012 (Figure 20.24)**

Kareninoides lii Yang, Makarkin & Ren, 2012: *Zootaxa*, 3597, 7.

Locality and horizon: Liulongtai, Yixian, Liaoning, China; Lower Cretaceous, Yixian Formation.

The specific epithet is dedicated to Mr. Yanjun Li for his donation of this specimen to the fossil collection of CNUB. Head rather elongate; vertex with indistinct longitudinal coronal (epicranial) suture; antenna poorly preserved, only visible as impression on matrix, filiform or moniliform, with scapus, pedicel slightly broader than segments of flagellum; mandibles apparently quite large (poorly preserved). Prothorax long, slightly longer than maximal length of head. Mesothorax only fragmentarily preserved. Metathorax nearly complete, but very poorly preserved. Abdomen about 21.8 mm long; the first abdominal tergite possesses distinct medio longitudinal suture distally bifid into parts of transverse suture. Second abdominal tergite long, slightly longer than its width; acrotergite of second tergite not identified or absent. Forewing elongate, rather narrow, with rounded apex. Costal space only slightly expanded in proximal part, equally narrow before fusion of ScP with RA. Pterostigma not preserved. Veinlets of ScP + RA poorly preserved, simple; two connected by cross-vein. Subcostal space very narrow, cross-veins not detected. Stem of RP smooth, with 15 regularly pectinate, slightly zigzagged branches. Cross-veins in radial space numerous, mostly not forming gradate series. Two short Banksian lines present, converging apically; Banksian folds not detected. M basally fused with R for considerable distance; divided into MA and MP well-proximal to origin of RP_1. MA arched, slightly zigzagged, MP strongly zigzagged [116].

Family Myrmeleontidae Latreille, 1802

The Myrmeleontidae, commonly known as "antlions", are the most diverse family of Neuroptera, with about 2000 described species worldwide. Antlions are typically large and conspicuous insects, predominantly found in arid and semiarid regions. Nevertheless, the Myrmeleontid fossils are relatively sparse comparing to the extant antlion species. Rice, in 1969, was the first to report a fossil myrmeleontid *Palaeoleon ferrogeneticus* from the Late Cretaceous deposits of Labrador, Canada [20, 120]. Although *Palaeoleon* has at times been re-assigned into a related but separate family, the Palaeoleontidae [121, 122], the genus and its species are most conservatively considered to be in a basal subfamily of Myrmeleontidae. The Lower Cretaceous Crato Formation in Brazil had yielded a diverse fauna of myrmeleontids represented by nine genera [1]. Hong, in 1988, described a myrmeleontid-like fossil *Liaoximyia* from the Liaoning of the Cretaceous [123]. Makarkin and Menon, 2005 [53, 123] pointed the type specimen of *Liaoximyia* is too incomplete to determine it systematic status, and suggested it possibly belongs to a mesochrysopid affinity. The true fossil antlion from China, *Choromyrmeleon* was described by Ren & Guo, 1996 [49] based on the specimens from the Yixian Formation near Beipiao City, Liaoning.

Only one genus included from the Cretaceous of Northern China: *Choromyrmeleon* Ren & Guo, 1996.

Choromyrmeleon Ren & Guo, 1996

Choromyrmeleon Ren & Guo, 1996, *Acta Zootaxon. Sin.*, 21 (4), 472 [49] (original designation).

Figure 20.25 *Choromyrmeleon aspoeckorum* Ren & Engel 2008 (Holotype, LB20003) [124].

Type species: *Choromyrmeleon othneius* Ren & Guo, 1996.

In both fore and hind wings, presectoral cross-veins absent. Sc and R coalesced at base for a considerable distance, Rs arising well before CuA fork in the forewing and posterior medial fork in the hind wing. CuP elongate, with many pectinate long branches [49].

Distribution and age: Liaoning; Early Cretaceous.

Two species included from the Cretaceous of Northern China (see Table 20.1).

Choromyrmeleon aspoeckorum Ren & Engel, 2008 (Figure 20.25)

Choromyrmeleon aspoeckorum Ren & Engel, 2008: *Alavesia*, 2, 184.

Locality and horizon: Huangbanjigou, Beipiao, Liaoning, China; Lower Cretaceous, Yixian Formation.

The specific name is dedicated to Drs. Horst and Ulrike Aspöck for their enormous contributions to the studies of the Neuropterida. It can be separated from *C. othneius* by the apical forking of Rs, i.e. beyond termination of R_1, Rs does not fork until marginal twigging in the new species, (vs. Rs continues to fork beyond termination of R_1 in *C. othneius*); apically R_1 and Rs more strongly arch posteriorly leaving a wide space between them and the wing apex, many times broader than C-Sc separation in the new species (vs. apically R_1 and Rs continue along wing apex separated by distance slightly less than C-Sc separation in *C. othneius*); and the location of the oblique vein distinctly before CuA fork in the new species (vs. confluent with CuA fork in *C. othneius*). Moreover, the head of this species is quite unusual in that it is narrow and prolonged rather than the more typical transverse form of extant myrmeleontids [124].

Family *Incertae sedis*

Guyiling Shi, Béthoux, Shih & Ren, 2012

Guyiling Shi, Béthoux, Shih & Ren, 2012, *Syst. Entomol.*, 37 (3), 618 [62] (original designation).

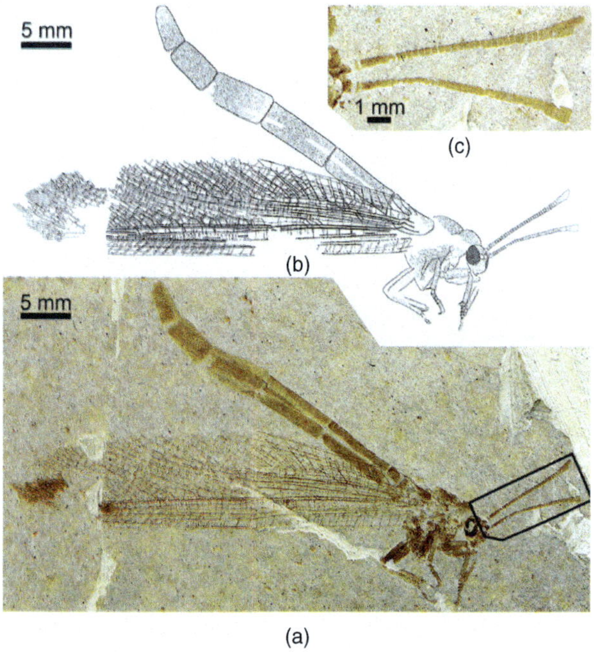

Figure 20.26 *Guyiling jianboni* Shi, Béthoux, Shih & Ren, 2012 (Holotype, CNU-NEU-LB2011014). (a), Photograph (composite, flipped horizontally); (b), Line drawing of habitus; (c), Detail of antennae, as located on A (composite) [62]. Source: Donated by Dr. Chungkun Shih.

Type species: *Guyiling jianboni* Shi, Béthoux, Shih & Ren, 2012.

The genus is only recorded by a single specimen from the Lower Cretaceous Yixian Formation. The specific epithet is dedicated to Mr. Jian Bon Shih, for demonstrating initiatives, creativity and innovation in his study and work, and for providing inspiration and support to Dr. Shih. The specimen exhibits derived traits such as distally dilated antennae and well-developed anterior Banksian line (known in Myrmeleontidae), but also genuine plesiomorphies (at the level of Myrmeleontiformia), such as the divergence of a distinct CuA_1 stem from $MP_2 + CuA_1$ (forewing), and a basal origin of MA (diverging from RP + MA; both wing pairs). This combination is unique among the "Palaeoleontidae", a group better considered as a paraphyletic assemblage of various stem-Myrmeleontiformia. Due to these inconsistent characters, Shi *et al.* deemed *G. jianboni* as an antlion-like lacewing, but in Family *Incertae sedis* [62].

Distribution and age: Liaoning; Early Cretaceous.

Only one species included from Cretaceous of Northern China (see Table 20.1).

***Guyiling jianboni* Shi, Béthoux, Shih & Ren, 2012 (Figure 20.26)**

Guyiling jianboni Shi, Béthoux, Shih & Ren, 2012: *Syst. Entomol.*, 37 (3), 618.

Locality and horizon: Huangbanjigou, Beipiao, Liaoning, China; Lower Cretaceous, Yixian Formation.

Antenna shorter than half of forewing length, gradually enlarged distally; wings: RP + MA arising close to wing base; anterior Banksian line present; elongate shape; forewing: no cross-veins in the area between ScP and RA; MP_2 fused with CuA for a short distance; a single main stem of CuA_1 diverging from $MP_2 + CuA_1$. The genus exhibits several characters diagnostic of the Myrmeleontidae, namely distally dilated antennae and well-developed anterior Banksian line. However, the genus also possesses the plesiomorphic traits such as the divergence of a distinct CuA_1 stem from $MP_2 + CuA_1$, and a basal origin of MA which are unique among "Palaeoleontidae". Due to these inconsistent characters, *G. jianboni* was considered as an antlion-like lacewing, but in Family *Incertae sedis* [62].

Family Nymphidae Rambur, 1842

Nymphidae are generally accepted to be the most basal family in the superfamily of Myrmeleontoidea, and the only myrmeleontoid family known from the Jurassic [125]. Diagnostic features of nymphids include elongated filiform antennae, ocelli absent, legs with a bifid arolium, wings with trichosors present, nygmata absent and thyridiate (incomplete) cross-veins present of varying lengths along the subcostal space [126]. The oldest fossil nymphid representatives are known from the Middle Jurassic of China [125–128]. To date, there are 14 extinct genera containing 23 species described from the Mesozoic and the Palaeogene deposits [126]. Eight extant genera containing about 35 described species are restricted to the Australian and Oriental regions, and they appear to be most diverse in Australia [128, 129].

Phylogenetic Research on Nymphidae

In 2015, Shi *et al.* first reconstructed the phylogeny of Nymphidae including both extant and fossil genera based on 68 morphological characters, as well as DNA molecular sequences [126] (Figure 20.27). According to the phylogenetic results, the monophyly of Nymphidae was supported to a high degree by a number of synapomorphies: a bifid arolium and wing subcostal thyridiate cross-veins present in the adults, and larval characters such as stemmata number, antennal flagellum shape, number of segments of the labial palp, a single large tooth on the jaw, and presence of thoracic and abdominal scoli. Nymphidae are distinctively divided into three clades: the basalmost Jurassic genus *Liminympha*, and two well-defined extant subfamilies Nymphinae and Myiodactylinae. The clade of Nymphinae includes three extant genera of *Nesydrion*, *Austronymphes*, and *Nymphes*, and three Jurassic and Cretaceous fossil genera of *Baissoleon*, *Sialium*, and *Nymphites* for sharing the presence of tibial spurs at least on hind legs,

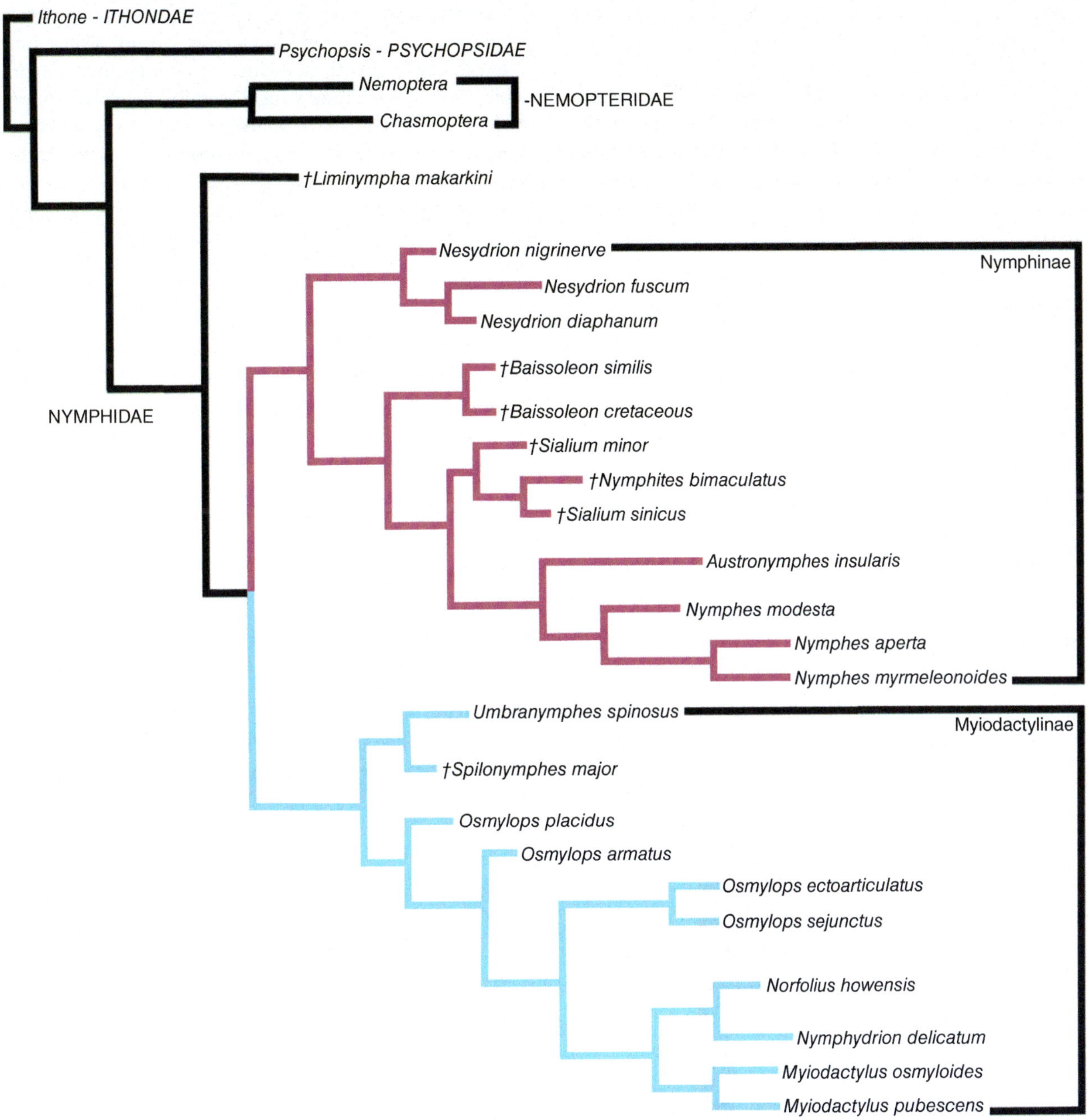

Figure 20.27 The phylogenetic relationships of extant and extinct insects in Nymphidae based on system morphology and DNA molecular data. Source: Modified from [126].

wing with narrow costal area and cross-veins present between branches of CuA, male genitalia with arms of gonarcus far separated medially and mediuncus having large lateral lobes. All extant nymphines are restricted to Australia, but it is clear that the subfamily was more widely distributed during the Mesozoic than at present, with fossil genera known from the Palearctic and Nearctic [125, 127, 130]. Myiodactylinae includes the extant genera of *Umbranymphes, Osmylops, Myiodactylus, Norfolius,* and *Nymphydrion,* and the Cretaceous genus of *Spilonymphes,* sharing absence of tibial spurs, wing costal area broad (often with multiple forked costal cross-veins), no cross-veins between forewing CuA branches; dorsal ends of gonarcus arms approximating each other (sometimes connected by a membrane); mediuncus small, ball-like or trilobate ventrally with varying sized teeth and no obvious lateral lobes. The extant Myiodactylinae are geographically more widely distributed than their sister clades, being found throughout New Guinea and Australia, including Norfolk Island [131–133]. The Cretaceous genus *Spilonymphes* from China indicates the clade had a more complicated scenario of evolution.

Genera included from the Jurassic and Cretaceous of Northern China: *Sialium* Westwood, 1854, *Nymphites* Haase, 1890, *Baissoleon* Makarkin, 1990, *Liminympha*, Ren & Engel, 2007, *Daonymphes* Makarkin, Yang, Shi & Ren, 2013 and *Spilonymphes* Shi, Winterton & Ren, 2015.

Sialium Westwood, 1854

Sialium Westwood, 1854, *Quart. J. Geol. Soc.*, 10, 378–396 [134] (original designation).

Type species: *Sialium sipylus* Westwood, 1854.

Tibial spurs present at least on hind legs; intracubital space dilated before CuA forked; two or more rows of cross-veins present between CuA branches; hind wing MP pectinately branched with more than eight branches; CuP with many branches [126].

Distribution and age: Inner Mongolia; Early Cretaceous.

Two species included from the Cretaceous of Northern China (see Table 20.1).

Sialium sinicus Shi, Winterton & Ren, 2015 (Figure 20.28)

Sialium sinicus Shi, Winterton & Ren, 2015: *Cladistics*, 31 (5), 470.

Locality and horizon: Liutiaogou, Ningcheng, Inner Mongolia, China; Lower Cretaceous, Yixian Formation.

Large lacewing with forewing 41 mm long and 11.8 mm wide as preserved; hind wing 37.5 mm long and 10.5 mm wide as preserved. Wing markings present in subcostal area and entire apical area in both wings, markings sparsely distributed on posterior part of forewing, one large spot present near distal one-third of wing length on hind wing. Forewing costal cross-veins seldom forked before pterostigma area. Humeral veinlet recurrent. Sc and R_1 closely parallel. Rs separated from R_1 at basal one-tenth of wing length. MA separated from Rs stem at basal one-fifth of wing length. MP forked between the origin of Rs and the separation of MA. MP_1 straight and unbranched. MP_2 pectinately branched with numerous branches terminating at outer margin. CuA pectinately branched with four branches terminating at the midway of wing margin. The space between CuA and CuP dilated before CuA forked [126].

Nymphites Haase, 1890

Nymphites Haase, 1890, *Neues Jahrb. Mineral. Geol.Paläontol.*, 2 (1), 1–33 [135] (original designation).

Type species: *Hemerobius priscus* Weyenbergh, 1869 [136].

Forewing elongate, broad proximally; all distal veinlets of ScP once or twice forked; MP with four to five pectinate branches, each connected by one cross-vein;

Figure 20.28 *Sialium sinicus* Shi, Winterton & Ren, 2015 (Holotype, CNU-NEU-NN2014002). (a), Photograph; (b) Line drawing [126].

CuP space broad, more than twice as wide as intracubital space. Hind wing MP with four to five pectinate branches; at least proximal branches of CuA connected by cross-veins; CuP short, not pectinately branched. Legs (at least hind) with tibial spurs [128].

Distribution and age: Inner Mongolia; Middle Jurassic.

Only one species included from the Jurassic of Northern China (see Table 20.1).

Baissoleon Makarkin, 1990

Baissoleon Makarkin, 1990, *Ann. Soc. Entomol. Fr. (N.S.)*, 26, 125 [97] (original designation).

Type species: *Baissoleon cretaceous* Makarkin, 1990.

Prothorax longer than wide; tibial spurs present at least on hind legs; wings slender; pterostigma distinct; Sc + R_1 entering margin before wing apex; costal area not dilated after the fusion of Sc and R_1; forewing CuP dichotomously forked [126].

Distribution and age: Liaoning; Early Cretaceous.

Only one species included from Cretaceous of Northern China (see Table 20.1).

Liminympha Ren & Engel, 2007

Liminympha Ren & Engel, 2007, *Ann. Zool.*, 57 (2), 212 [127] (original designation).

Type species: *Liminympha makarkini* Ren & Engel, 2007.

The specific epithet is dedicated to Dr. Makarkin for his authority on extant and fossil Neuroptera. Forewing costal space with numerous c-sc cross-veins; Sc and R_1 fused apically and then curving slightly posteriorly to terminate at wing margin beyond apex; Rs with 17 pectinate branches; hind wing with stem of MA absent [127].

Distribution and age: Inner Mongolia; Middle Jurassic.

Only one species included from the Jurassic of Northern China (see Table 20.1).

Daonymphes Makarkin, Yang, Shi & Ren, 2013

Daonymphes Makarkin, Yang, Shi & Ren, 2013, *ZooKeys*, 325, 3 [125] (original designation).

Type species: *Daonymphes bisulca* Makarkin, Yang, Shi & Ren, 2013.

Forewing broad proximally, vast majority of proximal subcostal veinlets forked, cross-veins between branches of MP absent, CuP space broad, nearly two times as wide as intracubital space [125].

Distribution and age: Inner Mongolia; Middle Jurassic.

Only one species included from the Jurassic of Northern China (see Table 20.1).

Spilonymphes Shi, Winterton & Ren, 2015

Spilonymphes Shi, Winterton & Ren, 2015, *Cladistics*, 31 (5), 464 [126] (original designation).

Type species: *Spilonymphes major* Shi, Winterton & Ren, 2015.

The new genus has a forewing costal space slightly broader than typical Nymphinae genera, and is more similar to recent myiodactyline genera. Forewing basal rs-m present, proximal to the oblique vein; MA separated from Rs close to the origin of Rs; CuP space broad, nearly three times as wide as intracubital space; hind wing CuA space broad [126].

Distribution and age: Liaoning; Early Cretaceous.

Only one species included from the Cretaceous of Northern China (see Table 20.1).

Spilonymphes major Shi, Winterton & Ren, 2015 (Figure 20.29)

Spilonymphes major Shi, Winterton & Ren, 2015: *Cladistics*, 31 (5), 464.

Locality and horizon: Huangbanjigou, Chaomidian, Liaoning, China; Lower Cretaceous, Yixian Formation.

Medium-sized lacewing. Forewing, costal cross-veins seldom forked and more numerous toward apex. Sc and R_1 fused at approximately distal one-quarter of wing length. Seven cross-veins present between R_1 and Rs. MA separated from Rs stem around the basal one-quarter of wing length. Rs with 17 branches. Outer gradate cross-veins and proximal irregular cross-veins present among Rs branches. MP forked before the separation of MA from Rs. MP_1 forked until near wing margin, with few branches. MP_2 straight and pectinately branched from the midway of its length, with numerous branches terminating at outer margin. CuA parallel to posterior margin with broad space between them. CuA pectinately branched with four or five branches midway along wing length. CuP space broad, about three times as broad as intracubital space. CuP pectinately branched with about six branches terminating at posterior margin. CuA and CuP branches continuously forked before terminating [126].

Family Osmylidae Leach, 1815

Osmylidae are a small but relatively basal family in Neuroptera, comprising *ca.* 225 extant species [137]. The extant osmylids are recorded in most biogeographical regions except for the Nearctic region. Adult osmylids are easily recognized by the brown to dark brown body and veins, presence of three distinct ocelli (absent in subfamily Gumillinae), well-developed pterostigma and nygmata, and fairly complicated venation. The larvae of osmylids are commonly found near the river or stream, which are considered to retain both aquatic and terrestrial habits. Osmylidae fossils have represented a basal

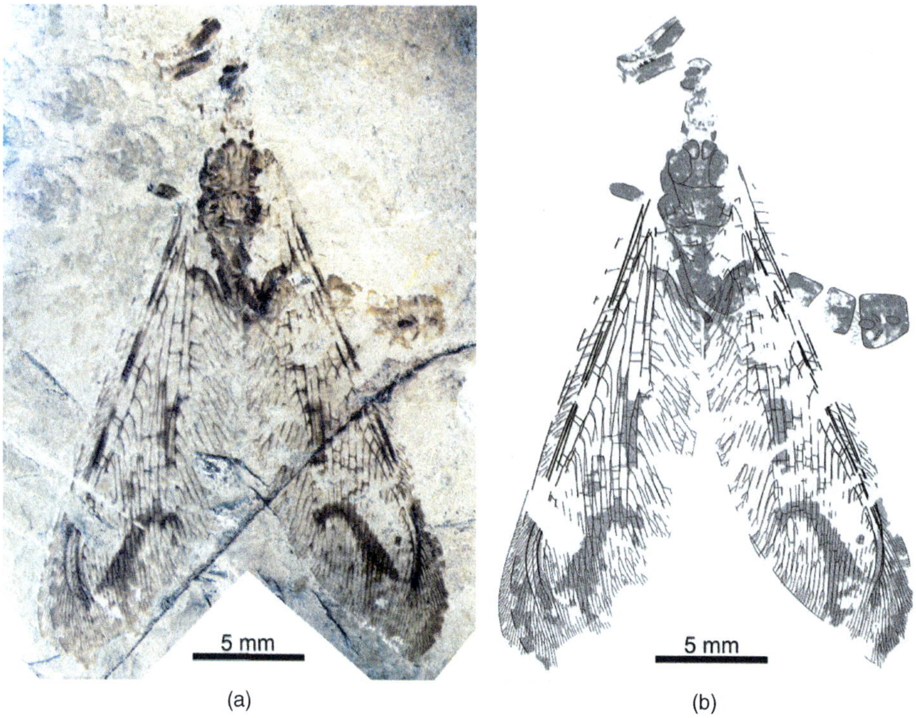

Figure 20.29 *Spilonymphes major* Shi, Winterton & Ren, 2015 (Holotype, CNU-NEU-LB2014001). (a), Photograph; (b), Line drawing [126].

branch within Neuroptera [9, 11, 16–19]. The analysis of molecular clock indicated the origin time of Osmylidae should be no later than the Later Permian [11, 12] and the divergence time of the major subfamilies should be no later than the end of the Triassic. This dating is in agreement with the rich Mesozoic fossil records for the Osmylidae [7]. The Mesozoic is a golden time during the evolution of Osmylidae, and about 30 genera have been recorded worldwide. In China, fossil osmylids are remarkably diverse, and hitherto 10 genera and 16 species have been described across five major extant subfamilies, i.e. Gumillinae, Osmylinae, Spilosmylinae, Protosmylinae, and Kempyninae.

Genera included from the Jurassic and Cretaceous of Northern China: *Epiosmylus* Panfilov,1980, *Yanosmylus* Ren, 1995, *Laccosmylus* Ren & Yin, 2003, *Enodinympha* Ren & Engel, 2007, *Nilionympha* Ren & Engel, 2007, *Palaeothyridosmylus* Wang, Liu & Ren, 2003, *Tenuosmylus* Wang, Liu & Ren, 2009, *Allotriosmylus* Yang, Makarkin & Ren, 2010, *Juraheterosmylus* Wang, Liu, Ren & Shih, 2010, *Archaeosmylidia* Makarkin, Yang & Ren, 2014 and *Jurakempynus* Wang, Liu, Ren & Shih, 2011.

Epiosmylus Panfilov, 1980

Epiosmylus Panfilov, 1980, *Mesozoic fossil insects*, 100 [56] (original designation).

Type species: *Epiosmylus longicornis* Panfilov, 1980.

Forewing slender, apex slightly pointed. The first branch of Rs far from the base of the wing. Veins Rs and MA separated at the middle of the wing. CuA is longer than CuP. Hind wing CuA is short, less than half the length of MP_2, CuP single [138].

Distribution and age: Inner Mongolia; Middle Jurassic.

Only one species included from the Jurassic of Northern China (see Table 20.1).

Yanosmylus Ren, 1995

Yanosmylus Ren, 1995, *Faunae and stratigraphy of Jurassic-Cretaceous in Beijing and the adjacent areas*, 191 [117] (original designation).

Type species: *Yanosmylus rarivenatus* Ren, 1995.

Anterior margin straight. Costal veins not forked. Sc area as wide as R_1. Some cross-veins present at base of branches of Rs. MA and MP branched three times at wing margin. 1A branched pectinately [117].

Distribution and age: Hebei; Early Cretaceous.

Only one species included from the Cretaceous of Northern China (see Table 20.1).

Enodinympha Ren & Engel, 2007

Enodinympha Ren & Engel, 2007, *Ann. Zool.*, 57 (2), 214 [127] (original designation).

Type species: *Enodinympha translucida* Ren & Engel, 2007.

Forewing costal space c-sc cross-veins becoming more dense at wing apex; subcostal cell without cross-veins; Rs with 11 pectinate branches; five basal rs-m cross-veins; 19 rs-ma cross-veins; 2A and 3A not forming a closed anal loop [127].

Distribution and age: Inner Mongolia; Jurassic.

Only one species included from the Jurassic of Northern China (see Table 20.1).

Nilionympha Ren & Engel, 2007

Nilionympha Ren & Engel, 2007, *Ann. Zool.*, 57 (2), 216 [126] (original designation).

Type species: *Nilionympha pulchella* Ren & Engel, 2007.

Hind wing c-sc cross-veins becoming denser toward wing apex; subcostal cell apparently without cross-veins; 15 r_1-rs cross-veins; Rs originating at wing base, stem of MA absent; anal field with a relatively small area in basal fifth of wing [126].

Distribution and age: Inner Mongolia; Middle Jurassic.

Two species included from the Jurassic of Northern China (see Table 20.1).

Palaeothyridosmylus Wang, Liu & Ren, 2009

Palaeothyridosmylus Wang, Liu & Ren, 2009, *Zootaxa*, 2034, 65 [139] (original designation).

Type species: *Palaeothyridosmylus septemaculatus* Wang, Liu & Ren, 2009.

Membrane fuscous, with some fenestrated spots. Rs with numerous branches, cross-veins in radial sector arranged irregularly, and not forming the gradate cross-veins. Cu forked at wing base. Cross-veins between MP and Cu more than one [139].

Distribution and age: Inner Mongolia; Middle Jurassic.

Only one species included from the Jurassic of Northern China (see Table 20.1).

Tenuosmylus Wang, Liu & Ren, 2009

Tenuosmylus Wang, Liu & Ren, 2009, *Acta Palaeontol. Pol.*, 54 (3), 557 [140] (original designation).

Type species: *Tenuosmylus brevineurus* Wang, Liu & Ren, 2009.

Forewing radial cross-veins numerous, arranged irregularly and not forming gradate cross-veins. Separation of Ma at one-third of wing length with first branch of Rs nearly at middle of wing. Rs with six to seven branches. Cu_2 has four pectinate branches [140].

Distribution and age: Inner Mongolia; Middle Jurassic.

Only one species included from the Jurassic of Northern China (see Table 20.1).

Allotriosmylus Yang, Makarkin & Ren, 2010

Allotriosmylus Yang, Makarkin & Ren, 2010, *Ann. Entomol. Soc. Am.*, 103 (6), 856 [141] (original designation).

Type species: *Allotriosmylus uniramosus* Yang, Makarkin & Ren, 2010.

Forewing elongate ovoid with rather rounded apex. Costal space relatively narrow, only slightly dilated at one fifth of its length. Subcostal veinlets simple, straight, rather widely spaced. Pterostigma distinct, elongate. Rs originates far from wing base, strongly zigzagged, with one long zigzagged branch. M appears fused basally with R for a rather long distance, dividing into MA and MP far distal from origin of Rs. MP zigzagged, with seven widely spaced pectinate branches, proximalmost deeply forked. CuA relatively short, straight, with two apical branches. CuP very thin basally, somewhat stouter distally; pectinate with two branches [141].

Distribution and age: Inner Mongolia; Middle Jurassic.

Only one species included from the Jurassic of Northern China (see Table 20.1).

Juraheterosmylus Wang, Liu, Ren & Shih, 2010

Juraheterosmylus Wang, Liu, Ren & Shih, 2010, *Zootaxa*, 2480, 46 [142] (original designation).

Type species: *Juraheterosmylus antiquatus* Wang, Liu, Ren & Shih, 2010.

Forewing with few spots, r_1-rs cross-veins no more than 10 branches. Cross-veins in radial sector arranged regularly, forming three rows of gradate series. Cu forked at the wing base, both branches approximately with equal lengths [142].

Distribution and age: Inner Mongolia; Middle Jurassic.

Three species included from the Jurassic of Northern China (see Table 20.1).

Jurakempynus Wang, Liu, Ren & Shih, 2011

Jurakempynus Wang, Liu, Ren & Shih, 2011, *Acta Palaeontol. Pol.*, 56 (4), 866 [143] (original designation).

Type species: *Jurakempynus sinensis* Wang, Liu, Ren & Shih, 2011.

Forewing membrane with many fuscous fragmentary spots; CuA forked at middle, with complicated branches; CuP forming some distal complicated branches. Hind wing with fuscous spots on the outer margin; MP forming two rows of cells. The genus is most related to the extant Australian subfamily Kempyninae, indicating that ancestors of kempynines were present in the Northern Hemisphere during the Middle Jurassic (165 Mya) and were apparently much more widely distributed than at present [143].

Distribution and age: Inner Mongolia; Middle Jurassic.

Figure 20.30 *Jurakempynus sinensis* Wang, Liu, Ren & Shih, 2011 (Holotype, CNU-NEU-NN2010204-1) [143].

Three species included from the Jurassic of Northern China (see Table 20.1).

***Jurakempynus sinensis* Wang, Liu, Ren & Shih, 2011 (Figure 20.30)**

Jurakempynus sinensis Wang, Liu, Ren & Shih, 2011: *Acta Palaeontol. Pol.*, 56 (4), 867.

Locality and horizon: Daohugou, Ningcheng, Inner Mongolia, China; Middle Jurassic, Jiulongshan Formation.

Forewing membrane with many fragmentary fuscous spots; some fuscous spots present at the anterior margin; five to seven spots resembling cross-veins present between Sc and R_1; nygmata not detected; pterostigma fuscous, with hyaline patches; trichosors well-defined on the margin; costal field relatively narrow, cross-veins forked distally; Rs branches numerous, each forming deep distal forks; cross-veins in radial area arranged irregularly, not forming the gradate series; MP forked with a short distance to the wing base, the both branches with simple distal forks; Cu forked at wing base; CuA long, biforked in middle, each branch forming complicated pectinate branches; CuP as long as the half of CuA, forming simple distal forks. Hind wing membrane with large fuscous markings close to the outer margin; nygmata present; trichosors well-defined on the posterior margin; costal field narrow, cross-veins simple; base of MA fused with Rs; MP field forming two rows of cells, but the basal cells arranged irregularly; MP_1 forming simple distal dichotomies, MP_2 forming some pectinate branches; Cu forked at wing base, CuA long, forming many pectinate branches, CuP as long as the half of CuA, with four to five pectinate branches [143].

***Archaeosmylidia* Makarkin, Yang & Ren, 2014**

Archaeosmylidia Makarkin, Yang & Ren, 2014, *Acta Palaeontol. Pol.*, 59 (1), 210 [144] (original designation).

Type species: *Archaeosmylidia fusca* Makarkin, Yang & Ren, 2014.

Archaeosmylidia differs from other osmylid genera by a combination of the following characters: subcostal cross-veins numerous, all longitudinal veins not zigzagged, CuP with a few branches, non-pectinate. *Archaeosmylidia* has numerous sc-r_1 cross-veins resembling Porisminae, but configuration of MP is distinctly different from both subfamilies. The venation of *Archaeosmylidia* is also similar to the extant Protosmylinae and Spilosmylinae, and accordingly the obliquely branched CuA and simply distal-branched CuP also indicate that it is not proper to assign it to any known subfamilies. The genus is a particular lineage within Osmylidae whose systematic status is to be determined [144].

Distribution and age: Inner Mongolia; Middle Jurassic.

Only one species included from the Jurassic of Northern China (see Table 20.1).

Family Osmylopsychopidae Martynova, 1949

Osmylopsychopidae was first set up by Martynova in 1949 as "Osmylopsychopsidae" [145] based on the genus *Osmylopsychops* Tillyard, 1923 [146] from the Late Triassic of Australia. Subsequently, Riek, in 1955, erected the new family Osmylopsychopidae [147] according to the same genus *Osmylopsychops*, and assigned two other genera *Archepsychops* Tillyard, 1919 [148] from the Late Triassic of Australia and *Mesopolystoechus* Martynov, 1937 [21] from the Early/Middle Jurassic of Tajikistan to this family. Makarkin and Archibald, in 2005, clarified the name of the family and pointed the right family name should be Osmylopsychopidae [148]. Jepson *et al.*, in 2009, assigned three Cretaceous genera of *Grammapsychops* Martynova, 1954 [149], *Embaneura* Zalessky, 1953 [150], and *Pulchroptionia* Martins-Neto, 1997 [151] to Osmylopsychopidae [152]. Lambkin, in 2014, reviewed all osmylopsychopid species from the Australian Triassic, and described the genus *Gayndahpsychops* from the Middle Triassic of Gayndah [153]. Peng *et al.* reviewed the psychopsoid fossil and pointed that Brongniartiellidae should be a synonym of Osmylopsychopidae [154]. They believed that Osmylopsychopidae was a dominant psychopsoid group during the Mesozoic, and described five genera and 10 species of Osmylopsychopidae from the Middle Jurassic Daohugou, China. Therefore, up to 21 genera from the Middle Triassic to the Late Cretaceous have been hitherto assigned to Osmylopsychopidae [47].

Genera included from the Jurassic of Northern China: *Undulopsychopsis* Peng, Makarkin, Wang & Ren, 2011, *Daopsychops* Peng, Makarkin & Ren, 2015, *Eupypsychops* Peng, Makarkin & Ren, 2015, *Nematopsychops* Peng, Makarkin & Ren, 2015, *Ochropsychops* Peng,

(a)

(b)

Figure 20.31 *Undulopsychopsis alexi* Peng, Makarkin & Ren, 2011 (Holotype, CYNB044p). (a), Photograph; (b), Line drawing [155].

Makarkin & Ren, 2015 and *Stenopteropsychops* Peng, Makarkin & Ren, 2015.

Undulopsychopsis Peng, Makarkin & Ren, 2011

Undulopsychopsis Peng, Makrkin, Wang & Ren, 2011, *ZooKeys*, 130, 221 [155] (original designation).

Type species: *Undulopsychopsis alexi* Peng, Makarkin & Ren, 2011.

The specific name is in honor of the distinguished Russian paleoentomologist Dr. Alexandr ("Alex") Rasnitsyn. Forewing costal gradate series absent; branches of Rs dichotomously branched; Rs_1 multi-branched, pectinate with branches directed anteriorly; M forked far distal to origin of Rs; CuP dichotomously branched. Posterior and outer margins of both wings undulate [155].

Distribution and age: Liaoning; Early Cretaceous.

Only one species included from the Cretaceous of Northern China (see Table 20.1).

Undulopsychopsis alexi Peng, Makarkin & Ren, 2011 (Figure 20.31)

Undulopsychopsis alexi Peng, Makarkin & Ren, 2011: *ZooKeys*, 130, 221.

Locality and horizon: Huangbanjigou, Beipiao, Liaoning, China; Lower Cretaceous, Yixian Formation.

Pronotum sub-rectangular, suffused with many long setae. Mesonotum with some long setae laterally. Forewing subtriangular. Costal space broad throughout; humeral veinlet slightly recurrent, branched. Sc and R_1 close distally but not fused. Rs_1 pectinately branched with branches directed anteriorly. MA and MP probably simple. CuA pectinately branched distal to fork of M. CuP multi-branched, dichotomous. 1A long, dichotomously branched. 2A multi-branched. Costal space basally with scarce cross-veins, not forming gradate series. Veins covered with dense setae, particularly long basally. Trichosors distinct. Wing membrane in general brownish; color pattern consists mainly of two pale transverse zigzagged bands which are proximally darker than other portions of wing. Wing margin with setae; posterior and outer margins undulate. Hind wings, outer margin undulate [155].

Daopsychops Peng, Makarkin & Ren, 2015

Daopsychops Peng, Makarkin & Ren, 2015, *J. Syst. Palaeontol.*, 14, 3 [154] (original designation).

Type species: *Daopsychops dissectus* Peng, Makarkin & Ren, 2015.

Forewing relatively elongate with rather acute apex possessing at least several costal cross-veins; distal nygmata; complete outer (distal most) series of cross-veins; Sc and R_1 fused distally; posterior and outer margins undulate; M forked much proximad origin of basalmost branch of CuA; MP with a few dichotomous branches; CuP relatively long, entering posterior wing margin approximately at mid-point [154].

Distribution and age: Inner Mongolia; Middle Jurassic.

Five species included from the Jurassic of Northern China (see Table 20.1).

Eupypsychops Peng, Makarkin & Ren, 2015

Eupypsychops Peng, Makarkin & Ren, 2015, *J. Syst. Palaeontol.*, 14, 16 [154] (original designation).

Type species: *Eupypsychops confinis* Peng, Makarkin & Ren, 2015.

Forewing relatively elongate with rather acute apex possessing gradate series of costal cross-veins; distal nygmata; complete outer (distalmost) series of cross-veins; undulate posterior and outer margins; Sc and R_1 not fused distally; M forked only slightly proximad origin of basalmost branch of CuA; MP with a few dichotomously branches; CuP relatively long, entering posterior wing margin approximately at mid-point [154].

Distribution and age: Inner Mongolia; Middle Jurassic.

Two species included from the Jurassic of Northern China (see Table 20.1).

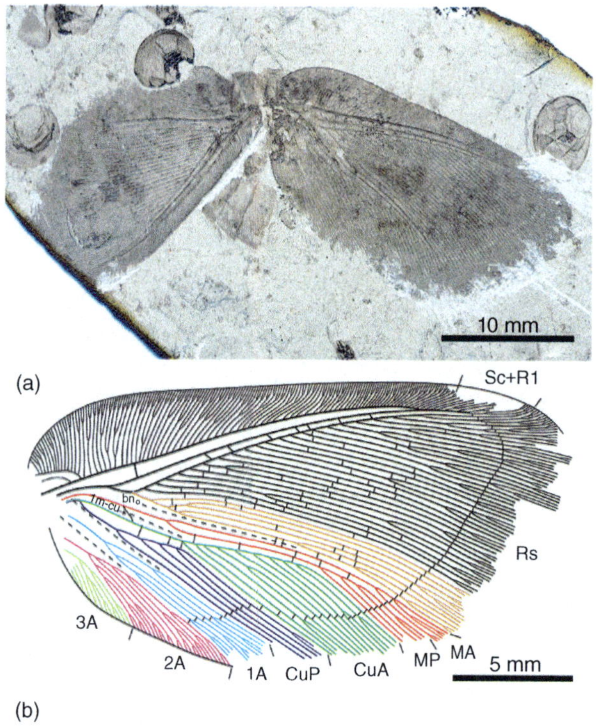

Figure 20.32 *Nematopsychops unicus* Peng, Makarkin & Ren, 2015 (Holotype, CNU-NEU-NN2011086p). (a), Photograph; (b), Line drawing [154].

Nematopsychops Peng, Makarkin & Ren, 2015

Nematopsychops Peng, Makarkin & Ren, 2015, *J. Syst. Palaeontol.*, 14, 18 [154] (original designation).

Type species: *Nematopsychops unicus* Peng, Makarkin & Ren, 2015.

In both wings, no costal cross-veins; Sc and R_1 fused distally; only one (basal) subcostal cross-vein; distal nygmata absent; complete outer (distalmost) series of cross-veins; MP with a few pectinate branches. In forewing, CuP pectinate, relatively long entering posterior wing margin approximately at mid-point [154].

Distribution and age: Inner Mongolia; Middle Jurassic.

Only one species included from the Jurassic of Northern China (see Table 20.1).

Nematopsychops unicus Peng, Makarkin & Ren, 2015 (Figure 20.32)

Nematopsychops unicus Peng, Makarkin & Ren, 2015: *J. Syst. Palaeontol.*, 14, 18.

Locality and horizon: Daohugou, Ningcheng, Inner Mongolia, China; Middle Jurassic, Jiulongshan Formation.

Forewing broadly ovate. Costal space very broad basally, narrowed toward apex; humeral veinlets strongly recurrent with 14 mostly forked branches. Sc and R_1 approaching toward apex, fused near wing apex; Sc, C and R_1 straight, not curved posteriad, with seven veinlets. Distal nygmata absent. M not fused basally with R; forked closer to wing base than in other genera. MA slightly arched before end-twigging; MP slightly sinuate, few pectinately branched. CuA regularly pectinately branched distally. Vein 1A long, dichotomously branched; 2A more dichotomously branched than 1A; 3A few branched. Costal cross-veins absent; radial cross-veins arranged in one complete long outer series from R_1 to 1A, other radial cross-veins irregularly spaced. Macrotrichia on longitudinal veins dense, moderately long. Trichosors prominent along preserved margins. Wing membrane dark brownish; color pattern indistinct, except for white cross-veins of outer gradate series. Hind wing sub-triangular, slightly smaller than forewing; costal space relatively broad for entire length, slightly dilated toward apex; humeral veinlets recurrent with three preserved branches. Sc and R_1 approaching toward apex, fused apically; Sc, C, and R_1 straight, extending slightly posteriad for long distance, with eight preserved veinlets. Distal nygmata absent. M not fused with R basally; MA straight before end-twigging; MP slightly incurved, few pectinately branched. CuA pectinately branched distally; CuP pectinately branched. Veins 1A and 2A strongly dichotomously branched; 3A few branched. Costal cross-veins absent; most radial cross-veins arranged in one complete long outer series from R_1 to 1A, few others irregularly spaced. Macrotrichia as in forewing. Trichosors prominent along preserved margins. Wing membrane dark brownish, color pattern indistinct, except pale cross-veins of outer gradate series [154].

Ochropsychops Peng, Makarkin & Ren, 2015

Ochropsychops Peng, Makarkin & Ren, 2015, *J. Syst. Palaeontol.*, 14, 20 [154] (original designation).

Type species: *Ochropsychops multus* Peng, Makarkin & Ren, 2015.

Forewing broad with obtuse apex possessing one gradate series of costal cross-veins; distal nygmata; complete outer (distalmost) series of cross-veins; numerous radial cross-veins proximad outer series arranged in several short series; Sc and R_1 fused distally; Sc, C and R_1 strongly curved posteriorly; posterior and outer margins not undulate; M forked at level of origin of basalmost branch of CuA; CuP relatively short, entering posterior wing margin well before mid-point [154].

Distribution and age: Inner Mongolia; Middle Jurassic.

Only one species included from the Jurassic of Northern China (see Table 20.1).

Stenopteropsychops Peng, Makarkin & Ren, 2015

Stenopteropsychops Peng, Makarkin & Ren, 2015, *J. Syst. Palaeontol.*, 14, 23 [154] (original designation).

Type species: *Stenopteropsychops trifasciatus* Peng, Makarkin & Ren, 2015.

Forewing relatively narrow with pale transverse fasciae; costal space relatively narrow basally; Rs_1 originated from R before Rs proper; M forked relatively close to wing base; CuP simple, only with marginal fork; 1A relatively few branched [154].

Distribution and age: Inner Mongolia; Middle Jurassic.

Only one species included from the Jurassic of Northern China (see Table 20.1).

Family Panfiloviidae Makarkin, 1990

The Mesozoic neuropteran family Panfiloviidae are very poorly known. It currently includes three genera: *Panfilovia* Makarkin, 1990 [97], *Osmylogramma* Ponomarenko, 1992 [113], and *Epipanfilovia* Yang, Makarkin & Ren, 2013 [156]. The *Panfilovia* was originally described as the monotypic genus *Grammosmylus* Panfilov, 1980 [56] from the Upper Jurassic Karatau locality, South Kazakhstan. It was the type and only genus of the family Grammosmylidae Panfilov, 1980 [56]. This generic name, however, turned out to be a homonym preoccupied by *Grammosmylus* Krüger, 1914 [157] (Neuroptera: Osmylidae). Makarkin, in 1990, proposed the replacement names *Panfilovia* and Panfiloviidae, respectively [97]. Later, Ponomarenko, in 1996, described another species in this genus, *Panfilovia fasciata* Ponomarenko, 1996 from the Lower Jurassic (Lower Toarcian) of Germany based on a fragmentary wing [158]. This species is here transferred to *Epipanfilovia* Yang, Makarkin & Ren, 2013 [156]. The systematic position of this family has been confused. Panfiloviidae was shown in the phylogenetic tree of Neuroptera as the only family whose relationships with other families are entirely unknown [74]. Makarkin and Archibald, in 2003, discussed a possible synonymy of Panfiloviidae with Kalligrammatidae [159]. But Martins-Neto (1997) [151] and Engel and Grimaldi (2008) [160] considered this family as belonging to the superfamily Osmyloidea of the suborder Hemerobiiformia. Ponomarenko, in 2002, placed it amongst the families related to the extant Psychopsidae [161], and Ren, in 2002, assigned it to the superfamily Myrmeleontoidea [58]. However, there is still no consistent agreement on the placement of the Panfiloviidae.

Only one genus included from the Jurassic of Northern China: *Epipanfilovia oviformis* Yang, Makarkin & Ren, 2013.

Epipanfilovia Yang, Makarkin & Ren, 2013

Epipanfilovia Yang, Makarkin & Ren, 2013, *Palaeontology*, 56 (1), 52 [156] (original designation).

Type species: *Epipanfilovia oviformis* Yang, Makarkin & Ren, 2013.

Sc and R_1 distally separated; M forked distal to origin of Rs_1; proximal branches of CuA originate at an obtuse angle, distal branches at acute angle; CuP incurved proximally, strongly arched distally [156].

Distribution and age: Inner Mongolia; Middle Jurassic.

Only one species included from the Jurassic of Northern China (see Table 20.1).

Family Parakseneuridae Yang, Makarkin & Ren, 2012

Parakseneuridae are an extinct family of Neuroptera with characteristic large wing size (forewing 50–75 mm long) and complicated venation. It is easily identified by the following characters: labial palpi stout, relatively short; antenna stout, filiform, apparently much shorter than forewing length; two tibial spurs straight, shorter than basitarsus; claws big, strongly curved; in both wings, humeral veinlet well-developed, strongly recurrent, branched; presumed ScA short, fused with ScP within humeral area; membrane covered with dense, long setae; RA (or ScP + RA) entering margin well before wing apex; subcostal cross-veins numerous; radial cross-veins irregularly spaced, not forming gradate series; in forewing, MP, CuA, CuP dichotomously branched; presumed AA_{1+2} very short (found in *Parakseneura*); AA_{3+4}, AP_{1+2}, AP_{3+4} deeply forked; in hind wing, presumed AA_{1+2} very short (found in *Pseudorapisma*); proximal half of hind wings considerably wider than distal half. Yang *et al.* investigated the phylogenetic position of Parakseneuridae by conducting the phylogenetic analysis of morphological character states for 33 families of extinct and extant Neuropterida combined with DNA sequence data for representatives of all extant families [160]. Parakseneuridae were recovered in a clade with Osmylopsychopidae, Prohemerobiidae, and Ithonidae. The family has three genera and 15 species from the Middle Jurassic of Daohugou (Inner Mongolia, China) and from the Early/Middle Jurassic of Sai-Sagul, Kyrgyzstan, i.e. *Shuraboneura* Khramov & Makarkin, 2012 [32].

Genera included from the Jurassic of Northern China: *Parakseneura* Yang, Makarkin & Ren, 2012 and *Pseudorapisma* Yang, Makarkin & Ren, 2012.

Parakseneura Yang, Makarkin & Ren, 2012

Parakseneura Yang, Makarkin & Ren, 2012, *PLoS ONE*, 7 (9), 4 [32] (original designation).

Type species: *Parakseneura nigromacula* Yang, Makarkin & Ren, 2012.

In forewing, outer margin undulate (smooth in *Shuraboneura* and *Pseudorapisma*); ScP, RA distally fused (separated in *Shuraboneura* and *Pseudorapisma*); presumed AA_{1+2} present, very short, fused with AA_{3+4}

Figure 20.33 *Parakseneura metallica* Yang, Makarkin & Ren, 2012 (Holotype, CNU-NEU-NN-2011019p) [32].

forming basal "loop" (absent in *Pseudorapisma*); in hind wing, basal sinuate cross-veins between R and M systems present (absent in *Pseudorapisma*) [32].

Distribution and age: Inner Mongolia; Middle Jurassic.

Eleven species included from the Jurassic of Northern China (see Table 20.1).

Parakseneura metallica Yang, Makarkin & Ren, 2012 (Figure 20.33)

Parakseneura metallica Yang, Makarkin & Ren, 2012: *PLoS ONE*, 7 (9), 13.

Locality and horizon: Daohugou, Ningcheng, Inner Mongolia, China; Middle Jurassic, Jiulongshan Formation.

Forewing unknown; hind wing somewhat metallic in color, shining with a faint tinge of blue; costal margin slightly convex; outer margin excised immediately posterior to apex, posteriorly slightly undulate. Trichosors prominent along outer margin, costal margins. Humeral plate well-developed, covered with many fine setae. ScA not preserved. Costal space equally moderately broad. All preserved subcostal veinlets forked once, few dichotomously branched; humeral veinlet recurrent, branched. Costal cross-veins forming one series in proximal half of wing. ScP and RA fused. ScP + RA relatively short, proximally bent toward RP, in general incurved; enter margin well before wing apex, with two long branched veinlets. Subcostal space moderately broad, with very scarcely-preserved cross-veins. RA space broader than subcostal space, with scarce irregularly spaced cross-veins. RP with seven widely spaced branches proximal to pterostigmal region; each branch profusely dichotomously branched distally; RP_1 originated near origin of RP. Basal r-m brace between R and M systems long, strongly sinuous. Medial fold distinct in proximal part of wing. M forked slightly distal to origin of RP_1 and MA dichotomously branched distally. MP pectinately branched, its anterior trace and two branches dichotomously branched distally. Cu forked near wing base. CuA and CuP relatively shallowly, dichotomously branched. Claval fold distinct. Color pattern in general dark brown, with paler regions; veins appear dark bluish [32].

Pseudorapisma Yang, Makarkin & Ren, 2012

Pseudorapisma Yang, Makarkin & Ren, 2012, *PLoS ONE*, 7 (9), 21 [32] (original designation).

Type species: *Pseudorapisma jurassicum* Yang, Makarkin & Ren, 2012.

Forewing elongate (broadly-ovate in *Parakseneura* and *Shuraboneura*), about 50–70 mm long, with large humeral plate; ScP and RA apically separated (fused in *Parakseneura*); basal anal "loop" formed by fusion of presumed AA_{1+2} and AA_{3+4} absent (present in *Parakseneura*). In hind wing, ScP and RA apically separated (fused in *Parakseneura*); CuA pectinately branched well proximal to branching of MP (distal in *Parakseneura*); basal sinuate cross-veins between R and M systems absent (present in *Parakseneura*) [32].

Distribution and age: Inner Mongolia; Middle Jurassic.

Three species included from the Jurassic of Northern China (see Table 20.1).

Family Psychopsidae Handlirsch, 1906

Psychopsidae, known as "silky lacewings", are a small relic family within Neuroptera, comprising 27 extant species in five genera, which are restricted in southern Africa, southeastern Asia and Australia [162, 163]. Adult psychopsids can be recognized by the broad and usually triangular forewing, a broad costal space, presence of a *vena triplica* (formed by specialized distal connection of ScP, RA, and RP), and densely setose on wings. Fossil psychopsids are diverse, with more than 40 species described up to date. However, the psychopsid affinity of these fossil taxa was often questionable, due to the lack of the synapomorphies with extant lineages (e.g. the pattern of Rs branches, the configuration of M and Cu in the forewing, the structure of *vena triplica*). According to Peng *et al.* [154], most Mesozoic psychosis should belong to Osmylopsychopidae, and the typical psychopsids are relatively rare as the extant. Therefore, the current psychopsid fossil are strongly in need of revision [164].

Genera included from the Jurassic and Cretaceous of Northern China: *Cretapsychops* Jepson, Makarkin & Jarzembowski, 2009 and *Alloepipsychopsis* Makarkin, Yang, Peng & Ren, 2012.

Cretapsychops Jepson, Makarkin & Jarzembowski, 2009

Cretapsychops Jepson, Makarkin & Jarzembowski, 2009, *Cretac. Res.*, 30, 1329 [152] (original designation).

Type species: *Cretapschops corami* Jepson, Makarkin & Jarzembowski, 2009.

Figure 20.34 *Cretapsychops decipiens* Peng, Makarkin, Yang & Ren, 2010 (Holotype, CNU-NEU-NN2010700p). (a), Photograph; (b), Line drawing [164].

In forewing, prominent color pattern present; costal gradate series present; R_1 distally with many long anteriorly-directed pectinate branches (veinlets); branches of Rs almost non-dichotomous; cross-veins in radial space form four transverse gradate series; CuA peculiarly branched; CuP non-pectinate, with few long branches. 1A dichotomously branched; 2A pectinate. In hind wing, costal gradate series present; branches of Rs non-dichotomous; cross-veins in radial space form several (two preserved) gradate series [164].

Distribution and age: Inner Mongolia; Middle Jurassic.

Only one species included from the Jurassic of Northern China (see Table 20.1).

Cretapsychops decipiens Peng, Makarkin, Yang & Ren, 2010 (Figure 20.34)

Cretapsychops decipiens Peng, Makarkin, Yang & Ren, 2010: *Zootaxa*, 2663, 60.

Locality and horizon: Daohugou, Ningcheng, Inner Mongolia, China; Middle Jurassic, Jiulongshan Formation.

Pronotum subrectangular, with several dark spots, and covered with many long setae on posterior and lateral edges. Mesonotum almost entirely dark, but pale in lateral edges, middorsal suture on prescutum, two lateral sides of scutellum; covered with long setae in lateral edges. Forewing broadly triangular, with rounded apex. Costal space broad throughout; humeral veinlet strongly recurrent, pectinately branched. Sc, R_1, and Rs strong, gradually converging toward apex. Rs with 34 branches, not forked before outer gradate series. Cross-veins in radial space form four transverse gradate series; R_1 and Rs connected by two cross-veins belonging to two distal-most gradate series. M forked far distal to origin of Rs. MA clearly concave; MP slightly sinuous. Cu forking near wing base. CuA peculiarly branched distally; CuP few branched. Anal space broad, nearly as wide as costal space. Trichosors present. All veins and wing margins with dense macrotrichia. Color pattern consists of a row of deep brown maculae along C in costal space, becoming lighter, smaller toward apex; two deep brown mottled regions in anal space; three patches with irregular wavy margins spreading on pale brown ground in radio-medial space. Hind wing costal space broad throughout, width over four times combined width of adjacent subcostal and R_1 spaces. Costal gradate series parallel to C then gradually curved to Sc, terminating at Sc near wing apex. Sc and R_1 gradually converging toward apex, but not fused. Two transverse gradate series of cross-veins in distal portion of radial space including cross-veins between R_1 and Rs. Trichosors present. All veins and wing margins with dense macrotrichia [164].

Alloepipsychopsis Makarkin, Yang, Peng & Ren, 2012

Alloepipsychopsis Makarkin, Yang, Peng, & Ren, 2012, *Cretac. Res.*, 35, 59 [71] (original designation).

Type species: *Alloepipsychopsis lata* Makarkin, Yang, Peng & Ren, 2012.

Only fragmentary forewing preserved. Large body size (preserved forewing length 32 mm, estimated length about 50 mm). Humeral veinlet recurrent, branched, with at least three distal branches. Costal cross-veins distally branched and interlinked by numerous veinlets. Dense cross-veins along the wing. Rs with eight preserved branches, deeply branched. M diverging at the separating point of Rs from R_1; M with four anteriorly directed pectinate branches in basal half of wing, all parallel to each other. CuA and CuP obliquely branched, forming complicated secondary branches [71].

Distribution and age: Liaoning; Early Cretaceous.

Only one species included from the Cretaceous of Northern China (see Table 20.1).

Family Saucrosmylidae Ren and Yin, 2003

Saucrosmylidae were initially established as a subfamily of Osmylidae for sharing the characters: presence of nygmata; presence of trichosors; Sc and R_1 fused apically and entering margin before wing apex [165]. Fang *et al.*, in 2015, considered the assignment to Osmylidae was primarily based on the synapomorphies, and elevated it to family rank [166] proposing the distinctive apomorphic characters, i.e. notably dense venation and two to seven rows of cells present between R_1 and Rs. Indeed saucrosmylids show more close relationships to the contemporaneous relative families (e.g. Kalligrammatidae, Grammolingiidae, Aetheogrammatidae, Panfiloviidae, and Parakseneuridae) instead of Osmylidae, which are characteristic of the typically large body size and complicated venation. Yang *et al.*, in 2012, also indicated that Saucrosmylidae should be allied to Myrmeleontiformia in the phylogeny of Neuroptera [32]. However the systematic position of Saucrosmylidae within Neuroptera is still not determined. The fossil specimens with unique gymnospermous leaf mimicry represented the particular lineages of the Neuroptera in adapting the Mesozoic environments (Section 29.2.1). To date, saucrosmylids have only been described from the Middle Jurassic of Daohugou, including seven genera and eight species.

Genera included from the Jurassic of Northern China: *Laccosmylus* Ren & Yin, 2003, *Rudiosmylus* Ren & Yin, 2003, *Saucrosmylus* Ren & Yin, 2003, *Bellinympha* Wang, Ren, Liu, Shih & Engel, 2010, *Huiyingosmylus* Liu, Zhang, Wang, Fang, Zheng, Zhang & Jarzembowski, 2013, *Daohugosmylus* Liu, Zhang, Wang, Fang, Zheng, Zhang & Jarzembowski, 2014 and *Ulrikezza* Fang, Ren & Wang, 2015.

Laccosmylus Ren & Yin, 2003

Laccosmylus Ren & Yin, 2003, *J. N.Y. Entomol. Soc.*, 111 (1), 7 [165] (original designation).

Type species: *Laccosmylus calophlebius* Ren & Yin, 2003.

Hind wing with distinct color markings; R_1 area at most with six to seven rows of cells; MA forked very earlier, before the middle point of the wing; two row of cells between MP_1 and MP_2; Rs with at least six main branches before Rs bent anteriorly [165].

Distribution and age: Inner Mongolia; Middle Jurassic.

Only one species included from the Jurassic of Northern China (see Table 20.1).

Rudiosmylus Ren & Yin, 2003

Rudiosmylus Ren & Yin, 2003, *J. N.Y. Entomol. Soc.*, 111 (1), 5 [165] (original designation).

Type species: *Rudiosmylus ningchengensis* Ren & Yin, 2003.

Hind wing without distinct color markings; R_1 area at most with four rows of cells; two row of cells between MP_1 and MP_2, Rs with at least six main branches before Rs bent anteriorly, all forked distally toward wing apex [165].

Distribution and age: Inner Mongolia; Middle Jurassic.

Figure 20.35 *Bellinympha filicifolia* Wang, Ren, Liu & Engel, 2010 (Holotype, CNU-NEU-NN2010240) [28].

Only one species included from the Jurassic of Northern China (see Table 20.1).

Saucrosmylus Ren & Yin, 2003

Saucrosmylus Ren & Yin, 2003, *J. N.Y. Entomol. Soc.*, 111 (1), 3 [165] (original designation).

Type species: *Saucrosmylus sambneurus* Ren & Yin, 2003.

Both fore and hind wing with distinct color markings, R_1 area with three rows of cells; hind wing with single row of cells between MP_1 and MP_2, Rs with only three main branches, two usually forked very basally, before Rs bent anteriorly [165].

Distribution and age: Inner Mongolia; Middle Jurassic.

Only one species included from the Jurassic of Northern China (see Table 20.1).

Bellinympha Wang, Ren, Liu, Shih & Engel, 2010

Bellinympha Wang, Ren, Liu, Shih & Engel, 2010, *Proc. Natl. Acad. Sci.*, 107 (37), 16212 [28] (original designation).

Type species: *Bellinympha filicifolia* Wang, Ren, Liu & Engel, 2010.

Large body size, forewing membrane with distinct pinna-like markings, MP branches resembling a leaf rachis, nygmata present in center of wing. Undulating margins to the forewings and hind wing. *Bellinympha* is the first and the earliest insect having the pinnate leaf mimesis in the Mesozoic, representing a 165-million-year-old specialization between insects and contemporaneous gymnosperms of the Cycadales or Bennettitales. The discovery of *Bellinympha* is of great significance to understanding the Mesozoic evolution and diversification of insects, particularly in the context of early co-evolution between plants and pollinators before angiosperms [28].

Distribution and age: Inner Mongolia; Middle Jurassic.

Two species included from the Jurassic of Northern China (see Table 20.1).

Bellinympha filicifolia **Wang, Ren, Liu & Engel, 2010 (Figure 20.35)**
Bellinympha filicifolia Wang, Ren, Liu & Engel, 2010: *Proc. Natl. Acad. Sci.*, 107 (37), 16213.

Locality and horizon: Daohugou, Ningcheng, Inner Mongolia, China; Middle Jurassic, Jiulongshan Formation.

Forewing with numerous pinnalike markings. Trichosors and nygmata present. Outer margin with conspicuous undulating projections. Costal cross-veins forked distally, interlinked by numerous veinlets. Rs expanded in middle and bent anteriorly sharply toward R_1. Space between Rs and R_1 broad, forming two to five rows of cells. Rs with 19 branches, each with complicated distal forks and angled toward posterior apical margin. Cross-veins in radial sector numerous, arranged irregularly. MA coalescent with Rs. MP forked close to wing base, MP_2 with eight to nine distal pectinate branches. CuA with numerous oblique pectinate branches medially, forming a large triangular region; CuP deeply dark; long, extending beyond midlength of CuA. Two rows of cells between A_1 and A_2. Hind wing, poorly preserved. Membrane with numerous oblique stripes close to apex, resembling the pinna-like forewings. Trichosors well spread across outer margin and part of anterior margin. Venation is similar to that of the forewing: costal cross-veins forked distally, interconnected by veinlets; Rs bent anteriorly toward R_1 distally, forming a large space between R_1 and Rs; cross-veins in radial sector numerous, arranged irregularly [28]. Details of ancient pinnate leaf mimicry by *Bellinympha*, please see Section 29.2.1.

Huiyingosmylus **Liu, Zhang, Wang, Fang, Zheng, Zhang & Jarzembowski, 2013**
Huiyingosmylus Liu, Zhang, Wang, Fang, Zheng, Zhang & Jarzembowski, 2013, *Zootaxa*, 3736, 388 [167] (original designation).

Type species: *Huiyingosmylus bellus* Liu, Zhang, Wang, Fang, Zheng, Zhang & Jarzembowski, 2013.

Forewing broad, decorated with fuscous spots. Outer margin strongly undulate. Cross-veins dense on the entire wing. Costal space moderately broad; costal veinlets dense and forked; Sc close to R_1; with at most six rows of cells [167].

Distribution and age: Inner Mongolia; Middle Jurassic.

Only one species included from the Jurassic of Northern China (see Table 20.1).

Daohugosmylus **Liu, Zhang, Wang, Fang, Zheng, Zhang & Jarzembowski, 2014**
Daohugosmylus Liu, Zhang, Wang, Fang, Zheng, Zhang & Jarzembowski, 2014, *Alcheringa*, 38, 302 [168] (original designation).

Type species: *Daohugosmylus castus* Liu, Zhang, Wang, Fang, Zheng, Zhang & Jarzembowski, 2014.

Hind wing broad. Outer margin smooth. Cross-veins dense over the entire wing. Costal space narrow; Sc very close to R_1; R_1 space wide with at most six rows of cells; Rs sharply bent anteriorly toward R_1; MP forked basally; CuA and CuP deeply forked [168].

Distribution and age: Inner Mongolia; Middle Jurassic.

Only one species included from the Jurassic of Northern China (see Table 20.1).

Ulrikezza **Fang, Ren & Wang, 2015**
Ulrikezza Fang, Ren & Wang, 2015, *PLoS ONE*, 10 (10), 5 [166] (original designation).

Type species: *Ulrikezza aspoeckae* Fang, Ren & Wang, 2015.

The epithets of the genus and type species are dedicated to Dr. Ulrike Aspöck for her outstanding contribution to the research of Neuroptera (Figure 20.36). Large insect, forewing slender and membrane decorated with seven fuscous eyespots and some small spotted markings; outer margin smooth, decorated with trichosors and microtrichia; costal cross-veins relatively simple, with shallow distal forks; space between R_1 and Rs with four to five rows of irregular cells; MA originated from Rs close to wing base, MP forked between separation of MA and Rs_1 from Rs; CuA and CuP deeply forked, CuA forked at midwing, forming complicated pectinate branches, CuP short and relatively simple, formed two to three distal dichotomous branches. Hind wing distinctively broader comparing to the forewing; costal cross-veins single branch, or occasionally with distal forks [166].

Distribution and age: Inner Mongolia; Middle Jurassic.

Only one species included from the Jurassic of Northern China (see Table 20.1).

Ulrikezza aspoeckae **Fang, Ren & Wang, 2015 (Figure 20.37)**
Ulrikezza aspoeckae Fang, Ren & Wang, 2015: *PLoS ONE*, 10 (10), 5.

Locality and horizon: Daohugou, Ningcheng, Inner Mongolia, China; Middle Jurassic, Jiulongshan Formation.

Forewing about 57 mm long and 22 mm wide as preserved. Trichosors and nygmata present. Costal space basally very narrow, then strongly expanded. Costal cross-veins forked distally, interlinked by numerous veinlets. Sc fused with R_1 apically and ending on costal margin before wing apex. Rs slightly bent anteriorly toward R_1; space between Rs and R_1 broad, forming four to five rows of irregular cells; Rs with six main branches before Rs bent anteriorly, each with complicated distal forks and angled toward posterior apical margin. In right forewing, MA coalescent with Rs_1; while MA is connected with Rs_1 through a thick vein in left forewing. CuA with numerous oblique pectinate branches forked

Figure 20.36 Dr. Ulrike Aspöck and the co-authors of the paper [166].

Figure 20.37 *Ulrikezza aspoeckae* Fang, Ren & Wang, 2015 (Holotype, CNU-NEU-NN2015001p) [166].

distally at about one fourth length of wing, forming large triangular area. CuA forked nine branches in right forewing while six branches in left forewing. CuP short and simple with only two main branches, forked slightly basad of CuA. Hind wing somewhat shorter than forewing, about 48 mm long and 21 mm wide. Venation is similar to the forewing. More than six rows of cross-veins between costal veinlets which are rather dense but only individual forked. MA barely forked tree branches distally. MP forked close to wing base. A couple rows of cells between MP_1 and MP_2. CuA with seven oblique pectinate branches forked distally at about one fourth length of wing, forming large triangular area. CuP short, with three oblique pectinate branches, forked slightly basad of Cu + A. Anal region is small, 1A curved [166].

Family *Incertae sedis*

Hongosmylites Makarkin and Archibald, 2005

Hongosmylites Makarkin and Archibald, 2005, *Zootaxa*, 1054, 18 [169] (original designation).

Type species: *Sinosmylites longus* Hong, 1996 [170].

The original name of *Hongosmylites* is actually a homonym of the other genus *Sinosmylites* Hong, 1983 [47], and Makarkin and Archibald, in 2005, clarified the mixtures of the both epithets and established *Hongosmylites* [169] based on the species *Sinosmylites longus* Hong, 1996 [170]. The genus *Hongosmylites* is diagnosed by these characters: medium-sized insect, forewing elongate, few cross-veins throughout wing except for a single r_1-rs cross-vein, costal cross-veins simple and dense, Sc and R_1 distally fused, Rs branches numerous, MA and MP strongly inclining and forming some distal branches, CuA and CuP well-developed and forming the oblique pectinate branches. Makarkin and Archibald indicated *Hongosmylites* was more similar to *Glottopteryx* for sharing the absence of end-twigging of the preserved veins, and suggested its psychopsoid affinity [169]. Nevertheless, according to the original description of *Hongosmylites*, it is distinctly different from *Glottopteryx*, i.e. extraordinarily scarce cross-veins, dense and simple costal cross-veins in *Hongosmylites* (vs. *Glottopteryx*, Bode 1953 [25]. Therefore the placement of *Hongosmylites* is still not determined [169].

Distribution and age: Shandong; Early Cretaceous.

Only one species included from the Cretaceous of Northern China (see Table 20.1).

Table 20.1 A list of fossil Neuroptera from the Jurassic and Cretaceous of China.

Family	Species	Locality	Horizon/Age	Citation
Aetheogrammatidae	*Aetheogramma bistriatum* Yang, Makarkin, Shih & Ren, 2015	Beipiao, Liaoning	Yixian Fm., K_1	Yang *et al.* [38]
	Aetheogramma speciosa Ren & Engel, 2008	Beipiao, Liaoning	Yixian Fm., K_1	Ren and Engel [36]
	Curtogramma ovatum Yang, Makarkin, Shih & Ren, 2015	Lingyuan, Liaoning	Yixian Fm., K_1	Yang *et al.* [38]
	Cyclicogramma rotundum Yang, Makarkin, Shih & Ren, 2015	Beipiao, Liaoning	Yixian Fm., K_1	Yang *et al.* [38]
	Ectopogramma kalligrammoides Engel, Huang, & Lin, 2011	Ningcheng, Inner Mongolia	Jiulongshan Fm., J_2	Engel *et al.* [37]
Ascalochrysidae	*Ascalochrysa megaptera* Ren & Makarkin, 2009	Beipiao, Liaoning	Yixian Fm., K_1	Ren and Makarkin [39]
Berothidae	*Oloberotha sinica* Ren & Guo, 1996	Beipiao, Liaoning	Yixian Fm., K_1	Ren and Guo [49]
	Sinosmylites fumosus Makarkin, Yang & Ren, 2011	Ningcheng, Inner Mongolia	Jiulongshan Fm., J_2	Makarkin *et al.* [46]
	Sinosmylites pectinatus Hong, 1983	Beipiao, Liaoning	Jiulongshan Fm., J_2	Hong [48]
	Sinosmylites rasnitsyni Makarkin, Yang & Ren, 2011	Ningcheng, Inner Mongolia	Jiulongshan Fm., J_2	Makarkin *et al.* [46]
Chrysopidae	*Drakochrysa sinica* Yang & Hong, 1990	Laiyang, Shandong	Laiyang Fm., K_1	Yang and Hong [57]
	Lembochrysa miniscula Ren & Guo, 1996	Beipiao, Liaoning	[b]Yixian Fm., K_1	Ren and Guo [49]
	Lembochrysa polyneura Ren & Guo, 1996	Beipiao, Liaoning	[b]Yixian Fm., K_1	Ren and Guo [49]
	Mesypochrysa sinica Khramov, Liu, Zhang & Jarzembowski, 2015	Ningcheng, Inner Mongolia	[b]Jiulongshan Fm., J_2	Khramov *et al.* [54]
Dipteromantispidae	*Dipteromantispa brevisubcosta* Makarkin, Yang & Ren, 2013	Beipiao, Liaoning	Yixian Fm., K_1	Makarkin *et al.* [30]
Grammolingiidae	*Chorilingia euryptera* Shi, Wang, Yang & Ren, 2012	Ningcheng, Inner Mongolia	Jiulongshan Fm., J_2	Shi *et al.* [65]
	Chorilingia parvica Shi, Wang, Yang & Ren, 2012	Ningcheng, Inner Mongolia	Jiulongshan Fm., J_2	Shi *et al.* [65]
	Chorilingia peregrina Shi, Wang, Yang & Ren, 2012	Ningcheng, Inner Mongolia	Jiulongshan Fm., J_2	Shi *et al.* [65]
	Chorilingia translucida Shi, Wang, Yang & Ren, 2012	Ningcheng, Inner Mongolia	Jiulongshan Fm., J_2	Shi *et al.* [65]
	Grammolingia binervis Shi, Wang & Ren, 2013	Ningcheng, Inner Mongolia	Jiulongshan Fm., J_2	Shi *et al.* [64]
	Grammolingia boi Ren, 2002	Ningcheng, Inner Mongolia	Jiulongshan Fm., J_2	Ren [58]
	Grammolingia sticta Shi, Wang & Ren, 2013	Ningcheng, Inner Mongolia	Jiulongshan Fm., J_2	Shi *et al.* [64]
	Grammolingia uniserialis Shi, Wang & Ren, 2013	Ningcheng, Inner Mongolia	Jiulongshan Fm., J_2	Shi *et al.* [64]
	Leptolingia calonervis Shi, Yang & Ren, 2011	Ningcheng, Inner Mongolia	Jiulongshan Fm., J_2	Shi *et al.* [59]
	Leptolingia imminuta Liu, Shi & Ren, 2011	Ningcheng, Inner Mongolia	Jiulongshan Fm., J_2	Liu *et al.* [61]

(Continued)

Table 20.1 (Continued)

Family	Species	Locality	Horizon/Age	Citation
	Leptolingia jurassica Ren, 2002	Ningcheng, Inner Mongolia	Jiulongshan Fm., J_2	Ren [58]
	Leptolingia tianyiensis Ren, 2002	Ningcheng, Inner Mongolia	Jiulongshan Fm., J_2	Ren [58]
	Litholingia eumorpha Ren, 2002	Ningcheng, Inner Mongolia	Jiulongshan Fm., J_2	Ren [58]
	Litholingia polychotoma Ren, 2002	Ningcheng, Inner Mongolia	Jiulongshan Fm., J_2	Ren [58]
	Litholingia ptesa Shi, Yang & Ren, 2011	Ningcheng, Inner Mongolia	Jiulongshan Fm., J_2	Shi *et al.* [59]
	Litholingia rhora Ren, 2002	Ningcheng, Inner Mongolia	Jiulongshan Fm., J_2	Ren [58]
Ithonidae	*Guithone bethouxi* Zheng, Ren & Wang, 2016	Ningcheng, Inner Mongolia	Jiulongshan Fm., J_2	Zheng *et al.* [73]
	Jurapolystoechotes melanolomus Ren, Engel & Lü, 2002	Ningcheng, Inner Mongolia	Jiulongshan Fm., J_2	Ren *et al.* [72]
	Lasiosmylus longus Zheng, Ren & Wang, 2016	Beipiao, Liaoning	Yixian Fm., K_1	Zheng *et al.* [70]
	Lasiosmylus newi Ren & Guo, 1996	Beipiao, Liaoning	[b]Yixian Fm., K_1	Ren and Guo [49]
	Mesopolystoechus wangyingziensis Hong, 1983	Luanping, Hebei	Jiulongshan Fm., J_2	Hong [48]
Kalligrammatidae	*Abrigramma calophleba* Yang, Wang, Labandeira, Shih & Ren, 2014	Pingquan, Hebei	Yixian Fm., K_1	Yang *et al.* [27]
	Affinigramma myrioneura Yang, Wang, Labandeira, Shih & Ren, 2014	Ningcheng, Inner Mongolia	Jiulongshan Fm., J_2	Yang *et al.* [27]
	Apochrysogramma rotundum Yang, Makarkin & Ren, 2011	Ningcheng, Inner Mongolia	Jiulongshan Fm., J_2	Yang *et al.* [92]
	Huiyingogramma formosum Liu, Zheng, Zhang, Wang, Fang & Zhang, 2014	Ningcheng, Inner Mongolia	Haifanggou Fm., J_2	Liu *et al.* [91]
	Ithigramma multinervium Yang, Wang, Labandeira, Shih & Ren, 2014	Ningcheng, Inner Mongolia	Yixian Fm., K_1	Yang *et al.* [27]
	Kalligramma albifasciatus Yang, Makarkin & Ren, 2014	Ningcheng, Inner Mongolia	Jiulongshan Fm., J_2	Yang *et al.* [92]
	Kalligramma brachyrhyncha Yang, Wang, Labandeira, Shih & Ren, 2014	Ningcheng, Inner Mongolia	Jiulongshan Fm., J_2	Yang *et al.* [27]
	Kalligramma circularium Yang, Wang, Labandeira, Shih & Ren, 2014	Ningcheng, Inner Mongolia	Jiulongshan Fm., J_2	Yang *et al.* [27]
	Kalligramma delicatum Liu, Khramov & Zhang, 2015	Ningcheng, Inner Mongolia	Jiulongshan Fm., J_2	Liu *et al.* [171]
	Kalligramma elegans Yang, Makarkin & Ren, 2014	Ningcheng, Inner Mongolia	Jiulongshan Fm., J_2	Yang *et al.* [92]
	Kalligramma jurarchegonium Zhang, 2003	Beipiao, Liaoning	Haifanggou Fm., J_2	Zhang [85]
	Kalligramma liaoningensis Ren & Guo, 1996	Beipiao, Liaoning	[b]Yixian Fm., K_1	Ren and Guo [49]
	Kalligramma paradoxum Liu, Zheng, Zhang, Wang, Fang & Zhang, 2014	Ningcheng, Inner Mongolia	Haifanggou Fm., J_2	Liu *et al.* [93]
	Kalligrammula lata Liu, Khramov, Zhang & Jarzembowski, 2015	Ningcheng, Inner Mongolia	Jiulongshan Fm., J_2	Liu *et al.* [91]

Table 20.1 (Continued)

Family	Species	Locality	Horizon/Age	Citation
	Kallihemerobius pleioneurus Ren & Oswald, 2002	Ningcheng, Inner Mongolia	Jiulongshan Fm., J_2	Ren and Oswald [82]
	Kallihemerobius aciedentatus Yang, Wang, Labandeira, Shih & Ren, 2014	Ningcheng, Inner Mongolia	Jiulongshan Fm., J_2	Yang *et al.* [27]
	Kallihemerobius almacellus Yang, Wang, Labandeira, Shih & Ren, 2014	Ningcheng, Inner Mongolia	Jiulongshan Fm., J_2	Yang *et al.* [27]
	Kallihemerobius feroculus Yang, Wang, Labandeira, Shih & Ren, 2014	Ningcheng, Inner Mongolia	Jiulongshan Fm., J_2	Yang *et al.* [27]
	Limnogramma hani Makarkin, Ren & Yang, 2009	Ningcheng, Inner Mongolia	Jiulongshan Fm., J_2	Makarkin *et al.* [87]
	Limnogramma mira Ren, 2003	Beipiao, Liaoning	Yixian Fm., K_1	Ren [83]
	Limnogramma mongolicum Makarkin, Ren & Yang, 2009	Ningcheng, Inner Mongolia	Jiulongshan Fm., J_2	Makarkin *et al.* [87]
	Oregramma aureolusa Yang, Wang, Labandeira, Shih & Ren, 2014	Ningcheng, Inner Mongolia	Yixian Fm., K_1	Yang *et al.* [27]
	Oregramma gloriosa Ren, 2003	Beipiao, Liaoning	Yixian Fm., K_1	Ren [83]
	Oregramma illecebrosa Yang, Wang, Labandeira, Shih & Ren, 2014	Beipiao, Liaoning	Yixian Fm., K_1	Yang *et al.* [27]
	Protokalligramma bifasciatum Yang, Makarkin & Ren, 2011	Ningcheng, Inner Mongolia	Jiulongshan Fm., J_2	Yang *et al.* [92]
	Sinokalligramma jurassicum Zhang, 2003	Ningcheng, Inner Mongolia	[b]Jiulongshan Fm., J_2	Zhang [84]
	Sophogramma eucalla Ren & Guo, 1996	Beipiao, Liaoning	Yixian Fm., K_1	Ren and Guo [49]
	Sophogramma lii Yang, Zhao & Ren, 2009	Beipiao, Liaoning	Yixian Fm., K_1	Yang *et al.* [88]
	Sophogramma papilionacea Ren & Guo, 1996	Beipiao, Liaoning	[b]Yixian Fm., K_1	Ren and Guo [49]
	Sophogramma pingquanica Yang, Wang, Labandeira, Shih & Ren, 2014	Chengde, Hebei	Yixian Fm., K_1	Yang *et al.* [27]
	Sophogramma plecophlebia Ren & Guo, 1996	Beipiao, Liaoning	Yixian Fm., K_1	Ren and Guo [49]
	Stelligramma allochroma Yang, Wang, Labandeira, Shih & Ren, 2014	Ningcheng, Inner Mongolia	Jiulongshan Fm., J_2	Yang *et al.* [27]
Mantispidae	*Archaeodrepanicus acutus* Jepson, Heads, Makarkin & Ren, 2013	Beipiao, Liaoning	Yixian Fm., K_1	Jepson *et al.* [99]
	Archaeodrepanicus nuddsi Jepson, Heads, Makarkin & Ren, 2013	Beipiao, Liaoning	Yixian Fm., K_1	Jepson *et al.* [99]
	Clavifemora rotundata Jepson, Heads, Makarkin & Ren, 2013	Ningcheng, Inner Mongolia	Jiulongshan Fm., J_2	Jepson *et al.* [99]
	Sinomesomantispa microdentata Jepson, Heads, Makarkin & Ren, 2013	Beipiao, Liaoning	Yixian Fm., K_1	Jepson *et al.* [99]
	Allopterus luianus Zhang, 1991	Laiyang, Shandong	[b]Laiyang Fm., K_1	Zhang [113]
	Kareninoides lii Yang, Makarkin & Ren, 2012	Beipiao, Liaoning	Yixian Fm., K_1	Yang *et al.* [116]
	Longicellochrysa yixiana Ren, Makarkin & Yang, 2010	Beipiao, Liaoning	Yixian Fm., K_1	Ren *et al.* [119]

(Continued)

Table 20.1 (Continued)

Family	Species	Locality	Horizon/Age	Citation
	Mesascalaphus yangi Ren, 1995	Beipiao, Liaoning	Yixian Fm., K_1	Ren *et al.* [117]
	Protoaristenymphes daohugouensis Yang, Makarkin & Ren, 2012	Ningcheng, Inner Mongolia	Jiulongshan Fm., J_2	Yang *et al.* [116]
	Siniphes delicatus Ren & Yin, 2002	Beipiao, Liaoning	Yixian Fm., K_1	Ren and Yin [118]
	Tachinymphes delicata Ren & Yin, 2002	Beipiao, Liaoning	Yixian Fm., K_1	Ren and Yin [118]
	Tachinymphes magnifica Nel, Delclos & Hutin, 2005	Beipiao, Liaoning	Yixian Fm., K_1	Nel *et al.* [52]
Myrmeleontidae	*Choromyrmeleon aspoeckorum* Ren & Engel, 2008	Beipiao, Liaoning	Yixian Fm., K_1	Ren and Engel [124]
	Choromyrmeleon othneius Ren & Guo, 1996	Beipiao, Liaoning	[b]Yixian Fm., K_1	Ren and Guo [49]
	[a] *Liaoximyia sinica* Hong, 1988	Kezuo, Liaoning	Sahai Fm., K_1	Hong [123]
Nymphidae	*Baissoleon similis* Shi, Winterton & Ren, 2015	Ningcheng, Inner Mongolia	Yixian Fm., K_1	Shi *et al.* [126]
	Daonymphes bisulca Makarkin, Yang, Shi, & Ren, 2013	Ningcheng, Inner Mongolia	Jiulongshan Fm., J_2	Makarkin*et al.* [125]
	Liminympha makarkini Ren & Engel, 2007	Ningcheng, Inner Mongolia	Jiulongshan Fm., J_2	Ren and Engel [127]
	Nymphites bimaculatus Shi, Makarkin, Yang, Archibald & Ren, 2013	Ningcheng, Inner Mongolia	Jiulongshan Fm., J_2	Shi *et al.* [128]
	Sialium minor Shi, Winterton & Ren, 2015	Beipiao, Liaoning	Yixian Fm., K_1	Shi *et al.* [126]
	Sialium sinicus Shi, Winterton & Ren, 2015	Ningcheng, Inner Mongolia	Yixian Fm., K_1	Shi *et al.* [126]
	Spilonymphes major Shi, Winterton & Ren, 2015	Beipiao, Liaoning	Yixian Fm., K_1	Shi *et al.* [126]
Osmylidae	*Allotriosmylus uniramosus* Yang, Makarkin & Ren, 2010	Ningcheng, Inner Mongolia	Jiulongshan Fm., J_2	Yang *et al.* [141]
	Archaeosmylidia fusca Makarkin, Yang & Ren, 2014	Ningcheng, Inner Mongolia	Jiulongshan Fm., J_2	Makarkin *et al.* [144]
	Epiosmylus panfilov Ren & Yin, 2002	Ningcheng, Inner Mongolia	Jiulongshan Fm., J_2	Ren *et al.* [138]
	Enodinympha translucida Ren & Engel, 2007	Ningcheng, Inner Mongolia	Jiulongshan Fm., J_2	Ren *et al.* [127]
	Juraheterosmylus antiquates Wang, Liu, Ren & Shih, 2010	Ningcheng, Inner Mongolia	Jiulongshan Fm., J_2	Wang *et al.* [142]
	Juraheterosmylus astictus Wang, Liu, Ren & Shih, 2010	Ningcheng, Inner Mongolia	Jiulongshan Fm., J_2	Wang *et al.* [142]
	Juraheterosmylus minor Wang, Liu, Ren & Shih, 2010	Ningcheng, Inner Mongolia	Jiulongshan Fm., J_2	Wang *et al.* [142]
	Jurakempynus bellatulus Wang, Liu, Ren & Shih, 2011	Ningcheng, Inner Mongolia	Jiulongshan Fm., J_2	Wang *et al.* [143]
	Jurakempynus epunctatus Wang, Liu, Ren & Shih, 2011	Ningcheng, Inner Mongolia	Jiulongshan Fm., J_2	Wang *et al.* [143]
	Jurakempynus sinensis Wang, Liu, Ren & Shih, 2011	Ningcheng, Inner Mongolia	Jiulongshan Fm., J_2	Wang *et al.* [143]
	Nilionympha imperfecta Ren & Engel, 2007	Ningcheng, Inner Mongolia	Jiulongshan Fm., J_2	Ren *et al.* [127]

Table 20.1 (Continued)

Family	Species	Locality	Horizon/Age	Citation
	Nilionympha pulchella Ren & Engel, 2007	Ningcheng, Inner Mongolia	Jiulongshan Fm., J_2	Ren *et al.* [127]
	Palaeothyridosmylus septemaculatus Wang, Liu & Ren, 2009	Ningcheng, Inner Mongolia	Jiulongshan Fm., J_2	Wang *et al.* [139]
	Tenuosmylus brevineurus Wang, Liu & Ren, 2009	Ningcheng, Inner Mongolia	Jiulongshan Fm., J_2	Wang *et al.* [140]
	Yanosmylus rarivenatus Ren, 1995	Chengde, Hebei	Yixian Fm., K_1	Ren [117]
Osmylitidae	[a] *Idiastogyia fatisca* Lin, 1986	Jiexi, Guangdong	Jinji Fm., J_1	Lin [35]
Osmylopsychopidae	*Daopsychops bifasciatus* Peng, Makarkin & Ren, 2015	Ningcheng, Inner Mongolia	Jiulongshan Fm., J_2	Peng *et al.* [154]
	Daopsychops clausus Peng, Makarkin & Ren, 2015	Ningcheng, Inner Mongolia	Jiulongshan Fm., J_2	Peng *et al.* [154]
	Daopsychops cubitalis Peng, Makarkin & Ren, 2015	Ningcheng, Inner Mongolia	Jiulongshan Fm., J_2	Peng *et al.* [154]
	Daopsychops dissectus Peng, Makarkin & Ren, 2015	Ningcheng, Inner Mongolia	Jiulongshan Fm., J_2	Peng *et al.* [154]
	Daopsychops inanis Peng, Makarkin & Ren, 2015	Ningcheng, Inner Mongolia	Jiulongshan Fm., J_2	Peng *et al.* [154]
	Eupypsychops confinis Peng, Makarkin & Ren, 2015	Ningcheng, Inner Mongolia	Jiulongshan Fm., J_2	Peng *et al.* [154]
	Eupypsychops ferox Peng, Makarkin & Ren, 2015	Ningcheng, Inner Mongolia	Jiulongshan Fm., J_2	Peng *et al.* [154]
	Nematopsychops unicus Peng, Makarkin & Ren, 2015	Ningcheng, Inner Mongolia	Jiulongshan Fm., J_2	Peng *et al.* [154]
	Ochropsychops multus Peng, Makarkjin & Ren, 2015	Ningcheng, Inner Mongolia	Jiulongshan Fm., J_2	Peng *et al.* [154]
	Stenopteropsychops trifasciatus Peng, Makarkin & Ren, 2015	Ningcheng, Inner Mongolia	Jiulongshan Fm., J_2	Peng *et al.* [154]
	Undulopsychopsis alexi Peng, Makarkin, Wang & Ren, 2011	Beipiao, Liaoning	Yixian Fm., K_1	Peng *et al.* [155]
Panfiloviidae	*Epipanfilovia oviformis* Yang, Makarkin & Ren, 2013	Ningcheng, Inner Mongolia	Jiulongshan Fm., J_2	Yang *et al.* [156]
Parakseneuridae	*Parakseneura albadelta* Yang, Makarkin & Ren, 2012	Ningcheng, Inner Mongolia	Jiulongshan Fm., J_2	Yang *et al.* [32]
	Parakseneura albomacula Yang, Makarkin & Ren, 2012	Ningcheng, Inner Mongolia	Jiulongshan Fm., J_2	Yang *et al.* [32]
	Parakseneura cavomacula Yang, Makarkin & Ren, 2012	Ningcheng, Inner Mongolia	Jiulongshan Fm., J_2	Yang *et al.* [32]
	Parakseneura curvivenis Yang, Makarkin & Ren, 2012	Ningcheng, Inner Mongolia	Jiulongshan Fm., J_2	Yang *et al.* [32]
	Parakseneura directa Yang, Makarkin & Ren, 2012	Ningcheng, Inner Mongolia	Jiulongshan Fm., J_2	Yang *et al.* [32]
	Parakseneura emarginata Yang, Makarkin & Ren, 2012	Ningcheng, Inner Mongolia	Jiulongshan Fm., J_2	Yang *et al.* [32]
	Parakseneura inflata Yang, Makarkin & Ren, 2012	Ningcheng, Inner Mongolia	Jiulongshan Fm., J_2	Yang *et al.* [32]
	Parakseneura metallica Yang, Makarkin & Ren, 2012	Ningcheng, Inner Mongolia	Jiulongshan Fm., J_2	Yang *et al.* [32]
	Parakseneura nigrolinea Yang, Makarkin & Ren, 2012	Ningcheng, Inner Mongolia	Jiulongshan Fm., J_2	Yang *et al.* [32]

(Continued)

Table 20.1 (Continued)

Family	Species	Locality	Horizon/Age	Citation
	Parakseneura nigromacula Yang, Makarkin & Ren, 2012	Ningcheng, Inner Mongolia	Jiulongshan Fm., J_2	Yang *et al.* [32]
	Parakseneura undula Yang, Makarkin & Ren, 2012	Ningcheng, Inner Mongolia	Jiulongshan Fm., J_2	Yang *et al.* [32]
	Pseudorapisma angustipenne Yang, Makarkin & Ren, 2012	Ningcheng, Inner Mongolia	Jiulongshan Fm., J_2	Yang *et al.* [32]
	Pseudorapisma jurassicum Yang, Makarkin & Ren, 2012	Ningcheng, Inner Mongolia	Jiulongshan Fm., J_2	Yang *et al.* [32]
	Pseudorapisma maculatum Yang, Makarkin & Ren, 2012	Ningcheng, Inner Mongolia	Jiulongshan Fm., J_2	Yang *et al.* [32]
Psychopsidae	*Alloepipsychopsis lata* Makarkin, Yang, Peng & Ren, 2012	Beipiao, Liaoning	Yixian Fm., K_1	Makarkin *et al.* [71]
	[a] *Angaropsychops sinicus* Wang, 1980	Ningcheng, Inner Mongolia	Yixian Fm., K_1	Wang [172]
	[a] *Beipiaopsychops triangulates* Hong, 1983	Beipiao, Liaoning	Yixian Fm., K_1	Hong [48]
	Cretapsychops decipiens Peng, Makarkin, Yang & Ren, 2010	Ningcheng, Inner Mongolia	Jiulongshan Fm., J_2	Peng *et al.* [164]
	[a] *Liutaipsychops borealis* Lin, 1994	Liutai, Jilin	Yincheng Fm., J_1	Lin [173]
	[a] *Sinopsychops chengdeensis* Hong, 1982	Chengde, Hebei	Jiulongshan Fm., J_2	Hong [174]
Saucrosmylidae	*Bellinympha dancei* Wang, Ren, Shih & Engel, 2010	Ningcheng, Inner Mongolia	Jiulongshan Fm., J_2	Wang *et al.* [28]
	Bellinympha filicifolia Wang, Ren, Liu & Engel, 2010	Ningcheng, Inner Mongolia	Jiulongshan Fm., J_2	Wang *et al.* [28]
	Daohugosmylus castus Liu, Zhang, Wang, Fang, Zheng, Zhang & Jarzembowski, 2014	Ningcheng, Inner Mongolia	Haifanggou Fm., J_2	Liu *et al.* [168]
	Huiyingosmylus bellus Liu, Zhang, Wang, Fang, Zheng, Zhang & Jarzembowski 2013	Ningcheng, Inner Mongolia	Jiulongshan Fm., J_2	Liu *et al.* [167]
	Laccosmylus calophlebius Ren & Yin, 2003	Ningcheng, Inner Mongolia	Jiulongshan Fm., J_2	Ren *et al.* [165]
	Rudiosmylus ningchengensis Ren & Yin, 2003	Ningcheng, Inner Mongolia	Jiulongshan Fm., J_2	Ren *et al.* [165]
	Saucrosmylus sambneurus Ren & Yin, 2003	Ningcheng, Inner Mongolia	Jiulongshan Fm., J_2	Ren *et al.* [165]
	Ulrikezza aspoeckae Fang, Ren & Wang, 2015	Ningcheng, Inner Mongolia	Jiulongshan Fm., J_2	Fang *et al.* [166]
Family *incertae sedis*	*Guyiling jianboni* Shi, Béthoux, Shih & Ren, 2012	Beipiao, Liaoning	Yixian Fm., K_1	Shi *et al.* [62]
	Hongosmylites longus Hong, 1996	Laiyang, Shandong	Laiyang Fm., K_1	Hong [170]
	a) *Mesohemerobius jeholensis* Ping, 1928	Beipiao, Liaoning	Yixian Fm., K_1	Ping [33]

a) The species is not present in the main text because original description, photos and line-drawings are not precise and the holotype cannot be rechecked.

b) Horizon/Age revised from the original paper based on updated information and data.

References

1 Oswald, J.D. (2017). *Neuropterida Species of the World. Version 5.0*, Lacewing Digital Library http://lacewing.tamu.edu/SpeciesCatalog/Main. TX, USA: Texas A&M University (Accessed 19 September 2018).

2 Mansell, M.W. (1983). A revision of the Australian Crocinae (Neuroptera: Nemopteridae). *Australian Journal of Zoology* **31**: 607–627.

3 Tjeder, B. (1967). Neuroptera-Planipennia. The lace-wings of Southern Africa. 6. Family Nemopteridae. *South African Animal Life* **13**: 290–501.

4 New, T.R. (1989). Planipennia (Lcewings). In: *Handbuch der Zoologie. Band IV. Arthropoda: Insecta. Part 30* (ed. M. Fischer), 1–132. Berlin: Walter de Gruyter.

5 Gallard, L. (1932). Notes on the feeding habits of the brown moth-lacewing, *Ithone fusca*. *Australian Naturalist* **8**: 168–170.

6 Hagen, K.S. and Tassan, R.L. (1970). The influence of food Wheast and related *Saccharomyces fragilis* yeast products on the fecundity of *Chrysopa carnea* (Neuroptera: Chrysopidae). *Canadian Entomologist* **102**: 806–811. https://doi.org/10.4039/Ent102806-7.

7 Winterton, S.L., Zhao, J., Garzón-Orduña, I.J. *et al.* (2017). The phylogeny of lance lacewings (Neuroptera: Osmylidae). *Systematic Entomology* **42** (3): 555–574.

8 Hagen, H.A. (1861). *Synopsis of the Neuroptera of North America, with a List of the South American Species*, Smithsonian Miscellaneous Collections, 4 (1), 210–211. Smithsonian Institution.

9 Withycombe, C.L. (1925). Some aspects of the biology and morphology of the Neuroptera. With special reference to the immature stages and their possible phylogenetic significance. *Transactions of the Entomological Society of London* **72**: 303–411.

10 Aspöck, U., Plant, J.D., and Nemeschkal, H.L. (2001). Cladistic analysis of Neuroptera and their systematic position within Neuropterida (Insecta: Holometabola: Neuroptera). *Systematic Entomology* **26**: 73–86. https://doi.org/10.1046/j.1365–3113.2001.00136.x.

11 Winterton, S.L., Hardy, N.B., and Wiegmann, B.M. (2010). On wings of lace: phylogeny and bayesian divergence time estimates of Neuropterida (Insecta) based on morphological and molecular data. *Systematic Entomology* **35** (3): 349–378.

12 Wang, Y.Y., Liu, X.Y., Garzón-Orduña, I.J. *et al.* (2016). Mitochondrial phylogenomics illuminates the evolutionary history of Neuropterida. *Cladistics* **33** (6): 617–636.

13 Withycombe, C.L. (1924). Further notes on the biology of some British Neuroptera. *The Entomologist* **57**: 145–152.

14 MacLeod, E.G. (1964) A comparative morphological study of the head capsule and cervix of larval Neuroptera (Insecta), PhD Thesis, Harvard University.

15 Beutel, R.G., Friedrich, F., and Aspöck, U. (2010). The larval head of Nevrorthidae and the phylogeny of Neuroptera (Insecta). *Zoological.l Journal of the Linnean Society* **158**: 533–562. https://doi.org/10.1111/j.1096–3642.2009.00560.x.

16 Beutel, R.G., Zimmermann, D., Kraus, M. *et al.* (2010). Head morphology of *Osmylus fulvicephalus* (Osmylidae, Neuroptera) and its phylogenetic implications. *Organisms Diversity & Evolution* **10**: 311–329. https://doi.org/10.1007/s13127–010–0024-0.

17 Haring, E. and Aspöck, U. (2004). Phylogeny of the Neuropterida: a first molecular approach. *Systematic Entomology* **29** (3): 415–430. https://doi.org/10.1111/j.0307–6970.2004.00263.x.

18 Aspöck, H. and Aspöck, U. (2008). Phylogenetic relevance of the genital sclerites of Neuropterida (Insecta: Holometabola). *Systematic Entomology* **33**: 97–127. https://doi.org/10.1111/j.1365–3113.2007.00396.x.

19 Randolf, S., Zimmermann, D., and Aspöck, U. (2014). Head anatomy of adult *Nevrorthus apatelios* and basal splitting events in Neuroptera (Neuroptera: Nevrorthidae). *Arthropod Systematics & Phylogeny* **72**: 111–136.

20 Carpenter, F.M. (1992). Superclass Hexapoda. In: *Treatise on Invertebrate Paleontology, Part R, Arthropoda 4*, vol. **3–4** (ed. R.C. Moore and R.L. Kaesler), 1–655. Boulder, CO: Geological Society of America.

21 Martynov, A.V. (1937). Liassic insects of Shurab and Kizil-Kija. *Trudy Paleontologicheskogo Instituta Akademiya Nauk SSSR* **7** (1): 1–232. (in Russian).

22 Riek, E.F. (1956). A re-examination of the mecopteroid and orthopteroid fossils (Insecta) from the Triassic beds at Denmark Hill, Queensland, with descriptions of further specimens. *Australian Journal of Zoology* **4** (1): 98–110.

23 Tillyard, R.J. (1922). Mesozoic insects of Queensland. No.9. Orthoptera, and additions to the Protorthoptera, Odonata, Hemiptera and Planipennia. *Proceedings of the Linnean Society of New South Wales* **47**: 447–470.

24 Handlirsch, A. (1906–1908). *Die fossilen Insekten und die Phylogenie der rezenten Formen: ein Handbuch für Paläontologen und Zoologen*, 1–1430. Leipzig: Wilhelm Engelmann.

25 Bode, A. (1953). Die insektenfauna des ostniedersächsischen oberen lias. *Palaeontographica Abteilung A* **103**: 1–375.

26 Labandeira, C.C., Yang, Q., Santiago-Blay, J.A. *et al.* (2016). The evolutionary convergence of mid-Mesozoic lacewings and Cenozoic butterflies. *Proceedings of the Royal Society B* **283** (1824): 20152893. https://doi.org/10.1098/rspb.2015.2893.

27 Yang, Q., Wang, Y.J., Labandeira, C.C. *et al.* (2014). Mesozoic lacewings from China provide phylogenetic insight into evolution of the Kalligrammatidae (Neuroptera). *BMC Evolutionary Biology* **14** (126): 1–30.

28 Wang, Y.J., Liu, Z.Q., Wang, X. *et al.* (2010). Ancient pinnate leaf mimesis among lacewings. *Proceedings of the National Academy of Sciences* **107** (37): 16212–16215.

29 Grimaldi, D.A. (2000). A diverse fauna of Neuropterodea in amber from the Cretaceous of New Jersey. In: *Studies on Fossil in Amber, with Particular Reference to the Cretaceous of New Jersey* (ed. D.A. Grimaldi), 259–303. Leiden: Backhuys Publishers.

30 Makarkin, V.N., Yang, Q., and Ren, D. (2013). A new Cretaceous family of enigmatic two-winged lacewings (Neuroptera). *Fossil Record* **16** (1): 67–75. https://doi.org/10.1002/mmng.201300002.

31 Liu, X.Y., Zhang, W.W., Winterton, S.L. *et al.* (2016). Early morphological specialization for insect-spider associations in Mesozoic lacewings. *Current Biology* **26** (12): 1590–1594. https://doi.org/10.1016/j.cub.2016.04.039.

32 Yang, Q., Makarkin, V.N., Winterton, S.L. *et al.* (2012). A remarkable new family of Jurassic insects (Neuroptera) with primitive wing venation and its phylogenetic position in Neuropterida. *PLoS One* **7** (9): 1–38.

33 Ping, C. (1928). Cretaceous fossil insects of China. *Palaeontologia Sinica* **13**: 1–47.

34 Hong, Y.C. (1980). *Mesozoic Stratigraphy and Paleontology of Shaanxi-Gangsu-Ningxia Basin*, 1–188. Beijing: Geological Publishing House.

35 Lin, Q.B. (1986). Early Mesozoic fossil insects from South China. *Palaeontologia Sinica. NSB* **170** (21): 67–68.

36 Ren, D. and Engel, M.S. (2008). Aetheogrammatidae, a new family of lacewings from the Mesozoic of China (Neuroptera: Myrmeleontiformia). *Journal of the Kansas Entomological Society* **81** (3): 161–167.

37 Engel, M.S., Huang, D.Y., and Lin, Q.B. (2011). A new genus and species of Aetheogrammatidae from the Jurassic of Inner Mongolia, China (Neuroptera). *Journal of the Kansas Entomological Society* **84**: 315–319. https://doi.org/10.2317/JKES110623.1.

38 Yang, Q., Makarkin, V.N., Shih, C.F., and Ren, D. (2015). New Aetheogrammatidae (Insecta: Neuroptera) from the Lower Cretaceous Yixian Formation, China. *Cretaceous Research* **55**: 25–31.

39 Ren, D. and Makarkin, V.N. (2009). Ascalochrysidae–a new lacewing family from the Mesozoic of China (Insecta: Neuroptera: Chrysopoidea). *Cretaceous Research* **30** (5): 1217–1222.

40 Nakahara, W. (1958). *The Neurorthinae, a New Subfamily of the Sisyridae (Neuroptera)*, vol. **32**, 19–32. Mushi.

41 Aspöck, U. (1986). The present state of knowledge of the family Berothidae (Neuropteroidea: Planipennia). In: *Recent Research in Neuropterology* (ed. J. Gepp, H. Aspöck and H. Holzel), 87–101. Graz: Privately printed.

42 Aspöck, H. and Aspöck, U. (1997). Studies on new and poorly-known Rhachiberothidae (Insecta: Neuroptera) from subsaharan Africa. *Annalen des Naturhistorischen Museums in Wien* **99**: 1–20.

43 Aspöck, U. and Nemeschkal, H.L. (1998). A cladistic analysis of the Berothidae (Neuroptera). *Acta Zoologica Fennica* **209**: 45–63.

44 Tjeder, B. (1959). Neuroptera-Planipennia. The Lace-wings of Southern Africa. 2. Family Berothidae. *South African Animal Life* **6**: 256–314.

45 Willmann, R. (1990). The phylogenetic position of the Rhachiberothinae and the basal sister-group relationships within the Mantispidae (Neuroptera). *Systematic Entomology* **15** (2): 253–265.

46 Makarkin, V.N., Yang, Q., and Ren, D. (2011). Two new species of *Sinosmylites* Hong (Neuroptera, Berothidae) from the Middle Jurassic of China, with notes on Mesoberothidae. *ZooKeys* **130**: 199–215. https://doi.org/10.3897/zookeys.130.1418.

47 Khramov, A.V. and Makarkin, V.N. (2015). New fossil Osmylopsychopidae (Neuroptera) from the Early/Middle Jurassic of Kyrgyzstan, Central Asia. *Zootaxa* **4059** (1): 115–132. https://doi.org/10.11646/zootaxa.4059.1.6.

48 Hong, Y.C. (1983). *Middle Jurassic Fossil Insects in North China*, 1–223. Beijing: Geological Publishing House.

49 Ren, D. and Guo, Z.G. (1996). On the new fossil genera and species of Neuroptera (Insecta) from the Late Jurassic of Northeast China. *Acta Zootaxonomica Sinica* **21** (4): 461–479.

50 Ansorge, J. and Schlüter, T. (1990). The earliest chrysopid: *Liassochrysa stigmatica* n. g., n. sp. from the Lower Jurassic of Dobbertin, Germany. *Neuroptera International* **6**: 87–93.

51 Wedmann, S. and Makarkin, V.N. (2007). A new genus of Mantispidae (Insecta: Neuroptera) from

the Eocene of Germany, with a review of the fossil record and palaeobiogeography of the family. *Zoological Journal of the Linnean Society* **149** (4): 701–716.

52 Nel, A., Delclos, X., and Hutin, A. (2005). Mesozoic chrysopid-like Planipennia: a phylogenetic approach (Insecta: Neuroptera). *Annales de la Société Entomologique de France* **41** (1): 29–69. https://doi.org/10.1080/00379271.2005.10697440.

53 Makarkin, V.N. and Menon, F. (2005). New species of the Mesochrysopidae (Insecta, Neuroptera) from the Crato Formation of Brazil (Lower Cretaceous), with taxonomic treatment of the family. *Cretaceous Research* **26** (5): 801–812. https://doi.org/10.1016/j.cretres.2005.05.009.

54 Khramov, A.V., Liu, Q., Zhang, H.C., and Jarzembowski, E.A. (2015). Early green lacewings (Insecta: Neuroptera: Chrysopidae) from the Jurassic of China and Kazakhstan. *Papers in Palaeontology* **2** (1): 25–39. https://doi.org/10.1002/spp2.1024.

55 Martynov, A.V. (1927). Jurassic fossil insects from Turkestan. Order Neuroptera. *Izvestia Akademii Nauk SSSR* **21** (5): 757–768.

56 Panfilov, D.V. (1980). New representatives of lacewings (Neuroptera) from the Jurassic of Karatau. In: *Mesozoic Fossil Insects* (ed. V.G. Dolin, D.V. Panfilov, A.G. Ponomarenko and L.N. Pritikina), 82–111. Kiev: Naukova Dumka (In Russian).

57 Yang, J.K. and Hong, Y.C. (1990). *Drakochrysa*, an Early Cretaceous new genus of Chrysopidae (Insecta Neuroptera) from Laiyang Basin, Shandong Province. *Geoscience* **4** (4): 15–26. (in Chinese with English abstract).

58 Ren, D. (2002). A new lacewing family (Neuroptera) from the Middle Jurassic of Inner Mongolia, China. *Entomologia Sinica* **9** (4): 53–67.

59 Shi, C.F., Yang, Q., and Ren, D. (2011). Two new fossil lacewing species from the Middle Jurassic of Inner Mongolia, China (Neuroptera: Grammolingiidae). *Acta Geologica Sinica* **85** (2): 482–489. https://doi.org/10.1111/j.1755–6724.2011.00416.x.

60 Khramov, A.V. (2010). A new lacewing (Insecta: Neuroptera: Grammolingiidae) from the Upper Jurassic of Mongolia. *Paleontological Journal* **44** (2): 188–191. https://doi.org/10.1134/S0031030110020103.

61 Liu, Y.S., Shi, C.F., and Ren, D. (2011). A new lacewing (Insecta: Neuroptera: Grammolingiidae) from the Middle Jurassic of Inner Mongolia, China. *Zootaxa* **2897**: 51–56.

62 Shi, C.F., Bethoux, O., Shih, C.K., and Ren, D. (2012). *Guyiling jianboni* gen. et sp. n, an antlion-like lacewing, illuminating homologies and transformations in Neuroptera wing venation. *Systematic Entomology* **37** (3): 617–631.

63 Khramov, A.V. (2012). The new fossil lacewings of Grammolingiidae (Neuroptera) from the Jurassic of Central Asia and Mongolia, with notes on biogeography of the family. *Zootaxa* **3478** (1): 297–308.

64 Shi, C.F., Wang, Y.J., and Ren, D. (2013). New species of *Grammolingia* Ren, 2002 from the Middle Jurassic of Inner Mongolia, China (Neuroptera: Grammolingiidae). *Fossil Record* **16** (2): 171–178.

65 Shi, C.F., Wang, Y.J., Yang, Q., and Ren, D. (2012). *Chorilingia* (Neuroptera: Grammolingiidae): a new genus of lacewings with four species from the Middle Jurassic of Inner Mongolia, China. *Alcheringa: An Australasian Journal of Palaeontology* **36** (3): 309–318.

66 Winterton, S.L. and Makarkin, V.N. (2010). Phylogeny of moth lacewings and giant lacewings (Neuroptera: Ithonidae, Polystoechotidae) using DNA sequence data, morphology and fossils. *Annals of the Entomological Society of America* **103** (4): 511–522.

67 Martynov, A.V. (1925). To the knowledge of fossil insects from Jurassic beds in Turkestan. 2. Raphidioptera (continued), Orthoptera (s.l.), Odonata, Neuroptera. *Izvestia Rossiiskoi Akademii Nauk* **19**: 569–598.

68 Makarkin, V.N. and Menon, F. (2007). First record of fossil 'rapismatid-like' Ithonidae (Insecta, Neuroptera) from the Lower Cretaceous Crato Formation of Brazil. *Cretaceous Reaearch* **28** (5): 743–753. https://doi.org/10.1016/j.cretres.2006.11.003.

69 Makarkin, V.N. and Archibald, S.B. (2009). A new genus and first Cenozoic fossil record of moth lacewings (Neuroptera: Ithonidae) from the Early Eocene of North America. *Zootaxa* **2063**: 55–63.

70 Zheng, B.Y., Ren, D., and Wang, Y.J. (2016). A new species of *Lasiosmylus* from the Early Cretaceous, China clarifies its genus-group placement in Ithonidae (Neuroptera). *ZooKeys* **636**: 1–41.

71 Makarkin, V.N., Yang, Q., Peng, Y.Y., and Ren, D. (2012). A comparative overview of the neuropteran assemblage of the Lower Cretaceous Yixian Formation (China), with description of a new genus of Psychopsidae (Insecta: Neuroptera). *Cretaceous Research* **35**: 57–68. https://doi.org/10.1016/j.cretres.2011.11.013.

72 Ren, D., Engel, M.S., and Lu, W. (2002). New giant lacewings from the Middle Jurassic of Inner Mongolia, China (Neuroptera, Polystoechotidae). *Journal of the Kansas Entomological Society* **75** (3): 188–193.

73 Zheng, B.Y., Ren, D., and Wang, Y.J. (2016). Earliest true moth lacewing from the Middle Jurassic

of Inner Mongolia, China. *Acta Palaeontologica Polonica* **61** (4): 847–851.
74 Grimaldi, D.A. and Engel, M.S. (eds.) (2005). *Evolution of the Insects*, 1–755. Cambridge: Cambridge University Press.
75 Walther, J. (1904). Die Fauna der Solnhofener Plattenkalke. Bionomisch betrachtet Denkschriften der Medizinisch-Naturwissenschaftlichen. *Denkschriften der Medizinisch-Naturwissenschaftlichen Gesellschaft zu Jena* **11**: 133–214.
76 Handlirsch, A. (1919). Eine neue Kalligrammide (Neuroptera) aus dem Solnhofener Plattenkalke. *Senckenbergiana* **1** (3): 61–63.
77 Martynova, O.M. (1947). Kalligrammatidae (Neuroptera) from the Jurassic shales of Karatau (Kazakhstanian SSR). *Doklady Akademii Nauk SSSR (N.S.)* **58**: 2055–2058. (in Russian).
78 Panfilov, D.V. (1968). Kalligrammatidae (Neuroptera) from the Jurassic deposits of Karatau. In: *Jurassic Insects of Karatau* (ed. B.B. Rohdendorf), 269–282. Moscow: Nauka Press (In Russian).
79 Ponomarenko, A.G. (1984). Neuroptera from the Jurassic of eastern Asia. *Paleontologicheskii Zhurnal* **1984**: 64–73. (In Russian).
80 Ponomarenko, A.G. (1992). New lacewings (Insecta, Neuroptera) from the Mesozoic of Mongolia. *Transactions of the Joint Soviet-Mongolian Paleontological Expedition* **41**: 101–111. (In Russian).
81 Jarzembowski, E.A. (2001). A new Wealden fossil lacewing. In: *Tunbridge Wells and Rusthall Commons: A History and Natural History* (ed. M.L.J. Rowlands), 56–58. Tunbridge Wells: Tunbridge Wells Museum and Art Gallery.
82 Ren, D. and Oswald, J.D. (2002). A new genus of kalligrammatid lacewings from the Middle Jurassic of China (Neuroptera: Kalligrammatidae). *Stuttgarter Beitrage zur Naturkunde, Serie B* **33**: 1–8.
83 Ren, D. (2003). Two new Late Jurassic genera of kalligrammatids from Beipiao, Liaoning (Neuroptera: Kalligrammatidae). *Acta Zootaxonomica Sinica* **28** (1): 105–109. (in Chinese with English abstract).
84 Zhang, J.F. (2003). Kalligrammatid lacewings from Upper Jurassic of Daohugou Formation in Inner Mongolia, China. *Acta Geologica Sinica* **77** (2): 141–147.
85 Zhang, J.F. and Zhang, H.C. (2003). *Kalligramma jurarchegonium* sp. nov. (Neuroptera: Kalligrammatidae) from the Middle Jurassic of Northeastern China. *Oriental Insects* **37** (1): 301–308.
86 Engel, M.S. (2005). A remarkable kalligrammatid lacewing from the Upper Jurassic of Kazakhstan (Neuroptera: Kalligrammatidae). *Transactions of the Kansas Academy of Science* **108** (1): 59–62. https://doi.org/10.1660/0022–8443(2005)108.
87 Makarkin, V.N., Ren, D., and Yang, Q. (2009). Two new species of Kalligrammatidae (Neuroptera) from the Jurassic of China, with comments on venational homologies. *Annals of the Entomological Society of America* **102** (6): 964–969. https://doi.org/10.1603/008.102.0606.
88 Yang, Q., Zhao, Y.Y., and Ren, D. (2009). An exceptionally well-preserved fossil kalligrammatid from the Jehol Biota. *Chinese Science Bulletin* **54** (10): 1732–1737.
89 Makarkin, V.N. (2010). New psychopsoid Neuroptera from the Lower Cretaceous of Baissa, Transbaikalia. *Annales de la Société entomologi.ue de France* **46** (1–2): 254–261. https://doi.org/10.1080/00379271.2010.10697666.
90 Yang, Q., Makarkin, V.N., and Ren, D. (2014). Two new species of *Kalligramma* Walther (Neuroptera: Kalligrammatidae) from the Middle Jurassic of China. *Annals of the Entomological Society of America* **107** (5): 917–925.
91 Liu, Q., Khramov, A.V., Zhang, H.C., and Jarzembowski, E.A. (2015). Two new species of *Kalligrammula* Handlirsch, 1919 (Insecta, Neuroptera, Kalligrammatidae) from the Jurassic of China and Kazakhstan. *Journal of Paleontology* **89**: 405–410.
92 Yang, Q., Makarkin, V.N., and Ren, D. (2011). Two interesting new genera of Kalligrammatidae (Neuroptera) from the Middle Jurassic of Daohugou, China. *Zootaxa* **2873** (6): 60–68.
93 Liu, Q., Zheng, D.R., Zhang, Q. *et al.* (2014). Two new kalligrammatids (Insecta, Neuroptera) from the Middle Jurassic of Daohugou, Inner Mongolia, China. *Alcheringa: An Australasian Journal of Palaeontology* **38** (1): 65–69. https://doi.org/10.1080/03115518.2013.828251.
94 Lambkin, K.J. (1986). A revision of the Australian Mantispidae (Insecta: Neuroptera) with a contribution to the classification of the family. I. General and Drepanicinae. *Australian Journal of Zoology Supplementary Series* **34** (116): 1–142. https://doi.org/10.1071/AJZS116.
95 Ohl, M. (2004). Annotated catalog of the Mantispidae of the world (Neuroptera). *Contributions on Entomology International ISSN* **5** (3): 129–264.
96 Makarkin, V.N. (1997). Fossil Neuroptera of the Lower Cretaceous of Baisa, East Siberia. Part 4: Psychopsidae. *Beitrage Zur Entomologie, Berlin* **47**: 489–492.
97 Makarkin, V.N. (1990). *Baissoleon cretaceus* gen. and sp. nov. Fossil Neuroptera from the Lower Cretaceous of Baisa, East Siberia. 2. Nymphitidae. *Annales de la Société Entomologique de France (N.S.)* **26** (1): 125–126.

98 Poinar, G.O. Jr., and Buckley, R. (2011). *Doratomantispa burmanica* n. gen., n. sp. (Neuroptera: Mantispidae), a new genus of mantidflies in Burmese amber. *Historical Biology* **23**: 169–176.

99 Jepson, J.E., Heads, S.W., Makarkin, V.N., and Ren, D. (2013). New fossil mantidflies (Insecta: Neuroptera: Mantispidae) from the Mesozoic of North-Eastern China. *Palaeontology* **56** (3): 603–613. https://doi.org/10.1111/pala.12005.

100 Ohl, M. (2011). Aboard a spider – a complex developmental strategy fossilized in amber. *Naturwissenschaften* **98** (5): 453–456. https://doi.org/10.1007/s00114–011–0783–2.

101 Jarzembowski, E.A. (1980). Fossil insects from the Bembridge Marls, Paleogene of the Isle of Wight, southern England. *Bulletin of the British Museum (Natural History) Geology* **33**: 237–293.

102 Cockerell, T.D.A. (1921). Fossil Arthropoda in the British Museum. Oligocene Hymenoptera from the Isle of Wight. *Annals and Magazine of Natural History* 7 (9): 13.

103 Nel, A. (1989). Deux nouveaux Mantispidae (Planipennia) fossiles de l'Oligocene du sud-est de la France. *Neuroptera International* **5**: 103–109.

104 Poinar, G. (2006). *Feroseta priscus* (Neuroptera: Mantispidae), a new genus and species of mantidflies in Dominican amber. *Proceedings of the Entomological Society of Washington* **108**: 411–417.

105 Engel, M.S. and Grimaldi, D.A. (2007). The neuropterid fauna of Dominican and Mexican amber (Neuropterida: Megaloptera, Neuroptera). *American Museum Novitates* **3587**: 1–58.

106 Adams, P.A. (1967). A review of the Mesochrysinae and Nothochrysinae (Neuroptera: Chrysopidae). *Bulletin of the Museum of Comparative Zoology* **135**: 215–238.

107 Schlüler, T. (1984). Phylogeny of Chrysopidae. In: *Biology of Chrysopidae* (ed. M. Canard, Y. Séméria and T.R. New), 1–8. Hague: Dr. W. Junk Publishers.

108 Séméria, Y. and Nel, A. (1990). *Paleochrysopa monteilsensis* gen. et sp. nov, a new fossil of Chrysopidae from the Upper Eocene Formation of Monteils (France), with a review of the known chrysopid fossils (Insecta: Neuroptera). In: *Advances in Neuropterology: Proceedings of the Third International Symposium on Neuropterology* (ed. M.W. Mansell and H. Aspöck), 27–32. Pretoria: South African Department of Agricultural Development.

109 Martins-Neto, R.G. (2003). The Santana Formation paleoentomofauna reviewed. Part I – Neuropteroida (Neuroptera and Raphidioptera): systematic and phylogeny, with description of new taxa. *Acta Geologica Leopoldensia (RS)* **25** (55): 35–66.

110 Ponomarenko, A.G. (2003). On some Neuroptera (Insecta) from Upper Jurassic Solnhofen Limestone. *Annals of the Upper Silesian Museum (Entomology)* **12**: 87–100.

111 Menon, F. and Makarkin, V.N. (2008). New fossil lacewings and antlions (Insecta, Neuroptera) from the Lower Cretaceous Crato Formation of Brazil. *Palaeontology* **51** (1): 149–162. https://doi.org/10.1111/j.1475-4983.2007.00740.x.

112 Martins-Neto, R.G. and Rodrigues, V.Z. (2009). New Neuroptera (Insecta, Osmylidae and Mesochrysopidae) from the Santana Formation, Lower Cretaceous of northeast Brazil. *Gaea: Journal of Geoscience* **5** (1): 1–15. https://doi.org/10.4013/gaea.2009.51.02.

113 Zhang, J.F. (1991). A new family of Neuroptera (Insecta) from the late Mesozoic of Shandong. *Science in China Series. B* **34**: 1105–1111.

114 Ponomarenko, A.G. (1992). Neuroptera (Insecta) from the Cretaceous of Transbaikalia. *Paleontological Journal* **26** (3): 43–50.

115 Nel, A. and Henrotay, M. (1994). Les Chrysopidae mésozoïques. État actuel des connaissances. Description d'un nouveau genre et nouvelle espèce dans le Jurassique inférieur (Lias) (Insecta, Neuroptera). *Annales de la Société Entomologique de France* **30**: 293–318.

116 Yang, Q., Makarkin, V.N., and Ren, D. (2012). New fossil Mesochrysopidae (Neuroptera) from the Mesozoic of China. *Zootaxa* **3597**: 1–14.

117 Ren, D. (1995). *Faunae and Stratigraphy of Jurassic-Cretaceous in Beijing and the Adjacent Areas*, 1–222. Beijing: Seismic Publishing House.

118 Ren, D. and Yin, J.C. (2002). A new genus and new species of lacewings in the Jurassic of China (Neuroptera: Myrmeleotoidea). *Acta Zootaxonomica Sinica* **27** (2): 269–273.

119 Ren, D., Makarkin, V.N., and Yang, Q. (2010). A new fossil genus of Mesochrysopidae (Neuroptera) from the Early Cretaceous Yixian Formation of China. *Zootaxa* **2523**: 50–56.

120 Rice, H.M.A. (1969). An antlion (Neuroptera) and a stonefly (Plecoptera) of Cretaceous age from Labrador, Newfoundland. *Geological Survey of Canada* **68–65**: 1–12.

121 Dobruskina, I.A., Ponomarenko, A.G., and Rasnitsyn, A.P. (1997). Fossil insects found in Israel. *Paleontologicheskii Zhurnal* **5**: 91–95.

122 Martins-neto, R.G. (2000). Remarks on the neuropterofauna (Insecta, Neuroptera) from the Brazilian Cretaceous, with keys for the identification of the known taxa. *Acta Geológica Hispánica* **35** (1): 97–118.

123 Hong, Y.C. (1988). Early Cretaceous Orthoptera, Neuroptera, Hymenoptera (Insecta) of Kezuo in

western Liaoning Province. *Entomotaxonomia* **10**: 119–130.

124 Ren, D. and Engel, M.S. (2008). A second antlion from the Mesozoic of Northeastern China (Neuroptera: Myrmeleontidae). *Alavesia* **2**: 183–186.

125 Makarkin, V.N., Yang, Q., Shi, C., and Ren, D. (2013). The presence of the recurrent veinlet in the Middle Jurassic Nymphidae (Neuroptera): a unique character condition in Myrmeleontoidea. *ZooKeys* **325**: 1–20. https://doi.org/10.3897/zookeys.325.5453.

126 Shi, C.F., Winterton, S.L., and Ren, D. (2015). Phylogeny of split-footed lacewings (Neuroptera, Nymphidae), with descriptions of new Cretaceous fossil species from China. *Cladistics* **31** (5): 455–490.

127 Ren, D. and Engel, M.S. (2007). A split-footed lacewing and two episomylines from the Jurassic of China (Neuroptera). *Annales Zoologici* **57** (2): 211–219.

128 Shi, C.F., Makarkin, V.N., Yang, Q. *et al.* (2013). New species of *Nymphites* Haase (Neuroptera: Nymphidae) from the Middle Jurassic of China, with a redescription of the type species of the genus. *Zootaxa* **3700** (3): 393–410.

129 New, T.R. (2003). *The Neuroptera of Malesia*. Leiden: Brill.

130 Archibald, S.B., Makarkin, V.N., and Ansorge, J. (2009). New fossil species of Nymphidae (Neuroptera) from the Eocene of North America and Europe. *Zootaxa* **2157**: 59–68.

131 New, T.R. (1981). A revision of the Australian Nymphidae (Insecta: Neuroptera). *Australian Journal of Zoology* **29** (5): 707–750. https://doi.org/10.1071/ZO9810707.

132 New, T.R. (1987). Nymphidae (Insecta: Neuroptera) from New Guinea. *Invertebrate Taxonomy* **1**: 807–815.

133 Oswald, J.D. (1998). *Osmylops* Banks (Neuroptera: Nymphidae): generic review and revision of the armatus species group. *Journal of Neuropterology* **1**: 79–108.

134 Westwood, J. (1854). Contributions to fossil entomology. *Quarterly Journal of the Geological Society* **10** (1–2): 378–396.

135 Haase, E. (1890). Remarks on the palaeontology of insects. *New Year's Book for Mineralogy, Geology and Paleontology, Stuttgart* **2** (1): 1–33.

136 Weyenbergh, H. Jr., (1869). Sur les insectes fossiles du calcaire lithographique de la Bavière, qui se trouvent au Musée Teyler. *Archives du Musée Teyler* **1** (2): 247–294.

137 Badano, D. and Winterton, S.L. (2017). New Philippine species of *Spilosmylus* Kolbe (Neuroptera, Osmylidae). *ZooKeys* **712** (1): 29–42. https://doi.org/10.3897/zookeys.712.19883.

138 Ren, D. and Yin, J.C. (2002). A new Middle Jurassic species of *Epioosmylus* from Inner Mongolia, China (Neuroptera: Osmylidae). *Acta Zootaxonomica Sinica* **27** (2): 274–277. (in Chinese with English abstract).

139 Wang, Y.J., Liu, Z.Q., and Ren, D. (2009). A new fossil lacewing genus from the Middle Jurassic of Inner Mongolia, China (Neuroptera: Osmylidae). *Zootaxa* **2034**: 65–68.

140 Wang, Y.J., Liu, Z.Q., and Ren, D. (2009). A new fossil lacewing genus and species from the Middle Jurassic of Inner Mongolia, China. *Acta Palaeontologica Polonica* **54** (3): 557–560.

141 Yang, Q., Makarkin, V.N., and Ren, D. (2010). Remarkable new genus of Gumillinae (Neuroptera: Osmylidae) from the Jurassic of China. *Annals of the Entomological Society of America* **103** (6): 855–859.

142 Wang, Y.J., Liu, Z.Q., Ren, D., and Shih, C.K. (2010). A new genus of Protosmylinae from the Middle Jurassic of China (Neuroptera: Osmylidae). *Zootaxa* **2480** (1): 45–53.

143 Wang, Y.J., Liu, Z.Q., Ren, D., and Shih, C.K. (2011). New Middle Jurassic kempynin osmylid lacewings from China. *Acta Palaeontologica Polonica* **56** (4): 865–869.

144 Makarkin, V.N., Yang, Q., and Ren, D. (2014). A new basal osmylid neuropteran insect from the Middle Jurassic of China linking Osmylidae to the Permian–Triassic Archeosmylidae. *Acta Palaeontologica Polonica* **59** (1): 209–214. https://doi.org/10.4202/app.2011.0018.

145 Martynova, O.M. (1949). Mesozoic lacewings (Neuroptera) and their bearing on concepts of phylogeny and systematics of the order. *Psychopharmacologia* **7** (3): 175–181.

146 Tillyard, R.J. (1923). Mesozoic insects of Queensland. No. 10. Summary of the Upper Triassic insect fauna of Ipswich, Queensland (with an appendix describing new Hemiptera and Planipennia). *Proceedings of the Linnean Society of New South Wales* **48**: 481–498.

147 Riek, E.F. (1955). Fossil insects from the Triassic beds at Mt. Crosby, Queensland. *Australian Journal of Zoology* **3** (4): 654–691.

148 Tillyard, R.J. (1919). Studies in Australian Neuroptera. No. 7. The life-history of Psychopsis elegans

(Guerin). *Proceedings of the Linnean Soceety of New South Wales* **43**: 787–818.
149 Martynova, O.M. (1954). Neuropterous insects from the Cretaceous deposits of Siberia. *Doklady Akademii Nauk SSSR* **94**: 1167–1169. (in Russian).
150 Zalessky, G. (1953). New localities of Cretaceous insects of the Volga, Kazakhstan, and Transbaikalia regions. *Antimicrobial Agents & Chemotherapy* **56** (6): 3114–3120.
151 Martins-Neto, R.G. (1997). Neurópteros (Insecta, Planipennia) da Formação Santana (Cretáceo Inferior), Bacia do Araripe, nordeste do Brasil. X-Descrição de novos táxons (Chrysopidae, Babinskaiidae, Myrmeleontidae, Ascalaphidae e Psychopsidae). *Revista Uniersidade de Guarulhos, Série Ciências Exatas e Technológicas* **2** (4): 68–83.
152 Jepson, J.E., Makarkin, V.N., and Jarzembowski, E.A. (2009). New lacewings (Insecta: Neuroptera) from the Lower Cretaceous Wealden supergroup of southern England. *Cretaceous Research* **30** (5): 1325–1338. https://doi.org/10.1016/j.cretres.2009.07.012.
153 Lambkin, K.J. (2014). Psychopsoid Neuroptera (Psychopidae, Osmylopsychopidae) from the Queensland Triassic. *Australian Entomologist* **41** (1): 57–76.
154 Peng, Y.Y., Makarkin, V.N., and Ren, D. (2015). Diverse new Middle Jurassic Osmylopsychopidae (Neuroptera) from China shed light on the classification of psychopsoids. *Journal of Systematic Palaeontology* **14** (4): 261–295. https://doi.org/10.1080/14772019.2015.1042080.
155 Peng, Y.Y., Makarkin, V.N., Wang, X.D., and Ren, D. (2011). A new fossil silky lacewing genus (Neuroptera, Psychopsidae) from the Early Cretaceous Yixian Formation of China. *ZooKeys* **130**: 217–228. https://doi.org/10.3897/zookeys.130.1576.
156 Yang, Q., Makarkin, V.N., and Ren, D. (2013). A new genus of the family Panfiloviidae (insect, neuroptera) from the Middle Jurassic of China. *Palaeontology* **56** (1): 49–59.
157 Krüger, L. (1914). Osmylidae. Beiträge zu einer monographie der neuropteren-familie der Osmyliden. *Entomologische Zeitung Stettin* **75**: 9–113.
158 Ponomarenko, A.G. (1996). Upper Liassic neuropterans (Insecta) from Lower Saxony, Germany. *Russian Entomological Journal* **4**: 73–89.
159 Makarkin, V.N. and Archibald, S.B. (2003). Family affinity of the genus *Palaeopsychops* Andersen with description of a new species from the Early Eocene of British Columbia (Neuroptera: Polystoechotidae). *Annals of the Entomological Society of America* **96** (3): 171–180. https://doi.org/10.1603/0013-8746(2003)096.
160 Engel, M.S. and Grimaldi, D.A. (2008). Diverse Neuropterida in Cretaceous amber, with particular reference to the paleofauna of Myanmar (Insecta). *Nova Supplementa Entomologica* **20**: 1–86.
161 Ponomarenko, A.G. (2002). Superorder Myrmeleontidea Latreille, 1802. In: *History of Insects* (ed. A.P. Rasnitsyn and D.L.J. Quicke), 176–192. Dordrecht: Kluwer Academic Publishers.
162 New, T.R. (1988). The Psychopsidae (Insecta: Neuroptera) of Australia and the oriental region. *Invertebrate Systematics* **2** (7): 841–883. https://doi.org/10.1071/IT9880841.
163 Wang, X.L. and Bao, R. (2006). A taxonomic study on the genus *Balmes* Navás from China (Neuroptera, Psychopsidae). *Acta Zootaxonomica Sinica* **31**: 846–850.
164 Peng, Y.Y., Makarkin, V.N., Yang, Q., and Ren, D. (2010). A new silky lacewing (Neuroptera: Psychopsidae) from the Middle Jurassic of Inner Mongolia, China. *Zootaxa* **2663**: 59–67.
165 Ren, D. and Yin, J.C. (2003). New 'osmylid-like' fossil Neuroptera from the Middle Jurassic of Inner Mongolia, China. *Journal of the New York Entomological Society* **111** (1): 1–11. https://doi.org/10.1664/0028-7199(2003)111[0001:NOFNFT]2.0.CO;2.
166 Fang, H., Ren, D., and Wang, Y.J. (2015). Familial clarification of Saucrosmylidae stat. nov. and new saucrosmylids from Daohugou, China (Insecta, Neuroptera). *PLoS One* **10** (10): e0141048. https://doi.org/10.1371/journal.pone.0141048.
167 Liu, Q., Zhang, H.C., Wang, B. *et al.* (2013). A new genus of Saucrosmylinae (Insecta, Neuroptera) from the Middle Jurassic of Daohugou, Inner Mongolia, China. *Zootaxa* **3736** (4): 387–391. https://doi.org/10.11646/zootaxa.3736.4.6.
168 Liu, Q., Zhang, H.C., Wang, B. *et al.* (2014). A new saucrosmylid lacewing (Insecta, Neuroptera) from the Middle Jurassic of Daohugou, Inner Mongolia, China. *Alcheringa: An Australasian Journal of Palaeontology* **38** (2): 301–304. https://doi.org/10.1080/03115518.2014.886849.
169 Makarkin, V.N. and Archibald, S.B. (2005). Substitute names for three genera of fossil Neuroptera, with taxonomic notes. *Zootaxa* **1054** (1): 15–23.
170 Hong, Y.C. (1996). A fossil new genus *Sinosmylites* (Insecta: Neuroptera) from Laiyang Basin, Shandong Province. *Memoirs of Beijing Natural History Museum* **55**: 55–62.

171 Liu, Q., Khramov, A.V., and Zhang, H.C. (2015). A new species of *Kalligramma* Walther, 1904 (Insecta, Neuroptera, Kalligrammatidae) from the Middle–Upper Jurassic of Daohugou, Inner Mongolia, China. *Alcheringa: an Australasian Journal of Palaeontology* **39** (3): 438–442. https://doi.org/10.1080/03115518.2015.1021564.

172 Wang, W.L. (1980). Mesozoic and Cenozoic. In: *Paleontological Atlas of Northeast China*, vol. **2** (ed. Shenyang Institute of Geology and Mineral Resources), 1–403. Beijing: Geological Publishing House.

173 Lin, Q.B. (1994). Cretaceous insects of China. *Cretaceous Research* **15** (3): 305–316. https://doi.org/10.1006/cres.1994.1019.

174 Hong, Y.C. (1982). *Mesozoic Fossil Insects of Jiuquan Basin in Gansu Province*, 155–156. Beijing: Geological Publishing House.

21

Coleoptera – Beetles

Yali Yu[1,2], *Zhenhua Liu*[1,2], *Chungkun Shih*[2,3], *and Dong Ren*[2]

[1] *Sun Yat-sen University, Guangzhou, Guangdong, China*
[2] *Capital Normal University, Haidian District, Beijing, China*
[3] *National Museum of Natural History, Smithsonian Institution, Washington, DC, USA*

21.1 Introduction to Coleoptera

Beetles are the most diverse group of organisms on Earth with about 420 000 described species from around the world. Statistically, one out of every five species of plants and animals is a beetle. An interesting observation by Dr. John B.S. Haldane, an English biologist, is "God has an inordinate fondness for stars and beetles."

The body sizes of beetles vary significantly, the smallest species, *Scydosella musawasensis* Hall, 1999 belonging to Ptiliidae, is about 0.33 mm in length [1]; while *Titanus giganteus* (Linnaeus, 1771) from South America and *Xixuthrus heros* Heer, 1868 from Fiji, both in Prioninae of Cerambycidae, can reach body lengths of 200 mm. Also, beetles are highly diverse in their different colors, forms and behaviors. However, they can be distinguished from other insect Orders by their chewing mouthparts (see Figure 3.2), but with some exceptional siphonate mouthparts (see Figure 3.13), holometabolous development and hardened forewings called elytra, which are the origin of their Order name Coleoptera, which means "sheath-winged" (Figure 21.1).

Similar to other insects, beetles have three main body parts: head, thorax and abdomen. In most beetles when at rest, the abdomens on the dorsal side are not observable due to coverage by elytra, while the membranous flight hind wings are folded under the elytra. The elytra protect the hind wings and internal organs in abdomen, help stabilize beetles in flight, and conserve precious bodily fluids. Some beetles, including most of the Staphylinidae (Figure 21.2a), Histeridae, Nitidulidae and some Melyridae, Cerambycidae and Curculionidae, have shortened elytra, with apical tergites of abdomen more sclerotized than those with complete elytra. Beetles usually have 11-segmented antennae, but may have fewer segments, which can be filiform, moniliform, capitate, clavate, lamellate, geniculate, serrate or pectinate. For weevils, in the families of Curculionidae, Dryophthoridae (Figure 21.2b), Brentidae and Nemonychidae, elbowed and non-elbowed antennae are inserted to the rostrum at various positions. Secondary sexual characters may present on mandibles, head, antennae (Figure 21.2c), pronotum, elytra or legs. One of the popular beetles, *Trypoxylus (Allomyrina) dichotomus* (Linnaeus, 1771) ("Japanese rhinoceros beetle", in Dynastinae, Scarabaeidae), has long and bifurcate projections on head and pronotum (Figure 21.2d) for the males, used to fight with other males for access to the females. On account of the combative and tenacious nature of this beetle, the Japanese made the samurai helmet based on the head of this rhinoceros beetle. Beetles, living in highly diverse habitats, have different types of legs to adapt to different environments and life styles, with the capabilities of swimming, digging, jumping, grasping or climbing.

Coleoptera are divided into four suborders: Archostemata, Myxophaga, Adephaga and Polyphaga. Although the four suborders are well-supported by phylogenetic analyses, the relationships among these suborders are still uncertain. The traditional relationship of Archostemata (Adephaga [Myxophaga + Polyphaga]) was proposed by Crowson [2] and supported by both morphological and molecular evidence [3, 4]. The phylogenetic analysis, based on 516 adult and larva characters by Lawrence *et al.* [5], revealed the systematic relationship as (Archostemata + Adephaga) + (Myxophaga + Polyphaga), which has not received significant support. For now, the most popular relationship was proposed by Kukalová-Peck and Lawrence [6, 7] as Polyphaga (Archostemata [Myxophaga + Adephaga]) based on hind wings alone (Figure 21.3). Of course, some other results have also been proposed by different research teams based on different data,

Rhythms of Insect Evolution: Evidence from the Jurassic and Cretaceous in Northern China, First Edition.
Edited by Dong Ren, Chungkun Shih, Taiping Gao, Yongjie Wang, and Yunzhi Yao.

Figure 21.1 A beetle ready to fly with opened elytra and extended hind wings. Source: Photo by Jason Shih.

for examplec Archostemata + Myxophaga + (Adephaga+ Polyphaga) [8], (Archostemata + Myxophaga) (Adephaga + Polyphaga) [9, 10], (Myxophaga + Adephaga) (Archostemata + Polyphaga) [11, 12] and Polyphaga (Adephaga [Archostemata + Myxophaga]) [13, 14], making the phylogenetic relationships still in contention.

Archostemata are the smallest suborder in the Coleoptera, with five extant and several extinct families, among which Jurodidae are doubtful with the only extant species *Sikhotealinia zhiltzovae* Lafer, 1996 described based on a single specimen from Russian Far East [15]. This suborder are usually treated as the basal group of beetles, mainly because of their resemblance to the oldest beetle fossils from the Permian and the Early Mesozoic as they all have clathrate and lattice-like elytral punctuation. Archostemata can be easily distinguished from the other three suborders by: exposed metatrochantin; the mesocoxal cavity broadly in contact with the metanepisternum (vs. never observed to this extent in other beetle suborders) and the prothorax with large and external

Figure 21.2 (a) *Stenus* spp. (Staphylinidae); (b) *Cyrtotrachelus thompsoni* Alono-Zarazaga & Lyal, 1999 (Dryophthoridae); (c) *Agapanthia amurensis* Kraatz, 1879 (Cerambycidae); (d) *Trypoxylus (Allomyrina) dichotomus* (Linnaeus, 1771) (Scarabaeidae). Source: Photos a and c by Zhenhua Liu, b and d by Jason Shih.

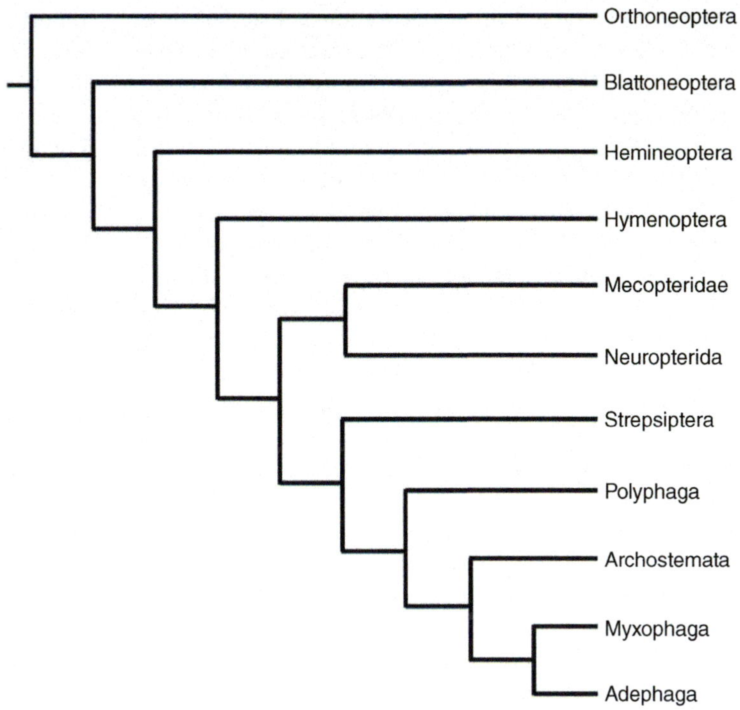

Figure 21.3 Phylogenetic relationships: the single most parsimonious tree based on 63 hind wing characters (tree length = 99, CI & RI = 0.96). Source: Modified from [7].

propleuron (similar to Adephaga, but different from Myxophaga and Polyphaga). Ommatidae and Cupedidae comprise most of the species of Archostemata with typical clathrate and lattice-like elytra which are usually covered by scales. While Crowsoniellidae and Micromalthidae, each with a single species *Crowsoniella relicta* Pace, 1975 and *Micromalthus debilis* LeConte, 1878 respectively, were once suggested to be in Polyphaga for their unique body shape or lacking some typical characters of Archostemata, but finally have been included in Archostemata for characters of the thorax, male genitalia and larvae.

Myxophaga contain small to tiny semiaquatic beetles with four small families: Lepiceridae, Torridincolidae, Hydroscaphidae and Sphaeriusidae. The propleuron of Myxophaga is well-developed and completely separating the notum and sternum or only extending behind the coxa, and notum and sternum are broadly joined anteriorly. Trochantin is apparently fused with the pleuron, forming the trochantin-pleuron as in Polyphaga.

Adephaga comprise 10 extant families: Gyrinidae, Haliplidae, Meruidae, Noteridae, Aspidytidae, Amphizoidae, Hygrobiidae, Dytiscidae, Trachypachidae, and Carabidae. However, some subfamilies in Carabidae have been treated as independent families sometimes. The propleuron in Adephaga is also largely external, but not extending to the anterior edge of the prothorax, resulting in narrow connection of notum and sternum. Adephaga can be separated from other suborders by sternite II divided by the metacoxae. They also have some synapomorphies, e.g. maxilla 2-segmented, palpiform galea; mentum with distinct lateral lobes enclosing the prementum; presence of ventral procoxal condyle; restriction in the mobility of the metacoxae and fusion of the abdominal sternites II–V.

Polyphaga contain most of the families and species of beetles with seven major infraorders, Staphyliniformia, Scarabaeiformia, Scirtiformia, Elateriformia, Derodontiformia, Bostrichiformia, and Cucujiformia. While positions of some taxa are still doubtful, some subfamilies of Hydrophilidae, Ptinidae, Melyridae, Erotylidae, Chrysomelidae, Curculionidae, etc. were once treated as families. There are still uncertainties due to different researchers currently using different systems. Derodontidae and Scirtoidea from different infraorders together were suggested to be a basal split from the remaining Polyphaga by Bocak *et al.* [4] based on molecular data. This suborder can be diagnosed by the propleuron being highly reduced and fused with trochantin, and it is internalized which makes it invisible on the prothorax. Another important character to separate Polyphaga from the other suborders is the presence of cervical sclerites which connect the head with the prothorax. Polyphaga include most of the beetles, among which some groups have important artistic or economic values. Apart from the *T. dichotomus* mentioned before, the Sacred Scarabs (*Scarabaeus sacer* Linnaeus, 1758) have been highly valued for their important religious meaning by ancient Egyptians, and people now use beautiful beetles of the Buprestidae and Elateridae to make necklaces and other decorations. Also, larvae of some Curculionidae, Scarabaeidae and Cerambycidae are traditional food in some regions, while adults of the giant water scavenger beetle (*Hydophilus triangularis* Say, 1823) are popular snacks in southern China. Also, some groups of Polyphaga are serious pests of crops and forest, causing huge economic losses. For example, *Monochamus alternatus* Hope, 1842 (Cerambycidae) is the major pests for pine forests in Asia as it's an efficient vector of pine wood nematode, which can cause pine wilt disease [16, 17]. *Lasioderma serricorne* (Fabricius, 1792) (Ptinidae, Anobiinae), known as the cigarette beetle, is one of the major pests for stored products such as tobacco, tea, beans, etc. all over the world [18, 19]. Many other beetles of Tenebrionidae, Chrysomelidae and Curculionidae cause many problems to stored grains or forests [20]. In contrast, some of the beetles are used as predators of other pest insects in biocontrol. *Cryptolaemu montrouzieri* (Mulsant, 1850), (Coccinellidae) from Australia is a famous predator of mealybugs and is widely used as biocontrol agent in many countries [21, 22]. *Tenebrio molitor* L., 1758 which are commonly used as animal and bird food and insect protein, is reported to biodegrade polystyrene and convert it to CO_2 by up to 48% within 12–14 hours in mealworm gut [23]. Furthermore, because of their large size, bright colors and huge mandibles or horns, some species of Lucanidae and Scarabaeidae are popular pets for insect hobbyists.

A key reason for beetles to be so diverse and speciose on earth is their capabilities of adaptation, not only by their various body structures enabling them to live in diverse habitats, but also in their wide range of feeding behaviors to gain advantages from the ecosystems, namely: mycophagy, phytophagy, saprophagy, predation, paratisism, and symbiosis.

Mycophagy is quite common for beetles like members of Staphylinidae, Lymexylidae, Erotylidae, Silvanidae (Figure 21.4a), Tenebrionidae, Curculionidae, Endomychidae, etc. Adults and larvae feed on spores, hymenia or other parts of fungi. Some groups like Lymexylidae and Scolytinae (Curulionidae) even have modified structures to carry the spores and raise the fungi to feed the larvae [24]. Phytophagous beetles are less populous compared to mycophagous ones; most of the beetles feeding on leaves belong to Chrysomeloidea and Curculionoidea.

Figure 21.4 (a) *Macrohyliota militaris* (Erichson, 1842) (Silvanidae); (b) *Henosepilachna vigintioctopunctata* (Fabricius, 1775) (Coccinellidae); (c) *Idgia* sp. (Prionoceridae); (d) *Tranes* spp. (Curculionidae). Source: Photos by Zhenhua Liu.

Some groups of Scarabaeidae, Coccinellidae, etc. are also phytophagous, among which some are even agricultural pests like *Henosepilachna vigintioctopunctata* (Fabricius, 1775) (Coccinellidae) (Figure 21.4b). In addition, many beetles feed on pollen while serving as pollinators, such as Lycidae, Elateriae, Prionoceridae (Figure 21.4c), Boganiidae, Erotylidae, Cerambycidae, Curculionidae etc., among which some members of Boganiidae, Erotylidae and Curulionidae (Figure 21.4d) are major pollinators of cycads [25–27].

In addition, there are some saprophagous beetles feeding on dead or decaying organic matter. For example, larvae of Lucanidae and some Scarabaeidae feed on rotten wood; adults of some Lucanidae and Helotidae feed on the sap of trees (Figure 21.5a). The most famous members of saprophagous beetles are dung beetles (Figure 21.5b), which serve as very important decomposers in the ecosystem. Also, beetles such as Silphidae and Dermestidae feeding on dead animals are also sapophagous. Furthermore, there are truly carnivorous beetles which can be predators or parasites. Many groups of Adephaga are terricolous or aquatic predators such as ground beetles (Carabidae) and diving beetles (Dytiscidae). For example, some species of *Carabus*, such as *Carabus ignimetalla* Bates, 1888, can feed on snails (Figure 21.5c). Also, there are other beetle predators like some Histeridae, Staphylinidae, etc. Parasitic behavior is not common for beetles compared to flies and wasps, but some of the parasitic beetles can be important biocontrol agents. For example, *Dastarcus helophroides* (Fairmaire, 1881) of Bothrderidae is very useful for controlling some Cerabycidae like *Apriona swainsoni* (Hope, 1840) [28]. Platypsyllinae of Leiodidae comprise some highly modified members which are ectoparasites of some rodents or insectivores [29]. Other famous parasitic beetles in the Rhipiceridae are known as "cicada parasite beetles" [30, 31] or others parasitizing different groups of Hymenoptera [32–34]. The most advanced parasitic behaviors are the social parasites, which occur in some groups of Paussinae (Carabidae), Pselaphinae (Staphylinidae) (Figure 21.5d) and Dermestidae.

Figure 21.5 (a) *Aegus dispar* Didier, 1931 (Lucanidae) feeding on dead or decaying organic matter; (b) *Onthophagus gazella* (Scarabaeidae, Scarabaeinae) as a dung recycler at the Savannah Oaks Ranch, Texas; (c) A predatory *Carabus ignimetalla* Bates, 1888 (Carabidae) feeding on snails; (d) Pselaphinae (Staphylinidae) as a social parasites. Source: Photos a, c and d by Zhenhua Liu, b by Dr. Chungkun Shih.

21.2 Progress in the Studies of Fossil Coleoptera

As one of the oldest group of holometabolous insects, fossil beetles have attracted a lot of interest from researchers, among whom A.G. Ponomarenko, A.G. Kirejtshuk, and R.A. Crowson have contributed significantly to the studies of fossil beetles.

Generally, beetles were thought to have had common ancestors with Megaloptera (see Chapter 18) and separated from them in the Early Permian [35, 36]. The oldest fossil of Coleoptera, *Coleopsis archaica* Kirejtshuk, Poschmann & Nel, 2014, was described from the earliest Permian of Germany [37]. This fossil beetle together with other Early Permian beetles traditionally belong to Tshekardocoleidae, which were suggested to be Protocoleoptera by Crowson [36], show some similar characters as extant Archostemata. However, they are distinctly differentiated from all the extant beetles based on their characters such as the 13-segmented antenna, forewings with well-defined veins and elytra extending beyond apex of abdomen. The Early Permian family Oborocoleidae described by Kukalová [38] with two species and the Cretaceous family Labradorocoleidae described by Ponomarenko [39] with only one species were both included in the superfamily of Tshekardocoleoidea by Bouchard *et al.* [40], which may need to be confirmed in the future. Interestingly, the Cretaceous family Umenocoleidae Chen and Tan, 1973, which were first thought to be beetles closely related to Tshekardocoleidae, was transferred to Protelytroptera [41] or even Blattodea [42]. Then, Kirejtshuk *et al.* [37] suggested it as a sister group of all other Coleoptera based on the venation. We place Umenocoleidae in Blattaria in Chapter 7. The family Moravocoleidae described by Kukalová-Peck and Beutel [43] for some described genera were included in Protocoleoptera by them, but was synonymized with Tshekardocoleidae later by Kirejtshuk *et al.* [37]. Kirejtshuk and Nel [44] separated *Coleopsis archaica* from Tshekardocoleidae and erected a new family Coleopsidae. Fossils of Tshekardocoleoidea were first found in the Early Permian in Russian Federation [45–47] and former Czechoslovakia [38], but with an unusual Middle Jurassic *Dictycoleus jurassicus* Hong, 1982 from Subei County of China, which may need further revision. There are also some Early Permian fossil beetles of *Permocoleus* Lubkin & Engel, 2005 from Wellington Formation of North America [48, 49], but these have not been included in any family. In the Late Permian, some more-developed fossil beetles were collected from all over the world, which have characters more similar to extant beetles, e.g. more reduced veins on forewings and 11-segmented antennae, while at least some of them still have 13-sgmented antennae like Permocupedidae. Crowson [35, 36] introduced a new suborder Archecoleoptera formerly to include these Late Permian beetles of Asiocoleidae, Schizocoleidae, Schizophoridae, Tladycupedidae, Permosynidae, Permocupedidae and Rhombocoleidae, while members of Permocupedidae were also described from the Early Permian of Brazil later [50]. Some families first described from the Triassic were also found from the Late Permian such as Triaplidae and Ademosynidae [51, 52]. Furthermore, Rhombocoleidae were included in the suborder Archostemata and Triaplidae were included in the suborder Adephaga by Carpenter [41], which were accepted by researchers later. Thus Archecoleoptera are apparently paraphyletic and together with Protocoleoptera were suggested to be invalid names [37]. More interestingly, members of Trachypachidae were described from Upper Permian Kedrovka beds [53], making them the earliest known members of an extant beetle family. Also, both Asiocoleoidea and Shizophoroidea were included in the suborder Myxophaga [40], and genera *Polysitum* Dunstan, 1923 and *Hydrobiites* Heer, 1865 were transferred to Polyphaga [54], within which *Polysitum kuznetskiense* Rohdendorf, 1961, *Hydrobiites tillyardi* Ponomarenko, 2011 and *Hydrobiites vladimiri* Ponomarenko, 2011 were described from the Late Permian of Russian Federation. Thus, Archostemata, Myxophga, Adephaga and Polyphaga all existed at least in the Late Permian, which makes the establishment time of four extant suborders much older than that suggested by Crowson [36].

Archostemata have been traditionally treated as the oldest group of the extant four suborders [2], mainly due to the limited and poorly recognized fossil record of other suborders and similarities of extant Archostemata with the Permian fossils. The basal position of Archostemata was also postulated by some phylogenetic researches [3, 4]. Other studies based on morphological and molecular characters [6, 7, 14, 55] have revealed that Polyphaga is probably the most basal lineage, and Derodontidae combined with Scirtidae and Eucinetidae form basal clades of Polyphaga. The recent fossil records show earlier origination of Polyphaga than generally accepted, but the relationships among the other three suborders are still in contention. The existence of four extant suborders in the Permian also provide more fossil evidence. A recently described fossilized wood of the conifer *Ningxiaites specialis* with unique insect borings from the Upper Permian (Changhsingian stage, *ca.* 254–252 Mya) attributed to extinct lineage of Polyphaga [56], seem to confirm that Polyphaga were very advanced at that stage, but did not survive the Permian-Triassic extinction. Thus, more fossil evidence and more studies are needed to address and elucidate which suborder is the most basal one.

The biota on earth experienced a severe ecological crisis at the end of Permian, also known as "Great Extinction", with up 90% of the species dying out during this extinction event. This "Largest Extinction of History" is not only the boundary between Permian and Triassic, but also the beginning of Mesozoic era. As a result of the great extinction, the records of fossil beetles during the Early Triassic were scarce, with only some beetles of Asiocoleidae, Schizocoleidae, Permosynidae and Coptoclavidae described from Russia and Angola [57–60], which might have been all aquatic [61]. No characters of Tshekardocoleidae have been found since the Early Triassic, while almost all later fossils had closely-fitting regularly striate elytra and 11-segmented antennae. Compared to the low proportion of Coleoptera during the Permian, beetles were much more abundant during the Triassic and found all over the world except for Antarctica. Crowson [36] suggested that the establishment of four extant suborders should have been achieved before the end of the Triassic. Real Archostemata-like members of the extant family Cupedidae were first found from the Triassic of Kyrgyzstan, Russia, Australia and South Africa [47, 62–64]. Also, some of the Archostemata described by Ponomarenko [47], e.g. *Hadeocoleus* Ponomarenko, 1969, were suggested to be Adephaga and some Ademosynidae described by Ponomarenko [47] were also suggested to be Polyphaga [36], mainly depending on the very short prosternum and reflexible head. Then, Crowson also suggested that the Coelocatiniidae might be the precursors of Myxophaga for its potential aquatic habits, and Myxophaga should have also appeared by the Triassic as the other three suborders were all recognized during this era. As mentioned before, members of all four extant suborders have been all recognized from the Permian recently. It seems that the differentiation and origination of these four extant suborders should have been much older, at least by the mid Permian or even earlier. What's more, even the major series within Adephaga should have been developed before the end of the Permian as one species of *Triaplus* Ponomarenko has been described from the Permian of Russia [51]; this was suggested to represent the ancestral stock of Hydradephaga [65]. All of these estimates are in contrast to the dating work by Toussaint *et al.* in 2017, and even the establishment time of some superfamilies of Polyphaga, e.g. Scirtoidea and Staphylinoidea, can be dated back to the Permian [66]. Evidence of the establishment of major series within the Polyphaga were still only found from the Triassic, as members of certain groups of Staphylinidae [67] and Elateridae [68–70] were described from the Triassic.

Fossil beetles from the Jurassic are much more diverse compared to those from the Triassic, especially for Polyphaga, while Archostemata are still quite common during the Jurassic. Distinct major series divisions are shown for this suborder; almost all the superfamilies should have occurred before the end of the Jurassic. Members of Polyphaga families, e.g. Liadytidae [65, 71], Hydrophilidae [71–73], Leiodidae [74], Hydraenidae [71, 75], Silphidae [76], Agyrtidae [71, 77], Staphylinidae [78, 79], Glaresidae [80, 81], Geotrupidae [82], Jurodidae [75, 83, 84], Lucanidae [85], Hybosoridae [86–88], Ochodaeidae [89], Mesocinetidae [90], Buprestidae [75, 91], Byrrhidae [80, 91], Eulichadidae [92–95], Lasiosynidae [96, 97], Artematopodidae [98], Cerophytidae [99, 100], Elateridae [101, 102], Eucnemidae [100], Derodontidae [103], Dermestidae [104], Trogossitidae [91, 92, 105, 106], Cleridae [107], Thanerocleridae [107], Prionoceridae [108], Paradrexidae [92, 93, 109, 110], Monotomidae [111], Nitidulidae [80, 91, 92], Mycetophagidae [91, 112], Mordellidae [113], Ripiphoridae [114], Tenebrionidae [80, 115, 116], Chrysomelidae [92, 117, 118], Nemonychidae [119–121], Anthribidae [122–124], Ithyceridae [124–126] have been documented from the Jurassic of Mongolia, Russia, China, Switzerland, Germany, Kazakhstan; most of them are extant families. Cupedidae and Ommatidae of Archostemata have also been described from the Jurassic, even comprising some extant genera like *Tetraphalerus* Waterhouse [47, 127–129], *Omma* Newman [129–132]. Adephaga families Triaplidae [93], Coptoclavidae [71, 72, 133], Gyrinidae [80, 91, 134], Dytiscidae [41, 46, 135, 136], Carabidae [75, 91, 137], and Trachypachidae [71, 93, 138] also show a long existence from the Jurassic until now. Some old groups which did not belong to the four extant suborders were also extended to the Jurassic, such as Asiocoleidae [139, 140], Schizocoleidae [75, 141], Schizophoridae [47, 92, 130, 142, 143], Coelocatiniidae [130], Taldycupedidae [93, 144, 145], Ademosynidae [47], Permosynidae [54, 75, 146] and Tshekardocoleidae mentioned before. More interestingly, some advanced biological habits of beetles were first recorded from the Jurassic. Crowson [36] suspected that the Cycadeoidea of the Jurassic should be pollinated by insects, most possibly by Boganiidae which was not described at that time, but has been reported recently by Liu *et al.* [147].

The beetles of the Early Cretaceous were similar to those of the Jurassic [61], but the diversity of Coleoptera became much broader after the Early Cretaceous, especially for Polyphaga. More than half of the extant families have been recorded by the Early Cretaceous with some extinct groups, and the fauna of beetles became more similar to recent species compared with the Jurassic beetles. For example, the reduction of the Archostemata and the increase of Polyphaga, the abundance of Carabidae rather than Trachypachidae. Furthermore, almost none of the extinct families of Archostemata, Myxophaga, Adephaga and those which haven't been included in any suborders, extended to the later part of Cretaceous, except for a species in the family Tritarsidae described from the Eocene of China [148]. *Hydroscapha jeholensis* Cai, Short & Huang 2012, the first example of extant family of the Myxophaga was described from the Lower Cretaceous Yixian Formation of China [149]. Some other particular changes in beetle fauna during the Cretaceous should be the abundant Staphylinidae and the relatively diverse Scarabaeoidea. Crowson [35, 36] suggested that the sudden rise of true birds, the appearance of Teleostean fish in fresh water and the rapid rise of Hymenoptera Parasitica were factors which had an important influence on beetles during the Cretaceous. Chrysomelidae, Cerambycidae and Cuculionidae, which are quite abundant nowadays, were still not common during the Cretaceous. What makes the Cretaceous fossils more interesting and important is that abundant fauna and flora were well-preserved in amber during this period, which can provide more information to support taxonomic and phylogenetic studies. Cretaceous amber have been found from different parts of the world such as the Early Cretaceous Lebanese and Spanish amber, the mid-Cretaceous Myanmar (Burmese) amber and Late Cretaceous New Jersey amber. Particularly, the mid-Cretaceous Myanmar amber have become a real research treasure recently for its rich biodiversity, not only for insects, but also including many plants, reptiles and even birds [150–152]. To date, 33 families of Coleoptera have been recorded from the Myanmar amber [153], while many more species, genera and families are still being studied and described. Similar to the Jurassic, some highly developed behaviors were reported from the Cretaceous, such as the termitophilous rove beetles [154, 155] and the parental care of Silphidae [76].

Cenozoic fossil beetles have been found all over the world and they were more and more similar to the extant groups. Similar to the Cretaceous, amber inclusions with abundant fauna and flora were also found in many localities, among which the Eocene Baltic amber is the most famous one, containing a great diversity of well-preserved creatures. Families of Chrysomelidae and Cerambycidae became more common during the Eocene, and nearly half of the genera are extant. In the Miocene Dominican amber, almost all the species belong to the extant genera and even some of the species still exist now.

In China, the studies of fossil beetles started much later than the western countries. In 1928, C. Ping [156] made the earliest contributions by working with the Cretaceous beetles. Until 1970s, Y.C. Hong and Q.B. Lin described many fossil beetles of different groups from the Permian to the Cenozoic [77, 93, 137, 141, 157–172]. H.C. Zhang and W.L. Wang continued to work on fossil beetles of China [101, 173–177]. After 2000, the studies of Chinese fossil beetles picked up speed, with most of the work done by the teams of Capital Normal University (CNU) and the Nanjing Institute of Geology and Palaeontology, Chinese Academy of Sciences (NIGPAS), in collaboration with Ponomarenko, Kirejtshuk, Jarzembowski, Slipinski, etc. In CNU, the studies of J.J. Tan, C.K. Shih, and D. Ren mainly focused on the Mesozoic Archostemata [96, 128, 129, 139, 143, 178–187], while the teams of H.L. Chang and D. Ren concentrated on the Elateriformia, e.g. Elateridae and Lasiosynidae [97, 188–193]. The work by M. Bai and D. Ren has contributed significantly to knowledge of the fossil Scarabaeoidea [194–203]. The teams of D.-Y. Huang in NIGPAS continued important research on taxonomy of various Mesozoic beetles [98, 103, 111, 132, 149, 155, 204–219] and biology of fossil beetles [76]. Also, B. Wang [116, 133, 138, 220–222], M. Liu [223–228], Y.L. Yue [229–234], Y.L. Yu [106, 235–239], Z.H. Liu [108, 147], Y. Hsiao [114], C.S. Deng [104] have made contributions to knowledge of fossil beetles from China. Also, E.V. Yan contributed important papers on fossil beetles from the Permian and Triassic in China, Russia and Australia [52, 140].

21.3 Representative Fossils of Coleoptera from Northern China

Suborder Archostemata Kolbe, 1908

Family Ademosynidae Ponomarenko, 1968

Ademosynidae were first described by Ponomarenko [130] from the Jurassic of Karatau, Kazakhstan, and later on many other species and genera have been reported on all the continents except Antarctica from the Permian to the Early Cretaceous, including 10 genera now. This family can be recognized mainly by the presence of prosternal sutures independently reaching into the anterior margin of the prosternum, metepisternum reaching the mesocoxal cavity, trochantinal suture of metaventrite and elytra without schiza and with puncture grooves.

Only one genus included from the Cretaceous of Northern China: *Atalosyne* Ren, Lu, Ji & Guo, 1995.

Atalosyne Ren, Lu, Ji & Guo, 1995

Atalosyne Ren, Lu, Ji & Guo, 1995, Fauna and Stratigraphy of Jurassic-Cretaceous in Beijing and the Adjacent Areas, 83 [240] (original designation).

Type species: *Atalosyne sinuolata* Ren, Lu, Ji & Guo, 1995.

Head wider than length, labrum broad; pronotum wide, with anterior angles acute and distinctly projecting;

prosternum slender, projecting between procoxae; metaventrite wider than length, discrimen distinct; elytra not extending over the apex of abdomen, with 10 rows of striae; abdominal ventrites with the first segment longest.

Distribution and age: Beijing; Early Cretaceous.

Only one species included from the Cretaceous of Northern China (see Table 21.1).

Family Asiocoleidae Rohdendorf, 1961

Asiocoleidae are a distinct beetle family including 11 genera described from the Permian to the Jurassic of China, Russia and Mongolia. The family Tricoleidae erected by Ponomarenko [47] has been transferred to Asiocoleidae. The Asiocoleidae are thought to be evolved from tshekardocoleids at early stages of their evolution [241], and can be characterized by the elytra having only two or three longitudinal veins or ribs with the spaces between them occupied by numerous rounded cells.

Only one genus included from the Jurassic of Northern China: *Loculitricoleus* Tan & Ren, 2009.

Loculitricoleus Tan & Ren, 2009

Loculitricoleus Tan & Ren, 2009, Mesozoic Archostematan Fauna from China, 144 [139] (original designation).

Type species: *Loculitricoleus tenuatus* Tan & Ren, 2009.

Head transverse, nearly rectangular, dorsal side with two indistinct protuberances, epicranial suture Y-shaped; mandibles medium-sized with basal area broaden, inner side with one tooth; pronotum with protuberances on dorsal side, posterior margin long and straight; elytra with three main veins, Rs, M + Cu vein and 2A are apparently different from the intermediate veins, dorsal side with more than 14 rows of cells which are small and rectangular, without black punctures around; abdominal ventrites with the first segment longest.

Distribution and age: Inner Mongolia of China, Middle Jurassic; Mongolia, Late Jurassic.

Two species included from the Jurassic of Northern China (see Table 21.1).

Family Cupedidae Laporte, 1836

The Cupedidae are the largest archostematan family. Cupedids are elongate, slender, parallel-sided, mostly distinctly flattened; body covered with scales. Head prognathous, dorsal surface of head with one or two pairs of macroscopic tubercles. Antenna inserted between and in front of eyes, 11-segmented, filiform. Prothorax mostly with tarsal grooves on ventral side of prothorax. Pronotum widest in anterior half, with acute anterior corners. Procoxal cavities separated. Elytra with 9–10 rows of window punctures. Tarsal formula 5-5-5; claws simple. Abdomen with five overlapping ventrites.

Genera included from the Jurassic and Cretaceous of Northern China: *Ensicupes* Hong, 1976, *Anthocoleus* Hong, 1983, *Celocoleus* Hong, 1983, *Euteticoleus* Hong, 1983, *Tetrocupes* Hong, 1983, *Hebeicupes* Zhang, 1986, *Longaevicupes* Ren, Lu, Ji & Guo, 1995, *Latocupes* Ren & Tan, 2006, *Furcicupes* Tan & Ren, 2006, *Gracilicupes* Tan, Ren & Shih, 2006, *Apriacma* Kirejtshuk, Nel & Kirejtshuk, 2016 and *Cretomerga* Kirejtshuk, Nel & Kirejtshuk, 2016.

Ensicupes Hong, 1976

Ensicupes Hong, 1976, Inner Mongolia, Volume 2, In: Palaeontological Atlas of the North China Region, 81–87 [157] (original designation).

Type species: *Ensicupes guyanensis* Hong, 1976.

Polygonal cells arranged into ambiguous and long longitudinal rows and weakly raised primary veins with the fusion of A1 and CuA.

Distribution and age: Inner Mongolia and Jilin; Early Cretaceous.

Two species included from the Cretaceous of Northern China (see Table 21.1).

Anthocoleus Hong, 1983

Anthocoleus Hong, 1983, Middle Jurassic Fossil Insects in North China, 85 [93] (original designation).

Type species: *Anthocoleus hebeiensis* Hong, 1983.

Elytron narrow with about 10 long and clearly longitudinal rows of irregular large cells and un-raised primary veins.

Distribution and age: Hebei; Middle Jurassic.

Only one species included from the Jurassic of Northern China (see Table 21.1).

Celocoleus Hong, 1983

Celocoleus Hong, 1983, Middle Jurassic Fossil Insects in North China, 87 [93] (original designation).

Type species: *Celocoleus densus* Hong, 1983.

Elytron narrow (about 11 mm long) with more than 10 long longitudinal rows of irregular large cells and un-raised primary veins.

Distribution and age: Hebei; Middle Jurassic.

Only one species included from the Jurassic of Northern China (see Table 21.1).

Euteticoleus Hong, 1983

Euteticoleus Hong, 1983, Middle Jurassic Fossil Insects in North China, 86 [93] (original designation).

Type species: *Euteticoleus radiatus* Hong, 1983.

Elytron comparatively large and narrow, with about 10 long longitudinal rows of irregular large cells and un-raised primary veins.

Distribution and age: Hebei; Middle Jurassic.

Only one species included from the Jurassic of Northern China (see Table 21.1).

Tetrocupes Hong, 1983

Tetrocupes Hong, 1983, Middle Jurassic Fossil Insects in North China, 84 [93] (original designation).

Type species: *Tetrocupes cavernasus* Hong, 1983.

Elytra length about 15 mm, slender, three to four times as long as wide; R, Rs, M and Cu longitudinal, straight and parallel, but absent posteriorly; with five to six rows of longitudinal cavities between veins, chaotic apically.

Distribution and age: Hebei; Middle Jurassic.

Only one species included from the Jurassic of Northern China (see Table 21.1).

Hebeicupes Zhang, 1986

Hebeicupes Zhang, 1986, Some fossil insects from the Jurassic of northern Hebei, China, In: The Paleontology and Stratigraphy of Shandong, Haiyang Publishing House, Beijing, 74–84 [118] (original designation).

Type species: *Hebeicupes formidabilis* Zhang, 1986.

Body robust, with the rather wide pronotum and the short elytra. Cells oval arranged into ambiguous and long longitudinal rows but clearly raised veins absent.

Distribution and age: Hebei; Middle Jurassic.

Only one species included from the Jurassic of Northern China (see Table 21.1).

Longaevicupes Ren, Lu, Ji & Guo, 1995

Longaevicupes Ren, Lu, Ji & Guo, 1995, Fauna and Stratigraphy of Jurassic-Cretaceous in Beijing and the Adjacent Areas, 82 [240] (original designation).

Type species: *Longaevicupes macilentus* Ren, Lu, Ji & Guo, 1995.

The elytron with nine long longitudinal rows of cells, the pterothorax and the abdomen are similar to those of the slender cupedines, but unlike the slender ommatines. The weak primary veins are combined with the fusion of A1 and CuA at elytral apex, metepisterna very wide.

Distribution and age: Beijing; Early Cretaceous.

Only one species included from the Cretaceous of Northern China (see Table 21.1).

Latocupes Ren & Tan, 2006

Latocupes Ren & Tan, 2006, *Ann. Zool. (Warszawa)*, 56 (3), 457–464 [178] (original designation).

Type species: *Latocupes fortis* Ren & Tan, 2006.

Head with two pairs of obvious tubercles; antennae reaching beyond the base of the prothorax; scape thicker than other antennomeres, pedicel slightly shorter than others, the third antennomere about 1.4 times as long as the length of pedicel, following antennomeres subequal in length. Pronotum slightly pentagonal or rectangular, anterior angles moderately orthogonal; prosternum without tarsal groove; prosternal process only shortly extending behind coxae. Basal four tarsomeres of hind leg with bilobed and gradually deepened. Elytra covered with brown scales.

Distribution and age: Inner Mongolia, Middle Jurassic; Beijing and Liaoning, Early Cretaceous.

Six species included from the Jurassic and Cretaceous of Northern China (see Table 21.1).

Furcicupes Tan & Ren, 2006

Furcicupes Tan & Ren, 2006, *J. Nat. Hist.*, 40 (47–48), 2653–2661 [182] (original designation).

Type species: *Furcicupes raucus* Tan & Ren, 2006.

Head with clear Y-shaped depression and two pairs of obvious tubercles. Neck narrow. Antennae reaching beyond the base of prothorax, scape thicker than others, pedicel slightly shorter than other antennomeres. Pronotum rectangle; anterior angles of prosternum bifid, prosternum before the procoxae short, prosternum without tarsal groove; disc without elevations; prosternal process only shallowly extending behind coxae. Elytron with nine rows of cells. Tarsi 5-segmented, the first and last segments of foreleg equal in length, both of them longer than other tarsomeres, two to four tarsomeres short, equal in length.

Distribution and age: Liaoning; Early Cretaceous.

Only one species included from the Cretaceous of Northern China (see Table 21.1).

Gracilicupes Tan, Ren & Shih, 2006

Gracilicupes Tan, Ren & Shih, 2006, *Ann. Zool. (Warszawa)*, 56 (1), 1–6 [185] (original designation).

Type species: *Gracilicupes crassicruralis* Tan, Ren & Shih, 2006.

Head sub-rhombic, broadest at eyes, bearing two pairs of unconspicuous tubercles, without antennal groove ventrally; with the comparatively narrow neck, very large eyes, long antennae; antennae more than half as long as entire insect, the third antennomere shorter than the first and second segments combined in length; pronotum very narrow, nearly quadrate, as wide as head, angles rounded, with two large elevations, sides curved (markedly narrower than head), anterior and posterior angles not projecting; prosternum in front of procoxae very short; the elytra with 10 long longitudinal rows of cells, the well-raised primary veins different in expression from the secondary veins, and the steeply sloping sides.

Distribution and age: Inner Mongolia; Middle Jurassic.

Two species included from the Jurassic of Northern China (see Table 21.1).

Gracilicupes crassicruralis **Tan, Ren & Shih, 2006 (Figure 21.6)**
Gracilicupes crassicruralis Tan, Ren & Shih, 2006: *Ann. Zool. (Warszawa)*, 56 (1), 1–6 [185].

Locality and horizon: Daohugou, Ningcheng, Inner Mongolia, China; Middle Jurassic, Jiulongshan Formation.

Body medium-sized, slender, covered with tubercles. Antenna filiform and 11-segmented. Pronotum transverse, 0.6 times as long as wide, anterior and posterior edges straight, the sides slightly curved, without propleuron; procoxal cavities separated, circular, small. Elytra about 1.5 times as wide as prothorax, longitudinal ridges with small tubercles, four times as long as wide; elytral cells quadrate, with one to two black-maculate round its margin elongated in the distal part of the elytron, approximately 53 cells formed in row [185].

Apriacma **Kirejtshuk, Nel & Kirejtshuk, 2016**
Apriacma Kirejtshuk, Nel & Kirejtshuk, 2016, *Invertebrate Zoology*, 13 (2), 61–190 [242] (original designation).

Type species: *Priacma tuberculosa* Tan, Ren & Shih, 2006.

Head with somewhat expressed Y- or V-shaped depression, well-projecting temples; dorsal surface with weakly raised tubercles, antennal insertions moderately widely separated and apparently not covered by tubercles. Scape subequal in length with flagellomeres or slightly longer than pedicel and somewhat thicker. Pronotum subtrapezoid and rectilinearly widening anteriorly, with widely explanate sides, projecting and subacute anterior angles, usually tuberculate surface and weak median elevated stripe. Part of prosternum before procoxae markedly longer than procoxae. Elytron with nine long longitudinal rows and a short prescutellar row interspaced by weak secondary and more raised primary veins; A1 straight, almost reaching elytral apex and fusing with CuA before apex; lateral edges subrectiliner, subrounded to subacute apices and apparently steeply (subvertically) sloping sides.

Distribution and age: Liaoning; Early Cretaceous.

Four species included from the Jurassic and Cretaceous of Northern China (see Table 21.1).

Cretomerga **Kirejtshuk, Nel & Kirejtshuk, 2016**
Cretomerga Kirejtshuk, Nel & Kirejtshuk, 2016, *Invertebrate Zoology*, 13 (2), 61–190 [242] (original designation).

Type species: *Priacmopsis subtilis* Tan & Ren, 2006.

Head with expressed Y-shaped depression, well-projecting temples; dorsal surface with weakly raised tubercles, antennal insertions moderately widely separated and apparently uncovered by tubercles. Scape subequal in length with flagellomeres or much longer than pedicel. Pronotum subpentagonal and rectilinearly widened anteriad, with smooth disc and apparently

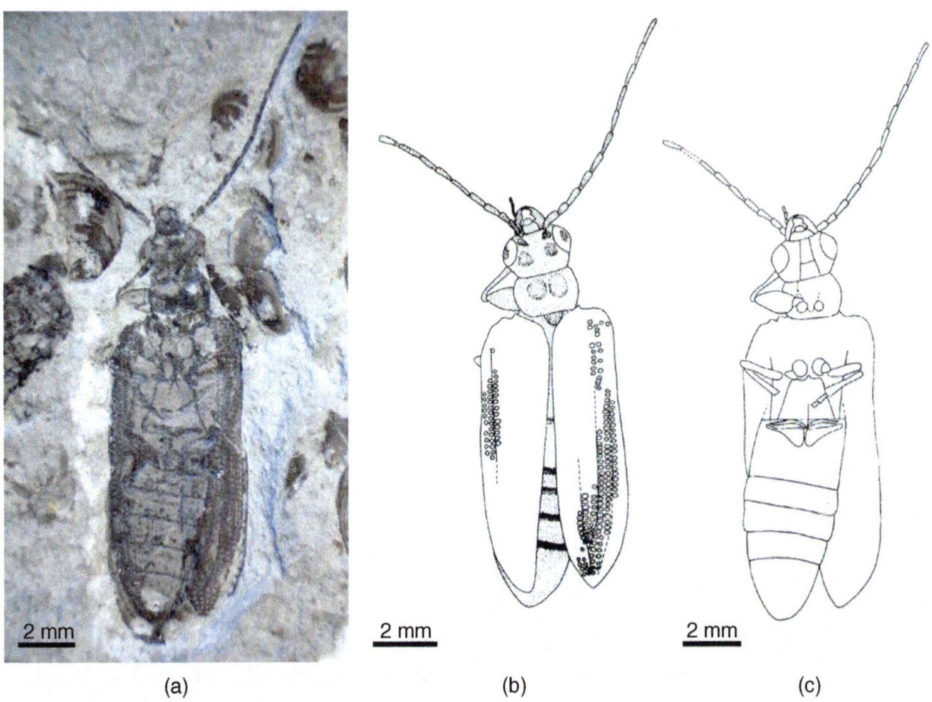

Figure 21.6 *Gracilicupes crassicruralis* Tan, Ren & Shih, 2006, (Holotype, CNU-COL-NN2005001). (a) Photograph; (b) Line drawing of dorsal view; (c) Line drawing of ventral view [185]. Source: Donated by Dr. Chungkun Shih.

widely explanate sides, arcuate anterior angles and strongly convex anterior edge. Part of prosternum before procoxae shorter than procoxae. Elytra with nine long longitudinal rows interspaced by weak secondary and well-raised primary veins; A1 slightly deviating at scutellum and almost completely rectilinearly reaching elytral apex, unclearly fusing with CuA and then with M before apex; broadly arcuate lateral edges, subrounded to subacute apices and apparently steeply (subvertically) sloping sides.

Distribution and age: Liaoning; Early Cretaceous.

Only one species included from the Cretaceous of Northern China (see Table 21.1).

Family Jurodidae Ponomarenko, 1985

Jurodids are amazing examples of "living fossils"; both extant and fossil jurodids are very rare. They are the "most mysterious representatives of beetles" [243] and "a phantom in beetle evolution" [244]. Jurodids have a mixture of morphological characters such as head slightly wider than long, dense pubescence; antennae inserted dorsally on the frons, directly behind the epistomal suture; antenna modified moniliform with 11 segments; pronotum distinctly narrower than elytra; prosternal process absent and abdomen with six visible ventrites. It may represent an important piece in the subordinal puzzle [84].

Only one genus included from the Jurassic of Northern China: *Jurodes* Ponomarenko, 1985.

Jurodes Ponomarenko, 1985

Jurodes Ponomarenko, 1985, Coleoptera, In: Rasnitsyn, A.P. (ed.), Jurassic Insects of Siberia and Mongolia, *Tr. Paleontol. Inst. Akad. Nauk SSSR*, 47–87 [75].

Type species: *Jurodes ignoramus* Ponomarenko, 1985.

Head quadrangular, length equal to maximal width at eye midlength, narrowing anteriorly and posteriorly. Eyes large, oval, weakly protruding laterally. Frons with wide, flat median elevation. Antennal insertions located on genae, in front of eyes; antennae not reaching base of elytra, apical antennomere egg-shaped, two times as long and wide as the penultimate one. Pronotum shorter than head, with rounded sides, narrowing anteriorly and posteriorly. Prosternum shorter than procoxae. Scutellum small, narrowing posteriorly to widely rounded apex. Elytron with more than 15 rows of punctures, distance between two nearest punctures in a row equal to diameter of one puncture. Metaventrite almost twice as long as mesoventrite, paracoxal suture strongly shifted to metaventrite posterior margin. Metanepisterna big, roundly widening before anterior margin and narrowing posteriorly. Abdomen egg-shaped, narrowing posteriorly from base.

Distribution and age: Inner Mongolia of China, Middle Jurassic; Mongolia, Late Jurassic; Russia, Late Jurassic.

Two species included from the Jurassic of Northern China (see Table 21.1).

Family Ommatidae Lawrence, 1999

Ommatidae are one of the "most ancestral" extant beetle families and a very small group. In contrast to the very limited range of the genera today, the recorded distribution of Ommatidae in the Mesozoic was much broader [129]. Ommatids have head prognathous; antennal insertions lateral; subantennal groove absent; antenna usually short, 11-segmented; pronotum widest anteriorly, procoxal cavities contiguous; abdominal ventrites more or less abutting one another and tarsi simple.

Genera included from the Jurassic and Cretaceous of Northern China: *Omma* Newman, 1839, *Tetraphalerus* Waterhouse, 1901, *Zygadenia* Handlirsch, 1906, *Notocupes* Ponomarenko, 1964, *Brochocoleus* Hong, 1982, *Fuscicupes* Hong & Wang, 1990, *Cionocoleus* Ren, Lu, Ji & Guo, 1995, *Monticupes* Ren, Lu, Ji & Guo, 1995 and *Pareuryomma* Tan, Wang, Ren & Yang, 2012.

Omma Newman, 1839

Omma Newman, 1839, *Ann. Mag. Nat. Hist.*, 3, 303 [245] (original designation).

Procarabus Oppenheim, 1888, *Paleontographica*, 34, 215–247 [246]; Syn. by Ponomarenko, 1971, *Paleontol. J.*, 1, 62–75 [247].

Ommomima Ponomarenko, 1964, *Paleontol. J.*, 2, 49–62 [127]; Syn. by Ponomarenko, 1969, *Trudy Paleontol. Inst.*, 125, 1–240 [47].

Type species: *Omma stanleyi* Newman, 1839.

Body elongate, moderately flattened or almost cylindrical. Surface often tuberculate or rarely spinose. Head subquadrate to slightly elongate, always with distinct neck region; temples usually shorter than eyes. Antennae filiform or somewhat moniliform, as long as or slightly shorter than head and prothorax together; antennomere 3 longest, as long as or longer than pedicel and antennomere 4 combined. Mandibles prominent, incurved. Labrum, clypeus, and frons fused. Pronotum subquadrate or slightly transverse, without lateral pronotal carinae. Legs moderately long and slender; tibiae usually not much longer than femora; mesofemora extending beyond side margins of body. Abdominal sternites arranged in one plane.

Distribution and age: Inner Mongolia of China, Middle Jurassic; Beijing of China, Late Jurassic; Germany, Late Jurassic; Kazakhstan, Middle Jurassic; Kyrgyzstan, Early Jurassic; Mongolia, Middle Jurassic, Early Cretaceous; Myanmar, Late Cretaceous; the United Kingdom, Early Jurassic; Russia, Early Cretaceous.

Two species included from the Jurassic and Cretaceous of Northern China (see Table 21.1).

Tetraphalerus Waterhouse, 1901

Tetraphalerus Waterhouse, 1901, *Ann. Mag. Nat. Hist.*, 7, 520–523 [248] (original designation).

Type species: *Tetraphalerus wagneri* Waterhouse, 1901.

Medium-sized, flat beetles. Head long, distance between base of head and eyes not shorter than half of diameter of eyes. Procoxae contiguous. Elytra usually with rows of small cells; disc of elytra smooth; cells possibly visible on outer surface of elytra. Middle and hind femora not extending laterally to body margins. Abdomen with five flat ventrites.

Distribution and age: Liaoning of China, Early Cretaceous; Inner Mongolia of China, Middle Jurassic; Hebei of China, Early Cretaceous; Hunan of China, Early Jurassic; Mongolia, Middle Jurassic, Early Cretaceous; Myanmar, Late Cretaceous; Australia, Early Jurassic; Kazakhstan, Middle Jurassic; Kyrgyzstan, Early Jurassic; Russia, Early Cretaceous, Late Cretaceous; Spain, Early Cretaceous.

Seven species included from the Jurassic and Cretaceous of Northern China (see Table 21.1).

Zygadenia Handlirsch, 1906

Zygadenia Handlirsch, 1906, *Die Fossilen Insekten und die Phylogenie der Rezenten Formen, parts I–IV*, Ein Handbuch fur Palaontologen und Zoologen, 1–640 [91].

Kakoselia Handlirsch, 1906, *Ein Handbuch fur Palaontologen und Zoologen*, 1–640 [91]; Syn. by Ponomarenko, 2006, *Paleontol. J.*, 40 (1), 90–99 [142].

Forticupes Hong and Wang, 1990, *Stratigraphy and Palaeontology of Laiyang Basin, Shandong Province*, 44–189 [173]; Syn. by Ponomarenko, 2006, *Paleontol. J.*, 40 (1), 90–99 [142].

Picticupes Hong and Wang, 1990, *Stratigraphy and Palaeontology of Laiyang Basin, Shandong Province*, 44–189 [173]; Syn. by Ponomarenko, 2006, *Paleontol. J.*, 40 (1), 90–99 [142].

Sinocupes Lin 1976, *Acta Palaeontol. Sin.*, 15 (1), 97–116 [168]; Syn. by Ponomarenko, 2006, *Paleontol. J.*, 40 (1), 90–99 [142].

Conexicoxa Lin, 1986, *Palaeontol. Sin., Series B*, 170 (21), 69–82 [141]; Syn. by Ponomarenko, 2006, *Paleontol. J.*, 40 (1), 90–99 [142].

Lupicupes Ren, Lu, Ji & Guo, 1995, *Fauna and Stratigraphy of Jurassic-Cretaceous in Beijing and the Adjacent Areas*, 73–90 [240]; Syn. by Ponomarenko, 2006, *Paleontol. J.*, 40 (1), 90–99 [142].

Type species: *Curculionites tuberculatus* Giebel, 1856.

Elytra comparatively broad, convex; four main veins present usually well-differentiated from intercalaries (intermediate veins); A2 and Cu fused before elytral apex, A2 + Cu ending on sutural margin; two longitudinal rows of window cells present between main veins, 20–30 cells per row; epipleural rim moderately wide with or without a row of cells; elytral apex may have a tail-like process.

Distribution and age: Liaoning of China, Early Cretaceous; Shandong of China, Early Cretaceous; Beijing of China, Early Cretaceous; Argentina, Late Cretaceous; Russia, Early Cretaceous; Spain, Early Cretaceous; the United Kingdom, Early Cretaceous; Australia, Early Jurassic; Germany, Late Jurassic; Kazakhstan, Middle Jurassic; Mongolia, Late Jurassic.

Four species included from the Cretaceous of Northern China (see Table 21.1).

Notocupes Ponomarenko, 1964

Notocupes Ponomarenko, 1964, *Paleontologicheskii Zhurnal*, 2, 49–62 [127] (original designation).

Type species: *Notocupes picturatus* Ponomarenko, 1964.

Medium-sized, flat beetles. Head long, with ridges under eyes. Procoxae contiguous, covered by short prosternal process. Elytra with rows of large cells, two veins nearest to sutural margin fused before reaching elytral apex. Abdomen with five convex ventrites; anterior margin of ventrites overlapped by hind part of previous one.

Distribution and age: Inner Mongolia of China, Middle Jurassic; Hebei of China, Early Cretaceous; Liaoning of China, Early Cretaceous; Shandong of China, Early Cretaceous; Hunan of China, Early Jurassic; Kazakhstan, Middle Jurassic, Late Cretaceous; Mongolia, Middle Jurassic, Late Jurassic, Early Cretaceous; Russia, Middle Jurassic, Late Jurassic, Early Cretaceous, Late Cretaceous; Kyrgyzstan, Early Jurassic, Middle Jurassic; Poland, Early Jurassic; Tajikistan, Early Jurassic.

Sixteen species included from the Jurassic and Cretaceous of Northern China (see Table 21.1).

Brochocoleus Hong, 1982

Brochocoleus Hong, 1982, Mesozoic Fossil Insects of Jiuquan Basin in Gansu Province, 103 [160] (original designation).

Diluticupes Ren, Lu, Ji & Guo, 1995, *Fauna and Stratigraphy of Jurassic-Cretaceous in Beijing and the Adjacent Areas*, 73–90 [240]; Syn. by Kirejtshuk, Ponomarenko, Prokin, Chang, Nikolajev & Ren, 2010, *Acta Geol. Sin.*, 84, 783–792 [249].

Type species: *Brochocoleus punctatus* Hong, 1982.

Body flattened and evenly covered with tubercles; head slightly longer than wide, not narrowed forwards;

mandibles comparatively short; antennae slightly moniliform, short, not reaching posterior margin of prothorax, 11-segmented, the third segment longer than other antennameres. Prothorax transverse, wider than head. Elytral epipleural space wide with four rows of cells in the basal half, all principal longitudinal veins parallel to elytral margin; or without distinct punctuation. Abdomen with five visible ventrites; all ventrites in one plane, the last ventrite about 1.5–2 times as long as the former one.

Distribution and age: Inner Mongolia of China, Middle Jurassic, Early Cretaceous; Gansu of China, Early Cretaceous; Beijing of China, Early Cretaceous; Liaoning of China, Early Cretaceous; Kazakhstan, Late Cretaceous; Mongolia, Early Cretaceous; Myanmar, Late Cretaceous; Russia, Late Cretaceous; Spain, Late Cretaceous; the United Kingdom, Early Jurassic, Late Cretaceous; Kyrgyzstan, Early Jurassic.

Seven species included from the Jurassic and Cretaceous of Northern China (see Table 21.1).

Fuscicupes Hong & Wang, 1990

Fuscicupes Hong & Wang, 1990, *Insect fossils of Laiyang Formation, In: The Stratigraphy and Palaeontology of Laiyang Basin, Shandong Province*, 44–189 [173] (original designation).

Type species: *Fuscicupes parvus* Hong & Wang, 1990.

Body slender and elongate, with two pairs of distinct tubercles on vertex; mandibles broad and strong, with at least three teeth in horizontal section; antenna filiform, 11-segmented, reaching posterior of pronotum, with the scape longest, pedicel slightly shorter than other segments. Pronotum rectangular, disc without tubercles, anterior angle biobular; prosternum without tibial groove, prosternal process slightly extending inferior margin of procoxae; elytra with eight rows of wing cells; five abdominal ventrites visible, the first ventrite as long as or slightly shorter than the last ventrite, the last ventrite more than twice as broad as the previous segment; tarsi with five segments, the first tarsomere of front tarsi as long as the fifth segment, which are longer than other tarsomeres, second to fourth segments equal in length.

Distribution and age: Shandong; Early Cretaceous.

Only one species included from the Cretaceous of Northern China (see Table 21.1).

Cionocoleus Ren, Lu, Ji & Guo, 1995

Cionocoleus Ren, Lu, Ji & Guo, 1995, Fauna and Stratigraphy of Jurassic-Cretaceous in Beijing and the Adjacent Areas, 73 [240] (original designation).

Type species: *Cionocoleus magicus* Ren, Lu, Ji & Guo, 1995.

Quite large and flattened. Head transverse, antennae filiform or somewhat moniliform, as long as or shorter than head and prothorax combined, the third antennomere longest; mandibles prominent, incurved, temples shorter than eyes. Pronotum nearly transverse, without distinct side edges, procoxal cavities contiguous. Elytra flattened on disc, without trace of cells.

Distribution and age: Beijing of China, Early Cretaceous; Inner Mongolia of China, Early Cretaceous; Liaoning of China, Early Cretaceous; Mongolia, Early Cretaceous; Russia, Early Cretaceous; Spain, Early Cretaceous; the United Kingdom, Early Cretaceous; Kazakhstan, Middle Jurassic.

Five species included from the Cretaceous of Northern China (see Table 21.1).

Monticupes Ren, Lu, Ji & Guo, 1995

Monticupes Ren, Lu, Ji & Guo, 1995, Fauna and Stratigraphy of Jurassic-Cretaceous in Beijing and the Adjacent Areas, 75 [240] (original designation).

Type species: *Monticupes surrectus* Ren, Lu, Ji & Guo, 1995.

Head longer than width, anterior angles extremely protruding, postocular constriction absent; mandibles distinct, with teeth; pronotum nearly quadratic, wider than head, pleurosternal suture curved; metaventrite without discrimen; elytra long with apex dehiscent, elytral epipleuron distinct, dorsal side with four coarse longitudinal ridges and two rows of angular cells.

Distribution and age: Beijing; Early Cretaceous.

Only one species included from the Cretaceous of Northern China (see Table 21.1).

Pareuryomma Tan, Wang, Ren & Yang, 2012

Pareuryomma Tan, Wang, Ren & Yang, 2012, *BMC Evol. Biol.*, 12 (113), 1–19 [129].

Euryomma Tan, Ren, Shih & Ge, 2006, *Acta Geol. Sin-Engl.*, 80 (4), 474–485 [187]; replaced by Tan, Wang, Ren & Yang, 2012, *BMC Evol. Biol.*, 12 (113), 1–19 [129].

Type species: *Euryomma tylodes* Tan, Ren, Shih & Ge, 2006.

Body surface with large round tubercles, especially on head and prothorax; dorsal head surface without macroscopic tubercles. Mandibles prominent, incurved, with vertically arranged teeth. Antennae not reaching posterior margin of prothorax, antennomeres 7–10 very slightly extended; the apical flagellomere short and wide, apically rounded, appearing inflated. Pronotum widest anteriorly, narrowing posteriorly; anterior angles moderately produced. Elytron very distinctly convex basally, flattened toward apex, with eight rows of cells on disc; epipleural space wide anteriorly, narrowing toward apex; with two rows of cells in the proximal half and one row in the apical half; principal longitudinal veins parallel to the sutural margin.

Distribution and age: Liaoning, Early Cretaceous; Inner Mongolia, Middle Jurassic.

Three species included from the Jurassic and Cretaceous of Northern China (see Table 21.1).

***Pareuryomma tylodes* (Tan, Ren, Shih & Ge, 2006) (Figure 21.7)**

Pareuryomma tylodes (Tan, Ren, Shih & Ge, 2006): *Acta Geol. Sin.-Engl.*, 80 (4), 474–485.

Locality and horizon: Huangbanjigou, Beipiao, Liaoning, China; Lower Cretaceous, Yixian Formation.

Body flattened, small-sized, rather evenly covered with large tubercles. Head longer than width, sub-rectangular, upper surface more or less flattened; eyes medium. Antenna filiform, 11-segmented, short. Pronotum trapezoid, 0.7 times as long as wide, obviously wider than head. Procoxal cavities contiguous, prosternal process not extending behind coxae. Abdomen with five visible ventrites.

Phylogenetic Relationships of the Ommatidae

The results of the phylogenetic analysis suggest six monophyletic groups (tribes): Pronotocupedini, Notocupedini, Lithocupedini, Brochocoleini, Ommatini and Tetraphalerini (Figure 21.8). Only the two subfamilies Lithocupedinae and Tetraphalerinae were recovered, while the other two, Notocupedinae and Ommatinae, turned out as paraphyletic [129].

Figure 21.7 *Pareuryomma tylodes* (Tan, Ren, Shih & Ge, 2006), (Holotype, CNU-COL-LB2005003) [187].

Family Schizophoridae Ponomarenko, 1968

This family was firstly described by Ponomarenko [130] from Jurassic of Karatau, most of genera and species were described from Permian to Cretaceous of Eurasia with only one undescribed species from North America. Schizophorids are mainly characterized by an elytro-thoracic interlocking device ("schiza"), combined with the exposure of propleura, elytra lack clathrate cells and rounded posteriorly. It was suggested to be in the suborder Myxophaga by Bouchard *et al.* [40], which is not followed by all other researchers.

Genera included from the Jurassic and Cretaceous of Northern China: *Homocatabrycus* Tan, Ren & Shih, 2007, *Menopraesagus* Tan, Ren & Shih, 2007, *Sinorhombocoleus* Tan & Ren, 2009, *Sinoschizala* Jarzembowski, Yan, Wang & Zhang, 2012 and *Abrhadeocoleodes* Tan, Ren, Shih & Yang, 2013.

***Homocatabrycus* Tan, Ren & Shih, 2007**

Homocatabrycus Tan, Ren & Shih, 2007, *Ann. Zool. (Warszawa)*, 57 (2), 231–247 [96] (original designation).

Type species: *Homocatabrycus liui* Tan, Ren & Shih, 2007.

The specific epithet is dedicated to Dr. Jiaxi Liu, Jingjing Tan's Professor in Capital Normal University, who provided guidance and initiative for her to pursue her graduate study of paleoentomology. Body flattened, oval; head large; mandibles large, projecting, bidentate; eyes displaced laterally; antennae filiform, pedicel distinctly thinner than scape and flagellomeres, about as long as wide; pronotum transverse, less than 1.5 times as wide as head, anterior angles tapered, extending forward, with narrow paranotals; mesocoxae large, oblong and separated. (Holotype fossil donated by Dr. Chungkun Shih.)

Distribution and age: Inner Mongolia; Middle Jurassic.

Only one species included from the Jurassic of Northern China (see Table 21.1).

***Menopraesagus* Tan, Ren & Shih, 2007**

Menopraesagus Tan, Ren & Shih, 2007, *Ann. Zool. (Warszawa)*, 57 (2), 231–247 [96] (original designation).

Type species: *Menopraesagus explanatus* Tan, Ren & Shih, 2007.

Body flattened, oval; head slightly hypognathous; mandibles small; eyes displaced dorsally; antennae filiform, slightly shorter than half as long as entire insect, pedicel distinctly thinner than scape and flagellomeres, about as long as wide, the third antennomere shorter

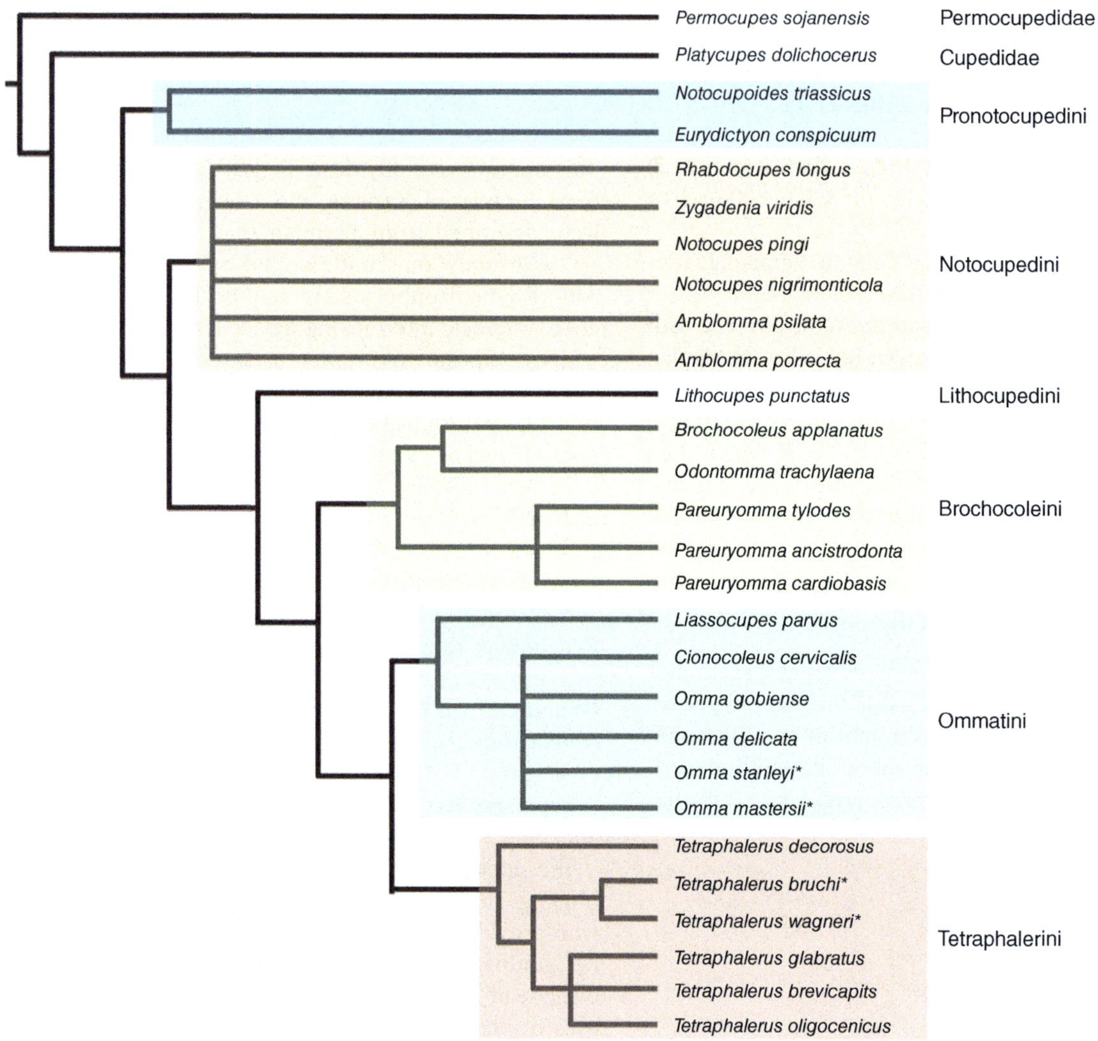

Figure 21.8 Results of the phylogenetic analyses as represented by strict consensus tree by NONA. Source: Modified from [129].

than scape and pedicel combined in length; pronotum less than 1.5 times as wide as head, anterior angles tapered, slightly extending forward, without markedly paranotals; mesocoxae slightly semicircular.

Distribution and age: Inner Mongolia; Middle Jurassic.

Four species included from the Jurassic of Northern China (see Table 21.1).

Sinorhombocoleus Tan & Ren, 2009

Sinorhombocoleus Tan & Ren, 2009, Mesozoic Archostematan Fauna from China, 158 [139].

Type species: *Sinorhombocoleus papposus* Tan & Ren, 2009.

Pronotum nearly trapezoid, lateral area of anterior margin slightly protruding, posterior margin straight; metasternal process extending to the middle of mesocoxae; elytra with at least 17 rows of cells, small and ovate; Rs, M + CuA and 2A distinct with RA and 3A also visible; anterior area with more than six rows of cells, areas between Rs and M + CuA with four rows of cells, areas between M + CuA and 2A with six rows of cells basally, reduced to three rows posteriorly.

Distribution and age: Liaoning; Early Cretaceous.

Only one species included from the Cretaceous of Northern China (see Table 21.1).

Sinoschizala Jarzembowski, Yan, Wang & Zhang, 2012

Sinoschizala Jarzembowski, Yan, Wang & Zhang, 2012, *Palaeoworld*, 21 (3–4), 160–166 [250] (original designation).

Type species: *Sinoschizala darani* Jarzembowski, Yan, Wang & Zhang, 2012.

The specific epithet is dedicated to Mr. Daran Zheng of NIGPAS (Nanjing Institute of Geology and Palaeontology, Chinese Academy of Sciences). Large schizophorid with elongate elytra and uniquely distinctive venation: 2A forked near base (with posterior branch strongest); RP2 and M + CuA converge distad of anal fork; RA simple, RP forked; short "schiza" along RP1 opposite basal part of concave indentation in anterior wing margin. Protarsomeres moderately elongate.

Distribution and age: Inner Mongolia; Middle Jurassic.

Only one species included from the Jurassic of Northern China (see Table 21.1).

Abrhadeocoleodes Tan, Ren, Shih & Yang, 2013

Abrhadeocoleodes Tan, Ren, Shih & Yang, 2013, *J. Syst. Palaeontol.*, 11 (1), 47–62 [143] (original designation).

Type species: *Abrohadeocoleodes eurycladus* Tan, Ren, Shih & Yang, 2013.

Body oval; head prognathous; mandibles compact, triangular, large, length of mandible nearly as long as half of head, with distinct bidentate apex; antennae inserted dorsally, above the dorsal joints of mandibles, linear space between antennal fossa slightly wider than width of labrum; antennae filiform, pedicel nearly as wide as scape, the third antennomere shorter than the first and second combined; procoxal cavities contiguous; anterior angles of metaventrite extending forward and exceeding metaventral process; end of elytra rounded, elytra not reaching posterior abdominal margin, part of sternite VII apparently exposed; legs thin and long, metatrochanter slightly triangular, distinctly elongated or normal.

Distribution and age: Inner Mongolia; Middle Jurassic.

Four species included from the Jurassic of Northern China (see Table 21.1).

Abrohadeocoleodes eurycladus Tan, Ren, Shih & Yang, 2013 (Figure 21.9)

Abrohadeocoleodes eurycladus Tan, Ren, Shih & Yang, 2013: *J. Syst. Palaeontol.*, 11 (1), 47–62.

Locality and horizon: Daohugou, Ningcheng, Inner Mongolia, China; Middle Jurassic, Jiulongshan Formation.

Head very small in relation to prothorax; antennae filiform, scape and pedicel very short, wider than long, distinctly thicker than elongate flagellomeres; anterior margin of pronotum 2.5 times as wide as posterior margin of head; cervical sclerite very distinct; mesoventrite with longitudinal and transverse sutures; metaventrite with broad longitudinal suture transformed into an elongated groove; metaventral process broad and flat; metatrochanter slightly triangular, distinctly elongated as in Adephaga [143].

Abrohadeocoleodes eurycladus might have been able to run fast and hunt other small insects in open wetlands or lakeshores based on the morphological and environmental characters. It is probably a close relative of the Mesozoic adephagan beetles [143, 251].

Figure 21.9 *Abrohadeocoleodes eurycladus* Tan, Ren, Shih & Yang, 2013, (Holotype, CNU-COL-NN2010803) [143].

Family Tshekardocoleidae Rohdendorf, 1944

Main veins on elytra run more obliquely in regard to the long axis of elytra compared to Cupedidae and Permocupedidae, not being parallel with fore or hind margin; anterior margin sharply divided off the rest, broad and flat, with a meshwork of veins; RS branching off at the junction of the first and second third of wing; apex of wing is reached by SC, RS and sutural rib; M with two long branches; Cu and A1 run parallel to each other and end at hind wing margin, often uniting with suture rib.

Only one genus included from the Jurassic and Cretaceous of Northern China: *Dictycoleus* Hong, 1982.

Dictycoleus Hong, 1982

Dictycoleus Hong, 1982, Mesozoic Fossil Insects of Jiuquan Basin in Gansu Province, 98 [160] (original designation).

Type species: *Dictycoleus jurassicus* Hong, 1982.

Elytral cells on basal area slightly smaller than those on posterior area, cells irregular, net-shaped. Surface with 11 longitudinal veins, among which there are usually two rows of cells, but sometimes with three to five rows; anterior areas wide, with veins net-shaped, areas between veins with only two rows of cells, adding to a few rows at middle; veins long and straight, parallel; longitudinal veins always with small concavities.

Distribution and age: Gansu; Early-Middle Jurassic.

Only one species included from the Jurassic of Northern China (see Table 21.1).

Suborder Myxophaga Crowson, 1955

Family Hydroscaphidae LeConte, 1874

Hydroscaphids are minute beetles with shortened elytra and resemblance to some small staphylinids. The tarsi are 3-segmented, notopleural suture distinct and the short, glabrous antenna may be 5- or 8-segmented.

Only one genus included from the Cretaceous of Northern China: *Hydroscapha* LeConte, 1874.

Hydroscapha LeConte, 1874

Hydroscapha LeConte, 1874, *T. Am. Entomol. Soc.*, 5, 43–72 [252] (original designation).

Type species: *Hydroscapha natans* LeConte, 1874.

Body small, scaphiform, rounded in front, narrowed behind, convex, elongate and shining; head rather large; prothorax narrower in front, with deflexed angles, base truncate not margined; scutellum small; elytra without striae, slightly punctulate, broadly truncate apically; abdomen projecting somewhat behind the elytra, with three segments visible, conical, not margined at the sides.

Distribution and age: Liaoning; Early Cretaceous.

Only one species included from the Cretaceous of Northern China (see Table 21.1).

Suborder Adephaga Schellenberg, 1806

Family Carabidae Latreille, 1802

Carabidae are a speciose and widely distributed family including the majority of terrestrial Adephaga. They usually have filiform antennae with antennomeres beyond three or four densely pubescent; prognathous head with large, predaceous mandibles; a grooming organ or antennal cleaner on the protibia; and the presence of erect, fixed tactile setae on the body; transverse groove on the metaventrite present and metacoxae relatively narrow.

Genera included from the Jurassic and Cretaceous of Northern China: *Cretorabus* Ponomarenko, 1977, *Nebrorabus* Ponomarenko, 1989, *Aethocarabus* Ren, Lu, Ji & Guo, 1995, *Denudirabus* Ren, Lu, Ji & Guo, 1995 and *Penecupes* Ren, Lu, Ji & Guo, 1995.

Cretorabus Ponomarenko, 1977

Cretorabus Ponomarenko, 1977, *Tr. Paleontol. Inst. Akad. Nauk SSSR*, 161, 17–119 [71] (original designation).

Cretarabus Hong & Wang, 1990, *Stratigraphy and Palaeontology of Laiyang Basin, Shandong Province*, 44–189 [173].

Type species: *Cretorabus capitatus* Ponomarenko, 1977.

Body wide, small or medium-sized. Head large, strongly transverse. Pronotum transverse, widest at anterior or middle portions, constricted behind middle. Mesoventrite longer than mesocoxae. Metaepisterna narrowed posteriorly. Metacoxal plates strongly narrowed at lateral half, extending as a narrow tongue up to lateral margins of metacoxae. Abdomen short, with rounded apex; last ventrite long, its anterior margin two-thirds as broad as the base of abdomen. Legs short, femora slightly extending beyond body sides. Elytra smooth or with numerous rows of large punctures.

Distribution and age: Inner Mongolia of China, Mongolia, Russia and the United Kingdom; Early Cretaceous.

Only one species included from the Cretaceous of Northern China (see Table 21.1).

Nebrorabus Ponomarenko, 1989

Nebrorabus Ponomarenko, 1989, *Paleontol. J.*, 2, 52–63 [253] (original designation).

Type species: *Nebrorabus baculum* Ponomarenko, 1989.

Small beetles with elongated and flattened body. Head large, triangular, almost the same length as pronotum, and the width of occiput less than 1.5 times as wide as base of pronotum. Prothorax strongly narrowed backwards, strongly narrower than elytra at base. The hind legs are longer than the middle coxae. Metacoxae oblique, their femoral lateral covers are greatly shortened. Elytra about three times as long as wide, with fine punctures and furrows or smooth, femora almost more than one third as long as body length, hind tibiae are longer than the femora.

Distribution and age: Beijing of China, Mongolia and Russia; Early Cretaceous.

Only one species included from the Cretaceous of Northern China (see Table 21.1).

Aethocarabus Ren, Lu, Ji & Guo, 1995

Aethocarabus Ren, Lu, Ji & Guo, 1995, Fauna and Stratigraphy of Jurassic-Cretaceous in Beijing and the Adjacent Areas, 84 [240] (original designation).

Type species: *Aethocarabus levigata* Ren, Lu, Ji & Guo, 1995.

Head wider than length; gula wide anteriorly, gula suture distinctly curved, emarginate; pleurosternal suture distinctly curved, propleuron moderately large; metaventrite with metakatepisternal suture distinct, discrimen absent; elytra smooth, punctures indistinct.

Distribution and age: Beijing; Early Cretaceous.

Only one species included from the Cretaceous of Northern China (see Table 21.1).

Denudirabus Ren, Lu, Ji & Guo, 1995

Denudirabus Ren, Lu, Ji & Guo, 1995, Fauna and Stratigraphy of Jurassic-Cretaceous in Beijing and the Adjacent Areas, 86 [240] (original designation).

Type species: *Denudirabus exstrius* Ren, Lu, Ji & Guo, 1995.

Head wider than length, mandibles strong, gular extremely broad anteriorly and narrow posteriorly; pronotum apparently transverse, anterior angles protruding forwardly, notopleural sutures similarly curved to the lateral margins of pronotum, pleurosternal sutures obliquely extending to the procoxae; prosternal process not distinctly projecting between coxae; mesocoxae closely separated; metakatepisternal suture distinct, metacoxae nealy rounded, metanepisternum distinct; elytra smooth without sculpture.

Distribution and age: Beijing; Early Cretaceous.

Only one species included from the Cretaceous of Northern China (see Table 21.1).

Penecupes Ren, Lu, Ji & Guo, 1995

Penecupes Ren, Lu, Ji & Guo, 1995, Fauna and Stratigraphy of Jurassic-Cretaceous in Beijing and the Adjacent Areas, 79 [240] (original designation).

Type species: *Penecupes rapax* Ren, Lu, Ji & Guo, 1995.

Head wider than length, eyes large, mandibles distinctly large; pronotum wide with anterior part covering the base of head, notopleural suture straight, pleurosternal suture curved; procoxae well-separated with prosternal process projecting between the coxae; mesocoxae separated; metaventrite without discrimen or transverse suture; elytra extending over the abdomen, anterior and posterior epipleuron distinct but without sculpture, dorsal side with nine rows of longitudinal rounded cells, without fine punctures around the cells.

Distribution and age: Beijing; Early Cretaceous.

Only one species included from the Cretaceous of Northern China (see Table 21.1).

Superfamily Dytiscoidea Bell, 1966

Family Coptoclavidae Ponomarenko, 1961

Coptoclavids are extinct predaceous water beetles. Both larva and adult fossils are common, although the diversity of known larvae is considerably narrower than that of adults. Adult coptoclavids had two pairs of eyes, while most larvae with raptorial forelegs and natatorial mid- and hind legs [133].

Genera included from the Jurassic and Cretaceous of Northern China: *Coptoclava* Ping, 1928, *Coptoclavisca* Ponomarenko, 1987 and *Daohugounectes* Wang, Ponomarenko & Zhang, 2009.

Coptoclava Ping, 1928

Coptoclava Ping, 1928, *Palaeontologica Sin. B*, 13 (1), 1–56 [156] (original designation).

Coptolavia Teixeira, 1975, *Boletim da Sociedade Geologica de Portugal*, 19 (3), 131–134 [75].

Type species: *Coptoclava longipoda* Ping, 1928.

Length 15–45 mm. Head protruding; mandibles strong with two to three pairs of teeth; maxillary palps 4-segmented; eyes on the dorsal side of head, rounded and well-separated; legs with femora slightly longer than tibiae, tibiae with a pair of strong apical spurs, tarsi sharpened apically with a pair of well-developed claws; abdomen with six ventrites, each ventrite with a pair of lateral branchiae and spiracles; male with strong harpagones, female with three pairs of valvulae, beginning with the fourth ventrite; elytra not covering the apex of abdomen, pygidium exposed; elytra coarse and convex, anterior margin arcuate with a few longitudinal stripes, meeting areas of anterior and posterior margins distinctly angled.

Larvae have two types: (i) Mature larvae: length 20–45 mm, shape similar to the adults; head, legs have characters similar to adults, abdomen 6-segmented with a pair of long and forked urogomphi on apex. (ii) Immature larvae: length 10–35 mm; head small, mandibles strong with two to three pairs of teeth; hind legs are swimmeret, fore legs aciculiform, all covered with swimming hairs; abdomen with two thick longitudinal tracheae and a pair of forked urogomphi apically.

Distribution and age: Gansu, Hebei, Jilin, Liaoning, Shandong, Inner Mongolia of China, Early Cretaceous; Mongolia, Early Cretaceous; Russia, Early Cretaceous; South Korea, Middle Cretaceous.

Only one species included from the Jurassic and Cretaceous of Northern China (see Table 21.1)

Coptoclavisca Ponomarenko, 1987

Coptoclavisca Ponomarenko, 1987, *Paleontol. J.*, 21 (2), 79–92 [135] (original designation).

Type species: *Coptoclavisca nigricollinus* Ponomarenko, 1987.

Small, with metacoxae plates comparatively widened, all legs without noticcable swimming hairs.

Distribution and age: Shandong of China, Mongolia; Early Cretaceous.

Only one species included from the Cretaceous of Northern China (see Table 21.1).

Daohugounectes Wang, Ponomarenko & Zhang, 2009

Daohugounectes Wang, Ponomarenko & Zhang, 2009, *Paleontol. J.*, 43 (6), 652–659 [133] (original designation).

Type species: *Daohugounectes primitivus* Wang, Ponomarenko & Zhang, 2009.

For larva: Head transverse, narrowing anterior of the base; nasale with rounded lateral lobes. Mandible with retinaculum near its middle. Prothoracic tergite somewhat longer than meso- and metathoracic tergites. Legs relatively short; mid- and hind leg distinctly longer than foreleg, their femora, tibiae and tarsi subequal in length, tibae slightly dilated at apex, tarsi roundly dilated anteriorly. Abdomen with nine segments, seven of them transverse.

For adult: Body medium-sized and oval. Head transverse, laterally rounded, with eyes not protruding. Pronotum transverse; base much narrower than width of elytral base; anterior margin weakly concave; prosternal process elongated and parallel-sided. Metaventrite with raised triangular platform. Border between abdominal segments III and IV indistinct. Elytra with protruded longitudinal ridges, interstices smooth. All legs with long and dense natatory setae.

Daohugounectes larvae possess some primitive characters such as the present of nasale; the elongate mandible with only two teeth; the mid- and hind legs not so strong; the relatively weakly-developed helical thickening of the main tracheae; and the present vestige of abdominal segment IX. They can be regarded as the most basal position among the known coptoclavid larvae in the subfamily Timarchopsinae. However *Daohugounectes* adults show some advanced characters and can be regarded as among the most advanced forms in the subfamily Timarchopsinae. So, *Daohugounectes* probably represent a transition between the Timarchopsinae and the advanced subfamily Charanoscaphinae [221].

Distribution and age: Inner Mongolia; Middle Jurassic.

Only one species included from the Jurassic of Northern China (see Table 21.1).

Family Dytiscidae Leach, 1815

Dytiscidae are a major aquatic adephagan family with worldwide distribution. They have streamlined form and strong hind legs with swimming hairs, antennal club absent, external propleuron present and large metacoxae lacking plates.

Genera included from the Jurassic and Cretaceous of Northern China: *Liadroporus* Prokin & Ren, 2010, *Liadytiscus* Prokin & Ren, 2010, *Mesoderus* Prokin & Ren, 2010, *Sinoporus* Prokin & Ren, 2010, *Liadyxianus* Prokin, Petrov, Wang & Ponomarenko, 2013 and *Mesodytes* Prokin, Petrov, Wang & Ponomarenko, 2013.

Liadroporus Prokin & Ren, 2010

Liadroporus Prokin & Ren, 2010, *Paleontol. J.*, 44 (5), 526–533 [254] (original designation).

Type species: *Liadroporus elegans* Prokin & Ren, 2010.

Medium-sized beetles with parallel-sided elytra. Pronotum not narrower than both elytra taken together, forming with them gently curved lateral outline. Head partly sunk in pronotum. Procoxae and mesocoxae rounded, mesocoxae spread widely apart. Prosternum with carina. Metasternum with rounded raised area. Metacoxae somewhat longer than metasternal "wing". Metacoxal processes rounded apically. Metafemora narrow. Epipleuron wide basally, gently narrowing and disappearing at level of middle of abdominal segment 5.

Distribution and age: Liaoning; Early Cretaceous.

Only one species included from the Cretaceous of Northern China (see Table 21.1).

Liadytiscus Prokin & Ren, 2010

Liadytiscus Prokin & Ren, 2010, *Paleontol. J.*, 44 (5), 526–533 [254] (original designation).

Type species: *Liadytiscus cretaceous* Prokin & Ren, 2010.

Medium-sized ovate beetles. Pronotum not narrower than both elytra taken together, forming with them gently curved lateral outline. Procoxae and mesocoxae rounded. Pronotum with carina. Metacoxae somewhat longer than metasternal "wing". Metasternal processes with straightly truncated apex, with incision. Mesotibiae and mesotarsi with swimming hairs. Hind legs long, 1.5 times as long as abdomen. Abdominal ventrite one shorter medially than any other abdominal ventrites. Epipleura long. Scutellum equilaterally triangular. Elytron smooth, narrowing toward apex, without sulci or puncture rows.

Distribution and age: Liaoning; Early Cretaceous.

Three species included from the Cretaceous of Northern China (see Table 21.1).

Mesoderus Prokin & Ren, 2010

Mesoderus Prokin & Ren, 2010, *Paleontol. J.*, 44 (5), 526–533 [254] (original designation).

Type species: *Mesoderus magnus* Prokin & Ren, 2010.

Procoxae closely set. Mesotibiae distally with swimming setae. Maximal length of lateral lobe of metaventrite relatively great, but smaller than maximal length of metacoxal plate. Metacoxal lines strongly diverging anteriad; metacoxal processes rounded posteriorly, somewhat narrowed anteriad. Metaventrite with rounded elevated median area. Metatibia subequal in length to metafemur, flattened; metatarsomeres also flattened.

Distribution and age: Inner Mongolia, Liaoning; Early Cretaceous.

Four species included from the Cretaceous of Northern China (see Table 21.1).

Sinoporus Prokin & Ren, 2010

Sinoporus Prokin & Ren, 2010, *Paleontological Journal*, 44 (5), 526–533 [254] (original designation).

Type species: *Sinoporus lineatus* Prokin & Ren, 2010.

Small ovate beetles with lateral margins of pronotum and elytron curved separately. Maximal body width at middle. Mesocoxae rounded, closely positioned. Scutellum exposed. Anterior margin of eye without incision. Metasternal "wing" long, with almost straight posterior margin, not reaching body sides, shorter than metacoxa. Medial metacoxal processes straight laterally and straightly truncated apically. Metatibia straight, reaching posterior margin of abdominal ventrite 4.

Distribution and age: Liaoning; Early Cretaceous.

Only one species included from the Cretaceous of Northern China (see Table 21.1).

Liadyxianus Prokin, Petrov, Wang & Ponomarenko, 2013

Liadyxianus Prokin, Petrov, Wang & Ponomarenko, 2013, *Zootaxa*, 3666 (2), 137–159 [136] (original designation).

Type species: *Liadyxianus kirejtshuki* Prokin, Petrov, Wang & Ponomarenko, 2013.

The specific epithet is dedicated to Dr. Kirejtshuk for his significant contribution to the studies of fossil beetles. Metacoxal plate somewhat shorter than lateral lobe of metaventrite. Metaventrite with elevated rounded median area. Metacoxal lines weakly diverging anteriad. Metacoxal processes somewhat conjointly truncate posteriorly, with posterior margins appearing somewhat sinuate, with incision between them. Lateral lobe of metaventrite not reaching lateral margin of metepisternum and metacoxal plate. Hind leg long, equal in length to abdomen. Metafemur, when stretched, reaching posterior margin of abdominal ventrite 3; metatibia longer than metafemur. Abdominal ventrite 1 shorter medially than any other abdominal ventrites, ventrite 2 longer medially than any other abdominal ventrites.

Distribution and age: Liaoning; Early Cretaceous.

Only one species included from and Cretaceous of Northern China (see Table 21.1).

Mesodytes Prokin, Petrov, Wang & Ponomarenko, 2013

Mesodytes Prokin, Petrov, Wang & Ponomarenko, 2013, *Zootaxa*, 3666 (2), 137–159 [136] (original designation).

Type species: *Mesodytes rhantoides* Prokin, Petrov, Wang & Ponomarenko, 2013.

Elytra smooth, translucent, covered with matted rows of dark spots, separated by a distance greater than or equal to their diameters. Lobe of prosternal process flat, elongated, reaching mesocoxae. Maximal length of lateral lobe of metaventrite relatively great, but smaller than maximal length of metacoxal plate. Lateral lobe of metaventrite not reaching lateral margins of metepisternum and mesocoxal plate. Metaventrite without elevated laterally limited area medially. Metacoxal lines weakly diverging anteriad. Metacoxal processes rounded posteriorly, slightly narrowed anteriad. Metafemur 1.5 times as long as metatibia. Metafemur, when outstretched, reaching middle of abdominal ventrite 4. The first metatarsomere 1.5 times as long as metatarsomere 2.

Distribution and age: Inner Mongolia; Early Cretaceous.

Only one species included from the Cretaceous of Northern China (see Table 21.1).

Family Haliplidae Aubé, 1836

Haliplidae are a small family comprising five extant genera and over 200 species distributed all over the world except for Antarctica and New Zealand. One extinct genus *Cretihaliplus* Ren, Zhu & Lu, 1995 was described from China, while Prokin and Ponomarenko, in 2013, suggested it should not belong to Haliplidae [255]. We just treat it as Haliplidae because they did not move it to any other family or established a new family. Main characters for recognizing this family are the short and glabrous antennae, the huge metacoxal plate covering metafemora, basal abdominal ventrites combined with strongly convex dorsal side and finely punctured elytra with several striae.

Only one genus included from the Cretaceous of Northern China: *Cretihaliplus* Ren, Zhu & Lu, 1995.

Cretihaliplus Ren, Zhu & Lu, 1995

Cretihaliplus Ren, Zhu & Lu, 1995, *Acta Geoscienta Sin.*, 4, 432–439 [256] (original designation).

Type species: *Cretihaliplus chifengensis* Ren, Zhu & Lu, 1995.

Body with dense and small punctures. Head prognathous, clypeus and labrum large, head protruding. Pronotum narrowed anteriorly, scutellum absent. Elytra with nine rows of longitudinal striae, not covering the abdomen completely, apex dehiscent. Metacoxal plates not extending beyond the third segment of abdominal ventrites.

Distribution and age: Inner Mongolia; Early Cretaceous.

Two species included from the Cretaceous of Northern China (see Table 21.1).

Family Liadytidae Ponomarenko, 1977

Liadytidae were first described by Ponomarenko [71] with two species belong to a single genus *Liadytes* Ponomarenko. This family were suspected to be aquatic beetles with oval, biconvex bodies. They can be distinguished by external scutellum; metepisterna reaching mesocoxal cavities, metaventrite with median longitudinal protuberance, sharply delimited laterally; hind coxae with anterior margin almost straight with small central emargination, bounded in front by an extension of the transverse metasternal suture which forms a single straight line with the anterior margin of the coxae; legs slender and long with natatory hairs on tibiae and tarsi.

Only one genus included from the Cretaceous of Northern China: *Ovidytes* Ren, Zhu & Lu, 1995.

Ovidytes Ren, Zhu & Lu, 1995

Ovidytes Ren, Zhu & Lu, 1995, *Acta Geoscienta Sin.*, 4, 432–439 [256] (original designation).

Type species: *Ovidytes gaoi* Ren, Zhu & Lu, 1995.

The specific epithet is in honor of famous geologist Professor Zhenxi Gao who is the founder of the Geological Museum of China. Body elongated ovate. Pronotum wide, both anterior and posterior curved, toward the base, prosternal process distinct; hind legs with metacoxae slightly enlarged, anterior margin curved, femora and tibiae with swimming hair, tarsomere 1 long.

Distribution and age: Inner Mongolia; Early Cretaceous.

Only one species included from the Cretaceous of Northern China (see Table 21.1).

Family Trachypachidae Thomson, 1857

T rachypachidae, called "false ground beetles", is a small family comprising the living Trachypachinae and extinct Eodromeinae. Adult trachypachids with tactile setae and protibial grooming organs, metacoxae extend laterally to meet the elytra so that the metepimeron does not lie in between the metanepisternum and the first abdominal ventrite, antennal pubescence not becoming denser apically as in Carabidae.

Genera included from the Jurassic and Cretaceous of Northern China: *Eodromeus* Ponomarenko, 1977, *Unda* Ponomarenko, 1977, *Fortiseode* Jia & Ren, 2011 and *Sinodromeus* Wang, Zhang & Ponomerenko, 2012.

Eodromeus Ponomarenko, 1977

Eodromeus Ponomarenko, 1977, *Tr. Paleontol. Inst. Akad. Nauk SSSR*, 161, 17–119 [71] (original designation).

Type species: *Eodromeus antiquus* Ponomarenko, 1977.

Head triangular, its length (including mandibles) commonly shorter than width of occiput. Temples shorter than eyes. Antenna short, scarcely extending beyond pronotal base. Pronotum with anterior margin very weakly concave; anterior angles not sharp; base slightly narrower than elytra combined. Metacoxal plates with posterior and lateral parts emarginate.

Distribution and age: Inner Mongolia of China, Middle Jurassic; Jilin of China, Early Cretaceous; Mongolia, Early Cretaceous; Russia, Early Cretaceous; Kazakhstan, Middle Jurassic.

Three species included from the Jurassic and Cretaceous of Northern China (see Table 21.1).

Unda Ponomarenko, 1977

Unda Ponomarenko, 1977, *Tr. Paleontol. Inst. Akad. Nauk SSSR*, 161, 17–119 [71] (original designation).

Type species: *Unda microplata* Ponomarenko, 1977.

Head transverse and triangular. Eyes small, not longer than temples. Antennae rather long, slender, filiform. Pronotum transverse, slightly narrowed at anterior third; anterior margin weakly concave; base slightly narrower than combined elytral base. Mesoventrite nearly equal to mesocoxae in length, distance between meso- and metacoxae longer than length of mesocoxae. Metacoxae strongly oblique. Abdomen rather short, only slightly longer than meso- and metathoraces combined; last venrite longer than the others, its anterior margin much narrower than base of abdomen.

Distribution and age: Inner Mongolia of China, Middle Jurassic; Beijing of China, Early Cretaceous; Russia, Late Jurassic.

Three species included from the Jurassic and Cretaceous of Northern China (see Table 21.1).

Fortiseode Jia & Ren, 2011

Fortiseode Jia & Ren, 2011, *Zootaxa*, 2736, 63–68 [257] (original designation).

Type species: *Fortiseode pervalimand* Jia & Ren, 2011.

Head transverse, length of head capsule more than half its width. Mandibles with the same length as the head. Pronotum cordiform, widest at about anterior one-fifth of the pronotum, anterior edge of pronotum strongly concave; base of pronotum with similar width as base of elytra; length of head including mandibles nearly equal to width of head capsule. Hind coxae strong, transverse. Legs with strong femora and tibiae, metatibiae shorter than 1/2 of elytra, metatarsi short, with length nearly half of metatibiae; trochanters large, longer than 1/3 of metafemora.

Distribution and age: Liaoning; Early Cretaceous.

Only one species included from the Cretaceous of Northern China (see Table 21.1).

Sinodromeus Wang, Zhang & Ponomerenko, 2012

Sinodromeus Wang, Zhang & Ponomerenko, 2012, *Palaeontology*, 55 (2), 341–353 [138] (original designation).

Type species: *Sinodromeus liutiaogouensis* Wang, Zhang & Ponomerenko, 2012.

Head transverse and triangular. Eyes small, not longer than temples. Antennae rather long, slender, filiform. Pronotum transverse, slightly tapering in anterior third; anterior margin weakly concave; base slightly narrower than elytra combined at base. Distance between meso- and metacoxae longer than mesocoxal length. Metacoxae strongly oblique. Abdomen rather short, only slightly longer than meso- and metathoraces combined; the last ventirte longer than the others, its anterior margin much narrower than base of abdomen.

Distribution and age: Inner Mongolia; Early Cretaceous.

Only one species included from the Cretaceous of Northern China (see Table 21.1).

Suborder Polyphaga Emery, 1886

Infraorder Staphyliniformia Lameere, 1900

Superfamily Hydrophiloidea Latreille, 1802

Family Helophoridae Leach, 1815

Almost all adults are aquatic, while larvae are terrestrial. Helophoridae, containing the primarily Holarctic genus *Helophorus*, have a unique pattern of longitudinal grooves on the pronotum and the fringes of long setae on the apical labial palpomere.

Only one genus included from the Jurassic and Cretaceous of Northern China: *Helophorus* Fabricius, 1775.

Helophorus Fabricius, 1775

Helophorus Fabricius, 1775, *Systema Entomologiae, sistens insectorum classes, ordines, genera, species, adiectis synonymis, locis, descriptionibus, observationibus*, Officina Libraria Kortii, Flensburgi & Lipsiae, 1–832 [258] (original designation).

Type species: *Silpha aquatica* Linnaeus, 1758 [259].

Adult: Head and pronotum at least partly with setiferous granules in most species; frontoclypeal suture very distinct, its median portion grooved; mentum distinctly transverse or only slightly wider than long; gula usually strongly constricted; pronotum with 1–5 longitudinal grooves; lateral margin of pronotum at least slightly crenulate; anterolateral portion of hypomeron with antennal groove; mesanepisterna not meeting mesally; mesoventrite very narrow at anterior margin, bearing a transverse ridge posteromedially; elytra never pale with dark stripes along elytral series [260].

Larva: Head prognathous; nasale simply triangular or with lobate lateral margins; epistomal lobes large, slightly overlapping nasale, bearing series of stout setae; mandible with two retinacular teeth; labium without ligula; each parietale with six stemmata situated in area of darker cuticle; all thoracic segments with large dorsal tergite; abdominal segments one to eight each with a pair of dorsal sclerites and an additional large sclerite laterally on each side of them; tracheal system holopneustic, spiracular atrium absent; urogomphi large, three-segmented.

Distribution and age: Liaoning of China, Early Cretaceous; Mongolia, Late Jurassic, Early Cretaceous; Russia, Late Jurassic.

Two species included from the Cretaceous of Northern China (see Table 21.1).

Family Hydrophilidae Latreille, 1802

Hydrophilidae, called "water scavenger beetles", currently represent the largest lineage of the superfamily Hydrophiloidea. They are characterized by having relatively short antennae consisting of seven, eight or nine segments and a distinct pubescent club of three or rarely four segments, preceded by a transverse, glabrous antennomere.

Genera included from the Jurassic and Cretaceous of Northern China: *Alegorius* Fikáček, Prokin, Yan, Yue, Wang, Ren & Beattie, 2014 and *Hydroyixia* Fikáček, Prokin, Yan, Yue, Wang, Ren & Beattie, 2014.

Alegorius Fikáček, Prokin, Yan, Yue, Wang, Ren & Beattie, 2014

Alegorius Fikáček, Prokin, Yan, Yue, Wang, Ren & Beattie, 2014, *Zool. J. Linn. Soc-Lond*, 170, 710–734 [261] (original designation).

Type species: *Alegorius yixianus* Fikáček, Prokin, Yan, Yue, Wang, Ren & Beattie, 2014.

Body medium-sized; labrum partly exposed in dorsal view; anterior margin of clypeus weakly excised mesally; prosternum moderately long, weakly carinate/elevated medially; scutellar shield small, triangular; anapleural sutures well-developed, strongly curved, nearly parallel-sided and widely separate anteriorly; mesocoxae very narrowly separated; abdomen with five ventrites; apical abdominal ventrite with deep and narrow emargination; elytron with deeply impressed sutural stria; elytra with series of punctures; tarsi of mesothoracic leg with five tarsomeres; mesotarsomere 1 very short.

Distribution and age: Liaoning; Early Cretaceous.

Only one species included from the Cretaceous of Northern China (see Table 21.1).

***Hydroyixia* Fikáček, Prokin, Yan, Yue, Wang, Ren & Beattie, 2014**

Hydroyixia Fikáček, Prokin, Yan, Yue, Wang, Ren & Beattie, 2014, *Zool. J. Linn. Soc-Lond*, 170, 710–734 [261] (original designation).

Type species: *Hydroyixia elongata* Fikáček, Prokin, Yan, Yue, Wang, Ren & Beattie, 2014.

Body medium-sized; labrum partly exposed in dorsal view; anterior margin of clypeus deeply and widely excised mesally; prosternum moderately long; scutellar shield small, triangular; anapleural sutures well-developed, weakly curved, converging anteriad; mesoventrite very narrow anteriorly; mesocoxae very narrowly separated; abdomen with five ventrites; apical abdominal ventrite with shallow wide emargination; elytron with deeply impressed sutural stria; elytra with series of punctures; elytral trichobothria very distinct; metathoracic leg with five tarsomeres; mesotarsomere 1 very short.

Distribution and age: Liaoning; Early Cretaceous.

Two species included from the Cretaceous of Northern China (see Table 21.1).

Superfamily Staphylinoidea Latreille, 1802

Family Staphylinidae Latreille, 1802

Staphylinidae, called "rove beetles", are a very large and diverse family, now recognized as the most diverse of living organisms. They usually have very short elytra exposing three or more abdominal tergites and highly derived wing folding mechanisms [204].

Genera included from the Jurassic and Cretaceous of Northern China: *Oxyporus* Fabricius, 1775, *Quedius* Stephens, 1829, *Protostaphylinus* Lin, 1976, *Cretoquedius* Ryvkin, 1988, *Laostaphylinus* Zhang, 1988, *Mesostaphylinus* Zhang, 1988, *Sinostaphylina* Hong & Wang, 1990, *Sinostaphylius* Hong, 1992, *Hesterniasca* Zhang, Wang & Xu, 1992, *Glabrimycetoporus* Yue, Zhao & Ren, 2009, *Megolisthaerus* Solodovnikov & Yue, 2010, *Sinoxytelus* Yue, Zhao & Ren, 2010, *Juroglypholoma* Cai & Huang, 2012, *Mesocoprophilus* Cai & Huang, 2013, *Pseudanotylus* Cai & Huang, 2013, *Sinanthobium* Cai & Huang, 2013, *Protodeleaster* Cai, Thayer, Huang, Wang & Newton, 2013, *Cretoprosopus* Solodovnikov & Yue, 2013, *Durothorax* Solodovnikov & Yue, 2013, *Paleothius* Solodovnikov & Yue, 2013, *Paleowinus* Solodovnikov & Yue, 2013, *Thayeralinus* Solodovnikov & Yue, 2013, *Cretoxyporus* Cai & Huang, 2014, *Protoxyporus* Cai & Huang, 2014, *Mesapatetica* Cai, Huang, Newton & Thayer, 2014, *Protolisthaerus* Cai, Beattie & Huang, 2015 and *Paleosiagonium* Yue, Gu, Yang, Wang & Ren, 2016.

***Oxyporus* Fabricius, 1775**

Oxyporus Fabricius, 1775, *Systema Entomologiae, sistens insectorvm classes, ordines, genera, species, adiectis synonymis, locis, descriptionibvs, observationibvs.* Officina Libraria Kortii, Flensburgi & Lipsiae, 1–832 [258].

Type species: *Staphylinus rufus* Linnaeus, 1810.

Adults have large prognathous head with long, curved mandibles and apically expanded terminal segments of labial palpi, antennae short and moniliform. Larvae have a distinctive trilobed mala and stout mandibles that are deeply bifid.

Distribution and age: Liaoning of China, Early Cretaceous; Myanmar, Late Cretaceous.

Only one species included from the Cretaceous of Northern China (see Table 21.1).

***Oxyporus yixianus* Solodovnikov & Yue, 2011 (Figure 21.10)**

Oxyporus yixianus Solodovnikov & Yue, 2011: *J. Syst. Palaeontol.*, 9 (4), 467–471.

Locality and horizon: Huangbanjigou, Beipiao, Liaoning, China; Lower Cretaceous, Yixian Formation.

Bi-colored species with robust body shape, without neck constriction and with mandibles having pronounced subapical cutting edge and very distinct, dorsally positioned dorso-lateral groove. Yue *et al.* placed it in *Oxyporus* Fabricius, 1775, the extant and

Figure 21.10 *Oxyporus yixianus* Solodovnikov & Yue, 2011, (Holotype, CNU-COL-LB2008087) [232].

only genus of the subfamily Oxyporinae. In addition to the hitherto-known, much younger fossils of Oxyporinae, the significantly older age of this newly discovered species makes the currently accepted phylogenetic hypothesis that Oxyporinae are a basal lineage of the Staphylininae group more plausible. They deemed that the remarkably modern and specialization-rich morphology of *Oxyporus yixianus* leads to the assumption that this species exemplifies a rather derived stage of divergence of Oxyporinae from the stem of the Staphylininae group. Divergence of Oxyporinae from the rest of the Staphylininae group lineage may have begun significantly earlier than the Early Cretaceous, and the fungus-associated biology of Oxyporinae is also apparently older than documented by *Oxyporus yixianus* [232].

Quedius Stephens, 1829

Quedius Stephens, 1829, *The Nomenclature of British insects; being a compendious list of such species as are contained in the Systematic Catalogue of British Insects, and forming a guide to their classification. 68 columns.* [262] (original designation).

Type species: *Staphylinus levicollis* Brullé, 1832.

Tarsal formula 5-5-5; presence of variously developed infraorbital ridges; pronotum with more or less deflected hypomera and developed membranous postcoxal process, with dorsal rows of one to three punctures on disc; prosternum usually with median longitudinal carina; mesosternum with sharp intercoxal process (mesocoxae contiguous); aedeagus with parameres fused into a single lamella.

Distribution and age: Liaoning; Early Cretaceous.

Only one species included from the Cretaceous of Northern China (see Table 21.1).

Protostaphylinus Lin, 1976

Protostaphylinus Lin, 1976, *Acta Palaeontol. Sin.*, 15 (1), 97–116 [168] (original designation).

Prostaphylinus Hong, 1983, *Middle Jurassic Fossil Insects in North China*, 223 [93].

Type species: *Protostaphylinus mirus* Lin, 1976.

Antenna 11-segmented, the first three segments rather long, the last five segments gradually dilated, the last segment remarkable dilated and elongate, other segments moniliform; head triangle-shaped, posterior edge narrower, unconspicuous; mandibles not protrudent; eyes located at posterior-laterally area of head; prothorax larger than head, width twice as long as length, pronotum smooth; elytra short and smooth; five abdominal ventrites visible, with straight posterior margin, the last two segments smaller.

Distribution and age: Liaoning; Middle Jurassic.

Only one species included from the Cretaceous of Northern China (see Table 21.1).

Cretoquedius Ryvkin, 1988

Cretoquedius Ryvkin, 1988, *Paleontol. Zh.*, 4, 103–106 [263] (original designation).

Type species: *Cretoquedius oculatus* Ryvkin, 1988.

Head with distinct neck constriction; eyes very large to moderate in size. 11-segmented antenna inserted at frons, antennal insertions not concealed. Pronotum distinctly transverse to about as long as wide, but always with distinct anterior angles, broadly rounded, indistinct posterior angles, and sides broadly rounded; pronotosternal sutures well-developed; prosternum with sharp longitudinal carina along furcasternum. Mesocoxae almost contiguous; mesoscutellum with two basal carinae. Elytra moderately elongate, without distinct punctuation or striae, seemingly without epipleural ridge. Tarsal formula 5–5–5, anterior tarsi with dilated tarsomeres I–IV. Abdomen elongate, with two pairs of paratergites on tergites III–VII; paratergites on each side of a segment well-separated from each other; male sternite VIII simple, without medial concavity, male lateral tergal sclerites IX produced into apically inflated and obtuse processes, with bunches of long setae; male sternite IX usually entire; abdominal tergites III–VII with one basal carina each; abdominal segments without fringes of long stout setae, but covered with somewhat dense short setation.

Distribution and age: Liaoning of China, Early Cretaceous; Russia, Late Cretaceous.

Three species included from the Cretaceous of Northern China (see Table 21.1).

Laostaphylinus Zhang, 1988

Laostaphylinus Zhang, 1988, *Acta Entomol. Sin.*, 31, 79–84 [264] (original designation).

Type species: *Laostaphylinus nigritellu* Zhang, 1988.

Body slender, small in size. Head large, subquadrate. Pronotum nearly quadrate, without setae or punctures. Anterior margin of pronotum slightly broader than posterior, with the sharp anterior angles somewhat produced forwardly, and with round posterior angles. Elytra short, with truncate posterior margin. Legs short and stout, tarsi 5-segmented. Hind tarsi with the first segment short. Abdomen with seven ventrites, with long setae on terminal part. Cerci absent.

Distribution and age: Shandong; Early Cretaceous.

Two species included from the Cretaceous of Northern China (see Table 21.1).

Mesostaphylinus Zhang, 1988

Mesostaphylinus Zhang, 1988, *Acta Entomol. Sin.*, 31, 79–84 [264] (original designation).

Type species: *Mesostaphylinus laiyangensis* Zhang, 1988.

Head with distinct neck constriction and elongate neck; eyes moderately large, about as long as temples. Antenna 11-segmented, inserted in front of eyes. Pronotum apparently slightly oblong, with broad pronotal hypomeron extending into distinct postcoxal process. Elytra relatively short, as long as pronotum, with a structure that can be interpreted as epipleural ridge extending from humerus to about middle of elytral length. All legs moderately long, with large contiguous anterior coxae, femora rather broad anteriorly, tarsomeres I– IV of fore tarsi slightly dilated. Abdomen with two pairs of paratergites at least on tergites III–VI; each pair of paratergites well-separated from each other; lateral tergal sclerites IX dorsally contiguous or fused, apically produced into sharp processes; abdominal tergites III–VII with one basal carina each.

Distribution and age: Shandong, Liaoning; Early Cretaceous.

Four species included from the Cretaceous of Northern China (see Table 21.1).

Sinostaphylina Hong & Wang, 1990

Sinostaphylina Hong & Wang, 1990, Insect fossils of Laiyang Formation. In: *The Stratigraphy and Palaeontology of Laiyang Basin, Shandong Province.* 44–189 [173] (original designation).

Type species: *Sinostaphylina nanligezhuangensis* Hong & Wang, 1990.

Medium-sized species, body length 13 mm. Antenna with scape dilated, following segments reduced abruptly, the first segment about three times as wide as the second segment; pronotum bowl-shaped, anterior margin obviously broader than posterior, posterior angles round; apical of elytral suture form a semilunar gap; legs wide and short, clavate. Abdomen with eight rectangular ventrites, each ventrite with a long hair along the posterior angle; apex of abdomen with fine setae.

Distribution and age: Shandong; Early Cretaceous.

Only one species included from the Cretaceous of Northern China (see Table 21.1).

Sinostaphylius Hong, 1992

Sinostaphylius Hong, 1992, Middle and Late Jurassic insects. In: *Jilin Province Bureau of Geology and Mineral Resource (ed.), Palaeontological Atlas of Jilin Province.* Jilin Science & Technology Press, Changchun, 410–425 [145] (original designation).

Type species: *Sinostaphylius xiejiajieensis* Hong, 1992.

Head nearly triangular with anterior margin rounded; antenna 11-segmented, filiform; pronotum nearly flat persimmon-shaped, lateral suture deep; hind legs strong, tibiae slender with spines on dorsal margin and two apical spurs; tarsi 5-segmented with one pair of claws; abdomen with seven ventrites and a pair of long filum terminale, the sixth and seventh segments with long posterolateral angles and long setae; elytra hard and leathery with truncate posterior margin, dorsal side with dense punctures.

Distribution and age: Jilin; Early Cretaceous.

Only one species included from the Cretaceous of Northern China (see Table 21.1).

Hesterniasca Zhang, Wang & Xu, 1992

Hesterniasca Zhang, Wang & Xu, 1992, *Entomotaxonomia*, 14, 277–281 [265] (original designation).

Type species: *Hesterniasca obesa* Zhang, Wang & Xu, 1992.

Body robust; head small, transverse and triangular-shaped; pronotum rather large, transverse and subquadrate; elytra short and wide; abdomen short and stout with a pair of broad gonocoxites; abdominal ventrites 3–7 with a pair of paratergites.

Distribution and age: Shandong, Liaoning; Early Cretaceous.

Two species included from the Cretaceous of Northern China (see Table 21.1).

Glabrimycetoporus Yue, Zhao & Ren, 2009

Glabrimycetoporus Yue, Zhao & Ren, 2009, *Zootaxa*, 2225, 63–68 [233] (original designation).

Type species: *Glabrimycetoporus amoenus* Yue, Zhao & Ren, 2009.

Head longer than wide, widest near base, slightly retracted under pronotum, without neck. Antenna long, with 11 antennomeres. Pronotum transverse, length/width ratio 0.60, sides broadly arcuate, widest just before posterior angles. Legs long, meso-and metatibiae with two long spurs; protarsi with the first four segments dilated, the last segment longest, slightly shorter than the first three segments combined. Elytra 1.4 times as long as pronotum at midline. Abdomen elongate, sides evenly tapering from base to acute apex and all abdominal segments covered with dense microsetae.

Distribution and age: Liaoning; Early Cretaceous.

Only one species included from the Cretaceous of Northern China (see Table 21.1).

Megolisthaerus Solodovnikov & Yue, 2010

Megolisthaerus Solodovnikov & Yue, 2010, *Insect Syst. Evol.*, 41 (4), 317–327 [231] (original designation).

Type species: *Megolisthaerus chinensis* Solodovnikov & Yue, 2010.

Body relatively large; mandibles elongate and apically pointed, with one or two preapical teeth; gular sutures separated, diverging anteriorly and posteriorly;

antennomere 1 very elongate, more than twice as long as antennomere 2; antennal insertions located at anterior margin of head capsule, with distance between antennal insertions slightly longer than distance from antennal insertion to eye; infraorbital ridges present; neck absent; scutellum large, with concave anterior margin; pro- and mesotarsi 5-segmented, elongate; abdominal tergites III–VI with small arcuate basolateral ridges; abdominal intersegmental membranes with fine and less rectangular sclerites.

Distribution and age: Liaoning; Early Cretaceous.

Two species included from the Cretaceous of Northern China (see Table 21.1).

Sinoxytelus Yue, Zhao & Ren, 2010

Sinoxytelus Yue, Zhao & Ren, 2010, *Cretac. Res.*, 31, 61–70 [234] (original designation).

Type species: *Sinoxytelus euglypheus* Yue, Zhao & Ren, 2010.

Apical maxillary palpi nearly as wide as, and longer than penultimate one; longitudinal grooves along and behind internal margin of eyes; gular suture completely separated; mesocoxae moderately separated by internal process; trochantin exposed; abdominal tergites 3–7 with curved basolateral ridge.

Distribution and age: Liaoning; Early Cretaceous.

Three species included from the Cretaceous of Northern China (see Table 21.1).

Juroglypholoma Cai & Huang, 2012

Juroglypholoma Cai & Huang, 2012, *J. Kansas Entomol. Soc.*, 85 (3), 239–244 [206] (original designation).

Type species: *Juroglypholoma antiquum* Cai & Huang, 2012.

Head slightly longer than wide; antenna with basal two antennomeres normal, not dilated, and distal three slightly dilated to form a weak club; antenna slightly shorter than head and pronotum combined; pronotum transverse, broadened posteriorly; elytra relatively short (for the subfamily), exposing four abdominal segments (tergites V–VIII); metafemora relatively short; epipleural keel of elytra short (about 0.63 times as long as elytral length in lateral view).

Distribution and age: Inner Mongolia of China, Middle Jurassic; Australia, Late Jurassic.

Only one species included from the Jurassic of Northern China (see Table 21.1).

Mesocoprophilus Cai & Huang, 2013

Mesocoprophilus Cai & Huang, 2013, *Insect Syst. Evol.*, 44, 213–220 [211] (original designation).

Type species: *Mesocoprophilus clavatus* Cai & Huang, 2013.

Head relatively large; antennae short, with the apical three antennomeres distinctly dilated to form a distinct club; mesocoxae sub-contiguous, narrowly separated by sharp mesoventral process; abdominal sternite II poorly-sclerotized; intercoxal process between metacoxae absent; abdominal segments III–VII each with two pairs of paratergites; tergites III–VII without basolateral ridges; and metatarsi 5-segmented.

Distribution and age: Liaoning; Early Cretaceous.

Only one species included from the Cretaceous of Northern China (see Table 21.1).

Pseudanotylus Cai & Huang, 2013

Pseudanotylus Cai & Huang, 2013, *Insect Syst. Evol.*, 44 (2), 203–212 [210] (original designation).

Type species: *Pseudanotylus archaicus* (Yue, Makranczy & Ren, 2012).

Anotylus archaicus Yue, Makranczy & Ren, 2012, *Journal of Paleontology*, 86, 508–512 [230].

Body relatively large and elongate; antennae short, moniliform; gular sutures moderately separated, diverging anteriorly and posteriorly; mandibles relatively large and robust; pronotum transverse, narrowed posteriorly; procoxae contiguous; mesocoxae sub-contiguous; all tibiae robust, broadened to apex, each with one or two distinct longitudinal ridge(s); abdominal segment II very short, poorly sclerotized; abdominal segments III–VII each with a single pair of wide paratergites; and all tergites without basolateral ridges.

Distribution and age: Liaoning, Early Cretaceous.

Only one species included from the Cretaceous of Northern China (see Table 21.1).

Sinanthobium Cai & Huang, 2013

Sinanthobium Cai & Huang, 2013, *Can. Entomol.*, 145 (5), 496–500 [209] (original designation).

Type species: *Sinanthobium daohugouense* Cai & Huang, 2013.

Body oval-shaped and small-sized (shorter than 2 mm long); antennae moderate, extending to the posterior margin of pronotum; antennomeres 8–10 nearly quadrate, not transverse nor elongate; pronotum very transverse, having a well-developed pronotal hypomeron with a subtriangularly produced process; elytra relatively long, exposing apical four abdominal segments (tergites V–VIII).

Distribution and age: Inner Mongolia; Middle Jurassic.

Only one species included from the Jurassic of Northern China (see Table 21.1).

Protodeleaster Cai, Thayer, Huang, Wang & Newton, 2013

Protodeleaster Cai, Thayer, Huang, Wang & Newton, 2013, *Cr. Palevol*, 12, 159–163 [216] (original designation).

Type species: *Protodeleaster glaber* Cai, Thayer, Huang, Wang & Newton, 2013.

Body relatively large and elongate, glabrous, impunctate; antennae filiform to moniliform, with apical five antennomeres slightly dilated; gular sutures very narrowly separated at middle of head; epistomal suture present, straight and entire; pronotum well-margined laterally, transverse; exposed part of procoxae small and rounded, procoxal cavities open, procoxal fissure probably open; mesocoxal cavities contiguous; mesoscutellum relatively large and exposed; meso- and metatibiae armed with one or two distinct ridges; abdominal segment II relatively short, poorly sclerotized; abdominal segments III–VII each with a single pair of wide paratergites; and tergites without basolateral ridges.

Distribution and age: Liaoning; Early Cretaceous.

Only one species included from the Cretaceous of Northern China (see Table 21.1).

Cretoprosopus Solodovnikov & Yue, 2013

Cretoprosopus Solodovnikov & Yue, 2013, *Cladistics*, 29, 360–403 [266] (original designation).

Type species: *Cretoprosopus problematicus* Solodovnikov & Yue, 2013.

Frontoclypeal suture present. Antenna 11-segmented. Labrum seemingly bilobed, strongly transverse. Maxillary palps with apical segment fusiform, about 2.5 times as long as wide, and longer than penultimate segment. Prothorax with apparently slightly oblong pronotum. Elytra relatively short, about as long as pronotum, covered with short setae, seemingly without epipleural ridge. All legs moderately long, tarsal formula 5-5-5: anterior legs robust, with broad femora and tibiae, anterior tarsi with strongly dilated tarsomeres I–IV; femora and tibiae of middle legs more slender than those of anterior legs, the latter with pronounced spines; posterior coxae about as long as wide, otherwise shape of posterior legs similar to middle legs; middle and posterior tarsi not expanded. Abdomen with two pairs of paratergites at least on tergites III–VI, paratergites of each pair well-separated from each other.

Distribution and age: Liaoning; Early Cretaceous.

Only one species included from the Cretaceous of Northern China (see Table 21.1).

Durothorax Solodovnikov & Yue, 2013

Durothorax Solodovnikov & Yue, 2013, *Cladistics*, 29, 360–403 [266] (original designation).

Type species: *Durothorax creticus* Solodovnikov & Yue, 2013.

Head capsule slightly elongate, with distinct neck constriction. Eyes small, tempora nearly four times as long as eyes. Eleven-segmented antenna inserted at frons, anteromedially from eyes. Anterior angles of pronotum strongly produced over anterior margin of prosternum, latter with slightly concave anterior margin; notosternal suture well-developed. Elytra about as long as pronotum. Mesocoxae contiguous. All legs moderately long with coxae typical for the tribe Staphylinini. Abdomen elongate, with two pairs of paratergites on tergites III–VI; paratergites on each side of abdominal segment well-separated from each other; paratergites of segment VII obscured by state of preservation; apical margin of sternite VIII simple, without medial concavity; lateral tergal sclerites IX produced into apically inflated and sharp processes; abdominal tergites III–VII each with one basal carina.

Distribution and age: Liaoning; Early Cretaceous.

Only one species included from the Cretaceous of Northern China (see Table 21.1).

Paleothius Solodovnikov & Yue, 2013

Paleothius Solodovnikov & Yue, 2013, *Cladistics*, 29, 360–403 [266] (original designation).

Type species: *Paleothius gracilis* Solodovnikov & Yue, 2013.

Head with broad but distinct neck. Antennae inserted at anterior margin of frons rather close to each other. Prosternum with straight apical margin and sharp intercoxal process. Elytra with sub-basal ridge long, not adjacent to elytral articulation and extending from level of middle of scutellum to elytral humerus. Mesoscutellum with one transverse carina. Mesosternum with sharp intercoxal process; mesocoxae contiguous. Metathorax relatively long; posterior coxae slightly wider than long. Legs relatively short, all about same length, with broad femora and slender tibiae. Middle and posterior tarsi not dilated, 5-segmented. Abdomen parallel-sided along most of its length; segments III–VI each with two pairs of paratergites, segment VII with one pair of short paratergites; segments IX and X not visible, except protruding bunches of setae at apex of lateral tergal sclerites IX; abdominal tergites III–VII with one basal carina each.

Distribution and age: Liaoning; Early Cretaceous.

Only one species included from the Cretaceous of Northern China (see Table 21.1).

Paleowinus Solodovnikov & Yue, 2013

Paleowinus Solodovnikov & Yue, 2013, *Cladistics*, 29, 360–403 [266] (original designation).

Type species: *Paleowinus rex* Solodovnikov & Yue, 2013.

Head without neck constriction. Eyes moderate size. Eleven-segmented antenna inserted at frons, anteromedially from eyes; antennal insertions not concealed. Pronotal hypomera apparently somewhat strongly inflexed under pronotum; anterior angles of pronotum produced over straight anterior margin of prosternum;

notosternal suture well-developed. Elytra moderately elongate, without distinct punctuation or striae. Mesothorax with subcontiguous coxal cavities. Hind wings well-developed, with completely separate MP3, MP4 and CuA veins. Tibiae externally armed with strong setae and spines, with large apical spurs; all tarsi 5-segmented, anterior tarsi with widened tarsomeres I–IV; anterior and middle coxae elongate, somewhat conical, contiguous; posterior coxae about as long as wide. Abdomen elongate, with two pairs of paratergites on segments III–VII; tergite I with prototergal glands; male sternite VIII simple, without medial concavity, male and female lateral tergal sclerites IX produced into apically inflated and sharp processes, with bunches of long setae; male and female sternite IX entire; in females ovipositor consisting of paired structures; abdominal tergite III always with two basal carinae, tergites IV–VII with one or two basal carinae each; apical abdominal segments with long stout setae.

Distribution and age: Liaoning; Early Cretaceous.

Five species included from the Cretaceous of Northern China (see Table 21.1).

Thayeralinus Solodovnikov & Yue, 2013

Thayeralinus Solodovnikov & Yue, 2013, *Cladistics*, 29, 360–403 [266] (original designation).

Type species: *Thayeralinus fieldi* Solodovnikov & Yue, 2013.

Head without neck constriction. Diameter of eyes significantly shorter than length of tempora. Eleven-segmented antenna inserted laterally at frons. Pronotum transverse, somewhat trapezoidal. Elytra from relatively short to moderately elongate, without distinct punctuation or striae, with distinct epipleura separated by epipleural ridge. Mesothorax with contiguous to subcontiguous coxal cavities. All legs moderately long, with robust tibiae armed externally with thick setae and spines, and with large apical spurs; all tarsi 5-segmented, anterior tarsi with widened tarsomeres I–IV; anterior and middle coxae elongate, somewhat conical; posterior coxae about as wide as long, with slightly exposed laterodorsal portion. Abdomen elongate, with well-developed segment II, two pairs of paratergites on tergites III–VI; tergite I with prototergal glands; male sternite VIII with slight medial concavity, female sternite VIII simple, with smoothly rounded apical margin; male and female lateral tergal sclerites IX produced into apically inflated and sharp processes, with bunches of long setae; male sternite IX entire, female with ovipositor consisting of paired proximal and distal gonocoxites; at least abdominal tergites III–VI with one basal carina each; often abdominal segments, especially apical one, laterally and apically fringed with long stout setae.

Distribution and age: Liaoning; Early Cretaceous.

Five species included from the Cretaceous of Northern China (see Table 21.1).

Cretoxyporus Cai & Huang, 2014

Cretoxyporus Cai & Huang, 2014, *Syst. Entomol.*, 39 (3), 500–505 [212] (original designation).

Type species: *Cretoxyporus extraneus* Cai & Huang, 2014.

Body relatively small; antenna with antennomeres 7–10 distinctly transverse, antennomere 7, at least, asymmetrical; mandibles long and sharp, crossed near apex; neck constriction present; gular sutures broadly separated; infraorbital ridges distinctly present, long; mesoventrite very short; mesocoxae ovate; mesocoxae widely separated by a large metaventral anterior process; intercoxal process between metacoxae absent.

Distribution and age: Liaoning; Early Cretaceous.

Only one species included from the Cretaceous of Northern China (see Table 21.1).

Protoxyporus Cai & Huang, 2014

Protoxyporus Cai & Huang, 2014, *Syst. Entomol.*, 39 (3), 500–505 [212] (original designation).

Type species: *Protoxyporus grandis* Cai & Huang, 2014.

Body robust and large; antennae relatively long compared with those of extant oxyporines; antennomeres 6–9 slightly transverse to slightly longer than wide; neck constriction absent; gular sutures narrowly separated; procoxae contiguous, transverse; mesocoxae moderately separated; mesocoxal separation formed by mesoventral posterior and metaventral anterior processes; intermetacoxal carina present on sternites II and III; mesotarsi not bilobed.

Distribution and age: Inner Mongolia; Early Cretaceous.

Only one species included from the Cretaceous of Northern China (see Table 21.1).

Mesapatetica Cai, Huang, Newton & Thayer, 2014

Mesapatetica Cai, Huang, Newton & Thayer, 2014, *J. Kansas Entomol. Soc.*, 87 (2), 219–224 [213] (original designation).

Type species: *Mesapatetica aenigmatica* Cai, Huang, Newton & Thayer, 2014.

Body moderate in size; antennae clubbed, antennomere 8 distinctly smaller than 7 and 9, at least the last five or six antennomeres densely pubescent; epistomal suture V-shaped, apparently without stem; mandibles strong, with acute apices and without subapical teeth; maxillary palpi long, penultimate palpomere shorter than the preceding one and much shorter than the apical one; pronotum transverse, well-margined; elytra long, covering abdominal tergite III, each elytron with nine rows of punctures, with striae present within grooves; tibiae thin, with two longitudinal ridges; procoxae transverse, contiguous; protrochantins exposed; mesocoxae contiguous; mesotarsi 5-segmented, tarsomeres 1–4 short, tarsomere 5 elongate; abdominal segments III–VII apparently with one pair of paratergites each; abdominal sternites II and III with distinct intercoxal carina between metacoxae; tergites IV and V each with a pair of basolateral ridges.

Distribution and age: Inner Mongolia; Middle Jurassic.

Only one species included from the Jurassic of Northern China (see Table 21.1).

Protolisthaerus Cai, Beattie & Huang, 2015

Protolisthaerus Cai, Beattie & Huang, 2015, *Gondwana Res.*, 28 (1), 425–431 [205] (original designation).

Type species: *Protolisthaerus jurassicus* Cai, Beattie & Huang, 2015.

Body comparatively large; antennae elongate, with antennomere 2 short and antennomeres 4–10 each longer than wide; pronotum transverse; elytra each with eight somewhat regular rows of punctures (with some wrinkles near apex); and pronotal postcoxal process normal, not separated from the rest of hypomeron by a thin suture as in modern species.

Distribution and age: Inner Mongolia; Middle Jurassic.

Only one species included from the Jurassic of Northern China (see Table 21.1).

Paleosiagonium Yue, Gu, Yang, Wang & Ren, 2016

Paleosiagonium Yue, Gu, Yang, Wang & Ren, 2016, *Cretac. Res.*, 63, 63–67 [229] (original designation).

Type species: *Paleosiagonium adaequatum* Yue, Gu, Yang, Wang & Ren, 2016.

Antennae short; tempora longer than eyes; gular sutures very narrowly separated, distinctly diverging anteriorly and posteriorly; protibiae without thick spines along outside edge; elytra short, without row of fused punctures; abdomen with elongate longitudinal median carina on abdominal sternite III, tergites III–VII each with one pair of curved basolateral ridges and two basal carinae.

Distribution and age: Liaoning; Early Cretaceous.

Two species included from the Cretaceous of Northern China (see Table 21.1).

Phylogenetic Research on Staphylinidae

Based on the rigorous (maximum parsimony (Figure 21.11), maximum likelihood and Bayesian inference) phylogenetic analysis of both extinct and extant taxa, it confirmed the sister relationships of Paederinae and Staphylininae and the monophyly of both subfamilies. It is also demonstrated that by the Early Cretaceous, Paederinae and Staphylininae were already diversified into groups, some of which now represent extant tribes but not the branches dominating in the modern biota. While the study of the Early Cretaceous rove beetle fauna pushes the estimated divergence time between Paederinae and Staphylininae down into the Jurassic, it also suggests that presently hyperdiverse groups of Staphylininae originated some time later than the Early Cretaceous [266].

Superfamily Scarabaeoidea Latreille, 1802

Family Alloioscarabaeidae Bai, Ren & Yang, 2012

Alloioscarabaeidae are easily recognized by having antennae with lamellate club, canthi of eyes absent, outer margin of protibia with four teeth, preapical surfaces of mesotibiae with transverse or oblique ridges or combs, mesocoxal cavities at middle narrowly separated, radial cell of hind wing open, RP3+4 of hind wing well-developed and not fused or closed to MP1+2 and abdomen 6-segmented. Alloioscarabaeidae were very likely not good diggers and might have fed on decaying organic materials [194].

Only one genus included from the Jurassic of Northern China: *Alloioscarabaeus* Bai, Ren & Yang, 2012.

Alloioscarabaeus Bai, Ren & Yang, 2012

Alloioscarabaeus Bai, Ren & Yang, 2012, *Insect Science*, 19, 159–171 [194] (original designation).

Type species: *Alloioscarabaeus cheni* Bai, Ren & Yang, 2012.

The specific epithet is in honor of Sicien Chen (Professor Shixiang Chen) for his great contribution to the research of insects from China. Broadly oval and compact. Antenna with 3-segmented lamellate club, club loose; clypeal margins convex. Pronotum obviously wider than head, widest at base. Scutellum present, triangular. Abdomen apex slightly rounded. Procoxal cavity strongly transverse; lateral portion of prosternum in front of coxae as long as mid-length of procoxal cavity, mesocoxal cavities not or slightly oblique; metafemora slender; protibia dilated apically; meso- and metatibia slender, not strongly widened; mesotibia outer margin without longitudinal row of teeth.

Distribution and age: Inner Mongolia; Middle Jurassic.

Only one species included from the Jurassic of Northern China (see Table 21.1).

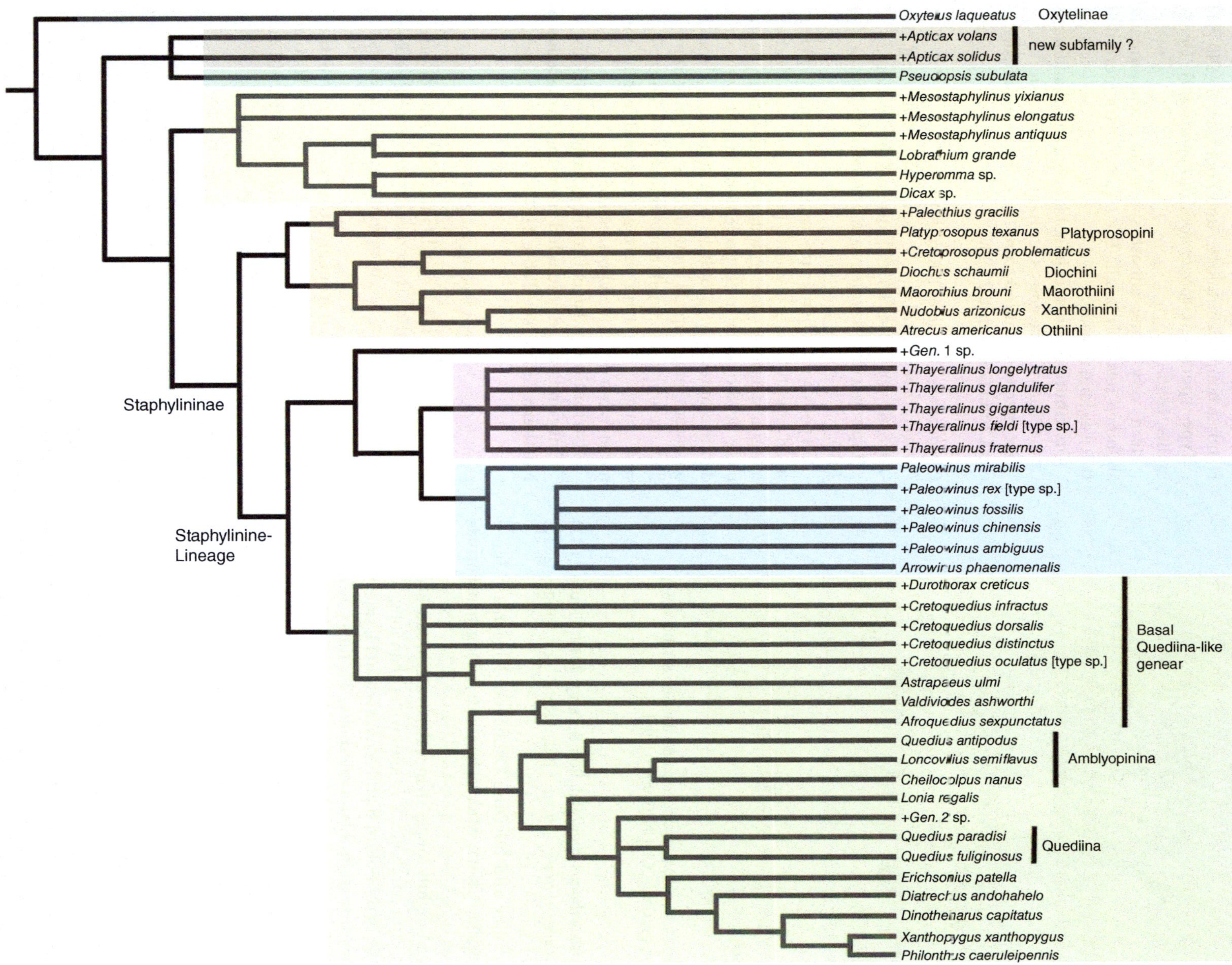

Figure 21.11 Results of maximal parsimony (MP) analyses. Majority rule consensus tree of analysis (MP, extant and fossil taxa together). Names of fossil species marked "+." Source: Modified from [266].

Figure 21.12 *Alloioscarabaeus cheni* Bai, Ren & Yang, 2012, (Holotype, CNU-COL-NN2010603) [194].

***Alloioscarabaeus cheni* Bai, Ren & Yang, 2012 (Figure 21.12)**
Alloioscarabaeus cheni Bai, Ren & Yang, 2012: *Insect Science*, 19, 159–171.

Locality and horizon: Daohugou, Ningcheng, Inner Mongolia, China; Middle Jurassic, Jiulongshan Formation.

Head slightly longer than wide, ovate, broadest at eyes. Pronotum parabolic, 2.2 times wider than long; anterior margin of pronotum slightly concave, lateral sides convex and with margin. Elytra parabolic, 3.1 times as long as pronotum; striae strongly and rather widely impressed, strial punctures distinct. Meso- and metatibiae with two spurs at the end subcqual in length and form. The first abdominal ventrite not completely divided by metacoxae, apex slightly rounded [194].

Family Geotrupidae Latreille, 1802

The Geotrupidae, a small family, comprise a number of relatively large and robust beetles. They are distributed worldwide except Madagascar and New Zealand [267].

Genera included from the Cretaceous of Northern China: *Geotrupoides* Handlirsch, 1906 and *Parageotrupes* Nikolajev & Ren, 2010.

***Geotrupoides* Handlirsch, 1906**
Geotrupoides Handlirsch, 1906, *Die Fossilen Insekten und die Phylogenie der Rezenten Formen, parts I–IV*. Ein Handbuch fur Palaontologen und Zoologen, 1–640 [91].

Type species: *Geotrupoides lithographicus* Deichmüller, 1886.

Head very large, almost spherical. Pronotum and abdomen relatively small. The striped elytra more than twice as long as wide.

Distribution and age: Jilin, Liaoning and Shandong of China, Early Cretaceous; Germany, Late Jurassic.

Six species included from the Cretaceous of Northern China (see Table 21.1).

***Parageotrupes* Nikolajev & Ren, 2010**
Parageotrupes Nikolajev & Ren, 2010, *Acta Geol. Sin. -Engl.*, 84 (4), 673–675 [267] (original designation).

Type species: *Parageotrupes incanus* Nikolajev & Ren, 2010.

Clypeus with tubercle; frontoclypeal suture V-shaped; small eye tubercle present; pronotum with leathery margin anteriorly, not armed; elytra distinctly denticulate at shoulder; male profemora armed with small denticle anteriorly; protibiae with numerous denticles on outer margin; apical tooth of male protibiae unmodified; mesotibiae and metatibiae transversely unicarinate, with single transverse carina near mid-length; male metafemora not armed posteriorly.

Distribution and age: Liaoning; Early Cretaceous.

Only one species included from the Cretaceous of Northern China (see Table 21.1).

Family Glaphyridae Macleay, 1819

The Glaphyridae are a small family and most diverse in the Palearctic realm. These beetles are brightly colored and densely hairy, often resembling bees and visiting flowers.

Genera included from and Cretaceous of Northern China: *Glaphyrus*, Latreille, 1807, *Lithohypna* Nikolajev & Ren, 2011, *Cretohypna* Yan, Nikolajev & Ren, 2012 and *Mesohypna* Nikolajev & Ren, 2013.

***Glaphyrus* Latreille, 1807**
Glaphyrus Latreille, 1807, *Genera crustaceorum et insectorum: secundum ordinem naturalem in familias disposita, iconibus exemplisque plurimis explicata.* Tomus Secundus. Amand Koenig, Parisiis et Argentorati, 1–280 [268].

Type species: *Scarabaeus maurus* Linnaeus, 1758.

Body oblong oval, small to fairly large size. Mandibles and labrum are clearly visible from above; mandibles extending beyond the apex of the labrum. Labrum narrow. Clypeus with two to three teeth on the anterior margin. Eyes small, partially separated by buccal protuberances. Scutellum is visible between the pronotum and the elytra from above. Elytra smooth or with four distinct longitudinal ribs; apexes of the elytra are often pointed. The coxae of the mid legs are not widely separated. Femora of hind legs are often very dilated. Front tibiae with three prongs.

Distribution and age: Liaoning; Early Cretaceous.

Only one species included from the Cretaceous of Northern China (see Table 21.1).

Lithohypna Nikolajev, Wang & Zhang, 2011

Lithohypna Nikolajev, Wang & Zhang, 2011, *Zootaxa*, 2811 (1), 47–52 [269] (original designation).

Fortishybosorus Yan, Bai & Ren, 2013, *Alcheringa*, 37 (2), 139–145 [201]; Syn. by Nikolajev, 2014, *Eur. Entomol. J.*, 13 (3), 253–256 [270].

Type species: *Lithohypna chifengensis* Nikolajev, Wang & Zhang, 2011.

Body elongate. Labrum large, approximately twice as wide as long, somewhat bilobed, sinuate at the middle. Clypeus with truncate, straight apical margin. Scutellum exposed, U-shaped. Elytra elongate, without striae, dehiscent at apex. Elytral lateral margin arcuate. Pygidium visible beyond elytra. Mesocoxae moderately separated. Mesofemora narrow. Protibia with three large denticles on the outer margin. Protarsi longer than protibiae, non-lamellate medially. Metatibiae with single, transverse carina on the outer margin.

Distribution and age: Inner Mongolia, Liaoning; Early Cretaceous.

Seven species included from the Cretaceous of Northern China (see Table 21.1).

Cretohypna Yan, Nikolajev & Ren, 2012

Cretohypna Yan, Nikolajev & Ren, 2012, *ZooKeys*, 241, 67–75 [202] (original designation).

Type species: *Cretohypna cristata* Yan, Nikolajev & Ren, 2012.

Large elongate oval and compact scarab beetle with head, pronotum and mesothorax very close to each other. Mandibles and labrum exposed beyond apex of clypeus and clearly visible in dorsal view of head, labrum approximately five times as wide as long. Pronotum subquadrate-shaped with concave anterior margin and slightly convex lateral and posterior margins. Scutellum triangular. Mesoepimeron clearly visible from above between pronotum and elytron. Elytra convex and thin, without longitudinal carina; hind wings well-developed. Legs short and strong, mesocoxae moderately separated, protibia with three large teeth on outer margin, apex of male mesotibia lamellate; mesotibia and metatibia with two apical spurs; male metatarsus shorter than corresponding tibia. Abdomen with six visible sternites, the first sternites not obscured by hind coxae. Pygidium exposed beyond apices of elytra.

Distribution and age: Inner Mongolia, Liaoning; Early Cretaceous.

Four species included from the Cretaceous of Northern China (see Table 21.1).

Mesohypna Nikolajev & Ren, 2013

Mesohypna Nikolajev & Ren, 2013, *Caucasian Entomological Bulletin*, 9 (1), 62–64 [271] (original designation).

Type species: *Mesohypna lopatini* Nikolajev & Ren, 2013.

The specific epithet is dedicated to the memory of the outstanding coleopterist I.K. Lopatin. Body medium-sized, oblong-oval. Mandibles and labrum visible on dorsal side; mandibles protrude forwardly and extending beyond the labrum. Labrum relatively long with strong anterior edge. Clypeus without projection along the anterior edge, lateral edge slightly convex, widest at about middle. Eyes small, partially divided by buccal protuberances. Antennae short. Epimers of middle thorax visible on dorsal side between pronotum and elytra. Scutellum triangular, short and wide. Elytra smooth without ridges, completely cover the propigidium and pygidium. Mesocoxae not widely separated. Hind femora relatively narrow, about 2.5 times as long as wide. Fore tibiae with tree spines along the outer margin. Fore tarsi of the male approximately equal to the hind leg; legs without blades or comb-like projections along the inner side. Meso- and metatarsi are noticeably longer than the corresponding tibiae. Mesotibiae with two transverse carinae on the outside. Mesotaris with the first tarsomere longest, somewhat longer than the apical segment, tarsomeres 2–4 almost equal in length. Hind legs with segments of almost the same ratio.

Distribution and age: Liaoning; Early Cretaceous.

Two species included from the Cretaceous of Northern China (see Table 21.1).

Family Glaresidae Kolbe, 1905

Glaresidae are a small and widely distributed family except Australia and Antarctica with only one extant genus. They usually occur in sandy areas of semi-arid regions, but little is known of their biology. Glaresids are characterized by head deflexed, eyes with a canthus and with euconeommatidia, Antenna 10-segmented with 3-segmented club, wings with veins RP1 and RP3+4 reduced and abdomen with five visible ventrites.

Only one genus included from the Cretaceous of Northern China: *Glaresis* Erichson, 1848.

Glaresis Erichson, 1848

Glaresis Erichson, 1848, *Naturgeschichte der Insecten Deutschlands. Erste Abtheilung. Coleoptera, Bd. 3.* [272] (original designation).

Type species: *Glaresis rufa* Erichson, 1848.

Antennae with basal segment largest, usually cupuliform and partially enclosing segment 2; eyes divided by large canthus, dorsal portion of eyes small, ventral portion large; mandibles heavily sclerotized, molar region varies from having three blunt teeth to having a plate-like surface, apex may be asymmetrically dentate with more teeth on one mandible than the other. Elytra with 8–10 costae. Thorax long, with well-developed wings, metasternum may have deep metasternal groove for reception of mesotarsi. Metipemeron with apical projection extended over lateral margin of elytra, locking elytra in place. Protibiae strongly developed for digging with three teeth; meso- and metatibia armed with dentiform processes and various ridges and setae; metatibiae broad, spatulate, metafemora and metatibia enlarged to cover abdomen in retracted position.

Distribution and age: Liaoning of China and Russia; Early Cretaceous.

Two species included from the Cretaceous of Northern China (see Table 21.1).

Family Hybosoridae Erichson, 1847

Hybosorids usually are ovate, convex and glabrous scarabs. It is a small family and widely distributed but most common and diverse in tropical regions. They can be distinguished by a combination of functional eighth spiracles, visible labrum and mandibles and cupuliform antennal club.

Genera included from the Cretaceous of Northern China: *Leptosorus* Nikolajev, 2006, *Mesoceratocanthus* Nikolajev, Wang, Liu & Zhang, 2010, *Crassisorus* Nikolajev, Wang & Zhang, 2012, *Pulcherhybosorus* Yan, Bai, & Ren, 2012, *Sinohybosorus* Nie, Bai, Ren & Yang, 2018 and *Sinochaetodus* Lu, Bai, Ren & Yang, 2018.

Leptosorus Nikolajev, 2006

Leptosorus Nikolajev, 2006, *Evraziatskii Entomologicheskii Zhurnal*, 5 (1), 12–13 [273] (original designation).

Type species: *Leptosorus zherikhini* Nikolajev, 2006.

The specific epithet is in honor of the famous Russian paleoentomologist V.V. Zherikhin. Body oblong-oval. Mandibles not hidden under the clypeus with a slightly convex anterior margin. Eyes partially separated by the buccal projection. Coxae of all legs in contact. The coxae of the middle legs are at right angles. Tibiae of the middle and hind legs are relatively thin, without transverse keels along the outer side. Tarsi of the middle and hind legs are slightly longer than the tibiae. The abdomen is completely hidden under the elytra.

Distribution and age: Inner Mongolia of China, Russia; Early Cretaceous.

Only one species included from the Cretaceous of Northern China (see Table 21.1).

Mesoceratocanthus Nikolajev, Wang, Liu & Zhang, 2010

Mesoceratocanthus Nikolajev, Wang, Liu & Zhang, 2010, *Acta Palaeontol. Sin.*, 49, 443–447 [274] (original designation).

Type species: *Mesoceratocanthus tuberculifrons* Nikolajev, Wang, Liu & Zhang, 2010.

Body large. Labrum narrow, oval, about 1/4 as wide as head. Anterior margin of clypeus serrated. Pronotum posteriorly with a deep V-shaped; lateral margin groove sinuate. Abdomen with six visible ventrites. Pygidium not covered by elytra. Profemora wide, almost as wide as long. Metatibiae slender, not distinctly widened.

Distribution and age: Inner Mongolia; Early Cretaceous.

Only one species included from the Cretaceous of Northern China (see Table 21.1).

Crassisorus Nikolajev, Wang & Zhang, 2012

Crassisorus Nikolajev, Wang & Zhang, 2012, *Euroasian Entomological Journal*, 11 (6), 503–506 [275] (original designation).

Type species: *Crassisorus fractus* Nikolajev, Wang & Zhang, 2012.

Winged. Body prominent, large (about 25 mm) and wide (about 12 mm). The first segment of the club of the antenna is enveloping, partially (possibly completely) hides the second and third segments. Scutellum triangular, short, wide. Pygidium is completely hidden under the elytra. Anterior tibiae with three large prongs along the outer margin. Middle legs without transverse carina, hind tibiae, possibly with one transverse carina. The apex of the middle legs with two spurs, approximately equal in length. The apex of the posterior tibiae is elongated into a long curved, hook-shaped process and, possibly, devoid of spurs. All tibiae are five-segmented, claws simple.

Distribution and age: Inner Mongolia; Early Cretaceous.

Only one species included from the Cretaceous of Northern China (see Table 21.1).

Pulcherhybosorus Yan, Bai & Ren, 2012

Pulcherhybosorus Yan, Bai & Ren, 2012, *Zootaxa*, 3478 (1), 201–204 [200] (original designation).

Type species: *Pulcherhybosorus tridentatus* Yan, Bai & Ren, 2012.

Elongate oval. Mandibles and labrum prominent, clearly visible in dorsal view of head, labrum and mandibles exposed in front of clypeus, anterior margin of clypeus moderately emarginated. Antenna with a 3-segmented lamellate club, club compact. Eyes large. Pronotum slightly wider than elytra, nearly trapezoidal, widest at base. Elytra convex, striae well-defined and without tubercles; hind wings well- developed, the MP3 vein present and MP4 vein absent. Abdomen with five visible ventrites. Protibiae with three teeth on outer margin, mesotibiae and metatibiae without a transverse carina, two spurs at the end of mesotibiae subequal in length, and two symmetrical claws at the end of mesotarsi.

Distribution and age: Inner Mongolia; Early Cretaceous.

Only one species included from the Cretaceous of Northern China (see Table 21.1).

Sinohybosorus Nie, Bai, Ren & Yang, 2018

Sinohybosorus Nie, Bai, Ren & Yang, 2018, *Cretac. Res.*, 86, 53–59 [276] (original designation).

Type species: *Sinohybosorus cheni* Nie, Bai, Ren & Yang, 2018.

The specific epithet is in honor of Sicien Chen (Professor Shixiang Chen) for his great contribution to the research of insects from China. Scutellum bigger, the length of elytron 5.3 times as long as scutellum; pronotum widest near 1/4 to base; legs robust, mesotibiae 2.2 times as long as wide, metatibiae 3.5 times as long as wide.

Distribution and age: Liaoning; Early Cretaceous.

Only one species included from the Cretaceous of Northern China (see Table 21.1).

Sinochaetodus Lu, Bai, Ren & Yang, 2018

Sinochaetodus Lu, Bai, Ren & Yang, 2018, *Cretac. Res.*, 86, 53–59 [276] (original designation).

Type species: *Sinochaetodus tridentatus* Lu, Bai, Ren & Yang, 2018.

Pronotum without areolateocellate, meso- and metatibiae slender, outer margin of mesotibiae without longitudinal row of teeth, outer margin of metatibiae with one longitudinal row of teeth.

Distribution and age: Liaoning; Early Cretaceous.

Only one species included from the Cretaceous of Northern China (see Table 21.1).

Family Lithoscarabaeidae Nikolajev, 1992

Lithoscarabaeidae was established by Nikolajev based on a single species *Proteroscarabaeus baissensis* Nikritin, 1977, and a genus *Lithoscarabaeus* described in 1992 [277]. Then, the second genus of this family, *Baisarabaeus* Nikolajev, 2005, was described based on several fossils with only part of the body and elytra [278]. Nikolajev *et al.* described a new species of *Baisarabaeus* from China based on a single elytron in 2013 [279].

Only one genus included from the Cretaceous of Northern China: *Baiscarabaeus* Nikolajev, 2005.

Baisarabaeus Nikolajev, 2005

Baisarabaeus Nikolajev, 2005, *Biologicheskie Nauki Kazakhstana*, 1, 117–120 [278] (original designation).

Type species: *Baisarabaeus rugosus* Nikolajev, 2005.

Relatively large beetle. The coxae of the middle legs are located at a right angle, not very wide, but clearly separated by a protuberance of the middle thorax; hind tibiae with no more than one transverse carina on the outside before the apex.

Distribution and age: Inner Mongolia; Early Cretaceous.

Only one species included from the Cretaceous of Northern China (see Table 21.1).

Family Lucanidae Latreille, 1804

Lucanids only have three abdominal ventrites and are usually recognized by their pectinate antennal club. In many cases, the mandibles are large and ornately toothed in the males which clearly indicate their sexual dimorphism.

Genera included from the Jurassic and Cretaceous of Northern China: *Juraesalus* Nikolajev, Wang, Liu & Zhang, 2011, *Sinaesalus* Nikolajev, Wang, Liu & Zhang, 2011, *Prosinodendron* Bai, Ren & Yang, 2012 and *Litholamprima* Nikolajev & Ren, 2015.

Juraesalus Nikolajev, Wang, Liu & Zhang, 2011

Juraesalus Nikolajev, Wang, Liu & Zhang, 2011, *Acta Palaeontologica Sin.*, 50, 41–47 [85] (original designation).

Type species: *Juraesalus atavus* Nikolajev, Wang, Liu & Zhang, 2011.

Body elongate and large. Eyes without canthi. Mandibles short. Antennae non-geniculate. Mesocoxae moderately separated. Protibia with two or three large denticles on the outer margin. Meso- and metatibiae with numerous denticles on the outer margin. Meso- and metatarsi nearly as long as meso- and metatibiae respectively. The first metatarsomere almost as long as the second tarsomere. Abdomen with five visible ventrites; the first four ventrites subequal in length.

Distribution and age: Inner Mongolia; Middle Jurassic.

Only one species included from the Jurassic of Northern China (see Table 21.1).

Sinaesalus Nikolajev, Wang, Liu & Zhang, 2011

Sinaesalus Nikolajev, Wang, Liu & Zhang, 2011, *Acta Palaeontol. Sin.*, 50, 41–47 [85] (original designation).

Type species: *Sinaesalus tenuipes* Nikolajev, Wang, Liu & Zhang, 2011.

Body large. Eyes without canthi. Mandibles short. Antenna non-geniculate. Scutellum about as long as wide. Mesocoxae narrowly separated. Protibia with four or five large denticles on the outer margin. Mesotarsus longer than mesotibia; metatarsus almost as long as metatibia. The first mesotarsomere longer than combined length of the second and third tarsomeres. The first metatarsomere approximately as long as the second–fourth tarsomeres combined. Abdomen with five visible ventrites; the first four ventrites subequal in length.

Distribution and age: Inner Mongolia; Early Cretaceous.

Three species included from the Cretaceous of Northern China (see Table 21.1).

Prosinodendron Bai, Ren & Yang, 2012

Prosinodendron Bai, Ren & Yang, 2012, *Cretac. Res.*, 34 (3), 334–339 [199] (original designation).

Type species: *Prosinodendron krelli* Bai, Ren & Yang, 2012.

The specific epithet is in honor of Dr. Krell for his great contribution to research on scarab fossils. Body elongate, antennae lamellate, club loose, incisor edge of mandible with two or three teeth, mandibular apex bidentate, anterior or mesal edge of eye not emarginated, metacoxae extending laterally to meet sides of body, scutellum well-developed.

Distribution and age: Liaoning; Early Cretaceous.

Only one species included from the Cretaceous of Northern China (see Table 21.1).

Litholamprima Nikolajev & Ren, 2015

Litholamprima Nikolajev & Ren, 2015, *Caucasian Entomol. Bull.*, 11, 15–18 [280] (original designation).

Type species: *Litholamprima longimana* Nikolajev & Ren, 2015.

Body elongated, relatively large, about 17 mm (without mandibles). The antennas are not geniculate, with a short basal segment. Eyes are whole, not separated by buccal protuberances. Pronotum of trapezoidal shape, slightly wider than length, without leathery border along anterior margin. Scutellum triangular, short, markedly wider than length. Elytra elongated, with slightly expressed grooves. Wings are well-developed. The anterior coxae are widely separated by an anterolateral process. The mesosternum is narrow, hidden under the coxae. Sexual dimorphism is evident in a strong elongation in males of mandibles and protibia; apical spur of protibia of the male modified - expanded and flattened, with a widely rounded apex.

Distribution and age: Liaoning; Early Cretaceous.

Only one species included from the Cretaceous of Northern China (see Table 21.1).

Family Ochodaeidae Mulsant & Rey, 1871

Ochodaeidae are a small and cosmopolitan family, occurring mainly in the Holarctic region, southern Africa and Madagascar. Adults are usually collected under lights in semi-arid, sandy areas.

Genera included from the Jurassic and Cretaceous of Northern China: *Mesochodaeus* Nikolajev & Ren, 2010 and *Yixianochodaeus* Nikolajev, 2015.

Mesochodaeus Nikolajev & Ren, 2010

Mesochodaeus Nikolajev & Ren, 2010, *Zootaxa*, 2553 (1), 65–68 [89] (original designation).

Type species: *Mesochodaeus daohugouensis* Nikolajev & Ren, 2010.

Mandibles externally rounded, visible beyond labrum in dorsal view; mesocoxae appear to be contiguous (but probably very narrowly separated in life); mesotibiae with two or three transverse carinae along outer edge.

Distribution and age: Inner Mongolia; Middle Jurassic.

Only one species included from the Jurassic of Northern China (see Table 21.1).

Yixianochodaeus Nikolajev, 2015

Yixianochodaeus Nikolajev, 2015, *Euroasian Entomological Journal*, 14, 21–26 [281] (original designation).

Type species: *Yixianochodaeus horridus* Nikolajev, 2015.

Body relatively small, oblong-oval; the color of the underside is two-colored-light, with dark spots. Mandibles and labrum are not hidden under the clypeus; the eyes are not separated by the buccal protuberances; on propigidia, a mechanism is developed to fix the wings when they move back; mesocoxae seem to be contiguous (it is possible that in a living beetle they were not widely separated); tibiae of the front legs with three teeth along the outer edge; tibiae of middle and hind legs, with at least two transverse carinae; tibiae of the hind legs are not flattened-round in cross section; at least one of the spurs of hind tibiae is comb-shaped.

Distribution and age: Liaoning; Early Cretaceous.

Only one species included from the Cretaceous of Northern China (see Table 21.1).

Family Pleocomidae LeConte, 1861

Pleocomidae contains only one extant genus *Pleocoma* occurring along the west coast of North America from

Washington to Baja California. Adult males are winged and fly during the first spring rain. Females are wingless and remain in burrows in the soil except to mate.

Genera included from the Cretaceous of Northern China: *Pleocoma* LeConte, 1856 and *Proteroscarabaeus* Grabau, 1923.

Pleocoma LeConte, 1856

Pleocoma LeConte, 1856, *P. Acad. Nat. Sci. Phila.*, 8, 19–25 [282] (original designation).

Type species: *Pleocoma fimbriata* LeConte, 1856.

Elongate oval, rather large beetles, displaying sexual dimorphism (males smaller and winged; females with reduced wings). Mandibles and labrum hidden under clypeus. Anterior margin of clypeus with high process directed dorsally or anterodorsally. Antenna 11-segmented, with 4–8-segmented lamellate club. Eyes not divided by genal canthus. Procoxal cavities (with clearly visible trochantin), open externally; mesocoxal cavities separated by very narrow process. Meso and metatibiae with one transverse carina externally.

Distribution and age: Liaoning; Early Cretaceous.

Only one species included from the Cretaceous of Northern China (see Table 21.1).

Proteroscarabaeus Grabau, 1923

Proteroscarabaeus Grabau, 1923, *Bulletin of the Geological Survey of China*, 5, 148–181 [283] (original designation).

Type species: *Proteroscarabaeus yeni* Grabau, 1923.

Large and medium-sized, broad, flattened beetles. Labrum and mandibles not fully concealed under clypeus. Pronotum transverse, laterally rounded. Spiracles of abdominal segments large, present on membrane. Elytra not covering apex of abdomen, without distinct punctate grooves. Wings with two veins between Cubitus and the basally joining anal.

Distribution and age: Jilin and Shandong of China and Russia; Early Cretaceous.

Two species included from the Cretaceous of Northern China (see Table 21.1).

Family Scarabaeidae Latreille, 1802

Scarabaeidae are a very large, diverse and cosmopolitan family. But their abundance and diversity drop off considerably in cooler climates.

Only one genus included from the Cretaceous of Northern China: *Prophaenognatha robusta* Bai, Ren & Yang, 2011.

Prophaenognatha Bai, Ren & Yang, 2011

Prophaenognatha Bai, Ren & Yang, 2011, *Acta Geol. Sin-Engl.*, 85 (5), 984–993 [198].

Type species: *Prophaenognatha robusta* Bai, Ren & Yang, 2011.

Medium-sized aclopines, labrum and mandibles prominent, clypeus with transverse ridge, protibiae with four teeth on outer margin, elytra with distinct punctuate striae, meso- and metatibiae slender, outer margin of mesotibiae without transverse ridge or tooth.

Distribution and age: Liaoning; Early Cretaceous.

Only one species included from the Cretaceous of Northern China (see Table 21.1).

Family Septiventeridae Bai, Ren, Shih & Yang, 2013

Septiventeridae are easily recognized by the abdomen with seven ventrites, mesocoxal cavities contiguous, antenna with 3-segmented lamellate club, radial cell of hind wing well-developed [196].

Only one genus included from the Cretaceous of Northern China: *Septiventer* Bai, Ren, Shih & Yang, 2013.

Septiventer Bai, Ren, Shih & Yang, 2013

Septiventer Bai, Ren, Shih & Yang, 2013, *J. Syst. Palaeontol.*, 11 (3), 359–374 [196] (original designation).

Type species: *Septiventer quadridentatus* Bai, Ren, Shih & Yang, 2013.

Broadly oval, compact and pilose. Head roughly trapezoid, widening behind small compound eyes; posterior edge almost straight; clypeal region long and broad, anteriorly rounded, apparently covering labrum. Antenna inserted on ventral side of head; with slender, elongate segments proximad 3-segmented, lamellate club. Mandibles completely concealed. Maxillary palp 3-segmented. Pronotum anteriorly slightly wider than head, widest at base. Scutellar shield triangular. Elytra shorter than abdomen. Radial cell of hind wing well-developed. Procoxal cavities strongly transverse; protibiae widening apically, with four teeth on outer margin; mesocoxal cavities contiguous, not or very slightly oblique; meso- and metatibiae slender; preapical surfaces of meso- and metatibia with one transverse ridge. Abdomen with seven ventrites.

Distribution and age: Liaoning; Early Cretaceous.

Only one species included from the Cretaceous of Northern China (see Table 21.1).

Septiventer quadridentatus Bai, Ren, Shih & Yang, 2013 (Figure 21.13)

Septiventer quadridentatus Bai, Ren, Shih & Yang, 2013: *J. Syst. Palaeontol.*, 11 (3), 359–374.

Locality and horizon: Huangbanjigou, Beipiao, Liaoning, China; Lower Cretaceous, Yixian Formation.

The first segment of antennal club is not cupuliform; all three club segments about equally long, longer than two antennomeres proximad of club combined. Pronotum

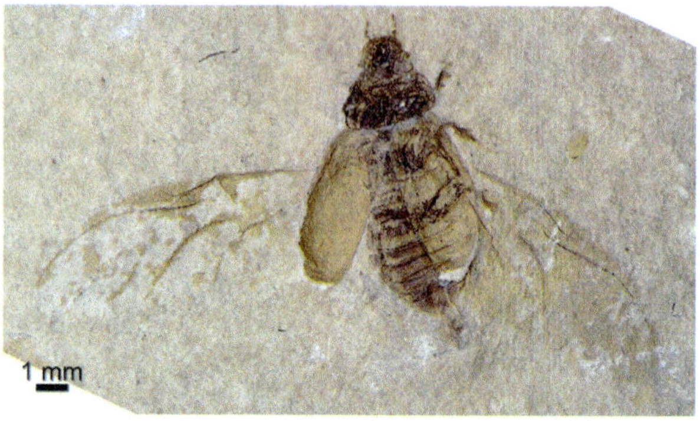

Figure 21.13 *Septiventer quadridentatus* Bai, Ren, Shih & Yang, 2013, (Holotype, CNU-COL-LB-2010-624) [196]. Source: Donated by Dr. Chungkun Shih.

2.2 times wider than long at base; anterior margin slightly concave; lateral margin rounded and marginate. Elytra subparallel, rounded anterolaterally, anteromesally and posteriorly; 2.2 times as long as wide at midlength; 0.8 times as wide as prothorax. Radial cell of hind wing with acute inner posterior angle at base; apical area 0.43 times as long as total wing length. Protarsi present and slender; contiguous mesocoxal cavities 1.5 times as long as wide; meso- and metatibia with long, shallow concavity on distal, external margin; meso- and metatibiae not distinctly widened; mesotibial outer margin without longitudinal row of teeth. Apical sternite VIII rounded. Aedeagus blade-like, asymmetric, with oblique, slightly concave distal edges; apex sharp; paramere possibly represented by a sharp triangular element [196].

Systematic Position of Septiventeridae Bai, Ren, Shih & Yang, 2013 (Cladistic evidence)

Septiventeridae are part of a largely unresolved polytomy in the strict consensus of the three cladograms obtained with all characters unweighted. They are the sister group of the remaining Scarabaeoidea in one of these trees (Figure 21.14). In this tree Glaresidae and Trogidae form two successive branches. This pattern suggests that Septiventeridae are a key taxon for understanding the early evolution of Scarabaeoidea, and that characters found in the three families may largely reflect the groundplan of the entire superfamily [196].

Infraorder Elateriformia Crowson, 1960

Superfamily Dascilloidea Guérin-Méneville, 1843

Family Dascillidae Guérin-Méneville, 1843

The beetle family Dascillidae are divided into two subfamilies, a free-living Dascillinae and often morphologically modified Karumiinae, some of which are apparently associated with subterranean termites [284, 285].

Only one genus included from the Cretaceous of Northern China: *Cretodascillus* Jin, Ślipiński, Pang & Ren, 2013.

Cretodascillus Jin, Ślipiński, Pang & Ren, 2013

Cretodascillus Jin, Ślipiński, Pang & Ren, 2013, *Ann. Zool. (Warszawa)*, 63 (3), 501–509 [285] (original designation).

Type species: *Cretodascillus sinensis* Jin, Ślipiński, Pang & Ren, 2013.

Elytra without distinct rows of punctures or striae, comparatively broader prosternal process and the weakly oval mesocoxae.

Distribution and age: Inner Mongolia; Early Cretaceous.

Only one species included from the Cretaceous of Northern China (see Table 21.1).

Superfamily Buprestoidea Leach, 1815

Family Buprestidae Leach, 1815

Buprestidae are a large cosmopolitan family of small to relatively large "jewel beetles", often bicolored with light and dark pattern or exhibiting bright metallic colors. They usually can be distinguished by short, hypognathous head with vertically oval, finely facetted eyes and a median endocarina; slender, more or less serrate antennae; broad prosternal process fitting into a large mesoventral cavity and often concealing most of the mesoventrite; exposed protrochantins which fit tightly against the prothorax and are not freely movable; well-defined transverse metaventral suture and solidly fused ventrites 1 and 2 [286].

Genera included from the Jurassic and Cretaceous of Northern China: *Planocoleus* Hong, 1982, *Sinoparathyrea* Pan, Chang & Ren, 2011, *Trapezitergum* Yu, Ślipiński & Shih, 2013 and *Mongoligenula* Yu, Ślipiński, Pang & Ren, 2015.

Planocoleus Hong, 1982

Planocoleus Hong, 1982, Mesozoic Fossil Insects of Jiuquan Basin in Gansu Province, 149 [160] (original designation).

Type species: *Planocoleus glabratus* Hong, 1982.

Elytra cutellate and broad, flat or slightly convex, anterior and posterior margins with partly parallel and straight; surfaces smooth or covered with weak striae, without punctate ornaments; anterior margin straight anteriorly and slightly oblique posteriorly, base narrow with apex distinctly constricted; posterior epipleuron with granular ornaments; surfaces simple, without complicated ornaments, radial field without short suture.

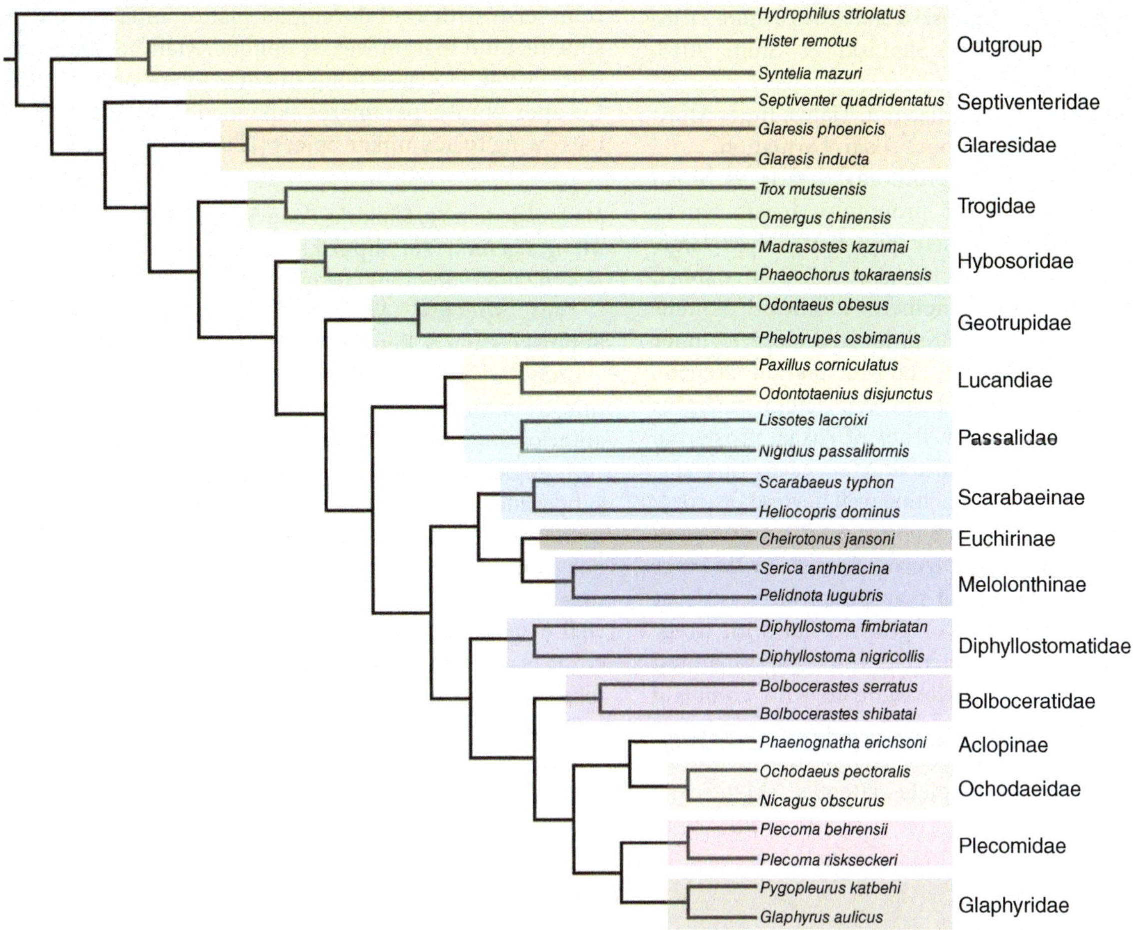

Figure 21.14 Preferred cladogram of three most parsimonious trees from the parsimony analysis of 27 species of extant Scarabaeoidea, one fossil species, and three outgroups (tree length = 270 steps, CI = 0.38, RI = 0.64). Source: Modified from [196].

Distribution and age: Gansu of China and Mongolia; Early Cretaceous.

Two species included from the Cretaceous of Northern China (see Table 21.1).

***Sinoparathyrea* Pan, Chang & Ren, 2011**

Sinoparathyrea Pan, Chang & Ren, 2011, *Zootaxa*, 2745, 53–62 [287] (original designation).

Type species: *Sinoparathyrea bimaculata* Pan, Chang & Ren, 2011.

Body medium-sized, about 13–18 mm long, subcylindrical, with head nearly as wide as anterior margin of pronotum. Pronotum about 1.6 times as wide as long, lateral margins weakly arched at anterior 3/4, then arcuately expanded to the posterior margin, posterior angles acute, widest at base. Elytra more than 3.0 times as long as wide, widest at middle; lateral margins weakly arcuate in anterior 2/3, then somewhat arcuately narrowed to obtuse apex, the last ventrite of abdomen broadly rounded at apex.

Distribution and age: Inner Mongolia; Middle Jurassic.

Three species included from the Jurassic of Northern China (see Table 21.1).

***Trapezitergum* Yu, Ślipiński & Shih, 2013**

Trapezitergum Yu, Ślipiński & Shih, 2013, *Zootaxa*, 3637 (3), 355–360 [239] (original designation).

Type species: *Trapezitergum grande* Yu, Ślipiński & Shih, 2013.

Large body, prothorax twice as wide as long with pronotum trapezoidal and the prosternum in front of procoxae much longer than procoxal diameter, the metaventral transverse suture distinctly bisinuate medially and the abdominal ventrite five triangular and weakly tridentate apically.

Distribution and age: Inner Mongolia; Early Cretaceous.

Only one species included from the Cretaceous of Northern China (see Table 21.1).

Trapezitergum grande **Yu, Ślipiński & Shih, 2013 (Figure 21.15)**
Trapezitergum grande Yu, Ślipiński & Shih, 2013: *Zootaxa*, 3637 (3), 355–360.

Locality and horizon: Liutiaogou, Tianyi, Inner Mongolia, China; Lower Cretaceous, Yixian Formation.

Head retracted into the prothorax, nearly as broad as the anterior portion of pronotum; frontoclypeus broad, feebly convex, densely punctuate; eyes large, oval, inner margins converging above. Antenna short, weakly serrate from antennomere 4 onward. Antennal sockets distinctly separated, located close to inner margin of eyes. Pronotum anterior margin arcuate, posterior margin slightly bisinuate; lateral margins straight. Prosternum long in front of coxae; prosternal process narrower than coxal diameter, narrowing apically, apex feebly pointed, reaching well beyond procoxa; postcoxal projections absent; notosternal sutures well-developed. Elytra with epipleuron narrow, with inner margin subparallel to elytral margin, surfaces without obvious rows of punctures or grooves but with irregular micropunctures. Mesocoxae rounded, separated by about coxal diameter. Metaventrite with complete discrimen; metanepisternum triangular, converging posteriorly. Transverse metaventral suture distinctly bisinuate medially, incomplete laterally. Metacoxae transverse with well-developed plates; anterior margin sinuate, hind margin weakly oblique. Male genitalia partially visible, elongate, penis (median lobe) parallel-sided, the apex obtuse; parameres expanded preapically before narrowing to acuminate apices [239].

Mongoligenula **Yu, Ślipiński, Pang & Ren, 2015**
Mongoligenula Yu, Ślipiński, Pang & Ren, 2015, *Cretac. Res.*, 52, 480–489 [237] (original designation).

Type species: *Mongoligenula altilabdominis* Yu, Ślipiński, Pang & Ren, 2015.

Body small. Head short and hypognathous. Pronotum transverse, roundly trapezoidal and widest at base; anterior and posterior margins slightly bisinuate; posterior angles acute. Prosternal process well-developed, subparallel and apex obtuse. Metaventrite wider than long with complete discrimen; metakatepisternal suture sinuate. Elytra with irregular spots and nine distinct longitudinal striae. The radial cell is well-developed. Pro- and mesocoxae separated; metacoxae transverse and contiguous. Ventrites 1 and 2 solidly fused.

Distribution and age: Liaoning; Early Cretaceous.

Two species included from the Cretaceous of Northern China (see Table 21.1).

Figure 21.15 *Trapezitergum grande* Yu, Ślipiński & Shih, 2013, (Holotype, CNU-COL-NN-2010411). (a) Photograph; (b) Line drawing [239]. Source: Donated by Dr. Chungkun Shih.

Family Schizopodidae LeConte, 1861

Schizopodidae, called "false jewel beetles", are a small group of beetles endemic to the western North America. They are very similar to, but can be separated from Buprestidae by the wide metanepisternum, deeply bilobed fourth tarsomere, and peculiar wing venation and folding [215, 286, 288].

Only one genus included from the Cretaceous of Northern China: *Mesoschizopus* Cai, Ślipiński & Huang, 2015.

Mesoschizopus Cai, Ślipiński & Huang, 2015

Mesoschizopus Cai, Ślipiński & Huang, 2015, *Cretac. Res.*, 52, 490–494 [215] (original designation).

Type species: *Mesoschizopus elegans* Cai, Ślipiński & Huang, 2015.

Body robust, moderately oval. Head slightly declined. Pronotum transverse. Prosternum short; prosternal process short and narrow. Metaventrite wider than long; discrimen complete; katepisternal suture complete and sinuate. Metanepisternum very broad. Elytron with color pattern and round-window punctures. Elytron with a long, slightly sinuate carina near outer edge. Hind wing with radial cell relatively wide; wedge cell fusiform, relatively narrow. Mesocoxae moderately separated. Metacoxae large, contiguous. Abdomen broad, with five ventrites.

Distribution and age: Liaoning; Early Cretaceous.

Only one species included from the Cretaceous of Northern China (see Table 21.1).

Superfamily Byrrhoidea Latreille, 1804

Family Byrrhidae Latreille, 1804

Byrrhids have highly convex and compact body, reduced clypeus and lack a transverse metaventral suture. They usually also have a deep transverse impression between the labrum and an abrupt frontal ridge and a series of compaction mechanisms [289].

Genera included from the Jurassic and Cretaceous of Northern China: *Fangshanella* Huang & Zhang, 1997 and *Mesobyrrhus* Huang & Zhang, 1997.

Fangshanella Huang & Zhang, 1997

Fangshanella Huang & Zhang, 1997, *J. Nanjing University (Natural Sciences)*, 33 (4), 562–569 [290] (original designation).

Type species: *Fangshanella stolida* Huang & Zhang, 1997.

Body slightly large in size and rather long, with ventral surface and legs chagrined. Head long and decorated with tuberculi. Prothorax rather long with sides inclined. Metathorax with fine setae on surface. Hind coxae convex. Tarsi with tiny lobes. Abdominal sutures apparently curved. Elytra with longitudinal striae not parallel to each other, and undeveloped epipleura.

Distribution and age: Beijing; Early Cretaceous.

Only one species included from the Cretaceous of Northern China (see Table 21.1).

Mesobyrrhus Huang & Zhang, 1997

Mesobyrrhus Huang & Zhang, 1997, *J. of Nanjing University (Natural Sciences)*, 33 (4), 562–569 [290] (original designation).

Type species: *Mesobyrrhus tanae* Huang & Zhang, 1997.

The specific epithet is dedicated to famous entomologist Mr. Juanjie Tan, a researcher in the Institute of Zoology, Chinese Academy of Sciences. Body oval in form, convex with ventral surface and legs chagrined. Vertex rather wide, lavigate. Pronotum with anterior margin intensely narrowed; lateral and anterior margins difficult to distinguish, arched; its width much longer than length. Tibiae gradually widened, not longer than femora, with apex not apparently enlarged. Tarsi with tiny lobes. Abdomen with apex rounded. Elytra decorated with several weak longitudinal striae parallel to each other.

Distribution and age: Beijing; Early Cretaceous.

Two species included from the Cretaceous of Northern China (see Table 21.1).

Family Eulichadidae Crowson, 1973

Eulichadidae are a small family containing two genera: *Eulichas* in Asia and the East Indies and *Stenocolus* in North America. They are relatively large, elaterid-like beetles (13–35 mm), clothed with short setae, which may form a pattern on the elytra [291].

Genera included from the Cretaceous of Northern China: *Mesaplus* Hong, 1983 and *Cretasyne* Yan, Wang & Zhang, 2013.

Mesaplus Hong, 1983

Mesaplus Hong, 1983, Middle Jurassic Fossil Insects in North China, 79 [93] (original designation).

Type species: *Mesaplus beipiaoensis* Hong, 1983.

Body medium-sized and slender. Head small, concealed under the pronotum; pronotum transverse, half-moon shaped, posterolateral areas tilt toward central, distinctly narrower than the base of elytra; scutellum small and triangular; mesoventrite not extending beyond metaventrite, metanepisternum long, reaching to the first abdominal ventrite; pro- and mesocoxae rounded, metacoxae transverse, triangular and contiguous, not forming right angle with apical areas separated;

metafemora narrower than metatibiae, metatibiae with one spur; abdomen longer than pterothorax; elytra long, three to four times as long as wide, apical acute, dehiscent about one half, with 11 rows of longitudinal ridges, areas between ridges with dense punctures arranged in a horizontal line.

Distribution and age: Liaoning; Middle Jurassic.

Only one species included from the Jurassic of Northern China (see Table 21.1).

Cretasyne Yan, Wang & Zhang, 2013

Cretasyne Yan, Wang & Zhang, 2013, *Cretac. Res.*, 40 (1), 43–50 [292] (original designation).

Type species: *Cretasyne lata* Yan, Wang & Zhang, 2013.

The suture between gular plate and submentum is obliterated, trapezoid pronotum almost absent, anterior angles rounded and posterior angles weakly protruded lateroposteriorly; intercoxal process of prosternum very wide, only half as wide as procoxae; elytral striae shallow, shortened striae absent.

Distribution and age: Inner Mongolia; Early Cretaceous.

Two species included from the Cretaceous of Northern China (see Table 21.1).

Family Heteroceridae MacLeay, 1855

Heteroceridae are a distinctive and invariable family worldwide. Their antennae are short and club-like and the legs strongly spiny. They are classified in two subfamilies and usually live close to water in habitats consisting of mud or very fine sand [293].

Only one genus included from the Cretaceous of Northern China: *Heterocerites* Ponomarenko, 1986.

Heterocerites Ponomarenko, 1986

Heterocerites Ponomarenko, 1986, Coleoptera. Scarabaeida (=Coleoptera), in Nasekomye v rannemelovykh ekosistemakh Zapadnoi Mongolii (*Insects in the Early Cretaceous Ecosystems of Western Mongolia)*, Moscow, Nauka, 84–105 [294] (original designation).

Type species: *Heterocerites kobdoensis* Ponomarenko, 1986.

This genus doesn't have distinct diagnostic character, which was suggested to describe Mesozic Heteroceridae of unclear generic position [294].

Distribution and age: Liaoning of China, Mongolia; Early Cretaceous.

Only one species included from the Cretaceous of Northern China (see Table 21.1).

Family Lasiosynidae Kirejtshuk, Chang, Ren & Shih, 2010

The extinct family Lasiosynidae includes numerous and morphologically-diverse Mesozoic elateriform beetles. They possess transverse coxae on all legs, exposed trochantins on the pro- and mesocoxae, retained discrimen on metaventrites, prognathous head, elongate radial cell and short apical field [97, 292].

Genera included from the Jurassic and Cretaceous of Northern China: *Lasiosyne* Tan, Ren & Shih, 2007, *Anacapitis* Yan, 2009, *Bupredactyla* Kirejtshuk, Chang, Ren & Shih, 2010 and *Parelateriformius* Yan & Wang, 2010.

Lasiosyne Tan, Ren & Shih, 2007

Lasiosyne Tan, Ren & Shih, 2007, *Ann. Zool. (Warszawa)*, 57 (2), 231–247 [96] (original designation).

Type species: *Lasiosyne euglyphea* Tan, Ren & Shih, 2007.

Body subparallel-sided to subcylindrical; head (sub) prognathous, antennal grooves usually expressed on ventral side; mandibles large and with acute apex; eyes large; labrum moderately short and transverse; cervical constriction distinct; antenna rather long and with 11 segments, antennomere two short and frequently transverse, remaining antennomeres elongate subcylindrical or subconical and usually rather long, sometimes widened apically; pronotum much narrower than elytral base, subquadrate to slightly transverse, not crenulate along base, anterior angles rounded and not projecting, anterior margin nearly as wide as head, posterior angles sharp, extending lateroposteriorly; metepisterna 2.5–3.0 times as long as wide; elytra with subacute apices and 11 striae (2 and 3 not complete); prosternal process moderately narrow; mesocoxae narrowly separated to subcontigous; metacoxal plates slightly to moderately developed only in median part of coxae; tarsi 5-segmented with moderately lobed tarsomeres 1–4.

Distribution and age: Inner Mongolia of China, Middle Jurassic; Mongolia, Late Jurassic, Early Cretaceous; Russia, Early Cretaceous.

Five species included from the Jurassic of Northern China (see Table 21.1).

Lasiosyne fedorenkoi Kirejtshuk, Chang & Ren, 2010 (Figure 21.16)

Lasiosyne fedorenkoi Kirejtshuk, Chang & Ren, 2010: *Ann. Soc. Entomol. Fr.*, 46 (1–2), 67–87.

Locality and horizon: Daohugou, Ningcheng, Inner Mongolia, China; Middle Jurassic, Jiulongshan Formation.

The specific epithet is devoted to D.N. Fedorenko who helped to correctly reconstruct venation and folding of hind wing of *Lasiosyne*. Anterior edge of pronotum

Figure 21.16 *Lasiosyne fedorenkoi* Kirejtshuk, Chang, Ren & Shih, 2010, (Holotype, CNU-COL-NN2006015) [97]. Source: Donated by Dr. Chungkun Shih.

subtruncated to slightly convex and somewhat narrower than posterior ones, its lateral sides slightly arcuate, anterior edge prosternum rather concave disposed markedly behind anterior edge of pronotum; metacoxae plates moderately raised. Among many beetle fossils collected in Northeastern China, only very few specimens have well-preserved hind wings in stretched-out and unfolded position. This specimen has a detailed imprint of pterothorax with elytra and perfectly preserved hind wing in unfolded condition exposing all veins and trace of many folds. This provided a rare opportunity for studying hind wing venation of *Lasiosyne*.

Anacapitis Yan, 2009

Anacapitis Yan, 2009, *Paleontol. J.*, 43, 78–82 [295] (original designation).

Brachysyne Tan & Ren, 2009, 347 [139]; Syn. by Kirejtshuk, Chang, Ren & Shih, 2010, *Ann. Soc. Entomol. Fr. (n.s.)*, 46, 67–87 [97].

Type species: *Anacapitis oblongus* Yan, 2009.

Head prognathous; labrum well-developed, broadly attached, visible from above; clypeus and frons separated by suture. Gular plate wide. Antennal bases situated close to anterior margins of eyes. Eyes large, ovate. Antennae filiform, not received in grooves on prosternum, second antennomere very small. Pronotum transverse, weakly emarginate anteriorly, rounded laterally, with pointed anterior angles and posterior angles extending posterad and slightly laterad. Procoxae spherical, not touching, posteriorly open; protrochantins concealed. Prosternal process triangular, with apex rounded, fitting in mesosternal groove. Mesocoxae large, slightly transversely ovate, with external trochantin, separated at midlength by short metasternal process. Metasternum trapeziform; paracoxal suture interrupted before longitudinal suture at distance of one-third of its width; laterally, terminating short of reaching border of metasternum. Metepisterna large, not reaching mesocoxal cavities. Metacoxae transverse, touching, with well-developed metacoxal plates. Femora usually twice as wide as tibiae. Tibiae slightly longer than femora, gradually widening toward apex. Metatarsi five-segmented; tarsomeres cylindrical, almost as wide as tibiae, without lobes or spines; claws simple. Elytra with deep, thin striae, fused in pairs near base, without rows of punctures. Abdomen with five visible segments.

Distribution and age: Inner Mongolia of China and Kazakhstan; Middle Jurassic.

Only one species included from the Jurassic of Northern China (see Table 21.1).

Bupredactyla Kirejtshuk, Chang, Ren & Shih, 2010

Bupredactyla Kirejtshuk, Chang, Ren & Shih, 2010, *Ann. Soc. Entomol. Fr.*, 46 (1–2), 67–87 [97] (original designation).

Type species: *Bupredactyla magna* Kirejtshuk, Chang, Ren & Shih, 2010.

Body apparently elongate and oval, pubescence fine and short; mandibles moderately developed and with acute apex; eyes moderately large; antenna rather long and apparently with 11 segments, most antennomeres elongate subcylindrical or subconical (thickened apically); pronotum with most width comparable with that of elytral base, transverse and subarcuate at sides, anterior angles rounded and not projecting, posterior angles sharp, extending lateroposteriorly; metepisterna slightly more than twice as long as wide; elytra with blunt apices and 11 striae (2 and 3 not complete); mesocoxae apparently narrowly separated to subcontigous; metacoxal plates slightly to moderately developed only in median part of coxae; tarsi four-segmented with moderately lobed tarsomeres 1–3 (indeed tarsi pseudotetramerous with a very small tarsomere 4 and very long tarsomere 5).

Distribution and age: Inner Mongolia; Middle Jurassic.

Only one species included from the Jurassic of Northern China (see Table 21.1).

Parelateriformius Yan & Wang, 2010

Parelateriformius Yan & Wang, 2010, *Paleontol. J.*, 44 (3), 297–302 [296] (original designation).

Type species: *Parelateriformius communis* Yan & Wang, 2010.

Antennal cavities large. Compound eyes large, oval; ocelli absent. Pronotum approximately 2–2.5 times as wide as head, 0.25 times as long as elytra, trapeziform, with bead, convex on disc and flattened laterally; sides of pronotum moderately curved; front angles not protruding anteriad, only slightly pointed; hind angles acute, protruding posteriad. Anterior margin of pronotum weakly concave; posterior margin rounded, curved posteriad, indented. Entire surface of pronotum densely and deeply punctured. Elytra elongate, 3.1 times as long as wide together; each elytron smoothly dilating from middle of its length, forming a curve at exterior margin, then again narrowing toward rounded apex. Epipleural bead broad; elytra with bead. Elytra with 10 deep grooves with punctures. Second groove (from suture) interrupted at three fourth of elytron length; other grooves reaching apex of elytron; sutural furrow absent. Surface of elytra covered with dense punctures forming parallel rugae. Metacoxae separated by short triangular process of metaventrite, slightly oblique, with rather small coxal plates.

Distribution and age: Inner Mongolia; Middle Jurassic.

Four species included from the Jurassic of Northern China (see Table 21.1).

Superfamily Elateroidea Leach, 1815

Family Artematopodidae Lacordaire, 1857

Artematopodidae are a small family of beetles forming one of the basal lineages of Elateroidea. They usually can be distinguished by the paired carinae on the prosternum and the apical interlocking tongue on the elytra [98, 297].

Genera included from the Jurassic of Northern China: *Tarsomegamerus* Zhang, 2005 and *Sinobrevipogon* Cai, Lawrence, Ślipiński & Huang, 2015.

Tarsomegamerus Zhang, 2005

Tarsomegamerus Zhang, 2005, *Géobios*, 38 (6), 865–871 [298] (original designation).

Type species: *Tarsomegamerus mesozoicus* Zhang, 2005.

Body distinctly convex and somewhat elongated. Head short and wide, distinctly transverse. Pronotum large, posterior angles rounded, and median longitudinal furrow present. Scutellum small, semicircular. Elytral epipleura narrow. Elytra with puncta arranged in nine striae. Pro- and mesocoxae ovate, and separated from each other, metacoxae elongate-triangular and somewhat oblique. Tibiae of all legs armed with median carinae. Claws simple, thick and long, obviously divergent. Abdomen with five ventrites, the first four ventrites subequal in length, with the fifth ventrite slightly longer than others.

Distribution and age: Inner Mongolia; Middle Jurassic.

Only one species included from the Jurassic of Northern China (see Table 21.1).

Sinobrevipogon Cai, Lawrence, Ślipiński & Huang, 2015

Sinobrevipogon Cai, Lawrence, Ślipiński & Huang, 2015, *Syst. Entomol.*, 40 (4), 779–788 [98] (original designation).

Type species: *Sinobrevipogon jurassicus* Cai, Lawrence, Ślipiński & Huang, 2015.

Moderately large, elongate oval, densely setose. Eyes relatively large, laterally protruding. Antenna long, 11-segmented, slightly serrate. Pronotum transverse, with complete lateral carinae; prosternum with a pair of longitudinal carinae in front of procoxae. Elytra striate, densely setose, with apical interlocking tongue on the ventral side of each elytron. Prosternal process relatively wide, subparallel-sided. Protrochantins exposed. Mesocoxae moderately widely separated; mesocoxal cavity closed by the mesepimeron and the anteromesal edge of metanepisternum. Metacoxae excavate, with narrow but complete coxal plates. Meso- and metatarsi five-segmented, metatarsomeres 1 and 2 elongate, tarsomeres 3 and 4 lobed. Abdominal ventrite 1 short, much shorter than ventrite 2; ventrite 5 very long, longer than ventrites 3 and 4 combined. Sutures between all abdominal ventrites somewhat curved; suture between ventrites 4 and 5 very strongly curved anteriorly.

Distribution and age: Inner Mongolia; Middle Jurassic.

Only one species included from the Jurassic of Northern China (see Table 21.1).

Phylogenetic Reconstructions of Artematopodidae

Eleven in-group taxa and two out-groups were included in a cladistic analysis based on 30 adult characters; the resulting tree, as shown in Figure 21.17, recovered the family Artematopodidae is monophyletic, including the two fossil genera *Sinobrevipogon* and *Tarsomegamerus*, but excluding the genus *Eulichas* (Eulichadidae). Within the family, three clades are supported: (i) *Electribius* authority; (ii) *Ctesibius* authority + *Brevipogon* authority + *Sinobrevipogon* + *Tarsomegamerus*; and (iii) the remaining extant genera, including *Allopogonia* authority [98].

Family Cerophytidae Latreille, 1834

Cerophytidae, a small family, can be distinguished by the complete absence of metacoxal plates and the presence of pectinate tarsal claws.

Only one genus included from the Cretaceous of Northern China: *Necromera* Martynov, 1926.

Necromera Martynov, 1926

Necromera Martynov, 1926, *Ezhegodnik Russkogo Paleontologicheskogo Obshestva*, 5 (1), 1–39 [92] (original designation).

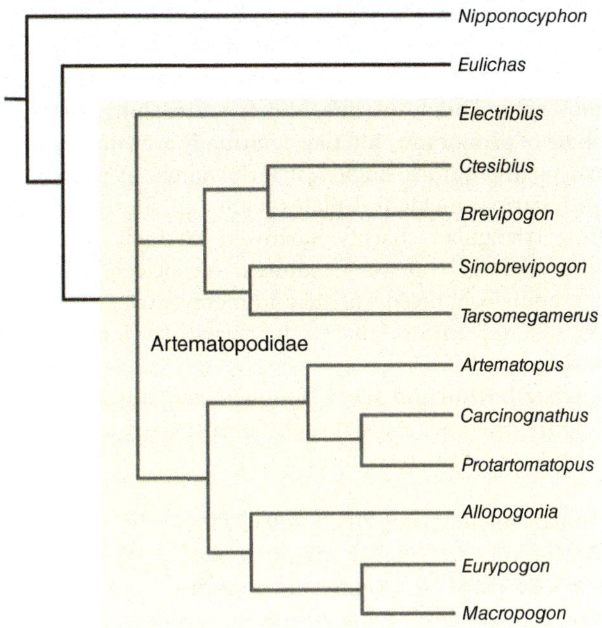

Figure 21.17 Phylogenetic reconstructions of Artematopodidae: a tree with characters mapped on branches using unambiguous optimization in Winclada. Source: modified from [98].

Idiomerus Dolin, 1980, *Naukova Dumka, Kiev*, 17–81 [299]; Syn. by Chang, Kirejtshuk & Ren, 2011, *Ann. Soc. Entomol. Fr. (n.s.)*, 2011, 47 (1–2), 33–44 [99].

Leptocnemus Hong & Wang, 1990, *Insect fossils of Laiyang Formation. In: The Stratigraphy and Palaeontology of Laiyang Basin, Shandong Province.* 105–120 [173]; Syn. by Chang, Kirejtshuk & Ren, 2011, *Ann. Soc. Entomol. Fr. (n.s.)*, 2011, 47 (1–2), 33–44 [99].

Type species: *Necromera baeckmanni* Martynov, 1926.

Length 4.5–10.0 mm. Head not retracted into prothorax, transverse with posterior constriction; frons projecting medially with median carina; antennal insertions approximate. Eyes very large; apparently finely facetted. Antenna 11-segmented, filiform or weakly serrate; pedicel as long as wide or weakly elongate, 0.5–0.8 times as long as antennomere 3. Terminal maxillary palpomere ovoid. Pronotum convex with anterior angles obtuse, posterior angles angulate, not prominent; lateral carina complete. Prosternum with prominent, apically rounded chin piece. Prosternal process angulate at level of procoxae, blade like posteriorly, extending to mesoventral cavity. Elytra without projecting shoulders; with at least eight impressed and punctate elytral striae. Mesocoxae separated by less than 0.5 diameter of coxa. Hind coxae transverse, contiguous with strongly reduced mesal coxal plates. Abdomen with five ventrites, ventrites 1–3 connate. Hind trochanter length about half length of hind femur. Tarsi with tarsomere 4 lobed; tarsomere 1 slightly longer than tarsomeres 2 and 3 combined; pretarsal claws narrow.

Distribution and age: Shandong and Liaoning of China, Early Cretaceous; Kazakhstan, Middle Jurassic.

Two species included from the Cretaceous of Northern China (see Table 21.1).

Family Elateridae Leach, 1815

Elateridae are an abundant and diverse family, commonly called "click-beetles". They are characterized by exposed labrum, projecting hind pronotal angles, long prosternum, globular procoxae with concealed trochantins, well-developed metacoxal plates, four connate ventrites, the clicking mechanism and lack of a transverse metaventral suture [300].

Genera included from the Jurassic and Cretaceous of Northern China: *Protagrypnus* Dolin, 1973, *Crytocoelus* Dolin & Nel, 2002, *Lithomerus* Dolin & Nel, 2002, *Bilineariselater* Chang & Ren, 2008, *Curtelater* Chang & Ren, 2008, *Paralithomerus* Chang, Zhang & Ren, 2008, *Paradesmatus* Chang, Kirejtshuk & Ren, 2009, *Paraprotagrypnus* Chang, Zhao & Ren, 2009, *Sinolithomerus* Dong & Huang, 2009, *Anoixis* Chang, Kirejtshuk & Ren, 2010, *Apoclion* Chang, Kirejtshuk & Ren, 2010, *Desmatinus* Chang, Kirejtshuk & Ren, 2010 and *Clavelater* Dong & Huang, 2011.

Protagrypnus Dolin, 1973

Protagrypnus Dolin, 1973, Fossil forms of click-beetles (Elateridae, Coleoptera) from Lower Jurassic of Middle Asia, In: Fauna and biology of Insects of Moldavia, "Shtiintsa", Kishinev, 72–82 [301] (original designation).

Type species: *Protagrypnus exoletus* Dolin, 1973.

Small to comparatively large elongate body; head transversal to subtriangular; prothorax with pointed acute posterior angles, prosternum rounded anteriorly, pronotosternal sutures apparently associated with sulciform grooves, procoxae opened externally, mesoventrite short with many transverse sutures divided into small "praeepisterna" and larger "postepisterna", mesepisterna with transverse suture, metacoxal plates weakly narrowed outwardly.

Distribution and age: Inner Mongolia of China, Middle Jurassic; Kyrgyzstan, Early Jurassic.

Only one species included from the Jurassic of Northern China (see Table 21.1).

Crytocoelus Dolin & Nel, 2002

Crytocoelus Dolin & Nel, 2002, *B. Soc. Entomol. Fr.*, 107 (4), 341–346 [302] (original designation).

Type species: *Crytocoelus major* Dolin & Nel, 2002.

Body elongate, posterior angles of pronotum obviously prolonged backward, with distinct short carina; antennae short, not reaching posterior angle of pronotum, scape robust, pedicel much shorter than scape and antennomere 3, antennomeres 3 and 4 distinctly elongate; chin piece moderately or strongly arcuate; scutellum rounded, semi-oval, or subtriangular, never cordate; elytra with faintly longitudinal striae; mesocoxae open to mesepimeron; metacoxal plates obtusely long triangular, evenly narrowed laterally; tarsi with five tarsomeres, tarsomeres 1–4 succeedingly shorter.

Distribution and age: Liaoning; Early Cretaceous.

Three species included from the Cretaceous of Northern China (see Table 21.1).

Lithomerus Dolin, 1980

Lithomerus Dolin, 1980, *Naukova Dumka, Kiev*, 17–81 [299] (original designation).

Type species: *Lithomerus cockerelli* Dolin, 1980.

Body elongate, elytra 2.5–4 times as long as prothorax. Prothorax transverse, narrowly rounded anteriorly; prosternal suture extended to procoxae. Metacoxae cavities laterally open to mesepimeron. Metacoxae short, not longer than diameter of mesocoxae, more or less narrowed laterally; metacoxal plates strongly narrowed laterally, almost disappeared near the outer edge. Metaventrite longer than prothorax and not less than 1.5 times shorter than the abdomen.

Distribution and age: Liaoning of China, Early Cretaceous; Australia, Early Jurassic; Kazakhstan, Middle Jurassic.

Only one species included from the Cretaceous of Northern China (see Table 21.1).

Bilineariselater Chang & Ren, 2008

Bilineariselater Chang & Ren, 2008, *Acta Geol. Sin-Engl.*, 82 (2), 236–243 [190] (original designation).

Type species: *Bilineariselater foveatus* Chang & Ren, 2008.

Body elongated; head sub-triangular, frons with ridge; antenna filiform, pedicel much shorter than scape and antennomere 3; prosternal pleural suture double, closed anteriorly; metaventrite without transverse sutures; mesocoxae open to mesepimeron; mesoventrite and metaventrite separated by distinct sutures; metacoxal plates slender and short, obtusely triangular, narrowed laterally.

Distribution and age: Liaoning; Early Cretaceous.

Only one species included from the Cretaceous of Northern China (see Table 21.1).

Curtelater Chang & Ren, 2008

Curtelater Chang & Ren, 2008, *Acta Geol. Sin. -Engl.*, 82 (2), 236–243 [190] (original designation).

Type species: *Curtelater wui* Chang & Ren, 2008.

The specific epithet is in honor of Qicheng Wu for his practical help for this study. Body elongated; mandibles toothed and bifid; antennae short, not reaching posterior angle of pronotum; chin piece normally arcuate; prosternal pleural suture single; posterior angle of pronotum with carina; metacoxal plates wide and short, obtusely long triangular, sharply narrowed laterally, metaventrite without transverse sutures; mesocoxae open to mesepimeron; mesoventrite and metaventrite separated by distinct suture; metaventrite with a longitudinal suture.

Distribution and age: Liaoning; Early Cretaceous.

Only one species included from the Cretaceous of Northern China (see Table 21.1).

Paralithomerus Chang, Zhang & Ren, 2008

Paralithomerus Chang, Zhang & Ren, 2008, *Zootaxa*, 1785, 54–62 [192] (original designation).

Type species: *Paralithomerus exquisitus* Chang, Zhang & Ren, 2008.

Metaventrite with transverse sutures, but prosternum without longitudinal furrows.

Distribution and age: Liaoning; Early Cretaceous.

Two species included from the Cretaceous of Northern China (see Table 21.1).

Paradesmatus Chang, Kirejtshuk & Ren, 2009

Paradesmatus Chang, Kirejtshuk & Ren, 2009, *Ann. Zool.* (Warszawa), 59 (1), 7–14 [189] (original designation).

Type species: *Paradesmatus baiae* Chang, Kirejtshuk & Ren, 2009.

The specific epithet is dedicated to Madam Li Bai for her contribution to the fossils collected and studied and her support for Dr. Chungkun Shih. Prosternum arcuate anteriorly, short, pronotum with trisinuate posterior edge, prosternum without median plate; mesocoxae rather narrowly separated, metacoxal femoral plates very large and triangular.

Distribution and age: Inner Mongolia, Middle Jurassic; Liaoning, Early Cretaceous.

Three species included from the Jurassic of Northern China (see Table 21.1).

Paraprotagrypnus Chang, Zhao & Ren, 2009

Paraprotagrypnus Chang, Zhao & Ren, 2009, *Prog. Nat. Sci. – Mater.*, 19 (10), 1433–1437 [193] (original designation).

Type species: *Paraprotagrypnus superbus* Chang, Zhao & Ren, 2009.

Metacoxal femoral plates short, sharply narrowed laterally; legs very slender.

Distribution and age: Inner Mongolia; Middle Jurassic.

Only one species included from the Jurassic of Northern China (see Table 21.1).

Sinolithomerus Dong & Huang, 2009

Sinolithomerus Dong & Huang, 2009, *Acta Palaeontol. Sin.*, 48 (1), 102–108 [303] (original designation).

Type species: *Sinolithomerus dolini* Dong & Huang, 2009.

The specific epithet is dedicated to entomologist V.G. Dolin from Ukraine for his great contribution in studying fossil and extant click beetles. Body very narrow, length about three times as long as width. Pronotum long and narrow, lateral edges straight; prosternal process behind coxae very narrow. Scutellum pentagonal. Mesocoxal cavities closed laterally by mesoventrite, metaventrite and mesepimeron. Hind coxae oblique, length of mesocoxal plates laterally about half of the length of coxae.

Distribution and age: Liaoning; Middle Jurassic.

Only one species included from the Jurassic of Northern China (see Table 21.1).

Anoixis Chang, Kirejtshuk & Ren, 2010

Anoixis Chang, Kirejtshuk & Ren, 2010, *Ann. Entomol. Soc. Am.*, 103 (6), 866–874 [188] (original designation).

Type species: *Anoixis complanus* Chang, Kirejtshuk & Ren, 2010.

Body robust, head comparatively short, pronotum subtrapeziform and with comparatively shortly projecting posterior angles, scutellum slightly transverse, prosternum with partly and comparatively deeply open pronotosternal sutures, rather narrow and very long intercoxal process, apparently somewhat explanate pronotal sides, slightly emarginated sides of elytra, metacoxae strongly oblique and with large subtriangular metacoxal plates demonstrating a clear curve at the middle of their inner edge.

Distribution and age: Liaoning; Early Cretaceous.

Only one species included from the Cretaceous of Northern China (see Table 21.1).

Apoclion Chang, Kirejtshuk & Ren, 2010

Apoclion Chang, Kirejtshuk & Ren, 2010, *Ann. Entomol. Soc. Am.*, 103 (6), 866–874 [188] (original designation).

Type species: *Apoclion clavatus* Chang, Kirejtshuk & Ren, 2010.

Small beetles with body moderately slender, head subtriangular, antenna with apical segment somewhat enlarged, pronotum subtrapeziform and with clear lateral carina, posterior angles of pronotum rather long projecting and sharply acute, scutellum somewhat elongate to slightly transverse, prosternum with moderately wide process, metacoxae strongly oblique and with metacoxal plates subtriangular and very larger, demonstrating subrectilinear inner edge or with a weak curve at their middle, elytra with a characteristic thickening in shape of the Y-network.

Distribution and age: Liaoning; Early Cretaceous.

Three species included from the Cretaceous of Northern China (see Table 21.1).

Desmatinus Chang, Kirejtshuk & Ren, 2010

Desmatinus Chang, Kirejtshuk & Ren, 2010, *Ann. Entomol. Soc. Am.*, 103 (6), 866–874 [188] (original designation).

Type species: *Desmatinus cognatus* Chang, Kirejtshuk & Ren, 2010.

Body moderately slender, head moderately short, pronotum arcuate at sides and rather narrowing to posterior angles, scutellum strongly transverse, prosternum with pronotosternal sutures apparently not open and rather wide intercoxal process, metacoxae strongly oblique and with subtriangular metacoxal plates.

Distribution and age: Liaoning; Early Cretaceous.

Only one species included from the Cretaceous of Northern China (see Table 21.1).

Desmatinus cognatus Chang, Kirejtshuk & Ren, 2010 (Figure 21.18)

Desmatinus cognatus Chang, Kirejtshuk & Ren, 2010: *Annals of the Entomological Society of America*, 103 (6), 866–874.

Locality and horizon: Huangbanjigou, Beipiao, Liaoning, China; Lower Cretaceous, Yixian Formation.

Head subtriangular and rather short behind eyes, apparently slightly convex; eyes comparatively large. Pronotum transverse, with lateral edges slightly arcuate, anterior edge shallowly emarginate, posterior edge trisinuate and without basal furrows, posterior angles comparatively short, without carinae. Scutellum strongly transverse (1.5 times as wide as long) and subtrapezoidal. Elytra somewhat wider than prothorax, approximately twice as long as wide combined, subacuminate at apex. Underside with uniform, rather sparse and moderately small punctures. Mentum subtrapezoidal. Procoxal cavities oval and small. Mesocoxal cavities suboval and open to mesepimera, closed to mesepisterna, Mesepimera subtriangular. Metaventrite apparently subflattened, with a longitudinal median suture. Metacoxae strongly oblique and with femoral plates demonstrating contiguous inner edges and strongly arcuate outer edges with a deep sinuation at outer end. Metatibiae with an apical spur, tarsomere 1 much longer than others, tarsomeres 2–4 progressively shortening distally, claws moderately long and simple [188].

Clavelater Dong & Huang, 2011

Clavelater Dong & Huang, 2011, *Acta Geol. Sin.*, 85 (6), 1224–1230 [102] (original designation).

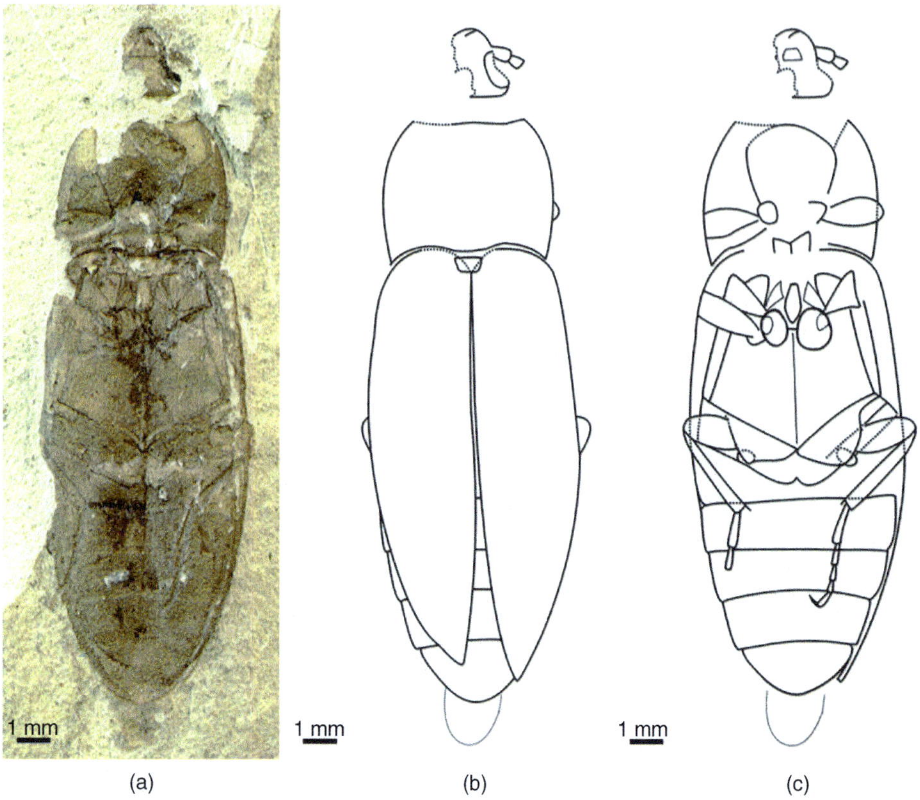

Figure 21.18 *Desmatinus cognatus* Chang, Kirejtshuk & Ren, 2010, (Holotype, CNU-COL-LB2008836). (a) Photograph, dorsal view; (b) Line drawing, dorsal view; (c) Line drawing, ventral view [188].

Type species: *Clavelater ningchengensis* Dong & Huang, 2011.

Body length 3.7 times as long as width; head prognathous; antenna reach to posterior edge of pronotum, length of antennomere 3 about 1.8 times as long as pedicel, the last six segments singularly clubbed; posterior angles of pronotum armed with short carina; pronotum as long as wide; chin piece short; median plate of prosternum narrowed backwards gradually; pronotosternal sutures open to procoxal cavities; mesocoxal cavities close by meso-, metaventrite and mesepimeron; metacoxal plates sharply narrowed laterally.

Distribution and age: Inner Mongolia; Middle Jurassic.

Only one species included from the Jurassic of Northern China (see Table 21.1).

Family Eucnemidae Eschscholtz, 1829

Eucnemidae, called "false click-beetles", are a large and cosmopolitan family. They can be distinguished by the labrum attached beneath the clypeus, usually more or less membranous and not visible externally and all five ventrites connate, and larvae often with reduced legs and mouthparts [304].

Only one genus included from the Cretaceous of Northern China: *Palaeoxenus* Horn, 1891.

Palaeoxenus Horn, 1891

Palaeoxenus Horn, 1891, *Trans. of the Am. Entomol. Soc.*, 18, 32–48. [305] (original designation).

Type species: *Palaeoxenus dobrni* Horn, 1878.

Antenna simple in both sexes, the last two segments shorter, the terminal one broader than long, squarely truncate, but chisel-shaped on its terminal edge. Mesosternum oblique, broadly grooved. Eyes longer in their vertical diameter and without distinct supra-orbital ridge.

Distribution and age: Liaoning; Early Cretaceous.

Only one species included from the Cretaceous of Northern China (see Table 21.1).

Palaeoxenus sinensis Chang, Muona & Teräväinen, 2016 (Figures 21.19 and 21.20)

Palaeoxenus sinensis Chang, Muona & Teräväinen, 2016: *Cladistics*, 32 (2), 211–214.

Locality and horizon: Huangbanjigou, Beipiao, Liaoning, China; Lower Cretaceous, Yixian Formation.

Larva body large, head shorter and wider, less elongated presternum, slightly larger presternal sclerite,

Figure 21.19 *Palaeoxenus sinensis* Chang, Muona & Teräväinen, 2016, (Holotype, 41HIII0181). Source: Photo provided by Dr. Huali Chang [307].

Figure 21.20 Ecological reconstruction of *Palaeoxenus sinensis* Chang, Muona & Teräväinen, 2016. Source: Artwork by Dr. Chen Wang.

slightly differently shaped median setal patches and slightly more widely separated urogomphi. The fossil larva shares the diagnostic features and defining apomorphies of the genus *Palaeoxenus* [306], the only living member of the false click-beetle subfamily Palaeoxeninae. *Palaeoxenus dohrni*, the extant species, is endemic to the southern Californian mountain ranges [304]. According to collection label data, it has been associated with both cedars and sugar pines [304]. The new finding proves that the highly specialized main eucnemid lineages had evolved 125 Mya, before the main radiation of the angiosperms and probably as an adaptation to development in gymnosperms [307].

Infraorder Derodontiformia LeConte, 1861

Superfamily Derodontoidea LeConte, 1861

Family Derodontidae LeConte, 1861

Derodontidae are a small family and characterized by several primitive coleopteran features, such as a membranous joint between the meso- and metathorax, the partial closure of the mesocoxal cavities by the metepisternum, the well-developed metacoxal plates, and the male abdomen tergite 10 distinct and separated from the ninth [308].

Only one genus included from the Jurassic of Northern China: *Juropeltastica* Cai, Lawrence, Ślipiński & Huang, 2014.

Juropeltastica Cai, Lawrence, Ślipiński & Huang, 2014

Juropeltastica Cai, Lawrence, Ślipiński & Huang, 2014, *Eur. J. Entomol.*, 111 (2), 299–302 [103] (original designation).

Type species: *Juropeltastica sinica* Cai, Lawrence, Ślipiński & Huang, 2014.

Body small. Head with frontal raised area delimited laterally and posteriorly by grooves. Antenna 11-segmented, with a loose 3-segmented club. Pronotum weakly explanate; lateral margins toothed/crenulate. Procoxae distinctly separated, transverse; protrochantins exposed.

Procoxal cavities externally closed. Elytra with one short mesobasal and three complete longitudinal carinae. Epipleural rim with a row of relatively large round-window punctures. Mesocoxal cavities transverse, broadly open laterally, closed by mesepimeron and metanepisternum. Metaventral discrimen present. Metacoxae contiguous, excavate, extending laterally to rim of elytra. Abdominal ventrite 1 very short, without parallel ridges. Ventrites 2–5 each with a pair of small curved basolateral ridges.

Distribution and age: Inner Mongolia; Middle Jurassic.

Only one species included from the Jurassic of Northern China (see Table 21.1).

Infraorder Bostrichiformia Forbes, 1926

Superfamily Bostrichoidea Latreille, 1802

Family Dermestidae Latreille, 1804

Dermestids are cosmopolitan bostrichiform beetles. They usually can be distinguished by compact body, covering of stout setae or scales forming a pattern, single and median ocellus on the head present, and having excavations and cavities for housing antennae and legs [309].

Only one genus included from the Jurassic of Northern China: *Paradermestes* Deng, Ślipiński, Ren & Pang, 2017.

Paradermestes Deng, Ślipiński, Ren & Pang, 2017

Paradermestes Deng, Ślipiński, Ren & Pang, 2017, *Ann. Zool. (Warszawa)*, 67 (1), 109–112 [104] (original designation).

Type species: *Paradermestes jurassicus* Deng, Ślipiński, Ren & Pang, 2017.

Broadly oval body, the lack of the median ocellus, 11-segmented antenna with 4-segmented antennal club, procoxae not projecting and separated by a complete prosternal process, the hind coxae bearing only mesally developed plates and extending laterally beyond metaventrite to elytral epipleura.

Distribution and age: Inner Mongolia; Middle Jurassic.

Only one species included from the Jurassic of Northern China (see Table 21.1).

Infraorder Cucujiformia Lameere, 1938

Superfamily Cleroidea Latreille, 1802

Family Cleridae Latreille, 1802

Cleridae, called "checkered beetles", are a large and diverse family within Cucujiformia: Cleroidea. They are distributed worldwide, although the majority of species inhabit the tropical and subtropical belts. Most of them are easily distinguished by the setose, often colorful body, clavate or capitate antennae, distinctly lobed tarsi, and projecting procoxae with more or less concealed trochantins [310].

Genera included from the Jurassic of Northern China: *Protoclerus* Kolibáč & Huang, 2016 and *Wangweiella* Kolibáč & Huang, 2016.

Protoclerus Kolibáč & Huang, 2016

Protoclerus Kolibáč & Huang, 2016, *Syst. Entomol.*, 41 (4), 808–823 [107] (original designation).

Type species: *Protoclerus korynetoides* Kolibáč & Huang, 2016.

Body broadly oval; compact antennal club composed of three symmetrical antennomeres; clypeus protracted; frontoclypeal suture distinctly emarginate; prothorax not constricted at base but widely connected with mesothorax (= pronotum in line with base of elytra); pronotum wider than long; sculpture of pronotum composed of granular punctures with setae at center; prothorax with distinct lateral carina; tarsomeres 1–4 not shortened or hidden (tarsomeres approximately the same in size); tarsomeres 1–4 with pulvilli; six abdominal ventrites.

Distribution and age: Inner Mongolia; Middle Jurassic.

Only one species included from the Jurassic of Northern China (see Table 21.1).

Wangweiella Kolibáč & Huang, 2016

Wangweiella Kolibáč & Huang, 2016, *Syst. Entomol.*, 41 (4), 808–823 [107] (original designation).

Type species: *Wangweiella calloviana* Kolibáč & Huang, 2016.

The generic name is in honor of Wei Wang (CE 699–759), a Chinese poet and painter of the Tang Dynasty. Body elongate, elytra three times as long as prothorax; antenna longer than head, with simple, loose, nearly symmetrical, 3-segmented club; eyes large but not extremely elevated (width of head not exceeding pronotum); prothorax with lateral carina, indistinctly narrowing at base and humeral portion of elytra wider than pronotum but mesosternum and mesonotum do not form a distinct "neck"; metacoxae reaches metepisternum; protarsi with basitarsus shortened but not covered from above by tarsomere 2; tarsomere 4 approximately as long as tarsomere 3; parameral portion of tegmen separated from phallobase.

Distribution and age: Inner Mongolia; Middle Jurassic.

Only one species included from the Jurassic of Northern China (see Table 21.1).

Family Prionoceridae Lacordaire, 1857

Prionocerids are brightly colored beetles, possessing the modified apical antennomere, head usually rostrate with large eyes, protibiae with a single spur and two or three protarsomeres in the male with a comb. The family live thoughout the Palearctic, Indian/Indomalayan and Afrotropic realms. Adult prionocerids are pollen feeders, while larvae are predaceous [311].

Only one genus included from the Jurassic of Northern China: *Idgiaites* Liu, Ślipiński, Leschen, Ren & Pang, 2015.

Idgiaites Liu, Ślipiński, Leschen, Ren & Pang, 2015

Idgiaites Liu, Ślipiński, Leschen, Ren & Pang, 2015, *Ann. Zool. (Warszawa)*, 65 (1), 41–52 [108] (original designation).

Type species: *Idgiaites jurassicus* Liu, Ślipiński, Leschen, Ren & Pang, 2015.

Body large. Head relatively small and prolonged beyond eyes, frontoclypeal region compressed but probably not rostrate. Antenna 11-segmented, longer than head, filiform and pubescent; antennomere 3 distinctly longer than other segments; terminal three antennomeres slightly expanded and flattened. Terminal segment of maxillary palp long, weakly expanded. Eyes medium-sized and rounded. Prothorax longer than head; lateral margins not explanate with smooth edges. Procoxal cavities large and rounded; procoxae conical, protrochantin exposed. Mesocoxal cavities large and oval, laterally open to mesepimeron. Profemora strongly broadened medially, meso- and metafemora slightly widened medially. Tibiae parallel-sided and slender. Metacoxae transverse; metacoxal cavities laterally open. Abdomen with six freely articulated ventrites, ventrite 1 longer than the others.

Distribution and age: Inner Mongolia; Middle Jurassic.

Only one species included from the Jurassic of Northern China (see Table 21.1).

Family Trogossitidae Fabricius, 1801

Trogossitidae, called "bark-gnawing beetles," are distributed worldwide and highly diverse, comprising at least 600 extant species [312]. Kolibáč, in 2006, studied the phylogenetic relationships of the family based on larval and adult characters [313], and proposed a 2-subfamily classification (Peltinae and Trogossitinae) that was revised by Kolibáč and Zaitsev [314], by elevating Lophocaterinae to the subfamily status [235].

Genera included from the Jurassic and Cretaceous of Northern China: *Palaeoendomychus* Zhang, 1992, *Sinosoronia* Zhang, 1992, *Eotenebroides* Ren, Lu, Ji & Guo, 1995, *Sinopeltis* Yu, Leschen, Ślipiński, Ren & Pang, 2012, *Latitergum* Yu, Ślipiński, Leschen, Ren & Pang, 2014, *Marginulatus* Yu, Ślipiński, Leschen, Ren & Pang, 2014, *Paracretocateres* Yu, Ślipiński, Leschen, Ren & Pang, 2015 and *Yixianteres* Yu, Ślipiński, Leschen, Ren & Pang, 2015.

Palaeoendomychus Zhang, 1992

Palaeoendomychus Zhang, 1992, *Acta Entomol. Sin.*, 35, 331–338 [315] (original designation).

Type species: *Palaeoendomychus gymnus* Zhang, 1992.

Body minute and compact, oval, without setae. Head deeply sunk into pronotum. Eyes rather large, widely separated, slightly prominent. Clypeus protracted. Antennae short, clubbed, flagellum cylindrical. Pronotum transverse, short, without membrane (sic) in front, but with lateral sides broadly flattened, posterior angles prominent. Scutellum small, narrowly triangular, distinctly longer than broad. Elytra with carinae, wide, dehiscent, as wide as pronotum at base, and both closely connected to one another, humeral angles rounded, terminal angles prominent, surface with striae. All pairs of coxae widely distant, hind tarsomeres widened or lobate. Legs normal, with tarsi short and narrow, the first and second tarsal segments triangular, each about as long as wide, the third cylindrical, noticeably longer than wide.

Distribution and age: Shandong; Early Cretaceous.

Only one species included from the Cretaceous of Northern China (see Table 21.1).

Sinosoronia Zhang, 1992

Sinosoronia Zhang, 1992, *Acta Entomol. Sin.*, 35, 331–338 [315] (original designation).

Type species: *Sinosoronia longiantennata* Zhang, 1992.

Body minute, oval. Head small, triangular, deeply sunk into pronotum. Eyes large. Antenna elongate, with loose club. Pronotum short, wide, anterior angles acute, clearly expanded forwards, posterior angles prominent, posterior margin as wide as base of elytra. Scutellum small, semicircular. Elytra long, dehiscent, covering apex of abdomen, shoulder prominent, terminal angles acute.

Distribution and age: Shandong; Early Cretaceous.

Only one species included from the Cretaceous of Northern China (see Table 21.1).

Eotenebroides Ren, Lu, Ji & Guo, 1995

Eotenebroides Ren, Lu, Ji & Guo, 1995, Fauna and Stratigraphy of Jurassic-Cretaceous in Beijing and the Adjacent Areas, 73–90 [240] (original designation).

Type species: *Eotenebroides tumoculus* Ren, Lu, Ji & Guo, 1995.

Body elongate. Head triangular, longer than wide, distinctly narrower than the pronotum. Eyes distinct, large, situated laterally at median. Antenna is not clubbed but segments probably successively widening from about segment 5 to 10; moreover, these segments are perhaps weakly asymmetrical. Pronotum trapezoidal or possibly weakly cordate, without distinctly projecting anterior corners. Scutellum large. Femora non-clavate, front coxae transverse, mesocoxae and metacoxae very close. Elytra not longer than abdomen, with six to eight carinae but imperceptible sculpture.

Distribution and age: Beijing; Early Cretaceous.

Only one species included from the Cretaceous of Northern China (see Table 21.1).

Sinopeltis Yu, Leschen, Ślipiński, Ren & Pang, 2012

Sinopeltis Yu, Leschen, Ślipiński, Ren & Pang, 2012, *Ann. Zool. (Warszawa)*, 62 (2), 245–252 [235] (original designation).

Type species: *Sinopeltis jurrasica* Yu, Leschen, Ślipiński, Ren & Pang, 2012.

Body broadly-ovoid and parallel-sided. Antennae distinctly clubbed with antennomeres symmetrical; antennal insertions visible in dorsal view. Eyes convex. Frontoclypeal suture present. Antennal grooves present and parallel. Anterior pronotal angles well-developed and subrounded. Mesoventrite not vaulted. Mesocoxae widely separated. Metaventrite lacking axillary space and metakatepisternal suture. Metacoxae not excavate and narrowly separated. Elytra with seriate punctuation. Abdominal ventrite 1 about as long as 2, intercoxal process narrow.

Distribution and age: Inner Mongolia; Middle Jurassic.

Two species included from the Jurassic of Northern China (see Table 21.1).

Latitergum **Yu, Ślipiński, Leschen, Ren & Pang, 2014**

Latitergum Yu, Ślipiński, Leschen, Ren & Pang, 2014, *Ann. Zool. (Warszawa)*, 64 (4), 667–676 [106] (original designation).

Type species: *Latitergum glabrum* Yu, Ślipiński, Leschen, Ren & Pang, 2014.

Body elongate and parallel-sided. Head relatively large only slightly narrower than pronotum. Antenna 11-segmented with loose 3-segmented, weakly asymmetrical club. Frontoclypeal suture indistinct. Pronotum widest near posterior margin; anterior pronotal angles not produced forward, angulate. Prosternal process not expanded apically; procoxal cavities narrowly open externally; procoxae transverse with exposed trochantins. Mesoventrite not vaulted. Mesocoxae moderately separated; their cavities externally open. Metaventrite lacking axillary space and metakatepisternal suture. Metacoxae flat and narrowly separated. Abdomen with five visible ventrites of approximately similar lengths.

Distribution and age: Inner Mongolia; Middle Jurassic.

Only one species included from the Jurassic of Northern China (see Table 21.1).

Marginulatus **Yu, Ślipiński, Leschen, Ren & Pang, 2014**

Marginulatus Yu, Ślipiński, Leschen, Ren & Pang, 2014, *Ann. Zool. (Warszawa)*, 64 (4), 667–676 [106] (original designation).

Type species: *Marginulatus venustus* Yu, Ślipiński, Leschen, Ren & Pang, 2014.

Body elongate and parallel-sided. Head relatively large, prognathous. Antenna short, 11-segmented with three terminal antennomeres forming tight club. Eyes convex. Pronotum widest at middle, anterior pronotal angles weakly prominent, blunt. Procoxae transverse with exposed trochantins. Procoxal cavities externally closed, with postcoxal hypomeral projections strongly overlapping apex of prosternal process. Mesoventrite not vaulted. Mesocoxae narrowly separated. Metaventrite with complete discrimen, lacking axillary spaces and metakatepisternal suture. Metacoxae flat, not excavate and very narrowly separated. Elytra with fine punctuation. Abdominal ventrites 1–4 somewhat equal in length, intercoxal process very narrow, apically pointed.

Distribution and age: Inner Mongolia; Middle Jurassic.

Only one species included from the Jurassic of Northern China (see Table 21.1).

Paracretocateres **Yu, Ślipiński, Leschen, Ren & Pang, 2015**

Paracretocateres Yu, Ślipiński, Leschen, Ren & Pang, 2015, *Cretac. Res.*, 53, 89–97 [236] (original designation).

Type species: *Paracretocateres bellus* Yu, Ślipiński, Leschen, Ren & Pang, 2015.

Body moderately broad and almost parallel-sided. Head prognathous. Dorsal body surface nearly bare. Antenna 11-segmented with distinct 3-segmented club. Eyes convex, medium-sized. Gular sutures conspicuous and widely separate. Anterior pronotal angles weakly projecting. Procoxae transverse; protrochantin exposed; procoxal cavities externally open. Meso- and metacoxae narrowly separated. Elytra with regular rows of punctures and weakly convex intervals.

Distribution and age: Liaoning; Early Cretaceous.

Only one species included from the Cretaceous of Northern China (see Table 21.1).

Yixianteres **Yu, Ślipiński, Leschen, Ren & Pang, 2015**

Yixianteres Yu, Ślipiński, Leschen, Ren & Pang, 2015, *Cretac. Res.*, 53, 89–97 [236] (original designation).

Type species: *Yixianteres beipiaoensis* Yu, Ślipiński, Leschen, Ren & Pang, 2015.

Body elongate and parallel-sided. Antenna 11-segmented with three terminal antennomeres forming asymmetrical club; subantennal groove probably present. Mandibles strongly developed, pointed apically. Frontoclypeal suture not visible. Eyes convex. Anterior pronotal angles prominent. Procoxae transverse, widely separated; procoxae with exposed protrochantins. Mesocoxae and metacoxae narrowly separated; mesocoxal cavities oval, externally open to mesepimeron; mesotrochantin exposed. Elytra with regular rows of punctures.

Distribution and age: Liaoning; Early Cretaceous.

Only one species included from the Cretaceous of Northern China (see Table 21.1).

Yixianteres beipiaoensis **Yu, Ślipiński, Leschen, Ren & Pang, 2015 (Figure 21.21)**

Yixianteres beipiaoensis Yu, Ślipiński, Leschen, Ren & Pang, 2015: *Cretac. Res.* 53, 89–97.

Locality and horizon: Huangbanjigou, Beipiao, Liaoning, China; Lower Cretaceous, Yixian Formation.

Head relatively large, prognathous, vertex simple. Eyes medium-sized, convex. Scape robust. Prothorax distinctly transverse, about 1.7 times broader than long, widest at base. Lateral margins slightly explanate; anterior angles projecting and acute; hind angles blunt or rounded. Procoxal cavities transverse, broadly open posteriorly; notosternal sutures visible. Prosternal process parallel-sided, relatively broad, about 0.5 times coxal width, straight or slightly expanding apically. Mesoventrite broad, arcuate anteriorly; mesanepisternum and mesepimeron trapezoidal. Mesocoxal cavities slightly oblique; mesometaventral process very narrow. Metaventrite trapezoidal, its base wider than anterior part, with complete discrimen. Metanepisternum moderately broad; metepimeron not visible. Metacoxae transverse, flat, very narrowly separated. Scutellum transverse, not clearly visible. Elytra oval, widest at

Figure 21.21 *Yixianteres beipiaoensis* Yu, Ślipiński, Leschen, Ren & Pang, 2015, (Holotype, CNU-COL-LB2011113) [236].

2/3 of length; dorsal surfaces with conspicuous regular punctuation. Abdomen with five freely articulated ventrites of more or less equal lengths [236].

Superfamily Cucujoidea Latreille, 1802

Family Boganiidae Sen Gupta & Crowson, 1966

Boganiidae are a very interesting and isolated family of basal cucujoids. Their adults and larvae all appear to be pollenophagous [316]. It includes two newly established Australian genera, *Boganium* and *Paracucujus*, along with three other genera subsequently removed to Phloeostichidae, Hobartiidae and Cavognathidae [147, 317, 318].

Only one genus included from the Jurassic of Northern China: *Palaeoboganium* Liu, Ślipiński, Lawrence, Ren & Pang, 2017.

Palaeoboganium Liu, Ślipiński, Lawrence, Ren & Pang, 2017

Palaeoboganium Liu, Ślipiński, Lawrence, Ren & Pang, 2017, *J. Syst. Palaeontol.*, 10, 1–10 [147] (original designation).

Type species: *Palaeoboganium jurassicum* Liu, Ślipiński, Lawrence, Ren & Pang, 2017.

Body size large. Head with strongly arcuate frontoclypeal suture; labrum not visible externally; mandible with dorsal tubercle fitting into depression on lateral edge of clypeus; subantennal groove absent; antennae short and weakly capitate; strongly transverse pronotum without glandular callosities; pro- and mesocoxae with exposed trochantins, metacoxae narrowly separated, not extending laterally to meet distinctly striate elytra; abdomen with five freely articulated ventrites.

Distribution and age: Inner Mongolia; Middle Jurassic.

Only one species included from the Jurassic of Northern China (see Table 21.1).

Palaeoboganium jurassicum Liu, Ślipiński, Lawrence, Ren & Pang, 2017 (Figure 21.22)

Palaeoboganium jurassicum Liu, Ślipiński, Lawrence, Ren & Pang, 2017: *J. Syst. Palaeontol.*, 10, 1–10.

Locality and horizon: Daohugou, Ningcheng, Inner Mongolia, China; Middle Jurassic, Jiulongshan Formation.

It is hard to prove that the extinct *Palaeoboganium jurassicum* was an ancient pollinator of cycads, but it was a good candidate based on its occurrence in Jurassic deposits that also contain cycads and on its phylogenetic placement near the genera that currently pollinate cycads in Africa and Australia. This suggests that the family Boganiidae may have been more diverse and widely distributed in the Mesozoic, and that a boganiid–cycad association may be older than indicated by the two extant genera associated with these plants [147].

Figure 21.22 *Palaeoboganium jurassicum* Liu, Ślipiński, Lawrence, Ren & Pang, 2017, (Holotype, CNU-COL-NN201601) [147].

Phylogenetic Relationship for the Boganiidae

Based on phylogenetic analysis by Liu *et al.* in 2017, the majority consensus tree of two maximal parsimony trees (length = 46; CI = 70; RI = 66) is shown in Figure 21.23. All trees support a monophyletic family Boganiidae with *Palaeoboganium* well-embedded in it either as the sister group of the clade (*Metacucujus* + *Paracucujus*) or at the base of Paracucujinae. *Palaeoboganium* was found to belong to Boganiidae based on at least three unique apomorphies (4,0 – lack of ventral subantennal grooves; 6,1 – labrum not visible dorsally, and 8,0 – simple mandibular apex). The Boganiidae clade is supported by four further potential apomorphies (10,2 – incisor edge of mandible with bifid or two fixed lobes; 15,0 – lacinia with bidentate uncus; 29,2 – penultimate mesotarsomere simple and much shorter than preceding tarsomere; 32,1 – phallobase of aedeagus asymmetrical) mapped by the algorithm but not observable on *Palaeoboganium* fossil impressions and scored as missing ("?") in the data matrix [147].

Family Monotomidae Laporte, 1840

The so-called "root-eating beetles" (Monotomidae) are typically small (1.5–6.0 mm) and slender predatory or mycophagous beetles found in all zoogeographical regions of the world [319].

Only one genus included from the Jurassic of Northern China: *Jurorhizophagus* Cai, Ślipiński & Huang, 2015.

Jurorhizophagus Cai, Ślipiński & Huang, 2015

Jurorhizophagus Cai, Ślipiński & Huang, 2015, *Alcheringa*, 39 (4), 488–493 [111] (original designation).

Type species: *Jurorhizophagus alienus* Cai, Ślipiński & Huang, 2015.

Relatively large for Monotomidae (*ca* 5 mm long), elongate, glabrous. Eyes lateral, finely faceted. Antenna short, 11-segmented, with a distinct 3-segmented club. Frontoclypeal suture present, straight. Pronotum transverse, with lateral carinae and a median longitudinal groove; Elytra long (covering first three abdominal ventrites), somewhat truncate apically, exposing two abdominal tergites; each elytron with about eight longitudinal lines. Prosternal process narrow and sub-parallel, not dilated at apex. Protrochantins exposed. Metatarsi 5-segmented. Abdominal ventrite 1 slightly shorter than ventrites 2 and 3 combined; postcoxal lines absent.

Distribution and age: Inner Mongolia; Middle Jurassic.

Only one species included from the Jurassic of Northern China (see Table 21.1).

Family Nitidulidae Latreille, 1982

Nitidulidae are a relatively large and very diverse family; they are usually characterized by having capitate antennae and strongly transverse procoxae with exposed trochantins and they always lack a maxillary galea [320].

Only one genus included from the Jurassic of Northern China: *Sinonitidulina* Hong, 1983.

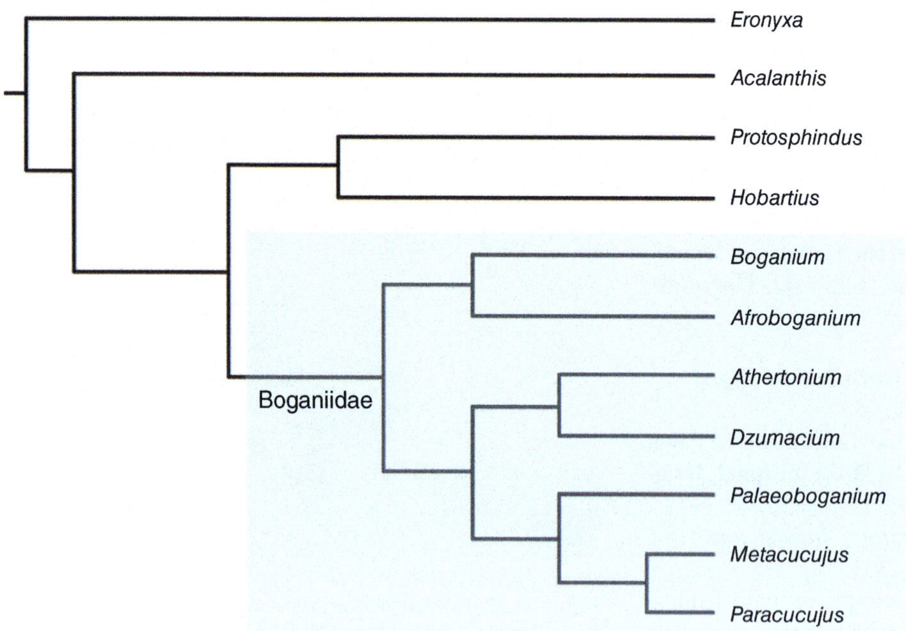

Figure 21.23 Majority consensus tree of two maximal parsimony trees (length = 46; CI = 70; RI = 66). Source: Modified from [147].

Sinonitidulina Hong, 1983

Sinonitidulina Hong, 1983, Middle Jurassic Fossil Insects in North China, 1–223 [93] (original designation).

Type species: *Sinonitidulina luanpingensis* Hong, 1983.

Body black; antennae short, slightly longer than head, never longer than head and pronotum combined, widened apically; pronotum slightly wider than length, covered with punctures; elytra very long, three times as long as wide, smooth or with striae and punctures. Length 9.2–10 mm, or longer.

Distribution and age: Hebei; Middle Jurassic.

Three species included from the Jurassic of Northern China (see Table 21.1).

Family Parandrexidae Kirejtshuk, 1994

Parandrexidae are a small extinct family of beetles comprising two genera *Parandrexis* Martynov, 1926 [92] and *Martynopsis* Soriano, Kirejtshuk & Delclòs, 2006 [112]. Originally *Parandrexis* was assigned to the subfamily Prioninae (Chrysomeloidea: Cerambycidae) [92], Kirejtshuk (1994) [109] proposed a new family Parandrexidae and assigned it to the infraorder Cucujiformia (Coleoptera: Polyphaga). Soriano *et al.*, in 2006, accepted this systematic placement and established *Martynopsis* Soriano, Kirejtshuk & Delclòs, 2006 in this family and put this family in the superfamily Cucujoidea [112].

Only one genus included from the Jurassic and Cretaceous of Northern China: *Parandrexis* Martynov, 1926.

Parandrexis Martynov, 1926

Parandrexis Martynov, 1926, *Ezhegodnik Russkogo Paleontologicheskogo Obshestva*, 5 (1), 1–39 [92] (original designation).

Type species: *Parandrexis parvula* Martynov, 1926.

Body slender or oblong. Antenna filiform, 11–12 segmented. Scape forming a clavate structure. Mandibles very long and narrow in males, but much shorter than head in females. Maxillary palp 3–4 segmented, thin and long. Mesothorax somewhat short and metathorax longer than mesothorax. Scutellum triangular. Prosternal lobe not projecting to mesocoxae. Femora fusiform but not strong. Tibiae much thinner than femora, as long as femur and slightly widening toward apex. Tarsi 4-segmented, short and thin, the distal tarsomere longer than others, claws simple. Females having approximately the same outlines as males, but with larger body size, shorter antennae and shorter mandibles.

Distribution and age: Inner Mongolia of China, Middle Jurassic; Liaoning of China, Early Cretaceous; Kazakhstan, Middle Jurassic.

Five species included from the Jurassic and Cretaceous of Northern China (see Table 21.1).

Superfamily Tenebrionoidea Latreille, 1802

Family Mordellidae Latreille, 1802

Mordellidae, called "tumbling flower beetles" are cosmopolitan in distribution. They can be distinguished by the wedge-like body and the presence of a spine-like posterior abdominal process formed by the seventh tergite. In addition, the metacoxae are very large and more or less fused to the metaventrite, the prothorax is reduced especially ventrally, the flattened head is opisthognathous that extending posteriorly over the anterior edge of the metaventrite [321].

Genera included from the Cretaceous of Northern China: *Cretanaspis* Huang & Yang, 1999, *Mirimordella* Liu, Lu & Ren, 2007 and *Bellimordella* Liu, Zhao & Ren, 2008.

Cretanaspis Huang & Yang, 1999

Cretanaspis Huang & Yang, 1999, *Acta Palaeontol. Sin.*, 38 (1), 125–132 [322] (original designation).

Type species: *Cretanaspis lushangfenensis* Huang & Yang, 1999.

Body small; head much flexed; eyes large, ovoid, without concave behind antenna insertions; antenna somewhat clubbed, middle segments small and rather serrate, the last segment markedly smaller than the penultimate segment; pronotum flexed, arcuate at base; sheath gradually narrowing toward tip; mesocoxae not contiguous, the fourth of middle tarsi extremely small; inlaying the third one which is lobed slightly; metacoxae very large, hind tibiae longer than the hind femora or tarsi, gradually broadening toward apex, transversely truncated at apex; in the middle and hind tibiae, the first two segments of hind tarsi with a very fine longitudinal ridge at apex; hind tibial spurs very developed, longer than the first segment of hind tarsi; claws developed; abdomen with six visible ventrites, attenuating backwards, the sixth rather pointed but not elongating; apex of the third abdominal segment with a filar appendage.

Distribution and age: Beijing; Early Cretaceous.

Only one species included from the Cretaceous of Northern China (see Table 21.1).

Mirimordella Liu, Lu & Ren, 2007

Mirimordella Liu, Lu & Ren, 2007, *Zootaxa*, 1415, 49–56 [223] (original designation).

Type species: *Mirimordella gracilicruralis* Liu, Lu & Ren, 2007.

Maxillary palpi linear, last segment without distinct enlargement. Scutellum long, triangular or rectangular. Elytra arched, sharply tapering on apical 1/3; apex pointed. Epicoxa present in front of metacoxal plate and immediately juxtaposed to metepisternum. Hind tibiae expanded apically, with oblique truncate apex, as long as femora; hind tarsi longer than tibiae. Abdomen with five visible abdominal segments.

Distribution and age: Liaoning; Early Cretaceous.

Only one species included from the Cretaceous of Northern China (see Table 21.1).

Bellimordella Liu, Zhao & Ren, 2008

Bellimordella Liu, Zhao & Ren, 2008, *Cretac. Res.*, 29 (3), 445–450 [228] (original designation).

Type species: *Bellimordella capitulifera* Liu, Zhao & Ren, 2008.

Maxillary palpi linear, the last segment without distinct enlargement. Antenna filiform, inserted in front of eyes, shorter than pronotum. Elytra flat, gradually narrowing caudad; apex rounded slightly. Epicoxa present in front of metacoxal plate and immediately juxtaposed to metepisternum. Middle and hind tibiae with one apical spur respectively, with straight truncate apex, much shorter than their tarsi. Abdomen with five visible abdominal segments, the last two segments exceed elytral apices.

Distribution and age: Liaoning; Early Cretaceous.

Three species included from the Cretaceous of Northern China (see Table 21.1).

Family Ripiphoridae Gemminger & Harold, 1870

Ripiphoridae are known as "wedge-shaped beetles" because of the wedge-shaped, humped and tapered body. They are highly derived group of predators and parasites, which may have evolved from melandryid-like ancestors. They are traditionally divided into six subfamilies: Pelecotominae, Micholaeminae, Ptilophorinae, Hemirhipidiinae, Ripidiinae, and Ripiphorinae [114, 323, 324].

Only one genus included from the Jurassic of Northern China: *Archaeoripiphorus* Hsiao, Yu & Deng, 2017.

Archaeoripiphorus Hsiao, Yu & Deng, 2017

Archaeoripiphorus Hsiao, Yu & Deng, 2017, *Eur. J. Taxon.*, 277, 1–13 [114] (original designation).

Type species: *Archaeoripiphorus nuwa* Hsiao, Yu & Deng, 2017.

Body large-sized (about 15–16 mm); head elongate, abruptly constricted posteriorly to form broad neck; eyes oval, shallowly emarginate, distinctly separated from each other; antenna 11-segmented, antennomeres IV to X rectangular or trapezoidal and antennomere XI with pointed apex; terminal maxillary palpomere elongate securiform, not modified, about four times as long as minimum width; pronotum almost triangular, trilobed at base; elytra complete, covering entire abdomen; abdomen with five ventrites; protibiae nearly as long as protarsi; apices of tibiae without spiniform seta; claws pectinate, at least in mid and hind legs. (Holotype fossil donated by Dr. Chungkun Shih.)

Distribution and age: Inner Mongolia; Middle Jurassic.

Only one species included from the Jurassic of Northern China (see Table 21.1).

Family Tenebrionidae Latreille, 1802

Tenebrionidae, a very large family, are usually characterized by the concealed antennal insertions, eyes more or less elongate-oval and slightly emarginated, procoxal cavities internally and externally closed, prosternal process convex, abdominal ventrites 1–3 connate, abdominal apex with paired defensive glands, tarsi 5-5-4 [325].

Genera included from the Cretaceous of Northern China: *Alphitopsis* Kirejtshuk, Nabozhenko & Nel, 2011 and *Platycteniopus* Chang, Nabozhenko, Pu, Xu, Jia & Li, 2016.

Alphitopsis Kirejtshuk, Nabozhenko & Nel, 2011

Alphitopsis Kirejtshuk, Nabozhenko & Nel, 2011, *Entomologicheskoe Obozrenie*, 90, 548–552 [326] (original designation).

Type species: *Alphitopsis initialis* Kirejtshuk, Nabozhenko & Nel, 2011.

Body comparatively large. Head moderately transverse, with eyes rounded oval from below. Antennae, beginning with antennomere 6, moniliform; each antennomere with narrow stick-shaped base. Apex of apical antennomere acuminate. Apical maxillary palpomere narrow, obliquely truncate at apex, not dolabriform. Mentum comparatively narrow, not narrowing anteriorly. Penultimate maxillary palpomere markedly elongate (considerably longer than wide), slightly widening from base to apex. Pronotum transverse, its posterior edge slightly bi-emarginate and anterior edge widely emarginate. Sides of elytra regularly rounded, not parallel-sided. Epipleura of elytra gradually (not sharply) narrowed to apex, however, they do not reaching sutural angle forming sharp edge at apex. Width of metacoxae about subequal to length of metaventrite.

Distribution and age: Liaoning; Early Cretaceous.

Only one species included from the Cretaceous of Northern China (see Table 21.1).

Platycteniopus Chang, Nabozhenko, Pu, Xu, Jia & Li, 2016

Platycteniopus Chang, Nabozhenko, Pu, Xu, Jia & Li, 2016, *Cretac. Res.*, 57, 289–293 [327] (original designation).

Type species: *Platycteniopus diversoculatus* Chang, Nabozhenko, Pu, Xu, Jia & Li, 2016.

Body large, elongate. Head nearly oval. Eyes round dorsally and transverse ventrally. Antennae serrate, first four antennomeres weakly transverse or with subequal width and length. Mandibules bifid. Pronotum bell-shaped, not narrower than humeral angles of elytra; posterior angles distinct, almost straight. Elytra wide, with broadly rounded lateral margins, narrowed to apex, densely and finely punctate. Mesoventrite process between mesocoxal cavities apically acute. Abdominal ventrite 6 (sternite VIII) with wide deep emargination. Abdominal ventrite 5 with straight apical margin. Fore and mid legs not long, profemora and mesofemora not protruding beyond pronotum and elytra. Hind legs longer, visibly protruding beyond elytra. Protibiae straight, metatibiae weakly bent. Tarsi narrow, filiform, their penultimate segment without membranous lobe. Protarsomere 1 equal in length to protarsomeres 2–4. Metatrochanters small, much shorter than metafemora. Male genitalia: basal piece long, apical piece short, outwardly curved.

Distribution and age: Liaoning; Early Cretaceous.

Only one species included from the Cretaceous of Northern China (see Table 21.1).

Superfamily Chrysomeloidea Latreille, 1802

Family Cerambycidae Latreille, 1802

Cerambycidae, known as "longhorn beetles", are a large and highly diverse family, constituting one of the largest groups of beetles, distributed worldwide excluding Antarctica [328]. They are usually characterized by the pseudotetramerous tarsi, the raised antennal insertions and long antennae, usually more than half as long as the body and sometimes much longer.

Genera included from the Cretaceous of Northern China: *Cretoprionus* Wang, Ma, McKenna, Yan, Zhang & Jarzembowski, 2014 and *Sinopraecipuus* Yu, Slipiński, Reid, Shih, Pang & Ren, 2015.

Cretoprionus Wang, Ma, McKenna, Yan, Zhang & Jarzembowski, 2014

Cretoprionus Wang, Ma, McKenna, Yan, Zhang & Jarzembowski, 2014, *J. Syst. Palaeontol.*, 12 (5), 565–574 [220] (original designation).

Type species: *Cretoprionus liutiaogouensis* Wang, Ma, McKenna, Yan, Zhang & Jarzembowski, 2014.

Body large; antennae stout and serrate, almost as long as body; pronotum broad, with lateral carinae and four teeth on each lateral side; elytron clearly wider than pronotum, with a pointed tip.

Distribution and age: Inner Mongolia; Early Cretaceous.

Only one species included from the Cretaceous of Northern China (see Table 21.1).

Phylogeny for the Major Lineages of Chrysomeloidea

Wang *et al.* (2014) incorporated two new calibration points as minimum constraints on the age of (1) Prioninae + Parandrinae and (2) Bruchinae using data from a recent molecular phylogenetic study of Chrysomeloidea, to reconstruct divergence times among the major lineages of Chrysomeloidea (Figure 21.24). Their analyses suggest that most chrysomeloid families appeared in the Jurassic and diversified over the course of the Cretaceous, a scenario consistent with the codiversification of Chrysomeloidea and their (predominantly) angiospermous hosts; however, the phylogeny of Chrysomeloidea remains incompletely resolved, and further elucidation of timing and patterns of chrysomeloid macroevolution will require additional studies [220].

Sinopraecipuus Yu, Ślipiński, Reid, Shih, Pang & Ren, 2015

Sinopraecipuus Yu, Ślipiński, Reid, Shih, Pang & Ren, 2015, *Cretac. Res.*, 52, 453–460 [238] (original designation).

Type species: *Sinopraecipuus bilobatus* Yu, Ślipiński, Reid, Shih, Pang & Ren, 2015.

Body elongate, somewhat oval. Antennae filiform with very short scape and relatively long pedicel. Pronotum transverse with margins finely crenulate, Procoxal cavities with hypomeral lobes posteriorly closing or almost closing the cavities. Mesocoxae apparently rather narrowly separated. Elytra with very pointed apices and dense punctures. Protibiae and mesotibiae with paired apical spurs. Tarsi pseudotetramerous.

Distribution and age: Liaoning; Early Cretaceous.

Only one species included from the Cretaceous of Northern China (see Table 21.1).

Sinopraecipuus bilobatus Yu, Ślipiński, Reid, Shih, Pang & Ren, 2015 (Figure 21.25)

Sinopraecipuus bilobatus Yu, Ślipiński, Reid, Shih, Pang & Ren, 2015, *Cretac. Res.*, 52, 453–460.

Locality and horizon: Dawangzhangzi, Lingyuan, Liaoning, China; Lower Cretaceous, Yixian Formation.

Sinopraecipuus bilobatus has many characters the same as the extant cerambycids to support its placement in Cerambycidae. However, *S. bilobatus* has several different characters when compared with extant cerambycids. Furthermore, placement of *S. bilobatus* with confidence in a specific subfamily of Cerambycidae is not possible, because of insufficient preserved evidence and different characters of various subfamilies present in *S. bilobatus*. This unique combination of characters provides us a glimpse of the early evolutionary history of cerambycids. It is proposed that some of these characters might have been plesiomorphies for ancient Cerambycidae, while changes and development of these and other characters might have resulted in taxa of various subfamilies in the subsequent evolutionary process [238].

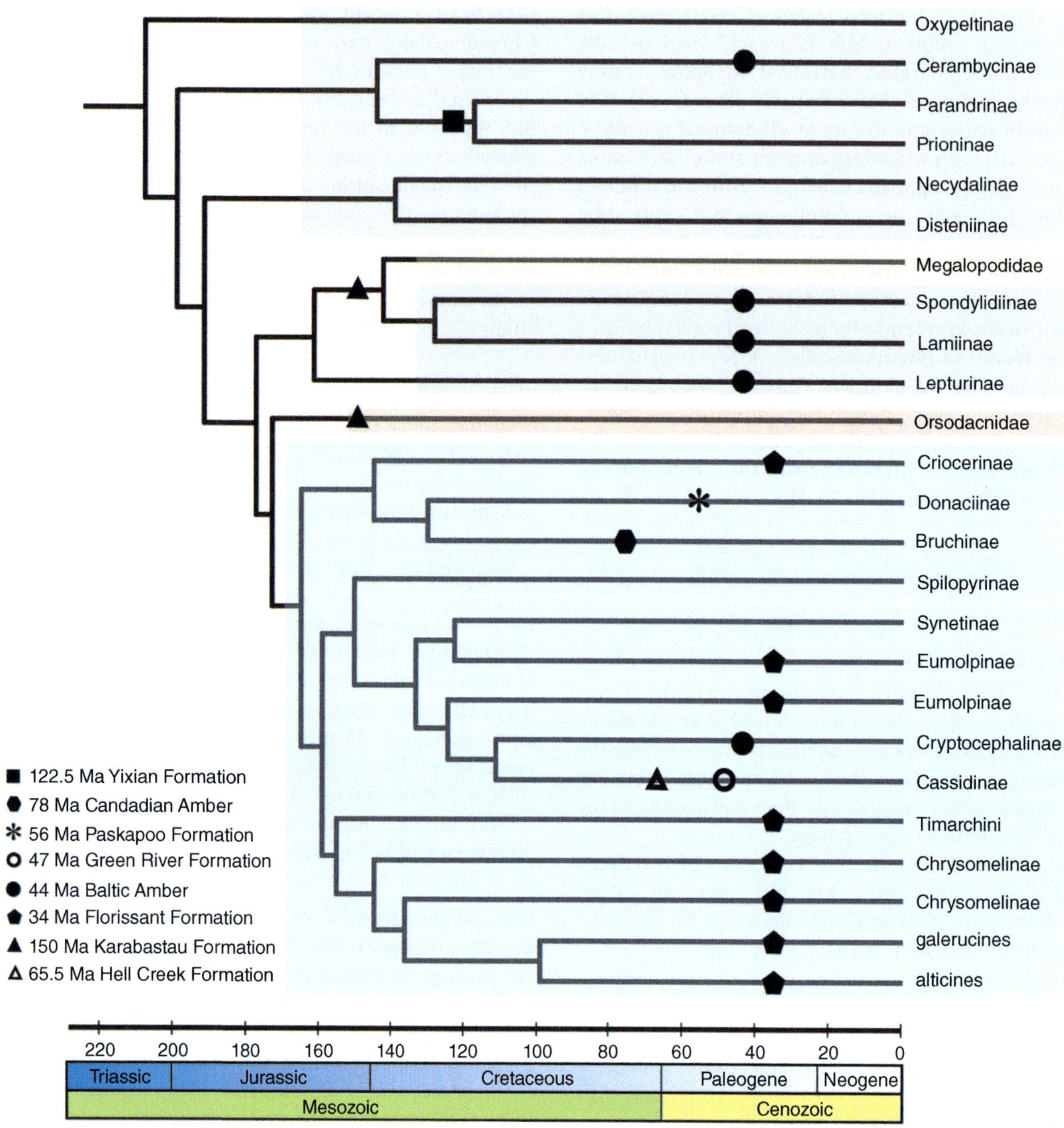

Figure 21.24 Simplified chronogram for higher-level groups of Chrysomeloidea. Branches representing Chrysomelidae are blue. Branches representing Cerambycidae are gray. Other families (Megalopodidae, Orsodacnidae) are yellow and orange. Black circles along branches represent the approximate placement of the earliest known fossils for each group [335–337]. Black triangles represent the placement of the earliest known, but questionable, fossils for each group. Source: Modified from [220, 338].

Family Chrysomelidae Latreille, 1802

Chrysomelidae, commonly called "leaf beetles", are a large and cosmopolitan family. Among the phytophagous beetle families, Chrysomelidae are second in number of species only to the Curculionidae. They are generally recognized by short to medium length antennae, antennal insertions not on prominences and not partly encircled by the eyes, general reduction of tibial spurs, highly variable but generally elongate-oval body form, leaf-feeding habits of adults, and larvae which are not true borers [329].

Only one genus included from the Cretaceous of Northern China: *Mesolaria* Zhang, 1986.

Mesolaria Zhang, 1986

Mesolaria Zhang, 1986: Some fossil insects from the Jurassic of northern Hebei, China. In: The Paleontology

Figure 21.25 *Sinopraecipuus bilobatus* Yu, Ślipiński, Reid, Shih, Pang & Ren, 2015, (Holotype, CNU-COL-LB20111101) [238]. Source: Donated by Dr. Chungkun Shih.

and Stratigraphy of Shandong, Haiyang Publishing House, Beijing, 74–84 [118] (original designation).

Type species: *Mesolaria longala* Zhang, 1986.

Body size medium; head small and perfectly round; eyes large; antennae serrated; pronotum posterolateral acute; elytra short and not covering all segments of abdomen, surface with fine and longitudinal striae; apex acute; legs slender.

Distribution and age: Hebei; Middle Jurassic.

Only one species included from the Jurassic of Northern China (see Table 21.1).

Superfamily Curculionoidea Latreille, 1802

Family Brentidae Billberg, 1820

Brentidae, called "straight-snouted weevils", are a cosmopolitan family of primarily xylophagous weevils in the tropics and the termperate regions. Recently, three subfamilies Apioninae, Cyladinae, and Nanophyinae in the Curculionidae, as well as the Ithycerinae, previously considered a separate family have been transferred to the Brentidae. They are characterized by having non-elbowed antennae.

Genera included from the Cretaceous of Northern China: *Cretonanophyes* Zherikhin, 1977 and *Abrocar* Liu & Ren, 2006.

Cretonanophyes Zherikhin, 1977

Cretonanophyes Zherikhin, 1977, Rhynchophora. Family Attelabidae. Family Curculionidae. In: Arnoldi, L.V., Zherikhin, V.V., Nikritin, L.M. *et al.* (eds.) Mezozoiskie Zhestkokrylye. *Tr. Paleontol. Inst. Akad. Nauk SSSR*, Volume 161, 176–180 [330] (original designation).

Leptocar Liu & Ren, 2007, *Sci. China, Ser. D. Earth Sci.*, 50 (5), 641–648 [225]; Syn. by Davis, Engel, Legalov & Ren, 2013, *J. of Syst. Palaeontol.*, 11 (4), 399–429 [331].

Rugosocar Legalov, 2009, *Amur. Zool. Zh.*, 1, 283–295 [332]; Syn. by Davis, Engel, Legalov & Ren, 2013, *J. of Syst. Palaeontol.*, 11 (4), 399–429 [331].

Type species: *Cretonanophyes longirostris* Zherikhin, 1977.

Rostrum elongate, slightly more than two times as long as pronotum, nearly straight. Pronotum and elytra with shallow punctures. Legs often densely covered with setae; forelegs often slightly longer than mid and hind legs. Elytra covered with elongate setae (not always visible in specimens), particularly near apex.

Distribution and age: Liaoning of China, Russia and Spain; Early Cretaceous.

Two species included from the Cretaceous of Northern China (see Table 21.1).

Abrocar Liu & Ren, 2006

Abrocar Liu & Ren, 2006, *Zootaxa*, 1176, 59–68 [224] (original designation).

Type species: *Abrocar brachyorhinos* Liu & Ren, 2006.

Rostrum fairly short, slightly more than length of pronotum; gula apparently present, narrow (gular [postoccipital] sutures paired); shallow scrobe present along basal half of rostrum. Compound eyes laterally elongated. Pronotum with acute lateral margin, narrower than anterior of elytra at shoulders. Pronotum and elytra with fairly dense covering of punctures. Body with short, sparse setae. The first probasitarsomere not to slightly expand.

Distribution and age: Liaoning; Early Cretaceous.

Four species included from the Cretaceous of Northern China (see Table 21.1).

Family Nemonychidae Bedel, 1882

The family Nemonychidae are generally considered to the most basal lineage of weevils, and many of the characters defining the family are plesiomorphic. This includes the relatively well-developed mouthparts with a free labrum, separate gular sutures, straight antenna, and free abdominal ventrites.

Genera included from the Jurassic and Cretaceous of Northern China: *Microprobelus* Liu, Ren & Shih, 2006, *Leptocar* Liu & Ren, 2007, *Chinocimberis* Legalov, 2009 and *Renicimberis* Legalov, 2009.

Microprobelus Liu, Ren & Shih, 2006

Microprobelus Liu, Ren & Shih, 2006, *Prog. Nat. Sci.*, 16, 885–888 [226] (original designation).

Type species: *Microprobelus liuae* Liu, Ren & Shih, 2006.

The specific epithet is in honor of Madam Xiuli Liu for her assistance and contribution in collecting Liaoning fossils, including this specimen. Rostrum projecting anteroventrally and fairly straight, only slightly curved. Antenna inserted slightly beyond middle of rostrum; antennal club weak, club articles barely larger than flagellar articles. Compound eyes round. Pronotum with acute lateral margin, extending dorsally toward posterior margin. Pronotum and elytra with dense covering of elongate setae. Tibiae with longitudinal, crenulated carina along outer margin. Tarsomeres slightly lobed.

Distribution and age: Liaoning; Early Cretaceous.

Only one species included from the Cretaceous of Northern China (see Table 21.1).

Chinocimberis Legalov, 2009

Chinocimberis Legalov, 2009, *Amurian Zoological Journal*, 1 (3), 200–213 [333] (original designation).

Chinabrenthorrhinus Legalov, 2009, *Amur. Zool. Zh.*, 1, 283–295 [332]; Syn. by Davis, Engel, Legalov & Ren, 2013, *J. of Syst. Palaeontol.*, 11 (4), 399–429 [331].

Type species: *Brenthorrhinoides angustipecteris* Liu, Ren & Tan, 2006.

Body moderately compressed dorsoventrally. Rostrum projecting anteriad and fairly straight, only slightly curved; labrum small. Antenna inserted near apex of rostrum; antennal club weak, club articles slightly larger than flagellar articles. Head angular before base of rostrum. Compound eyes slightly elongate, ovoid, bulging. Pronotum with acute lateral margin; lateral margins nearly parallel, posterior of pronotum only slightly wider than anterior. Pronotum and elytra with dense covering of elongate setae. Elytra with small punctures, if any, sparsely arranged (not in striae).

Distribution and age: Beijing and Liaoning of China and Mongolia; Early Cretaceous.

Three species included from the Cretaceous of Northern China (see Table 21.1).

Renicimberis Legalov, 2009

Renicimberis Legalov, 2009, *Amurian Zoological Journal*, 1 (3), 200–213 [333] (original designation).

Type species: *Brenthorrhinoides latipecteris* Liu, Ren & Tan, 2006.

Brenthorrhinoides latipecteris Liu, Ren & Tan, 2006, *Annales Zoologici (Warszawa)*, 56 (4), 605–612 [227] (original designation).

Renicimberis latipecteris (Liu, Ren & Tan, 2006), *Amurian Zoological Journal*, 1 (3), 200–213 [333].

Body moderately compressed dorsoventrally. Rostrum projecting anteriad and fairly straight, only slightly curved; rostrum slightly longer than nearly 1.5 times of length of pronotum. Antenna inserted near apex of rostrum; antennal club weak, club articles barely larger than flagellar articles. Head angular before base of rostrum. Compound eyes slightly elongate, ovoid. Pronotum with acute lateral margins; slightly narrow than elytral shoulders; pronotum widest at middle. Pronotum and elytra with moderately dense covering of elongate setae. Elytra lacking punctures.

Distribution and age: Liaoning; Early Cretaceous.

Only one species included from the Cretaceous of Northern China (see Table 21.1).

Renicimberis latipecteris (Liu, Ren & Tan, 2006) (Figure 21.26)

Renicimberis latipecteris (Liu, Ren & Tan, 2006): *Amurian Zoological Journal*, 1 (3), 200–213.

Locality and horizon: Huangbanjigou, Beipiao, Liaoning, China; Lower Cretaceous, Yixian Formation.

Figure 21.26 *Renicimberis latipecteris* (Liu, Ren & Tan, 2006), (Holotype, CNU-COL-LB2005101) [227].

Rostrum slightly longer than pronotum, four times as long as broad, apex weakly widened. Antenna reaching hind margin of head; scape longest, arched; club indistinct. Eyes big, in same length with the basal width of rostrum. Pronotum oblate, widest behind midlength, lateral sides rounded, 1.7 times as broad as long. Prothoracic coxae postmedial in position of prothorax, close to each other [227].

Suborder *Incertae Sedis*

Family Permosynidae Tillyard, 1924

The family is considered as a form taxon for isolated beetle elytra with punctuated and unpunctuated striae. Most of these beetles belong to Polyphage in the natural system, and a few specimens probably fall into Archostemata-Ademosynidae or Adephaga [334].

Only one genus included from the Jurassic and Cretaceous of Northern China: *Artematopodites* Ponomarenko, 1990.

Artematopodites Ponomarenko, 1990

Artematopodites Ponomarenko, 1990, Beetles. Scarabaeida. In: Rasnitsyn, A.P. (ed.), Late Mesozoic Insects of Eastern Transbaikalia. *Tr. Paleontol. Inst. Akad. Nauk SSSR*, 239, 39–82 [83] (original designation).

Type species: *Artematopodites latus* Ponomarenko, 1990.

Elongated elytra of medium or large size, with symmetrical apex; humeral angles indistinct, rounded. Elytral base slightly oblique, with flat, rather narrow border not covered by elytral disc, and with scutellar incision. Outer margin often with arcuation at elytral midlength or its basal third, with distinct, flat epipleural edge. Sutural margin with flat border, slightly curved, almost straight. Elytra edged, their surface covered with punctures or rugose sculpture. Elytra with 11 striae of punctures. The first stria lying very close to sutural border, reaching elytral apex; the second and third striae shortened; in most species, more than half as long as elytra; striae 4–11 reaching apex.

Distribution and age: Xinjiang of China, Early Jurassic, Middle Jurassic; Hebei of China, Middle Jurassic; Shanxi of China, Early Cretaceous; Russia, Late Jurassic, Early Cretaceous; Mongolia, Late Jurassic; Poland, Early Jurassic.

Five species included from the Jurassic and Cretaceous of Northern China (see Table 21.1).

Table 21.1 A list of fossil Coleoptera from the Jurassic and Cretaceous of China.

Family	Species	Locality	Horizon/Age	Citation
		Suborder Archostemata Kolbe, 1908		
Ademosynidae	*Atalosyne sinuolata* Ren, Lu, Ji & Guo, 1995	West hills, Beijing	Lushangfen Fm., K_1	Ren *et al.* [240]
	[a)]*Eremisyne xiazhuangensis* Wang, 1998	Fangshan, Beijing	Xiazhuang Fm., K_1	Wang [175]
Asiocoleidae	*Loculitricoleus flatus* Tan & Ren, 2009	Ningcheng, Inner Mongolia	Jiulongshan Fm., J_2	Tan and Ren [139]
	Loculitricoleus tenuatus Tan & Ren, 2009	Ningcheng, Inner Mongolia	Jiulongshan Fm., J_2	Tan and Ren [139]
[a)]Coelocatiniidae	[a)]*Catinius artus* Zhang, 1997	Zhixin Forest Farm, Jilin	Dalazi Fm., K_1	Zhang [177]
	[a)]*Catinius ovatus* Zhang, 1997	Zhixin Forest Farm, Jilin	Dalazi Fm., K_1	Zhang [177]
Cupedidae	*Ensicupes guyanensis* Hong, 1976	Guyang, Inner Mongolia	Guyang Fm., K_1	Hong [157]
	Ensicupes obtusus Zhang, 1997	Zhixin Forest Farm, Jilin	Dalazi Fm., K_1	Zhang [177]
	Anthocoleus hebeiensis Hong, 1983	Luoping, Hebei	Jiulongshan Fm., J_2	Hong [93]
	Celocoleus densus Hong, 1983	Luoping, Hebei	Jiulongshan Fm., J_2	Hong [93]
	Euteticoleus radiatus Hong, 1983	Chengde, Hebei	Jiulongshan Fm., J_2	Hong [93]
	Tetrocupes cavernasus Hong, 1983	Chengde, Hebei	Jiulongshan Fm., J_2	Hong [93]
	Hebeicupes formidabilis Zhang, 1986	Luoping, Hebei	Jiulongshan Fm., J_2	Zhang [118]

(Continued)

Table 21.1 (Continued)

Family	Species	Locality	Horizon/Age	Citation
	Longaevicupes macilentus Ren, Lu, Ji & Guo, 1995	West hills, Beijing	Lushangfen Fm., K_1	Ren *et al.* [240]
	Latocupes jiensis (Ren, Lu, Ji & Guo, 1995)	West hills, Beijing	Lushangfen Fm., K_1	Ren *et al.* [240]
	Latocupes bellus Ren & Tan, 2006	Beipiao, Liaoning	Yixian Fm., K_1	Ren and Tan [178]
	Latocupes fortis Ren & Tan, 2006	Beipiao, Liaoning	Yixian Fm., K_1	Ren and Tan [178]
	Latocupes angustilabialis (Tan, Huang & Ren, 2007)	Ningcheng, Inner Mongolia	Jiulongshan Fm., J_2	Tan *et al.* [180]
	Latocupes collaris (Tan, Huang & Ren, 2007)	Ningcheng, Inner Mongolia	Jiulongshan Fm., J_2	Tan *et al.* [180]
	Latocupes latilabialis (Tan, Huang & Ren, 2007)	Ningcheng, Inner Mongolia	Jiulongshan Fm., J_2	Tan *et al.* [180]
	Furcicupes raucus Tan & Ren, 2006	Beipiao, Liaoning	Yixian Fm., K_1	Tan and Ren [182]
	Gracilicupes crassicruralis Tan, Ren & Shih, 2006	Ningcheng, Inner Mongolia	Jiulongshan Fm., J_2	Tan *et al.* [185]
	Gracilicupes tenuicruralis Tan, Ren & Shih, 2006	Ningcheng, Inner Mongolia	Jiulongshan Fm., J_2	Tan *et al.* [185]
	Apriacma clavata (Tan, Ren & Shih, 2006)	Beipiao, Liaoning	Yixian Fm., K_1	Tan *et al.* [186]
	Apriacma latidentata (Tan, Ren & Shih, 2006)	Beipiao, Liaoning	Yixian Fm., K_1	Tan *et al.* [186]
	Apriacma renaria (Tan, Ren & Shih, 2006)	Beipiao, Liaoning	Yixian Fm., K_1	Tan *et al.* [186]
	Apriacma tuberculosa (Tan, Ren & Shih, 2006)	Beipiao, Liaoning	Yixian Fm., K_1	Tan *et al.* [186]
	Cretomerga subtilis (Tan & Ren, 2006)	Beipiao, Liaoning	Yixian Fm., K_1	Tan and Ren [182]
Jurodidae	*Jurodes daohugouensis* Yan, Wang, Ponomarenko & Zhang, 2014	Ningcheng, Inner Mongolia	Jiulongshan Fm., J_2	Yan *et al.* [84]
	Jurodes pygmaeus Yan, Wang, Ponomarenko & Zhang, 2014	Ningcheng, Inner Mongolia	Jiulongshan Fm., J_2	Yan *et al.* [84]
Ommatidae	*Omma delicata* Tan, Wang, Ren & Yang, 2012	Ningcheng, Inner Mongolia	Jiulongshan Fm., J_2	Tan *et al.* [129]
	Omma daxishanense Cai & Huang, 2017	Linglongta, Beijing	Tiaojishan Fm., J_3	Cai and Huang [132]
	Tetraphalerus laetus Lin, 1976	Shangyuan, Liaoning	Jianshangou Fm., K_1	Lin [168]
	[a]*Tetraphalerus largicoxa* Lin, 1986	Qiyang, Hunan	Guanyintan Fm., J_1	Lin [141]
	Tetraphalerus lentus Ren, Lu, Ji & Guo, 1995	Luanping, Hebei	Dabeigou Fm., K_1	Ren *et al.* [240]
	Tetraphalerus trachylaena (Ren, Tan & Ge, 2006)	Beipiao, Liaoning	Yixian Fm., K_1	Ren et al [179]
	Tetraphalerus curtinervis Tan, Ren & Shih, 2007	Beipiao, Liaoning	Yixian Fm., K_1	Tan *et al.* [96]
	Tetraphalerus latus Tan, Ren & Shih, 2007	Beipiao, Liaoning	Yixian Fm., K_1	Tan *et al.* [96]
	Tetraphalerus decorosus Tan, Wang, Ren & Yang, 2012	Ningcheng, Inner Mongolia	Jiulongshan Fm., J_2	Tan *et al.* [129]
	[a]*Memptus handlirschi* Ponomarenko, 1985	Qiyang, Hunan	Guanyintan Fm., J_1	Ponomarenko [75]
	Zygadenia laiyangensis Hong & Wang, 1990	Laiyang, Shandong	Laiyang Fm., K_1	Hong and Wang [173]
	Zygadenia liui Jarzembowski, Wang, Zhang & Fang, 2015	Xisanjia, Inner Mongolia	Yixian Fm., K_1	Jarzembowski *et al.*, 2015 [339]

(Continued)

Table 21.1 (Continued)

Family	Species	Locality	Horizon/Age	Citation
	Zygadenia trachylenus Ren, Lu, Ji & Guo, 1995	West hills, Beijing	Lushangfen Fm., K_1	Ren *et al.* [240]
	Zygadenia tuanwangensis Hong & Wang, 1990	Laiyang, Shandong	Laiyang Fm., K_1	Hong and Wang [173]
	Notocupes (Amblomma) epicharis Tan, Ren & Liu, 2005	Beipiao, Liaoning	Yixian Fm., K_1	Tan *et al.* [184]
	Notocupes (Amblomma) psilata Tan, Ren & Liu, 2005	Beipiao, Liaoning	Yixian Fm., K_1	Tan *et al.* [184]
	Notocupes (Amblomma) rudis Tan, Ren & Liu, 2005	Beipiao, Liaoning	Yixian Fm., K_1	Tan *et al.* [184]
	Notocupes (Amblomma) stabilis Tan, Ren & Liu, 2005	Beipiao, Liaoning	Yixian Fm., K_1	Tan *et al.* [184]
	Notocupes (Amblomma) cyclodonta Tan, Ren, Shih & Ge, 2006	Beipiao, Liaoning	Yixian Fm., K_1	Tan *et al.* [187]
	Notocupes (Amblomma) eumeura Tan, Ren, Shih & Ge, 2006	Beipiao, Liaoning	Yixian Fm., K_1	Tan *et al.* [187]
	Notocupes (Amblomma) miniscula Tan, Ren, Shih & Ge, 2006	Beipiao, Liaoning	Yixian Fm., K_1	Tan *et al.* [187]
	Notocupes (Amblomma) porrecta Tan, Ren, Shih & Ge, 2006	Beipiao, Liaoning	Yixian Fm., K_1	Tan *et al.* [187]
	Notocupes (Amblomma) protensa Tan, Ren, Shih & Ge, 2006	Beipiao, Liaoning	Yixian Fm., K_1	Tan *et al.* [187]
	Notocupes (Notocupes) validus (Lin, 1976)	Shangyuan, Liaoning	Yixian Fm., K_1	Lin [168]
	[a]*Notocupes (Notocupes) homorus* (Lin, 1986)	Lanshan, Hunan	Guanyintan Fm., J_1	Lin [141]
	Notocupes (Notocupes) dischides Zhang, 1986	Luoping, Hebei	Xiahuayuan Fm., K_1	Zhang [118]
	Notocupes (Notocupes) ludongensis Wang & Liu, 1996	Laiyang, Shandong	Laiyang Fm., K_1	Wang and Liu [176]
	Notocupes (Notocupes) alienus (Tan & Ren, 2006)	Beipiao, Liaoning	Yixian Fm., K_1	Tan and Ren [181]
	Notocupes (Notocupes) pingi Ponomarenko & Ren, 2010	Ningcheng, Inner Mongolia	Jiulongshan Fm., J_2	Ponomarenko and Ren [340]
	[a]*Notocupes (Notocupes) lini* Ponomarenko, Yan, Wang & Zhang, 2012	Qiyang, Hunan	Guanyintan Fm., J_1	Ponomarenko *et al.* [334]
	Brochocoleus punctatus Hong, 1982	Yumen, Gansu	[b]Chijinqiao Fm., K_1	Hong [160]
	Brochocoleus impressus (Ren, Lu, Ji & Guo, 1995)	West hills, Beijing	Lushangfen Fm., K_1	Ren *et al.* [240]
	Brochocoleus sulcatus Tan, Ren & Shih, 2007	Beipiao, Liaoning	Yixian Fm., K_1	Tan *et al.* [128]
	Brochocoleus applanatus Tan & Ren, 2009	Ningcheng, Inner Mongolia	Jiulongshan Fm., J_2	Tan and Ren [139]
	Brochocoleus magnus Tan & Ren, 2009	Ningcheng, Inner Mongolia	Jiulongshan Fm., J_2	Tan and Ren [139]
	Brochocoleus validus Tan & Ren, 2009	Ningcheng, Inner Mongolia	Jiulongshan Fm., J_2	Tan and Ren [139]
	Brochocoleus yangshuwanziensis Jarzembowski, Yan, Wang & Zhang, 2013	Ningcheng, Inner Mongolia	Yixian Fm., K_1	Jarembowski *et al.* [341]
	Fuscicupes parvus Hong & Wang, 1990	Laiyang, Shandong	Laiyang Fm., K_1	Hong and Wang [173]
	Cionocoleus magicus Ren, Lu, Ji & Guo, 1995	West hills, Beijing	Lushangfen Fm., K_1	Ren *et al.* [240]

(Continued)

Table 21.1 (Continued)

Family	Species	Locality	Horizon/Age	Citation
	Cionocoleus cervicalis Tan, Ren & Shih, 2007	Beipiao, Liaoning	Yixian Fm., K_1	Tan *et al.* [128]
	Cionocoleus planiusculus Tan, Ren & Shih, 2007	Beipiao, Liaoning	Yixian Fm., K_1	Tan *et al.* [128]
	Cionocoleus olympicus Jarzembowski, Yan, Wang & Zhang, 2013	Ningcheng, Inner Mongolia	Yixian Fm., K_1	Jarembowski *et al.* [341]
	Cionocoleus tanae Jarzembowski, Yan, Wang & Zhang, 2013	Ningcheng, Inner Mongolia	Yixian Fm., K_1	Jarembowski *et al.* [341]
	Monticupes surrectus Ren, Lu, Ji & Guo, 1995	West hills, Beijing	Lushangfen Fm., K_1	Ren *et al.* [240]
	Pareuryomma tylodes (Tan, Ren, Shih & Ge, 2006)	Beipiao, Liaoning	Yixian Fm., K_1	Tan *et al.* [187]
	Pareuryomma ancistrodonta Tan, Wang, Ren & Yang, 2012	Beipiao, Liaoning	Yixian Fm., K_1	Tan *et al.* [129]
	Pareuryomma cardiobasis Tan, Wang, Ren & Yang, 2012	Ningcheng, Inner Mongolia	Jiulongshan Fm., J_2	Tan *et al.* [129]
Schizophoridae	[a]*Metrorhynchites putativus* (Lin, 1986)	Liuyang, Hunan	Zaoshang Fm., J_1	Lin [141]
	Homocatabrycus liui Tan, Ren & Shih, 2007	Ningcheng, Inner Mongolia	Jiulongshan Fm., J_2	Tan *et al.* [96]
	Menopraesagus explanatus Tan, Ren & Shih, 2007	Ningcheng, Inner Mongolia	Jiulongshan Fm., J_2	Tan *et al.* [96]
	Menopraesagus grammicus Tan, Ren & Shih, 2007	Ningcheng, Inner Mongolia	Jiulongshan Fm., J_2	Tan *et al.* [96]
	Menopraesagus oryziformis Tan, Ren, Shih & Yang, 2013	Ningcheng, Inner Mongolia	Jiulongshan Fm., J_2	Tan *et al.* [143]
	Menopraesagus oxycerus Tan, Ren & Shih, 2007	Ningcheng, Inner Mongolia	Jiulongshan Fm., J_2	Tan *et al.* [96]
	Sinorhombocoleus papposus Tan & Ren, 2009	Beipiao, Liaoning	Yixian Fm., K_1	Tan and Ren [139]
	Sinoschizala darani Jarzembowski, Yan, Wang & Zhang, 2012	Ningcheng, Inner Mongolia	Jiulongshan Fm., J_2	Jarzembowski *et al.* [250]
	Abrohadeocoleodes amoenus Tan, Ren, Shih & Yang, 2013	Ningcheng, Inner Mongolia	Jiulongshan Fm., J_2	Tan *et al.* [143]
	Abrohadeocoleodes eurycladus Tan, Ren, Shih & Yang, 2013	Ningcheng, Inner Mongolia	Jiulongshan Fm., J_2	Tan *et al.* [143]
	Abrohadeocoleodes ooideus Tan, Ren, Shih & Yang, 2013	Ningcheng, Inner Mongolia	Jiulongshan Fm., J_2	Tan *et al.* [143]
	Abrhadeocoleodes patefactus Tan, Ren, Shih & Yang, 2013	Ningcheng, Inner Mongolia	Jiulongshan Fm., J_2	Tan *et al.* [143]
Taldycupedidae	[a]*Yuxianocoleus hebeiense* Hong, 1985	Yuxian, Hebei	Xiahuayuan Fm., J_2	Hong [144]
	[a]*Wuchangia latilimbata* Hong, 1986	Daye, Hubei	Wuchang Fm., J_1	Hong [163]
	[a]*Yiyangicupes huobashanense* Hong, 1988	Yiyang, Jiangxi	Lengshuiwu Fm., K_1	Hong [164]
Tshekardocoleidae	*Dictycoleus jurassicus* Hong, 1982	Hongliugeda, Gansu	Dashankou Fm., J_1-J_2	Hong [160]
[a]Cupedidae or Ommatidae	[a]*Gansucupes attenuatus* Hong, 1982	Yumen, Gansu	[b]Chijinqiao Fm., K_1	Hong [160]
[a]Cupedidae or Ommatidae	[a]*Chengdecupes baojiatunensis* Hong, 1992	Jiaohe, Jilin	Baojiatun Fm., J_3	Hong [145]

(Continued)

Table 21.1 (Continued)

Family	Species	Locality	Horizon/Age	Citation
[a]Cupedidae or Ommatidae	[a]*Chengdecupes jurassicus* Hong, 1983	Chengde, Hebei	Jiulongshan Fm., J_2	Hong [93]
[a]Cupedidae or Ommatidae	[a]*Chengdecupes kezuoense* Hong, 1987	Kezuo, Liaoning	Shahai Fm., K_1	Hong [342]
[a]Cupedidae or Taldycupedidae	[a]*Hebeicoleus sertulatus* Hong, 1992	Chengde, Hebei	Jiulongshan Fm., J_2	Hong [167]
		Suborder Myxophaga Crowson, 1955		
Hydroscaphidae	*Hydroscapha jeholensis* Cai, Short & Huang, 2012	Beipiao, Liaoning	Yixian Fm., K_1	Cai *et al.* [149]
		Suborder Adephaga Schellenberg, 1806		
Carabidae	[a]*Yunnanocarabus litus* Lin, 1977	Xiangyun, Yunnan	Zhanghe Fm., K_2	Lin [169]
	[a]*Conjunctia longa* Zhang, 1997	Zhixin Forest Farm, Jilin	Dalazi Fm., K_1	Zhang [177]
	Cretorabus rasnitsyni Wang & Zhang, 2011	Ningcheng, Inner Mongolia	Yixian Fm., K_1	Wang and Zhang [116]
	[a]*Protorabus polyphlebius* Ren, Lu, Ji & Guo, 1995	West hills, Beijing	Lushangfen Fm., K_1	Ren *et al.* [240]
	[a]*Protorabus minisculus* Zhang, 1997	Zhixin Forest Farm, Jilin	Dalazi Fm., K_1	Zhang [177]
	[a]*Fangshania punctata* Hong, 1981	Fangshan, Beijing	Lushangfen Fm., K_1	Hong [159]
	Nebrorabus tumoculus (Ren, Lu, Ji & Guo, 1995)	West hills, Beijing	Lushangfen Fm., K_1	Ren *et al.* [240]
	[a]*Atrirabus shandongensis* Hong & Wang, 1990	Laiyang, Shandong	Laiyang Fm., K_1	Hong and Wang [173]
	[a]*Atrirabus tuanwangensis* Hong & Wang, 1990	Laiyang, Shandong	Laiyang Fm., K_1	Hong and Wang [173]
	[a]*Magnirabus furvus* Hong & Wang, 1990	Laiyang, Shandong	Laiyang Fm., K_1	Hong and Wang [173]
	[a]*Cavicarabus lucensua* Hong, 1991	Daye, Hubei	Wuchang Fm., J_1	Hong [165]
	[a]*Wuchangicarabus latus* Hong, 1991	Daye, Hubei	Wuchang Fm., J_1	Hong [165]
	[a]*Lirabus granulatus* Hong, 1992	Kezuo, Liaoning	Shahai Fm., K_1	Hong [166]
	Aethocarabus levigata Ren, Lu, Ji & Guo, 1995	West hills, Beijing	Lushangfen Fm., K_1	Ren *et al.* [240]
	Denudirabus exstrius Ren, Lu, Ji & Guo, 1995	West hills, Beijing	Lushangfen Fm., K_1	Ren *et al.* [240]
	Penecupes rapax Ren, Lu, Ji & Guo, 1995	West hills, Beijing	Lushangfen Fm., K_1	Ren *et al.* [240]
[a]Caraboidea *Incertae sedis*	[a]*Eurycoleus arcuatus* Hong, 1982	Yumen, Gansu	Zhonggou Fm., K_1	Hong [160]
[a]Caraboidea *Incertae sedis*	[a]*Eurycoleus clypeolatus* Hong, 1982	Yumen, Gansu	Zhonggou Fm., K_1	Hong [160]
[a]Caraboidea *Incertae sedis*	[a]*Eurycoleus dimorphocellatus* Hong, 1982	Yumen, Gansu	Zhonggou Fm., K_1	Hong [160]
[a]Caraboidea *Incertae sedis*	[a]*Eurycoleus foveolatus* Hong, 1984	Fangshan, Beijing	Lushangfen Fm., K_1	Hong [162]
[a]Caraboidea *Incertae sedis*	[a]*Eurycoleus parvus* Hong, 1982	Yumen, Gansu	Zhonggou Fm., K_1	Hong [160]

(Continued)

Table 21.1 (Continued)

Family	Species	Locality	Horizon/Age	Citation
[a]Caraboidea *Incertae sedis*	[a]*Glottocoleus lenticulatus* Hong, 1982	Yumen, Gansu	Zhonggou Fm., K_1	Hong [160]
[a]Caraboidea *Incertae sedis*	[a]*Glyptocoleus stellatus* Hong, 1982	Yumen, Gansu	Zhonggou Fm., K_1	Hong [160]
[a]Caraboidea *Incertae sedis*	[a]*Grypocoleus fornicatus* Hong, 1982	Yumen, Gansu	Zhonggou Fm., K_1	Hong [160]
[a]Caraboidea *Incertae sedis*	[a]*Leptocoleus lenis* Hong, 1982	Yumen, Gansu	[b]Chijinqiao Fm., K_1	Hong [160]
[a]Caraboidea *Incertae sedis*	[a]*Mesocoleus zhonggouense* Hong, 1982	Yumen, Gansu	Zhonggou Fm., K_1	Hong [160]
[a]Caraboidea *Incertae sedis*	[a]*Obesofemoria chijinqiaoensis* Hong, 1982	Yumen, Gansu	[b]Chijinqiao Fm., K_1	Hong [160]
[a]Caraboidea *Incertae sedis*	[a]*Phyllocoleus striolatus* Hong, 1982	Yumen, Gansu	Zhonggou Fm., K_1	Hong [160]
[a]Caraboidea *Incertae sedis*	[a]*Pleurocoleus catenatus* Hong, 1982	Yumen, Gansu	Zhonggou Fm., K_1	Hong [160]
[a]Caraboidea *Incertae sedis*	[a]*Vago oblonga* Hong, 1982	Yumen, Gansu	[b]Chijinqiao Fm., K_1	Hong [160]
[a]Caraboidea *Incertae sedis*	[a]*Yumenocoleus intermedius* Hong, 1982	Yumen, Gansu	Zhonggou Fm., K_1	Hong [160]
[a]Caraboidea *Incertae sedis*	[a]*Yumenocoleus tenuis* Hong, 1988	Yiyang, Jiangxi	Lengshuiwu Fm., K_1	Hong [164]
[a]Caraboidea *Incertae sedis*	[a]*Yumenocoleus lineatus* Hong, 1982	Yumen, Gansu	Xiagou Fm., K_1	Hong [160]
[a]Caraboidea *Incertae sedis*	[a]*Yumenocoleus longus* Hong, 1982	Yumen, Gansu	Chijinpu Fm., K_1	Hong [160]
[a]Caraboidea *Incertae sedis*	[a]*Yumenocoleus nantianmenensis* Hong, 1984	Zhangjiakou, Hebei	Qingshila Fm., K_1	Hong [162]
[a]Caraboidea *Incertae sedis*	[a]*Jiangxicoleus guixiense* Hong, 1988	Yiyang, Jiangxi	Lengshuiwu Fm., K_1	Hong [164]
		Superfamily Dytiscoidea Bell, 1966		
[a]Caraboidea *Incertae sedis*	[a]*Jiangxicoleus yiyangense* Hong, 1988	Yiyang, Jiangxi	Lengshuiwu Fm., K_1	Hong [164]
Coptoclavidae	*Coptoclava longipoda* Ping, 1928	Jiuquan, Gansu; Yumen, Gansu; Luanping, Hebei; Yanji, Jilin; Damiaotazi hill, Liaoning; Kazuo, Liaoning;	[b]Chijinqiao Fm., K_1 [b]Chijinpu Fm., K_1; Dabeigou Fm., K_1; Dalazi Fm., K_1; [b]Jingangshan Fm., K_1; Jiufotang Fm., K_1;	Ping [156]

(Continued)

Table 21.1 (Continued)

Family	Species	Locality	Horizon/Age	Citation
		Laiyang, Shandong; Duolun, Inner Mongolia	Laiyang Fm., K_1; Jianchang Fm., K_1	
	Coptoclavisca grandioculus Zhang, 1992	Laiyang, Shandong	Laiyang Fm., K_1	Zhang [315]
	Daohugounectes primitives Wang, Ponomarenko & Zhang, 2009	Ningcheng, Inner Mongolia	Jiulongshan Fm., J_2	Wang *et al.* [133]
Dytiscidae	*Liadroporus elegans* Prokin & Ren, 2010	Beipiao, Liaoning	Yixian Fm., K_1	Prokin and Ren [254]
	Liadytiscus cretaceous Prokin & Ren, 2010	Beipiao, Liaoning	Yixian Fm., K_1	Prokin and Ren [254]
	Liadytiscus latus Prokin & Ren, 2010	Beipiao, Liaoning	Yixian Fm., K_1	Prokin and Ren [254]
	Liadytiscus longitibialis Prokin & Ren, 2010	Beipiao, Liaoning	Yixian Fm., K_1	Prokin and Ren [254]
	Mesoderus magnus Prokin & Ren, 2010	Beipiao, Liaoning	Yixian Fm., K_1	Prokin and Ren [254]
	Mesoderus ovatus Prokin, Petrov, Wang & Ponomarenko, 2013	Dashuangmiao, Inner Mongolia	Yixian Fm., K_1	Prokin *et al.* [136]
	Mesoderus punctatus Prokin, Petrov, Wang & Ponomarenko, 2013	Dashuangmiao, Inner Mongolia	Yixian Fm., K_1	Prokin *et al.* [136]
	Mesoderus ventralis Prokin & Ren, 2010	Beipiao, Liaoning	Yixian Fm., K_1	Prokin and Ren [254]
	Sinoporus lineatus Prokin & Ren, 2010	Beipiao, Liaoning	Yixian Fm., K_1	Prokin and Ren [254]
	Liadyxianus kirejtshuki Prokin, Petrov, Wang & Ponomarenko, 2013	Beipiao, Liaoning	Yixian Fm., K_1	Prokin *et al.* [136]
	Mesodytes rhantoides Prokin, Petrov, Wang & Ponomarenko, 2013	Dashuangmiao, Inner Mongolia	Yixian Fm., K_1	Prokin *et al.* [136]
Haliplidae	*Cretihaliplus chifengensis* Ren, Zhu & Lu, 1995	Chifeng, Inner Mongolia	Jiufotang Fm., K_1	Ren *et al.* [256]
	Cretihaliplus sidaojingensis Ren, Zhu & Lu, 1995	Chifeng, Inner Mongolia	Jiufotang Fm., K_1	Ren *et al.* [256]
Liadytidae	*Ovidytes gaoi* Ren, Zhu & Lu, 1995	Chifeng, Inner Mongolia	Jiufotang Fm., K_1	Ren *et al.* [256]
Trachypachidae	*Eodromeus daohugouensis* Wang, Zhang & Ponomerenko, 2012	Ningcheng, Inner Mongolia	Jiulongshan Fm., J_2	Wang *et al.* [138]
	Eodromeus robustus Wang, Zhang & Ponomerenko, 2012	Ningcheng, Inner Mongolia	Jiulongshan Fm., J_2	Wang *et al.* [138]
	Eodromeus viriosus Zhang, 1997	Zhixin Forest Farm, Jilin	Dalazi Fm., K_1	Zhang [177]
	Unda chifengensis Wang, Zhang & Ponomerenko, 2012	Ningcheng, Inner Mongolia	Jiulongshan Fm., J_2	Wang *et al.* [138]
	Unda chonggezhuangensis Wang, 1998	Fangshan, Beijing	Lushangfen Fm., K_1	Wang [175]
	Unda pandurata Ren, Lu, Ji & Guo, 1995	West hills, Beijing	Lushangfen Fm., K_1	Ren *et al.* [240]

(Continued)

Table 21.1 (Continued)

Family	Species	Locality	Horizon/Age	Citation
	Fortiseode pervalimand Jia & Ren, 2011	Beipiao, Liaoning	Yixian Fm., K_1	Jia and Ren [257]
	Sinodromeus liutiaogouensis Wang, Zhang & Ponomerenko, 2012	Dashuangmiao, Inner Mongolia	Yixian Fm., K_1	Wang *et al.* [138]
Triaplidae	[a]*Clypeus korlaensis* Hong, 1983	Tashidian, Xinjiang	Kezilenuer Fm., J_2	Hong [93]
		Suborder Polyphaga Emery, 1886		
		Superfamily Hydrophiloidea Latreille, 1802		
Helophoridae	*Helophorus (Mesosperchus) gracilis* (Prokin, Ren & Fikáček, 2010)	Beipiao, Liaoning	Yixian Fm., K_1	Prokin *et al.* [343]
	Helophorus (Mesosperchus) yixianus Fikáček, Prokin, Angus, Ponomarenko, Yue, Ren & Prokop, 2012	Beipiao, Liaoning	Yixian Fm., K_1	Fikáček *et al.* [344]
Hydrophilidae	*Alegorius yixianus* Fikáček, Prokin, Yan, Yue, Wang, Ren & Beattie, 2014	Beipiao, Liaoning	Yixian Fm., K_1	Fikáček *et al.* [261]
	Hydroyixia elongata Fikáček, Prokin, Yan, Yue, Wang, Ren & Beattie, 2014	Beipiao, Liaoning	Yixian Fm., K_1	Fikáček *et al.* [261]
	Hydroyixia latissima Fikáček, Prokin, Yan, Yue, Wang, Ren & Beattie, 2014	Beipiao, Liaoning	Yixian Fm., K_1	Fikáček *et al.* [261]
[a]Hydrophiloidea *Incertae sedis*	[a]*Sinosperchopsis silinae* Prokin, Ren & Fikáček, 2010	Beipiao, Liaoning	Yixian Fm., K_1	Prokin *et al.* [343]
[a]Hydrophiloidea *Incertae sedis*	[a]*Laetopsia hydraenoides* (Prokin, Ren & Fikáček, 2010)	Beipiao, Liaoning	Yixian Fm., K_1	Prokin *et al.* [343]
[a]Hydrophiloidea *Incertae sedis*	[a]*Laetopsia shatrovskiyi* (Prokin, Ren & Fikáček, 2010)	Beipiao, Liaoning	Yixian Fm., K_1	Prokin *et al.* [343]
Agyrtidae	[a]*Mesecanus lintouensis* (Lin, 1985)	Hanshan, Anhui	Hanshan Fm., J_2	Lin [77]
	[a]*Sinosilphia punctata* Hong & Wang, 1990	Laiyang, Shandong	Laiyang Fm., K_1	Hong and Wang [173]
		Superfamily Staphylinoidea Latreille, 1802		
Staphylinidae	*Oxyporus yixianus* Solodovnikov & Yue, 2011	Beipiao, Liaoning	Yixian Fm., K_1	Yue *et al.* [232]
	Quedius cretaceus Cai & Huang, 2013	Beipiao, Liaoning	Yixian Fm., K_1	Cai and Huang [208]
	Protostaphylinus mirus Lin, 1976	Beipiao, Liaoning	Haifanggou Fm., J_2	Lin [168]
	Cretoquedius distinctus Solodovnikov & Yue, 2013	Beipiao, Liaoning	Yixian Fm., K_1	Solodovnikov *et al.* [266]
	Cretoquedius dorsalis Solodovnikov & Yue, 2013	Beipiao, Liaoning	Yixian Fm., K_1	Solodovnikov *et al.* [266]
	Cretoquedius infractus Solodovnikov & Yue, 2013	Beipiao, Liaoning	Yixian Fm., K_1	Solodovnikov *et al.* [266]
	Laostaphylinus fuscus Zhang, 1988	Laiyang, Shandong	Laiyang Fm., K_1	Zhang [264]
	Laostaphylinus nigritellu Zhang, 1988	Laiyang, Shandong	Laiyang Fm., K_1	Zhang [264]
	Mesostaphylinus antiquus Solodovnikov & Yue, 2013	Beipiao, Liaoning	Yixian Fm., K_1	Solodovnikov *et al.* [266]

(Continued)

Table 21.1 (Continued)

Family	Species	Locality	Horizon/Age	Citation
	Mesostaphylinus elongates Solodovnikov & Yue, 2013	Beipiao, Liaoning	Yixian Fm., K_1	Solodovnikov *et al.* [266]
	Mesostaphylinus laiyangensis Zhang, 1988	Laiyang, Shandong	Laiyang Fm., K_1	Zhang [264]
	Mesostaphylinus yixianus Solodovnikov & Yue, 2013	Beipiao, Liaoning	Yixian Fm., K_1	Solodovnikov *et al.* [266]
	Sinostaphylina nanligezhuangensis Hong & Wang, 1990	Laiyang, Shandong	Laiyang Fm., K_1	Hong and Wang [173]
	Sinostaphylius xiejiajieensis Hong, 1992	Liaoyuan, Jilin	Yixian Fm., K_1	Hong [145]
	Hesterniasca lata Cai, Huang & Solodovnikov, 2011	Beipiao, Liaoning	Yixian Fm., K_1	Cai *et al.* [214]
	Hesterniasca obesa Zhang, Wang & Xu, 1992	Laiyang, Shandong	Laiyang Fm., K_1	Zhang *et al.* [265]
	Glabrimycetoporus amoenus Yue, Zhao & Ren, 2009	Beipiao, Liaoning	Yixian Fm., K_1	Yue *et al.* [233]
	Megolisthaerus chinensis Solodovnikov & Yue, 2010	Beipiao, Liaoning	Yixian Fm., K_1	Yue *et al.* [231]
	Megolisthaerus minor Cai & Huang, 2013	Beipiao, Liaoning	Yixian Fm., K_1	Cai and Huang [207]
	Sinoxytelus breviventer Yue, Zhao & Ren, 2010	Beipiao, Liaoning	Yixian Fm., K_1	Yue *et al.* [234]
	Sinoxytelus euglypheus Yue, Zhao & Ren, 2010	Beipiao, Liaoning	Yixian Fm., K_1	Yue *et al.* [234]
	Sinoxytelus longisetosus Yue, Zhao & Ren, 2010	Beipiao, Liaoning	Yixian Fm., K_1	Yue *et al.* [234]
	Juroglypholoma antiquum Cai & Huang, 2012	Ningcheng, Inner Mongolia	Jiulongshan Fm., J_2	Cai and Huang [206]
	Mesocoprophilus clavatus Cai & Huang, 2013	Beipiao, Liaoning	Yixian Fm., K_1	Cai and Huang [211]
	Pseudanotylus archaicus (Yue, Makranczy & Ren, 2012)	Beipiao, Liaoning	Yixian Fm., K_1	Yue *et al.* [210]
	Sinanthobium daohugouense Cai & Huang, 2013	Ningcheng, Inner Mongolia	Jiulongshan Fm., J_2	Cai and Huang [209]
	Protodeleaster glaber Cai, Thayer, Huang, Wang & Newton, 2013	Beipiao, Liaoning	Yixian Fm., K_1	Cai *et al.* [216]
	Cretoprosopus problematicus Solodovnikov & Yue, 2013	Beipiao, Liaoning	Yixian Fm., K_1	Solodovnikov *et al.* [266]
	Durothorax creticus Solodovnikov & Yue, 2013	Beipiao, Liaoning	Yixian Fm., K_1	Solodovnikov *et al.* [266]
	Paleothius gracilis Solodovnikov & Yue, 2013	Beipiao, Liaoning	Yixian Fm., K_1	Solodovnikov *et al.* [266]
	Paleowinus ambiguus Solodovnikov & Yue, 2013	Beipiao, Liaoning	Yixian Fm., K_1	Solodovnikov *et al.* [266]
	Paleowinus chinensis Solodovnikov & Yue, 2013	Beipiao, Liaoning	Yixian Fm., K_1	Solodovnikov *et al.* [266]
	Paleowinus fossilis Solodovnikov & Yue, 2013	Beipiao, Liaoning	Yixian Fm., K_1	Solodovnikov *et al.* [266]
	Paleowinus mirabilis Solodovnikov & Yue, 2013	Beipiao, Liaoning	Yixian Fm., K_1	Solodovnikov *et al.* [266]
	Paleowinus rex Solodovnikov & Yue, 2013	Beipiao, Liaoning	Yixian Fm., K_1	Solodovnikov *et al.* [266]

(Continued)

Table 21.1 (Continued)

Family	Species	Locality	Horizon/Age	Citation
	Thayeralinus fieldi Solodovnikov & Yue, 2013	Beipiao, Liaoning	Yixian Fm., K_1	Solodovnikov *et al.* [266]
	Thayeralinus fraternus (Zhang, Wang & Xu, 1992)	Beipiao, Liaoning	Yixian Fm., K_1	Zhang *et al.* [265]
	Thayeralinus giganteus Solodovnikov & Yue, 2013	Beipiao, Liaoning	Yixian Fm., K_1	Solodovnikov *et al.* [266]
	Thayeralinus glandulifer Solodovnikov & Yue, 2013	Beipiao, Liaoning	Yixian Fm., K_1	Solodovnikov *et al.* [266]
	Thayeralinus longelytratus Solodovnikov & Yue, 2013	Beipiao, Liaoning	Yixian Fm., K_1	Solodovnikov *et al.* [266]
	Cretoxyporus extraneus Cai & Huang, 2014	Beipiao, Liaoning	Yixian Fm., K_1	Cai and Huang [212]
	Protoxyporus grandis Cai & Huang, 2014	Dashuangmiao, Inner Mongolia	Yixian Fm., K_1	Cai and Huang [212]
	Mesapatetica aenigmatica Cai, Huang, Newton & Thayer, 2014	Ningcheng, Inner Mongolia	Jiulongshan Fm., J_2	Cai *et al.* [213]
	Protolisthaerus jurassicus Cai, Beattie & Huang, 2015	Ningcheng, Inner Mongolia	Haifanggou Fm., J_2	Cai *et al.* [205]
	Paleosiagonium adaequatum Yue, Gu, Yang, Wang & Ren, 2016	Beipiao, Liaoning	Yixian Fm., K_1	Yue *et al.* [229]
	Paleosiagonium brevelytratum Yue, Gu, Yang, Wang & Ren, 2016	Beipiao, Liaoning	Yixian Fm., K_1	Yue *et al.* [229]
Superfamily Scarabaeoidea Latreille, 1802				
Alloioscarabaeidae	*Alloioscarabaeus cheni* Bai, Ren & Yang, 2012	Ningcheng, Inner Mongolia	Jiulongshan Fm., J_2	Bai *et al.* [194]
Geotrupidae	*Geotrupoides jiaoheensis* Hong, 1992	Jiaohe, Liaoning	Baojiatun Fm., K_1	Hong [166]
	Geotrupoides kezuoensis Hong, 1992	Kezuo, Liaoning	Shahai Fm., K_1	Hong [166]
	Geotrupoides lithographicus Deichmüller, 1886	Laiyang, Shandong	Laiyang Fm., K_1	Deichmüller [82]
	Geotrupoides nodosus Hong & Wang, 1990	Laiyang, Shandong	Laiyang Fm., K_1	Hong and Wang [173]
	Geotrupoides saxosus Zhang, 1997	Zhixin Forest Farm, Jilin	Dalazi Fm., K_1	Zhang [177]
	Geotrupoides songyingziense Hong, 1984	Jianping, Liaoning	Yixian Fm., K_1	Hong [162]
	Parageotrupes incanus Nikolajev & Ren, 2010	Beipiao, Liaoning	Yixian Fm., K_1	Nikolajev and Ren [267]
Glaphyridae	*Glaphyrus ancestralis* Nikolajev & Ren, 2011	Beipiao, Liaoning	Yixian Fm., K_1	Nikolajev and Ren [345]
	Lithohypna chifengensis Nikolajev, Wang & Zhang, 2011	Dashuangmiao, Inner Mongolia	Yixian Fm., K_1	Nikolajev *et al.* [269]
	Lithohypna ericeusicus (Yan, Bai & Ren, 2013)	Beipiao, Liaoning	Yixian Fm., K_1	Yan *et al.* [201]
	Lithohypna laoningensis Nikolajev, 2014	Beipiao, Liaoning	Yixian Fm., K_1	Nikolajev [270]
	Lithohypna lepticephala Nikolajev & Ren, 2012	Beipiao, Liaoning	Yixian Fm., K_1	Nikolajev and Ren [346]
	Lithohypna longula Nikolajev & Ren, 2012	Beipiao, Liaoning	Yixian Fm., K_1	Nikolajev and Ren [346]

(Continued)

Table 21.1 (Continued)

Family	Species	Locality	Horizon/Age	Citation
	Lithohypna tuberculata Nikolajev & Ren, 2012	Beipiao, Liaoning	Yixian Fm., K_1	Nikolajev and Ren [346]
	Lithohypna yuxiana Nikolajev & Ren, 2012	Beipiao, Liaoning	Yixian Fm., K_1	Nikolajev and Ren [346]
	Cretohypna cristata Yan, Nikolajev & Ren 2012	Dashuangmiao, Inner Mongolia	Yixian Fm., K_1	Yan *et al.* [202]
	Cretohypna puncta Zhao, Bai, Shih & Ren, 2016	Dashuangmiao, Inner Mongolia	Yixian Fm., K_1	Zhao *et al.* [203]
	Cretohypna robusta Zhao, Bai, Shih & Ren, 2016	Dashuangmiao, Inner Mongolia	Yixian Fm., K_1	Zhao *et al.* [203]
	Cretohypna yixianensis Nikolajev & Ren, 2015	Beipiao, Liaoning	Yixian Fm., K_1	Nikolajev and Ren [347]
	Mesohypna lopatini Nikolajev & Ren, 2013	Beipiao, Liaoning	Yixian Fm., K_1	Nikolajev and Ren [271]
	Mesohypna probata Nikolajev, 2014	Beipiao, Liaoning	Yixian Fm., K_1	Nikolajev [270]
Glaresidae	*Glaresis orthochilus* Bai, Krell & Ren, 2010	Beipiao, Liaoning	Yixian Fm., K_1	Bai *et al.* [197]
	Glaresis tridentata Bai, Beutel & Ren, 2014	Beipiao, Liaoning	Yixian Fm., K_1	Bai *et al.* [195]
Hybosoridae	*Leptosorus fortus* (Ren, Zhu & Lu, 1995)	Chifeng, Inner Mongolia	Yixian Fm., K_1	Ren *et al.* [256]
	Mesoceratocanthus tuberculifrons Nikolajev, Wang, Liu & Zhang, 2010	Ningcheng, Inner Mongolia	Yixian Fm., K_1	Nikolajev *et al.* [274]
	Crassisorus fractus Nikolajev, Wang & Zhang, 2012	Dashuangmiao, Inner Mongolia	Yixian Fm., K_1	Nikolajev *et al.* [275]
	Pulcherhybosorus tridentatus Yan, Bai & Ren, 2012	Dashuangmiao, Inner Mongolia	Yixian Fm., K_1	Yan *et al.* [200]
	Sinohybosorus cheni Nie, Bai, Ren & Yang, 2018	Beipiao, Liaoning	Yixian Fm., K_1	Lu *et al.* [276]
	Sinochaetodus tridentatus Lu, Bai, Ren & Yang, 2018	Beipiao, Liaoning	Yixian Fm., K_1	Lu *et al.* [276]
Lithoscarabaeidae	*Baiscarabaeus yixianensis* Nikolajev, Wang & Zhang, 2013	Chifeng, Inner Mongolia	Yixian Fm., K_1	Nikolajev *et al.* [279]
Lucanidae	*Juraesalus atavus* Nikolajev, Wang, Liu & Zhang, 2011	Ningcheng, Inner Mongolia	Jiulongshan Fm., J_2	Nikolajev *et al.* [85]
	Sinaesalus curvipes Nikolajev, Wang, Liu & Zhang, 2011	Ningcheng, Inner Mongolia	Yixian Fm., K_1	Nikolajev *et al.* [85]
	Sinaesalus longipes Nikolajev, Wang, Liu & Zhang, 2011	Ningcheng, Inner Mongolia	Yixian Fm., K_1	Nikolajev *et al.* [85]
	Sinaesalus tenuipes Nikolajev, Wang, Liu & Zhang, 2011	Ningcheng, Inner Mongolia	Yixian Fm., K_1	Nikolajev *et al.* [85]
	Prosinodendron krelli Bai, Ren & Yang, 2012	Beipiao, Liaoning	Yixian Fm., K_1	Bai *et al.* [199]
	Litholamprima longimana Nikolajev & Ren, 2015	Beipiao, Liaoning	Yixian Fm., K_1	Nikolajev and Ren [280]
Ochodaeidae	*Mesochodaeus daohugouensis* Nikolajev & Ren, 2010	Ningcheng, Inner Mongolia	Jiulongshan Fm., J_2	Nikolajev & Ren [89]
	Yixianochodaeus horridus Nikolajev, 2015	Beipiao, Liaoning	Yixian Fm., K_1	Nikolajev [281]

(Continued)

Table 21.1 (Continued)

Family	Species	Locality	Horizon/Age	Citation
Pleocomidae	*Pleocoma dolichophylla* Nikolajev & Ren, 2012	Beipiao, Liaoning	Yixian Fm., K_1	Nikolajev & Ren [348]
	Proteroscarabaeus robustus Zhang, 1997	Zhixin Forest Farm, Jilin	Dalazi Fm., K_1	Zhang [177]
	Proteroscarabaeus yeni Grabau, 1923	Laiyang, Shandong	Laiyang Fm., K_1	Grabau [283]
Scarabaeidae	[a]*Prionocephale deplanate* Lin, 1980	Lanxi, Zhejiang	[b]Fangyuan Fm., K_2	Lin [170]
	Prophaenognatha robusta Bai, Ren & Yang, 2011	Beipiao, Liaoning	Yixian Fm., K_1	Bai *et al.* [198]
Septiventeridae	*Septiventer quadridentatus* Bai, Ren, Shih & Yang, 2013	Beipiao, Liaoning	Yixian Fm., K_1	Bai *et al.* [196]
[a]Scarabaeoidea *Incertae sedis*	[a]*Mesoscarabaeus corneus* Hong, 1982	Yumen, Gansu	[b]Chijinqiao Fm., K_1	Hong [160]
[a]Scarabaeoidea *Incertae sedis*	[a]*Mesoscarabaeus morulosus* Hong, 1982	Yumen, Gansu	[b]Chijinqiao Fm., K_1	Hong [160]
[a]Scarabaeoidea *Incertae sedis*	[a]*Hongscarabaeus brunneus* (Hong, 1982)	Yumen, Gansu	[b]Chijinqiao Fm., K_1	Hong [160]
		Superfamily Dascilloidea Guérin-Méneville, 1843		
Dascillidae	*Cretodascillus sinensis* Jin, Ślipiński, Pang & Ren, 2013	Dashuangmiao, Inner Mongolia	Yixian Fm., K_1	Jin *et al.* [285]
		Superfamily Buprestoidea Leach, 1815		
Buprestidae	*Planocoleus ensatus* Hong, 1982	Yumen, Gansu	Zhonggou Fm., K_1	Hong [160]
	Planocoleus glabratus Hong, 1982	Yumen, Gansu	Zhonggou Fm., K_1	Hong [160]
	[a]*Macrotonus tuanwangensis* Hong & Wang, 1990	Laiyang, Shandong	Laiyang Fm., K_1	Hong and Wang [173]
	Sinoparathyrea bimaculata Pan, Chang & Ren, 2011	Ningcheng, Inner Mongolia	Jiulongshan Fm., J_2	Pan *et al.* [287]
	Sinoparathyrea gracilenta Pan, Chang & Ren, 2011	Ningcheng, Inner Mongolia	Jiulongshan Fm., J_2	Pan *et al.* [287]
	Sinoparathyrea robusta Pan, Chang & Ren, 2011	Ningcheng, Inner Mongolia	Jiulongshan Fm., J_2	Pan *et al.* [287]
	Trapezitergum grande Yu, Ślipiński & Shih, 2013	Dashuangmiao, Inner Mongolia	Yixian Fm., K_1	Yu *et al.* [239]
	Mongoligenula altilabdominis Yu, Ślipiński, Pang & Ren, 2015	Beipiao, Liaoning	Yixian Fm., K_1	Yu *et al.* [237]
	Mongoligenula gracilis Yu, Ślipiński, Pang & Ren, 2015	Beipiao, Liaoning	Yixian Fm., K_1	Yu *et al.* [237]
Schizopodidae	*Mesoschizopus elegans* Cai, Ślipiński & Huang, 2015	Beipiao, Liaoning	Yixian Fm., K_1	Cai *et al.* [215]
		Superfamily Byrrhoidea Latreille, 1804		
Byrrhidae	*Fangshanella stolida* Huang & Zhang, 1997	Fangshan, Beijing	Lushangfen Fm., K_1	Huang and Zhang [290]
	Mesobyrrhus parvus Huang & Zhang, 1997	Fangshan, Beijing	Lushangfen Fm., K_1	Huang and Zhang [290]
	Mesobyrrhus tanae Huang & Zhang, 1997	Fangshan, Beijing	Lushangfen Fm., K_1	Huang and Zhang [290]

(Continued)

Table 21.1 (Continued)

Family	Species	Locality	Horizon/Age	Citation
Eulichadidae	*Mesaplus beipiaoensis* Hong, 1983	Beipiao, Liaoning	Haifanggou Fm., J_2	Hong [93]
	Cretasyne lata Yan, Wang & Zhang, 2013	Ningcheng, Inner Mongolia	Yixian Fm., K_1	Yan *et al.* [292]
	Cretasyne longa Yan, Wang & Zhang, 2013	Ningcheng, Inner Mongolia	Yixian Fm., K_1	Yan *et al.* [292]
Heteroceridae	*Heterocerites magnus* Prokin & Ren, 2011	Beipiao, Liaoning	Yixian Fm., K_1	Prokin and Ren [349]
[a]Byrrhoidea *Incertae sedis*	[a]*Serecoleus nadbitovae* Yan, Wang, Jarzembowski & Zhang, 2015	Ningcheng, Inner Mongolia	Jiulongshan Fm., J_2	Yan *et al.* [350]
[a]Dryopoidea *Incertae sedis*	[a]*Jiuquanocoleus punctatus* Hong, 1982	Yumen, Gansu	[b]Chijinqiao Fm., K_1	Hong [160]
Lasiosynidae	*Lasiosyne daohugouensis* Kirejtshuk, Chang, Ren & Shih, 2010	Ningcheng, Inner Mongolia	Jiulongshan Fm., J_2	Kirejtshuk *et al.* [97]
	Lasiosyne euglyphea Tan, Ren & Shih, 2007	Ningcheng, Inner Mongolia	Jiulongshan Fm., J_2	Tan *et al.* [96]
	Lasiosyne fedorenkoi Kirejtshuk, Chang, Ren & Shih, 2010	Ningcheng, Inner Mongolia	Jiulongshan Fm., J_2	Kirejtshuk *et al.* [97]
	Lasiosyne gratiosa Kirejtshuk, Chang, Ren & Shih, 2010	Ningcheng, Inner Mongolia	Jiulongshan Fm., J_2	Kirejtshuk *et al.* [97]
	Lasiosyne quadricollis Kirejtshuk, Chang, Ren & Shih, 2010	Ningcheng, Inner Mongolia	Jiulongshan Fm., J_2	Kirejtshuk *et al.* [97]
	Anacapitis plata Tan & Ren, 2009	Ningcheng, Inner Mongolia	Jiulongshan Fm., J_2	Tan and Ren [139]
	Bupredactyla magna Kirejtshuk, Chang, Ren & Shih, 2010	Ningcheng, Inner Mongolia	Jiulongshan Fm., J_2	Kirejtshuk *et al.* [97]
	Parelateriformius capitifossus Yan & Wang, 2010	Ningcheng, Inner Mongolia	Jiulongshan Fm., J_2	Yan and Wang [296]
	Parelateriformius communis Yan & Wang, 2010	Ningcheng, Inner Mongolia	Jiulongshan Fm., J_2	Yan and Wang [296]
	Parelateriformius mirabdominis Yan & Wang, 2010	Ningcheng, Inner Mongolia	Jiulongshan Fm., J_2	Yan and Wang [296]
	Parelateriformius villosus Yan & Wang, 2010	Ningcheng, Inner Mongolia	Jiulongshan Fm., J_2	Yan and Wang [296]
		Superfamily Elateroidea Leach, 1815		
Artematopodidae	*Tarsomegamerus mesozoicus* Zhang, 2005	Ningcheng, Inner Mongolia	Jiulongshan Fm., J_2	Zhang [298]
	Sinobrevipogon jurassicus Cai, Lawrence, Ślipiński & Huang, 2015	Ningcheng, Inner Mongolia	Jiulongshan Fm., J_2	Cai *et al.* [98]
Cerophytidae	*Necromera admiranda* Chang & Kirejtshuk, 2011	Beipiao, Liaoning	Yixian Fm., K_1	Chang and Kirejtshuk [99]
	Necromera longa (Hong & Wang, 1990)	Laiyang, Shandong	Laiyang Fm., K_1	Hong and Wang [173]
Elateridae	[a]*Sinoelaterium melanocolor* Ping, 1928	Beipiao, Liaoning	Yixian Fm., K_1	Ping [156]
	Protagrypnus robustus Chang, Kirejtshuk & Ren, 2009	Ningcheng, Inner Mongolia	Jiulongshan Fm., J_2	Chang *et al.* [189]
	[a]*Fengningia punctata* Hong, 1984	Fengning, Hebei	Jiufotang Fm., K_1	Hong [162]
	[a]*Archaeolus funestus* Lin, 1986	Hezhou, Guangxi	Shiti Fm., J_1	Lin [141]

(Continued)

Table 21.1 (Continued)

Family	Species	Locality	Horizon/Age	Citation
	[a)]*Artinama qinghuoensis* Lin, 1986	Liuyang, Hunan	Zaoshang Fm., J_1	Lin [141]
	[a)]*Gripecolous enallus* Lin, 1986	Hezhou, Guangxi	Shiti Fm., J_1	Lin [141]
	[a)]*Mercata festira* Lin, 1986	Hezhou, Guangxi	Shiti Fm., J_1	Lin [141]
	[a)]*Ovivagina longa* Zhang, 1997	Shawan, Xinjiang	Badaowan Fm., J_1	Zhang [101]
	Crytocoelus buffoni Dolin & Nel, 2002	Beipiao, Liaoning	Yixian Fm., K_1	Dolin and Nel [302]
	Crytocoelus gianteus Chang, Ren & Shih, 2007	Beipiao, Liaoning	Yixian Fm., K_1	Chang *et al.* [191]
	Crytocoelus major Dolin & Nel, 2002	Beipiao, Liaoning	Yixian Fm., K_1	Dolin and Nel [302]
	Lithomerus buyssoni Dolin & Nel, 2002	Beipiao, Liaoning	Yixian Fm., K_1	Dolin and Nel [302]
	Bilineariselater foveatus Chang & Ren, 2008	Beipiao, Liaoning	Yixian Fm., K_1	Chang and Ren [190]
	Curtelater wui Chang & Ren, 2008	Beipiao, Liaoning	Yixian Fm., K_1	Chang and Ren [190]
	Paralithomerus exquisitus Chang, Zhang & Ren, 2008	Beipiao, Liaoning	Yixian Fm., K_1	Chang *et al.* [192]
	Paralithomerus parallelus Chang, Zhang & Ren, 2008	Beipiao, Liaoning	Yixian Fm., K_1	Chang *et al.* [192]
	Paradesmatus baiae Chang, Kirejtshuk & Ren, 2009	Ningcheng, Inner Mongolia	Jiulongshan Fm., J_2	Chang *et al.* [189]
	Paradesmatus dilatatus Chang, Kirejtshuk & Ren, 2010	Beipiao, Liaoning	Yixian Fm., K_1	Chang *et al.* [188]
	Paradesmatus ponomarenkoi Chang, Kirejtshuk & Ren, 2009	Ningcheng, Inner Mongolia	Jiulongshan Fm., J_2	Chang *et al.* [189]
	Paraprotagrypnus superbus Chang, Zhao & Ren, 2009	Ningcheng, Inner Mongolia	Jiulongshan Fm., J_2	Chang *et al.* [193]
	Sinolithomerus dolini Dong & Huang, 2009	Beipiao, Liaoning	Haifanggou Fm., J_2	Dong and Huang [303]
	Anoixis complanus Chang, Kirejtshuk & Ren, 2010	Beipiao, Liaoning	Yixian Fm., K_1	Chang *et al.* [188]
	Apoclion antennatus Chang, Kirejtshuk & Ren, 2010	Beipiao, Liaoning	Yixian Fm., K_1	Chang *et al.* [188]
	Apoclion clavatus Chang, Kirejtshuk & Ren, 2010	Beipiao, Liaoning	Yixian Fm., K_1	Chang *et al.* [188]
	Apoclion dolini Chang, Kirejtshuk & Ren, 2010	Beipiao, Liaoning	Yixian Fm., K_1	Chang *et al.* [188]
	Desmatinus cognatus Chang, Kirejtshuk & Ren, 2010	Beipiao, Liaoning	Yixian Fm., K_1	Chang *et al.* [188]
	Clavelater ningchengensis Dong & Huang, 2011	Ningcheng, Inner Mongolia	Jiulongshan Fm., J_2	Dong and Huang [102]
Eucnemidae	*Palaeoxenus sinensis* Chang, Muona & Teräväinen, 2016	Beipiao, Liaoning	Yixian Fm., K_1	Chang *et al.* [307]
[a)]Elateroidea *Incertae sedis*	[a)]*Microcoleus brunneus* Hong, 1982	Yumen, Gansu	Chijinpu Fm., K_1	Hong [160]
		Superfamily Derodontoidea LeConte, 1861		
Derodontidae	*Juropeltastica sinica* Cai, Lawrence, Ślipiński & Huang, 2014	Ningcheng, Inner Mongolia	Jiulongshan Fm., J_2	Cai *et al.* [103]

(Continued)

Table 21.1 (Continued)

Family	Species	Locality	Horizon/Age	Citation
		Superfamily Bostrichoidea Latreille, 1802		
Dermestidae	*Paradermestes jurassicus* Deng, Ślipiński, Ren & Pang, 2017	Ningcheng, Inner Mongolia	Jiulongshan Fm., J_2	Deng *et al.* [104]
		Superfamily Cleroidea Latreille, 1802		
Cleridae	*Protoclerus korynetoides* Kolibáč & Huang, 2016	Ningcheng, Inner Mongolia	Haifanggou Fm., J_2	Kolibáč and Huang [107]
	Wangweiella calloviana Kolibáč & Huang, 2016	Ningcheng, Inner Mongolia	Haifanggou Fm., J_2	Kolibáč and Huang [107]
Prionoceridae	*Idgiaites jurassicus* Liu, Ślipiński, Leschen, Ren & Pang, 2015	Ningcheng, Inner Mongolia	Jiulongshan Fm., J_2	Liu *et al.* [108]
Trogossitidae	*Palaeoendomychus gymnus* Zhang, 1992	Laiyang, Shandong	Laiyang Fm., K_1	Zhang [315]
	Sinosoronia longiantennata Zhang, 1992	Laiyang, Shandong	Laiyang Fm., K_1	Zhang [315]
	Eotenebroides tumoculus Ren, Lu, Ji & Guo, 1995	West hills, Beijing	Lushangfen Fm., K_1	Ren *et al.* [240]
	Sinopeltis amoena Yu, Leschen, Ślipiński, Ren & Pang, 2012	Ningcheng, Inner Mongolia	Jiulongshan Fm., J_2	Yu *et al.* [235]
	Sinopeltis jurrasica Yu, Leschen, Ślipiński, Ren & Pang, 2012	Ningcheng, Inner Mongolia	Jiulongshan Fm., J_2	Yu *et al.* [235]
	Latitergum glabrum Yu, Ślipiński, Leschen, Ren & Pang, 2014	Ningcheng, Inner Mongolia	Jiulongshan Fm., J_2	Yu *et al.* [106]
	Marginulatus venustus Yu, Ślipiński, Leschen, Ren & Pang, 2014	Ningcheng, Inner Mongolia	Jiulongshan Fm., J_2	Yu *et al.* [106]
	Paracretocateres bellus Yu, Ślipiński, Leschen, Ren & Pang, 2015	Beipiao, Liaoning	Yixian Fm., K_1	Yu *et al.* [236]
	Yixianteres beipiaoensis Yu, Ślipiński, Leschen, Ren & Pang, 2015	Beipiao, Liaoning	Yixian Fm., K_1	Yu *et al.* [236]
[a]Cleroidea *Incertae sedis*	[a]*Cervicatinius complanus* Tan & Ren, 2007	Ningcheng, Inner Mongolia	Jiulongshan Fm., J_2	Tan and Ren [183]
[a]Cleroidea *Incertae sedis*	[a]*Forticatinius elegans* Tan & Ren, 2007	Beipiao, Liaoning	Yixian Fm., K_1	Tan and Ren [183]
[a]Cleroidea *Incertae sedis*	[a]*Mathesius liaoningensis* Kolibáč & Huang, 2011	Beipiao, Liaoning	Yixian Fm., K_1	Kolibáč and Huang [351]
		Superfamily Cucujoidea Latreille, 1802		
Boganiidae	*Palaeoboganium jurassicum* Liu, Ślipiński, Lawrence, Ren & Pang, 2017	Ningcheng, Inner Mongolia	Jiulongshan Fm., J_2	Liu *et al.* [147]
Cryptophagidae	[a]*Atomaria cretacea* Cai & Wang, 2013	Shangrao, Jiangxi	Shixi Fm., K_1	Cai & Wang [217]
Monotomidae	*Jurorhizophagus alienus* Cai, Ślipiński & Huang, 2015	Ningcheng, Inner Mongolia	Jiulongshan Fm., J_2	Cai *et al.* [111]
Nitidulidae	*Sinonitidulina liugouensis* Hong, 1983	Luoping, Hebei	Jiulongshan Fm., J_2	Hong [93]
	Sinonitidulina luanpingensis Hong, 1983	Luoping, Hebei	Jiulongshan Fm., J_2	Hong [93]
	Sinonitidulina punctata Hong, 1983	Luoping, Hebei	Jiulongshan Fm., J_2	Hong [93]

(Continued)

Table 21.1 (Continued)

Family	Species	Locality	Horizon/Age	Citation
Parandrexidae	*Parandrexis parvula* Martynov, 1926	Ningcheng, Inner Mongolia	Jiulongshan Fm., J_2	Martynov [92]
	Parandrexis beipiaoensis Hong, 1983	Beipiao, Liaoning	Yixian Fm., K_1	Hong [93]
	Parandrexis agilis Lu, Shih & Ren, 2015	Ningcheng, Inner Mongolia	Jiulongshan Fm., J_2	Lu *et al.* [352]
	Parandrexis longicornis Lu, Shih & Ren, 2015	Ningcheng, Inner Mongolia	Jiulongshan Fm., J_2	Lu *et al.* [352]
	Parandrexis oblongis Lu, Shih & Ren, 2015	Ningcheng, Inner Mongolia	Jiulongshan Fm., J_2	Lu *et al.* [352]
		Superfamily Tenebrionoidea Latreille, 1802		
Mordellidae	[a]*Liaoximordella hongi* Wang, 1993	Linyuan, Liaoning	Yixian Fm., K_1	Wang [174]
	Cretanaspis lushangfenensis Huang & Yang, 1999	Fangshan, Beijing	Lushangfen Fm., K_1	Huang and Yang [322]
	Mirimordella gracilicruralis Liu, Lu & Ren, 2007	Beipiao, Liaoning	Yixian Fm., K_1	Liu *et al.* [223]
	Bellimordella capitulifera Liu, Zhao & Ren, 2008	Beipiao, Liaoning	Yixian Fm., K_1	Liu *et al.* [228]
	Bellimordella longispina Liu, Zhao & Ren, 2008	Beipiao, Liaoning	Yixian Fm., K_1	Liu *et al.* [228]
	Bellimordella robusta Liu, Zhao & Ren, 2008	Beipiao, Liaoning	Yixian Fm., K_1	Liu *et al.* [228]
Ripiphoridae	*Archaeoripiphorus nuwa* Hsiao, Yu & Deng, 2017	Ningcheng, Inner Mongolia	Jiulongshan Fm., J_2	Hsiao *et al.* [114]
Tenebrionidae	*Alphitopsis initialis* Kirejtshuk, Nabozhenko & Nel, 2011	Beipiao, Liaoning	Yixian Fm., K_1	Kirejtshuk *et al.* [326]
	Platycteniopus diversoculatus Chang, Nabozhenko, Pu, Xu, Jia & Li, 2016	Beipiao, Liaoning	Yixian Fm., K_1	Chang *et al.* [327]
[a]Tenebrionoidea *Incertae sedis*	[a]*Glypta longa* Hong, 1984	Qingshila, Hebei	Qingshila Fm., K_1	Hong [162]
[a]Tenebrionoidea *Incertae sedis*	[a]*Glypta qingshilaensis* Hong, 1984	Qingshila, Hebei	Qingshila Fm., K_1	Hong [162]
[a]Tenebrionoidea *Incertae sedis*	[a]*Wuhua jurassica* Wang & Zhang, 2011	Ningcheng, Inner Mongolia	Jiulongshan Fm., J_2	Wang and Zhang [116]
		Superfamily Chrysomeloidea Latreille, 1802		
Cerambycidae	*Cretoprionus liutiaogouensis* Wang, Ma, McKenna, Yan, Zhang & Jarzembowski, 2014	Dashuangmiao, Inner Mongolia	Yixian Fm., K_1	Wang *et al.* [220]
	Sinopraecipuus bilobatus Yu, Ślipiński, Reid, Shih, Pang & Ren, 2015	Lingyuan, Liaoning	Yixian Fm., K_1	Yu *et al.* [238]
Chrysomelidae	*Mesolaria longala* Zhang, 1986	Luoping, Hebei	Jiulongshan Fm., J_2	Zhang [118]
Anthribidae	[a]*Protoscelis tuanwangensis* Hong & Wang, 1990	Laiyang, Shandong	Laiyang Fm., K_1	Hong and Wang [173]

(Continued)

Table 21.1 (Continued)

Family	Species	Locality	Horizon/Age	Citation
	Superfamily Curculionoidea Latreille, 1802			
Brentidae	*Cretonanophyes punctatus* Liu & Ren, 2007	Beipiao, Liaoning	Yixian Fm., K_1	Liu and Ren [225]
	Cretonanophyes zherikhini (Liu & Ren, 2006)	Beipiao, Liaoning	Yixian Fm., K_1	Liu and Ren [224]
	Abrocar brachyorhinos Liu & Ren, 2006	Beipiao, Liaoning	Yixian Fm., K_1	Liu and Ren [224]
	Abrocar concavus Davis, Engel, Legalov & Ren, 2013	Beipiao, Liaoning	Yixian Fm., K_1	Davis *et al.* [331]
	Abrocar macilentus Liu & Ren, 2007	Beipiao, Liaoning	Yixian Fm., K_1	Liu and Ren [225]
	Abrocar relicinus Davis, Engel, Legalov & Ren, 2013	Beipiao, Liaoning	Yixian Fm., K_1	Davis *et al.* [331]
Nemonychidae	[a]*Longidorsum generale* Zhang, 1997	Zhixin Forest Farm, Jilin	Dalaizi Fm., K_1	Zhang [177]
	Microprobelus liuae Liu, Ren & Shih, 2006	Beipiao, Liaoning	Yixian Fm., K_1	Liu *et al.* [226]
	Chinocimberis angustipecteris (Liu, Ren & Tan, 2006)	Beipiao, Liaoning	Yixian Fm., K_1	Liu *et al.* [227]
	Chinocimberis longidigitus (Ren, Lu, Ji & Guo, 1995)	West hills, Beijing	Lushangfen Fm., K_1	Ren *et al.* [240]
	Chinocimberis magnoculi (Liu, Ren & Tan, 2006)	Beipiao, Liaoning	Yixian Fm., K_1	Liu *et al.* [227]
	Renicimberis latipecteris (Liu, Ren & Tan, 2006)	Beipiao, Liaoning	Yixian Fm., K_1	Liu *et al.* [227]
Suborder *Incertae sedis*, Permosynidae	*Artematopodites insculptus* (Zhang, 1997)	Shawan, Xinjiang	Sangonghe Fm., J_1 Xishanyao Fm., J_2	Zhang [101]
	Artematopodites longus (Hong, 1983)	Luoping, Hebei	Jiulongshan Fm., J_2	Hong [93]
	Artematopodites prolixus (Zhang, 1997)	Shawan, Xinjiang	Badaowan Fm., J_1	Zhang [101]
	Artematopodites propinquus (Zhang, 1997)	Shawan, Xinjiang	Badaowan Fm., J_1	Zhang [101]
	Artematopodites shaanbeiensis Hong, 1995	Wuqi, Shanxi	Huachi-Huanhe Fm., K_1	Hong [353]
[a]Coleoptera *Incertae sedis*	[a]*Polysitum wudenghaoensis* Hong, 1976	Urad Qianqi, Inner Mongolia	Guyang Fm., K_1	Hong [157]
[a]Coleoptera *Incertae sedis*	[a]*Mesotricupes lineatus* Hong, 1982	Yumen, Gansu	Chijinpu Fm., K_1	Hong [160]
[a]Coleoptera *Incertae sedis*	[a]*Mesotricupes reticulatus* Hong, 1982	Yumen, Gansu	Xiagou Fm., K_1	Hong [160]
[a]Coleoptera *Incertae sedis*	[a]*Petalocupes arcus* Hong, 1982	Yumen, Gansu	Zhonggou Fm., K_1	Hong [160]
[a]Coleoptera *Incertae sedis*	[a]*Prosilpha nigrita* Hong, 1982	Yumen, Gansu	[b]Chijinqiao Fm., K_1	Hong [160]
[a]Coleoptera *Incertae sedis*	[a]*Sinocarabus longicornutus* Hong, 1982	Yumen, Gansu	[b]Chijinqiao Fm., K_1	Hong [160]

(Continued)

Table 21.1 (Continued)

Family	Species	Locality	Horizon/Age	Citation
[a]Coleoptera *Incertae sedis*	[a]*Synodus changmaensis* Hong, 1982	Yumen, Gansu	Chijinpu Fm., K_1	Hong [160]
[a]Coleoptera *Incertae sedis*	[a]*Tetillopsis parvula* Hong, 1982	Yumen, Gansu	Zhonggou Fm., K_1	Hong [160]
[a]Coleoptera *Incertae sedis*	[a]*Trypocoleus ramulosus* Hong, 1982	Yumen, Gansu	Zhonggou Fm., K_1	Hong
[a]Coleoptera *Incertae sedis*	[a]*Beipiaocarabus oblonga* Hong, 1983	Beipiao, Liaoning	Haifanggou Fm., J_2	Hong [93]
[a]Coleoptera *Incertae sedis*	[a]*Xinbinia foveolata* Hong, 1983	Xinbin, Liaoning	Houjiatun Fm., J_1-J_2	Hong [93]
[a]Coleoptera *Incertae sedis*	[a]*Anhuistoma hyla* Lin, 1985	Hanshan, Anhui	Hanshan Fm., J_2	Lin [77]
[a]Coleoptera *Incertae sedis*	[a]*Grammocolous arcuatus* Lin, 1986	Hezhou, Guangxi	Shiti Fm., J_1	Lin [141]
[a]Coleoptera *Incertae sedis*	[a]*Cretohelophorus yanensis* Ren, Lu, Ji & Guo, 1995	West hills, Beijing	Lushangfen Fm., K_1	Ren *et al.* [240]
[a]Coleoptera *Incertae sedis*	[a]*Jinxidiscus lushangfenensis* (Hong, 1981)	Fangshan, Beijing	Lushangfen Fm., K_1	Hong [159]
[a]Coleoptera *Incertae sedis*	[a]*Hubeicoleus tenuis* Hong, 1991	Daye, Hubei	Wuchang Fm., J_1	Hong [165]
[a]Coleoptera *Incertae sedis*	[a]*Orphnospercheus longjingensis* Hong, 1992	Yanji, Jilin	Dalazi Fm., K_1	Hong [145]

a) The species is not present in the main text because the original description, photos and line-drawings are not precise and the holotype cannot be rechecked.

b) Horizon/Age revised from the original paper based on updated information and data.

References

1 Polilov, A.A. (2015). How small is the smallest? New record and remeasuring of *Scydosella musawasensis* Hall, 1999 (Coleoptera, Ptiliidae), the smallest known free-living insect. *ZooKeys* **526**: 61–64.

2 Crowson, R.A. (1960). The phylogeny of Coleoptera. *Annual Review of Entomology* **5**: 111–134.

3 Beutel, R.G. and Haas, F. (2000). Phylogenetic relationships of the suborders of Coleoptera (Insecta). *Cladistics* **16** (1): 103–141.

4 Bocak, L., Barton, C., Crampton-Platt, A. *et al.* (2014). Building the Coleoptera tree-of-life for >8000 species: composition of public DNA data and fit with Linnaean classification. *Systematic Entomology* **39** (1): 97–110.

5 Lawrence, J.F., Ślipiński, A., Seago, A.E. *et al.* (2011). Phylogeny of the Coleoptera based on morphological characters of adults and larvae. *Annales Zoologici (Warszawa)* **61**: 1–217.

6 Kukalová-Peck, J. and Lawrence, J.F. (1993). Evolution of the hind wing in Coleoptera. *The Canadian Entomologist* **125** (2): 181–258.

7 Kukalová-Peck, J. and Lawrence, J.F. (2004). Relationships among coleopteran suborders and major endoneopteran lineages, evidence from hind wing characters. *European Journal of Entomology* **101** (1): 95–144.

8 Shull, V.L., Vogler, A.P., Baker, M.D. *et al.* (2001). Sequence alignment of 18S ribosomal RNA and the basal relationships of adephagan beetles: evidence for monophyly of aquatic families and the placement of Trachypachidae. *Systematic Biology* **50**: 945–969.

9 Hunt, T., Bergsten, J., Levkanicova, Z. *et al.* (2007). A comprehensive phylogeny of beetles reveals the evolutionary origins of a superradiation. *Science* **318**: 1913–1916.

10 Maddison, D.R., Moore, W., Baker, M.D. *et al.* (2009). Monophyly of terrestrial adephagan beetles as indicated by three nuclear genes (Coleoptera: Carabidae and Trachypachidae). *Zoologica Scripta* **38**: 43–62.

11 Pons, J., Ribera, I., Bertranpetit, J. *et al.* (2010). Nucleotide substitution rates for the full set of mitochondrial protein-coding genes in Coleoptera. *Molecular Phylogenetics and Evolution* **56** (2): 796–807.

12 Song, H., Sheffield, N.C., Cameron, S.L. *et al.* (2010). When phylogenetic assumptions are violated: base compositional heterogeneity and among-site rate variation in beetle mitochondrial phylogenomics. *Systematic Entomology* **35** (3): 429–448.

13 McKenna, D.D. and Farrell, B.D. (2009). Beetles (Coleoptera). In: *The Time Tree of Life* (ed. S.B. Hedges and S. Kumar), 278–289. Oxford: Oxford University Press.

14 Mckenna, D.D., Wild, A.L., Kanda, K. *et al.* (2015). The beetle tree of life reveals that Coleoptera survived end-Permian mass extinction to diversify during the Cretaceous terrestrial revolution. *Systematic Entomology* **40** (4): 835–880.

15 Lafer, G.S. (1996). Family sikhotealiniidae. In: *Key to the Insects of the Russian Far East, Volume III, Part 3* (ed. P.A. Ler), 390–396. Vladivostok: Dal'nauka.

16 Ning, T., Fang, L.Y., Tang, J. *et al.* (2004). Advances in research on *Bursaphelenchus xylophilus* and its key vector *Monochamus* spp. *Chinese Bulletin of Entomology* **41**: 97–104.

17 Teale, S.A., Wickham, J.D., Zhang, F.P. *et al.* (2011). A male-produced aggregation pheromone of *Monochamus alternatus* (Coleoptera: Cerambycidae), a major vector of pine wood nematode. *Journal of Economic Entomology* **104** (5): 1592–1598.

18 Ashworth, J.R. (1993). The biology of *Lasioderma serricorne. Journal of Stored Products Research* **29** (4): 291–303.

19 Hori, M., Miwa, M., and Iizawa, H. (2011). Host suitability of various stored food products for the cigarette beetle, *Lasioderma serricorne* (Coleoptera: Anobiidae). *Applied Entomology and Zoology* **46** (4): 463–469.

20 Padin, S., Bello, D.G., and Fabrizio, M. (2002). Grain loss caused by *Tribolium castaneum, Sitophilus oryzae* and *Acanthoscelides obtectus* in stored durum wheat and beans treated with *Beauveria bassiana. Journal of Stored Products Research* **38** (1): 69–74.

21 Kairo, M.T.K., Paraiso, O., Gautam, R.D. *et al.* (2013). *Cryptolaemus montrouzieri* (Mulsant) (Coccinellidae: Scymninae): a review of biology, ecology, and use in biological control with particular reference to potential impact on non-target organisms. *CAB Reviews: Perspectives in Agriculture, Veterinary Science, Nutrition and Natural Resources* **8** (005): 1–20.

22 Wu, H.S., Zhang, Y.H., Liu, P. *et al.* (2014). *Cryptolaemus montrouzieri* as a predator of the striped mealybug, *Ferrisia virgata*, reared on two hosts. *Journal of Applied Entomology* **138** (9): 662–669.

23 Wu, W., Yang, S., Brandon, A.M. *et al.* (2016) Rapid biodegradation of plastics by mealworms (larvae of Tenebrio molitor) brings hope to solve wasteplastic pollution. AGU Fall Meeting Abstracts. American Geophysical Union, Washington, USA.

24 Wheeler, Q.D. (1986). Revision of the genera of Lymexylidae (Coleoptera: Cucujiformia). *Bulletin of the American Museum of Natural History* **183** (2): 113–210.

25 Crowson, R.A. (1990). A new genus of Boganiidae (Coleoptera) from Australia, with observations on glandular openings, cycad associations and geographical distribution in the family. *Austral Entomology* **29** (2): 91–99.

26 EndrÃ, S. (1991). Boganiidae (Coleoptera: Cucujoidea) associated with cycads in South Africa: two new species and a new synonym. *Annals of the Transvaal Museum* **35** (20): 285–293.

27 Suinyuy, T.N., Donaldson, J.S., and Johnson, S.D. (2009). Insect pollination in the African cycad Encephalartos friderici-guilielmi Lehm. *South African Journal of Botany* **75** (4): 682–688.

28 Lu, X., Yang, Z., Sun, X. *et al.* (2011). Biological control of *Apriona swainsoni* (Coleoptera: Cerambycidae) by releasing the parasitic beetle *Dastarcus helophoroides* (Coleoptera: Bothrideridae). *Scientia Silvae Sinicae* **47** (10): 116–121.

29 Peck, S.B. (2006). Distribution and biology of the ectoparasitic beaver beetle *Platypsyllus castoris* Ritsema in North America (Coleoptera: Leiodidae: Platypsyllinae). *Insecta Mundi* **20** (1–2): 85–94.

30 Elzinga, R.J. (1977). Observations on *Sandalus niger* Knoch (Coleoptera: Sandalidae) with a description of the triungulin larva. *Journal of the Kansas Entomological Society* **50** (3): 324–328.

31 Lee, C.F., Satô, M., and Sakai, M. (2005). The Rhipiceridae of Taiwan and Japan (Insecta: Coleoptera: Dascilloidea). *Zoological Studies* **44** (4): 437–444.

32 Falin, Z.H., Arneson, L.C., and Wcislo, W.T. (2000). Night-flying sweat bees *Megalopta genalis* and *Me. ecuadoria* (Hymenoptera: Halictidae) as hosts of the parasitoid beetle *Macrosiagon gracilis* (Coleoptera: Rhipiphoridae). *Journal of the Kansas Entomological Society* **73** (3): 183–185.

33 Batelka, J. (2013). A review of the genus *Macrosiagon* in Laos (Coleoptera: Ripiphoridae). *Entomologica Basiliensia et Collectionis Frey* **34**: 319–325.

34 Auko, T.H., Trad, B.M., and Silvestre, R. (2014). Five new associations of parasitoids in potter wasps (Vespidae, Eumeninae). *Revista Brasileira de Entomologia* **58** (4): 376–378.

35 Crowson, R.A. (1975). The evolutionary history of Coleoptera, as documented by fossil and comparative evidence. In: *Atti del X Congresso Nazionale Italiano di Entomologia. Firenze*, 47–90. Italy: Accademia Nazionale Italian di Entomologia Palermo.

36 Crowson, R.A. (1981). *The Biology of the Coleoptera*. London: Academic Press, xii + 802 pp.

37 Kirejtshuk, A.G., Poschmann, M., Prokop, J. *et al.* (2014). Evolution of the elytral venation and structural adaptations in the oldest Palaeozoic beetles (Insecta: Coleoptera: Tshekardocoleidae). *Journal of Systematic Palaeontology* **12** (5): 575–600.

38 Kukalová, J. (1969). On the systematic position of the supposed Permian beetles, Tshecardocoleidae, with a description of a new collection from Moravia. *Sbornik Geologickych Ved, Paleontologie* **11**: 139–162.

39 Ponomarenko, A.G. (1969). Cretaceous insects from Labrador. 4. A new family of beetles (Coleoptera: Archostemata). *Psyche* **76** (3): 306–310.

40 Bouchard, P., Bousquet, Y., Davies, A.E. *et al.* (2011). Family-group names in Coleoptera (Insecta). *ZooKeys* **30** (88): 1–972.

41 Carpenter, F.M. (1992). Superclass Hexapoda. In: *Treatise on Invertebrate Paleontology, Arthropoda 4, 3/4* (ed. R.C. Moore and R.L. Kaesler). Boulder and Lawrence: The Geological Society of America and the University of Kansas, xii + 655 pp.

42 Vršanský, P. (2003). Umenocoleoidea—an amazing lineage of aberrant insects (Insecta, Blattaria). *AMBA Projekty* **7** (1): 1–32.

43 Kukalová-Peck, J. and Beutel, R.G. (2012). Is the Carboniferous a *Adiphlebia lacoana* really the "oldest beetle"? Critical reassessment and description of a new Permian beetle family. *Eropean Jounal of Entomology* **109**: 633–645.

44 Kirejtshuk, A.G. and Nel, A. (2016). Origin of the Coleoptera and significance of the fossil record. *Euroasian Entomological Journal* **15** (S1): 66–73.

45 Rohdendorf, B.B. (1944). A new family of Coleoptera from the Permian of the Urals. *Comptes Rendus (Doklady) de l'Académie des Sciences de l'URSS* **44** (6): 252–253.

46 Ponomarenko, A.G. (1963). Paleozoyskie zhuki Cupididea evropeyskoy chasti SSSR. *Paleontologicheskii ZhurnalI* **1963** (1): 70–85.

47 Ponomarenko, A.G. (1969). Istoricheskoe razvitie zhestkokrylykh-arkhostemat [Historical development of the Coleoptera–Archostemata]. *Trudy Paleontologicheskogo Instituta Akademiya Nauk SSSR* **125**: 1–240. (in Russian).

48 Lubkin, S.H. and Engel, M.S. (2005). *Permocoleus*, new genus, the first Permian beetle (Coleoptera) from North America. *Annals of the Entomological Society of America* **98** (1): 73–76.

49 Beckemeyer, R.J. and Engel, M.S. (2008). A second specimen of *Permocoleus* (Coleoptera) from the lower Permian Wellington Formation of Noble County, Oklahoma. *Journal of the Kansas Entomological Society* **81** (1): 4–7.

50 Pinto, I.D. (1987). Permian insects from Paraná Basin, South Brazil. 4. Coleoptera. *Pesquisas (Zoologia)* **19**: 5–12.

51 Volkov, A.N. (2013). New species of Triaplidae from the Babii Kamen'locality (Kuznetsk Basin). *Paleontological Journal* **47** (1): 94–97.

52 Yan, E.V., Beutel, R.G., and Ponomarenko, A.G. (2018). † Peltosynidae, a new beetle family from the Middle–Late Triassic of Kyrgyzstan: its affinities with Polyphaga (Insecta, Coleoptera) and the groundplan of this megadiverse suborder. *Journal of Systematic Palaeontology* **16** (6): 515–530.

53 Ponomarenko, A.G. and Volkov, A.N. (2013). *Ademosynoides asiaticus* Martynov, 1936, the earliest known member of an extant beetle family (Insecta, Coleoptera, Trachypachidae). *Paleontological Journal* **47** (6): 601–606.

54 Ponomarenko, A.G. (2014). Coleoptera, in Upper Jurassic Lagerstätte Shar Teg, southwestern Mongolia. *Paleontological Journal* **48**: 1658–1671.

55 Zhang, S.Q., Che, L.H., Li, Y. *et al.* (2018). Evolutionary history of Coleoptera revealed by extensive sampling of genes and species. *Nature Communications* **9** (1): 205.

56 Feng, Z., Wang, J., Rößler, R. *et al.* (2017). Late Permian wood-borings reveal an intricate network of ecological relationships. *Nature Communications* **8** (1): 556.

57 Teixeira, C. (1975). Sur une larve d'insecte fossile (Coleoptere) du "Karroo" de l'Angola. *Boletim da Sociedade Geologica de Portugal* **19** (3): 131–134.

58 Ponomarenko, A.G. (2004). Beetles (Insecta, Coleoptera) of the Late Permian and Early Triassic. *Paleontological Journal* **38** (Suppl 2): S185–S196.

59 Ponomarenko, A.G. (2008). New Triassic beetles (Coleoptera) from northern European Russia. *Paleontological Journal* **42** (6): 600–606.

60 Ponomarenko, A.G. (2016). New beetles (Insecta, Coleoptera) from the Lower Triassic of European Russia. *Paleontological Journal* **50** (3): 286–292.

61 Ponomarenko, A.G. (2002). Superorder Scarabaeidea Laicharting, 1781. Order Coleoptera Linné, 1758. The beetles. In: *History of Insects* (ed. A.P. Rasnitsyn and D.L.J. Quicke), 164–176. Dordrecht: Kluwer Academic Publishers.

62 Dunstan, B. (1923). Mesozoic insects of Queensland. Part 1, introduction and Coleoptera. *Geological Survey of Queensland Publication* **273**: 1–89.

63 Zeuner, F.E. (1960) A Triassic insect fauna from the Molteno beds of South Africa. XI Internationaler Kongress für Entomologie, Verhandlungen, 1, 304–306.

64 Riek, E.F. (1974). Upper Triassic insects from the Molteno "Formation", South Africa. *Palaeontologia Africana* **17**: 19–31.

65 Arnoldi, L.V., Zherikhin, V.V., Nikritin, L.M. *et al.* (1977). Mezozoiskie zhestkokryiye. *Akademiya Nauk SSSR, Trudy Paleontologicheskogo Instituta* **161**: 1–204.

66 Toussaint, E.F., Seidel, M., Arriaga-Varela, E. *et al.* (2017). The peril of dating beetles. *Systematic Entomology* **42** (1): 1–10.

67 Chatzimanolis, S., Grimaldi, D.A., Engel, M.S. *et al.* (2012). *Leehermania prorova*, the earliest staphyliniform beetle, from the Late Triassic of Virginia (Coleoptera: Staphylinidae). *American Museum Novitates* **3761**: 1–28.

68 Tillyard, R.J. (1918). Permian and Triassic insects from New South Wales, in the collection of Mr. John Mitchell. *The Proceedings of the Linnean Society of New South Wales* **42**: 720–756.

69 Martins-Neto, R.G., Gallego, O.F., and Mancuso, A. (2006). The Triassic insect fauna from Argentina Coleoptera from the Los Rastros Formation (Bermejo Basin), La Rioja Province. *Ameghiniana (Argentinian Association of Paleontology)* **43** (3): 1–20.

70 Martins-Neto, R.G. and Gallego, O.F. (2009). The Triassic insect fauna from Argentina. Blattoptera and Coleoptera from the Ischichuca Formation (Bermejo Basin), La Rioja Province. *Ameghiniana (Argentinian Association of Paleontology)* **46** (2): 361–372.

71 Ponomarenko, A.G. (1977). Suborder Adephaga, Polyphaga *Incertae Sedis*. In: *Mesozoic Coleoptera*, vol. **161** (ed. L.V. Arnoldi, V.V. Zherikhin, L.M. Nikritin, *et al.*), 17–119. Trudy Paleontologicheskogo Instituta Akadamii Nauk SSSR.

72 Bode, A. (1953). Die Insektenfauna des Ostniedersachsischen Oberen Lias. *Palaeontographica Abteilung A* **103**: 1–375.

73 Prokin, A.A. (2009). New water scavenger beetles (Coleoptera: Hydrophilidae) from the Mesozoic of Mongolia. *Paleontological Journal* **43**: 660–663.

74 Perkovsky, E.E. (1999). Novyy pozdnejurskiy rod i vid Leiodinae (Coleoptera, Leiodidae) iz Mongolii. *Vestnik Zoologii* **33**: 77–79.

75 Ponomarenko, A.G. (1985). Coleoptera. In: *Jurassic Insects of Siberia and Mongolia* (ed. A.P. Rasnitsyn), 47–87. Nauka, Moscow: Trudy paleontologicheskogo instituta akadamii nauk SSSR.

76 Cai, C.Y., Thayer, M.K., Engel, M.S. *et al.* (2014). Early origin of parental care in Mesozoic carrion beetles. *Proceedings of the National Academy of Sciences of the United States of America* **111** (39): 14170–14174.

77 Lin, Q.B. (1985). Insect fossils from the Hanshan Formation at Hanshan County, Anhui Province. *Acta Palaeontologica Sinica* **24** (3): 300–315.

78 Tikhomirova, A.L. (1968). Staphylinid beetles from the Jurassic of Karatau (Coleoptera, Staphylinidae). In: *Jurassic Insects of Karatau* (ed. B.B. Rohdendorf), 139–154. Moscow: Academiya Nauk SSSR.

79 Grebennikov, V.V. and Newton, A.F. (2012). Detecting the basal dichotomies in the monophylum of carrion and rove beetles (Insecta: Coleoptera: Silphidae and Staphylinidae) with emphasis on the Oxyteline group of subfamilies. *Arthropod Systematics & Phylogeny* **70**: 133–165.

80 Heer, O. (1865) Die Urwelt der Schweiz, pp. 1–622.

81 Krell, F.T. (2000). The fossil record of Mesozoic and Tertiary Scarabaeoidea (Coleoptera: Polyphaga). *Invertebrate Taxonomy* **14**: 871–905.

82 Deichmüller, J.V. (1886) Die Insecten aus dem lithographischen Schiefer. Mitteilungen, Koniglichen Mineralogisch-Geologischen und Prehistorischen Museum in Dresden, 7, 1–84.

83 Ponomarenko, A.G. (1990). Beetles. Scarabaeida. In: *Late Mesozoic Insects of Eastern Transbaikalia*, vol. **239** (ed. A.P. Rasnitsyn), 39–82. Nauka, Moscow: Trudy paleontologicheskogo instituta akadamii nauk SSSR.

84 Yan, E.V., Wang, B., Ponomarenko, A.G. *et al.* (2014). The most mysterious beetles: Jurassic Jurodidae (Insecta: Coleoptera) from China. *Gondwana Research* **25** (1): 214–225.

85 Nikolajev, G.V., Wang, B., Liu, Y. *et al.* (2011). Stag beetles from the Mesozoic of Inner Mongolia, China (Scarabaeoidea: Lucanidae). *Acta Palaeontologica Sinica* **50**: 41–47.

86 Nikolajev, G.V. (2005). Novyy rod triby Hybosorini (Coleoptera, Scarabaeidae) iz Mezozoya Azii. *Tethys Entomological Research* **11**: 27–28.

87 Nikolajev, G.V. (2008). Novyy vid roda *Jurahybosorus* Nikolajev (Coleoptera, Scarabaeoidea: Hybosoridae) iz verkhney Jury Kazakhstana. *Tethys Entomological Research* **16**: 27–30.

88 Nikolajev, G.V. (2010). On the Mesozoic taxa of scarabaeoid beetles of the Family Hybosoridae (Coleoptera: Scarabaeoidea). *Paleontological Journal* **44** (6): 649–653.

89 Nikolajev, G.V. and Ren, D. (2010). The oldest fossil Ochodaeidae (Coleoptera: Scarabaeoidea) from the Middle Jurassic of China. *Zootaxa* **2553** (1): 65–68.

90 Kirejtshuk, A.G. and Ponomarenko, A.G. (2010). A new coleopterous family Mesocinetidae fam. nov. (Coleoptera: Scirtoidea) from Late Mesozoic and notes on fossil remains from Shar-Teg (Upper Jurassic, South-Western Mongolia). *Zoosystematica Rossica* **19**: 301–325.

91 Handlirsch, A. (1906). Die Fossilen Insekten und die Phylogenie der Rezenten Formen, parts I–IV. In: *Ein Handbuch fur Palaontologen und Zoologen*, 1–640. Leipzig: Wilhelm Engelmann.

92 Martynov, A.V. (1926). K Poznaniyu Iskopaemykh Nasekomykh Yurskikh Slantsev Turkestana. 5. O Nekotorykh Formakh Zhukov (Coleoptera) [To the

knowledge of fossil insects from Jurassic beds in Turkestan 5. On some interesting Coleoptera]. *Ezhegodnik Russkogo Paleontologicheskogo Obshestva* **5** (1): 1–39.
93 Hong, Y.C. (1983). *Middle Jurassic Fossil Insects in North China*, 1–223. Beijing: Geological Publishing House.
94 Kirejtshuk, A.G. (2013). Taxonomic names, in Current knowledge of Coleoptera (Insecta) from the Lower Cretaceous Lebanese amber and taxonomical notes for some Mesozoic groups. *Terrestrial Arthropod Reviews* **6**: 103–134.
95 Kirejtshuk, A.G. (2017). Taxonomic notes on fossil beetles (Insecta: Coleoptera). *Russian Entomological Journal* **26**: 35–36.
96 Tan, J.J., Ren, D., and Shih, C.K. (2007). New beetles (Insecta: Coleoptera: Archostemata) from the Late Mesozoic of North China. *Annales Zoologici (Warszawa)* **57** (2): 231–247.
97 Kirejtshuk, A.G., Chang, H.L., Ren, D. *et al.* (2010). Family Lasiosynidae n. fam. new palaeoendemic Mesozoic family from the infraorder Elateriformia (Coleoptera: Polyphaga). *Annales De La Société Entomologique De France* **46** (1–2): 67–87.
98 Cai, C.Y., Lawrence, J.F., Ślipiński, A. *et al.* (2015). Jurassic artematopodid beetles and their implications for the early evolution of Artematopodidae (Coleoptera). *Systematic Entomology* **40** (4): 779–788.
99 Chang, H.L., Kirejtshuk, A.G., and Ren, D. (2011). On taxonomy and distribution of fossil Cerophytidae (Coleoptera, Polyphaga, Elateriformia) with description of a new species of *Necromera* Martynov, 1926. *Annales de la Société entomologique de France* **47** (1–2): 33–44.
100 Oberprieler, R.G., Ashman, L.G., Frese, M. *et al.* (2016). The first elateroid beetles (Coleoptera: Polyphaga: Elateroidea) from the Upper Jurassic of Australia. *Zootaxa* **4147**: 177–191.
101 Zhang, H.C. (1997). A new genus of Elateridae (Insecta, Coleoptera) from lower Middle Jurassic of Junggar Basin, Xinjiang, China. *Acta Micropalaeontologica Sinica* **14** (1): 71–77.
102 Dong, F.B. and Huang, D.-Y. (2011). A new elaterid from the Middle Jurassic Daohugou biota (Coleoptera: Elateridae: Protagrypninae). *Acta Geologica Sinica (English Edition)* **85** (6): 1224–1230.
103 Cai, C.Y., Lawrence, J.F., Ślipiński, A. *et al.* (2014). First fossil tooth-necked fungus beetle (Coleoptera: Derodontidae): *Juropeltastica sinica* gen. n. sp. n. from the Middle Jurassic of China. *European Journal of Entomology* **111** (2): 299–302.
104 Deng, C.S., Ślipiński, A., Ren, D. *et al.* (2017). The oldest dermestid beetle from the Middle Jurassic of China (Coleoptera: Dermestidae). *Annales Zoologici (Warszawa)* **67** (1): 109–112.
105 Kolibáč, J. (2013). Trogossitidae: A review of the beetle family, with a catalogue and keys. *Zookeys* **366**: 1–194.
106 Yu, Y.L., Ślipiński, A., Leschen, R.A.B. *et al.* (2014). Enigmatic Mesozoic Bark-Gnawing Beetles (Coleoptera: Trogossitidae) from the Jiulongshan Formation in China. *Annales Zoologici (Warszawa)* **64** (4): 667–676.
107 Kolibáč, J. and Huang, D.-Y. (2016). The oldest known clerid fossils from the Middle Jurassic of China, with a review of Cleridae systematics (Coleoptera). *Systematic Entomology* **41** (4): 808–823.
108 Liu, Z.H., Ślipiński, A., Leschen, R.A.B. *et al.* (2015). The oldest Prionoceridae (Coleoptera: Cleroidea) from the Middle Jurassic of China. *Annales Zoologici* **65** (1): 41–52.
109 Kirejtshuk, A.G. (1994). Parandrexidae fam. nov., Jurassic beetles of the infraorder Cucujiformia (Coleoptera, Polyphaga). *Paleontological Journal* **28** (1): 57–64.
110 Soriano, C., Kirejtshuk, A.G., and Delclos, X. (2006). The Mesozoic Laurasian family Parandrexidae (Insecta: Coleoptera), new species from the Lower Cretaceous of Spain. *Comptes Rendus Palevol* **5** (6): 779–784.
111 Cai, C.Y., Ślipiński, A., and Huang, D.-Y. (2015). The oldest root-eating beetle from the Middle Jurassic of China (Coleoptera, Monotomidae). *Alcheringa: An Australasian Journal of Palaeontology* **39** (4): 488–493.
112 Heer, O. (1852). Die Lias-Insel des Aargau's. *Zwei Geologische Vorträge Gehalten im März* **1**: 1–15.
113 Shchegoleva-Barovskaya, T.I. (1929). Der erste Vertreter der Familie Mordellidae (Coleoptera) aus der Juraformation Turkestans. *Doklady Akademii Nauk SSSR* **1929**: 27–29.
114 Hsiao, Y., Yu, Y.L., Deng, C.S. *et al.* (2017). The first fossil wedge-shaped beetle (Coleoptera, Ripiphoridae) from the Middle Jurassic of China. *European Journal of Taxonomy* **277**: 1–13.
115 Kirejtshuk, A.G., Merkl, O., and Kernegger, F. (2008). A new species of the genus *Pentaphyllus* Dejean, 1821 (Coleoptera, Tenebrionidae, Diaperinae) from the Baltic amber and checklist of the fossil Tenebrionidae. *Zoosystematica Rossica* **17**: 131–137.
116 Wang, B. and Zhang, H.C. (2011). The oldest Tenebrionoidea (Coleoptera) from the Middle Jurassic of China. *Journal of Paleontology* **85**: 266–270.
117 Ponomarenko, A.G. (1962). Taxonomic names, in Order Coleoptera. In: *Osnovy Paleontologii. Tom 9:*

Chlenistonogie, Trakheinye i Khelitserovye (ed. B.B. Rohdendorf), 241–267.

118 Zhang, J.F. (1986). Some fossil insects from the Jurassic of northern Hebei, China. In: *The Paleontology and Stratigraphy of Shandong*, 74–84. Beijing: Haiyang Publishing House.

119 Gratshev, V.G. and Zherikhin, V.V. (1995). A revision of the nemonychid weevil subfamily Brethorrhininae (Insecta, Coleoptera: Nemonychidae). *Paleontological Journal* **29** (4): 112–127.

120 Gratshev, V.G. and Legalov, A.A. (2009). New taxa of the family Nemonychidae (Coleoptera) from Jurassic and Early Cretaceous. *Euroasian Entomological Journal* **8**: 411–416.

121 Oberprieler, R.G. and Oberprieler, S.K. (2012). *Talbragarus averyi* gen. et sp. n., the first Jurassic weevil from the southern hemisphere (Coleoptera: Curculionoidea: Nemonychidae). *Zootaxa* **3478**: 256–266.

122 Medvedev, L.N. (1968). Zhuki-Listoyedy Yury Karatau (Coleoptera, Chrysomelidae) [Jurassic leaf beetles of Karatau (Coleoptera, Chrysomelidae)]. In: *Yurskoy Nasekomiye Karatau* [Jurassic Insects of Karatau] (ed. B.B. Rohdendorf), 155–165. Moscow: Akademiya nauk SSSR Otdelenie Obshchej Biologii.

123 Legalov, A.A. (2011). The first record of anthribid beetle from the Jurassic of Kazakhstan (Coleoptera: Anthribidae). *Paleontological Journal* **45** (6): 629–633.

124 Legalov, A.A. (2015). Fossil Mesozoic and Cenozoic weevils (Coleoptera, Obrienioidea, Curculionoidea). *Paleontological Journal* **49**: 1442–1513.

125 Arnoldi, L.V. (1977). Rhynchopora, in Mezozoiskie zhestkokryiye. *Akademiya Nauk SSSR, Trudy Paleontologicheskogo Instituta* **161**: 142–176.

126 Gratshev, V.G. and Zherikhin, V.V. (2000). New Early Cretaceous weevil taxa from Spain (Coleoptera, Curculionoidea). *Acta Geologica Hispanica* **35**: 37–46.

127 Ponomarenko, A.G. (1964). New beetles of of the family Cupedidae from the Jurassic of Karatau. *Paleontologicheskii Zhurnal* **2**: 49–62.

128 Tan, J.J., Ren, D., and Shih, C.K. (2007). New ommatids of Ommatinae (Coleoptera: Archostemata: Ommatidae) from the Yixian Formation of western Liaoning, China. *Progress in Natural Science* **17**: 803–811.

129 Tan, J.J., Wang, Y.J., Ren, D. *et al.* (2012). New fossil species of ommatids (Coleoptera: Archostemata) from the Middle Mesozoic of China illuminating the phylogeny of Ommatidae. *BMC Evolutionary Biology* **12** (113): 1–19.

130 Ponomarenko, A.G. (1968). Archostemata beetles from the Jurassic of the Karatau. In: *Jurassic Insects of the Karatau*, 118–138. Moscow: Nauka.

131 Ponomarenko, A.G. (1997). New beetles of the family Cupedidae from the Mesozoic of Mongolia. Ommatini, Mesocupedini, Priacmini. *Paleontological Journal* **31** (4): 389–399.

132 Cai, C.Y. and Huang, D.-Y. (2017). *Omma daxishanense* sp. nov., a fossil representative of an extant Australian endemic genus recorded from the Late Jurassic of China (Coleoptera: Ommatidae). *Alcheringa: An Australasian Journal of Palaeontology* **41** (2): 277–283.

133 Wang, B., Ponomarenko, A.G., and Zhang, H.C. (2009). A new coptoclavid larva (Coleoptera: Adephaga: Dytiscoidea) from the Middle Jurassic of China, and its phylogenetic implication. *Paleontological Journal* **43** (6): 652–659.

134 Ponomarenko, A.G. (1973). Mesozoic Whirligig Beetles (Gyrinidae, Coleoptera). *Paleontological Journal* 7 (4): 499–506.

135 Ponomarenko, A.G. (1987). New Mesozoic Water Beetles (Insecta, Coleoptera) From Asia. *Paleontological Journal* **21** (2): 79–92.

136 Prokin, A.A., Petrov, P.N., Wang, B. *et al.* (2013). New fossil taxa and notes on the Mesozoic evolution of Liadytidae and Dytiscidae (Coleoptera). *Zootaxa* **3666**: 137–159.

137 Hong, Y.C. (1985). Early Jurassic fossil insects from Daye of Hubei Province. *Professional Papers of Stratigraphy and Palaeontology* **15**: 181–187.

138 Wang, B., Zhang, H.C., and Ponomarenko, A.G. (2012). Mesozoic Trachypachidae (Insecta: Coleoptera) from China. *Palaeontology* **55** (2): 341–353.

139 Tan, J.J. and Ren, D. (2009). *Mesozoic Archostematan Fauna from China*, 1–347. Beijing: Science Press.

140 Ponomarenko, A.G., Yan, E.V., and Huang, D.-Y. (2014). New beetles (Coleoptera) from the terminal Middle Permian of China. *Paleontological Journal* **48** (2): 191–200.

141 Lin, Q.B. (1986). Early Mesozoic fossil insects from South China. *Palaeontologica Sinica, Series B* **170** (21): 69–82. (in Chinese with English abstract).

142 Ponomarenko, A.G. (2006). On the types of Mesozoic archostematan beetles (Insecta, Coleoptera, Archostemata) in the Natural History Museum, London. *Paleontological Journal* **40** (1): 90–99.

143 Tan, J.J., Ren, D., Shih, C.K. *et al.* (2013). New schizophorid fossils from China and possible evolutionary scenarios for Jurassic archostematan beetles. *Journal of Systematic Palaeontology* **11** (1): 47–62.

144 Hong, Y.C. (1985). New fossil insects of Xiahuayuan Formation in Yuxian County, Hebei Province. *Bulletin Tianjin Institute of Geological and Mineral Resources* **13**: 131–138. (in Chinese with English abstract).

145 Hong, Y.C. (1992). Middle and Late Jurassic insects. In: Jilin Province Bureau of Geology and Mineral Resource, ed. In: *Palaeontological Atlas of Jilin Province*, 410–425. Changchun: Jilin Science & Technology Press.

146 Yan, E.V. (2010). New taxa, in New Beetle Species of the Formal Genus *Artematopodites* (Coleoptera: Polyphaga), with Remarks on the Taxonomic Position of the Genera *Ovivagina* and *Sinonitidulina*. *Paleontological Journal* **44**: 451–456.

147 Liu, Z.H., Ślipiński, A., Lawrence, J.F. *et al.* (2017). *Palaeoboganium* gen. nov. from the Middle Jurassic of China (Coleoptera: Cucujoidea: Boganiidae): the first cycad pollinators? *Journal of Systematic Palaeontology* **10**: 1–10.

148 Hong, Y.C. (2002). *Amber Insects of China*, 1–653. Beijing: Science and Technic Publishing House (in Chinese).

149 Cai, C.Y., Short, A.E., and Huang, D.-Y. (2012). The first skiff beetle (Coleoptera: Myxophaga: Hydroscaphidae) from Early Cretaceous Jehol Biota. *Journal of Paleontology* **86** (1): 116–119.

150 Poinar, G. Jr.,, Chambers, K.L., and Buckley, R. (2007). *Eoëpigynia burmensis* gen. and sp. nov., an Early Cretaceous eudicot flower (Angiospermae) in Burmese amber. *Journal of the Botanical Research Institute of Texas* **1** (1): 91–96.

151 Xing, L.D., McKellar, R.C., Wang, M. *et al.* (2016). Mummified precocial bird wings in mid-Cretaceous Burmese amber. *Nature Communications* **7**: 12089.

152 Xing, L.D., McKellar, R.C., Xu, X. *et al.* (2016). A feathered dinosaur tail with primitive plumage trapped in mid-Cretaceous amber. *Current Biology* **26** (24): 3352–3360.

153 Ross, A.J. (2017) Burmese (Myanmar) amber taxa, on-line checklist v. 2017.4. www.researchgate.net/publication/322332066

154 Yamamoto, S., Maruyama, M., and Parker, J. (2016). Evidence for social parasitism of early insect societies by Cretaceous rove beetles. *Nature Communications* **7**: 13658.

155 Cai, C.Y., Huang, D.-Y., Newton, A.F. *et al.* (2017). Early evolution of specialized termitophily in Cretaceous Rove beetles. *Current Biology* **27** (8): 1229–1235.

156 Ping, C. (1928). Cretaceous fossil insects of China. *Palaeontologica Sinica B* **13** (1): 1–56.

157 Hong, Y.C. (1976). Inner Mongolia, Volume 2. In: *Paleontological Atlas of the North China Region*, 81–87. Beijing: Geological Publishing House.

158 Hong, Y.C. (1980). Fossil Insects, Volume 6. In: *Mesozoic Stratigraphy and Paleontology of Shanxi-Gansu-Ningxia Basin*, 111–114. Beijing: Geological Publishing House (in Chinese).

159 Hong, Y.C. (1981). Discovery of new Early Cretaceous insects from Xishan, Beijing. *Bulletin Tianjin Institute of Geological and Mineral Resources* **4**: 87–96. (in Chinese with English abstract).

160 Hong, Y.C. (1982). *Mesozoic Fossil Insects of Jiuquan Basin in Gansu Province*, 98–152. Bejing: Geological Publishing House (in Chinese).

161 Hong, Y.C. (1984). New fossil insects of Laiyang Formation in Laiyang Group, Shandong Province. *Professional Papers of Stratigraphy and Palaeontology* **15**: 181–189. (in Chinese).

162 Hong, Y.C. (1984). *Mesozoic Volume. Palaeontological Atlas of North China*, 161–171. Beijing: Geological Publishing House (in Chinese).

163 Hong, Y.C. (1986). Early Jurassic fossil insects from Daye of Hubei Province. *Professional Papers of Stratigraphy and Palaeontology* **15**: 181–189. (in Chinese).

164 Hong, Y.C. (1988). New fossil insects of Lenshuiwu Formation, Northeastern Jiangxi. *Professional Papers of Stratigraphy and Paleontology* **21**: 172–179. (in Chinese).

165 Hong, Y.C. (1991). Early Jurassic fossil insects from daye of Hubei Province. *Professional Papers of Stratigraphy and Paleontology* **15**: 181–189. (in Chinese).

166 Hong, Y.C. (1992). The study of Early Cretaceous Coleoptera, Raphidioptera, Diptera (Insecta) of Kazuo, Western Liaoning Province. *Acta Geologica Gansu* **1** (1): 1–13. (in Chinese with English abstract).

167 Hong, Y.C. (1992). Taxonomic names, in a new Middle Jurassic insect genus — *Hebeicoleus* gen. nov. (Insecta: Coleoptera) of Chengde, Hebei Province. *Memoirs of Beijing Natural History Museum* **51**: 37–44. (in Chinese).

168 Lin, Q.B. (1976). The Jurassic fossil insects from Western Liaoning. *Acta Palaeontologica Sinica* **15** (1): 97–116. (in Chinese).

169 Lin, Q.B. (1977). *Mesozoic Fossils from Yunnan, China*, 373–382. Beijing: Science Press (in Chinese).

170 Lin, Q.B. (1980). Fossil Insects of Mesozoic from Zhejiang and Anhui provinces. In: *Division and Correlation of Mesozoic Deposits of Volcanic Origin in Zhejiang and Anhui Provinces*, 211–234. Beijing: Science Press (in Chinese).

171 Lin, Q.B. (1983). Two beetle fossils from the Sanqiutian Formation (Upper Triassic) in Liuyang County of Hunan Province. *Bulletin of the Nanjing Institute of Geology and Palaeontology, Academia Sinica* **6**: 297–307.

172 Lin, Q.B. (1992). Late Triassic insect fauna from Toksun, Xinjiang. *Acta Palaeontologica Sinica* **31** (3): 313–335.

173 Hong, Y.C. and Wang, W.L. (1990). Insect fossils of Laiyang Formation. In: *The Stratigraphy and Palaeontology of Laiyang Basin, Shandong Province*, 44–189. Beijing: Geological Publishing House (in Chinese).

174 Wang, W.L. (1993). Study of Liaoximordellidae fam. nov. (Coleoptera) from Jurassic, west Liaoning Province, China. *Acta Geologica Sinica (English Edition)* **67** (1): 86–94.

175 Wang, W.L. (1998). Some fossil insects (Coleoptera) of western Beijing with discussion of development and succession of the late Mesozoic insect fauna. *Memoirs of Beijing Natural History Museum* **56**: 199–206.

176 Wang, W.L. and Liu, M.W. (1996). A new species of *Notocupes* from the Cretaceous of Laiyang Basin, Shanfong Province. *Memoirs of Beijing Natural History Museum* **55**: 79–82.

177 Zhang, H.C. (1997). Early Cretaceous insects from the Dalazi Formation of the Zhixin basin, Jilin Province, China. *Palaeoworld* **7** (8): 75–103. (in Chinese with English abstract).

178 Ren, D. and Tan, J.J. (2006). A new cupedid genus (Coleoptera: Archostemata: Cupedidae) from Jehol Biota of Western Liaoning, China. *Annales Zoologici (Warszawa)* **56** (3): 457–464.

179 Ren, D., Tan, J.J., and Ge, S.Q. (2006). New fossil ommatid (Coleoptera: Archostemata: Ommatidae) from Jehol Biota of Western Liaoning, China. *Progress in Natural Science: Materials International* **16** (6): 639–643.

180 Tan, J.J., Huang, D.-Y., and Ren, D. (2007). First record of Fossil *Mesocupes* from China (Coleoptera: Archostemata: Cupedidae). *Acta Geologica Sinica* **81** (5): 688–696.

181 Tan, J.J. and Ren, D. (2006). *Ovatocupes*: A new Cupedid genus (Coleoptera: Archostemata: Cupedidae) from the Jehol Biota (Late Jurassic) of western Liaoning, China. *Entomological News* **117** (2): 223–232.

182 Tan, J.J. and Ren, D. (2006). New fossil Priacmini (Insecta: Coleoptera: Archostemata: Cupedidae) from the Jehol Biota of China. *Journal of Natural History* **40** (47–48): 2653–2661.

183 Tan, J.J. and Ren, D. (2007). Two exceptionally well-preserved catiniids (Coleoptera: Archostemata: Catiniidae) from the Late Mesozoic of Northeastern China. *Annals of the Entomological Society of America* **100** (5): 666–672.

184 Tan, J.J., Ren, D., and Liu, M. (2005). New ommatidis from the Late Jurassic of western Liaoning, China (Coleoptera: Archostemata). *Insect Science* **12**: 211–220.

185 Tan, J.J., Ren, D., and Shih, C.K. (2006). New cupedids from the Middle Jurassic of Inner Mongolia, China (Coleoptera: Archostemata). *Annales Zoologici (Warszawa)* **56** (1): 1–6.

186 Tan, J.J., Ren, D., and Shih, C.K. (2006). First record of fossil *Priacma* (Coleoptera: Archostemata: Cupedidae) from the Jehol Biota of western Liaoning, China. *Zootaxa* **1326**: 55–68.

187 Tan, J.J., Ren, D., Shih, C.K. *et al.* (2006). New fossil beetles of the family Ommatidae from Jehol Biota of China (Coleoptera: Archostemata). *Acta Geologica Sinica (English Edition)* **80** (4): 474–485.

188 Chang, H.L., Kirejtshuk, A.G., and Ren, D. (2010). New fossil elaterids (Coleoptera: Polyphaga: Elateridae) from the Jehol Biota in China. *Annals of the Entomological Society of America* **103** (6): 866–874.

189 Chang, H.L., Kirejtshuk, A.G., Ren, D. *et al.* (2009). First fossil click beetles from the Middle Jurassic of Inner Mongolia, China (Coleoptera: Elateridae). *Annales Zoologici (Warszawa)* **59** (1): 7–14.

190 Chang, H.L. and Ren, D. (2008). New fossil beetles of the family Elateridae from the Jehol Biota of China (Coleoptera: Polyphaga). *Acta Geologica Sinica (English edition)* **82** (2): 236–243.

191 Chang, H.L., Ren, D., and Shih, C.K. (2007). New fossil elaterid (Coleoptera: Polyphaga: Elateridae) from Yixian Formation of western Liaoning, China. *Progress in Natural Science (English edition)* **17** (10): 1244–1249.

192 Chang, H.L., Zhang, F., and Ren, D. (2008). A new genus and two new species of fossil elaterids from the Yixian Formation of Western Liaoning, China (Coleoptera: Elateridae). *Zootaxa* **1785**: 54–62.

193 Chang, H.L., Zhao, Y.Y., and Ren, D. (2009). New fossil elaterids (Insect: Coleoptera: Polyphaga: Elateridae) from the Middle Jurassic of Inner Mongolia, China. *Progress in Natural Science: Materials International* **19** (10): 1433–1437.

194 Bai, M., Ahrens, D., Yang, X.K. *et al.* (2012). New fossil evidence of the early diversification of scarabs: *Alloioscarabaeus cheni* (Coleoptera: Scarabaeoidea) from the Middle Jurassic of Inner Mongolia, China. *Insect Science* **19**: 159–171.

195 Bai, M., Beutel, R.G., Liu, W.G. *et al.* (2014). Description of a new species of Glaresidae (Coleoptera: Scarabaeoidea) from the Jehol Biota

of China with a geometric morphometric evaluation. *Arthropod Systematics & Phylogeny* **72** (3): 223–236.

196 Bai, M., Beutel, R.G., Shih, C.K. *et al.* (2013). Septiventeridae, a new and ancestral fossil family of Scarabaeoidea (Insecta: Coleoptera) from the Late Jurassic to Early Cretaceous Yixian Formation. *Journal of Systematic Palaeontology* **11** (3): 359–374.

197 Bai, M., Krell, F.T., Ren, D. *et al.* (2010). A new, well-preserved species of Glaresidae (Coleoptera: Scarabaeoidea) from the Jehol Biota of China. *Acta Geologica Sinica (English Edition)* **84** (4): 676–679.

198 Bai, M., Ren, D., and Yang, X.K. (2011). *Prophaenognatha*, a new Aclopinae genus from the Yixian Formation, China and its phylogenetic position based on morphological characters (Coleoptera: Scarabaeidae). *Acta Geologica Sinica (English Edition)* **85** (5): 984–993.

199 Bai, M., Ren, D., and Yang, X.K. (2012). *Prosinodendron krelli*, from the Yixian Formation, China: a missing link among Lucanidae, Diphyllostomatidae and Passalidae (Coleoptera: Scarabaeoidea). *Cretaceous Research* **34** (3): 334–339.

200 Yan, Z., Bai, M., and Ren, D. (2012). A new fossil Hybosoridae (Coleoptera: Scarabaeoidea) from the Yixian Formation of China. *Zootaxa* **3478** (1): 201–204.

201 Yan, Z., Bai, M., and Ren, D. (2013). A new genus and species of fossil Hybosoridae (Coleoptera: Scarabaeoidea) from the Early Cretaceous Yixian Formation of Liaoning, China. *Alcheringa: An Australasian Journal of Palaeontology* **37** (2): 139–145.

202 Yan, Z., Nikolajev, V.G., and Ren, D. (2012). A new, well-preserved genus and species of fossil Glaphyridae (Coleoptera, Scarabaeoidea) from the Mesozoic Yixian Formation of Inner Mongolia, China. *ZooKeys* **241**: 67–75.

203 Zhao, H.Y., Bai, M., Shih, C.K. *et al.* (2016). Two new glaphyrids (coleoptera, scarabaeoidea) from the jehol biota, china. *Cretaceous Research* **59** (5): 1–9.

204 Cai, C.Y. (2015) Middle-late Mesozoic staphylinids: systematics, paleoethology and early evolution. Doctoral dissertation, Nanjing Institute of Geology and Palaeontology, Chinese Academy of Sciences.

205 Cai, C.Y., Beattie, R., and Huang, D.-Y. (2015). Jurassic olisthaerine rove beetles (Coleoptera: Staphylinidae): 165 million years of morphological and probably behavioral stasis. *Gondwana Research* **28** (1): 425–431.

206 Cai, C.Y. and Huang, D.-Y. (2012). Glypholomatine rove beetles (Coleoptera: Staphylinidae): a southern hemisphere recent group recorded from the Middle Jurassic of China. *Journal of the Kansas Entomological Society* **85** (3): 239–244.

207 Cai, C.Y. and Huang, D.-Y. (2013). *Megolisthaerus*, interpreted as staphylinine rove beetle (Coleoptera: Staphylinidae) based on new Early Cretaceous material from China. *Cretaceous Research* **40**: 207–211.

208 Cai, C.Y. and Huang, D.-Y. (2013). A new species of small-eyed *Quedius* (Coleoptera: Staphylinidae: Staphylininae) from the Early Cretaceous of China. *Cretaceous Research* **44**: 54–57.

209 Cai, C.Y. and Huang, D.-Y. (2013). *Sinanthobium daohugouense*, a tiny new omaliine rove beetle from the Middle Jurassic of China (Coleoptera, Staphylinidae). *The Canadian Entomologist* **145** (5): 496–500.

210 Cai, C.Y. and Huang, D.-Y. (2013). Discussion on the systematic position of the oxyteline rove beetle *Anotylus archaicus* Yue, Makranczy & Ren, 2012 (Coleoptera: Staphylinidae). *Insect Systematics & Evolution* **44** (2): 203–212.

211 Cai, C.Y. and Huang, D.-Y. (2013). *Mesocoprophilus clavatus*, a new oxyteline rove beetle (Coleoptera: Staphylinidae) from the Early Cretaceous of China. *Insect Systematics & Evolution* **44**: 213–220.

212 Cai, C.Y. and Huang, D.-Y. (2014). Diverse oxyporine rove beetles from the Early Cretaceous of China (Coleoptera: Staphylinidae). *Systematic Entomology* **39** (3): 500–505.

213 Cai, C.Y., Huang, D.-Y., Newton, A.F. *et al.* (2014). *Mesapatetica aenigmatica*, a new genus and species of rove beetles (Coleoptera, Staphylinidae) from the Middle Jurassic of China. *Journal of the Kansas Entomological Society* **87** (2): 219–224.

214 Cai, C.Y., Huang, D.-Y., and Solodovnikov, A. (2011). A new species of *Hesterniasca* (Coleoptera: Staphylinidae: Tachyporinae) from the Early Cretaceous of China with discussion of its systematic position. *Insect Systematics & Evolution* **42** (2): 213–220.

215 Cai, C.Y., Ślipiński, A., and Huang, D.-Y. (2015). First false jewel beetle (Coleoptera: Schizopodidae) from the Lower Cretaceous of China. *Cretaceous Research* **52**: 490–494.

216 Cai, C.Y., Thayer, M.K., Huang, D.-Y. *et al.* (2013). A basal oxyteline rove beetle (Coleoptera: Staphylinidae) from the Early Cretaceous of China: oldest record for the tribe Euphaniini. *Comptes Rendus Palevol* **12**: 159–163.

217 Cai, C.Y. and Wang, B. (2013). The oldest silken fungus beetle from the Early Cretaceous of Southern China (Coleoptera: Cryptophagidae: Atomariinae). *Alcheringa* **37** (4): 452–455.

218 Cai, C.Y., Yan, E.V., and Vasilenko, D.V. (2013). First record of *Sinoxytelus* (Coleoptera: Staphylinidae) from the Urey locality of Transbaikalia, Russia, with discussion on its systematic position. *Cretaceous Research* **41**: 237–241.

219 Dong, F.B., Cai, C.Y., and Huang, D.-Y. (2011). Revision of five mesozoic beetles from southern china. *Acta Palaeontologica Sinica* **50** (4): 481–491.

220 Wang, B., Ma, J.Y., Mckenna, D.D. *et al.* (2014). The earliest known longhorn beetle (Cerambycidae: Prioninae) and implications for the early evolution of Chrysomeloidea. *Journal of Systematic Palaeontology* **12** (5): 565–574.

221 Wang, B., Ponomarenko, A.G., and Zhang, H.C. (2010). Middle Jurassic Coptoclavidae (Insecta: Coleoptera: Dytiscoidea) from China: a Good Example of Mosaic Evolution. *Acta Geological Sinica* **84** (4): 680–687.

222 Wang, B. and Zhang, H.C. (2011). A new ground beetle (Carabidae, Protorabinae) from the Lower Cretaceous of Inner Mongolia, China. *ZooKeys* **130**: 229–237.

223 Liu, M., Lu, W.H., and Ren, D. (2007). A new fossil mordellid (Coleoptera: Tenebrionoidea: Mordellidae) from the Yixian Formation of Western Liaoning Province, China. *Zootaxa* **1415**: 49–56.

224 Liu, M. and Ren, D. (2006). First fossil Eccoptarthridae (Coleoptera: Curculionoidea) from the Mesozoic of China. *Zootaxa* **1176**: 59–68.

225 Liu, M. and Ren, D. (2007). New fossil eccoptarthrids (Coleoptera: Curculionoidea) from the Yixian Formation of western Liaoning, China. *Science in China Series D: Earth Sciences* **50**: 641–648.

226 Liu, M., Ren, D., and Shih, C.K. (2006). A new fossil weevil (Coleoptera, Curculionoidea, Belidae) from the Yixian Formation of western Liaoning, China. *Progress in Natural Science* **16**: 885–888.

227 Liu, M., Ren, D., and Tan, J.J. (2006). New fossil weevils (Coleoptera: Curculionoidea: Nemonychidae) from the Jehol Biota of western Liaoning, China. *Annales Zoologici (Warszawa)* **56** (4): 605–612.

228 Liu, M., Zhao, Y.Y., and Ren, D. (2008). Discovery of three new mordellids (Coleoptera, Tenebrionoidea) from the Yixian Formation of Western Liaoning, China. *Cretaceous Research* **29** (3): 445–450.

229 Yue, Y.L., Gu, J.J., Yang, Q. *et al.* (2016). The first fossil species of subfamily Piestinae (Coleoptera: Staphylinidae) from the Lower Cretaceous of China. *Cretaceous Research* **63**: 63–67.

230 Yue, Y.L., Makranczy, G., and Ren, D. (2012). A Mesozoic species of *Anotylus* (Coleoptera, Staphylinidae, Oxytelinae) from Liaoning, China, with the earliest evidence of sexual dimorphism in rove beetles. *Journal of Paleontology* **86**: 508–512.

231 Yue, Y.L., Ren, D., and Solodovnikov, A. (2010). *Megolisthaerus chinensis* gen. et sp. n. (Coleoptera: Staphylinidae *incertae sedis*): an enigmatic rove beetle lineage from the Early Cretaceous. *Insect Systematics & Evolution* **41** (4): 317–327.

232 Yue, Y.L., Ren, D., and Solodovnikov, A. (2011). The oldest fossil species of the rove beetle subfamily Oxyporinae (Coleoptera: Staphylinidae) from the Early Cretaceous (Yixian Formation, China) and its phylogenetic significance. *Journal of Systematic Palaeontology* **9** (4): 467–471.

233 Yue, Y.L., Zhao, Y.Y., and Ren, D. (2009). *Glabrimycetoporus amoenus*, a new tachyporine genus and species of Mesozoic Staphylinidae (Coleoptera) from Liaoning, China. *Zootaxa* **2225**: 63–68.

234 Yue, Y.L., Zhao, Y.Y., and Ren, D. (2010). Three new Mesozoic Staphylinids (Coleoptera) from Liaoning, China. *Cretaceous Research* **31**: 61–70.

235 Yu, Y.L., Leschen, R.A.B., Slipinski, A. *et al.* (2012). The first fossil bark-gnawing beetle from the Middle Jurassic of Inner Mongolia, China (Coleoptera: Trogossitidae). *Annales Zoologici (Warszawa)* **62** (2): 245–252.

236 Yu, Y.L., Ślipiński, A., Leschen, R.A.B. *et al.* (2015). New genera and species of Bark-Gnawing beetles (Coleoptera: Trogossitidae) from the Yixian Formation (Lower Cretaceous) of Western Liaoning, China. *Cretaceous Research* **53**: 89–97.

237 Yu, Y.L., Ślipiński, A., Pang, H. *et al.* (2015). A new genus and two new species of Buprestidae (Insecta: Coleoptera) from the Yixian Formation (Lower Cretaceous), Liaoning, China. *Cretaceous Research* **52**: 480–489.

238 Yu, Y.L., Ślipiński, A., Reid, C.K. *et al.* (2015). A new longhorn beetle (Coleoptera: Cerambycidae) from the Early Cretaceous Jehol Biota of Western Liaoning in China. *Cretaceous Research* **52**: 453–460.

239 Yu, Y.L., Ślipiński, A., Shih, C.K. *et al.* (2013). A new fossil jewel beetle (Coleoptera: Buprestidae) from the Early Cretaceous of Inner Mongolia, China. *Zootaxa* **3637** (3): 355–360.

240 Ren, D., Lu, L.W., Ji, S.A. *et al.* (1995). *Fauna and Stratigraphy of Jurassic-Cretaceous in Beijing and the Adjacent Areas*, 73–90. Beijing: Seismic Publishing House (in Chinese with English abstract).

241 Ponomarenko, A.G. and Prokin, A.A. (2015). Review of paleontological data on the evolution of aquatic beetles (Coleoptera). *Paleontological Journal* **49** (13): 1383–1412.

242 Kirejtshuk, A.G., Nel, A., and Kirejtshuk, P.A. (2016). Taxonomy of the reticulate beetles of the subfamily Cupedinae (Coleoptera: Archostemata), with a review of the historical development. *Invertebrate Zoology* **13** (2): 61–190.

243 Kirejtshuk, A.G. (1999). *Sikhotealinia zhiltzovae* Lafer, 1996 — recent representative of the Jurassic

coleopterous fauna (Coleoptera, Archostemata, Jurodidae). *Proceedings of the Zoological Institute RAS* **281**: 21–26.
244 Beutel, R.G., Friedrich, F., and Leschen, R.A.B. (2009). Charles Darwin, beetles and phylogenetics. *Naturwissenschaften* **96**: 1293–1312.
245 Newman, E. (1839). Supplementary note to the synonymy of *Passandra. Annals and Magazine of Natural History* **3**: 303–304.
246 Oppenheim, P. (1888). Die Insektenwelt des lithographischen Schiefers in Bayern. *Palaeontographica* **34**: 215–247.
247 Ponomarenko, A.G. (1971). Systematic position of some beetles from the Solenhofen shales of Bavaria. *Paleontologicheskii Zhurnal* **1**: 67–81.
248 Waterhouse, C.O. (1901). Two new genera of Coleoptera belong to the Cupesida and Prionida. *Annals and Magazine of Natural History* **7**: 520–523.
249 Kirejtshuk, A.G., Ponomarenko, A.G., Prokin, A.A. *et al.* (2010). Current knowledge of Mesozoic Coleoptera from Daohugou and Liaoning (northeast China). *Acta Geologica Sinica* **84**: 783–792.
250 Jarzembowski, E.A., Yan, E.V., Wang, B. *et al.* (2012). A new flying water beetle (Coleoptera: Schizophoridae) from the Jurassic Daohugou Lagerstätte. *Palaeoworld* **21** (3–4): 160–166.
251 Ponomarenko, A.G. (2003). Ecological evolution of beetles (Insecta: Coleoptera). *Acta zoological cracoviensia* **46** (Supplement – Fossil Insects): 319–328.
252 LeConte, J.L. (1874). Descriptions of new Coleoptera chiefly from the Pacific slope of North America. *Transactions of the American Entomological Society* **5**: 43–72.
253 Ponomarenko, A.G. (1989). New Jurassic and Cretaceous ground beetles (Insecta, Coleoptera, Caraboidea) from Asia. *Paleontological Journal* **2**: 52–63. (in Russian).
254 Prokin, A.A. and Ren, D. (2010). New Mesozoic diving beetles (Coleoptera, Dytiscidae) from China. *Paleontological Journal* **44** (5): 526–533.
255 Prokin, A.A. and Ponomarenko, A.G. (2013). The first record of crawling water beetles (Coleoptera, Haliplidae) in the Lower Cretaceous of Mongolia. *Paleontological Journal* **47** (1): 89–93.
256 Ren, D., Zhu, H.Z., and Lu, Y.Q. (1995). New discovery of Early Cretaceous fossil insects from Chifeng City, Inner Mongolia. *Acta Geoscienta Sinica* **4**: 432–439. (in Chinese with English abstract).
257 Jia, T., Liang, H., Chang, H.L. *et al.* (2011). A new genus and species of fossil Eodromeinae from the Yixian Formation of Western Liaoning, China (Coleoptera: Adephaga: Trachypachidae). *Zootaxa* **2736**: 63–68.
258 Fabricius, J.C. (1775). *Systema Entomologiae, sistens insectorvm classes, ordines, genera, species, adiectis synonymis, locis, descriptionibvs, observationibvs*, 1–832. Officina Libraria Kortii, Flensburgi & Lipsiae.
259 Linnaeus, C. (1758). *Systema Naturae per Regna Tria Naturae, Secundum Classes, Ordines, Genera, Species, cum Characteribus, Differentiis, Synonymis, Locis. Editio Decima* **1**: 1–824.
260 Angus, R.B. (1973). Pleistocene *Helophorus* (Coleoptera, Hydrophilidae) from Borislav and Starunia in the Western Ukraine, with a Reinterpretation of M. Lomnicki's Species, Description of a New Siberian Species, and Comparison with British Weichselian Faunas. *Philosophical Transactions of the Royal Society B Biological Sciences* **265** (869): 299–326.
261 Fikáček, M., Prokin, A., Yan, E. *et al.* (2014). Modern hydrophilid clades present and widespread in the Late Jurassic and Early Cretaceous (Coleoptera: Hydrophiloidea: Hydrophilidae). *Zoological Journal of the Linnean Society* **170** (4): 710–734.
262 Stephens, J.F. (1829). *The Nomenclature of British insects; being a compendious list of such species as are contained in the Systematic Catalogue of British Insects, and forming a guide to their classification. 68 columns*. London: Baldwin & Cradock.
263 Ryvkin, A.B. (1988). *Novye melovye Staphylinidae (Insecta) s dal'nego vostoka*, vol. **4**, 103–106. Paleontologicheskii Zhurnal.
264 Zhang, J.F. (1988). The Late Jurassic fossil Staphylinidae (Coleoptera) of China. *Acta Entomologica Sinica* **31**: 79–84. (in Chinese with English abstract).
265 Zhang, J.F., Wang, X.H., and Xu, G.Z. (1992). A new genus and two new species of fossil Staphylinidae (Coleoptera) from Laiyang, Shandong Province, China. *Entomotaxonomia* **14**: 277–281. (in Chinese with English abstract).
266 Solodovnikov, A., Yue, Y.L., Tarasov, S. *et al.* (2013). Extinct and extant rove beetles meet in the matrix: Early Cretaceous fossils shed light on the evolution of a hyperdiverse insect lineage (Coleoptera: Staphylinidae: Staphylininae). *Cladistics* **29**: 360–403.
267 Nikolajev, G.V. and Ren, D. (2010). New genus of the subfamily Geotrupinae (Coleoptera: Scarabaeoidea: Geotrupindae) from the Jehol Biota. *Acta Geologica Sinica (English Edition)* **84** (4): 673–675.
268 Latreille, P.A. (1807). *Genera crustaceorum et insectorum: secundum ordinem naturalem in familias disposita, iconibus exemplisque plurimis explicata. Tomus Secundus*, 1–280. Parisiis et Argentorati: Amand Koenig.

269 Nikolajev, G.V., Wang, B., and Zhang, H.C. (2011). A new fossil genus of the family Glaphyridae (Coleoptera: Scarabaeoidea) from the Lower Cretaceous Yixian Formation. *Zootaxa* **2811** (1): 47–52.

270 Nikolajev, G.V. (2014). On the position of the fossil species *Fortishybosorus ericeusicus* Yan, Bai et Ren, 2012 within the superfamily scarabs (Coleoptera: Scarabaeoidea). *Euroasian Entomological Journal* **13**: 253–256.

271 Nikolajev, G.V. and Ren, D. (2013). A new Glaphyridae genus (Coleoptera: Scarabaeidae) from the Yixian Formation. *Caucasian Entomological Bulletin* **9** (1): 62–64.

272 Erichson, W.F. (1848). *Naturgeschichte der Insecten Deutschlands. Erste Abtheilung. Coleoptera, Bd. 3.* Berlin: Verlag der Nicolaischen Buchhandlung.

273 Nikolajev, G.V. (2006). A new genus in Hybosorinae (Coleoptera, Scarabaeidae) from the Lower Cretaceous of Transbaikalia. *Evraziatskii Entomologicheskii Zhurnal* **5** (1): 12–13.

274 Nikolajev, G.V., Wang, B., Liu, Y. *et al.* (2010). First record of Mesozoic Ceratocanthinae (Coleoptera: Hybosoridae). *Acta Palaeontologica Sinica* **49**: 443–447.

275 Nikolajev, G.V., Wang, B., and Zhang, H.C. (2012). A new genus of the family Hybosoridae (Coleoptera, Scarabaeoidea) from the Yixian Lower Cretaceous Formation in China. *Euroasian Entomological Journal* **11** (6): 503–506.

276 Lu, Y.Y., Nie, R., Shih, C.K. *et al.* (2018). New Scarabaeoidea (Coleoptera) from the Lower Cretaceous Yixian Formation, western Liaoning Province, China: Elucidating the systematics of Mesozoic Hybosoridae. *Cretaceous Research* **86**: 53–59.

277 Nikolajev, G.V. (1992). Taxonomic criteria and generic composition of Mesozoic lamellicorn beetles (Coleoptera, Scarabaeidae). *Paleontological Journal* **26** (1): 96–111. (in Russian with English abstract.

278 Nikolajev, G.V. (2005). Novyy rod plastinchatousykh zhukov (Coleoptera, Scarabaeidae) iz nizhnemelovogo mestonakhozhdeniya Baisa v Zabaykal'e. *Biologicheskie Nauki Kazakhstana* **1**: 117–120. (in Russian with English abstract.

279 Nikolajev, G.V., Wang, B., and Zhang, H.C. (2013). The presence of the family Lithoscarabaeidae (Coleoptera, Scarabaeoidea) in the Yixian geological formation. *Euroasian Entomological Journal* **12**: 559–560. (in Russian with English abstract.

280 Nikolajev, G.V. and Ren, D. (2015). A new fossil Lucanidae subfamily (Coleoptera) from the Mesozoic of China. *Caucasian Entomological Bulletin* **11**: 15–18. in Russian with English abstract.

281 Nikolajev, G.V. (2015). On the systematic position of the new scarab beetles genus (Coleoptera: Scarabaeoidea: Ochodaeidae) from the Mesozoic of China. *Euroasian Entomological Journal* **14**: 21–26. (in Russian with English abstract.

282 LeConte, J.L. (1856). Notice of three genera of Scarabaeidae found in the United States. *Proceedings of the Academy of Natural Sciences of Philadelphia* **8**: 19–25.

283 Grabau, A.W. (1923). Cretaceous fossils from Shantung. *Bulletin of the Geological Survey of China* **5**: 148–181.

284 Lawrence, J.F. (2005). 16.1 Dascillidae Guérin-Méneville, 1843. In: *Handbook of Zoology. Volume IV Arthropoda: Insecta Part 38. Coleoptera, Beetles. Volume 1. Morphology and Systematics (Archostemata, Adephaga, Myxophaga, Polyphaga partim)* (ed. R.G. Beutel and R.A.B. Leschen), 451–456. New York, Berlin: Walter DeGruyter.

285 Jin, Z.Y., Ślipiński, A., Pang, H. *et al.* (2013). A new Mesozoic species of soft-bodied plant beetle (Coleoptera: Dascillidae) from the Early Cretaceous of Inner Mongolia, China with a review of fossil Dascillidae. *Annales Zoologici (Warszawa)* **63** (3): 501–509.

286 Bellamy, C.L. and Volkovitsh, M.G. (2005). Chapter 17. Buprestoidea Crowson, 1955. In: *Handbook of Zoology. Volume IV Arthropoda: Insecta Part 38. Coleoptera, Beetles. Volume 1. Morphology and Systematics (Archostemata, Adephaga, Myxophaga, Polyphaga partim)* (ed. R.G. Beutel and R.A.B. Leschen), 461–468. New York, Berlin: Walter DeGruyter.

287 Pan, X.X., Chang, H.L., Ren, D. *et al.* (2011). The first fossil buprestids from the Middle Jurassic Jiulongshan Formation of China (Coleoptera: Buprestidae). *Zootaxa* **2745** (5): 53–62.

288 Nelson, G.H. and Bellamy, C.L. (2002). Chapter 44. Schizopodidae LeConte 1861. In: *Polyphaga: Scarabaeoidea through Curculionoidea. American Beetles, Volume 2* (ed. R.H. Arnett Jr.,, M.C. Thomas, P.E. Skelley, *et al.*), 95–97. Boca Raton: CRC Press.

289 Johnson, P. (2005). 18.1 Byrrhidae Latreille, 1804. In: *Coleoptera, Beetles. Volume I: Morphology and Systematics (Archostemata, Adephaga, Myxophaga, Polyphaga partim). Handbook of Zoology Volume IV Arthropoda: Insecta. Part 38* (ed. R.G. Beutel and R.A.B. Leschen), 469–471. New York: Walter de Gruyter.

290 Huang, D.-Y. and Zhang, H.C. (1997). Fossil Byrrhidae (Insect, Coleoptera) from the Early Cretaceous of Western Beijing. *Journal of Naijing University (Natural Sciences)* **33** (4): 562–569.

291 Ivie, M.A. (2005). 18.11 Eulichadidae Crowson, 1973. In: *Coleoptera, Beetles. Volume I: Morphology and Systematics (Archostemata, Adephaga, Myxophaga,*

Polyphaga partim). Handbook of Zoology Volume IV Arthropoda: Insecta. Part 38 (ed. R.G. Beutel and R.A.B. Leschen), 547–551. New York: Walter de Gruyter.

292 Yan, E.V., Wang, B., and Zhang, H.C. (2013). First record of the beetle family Lasiosynidae (Insecta: Coleoptera) from the Lower Cretaceous of China. *Cretaceous Research* **40** (1): 43–50.

293 Vanin, S.A., Costa, C., Ide, S. *et al.* (2005). 18.6 Heteroceridae MacLeay, 1825. In: *Coleoptera, Beetles. Volume I: Morphology and Systematics (Archostemata, Adephaga, Myxophaga, Polyphaga partim). Handbook of Zoology Volume IV Arthropoda: Insecta. Part 38* (ed. R.G. Beutel and R.A.B. Leschen), 518–521. New York: Walter de Gruyter.

294 Ponomarenko, A.G. (1986). Coleoptera. Scarabaeida (=Coleoptera). In: *Nasekomye v rannemelovykh ekosistemakh Zapadnoi Mongolii* (Insects in the Early Cretaceous Ecosystems of Western Mongolia), 84–105. Moscow: Nauka.

295 Yan, E.V. (2009). A new genus of elateriform beetles (Coleoptera, Polyphaga) from the Middle-Late Jurassic of Karatau. *Paleontological Journal* **43** (1): 78–82.

296 Yan, E.V. and Wang, B. (2010). A new genus of Elateriform beetles (Coleoptera, Polyphaga) from the Jurassic of Daohugou, China. *Paleontological Journal* **44** (3): 297–302.

297 Lawrence, J.F. (2010). 4.2. Artematopodidae Lacordaire, 1857. In: *Handbook of Zoology. Volume IV Arthropoda: Insecta Part 38. Coleoptera, Beetles. Volume 2. Morphology and Systematics (Elateroidea, Bostrichiformia, Cucujiformia partim)* (ed. R.A.B. Leschen, R.G. Beutel and J.F. Lawrence), 42–47. New York, Berlin: Walter DeGruyter.

298 Zhang, J.F. (2005). The first find of chrysomelids (Insecta: Coleopetra: Chrysomeloidea) from Callovian-Oxfordian Daohugou biota of China. *Géobios* **38** (6): 865–871.

299 Dolin, V.G. (1980). Fossil click-beetles (Coleoptera, Elateridae) from the Upper Jurassic of Karatau. In: *Fossil insects of the Mesozoic*, 17–81. Kiev: Naukova Dumka.

300 Costa, C., Vanin, S.A., Lawrence, J.F. *et al.* (2010). 4.4. Cerophytidae Latreille, 1834. In: *Handbook of Zoology. Volume IV Arthropoda: Insecta Part 38. Coleoptera, Beetles. Volume 2. Morphology and Systematics (Elateroidea, Bostrichiformia, Cucujiformia partim)* (ed. R.A.B. Leschen, R.G. Beutel and J.F. Lawrence), 54–61. New York, Berlin: Walter DeGruyter.

301 Dolin, V.G. (1973). Fossil forms of click-beetles (Elateridae, Coleoptera) from Lower Jurassic of Middle Asia. In: *Fauna and biology of Insects of Moldavia* (ed. V.G. Dolin, D.V. Panfilov, A.G. Ponomarenko and L.N. Pritykina), 72–82. Kishinev: 'Shtiintsa' (in Russian.

302 Dolin, V.G. and Nel, A. (2002). Three new fossil Elateridae from Superior Mesozoic in China (Coleoptera). *Bulletin de la Société entomologique de France* **107** (4): 341–346. (in French).

303 Dong, F.B. and Huang, D.-Y. (2009). A new click beetle (Coleoptera: Elateridae) from Middle Jurassic Haifanggou Formation of Western Liaoning, China. *Acta Palaeontologica Sinica* **48** (1): 102–108.

304 Muona, J. (2000). A revision of the Nearctic Eucnemidae (Coleoptera). *Acta Zoologica Fennica* **212**: 1–106.

305 Horn, G.H. (1891). New species and miscellaneous notes. *Transactions of the American Entomological Society* **18**: 32–48.

306 Lawrence, J.F., Muona, J., Teräväinen, M. *et al.* (2007). *Anischia, Perothops* and the phylogeny of Elateroidea. *Insect Systematic Evolution* **38**: 205–239.

307 Chang, H.L., Muona, J., Pu, H.Y. *et al.* (2016). Chinese Cretaceous larva exposes a Southern Californian living fossil (Insecta, Coleoptera, Eucnemidae). *Cladistics* **32** (2): 211–214.

308 Leschen, A.B. and Beutel, R.G. (2010). Derodontidae LeConte, 1861. In: *Handbook of Zoology. Volume IV Arthropoda: Insecta Part 38. Coleoptera, Beetles. Volume 2. Morphology and Systematics (Polyphaga partim)* (ed. R.A.B. Leschen, R.G. Beutel and J.F. Lawrence), 180–185. New York, Berlin: Walter DeGruyter.

309 Lawrence, J.F. and Ślipiński, A. (2010). Dermestidae Latreille, 1804. In: *Handbook of Zoology. Volume IV Arthropoda: Insecta Part 38. Coleoptera, Beetles. Volume 2. Morphology and Systematics (Polyphaga partim)* (ed. R.A.B. Leschen, R.G. Beutel and J.F. Lawrence), 198–206. New York, Berlin: Walter DeGruyter.

310 Kolibáč, J. (2010). Cleridae Latreille, 1802. In: *Handbook of Zoology. Volume IV Arthropoda: Insecta Part 38. Coleoptera, Beetles. Volume 2. Morphology and Systematics (Polyphaga partim)* (ed. R.A.B. Leschen, R.G. Beutel and J.F. Lawrence), 257–261. New York, Berlin: Walter DeGruyter.

311 Lawrence, J.F. and Leschen, R.A.B. (2010). Prionoceridae Lacordaire, 1857. In: *Handbook of Zoology. Volume IV Arthropoda: Insecta Part 38. Coleoptera, Beetles. Volume 2. Morphology and Systematics (Polyphaga partim)* (ed. R.A.B. Leschen, R.G. Beutel and J.F. Lawrence), 268–271. New York, Berlin: Walter DeGruyter.

312 Kolibáč, J. and Leschen, R.A.B. (2010). Trogossitidae Fabricius, 1801. In: *Handbook of Zoology. Volume IV Arthropoda: Insecta Part 38. Coleoptera, Beetles.*

Volume 2. Morphology and Systematics (Elateroidea, Bostrichformia, Cucujiformia partim) (ed. R.A.B. Leschen, R.G. Beutel and J.F. Lawrence), 241–247. Berlin: Walter de Gruyter.

313 Kolibáč, J. (2006). A review of the Trogositidae. Part 2: Larval morphology, phylogeny and taxonomy (Coleoptera, Cleroidea). *Entomologica Basiliensia et Collectionis Frey* **28**: 105–153.

314 Kolibáč, J. and Zaitsev, A.A. (2010). A description of a larva of *Ancyrona diversa* Pic, 1921 and its phylogenetic implications (Coleoptera: Trogossitidae). *Zootaxa* **2451**: 53–62.

315 Zhang, J.F. (1992). Fossil Coleoptera from Laiyang, Shandong Province, China. *Acta Entomologica Sinica* **35**: 331–338. (in Chinese with English abstract).

316 Lawrence, J.F. and Ślipiński, A. (2010). Boganiidae Sen Gupta and Crowson, 1966. In: *Handbook of Zoology. Volume IV Arthropoda: Insecta Part 38. Coleoptera, Beetles. Volume 2. Morphology and Systematics (Elateroidea, Bostrichformia, Cucujiformia partim)* (ed. R.A.B. Leschen, R.G. Beutel and J.F. Lawrence), 282–286. Berlin: Walter de Gruyter.

317 Leschen, R.A.B., Lawrence, J.F., and Ślipiński, A. (2005). Classification of basal Cucujoidea (Coleoptera: Polyphaga): cladistic analysis, keys and review of new families. *Invertebrate Systematics* **19**: 17–73.

318 Escalona, H.E., Lawrence, J.F., Wanat, M. *et al.* (2015). Phylogeny and placement of Boganiidae (Coleoptera, Cucujoidea) with a review of Australian and New Caledonian taxa. *Systematic Entomology* **40**: 628–651.

319 Bousquet, Y. (2010). 10.8. Monotomidae Laporte, 1840. In: *Handbook of Zoology. Volume IV Arthropoda: Insecta Part 38. Coleoptera, Beetles. Volume 2. Morphology and Systematics (Elateroidea, Bostrichformia, Cucujiformia partim)* (ed. R.A.B. Leschen, R.G. Beutel and J.F. Lawrence), 319–324. Berlin, New York: Walter de Gruyter.

320 Jelínek, J., Carlton, C., Cline, A.R. *et al.* (2010). Nitidulidae Latreille, 1802. In: *Handbook of Zoology. Volume IV Arthropoda: Insecta Part 38. Coleoptera, Beetles. Volume 2. Morphology and Systematics (Elateroidea, Bostrichformia, Cucujiformia partim)* (ed. R.A.B. Leschen, R.G. Beutel and J.F. Lawrence), 390–407. Berlin: Walter de Gruyter.

321 Lawrence, J.F. and Ślipiński, A. (2010). Mordellidae Latreille, 1802. In: *Handbook of Zoology. Volume IV Arthropoda: Insecta Part 38. Coleoptera, Beetles. Volume 2. Morphology and Systematics (Elateroidea, Bostrichformia, Cucujiformia partim)* (ed. R.A.B. Leschen, R.G. Beutel and J.F. Lawrence), 533–537. Berlin: Walter de Gruyter.

322 Huang, D.-Y. and Yang, J. (1999). Early Cretaceous fossil Mordellidae (Insecta, Coleoptera) from western Beijing. *Acta Palaeontologica Sinica* **38** (1): 125–132. (in Chinese with English abstract).

323 Lawrence, J.F. and Newton, A.F. (1995). Families and subfamilies of Coleoptera (with selected genera, notes, and references and data on family-group names). In: *Biology, Phylogeny, and Classification of Coleoptera, Papers Celebrating the 80th Birthday of Roy A. Crowson* (ed. J. Pakaluk and S.A. Ślipiński), 779–1006. Warszawa: Muzeum i Instytut Zoologii PAN.

324 Lawrence, J.F., Falin, Z.H., and Ślipiński, A. (2010). 11.8. Ripiphoridae Gemminger and Harold, 1870 (Gerstaecker, 1855). In: *Handbook of Zoology. Volume IV Arthropoda: Insecta Part 38. Coleoptera, Beetles. Volume 2. Morphology and Systematics (Polyphaga partim)* (ed. R.A.B. Leschen, R.G. Beutel and J.F. Lawrence), 538–548. New York, Berlin: Walter DeGruyter.

325 Matthews, E.G., Lawrence, J.F., Bouchard, P. *et al.* (2010). Tenebrionidae Latreille, 1802. In: *Handbook of Zoology. Volume IV Arthropoda: Insecta Part 38. Coleoptera, Beetles. Volume 2. Morphology and Systematics (Polyphaga partim)* (ed. R.A.B. Leschen, R.G. Beutel and J.F. Lawrence), 574–659. New York, Berlin: Walter DeGruyter.

326 Kirejtshuk, A.G., Nabozhenko, M.V., and Nel, A. (2011). First Mesozoic representative of the subfamily Tenebrioninae (Coleoptera, Tenebrionidae) from the Lower Cretaceous of Yixian (China, Liaoning). *Entomologicheskoe Obozrenie* **90**: 548–552. (in Russian).

327 Chang, H.L., Nabozhenko, M., Pu, H.Y. *et al.* (2016). First record of fossil comb-clawed beetles of the tribe Cteniopodini (Insecta: Coleoptera: Tenebrionidae) from the Jehol Biota (Yixian Formation of China), Lower Cretaceous. *Cretaceous Research* **57**: 289–293.

328 Ślipiński, A. and Escalona, H. (2013). *Australian Longhorn Beetles (Coleoptera: Cerambycidae) Volume 1, Introduction and Subfamily Lamiinae.* CSIRO Publishing.

329 Riley, E.G., Clark, S.M., Flowers, R.W. *et al.* (2002). Chrysomelidae Latreille 1802. In: *American beetles* (ed. R.H. Arnett and M.C. Thomas), 617–691. CRC press.

330 Zherikhin, V.V. (1977). Rhynchophora. Family Attelabidae. Family Curculionidae. In: *Mezozoiskie Zhestkokrylye. Akademiya Nauk SSSR, Trudy Paleontologicheskogo Instituta. Volume 161* (ed. L.V. Arnoldi, V.V. Zherikhin, L.M. Nikritin, *et al.*), 176–180. Moscow: Nauka Publishers (in Russian).

331 Davis, S.R., Engel, M.S., Legalov, A. *et al.* (2013). Weevils of the Yixian Formation, China (Coleoptera: Curculionoidea): phylogenetic considerations and comparison with other Mesozoic Faunas. *Journal of Systematic Palaeontology* **11** (4): 399–429.

332 Legalov, A.A. (2009). Contribution to the knowledge of the Mesozoic Curculionoidea (Coleoptera). *Amurian Zoological Journal* **1**: 283–295.

333 Legalov, A.A. (2009). Annotaed checklist of fossil and recent species of the family Nemonychidae (Coleoptera) from the World Fauna. *Amurian Zoological Journal* **1** (3): 200–213.

334 Ponomarenko, A.G., Yan, E.V., Wang, B. *et al.* (2012). Revision of some Early Mesozoic beetles from China. *Acta Paleontological sinica* **51** (4): 475–490.

335 Santiago-Blay, J.A. (1994). Paleontology of leaf beetles. In: *Novel aspects of the biology of Chrysomelidae* (ed. P.H. Jolivet, M.L. Cox and E. Petitpierre), 1–68. Dordrecht: Kluwer Academic Publishers.

336 Gómez-Zurita, J., Hunt, T., Kopliku, F. *et al.* (2007). Recalibrated tree of leaf beetles (Chrysomelidae) indicates independent diversification of angiosperms and their insect herbivores. *PLoS ONE* **2**: e360. https://doi.org/10.1371/journal.pone.0000360.

337 Kirejtshuk, A.G. and Ponomarenko, A.G. (2013) Catalogue of fossil Coleoptera. Beetles (Coleoptera) and Coleopterologists. Zoological Institute of the Russian Academy of Sciences, St. Petersburg. (updated at www.zin.ru/Animalia/Coleoptera/eng/paleosys.htm)

338 Peralta-Medina, E. and Falcon-Lang, H.J. (2012). Cretaceous forest composition and productivity inferred from a global fossil wood database. *Geology* **40**: 219–222.

339 Jarzembowski, E.A., Wang, B., Zhang, H.C. *et al.* (2015). Boring beetles are not necessarily dull: New notocupedins (Insecta: Coleoptera) from the Mesozoic of Eurasia and East Gondwana. *Cretaceous Research* **52**: 431–439.

340 Ponomarenko, A.G. and Ren, D. (2010). First record of *Notocupes* (Coleoptera: Cupedidae) in locality Daohugou, Middle Jurassic of Inner Mongolia, China. *Annales zoologici (Warszawa)* **60** (2): 169–171.

341 Jarzembowski, E.A., Yan, E.V., Wang, B. *et al.* (2013). Brochocolein beetles (Insecta: Coleoptera) from the Lower Cretaceous of northeast China and southern England. *Cretaceous Research* **44**: 1–11.

342 Hong, Y.C. (1987). The study of Early Cretaceous insects of Kezuo, west Liaoning. *Professional Papers of Stratigraphy and Palaeontology* **18**: 76–87.

343 Prokin, A., Ren, D., and Fikáček, M. (2010). New Mesozoic water scavenger beetles from the Yixian Formation in China (Coleoptera: Hydrophiloidea). *Annales Zoologici (Warszawa)* **60** (2): 173–179.

344 Fikáček, M., Prokin, A., Angus, R.B. *et al.* (2012). Revision of Mesozoic fossils of the helophorid lineage of the superfamily Hydrophiloidea (Coleoptera: Polyphaga). *Acta Entomologica Musei Natioalis Pragae* **52** (1): 89–127.

345 Nikolajev, G.V. and Ren, D. (2011). The oldest species of the genus *Glaphyrus* Latr. (Coleoptera: Scarabaeoidea: Glaphyridae) from the Mesozoic of China. *Paleontological Journal* (2): 179–182.

346 Nikolajev, G.V. and Ren, D. (2012). New species of the genus *Lithohypna* Nikolajev, Wang et Zhang, 2011 (Coleoptera, Scarabaeidae, Glaphyridae) from the Yixian Formation, China. *Evraziatskiy Entomologicheskiy Zhurnal* **11** (3): 209–211.

347 Nikolajev, G.V. and Ren, D. (2015). A second species of the fossil genus *Cretohypna* Yan, Nikolajev et Ren (Coleoptera: Glaphyridae) from the Mesozoic of China. *Euroasian Entomological Journal* **14**: 142–143.

348 Nikolajev, G.V. and Ren, D. (2012). The earliest known species of the genus *Pleocoma* LeConte (Coleoptera, Scarabaeoidea, Pleocomidae) from the Mesozoic of China. *Paleontological Journal* **46** (5): 495–498.

349 Prokin, A.A. and Ren, D. (2011). New species of variegated mud-loving beetles (Coleoptera: Heteroceridae) from Mesozoic deposits of China. *Paleontological Journal* **45**: 284–286.

350 Yan, E.V., Wang, B., Jarzembowski, E.A. *et al.* (2015). The earliest byrrhoids (Coleoptera, Elateriformia) from the Jurassic of China and their evolutionary implications. *Proceedings of the Geologists Association* **126** (2): 211–219.

351 Kolibáč, J. and Huang, D.-Y. (2011). *Mathesius liaoningensis* gen. et sp. nov. of Jehol Biota, a presumptive relative of the clerid or thaneroclerid branches of Cleroidea (Coleoptera). *Zootaxa* **2872**: 1–17.

352 Lu, T.M., Zhao, Y.Y., Shih, C.K. *et al.* (2015). Three new species of *Parandrexis* (Coleoptera: Parandrexidae) from the Middle Jurassic Jiulongshan Formation of Inner Mongolia, China. *Entomotaxonomia* **37** (2): 111–122.

353 Hong, Y.C. (1995). Fossil insects of the northern Ordos Basin. *Acta Geologica Gansu* **4** (2): 1–9.

22

Hymenoptera – Sawflies and Wasps

Mei Wang[1,2], Longfeng Li[2,3], Chungkun Shih[2,4], Taiping Gao[2], and Dong Ren[2]

[1] Research Institute of Forest Ecology, Environment and Protection, Chinese Academy of Forestry, Haidian District, Beijing, China
[2] Capital Normal University, Haidian District, Beijing, China
[3] Gansu Agricultural University, Lanzhou, Gansu, China
[4] National Museum of Natural History, Smithsonian Institution, Washington, DC, USA

22.1 Introduction to Hymenoptera

Hymenoptera, encompassing sawflies, wasps, bees and ants, are speciose with more than 150 000 known species worldwide today. Hymenopterans have complete metamorphosis. The body sizes of Hymenopterans range from 0.14 to 70.0 mm in length. The males of *Dicopomorpha echmepterygis* (Mymaridae), a fairyfly species found in Costa Rica with body length about 0.14 mm, are considered as the smallest insects in the world hitherto. Although the Asian giant hornet (*Vespa mandarinia* in Vespidae) and the tarantula hawk (in the genera of *Pepsis* and *Hemipepsis* in Pompilidae) are notorious for their dangerous stings and large body sizes (up to about 50 mm long), the elegant and gracious females of *Pelecinus polyturator* (in Pelecinidae) are the longest hymenopteran with body lengths up to 90 mm.

Like all other insects, hymenopterans have three body sections, the head, thorax and abdomen, and usually two pairs of membranous wings with relatively few veins. The forewings are slightly larger than the hind wings. Antennae, mostly geniculate (elbow-shaped) or filiform (thread-like) and rarely pectinate (comb-like), are usually fairly long with nine or more antennomeres. Most adults have chewing mouthparts with mandibles, especially for those which feed on plant matter or prey on spiders and other insects. For honey bees and some wasps, the mouthparts have further evolved into a proboscis for sipping nectar.

The order Hymenoptera, which means "membrane wings", are classified into two suborders, Symphyta and Apocrita. Symphyta are a basal group of extant and fossil taxa, consisting of 25 families in seven superfamilies, i.e. Cephoidea, Orussoidea, Pamphilioidea, Siricoidea, Tenthredinoidea, Xiphydrioidea, and Xyeloidea (Figure 22.1). The Symphyta, comprising about 8000 species in more than 800 genera, are characterized by relatively complex wing venation, the abdomen broadly attached to the thorax and a long and robust saw-like ovipositor.

The Symphyta are commonly divided into three groups: true sawflies (phyllophaga) with Tenthredinidae as the largest family having more than 7500 extant species worldwide (Figure 22.2), woodwasps or xylophaga (Siricidae), and Orussidae. The three groups are distinguished by the true sawflies' ventral serrated or saw-like ovipositors for making holes in vegetation to deposit eggs, while the woodwasps' ovipositors penetrate wood while the Orussidae are external parasitoids of wood-boring beetles. The body sizes of sawflies vary greatly ranging from small species with only 2.5 mm in body length [2] to *Uroceras gigas*, up to 20 mm long. However, Gao *et al.*, in 2013, reported the largest sawfly, *Hoplitolyda duolunica* (see Figure 22.28) in Praesiricidae, Rasnitsyn, 1968 from the Early Cretaceous (125 Mya), with a body length of 55 mm and a wingspan of 92 mm [3]. In contrast, *Syspastoxyela rhaphidia* Engel and Huang, 2016 in Syspastoxyelidae Engel and Huang, 2016, from the Late Cretaceous Myanmar amber (99 Mya), has a body length of 2.8 mm [4].

In general, adult sawflies feed on pollen and nectar, while their larvae feed on leaves, fruits or other plant matter, except for orussid larvae being parasitoid. On the

Rhythms of Insect Evolution: Evidence from the Jurassic and Cretaceous in Northern China, First Edition.
Edited by Dong Ren, Chungkun Shih, Taiping Gao, Yongjie Wang, and Yunzhi Yao.

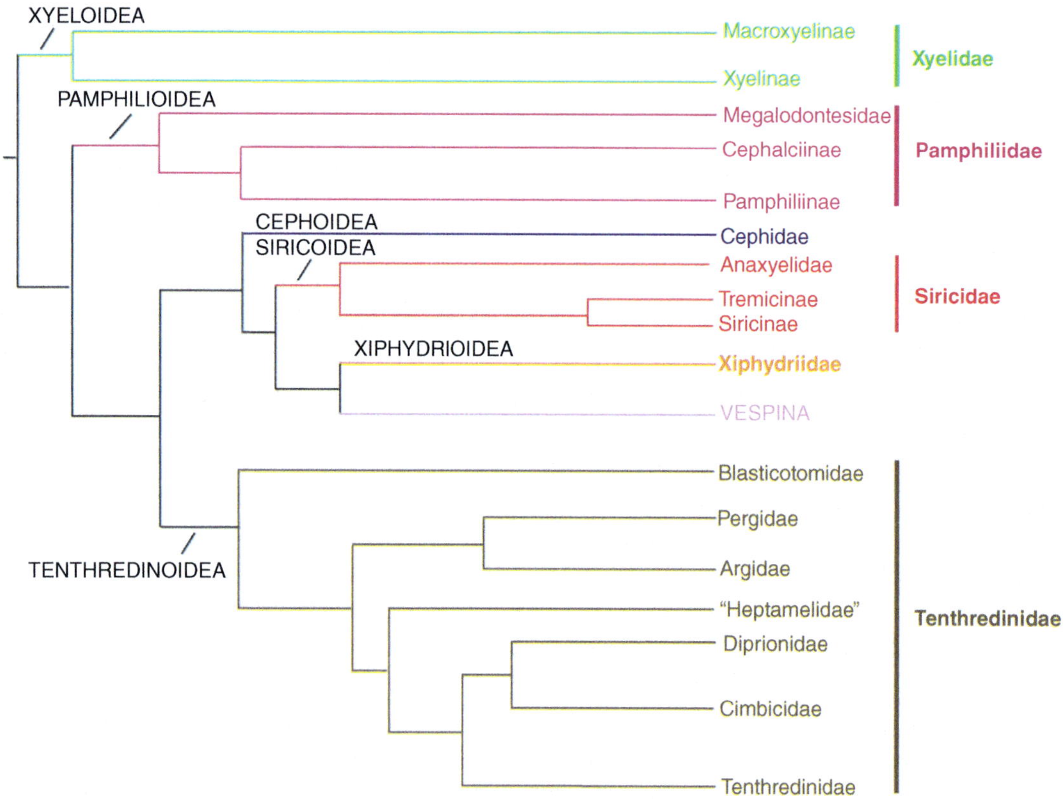

Figure 22.1 The phylogenetic hypothesis of the "basal" Hymenoptera and Unicalcarida, based on a Bayesian Inference (BI) analysis of the 12 + CAD3 + GLN3_hym data set. Source: modified from [1].

Figure 22.2 A sawfly, *Macrophy albicincta,* in the family of Tenthredinidae. Source: Photo by Dr. Chungkun Shih.

other hand, sawfly larvae are hosts of many parasitoid insects, especially parasitoid wasps. Some sawflies are considered as pests for the forestry and horticulture. For example, the larvae of pine sawflies, *Diprion pini* and *Neodiprion serifer* in Diprionidae, damage the pine forest, while iris sawfly larvae, *Rhadinoceraea micans* in Tenthredinidae, destroy the leaves of some iris plants.

Forest Damage by Sawflies

Pamphiliids, widely distributed in China, sometime break out, causing serious damage. For example, *Acantholyda posticalis* Matsumura, 1912 caused severe damage to pine secondary forest in Huguan County, Shanxi Province from 1967 to 1979, covering a forest area of about 2000 ha (1 ha = 10 000 m^2), and even gradually extending to the young plantations. The damaged forests appeared as if they had been burnt, and if the infestation continued for two to three years, pine trees could die. In 1988, the occurrence of *A. posticalis* reached to 734 ha in Lingbaochuankou forestry farms of Henan Province, resulting in a withered condition for the damaged pine trees [5].

In the 1990s, *Diprion* became one of the important forest insect pests, resulting in large economic debits for the forestry management and production in China. A particular pest was *Diprion jingyuanensis* Xiao & Zhang, 1994, which was found in forestry farms of Haas Mountain, Jingyuan County, Gansu Province in 1989 [6]. Later in 1990, it was discovered in Chishiqiao Village of Qinyuan County, Shanxi Province. The larvae can feed on needles of *Pinus tabulaeformis* Carr, and sometimes the whole plants can be completely eaten as if they had been burnt, which causes severe damage on forestry growth and development, even leading to death [6, 7]. Such severe damage caused by *Diprion* was rare previously in China, but the occurrence areas have increased rapidly year by year since 1990.

The Apocrita, including parasitoid wasps, wasps, bees, ants, etc., have simpler wing venation and a constricted narrow waist (petiole) between the first two abdominal segments. The first abdominal segment is fused to the thorax, which is called the propodeum. For apocritans, the term mesosoma represents the thorax and the fused propodeum, while metasoma is the abdomen excluding the first segment. The ovipositors of wasps and some bees are evolved to form stingers.

The Apocrita is further divided into two unranked divisions: the Parasitica, mainly parasitoid wasps; and the Aculeata, the true wasps, bees and ants (Figure 22.3). The Parasitica division comprise 11 superfamilies: i.e. Ceraphronoidea, Chalcidoidea, Cynipoidea, Evanioidea, Ichneumonoidea, Megalyroidea, Mymarommatoidea, Platygastroidea, Proctotrupoidea, Stephanoidea and Trigonaloidea [8]. Female parasitoid wasps, each equipped with a well-developed and specialized ovipositor of suitable length fit for purpose, lay eggs in their own unique way into or on hosts of insects or larvae or into plant crevices (Figure 22.4). Parasitoid wasps are also widely used in the biological control of pests, especially those in the Mymarommatoidea, Aphelinidae, Ichneumonidae, Braconidae and Encyrtidae.

The Aculeata division has three superfamilies: Apoidea, Chrysidoidea and Vespoidea. Most wasps and bees in the Aculeata have the capability of constructing nests (Figure 22.5). Some are simple nests just for storing food and providing protection for the brood. Others have a more complicated design and structure to serve the eusocial activities of the entire colony, including providing food for the brood (see box below). Contrary to common perceptions, many hymenopterans don't collect pollen and nectar; instead they prey on other insects or spiders to feed their next generation. Members of families Sphecidae (*sensu stricto*) and Crabronidae are well-known for their hunting skills. They construct nests in crevices in trees, rocks or underground in sand or soil and are solitary and usually found on flowers while preying on a wide range of spiders and insects, such as flies, weevils, bugs, crickets, cicadas and caterpillars (Figure 22.6). Vespidae wasps, such as spider wasps, yellow jackets, hornets, paper wasps, etc., are also known to provision their nests with caterpillars, spiders, etc. (Figure 22.7). Some even provide masticated insects to feed their larvae.

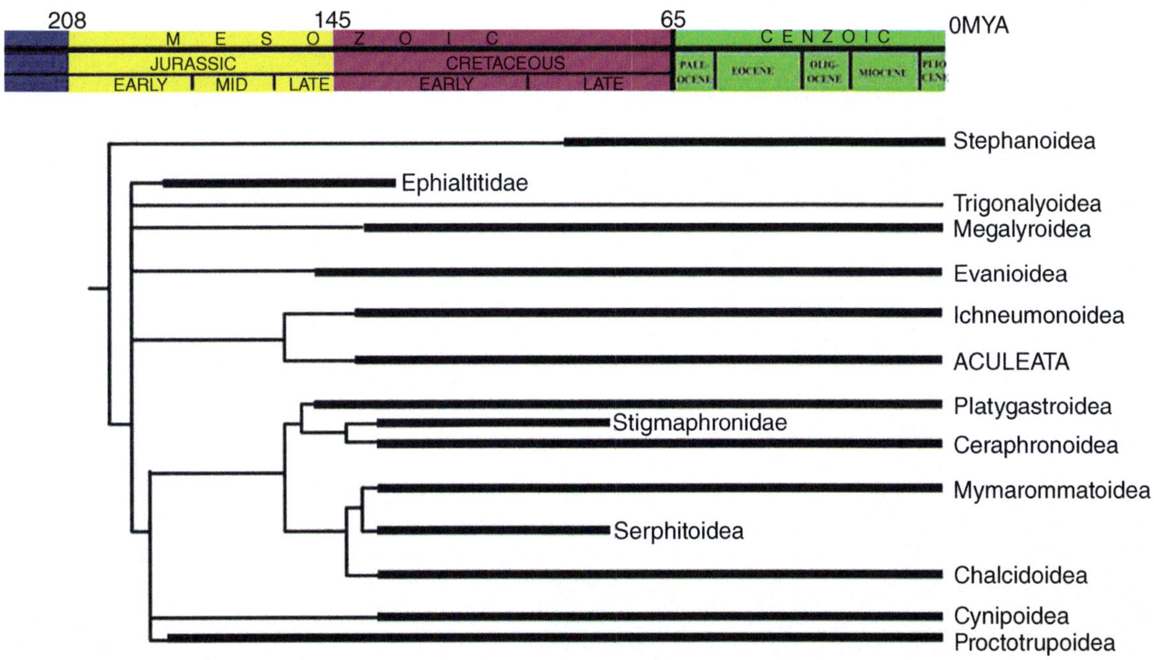

Figure 22.3 Phylogenetic hypothesis of the Apocritans in Hymenoptera. Source: modified from [8].

Figure 22.4 Parasitoid wasps deposit their eggs into the hosts. Source: Photos by Jason Shih.

Figure 22.5 A wasp building a nest to keep eggs. Source: Photo by Dr. Chungkun Shih.

Figure 22.6 Hunting wasps caught insects to feed their larvae in the nests Source: Photos by Jason Shih.

Figure 22.7 Vespidae wasps caught and brought various insects to their nests to feed their larvae. Source: Photos by Jason Shih.

Parasitoid Wasps

Parasitoid wasps deposit eggs on or into the body of hosts, which cover a wide ranges of insect larvae or adults such as moths, butterflies, flies, aphids, cicadas, beetles, or other hymenopterans, etc. (Figures 22.8 and 22.9). The wasp larvae grow while feeding on the non-vital tissues of the hosts without causing much damage; toward the end of their development, they feed voraciously and kill the hosts. After the final molt, the larva changes to a pupa (Figure 22.10). Some of the parasitoid wasps are used commercially for biological control of agricultural pests. For example, *Encarsia formosa* Gahan, 1924 (Aphelinidae, Chalcidoidea) are used to control whiteflies of various crops, e.g. tomato, cucumber, eggplant, strawberry etc., in the greenhouse or in the garden [9]; *Trichogramma galloi* Zucchi, 1988 (Trichogrammatidae) used to control the "sugarcane borers" (*Diatraea saccharalis* in Crambidae) which damage sugarcane, corn, rice, sorghum in Caribbean, Central America, the warmer parts of South America and gulf states of the USA [10]; *Aphidius matricariae* Haliday, 1834 (Braconidae, Ichneumonoidea) used to control various aphids [11]; and *Gonatocerus ashmeadi* Girault, 1915 (Mymaridae, Chalcidoidea) a very effective biocontrol agent to control the "glassy-winged sharpshooter" (*Homalodisca vitripennis* in Cicadellidae) in Florida, Hawaii, western Polynesia, and cooler climate in central and northern California [12, 13].

Figure 22.8 A "cicada killer wasp", *Sphecius speciosus,* caught a cicada to be used for feeding larvae. Source: Photo by Peter Jian Ming Shih.

Figure 22.9 Parasitoid wasps lay eggs into hosts via various modes. Source: Photos by Jason Shih.

Figure 22.10 Parasitoid braconid wasps attacked tobacco hornworm. Source: Photos by Dr. Chungkun Shih.

Insect Eusociality

Eusociality, the most sophisticated level of animal sociality, is characterized by cooperative brood care, a division of labor by reproductive and non-reproductive members and multiple generations of adults within a colony [14–16]. The division of labor creates various castes within a colony, e.g. queen, male drone, worker, soldier, etc. Eusocial behaviors, including a caste system and strict adhering to the individual's roles and functions, significantly enhance the survival and/or development of a colony. Judging by the vast speciosity, broad diversity, and high individual counts of bees, wasps, ants, and termites, it is clear that insect eusociality provides the colony competitive advantages in the ecosystem, which is one of the key factors leading to their successful evolution.

Only a limited number of insect species exhibit eusocial behaviors, e.g. four tribes of the bee family Apidae: Apini (e.g. honey bees – genus *Apis*), Bombini (e.g. bumble bees – genus *Bombus*), Euglossini (e.g. orchid bees – genus *Euglossa*), and Meliponini (e.g. stingless bees); two subfamilies of wasp family Vespidae: Polistinae (e.g. paper wasps) and Vespinae (e.g. hornets – genus *Vespa* and yellow jackets – genus *Vespula*); ants (family Formicidae), and termites (order Isoptera). In addition, *Austroplatypus incompertus*, a species of ambrosia beetle in the weevil family of Curculionidae native to Australia [17] and some species of gall-inducing insects, including the gall-forming aphid, *Pemphigus spyrothecae*, and thrips were also described as eusocial [18, 19].

The honey bee is an iconic example of having a caste system with complex eusocial behaviors. A typical colony consists of a queen bee with a body length 18–20 mm, 60 000–80 000 worker bees, 10–15 mm in body length and about 200 male drones, 15–17 mm long. Female bees are developed from fertilized eggs, while male drones from unfertilized eggs. Most fertilized eggs are hatched into worker bees. Only a few become "queen" larvae, which are fed with royal jelly, secreted by the glands of young worker bees, during their entire larval stages. The first queen bee emerging usually kills other queen pupae with her sting. Before the emergence of the new queen, the old queen leaves the hive with a swarm of workers to construct a new hive and to build a new colony.

The new queen bee takes to the air and mates with one to several male drones from other colonies. Then the queen returns to the hive to carry out her main duty of producing eggs and leading the colony. The queen bee may lay as many as 1500 eggs a day in spring and summer, producing up to one million eggs during her lifetime of up to four years. Pheromones are used by the queen to communicate with all members of the colony, including controlling the roles and functions of individuals, identifying friends or foes and defending against intruders. Drones, with a lifespan of up to four months, die after mating, while unmated drones are cast out or denied food when the honey inventory gets low.

Worker bees, with a lifespan of six to seven weeks in spring and summer, carry out various duties as they age, starting from handling eggs, secreting royal jelly, to feeding and caring for the larvae; then, duties change to maintaining or building honeycomb cells, cleaning the hive, adjusting the hive temperature, receiving nectar and pollen from foragers and defending the hive; and finally, they are upgraded to foraging for pollen and nectar. Honey bees are known to use dances to inform other foragers about the location (direction and distance), and quality and quantity of pollen and nectar sources.

Honey bees have the distinct honor for being the subject of research by Dr. Karl von Frisch for his Nobel Prize in Physiology or Medicine in 1973. Dr. von Frisch decoded and interpreted the honey bee's dance language which is used by returning bees loaded with food to inform other foragers about the location (direction and distance) and quality and quantity of pollen and nectar sources. Bee dance is a combination of various dancing movements, including vibrations of tail-wagging and frequencies of wing-beating to recruit and inform the outgoing forager bees [20, 21]. This is the only Nobel Prize awarded to research on entomological behavior, highlighting the importance and uniqueness of honey bees using intriguing dance language in communication.

22.2 Progress in the Studies of Fossil Hymenoptera

Rasnitsyn [22, 23] proposed Xyelidae as the most basal group, from which all other hymenopterans were derived. *Archexyela crosbyi* Riek, 1955 and *Archexyela ipswichensis* Engel, 2005 [24] in the family of Xyelidae, from the Late Triassic, Carnian of Mt. Crosby Formation in Queensland, Australia, are considered as two examples among the earliest fossil hymenopterans. The phylogeny tree of Xyelidae is composed of two branches, Macroxyelinae and Xyelinae + Madygellinae. It was proposed that Tenthredinoidea originated from Macroxyelinae, while other hymenopterans were all from the ancestor of Xyelinae + Madygellinae [22, 23]. In cladistic analysis, the Orussoidea within the Symphyta gave rise to the monophyletic suborder Apocrita (wasps, bees and ants) [25–27].

Undoubtedly, the most basal groups of Apocrita belong to superfamily Proctotrupoidea with 12 families. The earliest fossils of proctotrupoids are species in the families of Heloridae and Pelecinidae documented from the Middle Jurassic in China. In addition, Pelecinidae and Heloridae and Roproniidae have been reported from the Early Cretaceous of China, and Spathiopterygidae Engel and Ortega-Blanco 2013, from Cretaceous amber of Spain (Albian) and New Jersey (Turonian) [28]. Many other fossil families of Apocrita have been recorded from the Cretaceous.

Most fossil and amber Apidae have been described from the Cenozoic. The earliest described bee fossil to date is *Cretotrigona prisca* from the Late Cretaceous (*ca.* 65 Mya) in New Jersey amber [29, 30]. This amber species suggested that bee diversification had occurred by the Late Cretaceous, due to its derived position in a derived family of Apidae [30]. Furthermore, bee burrows, presumed halictine nests, were reported by Elliott and Nations [31] from the mid-Cretaceous (*ca.* 95 Mya) Dakota Formation in the eastern Arizona. Again, this suggested origination of bees may be even older, probably the late Early Cretaceous, *ca.* 120 Mya [8]. The fossil record of bees appears to be more abundant in the Cenozoic, especially in the Eocene and Oligocene (55 to 23 Mya), including a variety of fossil bees in the Baltic amber (*ca.* 44 Mya), in compression fossils from Florissant, Colorado (*ca.* 34 Mya), and in the Dominican amber (*ca.* 25 Mya). These fossil bees are very similar to the extant bees. The first *Apis* bees appeared in the fossil record at the Eocene – Oligocene boundary (*ca.* 34 Mya), in European deposits and Dominican amber.

In China, the study of hymenopteran fossils started over 40 years ago, with only a few reports published. In 1975, Youchong Hong found the first hymenopteran fossil *Sinosirex gigantea* Hong, 1975 from Weichang of Hebei [32]. In the following year, Qibin Lin reported an orussid fossil *Paroryssus suspectus* Lin, 1976 from Chaomidian Village of Liaoning [33]. From then on, more paleoentomologists have been involved in studies with new fossil findings, especially from Northeastern China. So far, there have been more than 130 species, 80 genera belonging to 34 families described in China. These specimens are well-preserved, many of which retain well-preserved morphological characters essential to the study of early evolution of Hymenoptera in distant times. During this period, significant contributions have been made by Youchong Hong, Qibin Lin, Junfeng Zhang, Haichun Zhang, Dong Ren, Yunyun Zhao, Chungkun Shih, Taiping Gao, Mei Wang, and Longfeng Li. We appreciate the cooperation and contribution from our collaborators, M. Buffington, M.S. Engel, C.C. Labandeira, H. Li, J. Huber, D.S. Kopylov, J. Ortega-Blanco, A.P. Rasnitsyn, M.J. Sharkey, M.J.H. Shih and J.A. Santiago-Blay; and our CNU Team past and present, Yaping Cai, Lichao Guo, Hua Feng, Chenxi Liu, Xiaoqing Shi and Chen Wang.

22.3 Representative Fossils of Hymenoptera from Northern China

Suborder "Symphyta" Gerstaecker, 1867

Superfamily Xyeloidea Newman, 1834

Family Xyelidae Newman, 1834

Xyelidae, considered as the most basal taxa in the evolution of Hymenoptera (Figure 22.1), are typified by having a long and enlarged first flagellomere followed by slender filament-like flagellomeres; forewing with Rs bifurcation; and the protruding ovipositor. The oldest fossil representatives are known from the Middle or Upper Triassic of Kyrgyzstan in Central Asia [22, 23, 34], the Upper Triassic of Australia [24, 35], South Africa [36], and Argentina [37]. Xyelidae, comprising four subfamilies: Macroxyelinae Ashmead, 1898, Xyelinae Newman, 1834, Archexyelinae Rasnitsyn, 1964 (only from the Triassic), and Madygellinae Rasnitsyn, 1969 (only from the Triassic), underwent massive diversifications during the Jurassic [23]. To date, there are 47 extinct genera containing 146 species described from the Mesozoic of Kyrgyzstan, China, Australia, Russia, Africa, and Argentina. But, since the Late Cretaceous, taxa of Xyelidae decreased rapidly. There are five extant genera in Xyelinae and Macroxyelinae containing about 65 described species [38]. Extant members of Xyelidae are mostly found in the Northern Hemisphere, especially in boreal regions, and their larvae typically feed on nectar and pollens, shoots of fir or on leaves of deciduous trees. Rasnitsyn once found pollens of Caytoniales or Pteridosperm in the intestine of *Anthoxyela anthophaga* Rasnitsyn, 1982, and considered *A. anthophaga* feeding on pollens [39].

Genera included from the Jurassic and Cretaceous of Northern China: *Angaridyela* Rasnitsyn, 1966, *Anthoxyela* Rasnitsyn, 1977, *Yanoxyela* Ren, Lu, Guo & Ji, 1995, *Ceratoxyela* Zhang & Zhang, 2000, *Heteroxyela* Zhang & Zhang, 2000, *Isoxyela* Zhang & Zhang, 2000, *Lethoxyela* Zhang & Zhang, 2000, *Liaoxyela* Zhang & Zhang, 2000, *Sinoxyela* Zhang & Zhang, 2000, *Abrotoxyela* Gao, Ren & Shih, 2009, *Brachyoxyela* Gao, Zhao & Ren, 2011, *Platyxyela* Wang, Shih & Ren, 2012, *Cathayxyela* Wang, Rasnitsyn & Ren, 2013 and *Aequixyela* Wang, Rasnitsyn & Ren, 2013

Angaridyela Rasnitsyn, 1966

Angaridyela Rasnitsyn, 1966, *Paleontol. Zhur.*, 1, 729 [40] (original designation).

Type species: *Angaridyela vitimica* Rasnitsyn, 1966.

Antenna with the first flagellomere almost equal to the remaining part of flagellomeres. Forewing with pterostigma sclerotized basally, and having a deep break; vein Sc connected to R before origin of Rs. Ovipositor with sheath thick and short.

Distribution and age: Liaoning; Early Cretaceous.

Four species included from the Cretaceous of Northern China (see Table 22.1).

Anthoxyela Rasnitsyn, 1977

Anthoxyela Rasnitsyn, 1977, *Paleontol. Zhur.*, 11 (3), 99 [41] (original designation).

Type species: *Anthoxyela baissensis* Rasnitsyn, 1977.

Antenna with the first flagellomere thicker and slightly longer than following flagellomeres. Forewing with pterostigma sclerotized basally only; vein Sc connected to R only for a very short distance beyond origin of Rs; hind wing with basal section of Rs longer, the cell 1Cu broad and basal part of vein cu-a extraordinarily bent.

Distribution and age: Liaoning; Early Cretaceous.

Only one species included from the Cretaceous of Northern China (see Table 22.1).

Yanoxyela Ren, Lu, Guo & Ji, 1995

Yanoxyela Ren, 1995, Faunae and Stratigraphy of Jurassic-Cretaceous in Beijing and the Adjacent Areas. Seismie Publishing House, Beijing. 107 [42] (original designation).

Type species: *Yanoxyela hongi* Ren, Lu, Guo & Ji, 1995.

The specific epithet is dedicated to Dr. Youchong Hong for being an excellent paleoentomology scientist and supervisor. He has made great contributions to the improvement of paleoentomology and provided guidance and inspiration to his student Dr. Dong Ren. Forewing with 1r-rs equal to 2r-rs in length; vein Sc close to R but not merging into it. Posterior margin of pronotum arched.

Distribution and age: Hebei; Early Cretaceous.

Only one species included from the Cretaceous of Northern China (see Table 22.1).

Ceratoxyela Zhang & Zhang, 2000

Ceratoxyela Zhang & Zhang, 2000, *Acta Palaeontol. Sin.*, 39 (4), 483 [43] (original designation).

Type species: *Ceratoxyela decorosa* Zhang & Zhang, 2000.

Antenna with terminal part of flagellum longer than basal three flagellomeres combined. Forewing with pterostigma greatly sclerotized basally; basal section of Rs slightly shorter than that of M. Ovipositor with sheath small and slightly longer than wide.

Distribution and age: Liaoning; Early Cretaceous.

Only one species included from the Cretaceous of Northern China (see Table 22.1).

Heteroxyela Zhang & Zhang, 2000

Heteroxyela Zhang & Zhang, 2000, *Acta Palaeontol. Sin.*, 39 (4), 484–485 [43] (original designation).

Type species: *Heteroxyela ignota* Zhang & Zhang, 2000.

Antenna with terminal part of flagellum shorter than the first flagellomere. Forewing with pterostigma sclerotized completely; basal section of Rs as long as that of M. Ovipositor short, sheath comparatively slender and twice as long as wide.

Distribution and age: Liaoning; Early Cretaceous.

Only one species included from the Cretaceous of Northern China (see Table 22.1).

Isoxyela Zhang & Zhang, 2000

Isoxyela Zhang & Zhang, 2000, *Acta Palaeontol. Sin.*, 39 (4), 487 [43] (original designation).

Type species: *Isoxyela rudis* Zhang & Zhang, 2000.

Antenna with terminal part of flagellum greatly shorter than the first flagellomere. Forewing with pterostigma sclerotized but membranous basally; basal section of Rs shorter than that of M. Ovipositor with sheath comparatively thick, gradually becoming thin toward its end and greatly longer than wide.

Distribution and age: Liaoning; Early Cretaceous.

Only one species included from the Cretaceous of Northern China (see Table 22.1).

Lethoxyela Zhang & Zhang, 2000

Lethoxyela Zhang & Zhang, 2000, *Acta Palaeontol. Sin.*, 39 (4), 481 [43] (original designation).

Type species: *Lethoxyela excurva* Zhang & Zhang, 2000.

Antenna with terminal part of flagellum distinctly longer than the first flagellomere. Forewing with pterostigma sclerotized basally and membranous distally; basal section of Rs slightly shorter than that of M. Ovipositor short with sheath slightly longer than wide.

Distribution and age: Liaoning; Early Cretaceous.

Two species included from the Cretaceous of Northern China (see Table 22.1).

Liaoxyela Zhang & Zhang, 2000

Liaoxyela Zhang & Zhang, 2000, *Acta Palaeontol. Sin.*, 39 (4), 484 [43] (original designation).

Type species: *Liaoxyela antiqua* Zhang & Zhang, 2000.

Antenna with terminal part of flagellum longer than the first flagellomere. Forewing with pterostigma sclerotized basally and on baso-foreside, with a shallow break; basal section of Rs slightly shorter than that of M. Ovipositor short, sheath broad and short, slightly longer than wide.

Distribution and age: Liaoning; Early Cretaceous.

Only one species included from the Cretaceous of Northern China (see Table 22.1).

Sinoxyela Zhang & Zhang, 2000

Sinoxyela Zhang & Zhang, 2000, *Acta Palaeontol. Sin.*, 39 (4), 485–486 [43] (original designation).

Type species: *Sinoxyela viriosa* Zhang & Zhang, 2000.

Antenna with terminal part of flagellum no shorter than the first flagellomere. Forewing with pterostigma sclerotized but membranous basally; basal section of Rs slightly shorter than that of M. Ovipositor with sheath comparatively thin, sub-uniform.

Distribution and age: Liaoning; Early Cretaceous.

Only one species included from the Cretaceous of Northern China (see Table 22.1).

Abrotoxyela Gao, Ren & Shih, 2009

Abrotoxyela Gao, Ren & Shih, 2009, *Zootaxa*, 2094, 53 [44] (original designation).

Type species: *Abrotoxyela lepida* Gao, Ren & Shih, 2009.

Antenna with the first flagellomere almost 2.5 times as long as the rest flagellomeres. Forewing with SC triple-branched; basal section of Rs approximately as long as that of M. Ovipositor short.

Distribution and age: Inner Mongolia; Middle Jurassic.

Two species included from the Jurassic of Northern China (see Table 22.1).

Brachyoxyela Gao, Zhao & Ren, 2011

Brachyoxyela Gao, Zhao & Ren, 2011, *Acta. Geol. Sin.-Engl.*, 85 (3), 529 [45] (original designation).

Type species: *Brachyoxyela brevinodia* Gao, Zhao & Ren, 2011.

Antenna with the first flagellomere nearly equal to the remaining flagellomeres combined; flagellum with at least 15 flagellomeres. Forewing with pterostigma sclerotized completely; Sc entering into R beyond the origin of RS.

Distribution and age: Liaoning; Early Cretaceous.

Two species included from the Cretaceous of Northern China (see Table 22.1).

Platyxyela Wang, Shih & Ren, 2012

Platyxyela Wang, Shih & Ren, 2012, *Zootaxa*, 3456, 84 [46] (original designation).

Type species: *Platyxyela unica* Wang, Shih & Ren, 2012.

Antenna with the first flagellomere shorter than the following flagellomeres combined. Forewing with pterostigma sclerotized completely; vein Sc meeting R only a short distance before origin of Rs; 1-Rs shorter than 1-M. Ovipositor with sheath extending far beyond metasomal apex, and relatively wide, nearly as wide as the length of mid metasomal terga.

Distribution and age: Inner Mongolia; Middle Jurassic.

Only one species included from the Jurassic of Northern China (see Table 22.1).

Platyxyela unica Wang, Shih & Ren, 2012 (Figure 22.11)

Platyxyela unica Wang, Shih & Ren, 2012: *Zootaxa*, 3456, 84.

Locality and horizon: Daohugou, Ningcheng, Inner Mongolia, China; Middle Jurassic, Jiulongshan Formation.

The body length excluding ovipositor sheath is 13.2 mm, and forewing 12.3 mm long. Head nearly round; eyes narrow; and three ocelli small. Pronotum long, trapeziform, with shortly pilose. Forewing with

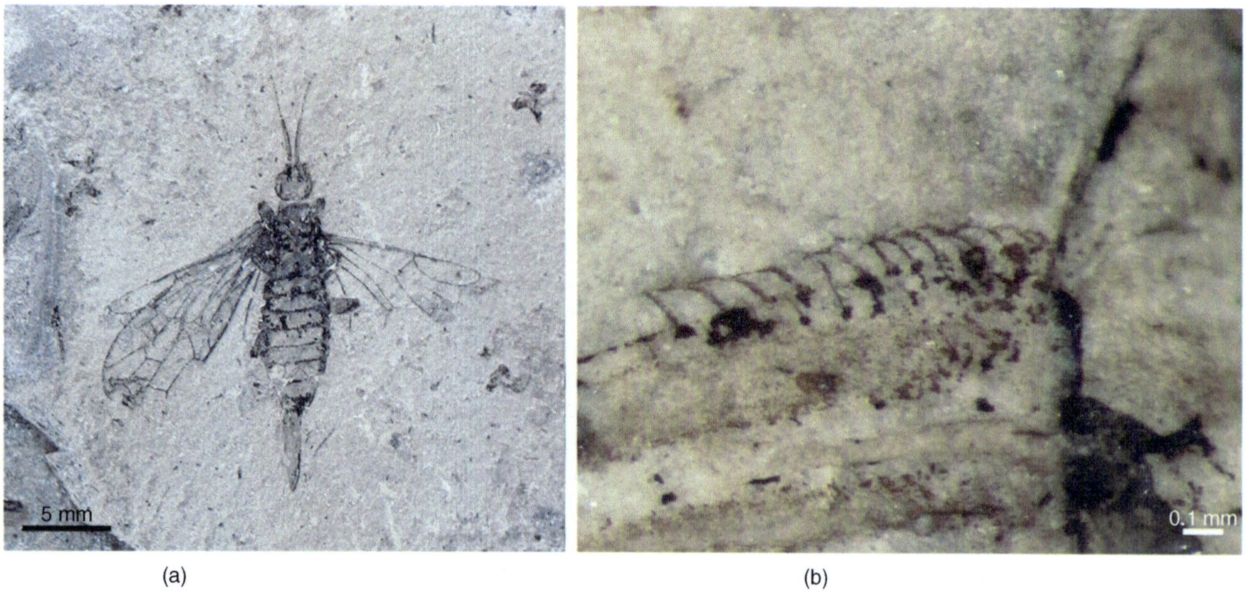

Figure 22.11 *Platyxyela unica* Wang, Shih & Ren, 2012 (Holotype, CNU-HYM-NN-2012100p) (a) Habitus; (b) Hamuli [46].

costal area obviously widen proximad of the base of Rs, posterior branch of Sc slightly longer than the basal section of Rs. Hind wing with 21 hamuli (Figure 22.11) arranged in two rows, 13 long and thick hamuli inserting at the anterior edge of C, and possibly eight shorter and thinner hamuli inserting slightly behind the first row [46].

***Cathayxyela* Wang, Rasnitsyn & Ren, 2014** *Cathayxyela* Wang, Rasnitsyn & Ren, 2014, *Acta Geol. Sin.-Engl.*, 88 (4), 1028–1029 [47] (original designation).

Type species: *Cathayxyela extensa* Wang, Rasnitsyn & Ren, 2014.

Antenna with the first flagellomere longer than head, with at least 12 antennomeres. Forewing with pterostigma pale and narrow; R almost straight; Sc touching R except apically, meeting C just before junction of 1-Rs and 1-M.

Distribution and age: Inner Mongolia; Middle Jurassic.

Only one species included from the Jurassic of Northern China (see Table 22.1).

***Cathayxyela extensa* Wang, Rasnitsyn & Ren, 2014 (Figure 22.12)**

Cathayxyela extensa Wang, Rasnitsyn & Ren, 2014: *Acta Geol. Sin.-Engl.*, 88 (4), 1029–1030.

Locality and horizon: Daohugou, Ningcheng, Inner Mongolia, China; Middle Jurassic, Jiulongshan Formation.

The body length excluding antennae is 7.5 mm, and forewing length up to end of cell 4r, 5.8 mm. Antenna with the first flagellomere straight, and nearly twice as long as the head. Mesopostnotum with a round foramen centrally; metanotum with cenchri comparatively long and wide. Forewing with the first segment of Rs (1-Rs) 0.5 (left wing) and 0.7 (right wing) times as long as the first segment of M (1-M) due to asymmetry; 1r-rs shorter than 2r-rs; 3-Cu at least 1.5 times as long as 1m-cu. The abdomen only slightly wider than the mesonotum, terga with no laterotergite separation apparent on the dorsal surface [47].

Figure 22.12 *Cathayxyela extensa* Wang, Rasnitsyn & Ren, 2014 (Holotype, CNU-HYM-NN-2012106p) [47].

Superfamily Tenthredinoidea Latreille, 1802

Family Xyelotomidae Rasnitsyn, 1968

Xyelotomidae, a small extinct family belonging to Tenthredinoidea, are reported to originate from Xyelidae [23]. They are typified by having xyelid-like antenna and simple vein Sc in the forewing. To date, 15 genera containing 21 species have been described from the Lower Jurassic to the Lower Cretaceous in Kazakhstan, France, England, Spain, and China [22, 23, 41, 42, 48–51].

Genera included from the Jurassic and Cretaceous of Northern China: *Xyelotoma* Rasnitsyn, 1968, *Xyelocerus* Rasnitsyn, 1968, *Liaotoma* Ren, Lu, Guo & Ji, 1995, *Abrotoma* Gao, Ren & Shih, 2009, *Paradoxotoma* Gao, Ren & Shih, 2009, *Synaptotoma* Gao, Ren & Shih, 2009 and *Aethotoma* Gao, Shih, Engel & Ren, 2016

***Xyelotoma* Rasnitsyn, 1968**

Xyelotoma Rasnitsyn, 1968, New Mesozoic sawflies (Hymenoptera, Symphyta). In Rohdendorf B. B. ed. Jurassic insects of Karatau. Nauka Press, Moscow, 225 [50] (original designation).

Type species: *Xyelotoma nigricornis* Rasnitsyn, 1968.

Antenna with the first flagellomere slightly shorter than head, almost twice as long as the rest of flagellum. Forewing with Sc present; R greatly thickening before the fully sclerotized pterostigma; the first free abscissa of Rs shorter than one sixth of that of M.

Distribution and age: Inner Mongolia; Middle Jurassic.

Only one species included from the Jurassic of Northern China (see Table 22.1).

***Xyelocerus* Rasnitsyn, 1968**

Xyelocerus Rasnitsyn, 1968, New Mesozoic sawflies (Hymenoptera, Symphyta). In Rohdendorf B. B. ed. Jurassic insects of Karatau. Nauka Press, Moscow, 226 [50] (original designation).

Type species: *Xyelocerus admirandus* Rasnitsyn, 1968.

Antenna with the first flagellomere nearly triple as long as the rest of flagellum. Forewing with pterostigma sclerotized but with a clear area at the center; Sc present and divided into two parts, with its basal section running into R distal to M separated from Cu.

Distribution and age: Inner Mongolia; Middle Jurassic.

Only one species included from the Jurassic of Northern China (see Table 22.1).

Liaotoma Ren, Lu, Guo & Ji, 1995

Liaotoma Ren, Lu, Guo & Ji, 1995, Faunae and Stratigraphy of Jurassic-Cretaceous in Beijing and the Adjacent Areas. Seismie Publishing House, Beijing. 110 [42] (original designation).

Type species: *Liaotoma linearis* Ren, Lu, Guo & Ji, 1995.

Forewing with pterostigma sclerotized basally; SC forming a cross-vein between C and R; 2r-rs present; first abscissa of Rs absent; and cell 1 mcu broad, almost equal to cell 2 mcu in size.

Distribution and age: Liaoning; Early Cretaceous.

Only one species included from the Cretaceous of Northern China (see Table 22.1).

Abrotoma Gao, Ren & Shih, 2009

Abrotoma Gao, Ren & Shih, 2009, *Ann. Entomol. Soc. Am.*, 102 (4), 592 [52] (original designation).

Type species: *Abrotoma robusta* Gao, Ren & Shih, 2009.

Forewing with pterostigma elongate, narrow but fully sclerotized. Forewing with Sc long, parallel to R, closer to R than to C, Sc2 lacking, Sc1 entering C just a short distance before Rs.

Distribution and age: Inner Mongolia; Middle Jurassic.

Only one species included from the Jurassic of Northern China (see Table 22.1).

Paradoxotoma Gao, Ren & Shih, 2009

Paradoxotoma Gao, Ren & Shih, 2009, *Ann. Entomol. Soc. Am.*, 102 (4), 591 [52] (original designation).

Type species: *Paradoxotoma tsaiae* Gao, Ren & Shih, 2009.

The specific epithet is dedicated to Ms. Nancy Chung-Nan Tsai, for her dedication and contribution to Chemical Engineering and for providing friendship and inspiration to Dr. Chungkun Shih. Forewing with pterostigma strongly sclerotized and relatively broad; Sc closer to C than to R, Sc2 longer, nearly vertical to R; first free abscissa of Rs preserved, nearly 1/4 as long as that of M.

Distribution and age: Inner Mongolia; Middle Jurassic.

Only one species included from the Jurassic of Northern China (see Table 22.1).

Synaptotoma Gao, Ren & Shih, 2009

Synaptotoma Gao, Ren & Shih, 2009, *Ann. Entomol. Soc. Am.*, 102 (4), 593 [52] (original designation).

Type species: *Synaptotoma limi* Gao, Ren & Shih, 2009.

The specific epithet is dedicated to John Lim, for providing guidance, friendship and inspiration to Dr. Chungkun Shih. Antenna with the first flagellomere nearly five times as long as the rest of flagellum. Forewing with Sc present as an intercostal cross-vein; pterostigma only sclerotized basally; abscissa of Rs missing, the first part of M converging with R, forming a uniquely combined vein.

Distribution and age: Inner Mongolia; Middle Jurassic.

Only one species included from the Jurassic of Northern China (see Table 22.1).

Aethotoma Gao, Shih, Engel & Ren, 2016

Aethotoma Gao, Shih, Engel & Ren, 2016, *BMC Evol. Biol.*, 16 (1), 155 [53] (original designation).

Type species: *Aethotoma aninomorpha* Gao, Shih, Engel & Ren, 2016.

Antenna with 10 antennomeres, enlarged first flagellomere about four times as long as remaining flagellomeres combined. Forewing with Sc simple, terminating into R, lacking a separate cross-vein like distal part; pterostigma fully sclerotized; first abscissa of Rs about 1/5 of that of M. Hind wing Sc with two branches.

Distribution and age: Inner Mongolia; Middle Jurassic.

Only one species included from the Jurassic of Northern China (see Table 22.1).

Aethotoma aninomorpha Gao, Shih, Engel & Ren, 2016 (Figure 22.13)

Aethotoma aninomorpha Gao, Shih, Engel & Ren, 2016: *BMC Evol. Biol.*, 16 (1), 155.

Locality and horizon: Daohugou, Ningcheng, Inner Mongolia, China; Middle Jurassic, Jiulongshan Formation.

The body length is 10.2 mm as preserved. Forewing with 2r-rs of left forewing forking into two short braches reaching Rs; right forewing with 2r-rs forming a small loop, then forking into two long braches reaching Rs in contrast to a linear 2r-rs in typical fossil and extant sawflies. Such rare asymmetrical venation found from fossil sawflies provides a glance at early occurrences of venational variability and instability, or possibly aberrational development, for insects in the latest Middle Jurassic. Hind wing has at least 12 basal hamuli visible at about 1/6 of hind wing length near basal part of vein C, traces of distal and intermediate group hamuli visible (but details not clear) [53].

Figure 22.13 *Aethotoma aninomorpha* Gao, Shih, Engel & Ren, 2016 (Holotype, CNU-HYM-NN-2012003) [53].

Asymmetrical Venation Indicating Variability and Instability or Possible Aberrational Development

The *Aethotoma aninomorpha* Gao, Shih, Engel & Ren, 2016 exhibits its unique asymmetrical venation, that is, the cross-vein 2r-rs of the right forewing forms a loop, then forks into two long branches reaching Rs (Figure 22.14c,f), while the 2r-rs of the left forewing forks into two short branches reaching Rs (Figure 22.14b,e), in contrast to a linear 2r-rs present in other fossil and extant sawflies. The previously described xyelotomid *Xyelocerus diaphanous* Gao, Ren & Shih, 2009, from the same locality, possesses a small cell of vein Rs between veins R and M in the left forewing (Figure 22.14h,i,k), but not in the right forewing (Figure 22.14h,j,l) [52].

A few other cases of clearly recognizable venational asymmetry on individual insects have been documented from the Mesozoic of China. *Paristopsyche angelineae* Qiao, Shih, Petrulevičius & Ren, 2013, a choristopsychid in Mecoptera, has vein MP_3 with two branches on the right forewing, instead of the typical one branch on the left forewing [54] (Figure 22.15a–c); the bittacid *Exilibittacus lii* Yang, Ren & Shih, 2012 has RP + MA and MP of its left hind wing bearing only three branches and RP_{1+2} and MP_3 not forking, even though RP + MA and MP of its left and right forewings has the typical four branches of most hangingflies [55] (Figure 22.15d–g); in the paratype of *Synapocossus sciacchitanoae* Wang, Shih & Ren, 2012, a palaeontinid in Homoptera, from Daohugou, China, the right forewing has 1 mm of coalescence of RP with branch M_1, but only a point contact on the left forewing [56] (Figure 22.15h–j); and differences between the left and right hind wings of the plecopteran *Sinosharaperla zhaoi* Liu, Sinitshenkova & Ren, 2007 [57] (Figure 22.15k–m).

Asymmetrical shape and size of the left and right wings on a specimen of *Epicharmesopsyche pentavenulosus* Shih, Qiao, Labandeira & Ren, 2013 seems to be a common condition for mesopsychid taxa (Mecoptera) from Northeastern China [58]. Notably, the asymmetrical wing shape and size of the left wings are broader than the right wings, which seems to be common for mesopsychids as reported by Ren *et al.* [59, 60] and in other specimens collected after the publication of these two papers. Asymmetrically unequal wings also occur in the dipteran family Ptychopteridae, as reported for *Eoptychopterina elenae* Ren & Krzemiński, 2002 [61], and for *E. postica* Liu, Shih & Ren, 2012 [62].

The finding of *Aethotoma aninomorpha* and other documented asymmetrical examples from the Mesozoic of Northeastern China provide a glimpse of early occurrences of venational variability and instability, or possibly aberrational development, for insects in the late Middle Jurassic. It is suggested that more attention should be paid to recognize and distinguish whether a venational character is an aberration or a diagnostic character in fossil insect classification, when we carry out taxonomy research based on a single wing of the aforementioned groups of insects.

Superfamily Pamphilioidea Cameron, 1890

Family Xyelydidae Rasnitsyn, 1968

Xyelydidae, a relatively little-known extinct family of sawflies, have long been regarded as a basal group within superfamily Pamphilioidea [1, 22, 23, 63]. However, the non-monophyly of Xyelydidae was confirmed based on phylogenetic analyses using morphological characters [64]. It has the following plesiomorphic characters: vein Sc with two branches and vein R being angular or bent near the origin of 1-Rs in the forewings and the mesopseudosternum positioned distantly from the fore-margin of the ventropleuron [63]. To date, 12 genera and 32 species of xyelydids have been reported [64], most of which are distributed in Kyrgyzstan, China, Kazakhstan and Russia. Most xyelydids have been reported from the Jurassic, with the oldest representatives *Sagulyda* spp. and *Ferganolyda* spp. from the Lower or Middle Jurassic Sogul Formation in Kyrgyzstan [63, 65, 66]. Only three genera (*Novalyda, Fissilyda* and *Rectilyda*) of xyelydids have been described in the Early

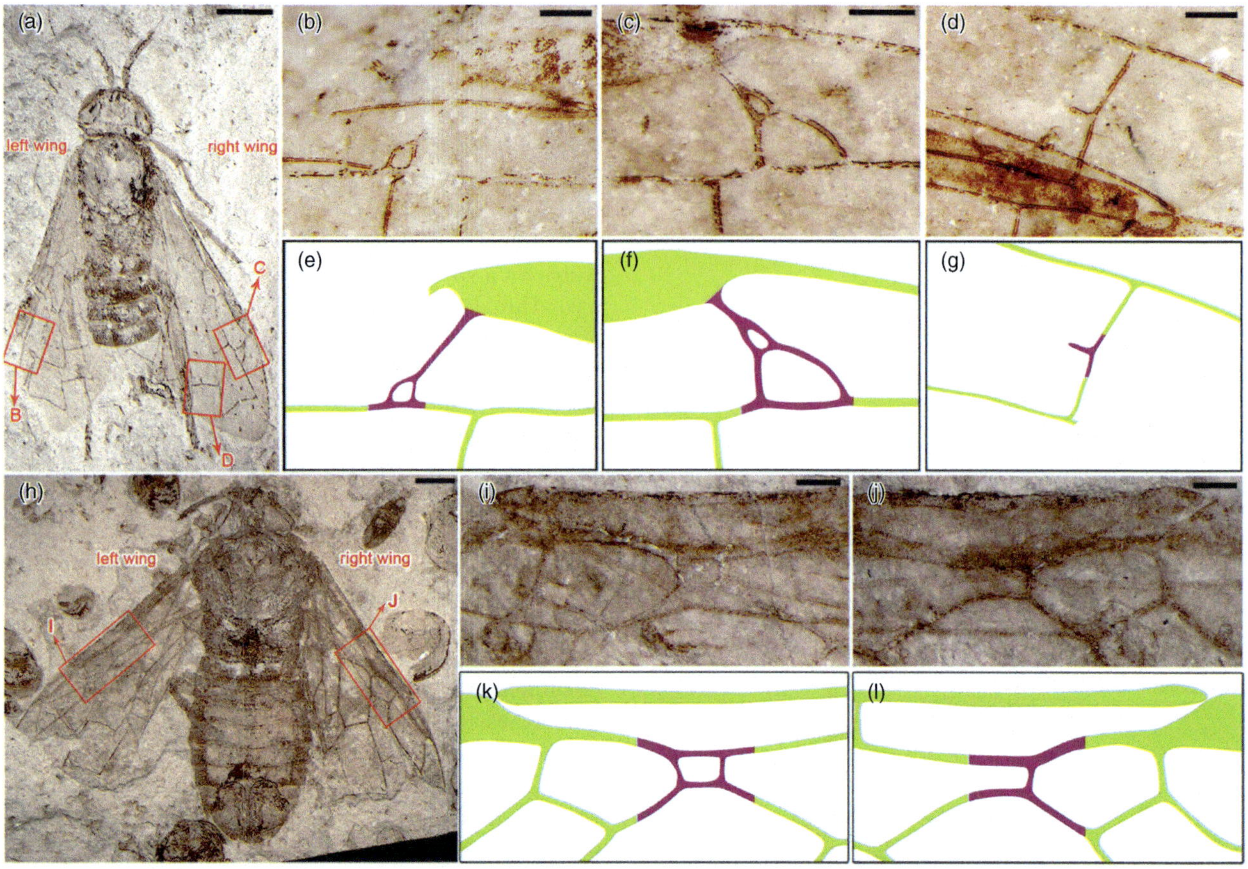

Figure 22.14 *Aethotoma aninomorpha* Gao, Shih, Engel & Ren, 2016 and *Xyelocerus diaphanous* Gao, Ren & Shih, 2009 display fluctuating asymmetry in the forewings, purple color highlights asymmetrical veins. (a) Holotype of *A. aninomorpha*; (b, e) Vein 2r-rs of the left forewing; (c, f) Vein 2r-rs of the right forewing of *A. aninomorpha*; (d, g) 2m-cu of the right forewing of *A. aninomorpha*; (h) Holotype of *X. diaphanus* from the same locality (No. CNU-HYM-NN-2008011p); (i, k) Part of the left forewing of *X. diaphanus*; (j, l) Part of right forewing of *X. diaphanus*, showing normal venation. Scale bars = 2 mm (a, h), 0.4 mm (b–d), 0.2 mm (i, j).

Cretaceous [67–69], which are the latest occurrences of xyelydids in the fossil records.

Genera included from the Jurassic and Cretaceous of Northern China: *Prolyda* Rasnitsyn, 1968, *Ferganolyda* Rasnitsyn, 1983, *Novalyda* Gao, Engel, Shih & Ren, 2013, *Rectilyda* Wang, Rasnitsyn, Shih & Ren, 2014, *Fissilyda* Wang, Rasnitsyn, Shih & Ren, 2015, *Medilyda* Wang & Rasnitsyn, 2016, *Brevilyda* Wang & Rasnitsyn, 2016 and *Strenolyda* Wang & Rasnitsyn, 2016.

Prolyda Rasnitsyn, 1968

Prolyda Rasnitsyn, 1968, New Mesozoic sawflies (Hymenoptera, Symphyta). In Rohdendorf B. B. ed. Jurassic insects of Karatau. Nauka Press, Moscow. 194 [50] (original designation).

Type species: *Prolyda karatavica* Rasnitsyn, 1968.

Antenna with the first flagellomere equal to head in length, but eight times as long as the second flagellomere. Forewing with pterostigma variable, completely sclerotized or partly sclerotized, or just membranous; M diverging from M+Cu at much larger angle than Cu; hind wing with 1r-m rather long, as long as or slightly shorter than 1-M.

Distribution and age: Inner Mongolia; Middle Jurassic.

Two species included from the Jurassic of Northern China (see Table 22.1, Figure 22.16).

Ferganolyda Rasnitsyn, 1983

Ferganolyda Rasnitsyn, 1983, *Paleontol. J.*, 2, 62 [63] (original designation).

Type species: *Ferganolyda cubitalis* Rasnitsyn, 1983.

Sexual dimorphism high in body size (female smaller) and in morphology of head and antenna. Head grotesquely widened, with mandible long, falcate, two-toothed as in Pamphiliidae; clypeus short, very wide; antenna with fourth segment enlarged, the same as the third one. Forewing with Sc having hind branch short, subvertical or weakly oblique; pterostigma variable (sclerotized on margins or entirely); first abscissa of Rs short or missing; cell 1mcu asymmetrical basally with fore margin straight.

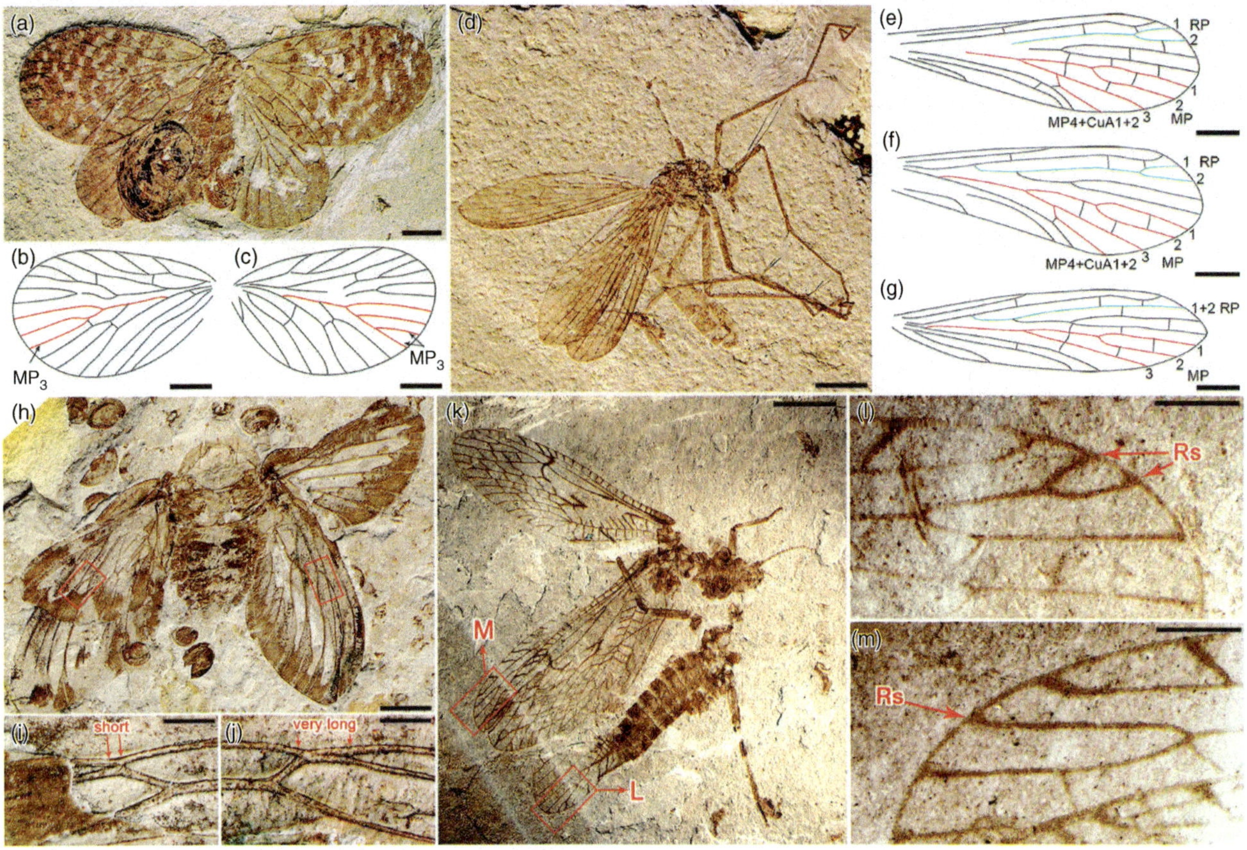

Figure 22.15 Several examples of fossil insects with wings possessing venational asymmetry reported from the Mesozoic of China. (a–c) *Paristopsyche angelineae* within Mecoptera showing the different branches of vein MP_3 between left and right forewings. (a) photograph of the specimen; (b) Line drawing of left forewing, vein MP_3 with typical one branch; (c) Line drawing of right forewing, vein MP_3 with two branches; (d–g) *Exilibittacus lii* within Mecoptera showing the different vein RP and MP on fore and hind wings. (d) Photograph of the specimen; (e, f) Right and left forewings with RP_{1+2} forking and MP with four branches; (g) Left hind wing with RP_{1+2} not forking and MP with three branches; (h–j) *Synapocossus sciacchitanoae* within Homoptera showing the RP coalesced with M_1 on right and left forewings. (h) Photograph of the specimen; (i) Left wing with the Rs coalesced with M_1 at a point; (j) Right wing with the RP coalesced with M_1 for about 1 mm; (k–m) *Sinosharaperla zhaoi* within Plecoptera showing different Rs forking on right and left forewings. (k) Photograph of the specimen; (l) Part of right hind wing showing Rs with two terminal branches; (m) Part of left hind wing, Rs with only one branch. Scale bar = 5 mm (H and K), 2 mm (a–d), 1 mm (e–g, i, j, l, and m).

Figure 22.16 A 3-D ecological reconstruction of *Prolyda* Rasnitsyn, 1968. Source: Modified from [64]. Artwork by Dr. Chen Wang.

Distribution and age: Inner Mongolia; Middle Jurassic.

Five species included from the Jurassic of Northern China (see Table 22.1).

Ferganolyda insolita **Wang, Rasnitsyn, Shih & Ren, 2015 (Figure 22.17)**

Ferganolyda insolita Wang, Rasnitsyn, Shih & Ren, 2015: *Alcheringa*, 39, 104–106.

Locality and horizon: Daohugou, Ningcheng, Inner Mongolia, China; Middle Jurassic, Jiulongshan Formation.

Body is large, 14.3 mm in length excluding antennae, forewing length up to end of cell 3r, 11.9 mm. Head capsule widest distal to midlength, with hind contour biconvex with deep central excision about 0.6 mm deep.

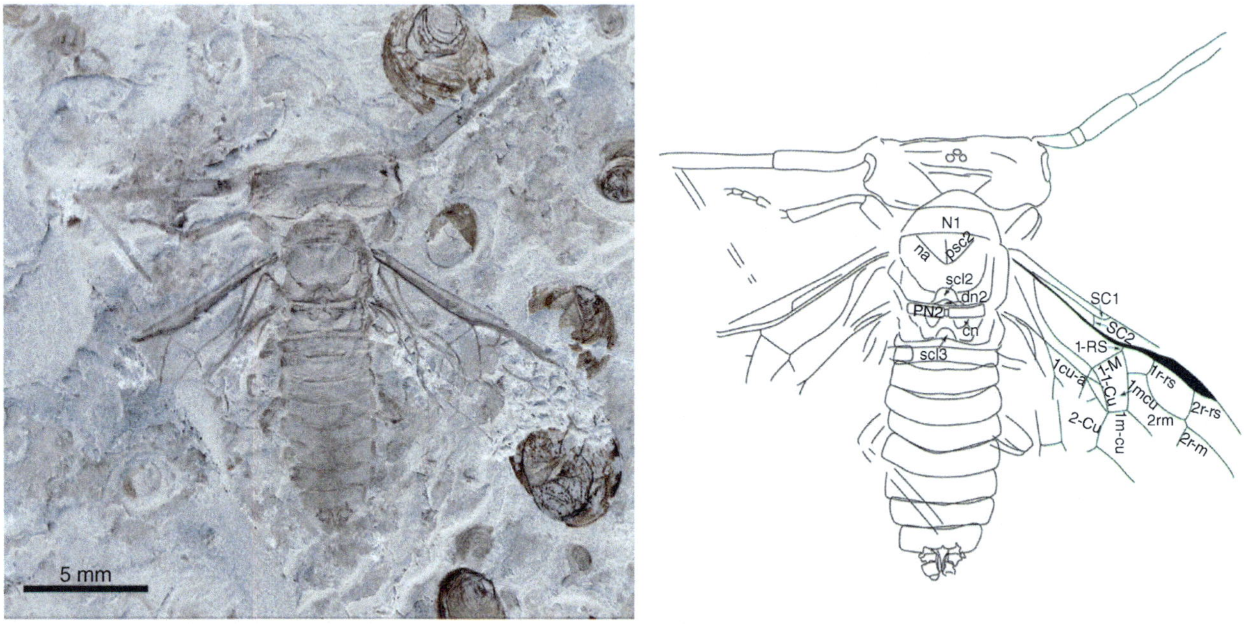

(a) (b)

Figure 22.17 *Ferganolyda insolita* Wang, Rasnitsyn, Shih & Ren, 2015 (Holotype, CNU-HYM-NN-2012136) (Male) (a) Habitus; (b) Line drawing [66].

Antenna with the second flagellomere nearly twice as long as, and 0.8 times as wide as, the first flagellomere. Forewing with pterostigma narrow and sclerotized completely; Sc2 subvertical and long, nearly equal to the length of Sc1; 1-Rs not distinct or only present as a very short stub; M+Cu with a S-like bend at base. Male genitalia distinct; gonocoxa trapezoidal, relatively small and short, about 0.7 times as long as wide, posterior margin curved inward; gonostyle quadrangular, large and long, 1.4 times as long as wide, with apical membranous disc; penial valve long, approaching subrectangular [66].

Various morphological characters of *Ferganolyda* from the Yanliao Biota, Jiulongshan Formation of Northeastern China during the latest Middle Jurassic highlight their broad diversity [65, 66]. On the size of the head and the shape of the fourth antennomere, Rasnitsyn *et al.* in 2006, attributed them to sexual dimorphism in Hymenoptera [65]. Based on our new specimens, we conclude that males of *Ferganolyda*, have greater ratios (3.1 to 3.6) of head width/length than those for females (1.9 to 2.1). This significant dimorphism can be used as a diagnostic character to distinguish the sexes when genitalia are not preserved or not discernible.

Novalyda Gao, Engel, Shih & Ren, 2013

Novalyda Gao, Engel, Shih & Ren, 2013, *J. Kansas Entomol. Soc.*, 86, 79 [67] (original designation).

Type species: *Novalyda cretacica* Gao, Engel, Shih & Ren, 2013.

Antenna with 15 antennomeres, first flagellomere just slightly shorter than following flagellomeres combined. Mandibles strong, possessing at least two teeth. Forewing with Sc two branches, Sc1 intersecting C before Rs divergence; first abscissa of Rs shorter than Sc2, about one-sixth of basal section of M.

Distribution and age: Liaoning; Early Cretaceous.

Two species included from the Cretaceous of Northern China (see Table 22.1).

Novalyda cretacica Gao, Engel, Shih & Ren, 2013 (Figure 22.18)

Novalyda cretacica Gao, Engel, Shih & Ren, 2013: *J. Kansas Entomol. Soc.*, 86, 79–82.

Locality and horizon: Huangbanjigou, Beipiao, Liaoning, China; Lower Cretaceous, Yixian Formation.

Body is 8.72 mm in length excluding antennae, forewing 6.83 mm long as preserved. Pronotum short; mesoscutum large, anterior edge nearly straight, with short notauli at an angle of 115°; mesoscutellum trapezoidal, nearly as large as metascutellum, but of different shape, the former oval but the latter trapezoidal [67].

Extant pamphilioids are reported to use their mandibles as weapons for defense or probably used in male-female interactions. However, the mandibles of *N. cretacica* are much shorter, broader, and seemingly flexible at the base. Given the general biology of pamphilioids and other sawflies, Gao *et al.* in 2013, concluded that it is likely that the mandibles in *N. cretacica* might have been used for boring out of a particular substrate,

Figure 22.18 *Novalyda cretacica* Gao, Engel, Shih & Ren, 2013 (Holotype, CNU-HYM-LB-2011016) [67].

such as wood or stems [67]. It was hypothesized that the larvae of *N. cretacica* might have lived inside contemporaneous gymnosperms, and thereby used their strong mandibles to excavate tunnels in wood, and escaped from these burrows after pupation. Certainly, this would not exclude their use of the same mandibles for male-female interactions.

Rectilyda Wang, Rasnitsyn, Shih & Ren, 2014

Rectilyda Wang, Rasnitsyn, Shih & Ren, 2014, *BMC Evol. Biol.*, 14, 131 [68] (original designation).

Type species: *Rectilyda sticta* Wang, Rasnitsyn, Shih & Ren, 2014.

Antenna with about 17 antennomeres; first flagellomere shorter than remaining flagellomeres combined. Forewing with Sc bifurcate; pterostigma narrow and long, sclerotized completely; 1-Rs reclival, as long as 1-M and linearly aligned with 1-M; 1-M nearly equal to 2-M in length. Hind wing with Sc1 and Sc2 present.

Distribution and age: Inner Mongolia; Early Cretaceous.

Only one species included from the Cretaceous of Northern China (see Table 22.1).

Rectilyda sticta Wang, Rasnitsyn, Shih & Ren, 2014 (Figure 22.19) *Rectilyda sticta* Wang, Rasnitsyn, Shih & Ren, 2014: *BMC Evol. Biol.*, 14, 131.

Locality and horizon: Nanpanying, Duolun, Inner Mongolia, China; Lower Cretaceous, Yixian Formation.

Length of body excluding antennae is 24.0 mm, forewing as preserved with length up to the end of cell 3r 17.5 mm. Entire body and all legs (excluding tarsi) are covered by dense, long and dark setae. Mandible small, sickle-shaped, with at least one inner tooth. Antenna with the remaining flagellomeres nearly as long as wide, gradually shorter and narrower toward the apex. Forewing with cell 3r at least 1.5 times as long as cells 1r and 2r. Hind wing with cross-vein m-cu distal to middle of cell rm; cu-a before the middle of cell mcu [68].

Enigmatic Nygmata

Nygmata, was once considered as the "facetic organs" in Trichoptera, and have sometimes been referred to as "presumed sensory spots" from 1989 to 1991 [70, 71]. The idea of nygmata as a groundplan character of Holometabola has been recently reiterated by Minet *et al.* (2010) [72].

Nygmata are only present in Symphyta among the extant hymenopterans. Our observations confirm that nygmata occur in almost every family of Symphyta, but not necessary in every genus. Generally, the families with the highest number of cells with nygmata are in the extant Pamphiliidae (up to five in the forewing and four in the hind wing) and Siricidae (up to five in the forewing and only two in the hind wing). The lowest number of cells with nygmata is recorded in the Cephidae (one and zero for the fore- and hind wings, respectively). Other families of Symphyta have intermediate numbers of cells. The nygmata of Symphyta are characteristically found in the forewing cells of 1rm and 2rm and in the hind wing cell of 2 + 3rm. Nygmata in the forewing 2rm and hind wing 2 + 3rm cells are the most stable (Figure 22.20).

Besides, there are seven forewing nygmata of *Parasialis latipennis* Ponomarenko, 1977 in Parasialidae in the Permian (taken as a rough model of a hymenopteran ancestor) (Figure 22.21a), six nygmata for *Asioxyela paurura* and *Madygenius primitives* in Xyelidae in the Triassic (Figure 22.21b), four nygmata for *Rectilyda sticta* Wang, Rasnitsyn, Shih & Ren, 2014 (Figure 22.21c), and one to three nygmata in most of the extant species among Symphyta (Figure 22.20a–e,g–j), except for Pamphiliidae and Siricidae, which have five (Figure 22.20f), thus we consider that there is a general decreasing trend for the number of forewing nygmata during the evolution of hymenopteran wings.

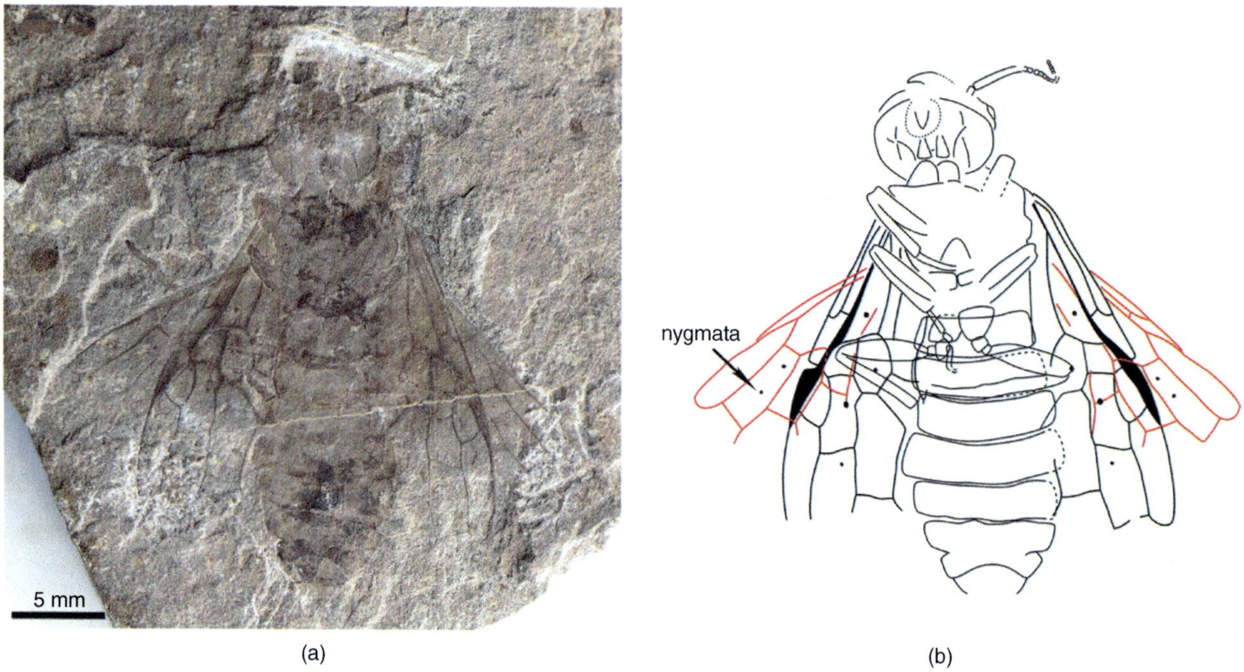

Figure 22.19 *Rectilyda sticta* Wang, Rasnitsyn, Shih & Ren, 2014 (Holotype, CNU-HYM-LB-2012125) (a) Habitus; (b) Line drawing [68].

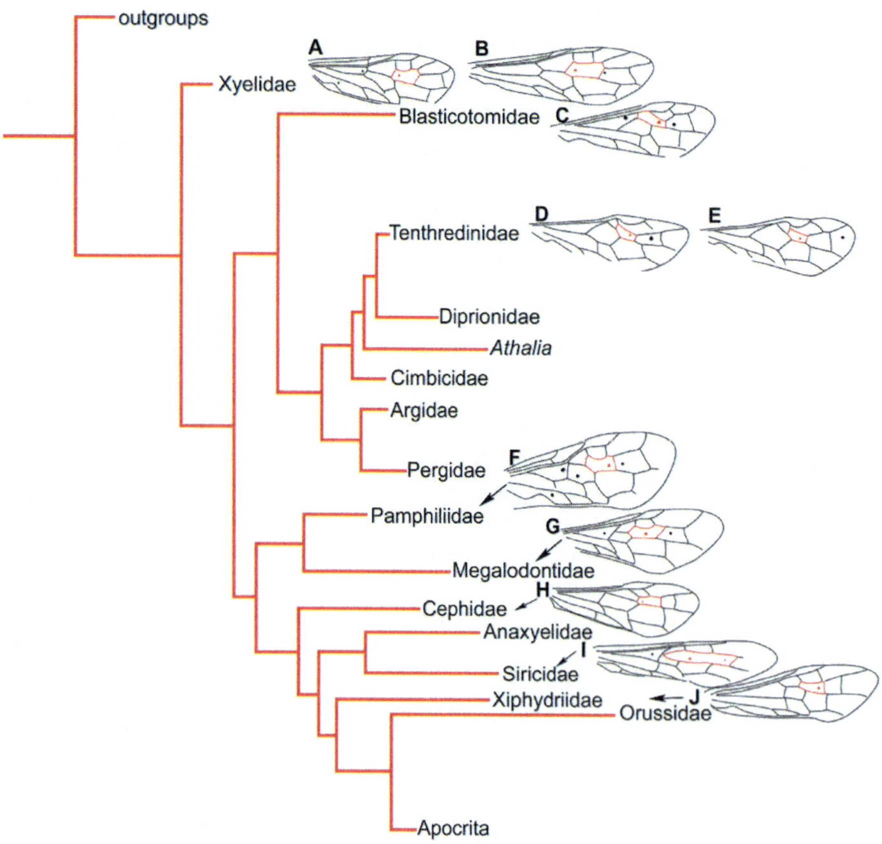

Figure 22.20 Cladogram of extant basal Hymenoptera after Ronquist *et al*. (figure 3 in [73], slightly modified), with relevant forewing venation containing nygmata among extant taxa mapped onto figure [68]. (a) *Macroxyela ferruginea* Say, 1824 (Xyelidae); (b) *Megaxyela major* Cresson, 1880 (Xyelidae); (c) *Blasticotoma filiceti* Klug, 1834 (Blasticotomidae); (d) *Empria candidata* Fallén, 1808 (Tenthredinidae); (e) *Empria formosana* Prous & Heidemaa, 2012 (Tenthredinidae); (f) *Onycholyda amplecta* Fabricius, 1804 (Pamphiliidae); (g) *Megalodontes cephalotes* Fabricius, 1781 (Megalodontidae); (h) *Cephus pygmeus* Linné, 1767 (Cephidae); (i) *Tremex columba* Linné, 1763 (Siricidae); (j) *Xiphydria camelus* Linné, 1758 (Xiphydriidae).

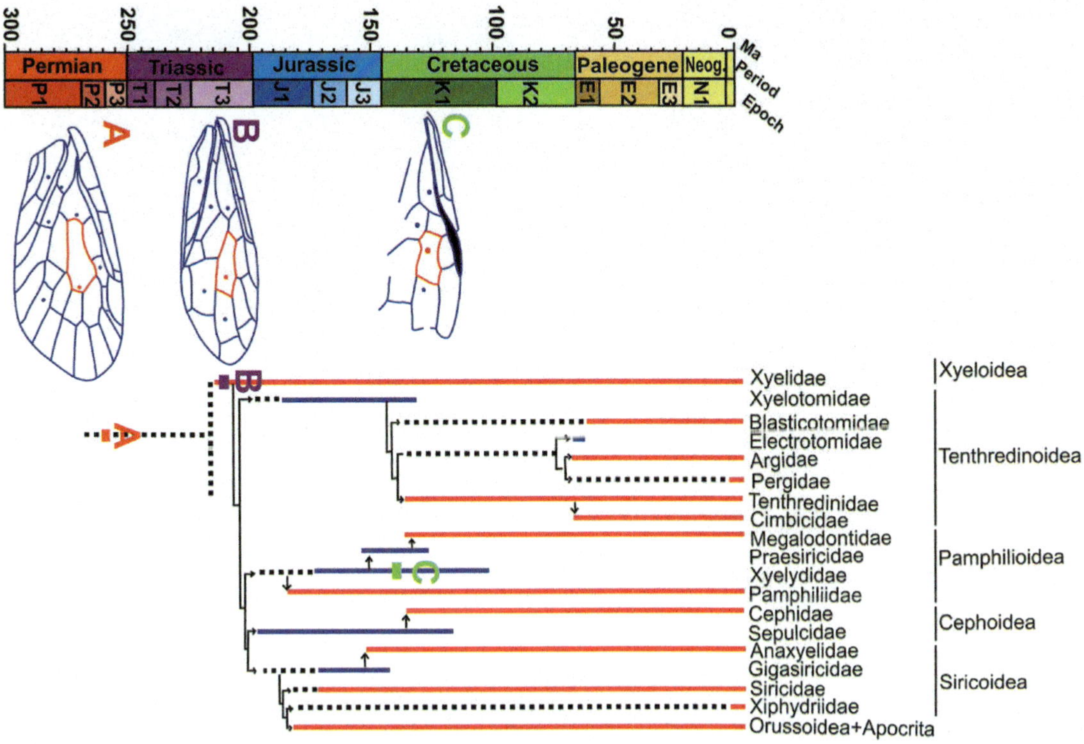

Figure 22.21 Noncladistic cladogram of basal Hymenoptera after Rasnitsyn (figure 331 in [74], partially modified), with forewing venation containing nygmata among fossil taxa mapped onto figure [68]. (a) *Parasialis latipennis* Ponomarenko, 1977 (Parasialidae); (b) Xyelidae (based on *Asioxyela paurura* Rasnitsyn, 1969 and *Madygenius primitivus* Rasnitsyn, 1969; after [75]); (c) *Rectilyda sticta* Wang, Rasnitsyn, Shih & Ren, 2015 (Xyelydidae).

***Fissilyda* Wang, Rasnitsyn, Shih & Ren, 2015**

Fissilyda Wang, Rasnitsyn, Shih & Ren, 2015, *Cretac. Res.*, 5, 171 [69] (original designation).

Type species: *Fissilyda compta* Wang, Rasnitsyn, Shih & Ren, 2015.

Antenna with at least 20 antennomeres. Forewing with pterostigma pale at base, and otherwise sclerotized completely; Sc two branched; 1-Rs short, at least 0.3 times as long as 1-M; 2r-m far from 2r-rs by at least 0.5 times of its own length; and cell 1mcu relatively large, almost from 1.35 times to 1.9 times as long as wide.

Distribution and age: Liaoning; Early Cretaceous.

Three species included from the Cretaceous of Northern China (see Table 22.1).

***Medilyda* Wang & Rasnitsyn, 2016**

Medilyda Wang & Rasnitsyn, 2016, *Cladistics*, 32 (3), 249 [64] (original designation).

Type species: *Medilyda procera* Wang & Rasnitsyn, 2016.

Head circular and massive, at least as wide as mesonotum. Forewing with pterostigma sclerotized completely, the latter narrow and long; Sc very close to R; posterior branch of Sc extremely short and almost diminished; 2r-rs meeting pterostigma almost at its mid-length; cell 1r at least as long as cell 2r. In hind wing, 1-Rs short, at most half of 1r-m.

Distribution and age: Inner Mongolia; Middle Jurassic.

Two species included from the Jurassic of Northern China (see Table 22.1).

***Medilyda procera* Wang & Rasnitsyn, 2016 (Figure 22.22)**

Medilyda procera Wang & Rasnitsyn, 2016, *Cladistics*, 32 (3), 250–251.

Locality and horizon: Daohugou, Ningcheng, Inner Mongolia, China; Middle Jurassic, Jiulongshan Formation.

Body is 15.0 mm in length excluding antennae, forewing length up to end of cell 4r, 11.1 mm. Head circular and massive, at least as wide as mesonotum. Forewing with 1cu-a located almost at the middle of cell 1mcu; RS+M slightly shorter than 2-M; 2r-m far from 2r-rs by longer than half of its own length; cell 2r rather small, almost one-third of cell 3r in length. In hind wing, 1-RS short, at most half of 1r-m [64].

Figure 22.22 *Medilyda procera* Wang & Rasnitsyn, 2016 (Holotype, CNU-HYM-NN-2012143) [64].

Figure 22.23 *Strenolyda marginalis* Wang & Rasnitsyn, 2016 (Holotype, CNU-HYM-NN-2012166) [64].

***Brevilyda* Wang & Rasnitsyn, 2016**

Brevilyda Wang & Rasnitsyn, 2016, *Cladistics*, 32 (3), 252 [64] (original designation).

Type species: *Brevilyda provecta* Wang & Rasnitsyn, 2016.

Forewing with pterostigma sclerotized completely, narrow and long, almost 4.5 times as long as wide; Sc1 almost equal to Sc2 in length; 1-Rs subvertical, more or less 0.4 times as long as 1-M; 2r-m antefurcal; cell 2r large, at least 1.5 times as long as cell 1r; cell 1mcu forming a hexagon, and at least 1.1 times as long as wide.

Distribution and age: Inner Mongolia; Middle Jurassic.

Only one species included from the Jurassic of Northern China (see Table 22.1).

***Strenolyda* Wang & Rasnitsyn, 2016**

Strenolyda Wang & Rasnitsyn, 2016, *Cladistics*, 32 (3), 254 [64] (original designation).

Type species: *Strenolyda marginalis* Wang & Rasnitsyn, 2016.

Forewing with pterostigma sclerotized completely or just around margins; Sc1 longer or almost as long as Sc2; 1-Rs short, shorter than half of 1-M; angle between 1-M and 1-Cu 100–140°; 2r-m antefurcal, sometimes just slightly postfurcal; cell 1mcu at least as long as wide, no more than 1.5 times as long as wide; cell 1r short, 0.6–0.7 times as long as cell 2r.

Distribution and age: Inner Mongolia; Middle Jurassic.

Two species included from the Jurassic of Northern China (see Table 22.1).

***Strenolyda marginalis* Wang & Rasnitsyn, 2016 (Figure 22.23)**

Strenolyda marginalis Wang & Rasnitsyn, 2016: *Cladistics*, 32 (3), 254–255.

Locality and horizon: Daohugou, Ningcheng, Inner Mongolia, China; Middle Jurassic, Jiulongshan Formation.

Length of body excluding antennae is 16.9 mm, forewing length up to end of cell 3r, 13.0 mm as preserved. Head massive and almost round; mandibles sickle-like, strong and long, reaching opposite side of head when closed, with apical tooth long, slanting sub-apical one placed basal of mandible. Hind wing with 19 hamuli present, and Sc well-developed with two branches [64].

Superfamily Pamphilioidea Cameron, 1890

Family "Praesiricidae" Rasnitsyn, 1968

Praesiricidae, an extinct family of sawflies, were first considered to be a subfamily of Siricoidea [22] based on only one compression fossil of a deformed insect, while some structures, e.g. mesothorax and ovipositor, were mistakenly interpreted before. After corrections of these mistakes, it was transferred from Siricoidea to Pamphilioidea by Rasnitsyn in 1983. There are 10 described genera and 20 species reported to date [22, 23, 50, 63, 76–79]. The earliest fossil record of Praesiricidae is *Aulidontes mandibulatus* Rasnitsyn, 1983 reported from the Upper Jurassic of Southern Kazakhstan [63].

Genera included from the Jurassic and Cretaceous of Northern China: *Rudisiricius* Gao, Rasnitsyn, Shih & Ren, 2010, *Archoxyelyda* Wang, Rasnitsyn & Ren,

2013, *Hoplitolyda* Gao, Shih, Rasnitsyn & Ren, 2013, *Decorisiricius* Wang, Rasnitsyn, Shih & Ren, 2016, *Limbisiricius* Wang, Rasnitsyn, Shih & Ren, 2016 and *Pallorisiricius* Wang, Rasnitsyn, Shih & Ren, 2016

Rudisiricius Gao, Rasnitsyn, Shih & Ren, 2010

Rudisiricius Gao, Rasnitsyn, Shih & Ren, 2010, *Ann. Soc. Entomol. Fr.*, 46 (1–2), 150 [76] (original designation).

Type species: *Rudisiricius belli* Gao, Rasnitsyn, Ren & Shih, 2010.

The specific epithet is dedicated to Mr. Nigel Bell for providing guidance, mentorship and inspiration to Dr. Chungkun Shih. Antenna with 17–24 antennomeres; scape long, thick or thickened apical, about as long as head and at least twice as long as wide; mandible sickle-shaped, robust. Forewing with 1-Rs at most half as long as Rs+M; M lacking free apex (Cu possessing it); cell 2a well-surpassing cross-vein cu-a. Hind wing with M and 1A lacking free apex (Cu possessing it); 1-M aligned with 1r-m.

Distribution and age: Liaoning; Early Cretaceous.

Nine species included from the Cretaceous of Northern China (see Table 22.1).

Rudisiricius tenellus Wang, Rasnitsyn, Shih & Ren, 2015 (Figure 22.24)

Rudisiricius tenellus Wang, Rasnitsyn, Shih & Ren, 2015: *Cretac. Res.*, 52, 574.

Locality and horizon: Huangbanjigou, Beipiao, Liaoning, China; Lower Cretaceous, Yixian Formation.

Figure 22.24 *Rudisiricius tenellus* Wang, Rasnitsyn, Shih & Ren, 2015 (Holotype, CNU-HYM-LB-2012114) [78].

Body length is 12.9 mm excluding antennae and forewing length up to the end of cell 3r, 8.9 mm. Antenna about 2.3 times as long as head width, with 20 antennomeres preserved; scape long, narrow, widened apically, more than four times (4.3–5.4 times) as long as wide; the first flagellomere about as long as following four combined, remaining flagellomeres elongate, growing shorter toward the apex. Clypeal margin slightly protruding. Forewing with pterostigma pale; R much thickening before pterostigma, about as wide as 1r-rs long; 1-Rs vertical to R (right wing) and nearly aligned with 1-M; 1-M vertical to RS+M and 1-Cu; 1cu-a interstitial [78].

An interesting feature of *Rudisiricius* is its high diversity and comparative abundance (14 specimens representing nine species) in the hymenopteran assemblage in the Early Cretaceous Jehol Biota, Yixian Formation, and nowhere else. The current data might suggest a narrow spatiotemporal distribution of the *Rudisiricius*. In addition, all reported specimens of *Rudisiricius* thus far are definite or possible males, while no definite females have been described. This is a very enigmatic case for Symphyta, because the vast majority of documented fossil symphytans are females. We are not aware of any other sufficiently abundant sawfly taxa with only male fossils preserved. We are not sure whether sedentary habits of females might be a plausible explanation of their lower probability of being fossilized at the bottom of a lake.

Archoxyelyda Wang, Rasnitsyn & Ren, 2013

Archoxyelyda Wang, Rasnitsyn & Ren, 2013, *Syst. Entomol.*, 38, 579 [77] (original designation).

Type species: *Archoxyelyda mirabilis* Wang, Rasnitsyn & Ren, 2013.

Head wider than long, with eyes about half as long as head. Mesopseudosternum triangular, wider than long, nearly as large as mesoscutum. Forewing with Rs+M almost vertical to 1-M and much longer than 1-Cu; 2-M short. Integument without coarse sculpture.

Distribution and age: Liaoning; Early Cretaceous.

Only one species included from the Cretaceous of Northern China (see Table 22.1).

Archoxyelyda mirabilis Wang, Rasnitsyn & Ren, 2013 (Figure 22.25)

Archoxyelyda mirabilis Wang, Rasnitsyn & Ren, 2013: *Syst. Entomol.*, 38, 579–581.

Locality and horizon: Huangbanjigou, Beipiao, Liaoning, China; Lower Cretaceous, Yixian Formation.

Body length is 9.4 mm excluding antennae and forewing length up to the end of cell 3r, 6.5 mm. Oral cavity (well traceable along its margins) isolated from mandibular orifices with lower clypeal parts, and lacking a sclerotized bridge between it and occipital foramen

Figure 22.25 *Archoxyelyda mirabilis* Wang, Rasnitsyn & Ren, 2013 (Holotype, CNU-HYM-LB-2012102) [77].

(postoccipital sclerites disconnected). Antennal flagellum with basal, thick portion multi-segmented and tightly connected, followed by 10–12 antennomeres less tightly connected, elongate or (subapical) subquadrate, growing gradually shorter and narrower apicad [77].

Early Development of Antenna in Basal Hymenoptera

Rasnitsyn has demonstrated that long and broad first flagellomere is a plesiomorphy of Hymenoptera [22, 23, 52, 74, 80, 81] and suggested that the xyelid-like antenna experienced transformation along different trajectories during the course of hymenopteran evolution, resulting in the first flagellomere that is not remarkably different from the remaining flagellomeres, as observed in most Hymenoptera.

Three pathways of transformation are as follows [77]:

1. Direct separation of the compound third antennomere (the first flagellomere) into its constituent antennomeres as a result of a heterochrony. This pathway is rather uncommonly realized (Figure 22.26a–d, i).
2. Simple size reduction resulted in the enlarged third antennomere looking like an ordinary flagellomere (Figure 22.26a–f,i). This trajectory seems to occur more frequently, as indicated by a number of examples [81].
3. The compound third antennomere is growing smaller, and simultaneously its constituent primary subunits become more and more clearly delimited (Figure 22.26a–e,g–i) until they are finally totally separated. This transformation series is rather uncommon [81].

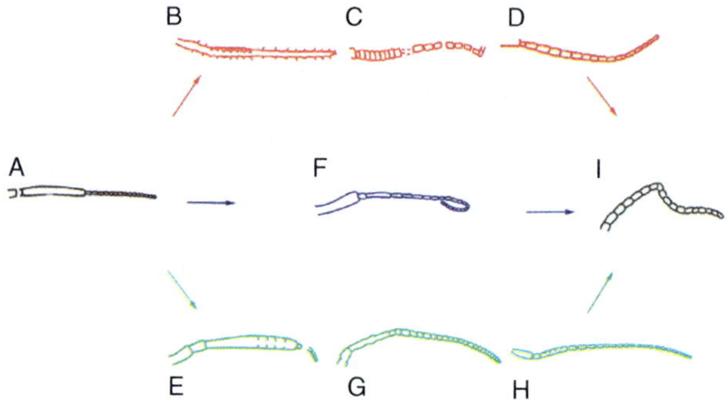

Figure 22.26 Antennae of various sawflies (modified from [77]). (a) *Platyxyela unica* Wang, Shih & Ren, 2012 (Xyelidae); (b) *Xyela julii* (Brébisson, 1818) (Xyelidae); (c) *Archoxyelyda mirabilis* (Praesiricidae); (d) *Brachysyntexis brachyuran* Rasnitsyn (1968) (Anaxyelidae); (e) *Xyelotoma macroclada* Gao, Ren & Shih, 2009 (Xyelotomidae); (f) *Rudisiricius belli* Gao, Rasnitsyn, Ren & Shih, 2010 (Praesiricidae); (g) *Cephalcia fasciipennis* (Cresson, 1880) (Pamphiliidae); (h) *Caenolyda reticulata* (Linnaeus, 1758) (Pamphiliidae); (i) *Microryssus minus* Rasnitsyn, 1968 (Paroryssidae).

Hoplitolyda Gao, Shih, Rasnitsyn & Ren, 2013

Hoplitolyda Gao, Shih, Rasnitsyn & Ren, 2013, *PLoS ONE*, 8, e62420 [3] (original designation).

Type species: *Hoplitolyda duolunica* Gao, Shih, Rasnitsyn & Ren, 2013.

Large hymenopteron with head subcircular, widest at mandibular base. Mandible sickle-shaped, with single preapical tooth. Antenna with scape moderately long and first flagellomere not disproportionally enlarged. Fore leg apparently resting on substrate with tibial apex and base of basitarsus flexed below tibia, with foretibial apical spur apparently lost. Forewing with Sc absent; R and pterostigma narrow; 1-Rs reclined, meeting 1-M at about 145°; RS+M very short; cells 2rm and 1mcu very long. Hind wing with Sc present.

Distribution and age: Inner Mongolia; Early Cretaceous.

Only one species included from the Cretaceous of Northern China (see Table 22.1).

Hoplitolyda duolunica Gao, Shih, Rasnitsyn & Ren, 2013 (Figure 22.27)

Hoplitolyda duolunica Gao, Shih, Rasnitsyn & Ren, 2013: *PLoS ONE*, 8, e62420.

Figure 22.27 *Hoplitolyda duolunica* Gao, Shih, Rasnitsyn & Ren, 2013 (Holotype, CNU-HYM-ND-2011016) [3].

Locality and horizon: Nanpanying, Duolun, Inner Mongolia, China; Lower Cretaceous, Yixian Formation.

This is the largest fossil hymenopteran hitherto with body length estimated >55.0 mm and a wingspan of >92.0 mm. Forewing with Sc absent; R and pterostigma narrow; 1-Rs reclined, meeting 1-M at about 145°, representing an intermediate in the transition from "Y" to "T" shapes; Rs+M very short; cells 2rm and 1mcu very long. *H. duolunica* is, to our knowledge, the only sawfly with Sc present in the hind wing but not in the forewing [3].

Early Evolution of Wing Venation in Basal Hymenoptera

Hymenoptera have a comparatively simple pattern of wing venation among many pterygote lineages, and it has been simplified from basal to derived taxa within the order. Gao *et al.*, in 2010, summarized the evolutionary patterns of changes observed in vein Sc across lower Hymenoptera based on fossil records [76]. Among early Xyelidae, such as *G. quadrifurcata*, the forewing Sc is four-branched, and this condition is reduced to three-branched in *Abrotoxyela* [44], *Xyelites* and *Shartexyela*, and two-branched in other genera. In Xyelotomidae, vein Sc varies widely from the groundplan of a free Sc with two branches ending on both C and R (in *Pseudoxyela)* to entirely lost (in *Leridotoma*), with various intermediate states. Nevertheless, Sc is more stable in the remaining sawflies, like pamphilioids, siricoids, cephoids, etc.

Furthermore, an additional evolutionary trend is revealed by the sections of Rs and M in the course of formation of the so-called basal vein as represented by a series of sawfly fossils. The direction of 1-Rs varies significantly in Pamphilioidea: 1-Rs is normally proclival (e.g. in *Prolyda*, *Xyelyda*, *Strophandria*, and *Ferganolyda*) or vertical (e.g. in *Sagulyda* and *Mesolyda*), resulting in the angles between 1-Rs and 1-M ranging from 86.8° to 140°. However, in the vast and extensive CNUB collections, *Rectilyda sticta* Wang, Rasnitsyn, Shih & Ren, 2014 [68] (Figure 22.19) and *Hoplitolyda duolunica* Gao, Shih, Rasnitsyn & Ren, 2013 [3] are the only two species to date with 1-Rs reclival, thus forming a "T" shape, which is normally present in the Apocrita of the Hymenoptera [53, 76].

Decorisiricius Wang, Rasnitsyn, Shih & Ren, 2016

Decorisiricius Wang, Rasnitsyn, Shih & Ren, 2016, *Syst. Entomol.*, 41, 44 [79] (original designation).

Type species: *Decorisiricius patulus* Wang, Rasnitsyn, Shih & Ren, 2016.

Head large, rounded, nearly as wide as mesothorax; antenna (incomplete) with 11–15 antennomeres, the first flagellomere at least 4.6 times as long as wide, thicker than following flagellomeres and about as long as the remaining three flagellomeres combined. Forewing with pterostigma completely sclerotized; 1-Rs about 0.4 times as long as 1-M; 1cu-a basal to the mid-length of cell 1mcu, nearly as long as 2-Cu.

Distribution and age: Liaoning; Early Cretaceous.

Two species included from the Cretaceous of Northern China (see Table 22.1).

Limbisiricius Wang, Rasnitsyn, Shih & Ren, 2016.

Limbisiricius Wang, Rasnitsyn, Shih & Ren, 2016, *Syst. Entomol.*, 41, 47 [79] (original designation).

Type species: *Limbisiricius aequalis* Wang, Rasnitsyn, Shih & Ren, 2016.

Head relatively large, rounded and flattened, nearly as wide as or slightly wider than mesothorax. Forewing with pterostigma sclerotized around margins, pale medially; 1-Rs 05–0.7 times as long as 1-M; RS+M as long as 2-M; 1r-rs and 2r-rs subvertical to wing anterior margin; cross-vein 1m-cu almost as long as 2-Cu; and 1cu-a originating near the middle of cell 1mcu.

Distribution and age: Inner Mongolia; Middle Jurassic.

Two species included from the Jurassic of Northern China (see Table 22.1).

Brevisiricius Wang, Rasnitsyn, Shih & Ren, 2016

Brevisiricius Wang, Rasnitsyn, Shih & Ren, 2016, *Syst. Entomol.*, 41, 51 [79] (original designation).

Type species: *Brevisiricius partialis* Wang, Rasnitsyn, Shih & Ren, 2016.

Forewing with pterostigma completely sclerotized; 1-Rs 0.65 times as long as 1-M; Rs+M short, about 0.62 times as long as 2-M; 1m-cu shorter than half of 2-Cu; 1cu-a located just slightly proximad of cell 1mcu.

Distribution and age: Inner Mongolia; Middle Jurassic.

Only one species included from the Jurassic of Northern China (see Table 22.1).

Superfamily Pamphilioidea Cameron, 1890

Family Pamphiliidae Cameron, 1890

Pamphiliidae have 10 extant genera comprising about 330 reported species [38], and four extinct genera containing eight species described from the Mesozoic, Paleogene and Neogene of China, Russia, USA and Spain [50, 63, 82, 83]. They are typified by having a medially split second tergum, lateral folding of the abdominal terga above the spiracle, and forewing with vein A markedly sinuate [84]. Extant members of Pamphiliidae are restricted to temperate regions of North America and Eurasia [85, 86], and their larvae typically feed on conifers, spinning silk to build tents or webs (Cephalciinae), or rolling angiosperm leaves to form tubes (Pamphiliinae) for feeding [84, 87, 88]. In addition to the two extant subfamilies, a third subfamily, Juralydinae, was proposed by Rasnitsyn [41], based on one genus, *Juralyda*, described from a single specimen of forewing.

Only one genus included from the Jurassic of Northern China: *Scabolyda* Wang, Rasnitsyn, Shih & Ren, 2014.

Scabolyda Wang, Rasnitsyn, Shih & Ren, 2014

Scabolyda Wang, Rasnitsyn, Shih & Ren, 2014, *Alcheringa*, 38, 392 [83] (original designation).

Type species: *Scabolyda orientalis* Wang, Rasnitsyn, Shih & Ren, 2014.

Antenna with the first flagellomere at least twice as long as the fourth. Mesoscutellum small, nearly as large as mesoprescutum. Forewing with pterostigma long and narrow, completely sclerotized; Sc well-developed and bifid; Rs+M at least 1.3 times as long as 1-M; 2-M about 0.8–0.9 times as long as Rs+M; angle between 1-M and 1-Cu over 90°.

Distribution and age: Inner Mongolia; Middle Jurassic.

Two species included from the Jurassic of Northern China (see Table 22.1).

Scabolyda orientalis Wang, Rasnitsyn, Shih & Ren, 2014 (Figure 22.28)

Scabolyda orientalis Wang, Rasnitsyn, Shih & Ren, 2014: *Alcheringa*, 38, 392–395.

Locality and horizon: Daohugou, Ningcheng, Inner Mongolia, China; Middle Jurassic, Jiulongshan Formation.

Length of body excluding antennae is 15.7 mm, forewing length up to end of cell 3r, 11.2 mm as preserved. Head massive and subcircular, length nearly equal to width; mandible long and narrow; oral cavity isolated from occipital foramen by a hypostomal bridge. Antenna 1.7 times as long as head width, densely covered with short setae; the first flagellomere nearly three times as long, and 1.5 times as wide, as the second; remaining cylindrical flagellomeres becoming shorter toward the apex [83].

Superfamily Pamphilioidea Cameron, 1890

Family Megalodontesidae Konow, 1897

Megalodontesidae are a small family with only one extant genus, *Megalodontes* Latreille, 1803, comprising about 88 described species, all of which are distributed in Eurasia [38]. In contrast to Pamphiliidae, Megalodontesidae are

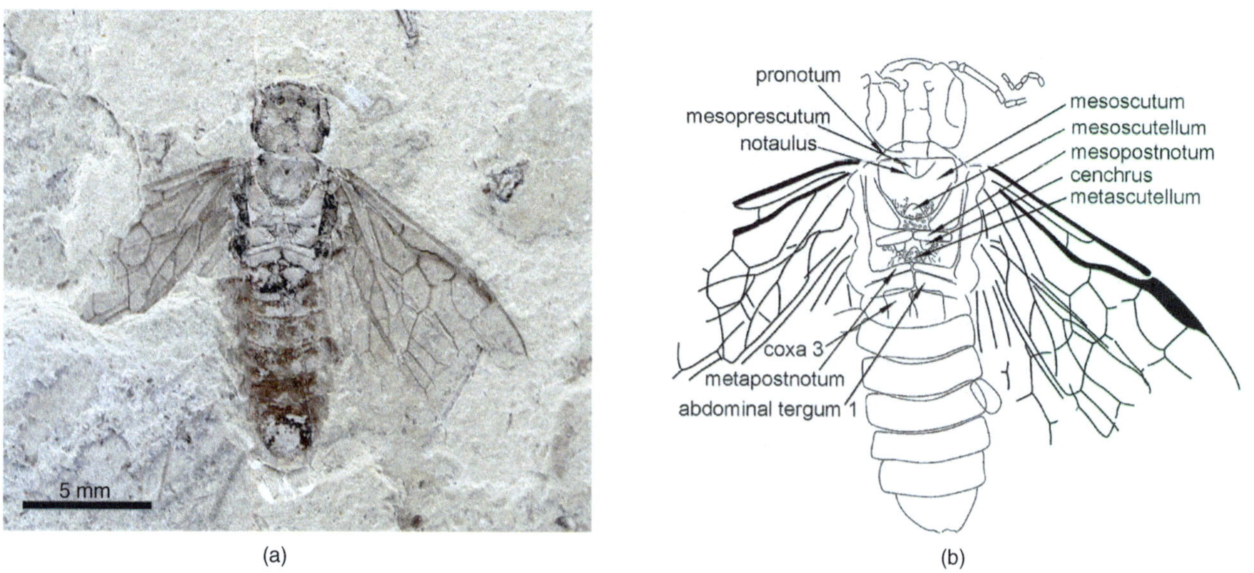

Figure 22.28 *Scabolyda orientalis* Wang, Rasnitsyn, Shih & Ren, 2014 (Holotype, CNU-HYM-NN-2012108) (a) Habitus; (b) Line drawing [83].

typified by having pectinate antennae and no lateral keels on the abdomen [88, 89]. The family has only one extinct representative, *Jibaissodes giganteus* from the Early Cretaceous of China [42], which was transferred from Baissodidae to Megalodontesidae by Rasnitsyn in 2000 [90].

Only one genus included from the Cretaceous of Northern China: *Jibaissodes* Ren, 1995.

Jibaissodes Ren, 1995

Jibaissodes Ren, 1995, Faunae and Stratigraphy of Jurassic-Cretaceous in Beijing and the Adjacent Areas. Seismie Publishing House, Beijing. 120 [42] (original designation).

Type species: *Jibaissodes giganteus* Ren, Lu, Guo & Ji, 1995.

Forewing with costal cell apically widened, broader than pterostigmal base at origin of Rs; first abscissa of M and that of 1-Rs forming a nearly straight line; basal vein confluent with 1cu-a; M+Cu straight with first free abscissa of Cu; marginal cell apex broadly rounded, reaching nearly to wing apex (rather than blunt, attributable to strong arching of apical Rs toward anterior wing margin). Abdominal terga without lateral creases.

Distribution and age: Liaoning; Early Cretaceous.

Two species included from the Cretaceous of Northern China (see Table 22.1).

Jibaissodes bellus Gao, Shih, Labandeira & Ren, 2016 (Figure 22.29)

Jibaissodes bellus Gao, Shih, Labandeira & Ren, 2016: *Proc. R. Soc. B*, 283 (1839), 20161448.

Locality and horizon: Huangbanjigou, Beipiao, Liaoning, China; Lower Cretaceous, Yixian Formation.

Length of body excluding antennae is 15.5 mm, forewing length up to end of cell 3r, 13.0 mm as preserved. Antenna plumose (feathery rather than serrate), with more than 30 flagellomeres, flagellum slightly longer than width of head; flagellomeres with plumose apical rami. Metanotal cenchri comparatively narrow and small [91].

(a)

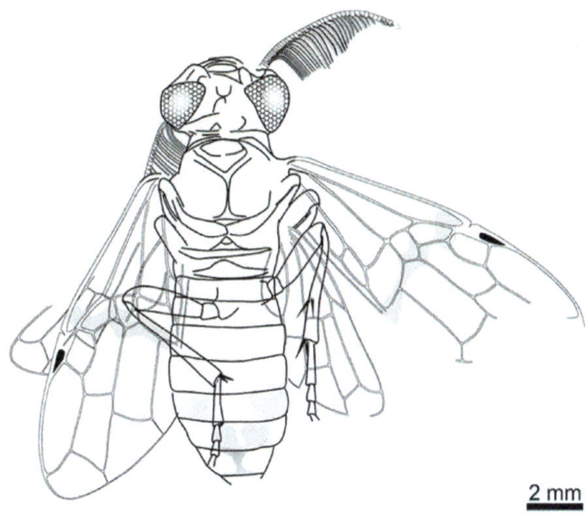

(b)

Figure 22.29 *Jibaissodes bellus* Gao, Shih, Labandeira & Ren, 2016 (Holotype, CNU-HYM-LB-2011009) (a) Habitus; (b) Line drawing [91].

Phylogenetic Research on Pamphilioidea

In order to evaluate the phylogenetic relationships among the genera of Pamphilioidea, we employed a morphological data set containing 44 species and 45 morphological characters; our phylogenetic analyses of the superfamily Pamphilioidea corroborated the monophyly of Pamphilioidea and Pamphiliidae, and also demonstrated that the extinct families Praesiricidae and Xyelydidae were not monophyletic (Figure 22.30). All members of Praesiricidae were therefore transferred to Megalodontesidae. Rudisiricinae, including *Megalodontes*, was renamed as Megalodontesinae. Based on fossil evidence, the origin and initial divergence of Pamphiliidae was pushed further back in time, possibly as early as the Early Jurassic.

As the inter-relationships among Pamphiliidae genera were poorly supported, a total-evidence data set including 17 species, 45 morphological characters, and seven genes was used in order to clarify the inner relationships of Pamphiliidae (Figure 22.31). This family was recovered as monophyletic with strong statistical support (Bayesian posterior probability: 0.99) and consisted of two major groups, i.e. three heterogeneous genera (*Scabolyda*, *Atocus*, and *Neurotoma*) as the sister-group to the other genera. Cephalciinae and Pamphiliinae were sisters with moderate posterior support (PP: 0.77) (see Figure 22.31).

Figure 22.30 Phylogeny of extant and extinct Pamphilioidea (modified from [64]). Strict consensus tree recovered from parsimony analyses of morphological characters. Character state changes are plotted on each node with bootstrap support values >50%. An asterisk indicates extant species.

Suborder Apocrita Gerstäcker, 1867

The Middle Jurassic Jiulongshan Formation and the Lower Cretaceous Yixian Formation, both located in Northeastern China, have yielded many well-preserved apocritan fossils. Based on morphologic characters of these apocritans, we have conducted taxonomic studies and phylogenetic analyses in order to provide insights of their evolutionary trends, originations and dispersals, and potential roles in the ecosystems. To date, we have described 23 genera, 67 species in 10 families: Pelecinidae, Heloridae, Mesoserphidae, Praeaulacidae, Anomopterellidae, Evaniidae, Scolebythidae, Ephialtitidae, Karatavitidae and Kuafuidae in the Suborder of Apocrita.

Superfamily Proctotrupoidea Latreille, 1802

Proctotrupoidea comprise 11 extant families, i.e. the diverse Diapriidae and Proctutrupidae, and the much less diverse Austroniidae, Heloridae, Maamingidae, Monomachidae, Pelecinidae, Peradeniidae, Proctorenyxidae, Roproniidae, and Vanhorniidae. Within Hymenoptera,

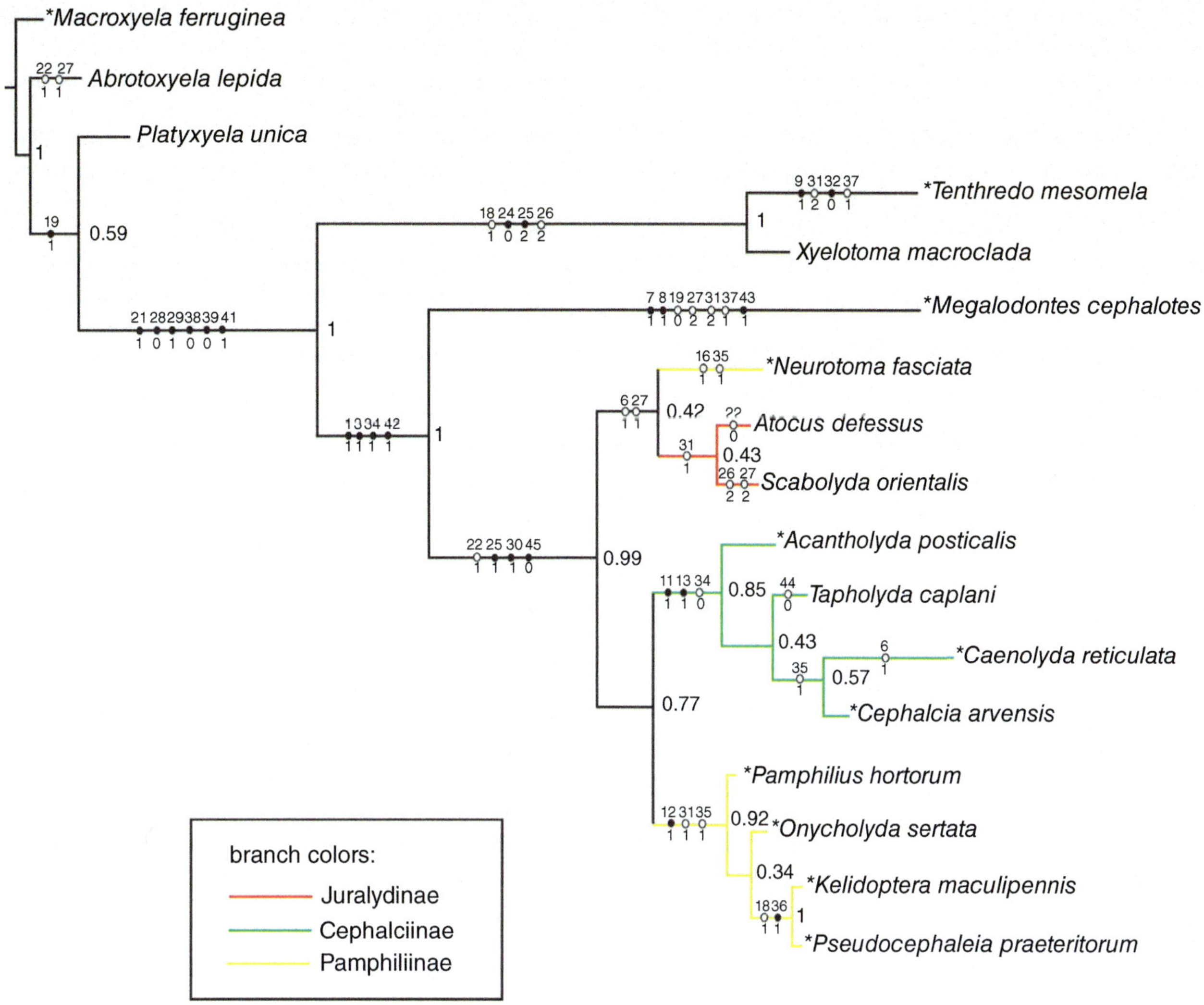

Figure 22.31 Phylogeny of extant and extinct Pamphiliidae (modified from [64]). Consensus tree recovered from Bayesian analysis of combined morphological characters and DNA sequence data. Bayesian posterior probability values are included at nodes. Synapomorphic characters mapped on each node are recovered from parsimony optimization of morphological characters alone. An asterisk indicates extant species.

this superfamily is significant and important for their long evolutionary history, special morphology and broad diversity [1]. Most extant species of Proctotrupoidea are small wasps except for the giant pelecinids [92, 93].

Family Pelecinidae Haliday, 1840

Pelecinids are elegant and beautiful extant wasps with unique metasoma and body sizes ranging from 20 to 90 mm [94]. Female pelecinids have elongate and exceptionally slender metasomas, while males have stout metasomal segments, the first metasomal segment strongly elongate [94]. In addition, a very interesting phenomenon was proposed by Brues [95] that the extant *P. polyturator* is a case of geographic parthenogenesis (asexual reproduction): parthenogenesis in temperate populations and amphigenesis (sexual reproduction) in tropical populations. It was confirmed that the populations of pelecinids in the USA and Canada are mainly thelytokous (females from unfertilized eggs) [94]. To date, the Pelecinidae comprise one extant genus *Pelecinus* and 17 extinct genera with 48 fossil species [96].

Genera included from the Jurassic and Cretaceous of Northern China: *Eopelecinus* Zhang, Rasnitsyn & Zhang, 2002, *Sinopelecinus* Zhang, Rasnitsyn & Zhang, 2002, *Scorpiopelecinus* Zhang, Rasnitsyn & Zhang, 2002, *Allopelecinus* Zhang & Rasnitsyn, 2006, *Archaeopelecinus* Shih, Liu & Ren, 2009, *Cathaypelecinus* Shih, Liu & Ren, 2009, *Shoushida* Liu, Shih & Ren, 2009, *Abropelecinus* Feng, Shih, Ren & Liu, 2010, *Azygopelecinus* Feng,

Shih, Ren & Liu, 2010, *Megapelecinus* Shih, Feng, Liu, Zhao & Ren, 2010 and *Stelepelecinus* Guo, Shih & Ren, 2016.

***Eopelecinus* Zhang, Rasnitsyn & Zhang, 2002**
Eopelecinus Zhang, Rasnitsyn & Zhang, 2002, *Cretac. Res.*, 23, 687 [97] (original designation).

Type species: *Eopelecinus vicinus* Zhang, Rasnitsyn & Zhang, 2002.

Antenna with 13–15 antennomeres. Forewing with two tubular veins present (C and R); pterostigma variable in shape and size. Pronotum short to long medially. Metasoma with one broad basal segment, succeeding segments usually tubular, but sometimes second metasomal segment club-shaped.

Distribution and age: Liaoning; Early Cretaceous.

Fourteen species included from the Cretaceous of Northern China (see Table 22.1).

***Sinopelecinus* Zhang, Rasnitsyn & Zhang, 2002**
Sinopelecinus Zhang, Rasnitsyn & Zhang, 2002, *Cretac. Res.*, 23, 88 [97] (original designation).

Type species: *Sinopelecinus delicatus* Zhang, Rasnitsyn & Zhang, 2002.

Male, antenna with 15 antennomeres. Forewing with two veins (C and R), R slightly bent; 2r-rs oblique apicad and slightly longer than width of pterostigma; Rs short, almost spectral; M+ Cu distinguishable and straight, bearing short rudiments of free M and 1cu-a. Metasoma with 1–6 segments nearly parallel-sided, rectangular, last segment triangular.

Distribution and age: Liaoning; Early Cretaceous.

Six species included from the Cretaceous of Northern China (see Table 22.1).

***Scorpiopelecinus* Zhang, Rasnitsyn & Zhang, 2002**
Scorpiopelecinus Zhang, Rasnitsyn & Zhang, 2002, *Cretac. Res.*, 23, 96 [97] (original designation).

Type species: *Scorpiopelecinus versatilis* Zhang, Rasnitsyn & Zhang, 2002.

Antenna with 14 antennomeres. Forewing with pterostigma narrow, acute apically, not widened beyond mid-length; 2r-rs distinct, vestigial Rs present distad 2r-rs, r open; M+Cu distinct. Hind wing with only C present. Pronotum comparatively long. Propodeum reticulate. Metasoma with three basal segments wide and remainder tubular.

Distribution and age: Liaoning; Early Cretaceous.

Two species included from the Cretaceous of Northern China (see Table 22.1).

***Allopelecinus* Zhang & Rasnitsyn, 2006**
Allopelecinus Zhang & Rasnitsyn, 2006, *Cretac. Res.*, 27, 687 [98] (original designation).

Type species: *Allopelecinus terpnus* Zhang & Rasnitsyn, 2006.

Antenna with 15 antennomeres. Pronotum long dorsally. Forewing with pterostigma narrow, acute apically, veins other than C, R lost. Propodeum reticulate. Metasoma with three basal segments wide, distinctly constricted at junctions, following two moderately narrowed, not strictly tubular, the sixth about as thick as the first. Ovipositor external, long.

Distribution and age: Shandong; Early Cretaceous.

Only one species included from the Cretaceous of Northern China (see Table 22.1).

***Archaeopelecinus* Shih, Liu & Ren, 2009**
Archaeopelecinus Shih, Liu & Ren, 2009, *Ann. Entomol. Soc. Am.*, 102, 25 [92] (original designation).

Type species: *Archaeopelecinus tebbei* Shih, Liu and Ren, 2009.

Antenna with 17 antennomeres or more, up to 23 antennomeres. Forewing with cell r closed; Rs_1 almost linear and reaching anterior margin of wing much before apex of wing; Rs_2 very short or vestige; M+Cu distinct; M and Cu distinct, almost straight, reaching outer margin of wing; cell 1mcu small and nearly triangular; The first abscissa of Rs shorter than or as long as that of M; cell 1 + 2r five-sided (surrounded by R, Rs, Rs+M, Rs, and 2r-rs); 1cu-a and 2cu-a distinct but not reaching posterior margin of wing. Mesosoma coarsely and irregularly reticulated. Metasoma with basal-most two segments broader than others and the first broader than the second, remaining tubular.

Distribution and age: Inner Mongolia; Middle Jurassic.

Two species included from the Jurassic of Northern China (see Table 22.1).

***Archaeopelecinus tebbei* Shih, Liu and Ren, 2009 (Figure 22.32)**
Archaeopelecinus tebbei Shih, Liu and Ren, 2009: *Ann. Entomol. Soc. Am.*, 102, 25.

Locality and horizon: Daohugou, Ningcheng, Inner Mongolia, China; Middle Jurassic, Jiulongshan Formation.

The specific epithet is dedicated to Stan Tebbe for being an excellent business leader, mentor, and role-model and providing guidance, motivation and inspiration to Dr. Chungkun Shih. Length of body excluding antennae is 22.7 mm. Head large, oval; eyes large, round, and spaced closely. Antenna with 23 antennomeres. Pronotum comparatively short, mesosoma nearly elongate-hexagonal. Forewing long with long and narrow pterostigma; cell r closed, Rs_1 straight, Rs_2 short and not vestige; cell 1mcu small and nearly triangular; cell 1 + 2r is five sided (surrounded by R, Rs, Rs+M, Rs and 2r-rs). Metasoma with basalmost two segments broader, remaining

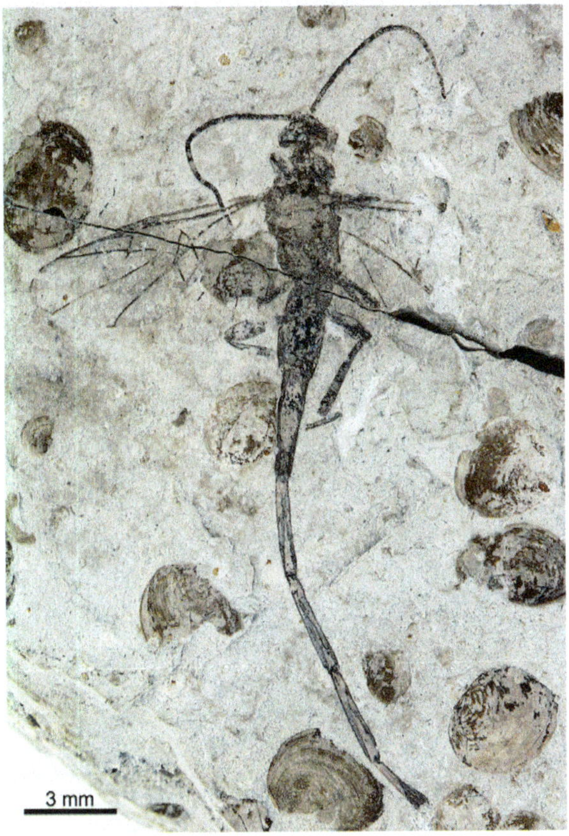

Figure 22.32 *Archaeopelecinus tebbei* Shih, Liu and Ren, 2009 (Holotype, CNU-HYM-NN-2006001p). Female. Source: Donated by Dr. Chungkun Shih [92].

tubular. This is the earliest pelecinid fossil documented hitherto and extending their existence to the Middle Jurassic [92].

Cathaypelecinus Shih, Liu & Ren, 2009

Cathaypelecinus Shih, Liu & Ren, 2009, *Ann. Entomol. Soc. Am.*, 102, 30 [92] (original designation).

Type species: *Cathaypelecinus daohugouensis* Shih, Liu & Ren, 2009.

Antenna with 17 antennomeres or more, up to 21 antennomeres, with relatively thin and long flagellomeres. Forewing with cell r closed; Rs_1 almost straight, reaching anterior margin of wing much before apex of wing; Rs_2 vestige; M+Cu distinct; M and Cu distinct, almost straight, reaching outer margin of wing; cell 1mcu small, four-sided, nearly trapezoid; first abscissa of Rs about the same length as that of M; cell 1+ 2r is five-sided (surrounded by R, Rs, Rs+M, Rs, and 2r-rs); 1cu-a and 2cu-a distinct and not reaching posterior margin of wing. Mesosoma coarsely and irregularly reticulate. Metasoma with two basalmost segments broader, especially the first segment, remaining tubular.

Distribution and age: Inner Mongolia; Middle Jurassic.

Only one species included from the Jurassic of Northern China (see Table 22.1).

Shoushida Liu, Shih & Ren, 2009

Shoushida Liu, Shih & Ren, 2009, *Zootaxa*, 2080, 48 [99] (original designation).

Type species: *Shoushida regilla* Liu, Shih & Ren, 2009.

The generic name is a Chinese pronunciation of Capital Normal University and the specific epithet is derived from the Latin *"regilla"*, meaning "honorable and distinguished", in praise of the Capital Normal University. Antenna with 14 antennomeres. Forewing Rs forking into two branches, Rs_1 and Rs_2; Rs_1 straight and extending to wing margin much before apex; Rs_2 a nebulous vein, long but not reaching the wing margin; Rs, Rs_1, Rs_2 and 2r–rs form an "X" pattern; M+Cu distinct; pterostigma long and narrow. Notauli reaching the pronotum, propodeum reticulate. Metasoma with two basalmost segments broader than others and the first broader than the second, the following three segments tubular, the last segment fusiform.

Distribution and age: Liaoning; Early Cretaceous.

Two species included from the Cretaceous of Northern China (see Table 22.1).

Shoushida infera Guo, Shih & Ren, 2016 (Figure 22.33)

Shoushida infera Guo, Shih & Ren, 2016: *Cretac. Res.*, 61, 154.

Locality and horizon: Huangbanjigou, Liaoning, China; Lower Cretaceous, Yixian Formation.

This is the second fossil record of pelecinid with "X" type forewing venation formed by Rs, Rs_1, Rs_2 and 2r-rs from this locality and from China. Length of body excluding antennae is 12.5 mm. Head medium-sized, transversely broad in lateral view, antenna with 14 antennomeres. Mesosoma elongate-oval in lateral view; pronotum comparatively short dorsally and laterally, mesoscutum medium-sized, comparatively flat; mesoscutellum slightly concave; propodeum coarsely and irregularly reticulate. Forewing with only two tubular veins (C and R); cell r nearly triangular; the first abscissa of Rs (1-Rs) as long as that of M (1-M); 1m-cu very short; shape of cell 1mcu trapezoid; Rs forking into Rs_1 and Rs_2; cu-a interstitial, remnant; 2cu-a distal of 1m-cu. Metasoma slender, the first segment relatively broad, nearly trapezoid, the second segment inversely trapezoid, other segments slender and tubular [96].

Abropelecinus Feng, Shih, Ren & Liu, 2010

Abropelecinus Feng, Shih, Ren & Liu, 2010, *Geolo. Carpath.*, 61, 464 [100] (original designation).

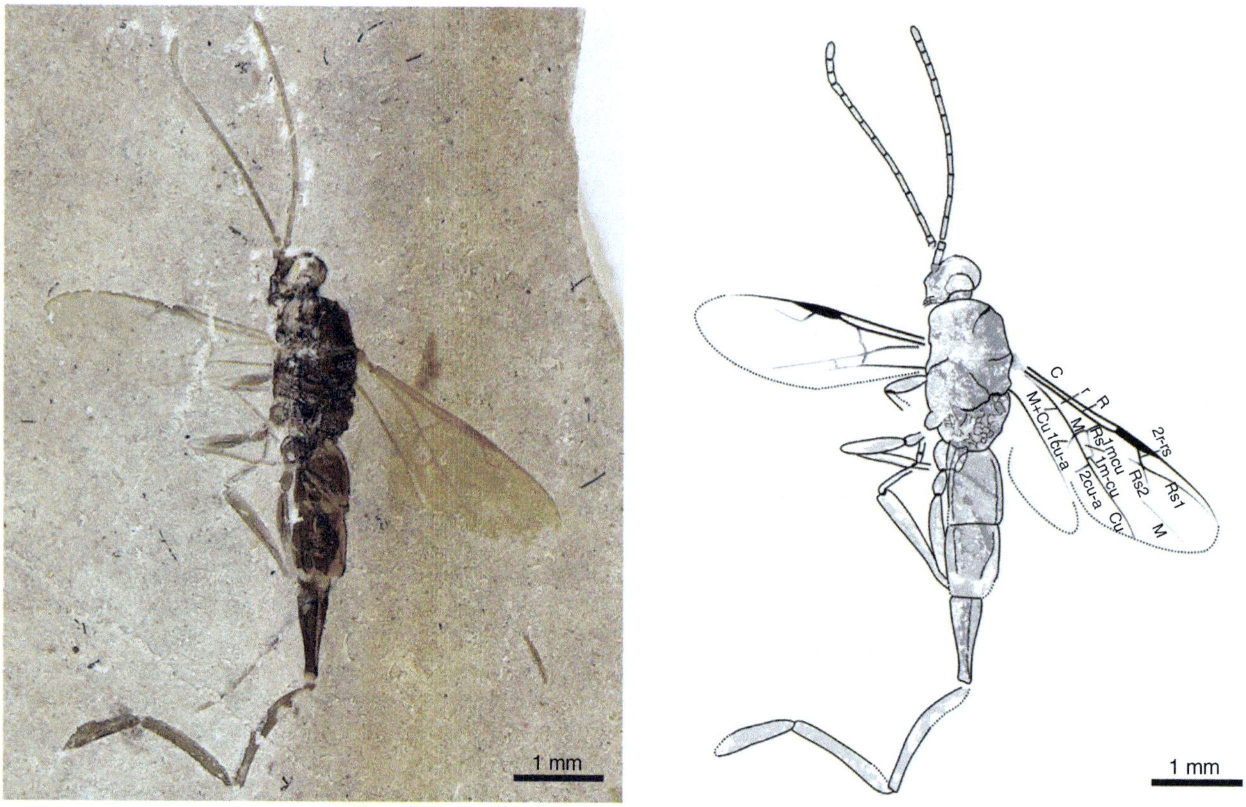

(a) (b)

Figure 22.33 *Shoushida infera* Guo, Shih & Ren, 2016 (Holotype CNU-HYM-LB2006105) (a) Habitus; (b) Line drawing [96].

Type species: *Abropelecinus annulatus* Feng, Shih, Ren & Liu, 2010.

Male, antenna with 13 antennomeres. Forewing with only two veins present (C and R). Metasoma with the seventh segment oblong, slightly acute apically (sometimes tergum and sternum separated to an extent).

Distribution and age: Liaoning; Early Cretaceous.

Only one species included from the Cretaceous of Northern China (see Table 22.1).

Azygopelecinus Feng, Shih, Ren & Liu, 2010

Azygopelecinus Feng, Shih, Ren & Liu, 2010, *Geolo. Carpath.*, 61, 465 [100] (original designation).

Type species: *Azygopelecinus clavatus* Feng, Shih, Ren & Liu, 2010.

Male, forewing with only two veins distinct (C and R); 2r-rs issuing medial 1/3 of pterostigma; Rs short; M+Cu discernible, straight. The first metasomal segment thin and rectangular, the second and third ones nearly triangular, very narrow basally, the fourth to sixth ones nearly trapeziform, the last metasomal segment short triangular.

Distribution and age: Liaoning; Early Cretaceous.

Only one species included from the Cretaceous of Northern China (see Table 22.1).

Megapelecinus Shih, Feng, Liu, Zhao & Ren, 2010

Megapelecinus Shih, Feng, Liu, Zhao & Ren, 2010, *Ann. Entomol. Soc. Am.*, 103, 877 [93] (original designation).

Type species: *Megapelecinus changi* Shih, Feng, Liu, Zhao & Ren, 2010.

The specific name of *changi* is in honor of Yung-Chang Chang for providing guidance and inspiration to Dr. Chungkun Shih. Antenna with 22 antennomeres or more, up to 26. Forewing with cell r closed; Rs_1 almost straight and reaching anterior margin of wing much before wing apex; Rs+M after cell 1mcu forking to Rs and M at a distal point making Rs only about 1/5 as long as Rs+M; Rs_2 absent; cell 1 + 2r six sided; the first abscissa of Rs two to three times as long as that of M; cell 1mcu small, narrow and nearly trapezoid with 1m-cu as long as or slightly shorter than that of M; and 1cu-a and 2cu-a distinct but very short, not reaching posterior margin. Metasoma has basal most one or two segments broader than others, and the first segment is broader than the second.

Distribution and age: Liaoning; Early Cretaceous.

Two species included from the Cretaceous of Northern China (see Table 22.1).

Megapelecinus changi **Shih, Liu & Ren, 2010 (Figure 22.34)**
Megapelecinus changi Shih, Liu & Ren, 2010: *Ann. Entomol. Soc. Am.*, 103, 877.

Locality and horizon: Huangbanjigou, Liaoning, China; Lower Cretaceous, Yixian Formation.

The largest pelecinid fossil hitherto with a total body length of 50.9 mm, similar to those of the extant female pelecinids. Head round and large, eyes small and widely separated. Antenna filiform and long with 26 antennomeres. Mesosoma poorly preserved with a length of 8 mm. Pronotum very short dorsally; mesosoma notauli distinctly and arched inwards; propodeal sculpturing reticulate. Forewing with pterostigma longer than wide and acute apically; M and Cu parallel; 2r-rs arises from about basal 1/3 of pterostigma, slightly oblique apicad, outer margin has straw yellow faint speckles. Hind wing has only C present along anterior margin and straw yellow faint speckles on outer margin. The first metasoma segment suboval, the second segment trapezoid with base wider than apex, the first and second segments broader than others, and the first broader than the second, the third through fifth slender; the sixth medial slightly expanded and blunt apically [93].

Stelepelecinus **Guo, Shih & Ren, 2016**
Stelepelecinus Guo, Shih & Ren, 2016, *Cretac. Res.*, 61, 156 [96] (original designation).

Type species: *Stelepelecinus longus* Guo, Shih & Ren, 2016.

Male, antenna with 14 antennomeres. Forewing with only two veins present (C and R). Metasoma clavate, narrowing caudad, six segments, the first metasomal segment noticeably long, petiole-like.

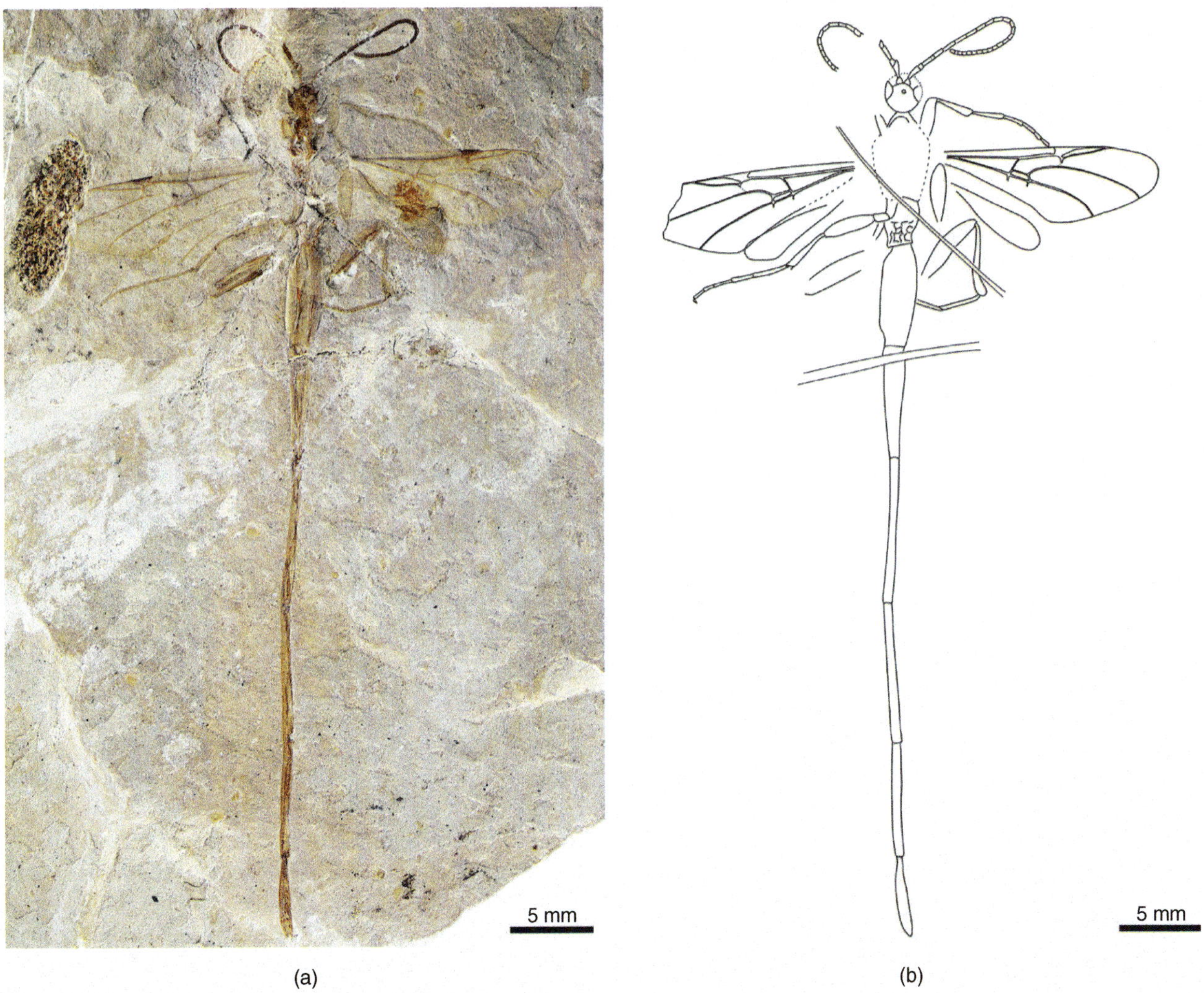

Figure 22.34 *Megapelecinus changi* Shih, Liu & Ren, 2010 (Holotype CNU-HY-LB2006036p). Female, (a) Habitus; (b) Line drawing. Source: Donated by Dr. Chungkun Shih [93].

Distribution and age: Liaoning; Early Cretaceous.

Only one species included from the Cretaceous of Northern China (see Table 22.1).

Stelepelecinus longus **Guo, Shih & Ren, 2016 (Figure 22.35)**
Stelepelecinus longus Guo, Shih & Ren, 2016: *Cretac. Res.*, 61, 156.

Locality and horizon: Huangbanjigou, Liaoning, China; Lower Cretaceous, Yixian Formation.

Figure 22.35 *Stelepelecinus longus,* Guo, Shih & Ren, 2016 (Holotype CNU-HYM-LB2015093) [96].

Figure 22.36 *Brachypelecinus eythyntus* Guo, Shih & Ren, 2016 (Holotype, male, CNU-HYM-MA2016001) [102].

Length of body excluding antennae is 9.2 mm. Head large, trapezoid; antenna with 14 antennomeres as preserved. Mesosoma elongate-oval in lateral view; pronotum comparatively long, mesoscutum apparently long, mesoscutellum distinctly short, convex; metanotum moderately short, comparatively flat, mesoscutellum, metanotum, metapleura and propodeum densely reticulate. The apex of the forewing somewhat acuminate, R is straight or slightly bending posteriad; the pterostigma is entirely sclerotized, thin, long, rectangular (truncate apically). Metasoma clavate, six segments visible, the first segment noticeably long, petiole-like [96].

Amber Pelecinid from Myanmar

Brachypelecinus eythyntus Guo, Shih & Ren, 2016, male, was collected from Kachin (Hukawng Valley) of northern Myanmar, mid-Cretaceous (98.79 ± 0.62 Mya [101]. Body size 5.1 mm, antenna with 13 antennomeres. Forewing with membrane hyaline with abundant setae, cell 1mcu small, four-sided, nearly trapezoid; M, Cu and A reaching wing apex, slightly more pigmented near wing margin than the distalmost abscissae of Rs_1, Rs_2; 1m-cu far basad apex of 2cu-a. Metasoma stick-like with six segments [102] (see Figure 22.36).

Morphology, Phylogeny, Evolution and Dispersal of Pelecinids over 165 Million Years

Large extant female pelecinid wasps, elegant and beautiful with uniquely elongated metasoma, are found only in North, Central, and South America. These distinctive wasps are "living fossils" in a relict family of Pelecinidae. Discoveries of well-preserved pelecinid fossils, e.g. *Archaeopelecinus tebbei, Shoushida regilla, Megapelecinus changi* and *Megapelecinus nashi*, from the Middle Jurassic (165 Mya) and the Early Cretaceous (125 Mya) of Northeastern China, provide strong evidence for morphological changes and evolution of pelecinids, such as body size, antenna and wing venation, over the past 165 million years. A phylogenetic analysis was carried out by Shih *et al.* [93], using 22 morphological characters, 12 extinct pelecinid taxa and one extant pelecinid taxon at the generic level. The results, as shown in Figure 22.37, suggest that *Megapelecinus* Shih, Liu & Ren, 2010 is the most basal pelecinid, whereas *Cathaypelecinus* Shih, Liu et Ren 2009, *Archaeopelecinus* Shih, Liu et Ren 2009, and *Iscopinus* Kozlov, 1974 forming a basal clade. Natural selection is demonstrated in pelecinids by the preference of larger female body size offering potentially better ovipositing capability and more efficient forewing venation with an X pattern providing

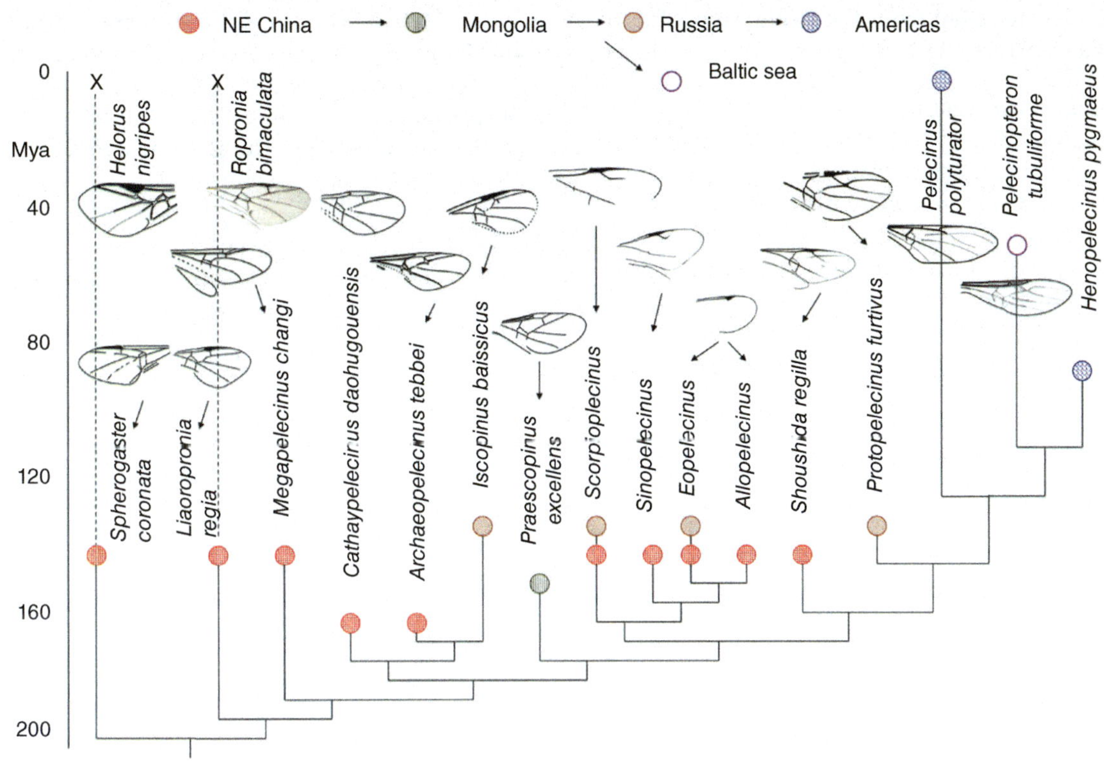

Figure 22.37 Pelecinidae and two outgroups with their respective characteristic forewing in geological context [93].

potentially stronger wing structure and better flight performance for large-sized pelecinids. Furthermore, temporal and spatial analyses indicate that the most parsimonious hypothesis is that pelecinid might have originated from Northeastern China, spread to Central and Eastern Asia, and then dispersed to Americas [93].

Family Heloridae Foerster, 1856

Heloridae, a very small extant family but distributed nearly worldwide, belong to the superfamily Proctotrupoidea [1]. The extant Heloridae comprise only one genus, *Helorus* Latreille, 1802, all species of which are parasitoids of Chrysopidae within Neuroptera [103]. *Helorus* is characterized by having antenna with 15 antennomeres, a distinct pterostigma, a subtriangular first discal cell (1mcu), vein Rs+M present, 1Rs absent, an elongate first metasomal segment, and metasomal tergites 2–4 fused into a syntergite [104]. However, the extinct helorid wasps have more diverse fossil records, and ten genera with 18 species have been reported from China, Russia and Mongolia from the Middle Jurassic to the Early Cretaceous [105–113].

Genera included from the Jurassic and Cretaceous of Northern China: *Protocyrtus* Rohdendorf, 1938, *Laiyanghelorus* Zhang, 1992, *Liaoropronia* Zhang & Zhang, 2001, *Spherogaster* Zhang & Zhang, 2001, *Archaeohelorus* Shih, Feng & Ren, 2011, *Sinohelorus* Shi, Zhao, Shih & Ren, 2013, *Bellohelorus* Li, Shih & Ren, 2017 and *Novhelorus* Li, Shih & Ren, 2017.

Protocyrtus Rohdendorf, 1938

Protocyrtus Rohdendorf, 1938, *Trudy Paleontologicheskogo Instituta Academii Nauk SSSR*, 7, 29–67 [105] (original designation).

Type species: *Protocyrtus jurassicus* Rohdendorf, 1938.

Forewing with 1-Rs & 1-M nearly straight; 1-M inclined; 1cu-a slightly postfurcal; 2cu-a interstitial; the position of Rs+M forking at about 40% of distance between 1m-cu and 2r-rs; closer to 1m-cu than to 2r-rs. The first metasomal segment broad.

Distribution and age: Liaoning; Early Cretaceous.

Two species included from the Cretaceous of Northern China (see Table 22.1).

Laiyanghelorus Zhang, 1992

Laiyanghelorus Zhang, 1992, *Entomotaxonomia*, 14, 1 [108] (original designation).

Type species: *Laiyanghelorus erymnus* Zhang, 1992.

Forewing long and broad; 2r-rs nearly twice as long as 3r is wide; 2Rs distinctly curved at the same level as the origin of 2r-rs; cell 1mcu markedly narrow, the length nearly four times as long as the width.

Distribution and age: Laiyang, Shandong; Early Cretaceous.

Only one species included from the Cretaceous of Northern China (see Table 22.1).

***Liaoropronia* Zhang & Zhang, 2001**

Liaoropronia Zhang & Zhang, 2001, *Acta Micropalaeontol. Sin.*, 18, 22 [109] (original designation).

Type species: *Liaoropronia leonine* Zhang & Zhang, 2001.

Head large with eyes large; antenna with at least 14 antennomeres with scape swollen. Propodeal tergum and pleura areolated. Petiole thin and distinctly longer than wide. Metasoma with the first segment bell-shaped; remaining part short. Forewing with pterostigma narrow and elongate; 2r-rs nearly perpendicular to anterior margin; the first abscissa of Rs shorter than basal section of M; 1 m-2cu greatly shorter than the first abscissa of M; cu-a interstitial. Hind wing with only anterior margin present.

Distribution and age: Liaoning; Early Cretaceous.

Two species included from theCretaceous of Northern China (see Table 22.1).

***Spherogaster* Zhang & Zhang, 2001**

Spherogaster Zhang & Zhang, 2001, *Acta Micropalaeontol. Sin.*, 18, 21 [109] (original designation).

Type species: *Spherogaster coronata* Zhang & Zhang, 2001.

Body large in size. Antenna elongate, 26 antennomeres and longer than body. Metasoma with the first segment thick and short and drum-shaped; remaining part small and spherical. Forewing with pterostigma narrow and elongate, triangular; 2r-rs longer than pterostigmal width, running toward posterior margin (nearly perpendicular to anterior margin); pterostigma issuing 2r-rs near its base; the first abscissa of Rs strongly inclined toward wing base, greatly longer than basal section of M; cell 1mcu greatly narrow and subtriangular.

Distribution and age: Liaoning; Early Cretaceous.

Two species included from the Jurassic and Cretaceous of Northern China (see Table 22.1).

***Archaeohelorus* Shih, Feng & Ren, 2011**

Archaeohelorus Shih, Feng & Ren, 2011, *Ann. Entomol. Soc. Am.*, 104, 1337 [110] (original designation).

Type species: *Archaeohelorus hoi* Shih, Feng & Ren, 2011.

The specific epithet is dedicated to Dr. Teh Chung Ho for dedication and contribution to science and engineering, especially on catalysis; for providing guidance, mentorship, and inspiration to Dr. Chungkun Shih. Body small in size. Antenna with 16 or 17 antennomeres, excluding a ring-like anellus on the first flagellomere, longer than head and mesosoma combined. Metasoma with six or seven segments, the first segment narrow, transverse or slightly conical. Forewing with the first abscissa of Rs (1-Rs) as long as or slightly longer than that of M (1-M) and slightly inclined toward wing base; cell 1mcu small and subtriangular; and 1cu-a in line with the basal section of M (1-M); and 2cu-a intersecting Cu and Rs+M at the same point.

Distribution and age: Liaoning; Early Cretaceous.

Three species included from the Jurassic of Northern China (see Table 22.1).

***Archaeohelorus hoi* Shih, Feng & Ren, 2011 (Figure 22.38)**

Archaeohelorus hoi Shih, Feng & Ren, 2011: *Ann. Entomol. Soc. Am.*, 104, 1338.

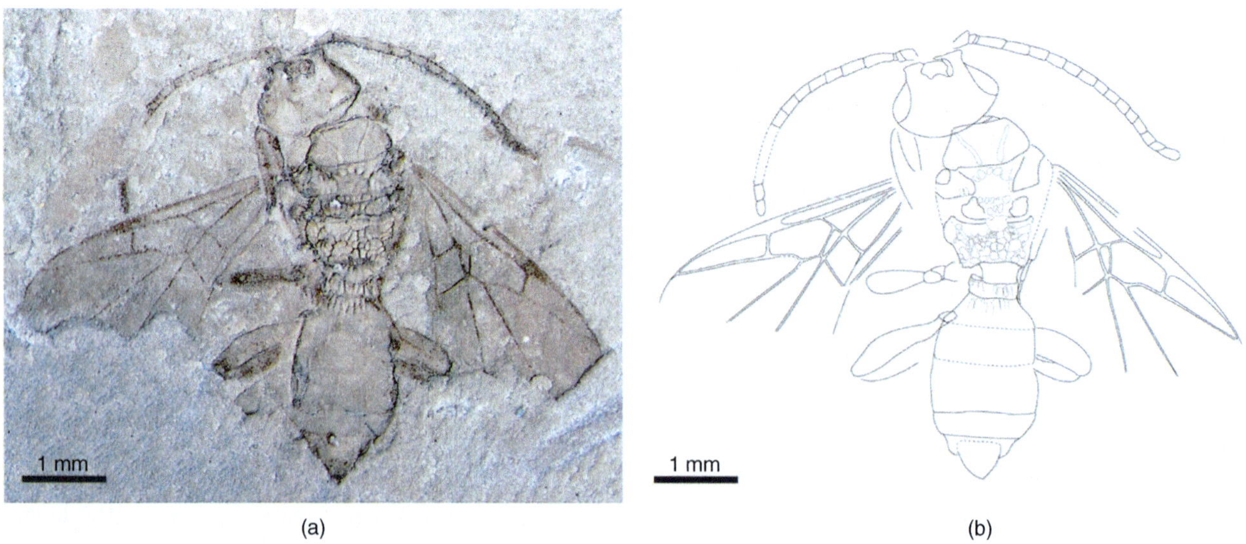

Figure 22.38 *Archaeohelorus hoi* Shih, Feng & Ren, 2011, (Holotype CNU-HYNN2008005). (a) Habitus; (b) Line drawing. Source: Donated by Dr. Chungkun Shih [110].

Figure 22.39 *Sinohelorus elegans* Shi, Shih & Ren, 2013 (Holotype CNU-HYM-LB-2012001) (a) Habitus; (b) Line drawing [111].

Locality and horizon: Daohugou, Inner Mongolia, China; Middle Jurassic, Jiulongshan Formation.

Length of body excluding antennae is 5.12 mm. Head and eyes large, antenna with 16 antennomeres. Mesosoma suboval, pronotum narrow anteriorly, mesoscutum trapezoidal with notauli robust and concave, metanotum relatively wide; propodeum areolated. Forewing broad, short and subtriangular; pterostigma long and acute apically; C robust and extending to apex of forewing; 2r-rs arising from basal 1/3 of pterostigma; R robust; Rs straight; cell r closed and triangular; cell 1 + 2r six-sided and surrounded by R, 1-Rs, 1-Rs+M, 2-Rs+M, 2-Rs, and 2r-rs; cell 1mcu small and subtriangular. Metasoma suboval with seven segments. This is the earliest fossil record hitherto of helorid and extending their existence to the Middle Jurassic [110].

***Sinohelorus* Shi, Zhao, Shih & Ren, 2013**

Sinohelorus Shi, Zhao, Shih & Ren, 2013, *Cretac. Res.*, 41, 137 [111] (original designation).

Type species: *Sinohelorus elegans* Shi, Shih & Ren, 2013.

Metasoma with seven segments, the first segment elongate, the second segment long and compound (fusion of several ones), others short, and all apparently smooth. Forewing with the first abscissa of Rs (1-Rs) as long as or slightly shorter than that of M (1-M) and slightly inclined toward wing base while 1-M inclined toward wing apex; 1-Rs together with the 1-M and RS+M form a "Y" configuration; cell 1mcu small and subtriangular; 1cu-a in line with the basal section of M (1-M); 2cu-a intersecting with Cu and Rs+M at the same point or slightly basal; M and Rs branching at about 30% from 1m-cu of the length between 1m-cu and 2r-rs.

Distribution and age: Liaoning; Early Cretaceous.

Only one species included from the Cretaceous of Northern China (see Table 22.1).

***Sinohelorus elegans* Shi, Shih & Ren, 2013 (Figure 22.39)**

Sinohelorus elegans Shi, Shih & Ren, 2013: *Cretac. Res.*, 41, 137.

Locality and horizon: Huangbanjigou, Liaoning, China; Lower Cretaceous, Yixian Formation.

Length of body excluding antennae is 7.65 mm. Head partially preserved, antenna with 24 antennomeres preserved. Forewing with 2r-rs arising from basal 1/3 of pterostigma; R robust; Rs straight; cell r four-sided; cell 1 + 2r six-sided; 1-Rs and 1-M forming an angle of about 135^0 and combined with Rs+M forming a "Y" configuration; cell 1mcu small and subtriangular;cross-veins of 1cu-a and 2cu-a distinct and 1cu-a in line with 1-M, and 2cu-a slightly basal with Cu and Rs+M. Metasoma spindly with seven segments; the first segment elongate [111].

***Bellohelorus* Li, Shih & Ren, 2017**

Bellohelorus Li, Shih & Ren, 2017, *Alcheringa*, 17, 3 [113] (original designation).

Type species: *Bellohelorus fortis* Li, Shih & Ren, 2017.

Head transversely broad, antenna with 19 antennomeres. Metasoma with the first segment elongate and gradually wider from base to apex, the second segment long and compound (fusion of several segments), with remaining segments distinctly shorter than the first segment. Forewing with 1-Rs longer than 1-M, 1-Rs, and 1-M straight, inclined toward pterostigma; 1cu-a slightly postfurcal; 2cu-a interstitial; cell 1mcu long and narrow; the position of Rs+M forking at nearly 40% of distance between 1m-cu and 2r-rs, closer to 1m-cu than to 2r-rs; 2r-rs long and straight, about three times as long as pterostigma wide.

Distribution and age: Liaoning; Early Cretaceous.

Only one species included from the Cretaceous of Northern China (see Table 22.1).

***Bellohelorus fortis* Li, Shih & Ren, 2017 (Figure 22.40)**

Bellohelorus fortis Li, Shih & Ren, 2017: *Alcheringa*, 17, 9.

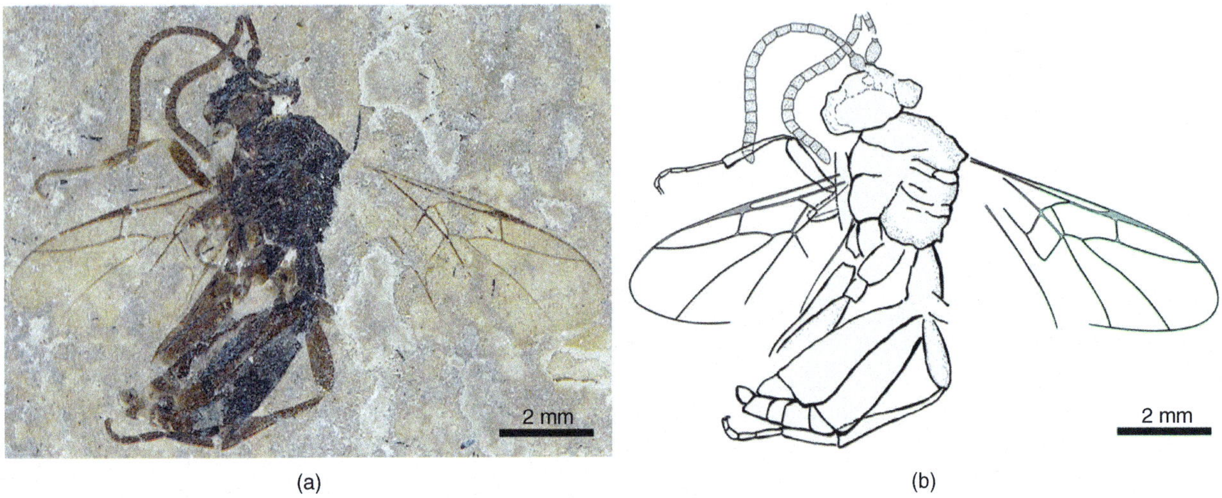

Figure 22.40 *Bellohelorus fortis* Li, Shih & Ren, 2017 (Holotype CNU-HYM-LB-2016001p). (a) Habitus; (b) Line drawing [113].

Locality and horizon: Huangbanjigou, Liaoning, China; Lower Cretaceous, Yixian Formation.

Length of body excluding antennae is 8.30 mm. Head twice as wide as long, antenna filiform, thick, curved and long. Mesosoma distinctly shorter than the metasoma, metanotum distinctly short, but nearly as wide as the base of the propodeum. The first metasoma segment elongate and the second long and compound. Forewing is 5.5 mm long, with 1-Rs longer than 1-M; 2r-rs long and issues from the basal 1/3 of the pterostigma, both 1cu-a and 2cu-a reaching A; cell 1mcu narrowly triangular, nearly three times as long as wide; cell 1 + 2r pentagonal and surrounded by R, 1-Rs, 1Rs+M, (2Rs+M) + (2-RS), and 2r-rs shorter than 3r, both 1Rs+M and 2Rs distinctly longer than 2Rs+M [113].

***Novhelorus* Li, Shih & Ren, 2017**

Novhelorus Li, Shih & Ren, 2017, *Alcheringa*, 17, 6 [113] (original designation).

Type species: *Novhelorus macilentus* Li, Shih & Ren, 2017

Body slender. Forewing with 1-Rs longer than 1-M; the distance between the base of pterostigma and the origin of 1-Rs nearly as long as pterostigma length; 1Rs+M nearly as long as 2Rs+M; the position of Rs+M forking nearly at half distance between 1m-cu and 2r-rs, both 1cu-a and 2cu-a postfurcal.

Distribution and age: Liaoning; Early Cretaceous.

Two species included from the Cretaceous of Northern China (see Table 22.1).

Family Mesoserphidae Kozlov, 1970

Mesoserphidae, the only extinct family of Proctotrupoidea, comprise two subfamilies: Mesoserphinae Kozlov, 1970 and Karataoserphinae Rasnitsyn, 1994 [1]. Mesoserphids are small to medium-sized wasps (1.7–8.0 mm long as reported, some incomplete specimens may be slightly larger) with the following key characters: 11–18 (25 in one reported case) antennomeres; forewing with the first abscissa Rs (1-Rs) and M (1-M) combined longer than 1cu-a; 2r-rs approximately perpendicular to anterior wing margin; cell 1mcu usually rectangular; and venation likely reduced significantly. Metasoma short, ovoid or spindle-shaped, weakly sclerotized and ovipositor external with different length and forms, sheath not modified [114]. The earliest fossil records of Mesoserphidae are described from the Middle Jurassic Jiulongshan Formation of Northeastern China [110]. To date, 22 genera with 53 species in this family have been described [114–120]. The majority of mesoserphids have been reported from Kazakhstan and China from the Middle Jurassic to the Early Cretaceous [120].

Genera included from the Jurassic and Cretaceous of Northern China: *Mesoserphus* Kozlov, 1968, *Karataoserphus* Rasnitsyn, 1994, *Beipiaoserphus* Zhang & Zhang, 2000, *Sinoserphus* Shih, Feng & Ren, 2011, *Yanliaoserphus* Shih, Feng & Ren, 2011, *Amboserphus* Li, Rasnitsyn, Shih & Ren, 2017, *Apiciserphus* Li, Rasnitsyn, Shih & Ren, 2017, *Basiserphus* Li, Rasnitsyn, Shih & Ren, 2017, *Choriserphus* Li, Rasnitsyn, Shih & Ren, 2017, *Novserphus* Li, Rasnitsyn, Shih & Ren, 2017 and *Ozososerphus* Li, Rasnitsyn, Shih & Ren, 2017.

***Mesoserphus* Kozlov, 1968**

Mesoserphus Kozlov, 1968, *Academiya Nauk SSSR Otdelenie Obshchej Biologii, Moscow* [115] (original designation).

Type species: *Mesoserphus karatavicus* Kozlov, 1968.

Antenna with 16 antennomeres. Forewing with both 1cu-a and 2cu-a interstitial; the length of 1-M as long as 1m-cu; the forking of Rs+M located approximately 1/5 between 1m-cu and 2r-rs. Slender body with elongated elliptical metasoma, ovipositor longer than metasoma.

Distribution and age: Inner Mongolia; Middle Jurassic.

Only one species included from the Jurassic of Northern China (see Table 22.1).

Karataoserphus Rasnitsyn, 1994

Karataoserphus Rasnitsyn, 1994, *Paleontol. Zhur.*, 2, 115–119 [117] (original designation).

Type species: *Karataoserphus dorsoniger* Rasnitsyn, 1994.

Antenna with ≥20 antennomeres. Forewing with four enclosed cells, 1-Rs longer than 1-M; pterostigma narrower and long; 1cu-a postfurcal, 2cu-a postfurcal; the forking of Rs+M located closer to 1m-cu than to 2r-rs. Hind wing with enclosed cell r. Short ovipositor surpassing abdominal apex.

Distribution and age: Inner Mongolia; Middle Jurassic.

Three species included from the Jurassic and Cretaceous of Northern China (see Table 22.1).

Beipiaoserphus Zhang & Zhang, 2000

Beipiaoserphus Zhang & Zhang, 2000, *Entomotaxomomia*, 22, 279 [121] (original designation).

Type species: *Beipiaoserphus elegans* Zhang & Zhang, 2000.

Antenna with ≥20 antennomeres. Forewing with 1cu-a antefurcal; 2cu-a antefurcal; the forking of Rs+M located approximately the middle of the distance between 1m-cu and 2r-rs. Hind wing with enclosed cell r. Short ovipositor not or slightly surpassing abdominal apex.

Distribution and age: Liaoning; Early Cretaceous.

Only one species included from the Cretaceous of Northern China (see Table 22.1).

Sinoserphus Shih, Feng & Ren, 2011

Sinoserphus Shih, Feng & Ren, 2011, *Ann. Entomol. Soc. Am.*, 104, 1340 [110] (original designation).

Type species: *Sinoserphus wui* Shih, Feng & Ren, 2011.

The specific epithet is dedicated to James Wu for providing guidance, friendship, and inspiration to Dr. Chungkun Shih. Antenna with >19 antennomeres. Forewing with four enclosed cells: 3r, 1 + 2r, 1mcu and cua; both 1cu-a and 2cu-a interstitial; the position of Rs+M forking from 1/5 or 1/3 of the distance between 1m-cu and 2r-rs, closer to 1m-cu than to 2r-rs. Hind wing, if well-preserved, with the following tubular veins: C, R, M+Cu, Cu, M, r-m; cell r absent.

Distribution and age: Inner Mongolia; Middle Jurassic.

Six species included from the Jurassic and Cretaceous of Northern China (see Table 22.1).

Sinoserphus shihae Shih, Feng & Ren, 2011 (Figure 22.41)

Sinoserphus shihae Shih, Feng & Ren, 2011: *Ann. Entomol. Soc. Am.*, 104, 1341.

Locality and horizon: Daohugou, Inner Mongolia, China; Middle Jurassic, Jiulongshan Formation.

The specific name is dedicated to Dr. Zong Huai Shih for dedication and contribution to nanotechnology and for providing guidance and inspiration to Dr. Chungkun Shih. Length of body excluding antennae is 12.8 mm, head oval and large, eyes large. Antenna filiform, thick, and long with 22 antennomeres. Forewing with 2r-rs arising from about middle of pterostigma; C long and extending to apex of forewing; R robust; cell r closed and triangular; Rs straight reaching distal C. Cell 1+

(a)

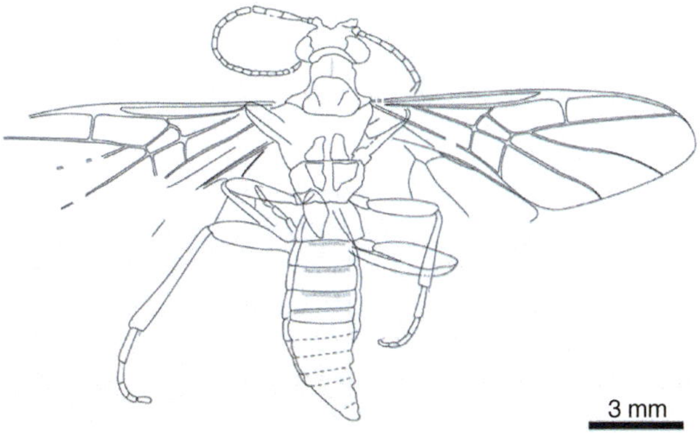

(b)

Figure 22.41 *Sinoserphus shihae* Shih, Feng & Ren, 2011 (Holotype CNU-HY-NN2008001) (a) Habitus; (b) Line drawing. Source: Donated by Dr. Chungkun Shih [110].

2r five-sided; M+Cu straight and distinct; M and Cu distinct; 1-Rs 1/3 as long as 1-M; 1cu-a in line with 1-M; and 2cu-aintersecting Cu in line (on right wing) with and slightly basal (on left wing) than 1m-cu. Hind wing with tubular veins C, R, M+Cu, Cu, M, and r-m; cell r absent. Metasoma long oval with eight terga discernible [110].

Yanliaoserphus Shih, Feng & Ren, 2011

Yanliaoserphus Shih, Feng & Ren, 2011, *Ann. Entomol. Soc. Am.*, 104, 1344 [110] (original designation).

Type species: *Yanliaoserphus jurassicus* Shih, Feng & Ren, 2011.

Forewing with pterostigma long and narrow; 2r-rs 1.3 times as long as pterostigmal width; first abscissa of Rs (1-Rs) as long as that of M (1-M); cell 1mcu rectangular with basal section of M (1-M) as long as 1m-cu; 1cu-a and 2cu-a distinct; 1m-cu distal; 1cu-a distal and both are short; forking of Rs+M located approximately 1/5 of the distance between the 1m-cu and 2r-rs. Hind wing with the following tubular veins: C, R, M+Cu, Cu, M, and r-m; cell r absent.

Distribution and age: Inner Mongolia; Middle Jurassic.

Only one species included from the Jurassic of Northern China (see Table 22.1).

Amboserphus Li, Rasnitsyn, Shih & Ren, 2017

Amboserphus Li, Rasnitsyn, Shih & Ren, 2017, *J. Syst. Palaeontol.*, 15, 623 [120] (original designation).

Type species: *Amboserphus tumidus* Li, Rasnitsyn, Shih & Ren, 2017.

Antenna with >15 antennomeres. Forewing with 1cu-a postfurcal, 2cu-a antefurcal; the forking of Rs+M located approximately one half of the distance between 1m-cu and 2r-rs. Ovipositor not surpassing abdominal apex.

Distribution and age: Liaoning; Early Cretaceous.

Three species included from the Cretaceous of Northern China (see Table 22.1).

Apiciserphus Li, Rasnitsyn, Shih & Ren, 2017

Apiciserphus Li, Rasnitsyn, Shih & Ren, 2017, *J. Syst. Palaeontol.*, 15, 623 [120] (original designation).

Type species: *Apiciserphus augustus* Li, Rasnitsyn, Shih & Ren, 2017.

Antenna with ≥20 antennomeres. Forewing with 1-Rs about 1.5 times as long as 1-M; 1-M longer than 1m-cu; 2r-rs arising from the distal 1/3 of pterostigma; 1cu-a postfurcal; 2cu-a postfurcal; the forking of Rs+M located approximately 1/3 between 1m-cu and 2r-rs, closer to 1m-cu. Hind wing with cell r open. Short ovipositor hardly surpassing abdominal apex.

Distribution and age: Inner Mongolia; Middle Jurassic.

Only one species included from the Jurassic of Northern China (see Table 22.1).

Basiserphus Li, Rasnitsyn, Shih & Ren, 2017

Basiserphus Li, Rasnitsyn, Shih & Ren, 2017, *J. Syst. Palaeontol*, 15, 627 [120] (original designation).

Type species: *Basiserphus loculatus* Li, Rasnitsyn, Shih & Ren, 2017.

Forewing with 1cu-a antefurcal; 2cu-a antefurcal; the forking of Rs+M located approximately 1/6 or 1/4 of the distance between 1m-cu and 2r-rs, closer to 1m-cu. Hind wing with following tubular veins: C, R, M+Cu, Cu, M, r-m; cell r absent.

Distribution and age: Inner Mongolia; Middle Jurassic; Liaoning; Early Cretaceous.

Two species included from the Jurassic and Cretaceous of Northern China (see Table 22.1).

Choriserphus Li, Rasnitsyn, Shih & Ren, 2017

Choriserphus Li, Rasnitsyn, Shih & Ren, 2017, *J. Syst. Palaeontol*, 15, 626 [120] (original designation).

Type species: *Choriserphus bellus* Li, Rasnitsyn, Shih & Ren, 2017.

Antenna with more than 15 antennomeres. Forewing with 1cu-a interstitial; 2cu-a postfurcal; the forking of Rs+M located approximately 1/5 or 1/3 of the distance between 1m-cu and 2r-rs. Hind wing, if well-preserved, with the following tubular veins: C, R, M+Cu, Cu, M, r-m; cell r absent. Short ovipositor hardly surpassing abdominal apex.

Distribution and age: Inner Mongolia; Middle Jurassic.

Two species included from the Jurassic of Northern China (see Table 22.1).

Novserphus Li, Rasnitsyn, Shih & Ren, 2017

Novserphus Li, Rasnitsyn, Shih & Ren, 2017, *J. Syst. Palaeontol*, 15, 628 [120] (original designation).

Type species: *Novserphus ningchengensis* Li, Rasnitsyn, Shih & Ren, 2017.

Forewing with four enclosed cells; 1-Rs longer than 1-M; 1-M nearly as long as 1m-cu; 2r-rs arising from the middle of pterostigma; 1cu-a postfurcal; 2cu-a interstitial; the forking of Rs+M located approximately 1/3 of the distance between 1m-cu and 2r-rs. Hind wing with enclosed cell r. Short ovipositor surpassing abdominal apex.

Distribution and age: Inner Mongolia; Middle Jurassic.

Only one species included from the Jurassic of Northern China (see Table 22.1).

Ozososerphus Li, Rasnitsyn, Shih & Ren, 2017

Ozososerphus Li, Rasnitsyn, Shih & Ren, 2017, *J. Syst. Palaeontol*, 15, 629 [120] (original designation).

Type species: *Ozososerphus lepidus* Li, Rasnitsyn, Shih & Ren, 2017.

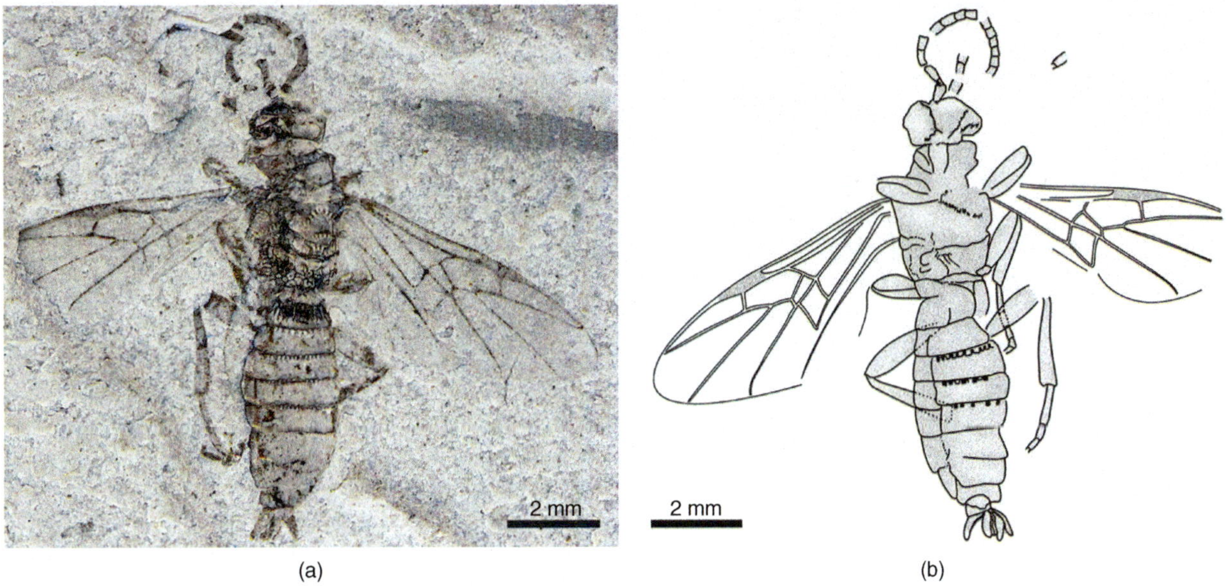

Figure 22.42 *Ozososerphus lepidus* Li, Rasnitsyn, Shih & Ren, 2017 (Holotype CNU-HYM-NN-2014013p) (a) Habitus; (b) Line drawing [120].

Antenna with >20 antennomeres. Forewing with four enclosed cells; 1cu-a interstitial; 2cu-a antefurcal; the forking of Rs+M located approximately from 1/4 to half of the distance between 1m-cu and 2r-rs. Hind wing, if well-preserved, with the following tubular veins: C, R, M+Cu, Cu, M, r-m; cell r absent. Short ovipositor surpassing abdominal apex.

Distribution and age: Inner Mongolia; Middle Jurassic.

Three species included from the Jurassic of Northern China (see Table 22.1).

***Ozososerphus lepidus* Li, Rasnitsyn, Shih & Ren, 2017 (Figure 22.42)**

Ozososerphus lepidus Li, Rasnitsyn, Shih & Ren, 2017: *J. Syst. Palaeontol*, 2017, 629.

Locality and horizon: Daohugou, Inner Mongolia, China; Middle Jurassic, Jiulongshan Formation.

Body length 9.58 mm. Head rounded (length 1.06 mm and width 1.71 mm); antenna partially preserved with 8 antennomeres. Mesosoma 1.6 times as long as wide. Metasoma long oval with eight segments, all segments nearly equal in length. Genitalia extruded, basal ring, parameres and aedeagus discernible. Forewing length 6.53 mm and width 2.45 mm, with 1-Rs as long as 1-M; 1-M nearly 2 times as long as 1m-cu; 2r-rs arising from the middle of pterostigma, straight, length slightly wider than pterostigma; cell 1 + 2r five-sided; 1-Rs origin at a distance from pterostigma (the distance nearly as long as 1-Rs + 1-M); the forking of Rs+M located approximately half of the distance between 1m-cu and 2r-rs [120].

Morphological Characters for Classifying Mesoserphidae

After surveying 62 Mesoserphidae specimens, we concluded that the following morphological diagnostic characters are necessary for identifying and placing specimens in the genera of Mesoserphidae.

1. Forewing with cell 1mcu four-sided.
2. Seven different types of pa, pp, aa, ip, pi, ia and ii formed by 1-M vs. 1cu-a, and 1m-cu vs. 2cu-a in the forewing (see Figure 22.43).
3. Three location patterns for the position of Rs+M forking in the forewing: 1), forking of Rs+M located approximately 1/5 of the distance between 1m-cu and 2r-rs, closer to 1m-cu than 2r-rs; 2), forking of Rs+M located approximately 1/4 or 1/3 of the distance between 1m-cu and 2r-rs, closer to 1m-cu than 2r-rs; and 3), forking of Rs+M located nearly 1/2 of the distance between 1m-cu and 2r-rs.
4. Free M in the forewing always far disconnected from 2-Rs+M or free M connected with 2-Rs+M.
5. Hind wing venation, if preserved, shows that M lost or present, cell r open or closed.
6. Ovipositor is diverse and variable, e.g. very long in *Mesoserphus*, but usually short in the remaining genera.
7. The numbers of antennomeres, most species have antenna with less than 20 antennomeres, ranging from 13 to 19, only species in *Beipiaoserphus*, *Sinoserphus* and *Ozososerphus* with more than 20 antennomeres [120].

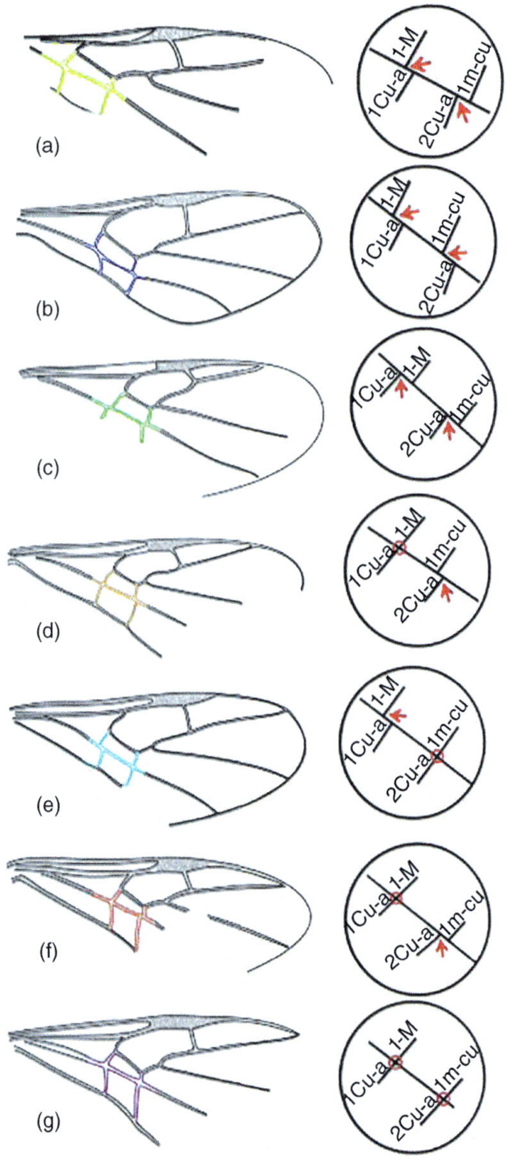

Figure 22.43 Line drawings of forewings in Mesoserphidae highlighting seven different types for the relative positions of forewing veins 1-M vs. 1cu-a, and 1m-cu vs. 2cu-a, which are designated as pa, pp, aa, ip, pi, ia and ii. (i, interstitial; a, antefurcal; p, postfurcal). (a) *Amboserphus tumidus*. (b) *Apiciserphus tenuis*. (c) *Basiserphus longa*. (d) *Choriserphus bellus*. (e) *Novserphus ningchengensis*. (f) *Ozososerphus lepidus*. (g) *Sinoserphus wui*.

Family Scolebythidae Evans, 1963

Scolebythidae are a very small family within the basal aculeate superfamily Chrysidoidea. Only four extant genera with six species and seven extinct genera with nine species have been described so far. For extinct species, only one genus with two species have been described based on compression fossils from the Yixian Formation of Northeastern China, the others are from amber [122]. The scolebythids have a primitive biological strait, like most other apocritans, that they are gregarious ectoparasitoids of wood-boring beetles in Cerambycidae and Anobiidae [123, 124]. The Scolebythidae are readily and easily identified by their unique apomorphy of large, exposed, diamond-shaped prosternum [125–127].

Only one genus included from the Jurassic and Cretaceous of Northern China: *Mirabythus* Cai, Shih & Ren, 2012.

Mirabythus Cai, Shih & Ren, 2012

Mirabythus Cai, Shih & Ren, 2012, *Zootaxa*, 3504, 58 [122] (original designation).

Type species: *Mirabythus lechrius* Cai, Shih & Ren, 2012.

Frontal prominence absent, clypeus short, conspicuously transverse. Mandible short and broad with notches on the anterior margin. Occipital carina present. Pronotal collar absent, dorsal surface of pronotum shortened. Propleura well-developed, anteriorly forming a short neck. Notauli present. Prosternum large, exposed, diamond-shaped. Parapsidal lines absent. Metanotum very short. Forewing cells 1cu, 2cu, r, 1 m, 2r1, and 1r1 closed; R_1 present and fusing with Rs apically; pterostigma relatively small, slightly convex on costal margin; r-m cross-vein present, nebular; 2cu-a cross-vein present, nebular; Cu vein present beyond 2cu-a cross-vein.

Distribution and age: Liaoning; Early Cretaceous.

Two species included from the Cretaceous of Northern China (see Table 22.1).

Mirabythus lechrius Cai, Shih & Ren, 2012 (Figure 22.44)

Mirabythus lechrius Cai, Shih & Ren, 2012: *Zootaxa*, 3504, 59.

Locality and horizon: Huangbanjigou, Liaoning, China; Lower Cretaceous, Yixian Formation.

Length of body excluding antennae is 9.2 mm, head rounded. Frontal prominence absent. Clypeus short, conspicuously transverse. Mandible with notches on the anterior margin. Ocelli apparently surpassing upper tangent of compound eyes. Occipital carina present. Pronotal collar absent, dorsal surface of pronotum shortened. Prosternum large, exposed, diamond-shaped. Parapsidal lines absent. Mesoscutal-mesoscutellar sulcus transverse. Metanotum very short, with a narrow transverse band, slightly prolonged posteriorly. Propodeum with median longitudinal groove extending from anterior margin to near posterior rim. Metasoma large and swollen, length 5.2 mm, slight sidewall compression [122].

Superfamily Evanioidea Latreille, 1802

The superfamily Evanioidea are supported by the articulation of the metasoma in a strongly dorsal position,

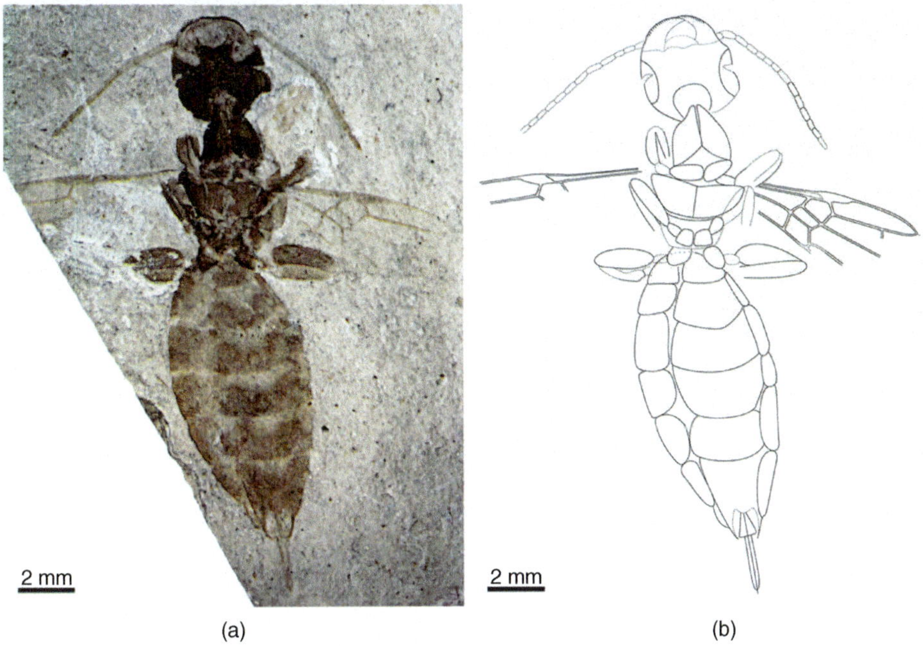

Figure 22.44 *Mirabythus lechrius* Cai, Shih & Ren, 2012 (Holotype CNU-HYM-LB2012105p) (a) Habitus; (b) Line drawing [122].

well above and separated from the metacoxal articulations, and other traits largely from the endoskeletal and musculature as well as molecular data [25]. Evanioidea comprise three extant families of distinctive wasps, i.e. Aulacidae, Gasteruptiidae and Evaniidae [1]. A few extinct families, e.g. Praeaulacidae, Anomopterellidae, Baissidae, Andreneliidae and Othniodellithidae, from the Mesozoic have been proposed to be within Evanioidea [128–130].

Family Anomopterellidae Rasnitsyn, 1975

Anomopterellidae, an extinct family of wasps, were originally classified as a family of the Evanioidea, and later lowered to a subfamily of Praeaulacidae. As mentioned before, Li *et al.*, in 2013, restored Anomopterellinae as a family with *Anomopterella* Rasnitsyn, 1975, and two additional genera *Synaphopterella* and *Choristopterella*, based on new fossil specimens from Daohugou and phylogenetic analyses of Evanioidea. The phylogenetic analyses also suggest that the extinct family Praeaulacidae are the most basal group of Evanioidea, and Anomopterellidae are the second family of Evanioidea described from the Middle Jurassic of China [131].

Genera included from the Jurassic and Cretaceous of Northern China: *Anomopterella* Rasnitsyn, 1975 and *Synaphopterella* Li, Rasnitsyn, Shih & Ren, 2013.

Anomopterella Rasnitsyn, 1975

Anomopterella Rasnitsyn, 1975, *Academy of Science of the USSR*, 90 [129] (original designation).

Type species: *Anomopterella mirabills* Rasnitsyn, 1975.

Forewing venation with 1r-rs absent or rudimentary like a stub; M+Cu not aligned with RS+M; 1-M present; 2r-rs and 2m-cu basad of 3r-m; vein cu-a interstitial or slightly postfurcal. Metasoma broadest beyond its mid-length, with the first segment gradually broadening distally and longer than any other segments. Ovipositor short and only slightly extending beyond metasomal apex, with sheaths shorter than basal sclerite.

Distribution and age: Inner Mongolia; Middle Jurassic.

Eight species included from the Jurassic of Northern China (see Table 22.1).

Anomopterella pygmea Li, Shih & Ren, 2014 (Figure 22.45)

Anomopterella pygmea Li, Shih & Ren, 2014: *Geolo. Carpath.*, 65, 366.

Locality and horizon: Daohugou, Inner Mongolia, China; Middle Jurassic, Jiulongshan Formation.

Head round, compound eyes large, antenna with 25 antennomeres. Wings well-preserved, with 1-Rs longer than 1-M; 1r-rs absent; 2r-rs issuing from pterostigma nearly at the apex of it; 3-Rs longer than 3r-m; cell 3r broad; cell 2 + 3rm wider than 2mcu, in contact with 1mcu by a point; cu-a interstitial and curved, nearly as long as 1-Rs. Hind wing with 1-Rs, 1-M and Rs + Mforming "Y-shape"; 1-Rs nearly as long as 1-M; cu-a present and curved. Metasomal short with six segments, the first metasomal segment particularly narrow like a distinct petiole [132].

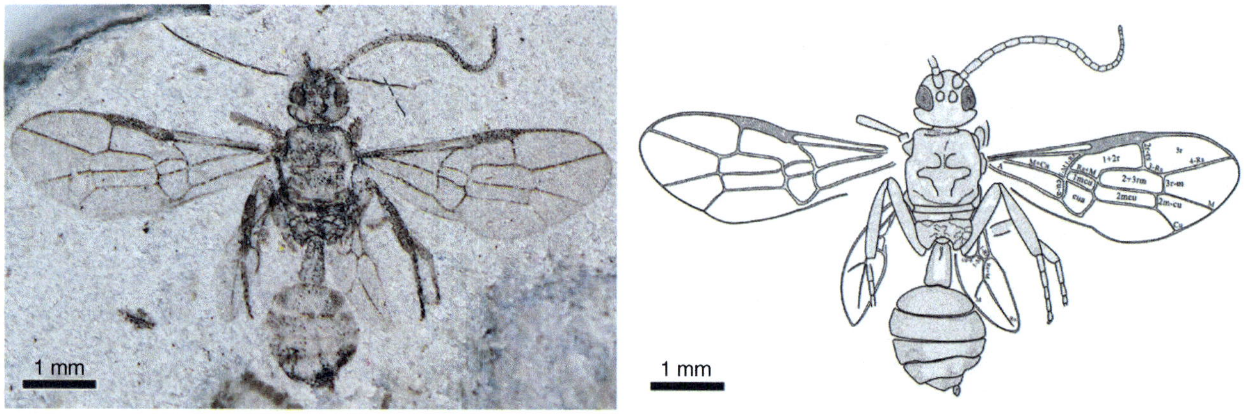

(a) (b)

Figure 22.45 *Anomopterella pygmea* Li, Shih & Ren, 2014 (Holotype CNU-HYM-NN-2012045) (a) Habitus; (b) Line drawing [132].

Synaphopterella Li, Rasnitsyn, Shih & Ren, 2013

Synaphopterella Li, Rasnitsyn, Shih & Ren, 2013, *PLoS ONE*, 8, 4 [131] (original designation).

Type species: *Synaphopterella patula* Li, Rasnitsyn, Shih & Ren, 2013.

Antenna moderately thick. Mesosoma short, pronotum and propodeum both short. Forewing with 2r-rs meeting pterostigma near apex; 1r-rs rudimentary; Rs+M aligned with M+Cu (1-M lost); 3r-m and 2m-cu coincide; 1-Cu subvertical and about as long as 1cu-a.

Distribution and age: Inner Mongolia; Middle Jurassic.

Only one species included from the Jurassic of Northern China (see Table 22.1).

Family Evaniidae Latreille, 1802

Evaniidae, a family of parasitoid wasps, are commonly called "ensign wasps" due to their long and thin petiole and shape of metasoma. This family, distributed all over the world except for the Polar regions, comprise 20 extant genera with over 400 described species [133]. Evaniidae are distinguished from other hymenopterans by two possible synapomorphies: (i) loss of all functional metasomal spiracles except those on the seventh segments and (ii) the articulation of the metasoma and mesosoma arising from near the dorsal-most surface of the mesosoma [134].

Only one genus included from the Jurassic and Cretaceous of Northern China: *Procretevania* Zhang & Zhang, 2000.

Procretevania Zhang & Zhang, 2000

Procretevania Zhang & Zhang, 2000, *Acta Micropalaeontol. Sin.* 47, 49 [135] (original designation).

Type species: *Procretevania pristina* Zhang and Zhang, 2000.

Maxillary palpi 4-segmented. Mesosoma high with pronotum short; mesonotum long with scutellum broad, short and slightly convex; propodeum, mesopleura and metapleura areolated. Forewing with basal section of Rs short and inclined toward wing base; cell rm open. Petiole slightly shorter than mesosoma. Metasoma with the first segment short and semiovoid; other part spherical, nearly as long as mesosoma. Ovipositor comparatively short.

Distribution and age: Liaoning; Early Cretaceous.

Four species included from the Cretaceous of Northern China (see Table 22.1).

Procretevania mitis Li, Shih & Ren, 2014 (Figure 22.46)

Procretevania mitis Li, Shih & Ren, 2014: *Cretac. Res.*, 47, 49.

Locality and horizon: Huangbanjigou, Liaoning, China; Lower Cretaceous, Yixian Formation.

Length of body excluding antennae is 11.3 mm. Pronotum very short; propleuron transversely wide; mesopleuron areolate; metanotum very short. Propodeum arched toward the attachment with petiole. Forewing with 1-Rs distinctly shorter than 1-M; pterostigma narrow, long, parallel-sided; 1 + 2r cell slightly longer than 3r cell; 3Rs slightly bent posteriorly a little beyond 2r-rs; 2 M geniculate anteriorly slightly basad of 2r-rs, and 3 M slightly bent; 1m-cu subparallel to 1-M; cu-a postfurcal; A partially discernible. Metasoma with the first segment short, half as long as petiole, triangle-shaped in lateral aspect, remaining part ovate [136].

Family Praeaulacidae Rasnitsyn, 1972

Praeaulacidae, an extinct family, have been proposed as the ancestral group of the Evanioidea [137]. Up to date, 20 extinct genera belonging to three subfamilies have been reported: Cretocleistogastrinae from the Early Cretaceous of East Asia and Australia, Praeaulacinae from the

Figure 22.46 *Procretevania mitis* Li, Shih & Ren, 2014 (Holotype CNU-HYM-LB-2013001) [136].

Late Jurassic of South Kazakhstan and Nevaniinae from the Middle Jurassic of Northeastern China. As aforementioned, the previously included Anomopterellinae with a genus, *Anomopterella* Rasnitsyn, 1975, from the Late Jurassic of South Kazakhstan [127, 129], was restored to family status as Anomopterellidae to include two additional genera *Synaphopterella* and *Choristopterella* based on new fossil specimens from Daohugou and a phylogenetic analysis of Evanioidea [131].

Genera included from the Jurassic and Cretaceous of Northern China: *Praeaulacus* Rasnitsyn, 1972, *Aulacogastrinus* Rasnitsyn, 1983, *Sinowestratia* Zhang & Zhang, 2000, *Nevania* Zhang & Rasnitsyn, 2007, *Sinaulacogastrinus* Zhang & Rasnitsyn, 2008, *Eosaulacus* Zhang & Rasnitsyn, 2008, *Eonevania* Rasnitsyn & Zhang, 2010, *Sinevania* Rasnitsyn & Zhang, 2010 and *Archaulacus* Li, Shih & Ren, 2014.

Praeaulacus Rasnitsyn, 1972

Praeaulacus Rasnitsyn, 1972, *Paleontol. Zhur.*, 6, 427 [127] (original designation).

Type species: *Praeaulacus ramosus* Rasnitsyn, 1972.

Forewing with first abscissa of Rs (1-Rs) not longer than its distance to pterostigma; 2r-rs meeting Rs basad of 2r-m (or, unusually, at the point where 2r-m meets Rs, and on such a condition 3rm longer than 2rm); 2r-rs shorter or, unusually, slightly longer than maximal width of 2rm; 3rm slightly shorter than 2rm, with outer margin (3r-m) not strongly oblique; 2rm in contact with 1mcu by a short section of M (2 M). Hind wing with M and Cu possessing free ends; cu-a meeting Cu distad of M+Cu fork, and oblique. The first metasomal segment conical or slightly elongate basally. Ovipositor long with sheaths at least two-thirds length of forewing.

Distribution and age: Inner Mongolia; Middle Jurassic.

Eleven species included from the Jurassic of Northern China (see Table 22.1).

Praeaulacus obtutus Li, Shih & Ren, 2014 (Figure 22.47)

Praeaulacus obtutus Li, Shih & Ren, 2014: *J. Nat. Hist.*, 49, 831.

Locality and horizon: Daohugou, Inner Mongolia, China; Middle Jurassic, Jiulongshan Formation.

Length of body excluding antennae is 6.2 mm, head medium-sized with large eyes and irregular oval-shape, antenna with 16 antennomeres. Mesosoma stout, propodeum short, but broad in profile aspect. Forewing with 1-Rs shorter than 1-M; 2r-m nearly straight and 3r-m slightly bending; 2rm longer than 3rm; 1mcu nearly parallelogram, about twice as long as wide; cu-a interstitial and longer than 1-M. Hind wing with r-m straight, shorter than 1Rs and 1-M; cu-a strongly oblique; vein A present. Metasoma with the first segment petiole-like, propodeal foramen attached high on propodeum and much closer to metanotum than to hind coxa [138].

Amber Praeaulacid from Myanmar

Habraulacus zhaoi Li, Rasnitsyn, Shih & Ren, 2015 was collected from Kachin (Hukawng Valley) of northern Myanmar, mid-Cretaceous (98.79 ± 0.62 Mya [101]. The specific name is dedicated to Renhao Zhao for his generous donation of the amber fossil for scientific research. Body length about 5.6 mm. Head large, antenna filiform, thin and long, with more than 30 antennomeres. Mesosoma short and stout, only the contour lines visible. Metasoma elongate with the first segment about five times as long as its maximum width, apparently attached near propodeal posterior. Forewing with pterostigma long and narrow; 1-Rs arising at a distance from pterostigma; cu-a interstitial; 2r-rs arising from basal 1/3 of pterostigma, directed slightly oblique apicad, both 2rm and 3r-m present well; 2r-m slightly basal than 2m-cu; 3r-m longer than 2r-m, V-like bent; 2m-cu nearly in line with 2r-m; cell 2mcu long and narrow [139] (see Figure 22.48).

Aulacogastrinus Rasnitsyn, 1983

Aulacogastrinus Rasnitsyn, 1983, *Paleontol. J.*, 61, 156 [116] (original designation).

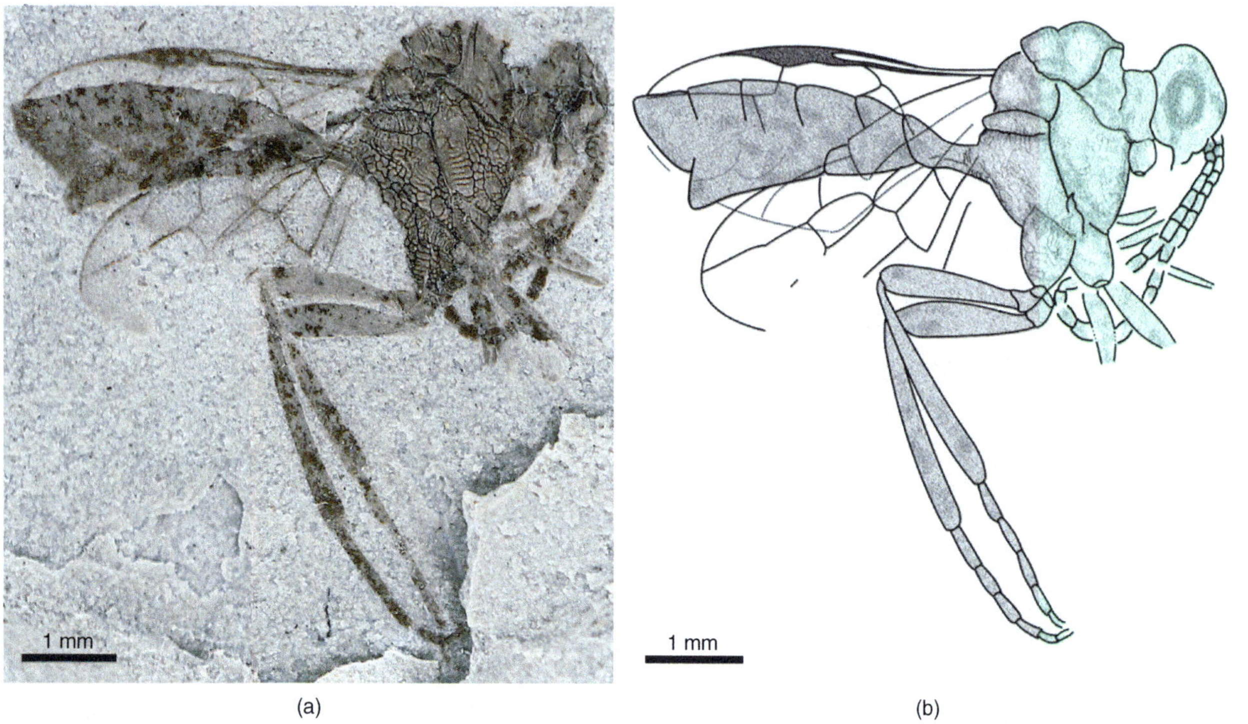

Figure 22.47 *Praeaulacus obtutus* Li, Shih & Ren, 2014 (Holotype CNU-HYM-NN-2012039p) (a) Habitus; (b) Line drawing [138].

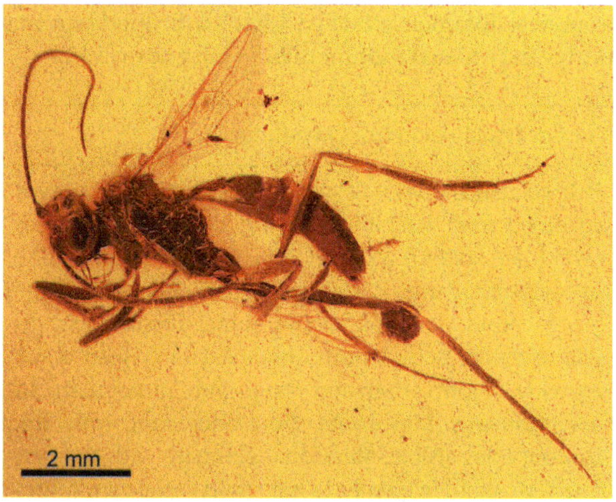

Figure 22.48 *Habraulacus zhaoi* Li, Rasnitsyn, Shih& Ren, 2015 (Holotype, CNU-HYM-MA-2014001) [139]. Source: Donated by Mr. Renhao Zhao.

Type species: *Aulacogastrinus ater* Rasnitsyn, 1983.

Head transverse, not elongate. Forewing with Rs origin far from pterostigma; 2r-rs longer than maximal width of 2rm, meeting Rs just at 2r-m or basad of that; 3rm not distinctly oblique; 2rm in contact with 1mcu by short 2 M. Hind wing with cu-a meeting M+Cu near forking of M+Cu into M and Cu. Metasoma thick with the first segment elongate, contracted and tubular basally, and bell-mouthed distally. Ovipositor sheaths at least as long as forewing.

Distribution and age: Inner Mongolia; Middle Jurassic.

Two species included from the Jurassic of Northern China (see Table 22.1).

Nevania Zhang & Rasnitsyn, 2007

Nevania Zhang & Rasnitsyn, 2007, *Insect Syst. Evol.*, 38, 2 [140] (original designation).

Type species: *Nevania robusta* Zhang & Rasnitsyn, 2007.

Medium sized, antenna inserted far above mandibles, usually consisting of over 28 segments. Mesosoma short and slightly high; pronotum and mesoscutum transversely, and mesopleura archedly ridged; metapostnotum very short; propodeum dorsally and laterally reticulate. Forewing venation complete with 11 closed cells; 1r-rs spectral or completely absent; cu-a interstitial or slightly postfurcal; 2A complete. Hind wing venation complete with C, Rs and M reaching wing margin; r closed; first section of Cu smoothly connected with cu-a, free Cu vestigial or completely absent; A distinctly angulate quite basad of cu-a. Metasoma with the first and second segments long tube-like, swollen apically, and subequal in length.

Distribution and age: Inner Mongolia; Middle Jurassic.

Eight species included from the Jurassic of Northern China (see Table 22.1).

Sinaulacogastrinus Zhang & Rasnitsyn, 2008

Sinaulacogastrinus Zhang & Rasnitsyn, 2008, *J. Syst. Palaeontol.*, 6, 481 [137] (original designation).

Type species: *Sinaulacogastrinus eucallus Zhang* & Rasnitsyn, 2008.

Head transversely broad. Forewing with Rs origin far from pterostigma; 2r-rs meeting Rs quite basad of 2r-m and about as long as maximal width of 2rm; 3r-m subvertical; 2rm about as long as 3rm; 2rm in contact with 1mcu by a point. Hind wing with M straight, not forming an obvious angle; free M and Cu distinct; cu-a meeting M+Cu basad of its forking into M and Cu. The first metasomal segment conical for basal half and rectangular for distal half. Ovipositor with sheaths longer than forewing.

Distribution and age: Inner Mongolia; Middle Jurassic.

Two species included from the Jurassic of Northern China (see Table 22.1).

Eosaulacus Zhang & Rasnitsyn, 2008

Eosaulacus Zhang & Rasnitsyn, 2008, *J. Syst. Palaeontol.*, 6, 481 [137] (original designation).

Type species: *Eosaulacus giganteus* Zhang & Rasnitsyn, 2008.

Body large, antenna filiform with over 30 antennomeres. Forewing with Rs origin at base of pterostigma; the first abscissa of Rs much shorter than that of M; 2r-rs originating basad of pterostigmal mid-length, meeting Rs just at the point where 2r-m joins, and greatly longer than maximal width of 2rm; 3r-m greatly oblique and sinuate; 2rm shorter than 3rm in terms of their posterior margins. Hind wing with M arched before r-m; cu-a meeting M+Cu basad of its forking into M and Cu, and obviously oblique; free M, Cu and A present. The first metasomal segment long conical and transversely ridged. Ovipositor with sheaths longer than hind wing.

Distribution and age: Inner Mongolia; Middle Jurassic.

Two species included from the Jurassic of Northern China (see Table 22.1).

Eonevania Rasnitsyn & Zhang, 2010

Eonevania Rasnitsyn & Zhang, 2010, *Acta Geol. Sin.-Engl.*, 84, 854 [141] (original designation).

Type species: *Eonevania robusta* Rasnitsyn & Zhang, 2010.

Forewing with area between 1A and 2A not enlarged and with 1a-2a present. Two basal metasomal segments are not really tube-like, even though distinctly thinner than remaining ones.

Distribution and age: Inner Mongolia; Middle Jurassic.

Only one species included from the Jurassic of Northern China (see Table 22.1).

Eonevania robusta Rasnitsyn & Zhang, 2010 (Figure 22.49)

Eonevania robusta Rasnitsyn & Zhang, 2010: *Acta Geol. Sin.-Engl.*, 84, 854.

Locality and horizon: Daohugou, Inner Mongolia, China; Middle Jurassic, Jiulongshan Formation.

Head small with eyes large and ovoid, antenna with 18 antennomeres. Pronotum very short. Mesonotum with notauli V-shaped and reaching transscutal suture. Forewing with first abscissa of Rs (1-Rs) 1.7 times as long as that of M (1-M); 2rm and 1mcu in contact by a point; cu-a meeting Cu quite beyond M+Cu forking; 2A complete; 1a-2a slightly shorter than cu-a, and 2a slightly longer than cua. Metasoma with the first segment slightly thinner than propodeal foramen, and gradually becoming thick toward its apex, nearly trapezoid, the second very similar to the first. Ovipositor short and slightly exerted [141].

Sinevania Rasnitsyn & Zhang, 2010

Sinevania Rasnitsyn & Zhang, 2010, *Acta Geol. Sin.-Engl.*, 84, 854 [141] (original designation).

Type species: *Sinevania speciosa* Rasnitsyn & Zhang, 2010.

The forewing basal vein (Rs and M) is slightly angular at junction of Rs+M. The first metasomal segment is almost tube-like except that it is slightly bulged at its basal third where at least one spiracle is observed.

Distribution and age: Inner Mongolia; Middle Jurassic.

Figure 22.49 *Eonevania robusta* Rasnitsyn & Zhang, 2010 (Holotype NND0029/NIGP151947) [141]. Source: Photo provided by Dr. Haichun Zhang.

Only one species included from the Jurassic of Northern China (see Table 22.1).

***Archaulacus* Li, Shih & Ren, 2014**

Archaulacus Li, Shih & Ren, 2014, *Zootaxa*, 3814, 433 [142] (original designation).

Type species: *Archaulacus probus* Li, Shih & Ren, 2014.

Forewing with Rs originating far from pterostigma, pterostigma narrow, long, parallel-sided; the first abscissa of Rs subvertical to R; 2r-rs meeting Rs quite basad of 2r-m and slightly longer than the maximal width of 2rm; 2m-cu slightly basad of 2r-m, 2rm slightly longer and narrower than 3rm; 2rm in contact with 1mcu by a point; cu-a postfurcal. Hind wing with C present. Metasomal long-ovoid with the first metasomal segment nearly triangular, ovipositor longer than forewing.

Distribution and age: Inner Mongolia; Middle Jurassic.

Only one species included from the Jurassic of Northern China (see Table 22.1).

***Archaulacus probus* Li, Shih & Ren, 2014 (Figure 22.50)**

Archaulacus probus Li, Shih & Ren, 2014: *Zootaxa*, 3814, 433.

Locality and horizon: Daohugou, Inner Mongolia, China; Middle Jurassic, Jiulongshan Formation.

Head medium-sized, antenna with 20 antennomeres as preserved. Forewing with Rs origin far from pterostigma; first abscissa of Rs (1-Rs) subvertical to R and Rs+M; 1-Rs shorter than that of 1-M; 2r-rs meeting Rs quite basad of 2r-m and slightly longer than the maximal width of 2rm; 2m-cu slightly basad of 2r-m; 2rm in contact with 1mcu by a point; cu-a postfurcal and slightly shorter than first abscissa of M (1-M). Metasoma long-ovoid, with seven segments. Ovipositor slightly longer than forewing [142].

Figure 22.50 *Archaulacus probus* Li, Shih & Ren, 2014 (Holotype CNU-HYM-NN-2012033) [142].

Family Ephialtitidae Handlirsh, 1906

The extinct Ephialtitidae are considered as the second most basal group in the suborder Vespina of Hymenoptera. Together with the most basal group of Karatavitidae, they jointly constitute the superfamily Ephialtitoidea [143]. In recent literature [131, 141], Karatavitidae have been transferred to Orussoidea to form the stem group of Vespina, while Ephialtitidae moved to Stephanoidea as the stem group of Apocrita. The Stephanoidea were also treated as basal to Evanioidea and (Ceraphronomorpha + Proctotrupomorpha + [Ichneumonomorpha + Vespomorpha]). Ephialtitidae comprise two subfamilies: Ephialtitinae Handlirsh, 1906 and Symphytopterinae Rasnitsyn, 1980. Up to date, 29 genera with 77 species of ephialtitids have been described [144].

Genera included from the Jurassic and Cretaceous of Northern China: *Asiephialtites* Rasnitsyn, 1975, *Proapocritus* Rasnitsyn, 1975, *Stephanogaster* Rasnitsyn, 1975, *Symphytopterus* Rasnitsyn, 1975, *Karataus* Rasnitsyn, 1977, *Sinephialtites* Zhang, 1985, *Crephanogaster* Rasnitsyn, 1990, *Praeproapocritus* Rasnitsyn & Zhang, 2010, *Acephialtitia* Li, Shih, Rasnitsyn & Ren, 2015 and *Proephialtitia* Li, Shih, Rasnitsyn & Ren, 2015.

***Asiephialtites* Rasnitsyn, 1975**

Asiephialtites Rasnitsyn, 1975, *Academy of Science of the USSR*, 30 [129] (original designation).

Type species: *Asiephialtites niger* Rasnitsyn, 1975.

Forewing with cu-a interstitial; 1-Rs distinctly subvertical; 3r-m and 2m-cu present; 1a-2a lost; cell 3rm much longer than 1mcu. Hind wing r cell closed and 1-M shorter and nearly straight. Ovipositor long.

Distribution and age: Inner Mongolia; Middle Jurassic.

Only one species included from the Jurassic of Northern China (see Table 22.1).

***Proapocritus* Rasnitsyn, 1975**

Proapocritus Rasnitsyn, 1975, *Academy of Science of the USSR*, 22 [129] (original designation).

Type species: *Proapocritus praecursor* Rasnitsyn, 1975.

Forewing with the first abscissa of Rs directed slightly postero distally; 1r-rs present or absent; 3r-m and 2m-cu tubular; cu-a interstitial or slightly postfurcal; 2A at least rudimentary; 2a closed. Hind wing with C present; Rs originating not based of M+Cu forking.

Distribution and age: Inner Mongolia; Middle Jurassic.

Eight species included from the Jurassic and Cretaceous of Northern China (see Table 22.1).

Symphytopterus Rasnitsyn, 1975

Symphytopterus Rasnitsyn, 1975, *Academy of Science of the USSR*, 42 [129] (original designation).

Type species: *Symphytopterus nigricornis* Rasnitsyn, 1975.

Forewing venation complete: 1r-rs absent or rudiment or distinctly present; cu-a postfurcal or interstitial; 2r-m, 3r-m, 2m-cu and a1-a2 present. Hind wing with cell r closed or not closed; r-m usually shorter than 1-Rs; M before r-m usually curved, less direct; Cu developed, reaching wing margin. Abdomen thickened, especially widest approximately midway. Ovipositor short.

Distribution and age: Inner Mongolia; Middle Jurassic.

Only one species included from the Jurassic of Northern China (see Table 22.1).

Symphytopterus graciler Wang, Li & Shih, 2015 (Figure 22.51)

Symphytopterus graciler Wang, Li & Shih, 2015: *Insect Syst. Evol.*, 46, 474.

Locality and horizon: Daohugou, Inner Mongolia, China; Middle Jurassic, Jiulongshan Formation.

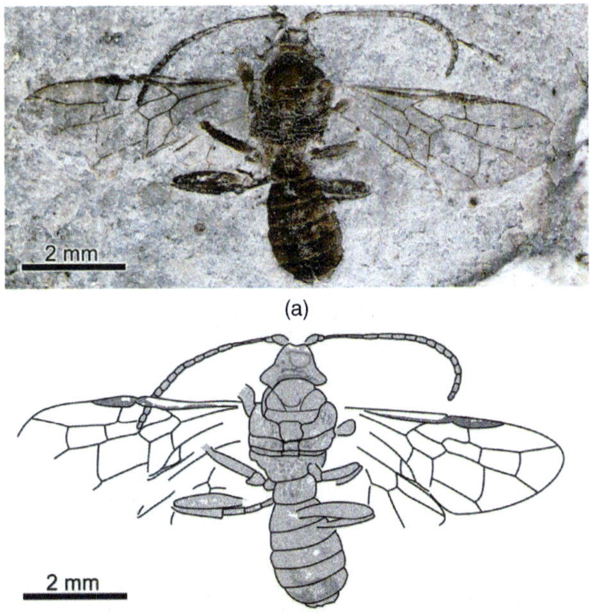

(a)

(b)

Figure 22.51 *Symphytopterus graciler* Wang, Li & Shih, 2015 (Holotype CNU-HYM-NN-2012036) (a) Habitus; (b) Line drawing [145].

Head medium-sized, antenna with 18 antennomeres. Pronotum short, mesonotum with notauli U-shaped, and reaching transscutal suture. Forewing with pterostigma acuminate, issuing 2r-rs at its 1/3 length; the first abscissa of Rs slightly longer than that of M; 2r-rs 1.5 times as long as pterostigma width and equal to maximal width of 2rm; 2r-m subvertical; 3r-m slightly bent; 1m-cu meeting 2rm beyond Rs + M fork; cu-a postfurcal; 1a-2a present; 2A nearly complete. Metasomal broad-ovoid, with six segments visible [145].

Karataus Rasnitsyn, 1977

Karataus Rasnitsyn, 1977, *Paleontol. J.*, 11, 103 [41] (original designation).

Type species: *Karataus pedalis* Rasnitsyn, 1977.

Antenna with at least 20 antennomeres. Hind femur greatly swollen in males but just incrassate in females. Forewing with basal section of Rs strongly oblique toward wing base; 1r-rs absent to complete; 2r-m closer to 2m-cu than to 2r-rs; 2rm longer than 3rm; 1m-cu joining M slightly beyond bifurcation of Rs+M; Rs+M forking nearly at the level of pterostigmal base; cu-a much distad of M+Cu forking; 2A present basad of a1-a2; a1-a2 complete and cell 2a closed. Hind wing with basal section of Rs much longer than r-m; r closed at wing anterior margin; r-m meeting r at its hind margin before its mid-length. Metasoma fusiform and widest beyond its mid-length. Ovipositor slightly extending beyond metasomal apex.

Distribution and age: Inner Mongolia; Middle Jurassic.

Six species included from the Jurassic of Northern China (see Table 22.1).

Crephanogaster Rasnitsyn, 1990

Crephanogaster Rasnitsyn, 1990, *Academy of Science of the USSR*, 239, 187 [107].

Type species: *Crephanogaster fomorata* Rasnitsyn, 1990.

Antenna with 26 antennomeres. Forewing venation complete; 1r-rs rudiment, 2A, a1–a2 present; basal section of Rs nearly perpendicular to R; cells 2rm and 3rm comparatively short; cu-a slightly postfurcal or interstitial. Hind wing with cell r closed.

Distribution and age: Liaoning; Early Cretaceous.

Only one species included from the Cretaceous of Northern China (see Table 22.1).

Praeproapocritus Rasnitsyn & Zhang, 2010

Praeproapocritus Rasnitsyn & Zhang, 2010, *Acta. Geol. Sin.-Engl.*, 84, 852 [141] (original designation).

Type species: *Praeproapocritus vugates* Rasnitsyn & Zhang, 2010.

Antenna usually has over 30 antennomeres. Forewing with the first abscissa of Rs directed slightly posterodistally; 1r-rs, 3r-m and 2m-cu present; 2A complete but slightly curved subbasally, and the metanotal cenchri very small but distinguishable. Hind wing with Rs origination distad of M+Cu forking.

Distribution and age: Inner Mongolia; Middle Jurassic.

Two species included from the Jurassic of Northern China (see Table 22.1).

Praeproapocritus flexus **Li, Shih & Ren, 2013 (Figure 22.52)**

Praeproapocritus flexus Li, Shih & Ren, 2013: *Acta. Geol. Sin.-Engl.*, 87, 1488.

Locality and horizon: Daohugou, Inner Mongolia, China; Middle Jurassic, Jiulongshan Formation.

Length of body excluding antennae is 14.7 mm. Head and eyes large, antenna with 33 antennomeres. Mesosoma much longer than wide; pronotum short; mesonotum with scutum slightly longer than scutellum, notauli V-shaped and reaching transscutal suture; cenchri very small; propodeum nearly trapezoid. Forewing with 1-Rs distinctly shorter than 1-M; 1r-rs complete and subparallel to 2r-rs; 2r-m shorter than 3r-m; 1mcu nearly rectangle, about twice as long as wide; 2m-cu slightly curved; cu-a interstitial; 2A complete; a_1-a_2 well-presented. Hind wing with 1Rs, r-m, 1 M, cu-a, M+Cu and A. Metasoma with eight segments visible [146].

Acephialtitia **Li, Shih, Rasnitsyn & Ren, 2015**

Acephialtitia Li, Shih, Rasnitsyn & Ren, 2015, *BMC Evol. Biol.*, 15, 3 [144] (original designation).

Type species: *Acephialtitia colossa* Li, Shih, Rasnitsyn & Ren, 2015.

Forewing with 1-Rs reclival, shorter than 1-M; 1r-rs, 2r-rs, 2r-m, 3r-m, 2 m-cu present; 1r-rs very long, subparallel to RS + M; 2r-m and 3r-m subvertical, distant for much more than their length; cu-a slightly postfurcal; 2r-rs issuing from the mid-length of pterostigma; cell 1mcu in contact with 2rm by a point; cell 2rm shorter than 3rm, both much shorter than 1mcu, 2rm base distal comparing pterostigmal base; hind wing with Rs, M, Cu and r-m; 1-M gently curved; cu-a slightly postfurcal. Mesosoma long, not specialized. Metasoma broadly attaching to propodeum, metasoma slightly, smoothly widening rearwards. Legs slender, ordinary, trochantellus present. Ovipositor much longer than body, gently bent downwards.

Distribution and age: Liaoning; Early Cretaceous.

Only one species included from the Jurassic of Northern China (see Table 22.1).

Acephialtitia colossa **Li, Shih, Rasnitsyn & Ren, 2015 (Figure 22.53)**

Acephialtitia colossa Li, Shih, Rasnitsyn & Ren, 2015: *BMC Evol. Biol.*, 15, 3.

Locality and horizon: Huangbanjigou, Liaoning, China; Lower Cretaceous, Yixian Formation.

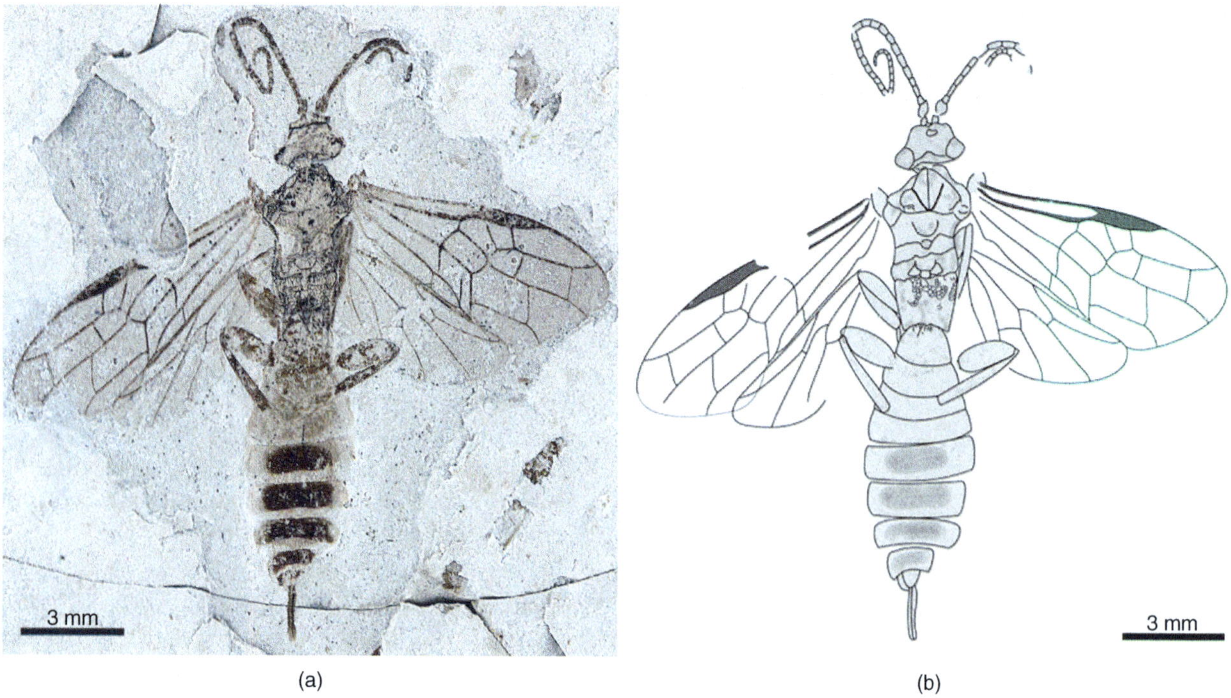

Figure 22.52 *Praeproapocritus flexus* Li, Shih & Ren, 2013 (Holotype CNU-HYM-NN-2012031p) (a) Habitus; (b) Line drawing [146].

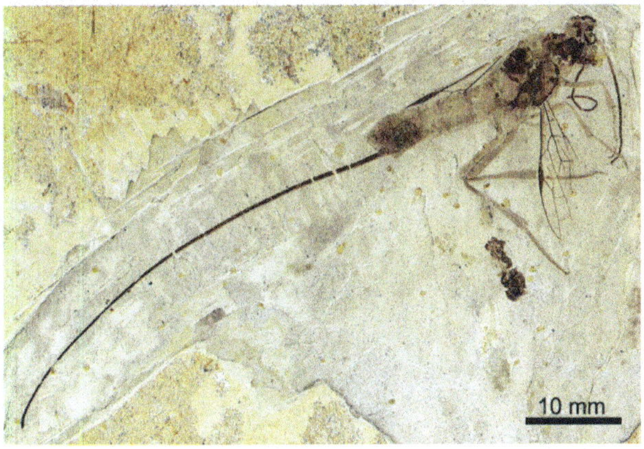

Figure 22.53 *Acephialtitia colossa* Li, Shih, Rasnitsyn & Ren, 2015 (Holotype CNU-HYM-LB-2013004) [144].

Figure 22.54 Reconstruction of *Acephialtitia colossa* Li, Shih, Rasnitsyn & Ren, 2015 (Holotype CNU-HYM-LB-2013004). Source: Artwork by Dr. Chen Wang.

Length of body excluding antennae and ovipositor is 28.3 mm. Head short with a large eye, antenna consisting of over 25 antennomeres. Mesosoma longer than high; pronotum moderately short; mesonotum transversely ridged; transverse suture straight; axillae evident; metapostnotum narrow but distinct. Forewing with 1-Rs shorter than 1-M; 2r-rs issuing from pterostigma slightly basad of its mid-length; 1r-rs long, directed to RS base; 2r-m shorter than 3r-m, both cross-veins slightly curved, subvertical; cu-a slightly postfurcal. Hind wing with Rs, M, Cu and r-m preserved; 1-M long, gently curved; cu-a slightly postfurcal. Metasoma broadly attaching to propodeum, with eight segments visible. Ovipositor slim and bent slightly downward with a sheath, approximately 50.6 mm long [144] (see Figure 22.54).

***Proephialtitia* Li, Shih, Rasnitsyn & Ren, 2015**

Proephialtitia Li, Shih, Rasnitsyn & Ren, 2015, *BMC Evol. Biol.*, 15, 6 [144] (original designation).

Type species: *Proephialtitia acantha* Li, Shih, Rasnitsyn & Ren, 2015.

Forewing with 1-Rs slightly reclival, shorter than 1-M and distinctly angular to it (for about 120°), and distant from pterostigma for almost three times its own length; 1r-rs short or absent; 2r-rs, 2rm, 3r-m, 2m-cu, cu-a and a1-a2 present; base of cell 2rm proximal in respect of pterostigma base; 2r-m and 3r-m subvertical, distant for much more than their own lengths; cu-a interstitial; hind wing with Rs, M, Cu and r-m, cu-a present; 1-M gently curved; cu-a slightly antefurcal. Ovipositor almost straight, much longer than body.

Distribution and age: Inner Mongolia; Middle Jurassic.

Two species included from the Jurassic of Northern China (see Table 22.1).

***Proephialtitia acantha* Li, Shih, Rasnitsyn & Ren, 2015 (Figure 22.55)**

Proephialtitia acantha Li, Shih, Rasnitsyn & Ren, 2015: *BMC Evol. Biol.*, 6.

Locality and horizon: Daohugou, Inner Mongolia, China; Middle Jurassic, Jiulongshan Formation.

Length of body excluding antennae and ovipositor is 23.7 mm. Forewing with the first abscissa of Rs shorter than that of M; 1-Rs vertical to Rs and forming an angle of about 120^0 with 1-M at Rs+M; 1r-rs absent; 2r-m and 2m-cu partially preserved; 3r-m complete and slightly curved; cu-a interstitial, slightly shorter than 1-M; 2A incompletely preserved, meeting 1A before level of 1m-cu; hind wing with r-m distant from both RS and M bases; 1-M gently curved; cu-a S-like bent, slightly antefurcal. Metasoma broadly attaching to propodeum, with eight segments visible. Ovipositor slim with a sheath, approximately 38.3 mm long as preserved [144].

Figure 22.55 *Proephialtitia acantha* Li, Shih, Rasnitsyn & Ren, 2015 (Holotype CNU-HYM-NN-2014004p). Source: Donated by Dr. Chungkun Shih [144].

The Origin and Transformation of the Propodeal-Metasomal Articulation in Basal Apocrita (Including Evanioidea)

After studying the wasp waists of described fossil species of Ephialtitidae, Li *et al.*, in 2015, identified three typical but different propodeal-metasomal articulations [144]: the shape of the first metasomal segment varies from narrow and greatly elongate with distal end broader than the proximal end (*Proapocritus elegans* Rasnitsyn & Zhang, 2010), to slightly broad with the sides straight and subparallel (*Proapocritus densipediculus* Rasnitsyn & Zhang, 2010), then, to transversally broad with the sides nearly straight and subparallel (*Acephialtitia colossa* Li, Shih, Rasnitsyn & Ren, 2015). Furthermore, Li *et al.* proposed a hypothesis that the broad articulation between the propodeum and metasoma in Ephialtitidae was passed on from a still more primitive family of Karatavitidae, and Ephialtitidae was interpreted as a stem group of Apocrita demonstrating the origin and transformation of the wasp waist characters of the Apocrita. This ground plan structure has gradually transformed into narrower and more mobile articulation mechanism during the evolution of basal apocritans. Different pathways of the transformation of the "wasp waist" have been observed in Figure 22.56 [144].

(1) A key character of Ephialtitidae, as well as of more primitive taxa (Symphyta, including Karatavitidae), is that the first abdominal segment (propodeum) is more or less convex (bent) transversally but even (little or not at all bent) longitudinally (Figure 22.56a). The first pathway of transformation of the metasomal attachment is demonstrated by the Jurassic Kuafuidae. This time both articulating foramens got narrower accompanied with the propodeum dorsum bent longitudinally (Figure 22.56b). As a result, the narrow and mobile metasomal articulation appeared low on the propodeum (close to hind coxae), as is typical of the remaining Apocrita, that is, the Aculeata, Ichneumonomorpha, Proctotrupomorpha, and Ceraphronomorpha (including Trigonalidae, Megalyridae and Ceraphronoidea).

(2) The second pathway of transformation is the pattern (Figure 22.56a) basically retained in the extant family Stephanidae (Figure 22.56c), except that the articulatory foramen is often narrow here. The propodeal dorsum persists flat but becomes slanting toward its apex rather than horizontal (a feature not uncommon in Ephialtitidae themselves, cf. *Proephialtitia acanthi* Li, Shih, Rasnitsyn & Ren, 2015).

(3) Some other Ephialtitidae (Figure 22.56d) demonstrate initial steps in a different direction. Their articulatory orifices transformed in a disparate way so as the propodeal one retained wide whilst the metasomal base became narrow and hinged to the hind upper rim of propodeum, leaving membranous most of the hind face of propodeum. Sclerotization of this membranous space from its lower sides toward midline is observable in some Jurassic evanioids of the extinct family Praeaulacidae (Figure 22.56e). This sclerotization resulted in closure of the propodeum under the metasomal attachment which is the key synapomorphy of Evanioidea (Figure 22.56f).

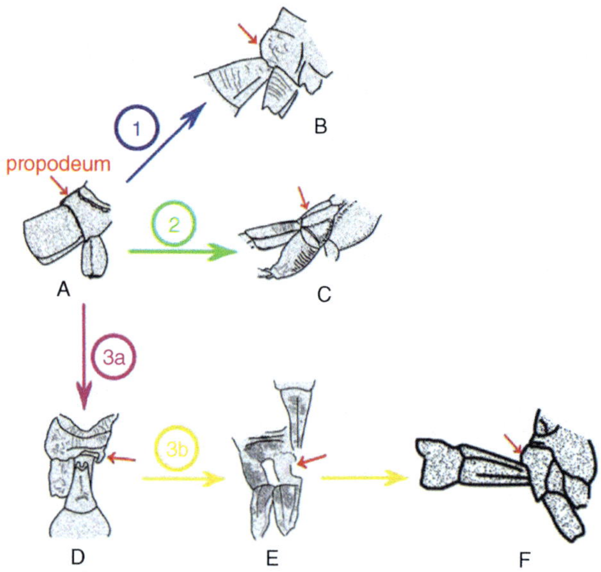

Figure 22.56 Transformation of the propodeum-metasomal articulation in basal Apocrita (including Evanioidea). (A) *Acephialtitia colossa* Li, Shih, Rasnitsyn & Ren, 2015; (B) *Kuafua polyneura* Rasnitsyn & Zhang, 2010 (Kuafuidae); (C) *Schlettererius cinctipes* (Cresson, 1880) (Stephanidae); (D) *Proapocritus sculptus* Rasnitsyn & Zhang, 2010; (E) *Eosaulacus giganteus* Zhang & Rasnitsyn, 2008 (Preaulacidae); (F) *Eosaulacus granulates* Zhang & Rasnitsyn, 2008 (Preaulacidae). Red arrows indicate propodeum, encircled numbers denote main transformation pathways, as follows: ① – pathway toward Kuafuidae and further to main taxa of Apocrita; ② – pathway toward Stephanidae; ③ – pathway toward basal Evanioidea. Source: modified from [144].

Family Karatavitidae Rasnitsyn, 1963

The extinct Karatavitidae Rasnitsyn, 1963 belong to the superfamily Karatavitoidea Rasnitsyn, 1963. Karatavitoidea also contain Orussidae Newman, 1834. So far, seven genera have been assigned to the Karatavitidae. The genera of Karatavitidae played important roles in the early evolution leading to Apocrita. Rasnitsyn *et al.*, in 2006, inferred a smooth transition from the genus of *Praeratavites*, with subbasal loop of vein 2A in the forewing, which resembles the common ancestor of all Vespina *s.l.* to *Karatavites*, without the loop of 2A, which resembles the common ancestor of Apocrita [81].

Genera included from the Jurassic and Cretaceous of Northern China: *Karatavites* Rasnitsyn, 1963; *Praeratavites* Rasnitsyn, Ansorge & Zhang, 2006, *Praporyssites* Rasnitsyn, Ansorge & Zhang, 2006, *Postxiphydria* Rasnitsyn & Zhang, 2010, *Postxiphydroides* Rasnitsyn & Zhang, 2010 and *Praeratavitioides* Rasnitsyn & Zhang, 2010.

Karatavites Rasnitsyn, 1963

Karatavites Rasnitsyn, 1963, *Paleontol. Zhur.*, 1, 97 [147] (original designation).

Type species: *Karatavites angustus* Rasnitsyn, 1963.

Forewing with pterostigma starting at lr-rs, lr-rs not longer than 2r-rs and well distant from it, 2r-rs joining pterostigma before its midlength; cu-a joining basal 0.3 of cell lmcu; 3r-m and 2m-cu flexible (annulated); 2A not looped subbasally. Propodeum with neither membraneous line nor carina along midline. Ovipositor flat (saw-like).

Distribution and age: Inner Mongolia; Middle Jurassic.

Two species included from the Jurassic of Northern China (see Table 22.1).

Praeratavites Rasnitsyn, Ansorge & Zhang, 2006

Praeratavites Rasnitsyn, Ansorge & Zhang, 2006, *Insect Syst. Evol.*, 37, 181 [148] (original designation).

Type species: *Praeratavites daohugou* Rasnitsyn, Ansorge & Zhang, 2006.

Forewing with pterostigma starting at 1r-rs; 1r-rs not longer than 2r-rs and well distant from it, 2r-rs joining pterostigma within its distal 0.3; cu-a joining mid 0.3 of cell 1mcu; 3r-m and 2m-cu tubular; 2A looped subbasally. Propodeum with neither membraneous line nor carina along midline. Ovipositor flat (saw-like).

Distribution and age: Inner Mongolia; Middle Jurassic.

Three species included from the Jurassic of Northern China (see Table 22.1).

Praeratavites rasnitsyni Shih, Li & Ren, 2017 (Figure 22.57)

Praeratavites rasnitsyni Shih, Li & Ren, 2017: *Alcheringa*, 2017, 3.

Locality and horizon: Daohugou, Ningcheng, Inner Mongolia, China; Middle Jurassic, Jiulongshan Formation.

The specific epithet is in honor of Dr. Alexandr Rasnitsyn for his significant contribution to palaeoentomology, especially to the early evolution and development of Hymenoptera, and for his erection of Karatavitidae in 1963. Body about 10.9 mm long, excluding antennae, including ovipositor. Head circular; antennae partly preserved, with scape long and thick; pedicel distinctly shorter and narrower than scape. Forewing with pterostigma large; 2r-rs issuing from nearly its apical 1/4; 1r-rs slightly shorter than 2r-rs, issuing from nearly its basal 1/4; R distinctly angular at Rs base; Rs gently arching basally; the first abscissa of Rs about twice as long as that of M; Rs+M not reaching the midlength of the anterior side of cell 1mcu; cu-a nearly reaching the midlength of posterior side of 1mcu; cross-vein 3r-m about twice as long as 2r-m; 2m-cu twice as long as 1m-cu; 1a-2a positioned much more basally than M+Cu forking; the sub-basal loop of 2A present [149].

Figure 22.57 *Praeratavites rasnitsyni* Shih, Li & Ren, 2017 (Holotype CNU-HYM-NN-2016001p, Female) (a) Habitus; (b) Line drawing [149].

Praeparyssites **Rasnitsyn, Ansorge & Zhang, 2006**

Praeparyssites Rasnitsyn, Ansorge & Zhang, 2006, *Insect Syst. Evol.*, 370, 186 [148] (original designation).

Type species: *Praeparyssites orientalis* Rasnitsyn, Ansorge & Zhang, 2006.

Forewing venation further reduced: 2r-rs, 3r-m, 2m-cu and Rs incomplete at best between Rs+M and 1r-rs; 2A lost between basal loop and 1a-2a; pterostigma extending basad far beyond 1r-rs, and in particularly long cell 2a; and in hind wing, in relatively basal position of cross-veins m-cu, cu-s and 1a-2a.

Distribution and age: Inner Mongolia; Middle Jurassic.

Four species included from the Jurassic of Northern China (see Table 22.1).

Postxiphydria **Rasnitsyn & Zhang, 2010**

Postxiphydria Rasnitsyn & Zhang, 2010, *Acta Geol. Sin.-Engl.*, 84, 841 [141] (original designation).

Type species: *Postxiphydria daohugouensis* Rasnitsyn & Zhang, 2010.

Metapostnotum forms two disconnected triangular plates similar to those in Xiphydriidae and other Siricina, except that their fore boundaries represent sutures and not narrow and deep gaps (entrances into metaphragma). Halves of the first abdominal tergum (propodeum) are apparently fused but not firmly, which resulted in easy breakage under fossilization, with uneven margins of the resulting cleft indicating its secondary nature. Cenchri and forewing 2A loop present. Forewing cross-veins 3r-m and 2m-cu present, tubular.

Distribution and age: Inner Mongolia; Middle Jurassic.

Two species included from the Jurassic of Northern China (see Table 22.1).

Postxiphydroides **Rasnitsyn & Zhang, 2010**

Postxiphydroides Rasnitsyn & Zhang, 2010, *Acta Geol. Sin.-Engl.*, 84, 843 [141] (original designation).

Type species: *Postxiphydroides strenuous* Rasnitsyn & Zhang, 2010.

Triangular metapostnotum and propodeum very short, propodeal halves completely fused forming medial suture. Forewing with cross-vein lr-rs distinctly longer than 2r-rs and running close to it; 3r-m and 2m-cu present, tubular; 2A loop present. Hind wing Cu base free.

Distribution and age: Inner Mongolia; Middle Jurassic.

Only one species included from the Jurassic of Northern China (see Table 22.1).

Praeratavitioides **Rasnitsyn & Zhang, 2010**

Praeratavitioides Rasnitsyn & Zhang, 2010, *Acta Geol. Sin.-Engl.*, 84, 843 [141] (original designation).

Type species: *Praeratavitioides amabilis* Rasnitsyn & Zhang, 2010.

Forewing with pterostigma starting at 1r-rs, 1r-rs not longer than 2r-rs and well distant from it; 2r-rs joining pterostigma within its distal 0.3; cu-a joining mid 0.3 of cell 1mcu; 3r-m and 2m-cu tubular; 2A looped subbasally. Propodeum with neither membraneous line nor carina along midline. Ovipositor thin needle-like.

Distribution and age: Inner Mongolia; Middle Jurassic.

Four species included from the Jurassic of Northern China (see Table 22.1).

Family Kuafuidae Rasnitsyn & Zhang, 2010

The family Kuafuidae was established in 2010 by the type genus *Kuafua* Rasnitsyn & Zhang, 2010 from the latest Middle Jurassic of Inner Mongolia, China. Kuafuidae comprise three genera [141]. The other two described genera, *Arthrogaster* Rasnitsyn, 1975 and *Leptogastrella* Rasnitsyn, 1975, are from the Upper Jurassic of Kazakhstan. Based on the intuitive cladogram of the suborder Vespina, with Xiphydriidae as a sister group [141], and the following characters: propodeum bent longitudinally (different from Orussomorpha); low position of the metasomal attachment (different from Evaniomorpha); 2A of forewing present, distal of la-2a; and marginal cell of hind wing closed at the anterior margin (different from Ceraphronomorpha, Proctotrupomorpha, Ichneumonomorpha and Vespomorpha), Rasnitsyn and Zhang considered Kuafuidae an intermediate position between Ephialtitidae and the higher Apocrita [141].

Only one genus included from the Jurassic and Cretaceous of Northern China: *Kuafua* Rasnitsyn & Zhang, 2010.

Figure 22.58 *Kuafua polyneura* Rasnitsyn & Zhang, 2010 (Holotype NND2005, 6/NIGP151949) [141]. Source: Photo provided by Dr. Haichun Zhang.

Kuafua **Rasnitsyn & Zhang, 2010**
Kuafua Rasnitsyn & Zhang, 2010, *Acta Geol. Sin.-Engl.*, 84, 855 [141] (original designation).

Type species: *Kuafua polyneura* Rasnitsyn & Zhang, 2010.

Forewing with the first abscissa of Rs subvertical to R; cu-a interstitial; 1r-rs, 3r-m and 1a-2a present; 2A nearly complete.

Distribution and age: Inner Mongolia; Middle Jurassic.

Only one species included from the Jurassic of Northern China (see Table 22.1).

Kuafua polyneura **Rasnitsyn & Zhang, 2010 (Figure 22.58)**
Kuafua polyneura Rasnitsyn & Zhang, 2010: *Acta Geol. Sin.-Engl.*, 84, 856.

Locality and horizon: Daohugou, Ningcheng, Inner Mongolia, China; Middle Jurassic, Jiulongshan Formation.

Head small, antenna with 21 antennomeres. Forewing with pterostigma long, acuminate, issuing 2r-rs at its mid-length. First abscissa of Rs as long as that of M; 1r-rs weak but complete, 2r-rs parallel to 1r-rs and 1Rs; cell 2rm and 3rm about of equal length; 2r-m slightly shorter than 3r-m; and both cross-veins subvertical; 1m-cu meeting 2rm slightly beyond Rs+M forking, and slightly shorter than section of Cu adjacent distally; 2A complete; 1a-2a distinguishable. Metasoma with basal half-cone-shaped and distal half nearly cylindrical in side view. Ovipositor long with sheaths [141].

Table 22.1 A list of fossil hymenoptera from the Jurassic and Cretaceous of China.

Family	Species	Locality	Horizon/Age	Citation
Suborder Symphyta Gerstaecker, 1867				
Megalodontesidae	*Jibaissodes giganteus* Ren, Lu, Guo & Ji, 1995	Beipiao, Liaoning	Yixian Fm., K_1	Ren *et al.* [42]
	Jibaissodes bellus Gao, Shih, Labandeira & Ren, 2016	Beipiao, Liaoning	Yixian Fm., K_1	Gao *et al.* [91]
Pamphiliidae	*Scabolyda orientalis* Wang, Rasnitsyn, Shih & Ren, 2014	Ningcheng, Inner Mongolia	Jiulongshan Fm., J_2	Wang *et al.* [83]
	Scabolyda incompleta Wang, Rasnitsyn, Shih & Ren, 2014	Ningcheng, Inner Mongolia	Jiulongshan Fm., J_2	Wang *et al.* [83]
Praesiricidae	*Rudisiricius belli* Gao, Rasnitsyn, Shih & Ren, 2010	Beipiao, Liaoning	Yixian Fm., K_1	Gao *et al.* [76]
	Rudisiricius crassinodus Gao, Rasnitsyn, Shih & Ren, 2010	Beipiao, Liaoning	Yixian Fm., K_1	Gao *et al.* [76]
	Rudisiricius scelsus Gao, Rasnitsyn, Shih & Ren, 2010	Beipiao, Liaoning	Yixian Fm., K_1	Gao *et al.* [76]
	Rudisiricius validus Wang, Rasnitsyn, Shih & Ren, 2015	Beipiao, Liaoning	Yixian Fm., K_1	Wang *et al.* [78]
	Rudisiricius ferox Wang, Rasnitsyn, Shih & Ren, 2015	Beipiao, Liaoning	Yixian Fm., K_1	Wang *et al.* [78]
	Rudisiricius ater Wang, Rasnitsyn, Shih & Ren, 2015	Beipiao, Liaoning	Yixian Fm., K_1	Wang *et al.* [78]
	Rudisiricius tenellus Wang, Rasnitsyn, Shih & Ren, 2015	Beipiao, Liaoning	Yixian Fm., K_1	Wang *et al.* [78]
	Rudisiricius membranaceous Wang, Rasnitsyn, Shih & Ren, 2015	Beipiao, Liaoning	Yixian Fm., K_1	Wang *et al.* [78]
	Rudisiricius parvus Wang, Rasnitsyn, Shih & Ren, 2014	Beipiao, Liaoning	Yixian Fm., K_1	Wang *et al.* [78]
	Archoxyelyda mirabilis Wang, Rasnitsyn & Ren, 2013	Beipiao, Liaoning	Yixian Fm., K_1	Wang *et al.* [77]
	Hoplitolyda duolunica Gao, Shih, Rasnitsyn & Ren, 2013	Duolun, Inner Mongolia	Yixian Fm., K_1	Gao *et al.* [3]

(Continued)

Table 22.1 (Continued)

Family	Species	Locality	Horizon/Age	Citation
	Decorisiricius patulus Wang, Rasnitsyn, Shih & Ren, 2016	Beipiao, Liaoning	Yixian Fm., K_1	Wang *et al.* [79]
	Decorisiricius latus Wang, Rasnitsyn, Shih & Ren, 2016	Beipiao, Liaoning	Yixian Fm., K_1	Wang *et al.* [79]
	Limbisiricius aequalis Wang, Rasnitsyn, Shih & Ren, 2016	Ningcheng, Inner Mongolia	Jiulongshan Fm., J_2	Wang *et al.* [79]
	Limbisiricius complanatus Wang, Rasnitsyn, Shih & Ren, 2016	Ningcheng, Inner Mongolia	Jiulongshan Fm., J_2	Wang *et al.* [79]
	Brevisiricius partialis Wang, Rasnitsyn, Shih & Ren, 2016	Ningcheng, Inner Mongolia	Jiulongshan Fm., J_2	Wang *et al.* [79]
Xyelidae	*Angaridyela robusta* Zhang & Zhang, 2000	Beipiao, Liaoning	Yixian Fm., K_1	Zhang and Zhang [43]
	Angaridyela exculpta Zhang & Zhang, 2000	Beipiao, Liaoning	Yixian Fm., K_1	Zhang and Zhang [43]
	Angaridyela suspecta Zhang & Zhang, 2000	Beipiao, Liaoning	Yixian Fm., K_1	Zhang and Zhang [43]
	Angaridyela endemica Zhang & Zhang, 2000	Beipiao, Liaoning	Yixian Fm., K_1	Zhang and Zhang [43]
	Anthoxyela orientalis Gao & Ren, 2008	Beipiao, Liaoning	Yixian Fm., K_1	Gao and Ren [150]
	Yanoxyela hongi Ren, Lu, Guo & Ji, 1995	Beipiao, Liaoning	Yixian Fm., K_1	Ren *et al.* [42]
	Ceratoxyela decorosa Zhang & Zhang, 2000	Beipiao, Liaoning	Yixian Fm., K_1	Zhang and Zhang [43]
	Heteroxyela ignota Zhang & Zhang, 2000	Beipiao, Liaoning	Yixian Fm., K_1	Zhang and Zhang [43]
	Isoxyela rudis Zhang & Zhang, 2000	Beipiao, Liaoning	Yixian Fm., K_1	Zhang and Zhang [43]
	Lethoxyela excurva Zhang & Zhang, 2000	Beipiao, Liaoning	Yixian Fm., K_1	Zhang and Zhang [43]
	Liaoxyela antiqua Zhang & Zhang, 2000	Beipiao, Liaoning	Yixian Fm., K_1	Zhang and Zhang [43]
	Sinoxyela viriosa Zhang & Zhang, 2000	Beipiao, Liaoning	Yixian Fm., K_1	Zhang and Zhang [43]
	Abrotoxyela lepida Gao, Ren & Shih, 2009	Ningcheng, Inner Mongolia	Jiulongshan Fm., J_2	Gao *et al.* [44]
	Abrotoxyela multiciliata Gao, Ren & Shih, 2009	Ningcheng, Inner Mongolia	Jiulongshan Fm., J_2	Gao *et al.* [44]
	Brachyoxyela brevinodia Gao, Zhao & Ren, 2011	Beipiao, Liaoning	Yixian Fm., K_1	Gao *et al.* [45]
	Brachyoxyela gracilenta Gao, Zhao & Ren, 2011	Beipiao, Liaoning	Yixian Fm., K_1	Gao *et al.* [45]
	Platyxyela unica Wang, Shih & Ren, 2012	Ningcheng, Inner Mongolia	Jiulongshan Fm., J_2	Wang *et al.* [46]
	Cathayxyela extensa Wang, Rasnitsyn & Ren, 2013	Ningcheng, Inner Mongolia	Jiulongshan Fm., J_2	Wang *et al.* [47]
	Aequixyela immensa Wang, Rasnitsyn & Ren, 2013	Ningcheng, Inner Mongolia	Jiulongshan Fm., J_2	Wang *et al.* [47]

(Continued)

Table 22.1 (Continued)

Family	Species	Locality	Horizon/Age	Citation
Xyelotomidae	*Xyelotoma macroclada* Gao, Ren & Shih, 2009	Ningcheng, Inner Mongolia	Jiulongshan Fm., J_2	Gao *et al.* [52]
	Xyelocerus diaphanous Gao, Ren & Shih, 2009	Ningcheng, Inner Mongolia	Jiulongshan Fm., J_2	Gao *et al.* [52]
	Liaotoma linearis Ren, Lu, Guo & Ji, 1995	Beipiao, Liaoning	Yixian Fm., K_1	Ren *et al.* [42]
	Abrotoma robusta Gao, Ren & Shih, 2009	Ningcheng, Inner Mongolia	Jiulongshan Fm., J_2	Gao *et al.* [52]
	Paradoxotoma tsaiae Gao, Ren & Shih, 2009	Ningcheng, Inner Mongolia	Jiulongshan Fm., J_2	Gao *et al.* [52]
	Synaptotoma limi Gao, Ren & Shih, 2009	Ningcheng, Inner Mongolia	Jiulongshan Fm., J_2	Gao *et al.* [52]
	Aethotoma aninomorpha Gao, Shih, Engel & Ren, 2016	Ningcheng, Inner Mongolia	Jiulongshan Fm., J_2	Gao *et al.* [53]
Xyelydidae	*Prolyda dimidia* Wang, Shih, Rasnitsyn & Wang, 2016	Ningcheng, Inner Mongolia	Jiulongshan Fm., J_2	Wang *et al.* [151]
	Prolyda elegantula Wang, Shih, Rasnitsyn & Wang, 2016	Ningcheng, Inner Mongolia	Jiulongshan Fm., J_2	Wang *et al.* [151]
	Ferganolyda scylla Rasnitsyn, Zhang & Wang, 2006	Ningcheng, Inner Mongolia	Jiulongshan Fm., J_2	Rasnitsyn *et al.* [65]
	Ferganolyda charybdis Rasnitsyn, Zhang & Wang, 2006	Ningcheng, Inner Mongolia	Jiulongshan Fm., J_2	Rasnitsyn *et al.* [65]
	Ferganolyda chungkuei Rasnitsyn, Zhang & Wang, 2006	Ningcheng, Inner Mongolia	Jiulongshan Fm., J_2	Rasnitsyn *et al.* [65]
	Ferganolyda eucalla Wang, Rasnitsyn, Shih & Ren, 2015	Ningcheng, Inner Mongolia	Jiulongshan Fm., J_2	Wang *et al.* [66]
	Ferganolyda insolita Wang, Rasnitsyn, Shih & Ren, 2015	Ningcheng, Inner Mongolia	Jiulongshan Fm., J_2	Wang *et al.* [66]
	Novalyda cretacica Gao, Engel, Shih & Ren, 2013	Beipiao, Liaoning	Yixian Fm., K_1	Gao *et al.* [67]
	Novalyda decora Wang, Rasnitsyn, Shih & Ren, 2015	Beipiao, Liaoning	Yixian Fm., K_1	Wang *et al.* [69]
	Rectilyda sticta Wang, Rasnitsyn, Shih & Ren, 2014	Duolun, Inner Mongolia	Yixian Fm., K_1	Wang *et al.* [68]
	Fissilyda compta Wang, Rasnitsyn, Shih & Ren, 2015	Beipiao, Liaoning	Yixian Fm., K_1	Wang *et al.* [69]
	Fissilyda alba Wang, Rasnitsyn, Shih & Ren, 2015	Beipiao, Liaoning	Yixian Fm., K_1	Wang *et al.* [69]
	Fissilyda parilis Wang, Rasnitsyn, Shih & Ren, 2015	Beipiao, Liaoning	Yixian Fm., K_1	Wang *et al.* [69]
	Medilyda procera Wang & Rasnitsyn, 2016	Ningcheng, Inner Mongolia	Jiulongshan Fm., J_2	Wang *et al.* [64]
	Medilyda distorta Wang & Rasnitsyn, 2016	Ningcheng, Inner Mongolia	Jiulongshan Fm., J_2	Wang *et al.* [64]
	Brevilyda provecta Wang & Rasnitsyn, 2016	Ningcheng, Inner Mongolia	Jiulongshan Fm., J_2	Wang *et al.* [64]
	Strenolyda marginalis Wang & Rasnitsyn, 2016	Ningcheng, Inner Mongolia	Jiulongshan Fm., J_2	Wang *et al.* [64]
	Strenolyda retrorsa Wang & Rasnitsyn, 2016	Ningcheng, Inner Mongolia	Jiulongshan Fm., J_2	Wang *et al.* [64]

(Continued)

Table 22.1 (Continued)

Family	Species	Locality	Horizon/Age	Citation
Suborder Apocrita Gerstäcker, 1867				
Anomopterellidae	*Anomopterella huang* Zhang & Rasnitsyn, 2008	Ningcheng, Inner Mongolia	Jiulongshan Fm., J_2	Zhang and Rasnitsyn [137]
	Anomopterella coalita Li, Rasnitsyn, Shih & Ren, 2013	Ningcheng, Inner Mongolia	Jiulongshan Fm., J_2	Li *et al.* [131]
	Anomopterella ampla Li, Rasnitsyn, Shih & Ren, 2013	Ningcheng, Inner Mongolia	Jiulongshan Fm., J_2	Li *et al.* [131]
	Anomopterella brachystelis Li, Rasnitsyn, Shih & Ren, 2013	Ningcheng, Inner Mongolia	Jiulongshan Fm., J_2	Li et al [131]
	Anomopterella divergens Li, Rasnitsyn, Shih & Ren, 2013	Ningcheng, Inner Mongolia	Jiulongshan Fm., J_2	Li *et al.* [131]
	Anomopterella ovalis Li, Rasnitsyn, Shih & Ren, 2013	Ningcheng, Inner Mongolia	Jiulongshan Fm., J_2	Li *et al.* [131]
	Anomopterella brevis Li, Shih & Ren, 2014	Ningcheng, Inner Mongolia	Jiulongshan Fm., J_2	Li *et al.* [132]
	Anomopterella pygmea Li, Shih & Ren, 2014	Ningcheng, Inner Mongolia	Jiulongshan Fm., J_2	Li *et al.* [132]
	Synaphopterella patula Li, Rasnitsyn, Shih & Ren, 2013	Ningcheng, Inner Mongolia	Jiulongshan Fm., J_2	Li *et al.* [131]
Ephialtitidae	*Acephialtitia colossa* Li, Shih, Rasnitsyn & Ren, 2015	Beipiao, Liaoning	Yixian Fm., K_1	Li *et al.* [144]
	Asiephialtites lini Rasnitsyn & Zhang, 2010	Ningcheng, Inner Mongolia	Yixian Fm., K_1	Rasnitsyn and Zhang [141]
	Crephanogaster rara Zhang, Rasnitsyn & Zhang, 2002	Beipiao, Liaoning	Yixian Fm., K_1	Zhang *et al.* [152]
	Karataus daohugouensis Zhang, Zhang & Rasnitsyn, 2014	Ningcheng, Inner Mongolia	Jiulongshan Fm., J_2	Zhang *et al.* [155]
	Karataus strenuus Zhang, Zhang & Rasnitsyn, 2014	Ningcheng, Inner Mongolia	Jiulongshan Fm., J_2	Zhang *et al.* [155]
	Karataus vigoratus Zhang, Zhang & Rasnitsyn, 2014	Ningcheng, Inner Mongolia	Jiulongshan Fm., J_2	Zhang *et al.* [155]
	Karataus exilis Zhang, Zhang & Rasnitsyn, 2014	Ningcheng, Inner Mongolia	Jiulongshan Fm., J_2	Zhang *et al.* [155]
	Karataus orientalis Zhang, Zhang & Rasnitsyn, 2014	Ningcheng, Inner Mongolia	Jiulongshan Fm., J_2	Zhang *et al.* [155]
	Praeproapocritus vulgates Rasnitsyn & Zhang, 2010	Ningcheng, Inner Mongolia	Jiulongshan Fm., J_2	Rasnitsyn and Zhang [141]
	Praeproapocritus flexus Li, Shih & Ren, 2013	Ningcheng, Inner Mongolia	Jiulongshan Fm., J_2	Li *et al.* [146]
	Proapocritus atropus Rasnitsyn & Zhang, 2010	Ningcheng, Inner Mongolia	Jiulongshan Fm., J_2	Rasnitsyn and Zhang [141]
	Proapocritus densipediculus Rasnitsyn & Zhang, 2010	Ningcheng, Inner Mongolia	Jiulongshan Fm., J_2	Rasnitsyn and Zhang [141]
	Proapocritus elegans Rasnitsyn & Zhang, 2010	Ningcheng, Inner Mongolia	Jiulongshan Fm., J_2	Rasnitsyn and Zhang [141]
	Proapocritus formosus Rasnitsyn & Zhang, 2010	Ningcheng, Inner Mongolia	Jiulongshan Fm., J_2	Rasnitsyn and Zhang [141]
	Proapocritus longatennatus Rasnitsyn & Zhang, 2010	Ningcheng, Inner Mongolia	Jiulongshan Fm., J_2	Rasnitsyn and Zhang [141]

(Continued)

Table 22.1 (Continued)

Family	Species	Locality	Horizon/Age	Citation
	Proapocritus sculptus Rasnitsyn & Zhang, 2010	Ningcheng, Inner Mongolia	Jiulongshan Fm., J_2	Rasnitsyn and Zhang [141]
	Proapocritus parallelus Rasnitsyn & Zhang, 2010	Ningcheng, Inner Mongolia	Jiulongshan Fm., J_2	Rasnitsyn and Zhang [141]
	Proapocritus parallelus Li, Shih & Ren, 2013	Ningcheng, Inner Mongolia	Jiulongshan Fm., J_2	Li *et al.* [146]
	Proephialtitia acanthi Li, Shih, Rasnitsyn & Ren, 2015	Ningcheng, Inner Mongolia	Jiulongshan Fm., J_2	Li *et al.* [144]
	Proephialtitia tenuata Li, Shih, Rasnitsyn & Ren, 2015	Ningcheng, Inner Mongolia	Jiulongshan Fm., J_2	Li *et al.* [144]
	Sinephialtites glyptus Zhang, 1986	Ningcheng, Inner Mongolia	Jiulongshan Fm., J_2	Zhang [159]
	Stephanogaster pristinus Rasnitsyn, 1975	Ningcheng, Inner Mongolia	Jiulongshan Fm., J_2	Rasnitsyn [130]
	Symphytopterus graciler Wang, Li & Shih, 2015	Ningcheng, Inner Mongolia	Jiulongshan Fm., J_2	Wang *et al.* [145]
	Tuphephialtites zherikhini Zhang, Rasnitsyn & Zhang, 2002	Beipiao, Liaoning	Yixian Fm., K_1	Zhang *et al.* [152]
Evaniidae	*Procretevania pristina* Zhang & Zhang, 2000	Beipiao, Liaoning	Yixian Fm., K_1	Zhang and Zhang [135]
	Procretevania vesca Zhang, Rasnitsyn & Wang, 2007	Beipiao, Liaoning	Yixian Fm., K_1	Zhang *et al.* [154]
	Procretevania exquisite Zhang, Rasnitsyn & Wang, 2007	Beipiao, Liaoning	Yixian Fm., K_1	Zhang *et al.* [154]
	Procretevania mitis Li, Shih & Ren, 2014	Beipiao, Liaoning	Yixian Fm., K_1	Li *et al.* [138]
Heloridae	*Archaeohelorus hoi* Shih, Feng & Ren, 2011	Ningcheng, Inner Mongolia	Jiulongshan Fm., J_2	Shih *et al.* [110]
	Archaeohelorus polyneurus Shi, Zhao, Shih & Ren, 2014	Ningcheng, Inner Mongolia	Jiulongshan Fm., J_2	Shi *et al.* [112]
	Archaeohelorus tensus Shi, Zhao, Shih & Ren, 2014	Ningcheng, Inner Mongolia	Jiulongshan Fm., J_2	Shi *et al.* [112]
	Bellohelorus fortis Li, Shih & Ren, 2017	Beipiao, Liaoning	Yixian Fm., K_1	Li *et al.* [113]
	Laiyanghelorus erymnus Zhang, 1992	Laiyang, Shandong Beipiao, Liaoning	Laiyang Fm., K_1; Yixian Fm., K_1	Zhang [108]
	Liaoropronia leonine Zhang & Zhang, 2001	Beipiao, Liaoning	Yixian Fm., K_1	Zhang and Zhang [109]
	Liaoropronia regia Zhang & Zhang, 2001	Beipiao, Liaoning	Yixian Fm., K_1	Zhang and Zhang [109]
	Novhelorus macilentus Li, Shih & Ren, 2017	Beipiao, Liaoning	Yixian Fm., K_1	Li *et al.* [113]
	Novhelorus saltatrix (Shi, Zhao, Shih & Ren) Li, Shih & Ren, 2017	Beipiao, Liaoning	Yixian Fm., K_1	Li *et al.* [113]
	Protocyrtus validus Zhang & Zhang, 2001	Beipiao, Liaoning	Yixian Fm., K_1	Zhang and Zhang [109]
	Protocyrtus parilis Li, Shih & Ren, 2017	Beipiao, Liaoning	Yixian Fm., K_1	Li *et al.* [113]

(Continued)

Table 22.1 (Continued)

Family	Species	Locality	Horizon/Age	Citation
	Sinohelorus elegans Shi, Zhao, Shih & Ren, 2013	Beipiao, Liaoning	Yixian Fm., K_1	Shi *et al.* [112]
	Spherogaster coronata Zhang & Zhang, 2001	Beipiao, Liaoning	Yixian Fm., K_1	Zhang and Zhang [109]
	Spherogaster beipiaoensis (Shi, Zhao, Shih & Ren) Li, Shih & Ren, 2017	Beipiao, Liaoning	Yixian Fm., K_1	Li *et al.* [113]
Karatavitidae	*Karatavites junfengi* Rasnitsyn & Zhang, 2010	Ningcheng, Inner Mongolia	Jiulongshan Fm., J_2	Rasnitsyn and Zhang [141]
	Karatavites ningchengensis M. Shih, Li & Ren, 2017	Ningcheng, Inner Mongolia	Jiulongshan Fm., J_2	Shih *et al.* [149]
	Praeratavites daohugou Rasnitsyn, Ansorge & Zhang, 2006	Ningcheng, Inner Mongolia	Jiulongshan Fm., J_2	Rasnitsyn *et al.* [148]
	Praeratavites wuhuaensis Rasnitsyn & Zhang, 2010	Ningcheng, Inner Mongolia	Jiulongshan Fm., J_2	Rasnitsyn and Zhang [141]
	Praeratavites perspicuus Rasnitsyn & Zhang, 2010	Ningcheng, Inner Mongolia	Jiulongshan Fm., J_2	Rasnitsyn and Zhang [141]
	Praeratavites rasnitsyni M. Shih, Li & Ren, 2017	Ningcheng, Inner Mongolia	Jiulongshan Fm., J_2	Shih *et al.* [149]
	Postxiphydria daohugouensis Rasnitsyn & Zhang, 2010	Ningcheng, Inner Mongolia	Jiulongshan Fm., J_2	Rasnitsyn and Zhang [141]
	Postxiphydria ningchengensis Rasnitsyn & Zhang, 2010	Ningcheng, Inner Mongolia	Jiulongshan Fm., J_2	Rasnitsyn and Zhang [141]
	Praeratavitioides amabilis Rasnitsyn & Zhang, 2010	Ningcheng, Inner Mongolia	Jiulongshan Fm., J_2	Rasnitsyn and Zhang [141]
Kuafuidae	*Kuafua polyneura* Rasnitsyn & Zhang, 2010	Ningcheng, Inner Mongolia	Jiulongshan Fm., J_2	Rasnitsyn and Zhang [141]
Mesoserphidae	*Amboserphus tumidus* Li, Rasnitsyn, Shih & Ren, 2017	Beipiao, Liaoning	Yixian Fm., K_1	Li *et al.* [120]
	Amboserphus beipiaoensis Li, Rasnitsyn, Shih & Ren, 2017	Beipiao, Liaoning	Yixian Fm., K_1	Li *et al.* [120]
	Amboserphus dimidius Li, Rasnitsyn, Shih & Ren, 2017	Beipiao, Liaoning	Yixian Fm., K_1	Li *et al.* [120]
	Apiciserphus augustus Li, Rasnitsyn, Shih & Ren, 2017	Ningcheng, Inner Mongolia	Jiulongshan Fm., J_2	Li *et al.* [120]
	Basiserphus loculatus Li, Rasnitsyn, Shih & Ren, 2017	Beipiao, Liaoning	Yixian Fm., K_1	Li *et al.* [120]
	Basiserphus longus Li, Rasnitsyn, Shih & Ren, 2017	Ningcheng, Inner Mongolia	Jiulongshan Fm., J_2	Li *et al.* [120]
	Choriserphus bellus Li, Rasnitsyn, Shih & Ren, 2017	Ningcheng, Inner Mongolia	Jiulongshan Fm., J_2	Li *et al.* [120]
	Choriserphus gigantus Li, Rasnitsyn, Shih & Ren, 2017	Ningcheng, Inner Mongolia	Jiulongshan Fm., J_2	Li *et al.* [120]
	Novserphus ningchengensis Li, Rasnitsyn, Shih & Ren, 2017	Ningcheng, Inner Mongolia	Jiulongshan Fm., J_2	Li *et al.* [120]
	Mesoserphus venustus Li, Rasnitsyn, Shih & Ren, 2017	Ningcheng, Inner Mongolia	Jiulongshan Fm., J_2	Li *et al.* [120]
	Ozososerphus lepidus Li, Rasnitsyn, Shih & Ren, 2017	Ningcheng, Inner Mongolia	Jiulongshan Fm., J_2	Li *et al.* [120]

(Continued)

Table 22.1 (Continued)

Family	Species	Locality	Horizon/Age	Citation
	Ozososerphus ovatus Li, Rasnitsyn, Shih & Ren, 2017	Ningcheng, Inner Mongolia	Jiulongshan Fm., J_2	Li *et al.* [120]
	Ozososerphus cuboidus Li, Rasnitsyn, Shih & Ren, 2017	Ningcheng, Inner Mongolia	Jiulongshan Fm., J_2	Li *et al.* [120]
	Sinoserphus wui Shih, Feng & Ren, 2011	Ningcheng, Inner Mongolia	Jiulongshan Fm., J_2	Shih *et al.* [110]
	Sinoserphus shihae Shih, Feng & Ren, 2011	Ningcheng, Inner Mongolia	Jiulongshan Fm., J_2	Shih *et al.* [110]
	Sinoserphus lillianae Shih, Feng & Ren, 2011	Ningcheng, Inner Mongolia	Jiulongshan Fm., J_2	Shih *et al.* [110]
	Sinoserphus flexilis Li, Rasnitsyn, Shih & Ren, 2017	Ningcheng, Inner Mongolia	Jiulongshan Fm., J_2	Li *et al.* [120]
	Sinoserphus grossus Li, Rasnitsyn, Shih & Ren, 2017	Ningcheng, Inner Mongolia	Jiulongshan Fm., J_2	Li *et al.* [120]
	Sinoserphus petilus Li, Rasnitsyn, Shih & Ren, 2017	Ningcheng, Inner Mongolia	Jiulongshan Fm., J_2	Li *et al.* [120]
	Yanliaoserphus jurassicus Shih, Feng & Ren, 2011	Ningcheng, Inner Mongolia	Jiulongshan Fm., J_2	Shih *et al.* [111]
	Beipiaoserphus elegans Zhang & Zhang, 2000	Beipiao, Liaoning	Yixian Fm., K_1	Zhang and Zhang [121]
	Karataoserphus adaequatus Li, Rasnitsyn, Shih & Ren, 2017	Ningcheng, Inner Mongolia	Jiulongshan Fm., J_2	Li *et al.* [120]
	Karataoserphus gracilentus Li, Rasnitsyn, Shih & Ren, 2017	Ningcheng, Inner Mongolia	Jiulongshan Fm., J_2	Li *et al.* [120]
	Karataoserphus sinicus (Ping) Li, Rasnitsyn, Shih & Ren, 2017	Ningcheng, Inner Mongolia	Jiulongshan Fm., J_2	Li *et al.* [120]
Pelecinidae	*Allopelecinus terpnus* Zhang & Rasnitsyn, 2006	Laiyang, Shandong	Laiyang Fm., K_1	Zhang and Rasnitsyn [98]
	Archaeopelecinus tebbei Shih, Liu & Ren, 2009	Ningcheng, Inner Mongolia	Jiulongshan Fm., J_2	Shih *et al.* [92]
	Archaeopelecinus jinzhouensis Shih, Liu & Ren, 2009	Ningcheng, Inner Mongolia	Jiulongshan Fm., J_2	Shih *et al.* [92]
	Abropelecinus annulatus Feng, Shih, Ren & Liu, 2010	Beipiao, Liaoning	Yixian Fm., K_1	Feng *et al.* [100]
	Azygopelecinus clavatus Feng, Shih, Ren & Liu, 2010	Beipiao, Liaoning	Yixian Fm., K_1	Feng *et al.* [100]
	Cathaypelecinus daohugouensis Shih, Liu & Ren, 2009	Ningcheng, Inner Mongolia	Jiulongshan Fm., J_2	Shih *et al.* [92]
	Eopelecinus eucallus Zhang, 2005	Laiyang, Shandong	Laiyang Fm., K_1	Zhang [155]
	Eopelecinus giganteus Zhang, 2005	Laiyang, Shandong	Laiyang Fm., K_1	Zhang [155]
	Eopelecinus hodoiporus Zhang, 2005	Laiyang, Shandong	Laiyang Fm., K_1	Zhang [155]
	Eopelecinus laiyangicus Zhang, 2005	Laiyang, Shandong	Laiyang Fm., K_1	Zhang [155]
	Eopelecinus leptaleus Zhang, 2005	Laiyang, Shandong	Laiyang Fm., K_1	Zhang [155]
	Eopelecinus mecometasomatus Zhang, 2005	Laiyang, Shandong	Laiyang Fm., K_1	Zhang [155]

(Continued)

Table 22.1 (Continued)

Family	Species	Locality	Horizon/Age	Citation
	Eopelecinus mesomicrus Zhang, 2005	Laiyang, Shandong	Laiyang Fm., K_1	Zhang [155]
	Eopelecinus pusillus Zhang, 2005	Laiyang, Shandong	Laiyang Fm., K_1	Zhang [155]
	Eopelecinus shangyuanensis Zhang, Rasnitsyn & Zhang, 2002	Beipiao, Liaoning	Yixian Fm., K_1	Zhang *et al.* [97]
	Eopelecinus similaris Zhang, Rasnitsyn & Zhang, 2002	Beipiao, Liaoning	Yixian Fm., K_1	Zhang *et al.* [97]
	Eopelecinus vicinus Zhang, Rasnitsyn & Zhang, 2002	Beipiao, Liaoning	Yixian Fm., K_1	Zhang *et al.* [97]
	Eopelecinus yuanjiawaensis Duan & Chen, 2006	Beipiao, Liaoning	Yixian Fm., K_1	Duan and Cheng [158]
	Eopelecinus huangi Liu, Gao, Shih & Ren, 2011	Beipiao, Liaoning	Yixian Fm., K_1	Liu *et al.* [156]
	Eopelecinus tumidus Liu, Gao, Shih & Ren, 2011	Beipiao, Liaoning	Yixian Fm., K_1	Liu *et al.* [156]
	Megapelecinus changi Shih, Feng, Liu, Zhao & Ren, 2010	Beipiao, Liaoning	Yixian Fm., K_1	Shih *et al.* [93]
	Megapelecinus nashi Shih, Feng, Liu, Zhao & Ren, 2010	Beipiao, Liaoning	Yixian Fm., K_1	Shih *et al.* [93]
	Scorpiopelecinus versatilis Zhang, Rasnitsyn & Zhang, 2002	Beipiao, Liaoning	Yixian Fm., K_1	Zhang *et al.* [97]
	Scorpiopelecinus laetus Zhang & Rasnitsyn, 2004	Beipiao, Liaoning	Yixian Fm., K_1	Zhang and Rasnitsyn [158]
	Shoushida regilla Liu, Shih & Ren, 2009	Beipiao, Liaoning	Yixian Fm., K_1	Liu *et al.* [99]
	Shoushida infera Guo, Shih & Ren, 2016	Beipiao, Liaoning	Yixian Fm., K_1	Guo *et al.* [96]
	Sinopelecinus daspletis Zhang & Rasnitsyn, 2006	Laiyang, Shandong	Laiyang Fm., K_1	Zhang and Rasnitsyn [98]
	Sinopelecinus delicatus Zhang, Rasnitsyn, & Zhang, 2002	Beipiao, Liaoning	Yixian Fm., K_1	Zhang *et al.* [97]
	Sinopelecinus epigaeus Zhang, Rasnitsyn & Zhang, 2002	Beipiao, Liaoning	Yixian Fm., K_1	Zhang *et al.* [97]
	Sinopelecinus hierus Zhang & Rasnitsyn, 2006	Laiyang, Shandong	Laiyang Fm., K_1	Zhang and Rasnitsyn [98]
	Sinopelecinus magicus Zhang, Rasnitsyn & Zhang, 2002	Beipiao, Liaoning	Yixian Fm., K_1	Zhang *et al.* [97]
	Sinopelecinus viriosus Zhang, Rasnitsyn & Zhang, 2002	Beipiao, Liaoning	Yixian Fm., K_1	Zhang *et al.* [97]
	Stelepelecinus longus Guo, Shih & Ren, 2016	Beipiao, Liaoning	Yixian Fm., K_1	Guo *et al.* [96]
Praeaulacidae	*Archaulacus probus* Li, Shih & Ren, 2014	Ningcheng, Inner Mongolia	Jiulongshan Fm., J_2	Li *et al.* [142]
	Aulacogastrinus hebeiensis Zhang & Rasnitsyn, 2008	Ningcheng, Inner Mongolia	Jiulongshan Fm., J_2	Zhang and Rasnitsyn [137]
	Anomopterella huangi Zhang & Rasnitsyn, 2008	Ningcheng, Inner Mongolia	Jiulongshan Fm., J_2	Zhang and Rasnitsyn [137]
	Aulacogastrinus insculptus Zhang & Rasnitsyn, 2008	Ningcheng, Inner Mongolia	Jiulongshan Fm., J_2	Zhang and Rasnitsyn [137]
	Aulacogastrinus longaciculatus Zhang & Rasnitsyn, 2008	Ningcheng, Inner Mongolia	Jiulongshan Fm., J_2	Zhang and Rasnitsyn [137]

(Continued)

Table 22.1 (Continued)

Family	Species	Locality	Horizon/Age	Citation
	Eosaulacus giganteus Zhang & Rasnitsyn, 2008	Ningcheng, Inner Mongolia	Jiulongshan Fm., J_2	Zhang and Rasnitsyn [137]
	Eosaulacus granulatus Zhang & Rasnitsyn, 2008	Ningcheng, Inner Mongolia	Jiulongshan Fm., J_2	Zhang and Rasnitsyn [137]
	Eonevania robusta Rasnitsyn & Zhang, 2010	Ningcheng, Inner Mongolia	Jiulongshan Fm., J_2	Rasnitsyn and Zhang [141]
	Nevania delicata Zhang & Rasnitsyn, 2007	Ningcheng, Inner Mongolia	Jiulongshan Fm., J_2	Zhang and Rasnitsyn [140]
	Nevania exquisita Zhang & Rasnitsyn, 2007	Ningcheng, Inner Mongolia	Jiulongshan Fm., J_2	Zhang and Rasnitsyn [140]
	Nevania ferocula Zhang & Rasnitsyn, 2007	Ningcheng, Inner Mongolia	Jiulongshan Fm., J_2	Zhang and Rasnitsyn [140]
	Nevania malleata Zhang & Rasnitsyn, 2007	Ningcheng, Inner Mongolia	Jiulongshan Fm., J_2	Zhang and Rasnitsyn [140]
	Nevania retenta Zhang & Rasnitsyn, 2007	Ningcheng, Inner Mongolia	Jiulongshan Fm., J_2	Zhang and Rasnitsyn [140]
	Nevania robusta Zhang & Rasnitsyn, 2007	Ningcheng, Inner Mongolia	Jiulongshan Fm., J_2	Zhang and Rasnitsyn [140]
	Nevania perbella Li, Shih & Ren, 2013	Ningcheng, Inner Mongolia	Jiulongshan Fm., J_2	Li *et al.* [157]
	Nevania aspectabilis Li, Shih & Ren, 2013	Ningcheng, Inner Mongolia	Jiulongshan Fm., J_2	Li *et al.* [157]
	Praeaulacus afflatus Zhang & Rasnitsyn, 2008	Ningcheng, Inner Mongolia	Jiulongshan Fm., J_2	Zhang and Rasnitsyn [137]
	Praeaulacus daohugouensis Zhang & Rasnitsyn, 2008	Ningcheng, Inner Mongolia	Jiulongshan Fm., J_2	Zhang and Rasnitsyn [137]
	Praeaulacus exquisitus Zhang & Rasnitsyn, 2008	Ningcheng, Inner Mongolia	Jiulongshan Fm., J_2	Zhang and Rasnitsyn [137]
	Praeaulacus orientalis Zhang & Rasnitsyn, 2008	Ningcheng, Inner Mongolia	Jiulongshan Fm., J_2	Zhang and Rasnitsyn [137]
	Praeaulacus robustus Zhang & Rasnitsyn, 2008	Ningcheng, Inner Mongolia	Jiulongshan Fm., J_2	Zhang and Rasnitsyn [137]
	Praeaulacus scabratus Zhang & Rasnitsyn, 2008	Ningcheng, Inner Mongolia	Jiulongshan Fm., J_2	Zhang and Rasnitsyn [137]
	Praeaulacus sculptus Zhang & Rasnitsyn, 2008	Ningcheng, Inner Mongolia	Jiulongshan Fm., J_2	Zhang and Rasnitsyn [137]
	Praeaulacus subrhombeus Li, Shih & Ren, 2014	Ningcheng, Inner Mongolia	Jiulongshan Fm., J_2	Li *et al.* [142]
	Praeaulacus tenellus Li, Shih & Ren, 2014	Ningcheng, Inner Mongolia	Jiulongshan Fm., J_2	Li *et al.* [142]
	Praeaulacus obtutus Li, Shih & Ren, 2014	Ningcheng, Inner Mongolia	Jiulongshan Fm., J_2	Li and Shih [138]
	Praeaulacus byssinus Wang, Li & Shih, 2015	Ningcheng, Inner Mongolia	Jiulongshan Fm., J_2	Wang *et al.* [145]
	Praeaulacon elegantulus Zhang & Rasnitsyn, 2008	Ningcheng, Inner Mongolia	Jiulongshan Fm., J_2	Zhang and Rasnitsyn [137]
	Praeaulacon ningchengensis Zhang & Rasnitsyn, 2008	Ningcheng, Inner Mongolia	Jiulongshan Fm., J_2	Zhang and Rasnitsyn [137]
	Sinaulacogastrinus eucallus Zhang & Rasnitsyn, 2008	Ningcheng, Inner Mongolia	Jiulongshan Fm., J_2	Zhang and Rasnitsyn [137]

(Continued)

Table 22.1 (Continued)

Family	Species	Locality	Horizon/Age	Citation
	Sinowestratia communicata Zhang & Zhang, 2000	Beipiao, Liaoning	Yixian Fm., K_1	Zhang and Zhang [135]
	Sinevania speciosa Rasnitsyn & Zhang, 2010	Ningcheng, Inner Mongolia	Jiulongshan Fm., J_2	Rasnitsyn and Zhang [141]
Scolebythidae	*Mirabythus lechrius* Cai, Shih & Ren, 2012	Beipiao, Liaoning	Yixian Fm., K_1	Cai *et al.* [122]
	Mirabythus liae Cai, Shih & Ren, 2012	Beipiao, Liaoning	Yixian Fm., K_1	Cai *et al.* [122]

References

1 Malm, T. and Nyman, T. (2015). Phylogeny of the symphytan grade of Hymenoptera: new pieces into the old jigsaw (fly) puzzle. *Cladistics* **31**: 1–17.

2 Benson, R.B. (1952). *Handbooks for the Identification of British Insects: VI Hymenoptera 2 Symphyta Section (b)*, 51. London: Royal Entomological Society of London.

3 Gao, T.P., Shih, C.K., Rasnitsyn, A.P., and Ren, D. (2013). *Hoplitolyda duolunica* gen. et sp. nov. (Insecta, Hymenoptera, Praesiricidae), the hitherto largest sawfly from the Mesozoic of China. *PLoS ONE* **8** (5): e62420.

4 Engel, M.S., Huang, D.-Y., Alqarni, A.S., and Cai, C.Y. (2016). An unusual new lineage of sawflies (Hymenoptera) in Upper Cretaceous amber from northern Myanmar. *Cretaceous Research* **60**: 281–286.

5 Xiao, G.R., Huang, X.Y., Zhou, S.Z. et al. (1991). *Economic Sawfly Fauna of China*. Shaanxi, China: Tianze Publishing House.

6 Xiao, G.R. and Zhang, Y. (1994). A new species of the genus *Diprion* (Hymenoptera: Diprionidae) from China. *Forest Research* **6**: 663–664.

7 Wang, L.Z., Zhao, J.L., Li, F.Y. et al. (1994). Research on biology of *Diprion* sp. and its biological control. *Forest Pest and Disease* **3**: 12–14.

8 Grimaldi, D. and Engel, M.S. (2005). *Evolution of the Insects*. New York: Cambridge University Press.

9 Hoddle, M.S., VanDriesche, R.G., and Sanderson, J.P. (1998). Biology and use of the Whitefly parasitoid *Encarsia formosa*. *Annual Review of Entomology* **43**: 645–669.

10 Rafikov, M. and de Holanda Limeira, E. (2012). Mathematical modelling of the biological pest control of the sugarcane borer. *International Journal of Computer Mathematics* **89** (3): 390–401.

11 Zamani, A.A., Talebi, A., Fathipour, Y., and Baniameri, V. (2007). Effect of temperature on life history of *Aphidius colemani* and *Aphidius matricariae* (Hymenoptera: Braconidae), two parasitoids of *Aphis gossypii* and *Myzus persicae* (Homoptera: Aphididae). *Environmental Entomology* **36** (2): 263–271.

12 Pilkington, L.J. and Hoddle, M.S. (2006). Reproductive and developmental biology of *Gonatocerus ashmeadi* (Hymenoptera: Mymaridae), an egg parasitoid of *Homalodisca coagulata* (Hemiptera: Cicadellidae). *Biological Control* **37**: 266–275. https://doi.org/10.1016/j.biocontrol.2006.02.006.

13 Charles, J.G. and Logan, D.P. (2013). Predicting the distribution of *Gonatocerus ashmeadi*, an egg parasitoid of glassy winged sharpshooter, in New Zealand. *New Zealand Entomologist* **36** (2): 73–81. https://doi.org/10.1080/00779962.2012.751776.

14 Batra, S.W.T. (1968). Behavior of some social and solitary Halictine bees within their nests: a comparative study (Hymenoptera: Halictidae). *Journal of the Kansas Entomological Society* **41** (1): 120–133.

15 Crespi, B.J. and Yanega, D. (1995). The definition of eusociality. *Behavior Ecology* **6**: 109–115. https://doi.org/10.1093/beheco/6.1.109.

16 Wilson, E.O. and Hölldobler, B. (2005). Eusociality: origin and consequences. *Proceedings of National Academy of Sciences* **102** (38): 13367–13371. https://doi.org/10.1073/pnas.0505858102.

17 Kent, D.S. and Simpson, J.A. (1992). Eusociality in the beetle *Austroplatypus incompertus* (Coleoptera: Curculionidae). *Naturwissenschaften* **79**: 86–87. https://doi.org/10.1007/BF01131810.

18 Stern, D.L. (1994). A phylogenetic analysis of soldier evolution in the aphid family Hormaphididae. *Proceedings of the Royal Society* **256**: 203–209. https://doi.org/10.1098/rspb.1994.0071.

19 Aoki, S. and Imai, M. (2005). Factors affecting the proportion of sterile soldiers in growing aphid

colonies. *Population Ecology* **47**: 127–136. https://doi.org/10.1007/s10144-005-0218-z.

20 von Frisch, K. (1967). *The Dance Language and Orientation of Bees*, 566. Cambridge, Massachusetts: The Belknap Press of Harvard University Press.

21 von Frisch, K. (1976). *Bees: Their Vision, Chemical Senses, and Language*, 157. Ithaca, N.Y: Cornell University Press, Revised Edition.

22 Rasnitsyn, A.P. (1969). *Origin and evolution of lower Hymenoptera*, vol. **123**, 1–196. Transactions of Paleontological Institution. Academy of Sciences, USSR.

23 Rasnitsyn, A.P. (1980). *Origin and evolution of Hymenoptera*, vol. **174**, 1–192. Transactions of Paleontological Institution. Academy of Sciences, USSR.

24 Engel, M.S. (2005). A new sawfly from the Triassic of Queensland (Hymenoptera: Xyelidae). *Memoirs of the Queensland Museum* **51** (2): 558.

25 Sharkey, M.J., Carpenter, J.M., Vilhelmsen, L. et al. (2012). Phylogenetic relationships among superfamilies of Hymenoptera. *Cladistics* **28** (1): 80–112. https://doi.org/10.1111/j.1096-0031.2011.00366.x.

26 Mao, M., Gibson, T., and Dowton, M. (2015). Higher-level phylogeny of the Hymenoptera inferred from mitochondrial genomes. *Molecular Phylogenetics and Evolution* **84**: 34–43. https://doi.org/10.1016/j.ympev.2014.12.009.

27 Song, S.N., Tang, P., Wei, S.J., and Chen, X.X. (2016). Comparative and phylogenetic analysis of the mitochondrial genomes in basal hymenopterans. *Scientific Reports* **6**: 20972.

28 Engel, M.S., Ortega-Blanco, J., Soriano, C. et al. (2013). A new lineage of enigmatic diaprioid wasps in Cretaceous amber (Hymenoptera, Diaprioidea). *American Museum Novitates* **3771**: 1–23.

29 Michener, C.D. and Grimaldi, D.A. (1988). A Trigona from Late Cretaceous Amber of New Jersey (Hymenoptera: Apidae: Meliponinae). *American Museum Novitates* **2917**: 1–10.

30 Engel, M.S. (2000). A new interpretation of the oldest fossil bee (Hymenoptera: Apidae). *American Museum Novitates* **3296**: 1–10.

31 Elliott, D.K. and Nations, J.D. (1998). Bee burrows in the Late Cretaceous (Late Cenomanian) Dakota Formation, northeastern Arizona. *Ichnos* **5**: 243–253. https://doi.org/10.1080/10420949809386423.

32 Hong, Y.C. (1975). A new family - Sinosiricidae (Hymenoptera: Siricoidea) in west Weichangder, Hebei Province. *Acta Entomologica Sinica* **18** (2): 235–241.

33 Lin, Q.B. (1976). The Jurassic fossil insects from western Liaoning. *Acta Palaeontologica Sinica* **15**: 97–116.

34 Rasnitsyn, A.P. (1964). New Triassic Hymenoptera of the Middle Asia. *Paleontologicheskiy Zhurnal* **1**: 88–97.

35 Riek, E.F. (1955). Fossil insects from the Triassic beds at Mt.Crosby, Queensland. *Australian Journal of Zoology* **3** (4): 654–657.

36 Schlüter, T. (2000). *Moltenia rieki* n. gen., n. sp. (Hymenoptera: Xyelidae?), a tentative sawfly from the Molteno Formation (Upper Triassic), South Africa. *Paläontologische Zeitschrift* **74**: 75–78.

37 Lara, M.B., Rasnitsyn, A.P., and Zavattieric, A.M. (2014). *Potrerilloxyela menendezi* gen. et sp. nov. from the late Triassic of Argentina: the oldest representative of Xyelidae (Hymenoptera: Symphyta) for Americas. *Paleontology Journal* **48** (2): 182–190.

38 Taeger, A., Blank, S.M., and Liston, A.D. (2010). World catalog of Symphyta (Hymenoptera). *Zootaxa* **2580**: 1–1064.

39 Krassilov, V.A. and Rasnitsyn, A.P. (1982). A unique finding: pollen in the intestine of Early Cretaceous sawflies. *Paleontologicheskij Zhurnal* **3**: 78–87.

40 Rasnitsyn, A.P. (1966). New Xyelidae (Hymenoptera) from the Mesozoic of Asia. *Paleontologicheskiy Zhurnal* **4**: 69–85.

41 Rasnitsyn, A.P. (1977). New Hymenoptera from the Jurassic and Cretaceous of Asia. *Paleontologicheskiy Zhurnal* **11** (3): 349–357.

42 Ren, D., Lu, L.W., Guo, Z.G., and Ji, S.A. (1995). *Faunae and Stratigraphy of Jurassic-Cretaceous in Beijing and the Adjacent Areas*, 106–112. Beijing: Seismie Publishing House.

43 Zhang, H.C. and Zhang, J.F. (2000). Xyelid sawflies (Insecta, Hymenoptera) from the Upper Jurassic Yixian Formation of Western Liaoning, China. *Acta Palaeontologica Sinica* **39** (4): 476–492.

44 Gao, T.P., Ren, D., and Shih, C.K. (2009). *Abrotoxyela* gen. nov. (Insecta, Hymenoptera, Xyelidae) from the Middle Jurassic of Inner Mongolia, China. *Zootaxa* **2094**: 52–59.

45 Gao, T.P., Zhao, Y.Y., and Ren, D. (2011). New fossil Xyelidae (Insecta, Hymenoptera) from the Yixian Formation of western Liaoning, China. *Acta Geologica Sinica (English Edition)* **85** (3): 528–532.

46 Wang, M., Shih, C.K., and Ren, D. (2012). *Platyxyela* gen. nov. (Hymenoptera, Xyelidae, Macroxyelinae) from the Middle Jurassic of China. *Zootaxa* **3456**: 82–88.

47 Wang, M., Rasnitsyn, A.P., and Ren, D. (2014). Two new fossil sawflies (Hymenoptera, Xyelidae, Xyelinae) from Middle Jurassic of China. *Acta Geologica Sinica – English Edition* **88** (4): 1027–1033.

48 Hong, Y.C. (1982). *Mesozoic Fossil Insects of Jiuquan Basin in Gansu Province*, 173–175. Beijing: Geological Publishing House.

49 Pagliano, G. and Scaramozzino, P. (1990). *Elenco dei generi di Hymenoptera del mondo. Members of the Entomological Society of Italy*, vol. **68**, 1–210.

50 Rasnitsyn, A.P. (1968). New Mesozoic sawflies (Hymenoptera, Symphyta). In: *Jurassic Insects of Karatau* (ed. B.B. Rohdendorf), 190–236. Mescow: Nauka Press.

51 Nel, A., Petrulevicius, J.F., and Henrotay, M. (2004). New early Jurassic sawflies from Luxembourg: the oldest record of Tenthredinoidea (Hymenoptera: "Symphyta"). *Acta Palaeontologica Sinica* **49**: 283–288.

52 Gao, T.P., Ren, D., and Shih, C.K. (2009). The first Xyelotomidae (Hymenoptera) from the Middle Jurassic in China. *Annals of the Entomological Society of America* **102** (4): 588–596.

53 Gao, T.P., Shih, C.K., Engel, M.S., and Ren, D. (2016). A new xyelotomid (Hymenoptera) from the Middle Jurassic of China displaying enigmatic venational asymmetry. *BMC Evolutionary Biology* **16** (1): 155.

54 Qiao, X., Shih, C.K., Petrulevičius, J.F., and Ren, D. (2013). Fossils from the Middle Jurassic of China shed light on morphology of Choristopsychidae (Insecta, Mecoptera). *ZooKeys* **318**: 91–111.

55 Yang, X.G., Ren, D., and Shih, C.K. (2012). New fossil hangingflies (Mecoptera, Raptipeda, Bittacidae) from the Middle Jurassic to Early Cretaceous of Northeastern China. *Geodiversitas* **34** (4): 785–799.

56 Wang, Y., Shih, C.K., Szwedo, J., and Ren, D. (2012). New fossil palaeontinids (Hemiptera, Cicadomorpha, Palaeontinidae) from the Middle Jurassic of Daohugou, China. *Alcheringa* **37** (1): 1–12.

57 Liu, Y.S., Sinitshenkova, N.D., and Ren, D. (2007). A new genus and species of stonefly (Insecta: Plecoptera) from the Yixian Formation, Liaoning Province, China. *Cretaceous Research* **28** (2): 322–326.

58 Shih, C.K., Qiao, X., Labandeira, C.C., and Ren, D. (2013). A new mesopsychid (Mecoptera) from the Middle Jurassic of Northeastern China. *Acta Geologica Sinica (English Edition)* **87** (5): 1235–1241.

59 Ren, D., Labandeira, C.C., Santiago-Blay, J.A. et al. (2009). A probable pollination mode before angiosperms: Eurasian, long-proboscid scorpionflies. *Science* **326** (5954): 840–847.

60 Ren, D., Labandeira, C.C., and Shih, C.K. (2010). New Mesozoic Mesopsychidae (Mecoptera) from Northeastern China. *Acta Geologica Sinica (English Edition)* **84** (4): 720–731.

61 Ren, D. and Krzemiński, W. (2002). Eoptychopteridae (Diptera) from the Middle Jurassic of China. *Annales Zoologici* **52** (2): 207–210.

62 Liu, L.X., Shih, C.K., and Ren, D. (2012). Two new species of Ptychopteridae and Trichoceridae from the Middle Jurassic of Northeastern China (Insecta: Diptera: Nematocera). *Zootaxa* **3501**: 55–62.

63 Rasnitsyn, A.P. (1983). Fossil Hymenoptera of the superfamily Pamphilioidea. *Paleontological Journal* **2**: 56–70.

64 Wang, M., Rasnitsyn, A.P., Li, H. et al. (2016). Phylogenetic analyses elucidate the inter-relationships of Pamphilioidea (Hymenoptera, Symphyta). *Cladistics* **32**: 239–260.

65 Rasnitsyn, A.P., Zhang, H.C., and Wang, B. (2006). Bizarre fossil insects: web-spinning sawflies of the genus *Ferganolyda* (Vespida, Pamphilioidea) from the Middle Jurassic of Daohugou, Inner Mongolia, China. *Paleontology* **49**: 907–916.

66 Wang, M., Rasnitsyn, A.P., Shih, C.K., and Ren, D. (2015). New fossil records of bizarre *Ferganolyda* (Hymenoptera: Xyelydidae) from the Middle Jurassic of China. *Alcheringa* **39**: 99–108.

67 Gao, T.P., Engel, M.S., Blanco, J.O. et al. (2013). A new xyelydid sawfly from the Early Cretaceous of China (Hymenoptera: Xyelydidae). *Journal of the Kansas Entomological Society* **86**: 78–83.

68 Wang, M., Rasnitsyn, A.P., Shih, C.K., and Ren, D. (2014). A new Cretaceous genus of xyelydid sawfly illuminating nygmata evolution in Hymenoptera. *BMC Evolutionary Biology* **14**: 131.

69 Wang, M., Rasnitsyn, A.P., Shih, C.K., and Ren, D. (2015). New xyelydid sawflies from the Lower Cretaceous of China. *Cretaceous Research* **5**: 169–178.

70 Nichols, S.W. and Schuh, R.T., managing ed. (1989). *The Torre-Bueno Glossary of Entomology (Revised Edition of a Glossary of Entomology by J. R. de la Torre-Bueno; Including Supplement a by George S. Tulloch)*. New York: New York Entomological Society & American Museum of Natural History. pp. xvii + 840.

71 New, T.R. (1991). *Neuroptera (Lacewings)*. In: *The Insects of Australia: A Textbook for Students and Research Workers*, 2e, 525–542. Melbourne: Melbourne University Press.

72 Minet, J., Huang, D.Y., Wu, H., and Nel, A. (2010). Early mecopterida and the systematic position of the microptysmatidae (insecta: endopterygota). *Annales de la Société entomologique de France* **46** (1–2): 262–270.

73 Ronquist, F., Klopfstein, S., Vilhelmsen, L. et al. (2012). A total-evidence approach to dating with fossils, applied to the early radiation of the Hymenoptera. *Systematic Biology* **61**: 973–999.

74 Rasnitsyn, A.P. (2002). Superorder Vespidea Laicharting, 1781. Order Hymenoptera Linné, 1758 (=Vespida Laicharting, 1781). In: *History of Insects*

(ed. A.P. Rasnitsyn and L.J.D. Quicke), 242–254. Netherlands: Kluwer Academic Publishers.

75 Shcherbakov, D.E. (2013). Permian ancestors of hymenoptera and raphidioptera. *ZooKey* **358**: 45–67.

76 Gao, T.P., Rasnitsyn, A.P., Ren, D., and Shih, C.K. (2010). The first Praesiricidae (Hymenoptera) from Northeast China. *Annales de la Société Entomologique de France* **46** (1–2): 148–153.

77 Wang, M., Rasnitsyn, A.P., and Ren, D. (2013). New sawfly fossil from the Lower Cretaceous of China elucidates the antennal evolution in lower Hymenoptera (Pamphilioidea: Praesiricidae: Archoxyelydinae subfam. n.). *Systematic Entomology* **38**: 577–584.

78 Wang, M., Rasnitsyn, A.P., Shih, C.K., and Ren, D. (2015). Revision of the genus *Rudisiricius* (Hymenoptera, Praesiricidae) with six new species from Jehol biota, China. *Cretaceous Research* **52**: 570–578.

79 Wang, M., Rasnitsyn, A.P., Shih, C.K. et al. (2016). New fossils from China elucidating the phylogeny of Praesiricidae (Insecta: Hymenoptera). *Systematic Entomology* **41**: 41–55.

80 Rasnitsyn, A.P. (1996). Conceptual issues in Phylogeny, taxonomy, and nomenclature. *Contributions to Zoology* **66** (1): 3–41.

81 Rasnitsyn, A.P. (2006). Ontology of evolution and methodology of taxonomy. *Paleontological Journal* **40** (6): 679–737.

82 Penãlver, E. and Arillo, A. (2002). Primer registro fósil del género *Acantholyda* (Insecta: Hymenoptera: Pamphiliidae), Mioceno inferior de Ribesalbes (Espanã). *Revista Espanõla de Paleontologia* **17**: 73–81.

83 Wang, M., Rasnitsyn, A.P., Shih, C.K., and Ren, D. (2014). A new fossil genus in Pamphiliidae (Hymenoptera) from China. *Alcheringa* **38**: 391–397.

84 Viitasaari, M. (ed.) (2002). *Sawflies (Hymenoptera, Symphyta), I: A Review of the Suborder, the Western Palaearctic Taxa of Xyeloidea and Pamphilioidea*, vol. **1**, 516. Tremex Press Ltd., Helsinki.

85 Eidt, C.D. (1969). The life histories, distribution, and immature forms of the North American sawflies of the genus Cephalcia (Hymenoptera: Pamphiliidae). *Members of the Entomological Society of Canada* **101**: 5–56.

86 Xiao, G.R. (1992). *Forest Insects of China*, 2e, 1144–1204. Beijing: Chinese Forestry Publishing House.

87 Benson, R.B. (1945). Classification of the Pamphiliidae (Hymenoptera Symphyta). *Proceedings of the Royal Entomological Society of London (B)* **14**: 25–33.

88 Benson, R.B. (1968). Hymenoptera from Turkey, Symphyta. *Bulletin of the British Museum (Natural History)* **22**: 111–207.

89 Goulet, H. (1993). Superfamilies Cephoidea, Megalodontoidea, Orussoidea, Siricoidea, Tenthredinoidea, and Xyeloidea. In: *Hymenoptera of the World: An Identification Guide to Families. Centre for Land and Biological Resources Research* (ed. H. Goulet and J.T. Huber), 101–129. Ottawa, Ontario: Research Branch Agriculture Canada.

90 Rasnitsyn, A.P. (2000). Testing cladograms by fossil record: the ghost range test. *Contributions to Zoology* **69**: 251–258.

91 Gao, T.P., Shih, C.K., Labandeira, C.C. et al. (2016). Convergent evolution of ramified antennae in insect lineages from the Early Cretaceous of Northeastern China. *Proceedings of the Royal Society of London B* **283** (1839): 20161448.

92 Shih, C.K., Liu, C.X., and Ren, D. (2009). The earliest fossil record of pelecinid wasps (Inseta: Hymenoptera: Proctotrupoidea: Pelecinidae) from Inner Mongolia, China. *Annals of the Entomological Society of America* **102** (1): 20–38.

93 Shih, C.K., Feng, H., Liu, C.X. et al. (2010). Morphology, phylogeny, evolution, and dispersal of pelecinid wasps (Hymenoptera: Pelecinidae) over 165 Million Years. *Annals of the Entomological Society of America* **103** (6): 875–885.

94 Johnson, N.F. and Musetti, L. (1999). Revision of the proctotrupoid genus *Pelecinus Latreille* (Hymenoptera: Pelecinidae). *Journal of Natural History* **33**: 1513–1543.

95 Brues, C.T. (1928). A note on the genus *Pelecinus. Psyche* **35**: 205–209.

96 Guo, L.C., Shih, C.K., Li, L.F., and Ren, D. (2016). New pelecinid wasps (Hymenoptera: Pelecinidae) from the Yixian Formation of western Liaoning, China. *Cretaceous Research* **61**: 151–160.

97 Zhang, H.C., Rasnitsyn, A.P., and Zhang, J.F. (2002). Pelecinid wasps (Insecta: Hymenoptera: Proctotrupoidea) from the Yixian Formation of western Liaoning, China. *Cretaceous Research* **23**: 87–98.

98 Zhang, J.F. and Rasnitsyn, A.P. (2006). New extinct taxa of Pelecinidae *sensu lato* (Hymenoptera: Proctotrupoidea) in the Laiyang Formation, Shandong, China. *Cretaceous Research* **27**: 684–688.

99 Liu, C.X., Shih, C.K., and Ren, D. (2009). The earliest fossil record of the wasp subfamily Pelecininae (Hymenoptera: Proctotrupoidea: Pelecinidae) from the Yixian Formation of China. *Zootaxa* **2080**: 7–54.

100 Feng, H., Shih, C.K., and Ren, D. (2010). New male pelecinid wasps (Hymenoptera: Pelecinidae) from the Yixian Formation of western Liaoning (China). *Geologica Carpathica* **61** (6): 463–468.

101 Shi, G.H., Grimaldi, D.A., Harlow, G.E. et al. (2012). Age constraint on Burmese amber based on U-Pb dating of zircons. *Cretaceous Research* **37**: 155–163.

102 Guo, L.C., Shih, C.K., Li, L.F., and Ren, D. (2016). New pelecinid wasps (Hymenoptera: Pelecinidae) from Upper Cretaceous Myanmar amber. *Cretaceous Research* **61**: 151–160.

103 Achterberg, C. (2006). European species of the genus *Helorus* Latreille (Hymenoptera: Heloridae), with description of a new species from Sulawesi (Indonesia). *Zoologische Mededelingen, Leiden* **80**: 1–12.

104 Kusigemati, K. (1987). The Heloridae (Hymenoptera: Proctotrupoidea) of Japan. *Kontyû* **55**: 477–485.

105 Rohdendorf, B.B. (1938). Mesozoic Diptera from Karatau. I Brachycera and some Nematocera. *Trudy Paleontologicheskogo Instituta Academii Nauk SSSR* 7: 29–67.

106 Rasnitsyn, A.P. (1986). Order Vespida (Hymenoptera). In: *Insects in the Early Cretaceous Ecosystem of the West Mongolia*, vol. **28**, 154–164. Transactions of the Joint Soviet- Mongolian Paleontological Expedition.

107 Rasnitsyn, A.P. (1990). Hymenopterans Vespida. In late Mesozoic insects of eastern Transbailalia. *Trudy Paleontologicheskogo Instituta Academii Nauk SSSR* **239**: 177–205.

108 Zhang, J.F. (1992). Two new genera and species of Heloridae (Hymenoptera) from late Mesozoic of China. *Entomotaxonomia* **14**: 222–228.

109 Zhang, H.C. and Zhang, J.F. (2001). Proctotrupoid wasps (Insecta, Hymenoptera) from the Yixian Formation of western Liaoning province. *Acta Micropalaeontologica Sinica* **18**: 11–28.

110 Shih, C.K., Feng, H., and Ren, D. (2011). New fossil Heloridae and Mesoserphidae wasps (Insecta, Hymenoptera, Proctotrupoidea) from the Middle Jurassic of China. *Annals of the Entomological Society of America* **104**: 1334–1348.

111 Shi, X.Q., Zhao, Y.Y., Shih, C.K., and Ren, D. (2013). New fossil helorid wasps (Insecta, Hymenoptera, Proctotrupoidea) from the Jehol Biota, China. *Cretaceous Research* **41**: 136–142.

112 Shi, X.Q., Zhao, Y.Y., Shih, C.K., and Ren, D. (2014). Two new species of *Archaeohelorus* (Hymenoptera, Proctotrupoidea, Heloridae) from the Middle Jurassic of China. *Zookeys* **369**: 49–59.

113 Li, L.F., Shih, C.K., and Ren, D. (2017). New fossil helorid wasps (Hymenoptera, Proctotrupoidea) from the Early Cretaceous of China. *Alcheringa* **17**: 1–13.

114 Rasnitsyn, A.P. (1986). New species of the Mesoserphidae hymenopteran family from Upper Jurassic Kara-Tau. *Vestnik Zoologii* **2**: 19–25.

115 Kozlov, M.A. (1968). Jurassic Proctotrupidea (Hymenoptera). In: *Jurassic Insects of Karatau* (ed. B.B. Rohdendorf), 237–240. Moscow, Russia: Academiya Nauk SSSR Otdelenie Obshchej Biologii.

116 Rasnitsyn, A.P. (1983). Hymenoptera from the Jurassic of East Siberia. *Bulletin of Moscow Society of Naturalists Biological Series* **58**: 85–94.

117 Rasnitsyn, A.P. (1994). *New Late Jurassic Mesoseprhidae (Vespida, Proctotrupoidea)*, 115–119. Paleontologicheskii Zhurnal.

118 Shi, X.Q., Zhao, Y.Y., Shih, C.K., and Ren, D. (2013). New fossil mesoserphid wasps (insect, Hymenoptera, Proctotrupoidea) from the Jehol Biota, China. *Zootaxa* **3710**: 591–599.

119 Zhang, H.C., Zheng, D.R., Zhang, Q. et al. (2013). Re-description and systematic of *Paraulacus sinicus* Ping, 1928 (Insecta, Hymenoptera). *Palaeoworld* **22**: 32–35.

120 Li, L.F., Rasnitsyn, A.P., Shih, C.K., and Ren, D. (2017). The Mesozoic family Mesoserphidae and its phylogeny (Hymenoptera, Apocrita, Proctotrupoidea). *Journal of Systematic Palaeontology* **15** (8): 617–639.

121 Zhang, H.C. and Zhang, J.F. (2000). A new genus of Mesoserphidae (Hymenoptera: Proctotrupoidea) from the Upper Jurassic of northeast China. *Entomotaxonomia* **22**: 279–282.

122 Cai, Y.P., Zhao, Y.Y., Shih, C.K., and Ren, D. (2012). A new genus of Scolebythidae (Hymenoptera: Chrysidoidea) from the Early Cretaceous of China. *Zootaxa* **3504**: 56–66.

123 Brothers, D.J. (1981). Note on the biology of *Ycaploca evansi* (Hymenoptera: Scolebythidae). *Journal of the Entomological Society of Southern Africa* **44**: 107–108.

124 Melo, G.A.R. (2000). Biology of an extant species of the scolebythid genus *Dominibythus* (Hymenoptera: Chrysidoidea: Scolebythidae), with description of its mature larva. In: *Hymenoptera: Evolution, Biodiversity and Biological Control* (ed. A.D. Austin and M. Dowton), 281–284. Collingwood: CSIRO.

125 Carpenter, J.M. (1986). Cladistics of the Chrysidoidea (Hymenoptera). *Journal of the New York Entomological Society* **94**: 303–330.

126 Prentice, M.A., Poinar, G.O. Jr.,, and Milki, R. (1996). Fossil scolebythids (Hymenoptera, Scolebythidae) from Lebanese and Dominican amber. *Proceedings of the Entomological Society of Washington* **98**: 802–811.

127 Rasnitsyn, A.P. (1972). Praeaulacidae (Hymenoptera) from the Upper Jurassic of Karatau. *Paleontologicheskiy Zhurnal* **1**: 72–87.

128 Engel, M.S., Huang, D.-Y., Alqarni, A.S., and Cai, C.Y. (2016). A remarkable evanioid wasp in

mid-Cretaceous amber from northern Myanmar (Hymenoptera: Evanioidea). *Cretaceous Research* **60**: 121–127.

129 Rasnitsyn, A.P. (1975). Hymenoptera Apocrita of Mesozoic. *Trudy Paleontologicheskogo Instituta Academii Nauk SSSR* **147**: 1–134.

130 Rasnitsyn, A.P. and Martínez-Delclòs, X. (2000). Wasps (Insecta: Vespida = Hymenoptera) from the Early Cretaceous of Spain. *Acta Geologica Hispanica* **35**: 65–95.

131 Li, L.F., Rasnitsyn, A.P., Shih, C.K., and Ren, D. (2013). Anomopterellidae restored with two new genera and its phylogeny in Evanioidea (Hymenoptera). *PLoS One* **8**: e82587.

132 Li, L.F., Shih, C.K., and Ren, D. (2014). Revision of *Anomopterella* Rasnitsyn, 1975 (Insecta, Hymenoptera, Anomopterellidae) with two new Middle Jurassic species from Northeastern China. *Geologica Carpathica* **65** (5): 365–374.

133 Ghahari, H. and Deans, A.R. (2010). A comment on Iranian Ensign wasps (Hymenoptera: Evanoidea: Evaniidae). *Munis Entomology and Zoology* **5**: 295–296.

134 Deans, A.R., Gillespie, J.J., and Yoder, M.J. (2006). An evaluation of ensign wasp classification (Hymenoptera: Evaniidae) based on molecular data and insights from ribosomal RNA secondary structure. *Systematic Entomology* **31**: 517–528.

135 Zhang, H.C. and Zhang, J.F. (2000). A new genus and two new species of Hymenoptera (Insecta) from the Upper Jurassic Yixian Formation of Beipiao, western Liaoning. *Acta Micropalaeontologica Sinica* **17**: 286–290. (in English, Chinese abstract).

136 Li, L.F., Shih, C.K., and Ren, D. (2014). New fossil evaniids (Hymenoptera, Evanioidea) from the Yixian Formation of western Liaoning, China. *Cretaceous Research* **47**: 48–55.

137 Zhang, H.C. and Rasnitsyn, A.P. (2008). Middle Jurassic Praeaulacidae (Insecta: Hymenoptera: Evanioidea) of Inner Mongolia and Kazakhstan. *Journal of Systematic Palaeontology* **6**: 463–487.

138 Li, L.F. and Shih, C.K. (2014). Two new fossil wasps (Insecta: Hymenoptera: Apocrita) from Northeastern China. *Journal of Natural History* **49** (13–14): 829–840.

139 Li, L.F., Rasnitsyn, A.P., Shih, C.K., and Ren, D. (2015). A new genus and species of Praeaulacidae (Hymenoptera: Evanioidea) from Upper Cretaceous Myanmar amber. *Cretaceous Research* **41** (2013): 136–142.

140 Zhang, H.C. and Rasnitsyn, A.P. (2007). Nevaniinae subfam. n., a new fossil taxon (Insecta: Hymenoptera: Evanioidea: Praeaulacidae) from the Middle Jurassic of Daohugou in Inner Mongolia, China. *Insect Systematics & Evolution* **38** (2): 149–166.

141 Rasnitsyn, A.P. and Zhang, H.C. (2010). Early evolution of Apocrita (Insect, Hymenoptera) as indicated by new findings in the Middle Jurassic of Daohugou, Northeast China. *Acta Geologica Sinica (English Edition)* **84**: 834–873.

142 Li, L.F., Shih, C.K., and Ren, D. (2014). New fossil Praeaulacinae wasps (Insect: Hymenoptera: Evanioidea: Praeaulacidae) from the Middle Jurassic of China. *Zootaxa* **3814** (3): 432–442.

143 Rasnitsyn, A.P. (1988). An outline of evolution of the hymenopterous insects (order Vespida). *Oriental Inssets* **22**: 115–145.

144 Li, L.F., Shih, C.K., Rasnitsyn, A.P., and Ren, D. (2015). New fossil ephialtitids elucidating the origin and transformation of the propodeal-metasomal articulation in Apocrita (Hymenoptera). *BMC Evolutionary Biology* **15** (1): 1–17.

145 Wang, M.L., Li, L.F., and Shih, C.K. (2015). New fossil wasps (Hymenoptera, Apocrita) from the Middle Jurassic of China. *Insect Systematics & Evolution* **46** (5): 471–484.

146 Li, L.F., Shih, C.K., and Ren, D. (2013). Two new wasps (Hymenoptera: Stephanoidea: Ephialtitidae) from the Middle Jurassic of China. *Acta Geologica Sinica (English Edition)* **87** (6): 1486–1494.

147 Rasnitsyn, A.P. (1963). Late Jurassic Hymenoptera of Karatau. *Paleontologicheskii Zhurnal* **1**: 86–99.

148 Rasnitsyn, A.P., Ansorge, J., and Zhang, H.C. (2006). Ancestry of the orussoid wasps, with description of three new genera and species of Karatavitidae (Hymenoptera = Vespida: Karatavitoidea stat. nov.). *Insect Systematics and Evolution* **37**: 179–190.

149 Shih, M.J.H., Li, L.F., and Ren, D. (2017). Application of geometric morphometric analyses to confirm two new species of Karatavitidae (Hymenoptera: Karatavitoidea) from Northeastern China. *Alcheringa* 1–10.

150 Gao, T.P. and Ren, D. (2008). Description of a new fossil Anthoxyela species (Hymenoptera, Xyelidae) from Yixian Formation of Northeast China. *Zootaxa* **1842**: 56–62.

151 Wang, C., Shih, C.K., Rasnitsyn, A.P., and Wang, M. (2016). Two new species of *Prolyda* from the Middle Jurassic of China (Hymenoptera, Pamphilioidea). *ZooKeys* **569**: 71.

152 Zhang, H.C., Rasnitsyn, A.P., and Zhang, J.F. (2002). Two ephialtitid wasps (Insecta, Hymenoptera, Ephialtitoidea) from the Yixian Formation of western Liaoning, China. *Cretaceous Research* **23**: 401–407.

153 Zhang, Q., Zhang, H.C., Rasnitsyn, A.P. et al. (2014). New ephialtitidae (Insecta: Hymenoptera) from the jurassic daohugou beds of Inner Mongolia, China. *Palaeoworld* **23** (3–4): 276–284.

154 Zhang, H.C., Rasnitsyn, A.P., Wang, D.J., and Zhang, Y.T. (2007). Some hatchet wasps (Hymenoptera, Evaniidae) from the Yixian Formation of western Liaoning, China. *Cretaceous Research* **28**: 310–316.

155 Zhang, J.F. (2005). Eight new species of the genus Eopelecinus (Hymenoptera: Prototrupoidea) from the Laiyang Formation, Shandong Province, China. *Paleontological Journal* **39**: 417–427.

156 Liu, C.X., Gao, T.P., Shih, C.K., and Ren, D. (2011). New Pelecinid wasps (Hymenoptera: Proctotrupoidea: Pelecinidae) from the Yixian Formation of Western Liaoning, China. *Acta Geologica Sinica (English Edition)* **85** (4): 749–757.

157 Li, L.F., Shih, C.K., and Ren, D. (2013). Two new species of Nevania (Hymenoptera: Evanioidea: Praeaulacidae: Nevaniinae) from the Middle Jurassic of China. *Alcheringa* **38**: 140–147.

158 Zhang, H.C. and Rasnitsyn, A.P. (2004). Pelecinid wasps (Insecta: Hymenoptera: Proctotrupoidea) from the Cretaceous of Russia and Mongolia. *Cretaceous Research* **25**: 807–825.

159 Zhang, J.F. (1986). A new Middle Jurassic insect genus Sinephialtites of Ephialtitidae discovered in China. *Acta Palaeontologica Sinica* **25**: 585–590.

160 Duan, Y. and Cheng, S.L. (2006). A new species of Pelecinidae (Hymenoptera: Proctotrupoidea) from the Lower Cretaceous Jiufotang Formation of western Liaoning. *Acta Palaeontologica Sinica* **45**: 393–398.

23

Diptera – True Flies with Two Wings

Ye Han[1], Xiuna Ye[1], Cuiping Feng[1], Kuiyan Zhang[1,2], Chungkun Shih[1,3], and Dong Ren[1]

[1] Capital Normal University, Haidian District, Beijing, China
[2] Institute of Zoology, Chinese Academy of Sciences; Chaoyang District, Beijing, China
[3] National Museum of Natural History, Smithsonian Institution, Washington, DC, USA

23.1 Introduction to Diptera

Flies have been used by people to denote various types of insects, such as mayflies, dragonflies, damselflies, butterflies, fireflies, caddisflies, stoneflies, fishflies, scorpionflies, etc. However, the true flies are in the order Diptera. Diptera can be easily distinguished from other insects, because dipterans have only two wings and the hind wings have been evolved into "halteres". That is why they are called "Diptera," "di" means "two", "ptera" means "wing" in Greek. Even with only a pair of wings, Diptera are one of the best fliers in the Class Insecta (Figure 23.1) with excellent capability of fast speed, hovering, backward flying and sudden direction changes. Typically, flies have a pair of highly developed compound eyes (Figure 23.2). For example, the compound eyes of house flies (*Musca domestica* Linnaeus, 1758) have 3500–4000 ommatidia, which occupy almost 60–90% of the whole head. Using the image inputs from thousands of ommatidia, dipterans possess a very wide view angle and can detect fast movement to escape predators or enemies and to find food or potential mates.

As highly evolved Holometabola insects, the life cycles of dipterans pass through four developmental stages: eggs, larvae, pupae and adults. Some flies (e.g. Muscidae, Calliphoridae, Tachinidae and Sarcophagidae) have ovoviviparous habitus, i.e. the females incubate the fertilized eggs and accommodate developing embryos inside the reproductive tracts for a prolonged duration, which is a successful strategy to protect and enhance the survival rate of offspring [1]. Diptera play a vital role in nature, and many species have close links with people's life. They are pollinators (e.g. hover flies), predators (e.g. robber flies, long-legged flies), scavengers (e.g. fruit flies, flesh flies), parasitoids (e.g. bee flies, flower flies) and hematophages (e.g. deer flies), etc. (Figures 23.3 and 23.4). Diptera can be found everywhere around the world except the polar regions and the distant seas.

As one of the four major insect orders, Diptera comprise about 10 000 genera in 150 families, at least 154 000 described species, representing 10–12% of animal species [2–4]. In the past, Diptera were classified into three suborders, Nematocera, Brachycera and Cyclorrhapha. Currently, based on both morphological and molecular data, "Nematocera" and "Cyclorrhapha" are not logical suborders any more (Figure 23.5). Thus, to be precise, there is only one suborder in Diptera now, i.e. Brachycera (including "Cyclorrhapha"). The former "Nematocera," which is paraphyletic, is called "Lower Diptera" temporarily.

The "Lower Diptera" comprise approximately 26–40 families and more than 52 000 species worldwide [5–7]. Previously, "Lower Diptera" were divided into 4–7 Infraorders: (Axymyiomorpha), (Ptychopteromorpha), Culicomorpha, (Blephariceromorpha), Bibionomorpha, Psychodomorpha and Tipulomorpha. In fact, these classifications have not been fully acceptable by all researchers. In particular, the phylogeny of Tipulomorpha is still confusing [8]. According to the latest research, only Culicomorpha and Bibionomorpha have been found to have significant support [4]. The "Lower Diptera" represent flies with long antennae. They are typically characterized by filamentous antenna with more than six antennomeres, for example, gnats, midges, mosquitoes, etc. Some of them are notorious disease-carriers or pests. Five families (Ceratopogonidae, Simuliidae, Culicidae, Corethrellidae and Psychodidae) contain blood-feeding species. Mosquitos, called "the Deadliest Animal in the World," can transmit extremely harmful infections, such as malaria, yellow fever, West Nile virus, dengue fever, encephalitis, Zika virus, etc.

Rhythms of Insect Evolution: Evidence from the Jurassic and Cretaceous in Northern China, First Edition.
Edited by Dong Ren, Chungkun Shih, Taiping Gao, Yongjie Wang, and Yunzhi Yao.

Figure 23.1 Flower flies (Syrphidae) mating while flying. Source: Photo by Jason Shih.

Figure 23.2 Houseflies (Muscidae) mating. Source: Photo by Jason Shih.

Figure 23.3 A long-legged fly (Dolichopodidae) feeding on a small insect. Source: Photo by Jason Shih.

Figure 23.4 A flower fly (Syrphidae) laying eggs on aphids. Source: Photo by Jason Shih.

Brachycera is a monophyletic group. This suborder includes four Infraorders: Xylophagomorpha, Stratiomyomorpha, Tabanomorpha and Muscomorpha. Lower Brachycera (previous "Orthorrhapha") comprise about 24 000 species in about 20 families assigned to three Infraorders of Stratiomyomorpha, Tabanomorpha, and Xylophagomorpha and two superfamilies of Asiloidea and Nemestrinoidea [9]. In this group, all hematophagous species come from Tabanomorpha. All or some species of Athericidae, Rhagionidae, and Tabanidae are well-known as blood-suckers [10]. Female horse flies of the species *Philoliche rostrata* (Family Tabanidae) have elongated proboscides and short biting mouthparts to carry out both nectar- and blood-feeding. Their mouthparts have two functional units, i.e. the proximal biting and piercing part comprising labrum, mandible blade, hypopharynx and maxilla, while distal siphoning part including a long prementum from the labial base and apical labella, to adapt to the different diets [11].

Eremoneura (belonging to Infraorder Muscomorpha, including Empidoidea and Cyclorrhapha) are also one of the best-supported higher-level brachyceran clades [12]. Empidoidea as an intermediate superfamily provide a vital link between Lower Brachycera and Cyclorrhapha. More than 12 000 species have been recorded in this superfamily (4900 species in Empididae and 7000 species in Dolichopodidae).

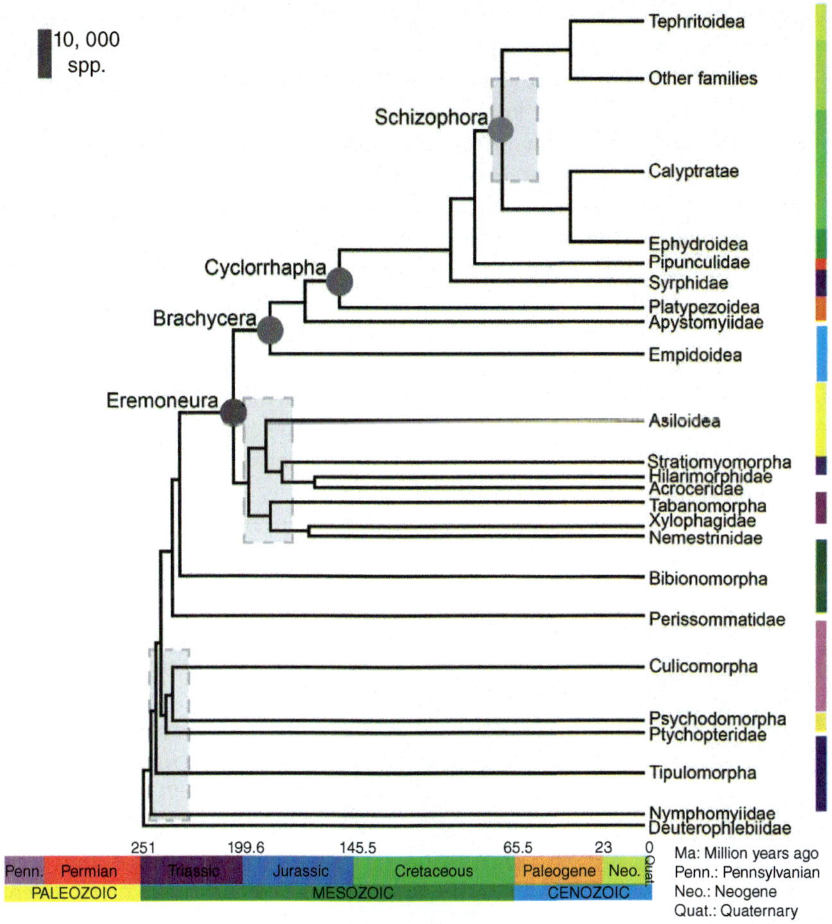

Figure 23.5 Chronogram depicting dipteran phylogeny and estimated age of clade divergences. Source: Modified from [5].

In the last few decades, Cyclorrhapha was considered to include "Aschiza" and Schizophora (Acalyptratae + Calyptratae). Now, we treat "Cyclorrhapha" as an unranked taxon within the Infraorder Muscomorpha. These are so-called "Higher flies", comprising more than 80 000 species. Adult "Higher flies" exit from their puparium by cracking a circular line of weakness. This is the reason why they are called "cyclorrhaphous flies". However, the systematic position and interrelationships of "Cyclorrhapha" are far from being resolved.

In Diptera, Schizophora have the highest diversity, comprising some 80 families in 12 superfamilies. As one of the best supported monophyletic groups, Schizophora are supported on the basis of a series of apomorphic character states, such as the inverted "U"-shaped ptilinum (above the base of antennae), unforked M_{1+2}, and shortened anal cell.

For blood-feeding, the following five families: Muscidae (Superfamily Muscoidea), Glossinidae, Hippoboscidae, Streblidae and Nycteribiidae (Superfamily Hippoboscoidea) have blood-feeders. Unlike other hematophagous Diptera, both sexes of the Calyptratae blood-feeding flies feed on blood. Undoubtedly, blood-feeding habit has independently evolved several times in Diptera. Different species have adapted to different vertebrate hosts, e.g. birds, bats, primates, etc., and developed individual mating systems/life cycles [13].

2015 Nobel Prize in Physiology or Medicine – A Novel Therapy against Malaria

Youyou Tu, Chief Professor at the China Academy of Traditional Chinese Medicine since 2000 (Figure 23.6), was awarded half of the Nobel Prize in Physiology or Medicine for her discoveries concerning a novel therapy against Malaria in 1970s [14].

Malaria has been with humankind for as long as we know. It is a mosquito-borne disease caused by single-cell parasites, which invade red blood cells, causing fever, and in severe cases brain damage and death. More than 3.4 billion of the world's most vulnerable citizens are at risk of contracting Malaria, and each year it claims more than 450 000 lives, predominantly among children.

Malaria was traditionally treated by chloroquine or quinine, but with declining success. By the late 1960s, efforts to eradicate Malaria had failed and the disease was on the rise. At that time, Youyou Tu in China turned to traditional herbal medicine to tackle the challenge of developing novel Malaria therapies. From a large-scale screen of herbal remedies in Malaria-infected animals, an extract from the plant *Artemisia annua* emerged as an interesting candidate. However, the results were inconsistent, so Tu revisited the ancient literature and discovered clues that guided her in her quest to successfully extract the active component from *A. annua*. Tu was the first to show that this component, later called Artemisinin, was highly effective against the Malaria parasite, both in infected animals and in humans. Artemisinin represents a new class of antimalarial agents that rapidly kill the Malaria parasites at an early stage of their development, which explains its unprecedented potency in the treatment of severe Malaria [15].

Figure 23.6 Nobel Laureate Youyou Tu [15].

23.2 Progress in the Studies of Fossil Diptera

The first fossil dipteran species was described by Scheuchzer in 1709 based on a female specimen of Tipulidae from the Eocene of Italy, even though it was originally assigned to Odonata [16]. As for the earliest dipterans, they were often mistakenly mixed with Mecoptera. Some described Permian "dipterans" have been eventually transferred to Mecoptera [17–20]. The earliest confirmed dipterans are from the Early Triassic of France, representing two isolated wings that belong to two families of Grauvogeliidae and Rhagionidae respectively [21]. In the Triassic, the dipterans had completed the worldwide distribution, recorded from Asia, Australia, North America and Europe. Some major lineages of Diptera, i.e. Bibionomorpha, Culicomorpha, Psychodomorpha, and Tipulomorpha, have been recorded. Therefore, it is suggested that the origination time of Diptera might be much earlier than the current fossil records.

In the history, many dipterists have made significant contributions to the research of Diptera fossils. The distinguished German dipterist Hermann Loew had described more than 4000 species of flies, mosquitoes, gnats and midges in his life. His excellent works on Baltic amber dipterans greatly promoted the research on Baltic amber. Scudder Samuel Hubbard, an American paleontologist, contributed two important monographs, *Fossil Insects of North America* and *The Pre-tertiary Insects and The Tertiary Insects of North America*, and described numerous fossil dipterans from America. In the twentieth century, the studies of dipteran fossils flourished, and many paleontologists have made significant contributions. Cockerell described plenty of dipterans from the Cenozoic (Tertiary) deposits worldwide. The monograph *Die Fossilen Insekten* by A. Handlirsch [22] is the first comprehensive summary on fossil insects in that many valid names of dipteran fossils were clarified. The Russian paleontologist B.B. Rohdendorf made significant contributions on the taxonomy and phylogeny of the fossil dipterans. Carpenter [23] and Evenhuis [24] respectively completed the world catalog of dipteran fossils and laid the foundation of the studies of fossil dipterans. Furthermore, V.G. Kovalev focused on the Brachycera and described many new taxa from the Jurassic to Cretaceous of Siberia. Later on, M.B. Mostovski, A. Nagatomi, W. Krzemiński and D.A. Grimaldi, as the representatives, have described numerous new taxa of fossil dipterans which have enhanced our understanding of Diptera.

Fossil Diptera studies in China started in 1923 when Grabau described four species from the Early Cretaceous of Laiyang, China, including the first new fossil dipteran species *Samarura gregaria* [25]. Then, in 1928, Ping reviewed the species and established the first new fossil dipteran genus of China, *Chironomaptera* Ping, 1928 [26]. After an interruption of 30 years, research on dipteran fossils developed rapidly, and many fossil dipterans have been described mainly from three important Mesozoic fossil localities: Daohugou in Ningcheng of Inner Mongolia, Huangbanjigou in Beipiao of Liaoning and Laiyang of Shandong. Besides, fossil dipterans were also found from many other deposits, e.g. Guangxi, Hunan, Zhejiang, and Gansu, etc. During this period, significant contributions have been made by Qibin Lin, Youchong Hong, Wenli Wang, Junfeng Zhang, Dong Ren, Ding Yang, Haichun Zhang, Diying Huang, Zhijun Zhang, Chungkun Shih and Kuiyan Zhang. Many particular dipteran fossils have been documented that greatly enhanced our understanding on the evolution and phylogeny of Diptera. To date, about 262 fossil species in 157 genera and 46 families have been described from China.

23.3 Representative Fossils of Diptera from Northern China

Suborder "Nematocera" Latreille, 1825

Family Antefungivoridae Rohdendorf, 1938

Antefungivoridae is an extinct family belonging to Bibionomorpha, Sciaroidea. It was established by Rohdendorf based on the genus *Antefungivora* from the Late Jurassic of Karatau. The family status of Antefungivoridae have not been well-documented, and Rohdendorf even assigned it to Pleciominidae [27]. Carpenter pointed out that cross-veins between M and Rs in *Antefungivora* is distant to the origin of Rs which is different to *Pleciomima*, but he accepted the assignment to Pleciominidae [23]. In fact, the family status of Antefungivoridae is still not clear, and the family rank was tentatively retained by some researchers [7, 24]. Up to date, there are 9 extinct genera containing 44 species described [7].

Genera included from the Jurassic and Cretaceous of Northern China: *Antefungivora* Rohdendorf, 1938, *Lycoriomimodes* Rohdendorf, 1946, *Mimallactoneura* Rohdendorf, 1946 and *Aortomima* Zhang, Zhang, Liu & Shangguan, 1986.

Antefungivora Rohdendorf, 1938

Antefungivora Rohdendorf, 1938, *Trudy Paleontol. Inst.*, 7 (3), 48 [28] (original designation).

Lycoriomima Rohdendorf, 1946, *Trudy Paleontol. Inst.*, 13 (2), 62 [27]. Syn. by Ansorge, 1996, *Neue Palaeontol. Abh.*, 2, 98 [29].

Lycorioplecia Rohdendorf, 1946, *Trudy Paleontol. Inst.*, 13 (2), 66 [27]. Syn. by Ansorge, 1996, *Neue Palaeontol. Abh.*, 2, 98 [29].

Paritonida Rohdendorf, 1946, *Trudy Paleontol. Inst.*, 13 (2), 74 [27]. Syn. by Ansorge, 1996, *Neue Palaeontol. Abh.*, 2, 98 [29].

Archilycoria Rohdendorf, 1962, *Fundamentals of Paleontology*, 328 [30]. Syn. by Ansorge, 1996, *Neue Palaeontol. Abh.*, 2, 98 [29].

Type species: *Antefungivora prima* Rohdendorf, 1938.

Body 3–4 mm long. R nearly straight and long, ending at 2/3 of length of the wing. The stem of Rs about three times as long as the length of cross-vein r-m, arched slightly. Sc not obvious. M short [27, 31].

Distribution and age: Zhejiang, Liaoning; Late Jurassic, Early Cretaceous.

Two species included from the Jurassic and Cretaceous of China (see Table 23.1).

Lycoriomimodes Rohdendorf, 1946

Lycoriomimodes Rohdendorf, 1946, *Trudy Paleontol. Inst.*, 13 (2), 67 [27] (original designation).

Lycoriomimella Rohdendorf, 1946, *Trudy Paleontol. Inst.*, 13 (2), 69 [27]. Syn. by Kovalev, 1990, *Trudy Paleontolo. Inst.*, 239, 160 [32].

Pleciomimella Rohdendorf, 1946, *Trudy Paleontol. Inst.*, 13 (2), 71 [27]. Syn. by Kovalev, 1990, *Trudy Paleontolo. Inst.*, 239, 160 [32].

Megalycoriomima Rohdendorf, 1962, *Fundamentals of Paleontology*, 326 [30]. Syn. by Kovalev, 1990, *Trudy Paleontolo. Inst.*, 239, 160 [32].

Sinemedia Rohdendorf, 1962, *Fundamentals of Paleontology*, 328 [30]. Syn. by Kovalev, 1990, *Trudy Paleontolo. Inst.*, 239, 160 [32].

Type species: *Lycoriomimodes deformatus* Rohdendorf, 1946.

Wing 1.6 mm long. R straight, just at terminal curved to C, about 2/3 of length of the wing. Cross-vein r-m slightly shorter than the stem of Rs. Rs curved backward. M and Cu not obviously weak. Pterostigma absent. CuP not obvious [27, 31].

Distribution and age: Liaoning, Shandong, Hebei; Middle Jurassic, Early Cretaceous.

Seven species included from the Jurassic and Cretaceous of Northern China (see Table 23.1).

Mimallactoneura Rohdendorf, 1946

Mimallactoneura Rohdendorf, 1946, *Trudy Paleontol. Inst.*, 13 (2), 72 [27] (original designation).

Type species: *Mimallactoneura vetusta* Rohdendorf, 1946.

Wing 3 mm long. R almost completely straight. The stem of Rs about five times as long as length of cross-vein r-m. Rs bends slightly back in the distance. Sc weak, not obvious. M and CuP remarkable [27, 31].

Distribution and age: Liaoning, Shandong; Middle Jurassic, Early Cretaceous.

Two species included from the Jurassic and Cretaceous of Northern China (see Table 23.1).

***Aortomima* Zhang, Zhang, Liu & Shangguan, 1986**
Aortomima Zhang, Zhang, Liu & Shangguan, 1986, *Geol. Shandong,* 2 (1), 27 [33] (original designation).

Type species: *Aortomima shandogensis* Zhang, Zhang, Liu & Shangguan, 1986.

Head large; antenna long, with 15 antennomeres. Thorax small. Sc absent; R short, near costal margin; Rs separated from R at the 2/3 of the wing, obvious bending. The stem of Rs as long as cross-vein r-m length; M anterior branch long, posterior branch weak; CuA thick. Legs slender; foreleg with one spur. Abdomen stout with eight segments.

Distribution and age: Shandong; Early Cretaceous.

Only one species included from the Cretaceous of Northern China (see Table 23.1).

Family Axymyiidae Shannon, 1921

Axymyiidae, a small family belonging to Axymyioidea, usually have body lengths from 11.4–16.5 mm. They are typified by having a pterostigma; cross-vein r-m strongly oblique; R_2 appearing as a short branch nearly perpendicular to R_{2+3}, ending in C at or just distal to ending of R_1; cross-vein r-m strongly oblique, appearing as a continuation of Rs, and M two-branched. Mamaev and Krivosheina thought Axymyiidae was very different from other Nematocera after study of larvae and ecological habits [34]. To date, three genera of extant Holarctic axymyiids are known from Canada, China, Hungary, Japan, Russia, and USA, and four extinct Middle Jurassic axymyiid genera and species have been described from China [35]. The history of studying of Axymyiidae fossils is more than 100 years. It can be roughly divided into two stages. The early twentieth century is an initial stage with the main feature of classifying fossils relying on simple morphological characteristics. The mid-twentieth century is the development stage with the main feature of extension from morphological studies to the system development, ecology and other fields [36].

Genera included from the Jurassic of Northern China: *Juraxymyia* Zhang, 2010, *Sinaxymyia* Zhang, 2010 and *Raraxymyia* Shi, Zhu, Shih & Ren, 2013.

***Juraxymyia* Zhang, 2010**
Juraxymyia Zhang, 2010, *Ann. Entomol. Soc. Am.*, 103, 460–461 [37] (original designation).

Type species: *Psocites fossilis* Zhang, 2004.

Sc relatively long; R_{2+3} furcated at an obtuse angle; cross-vein r-m drifted distally (situated beyond the middle of wing), slightly oblique and much shorter than dM_{1+2}; fork of Rs distad to level of fork of M; bM_{1+2} no more than 1.5 times as long as dM_{1+2}. CuP short.

Distribution and age: Inner Mongolia; Middle Jurassic.

Only one species included from the Jurassic of Northern China (see Table 23.1).

***Sinaxymyia* Zhang, 2010**
Sinaxymyia Zhang, 2010, *Ann. Entomol. Soc. Am.*, 103 (4), 462 [37] (original designation).

Type species: *Sinaxymyia rara* Zhang, 2010.

Sc relatively short; R_2 and R_3 forming an obtuse angle; R_2 ending in C distad to end of R_1; bR_{4+5} short and nearly as long as r-m; cross-vein r-m situated at the middle of wing, perpendicular to M_{1+2}; bM_{1+2} shorter than dM_{1+2}. CuP short.

Distribution and age: Inner Mongolia; Middle Jurassic.

Only one species included from the Jurassic of Northern China (see Table 23.1).

***Raraxymyia* Shi, Zhu, Shih & Ren, 2013**
Raraxymyia Shi, Zhu, Shih & Ren, 2013, *Acta. Geol. Sin. -Engl.*, 87 (5), 1229 [35] (original designation).

Type species: *Raraxymyia parallela* Shi, Zhu, Shih & Ren, 2013.

Pterostigma fusiform, reaching R_2. R_{2+3} branched to R_2 and R_3, both R_2 and R_3 forming an obtuse angle. R_2 like a cross-vein between C and the end of R_1, or at R_1 close to end of R_1. Vein bM_{1+2} shorter than dM_{1+2}. Vein dM_{1+2} much longer than r-m. Vein bM_{1+2} and r-m forming a sharp angle.

Distribution and age: Inner Mongolia; Middle Jurassic.

Two species included from the Jurassic of Northern China (see Table 23.1).

***Raraxymyia parallela* Shi, Zhu, Shih & Ren, 2013 (Figure 23.7)**
Raraxymyia parallela Shi, Zhu, Shih & Ren, 2013: *Acta. Geol. Sin.-Engl.*, 87 (5), 1229.

Locality and horizon: Daohugou, Ningcheng, Inner Mongolia, China; Middle Jurassic, Jiulongshan Formation.

Medium-sized fly. Body length 14.7–16.2 mm, antenna length 0.6–0.7 mm, width 0.2 mm; head length 1.0–1.6 mm, width 1.8 mm; wing length 12.8 mm. Pterostigma fusiform, reaching R_2. Sc relatively long. R furcated to four branches. Rs arising from less than 1/3 of the wing length, bR_{4+5} long and much longer than r-m, but shorter than Rs. M furcated to three branches. M_1, R_{4+5}, R_3 arched and subparallel. Vein bM_{1+2} shorter than dM_{1+2}. dM_{1+2} longer than r-m. Cup shorter. Halter short and massive [35].

Family Blephariceridae Loew, 1862

Blephariceridae, commonly called “net-winged midges”, constitute a very peculiar and isolated group of Nematocera. They are typified by having unobvious reticular fold

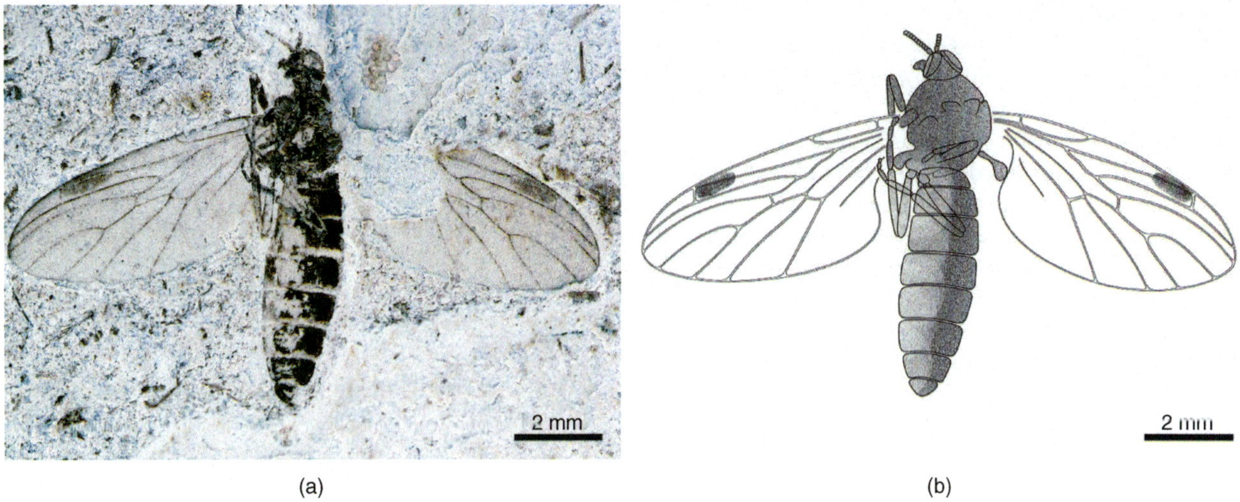

Figure 23.7 *Raraxymyia parallela* Shi, Zhu, Shih & Ren, 2013 (Holotype, CNU-DIP-NN2012874). (a), Photograph; (b), Line drawing.

on forewing. Mouthparts of adults are usually sexually dimorphic. Females have mandibles for sucking blood or eating insects, while males, no mandibles. Larvae are attached to submerged stones in swift-flowing water, usually in mountain streams. To date, there are five extinct genera containing seven species described [7]. In addition, more than 45 extant genera containing over 700 species are known around the world. Relationships of the genera among this family have been variously interpreted and are far from being resolved [38].

Genera included from the Jurassic of Northern China: *Megathon* Lukashevich & Shcherbakov, 1997 and *Brianina* Zhang & Lukashevich, 2007.

Megathon Lukashevich & Shcherbakov, 1997

Megathon Lukashevich & Shcherbakov, 1997, *Neues. Jahrb. Geol. P.-M.*, 11, 639–646 [39] (original designation).

Type species: *Megathon zwicki* Lukashevich & Shcherbakov, 1997.

The specific epithet is dedicated to Dr. Zwick for his contribution to fossil insects. Antenna not longer than head width; eyes bisected and dichoptic in female. Wing widest about midlength. Ninth abdominal tergum of female normal. Very long setae absent from both eyes and female terminalia. Fore femora straight [38, 39].

Distribution and age: Inner Mongolia; Middle Jurassic.

Only one species included from the Jurassic of Northern China (see Table 23.1).

Brianina Zhang & Lukashevich, 2007

Brianina Zhang & Lukashevich, 2007, *Cretac. Res.*, 28, 303 [38] (original designation).

Type species: *Brianina longitibialis* Zhang & Lukashevich, 2007.

Rs without backward spur at beginning; Rs forked into R_{2+3} and short R_{4+5} stalk; M_2 connected to M_1 without connection with M_{3+4}; desclerotized CuP not reaching wing margin. Hind tibiae curved; tibiae longer than femora.

Distribution and age: Inner Mongolia; Middle Jurassic.

Only one species included from the Jurassic of Northern China (see Table 23.1).

Family Bibionidae Kirby, 1837

Bibionids, belonging to Bibionomorpha, have the most extensive fossil records of any dipteran family. Bibionidae are medium-sized flies with body lengths from 4.0 to 10.0 mm. The antennae are moniliform. The front tibiae bear large strong spurs or a circle of spines. The tarsi are five-segmented and bear tarsal claws, pulvilli, and a well-developed empodium. The discoidal cell is absent; R_{4+5} is simple or branched; R developed with three branches [40]. Fossil bibionids are known questionably from the Jurassic, while some fossils from the early part of the Upper Cretaceous look quite similar to modern species. Most fossil species are easily identified with extant genera. Fossils from Europe include a large number of specimens of the mainly tropical genus *Plecia* which is now entirely absent from Europe, demonstrating a warmer climate during the Cenozoic. To date, there are nine extinct genera containing over 350 species described [24], but only one species in one genus reported from China. Over 1100 extant species belonging to 12 genera are documented around the world [7].

Only one genus included from the Cretaceous of Northern China: *Lichnoplecia* Ren, Lu, Guo & Ji, 1995.

Lichnoplecia Ren, 1995

Lichnoplecia Ren, 1995, *Faunae and Stratigraphy of Jurassic-Cretaceous in Beijing and the Adjacent Areas*, 102 [40] (original designation).

Type species: *Lichnoplecia kovalevi* Ren, 1995.

The specific epithet is dedicated to Dr. Kovalev for his focus on the Brachycera and his descriptions of many new taxa from the Jurassic to Cretaceous of Siberia. Rs fork after r-m; Rs_2 shorter than cross-vein r-m; the length of R_{4+5} 1.5 times as long as R_{2+3}, R_{2+3} nearly parallel to R_{4+5}; M_{1+2} fork behind the fork of Rs; Sc distant from R_1.

Distribution and age: Hebei; Early Cretaceous.

Only one species included from the Cretaceous of Northern China (see Table 23.1).

Family Chaoboridae Edwards, 1920

Chaoboridae, an extant family belonging to Culicomorpha, are diagnosed by having similar wing venation with Culicidae (i.e. R_1 ending in C anterior to R_2, R_3 and R_{4+5}; M with two branches); antenna with 13 flagellomeres, plumose in males and larvae with a pair of prominent air sacs in both thorax and abdominal segment seven [41]. Up to now, there are 20 extinct genera containing 47 species described [24]. Over 80 species belonging to 33 extant genera have been described around the world [7]. Larvae are aquatic, mostly transparent when alive. They feed mainly on small insects such as mosquito larvae and crustaceans such as *Daphnia*. Adult Chaoboridae are confused with some Chironomidae, but they are distinguished by wing venation. The oldest known fossils of the family are from the Middle Jurassic deposits of Siberia.

One genus included from the Cretaceous of Northern China: *Mesochaoborus* Zhang, Zhang, Liu & Shangguan, 1986.

Mesochaoborus Zhang, Zhang, Liu & Shangguan, 1986

Mesochaoborus Zhang, Zhang, Liu & Shangguan, 1986, *Geol. Shandong*, 2, 19 [33] (original designation).

Type species: *Leptoplecia zhangshanyingensis* Hong, 1984.

Body robust, medium sized. Eyes large and dichoptic. Antenna filiform with 14 antennomeres. Proboscis short with maxillary palpus 5-segmented. Thorax well-developed, scutum convex and scutellum obvious. Wing width/length (W/L) about 0.33; C straight and thick; Sc about 3/4 length of the wing; R long and straight; the stem of R_{2+3} longer than Rs stem; cross-veins m-cu and r-m at the middle of the wing. Abdomen strong with seven visible segments.

Distribution and age: Hebei, Shandong; Early Cretaceous.

Two species included from the Cretaceous of Northern China (see Table 23.1).

Family Chironomidae Macquart, 1838

Chironomidae is a large family belonging to Chironomoidea, Nematocera, Culicomorpha with extant and fossil taxa [42]. They are characterized by ocelli absent and unbranched M. Up to date, there are more than 100 extinct genera containing 287 species described [24], while over 500 extant genera including more than 7000 species have been documented globally [7].

Genera included from the Jurassic and Cretaceous of Northern China: *Manlayamyia* Kalugina, 1980, *Oryctochlus* Kalugina, 1985, *Coelochironoma* Zhang, Zhang, Liu & Shangguan, 1986 and *Sinoryctochlus* Zhang, 1991.

Manlayamyia Kalugina, 1980

Manlayamyia Kalugina, 1980, *Trudy Sovmest. Sov.-Mongol. Paleontol. Ekped.*, 13, 63 [43] (original designation).

Type species: *Manlayamyia litorina* Kalugina, 1980.

Body like mosquito; gonostyle evident, closed to gonocoxa, pincer-like. Antenna slender and short, length of every antennomere slightly longer than the width, with setae [42, 43].

Distribution and age: Hebei, Late Jurassic.

Only one species included from the Jurassic of Northern China (see Table 23.1).

Oryctochlus Kalugina, 1985

Oryctochlus, Kalugina, 1985, *Dipterous insects of Jurassic Siberia*, 159 [44] (original designation).

Type species: *Oryctochlus vulcanus* Kalugina, 1985.

Body small. Ommateum bare. Thorax short, tergum short; claws short; pulvillus and empodium not obvious. C terminating at the ending of Rs; Rs simple; the space between R and Rs wide; r-m longer than the base of Rs [42, 44].

Distribution and age: Shandong; Early Cretaceous.

Only one species included from the Cretaceous of Northern China (see Table 23.1).

Coelochironoma Zhang, Zhang, Liu & Shangguan, 1986

Coelochironoma Zhang, Zhang, Liu & Shangguan, 1986, *Geol. Shandong*, 2 (1), 24 [33] (original designation).

Type species: *Coelochironoma xantha* Zhang, Zhang, Liu & Shangguan, 1986.

Antenna long and thick with 12 antennomeres. Sc thick, as 2/3 of the length of wing; Rs two branches, the latter slightly deflexed; M, CuA, and A thin; M with two branches, r-m at the base of Rs; abdomen stout with eight segments.

Distribution and age: Shandong; Early Cretaceous.

Only one species included from the Cretaceous of Northern China (see Table 23.1).

Sinoryctochlus Zhang, 1991

Sinoryctochlus Zhang, 1991, *Acta Palaeontol. Sin.*, 30 (5), 563 [42] (original designation).

Type species: *Sinoryctochlus insolitus* Zhang, 1991.

Body small size. Head large. Antenna short, 11 antennomeres. Legs slender. Cercus uncinate. Wing short; Rs simple; cross-vein r-m nearly as long as the base of Rs; the base of M_{3+4} and cross-vein m-cu not obviously.

Distribution and age: Shandong; Early Cretaceous.

Only one species included from the Cretaceous of Northern China (see Table 23.1).

Family Eopleciidae Rohdendorf, 1945

Eopleciidae is an extinct family used to be known only from the Jurassic, until Hong and Wang documented *Gansuplecia triporata* Hong, 1990 from the Early Cretaceous [45]. This family is characterized by the presence of two sturdy anterior branches of Rs, of which the proximal one branches off from the main vein of Rs closer to the base of the wing than r-m and Sc long, longer than half of the wing [46]. To date, six genera including six species are included in Eopleciidae [24]. Kovalev, in 1982, implied that the venation of *Beipiaoplecia* Lin, 1976 might have been misinterpreted and the genus might belong to the Asilomorpha [47], but when Kovalev added the genus *Eomycetophila* Kovalev, 1990 and reviewed the generic composition of the Eopleciidae, *Beipiaoplecia* was retained [32]. It is provisionally retained here in the Eopleciidae pending further study [24].

Genera included from the Jurassic of Northern China: *Beipiaoplecia* Lin, 1976 and *Jurolaemargus* Evenhuis, 1994.

Beipiaoplecia Lin, 1976

Beipiaoplecia Lin, 1976, *Acta Palaeontol. Sin.*, 15 (1), 110 [48] (original designation).

Type species: *Beipiaoplecia malleformis* Lin, 1976.

Two branches of Rs short, the last one very long; main stem of M long, M_4 not co-terminous with CuA. The distal of abdomen tiny.

Distribution and age: Liaoning; Middle Jurassic.

Only one species included from the Jurassic of Northern China (see Table 23.1).

Jurolaemargus Evenhuis, 1994

Jurolaemargus Evenhuis, 1994, *Catalogue of the fossil flies of the world (Insecta: Diptera)*, 112 [24] (original designation).

Laemargus Hong, 1983, *Middle Jurassic fossil insects in North China*, 119 [31]. Syn. by Evenhuis, 1994, *Catalogue of the fossil flies of the world (Insecta: Diptera)*, 112 [24].

Type species: *Laemargus yujiagouensis* Hong, 1983.

Protibiae with two spurs, metatibiae with two to three spurs. Wing long, W/L about 0.25–0.29; Sc elongated and slanted, at the middle of wing; Rs branched late, arched base; M separated with CuA, M_{1+2} and M_4 before Rs; m-cu absent.

Distribution and age: Liaoning; Middle Jurassic.

Only one species included from the Jurassic of Northern China (see Table 23.1).

Family Limoniidae Speiser, 1909

The Limoniidae are one of the largest families of Tipuloidea, with approximately 11 000 described species worldwide [49]. They are subdivided into four extant subfamilies: Pediciinae, Eriopterinae, Hexatomonae, and Limoniinae; and two extinct subfamilies: Architipulinae and Eotipulinae [50]. Many fossil limoniids have been described since the Triassic and from the Mesozoic deposits of all continents except for Antarctica. They are the most ancient living family of the Diptera [51]. In China, some species have been described in Limoniidae or in a relative family of Tipulidae [30, 48]. The larvae of the family are found in aquatic or semiaquatic environments where they feed on decaying plant matter, fungi or algae, while adults, whether from terrestrial or aquatic habitats, generally congregate along aquatic environments and moist grounds [52].

Genera included from the Jurassic of Northern China: *Architipula* Handlirsch, 1906, *Mesotipula* Handlirsch, 1920, *Archimesotipula* Lin, 1986 and *Cretolimonia* Kalugina, 1986.

Architipula Handlirsch, 1906

Architipula Handlirsch, 1906, *Die fossilen Insekten und die Phylogenie der rezenten Formen. Ein Handbuch für Paläontologen und Zoologen*, 490 [22] (original designation).

Protipula Handlirsch, 1906, *Die fossilen Insekten und die Phylogenie der rezenten Formen. Ein Handbuch für Paläontologen und Zoologen*, 491 [22]. Syn. by Evenhuis, 1994, *Catalogue of the fossil flies of the world (Insecta: Diptera)*, 62 [24].

Liassotipula Tillyard, 1933, *The Panorpoid complex in the British Rhaetic and Lias. Fossil Insects no.3*, 74 [53]. Syn. by Kopec *et al.*, 2017, *Palaeontol. Electron.*, 20 (1), 2 [54].

Paratipula Bode, 1953, *Palaeontographica (A)*, 103, 305 [55]. Syn. by Evenhuis, 1994, *Catalogue of the fossil flies of the world (Insecta: Diptera)*, 62 [24].

Eoasilidea Bode, 1953, *Palaeontographica (A)*, 103, 315 [55]. Syn. by Krzemiński & Kovalev, 1988, *Syst. Entomol.*, 13, 55 [56].

Type species: *Protipula seebachiana* Handlirsch, 1906.

The wing has a narrow subcostal area; Sc nearly opposite the fork of Rs; R with four or five branches and the cross-vein r-m enters R_5 not far from the first fork of Rs. M forked into four branches. A_1 and A_2 reaching wing margin [22, 57].

Distribution and age: Inner Mongolia; Middle Jurassic.

Four species included from the Jurassic of Northern China (see Table 23.1).

Mesotipula Handlirsch, 1920

Mesotipula Handlirsch, 1920, *Handbuch der Entomologie*, 205 [58] (original designation).

Type species: *Mesotipula brachyptera* Handlirsch, 1920.

Long Sc terminating at C beyond wing midlength at the same level with Rs furcation; cross-vein sc-r terminal; Rs forking into very short R_{2+3+4} stem (shorter than bR_5) and R_5 or R_{2+3} and very short R_{4+5} with R_4 always alighting with Rs; R_3 tip closer to R_4 tip than to R_1; R_5 terminating at wing apex; long Rs originating in proximal half of the wing, proximal to A_1 tip; Cu almost straight at m-cu [58, 59].

Distribution and age: Inner Mongolia; Middle Jurassic.

Only one species included from the Jurassic of Northern China (see Table 23.1).

Mesotipula gloriosa Gao, Shih, Zhao & Ren, 2015 (Figure 23.8)

Mesotipula gloriosa Gao, Shih, Zhao & Ren, 2015: *Acta Geol. Sin.*, 89, 1792–1794.

Locality and horizon: Daohugou, Ningcheng, Inner Mongolia, China; Middle Jurassic, Jiulongshan Formation.

A large-sized female crane fly, body length 11.8 mm with well-preserved wings. Head oval; Palpi and antennae partly preserved; compound eyes with clear facets. Thorax well-preserved in dorsal view; the boundaries among prothorax, mesothorax and metathorax distinct; mesonotum robust and well-preserved; while some parts of mesopleura distinguishable; halteres not preserved. Legs long and slender. Abdomen relatively long. Female with ovipositor discernible. The right wing well-preserved. Length 9.3 mm and width 3.5 mm. Wing clear with light-colored pterostigma and pigmented region between distal part of R_1, R_2 and R_3. Venation: Sc long, 0.61 times of the wing length and terminating at C about the same level of branching of Rs; cross-vein sc-r terminal. R_1 long; Rs arising from 0.31 times as long as the wing; Rs, at 0.63 times length of wing, bifurcating into very short R_{2+3+4} and R_5; R_2 close to R_1 tip, about 3/5 its length before the tip of R_1, transverse; R_{2+3} 1.5 times as long as R_3 and 2.5 times as long as R_2. Stem of vein M bifurcating at the same level of the branching of Rs; discal cell broader distally, 0.12 times of the wing length; cell m_1 petiolate; cell m_1 1.1 times as long as petiole

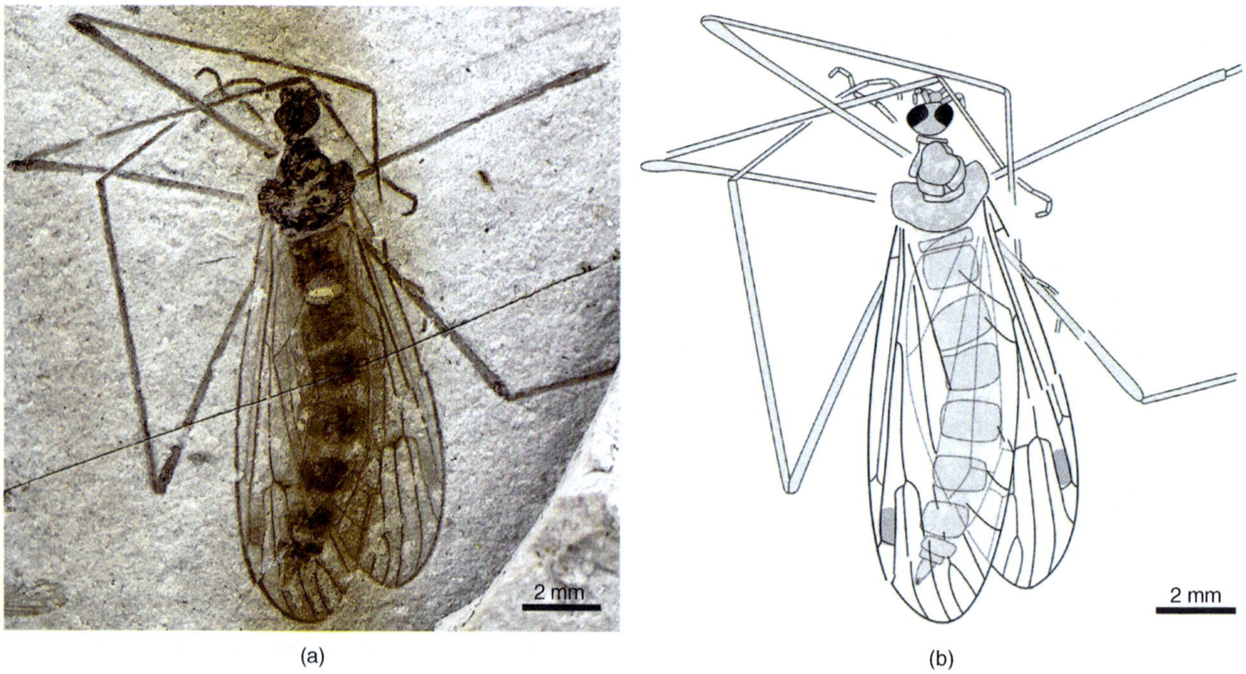

Figure 23.8 *Mesotipula gloriosa* Gao, Shih, Zhao & Ren, 2015 (Holotype, CNU-DIP-NN2015003). (a), Photograph; (b), Line drawing.

of cell m_1 (dM_{1+2}), mM_{1+2} nearly as long as petiole of cell m_1 (dM_{1+2}). Cross-vein m-cu far proximad of M_{3+4} bifurcation, bM_{3+4} 1.5 times as long as dM_{3+4}; Cu long. Anal veins long and divergent [59].

Archimesotipula Lin, 1986

Archimesotipula Lin, 1986, *Palaeontol. Sin.*, 170, 87 [60] (original designation).

Type species: *Archimesotipula antefortis* Lin, 1986.

Sc fine, located near wing anterior edge, R ending at the costal margin, Rs two branches, cross-vein r-m at the middle of the wing, M four-branched, CuA branched.

Distribution and age: Guangxi; Early Jurassic.

Only one species included from the Jurassic of Southern China (see Table 23.1).

Cretolimonia Kalugina, 1986

Cretolimonia Kalugina, 1986, *Trudy Sovmest. Sov.-Mongol. Paleontol. Ekped.*, 28, 115 [61] (original designation).

Type species: *Cretolimonia popovi* Kalugina, 1986.

Distinctly enlarged genitalia of male. Very thin and long leg in which the femur and tibia combined clearly longer than abdomen but shorter than wing. Cross-vein sc-r absent; Rs forking symmetrically into long R_{3+4} stem and R_5; R_2 absent; R_3 short, terminating at C closer to R_1 than to R_4 or equidistant [59, 61].

Distribution and age: Inner Mongolia; Middle Jurassic.

Only one species included from the Jurassic of Northern China (see Table 23.1).

Cretolimonia excelsa Gao, Shih, Zhao & Ren, 2015 (Figure 23.9)

Cretolimonia excelsa Gao, Shih, Zhao & Ren, 2015: *Acta Geol. Sin.*, 89, 1791–1792.

Locality and horizon: Daohugou, Ningcheng, Inner Mongolia, China; Middle Jurassic, Jiulongshan Formation.

A large-sized crane fly with well-preserved wings, body and head. Body length 12.3 mm; head 0.8–0.9 mm, oviform. The compound eyes with distinctly clear facets. Antenna 2.3 mm long, with 16 flagellomeres. Thorax length 2.1–2.2 mm. Prothorax barely visible; anterior part of scutum in mesothorax arched convex; the boundary among prescutum, scutum, scutellim and mediotertit is distinct in lateral view. Halteres not preserved. The legs slender; the femora much shorter than the abdomen. The abdomen relatively long and thin, length 9.4 mm, with eight clear segments as preserved. Female with ovipositor preserved. Wing: The right wing well-preserved, and the left wing is partly-folded and overlapped with abdomen. Length 10.6–11.1 mm and width 3.5 mm. Wing without any dark marks except for distinct large pterostigma along distal part of R_1 and R_3. Sc 0.67–0.69 times as long as the wing length, and terminating significantly distad of branching of Rs. Cross-vein sc-r absent. R_1 long; Rs arising from 0.37 times of length of wing, Rs bifurcating into R_{3+4} and R_5 at 0.63–0.64 times of length of wing and at the same level of bifurcation of stem of vein M; R_2 absent; R_3 short, terminating at C closer to R_1 than to R_4; R_{3+4} long, 0.75–0.81 times as long as R_4 and 3.28–3.60 times as long as R_3; R_5 straight, bR_5 short, subequal to the short r-m. Discal cell small, about 1/10 of the wing length; cell m_1 petiolate; mM_{1+2} 0.77–0.86 times as long as petiole of cell m_1 (dM_{1+2}); cell m_1 1.50–1.57 times as long as petiole of cell m_1 (dM_{1+2}). Cross-vein m-cu joining M_{3+4} at the middle of d cell. Cu distinctly bending at m-cu. Anal veins elongate and divergent; A_1 terminating at posterior margin the same level of Rs bifurcation [59].

Family Luanpingitidae Zhang, 1986

Luanpingitidae, a small extinct family, are characterized by body small-sized, anterior margin of wing straight; veins of C, R and Rs obviously more robust than M, Cu and A; vein C ending before the anal margin; Sc long; C, R and Rs stout, while others thin and not reaching posterior margin of wing; R long and simple and M with three branches. Because of these characters, it was set up as a family in the Chironomoidea Maequart, 1838. This family has only one genus with one species [62].

Only one genus included from the Jurassic of Northern China: *Luanpingites* Zhang, 1986.

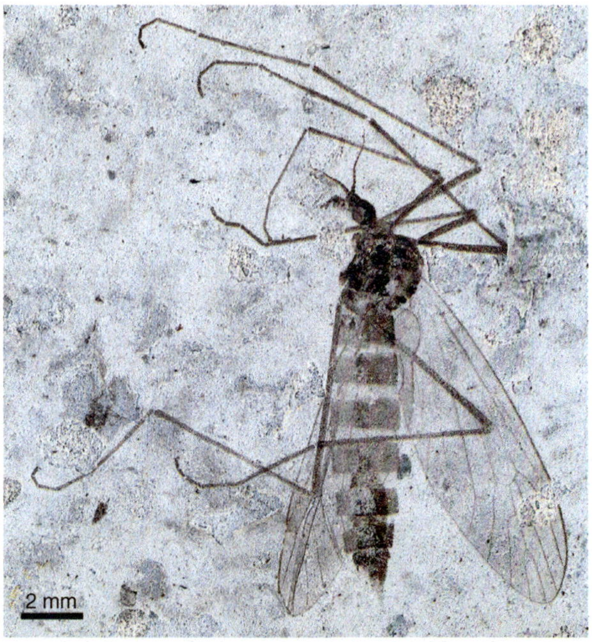

Figure 23.9 *Cretolimonia excelsa* Gao, Shih, Zhao & Ren, 2015 (Holotype, female, CNU-DIP-NN2015001).

***Luanpingites* Zhang, 1986**

Luanpingites Zhang, 1986, *Acta Palaeontol. Sin.*, 25 (1), 49 [62] (original designation).

Type species: *Luanpingites flavus* Zhang, 1986.

Wings extending beyond the apex of abdomen. C straight, ending near apex of wing; Sc long; R stout and long; Rs stout, weakly at base, with terminal part bent downwards. Cell cup narrowly.

Distribution and age: Hebei; Middle Jurassic.

Only one species included from the Jurassic of Northern China (see Table 23.1).

Family Mesosciophilidae Rohdendorf, 1946

Mesosciophilidae is one of the extinct families of Nematocera. In 1946, Rohdendorf first described the members of Mesosciophilidae as a subfamily Mesosciophilinae within the family of Allactoneuridae [27]. Rohdendorf, in 1964, renamed Allactoneuridae as Fungivoritidae [63]. In 1985, Kovalev thought that Fungivoritidae was a junior synonym of Pleciofungivoridae, and raised the subfamily Mesosciophilinae to the level of family Mesosciophilidae [44]. In 1993, Blagoderov emended the diagnosis of Mesosciophilidae, and erected the genus *Mesosciophilopsis* with three species from the Neocomian, Lower Cretaceous of Transbaykal, Baysa [64]. Fossils of Mesosciophilidae described previously have been mostly from the Middle and Late Jurassic. This family became very rare in the Early Cretaceous, apparently replaced by Mycetophilidae [64]. Our knowledge of the Mesosciophilidae is based mostly on impression fossils of wings from the Jurassic and Early Cretaceous [65]. Fossils retaining the entire body and wings are quite rare in this group. Up to date, after the corrections and transfers, 9 genera with 28 species of mesosciophilids have been described from the Jurassic of Siberia and Kazakhstan, the Middle Jurassic of Inner Mongolia, and the Lower Cretaceous of Transbaikalia and Beipiao, Liaoning [66].

Genera included from the Jurassic and Cretaceous of Northern China: *Mesosciophila* Rohdendorf, 1946, *Paramesosciophilodes* Zhang, 2007, *Jurasciophila* Li & Ren, 2009, *Similsciophila* Shi, Shih & Ren, 2014 and *Orentalphila* Lin, Shih & Ren, 2015.

***Mesosciophila* Rohdendorf, 1946**

Mesosciophila Rohdendorf, 1946, *Trudy Paleontol. Inst.*, 13 (2), 76 [27] (original designation).

Type species: *Mesosciophila venosa* Rohdendorf, 1946.

Antenna filiform with 16 antennomeres. Sc_1 ending distad to level of Rs origin, Sc_2 situated clearly basad to Rs origin; bRs longer than r-m; R_1 slightly curved; both R_1 and R_{4+5} divergent terminally; Rs furcated distad to fork of M_{1+2}; R_{2+3} oblique; cell r moderately large, 1/4–1/5 of length of wing; stem of M not developed [27, 65].

Distribution and age: Inner Mongolia; Middle Jurassic.

Two species included from the Jurassic of Northern China (see Table 23.1).

***Mesosciophilodes* Rohdendorf, 1946**

Mesosciophilodes Rohdendorf, 1946, *Trudy Paleontol. Inst.*, **13** (2), 76 [27] (original designation).

Type species: *Mesosciophilodes angustipennis* Rohdendorf, 1946.

Antenna filiform with 16 antennomeres. Sc_2 developed, but somewhat faint and slender; section of R much longer than bRs; bRs nearly as long as r-m; R_{4+5} alomost straight; M_{3+4} running close to CuA basally [27].

Distribution and age: Inner Mongolia; Middle Jurassic.

Only one species included from the Jurassic of Northern China (see Table 23.1).

***Paramesosciophilodes* Zhang, 2007**

Paramesosciophilodes Zhang, 2007, *Cretac. Res.*, 28, 298 [65] (original designation).

Type species: *Paramesosciophilodes ningchengensis* Zhang, 2007.

Sc_1 elongate; Sc_2 situated distinctly basad to Rs origin; both R_1 and R_{4+5} divergent terminally; Rs furcated distad or at level of fork of M_{1+2}; cell r 0.16–0.19 times as long as wing length; stem of M not developed; M_{1+2} furcated slightly distad, or basad, to level of Sc_1 ending.

Distribution and age: Inner Mongolia; Middle Jurassic.

Five species included from the Jurassic of Northern China (see Table 23.1).

***Jurasciophila* Li & Ren, 2009**

Jurasciophila Li & Ren, 2009, *Prog. Nat. Sci.*, 19, 1838 [67] (original designation).

Type species: *Jurasciophila curvula* Li & Ren, 2009.

Antenna usually with 15–16 antennomeres. Wing 2.3–2.6 times as long as wide. Sc_1 ending located at the level of or slightly exceeding the forking of bRs and r-m; both R_1 and R_{4+5} divergent terminally; M_{1+2} furcated distinctly at the level of R_{2+3} or earlier; M_{3+4} running close to CuA basally, but neither coalescent.

Distribution and age: Inner Mongolia; Middle Jurassic.

Two species included from the Jurassic of Northern China (see Table 23.1).

***Jurasciophila lepida* Li & Ren, 2009 (Figure 23.10)**

Jurasciophila lepida Li & Ren, 2009: *Prog. Nat. Sci.*, 19, 1838–1840.

Locality and horizon: Daohugou, Ningcheng, Inner Mongolia, China; Middle Jurassic, Jiulongshan Formation.

A female with dorsal-ventral aspect. Color gray, except head and metascutum are dark brownish. Head not well-preserved; antennae not preserved; eyes large, suboval. Thorax approximately elliptic. Wing 2.4 times as long as wide. Sc_1 0.38–0.40 times as long as wing, its ending beyond the level of the crossing of bRs and r-m, Sc_2 well-developed, R_1 straight, lacking sharp bend, section of R from Sc_2 to Rs origin about 1.5 times as long as bRs, bRs slightly longer than r-m, about 1.2 times, both R_1 and R_{4+5} divergent terminally, R_{4+5} tibiae and tarsi with one or two rows of short setae, tibial spur well-developed, claws very small. Abdomen broadest at midlength, narrowing from fifth segment to apex, eighth segment is smaller than preceding segments slightly, narrow apically. Female terminalia obtuse-triangular with short and cuspidate cerci [67].

***Similsciophila* Shi, Shih & Ren, 2014**

Similsciophila Shi, Shih & Ren, 2014, *J. Nat. Hist.*, 49 (19–20), 1149 [68] (original designation).

Type species: *Similsciophila singularis* Shi, Shih & Ren, 2014.

Sc elongate, shorter than half of wing length; sc-r situated distinctly basal to Rs origin, arising near midway between h to Sc ending at margin; cell r distinctly large; Rs furcated distad to fork of M_{1+2}; bRs longer than r-m; R_1 slightly curved; R_{4+5} arched near its mid-length; M_{1+2} furcated slightly distad to level of Sc ending.

Distribution and age: Inner Mongolia, Liaoning; Middle Jurassic, Early Cretaceous.

Three species included from the Jurassic and Cretaceous of Northern China (see Table 23.1).

***Similsciophila undulata* Lin, Shih & Ren, 2015 (Figure 23.11)**

Similsciophila undulata Lin, Shih & Ren, 2015: *Cretac. Res.*, 54, 96.

Locality and horizon: Huangbanjigou, Beipiao, Liaoning, China; Lower Cretaceous, Yixian Formation.

Body length about 8.1 mm. Antenna long, clearly exceeding head and thorax combined; scape and pedicel rounded; the length of middle flagellomeres about two times as long as wide. Maxillary palpi four segments visible as preserved. Wing 5 mm long as preserved; wing length about 2.1 times as long as wide. Sc_1 reaching C forming a narrow costal field. Vein bRs 1.2 times as long as cross-vein r-m, dRs clearly more than two times as long as bRs. Fore, mid and hind legs with coxae expanded; femur clearly thicker than tibia, tibia spur well-developed; claws very small. Abdomen with the fourth to seventh tergites wider than the others; the ninth tergite, very small, mostly hidden by the eighth. Genital, complex and slightly narrower than the eighth tergite, with gonocoxites massive and rounded, and gonostylus cylindrical, terminating with small apex, distinctly longer than the eighth and ninth tergites combined [66].

***Orentalphila* Lin, Shih & Ren, 2015**

Orentalphila Lin, Shih & Ren, 2015, *Cretac. Res.*, 54, 90 [66] (original designation).

Type species: *Orentalphila gravia* Lin, Shih & Ren, 2015.

Wing 2–2.4 times as long as wide. Sc 1/4–1/3 of wing length; Sc_1 reaching C slightly proximad or at same level of R forking; Sc_2 proximad of R forking. R_{2+3} sigmoidly curved, oblique to R_1. Vein bRs smoothly curved. R_{4+5} slightly curved, almost parallel with R_1, cell r small. M_{1+2} forking far distad or at same level of the position of R_{2+3} forking.

Figure 23.10 *Jurasciophila lepida* Li & Ren, 2009 (Holotype, female, CNU-DIPNN2008586).

Figure 23.11 *Similsciophila undulata* Lin, Shih & Ren, 2015 (Holotype, CNU-DIP-LB-2013402).

Figure 23.12 *Orentalphila gravia* Lin, Shih & Ren, 2015 (Holotype, CNU-DIP-LB-2013403).

Distribution and age: Liaoning; Early Cretaceous.

Two species included from the Cretaceous of Northern China (see Table 23.1).

Orentalphila gravia **Lin, Shih & Ren, 2015 (Figure 23.12)**
Orentalphila gravia Lin, Shih & Ren, 2015: *Cretac. Res.*, 54, 91–93.

Locality and horizon: Huangbanjigou, Beipiao, Liaoning, China; Lower Cretaceous, Yixian Formation.

Body length about 4.6–4.8 mm; body dark. Antenna with seven antennomeres as preserved; scape and pedicel rounded; the length of middle flagellomeres about two times as long as wide. Maxillary palpi segments not preserved. Wing length 2.9–3.7 mm, about 2.1–2.3 times as long as wide. Sc_1 reaching C slightly proximad of R forking; Sc_2 origination at 7/10–4/5 of Sc. Cross-vein r-m 1.2–1.3 times as long as bRs, bRs slightly shorter than dRs. Fore, mid and hind legs with coxae expanded; femur clearly thicker than tibia; tibia spur well-developed; claws very small. Abdomen dark, covered with dense pubescence; the fourth tergite widest; the eighth tergite connected with the ninth tightly. Halteres small and narrow, covered with dense pubescence. Cerci well-developed, two segmented, almost as long as the eighth and ninth tergites combined [66].

Family Mycetophilidae Johannesen, 1910

Mycetophilidae is a small family with extant and fossil taxa having body lengths from 2.2 to 13 mm. Mycetophilidae occur on all continental areas except Antarctica (from northern Greenland to Tierra del Fuego) and on most oceanic islands. About 3000 species have been described but the actual number of species is undoubtedly much greater. Rohdendorf has referred fossils of various periods from the Upper Triassic onward to several extinct genera and families of Fungivoroidea (Sciaroidea). But the oldest Diptera fossils are definitely referable to the mycetophilids, including some extant genera are well-represented in amber deposits. This family appears to have been well-established and diversified by the Cretaceous at the latest. In 1946, Rohdendorf described the genus *Mesosciophila* and other mesosciophiloides from the Karabastau Formation, Kazakhstan [27]. However, Rohdendorf designated them to the family Allactoneuridae. In 1985, because of the vein Rs_4 directly joining vein R_1, Rohdendorf transferred them to the family Mycetophilidae [44]. In addition, three extinct Mycetophilidae genera and species have been described from Chengde, Hebei, Beijing and Kazuo, Liaoning.

Only one genus included from the Cretaceous of Northern China: *Huaxiasciophilites* Zhang, Hong & Li, 2001.

Huaxiasciophilites **Zhang, Hong & Li, 2001**
Huaxiasciophilites Zhang, Hong & Li, 2001, *Entomol. Sin.*, 8 (3), 194 [69] (original designation).

Type species: *Huaxiasciophilites jingxiensis* Zhang, Hong & Li, 2001.

Sc short, not exceeding the forking of R and Rs, with branches; r-m longer than basal part of Rs; Rs very thick and wavy; and radial cell very short.

Distribution and age: Beijing; Early Cretaceous.

Only one species included from the Cretaceous of Northern China (see Table 23.1).

Family Pediciidae Osten Sacken, 1860

Pediciidae, widespread in the world except for the Afrotropical and Antarctic regions, is an extant family of the Tipuloidea [49]. The cosmopolitan family comprise about 500 species in over 10 genera [7]. To date, 23 fossil species belonging to seven genera have been reported, ranging from the Middle Jurassic to the Upper Oligocene [24]. The fossil taxonomic history of these flies is complicated. Four separate families, i.e. Archilimoniidae, Gracilitipulidae, Palaeolimnobiidae, and Zhangobiidae have been proposed for these fossil taxa, which have been all transferred to Pediciidae by Pape *et al.* in 2011

[7]. But it is still contentious. The Pediciidae currently are divided into two subfamilies of Pediciinae and Ulina [49]. The phylogenetic relationships among genera and subgenera of Pediciidae have not yet been investigated [70].

Only one genus included from the Jurassic of Northern China: *Praearchitipula* Kalugina, 1985.

Praearchitipula Kalugina, 1985

Praearchitipula Kalugina, 1985, *Dipterous insects of Jurassic Siberia,* 198 [44] [in Russian] (original designation).

Type species: *Praearchitipula notabilis* Kalugina, 1985.

Sc distad far from the bifurcation of Rs; cross-vein sc-r before middle of wing, proximad of or just distad of the bifurcation of R_b to R_1 and Rs; R_2 at the end of R_1; Rs forked to a long R_{2+3} and a short R_{4+5}; cross-vein m-cu approximately at the middle or distad of cell d base [44, 70].

Distribution and age: Inner Mongolia; Middle Jurassic.

Three species included from the Jurassic of Northern China (see Table 23.1).

Praearchitipula apprima Gao, Shih, Kopeć, Krzemiński & Ren, 2015 (Figure 23.13)

Praearchitipula apprima Gao, Shih, Kopeć, Krzemiński & Ren, 2015: *Zootaxa,* 3963 (2), 243–246.

Locality and horizon: Daohugou, Ningcheng, Inner Mongolia, China; Middle Jurassic, Jiulongshan Formation.

A large-sized crane fly with well-preserved wings, body and head. Body length 11.4–13 mm, head length 0.5–0.7 mm, subcircular. The compound eyes with distinct facets, no setae on eyes preserved. Antenna length 1.6 mm, 2.1 times as long as the head length; left antenna with 14 flagellomeres preserved, right antenna 16-segmented. Palpi with four segments, 1 mm long. Thorax length about 2 mm; prothorax barely visible; anterior part of scutum in mesothorax arched convex; the boundary between prescutum, scutum and scutellum distinct; mediotergite well-developed in lateral view. Legs long and slender. Abdomen well-preserved, relatively long and slender, length 8.9–10.3 mm, with nine segments. Female with ovipositor preserved. Wing: W/L about 0.32, no dark marks except for elongated pterostigma. Sc long, 0.74 times of the wing length and terminating significantly distad of the forking of R_{4+5}; cross-vein sc-r at middle of length of Sc. Rs of moderate length, strongly arcuated at origin; Rs arising from middle of length of wing; R_1 long; R_2 close to R_1 tip, transverse; R_{4+5} short, 0.2 times as long as R_{2+3}; stem of M bifurcating at the same level of branching of Rs; cell m_1 petiolate; dM_{1+2} half as long as cell m_1; r-m at 0.4 times of R_{4+5}; cell d broad distally, 0.11 times length of wing; bM_{3+4} 2.2 times as long as dM_{3+4}. Cu distinctly bending at cross-vein m-cu, straight beyond it. A elongate and divergent [70].

Figure 23.13 *Praearchitipula apprima* Gao, Shih, Kopeć, Krzemiński & Ren, 2015 (Holotype, CNU-DIP-NN2014001).

Family Pleciofungivoridae Rohdendprf, 1946

Pleciofungivoridae is an extinct family with venational similarities to some Bibionomorpha placed in Fungivoroidea, only reported in the Mesozoic. Up to now, there are 18 genera including 66 species described [24]. The name and the range of this family have been changed several times since this family was erected.

Genera included from the Jurassic and Cretaceous of Northern China: *Pleciofungivora* Rohdendorf, 1938, *Eohesperinus* Rohdendorf, 1946 and *Opiparifungivora* Ren, 1995.

Pleciofungivora Rohdendorf, 1938

Pleciofungivora Rohdendorf, 1938, *Trudy Paleontol. Inst.,* 7, 42–43 [28] (original designation).

Allactoneurites Rohdendorf, 1938, *Trudy Paleontol. Inst.,* 7, 43 [28]. Syn. by Kovalev, 1987, *Paleontol. Zh.,* 1987 (2), 69–82 [71].

Polyneurisca Rohdendorf, 1946, *Trudy Paleontol. Inst.,* 13 (2), 52 [27]. Syn. by Kovalev, 1987, *Paleontol. Zh.,* 1987 (2), 69–82 [71].

Transversiplecia Rohdendorf, 1946, *Trudy Paleontol. Inst.,* 13 (2), 53 [27]. Syn. by Kovalev, 1987, *Paleontol. Zh.,* 1987 (2), 69–82 [71].

Type species: *Pleciofungivora latipennis* Rohdendorf, 1938.

Antennomeres straight, compact and mostly longer than wide. Wing triangular.

Distribution and age: Hebei; Late Jurassic.

Only one species included from the Jurassic of Northern China (see Table 23.1).

Eohesperinus Rohdendorf, 1946

Eohesperinus Rohdendorf, 1946, *Trudy Paleontol. Inst.*, 13, 60 [27] (original designation).

Eopachyneura Rohdendorf, 1946, *Trudy Paleontol. Inst.*, 13, 57 [27]. Syn. by Kovalev, 1985, *Dvukrylye nasekomye Yury Sibiri*, 137 [44].

Type species: *Eohesperinus martynovi* Rohdendorf, 1946.

The specific epithet is dedicated to A.V. Martynov for his contribution to insect fossil research. Rs with a slant vein ended at the C; pterostigma subrotund; Rs slant, about 1.5 times as long as r-m; M branches less, far away from r-m [27, 31].

Distribution and age: Shandong, Liaoning; Early Cretaceous, Middle Jurassic.

Two species included from the Cretaceous of Northern China (see Table 23.1).

Opiparifungivora Ren, 1995

Opiparifungivora Ren, 1995, *Faunae and stratigraphy of Jurassic-Cretaceous in Beijing and the adjacent areas*, 104 [40] (original designation).

Type species: *Opiparifungivora aliena* Ren, 1995.

Rs_1 shorter than three times of cross-vein r-m; Sc ending before the beginning of Rs; Rs_2 long, about 2 times as long as Rs_1; M_1 and M_2 much longer than M_3; R_4 as long as Rs_1, forming an angle with R_5 larger than 45°.

Distribution and age: Hebei; Early Cretaceous.

Only one species included from the Cretaceous of Northern China (see Table 23.1).

Family Protobibionidae Rohdendorf, 1946

Protobibionidae is an extinct family of nematocerous flies known only from the Upper Jurassic deposits of the Karatua, Kazakhstan (Kimmeridgian Stage) and the Early Cretaceous Laiyang Basin, China [24]. There is only one species in one genus from China in the Early Cretaceous [33].

Only one genus included from the Cretaceous of Northern China: *Protobibio* Rohdendorf, 1946.

Protobibio Rohdendorf, 1946

Protobibio Rohdendorf, 1946, *Trudy Paleontol. Inst.*, 13 (2), 47 [27] (original designation).

Type species: *Protobibio jurassicus* Rohdendorf, 1946.

Head medium size, round; eyes not distinguishable; antennae thick and long, significantly longer than 1/2 of insect body length; the length of the wing 2.5 times the width. Wing slender and long. Vein Sc shorter than 0.5 times of the length of wing. Vein r-m slanting and long [27, 33].

Distribution and age: Shandong; Early Cretaceous.

Only one species included from the Cretaceous of Northern China (see Table 23.1).

Family Protopleciidae Rohdendorf, 1946

Protopleciidae is an extinct family known from the Jurassic of Europe and Asia [57]. The earliest record of the Protopleciidae is *Macropeza liasina* Geinitz 1884 from the Early Jurassic in Liassic Formation of Germany [72]. The Protopleciidae, erected by Rohdendorf in 1946, comprise originally three genera: *Protoplecia* Handlirsch, 1906, *Mesoplecia* Rohdendorf, 1938 and *Mesopleciella* Rohdendorf, 1946 [24, 27]. Zhang, in 2007, reported four species in two genera as members of the Protopleciidae, emended the diagnosis of *Mesoplecia* Rohdendorf, 1938 and excluded *Paraoligus exilus* Lin, 1976 and *Mesoplecia xinboensis* Hong, 1984 from this family, but stated that an alternative placement could not be suggested [73]. Up to date, more than 10 genera including 66 species have been described [24].

Genera included from the Jurassic and Cretaceous of Northern China: *Mesoplecia* Rohdendorf, 1938, *Sinoplecia*, Lin 1976 and *Epimesoplecia* Zhang, 2007.

Mesoplecia Rohdendorf, 1938

Mesoplecia Rohdendorf, 1938, *Trudy Paleontol. Inst.*, 7 (3), 49 [28] (original designation).

Type species: *Mesoplecia jurassica* Rohdendorf, 1938.

Head small. Vein bRs less than twice as long as dRs; M_1, M_2 at least five times as long as dM_{1+2}; bM_{1+2} longer than dM_{1+2}. Legs with coxae and femora thick [28, 73].

Distribution and age: Hebei, Inner Mongolia; Early Cretaceous, Middle Jurassic.

Eight species included from the Jurassic and Cretaceous of Northern China (see Table 23.1).

Mesoplecia fastigata Lin, Shih & Ren, 2014 (Figure 23.14)

Mesoplecia fastigata Lin, Shih & Ren, 2014: *Zootaxa*, 3838 (5), 545–556.

Locality and horizon: Daohugou, Ningcheng, Inner Mongolia, China; Middle Jurassic, Jiulongshan Formation.

The boundary among prothorax, mesothorax and metathorax not distinct in ventral view; the right halter small. Wing narrow and long, W/L = 0.4; vein h well-developed, straight and short; Sc well-developed, reaching anterior margin at the same level with cross-vein r-m; cross-vein sc-r present; R_1 straight; Rs arising from R_1 at basal 1/3 of wing length, furcating slightly distad of the level of M_{1+2} forking; pterostigma close to oval; bRs about 3.7 times as long as r-m; R_{4+5} weakly curved upward medially, approximately 1.8 times as long as R_{2+3}, and almost equal to the basal part of Rs; M_1 longer than M_2, almost equal to M_{3+4}; CuA strongly curved, ending at posterior margin. Vein A_1 reaching the posterior margin in the left wing. The coxae of forelegs and right hind leg expanded. Femur and tibia of forelegs well-preserved, femur distinctly thicker than tibia, covered with dense setae. Mid legs only right tibia well-preserved, tibia long and slender, tibial spur not preserved. Right hind leg well-preserved; femur cylindrical and thick; tibia slightly less than two times as long as femur; two tibial spurs well-developed; first to fifth tarsomeres well-developed with two pretarsal claws. Sternites of segments one to seven clearly visible in ventral view. The first sternite significantly narrower than the second; the third to fifth sternite widest, the eighth sternite combined with genitalia undivided, with cerci apically [74].

***Sinoplecia* Lin, 1976**

Sinoplecia Lin, 1976, *Acta Palaeontol. Sin.*, 15 (1), 109 [48] (original designation).

Type species: *Sinoplecia parvita* Lin, 1976.

Body small. Antenna filiform, two times as long as lengths of head and thorax combined. Tibiae longer than femora. Rs branched at the base of the wing; M_{1+2} separated after the middle of the wing; cross-vein m-cu between M_4 and CuA; CuA curved hard.

Distribution and age: Liaoning; Middle Jurassic.

Only one species included from the Jurassic of Northern China (see Table 23.1).

***Epimesoplecia* Zhang, 2007**

Epimesoplecia elenae Zhang, 2007, *Cretac. Res.*, 28, 292 [73] (original designation).

Type species: *Epimesoplecia shcherbakovi* Zhang, 2007.

The specific epithet is dedicated to Dr. Shcherbakov for his study of and contribution to fossil dipterans. Antenna filiform, at least longer than twice of head length. Wing narrow, long, with fore margin straight; Sc elongate, nearly half as long as wing; bRs more than 2.5 times of dRs length; R_{2+3} long; bM_{1+2} slightly longer, or a bit shorter than dM_{1+2}; bM_{3+4} longer, or slightly shorter than m-cu.

Distribution and age: Inner Mongolia; Middle Jurassic.

Seven species included from the Jurassic of Northern China (see Table 23.1).

***Epimesoplecia plethora* Lin, Shih & Ren, 2015 (Figure 23.15)**

Epimesoplecia plethora Lin, Shih & Ren, 2015: *ZooKeys*, 492, 127–129.

Locality and horizon: Daohugou, Ningcheng, Inner Mongolia, China; Middle Jurassic, Jiulongshan Formation.

Compound eyes crescent. Antennae moniliform. Sc very close to the level of r-m; fork of Rs distad of fork of M_{1+2}; Rs distad of cross-vein r-m; bRs less than two times (1.6–1.8) as long as dRs, the latter about three times (2.6–3) as long as r-m; R_{2+3}, sigmoidly curved, distinctly shorter than bRs and dRs combined; bM_{1+2}

Figure 23.14 *Mesoplecia fastigata* Lin, Shih & Ren, 2014 (Holotype, CNU-DIP-NN-2013101).

Figure 23.15 *Epimesoplecia plethora* Lin, Shih & Ren, 2015 (Holotype, CNU-DIP-NN2013202).

shorter than dM_{1+2}; bM_{3+4} shorter than m-cu; cell bp as wide as cell ba terminally [75].

Family Protorhyphidae Handlirsch 1906

Protorhyphidae is a small, extinct family belonging to Anisopodoidea [76]. The family was erected based on the genus *Protorhyphus* Handlirsch, 1906 from Lower Jurassic deposits of Germany [24]. Up to now, there are five genera including 17 species described from the Triassic to the Cretaceous. Rohdendorf, in 1964, placed the family between the Oligophrynidae and Olbiogastridae (both considered here in the family Anisopodidae) of his Superfamily Rhyphidea [63].

Only one genus included from the Jurassic and Cretaceous of Northern China: *Protorhyphus* Handlirsch 1906.

Protorhyphus Handlirsch, 1906

Protorhyphus Handlirsch, 1906, *Die fossilen Insekten und die Phylogenie der rezenten Formen. Ein Handbuch für Paläontologen und Zoologen*, 487 [22] (original designation).

Type species: *Phryganidium simplex* Geinitz, 1887.

Sc strong, less than half of wing length. R_4 longer than R_{4+5}. Stem of M thin, desclerotized. M_1 and M_2 diverging from cell d independently (im meeting bifurcation M_{1+2} or M_2) [22, 31].

Distribution and age: Inner Mongolia, Liaoning; Middle Jurassic, Early Cretaceous.

Three species included from the Jurassic and Cretaceous of Northern China (see Table 23.1).

Family Ptychopteridae Osten-Sacken, 1862

Ptychopteridae is a very small family but widespread in the world except the Australasian and the Antarctic regions [77]. Lukashevich [78], in 2008, synonymized Eoptychopteridae with Ptychopteridae. Ptychopteridae thus comprise two extant subfamilies of Ptychopterinae and Bittacomorphinae and three extinct subfamilies of Proptychopterininae, Eoptychopterinae and Eoptychopterininae [79]. To date, there are 27 genera and 156 species in this family [7].

Genera included from the Jurassic of Northern China: *Eoptychoptera* Handlirsch, 1906, *Crenoptychoptera* Kalugina, 1985 and *Eoptychopterina* Kalugina, 1985.

Eoptychoptera Handlirsch, 1906

Eoptychoptera Handlirsch, 1906, *Die fossilen Insekten und die Phylogenie der rezenten Formen.Ein Handbuch fur Palaontologen und Zoologen*, 489 [22] (original designation).

Proptychoptera Handlirsch, 1906, *Die fossilen Insekten und die Phylogenie der rezenten Formen. Ein Handbuch für Paläontologen und Zoologen*, 489 [22]. Syn. by Lukashevich *et al.*, 1998, *Polskie Pismo Ent.*, 67 [80].

Metaxybittacus Bode, 1953, *Palaeontographica (A)*, 103, 287 [55]. Syn. by Ansorge, 1996, *Neue Palaeontol. Abh.*, 2, 75 [29].

Acritorhyphus Bode, 1953, *Palaeontographica (A)*, 103, 299 [55]. Syn. by Ansorge, 1996, *Neue Palaeontol. Abh.*, 2, 81 [29].

Palaeolimnobia Bode, 1953, *Palaeontographica (A)*, 103, 301 [55]. Syn. by Lukashevich *et al.*, 1998, *Polskie Pismo Entomol.*, 338 [80].

Type species: *Eoptychoptera simplex* Handlirsch, 1906.

Body rather compact, wings with four medial veins, m-m beyond M_{1+2} furcation (connecting M_2 and M_3).

Distribution and age: Inner Mongolia; Middle Jurassic.

Two species included from the Jurassic of Northern China (see Table 23.1).

Crenoptychoptera Kalugina, 1985

Crenoptychoptera Kalugina, 1985, *Dipterous Insects of Jurassic Siberia*, 41 [46] (original designation).

Type species: *Crenoptychoptera antica* Kalugina, 1985.

Abdomen (male) slender, distally widened. Wings slender, narrowed at apex. Three medial veins (M_1, M_2, M_{3+4}); m-m long and curved, connecting M_2 near base; a cross-vein at basal part of M_2 [46, 81].

Distribution and age: Inner Mongolia; Middle Jurassic.

Three species included from the Jurassic of Northern China (see Table 23.1).

Eoptychopterina Kalugina, 1985

Eoptychopterina Kalugina, 1985, *Dipterous insects of Jurassic Siberia*, 37 [46] (original designation).

Type species: *Eoptychopterina rohdendorphi* Kalugina, 1985.

The specific epithet is dedicated to Dr. Rohdendorph for his contribution to studies of fossil dipterans. Antenna short, rounded; labrum strongly sclerotized; wide, slightly rounded stylets. Rs strongly kinked toward medial veins two-thirds along its length; R_{4+5} bent at r-m; im between M_{1+2} and unforked M_{3+4} [46, 82].

Distribution and age: Inner Mongolia; Middle Jurassic.

Five species included from the Jurassic of Northern China (see Table 23.1).

Eoptychopterina postica Liu, Shih & Ren, 2012 (Figure 23.16)

Eoptychopterina postica Liu, Shih & Ren, 2012: *Zootaxa*, **3501**, 55–62.

Figure 23.16 *Eoptychopterina postica* Liu, Shih & Ren, 2012 (Holotype, CNU-DIP-NN2011001).

Locality and horizon: Daohugou, Ningcheng, Inner Mongolia, China; Middle Jurassic, Jiulongshan Formation.

Body stout. Head round and significantly narrower than thorax. Left antenna short, with 11 antennomeres. Eyes large, ommatidia clearly visible. Thorax subovoid. Abdomen length 5.9 mm with eight visible segments. Male genitalia with harpagones preserved. Wing membrane densely pubescent, both wings with distinctly darker area along C. Wing longer than body. The left wing longer and narrower than the right wing. Asymmetric left and right wings are also shown on *Eoptychopterina elenae* Ren & Krzemiński, 2002 and a few other fossils in our CNU collection. Sc long, nearly 3/5 length of the wing, and terminating at C proximal to furcation of R_{4+5}. Stem R thick and strong; R_1 long and with a slight bend at R_2; R_3 longer than R_{2+3}, and shorter than R_4. The apical part of R_3 curving anteriorly. R_{4+5} bent at cross-vein r-m, cross-vein r-m longer than bR_{4+5}, dR_{4+5} two times as long as bR_{4+5}; R_{4+5} ramified symmetrically. Stem vein M weak; dM_{1+2} slightly longer than bM_{1+2}; M_1 smoothly curved at base, and obviously longer than M_2; cross-vein m-m connecting M_2 and M_{3+4}. CuA distinctly bent at m-cu, and straight beyond it. A_1 strongly curved to wing margin [79].

Family Strashilidae Rasnitsyn, 1992

In 1992, Rasnitsyn [83] described an enigmatic species, *Strashila incredibilis*, from the Upper Jurassic of Siberia and put it in a family of Strashilidae (Order *incertae sedis*), suggesting possible siphonapteran affinities. The *S. incredibilis* has very strange characters: segments of abdomen with long tubular appendages; chelate hind leg, formed by crassate femur, basitarsus, and tibial apical process, with tarsus otherwise normal; head hypognathous, with short moniliform antenna and possibly with short suctorial beak, and thorax completely wingless, lacking midventral suture. Vrsansky *et al.* [84] erected a family of Vosilidae Vršanský & Ren, 2010 and designated an insect order, Nakridletia, belonging to Papilionidea Laicharting, 1781 (= Mecopteroidea auct.). In 2013, Huang *et al.* [85] analyzed the morphology of strashilids from the Daohugou beds and documented that strashilids are highly specialized dipterans bearing large membranous wings and having substantial sexual dimorphism of the hind legs and abdominal extensions. Huang *et al.* reported that *Strashila daohugouensis* Huang, Nel, Cai, Lin & Engel, 2013 resembles the recent (extant) and relict members of the aquatic fly family Nymphomyiidae. Their ontogenies are distinguished by the persistence in adult males of larval abdominal respiratory gills, representing a unique case of pedomorphism among endopterygote insects. Adult strashilids were probably aquatic or amphibious, shedding their wings after emergence and mating in the water [85]. But the characteristics and classification of this group are still in contention.

Genera included from the Jurassic of Northern China: *Strashila* Rasnitsyn, 1992 and *Vosila* Vrsansky & Ren, 2010.

Strashila Rasnitsyn, 1992

Strashila Rasnitsyn, 1992, *Psyche*, 99 (4), 324 [83] (original designation).

Type species: *Strashila incredibilis* Rasnitsyn, 1992.

Fore and mid tibiae subclavate; hind ones inflated similar to the femora except having still shorter stalk, elongated apically at flexor side into slightly curved process with narrowly rounded apex. Tarsi subequal in length, thin; with basitarsus slightly or, in hind leg, conspicuously thicker than distal segments, with long apical segment and moderately long, gently curved claws.

Distribution and age: Inner Mongolia; Middle Jurassic.

Only one species included from the Jurassic of Northern China (see Table 23.1).

Vosila Vrsansky & Ren, 2010

Vosila Vrsansky & Ren, 2010, *Amba projekty*, 8 (1), 2 [84] (original designation).

Type species: *Vosila sinensis* Vrsansky & Ren, 2010.

Differing from *Strashila* in having more robust pronotum and meso- and metanotum, all femora more massive, hindleg nipper aperture round, and pretarsi very long. Generally, *Vosila* has fewer chaeta. Unlike *Strashila*, the abdominal appendages with large gill-like structures, but without lateral abdominal prolongations.

Distribution and age: Inner Mongolia; Middle Jurassic.

Only one species included from the Jurassic of Northern China (see Table 23.1).

Figure 23.17 *Vosila sinensis* Vršanský & Ren, 2010 (Holotype, CNU-PARA-001).

Vosila sinensis Vrsansky & Ren, 2010 (Figure 23.17)

Vosila sinensis Vrsansky & Ren, 2010: *Amba projekty,* 8 (1), 2.

Locality and horizon: Daohugou, Ningcheng, Inner Mongolia, China; Middle Jurassic, Jiulongshan Formation.

Head covered by slightly transversal, quadrate pronotum. Meso and metanotum extremely narrow. Abdomen almost lacking sclerotization, about 5 mm long and 1 mm wide as preserved. Abdominal appendages consisting of three distinct parts. The basalmost articulation is weakly sclerotized, narrow, with proximal part with an aperture. The central hardly sclerotized part consists of three lobes, two proximal parts closely associated. To the central part, plain, gill-like appendages are articulated. Three to the distal lobe, eight to the central lobe, and a single to the proximal lobe. Terminalia protruded from body, about 0.5 mm long. Foreleg femur robust, with numerous long sensilla, tibia only with five sensilla. Basitarsus with a terminal claw and a row of very small sensillar pitts. Segment 2 attached subapically. Midleg femur robust, with numerous spurs, tibia with two longitudinal ridges. Hind legs asymmetrical. Right coxa very long (1.3 mm), femur not very robust with numerous sensilla (1.8 mm), tibia very long (2.8 mm). Left femur extremely robust with two longitudinal ridges and numerous sensilla. Tibia with apical processus, very long, with two longitudinal ridges. Nipper aperture formed by the apical processus and basitarsus round and wide [84].

Family Tanyderidae Osten-Sacken, 1879

Tanyderidae are small to large extant basal crane flies and distributed worldwide, achieving their greatest species diversity in the southern hemisphere [86], including North and South America, Africa, Australia, New Zealand and various islands in the Pacific Ocean. This small recent family represented by 55 species in 12 genera [7]. Immature stages are aquatic, living in cobble or sand bottom streams. Tanyderids were more common in the Mesozoic. The most ancient fossils of Tanyderidae are known from the Toarcian of Germany [86, 87]. Based on examinations of the wing venation, the tanyderids are often considered to be the most ancient extant dipteran family with a plesiomorphic venation. Their wings correspond to the ground plan of dipteran wing, except for the reduction of A_2.

Genera included from the Jurassic of Northern China: *Praemacrochile* Kalugina, 1985 and *Protanyderus* Handlirsch, 1909.

Praemacrochile Kalugina, 1985

Praemacrochile Kalugina, 1985, *Dipterous insects of Jurassic Siberia*, 35 [44] (original designation).

Type species: *Praemacrochile stackelbergi* Kalugina, 1985.

Antenna long, scape long and cylindrical, pedicel short and oval, flagellomeres elongated. Tibial spurs absent. Long Sc terminates beyond wing midlength, distal to Rs furcation. Long Rs stem begins in proximal half of wing. Rs forking into R_{2+3} stem and R_{4+5} with long R_{4+5} in alignment with Rs. Fork R_{2+3} shorter than its petiole.

Distribution and age: Inner Mongolia; Middle Jurassic.

Four species included from the Jurassic of Northern China (see Table 23.1).

Praemacrochile ovalum Dong, Shih, Skibińska, Krzemiński & Ren, 2015 (Figure 23.18)

Praemacrochile ovalum Dong, Shih, Skibińska, Krzemiński & Ren, 2015: *Alcheringa*, 39, 498–501.

Locality and horizon: Daohugou, Ningcheng, Inner Mongolia, China; Middle Jurassic, Jiulongshan Formation.

Figure 23.18 *Praemacrochile ovalum* Dong, Shih, Skibińska, Krzemiński & Ren, 2015 (Holotype CNU-DIP-NN2014550p).

Thorax length 1.5 mm and width 1.3 mm, nearly circular, with robust and well-developed mesonotum. Wing length 7.5 mm, narrow and long with pterostigma; venation clear; Sc about 0.6 times as long as the wing and ending at the anterior margin basad to R_{4+5} furcation; Rs arising 1/3 distance from the base to apex of the wing, Rs furcation at 0.63 wing length; R_1 ending at the same level with furcation of R_{2+3}; R_{2+3} almost 4.2 times as long as R_2; R_4 6.0 times as long as R_{4+5}, R_{4+5} 2.0 times as long as r-m; M_1 1.45 times as long as M_{1+2}; cell d narrow and long and almost 0.25 times wing length; M_3 curved posteriorly and M_4 curved anteriorly, cross-vein m_3–m_4 between the M_3 and M_4; Cu long and reaching the posterior margin, 0.61 times wing length. Vein A long, slightly curving and reaching the posterior margin, 0.54 times wing length. Abdomen relatively long and thin, only three segments preserved. Without spurs; t_1 about twice as long as t_2 [88].

Protanyderus Handlirsch, 1909

Protanyderus Handlirsch, 1909, *Ann. K. K. Naturhist. Hofmus.*, 23, 270 [89] (original designation).

Type species: *Protanyderus vipio* Osten-Sacken, 1877.

Tibial spurs conspicuous. Wing spotted. Rs forking into R_{2+3} stem and R_{4+5} with very short R_{4+5} aligned with Rs. Fork R_{2+3} shorter than its petiole, furcating at or beyond midlength of distal section of R_1. Supernumerary veins absent from wing cells [82, 89].

Distribution and age: Inner Mongolia; Middle Jurassic.

Only one species included from the Jurassic of Northern China (see Table 23.1).

> **An Amber Tanyderid from the mid-Cretaceous of Myanmar**
>
> A tanyderid, *Similinannotanyderus lii*, is described from Myanmar (Burmese) amber (Figure 23.19). The specific epithet is to honor Mr. Jun Li for his generous donation of this Myanmar amber insect and other amber insects for our studies. It differs from other genera of Tanyderidae based on special characters of wing venation and male genitalia. Wing short (3.9 mm in length); Sc rather long about 0.6 times as long as the wing; R_2 long; R_{2+3} forking at 4/5 of wing length; cross-vein m-cu clearly longer than M_{3+4}; anal lobe reduced, rounded; R_1 short, 0.82 times as long as wing length; M_1 almost equal to M_{1+2}. Legs with two strong spurs; t1 clearly longer than t2 (t1/t2 = 4.1). Gonocoxite long and rather narrow, gonostylus with a big protuberance covered by strong and short bristles. This new taxon broadens the diversity of Tanyderidae in the Cretaceous, and its morphological characters enhance our understanding of the development and evolution of the basal crane flies [90].

Figure 23.19 *Similinannotanyderus lii* Dong, Shih & Ren, 2014 (Holotype, CNU-DIP-MA2014001), Source: Donated by Jun Li.

Family Tipulidae Latreille, 1802

Tipulidae, belonging to the Infraorder Tipulomorpha, are one of the most speciose extant families of the Diptera. To date, there are more than 4000 described extant and extinct species of Tipulidae, distributed almost all over the world [24, 91]. They are medium to large-sized insects, distinguished by their very long terminal segment of maxillary palpi, distinct nasus as most Tipulidae, antenna with 8–15 antennomeres, thorax with a V-shaped mesonotal suture, and the absence of ocelli. On their wings, the end of Sc is usually atrophied, and Sc joins through sc-r with the radial vein. Cross-vein m-cu is situated at or close to the forking of M_{3+4}. Tipulidae are divided into three subfamilies: Ctenophorinae Kertész, 1902, Dolichopezinae Kertész, 1902, and Tipulinae Latreille, 1802. More than 100 fossil species have been described from the USA, Russia, UK, France, Germany, Spain, Denmark, Czech Republic, Croatia, Canada, China, Dominica, Brazil, and Baltic Region. The earliest tipulid fossils reported so far are three species of *Leptotarsus* Guerin-Meneville, 1831 [92], from the Lower Cretaceous La Huerguina Formation in Spain, c. 126 Mya, not classified to definite subgenus [93] and *Leptotarsus (Longurio) primitivus* Shih, Dong, Kania, Liu, Krzemiński & Ren, 2015 from the Lower Cretaceous (c. 125 Mya) Yixian Formation of Huangbanjigou, Beipiao, Liaoning, China [92].

Only one genus included from the Cretaceous of Northern China: *Leptotarsus* Guérin-Méneville, 1831.

Leptotarsus Guérin-Méneville, 1831

Leptotarsus Guérin-Méneville, 1831, *Voyageautour du monde, execute par ordre du Roi, sur la corvette de samajeste La Coquille etc*, 20 [94] (original designation).

Type species: *Leptotarsus macquarti* Guérin-Méneville, 1831.

Large size; each respiratory horn elongated, nearly half body length, apex flat with rounded margin; dorsum of tergum VIII with two pairs of large spine-like curved lobes, all with smaller spines along length; above male genital sheaths, a pair of dome-like swellings with terminal papillae.

Distribution and age: Liaoning; Early Cretaceous.

Only one species included from the Cretaceous of Northern China (see Table 23.1).

Leptotarsus (Longurio) primitivus **Shih, Dong, Kania, Liu, Krzemiński & Ren, 2015 (Figure 23.20)**

Leptotarsus (Longurio) primitivus Shih, Dong, Kania, Liu, Krzemiński & Ren, 2015: *Cretac. Res.*, 54, 100.

Locality and horizon: Huangbanjigou, Beipiao, Liaoning, China; Lower Cretaceous, Yixian Formation.

Shih *et al.* described *Leptotarsus (Longurio) primitivus*, a representative of the family Tipulidae from the Lower Cretaceous (c. 125 Mya) Yixian Formation of Huangbanjigou, Beipiao, Liaoning, China [87]. This finding represents one of the earliest fossil records of the Tipulidae in the world. Among 20 subgenera from the genus *Leptotarsus* Guérin-Méneville, 1831, this species from China is classified to the subgenus *Longurio* Loew, 1869 based on characters of wing venation and morphology of rostrum and antenna. The *L. primitivus* has some of the most plesiomorphic features: the presence of well-developed vein Sc, connecting with costal (C) vein and the placement cross-vein m-cu distad of the bifurcation of vein M_{3+4} to M_3 and M_4. The species was placed in the subgenus *Longurio* by a combination of these characters: the two above-mentioned plesiomorphic features plus short rostrum and long antennae [92].

Family Trichoceridae Kertesz, 1902

Trichoceridae, commonly called winter crane flies, comprise 15 extant genera and 183 species [7]. The adults not only live in cold environment, but also mate and lay eggs under the snow cover in winter [95]. To date, there are 79 extinct species belonging to 12 genera recorded, and assigned into three subfamilies of Trichocerinae, Paracladurinae and Kovalevinae. They are typified by the following characters: Sc ending proximad of the forking of R_2, shape of d cell and A_2 rather short and bending sharply toward posterior margin. The oldest species of trichocerids has been described from Lower Jurassic of Germany [96].

Genera included from the Jurassic of Northern China: *Eotrichocera* Kalugina, 1985 and *Tanyochoreta* Zhang, 2006.

Eotrichocera **Kalugina, 1985**

Eotrichocera Kalugina, 1985, *Dipterous insects of Jurassic Siberia*, 47 [44] (original designation).

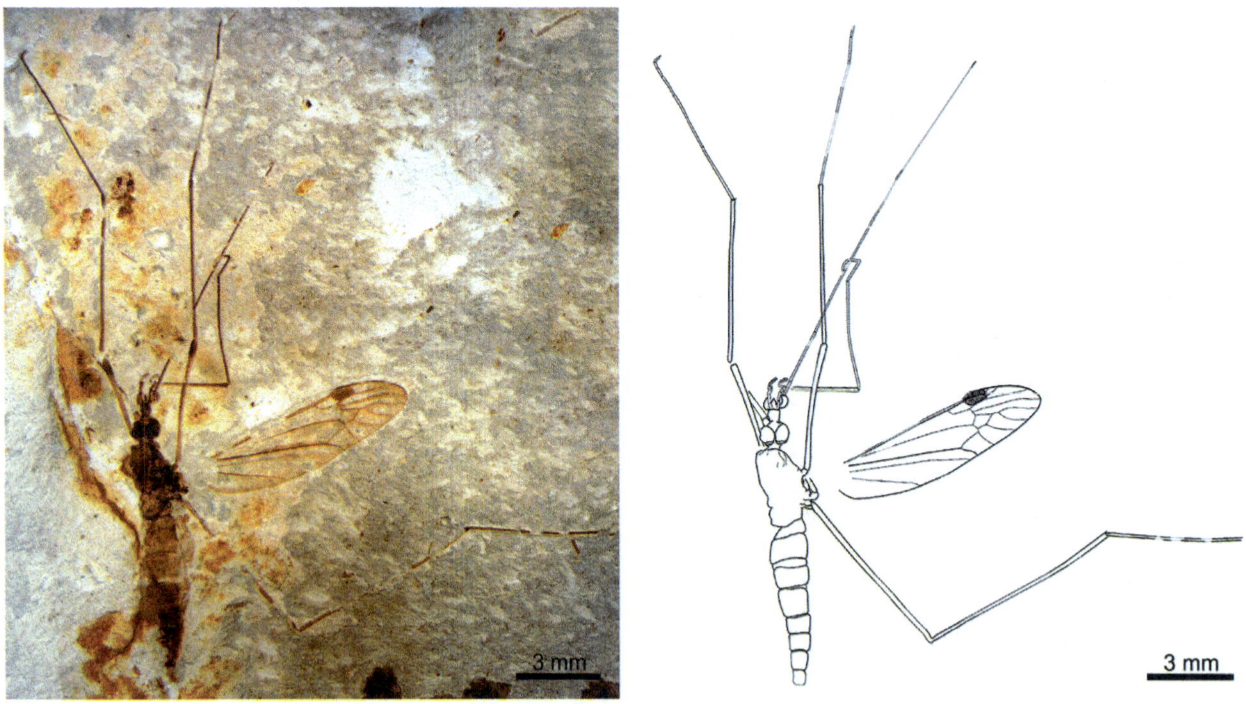

(a) (b)

Figure 23.20 *Leptotarsus (Longurio) primitivus* Shih, Dong, Kania, Liu, Krzemiński & Ren, 2015 (Holotype, CNU-DIP-NN2008108). (a) Photograph of habitus; (b) Line drawing. Source: Donated by Dr. Chungkun Shih.

Figure 23.21 *Eotrichocera (Archaeotrichocera) longensis* Dong, Shih & Ren, 2014 (Holotype, CNU-DIP-NN2013133).

Type species: *Eotrichocera (Eotrichocera) christinae* Kalugina, 1985.

Medium-sized winter crane flies. Sc_1 more than 3/4 length of wing; cell d long and wide; both bM_{1+2} and mM_{1+2} strongly flexed at cross-vein r-m. Legs with tibial spurs present; basitarsomere longer than the second tarsomere [44, 97].

Distribution and age: Inner Mongolia; Middle Jurassic.

Four species included from the Jurassic of Northern China (see Table 23.1).

Eotrichocera (Archaeotrichocera) longensis Dong, Shih & Ren, 2014 (Figure 23.21)

Eotrichocera (Archaeotrichocera) longensis Dong, Shih & Ren, 2014: *ZooKeys*, **411**, 145–160.

Locality and horizon: Daohugou, Ningcheng, Inner Mongolia, China; Middle Jurassic, Jiulongshan Formation.

Body length 13 mm (including head). Antenna very long, about 3.5 times as long as the head length, palpi about two times as long as the head length. Thorax higher in lateral view, subcircular in shape, mesonotum robust. Wing shorter than abdomen, length 9.0 mm, narrow and long. Sc rather short; ending at anterior margin proximad of R_2; cross-vein sc-r locating at 2/3 of Rs; Rs arising about 1/5 from the base of the wing; R_2 about 1/10 of length of R_3; bM_{1+2} nearly as long as the length of cross-vein r-m; d-cell narrow and long, almost 1/5 length of wing; both cross-veins m-m and m-cu intersecting with M_4 at the same point; Vein A divided into A_1 and A_2; A_1 long, slightly curving; A_2 short, almost 0.3 times as long as length of A_1, bending sharply. Abdomen long and thin, with 10 segments [98].

Tanyochoreta Zhang, 2006

Tanyochoreta Zhang, 2006, *Can. J. Earth. Sci.*, 43, 11 [97] (original designation).

Type species: *Tanyochoreta (Tanyochoreta) integera* Zhang, 2006.

Medium-sized winter crane flies. Mouthparts with labella large, surrounding tip of labrum and hyphopharynx. Sc_1 about 0.7 times length of wing; cell d large; both bM_{1+2} and mM_{1+2} almost in line at cross-vein r-m. Legs with tibial spurs present; basitarsomere clearly shorter than the second tarsomere.

Distribution and age: Inner Mongolia; Middle Jurassic.

Three species included from the Jurassic of Northern China (see Table 23.1).

Family Zhangobiidae Evenhuis 1994

Zhangobiidae are a small family in the superfamily Tipuloidea. In 1986, Zhang *et al.* erected the family Palaeolimnobiidae [33] for the genera of *Palaeolimnobia* Zhang, Zhang, Liu & Shangguan, 1986 and *Ceuthoneura* Zhang, Zhang, Liu & Shangguan, 1986 [33]. In 1994, Evenhuis [24] changed the family name to Zhangobiidae. Furthermore, Pape *et al.*, in 2011, proposed to transfer four separate families (i.e. Archilimoniidae, Gracilitipulidae, Palaeolimnobiidae, and Zhangobiidae) to Pediciidae [7]. Pediciidae are virtually cosmopolitan with some 500 species in over 10 genera [7]. The fossil taxonomic records of Pediciidae are complicated, and we need more evidence to support or justify the transfer of these four families to Pediciidae. The fossil records of Zhangobiidae have only two genera described from the Early Cretaceous of Shandong, China.

Genera included from the Cretaceous of Northern China: *Ceuthoneura* Zhang, Zhang, Liu & Shangguan, 1986 and *Zhangobia* Evenhuis 1994.

Ceuthoneura Zhang, Zhang, Liu & Shangguan, 1986

Ceuthoneura Zhang, Zhang, Liu & Shangguan, 1986, *Geol. Shandong*, 2, 19 [33] (original designation).

Type species: *Ceuthoneura dolichoptera* Zhang, Zhang, Liu & Shangguan, 1986.

Wing narrow and long, venation weak, not obvious. Eyes large. Antenna long, filiform. Thorax globose. Legs very long and slender, tibia longer than femur; tarsi long. Abdomen eight segments visible.

Distribution and age: Shandong; Early Cretaceous.

Only one species included from the Cretaceous of Northern China (see Table 23.1).

Zhangobia Evenhuis, 1994

Zhangobia Evenhuis, 1994, *Catalogue of the Fossil Flies of the World (Insecta: Diptera)*, 54 [24] (original designation).

Palaeolimnobia Zhang, Zhang, Liu & Shangguan, 1986, *Geol. Shandong*, 2, 16 [33].

Type species: *Palaeolimnobia laiyangensis* Zhang, Zhang, Liu & Shangguan, 1986.

Eyes large. Antenna long. Thorax globose. Wing slender; Sc relatively weak, Rs three branches and separated from R the base of the wing; M branches long, separated from CuA; M with cross-vein; CuA short. Abdomen nine segments.

Distribution and age: Shandong; Early Cretaceous.

Only one species included from the Cretaceous of Northern China (see Table 23.1).

Suborder "Brachycera" Schiner, 1862

Family Archisargidae Rohdendorf, 1962

Archisargidae, an extinct family, usually have these distinguished characters: head round, veins R and M usually with four branches, Sc and R_1 long and straight and discoidal cell distinctly shift distally. Archisargidae was erected by Rohdendorf in 1962 based on *Archirhagio obscures* Rohdendorf, 1938 and *Archisargus pulcher* Rohdendorf, 1938. Archisargidae have been documented from the Late Jurassic to the Early Cretaceous [30] and distributed in Laurasia and Gondwana, include China, Kazakhstan, Mongolia and Australia. Archisargidae include two subfamilies of Archisarginae Rohdendorf, 1962 and Mostovskisarginae Zhang, 2010. To date, two subfamilies, including 38 species in 12 genera, have been described in China.

Genera included from the Jurassic of Northern China: *Archirhagio* Rohdendorf, 1938, *Archisargus* Rohdendorf, 1938, *Mesosolva* Hong, 1983, *Ovisargus* Mostovski, 1996, *Sharasargus* Mostovski, 1996, *Calosargus* Mostovski, 1997, *Brevisolva* Zhang, Ren & Shih, 2010, *Mostovskisargus* Zhang, 2010, *Flagellisargus* Zhang, 2012, *Novisargus* Zhang, 2014 and *Tabanisargus* Zhang, 2014.

Archirhagio Rohdendorf, 1938

Archirhagio Rohdendorf, 1938, *Trudy Paleontol. Inst.*, 7 (3), 3 [28] (original designation).

Type species: *Archirhagio obscures* Rohdendorf, 1938.

Large size, relatively thick. Head subglobose, ommateum nudity, dichoptic; antenna shortwing long and narrow; Sc, R_1 and R_{2+3} long, cell d long and narrow, cell m_3 wide open, cell cup narrowly open [28, 99].

Distribution and age: Inner Mongolia; Middle Jurassic.

Five species included from the Jurassic of Northern China (see Table 23.1).

Archirhagio gracilentus Wang, Shih, Ren & Wang, 2017 (Figure 23.22)

Archirhagio gracilentus Wang, Shih, Ren & Wang, 2017: *Syst. Entomol.*, 42, 232.

Locality and horizon: Daohugou, Ningcheng, Inner Mongolia, China; Middle Jurassic, Jiulongshan Formation.

The body 15.2 mm long; wing 10.1 mm long and 2.3 mm wide. Head sub-elliptical, about as wide as thorax. Compound eye bare. Scape edged with small serrated ridges distally; pedicel short, first flagellomere swollen and concave. Thorax sub-ovate. Legs elongated. The wing elongated and hyaline. Costa terminating at apex of wing. Humeral vein present. Rs about twice as long as bR_{4+5}. R_{4+5} fork deep, R_4 nearly parallel to R_5 proximally. Abdomen slender and cylindrical. The first segment shorter than the second. Male: eight pregenital segments visible. Tergite ninth prominently concave posteriad, forming two triangular lobes in dorsal view. Gonocoxite well-developed, with the distinct inner appendix-like teeth caudally. Pair of parameres inserted in gonocoxite, bearing many long bristles at the tips. Female: only seven pregenital segments visible. Genitalia preserved partly clearly, tergite 8, 9 and sternite 8 small. Hypogynial valves hook-like, protruding out of the abdomen [100].

Archisargus Rohdendorf, 1938

Archisargus Rohdendorf, 1938, *Trudy Paleontol. Inst.*, 7 (3), 30 [28] (original designation).

Type species: *Archisargus pulcher* Rohdendorf, 1938.

Large and stout. Vein Sc long; vein R_5 ending behind wing tip; veins M_2, M_3, M_4 and CuA short. Vein CuP present. Cell sc wide open, mouth of cell sc wider than distance between ends of veins R_1 and R_5; mouth of cell r_1 narrowly open (or nearly closed); mouth of r_{2+3} narrower than that of cell r_4. Cell d distinctly enlarged, adjacent to wing margin. Cell cup closed, apical part truncate. Abdomen slender, almost parallel-sided [101].

Distribution and age: Inner Mongolia; Middle Jurassic.

Two species included from the Jurassic of Northern China (see Table 23.1).

Mesosolva Hong, 1983

Mesosolva Hong, 1983, *Middle Jurassic fossil insects in North China*, 133 [31] (original designation).

Prosolva Hong, 1983, *Middle Jurassic fossil insects in North China*, 135 [31]. Syn. by Zhang, 2012, *ZooKeys*, 238, 57–76 [102].

Type species: *Mesosolva parva* Hong, 1983.

R with four branches and M with three branches. Vein CuA_1 arising from cell bm. Mouth of cell sc wide open. Cell m_3 closed, with apical petiole. Cell cup always open [31, 103].

Distribution and age: Inner Mongolia; Middle Jurassic.

Five species included from the Jurassic of Northern China (see Table 23.1).

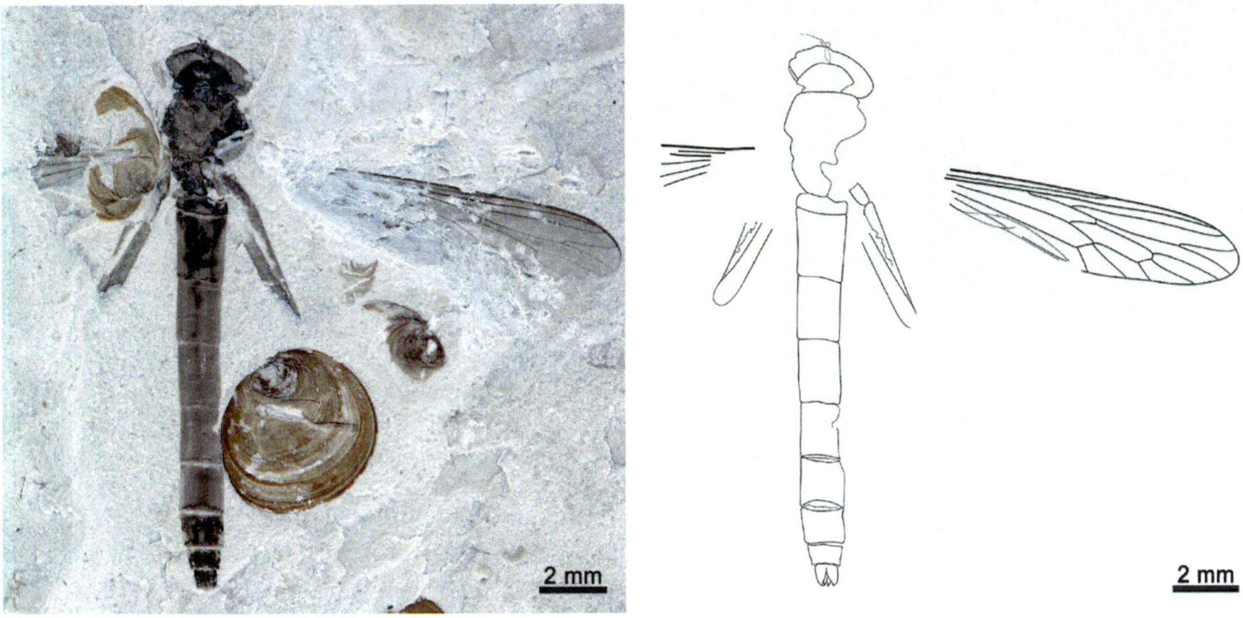

(a) (b)

Figure 23.22 *Archirhagio gracilentus* Wang, Shih, Ren & Wang, 2017 (Holotype, CNU-DIP-NN2016001). (a) Photograph; (b) Line drawing.

***Ovisargus* Mostovski, 1996**

Ovisargus Mostovski, 1996, *Russ. Ent. J.*, 5 (1–4), 121 [104] (original designation).

Type species: *Ovisargus gracilis* Mostovski, 1996.

Sc short, ending nearly at the middle of wing; section bR_{4+5} relatively long, dR_{4+5} long, the fork of R_{4+5} very short; M three branches; M_3 emit from d; cell m_1, m_2 and m_3 wide open; cell cup closed with short petiole; cell d short and wide [99, 104].

Distribution and age: Inner Mongolia; Middle Jurassic.

Only one species included from the Jurassic of Northern China (see Table 23.1).

***Sharasargus* Mostovski, 1996**

Sharasargus Mostovski, 1996, *Russ. Ent. J.*, 5 (1–4), 124 [104] (original designation).

Type species: *Sharasargus ruptus* Mostovski, 1996.

First flagellomere tapered to apex; Sc ending far beyond middle of wing; veins R_1 and R_{2+3} long; vein R_5 ending beyond wing apex; R four branches. Cell r_1 open; M three branches; m_1, m_2 and m_3 cell wide open; cup cell open [104, 105].

Distribution and age: Inner Mongolia; Middle Jurassic.

Three species included from the Jurassic and Cretaceous of Northern China (see Table 23.1).

***Calosargus* Mostovski, 1997**

Calosargus Mostovski, 1997, *Paleontol. J.*, 1, 74 [106] (original designation).

Type species: *Calosargus (Calosargus) tatianae* Mostovski, 1997.

Wing slender; vein Sc ending far beyond middle of wing; R_{2+3} short, R_{2+3} and R_1 intersect before the margin of the wing; some species M_1 and M_2 combined; M_4 growing out of cell d; cell cup narrowly open [106, 107].

Distribution and age: Inner Mongolia; Middle Jurassic.

Seven species included from the Jurassic of Northern China (see Table 23.1).

***Calosargus (Calosargus) bellus* Zhang, Yang & Ren, 2007 (Figure 23.23)**

Calosargus (Calosargus) bellus Zhang, Yang & Ren, 2007: *Zootaxa*, 1645, 15.

Locality and horizon: Daohugou, Ningcheng, Inner Mongolia, China; Middle Jurassic, Jiulongshan Formation.

Body length 17.3–17.6 mm; wing length 9.4–10.3 mm, wing width 2.8–2.9 mm. Head round, slightly narrower than thorax. Eyes large, bare, dichoptic; ommatidia visible clearly. Thorax black. Legs pubescent. Fore and mid tibiae slender; hind tibia with one spur, femur long, robust; the first tarsomere distinctly longer than following tarsomeres. Wing hyaline: vein Rs_1 about twice as long as vein Rs_2; vein R_{2+3} arising slightly distal of base of cell d; vein R_{2+3} ending at vein R_1 near wing margin; cross-vein r-m located at about basal 2/5 of cell d. Abdomen slender, narrowing toward tip, covered with setulae. Eight segments visible. Segment I evidently short. Segments I and II with wide black stripe at middle. Segments II to VII with distinct striae at side [107].

Figure 23.23 *Calosargus bellus* Zhang, Yang & Ren, 2007 (Holotype, CNU-DB-NN2007004). Donated by Dr. Chungkun Shih.

Brevisolva Zhang, Ren & Shih, 2010

Brevisolva Zhang, Ren & Shih, 2010, *Entomol. Sci.*, 13, 78 [103] (original designation).

Type species: *Brevisolva daohugouensis* Zhang, Ren & Shih, 2010.

Veins Sc, R_1 and R_{2+3} long and straight. Vein Sc distinctly distal of cross-vein r-m. Veins R_5 and Rs short. Cross-vein r-m close to the base of cell d. Four medial veins present. Vein CuA_1 arising from cell bm. Veins M_3 and M_4 meet before the margin. Cell cup open.

Distribution and age: Inner Mongolia; Middle Jurassic.

Only one species included from the Jurassic of Northern China (see Table 23.1).

Mostovskisargus Zhang, 2010

Mostovskisargus Zhang, 2010, *Palaeontology*, 53 (2), 310 [108] (original designation).

Type species: *Mostovskisargus portentosus* Zhang, 2010.

Antennal first flagellomere conical and tapered to apex. R_{2+3} arched sharply at basal. Rs_2 short. Rs fork shifted distally of m–cu. R_{4+5} fork deep. Veins M_3 and M_4 present, dM_{1+2} present. Cell m_3 open widely. Cell d relatively short. M_4 and CuA connected by cross-vein m–cu. CuA nearly parallel to CuP, and cell cup wide open.

Distribution and age: Inner Mongolia; Middle Jurassic.

Two species included from the Jurassic of Northern China (see Table 23.1).

Flagellisargus Zhang, 2012

Flagellisargus Zhang, 2012, *J. Paleontol.*, 86 (5), 879 [109] (original designation).

Type species: *Flagellisargus sinicus* Zhang, 2012.

R with four branches and M with three branches. Rs fork and d base nearly at the same level. The fork of R_{4+5} shallow. R_5 ending at wing apex. Cell r_1 open, cell cu narrowly open. Male genitalia large, with stout and long gonostyli.

Distribution and age: Inner Mongolia; Middle Jurassic.

Three species included from the Jurassic of Northern China (see Table 23.1).

Novisargus Zhang, 2014

Novisargus Zhang, 2014, *J. Syst. Palaeontol.*, 07, 12 [99] (original designation).

Type species: *Novisargus rarus* Zhang, 2014.

Body size large. Head large, globose, antenna short and slender. Wings slender, the length is longer than three times the width. Vein Sc long, extending far beyond the middle of wing. Vein R with three branches, Sc, R_1 and R_{2+3} long and straight, R_{4+5} single; M veins with four branches; r-m at the middle of the d cell. Cell cup narrowly open.

Distribution and age: Inner Mongolia; Middle Jurassic.

Only one species included from the Jurassic of Northern China (see Table 23.1).

Tabanisargus Zhang, 2014

Tabanisargus Zhang, 2014, *J. Syst. Palaeontol.*, 07, 12 [99] (original designation).

Type species: *Tabanisargus daohugous* Zhang, 2014.

Large-sized; head round; wing wide; R four branches, M four branches, Sc, R_1, R_{2+3} and R_{4+5} nearly parallel; spurious vein in br; Cell d small; R_{4+5} fork deep narrow. Cell cup narrowly open; M_4 emitting from cell d.

Distribution and age: Inner Mongolia; Middle Jurassic.

Only one species included from the Jurassic of Northern China (see Table 23.1).

Family Athericidae Stuckenberg, 1973

Athericidae, known as "water snipe flies", are a small, cosmopolitan family included in the superfamily Tabanoidea [110]. There are only 11 extant genera and about 121 species described around the world, most of which were previously placed in Rhagionidae. Now, they are known to be more closely related to Tabanidae. Although their wings are similar to those of Rhagionidae, athericids can be distinguished by the following characters: fuscous to black with fasciated abdomens; antenna with basal flagellomeres reniform, extending ventrally beyond pedicel and R_1 and R_{2+3} ending together in C above or distal to fork of R_{4+5}. The adults mostly feed on nectar but some species feed on blood of birds and mammals and some are the parasites of amphibians. The family is poorly represented in the fossil records, with only seven genera known to date, from Denmark, France, Russian, USA, Australia and China. There are only two fossil genera and species from the Middle Jurassic and the Early Cretaceous of China [111].

Genera included from the Jurassic and Cretaceous of Northern China: *Sinocretomyia* Zhang, 2012 and *Qiyia* Chen, Wang & Engel, 2014.

Sinocretomyia Zhang, 2012

Sinocretomyia Zhang, 2012, *Cretac. Res.*, 36, 3 [111] (original designation).

Type species: *Sinocretomyia minuscula* Zhang, 2012.

Flagellum long, not segmented. Arista slender and long. R_{2+3} short, strongly sinuous, and joining R_1 at costa. Basal cells comparatively long, their distal ends extending distal to level of Sc termination. M_1 weakly arched. Fourth posterior cell (m_3) slightly widened distally; M_3 and M_4 somewhat divergent. Cells m_2 and m_3 elongate. Cell cu closed or very narrowly opened. Pterostigma absent. Cercus one-segmented, oviform.

Distribution and age: Shandong; Early Cretaceous.

Only one species included from the Cretaceous of Northern China (see Table 23.1).

Qiyia Chen, Wang & Engel, 2014

Qiyia Chen, Wang & Engel, 2014, *eLife*: e02844, 2 [112] (original designation).

Type species: *Qiyia jurassica* Chen, Wang & Engel, 2014.

This is a bizarre larva from the Middle Jurassic which was a bloodsucking parasite of salamanders. Three thoracic segments fused, with a ventral sucker; two pairs of dorsal spines on abdominal segments one to seven; abdominal segments one to six with paired ventral prolegs bearing upward directed bristles and apical crochets; extended seventh proleg; two pairs of anal papillae; sclerotized terminal processes with stiff setae.

Distribution and age: Inner Mongolia; Middle Jurassic.

Only one species included from the Jurassic of Northern China (see Table 23.1).

Qiyia jurassica Chen, Wang & Engel, 2014 (Figure 23.24)

Qiyia jurassica Chen, Wang & Engel, 2014: *eLife*, e02844, 2.

Figure 23.24 *Qiyia jurassica* Chen, Wang & Engel, 2014, additional material. Source: Photo provided by Dr. Chungkun Shih.

Locality and horizon: Daohugou, Ningcheng, Inner Mongolia, China; Middle Jurassic, Jiulongshan Formation.

Qiyi is from the Chinese "qiyi" meaning bizarre. *Q. jurassica* exhibits adaptations to an aquatic habitat. Body elongate, 18–24 mm long. Head greatly reduced and partly retractile into thorax; antennae and eyes not visible; a pair of sclerotized tentorial rods. Sucker retractile, diameter about 2 mm, located ventrally on thoracic segment and consisting of a circular suction disc with central opening about one quarter of disc diameter; peripheral area of disc thin and flexible. Six robust, sclerotized ridges on sucker, radially arranged, covered by soft skin with small spines; distal part of each ridge thickened, probably with three processes embedded in musculature. Three pairs of small spines with simple shafts on dorsolateral margins of thorax, two pairs on dorsolateral margins of abdominal segments one to seven, and one pair on abdominal segment eight [112].

Family Eremochaetidae Ussatchov, 1968

Eremochaetidae is an endemic Mesozoic family, which has been found from the Late Jurassic to the Early Cretaceous. This family possesses a special combination of characters: the cross-vein r-m is absent, causing the vein R_{4+5} (sometimes veins R_{2+3} and R_{4+5}) to arise from cell d; the wings are weakly sclerotized; thus, the apex of the wings is always obscure and the ovipositor is needle-shaped. To date, two subfamilies, Eremochaetinae Ussatchov, 1968 (including six genera and seven species) and Eremomukhinae Mostovski, 1996 (including three genera and 10 species) have been described [113].

Genera included from the Cretaceous of Northern China: *Dissup* Evenhuis, 1994, *Alleremonomus* Ren & Guo, 1995, *Eremomukha* Mostovski, 1996 and *Lepteremochaetus* Ren, 1998.

Dissup Evenhuis, 1994

Dissup Evenhuis, 1994, *Catalogue of the fossil flies of the world (Insecta: Diptera)*, 316 [24] (original designation).

Type species: *Eremonomus irae* Kovalev, 1989.

Antennae invisible, arista slender. Sc short, ending at the middle of anterior margin of wing; cross-vein r-m absent, vein R_{4+5} arising from cell d; vein R_4 extremely short; cell sc wide open; cell br obviously longer and wider than cell bm [24, 113].

Distribution and age: Liaoning; Early Cretaceous.

Only one species included from the Cretaceous of Northern China (see Table 23.1).

Alleremonomus Ren & Guo, 1995

Alleremonomus Ren & Guo, 1995, *Entomologia Sin.*, 2 (4), 301 [114] (original designation).

Type species: *Alleremonomus xingi* Ren & Guo, 1995.

The specific epithet is dedicated to Dr. Xing for his outstanding contributions to Paleontology. Vein C ending before the tip of R_1; R_{2+3} originating from the base of Rs; R_{4+5} forked early; R_5 long, reaching the tip of wing; cell d located above the center of wing; cross-vein m-m obviously longer than M_3.

Distribution and age: Liaoning; Early Cretaceous.

Two species included from the Cretaceous of Northern China (see Table 23.1).

Eremomukha Mostovski, 1996

Eremomukha Mostovski, 1996, *Russ. Ent. J.*, 5, 118 [104] (original designation).

Type species: *Eremomukha (Eremomukha) tsokotukha* Mostovski, 1996.

Cell r_1 open. Fork R_{4+5} short and broad. Posterior cells 4 in number and vein M_4 (arising from the second basal cell) absent. Anal cell petiolate.

Distribution and age: Liaoning; Early Cretaceous.

Two species included from and Cretaceous of Northern China (see Table 23.1).

Lepteremochaetus Ren, 1998

Lepteremochaetus Ren, 1998, *Acta Zootaxonom. Sin.*, 23, 78 [115] (original designation).

Type species: *Lepteremochaetus lithoecius* Ren, 1998.

Vein C ending at wing tip; R_{2+3} short, originating from cell d basally, terminating on the R_1 far from its apex; with four branches. A_1 extending to wing margin.

Distribution and age: Liaoning; Early Cretaceous.

Two species included from the Cretaceous of Northern China (see Table 23.1).

Family Kovalevisargidae Mostovski, 1997

Kovalevisargidae is an extinct family in Archisargoidea [116]. The Kovalevisargidae, considered to be the sister taxon of Archisargidae, is a rather small family comprising only six species documented, respectively, to two genera [95]. They were distributed in China and Kazakhstan in the Middle to Late Jurassic [106, 117].

Genera included from the Jurassic of Northern China: *Kerosargus* Mostovski, 1997 and *Kovalevisargus* Mostovski, 1997.

Kerosargus Mostovski, 1997

Kerosargus Mostovski, 1997, *Paleontol. J.*, 31 (1), 76 [106] (original designation).

Type species: *Kerosargus argus* Mostovski, 1997.

Body short and thick. R with three branches. R_{2+3} short. Rs_1 shorter than Rs_2. Cross-vein r-m connecting Rs at the base of Rs fork. M_4 and CuA connected by cross-vein m-cu. Cell r_{2+3} widely open. Cell d elongate, longer than M_1 and M_2. Cell cu and m_3 open [106, 117].

Distribution and age: Inner Mongolia; Middle Jurassic.

Only one species included from the Jurassic of Northern China (see Table 23.1).

Kovalevisargus Mostovski, 1997

Kovalevisargus Mostovski, 1997, *Paleontol. J.*, 31 (1), 76 [106] (original designation).

Type species: *Kovalevisargus clarigenus* Mostovski, 1997.

Body slender in male, covered with dense setae. All the branches of R strongly costalized. R_{2+3} long. R_{4+5} close to R_{2+3}, ending in anterior margin of wing. Pt present. M furcated distad of origin of Rs. Cell d short. Cells m_3 and cu widely open [106, 117].

Distribution and age: Inner Mongolia; Middle Jurassic.

Three species included from the Jurassic of Northern China (see Table 23.1).

Family Nemestrinidae Macquart, 1834

Nemestrinidae, commonly called "tangle-veined flies," is a small group of brachycerous flies with about 300 extant and fossil species in about 30 genera. The adults are medium to large, stout-bodied flies, often characterized by a conspicuous dense pilosity; the wings usually longer than the body, venation complex with a compound diagonal vein; tibiae without apical spurs, and tarsi with an empodium [118, 119]. They have highly-developed flying abilities and are frequently observed on flowers. Larvae are endoparasitoids of either grasshoppers or scarab beetles. Fossil nemestrinids have been well-recorded since the Mesozoic and probably originated in the Late Triassic/Early Jurassic [120]. The most ancient and undoubted nemestrinid fly was reported from the Lower Jurassic of Germany. Recently, many Jurassic fossil nemestrinid flies have been reported in the Palearctic region. China is an important locality for fossil nemestrinid flies and two genera with three species have been described in China to date.

Genera included from the Jurassic and Cretaceous of Northern China: *Protonemestrius* Rohdendorf, 1968, *Florinemestrius* Ren, 1998 and *Ahirmoneura* Zhang, Yang & Ren, 2008.

Protonemestrius Rohdendorf, 1968

Protonemestrius Rohdendorf 1968, *Jurassic Insects of Karatau*, 182 [121] (original designation).

Type species: *Protonemestrius martynovi* Rohdendorf, 1968.

The specific epithet is dedicated to Martynov A.V. for his contribution to insect fossil research. *Protonemestrius* is similar to *Archinemestrius* by having no "phragma", having C and R_1 markedly thickened, and having Sc short, but different from having distinct cross-vein r-m, relatively long M_2 and the first basal cell longer than the second basal cell [121, 122].

Distribution and age: Liaoning; Early Cretaceous.

Two species included from the Cretaceous of Northern China (see Table 23.1).

Protonemestrius jurassicus Ren, 1998 (Figure 23.25)

Protonemestrius jurassicus Ren, 1998: *Acta Zootaxonom. Sin.*, 23 (1), 73.

Figure 23.25 *Protonemestrius jurassicus* Ren, 1998 (Holotype, LB97005).

Locality and horizon: Near Chaomidian, Beipiao, Liaoning, China; Lower Cretaceous, Yixian Formation.

Body (excluding proboscis) at least 13 mm long; proboscis (preserved part) 5 mm long; wing 10 mm long and 3 mm wide. Head large, hemispheric, as wide as thorax. Eyes bare, with distinct facets, dichoptic, covering most parts of head (presumably female). Antennae and ocelli unknown. Proboscis thin and long, about three times as long as head. Scutum divided into many parts by the sutures, its details obscure. Parts of three legs are preserved. The right fore leg almost complete, tibia longer than femur. Five tarsal segments distinct. Basitarsus longest, but shorter than the remainders combined. Two claws strong. Empodium pulvilli-form. All legs lacking bristles. Wing venation well-preserved. C ending at the tip of M_1. Sc long, slightly beyond the level of cross-vein r-m. R_{2+3} straight, originating from Rs slightly distad to fork of M. R_{4+5} forking deeply. M_1 terminating at wing tip. M_2 parallel to M_1, ending on margin below the apex of wing. Diagonal vein complete and reaching wing margin. CuA meeting 1A at wing margin. CuP present. Cross-vein m-cu connecting with M_4 distad to issuing point of M_4 from cell d. Abdomen with eight visible segments, the second one is longest and widest, tapering distally. Whole body lacking dense setae and bristles [115].

Protonemestrius jurassicus Ren, 1998 with long and thin proboscis implies that "flowers" or ovuliferous reproductive organs of gymnosperms with long tubular structure might have been present in the ecosystem. This species should have been pollinating insects highlighting the significant early interactions among flies and plants [123].

Florinemestrius Ren, 1998

Florinemestrius Ren, 1998, *Acta Zootaxonom. Sin.*, 23 (1), 74 [115] (original designation).

Type species: *Florinemestrius pulcherrimus* Ren, 1998.

Head almost as wide as thorax. Eyes bare. Proboscis elongate and stout. Sc distinctly beyond the level of fork of R_{4+5}. R_{2+3} originating from Rs distal to the level of fork of M. All branches of R and M terminating on anterior margin at apex. Diagonal vein complete, reaching posterior wing margin. A cross-vein present between M_1 and M_2.

Distribution and age: Liaoning; Early Cretaceous.

Only one species included from the Cretaceous of Northern China (see Table 23.1).

Florinemestrius pulcherrimus Ren, 1998 (Figure 23.26)

Florinemestrius pulcherrimus Ren, 1998: *Acta Zootaxonom. Sin.*, 23 (1), 74.

Locality and horizon: Near Chaomidian, Beipiao, Liaoning, China; Lower Cretaceous, Yixian Formation.

Figure 23.26 *Florinemestrius pulcherrimus* Ren, 1998 (Holotype, LB97007).

Body 26.0 mm long; proboscis (preserved part) 3.1 mm long; wing 18.5 mm long and 5.0 mm wide. The head obviously smaller than thorax. Eyes bare, occupying whole anterior part of head. Proboscis strong and straight, longer than head height. Antenna with three distinct segments, scape roundish oval, longer than pedicel, the third segment conicoval, with a long terminal style. Palpi well-developed and two-segmented. Thorax with a relatively convex scutum in lateral view. Legs slender. Both fore and mid femora almost as long as tibiae. Hind femur obviously shorter than hind tibia. Mid basitarsus somewhat shorter than the remainder combined, Hind basitarsus distinctly swollen, as long as the remainders combined. Claws strong. Empodium pulvilliform. Wing vein C ending at M_2 tip. Sc long, reaching in the C beyond R_{4+5} fork. R_{2+3} originating after M fork. R_{4+5} forking deeply. M_1 parallel to M_2, both terminating on anterior margin of wing apex. A cross-vein present between M_1 and M_2. CuA meeting 1A on wing margin. CuP absent. Abdomen consisting of eight visible pregenital segments, with dense setae, widest part lying in the third segment, tapering from segment four to apex, genital segment small and obscure. This holotype may be a female [115].

Entant tangle-veined flies visit flowers and are considered as important pollinators. *Florinemestrius pulcherrimus* Ren, 1998, with relatively short and stout proboscis, were suitably equipped for visiting and feeding on sugary fluids from open or short tubular "flowers" or ovuliferous reproductive organs of gymnosperms. This specimen, with well-preserved detailed structure of mouthparts, provides a valuable evidence in studying the feeding habit of nemestrinid flies and the co-evolution relationship among plants and pollinators [123].

Ahirmoneura Zhang, Yang & Ren, 2008

Ahirmoneura Zhang, Yang & Ren, 2008, *Acta Palaeontol. Pol.*, 53 (1), 162 [124] (original designation).

Type species: *Ahirmoneura neimengguensis* Zhang, Yang & Ren, 2008.

Base of vein C swollen and sclerotized with dense setulae. Vein Sc short, ending slightly beyond the middle of wing. Vein R_3 absent; vein R_4 cambered basally. Cross-vein r-m located at extreme base of cell d. Vein M_2 arising from apex of cell d; veins M_3 and CuA_1 fused before wing margin with a petiole. Vein CuA_1 arising from apex of cell d.

Distribution and age: Inner Mongolia; Middle Jurassic.

Only one species included from the Jurassic of Northern China (see Table 23.1).

Family Orientisargidae Zhang, 2012

Orientisargidae is a small extinct family under Archisargoidea, including only one genus and one species from the Middle Jurassic of Inner Mongolia, China. Orientisargidae have similar body structures and wing venation to those of the Archisargidae Rohdendorf, 1962. However, Orientisargidae can be differentiated by the simple R_{4+5}, the origin of Rs which is clearly proximal, the strongly flexed M_{3+4} stem and touching CuA instead of m-cu; R_{4+5} is simple and R furcated pectinately and the origin of Rs is proximal. Orientisargidae is also similar to the family Kovalevisargidae Mostovski, 1997. However, it is distinct from all kovalevisargids in having longer Sc, R_1 and R_{2+3}, the closed cell m_3 and the absence of m-cu. It is interesting that the long pedicel, the absence of arista or style on the tip of antenna and the absence of empodium may be the unique features of Orientisargidae, which are not present in Archisargidae nor in Kovalevisargidae [102].

Only one genus included from the Jurassic of Northern China: *Orientisargus* Zhang, 2012.

Orientisargus Zhang, 2012

Orientisargus Zhang, 2012, *ZooKeys*, 238, 60 [102] (original designation).

Type species: *Orientisargus illecebrosus* Zhang, 2012.

The first antennal flagellomere conical. R_{2+3} arched medially, ending in C before wing tip, and far from R_1 ending. R_{4+5} ending below wing tip. Rs stem and bR_{4+5} short. Rs fork shifted distally of M fork. Cross-vein r-m meeting R_{4+5} and M_{1+2}, near cell d base. Four medial veins present. Cells d and m_3 narrow and long, the latter cell with long petiole. Section bM_{3+4} shorter than r-m. CuA and CuP subparallel and cu cell wide open.

Distribution and age: Inner Mongolia; Middle Jurassic.

Only one species included from the Jurassic of Northern China (see Table 23.1).

Family Origoasilidae Zhang, Yang & Ren, 2011

Origoasilidae, with only one genus and one species, has been reported based on specimens collected from the Yixian Formation in Yangshuling Township, Pingquan County, northern Hebei Province. Because of the typical brachycerous venation and antennae, Origoasilidae is a family in Orthorrhapha [125].

Only one genus included from the Cretaceous of Northern China: *Origoasilus* Zhang, Yang & Ren, 2011.

Origoasilus Zhang, Yang & Ren, 2011

Origoasilus Zhang, Yang & Ren, 2011, *Acta Geol. Sin. – Engl.*, 85 (5), 995 [125] (original designation).

Type species: *Origoasilus pingquanensis* Zhang, Yang & Ren, 2011.

Head round. Wing much slender; Abdomen with nine segments visible. Veins Sc, R_1 and R_{2+3} long and straight, almost parallel. R_{4+5} fork deep and narrow. R_4 and R_5 long, much longer than Rs_3. R_4 nearly as long as M_1; Rs_1 long, longer than Rs_2 and Rs_3; br longer than bm; cross-vein r-m located at apex 1/4 of d cell; M_4 arising from the intersection of d and bm cell. Cell m_3 closed with stipe.

Distribution and age: Hebei; Early Cretaceous.

Only one species included from the Cretaceous of Northern China (see Table 23.1).

Origoasilus pingquanensis Zhang, Yang & Ren, 2011 (Figure 23.27)

Origoasilus pingquanensis Zhang, Yang & Ren, 2011: *Acta Geol. Sin.-Engl.*, 85 (5), 995.

Locality and horizon: Yangshuling, Pingquan, Chengde, Hebei, China; Lower Cretaceous, Yixian Formation.

Body large, stout and dark. Head slightly narrower than thorax in dorsal view. Eyes large, without setae; ommatidia visible clearly. Antennal flagellum distinctly elongated, with a tapering stylus. Thorax black, robust. Legs slender; hind legs partly visible with dense setae, basitarsus much longer and wider than succeeding segments. Wings rather narrow. Sc extremely elongated, ending far beyond the middle of wing, distal of cross-vein r-m. Vein R_1 long and straight, distal of origin vein R_4. Vein Rs_1 long, arising from basal 1/3 of wing; vein Rs_2 and Rs_3 subequal in length, shorter than vein Rs_1. Vein R_4 somewhat sinuate. Cross-vein r-m located at apical 1/4 of cell d. Vein M_1 subequal to vein R_5 in length; veins M_1 and M_2 converged at apex of cell d; veins M_3 and CuA_1 fused before wing margin with a petiole; base of vein CuA_1 curved upward, cross-vein m-cu absent. Mouth of cell sc narrowly open. Cell br slender, longer but distinctly narrower than cell bm. Cell d about twice as long as cell m_3. Mouth of cell m_1 widely open, longer than that of cell m_2; cell m_3 small and closed, triangular. Anal lobe well-developed. Alula broad. Halters developed, brown. Abdomen robust, curved with dense setulae, and nine segments present. Segment IX much longer than segment VIII, longitudinal fissure becoming two semilunar portions. Cerci visible [125].

Family Protapioceridae Ren, 1998

Protapioceridae, established by Ren in 1998 [115], belongs to the superfamily Asiloidea. They are a very rare but distinct family known only from the Early Cretaceous of China. Up to date, two genera of *Protapiocera* Ren, 1998 and *Pseudapiocera* Zhang, 2015 are known from the Yixian and Laiyang Formations in Liaoning and Shandong Provinces of China. They are characterized by the Rs originating from R_1 at midwing; R_5 and four branches of M are obsolete apically; all posterior cells open; 1A strongly curved before wing margin, almost at a right angle; spurs and bristles on the body and legs absent and hind basitarsi not extremely longer than other tarsal segments. The family is considered probably closely related to extant Apioceridae and Mydidae. So, it seems that Protapioceridae was a Mesozoic ancestor of the sister groups of Apioceridae and Mydidae [126].

Genera included from the Cretaceous of Northern China: *Protapiocera* Ren, 1998 and *Pseudapiocera* Zhang, 2015.

Protapiocera Ren, 1998

Protapiocera Ren, 1998, *Acta Zootaxonom. Sin.*, 23 (1), 76 [115] (original designation).

Type species: *Protapiocera megista* Ren, 1998.

Body robust with dense setulae. Eyes large, holoptic in male. Wing slender, with distinct venation on basal part, but absent or obscure on apical part of wing in natural state (not caused by preservation). Veins R_{2+3} and R_4 contiguous, both ending on vein R_1 apically. Vein M_3 present, vein CuA_1 arising from cell bm. Vein CuP present; cell cup closed, apical part truncate.

Distribution and age: Liaoning; Early Cretaceous.

Three species included from the Cretaceous of Northern China (see Table 23.1).

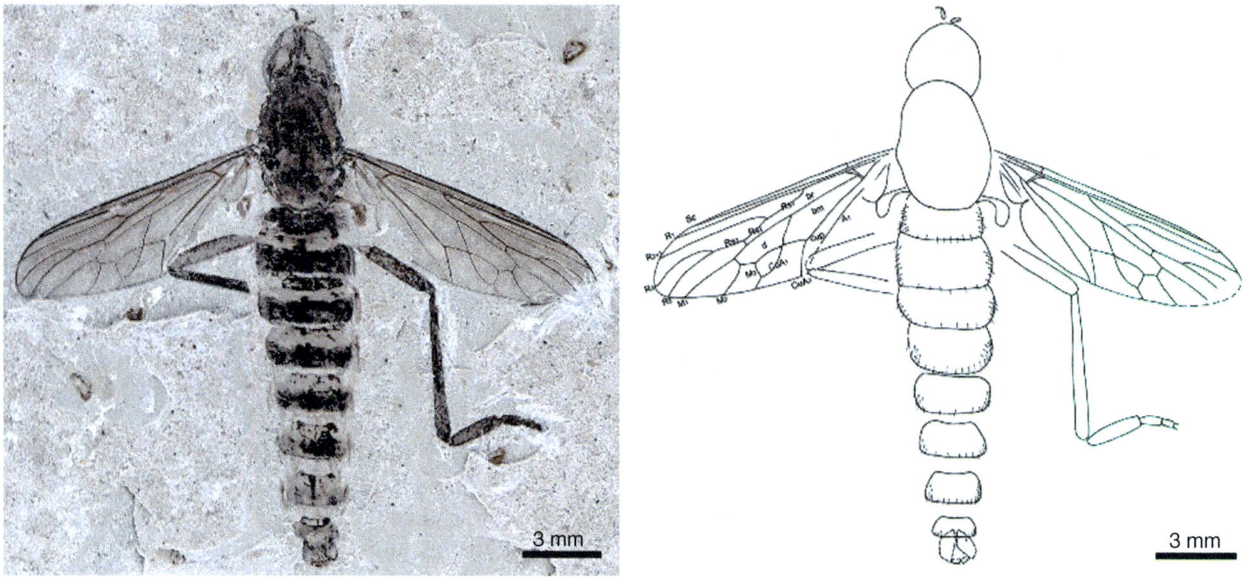

(a) (b)

Figure 23.27 *Origoasilus pingquanensis* Zhang, Yang & Ren, 2011 (Holotype, CNU-DIB-HC2008001). (a) Photograph; (b) Line drawing.

Pseudapiocera Zhang, 2015

Pseudapiocera Zhang, 2015, *Alcheringa*, 39 (4), 460 [127] (original designation).

Type species: *Pseudapiocera shandongensis* Zhang, 2015.

Antennal arista absent; scutum relatively small; scutellum very large, subtrapezoid; R_{2+3} joining R_4 before R_1, cell r_3 transversely oblong, $R_{2+3} + R_4$ joining R_1 before wing margin; origin of Rs slightly distad to level of M fork; M with three branches; petiole of CuA + CuP absent. Abdomen gradually swollen apically.

Distribution and age: Shandong; Early Cretaceous.

Only one species included from the Cretaceous of Northern China (see Table 23.1).

Family Protempididae Ussatchov, 1968

The family Protempididae was originally proposed by Ussatchev, in 1968, based on a single specimen of *Protempis antennata* Ussatchev, 1968 [128]. The wing has characters of basal Empididae, but also is reminiscent of such orthorrhaphous fly genera as *Apystomyia* Melander, 1950 and *Hilarimorpha* Schiner, 1860. Up to date, there are two extinct genera with four species described from China and Kazakhstan.

Genera included from the Cretaceous of Northern China: *Protempis* Ussatchov, 1968 and *Helempis* Ren, 1998.

Protempis Ussatchov, 1968

Protempis Ussatchev, 1968, *Ent. Obozr.*, 47, 623 [128] (original designation).

Type species: *Protempis antennata* Ussatchov, 1968.

R_1 ending on C slightly beyond middle of the wing. Pterostigma extending between Sc and R_1 distally. R_{4+5} forking distally. M with three long branches, issuing from cell d and ending at wing margin. CuA meeting 1A far from wing margin. CuP and 2A absent [122, 128].

Distribution and age: Liaoning; Early Cretaceous.

Only one species included from the Cretaceous of Northern China (see Table 23.1).

Helempis Ren, 1998

Helempis Ren, 1998, *Acta Zootaxonom. Sin.*, 23 (1), 80 [115] (original designation).

Type species: *Helempis yixianensis* Ren, 1998.

Sc ending on the C far beyond midwing. R_1 ending on the C near wing tip. R_{4+5} forking distally. M with three branches. Rs originating from R_1 at midwing, near the level of fork of M. Cross-vein r-m meeting Rs near the originating point of R_{2+3}. CuA meeting 1A before the wing margin. CuP distinctly present. The second abdominal segment obviously longer. Femora distinctly swollen.

Distribution and age: Liaoning; Early Cretaceous.

Two species included from the Cretaceous of Northern China (see Table 23.1).

Family Protobrachyceridae Rohdendorf, 1962

Protobrachyceridae is an archaic representative of the Brachycera with only one known genus and two species, *Protobrachyceron liasinum* Handlirsch, 1920 and *Protobrachyceron zessini* Krzemiński & Ansorge, 2000, described from the Early Jurassic (lower Toarcian) of Germany. The family can be distinguished by the following characters: vein Sc ending at midwing, both vein R_4 and cell d distinctly shorter than vein M_1 in length, vein CuP well visible, mouth of cell r_1 narrow, five posterior cells widely open, cells m_3 and cup narrowly open. Originally, the genus *Protobrachyceron* Handlirsch, 1920 was treated as a member of Xylophagidae [58]. Then, it was upgraded as a separate family Protobrachyceridae [30, 122]. Zhang *et al.* reported *Protobrachyceron sinensis* Zhang, Yang & Ren, 2008, from the Middle Jurassic Daohugou, Inner Mongolia, with good preservation of both wings and most of the body, which is the first record from China [129].

Only one genus included from the Jurassic of Northern China: *Protobrachyceron* Handlirsch 1920.

Protobrachyceron Handlirsch, 1920

Protobrachyceron Handlirsch, 1920, *Handbuch der Entomologie,* 205 [58] (original designation).

Type species: *Protobrachyceron liasinum* Handlirsch, 1920.

Body small. Vein Sc ending at midwing, slightly longer than cell br (basal radial); veins R_1 and R_{2+3} close to each other at wing margin; both vein R_4 and cell d distinctly shorter than vein M_1 in length; cross-vein r-m at basal 1/5–1/4 of cell d; five posterior cells widely open, mouth of cells m_3 and cup narrow [58, 129].

Distribution and age: Inner Mongolia; Middle Jurassic.

Only one species included from the Jurassic of Northern China (see Table 23.1).

Family Rhagionemestriidae Ussatchov, 1968

Rhagionemestriidae is a small extinct family belonging to the superfamily Nemestrinoidea. They are characterized by costal vein circumambient, becoming thinner beyond R_4 or wing apex, R rather long, medial veins of wing represented by either convex R_5 and fold running along R_S, R_{4+5} and M_1 beyond r-m (in this case R_{2+3} curved and terminate near R) or diagonal vein (R_{2+3} straight, otherwise tips of R), anal cell open or closed [130]. The family includes two subfamilies: Rhagionemestriinae and Heterostominae [108]. Up to now, there are five genera with eight species, documented from China, England, Kazakhstan, Mongolia and Spain.

Only one genus included from the Jurassic of Northern China: *Jurassinemestrinus* Zhang, 2010.

Jurassinemestrinus Zhang, 2010

Jurassinemestrinus Zhang, 2010, *Paleontology,* 52 (2), 313 [108] (original designation).

Type species: *Jurassinemestrinus orientalis* Zhang, 2010.

Antenna small, with first flagellomere cone-like, arista well-developed. Vein Rs fork basad to M fork; Rs_1 much shorter than Rs_2; R_{4+5} coalesced with M_{1+2} for a considerable distance. R_{4+5} fork shallow, nearly at the same level with the ending of R_1. Diagonal vein formed of Rs, R_{4+5}, $R_{4+5}+M_{1+2}$, M_{1+2} beyond fusion and M_2. Cell br longer and narrower than cell bm. Cell d elongate. Cell cup closed.

Distribution and age: Inner Mongolia; Middle Jurassic.

Only one species included from the Jurassic of Northern China (see Table 23.1).

Family Rhagionempididae Rohdendorf, 1938

Rhagionempididae are an extinct family from the Middle to Late Jurassic. Rhagionempididae can be distinguished with other families by antennal flagellum rounded triangular, with a style pointed at apex and no longer than preceding segment; mesonotum with paired longitudinal rows of setae, palpus two-segmented, tibial spurs absent, anal lobe of wing well-developed and alula well-developed [122]. Hitherto, five extinct genera with eight species from Russia and China are attributed to Rhagionempididae. The earliest record of this family is *Rhagionempis tabanicornis* Rohdendorf, 1938 [28].

Only one genus included from the Jurassic of Northern China: *Ussatchovia* Kovalev, 1982.

Ussatchovia Kovalev, 1982

Ussatchovia Kovalev, 1982, *Paleontol. J.,* 16 (3), 91 [47] (original designation).

Type species: *Ussatchovia jurassica* Kovalev, 1982.

Small, robust snipe flies. Antenna with long arista. Costal section R_1-R_{2+3} nearly as long as section Sc-R_1. Pterostigma present. Fork M_{1+2} basad to level of R_1 ending. Cross-vein m-m slightly arched downward. Cell m_1 longer than discal cell. Gonostyli of malegenitalia stout and large [47, 131].

Distribution and age: Inner Mongolia; Middle Jurassic.

Two species included from the Jurassic of Northern China (see Table 23.1).

Family Rhagionidae Latreille, 1802

Rhagionidae, commonly called "snipe flies", are one of the most ancient lower brachycera family of Tabanomorpha. They are medium or large flies with slender bodies and stilt-like legs, typically brown and yellow and lack bristles. Their adults have mouthparts adapted for piercing, and many species are bloodsuckers, while others prey on insects. Their larvae are also predatory and most are terrestrial, although some are aquatic. The family comprise four subfamilies: Arthrocerinae, Chrysopilinae, Rhagioninae and Spaniinae [132]. Up to now, 26 extant genera and more than 750 extant species have been described throughout the world. Fossil Rhagionidae have 35 genera and above 80 species. The oldest record of this family is the Early-Middle Triassic *Gallia alsatica* Krzemiński & Krzeminska, 2003 [133]. Recently, a large number of Mesozoic snipe flies have been described worldwide [134]. To date, many rhagionids have been described from the Middle Jurassic Jiulongshan Formation in Daohugou, Inner Mongolia, China. These fossils become a significant part of Rhagionidae for research and studies on the evolution of Diptera insects.

Genera included from the Jurassic and Cretaceous of Northern China: *Protorhagio* Rohdendorf, 1938, *Palaeobolbomyia* Kovalev, 1982, *Scelorhagio* Zhang, Zhang & Li, 1993, *Basilorhagio* Ren, 1995, *Palaeoarthroteles* Kovalev & Mostovski, 1997, *Oiobrachyceron* Ren, 1998, *Orsobrachyceron* Ren, 1998, *Sinorhagio* Zhang, Yang & Ren, 2006, *Achrysopilus* Zhang, Yang & Ren, 2008, *Lithorhagio* Zhang & Li, 2012, *Daohugorhagio* Zhang, 2013, *Parachrysopilus* Zhang, 2013 and *Trichorhagio* Zhang, 2013.

Protorhagio Rohdendorf, 1938

Protorhagio Rohdendorf, 1938, *Trudy Paleontol. Inst.*, 7 (3), 37 [28] (original designation).

Type species: *Protorhagio capitatus* Rohdendorf, 1938.

The first flagellomere rather large, other flagellomeres forming a kind of thickened arista, with distinct segmentation; flagellomeres longer than wide. Mesopleura setose. Mesonotal setae not in rows. Scutellum setose. R_1 with setae. R_{2+3} apically subparallel to R_1 or slightly curved, or cell r_1 markedly narrowing apically. R_5 ending below wing tip. Cross-vein r-m before or at midlength of discoidal cell (d). Four medial veins. Distal section of M_3 evenly arched anteriorly or almost straight, as long as, or shorter than, basal section; cell m_3 narrowing apically. Anal cell open. Pterostigma usually present. Tibial spur formula 0 : 2 : 2, spurs well-developed. Tarsomeres not dilated. Three spherical sclerotized spermathecae. Cerci double-segmented, basal segment broad with posteroventral projection, apical segment rounded elliptical [28, 135].

Distribution and age: Inner Mongolia, Middle Jurassic.

Only one species included from the Jurassic of Northern China (see Table 23.1).

Palaeobolbomyia Kovalev, 1982

Palaeobolbomyia Kovalev, 1982, *Paleontol. J.*, 16 (3), 94 [47] (original designation).

Type species: *Palaeobolbomyia sibirica* Kovalev, 1982.

Antennal flagellum five- to six-segmented; Costal section R_1–R_{2+3} longer than Sc–R_1. Vein R_{2+3} S-shaped, strongly curved under point of R_1 insertion. Fork R_{4+5} rather long. R_5 ends at wing tip. M_3 absent. Basal section of M_4 usually absent. Anal cell closed. Pterostigma present. Alula developed. Hind coxa bears blunt tubercle anteriorly. Hind tibia bears two short ventral spurs. Three (?) spherical spermathecae. Ovipositor short, weakly extended. Cerci 2-segmented, basal segment lacking large ventral lobe, distal segment short oval [47, 136].

Distribution and age: Inner Mongolia; Middle Jurassic.

Only one species included from the Jurassic of Northern China (see Table 23.1).

Scelorhagio Zhang, Zhang & Li, 1993

Scelorhagio Zhang, Zhang & Li, 1993, *Acta Palaeontol. Sin.*, 32 (6), 664 [137] (original designation).

Type species: *Scelorhagio mecomastigus* Zhang, Zhang & Li, 1993.

Small in size. The scape and pedicel wide flat, aristae elongated, segmentation not obvious. Wings long and narrow, leading veins with R_1 base merging, terminating in central part of wing at the anterior margin; Rs extremely short, R_{2+3} ending obviously curved; R_4 longer than M_1. The five posterior chambers are wide open and the breech chambers closed.

Distribution and age: Shandong; Early Cretaceous.

Only one species included from the Cretaceous of Northern China (see Table 23.1).

Basilorhagio Ren, 1995

Basilorhagio Ren, 1995, *Faunae and Stratigraphy of Jurassic-Cretaceous in Beijing and the Adjacent Areas*, 106 [40] (original designation).

Type species: *Basilorhagio venustus* Ren, 1995.

Small body, head large. Mesonotum obvious uplift. Coxae thick long, more than half the length of the femur. Lateral veins short, shorter than middle of wing; R_1 base thickened, R_1 and R_{2+3} long and straight; R_4 apparently short; disk near wing base; M_1 and M_2 base obviously converging.

Distribution and age: Hebei; Early Cretaceous.

Only one species included from the Cretaceous of Northern China (see Table 23.1).

Palaeoarthroteles Kovalev & Mostovski, 1997

Palaeoarthroteles Kovalev & Mostovski, 1997, *Paleontol. Zh.*, 1997 (5), 523–527 [138] (original designation).

Type species: *Palaeoarthroteles mesozoicus* Kovalev & Mostovski, 1997.

Face not produced conically forwards. Genae clearly visible in profile. Subsegments of third antennal segment (= flagellum) homonomous, not forming a type of arista beginning from the second subsegment. Proboscis long, labium fleshy, labella broad. Palps long, two-segmented, with distal segment conical. Mesopleura with setae. Laterotergites bare. Tibial spurs 0 : 2 : 2, well-developed. Hind basitarsi not widened. R_1 with setae. Cross-vein rm. before midlength of discal cell. Four medial veins. Distal section of M_3 nearly straight or weakly S-shaped, not longer than basal one; the latter curved so that discal cell is slightly broadened distally. Anal cell open. Pterostigma present. Cerci two-segmented, basal segment with a lobe directed ventrally.

Distribution and age: Inner Mongolia; Middle Jurassic.

Two species included from the Jurassic of Northern China (see Table 23.1).

Oiobrachyceron Ren, 1998

Oiobrachyceron Ren, 1998, *Acta Zootaxonom. Sin.*, 23 (1), 71 [115] (original designation).

Type species: *Oiobrachyceron limnogenus* Ren, 1998.

Head small. Eyes bare. Antenna stout, with seven flagellomeres. Proboscis stout, shorter than head height. Tibial spur absent. Sc terminating in the C at midwing. R_{4+5} forking at same level of fork of M_{1+2}. Both R_4 and R_5 curved. M with four branches. Cell m_3 open. CuA meeting 1A before its apex.

Distribution and age: Liaoning; Early Cretaceous.

Only one species included from the Cretaceous of Northern China (see Table 23.1).

Orsobrachyceron Ren, 1998

Orsobrachyceron Ren, 1998, *Acta Zootaxonom. Sin.*, 23 (1), 69 [115] (original designation).

Type species: *Orsobrachyceron chinensis* Ren, 1998.

Head small in size. Eyes bare. Antennae slender, the third segment annulated, with seven flagellomeres. Proboscis stout, almost as long as head height. Fore tibial spur absent. Sc terminating beyond the midwing slightly. R_{4+5} forking distad fork of M_{1+2}. R_4 and R_5 almost parallel. M with four branches. Cell m_3 closed at margin. CuA meeting 1A near its apex.

Distribution and age: Liaoning; Early Cretaceous.

Only one species included from the Cretaceous of Northern China (see Table 23.1).

Sinorhagio Zhang, Yang & Ren, 2006

Sinorhagio Zhang, Yang & Ren, 2006, *Zootaxa*, 1134, 53 [139] (original designation).

Type species: *Sinorhagio daohugouensis* Zhang, Yang & Ren, 2006.

Flagellum long, nearly spine-like, tapering toward tip. Mouths of cell sc, cell r_1 and cell r_{2+3} subequal in width; mouth of cell r_4 much narrower than that of cell r_{2+3}. Vein R_{2+3} long, straight apically. Rs nearly as long as trunk of $R_4 + R_5$. Veins M_1 and M_2 convergent basally at cell d; vein CuA_1 present and arising from cell bm. Cell cup closed with a short petiole apically.

Distribution and age: Inner Mongolia; Middle Jurassic.

Two species included from the Jurassic of Northern China (see Table 23.1).

Achrysopilus Zhang, Yang & Ren, 2008

Achrysopilus Zhang, Yang & Ren, 2008, *Biologia*, 63 (1), 114 [140] (original designation).

Type species: *Achrysopilus neimenguensis* Zhang, Yang & Ren, 2008.

Body with dense setluae. Pterostigma distinct. Sc ending at wing midlength. R_5 longer than distance between base of R_4 and cross-vein r-m. Cross-vein r-m at basal 1/4 of cell d. M_1 and M_2 bifurcating distad of cross-vein m-m. CuA_1 arising from base of cell d. CuA_2 and A_1 converged apically with a short petiole. Anal lobe distinct. Five posterior cells open wide. Abdomen robust, with nine segments.

Distribution and age: Inner Mongolia; Middle Jurassic.

Only one species included from the Jurassic of Northern China (see Table 23.1).

Achrysopilus neimenguensis Zhang, Yang & Ren, 2008 (Figure 23.28)

Achrysopilus neimenguensis Zhang, Yang & Ren, 2008: *Biologia*, 63 (1), 114–115.

Locality and horizon: Daohugou, Ningcheng, Inner Mongolia, China; Middle Jurassic, Jiulongshan Formation.

Figure 23.28 *Achrysopilus neimenguensis* Zhang, Yang & Ren, 2008 (Holotype, CNU-DB-NN2006–001).

Male body length 7.0 mm, wing length 5.5 mm and width 2.0 mm. Head with dense long setulae. Eyes large and round in lateral view. Visible part of proboscis fleshy, short and wide. Thorax black with long setulae. Mesonotum slight convex. Five legs partially preserved; femora slender, not swollen; the first tarsomere distinctly longer than following tarsomeres. Apical spurs not seen on tibiae. Wing membrane infuscated. Vein Sc ending at midwing. Pterostigma distinct, crossing apical half of vein R_1; all branches of vein R nearly straight, basal half of vein R_1 distinctly thickened; vein Rs short; vein R_5 longer than distance between base of vein R_4 and cross-vein r-m, slightly shorter than trunk of veins $R_4 + R_5$. Cross-vein r-m located at basal 1/4 of cell d. Veins M_1 and M_2 bifurcating distad of cross-vein m-m, fork of veins $M_1 + M_2$ slightly proximal of that of veins $R_4 + R_5$; vein CuA_1 arising from base of cell d. Veins CuA_2 and A_1 converged apically with a short petiole; vein CuP present, vein A_2 short and thin. Anal lobe large. Mouth of cell r_4 slight narrower than that of cell r_{2+3}. Five posterior cells open wide, mouth of cell m_2 about two times as wide as that of cell m_3. Cell br much narrower than cell bm. Cell d pentagonal. Cell cup close. Abdomen robust, covered with setulae, rounded apically. Nine segments visible, segment II widest, then narrowing toward tip, but boundaries of segments I and II, VIII and IV obscure in the specimen [140].

Lithorhagio Zhang & Li, 2012

Lithorhagio Zhang & Li, 2012, *Paleontol. J.*, 46 (2), 160 [135] (original designation).

Type species: *Lithorhagio megalocephalus* Zhang & Li, 2012.

Mouthparts forming proboscis for sucking. Antenna conical, flagellum with eight annular flagellomeres. Head, thorax, and legs lacking setae. Fork of R_{4+5} moderately long, widened distally. R_4 sinuate. Cross-vein r–m far before discoidal cell midlength. Cell r_5 narrowed apically. Cell br shorter than bm. Cell cu closed, with pointed stalk. Pterostigma blurry.

Distribution and age: Inner Mongolia; Middle Jurassic.

Only one species included from the Jurassic of Northern China (see Table 23.1).

Daohugorhagio Zhang, 2013

Daohugorhagio Zhang, 2013, *Palaeontology*, 56 (1), 221 [134] (original designation).

Type species: *Daohugorhagio elongatus* Zhang, 2013.

Antennal flagellum cornute with eight antenomeres. Costal vein encircling wing. R_1 upcurved near the terminus. Costal section of R_1–R_{2+3} very short, and markedly shorter than that of Sc–R_1 or R_{2+3}–R_4. R_{2+3} straight. Fork of R_{4+5} moderately narrow and long. Section of dM_3 as long as section of bM_3. Basal section of M_4 (bM_4) present. Cell cu closed and with a short petiole apically. Pterostigma present.

Distribution and age: Inner Mongolia; Middle Jurassic.

Only one species included from the Jurassic of Northern China (see Table 23.1).

Parachrysopilus Zhang, 2013

Parachrysopilus Zhang, 2013, *Palaeontology*, 56 (1), 225 [134] (original designation).

Type species: *Parachrysopilus jurassicus* Zhang, 2013.

Antenna rather short, antennal flagellum cornute, containing five or six heteronomous segments. Sc short, less than one half of wing length. Fork of R_{4+5} rather long (at level of end of R_1). M_{1+2} furcated basad to terminus of discal cell. Section of dM_3 longer than section of bM_3. Cell cu closed and with a short petiole apically. Alula not developed. Tibial spurs of hind legs equivocal.

Distribution and age: Inner Mongolia; Middle Jurassic.

Only one species included from the Jurassic of Northern China (see Table 23.1).

Trichorhagio Zhang, 2013

Trichorhagio Zhang, 2013, *Palaeontology*, 56 (1), 218 [134] (original designation).

Type species: *Trichorhagio gregarius* Zhang, 2013.

Body pubescent. Antennal flagellum cornute, eight- (in female) or nine-segmented (in male). R_1 straight. Costal section of R_1-R_{2+3} much shorter than that of Sc-R_1 or R_{2+3}-R_4. R_{2+3} downcurved at its midlength. Fork of R_{4+5} rather narrow and long. Section of dM_3 shorter than section of bM_3. Cell cu closed with a short petiole apically. Pterostigma present. Tibial spurs of hind leg well-developed. Empodium broad-ovoid.

Distribution and age: Inner Mongolia; Middle Jurassic.

Only one species included from the Jurassic of Northern China (see Table 23.1).

Family Tabanidae Latreille, 1802

Tabanidae, commonly known as "horseflies", belong to lower Brachycera of Tanbanomorpha and comprise three subfamilies: Chrysopsinae, Pangoniinae and Tabaninae. They are medium to large flies, without bristle, flagellum with annuli, empodia pulvilliform and calypter large. There are approximately 4500 living species known world-wide except at extreme northern and southern latitudes, over 1000 species of which have been assigned into the genus *Tabanus* Linnaeus, 1758. They are important pollinators of flowers but most of the females are also bloodsuckers and can attack reptiles, birds and mammals [111]. The oldest fossil horseflies are described as *Palaepangonius eupterus* Ren, 1998 and *Eopangonius pletud* Ren, 1998 from the Early Cretaceous in Beipiao of Liaoning. These fossils, with a complete body and a well-developed long proboscis, are closely associated with the pre-Cretaceous origin of angiosperms [123]. Adult tabanid flies are terrestrial and lay their eggs onto marsh plants near bodies of fresh water because the larvae develop in water or mud [141].

Genera included from the Cretaceous of Northern China: *Allomyia* Ren, 1998, *Eopangonius* Ren, 1998, *Palaepangonius* Ren, 1998, *Pauromyia* Ren, 1998 and *Laiyangitabanus* Zhang, 2012.

Allomyia Ren, 1998

Allomyia Ren, 1998, *Acta Zootaxonom. Sin.*, 23 (1), 68 [115] (original designation).

Type species: *Allomyia ruderalis* Ren, 1998.

Medium-sized flies, stout. Eyes bare. Ocelli absent. Sc ending in the C at midwing. R_1 obviously longer than Sc. The basal section of Rs very short. R_4 without an appendix. Cross-vein r-m at base of discal cell. Cell M_3 narrowed at margin.

Distribution and age: Liaoning; Early Cretaceous.

Only one species included from the Cretaceous of Northern China (see Table 23.1).

Eopangonius Ren, 1998

Eopangonius Ren, 1998, *Acta Zootaxonom. Sin.*, 23 (1), 68 [115] (original designation).

Type species: *Eopangonius pletus* Ren, 1998.

Proboscis well-developed, its trace slightly longer than head height. Eyes bare. Ocelli present, arranged in triangle. Rs originating basally from straight R_1. Cell sc not narrowed at margin. R_{4+5} forking after the level of fork of M_{1+2}. Cell r_4 long. R_4 without an appendix. Cell r_5 open. Abdomen distinctly stout.

Distribution and age: Liaoning; Early Cretaceous.

Only one species included from the Cretaceous of Northern China (see Table 23.1).

Palaepangonius Ren, 1998

Palaepangonius Ren, 1998, *Acta Zootaxonom. Sin.*, 23 (1), 66 [115] (original designation).

Type species: *Palaepangonius eupterus* Ren, 1998.

Proboscis well-developed, longer than head height. Eyes bare. Hind tibial spurs present. R_1 near Sc terminately, thus cell Sc distinctly narrowed at margin. Rs originating from R_1 basally. R_{4+5} forking before level of fork of M_{1+2}. R_4 and R_5 very long, divergent, with a more elongate cell R_4. R_4 with an obvious appendix. Cell r_5 open.

Distribution and age: Liaoning; Early Cretaceous.

Only one species included from the Cretaceous of Northern China (see Table 23.1).

Palaepangonius eupterus Ren, 1998 (Figure 23.29)

Palaepangonius eupterus Ren, 1998: *Acta Zootaxonom. Sin.*, 23 (1), 66.

Locality and horizon: Near Chaomidian, Beipiao, Liaoning, China; Lower Cretaceous, Yixian Formation.

Head subtriangular. Eyes bare, distinctly dichoptic, occupying most part of head (indicating this specimen is a female). Proboscis very long, preserved part

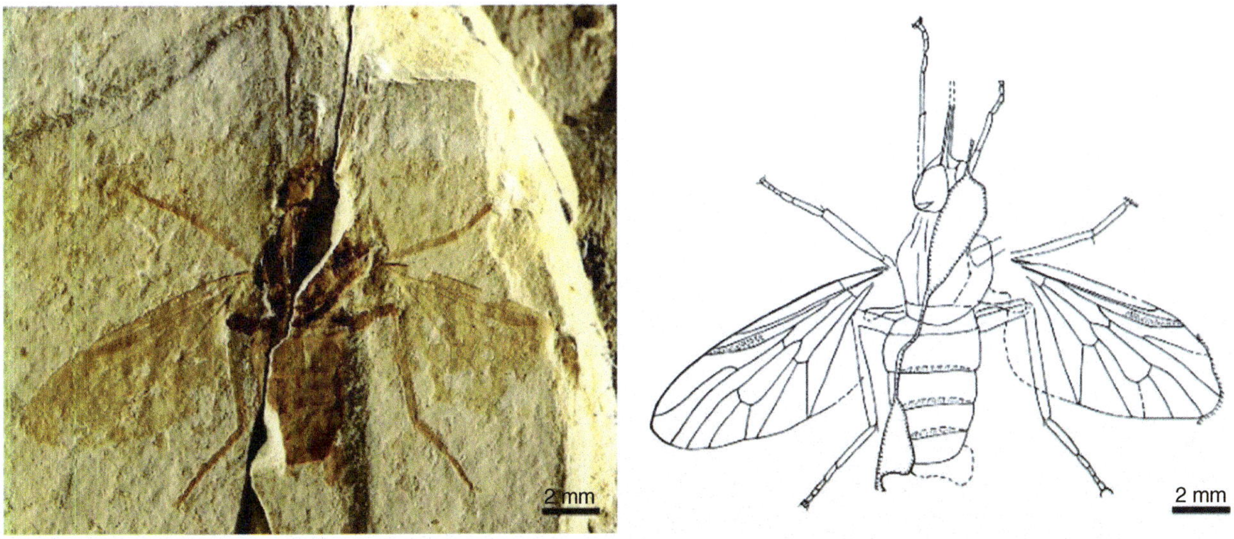

(a) (b)

Figure 23.29 *Palaepangonius eupterus* Ren, 1998 (Holotype, LB97017). (a) Photograph of habitus, (b) Line drawing.

about 1.5 times as long as head height. Antennae and ocelli unknown. Preservation of thorax not sufficiently detailed to describe. Legs clothed with tiny and dense setae, with well-developed mid and hind tibial spurs. Basitarsus nearly as long as remainders combined. Both pulvilli and empodia well-preserved. Empodia obviously pulvilliform. Wings lipochromous, with a distinct pterostigma. The vein C stout, ending at wing tip. Sc terminating at midwing. R_1 short, just beyond the Sc slightly in length. Rs originating from R_1 basally. R_{2+3} obviously curved upward terminally. R_4 with an appendix. Both R_4 and R_5 longer than M_1 and M_2. R_{4+5} forking before level of fork of M_{1+2}. Cross-vein r-m at base of discal cell (d). Cell d giving three veins (M_1, M_2 and M_3), the last one close to M_4 distally. Two halteres well-preserved. Abdomen with five visible segments. Terminalia unknown [115].

Pauromyia Ren, 1998

Pauromyia Ren, 1998, *Acta Zootaxonom. Sin.*, 23 (1), 72 [115] (original designation).

Type species: *Pauromyia oresbia* Ren, 1998.

Medium size. Mesonotum obvious uplift. Slender feet. Wing slender, subterminal veins terminating in central wing later. R_1 long, straight, R_{2+3} ends slightly curved; R_5 shorter than R_4 and median transverse pitch; median transverse vein at posterior disk center; M_1 and M_2 basal joining, opening of anal chamber wide.

Distribution and age: Liaoning; Early Cretaceous.

Only one species included from the Cretaceous of Northern China (see Table 23.1).

Laiyangitabanus Zhang, 2012

Laiyangitabanus Zhang, 2012, *Cretac. Res.*, 36, 2 [111] (original designation).

Type species: *Laiyangitabanus formosus* Zhang, 2012.

Eyes dichoptic. Ocelli absent. Wing somewhat narrowed and attenuated apically. C base thickened. Fork R_{4+5} broad, R_4 attached to R_5 almost at right angle. Cell br relatively long, and longer than cell bm. Discal cell short and wide. Pterostigma well-developed beneath R_1. Tibial spurs of hind leg short. Cerci one-segmented, short, oviform.

Distribution and age: Shandong; Early Cretaceous.

Only one species included from the Cretaceous of Northern China (see Table 23.1).

Family Uranorhagionidae Zhang, Yang & Ren, 2010

Uranorhagionidae, comprising two genera and four species, are reported in the Middle Jurassic of Daohugou Village, Inner Mongolia. Uranorhagionidae have a mixture of distinct characters of two families in two disparate superfamilies, Rhagionemestriidae (Nemestrinoidea) and Rhagionidae (Tabanoidea). Thus, this family might be in an inclusive position in dipteran phylogeny. We can differentiate Uranorhagionidae from Rhagionidae by the following characters: veins Sc and R_1 long; vein R_{2+3} fornical at base; vein CuA_2 short; cell d extending to near the wing margin (in the genus of *Uranorhagio*); hind femur swollen at apex; and nine abdominal segments. This family is tentatively placed as a member of Tabanoidea [142].

Genera included from the Jurassic of Northern China: *Strenorhagio* Zhang, Yang, Ren & Shih, 2010 and *Uranorhagio* Zhang, Yang & Ren, 2010.

Strenorhagio Zhang, Yang, Ren & Shih, 2010

Strenorhagio Zhang, Yang, Ren & Shih, 2010, *Zool. J. Linn. Soc.*, 158 (3), 566 [142].

Type species: *Strenorhagio deviatus* Zhang, Yang & Shih, 2010.

Body stout; hind tibia with a spur; veins R_1 and Rs_1 long; vein R_{2+3} strongly bent upwards at base; vein R_4 beginning distal to apex of cell d; cross-vein r-m at basal third of cell d.

Distribution and age: Inner Mongolia; Middle Jurassic.

Four species included from the Jurassic of Northern China (see Table 23.1).

Strenorhagio grimaldi Zhang, Ren & Shih, 2010 (Figure 23.30)

Strenorhagio grimaldi Zhang, Ren & Shih, 2010: *Zool. J. Linn. Soc.*, 158 (3), 566.

Locality and horizon: Daohugou, Ningcheng, Inner Mongolia, China; Middle Jurassic, Jiulongshan Formation.

The specific epithet is dedicated to Dr. D.A. Grimaldi for his contribution and dedication to Paleoentomology and evolution of the insects. Body large sized, stout, 17.0 mm long. Head round, slightly narrower than thorax. Eyes large, bare; ommatidia visible clearly, upper

Figure 23.30 *Strenorhagio grimaldi* Zhang, Ren & Shih, 2010 (Holotype, CNU-DIB-NN2007019) Source: Donated by Dr. Chungkun Shih.

area comprising of large facets. R_1 and Rs long; R_{2+3} arising from base of cell r_5, distinctly and strongly bended upward at base; R_5 ending slightly beyond wing apex. Cross-vein r-m at basal 1/3 of cell d. All branches of M long, M_1 and M_2 bifurcating distal of cross-vein m-m, basal trunk of M_1 and M_2 shorter than cross-vein m-m. CuA_1 arising from cell bm, cross-vein m-cu short. Cell bm about as wide as cell br; cell d slender, far from wing margin. Five posterior cells and cell cup wide open; mouth of cell m_1 slightly narrower than that of cell m_2; mouth of cell m_3 distinctly narrower than that of cell m_1, about 1/2 as wide as that of cell cua_1 [142].

Uranorhagio Zhang, Yang & Ren, 2010

Uranorhagio Zhang, Yang & Ren, 2010, *Zool. J. Linn. Soc.*, 158 (3), 563 [142].

Type species: *Uranorhagio daohugouensis* Zhang, Yang & Ren, 2010.

Body stout. Hind femur robust, strongly swollen at apical part, hind tibia with two spurs; empodium pulvilliform. Vein R_{2+3} distinctly strong and bent upwards at base, distal parts of veins R_{2+3}, R_4, and R_5, straight and parallel with each other, beginning of vein R_4 proximal to apex of cell d; distal to beginning of vein R_{2+3}; veins M_1 and M_2 bifurcating distal to cross-vein m–m, base of vein M_1 curved upwards; cell r_4 much narrower.

Distribution and age: Inner Mongolia; Middle Jurassic.

Only one species included from the Jurassic of Northern China (see Table 23.1).

Family Zhangsolvidae Nagatomi & Yang, 1998

The Zhangsolvidae is an extinct family of Stratiomyomorpha erected by Nagatomi and Yang in 1998 [122] based on fossils from the Early Cretaceous of Shandong, China. All zhangsolvids, possessing a long proboscis, were probably pollinators [143]. They are typified by having rather long proboscis with small labellum at the apex; vein C ending at or near apex of wing, M_1 strongly arched, M_3 fused to CuA_1 and CuA_2 fused to A_1 near wing margin [144]. Up to now, there are four genera containing seven species described. The monophyletic family Zhangsolvidae are reported in shale from China, limestones from Brazil and amber from Spanish and Myanmar.

Only one genus included from the Cretaceous of Northern China: *Zhangsolva* Nagatomi & Yang, 1998.

Zhangsolva Nagatomi & Yang, 1998

Zhangsolva Nagatomi & Yang, 1998, *Ent. Mon. Mag.*, 134, 139 [122] (original designation).

Type species: *Archisolva cupressa* Zhang, Zhang & Li, 1993.

Antennal flagellum 12 segmented; proboscis conspicuously long; long basal section of Rs; costal cell concave before Rs, possibly representing a sc-r cross-vein [122, 137].

Distribution and age: Shandong; Early Cretaceous.

Only one species included from the Cretaceous of Northern China (see Table 23.1).

Table 23.1 A list of fossil Diptera from the Jurassic and Cretaceous of China.

Family	Species	Locality	Horizon/age	Citation
		Suborder Nematocera Latreille, 1825		
Anisopodidae	[a]*Megarhyphus rarus* Zhang, 2007	Ningcheng, Inner Mongolia	[b]Jiulongshan Fm., J_2	Zhang [76]
	[a]*Mesobrachyopteryx shandongensis* Hong & Wang, 1990	Laiyang, Shandong	Laiyang Fm., K_1	Hong and Wang [45]
Antefungivoridae	*Antefungivora mictis* (Lin, 1980)	Jiande, Zhejiang	Laocun Fm., J_3	Lin [145]
	Antefungivora haifanggouensis (Hong, 1983)	Beipiao, Liaoning	Haifanggou Fm., J_2	Hong [31]
	Aortomima shandongensis Zhang, Zhang, Liu & Shangguan, 1986	Laiyang, Shandong	[b]Laiyang Fm., K_1	Zhang *et al.* [33]
	[a]*Baishuilingella micris* Lin, 1980	Jiande, Zhejiang	Laocun Fm., J_3	Lin [145]
	Lycoriomimodes oblongus (Hong, 1983)	Beipiao, Liaoning	Haifanggou Fm., J_2	Hong [31]
	Lycoriomimodes producopoda (Lin, 1976)	Beipiao, Liaoning	Haifanggou Fm., J_2	Lin [48]

Table 23.1 (Continued)

Family	Species	Locality	Horizon/age	Citation
	Lycoriomimodes perbella (Zhang, Zhang, Liu & Shangguan, 1986)	Laiyang, Shandong	[b]Laiyang Fm., K_1	Zhang *et al.* [33]
	Lycoriomimodes longiradiata (Hong & Wang, 1990)	Laiyang, Shandong	Laiyang Fm., K_1	Hong and Wang [45]
	Lycoriomimodes ovatus Hong & Wang, 1990	Laiyang, Shandong	Laiyang Fm., K_1	Hong and Wang [45]
	Lycoriomimodes parva (Hong & Wang, 1990)	Laiyang, Shandong	Laiyang Fm., K_1	Hong and Wang [45]
	Lycoriomimodes luanpingensis (Hong, 1983)	Luanping, Hebei	Jiulongshan Fm., J_2	Hong [31]
	Mimallactoneura lirata Hong, 1983	Beipiao, Liaoning	Haifanggou Fm., J_2	Hong [31]
	Mimallactoneura tuanwangensis Hong & Wang, 1990	Laiyang, Shandong	Laiyang Fm., K_1	Hong and Wang [45]
Asiochaoboridae	[a]*Asiochaoborus tenuous* Hong & Wang, 1990	Laiyang, Shandong	Laiyang Fm., K_1	Hong and Wang [45]
	[a]*Chaoboropsis longipedalis* Hong & Wang, 1990	Laiyang, Shandong	Laiyang Fm., K_1	Hong and Wang [45]
	[a]*Sinochaoborus dividus* Hong & Wang, 1990	Laiyang, Shandong	Laiyang Fm., K_1	Hong and Wang [45]
	[a]*Sunochaoborus laiyangensis* Hong & Wang, 1990	Laiyang, Shandong	Laiyang Fm., K_1	Hong and Wang [45]
Axymyiidae	*Juraxymyia fossilis* (Zhang, 2004)	Ningcheng, Inner Mongolia	[b]Jiulongshan Fm., J_2	Zhang [146]
	[a]*Psocites pectinatus* (Hong, 1983)	Beipiao, Liaoning	Haifanggou Fm., J_2	Hong [31]
	Raraxymyia parallela Shi, Zhu, Shih & Ren, 2013	Ningcheng, Inner Mongolia	Jiulongshan Fm., J_2	Shi *et al.* [35]
	Raraxymyia proxima Shi, Zhu, Shih & Ren, 2013	Ningcheng, Inner Mongolia	Jiulongshan Fm., J_2	Shi *et al.* [35]
	Sinaxymyia rara Zhang, 2010	Ningcheng, Inner Mongolia	[b]Jiulongshan Fm., J_2	Zhang [37]
Blephariceridae	*Brianina longitibialis* Zhang, & Lukashevich, 2007	Ningcheng, Inner Mongolia	Jiulongshan Fm., J_2	Zhang and Lukashevich [38]
	Megathon brodskyi Zhang, & Lukashevich, 2007	Ningcheng, Inner Mongolia	Jiulongshan Fm., J_2	Zhang and Lukashevich [38]
Bibionidae	*Lichnoplecia kovalevi* Ren, 1995	Chengde, Hebei	[b]Yixian Fm., K_1	Ren [40]
Chaoboridae	[a]*Chironomaptera robustus* (Lin, 1976)	Kazuo, Liaoning	[b]Jiufotang Fm., K_1	Lin [48]
	[a]*Chironomaptera gregaria* (Grabau, 1923)	Kazuo, Liaoning	[b]Jiufotang Fm., K_1	Grabau [25]
	[a]*Chironomaptera melanura* (Ping, 1928)	Zhuji, Zhejiang	Shouchang Fm., J_3	Ping [26]
	Mesochaoborus zhangshanyingensis (Hong, 1984)	Longhua, Hebei	[b]Yixian Fm., K_1	Hong [147]
	Mesochaoborus pallens Zhang, Zhang, Liu & Shangguan 1986	Laiyang, Shandong	[b]Laiyang Fm., K_1	Zhang *et al.* [33]

(Continued)

Table 23.1 (Continued)

Family	Species	Locality	Horizon/age	Citation
Chironomidae	*Coelochironoma xantha* Zhang, Zhang, Liu & Shangguan, 1986	Laiyang, Shandong	[b]Laiyang Fm., K_1	Zhang *et al.* [33]
	Manlayamyia dabeigouensis Zhang, 1991	Luanping, Hebei	Dabeigou Fm., J_3	Zhang [42]
	[a]*Orusa barba* Lin, 1980	Jiande, Zhejiang	Laocun Fm., J_3	Lin [145]
	Oryctochlus contiguus Zhang, 1991	Laiyang, Shandong	[b]Laiyang Fm., K_1	Zhang [42]
	Sinoryctochlus insolitus Zhang, 1991	Laiyang, Shandong	[b]Laiyang Fm., K_1	Zhang [42]
	[a]*Tendipopsis colorata* Hong & Wang, 1990	Laiyang, Shandong	Laiyang Fm., K_1	Hong and Wang [45]
	[a]*Tinactum solusum* Lin, 1980	Jiande, Zhejiang	Laocun Fm., J_3	Lin [145]
	[a]*Viduata otiosa* Lin, 1980	Jiande, Zhejiang	Laocun Fm., J_3	Lin [145]
Eopleciidae	*Beipiaoplecia malleformis* Lin, 1976	Beipiao, Liaoning	Haifanggou Fm., J_2	Lin [48]
	[a]*Gansuplecia triporata* Hong, 1990	Chongxin, Gansu	Huanhe Fm., K_1	Hong and Wang [45]
	Jurolaemargus yujiagouensis Hong, 1983	Beipiao, Liaoning	Haifanggou Fm., J_2	Hong [31]
	[a]*Leptoplecia laevis* Hong, 1983	Beipiao, Liaoning	Haifanggou Fm., J_2	Hong [31]
Gracilitipulidae	[a]*Gracilitipula asiatica* Hong & Wang, 1990	Laiyang, Shandong	Laiyang Fm., K_1	Hong and Wang [45]
Limoniidae	*Archimesotipula antefortis* Lin, 1986	Zhongshan, Guangxi	Shiti Fm., J_1	Lin [60]
	Architipula chinensis Zhang, 2004	Ningcheng, Inner Mongolia	Jiulongshan Fm., J_2	Zhang [148]
	Architipula conformis Hao & Ren, 2009	Ningcheng, Inner Mongolia	Jiulongshan Fm., J_2	Hao and Ren [149]
	Architipula insolita Zhang, 2006	Ningcheng, Inner Mongolia	[b]Jiulongshan Fm., J_2	Zhang [50]
	Architipula trichoclada Zhang, 2006	Ningcheng, Inner Mongolia	[b]Jiulongshan Fm., J_2	Zhang [50]
	Cretolimonia excelsa Gao, Shih, Zhao & Ren 2015	Ningcheng, Inner Mongolia	Jiulongshan Fm., J_2	Gao *et al.* [59]
	[a]*Eotipulina eximia* Zhang, 2006	Ningcheng, Inner Mongolia	[b]Jiulongshan Fm., J_2	Zhang [50]
	[a]*Eotipuloptera dignata* Zhang, 2006	Ningcheng, Inner Mongolia	[b]Jiulongshan Fm., J_2	Zhang [50]
	Mesotipula gloriosa Gao, Shih, Zhao & Ren 2015	Ningcheng, Inner Mongolia	Jiulongshan Fm., J_2	Gao *et al.* [59]
	[a]*Sinotipula huabeiensis* (Hong, 1983)	Beipiao, Liaoning	Haifanggou Fm., J_2	Hong [31]
	[a]*Xutipula longipetalis* Hong, 1983	Beipiao, Liaoning	Haifanggou Fm., J_2	Hong [31]
Limnorhyphidae	[a]*Limnorhyphus haifanggouensis* Hong, 1983	Beipiao, Liaoning	Haifanggou Fm., J_2	Hong [31]
Luanpingitidae	*Luanpingites flavus* Zhang, 1986	Luanping, Hebei	Xiahuayuan Fm., J_2	Zhang [62]

(Continued)

Table 23.1 (Continued)

Family	Species	Locality	Horizon/age	Citation
Mesosciophilidae	*Jurasciophila curvula* Li & Ren, 2009	Ningcheng, Inner Mongolia	Jiulongshan Fm., J_2	Li and Ren [67]
	Jurasciophila lepida Li & Ren, 2009	Ningcheng, Inner Mongolia	Jiulongshan Fm., J_2	Li and Ren [67]
	Mesosciophila abstracta Zhang, 2008	Ningcheng, Inner Mongolia	[b]Jiulongshan Fm., J_2	Zhang [150]
	Mesosciophila eucalla Zhang, 2007	Ningcheng, Inner Mongolia	[b]Jiulongshan Fm., J_2	Zhang [65]
	Mesosciophila sigmoidea Wang, Zhao & Ren, 2012	Ningcheng, Inner Mongolia	Jiulongshan Fm., J_2	Wang *et al.* [151]
	Mesosciophilodes synchrona Zhang, 2008	Ningcheng, Inner Mongolia	[b]Jiulongshan Fm., J_2	Zhang [150]
	Orentalphila caloa Lin, Shih & Ren, 2015	Beipiao, Liaoning	Yixian Fm., K_1	Lin *et al.* [66]
	Orentalphila gravia Lin, Shih & Ren, 2015	Beipiao, Liaoning	Yixian Fm., K_1	Lin *et al.* [66]
	Paramesosciophilodes aequus Wang, Zhao & Ren, 2012	Ningcheng, Inner Mongolia	Jiulongshan Fm., J_2	Wang *et al.* [151]
	Paramesosciophilodes bellus Gao, Shi, Shih & Ren, 2015	Ningcheng, Inner Mongolia	Jiulongshan Fm., J_2	Gao *et al.* [152]
	Paramesosciophilodes eximia Zhang, 2008	Ningcheng, Inner Mongolia	[b]Jiulongshan Fm., J_2	Zhang [150]
	Paramesosciophilodes ningchengensis Zhang, 2007	Ningcheng, Inner Mongolia	[b]Jiulongshan Fm., J_2	Zhang [65]
	Paramesosciophilodes rarissima Gao, Shi, Shih & Ren, 2015	Ningcheng, Inner Mongolia	Jiulongshan Fm., J_2	Gao *et al.* [152]
	Similsciophila singularis Shi, Shih & Ren, 2014	Ningcheng, Inner Mongolia	Jiulongshan Fm., J_2	Shi *et al.* [68]
	Similsciophila sinuata Shi, Shih & Ren, 2014	Ningcheng, Inner Mongolia	Jiulongshan Fm., J_2	Shi *et al.* [68]
	Similsciophila undulata Lin, Shih & Ren, 2015	Beipiao, Liaoning	Yixian Fm., K_1	Lin *et al.* [66]
	[a]*Sinosciophila angustia* Lin, Shih & Ren, 2015	Beipiao, Liaoning	Yixian Fm., K_1	Lin *et al.* [66]
	[a]*Sinosciophila meileyingziensis* Hong, 1992	Kazuo, Liaoning	Shahai Fm., K_1	Hong [153]
	[a]*Sinosciophila seboa* Lin, Shih & Ren, 2015	Beipiao, Liaoning	Yixian Fm., K_1	Lin *et al.* [66]
Mycetophilidae	*Huaxiasciophilites jingxiensis* Zhang, Hong & Li, 2001	Jingxi, Beijing	Lushangfen Fm., K_1	Zhang *et al.* [69]
	[a]*Liaoxifungivora simplicis* Hong, Wang & Sun, 1992	Kazuo, Liaoning	Shahai Fm., K_1	Hong *et al.* [154]
Parapleciidae	[a]*Paraplecia ovata* Hong, 1983	Beipiao, Liaoning	Haifanggou Fm., J_2	Hong [31]
Paraxymyiidae	[a]*Arcus beipiaoensis* Hong, 1983	Beipiao, Liaoning	Haifanggou Fm., J_2	Hong [31]
Pediciidae	*Praearchitipula abnormis* (Hao & Ren, 2009)	Ningcheng, Inner Mongolia	Jiulongshan Fm., J_2	Hao and Ren [149]
	Praearchitipula apprima Gao, Shih, Kopeć, Krzemiński & Ren, 2015	Ningcheng, Inner Mongolia	Jiulongshan Fm., J_2	Gao *et al.* [70]

(Continued)

Table 23.1 (Continued)

Family	Species	Locality	Horizon/age	Citation
	Praearchitipula mirabilis Gao, Shih, Kopeć, Krzemiński & Ren, 2015	Ningcheng, Inner Mongolia	Jiulongshan Fm., J_2	*et al.* [70]
Pleciofungivoridae	*Eohesperinus latus* Hong & Wang, 1990	Laiyang, Shandong	Laiyang Fm., K_1	Hong and Wang [45]
	Eohesperinus gracilis Hong, 1983	Beipiao, Liaoning	Haifanggou Fm., J_2	Hong [31]
	[a]*Fera jurassica* Hong, 1983	Beipiao, Liaoning	Haifanggou Fm., J_2	Hong [31]
	[a]*Fera parva* Hong, 1983	Beipiao, Liaoning	Haifanggou Fm., J_2	Hong [31]
	[a]*Mesopleciofungivora martynovae* Hong & Wang, 1990	Laiyang, Shandong	Laiyang Fm., K_1	Hong and Wang [45]
	Opiparifungivora aliena Ren, 1995	Chengde, Hebei	[b]Yixian Fm., K_1	Ren *et al.* [40]
	[a]*Parapleciofungivora triangulata* Hong & Wang, 1990	Laiyang, Shandong	Laiyang Fm., K_1	Hong and Wang [45]
	Pleciofungivora yangtianense (Hong, 1984)	Suocheng, Hebei	Dabeigou Fm., J_3	Hong [147]
Protendipedidae	[a]*Priscotendipes mirus* Zhang, 1986	Weichang, Hebei	Dabeigou Fm., J_3	Zhang [62]
	[a]*Protendipes huabensis* Zhang, 1986	Weichang, Hebei	Dabeigou Fm., J_3	Zhang [62]
Protobibionidae	*Protobibio orientalis* Zhang, Zhang, Liu & Shangguan, 1986	Laiyang, Shandong	[b]Laiyang Fm., K_1	Zhang *et al.* [33]
Protopleciidae	*Epimesoplecia ambloneura* Lin, Shih & Ren, 2015	Ningcheng, Inner Mongolia	Jiulongshan Fm., J_2	Lin *et al.* [75]
	Epimesoplecia elenae Zhang, 2007	Ningcheng, Inner Mongolia	[b]Jiulongshan Fm., J_2	Zhang [73]
	Epimesoplecia macrostrena Lin, Shih & Ren, 2015	Ningcheng, Inner Mongolia	Jiulongshan Fm., J_2	Lin *et al.* [75]
	Epimesoplecia plethora Lin, Shih & Ren, 2015	Ningcheng, Inner Mongolia	Jiulongshan Fm., J_2	Lin *et al.* [75]
	Epimesoplecia prosoneura Lin, Shih & Ren, 2015	Ningcheng, Inner Mongolia	Jiulongshan Fm., J_2	Lin *et al.* [75]
	Epimesoplecia shcherbakovi Zhang, 2007	Ningcheng, Inner Mongolia	[b]Jiulongshan Fm., J_2	Zhang [73]
	Epimesoplecia stana Lin, Shih & Ren, 2015	Ningcheng, Inner Mongolia	Jiulongshan Fm., J_2	Lin *et al.* [75]
	[a]*Hebeiplecia brunnea* Hong, 1983	Zhouyingzi, Hebei	Jiulongshan Fm., J_2	Hong [31]
	[a]*Huaxiaplecia zhongguanensis* Hong, 1992	Kazuo, Liaoning	Shahai Fm., K_1	Hong [153]
	Mesoplecia anfracta Hao & Ren, 2009	Ningcheng, Inner Mongolia	Jiulongshan Fm., J_2	Hao and Ren [57]
	Mesoplecia antiqua Hao & Ren, 2009	Ningcheng, Inner Mongolia	Jiulongshan Fm., J_2	Hao and Ren [57]
	Mesoplecia coadnata Hao & Ren, 2009	Ningcheng, Inner Mongolia	Jiulongshan Fm., J_2	Hao and Ren [57]
	Mesoplecia fastigata Lin, Shih & Ren, 2014	Ningcheng, Inner Mongolia	Jiulongshan Fm., J_2	Lin *et al.* [74]

(Continued)

Table 23.1 (Continued)

Family	Species	Locality	Horizon/age	Citation
	Mesoplecia mediana Zhang, 2007	Ningcheng, Inner Mongolia	[b]Jiulongshan Fm., J_2	Zhang [73]
	Mesoplecia plena Lin, Shih & Ren, 2014	Ningcheng, Inner Mongolia	Jiulongshan Fm., J_2	Lin *et al.* [74]
	Mesoplecia sinica Zhang, 2007	Ningcheng, Inner Mongolia	[b]Jiulongshan Fm., J_2	Zhang [73]
	Mesoplecia xinboensis Hong, 1984	Weichang, Hebei	[b]Dabeigou Fm., K_1	Hong [147]
	[a]*Palaeohesperinus guidongensis* Lin, 1986	Zhongshan, Guangxi	Shiti Fm., J_1	Lin [60]
	[a]*Pleciopsis longa* Hong, 1983	Zhouyingzi, Hebei	Jiulongshan Fm., J_2	Hong [31]
	[a]*Pseudoplecia ovata* Hong & Wang, 1990	Laiyang, Shandong	Laiyang Fm., K_1	Hong and Wang [45]
	Sinoplecia parvita Lin, 1976	Beipiao, Liaoning	Haifanggou Fm., J_2	Lin [48]
	[a]*Sunoplecia curvata* Hong & Wang, 1990	Laiyang, Shandong	Laiyang Fm., K_1	Hong and Wang [45]
	[a]*Sunoplecia liaoningensis* Hong, 1983	Beipiao, Liaoning	Haifanggou Fm., J_2	Hong [31]
	[a]*Sunoplecia longa* Hong, 1983	Beipiao, Liaoning	Haifanggou Fm., J_2	Hong [31]
Protorhyphidae	*Protorhyphus arcuatus* (Hong,1983)	Beipiao, Liaoning	Yixian Fm., K_1	Hong [31]
	Protorhyphus liaoningicus Zhang, 2007	Beipiao, Liaoning	Haifanggou Fm., J_2	Zhang [76]
	Protorhyphus neimonggolensis Zhang, 2007	Ningcheng, Inner Mongolia	[b]Jiulongshan Fm., J_2	Zhang [76]
Ptychopteridae	*Crenoptychoptera decorosa* Hao, Dong & Ren, 2009	Ningcheng, Inner Mongolia	Jiulongshan Fm., J_2	Hao *et al.* [81]
	Crenoptychoptera vicina Hao, Dong & Ren, 2009	Ningcheng, Inner Mongolia	Jiulongshan Fm., J_2	Hao *et al.* [81]
	Crenoptychoptera vulgaris Hao, Dong & Ren, 2009	Ningcheng, Inner Mongolia	Jiulongshan Fm., J_2	Hao *et al.* [81]
	Eoptychoptera ansorgei Ren & Krzemiński, 2002	Ningcheng, Inner Mongolia	Jiulongshan Fm., J_2	Ren and Krzemiński[155]
	Eoptychoptera jurassica Ren & Krzemiński, 2002	Ningcheng, Inner Mongolia	Jiulongshan Fm., J_2	Ren and Krzemiński[155]
	Eoptychopterina antica Hao, Ren & Shih, 2009	Ningcheng, Inner Mongolia	Jiulongshan Fm., J_2	Hao *et al.* [156]
	Eoptychopterina elenae Ren & Krzemiński, 2002	Ningcheng, Inner Mongolia	Jiulongshan Fm., J_2	Ren and Krzemiński[155]
	Eoptychopterina gigantea Zhang, 2004	Ningcheng, Inner Mongolia	[b]Jiulongshan Fm., J_2	Zhang [157]
	Eoptychopterina mediata Hao, Ren & Shih, 2009	Ningcheng, Inner Mongolia	Jiulongshan Fm., J_2	Hao *et al.* [156]
	Eoptychopterina postica Liu, Shih & Ren, 2012	Ningcheng, Inner Mongolia	Jiulongshan Fm., J_2	Liu *et al.* [79]
Serendipidae	[a]*Serendipa laiyangensis* (Hong & Wang, 1990)	Laiyang, Shandong	Laiyang Fm., K_1	Hong and Wang [45]
	[a]*Serendipa tuanwangensis* (Hong & Wang, 1990)	Laiyang, Shandong	Laiyang Fm., K_1	Hong and Wang [45]

(Continued)

Table 23.1 (Continued)

Family	Species	Locality	Horizon/age	Citation
	[a]*Thamnitendipes vegetabilis* Hong & Wang, 1990	Laiyang, Shandong	Laiyang Fm., K_1	Hong and Wang [45]
Sinotendipedidae	[a]*Sinotendipes tuanwangensis* Hong & Wang, 1990	Laiyang, Shandong	Laiyang Fm., K_1	Hong and Wang [45]
Strashilidae	*Strashila daohugouensis* Huang, Nel, Cai, Lin & Engel, 2013	Ningcheng, Inner Mongolia	Jiulongshan Fm., J_2	Huang *et al.* [85]
	Vosila sinensis Vršanský & Ren, 2010	Ningcheng, Inner Mongolia	Jiulongshan Fm., J_2	Vršanský *et al.* [84]
Tanyderidae	*Protanyderus astictum* Dong, Shih, Skibińska, Krzemiński & Ren, 2015	Ningcheng, Inner Mongolia	Jiulongshan Fm., J_2	Dong *et al.* [88]
	Praemacrochile chinensis Krzemiński & Ren, 2001	Ningcheng, Inner Mongolia	Jiulongshan Fm., J_2	Krzemiński and Ren [158]
	Praemacrochile dryasis Dong, Shih, Skibińska, Krzemiński & Ren, 2015	Ningcheng, Inner Mongolia	Jiulongshan Fm., J_2	Dong *et al.* [88]
	Praemacrochile ovalum Dong, Shih, Skibińska, Krzemiński & Ren, 2015	Ningcheng, Inner Mongolia	Jiulongshan Fm., J_2	Dong *et al.* [88]
	Praemacrochile vulcanium Zhang, 2004	Ningcheng, Inner Mongolia	Jiulongshan Fm., J_2	Zhang [148]
Tipulidae	*Leptotarsus (Longurio) primitivus* Shih, Dong, Kania, Liu, Krzemiński & Ren, 2015	Beipiao, Liaoning	Yixian Fm., K_1	Shih *et al.* [92]
Trichoceridae	*Eotrichocera (Archaeotrichocera) amabilis* Dong, Shih & Ren, 2014	Ningcheng, Inner Mongolia	Jiulongshan Fm., J_2	Dong *et al.* [98]
	Eotrichocera (Archaeotrichocera) ephemera (Zhang, 2006)	Ningcheng, Inner Mongolia	[b]Jiulongshan Fm., J_2	Zhang [97]
	Eotrichocera (Archaeotrichocera) longensis Dong, Shih & Ren, 2014	Ningcheng, Inner Mongolia	Jiulongshan Fm., J_2	Dong *et al.* [98]
	Eotrichocera (Archaeotrichocera) spatiosa (Liu, Shih & Ren, 2012)	Ningcheng, Inner Mongolia	Jiulongshan Fm., J_2	Liu *et al.* [79]
	[a]*Mesotrichocera laiyangensis* Hong & Wang, 1990	Laiyang, Shandong	Laiyang Fm., K_1	Hong and Wang [45]
	Tanyochoreta (Tanyochoreta) chifengica (Zhang, 2006)	Ningcheng, Inner Mongolia	[b]Jiulongshan Fm., J_2	Zhang [97]
	Tanyochoreta (Tanyochoreta) integera Zhang, 2006	Ningcheng, Inner Mongolia	[b]Jiulongshan Fm., J_2	Zhang [97]
	Tanyochoreta (Sinotrichocera) parva (Zhang, 2006)	Ningcheng, Inner Mongolia	[b]Jiulongshan Fm., J_2	Zhang [97]
Zhangobiidae	*Ceuthoneura dolichoptera* Zhang, Zhang, Liu & Shangguan, 1986	Laiyang, Shandong	Laiyang Fm., K_1	Zhang *et al.* [33]
	Zhangobia laiyangensis (Zhang, Zhang, Liu & Shangguan, 1986)	Laiyang, Shandong	Laiyang Fm., K_1	Zhang *et al.* [33]

(Continued)

Table 23.1 (Continued)

Family	Species	Locality	Horizon/age	Citation
Family *incertae sedis*	[a]*Brachyopteryx weichangensis* Hong, 1984	Weichang, Hebei	[b]Dabeigou Fm., K_1	Hong [147]
	[a]*Huaxiarhyphus chichengensis* Hong, 1996	Chicheng, Hebei	Houcheng Fm., K_1	Hong [159]
	[a]*Mesasimulium lahaigouense* Zhang, 1986	Weichang, Hebei	Dabeigou Fm., K_1	Zhang [62]
	[a]*Paucivena elongata* Lin, 1976	Beipiao, Liaoning	Haifanggou Fm., J_2	Lin [48]
	[a]*Platyplecia suni* Hong, 1983	Beipiao, Liaoning	Haifanggou Fm., J_2	Hong [31]
	[a]*Platyplecia parva* Hong, 1983	Beipiao, Liaoning	Haifanggou Fm., J_2	Hong [31]
	[a]*Raptatores erraticus* Hong, 1983	Zhouyingzi, Hebei	Jiulongshan Fm., J_2	Hong [31]
	Suborder Brachycera Schiner, 1862			
Archisargidae	*Archirhagio gracilentus* Wang, Shih, Ren & Wang, 2017	Ningcheng, Inner Mongolia	Jiulongshan Fm., J_2	Wang *et al.* [100]
	Archirhagio mostovskii Zhang, 2015	Ningcheng, Inner Mongolia	[b]Jiulongshan Fm., J_2	Zhang [99]
	Archirhagio striatus Zhang & Zhang, 2003	Ningcheng, Inner Mongolia	Jiulongshan Fm., J_2	Zhang and Zhang [160]
	Archirhagio varius Zhang, 2015	Ningcheng, Inner Mongolia	[b]Jiulongshan Fm., J_2	Zhang [99]
	Archirhagio zhangi Zhang, Yang & Ren, 2009	Ningcheng, Inner Mongolia	Jiulongshan Fm., J_2	Zhang *et al.* [161]
	Archisargus spurivenius Zhang, Yang & Ren, 2007	Ningcheng, Inner Mongolia	Jiulongshan Fm., J_2	Zhang *et al.* [101]
	Archisargus strigatus Zhang, Yang & Ren, 2007	Ningcheng, Inner Mongolia	Jiulongshan Fm., J_2	Zhang *et al.* [101]
	Brevisolva daohugouensis Zhang, Ren & Shih, 2010	Ningcheng, Inner Mongolia	Jiulongshan Fm., J_2	Zhang *et al.* [103]
	Calosargus (*Calosargus*) *antiquus* Zhang, Yang & Ren, 2007	Ningcheng, Inner Mongolia	Jiulongshan Fm., J_2	Zhang *et al.* [107]
	Calosargus (*Calosargus*) *bellus* Zhang, Yang & Ren, 2007	Ningcheng, Inner Mongolia	Jiulongshan Fm., J_2	Zhang *et al.* [107]
	Calosargus (*Calosargus*) *daohugouensis* Zhang, Yang & Ren, 2007	Ningcheng, Inner Mongolia	Jiulongshan Fm., J_2	Zhang *et al.* [107]
	Calosargus (*Calosargus*) *hani* Zhang, Yang & Ren, 2007	Ningcheng, Inner Mongolia	Jiulongshan Fm., J_2	Zhang *et al.* [107]
	Calosargus (*Calosargus*) *tenuicellus* Zhang, Yang & Ren, 2007	Ningcheng, Inner Mongolia	Jiulongshan Fm., J_2	Zhang *et al.* [107]
	Calosargus (*Calosargus*) *validus* Zhang, Yang & Ren, 2007	Ningcheng, Inner Mongolia	Jiulongshan Fm., J_2	Zhang *et al.* [107]
	Calosargus (*Pterosargus*) *sinicus* Zhang, 2010	Ningcheng, Inner Mongolia	[b]Jiulongshan Fm., J_2	Zhang [108]
	Flagellisargus robustus Zhang, 2012	Ningcheng, Inner Mongolia	[b]Jiulongshan Fm., J_2	Zhang [109]
	Flagellisargus sinicus Zhang, 2012	Ningcheng, Inner Mongolia	[b]Jiulongshan Fm., J_2	Zhang [109]

(Continued)

Table 23.1 (Continued)

Family	Species	Locality	Horizon/age	Citation
	Flagellisargus venustus Zhang, 2012	Ningcheng, Inner Mongolia	[b)]Jiulongshan Fm., J_2	Zhang [109]
	Mesosolva daohugouensis Zhang & Zhang, 2003	Ningcheng, Inner Mongolia	[b)]Jiulongshan Fm., J_2	Zhang and Zhang [160]
	Mesosolva huabeiensis (Hong, 1983)	Beipiao, Liaoning	Haifanggou Fm., J_2	Hong [31]
	Mesosolva jurassica Zhang, Yang & Ren, 2010	Ningcheng, Inner Mongolia	Jiulongshan Fm., J_2	Zhang *et al.* [103]
	Mesosolva parva Hong, 1983	Beipiao, Liaoning	Haifanggou Fm., J_2	Hong [31]
	Mesosolva sinensis Zhang, Yang & Ren, 2010	Ningcheng, Inner Mongolia	Jiulongshan Fm., J_2	Zhang *et al.* [103]
	Mostovskisargus portentosus Zhang, 2010	Ningcheng, Inner Mongolia	[b)]Jiulongshan Fm., J_2	Zhang [108]
	Mostovskisargus signatu Zhang, 2010	Ningcheng, Inner Mongolia	[b)]Jiulongshan Fm., J_2	Zhang [108]
	Novisargus rarus Zhang, 2015	Ningcheng, Inner Mongolia	[b)]Jiulongshan Fm., J_2	Zhang [99]
	Ovisargus (Ovisargus) singulus Zhang, 2015	Ningcheng, Inner Mongolia	[b)]Jiulongshan Fm., J_2	Zhang [99]
	Sharasargus eximius Zhang, Yang & Ren, 2008	Ningcheng, Inner Mongolia	Jiulongshan Fm., J_2	Zhang *et al.* [105]
	Sharasargus fortis Zhang, Yang & Ren, 2008	Ningcheng, Inner Mongolia	Jiulongshan Fm., J_2	Zhang *et al.* [105]
	Sharasargus maculus Zhang, 2015	Ningcheng, Inner Mongolia	[b)]Jiulongshan Fm., J_2	Zhang [99]
	Tabanisargus daohugous Zhang, 2015	Ningcheng, Inner Mongolia	[b)]Jiulongshan Fm., J_2	Zhang [99]
Athericidae	*Qiyia jurassica* Chen, Wang & Engel, 2014	Ningcheng, Inner Mongolia	[b)]Jiulongshan Fm., J_2	Chen *et al.* [112]
	Sinocretomyia minuscula Zhang, 2012	Laiyang, Shandong	Laiyang Fm., K_1	Zhang [111]
Eremochaetidae	*Alleremonomus liaoningensis* Ren & Guo, 1995	Beipiao, Liaoning	[b)]Yixian Fm., K_1	Ren and Guo [114]
	Alleremonomus xingi Ren & Guo, 1995	Beipiao, Liaoning	[b)]Yixian Fm., K_1	Ren and Guo [114]
	Dissup clausus Zhang, Yang & Ren, 2014	Beipiao, Liaoning	Yixian Fm., K_1	Zhang *et al.* [113]
	Eremomukha (Eremomukha) angusta Zhang, 2014	Beipiao, Liaoning	Yixian Fm., K_1	Zhang [162]
	Eremomukha (Eremomukha) tenuissima Zhang, 2014	Beipiao, Liaoning	Yixian Fm., K_1	Zhang [162]
	Lepteremochaetus elegans Zhang, 2014	Beipiao, Liaoning	Yixian Fm., K_1	Zhang [162]
	Lepteremochaetus lithoecius Ren, 1998	Beipiao, Liaoning	[b)]Yixian Fm., K_1	Ren [115]
Kovalevisargidae	*Kerosargus sororius* Zhang, 2011	Ningcheng, Inner Mongolia	[b)]Jiulongshan Fm., J_2	Zhang [117]
	Kovalevisargus brachypterus Zhang, 2011	Ningcheng, Inner Mongolia	[b)]Jiulongshan Fm., J_2	Zhang [117]
	Kovalevisargus haifanggouensis Zhang, 2015	Ningcheng, Inner Mongolia	[b)]Jiulongshan Fm., J_2	Zhang [99]

(Continued)

Table 23.1 (Continued)

Family	Species	Locality	Horizon/age	Citation
	Kovalevisargus macropterus Zhang, 2011	Ningcheng, Inner Mongolia	[b]Jiulongshan Fm., J_2	Zhang [117]
Nemestrinidae	*Ahirmoneura neimengguensis* Zhang, Yang & Ren, 2008	Ningcheng, Inner Mongolia	Jiulongshan Fm., J_2	Zhang *et al.* [124]
	Florinemestrius pulcherrimus Ren, 1998	Beipiao, Liaoning	[b]Yixian Fm., K_1	Ren [115]
	Protonemestrius beipiaoensis Ren, 1998	Beipiao, Liaoning	[b]Yixian Fm., K_1	Ren [115]
	Protonemestrius jurassicus Ren, 1998	Beipiao, Liaoning	[b]Yixian Fm., K_1	Ren [115]
Orientisargidae	*Orientisargus illecebrosus* Zhang, 2012	Ningcheng, Inner Mongolia	[b]Jiulongshan Fm., J_2	Zhang [102]
Origoasilidae	*Origoasilus pingquanensis* Zhang, Yang & Ren, 2011	Chengde, Hebei	[b]Yixian Fm., K_1	Zhang *et al.* [125]
Protapioceridae	*Protapiocera convergens* Zhang, Yang & Ren, 2007	Beipiao, Liaoning	[b]Yixian Fm., K_1	Zhang *et al.* [126]
	Protapiocera ischyra Ren, 1998	Beipiao, Liaoning	[b]Yixian Fm., K_1	Ren [115]
	Protapiocera megista Ren, 1998	Beipiao, Liaoning	[b]Yixian Fm., K_1	Ren [115]
	Pseudapiocera shandongensis Zhang, 2015	Laiyang, Shandong	Laiyang Fm., K_1	Zhang [127]
Protempididae	*Helempis eucalla* Ren, 1998	Beipiao, Liaoning	[b]Yixian Fm., K_1	Ren [115]
	Helempis yixianensis Ren, 1998	Beipiao, Liaoning	[b]Yixian Fm., K_1	Ren [115]
	Protempis minuta Ren, 1998	Beipiao, Liaoning	[b]Yixian Fm., K_1	Ren [115]
Protobrachyceridae	*Protobrachyceron sinensis* Zhang, Yang & Ren, 2008	Ningcheng, Inner Mongolia	Jiulongshan Fm., J_2	Zhang *et al.* [129]
Rhagionemestriidae	*Jurassinemestrinus orientalis* Zhang, 2010	Ningcheng, Inner Mongolia	[b]Jiulongshan Fm., J_2	Zhang [108]
Rhagionempididae	*Ussatchovia gracilenta* Zhang, 2010	Ningcheng, Inner Mongolia	[b]Jiulongshan Fm., J_2	Zhang [131]
	Ussatchovia robusta Zhang, 2010	Ningcheng, Inner Mongolia	[b]Jiulongshan Fm., J_2	Zhang [131]
Rhagionidae	*Achrysopilus neimenguensis* Zhang, Yang & Ren, 2008	Ningcheng, Inner Mongolia	Jiulongshan Fm., J_2	Zhang *et al.* [140]
	Basilorhagio venustus Ren, 1995	Chengde, Hebei	[b]Yixian Fm., K_1	Ren [40]
	Daohugorhagio elongatus Zhang, 2013	Ningcheng, Inner Mongolia	[b]Jiulongshan Fm., J_2	Zhang [134]
	Lithorhagio megalocephalus Zhang & Li, 2012	Ningcheng, Inner Mongolia	[b]Jiulongshan Fm., J_2	Zhang and Li [135]
	[a]*Longhuaia orientalis* Hong, 1992	Chengde, Hebei	[b]Yixian Fm., K_1	Hong [153]
	Oiobrachyceron limnogenus Ren, 1998	Beipiao, Liaoning	[b]Yixian Fm., K_1	Ren [115]
	Orsobrachyceron chinensis Ren, 1998	Beipiao, Liaoning	[b]Yixian Fm., K_1	Ren [115]

(Continued)

Table 23.1 (Continued)

Family	Species	Locality	Horizon/age	Citation
	Palaeoarthroteles jurassicus Zhang, 2011	Ningcheng, Inner Mongolia	[b]Jiulongshan Fm., J_2	Zhang [163]
	Palaeoarthroteles pallidus Zhang, 2011	Ningcheng, Inner Mongolia	[b]Jiulongshan Fm., J_2	Zhang [163]
	Palaeobolbomyia sinica Zhang, 2010	Ningcheng, Inner Mongolia	[b]Jiulongshan Fm., J_2	Zhang [131]
	Parachrysopilus jurassicus Zhang, 2013	Ningcheng, Inner Mongolia	[b]Jiulongshan Fm., J_2	Zhang [134]
	Protorhagio parvus Zhang & Li, 2012	Ningcheng, Inner Mongolia	[b]Jiulongshan Fm., J_2	Zhang and Li [135]
	Scelorhagio mecomastigus Zhang, Zhang & Li, 1993	Laiyang, Shandong	[b]Laiyang Fm., K_1	Zhang *et al.* [137]
	Sinorhagio daohugouensis Zhang, Yang & Ren, 2006	Ningcheng, Inner Mongolia	Jiulongshan Fm., J_2	Zhang *et al.* [139]
	Sinorhagio sinuatus Zhang, 2013	Ningcheng, Inner Mongolia	[b]Jiulongshan Fm., J_2	Zhang [134]
	Trichorhagio gregarius Zhang, 2013	Ningcheng, Inner Mongolia	[b]Jiulongshan Fm., J_2	Zhang [134]
Sinonemestriidae	[a]*Sinonemestrius completus* Zhang, 2017	Laiyang, Shandong	Laiyang Fm., K_1	Zhang [164]
	[a]*Sinonemestrius tuanwangensis* Hong & Wang, 1990	Laiyang, Shandong	Laiyang Fm., K_1	Hong and Wang [45]
Tabanidae	*Allomyia ruderalis* Ren, 1998	Beipiao, Liaoning	[b]Yixian Fm., K_1	Ren [115]
	Eopangonius pletus Ren, 1998	Beipiao, Liaoning	[b]Yixian Fm., K_1	Ren [115]
	Laiyangitabanus formosus Zhang, 2012	Laiyang, Shandong	Laiyang Fm., K_1	Zhang [111]
	Palaepangonius eupterus Ren, 1998	Beipiao, Liaoning	[b]Yixian Fm., K_1	Ren [115]
	Pauromyia oresbia Ren, 1998	Beipiao, Liaoning	[b]Yixian Fm., K_1	Ren [115]
Uranorhagionidae	*Strenorhagio asymmetricus* Zhang, Yang & Ren, 2010	Ningcheng, Inner Mongolia	Jiulongshan Fm., J_2	Zhang *et al.* [142]
	Strenorhagio deviatus Zhang, Yang & Shih, 2010	Ningcheng, Inner Mongolia	Jiulongshan Fm., J_2	Zhang *et al.* [142]
	Strenorhagio conjugovenius Zhang, Yang & Ren, 2010	Ningcheng, Inner Mongolia	Jiulongshan Fm., J_2	Zhang *et al.* [142]
	Strenorhagio grimaldi Zhang, Ren & Shih, 2010	Ningcheng, Inner Mongolia	Jiulongshan Fm., J_2	Zhang *et al.* [142]
	Uranorhagio daohugouensis Zhang, Yang & Ren, 2010	Ningcheng, Inner Mongolia	Jiulongshan Fm., J_2	Zhang *et al.* [142]
Zhangsolvidae	*Zhangsolva cupressa* (Zhang, Zhang & Li, 1993)	Laiyang, Shandong	[b]Laiyang Fm., K_1	Zhang *et al.* [137]
Family *incertae sedis*	[a]*Gigantoberis liaoningensis* Huang & Lin, 2007	Beipiao, Liaoning	Yixian Fm., K_1	Huang and Lin [165]
	[a]*Mesomphrale asiaticum* Hong & Wang, 1990	Laiyang, Shandong	Laiyang Fm., K_1	Hong and Wang [45]

(Continued)

Table 23.1 (Continued)

Family	Species	Locality	Horizon/age	Citation
	[a]*Mesorhagiophryne incerta* Hong & Wang, 1990	Laiyang, Shandong	Laiyang Fm., K_1	Hong and Wang [45]
	[a]*Mesorhagiophryne robusta* Hong & Wang, 1990	Laiyang, Shandong	Laiyang Fm., K_1	Hong and Wang [45]
	[a]*Mesostratiomyia laiyangensis* Hong & Wang, 1990	Laiyang, Shandong	Laiyang Fm., K_1	Hong and Wang [45]
	[a]*Stratiomyopsis robusta* Hong & Wang, 1990	Laiyang, Shandong	Laiyang Fm., K_1	Hong and Wang [45]

a) The species is not present in the main text because original description, photos and line-drawings are not precise and the holotype cannot be rechecked.
b) Horizon/Age revised from the original paper based on updated information and data.

References

1 Wiman, N.G. and Jones, V.P. (2013). Influence of oviposition strategy of *Nemorilla pyste* and *Nilea erecta* (Diptera: Tachinidae) on parasitoid fertility and host mortality. *Biological Control* **64** (3): 195–202. https://doi.org/10.1016/j.biocontrol.2012.12.008.

2 Yeates, D.K. and Wiegmann, B.M. (1999). Congruence and controversy: toward a higher-level phylogeny of the Diptera. *Annual Review of Entomology* **44**: 397–428.

3 Yeates, D.K. and Wiegmann, B.M. (2005). Phylogeny and evolution of Diptera: recent insights and new perspectives. In: *The Evolutionary Biology of Flies* (ed. D.K. Yeates and B.M. Wiegmann), 14–44. New York: Columbia University Press.

4 Lambkin, C., Pape, T., Sinclair, B.J. *et al.* (2013). The phylogenetic relationships among infraorders and superfamilies of Diptera based on morphological evidence. *Systematic Entomology* **38**: 164–179. https://doi.org/10.1111/j.1365-3113.2012.00652.x.

5 Wiegmann, B.M., Trautwein, M.D., Winkler, I.S. *et al.* (2011). Episodic radiations in the fly tree of life. *Proceedings of the National Academy of Sciences of the United States of America* **108** (14): 5690–5695. https://doi.org/10.1073/pnas.1012675108.

6 Mcalpine, J.F., Peterson, B.V., Shewell, G.E. *et al.* (eds.) (1981). *Manual of nearctic diptera. Monograph*, vol. **28**. Ottawa, ON: Biosystematics Reaserch Institute.

7 Pape, T., Blagoderov, V., and Mostovski, M. (2011). Order Diptera Linnaeus, 1758, in animal biodiversity: an outline of higher-level classification and survey of taxonomic richness. *Zootaxa* **3148**: 1–237.

8 Bertone, M.A., Courtney, G.W., and Wiegmann, B.M. (2008). Phylogenetics and temporal diversification of the earliest true flies (Insecta: Diptera) based on multiple nuclear genes. *Systematic Entomology* **33** (4): 668–687.

9 Yeates, D.K., Wiegmann, B.M., Courtney, G.W. *et al.* (2007). Phylogeny and systematics of Diptera: two decades of progress and prospects. *Zootaxa* **1668**: 565–590.

10 Nagatomi, A. and Soroida, K. (1985). The structure of the mouthparts of the orthorrhaphous Brachycera (Diptera) with special reference to blood-sucking. *Beiträge zur Entomologie* **35**: 263–368.

11 Karolyi, F., Colville, J.F., Handschuh, S. *et al.* (2014). One proboscis, two tasks: adaptations to blood-feeding and nectar-extracting in long-proboscid horse flies (Tabanidae, *Philoliche*). *Arthropod Structure & Development* **43**: 403–413.

12 Yeates, D.K. (2002). Relationships of the lower Brachycera (Diptera): a quantitative synthesis of morphological characters. *Zoologica Scripta* **31**: 105–121.

13 Yuval, B. (2006). Mating systems of blood-feeding Flies. *Annual Review of Entomology* **51**: 413–440.

14 Tu, Y.Y., Ni, M.Y., Zhong, Y.R. *et al.* (1981). Studies on the constituents of *Artemisia annua. Acta Pharmaceutica Sinica* **16**: 366–370. (in Chinese).

15 Nobelprize.org. Nobel Media AB (2014) The 2015 Nobel Prize in Physiology or Medicine. Press Release, 20 Mar.

16 Massalongo, A.B. (1856). Prodromo di un'entomologia fossile del *m. bolca*. In: *Studii Paleontologici* (ed. G. Antonelli), 11–21. Verona.

17 Tillyard, R.J. (1929). Permian Diptera from Warner's Bay. N.S.W. *Nature* **123**: 778–779.

18 Tillyard, R.J. (1937). The ancestors of the Diptera. *Nature* **139**: 66–67.

19 Riek, E.F. (1977). Four-winged Diptera from the Upper Permian of Australia. *Proceedings of the Linnean Society of New South Wales* **101** (4): 250–255.

20 Willmann, R. (1989). Rediscovered: *Pennotipula patricia*, the oldest known fly. *Naturwissenschaften* **76**: 375–377.

21 Krzemiński, W., Krzemińska, E., and Papier, F. (1994). *Grauvogelia arzvilleriana* sp. n. – the oldest Diptrea species (Lower/Middle Triassic of France). *Acta Zoologica Cracoviensia* 7 (2): 95–99.

22 Handlirsch, A. (1906–1908). *Die fossilen Insekten und die Phylogenie der rezenten Formen. Ein Handbuch für Paläontologen und Zoologen* (ed. W. Engelmann), 1–1430. Leipzig.

23 Carpenter, F.M. (1992). *Treatise on Invertebrate Paleontology, Part R, Arthropoda 4, vol. 3: Superclass Hexapoda*. Colorado and Kansas: Geological Society of America and The University of Kansas.

24 Evenhuis, N.L. (1994). *Catalogue of the fossil flies of the world (Insecta: Diptera)*. Leiden: Backhuys Publishers.

25 Grabau, A.W. (1923). Cretaceous fossils from Shantung. *ACM Communications in Computer Algebra* **48** (3/4): 78–89.

26 Ping, C. (1928). Study of the Cretaceous fossil insects of China. *Palaeontologia Sinica* **13** (1): 1–56.

27 Rohdendorf, B.B. (1946). Evolution of the wing and the phylogeny of oligoneura (Diptera, Nematocera). *Trudy Paleontologicheskogo Instituta, Academii Nauk SSSR* **13** (2): 1–108. (in Russian).

28 Rohdendorf, B.B. (1938). Dipterous insects of the Mesozoic of Karatau. I. Brachycera and part of the Nematocera. *Trudy Paleontologicheskogo Instituta, Akademii Nauk SSSR* **7** (3): 29–67. (in Russian).

29 Ansorge, J. (1996). The Upper Liassic insects of Grimmen (Pomerania, north Germany). *Neue Paläontologische Abhandlungen* **2**: 1–132.

30 Rohdendorf, B.B. (1962). Order Diptera. Flies. In: *Fundamentals of paleontology*, vol. **9** (ed. B.B. Rohdendorf), 307–345. Moscow: Izdatel'stvo Akad. Nauk.

31 Hong, Y.C. (1983). *Middle Jurassic Fossil Insects in North China*. Beijing: Geological Publishing House (in Chinese).

32 Kovalev, V.G. (1990). Diptera, Muscida, in Pozdne-Mezozoyskie Nasekomye Vostochnogo Zabaykal'ya. *Akademiya Nauk SSSR, Trudy Paleontologicheskogo Instituta* **239**: 123–177. (in Russian).

33 Zhang, J.F., Zhang, S., Liu, D.Z., and Shangguan, Y.N. (1986). Fossil insects (Diptera, Nematocera) of Laiyang basin in Shandong Province. *Geology of Shandong* **2** (1): 14–39. (in Chinese with English abstract).

34 Mamaev, B.M. and Krivosheina, N.P. (1966). New date on the taxonomy and biology of Diptera of the family Axymyiddae. *Entomologicheskoe Obozrenie* **45** (1): 168–180. (in Russian).

35 Shi, G.F., Zhu, Y., Shih, C.K., and Ren, D. (2013). A new axymyiid genus with two new species from the Middle Jurassic of China (Diptera: Nematocera: Axymyiidae). *Annals of The Entomological Society of America* **87** (5): 1228–1234.

36 Shi, G.F. and Ren, D. (2013). Progress in the research of fossil Axymyiidae (Insecta: Diptera). *Environmental Entomology* **35** (5): 650–655. (in Chinese with English abstract).

37 Zhang, J.F. (2010). Two new genera and one new species of Jurassic Axymyiidae (Diptera: Nematocera), with revision and redescription of the extinct taxa. *Annals of The Entomological Society of America* **103** (4): 455–464. https://doi.org/10.1603/AN09073.

38 Zhang, J.F. and Lukashevich, E.D. (2007). The oldest known net-winged midges (Insecta: Diptera: Blephariceridae) from the late Mesozoic of northeast China. *Cretaceous Research* **28** (2): 302–309. https://doi.org/10.1016/j.cretres.2006.05.012.

39 Lukashevich, E.D. and Shcherbakov, D.E. (1997). A first find of net-winged midges (Blephariceridae, Diptera) in the Mesozoic. *Neues Jahrbuch für Geologie und Paläontologie-Monatshefte* **11**: 639–646.

40 Ren, D., Lu, L.W., Guo, Z.G., and Ji, S.A. (eds.) (1995). *Faunae and Stratigraphy of Jurassic-Cretaceous in Beijing and the Adjacent Areas*. Beijing: Geological Publishing House.

41 Brown, B.V., Borkent, A., Cumming, J.M. *et al.* (eds.) (2009). *Manual of Central American Diptera*, vol. **1**. Ottawa: NRC Research Press.

42 Zhang, J.F. (1991). New genera and new species of Chironomidae (Diptera, Insecta) from Late Jurassic of China. *Acta Palaeontologica Sinica* **30**: 556–569. (in Chinese with English abstract)

43 Kalugina, N.S. (1980). Mosquitoes of the Chaoboridae and Chironimidae of the sediments of Lake Manlay. *Trudy Sovmestnaya Sovetsko–Mongol'skaya Paleontologicheskaya Ekpeditsiya* **13**: 61–65.

44 Kalugina, N.S. and Kovalev, V.G. (eds.) (1985). *Dipterous insects of Jurassic Siberia*. Moscow: Paleontological Institute, Akademia Nauk (in Russian).

45 Hong, Y.C. and Wang, W.L. (1990). Fossil insects from the Laiyang Basin, Shandong Province. In: *The stratigraphy and palaeontology of Laiyang Basin, Shandong Province* (ed. Regional Geological Surveying Team, Shandong Bureau of Geology and Mineral

Resources), 44–189. Beijing: Geological Publishing House (in Chinese).
46 Rohdendorf, B.B., Hocking, B., Oldroyd, H., and Ball, G.E. (eds.) (1974). *The Historical Development of Diptera*. Edmonton: University of Alberta.
47 Kovalev, V.G. (1982). Some Jurassic Diptera rhagionids (Muscida, Rhagionidae). *Paleontological Journal* **16** (3): 87–99.
48 Lin, Q.B. (1976). The Jurassic fossil insects from Western Liaoning. *Acta Palaeontologica Sinica* **15** (1): 86–116. (in Chinese)
49 Oosterbroek, P. (2018) *Catalogue of the Craneflies of the World (Diptera, Tipuloidea: Pediciidae, Limoniidae, Cylindrotomidae, Tipulidae).* http://nlbif.eti.uva.nl/ccw/index.phpn (accessed 16 November 2017).
50 Zhang, J.F. (2006). Jurassic limoniid dipterans from China (Diptera: Limoniidae). *Oriental Insects* **40** (1): 115–126. https://doi.org/10.1080/00305316.2006.10417463.
51 Lukashevich, E.D. (2009). Limoniidae (Diptera) in the Upper Jurassic of Shar Teg, Mongolia. *Zoosymposia* **3**: 131–145.
52 Pritchard, G. (1983). Biology of Tipulidae. *Annual Review of Entomology* **28** (122). https://doi.org/10.1146/annurev.en.28.010183.000245.
53 Tillyard, R.J. (1933). *The Panorpoid Complex in the British Rhaetic and Lias. Fossil Insects No. 3*. London: British Museum (Natural History).
54 Kopeć, K., Krzemiński, W., Skowron, K., and Coram, R. (2017). The genera *Architipula* Handlirsch, 1906 and *Grimmenia* Krzemiński and Zessin, 1990 (Diptera: Limoniidae) from the Lower Jurassic of England. *Palaeontologia Electronica* **20** (1): 1–7.
55 Bode, A. (1953). Die Insektenfauna des osterniedersächsischen Oberen Lias. *Palaeontographica (A)* **103**: 1–375.
56 Krzemiński, W. and Kovalev, V.G. (1988). The taxonomic status of *Architipula fragmentosa* (Bode) and the family Eoasilidae (Diptera) from the Lower Jurassic. *Systematic Entomology* **13**: 55–56.
57 Hao, J.Y. and Ren, D. (2009). Middle Jurassic Protopleciidae from Daohougou, Inner Mongolia, China (Insecta, Diptera). *Acta Zootaxonomica Sinica* **34** (3): 106–110.
58 Handlirsch, A. (1920). Palaeontologie. In: *Handbuch der Entomologie, Band III* (ed. C.W.M. Schroeder), 117–208. G. Fischer, Jena.
59 Gao, J.Q., Shih, C.K., Zhao, Y.Y., and Ren, D. (2015). New species of *Cretolimonia* and *Mesotipula* (Diptera: Limoniidae) from the Middle Jurassic of Northeastern China. *Acta Geologica Sinica* **89** (6): 1789–1796. https://doi.org/10.1111/1755-6724.12597.
60 Lin, Q.B. (1986). *Early Mesozoic Fossil Insects from South China*. Beijing: Palaeontologia Sinica.
61 Kalugina, N.S. (1986). Flies. Muscida (=Diptera), in Nasekomye v rannemelovykh ekosistemakh zapadnoy Mongolii. *The Joint Soviet-Mongolian Palaeontological Expedition* **28**: 112–125.
62 Zhang, J.F. (1986). Some fossil insects from the Jurassic of northern Hebei, China. In: *The Paleontology and Stratigraphy of Shandong* (ed. Paleontology Society of Shandong, China), 1–95. Beijing: Ocean Press (in Chinese).
63 Rohdendorf, B.B. (1964). Historical development of the Diptera. *Trudy Paleontologicheskogo Instituta, Akademii Nauk SSSR* **100**: 1–311. (in Russian).
64 Blagoderov, V.A. (1994). Dipterans (Mesosciophilidae) from the Lower Cretaceous of Transbaykal. *Paleontological Journal* **27**: 123–130.
65 Zhang, J.F. (2007). New mesosciophilid gnats (Insecta: Diptera: Mesosciophilidae) in the Daohugou biota of Inner Mongolia. China. *Cretaceous Research* **28** (2): 297–301. https://doi.org/10.1016/j.cretres.2006.05.007.
66 Lin, X.Q., Shih, C.K., and Ren, D. (2015). New fossil mesosciophilids (Diptera: Nematocera) from the Yixian Formation (Lower Cretaceous) of western Liaoning, China. *Cretaceous Research* **54**: 86–97. https://doi.org/10.1016/j.cretres.2014.11.007.
67 Li, T. and Ren, D. (2009). A new fossil genus of Mesosciophilidae (Diptera, Nematocera) with two new species from the Middle Jurassic of Inner Mongolia, China. *Progress in Natural Science* **19** (12): 1837–1841.
68 Shi, G.F., Shih, C.K., and Ren, D. (2015). A new genus with two new species of mesosciophilids from the Middle Jurassic of China (Diptera: Nematocera: Mesosciophilidae). *Journal of Natural History* **49**: 1147–1158. https://doi.org/10.1080/00222933.2014.951085.
69 Zhang, Z.J., Hong, Y.C., and Li, Z.Y. (2001). Description of a new fossil genus and species *Huaxiasciophilites jingxiensis* (Diptera: Mesosciophilidae) from Early Cretaceous Jingxi Basin of Beijing. *Entomologia Sinica* **8** (3): 193–198. https://doi.org/10.1111/j.1744-7917.2001.tb00440.x.
70 Gao, J.Q., Shih, C.K., Kopeć, K. *et al.* (2015). New species and revisions of Pediciidae (Diptera) from the Middle Jurassic of Northeastern China and Russia. *Zootaxa* **3963** (2): 240–249. https://doi.org/10.11646/zootaxa.3964.1.4.
71 Kovalev, V.G. (1987). The Mesozoic mycetophiloid Diptera of the family Pleciofungivoridae. *Paleontologicheskii Zhurnal* **1987** (2): 69–82.
72 Geinitz, F.E. (1884). Über die Fauna des Dobbertiner Lias. *Zeitschrift der Deutschen Geologischen Gesellschaft* **36**: 566–583.

73 Zhang, J.F. (2007). New Mesozoic Protopleciidae (Insecta: Diptera: Nematocera) from China. *Cretaceous Research* **28** (2): 289–296. https://doi.org/10.1016/j.cretres.2006.05.009.

74 Lin, X.Q., Shih, C.K., and Ren, D. (2014). Two new species of *Mesoplecia* (Insecta: Diptera: Protopleciidae) from the late Middle Jurassic of China. *Zootaxa* **3838** (5): 545–556. https://doi.org/10.11646/zootaxa.3838.5.3.

75 Lin, X.Q., Shih, C.K., and Ren, D. (2015). Revision of the genus *Epimesoplecia* Zhang, 2007 (Diptera, Nematocera, Protopleciidae) with five new species. *ZooKeys* **492**: 123–143.

76 Zhang, J.F. (2007). Some anisopodoids (Insecta: Diptera: Anisopodoidea) from late Mesozoic deposits of northeast China. *Cretaceous Research* **28** (2): 281–288. https://doi.org/10.1016/j.cretres.2006.05.008.

77 Krzemiński, W. and Prokop, J. (2011). Prokop. *Ptychoptera deleta* Novák, 1877 from the Early Miocene of the Czech Republic: redescription of the first fossil attributed to Ptychopteridae (Diptera). *ZooKeys* **130** (130): 299–305.

78 Lukashevich, E.D. (2008). Ptychopteridae (insecta: diptera): history of its study and limits of the family. *Paleontological Journal* **42** (1): 66–74.

79 Liu, L.X., Shih, C.K., and Ren, D. (2012). Two new species of Ptychopteridae and Trichoceridae from the Middle Jurassic of Northeastern China (Insecta: Diptera: Nematocera). *Zootaxa* (3501, 3501): 55–62.

80 Lukashevich, E.D., Ansorge, J., Krzemiński, W., and Krzemińska, E. (1998). Revision of Eoptychopterinae (Diptera: Eoptychopteridae). *Polskie Pismo Entomologiczne* **67** (3–4): 311–343.

81 Hao, J.Y., Dong, K.Q., and Ren, D. (2009). Middle Jurassic Eoptychopteridae from Daohugou, Inner Mongolia, China (Insecta, Diptera, Eoptychopteridae). *Acta Geologica Sinica* **34** (1): 106–110.

82 Lukashevich, E.D., Coram, R.A., and Jarzembowski, E.A. (2001). New true flies (Insecta: Diptera) from the lower Cretaceous of Southern England. *Cretaceous Research* **22** (4): 451–460.

83 Rasnitsyn, A.P. (1992). *Strashila incredibilis*, a new enigmatic mecopteroid insect with possible siphonapteran affinities from the Upper Jurassic of Siberia. *Psyche A Journal of Entomology* **99** (4): 323–333. https://doi.org/10.1155/1992/20491.

84 Vršanský, P., Ren, D., and Shih, C.K. (2010). Nakridletia ord. n. – enigmatic insect parasites support sociality and endothermy of pterosaurs. *Amba Projekty* **8** (1): 1–16.

85 Huang, D.-Y., Nel, A., Cai, C.Y. *et al.* (2013). Amphibious flies and paedomorphism in the Jurassic period. *Nature* **495** (7439): 94–97.

86 Ansorge, J. and Krzemiński, W. (2002). Lower Jurassic tanyderids (Diptera: Tanyderidae) from Germany. *Studia Dipterologica* **9**: 21–29.

87 Ansorge, J. (1994). Tanyderidae and Psychodidae (Insecta; Diptera) from the Lower Jurassic of Northeastern Germany. *Palaontologische Zeitschrift* **68** (12): 199–210.

88 Dong, F., Shih, C.K., Skibińska, K. *et al.* (2015). New species of Tanyderidae (Diptera) from the Jiulongshan Formation of China. *Alcheringa: An Australasian Journal of Palaeontology* **39** (4): 494–507.

89 Handlirsch, A. (1909). Zur phylogenie und flügelmorphologie der ptychopteriden (Dipteren). *Annalen des Kaiserlich-Königlichen Naturhistorischen Hofmuseums in Wien* **23**: 263–272.

90 Dong, F., Shih, C.K., and Ren, D. (2015). A new genus of Tanyderidae (Insecta: Diptera) from Myanmar amber, Upper Cretaceous. *Cretaceous Research* **54**: 260–265. https://doi.org/10.1016/j.cretres.2014.12.011.

91 Kania, I. and Nel, A. (2013). New fossil Tipulidae from the volcano-sedimentary Latest Oligocene of Bes-Konak (Turkey). *Polish Journal of Entomology* **82** (4): 327–338.

92 Shih, C.K., Dong, F., Kania, I. *et al.* (2015). A new species of Tipulidae (Diptera) from the Lower Cretaceous Yixian Formation of Liaoning, China – evolutionary implications. *Cretaceous Research* **54**: 98–105.

93 Ribeiro, G.C. and Lukashevich, E.D. (2014). New *Leptotarsus* from the Early Cretaceous of Brazil and Spain: the oldest members of the family Tipulidae (Diptera). *Zootaxa* **3753** (4): 347–363. https://doi.org/10.11646/zootaxa.3753.4.4.

94 Guérin-Méneville, F.E. (1831). Insectes. In: *Voyageautour du monde, execute par ordre du Roi, sur la corvette de samajeste La Coquille etc* (ed. L.I. Duperrey), 1–21. Atlas, Paris plates: Histoire Naturelle, Zoologie.

95 Hågvar, S. and Krzemińska, E. (2007). Contribution to the winter phenology of Trichoceridae (Diptera) in snow-covered southern Norway. *Studia Dipterologica* **14**: 271–283.

96 Krzemińska, E., Krzemiński, W., and Dahl, C. (eds.) (2009). *Monograph of Fossil Trichoceridae (Diptera) over 180 Million Years of Evolution*, 35–40. Krakow, Poland: Institute of Systematics and Evolution of Animals of the Polish Academy of Sciences Press.

97 Zhang, J.F. (2006). New winter crane flies (Insecta: Diptera: Trichoceridae) from the Jurassic Daohugou Formation (Inner Mongolia, China) and their associated biota. *Canadian Journal of Earthences* **43** (1): 9–22.

98 Dong, F., Shih, C.K., and Ren, D. (2014). Two new species of Trichoceridae from the Middle Jurassic Jiulongshan Formation of Inner Mongolia, China. *ZooKeys* **411** (411): 145–160.

99 Zhang, J.F. (2015). Archisargoid flies (Diptera, Brachycera, Archisargidae and Kovalevisargidae) from the Jurassic Daohugou biota of China, and the related biostratigraphic correlation and geological age. *Journal of Systematic Palaeontology* **13** (10): 857–881. https://doi.org/10.1080/14772019.2014.960902.

100 Wang, F.Y., Shih, C.K., Ren, D., and Wang, Y.J. (2017). Quantitative assessments and taxonomic revision of the genus *Archirhagio* with a new species from Daohugou, China (Diptera: Archisargidae). *Systematic Entomology* **42** (1): 230–239.

101 Zhang, K.Y., Yang, D., Ren, D., and Shih, C.K. (2007). The earliest species of the extinct genus *Archisargus* from China (Diptera: Brachycera: Archisargidae). *Annales Zoologici* **57** (4): 827–832. https://doi.org/10.3161/000345407783742006.

102 Zhang, J.F. (2012). Orientisargidae fam. n., a new Jurassic family of Archisargoidea (Diptera, Brachycera), with review of Archisargidae from China. *ZooKeys* **238** (238): 57–76. https://doi.org/10.3897/zookeys.238.3624.

103 Zhang, K.Y., Yang, D., Ren, D., and Shih, C.K. (2010). New archisargids from China (Insecta: Diptera). *Entomological Science* **13** (1): 75–80. https://doi.org/10.1111/j.1479-8298.2009.00350.x.

104 Mostovski, M.B. (1996). To the knowledge of Archisargoidea (Diptera, Brachycera). families Eremochaetidae and Archisargidae. *Russian Entomological Journal* **5** (1–4), 117–124.

105 Zhang, K.Y., Yang, D., and Ren, D. (2008). Middle Jurassic fossils of the genus *Sharasargus* from Inner Mongolia, China (Diptera: Archisargidae). *Entomological Science* **11**: 269–272. https://doi.org/10.1111/j.1479-8298.2008.00271.x.

106 Mostovski, M.B. (1997). To the knowledge of fossil dipterans of superfamily Archisargoidea (Diptera, Brachycera). *Paleontological Journal* **31** (1): 72–78. (in Russian with English abstract)

107 Zhang, K.Y., Yang, D., Ren, D., and Shih, C.K. (2007). The oldest *Calosargus* Mostovski, 1997 from the Middle Jurassic of China (Diptera: Brachycera: Archisargidae). *Zootaxa* **1645**: 1–17.

108 Zhang, J.F. (2010). Records of bizarre Jurassic brachycerans in the Daohugou biota, China (Diptera, Brachycera, Archisargidae and Rhagionemestriidae). *Palaeontology* **53** (2): 307–317. https://doi.org/10.1002/bip.20327.

109 Zhang, J.F. (2012). Distinct but rare archisargid flies from the Jurassic of China (Diptera, Brachycera, Archidargidae) with discussion of the systematic position of *Origoasilus pingquanensis* Zhang *et al.*, 2011. *Journal of Paleontology* **86** (5): 878–885. https://doi.org/10.1666/12-028.1.

110 Stuckenberg, B.R. (1973). The Athericidae, a new family in the lower Brachycera (Diptera). *Annals of the Natal Museum* **21**: 649–673.

111 Zhang, J.F. (2012). New horseflies and water snipe-flies (Diptera: Tabanidae and Athericidae) from the Lower Cretaceous of China. *Cretaceous Research* **36**: 1–5. https://doi.org/10.1016/j.cretres.2012.01.004.

112 Chen, J.B., Wang, M., Engel, M.S. *et al.* (2014). Extreme adaptations for aquatic ectoparasitism in a Jurassic fly larva. *eLife* **3** (3): e02844. https://doi.org/10.7554/eLife.02844.

113 Zhang, K.Y., Yang, D., and Ren, D. (2014). New short-horned flies (Diptera: Eremochaetidae) from the Early Cretaceous of China. *Zootaxa* **3760**: 479–486. https://doi.org/10.11646/zootaxa.3760.3.15.

114 Ren, D. and Guo, Z.G. (1995). A new genus and two new species of short-horned flies of Upper Jurassic from Northeast China (Diptera: Eremochaetidae). *Insect Science* **2** (4): 300–307. https://doi.org/10.1111/j.1744-7917.1995.tb00051.x.

115 Ren, D. (1998). Late Jurassic Brachycera from Northeastern China (Insecta: Diptera). *Acta Zootaxonomica Sinica* **23** (1): 65–83. (in Chinese with English abstract).

116 Grimaldi, D.A. and Barden, P. (2016). The Mesozoic family Eremochaetidae (Diptera: Brachycera) in Burmese amber and relationships of Archisargoidea: Brachycera, in Cretaceous amber, part VIII. *American Museum Novitates* **3865**: 1–29.

117 Zhang, J.F. (2011). Three distinct but rare kovalevisargid flies from the Jurassic Daohugou biota, China (Insecta, Diptera, Brachycera, Kovalevisargidae). *Palaeontology* **54** (1): 163–170. https://doi.org/10.1111/j.1475-4983.2010.01010.x.

118 Yeates, D.K. (1994). The cladistics and classification of the Bombyliidae (Diptera: Asiloidea). *Bulletin of the American Museum of Natural History* **219** (219): 1–191.

119 Wedmann, S. (2007). A nemestrinid fly (Insecta: Diptera: Nemestrinidae: cf. Hirmoneura) from the Eocene Messel pit (Germany). *Journal of Paleontology* **81** (5): 1114–1117.

120 Ansorge, J. and Mostovski, M.B. (2000). Redescription of *Prohirmoneura jurassica* Handlirsch 1906

(Diptera: Nemestrinidae) from the Lower Tithonian lithographic limestone of Eichstätt (Bavaria). *Neues Jahrbuch fur Geologie und Palaontologie-Monatshefte* **2000** (4): 235–243.

121 Rohdendorf, B.B. (1968). New Mesozoic nemestrinids (Diptera, Nemestrinidae). In: *Jurassic insects of the Karatau* (ed. B.B. Rohdendorf), 180–189. Moscow: Nauka (in Russian).

122 Nagatomi, A. and Yang, D. (1998). A review of extinct Mesozoic genera and families of Brachycera (Insecta, Diptera, Orthorrhapha). *Entomologist's Monthly Magazine* **134** (1998): 95–196.

123 Ren, D. (1998). Flower-associated Brachycera flies as fossil evidences for Jurassic angiosperm origins. *Science* **280** (5360): 85–88. https://doi.org/10.1126/science.280.5360.85.

124 Zhang, K.Y., Yang, D., Ren, D., and Ge, F.C. (2008). New Middle Jurassic tangle-veined flies from Inner Mongolia, China. *Acta Palaeontologica Polonica* **53** (1): 159–162.

125 Zhang, K.Y., Yang, D., and Ren, D. (2011). A new brachycerans family Origoasilidae fam. nov. from the Late Mesozoic of China (Insecta: Diptera). *Acta Geologica Sinica (English Edition)* **85** (5): 994–997. https://doi.org/10.1111/j.1755-6724.2011.00233.x.

126 Zhang, K.Y., Yang, D., and Ren, D. (2007). Notes on the extinct family Protapioceridae, with description of a new species from China (Insecta: Diptera: Asiloidea). *Zootaxa* **1530** (1): 27–32.

127 Zhang, J.F. (2015). *Pseudapiocera shandongensis* gen. et sp. nov., a protapiocerid fly (Diptera: Brachycera: Protapioceridae) from the Early Cretaceous Jehol Biota, China. *Alcheringa: An Australasian Journal of Palaeontology* **39** (4): 459–464. https://doi.org/10.1080/03115518.2015.1022335.

128 Ussatchev, D.A. (1968). New Jurassic Asilomorpha (Diptera) in Karatau. *Entomologicheskoe Obozrenie* **47**: 617–628.

129 Zhang, K.Y., Yang, D., and Ren, D. (2008). The first Middle Jurassic *Protobrachyceron* Handlirsch fly (Diptera: Brachycera: Protobrachyceridae) from Inner Mongolia (China). *Zootaxa* **1879**: 61–64. https://doi.org/10.5281/zenodo.184177.

130 Mostovski, M.B. and Martínez-Delclòs, X. (2000). New Nemestrinoidea (Diptera: Brachycera) from the Upper Jurassic-Lower Cretaceous of Eurasia, taxonomy. *Entomological Problems* **31** (2): 137–148.

131 Zhang, J.F. (2010). New species of *Palaeobolbomyia* Kovalev and *Ussatchovia* Kovalev (Diptera, Brachycera, Rhagionidae) from the Callovian-Oxfordian (Jurassic) Daohugou biota of China: Biostratigraphic and paleoecologic implications. *Geobios* **43** (6): 663–669. https://doi.org/10.1016/j.geobios.2010.06.004.

132 Kerr, P.H. (2010). Phylogeny and classification of Rhagionidae, with implications for *Tabanomorpha* (Diptera: Brachycera). *Zootaxa* **2592**: 1–133.

133 Krzemiński, W. and Krzeminska, E. (2003). Triassic Diptera: descriptions, revisions and phylogenetic relations. *Acta Zoologica Cracoviensia* **46** (Supplement – Fossil Insects: 153–184.

134 Zhang, J.F. (2013). Snipe flies (Diptera: Rhagionidae) from the Daohugou Formation (Jurassic), Inner Mongolia, and the systematic position of related records in China. *Palaeontology* **56** (1): 217–228. https://doi.org/10.1111/j.1475-4983.2012.01192.x.

135 Zhang, J.F. and Li, H.J. (2012). New taxa of snipe flies (Diptera: Brachycera: Rhagionidae) in the Daohugou biota, China. *Paleontological Journal* **46** (2): 157–163.

136 Mostovski, M.B. (2000). Contributions to the study of fossil snipe-flies (Diptera: Rhagionidae): the genus *Palaeobolbomyia*. *Paleontological Journal* **34** (Supplement 3): 360–366.

137 Zhang, J.F., Zhang, S., and Li, L.Y. (1993). Mesozoic gadflies (Insecta: Diptera). *Acta Palaeontologica Sinica* **32** (6): 662–672.

138 Kovalev, V.G. and Mostovski, M.B. (1997). A new genus of snipe-flies (Diptera, Rhagionidae) from the Mesozoic of eastern Transbaikalia. *Paleontologicheskii Zhurnal* **1997** (5): 86–90. (in Russian with English abstract).

139 Zhang, K.Y., Yang, D., and Ren, D. (2006). The first snipe fly (Diptera: Rhagionidae) from the Middle Jurassic of Inner Mongolia, China. *Zootaxa* **1134** (1134): 51–57.

140 Zhang, K.Y., Yang, D., and Ren, D. (2008). A new genus and species of Middle Jurassic rhagionids from China (Diptera, Rhagionidae). *Biologia* **63** (1): 113–116. https://doi.org/10.2478/s11756-008-0012-4.

141 Tashiro, H. and Schwardt, H.H. (1953). Biological studies of horseflies in New York. *Journal of Economic Entomology* **46** (5): 813–822.

142 Zhang, K.Y., Yang, D., Ren, D., and Shih, C.K. (2010). An evolutional special case in the Lower Orthorrhapha: some attractive fossil flies from the Middle Jurassic of China (Insecta: Diptera: Brachycera). *Zoological Journal of the Linnean Society* **158** (3): 563–572.

143 Grimaldi, D.A. (2016). Diverse orthorrhaphan flies (Insecta: Diptera: Brachycera) in amber from the Cretaceous of Myanmar: Brachycera in Cretaceous amber, part VII. *Bulletin of the American Museum of Natural History* **408** (1): 1–131.

144 Arillo, A., Peñalver, E., Pérezdelafuente, R. *et al.* (2015). Long-proboscid brachyceran flies in Cretaceous amber (Diptera: Stratiomyomorpha: Zhangsolvidae). *Systematic Entomology* **40** (1): 242–267. https://doi.org/10.1111/syen.12106.

145 Lin, Q.B. (1980). Fossil insects of the Mesozoic from Zhejiang and Anhui Provinces. In: *Division and Correlation of Mesozoic Deposits of Volcanic Origin in Zhejiang and Anhui Provinces* (ed. Nanjing Institute of Geology and Paleontology, Academia Sinica), 211–234. Beijing: Science Press (in Chinese).

146 Zhang, J.F. (2004). First description of axymyiid fossils (Insecta: Diptera: Axymyiidae). *Geobios* **37** (5): 687–694. https://doi.org/10.1016/j.geobios.2003.04.007.

147 Hong, Y.C. (1984). Class Insecta, Diptera. In: *Paleontological Atlas of North China (II) Mesozoic* (ed. Tianjin Institute of Geology and Mineral Resources), 175–179. Beijing: Geological Publishing House (in Chinese).

148 Zhang, J.F. (2004). Nematocerans dipterans from the Jurassic of China (Insecta: Diptera: Limoniidae, Tanyderidae). *Paleontologicheskil Zhurnal* **5**: 53–57. (in Russian with English abstract).

149 Hao, J.Y. and Ren, D. (2009). Two new fossil species of Limoniidae (Diptera: Nematocera) from the Middle Jurassic of northeast, China. *Entomological News* **120** (2): 171–178. https://doi.org/10.3157/021.120.0206.

150 Zhang, J.F. (2008). Three new species of Mesosciophilid gnats from the Middle-Late Jurassic of China (Insecta: Diptera: Nematocera: Mesosciophilidae). *Pakistan Journal of Biological Sciences* **11** (22): 2567. https://doi.org/10.3923/pjbs.2008.2567.2572.

151 Wang, Q., Zhao, Y.Y., and Ren, D. (2012). Two new species of Mesosciophilidae (Insecta: Diptera: Nematocera) from the Yanliao Biota of Inner Mongolia, China. *Alcheringa: An Australasian Journal of Palaeontology* **36** (4): 509–514.

152 Gao, J.Q., Shi, G.F., Shih, C.K., and Ren, D. (2015). Two new species of Paramesosciophilodes (Diptera, Nematocera, Mesosciophilidae) from the Middle Jurassic of China. *ZooKeys* **2015** (511): 117.

153 Hong, Y.C. (1992). The study of Early Cretaceous Coleoptera, Raphidioptera, Diptera (Insecta) of Kezuo, west Liaoning Province. *Acta Geologica Gansu* **1**: 1–11.

154 Hong, Y.C., Wang, Z.B., and Sun, W. (1992). The stratigraphy and fossil insects of Zhongguang basin, Hebei province. *Memoirs of the Beijing Natural History Museum* **51** (3): 20–36.

155 Ren, D. and Krzemiński, W. (2002). Eoptychopteridae (Diptera) from the Middle Jurassic of China. *Annales Zoologici* **52** (2): 207–210.

156 Hao, J.Y., Ren, D., and Shih, C.K. (2009). New fossils of Eoptychopteridae (Diptera) from the Middle Jurassic of Northeastern China. *Acta Geologica Sinica* **83** (2): 222–228. https://doi.org/10.1111/j.1755-6724.2009.00022.x.

157 Zhang, J.F. (2004). A new gigantic species of *Eoptychopterina* (Diptera: Eoptychopteridae) from Jurassic of Northeastern China. *Oriental Insects* **38** (1): 173–178. https://doi.org/10.1080/00305316.2004.10417384.

158 Krzemiński, W. and Ren, D. (2001). *Praemacrochile chinensis* sp. n. from the Middle Jurassic of China [Diptera: Tanyderidae]. *Polskie pismo entomologiczne* **70** (2): 127–129.

159 Hong, Y.C. (1996). A new genus *Huaxiarhyphus* gen. nov. (Diptera, Rhyphoidea) from the Honcheng formation, Hebei Province. *Memoirs of Beijing Natural History Museum* **55** (12): 47–53.

160 Zhang, J.F. and Zhang, H.C. (2003). Two new species of archisargids (Insecta: Diptera: Archisargidae) from the Upper Jurassic Daohugou Formation (Inner Mongolia, Northeastern China). *Paleontological Journal* **37** (4): 409–412.

161 Zhang, K.Y., Li, J.H., Yang, D., and Ren, D. (2009). A new species of *Archirhagio* Rohdendorf, 1938 from the Middle Jurassic of Inner Mongolia of China (Diptera: Archisargidae). *Zootaxa* **1984**: 61–65.

162 Zhang, J.F. (2014). New male eremochaetid flies (Diptera, Brachycera, Eremochaetidae) from the Lower Cretaceous of China. *Cretaceous Research* **49** (2): 205–213. https://doi.org/10.1016/j.cretres.2014.02.012.

163 Zhang, J.F. (2011). Two new species of *Palaeoarthroteles* Kovalev and Mostovski (Diptera, Rhagionidae) from the Callovian-Oxfordian (Jurassic) Daohugou biota of China. *Geobios* **44** (6): 635–639. https://doi.org/10.1016/j.geobios.2011.01.006.

164 Zhang, J.F. (2017). On the enigmatic *Sinonemestrius* Hong & Wang, 1990, with description of a new species based on a complete fossil fly (Diptera, Brachycera, Tabanomorpha, Heterostomidae). *Deutsche Entomologische Zeitschrift* **64**: 61–67. https://doi.org/10.3897/dez.64.11724.

165 Huang, D.-Y. and Lin, Q.B. (2007). A new soldier fly (Diptera, Stratiomyidae) from the Lower Cretaceous of Liaoning Province, Northeast China. *Cretaceous Research* **28** (2): 317–321. https://doi.org/10.1016/j.cretres.2006.05.006.

24

Mecoptera – Scorpionflies and Hangingflies

Xiaodan Lin[1], *Chungkun Shih*[1,2], *Sheng Li*[1], *and Dong Ren*[1]

[1] *Capital Normal University, Haidian District, Beijing, China*
[2] *National Museum of Natural History, Smithsonian Institution, Washington, DC, USA*

24.1 Introduction to Mecoptera

Mecoptera, from a Greek word meaning "long wings", are a small relict order, commonly called "scorpionflies" or "hangingflies". The name of scorpionfly is derived from the enlarged and uplifted male genitalia which look like stingers of scorpions. Extant scorpionflies comprise nine families, 38 genera with 650 species distributed worldwide. In contrast, extinct families are much more abundant, with a total of 39 families including 210 genera and about 710 species in the fossil records, especially speciose in the Mesozoic [1–3].

Mecoptera are one of the oldest groups of Holometabola, and the only one that have compound eyes in larval phase [4–6]. The latest molecular phylogenetic studies in Misof *et al.* 2014 indicate that extant Mecoptera are a monophyletic group, and Mecoptera have very close relationship with Siphonaptera (fleas) and Diptera (true flies) (Figure 24.1) [7]. The phylogenetic analysis, based on morphological characters of most extinct and extant families of Mecoptera, Siphonaptera and Diptera, revealed similar result of the monophyly of a clade including all extant and most geochronologically recent extinct mecopterans and phylogenetic relationships among extant mecopterans, siphonapterans and dipterans. However, the order Mecoptera including extinct and extant families are paraphyletic (Figure 24.2) [2].

Extant scorpionflies comprise nine families: Apteropanorpidae, Bittacidae, Boreidae, Choristidae, Eomeropidae, Meropeidae, Nannochoristidae, Panorpidae and Panorpodidae, most of which have chewing mouthparts with different degrees of extension. The morphology and habits of larvae are relatively diverse: the larvae of Nannochoristidae are aquatic and feeding on small arthropods; Boreidae larvae live near the bryophyte covered by snow and ice, and most are phytophagous; most larvae of Bittacidae and Panorpidae are necrophilous; the larval feeding habits of Panorpodidae, Choristidae and Apteropanorpidae are not clear yet [8, 9]. On the other hand, adults of extant Mecoptera often live in the cold shrub layer and feed on dead arthropods or rotten organic substances (Figure 24.3), but in some cases stealing caught insects on spider webs (Figure 24.4) [5, 10, 11]. Bittacidae hang on the stem or leaf edge and prey on other arthropods (Figure 24.5) [12]. Panorpodidae feed on leaves (or maybe some pollen) [13], and Eomeropidae have been readily attracted to oatmeal bait in Chile [14]. To date, there are only three extant families documented in China, Panorpidae Latreille, 1802, Bittacidae Handlirsch, 1906 and Panorpodidae Handlirsch, 1920 [15].

The body sizes of Mecoptera range from 2 to 35 mm in length. The largest species belong to Panorpidae and Bittacidae. Mecopterans usually have two pairs of membranous wings with venation, just like most basal Holometabola insects. The fore- and hind wings share similar shape, but hind wings slightly smaller than forewings [16, 17]. Like many other insects, heads of Mecoptera are nearly triangular in dorsal view, with large and separated compound eyes, and antennae mostly filiform. But in very rare fossil cases, antenna is comb-like with multiple antennomeres [5, 17–19].

The fossil Mecoptera are now classified into eight suborders: Aneuretopsychina Rasnitsyn & Kozlov, 1990; Eumecoptera Tillyard, 1919; Nannomecoptera Hinton, 1981; Neomecoptera Hinton, 1958; Paratrichoptera Tillyard, 1919; Pistillifera Willmann, 1987; Protodiptera Tillyard, 1937; and Protomecoptera Tillyard, 1917. In addition, 12 more extinct families are suborder *incertae sedis*: Anormochoristidae Tillyard, 1926; Apteropanorpidae Byers, 1965; Belmontiidae Tillyard, 1919; Choristopsychidae Martynov, 1937; Englathaumatidae Novokshonov, Ross, Krzemiński & Soszynska-Maj, 2016; Lithopanorpidae Carpenter, 1930; Mendozachoristidae Brauckmann, Gallego, Hauschke,

Rhythms of Insect Evolution: Evidence from the Jurassic and Cretaceous in Northern China, First Edition.
Edited by Dong Ren, Chungkun Shih, Taiping Gao, Yongjie Wang, and Yunzhi Yao.

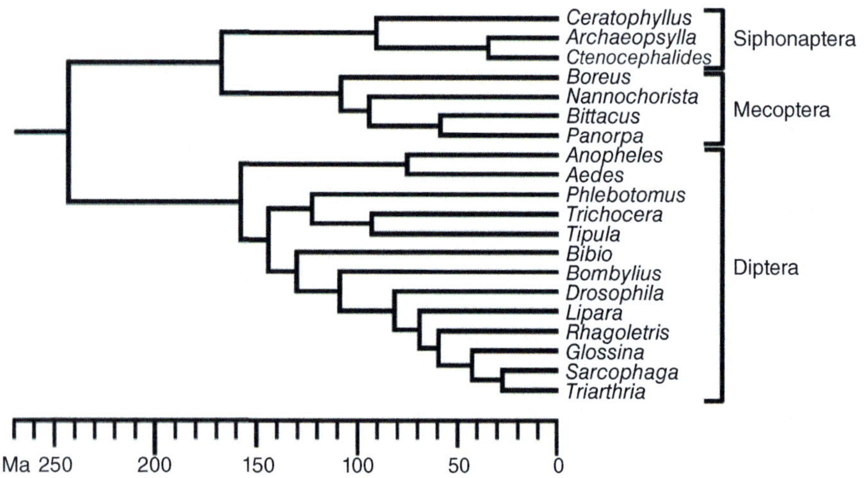

Figure 24.1 Dated phylogenetic tree inferred through a maximum-likelihood analysis of 413 459 amino acid sites divided into 479 metapartitions, the part of four extant Mecoptera families, Diptera and Siphonaptera. Source: Modified from [7].

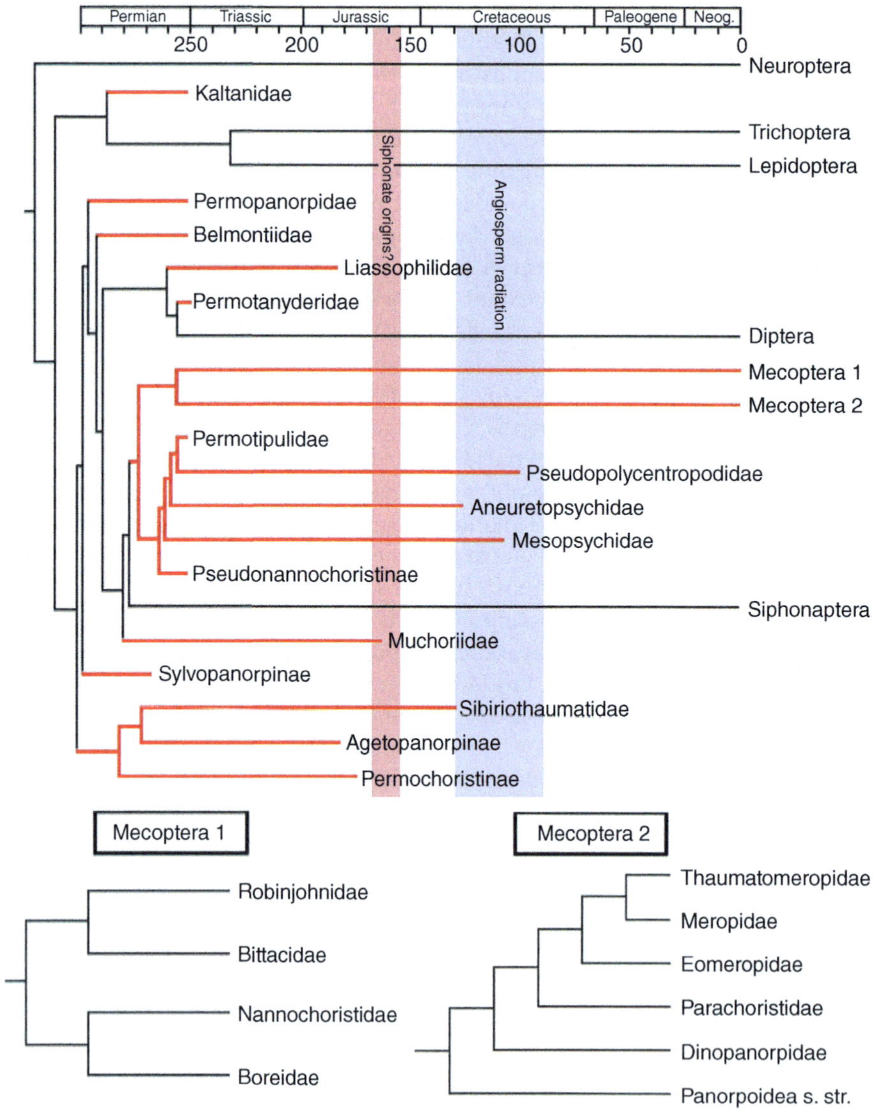

Figure 24.2 Phylogenetic results of Mecoptera, Diptera and Siphonaptera (modified from [2]). Top are geologic periods calibrated to a time scale in million years. Red bar denotes the possible originated time of long proboscis, the purple bar shows the transition from gymnosperm to angiosperm. The red branches denote the Mecoptera families and clades. Below are two trees indicating the details of Mecoptera 1 and 2.

Figure 24.3 Male scorpionflies feeding. Source: Photo by Jason Shih.

Figure 24.4 A female scorpionfly stealing caught insects on spider web. Source: Photo by Jason Shih.

Figure 24.5 Male (left) and female of *Bittacus planus* Cheng, 1949. Source: Photo provided by Dr. Baozhen Hua.

Martins-Neto, Groening, Ilger & Lara, 2010; Pseudonannochoristidae Novokshonov, 1994; Sibiriothaumatidae Sukatcheva & Novokshonov, 1998; Thaumatomeropidae Willmann, 1978; Tshekarchiereidae Novokshonov, 1997 and Volitorididae Hong, 1982. Four of the eight suborders, Aneuretopsychina, Eumecoptera, Nannomecoptera and Pistillifera, have fossil records in China.

Aneuretopsychina are an extinct suborder, now comprising only four families: Aneuretopsychidae Rasnitsyn & Kozlov, 1990, Mesopsychidae Tillyard, 1917, Nedubroviidae Bashkuev, 2011 and Pseudopolycentropodidae Handlirsch, 1920. The body sizes vary from only 1.79 mm long for *Parapolycentropus paraburmiticus* (Pseudopolycentropodidae) [20] to 28 mm long for *Lichnomesopsyche gloriae* (Mesopsychidae) [21]. There are no fossil larvae of Aneuretopsychina reported so far. In general, the adults of these scorpionflies have evolved long siphonate proboscides to feed on the pollination drops of gymnosperms, such as gingko, cycads or bennettitales [21–23] (details see Chapter 28). However, *Parapolycentropus* Grimaldi & Rasnitsyn, 2005 of Pseudopolycentropodidae, reported in Myanmar (Burmese) amber, have very small body sizes and only one pair of forewings, while the hind wings degraded as small lobes similar to haltere (Figure 24.6) [20, 24].

Eumecoptera include three families, Holcorpidae Willmann, 1989, Mesorthophlebiidae Hong, 1983 and Parachoristidae Handlirsch, 1937. Although Parachoristidae were considered to have close relationship with Permochoristidae and Orthophlebiidae, but distinguished from them by many obvious morphological characters of wings [25]. The enlarged and uplifted male genitalia are typical for many male Mecoptera, three families of which, the extant family Panorpidae, and

Figure 24.6 *Parapolycentropus paraburmiticus* Grimaldi & Rasnitsyn, 2005 (new material, male, CNU-MEC-MA-2015047).

the extinct families Orthophlebiidae and Holcorpidae, have extremely elongated abdominal segments for male genitalia. In particular, the sixth to eighth abdominal segments of Holcorpidae are extremely elongated, most likely for sexual display or fighting to compete for potential mates (see Figure 24.17 and Section 30.2).

Nannomecoptera have two families, Robinjohniidae and the extant Nannochoristidae. Robinjohniidae were established by Martynova in 1948, but the original type genus (*Robinjohnia* Martynova, 1948) was transferred to Nannochoristidae by Riek in 1968 [26]. Then Novokshonov described and revised *Minusinia* as the only genus of the family in 1994 and 1997 [27, 28]. Nannochoristidae play an important role in extant Mecoptera and also have abundant fossil records in the Mesozoic. The most common genus in the mid-Mesozoic of Asia is *Itaphlebia* Sukatsheva, 1985, which exhibit some similar characters as extant species [29].

Pistillifera are further divided into two infraorders, Opisthogonopora Willmann, 1987 and Raptipedia Willmann, 1987. Opisthogonopora have two clades, Meropomorpha and Panorpomorpha, and Raptipedia include two families, Cimbrophlebiidae and the extant Bittacidae. Meropomorpha only contain one family Meropeidae, but Panorpomorpha comprise two clades, Eomeropina (Eomeropidae) and Panorpina (Choristoidea and Panorpoidea). The Pistillifera have higher diversity than other suborders in the Mesozoic of China. To date, there are six reported families in this suborder: Bittacidae, Cimbrophlebiidae, Meropeidae, Eomeropidae, Orthophlebiidae and Panorpidae.

In recent years, the results of the phylogenetic analyses based on morphological characters and molecular data have deepened the understanding of various families. At the same time, the geometric morphometric analyses (GMA) also bring a breakthrough in confirming the genus–species relationship of many families. For instance, Mesopsychidae have already been determined by the classification position in Aneuretopsychina, nevertheless the relationships of different genus and species are still unclear, Lin *et al.* initiated both phylogenetic and GMA and obtained the interesting and important results [30]. In addition, as Cimbrophlebiidae are an extinct family with only a few morphological characters clearly identified, the results by using the phylogenetic analyses and GMA have proven the monophyly of this family [31].

Figure 24.7 Nuptial gifts and mating of *Dicerapanorpa magna* Chou, 1981. Source: Photo provided by Dr. Baozhen Hua.

Special Nuptial Gifts

Some male birds, insects and spiders adopt a mating strategy by providing nuptial gifts to potential mating partners before mating or courtship to increase the mating opportunity. The nuptial gifts may be food items or seminal material. For some katydids, the nuptial gifts are packaged with the sperm of the male, while for some scorpionflies and hangingflies, the gifts are captured prey like dead insects or fruits of plants (Figure 24.7). In addition, some scorpionflies may use "gland secretion", usually from the salivary glands or Needham's sac [32, 33]. For some species, using the salivary gland secretion as presents is more common than offering the prey [34]. On the other hand, *Panorpa liui* Hua, 1997 and *Panorpa jilinensis* Zhou, 1999 generally offer the dead insects as nuptial gifts and without well-developed salivary glands [35, 36].

Hangingflies (Bittacidae) are predatory, usually hanging their body in some low vegetation by their forelegs, and catching soft-bodied insects with their middle and hind legs [12, 37–40]. Male hangingflies usually catch small insects and release sex pheromone to attract females. If the prey insects are not acceptable, the male hangingfly will discard it and catch a new and better one to attract the potential mates. Such nuptial gift behavior is relatively rare in other orders of insects. If the "gift" is deemed to be acceptable, the female will accept the gift for food (Figure 24.7). While the female enjoys this offering, the male completes its task of gene propagation [41].

24.2 Progress in the Studies of Fossil Mecoptera

The fossil records indicate that the earliest Mecoptera occurred in the Early Permian, and then flourished in the Late Permian and the Mesozoic. However, after the Late Cretaceous, due to the environmental conditions, most of the fossil species died out in succession [42, 43].

Numerous cladistics analyses suggested that Kaltanidae are the most basal group in Mecoptera. Their fossil records, the same as Permochoristidae, can be found in the entire Permian [2, 27, 44]. To date, there are 10 described genera, but most of them with only incomplete wings preserved. Recently, much fossil evidence indicates the origination of Mecoptera was no later than Carboniferous/Permian boundary [28, 45–54].

The Aneuretopsychina, replacing Permochoristidae, became the most dominant group of Mecoptera from the Latest Permian to the Middle Triassic. In the Jurassic and the Early Cretaceous, Aneuretopsychina played the roles as important pollinators for a large proportion of gymnosperms [2]. According to the fossil records, the earliest record of this suborder is a mesopsychid, belonging to *Permopsyche* Bashkuev, 2011, from the Late Permian of Russia and Australia [54, 55].

In the Late Jurassic, Aneuretopsychina were gradually replaced by species of Parachoristidae and Orthophlebiidae. There is no doubt that the basal group of Eumecoptera belongs to Parachoristidae with nine genera and 23 species, which were close to Permochoristidae. Furthermore, the oldest described fossil is *Triasoparachorista huaxiaensis* Hong, 2009 from the Middle Triassic of China [55]. Other four genera, *Neoparachorista* Riek, 1954, *Triassochorista* Willmann, 1989, *Kirgizichorista* Novokshonov, 2001 and *Panorpaenigma* Novokshonov, 2001 have been recorded from the Late Triassic of Australia and Kyrgyzstan [50, 56, 57].

Only two families (Nannochoristidae and Robinjohniidae) are included in Nannomecoptera. Hitherto the earliest described fossil is *Afrochoristella maclachlani* Pinto & Ornellas, 1978 from the Early Permian of South Africa. However, *Minusinia martynovae* Novokshonov, 1994 is the oldest record of this suborder [27, 28, 58]. The earliest modern groups of the Mecoptera are derived from the Orthophlebiidae [53], belonging to Pistillifera which are the largest suborder up to date.

The research into Mecoptera fossils has a history of more than 100 years. Handlirsch and other researchers published many papers about mecopterans in the early twentieth century [47, 59–64]. However, the study of fossil Mecoptera started relatively late in China, and there were no formal taxonomic researches until the 1970s, when Dr. Hong and Dr. Lin initiated the early studies of mecopteran fossil in China [65, 66]. During the following 40 years, Chinese paleoentomologists published more than 50 papers and treatises. Up to now, 18 families and about 190 species, most from Northeastern China, have been reported with clear morphological features, such as wing and body characters.

During this time, many researchers have made outstanding contributions, especially Youchong Hong, Qibin Lin, Haichun Zhang, Junfeng Zhang, Dong Ren and Chungkun Shih. We also appreciate our collaborators, Conrad C. Labandeira, Jorge A. Santiago-Blay, Alexandr Rasnitsyn, Alexei Bashkuev, Carol L. Hotton, David Dilcher, M. Amelia V. Logan, Petrulevičius F. Julian and Matthew J. H. Shih, and also our CNU Team past and present, Yongjie Wang, Junxia Zhang, Nan Liu, Jianhai Sun, Xiaoguang Yang, Qi Wang, Xiao Qiao, Chen Wang, Sulin Liu, He Ding, Lei Li, Yizi Cao, Xiao Zhang and Xiaodan Lin.

24.3 Representative Fossils of Mecoptera from Northern China

Suborder Aneuretopsychina Rasnitsyn & Kozlov, 1990

Family Aneuretopsychidae Rasnitsyn & Kozlov, 1990

Aneuretopsychidae, with a distinctly long proboscis, are a small and mysterious family, comprising two genera and seven species hitherto. The type species is *Aneuretopsyche rostrata* Rasnitsyn & Kozlov, 1990 [67], from the Late Jurassic of Kazakhstan. Ren *et al.* amended the diagnosis of this family in 2011: a distinctly prolonged and siphonate proboscis present, probably opisthognathous or hypognathous, covered with well-developed annulated microtrichia, a lobed and fleshy pseudolabellum at the tip; antenna longer than proboscis and multi-articulated, antennomeres covered with annulated microtrichia; forewing Sc multiple-branched, Rs and MA branched, MP with four branches, CuA single or probably bifurcating; hind wing distinctly broader than forewing and setae on the legs formed distinctive rings [68]. In 2012, Qiao *et al.* further revised the diagnosis by adding: compound eyes big and oval, ocellus small and nearly round; CuA joining MP slightly basal to Rs + MA from R; the Rs + MA distinctly proximal to the first bifurcation of MP [69].

Only one genus included from the Cretaceous of Northern China: *Jeholopsyche* Ren, Shih & Labandeira, 2011.

Jeholopsyche Ren, Shih & Labandeira, 2011

Jeholopsyche Ren, Shih & Labandeira, 2011, *ZooKeys*, 129, 20 [68] (original designation).

Type species: *Jeholopsyche liaoningensis* Ren, Shih & Labandeira, 2011.

In forewing, Sc with three branches; R_1 single, MP from MP + CuA slightly proximal to Rs + MA from R; Rs + MA bifurcation distinctly basal to the first bifurcation of MP. Both fore- and mid basitarsus shorter than the rest of four tarsomeres combined, the basitarsus in hind legs almost as long as the rest four tarsomeres.

Distribution and age: Liaoning; Early Cretaceous.

Four species included from the Cretaceous of China (see Table 24.1).

Jeholopsyche liaoningensis **Ren, Shih & Labandeira, 2011 (Figure 24.8)**

Jeholopsyche liaoningensis Ren, Shih & Labandeira, 2011: *ZooKeys*, 129, 20.

Locality and horizon: Huangbanjigou, Beipiao, Liaoning, China; Lower Cretaceous, Yixian Formation.

The body length excluding antennae and proboscis at least 23 mm; proboscis length 6.8 mm, and antenna length at least 10 mm; Head oviform, compound eyes widely separated; Proboscis straight and siphonate, and a pseudolabellum present at the tip (Figure 24.8b); Pronotum much smaller than meso- and metanotum; All legs covered with annulated pubescence, Hind legs longer than fore- and mid legs, with at least one apical spur; All pretarsi with one pair of claws and reduced pulvillus; Forewing slender, length at least 21.5 mm, and width 6 mm; Sc long, almost at the same level as MA; Rs and MA with two branches, MP with four branches; Anal area distinctly broad, A_1 long; Pterostigma absent. Abdomen with 10 visible segments, 9–11 segments obscure but enlarged, indicating male sex [2, 68].

(a) (b)

Figure 24.8 *Jeholopsyche liaoningensis* Ren, Shih & Labandeira, 2011 (Holotype, CNU-MEC-LB-2005002p, male). (a) Photograph of habitus; (b) Detailed mouthparts in alcohol. Source: Donated by Dr. Chungkun Shih

Family Mesopsychidae Tillyard, 1917

Mesopsychidae is a small and extinct family, comprising 11 genera with 30 species from the Upper Permian to Lower Cretaceous hitherto. Similar to other members of Aneuretopsychina, they also have long siphonate proboscides used for feeding on pollination drops of gymnosperms [2, 22, 23]. The oldest fossil mesopsychids belong to *Mesopsyche* Tillyard, 1917, from the Uppermost Permian Russia [70, 71], and others described from Australia, Kyrgyzstan, South Africa, Ukraine, Tajikistan and China [30]. The diagnosis of this family was emended by Ren, Labandeira & Shih in 2010: adult with long and siphonate proboscis, antenna filiform; the setae on the legs not forming rings; fore- and hind wings nearly have the same size and shape; the vast majority of the taxa with wings emarginated at the CuP apex; MP with four branches, and CuA single [21]. Gao *et al.* reported one mesopsychid with ramified antennae, *Vitimopsyche pectinella* Gao, Shih, Labandeira, Santiago-Blay, Yao & Ren, 2016, from the Lower Cretaceous Yixian Formation of Northeastern China. These special antennae might have provided benefit of increased sensory capability [19, 72, 73], but extant mecopterans do not have such antennae.

Genera included from the Jurassic and Cretaceous of Northern China: *Vitimopsyche* Novokshonov & Sukatsheva, 2001, *Lichnomesopsyche* Ren, Labandeira & Shih, 2010 and *Epicharmesopsyche* Shih, Qiao, Labandeira & Ren, 2013.

Vitimopsyche **Novokshonov & Sukatsheva, 2001**

Vitimopsyche Novokshonov & Sukatsheva, 2001, *Paleontol. J.*, 35 (2), 179 [74] (original designation).

Type species: *Vitimopsyche torta* Novokshonov & Sukatsheva, 2001.

Wing is large and broad, distinctly emarginated at CuP apex. In forewing, Sc long and far distal to MA forking, Rs with two short veins, MP_1 distal to RS + MA. The stem of MA and MA_1 formed S-shape, MP from MP + CuA more proximal to Rs + MA from R in both fore- and hind wing. Hind wing Sc short, not distal to MA; MP from MP + CuA slightly distal or proximal to Rs + MA from R_1.

Distribution and age: Inner Mongolia, Middle Jurassic; Liaoning, Early Cretaceous.

Three species included from the Jurassic and Cretaceous of China (see Table 24.1).

Vitimopsyche pectinella **Gao, Shih, Labandeira, Santiago-Blay, Yao & Ren, 2016 (Figure 24.9)**

Vitimopsyche pectinella Gao, Shih, Labandeira, Santiago-Blay, Yao & Ren, 2016: *Proc. R. Soc. Lond. B*, 283 (20161448), 2–3.

(a) (b)

Figure 24.9 *Vitimopsyche pectinella* Gao, Shih, Labandeira, Santiago-Blay, Yao & Ren, 2016 (Holotype, CNU-MEC-LB-2012088p). (a) Photograph; (b) Line drawing. Source: Modified from [19].

Locality and horizon: Huangbanjigou, Beipiao, Liaoning, China; Lower Cretaceous, Yixian Formation.

The length of proboscis is at least 10 mm, proboscis straight and covered with dense setae. Antenna is pectinate, length about 7.8 mm. Forewing broad, Sc long, with only one anterior branch; R_1 single; Rs and MA with two branches each; MP with four long branches; two anal veins well-developed. Hind wing costal area covered with dense and short setae, Sc long and without anterior branch, reaching C distal to MP forking, MP from the stem of MP + CuA distal to Rs + MA from R_1 [19].

***Lichnomesopsyche* Ren, Labandeira & Shih, 2010**

Lichnomesopsyche Ren, Labandeira & Shih, 2010, *Acta Geol. Sin.-Engl.*, 84 (4), 721 [21] (original designation).

Type species: *Lichnomesopsyche gloriae* Ren, Labandeira & Shih, 2010.

The specific epithet is dedicated to Ms. Gloria A. Shih, spouse of Dr. Chungkun Shih, for her care and providing support and encouragement to him in collecting and studying insect and plant fossils. Wing is elongated, distinctly emarginated at CuP apex. In forewing, Sc proximal to MA forking, and with one anterior branch; Rs and MA both with two endings; MP originated from MP + CuA proximal or distal to Rs + MA from R_1; the base of MA straight. In hind wing, Sc distal to MA originated from Rs, and MP from MP + CuA distal to Rs + MA from R.

Distribution and age: Inner Mongolia; Middle Jurassic.

Three species included from the Jurassic of China (see Table 24.1).

***Lichnomesopsyche gloriae* Ren, Labandeira & Shih, 2010 (Figure 24.10)**

Lichnomesopsyche gloriae Ren, Labandeira & Shih, 2010: *Acta Geol. Sin.-Engl.*, 84 (4), 721–722.

Locality and horizon: Daohugou, Ningcheng, Inner Mongolia, China; Middle Jurassic, Jiulongshan Formation.

The length of forewing is 24 mm and width is 8 mm; proboscis length as least 9 mm, body length excluding antennae and proboscis is 23 mm. Head nearly triangular in dorsal view, compound eyes large. Proboscis long and straight, covered with dense setae, clypeus well-developed. Forewing slender, Sc distal to MA forking, with one anterior branch; Rs and MA with two branches each; Rs forking distal to MA; MP with four long branches, proximal to MA + Rs forking; anal area broad, A_1 and A_2 well-developed. Hind wing similar to forewing in size and shape, Sc short, without anterior branch, R_1 curved in pterostigma area. Wings lack of circular spots. Female abdomen with 10 segments, cerci with two segments (Figure 24.10). Male abdomen

Figure 24.10 *Lichnomesopsyche gloriae* Ren, Labandeira & Shih, 2010 (Holotype, CNU-MEC-NN-2005020p). Source: Donated by Dr. Chungkun Shih.

with 9 segments and the 9–11 segments enlarged and bulbous [21].

Lichnomesopsyche prochorista Lin, Shih, Labandeira & Ren, 2016 (Figure 24.11)

Lichnomesopsyche prochorista Lin, Shih, Labandeira & Ren, 2016: *BMC Evo. Bio.*, 16 (1), 8–11.

Locality and horizon: Daohugou, Ningcheng, Inner Mongolia, China; Middle Jurassic, Jiulongshan Formation.

The length of the body excluding antennae and proboscis is 22.2 mm; forewing length is 24.6 mm and width 7.5 mm. Head prognathous, compound eyes widely separated. Proboscis long and slightly curved, covered with dense setae. Forewing slender, Sc long with one anterior branch; MP forking slightly proximal to MA + Rs; two anal veins well-developed. Hind wing similar to forewing in shape but slightly smaller, Sc short, without anterior branch, and R_1 curved in pterostigma area. In both fore- and hind wing, Rs forking proximal to MA. Female abdomen with 10 visible segments, cerci with at least two segments and not fused with each other (Figure 24.11). Male abdomen with 9 visible segments, 9 and 10 segments bulbous [30].

Epicharmesopsyche Shih, Qiao, Labandeira & Ren, 2013

Epicharmesopsyche Shih, Qiao, Labandeira & Ren, 2013, *Acta Geol. Sin.-Engl.*, 87 (5), 1237 [75] (original designation).

Type species: *Epicharmesopsyche pentavenulosa* Shih, Qiao, Labandeira & Ren, 2013.

Antenna long and filiform. In both fore- and hind wings, Sc long and without anterior branches, almost at the same level of Rs forking; MP_1 forking proximal to Rs + MA; MP with five branches.

Distribution and age: Inner Mongolia; Middle Jurassic.

Only one species included from the Jurassic of China (see Table 24.1).

Epicharmesopsyche pentavenulosa Shih, Qiao, Labandeira & Ren, 2013 (Figure 24.12)

Epicharmesopsyche pentavenulosa Shih, Qiao, Labandeira & Ren, 2013: *Acta Geol. Sin.-Engl.*, 87 (5), 1237.

Locality and horizon: Daohugou, Ningcheng, Inner Mongolia, China; Middle Jurassic, Jiulongshan Formation.

The body length excluding the antennae is 20.8 mm, antenna very long, about 16.6 mm in length. Compound eyes oval and large, three ocelli present. Proboscis absent or missing in preservation. In left forewing, Sc with only one anterior branch; R_1 curved near the wing margin; Rs forking much distal to MA; the stem of MA and MA_1 forming a S-shape; CuA, CuP, A_1 and A_2 single and well-developed. The right forewing is just the same as the left in shape, but much narrower. There are dark and symmetrical spots on both forewings. In left hind wing, MP forking proximal to Rs + MA, and there are four long frenula bristles at the base of the costal margin. In contrast, right hind wing has three long frenula bristles. Abdomen with eight visible segments, male genitalia present as small lobe [75].

Phylogenetic and Geometric Morphometric Research on Mesopsychidae

In order to clarify the genus-level relationships of Mesopsychidae and explore the origin of the siphonate proboscis, a matrix of data including 16 taxa and 26 morphological characters were analyzed by NONA and PAUP. To supplement the phylogeny analysis, the GMA comprise the same taxa with 38 and 42 landmarks were produced. Both results support that Mesopsychidae are monophyletic group and two genera, *Vitimopsyche* and *Lichnomesopsyche* are also monophyletic, *Permopsyche* and *Mesopsyche* are paraphyletic (Figure 24.13). Moreover, it is hypothesized that long proboscis may have independently originated four or five times within early Mecoptera (Figure 24.13a), and the repeated rounds

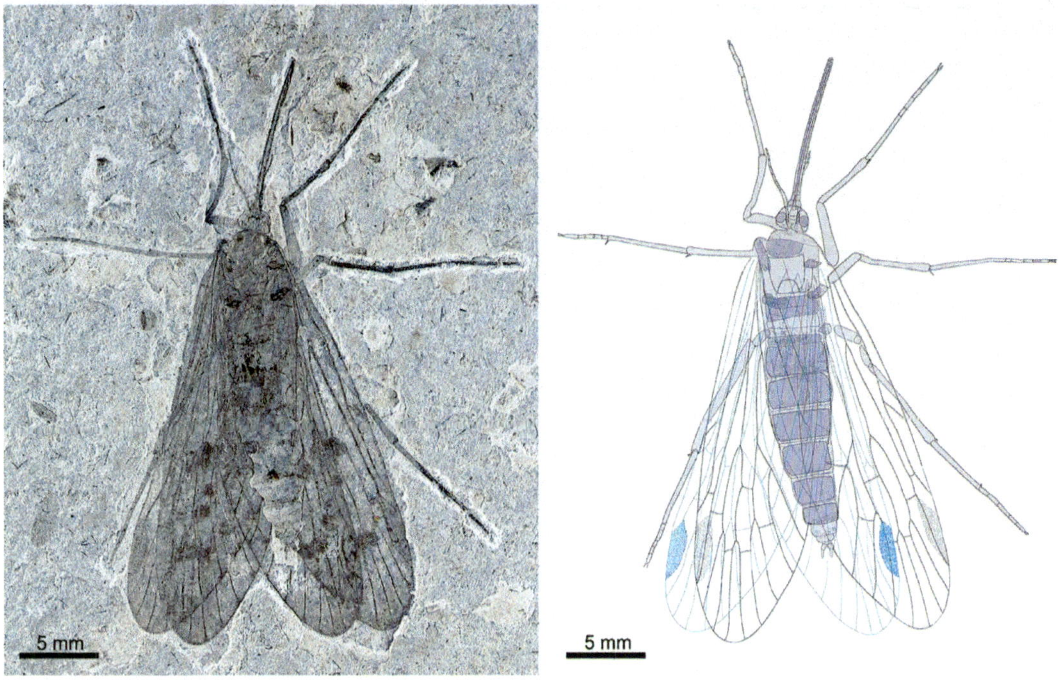

(a) (b)

Figure 24.11 *Lichnomesopsyche prochorista* Lin, Shih, Labandeira & Ren, 2016 (Holotype, CNU-MEC-NN-2015002p). (a) Photograph; (b) Overlay drawing. Source: Modified from [30].

Figure 24.12 *Epicharmesopsyche pentavenulosa* Shih, Qiao, Labandeira & Ren, 2013 (Holotype, CNU-MEC-NN-2011095p).

of suppression of *ext* and *hth* genes could explain the proboscis origin [30].

Family Pseudopolycentropodidae Handlirsch, 1925

Pseudopolycentropodidae, a small and extinct family of early Mecoptera, are considered to have close relationship with Mesopsychidae, Aneuretopsychidae and Nedubroviidae [54]. Pseudopolycentropodids are diagnosed by the following characters: forewing triangular or ovoid, hind wing with similar shape as forewing, but much smaller in size or degraded as a minute lobe similar to haltere such as *Parapolycentropus* Grimaldi & Rasnitsyn, 2005; Sc short in both fore- and hind wings; R_1 single, Rs with four branches; M with five branches; dc cell present; and CuA single. Up to date, there are four genera and 13 species described from Myanmar [24], USA [24], Kazakhstan [64], Kyrgyzstan [76], Germany [60, 77], United Kingdom [78], France [79] and China. The oldest fossil record, *Pseudopolycentropus triasicus* Papier, Nel & Grauvogel-Stamm, 1996, is from the Middle Triassic (Ladinian) of France. Interestingly, the hind wings of *Parapolycentropus* Grimaldi & Rasnitsyn, 2005 (Figure 24.6) are totally reduced to small haltere-like structures. Moreover, their wing shape, venation and other body features are also different from other pseudopolycentropodids. It is hypothesized that

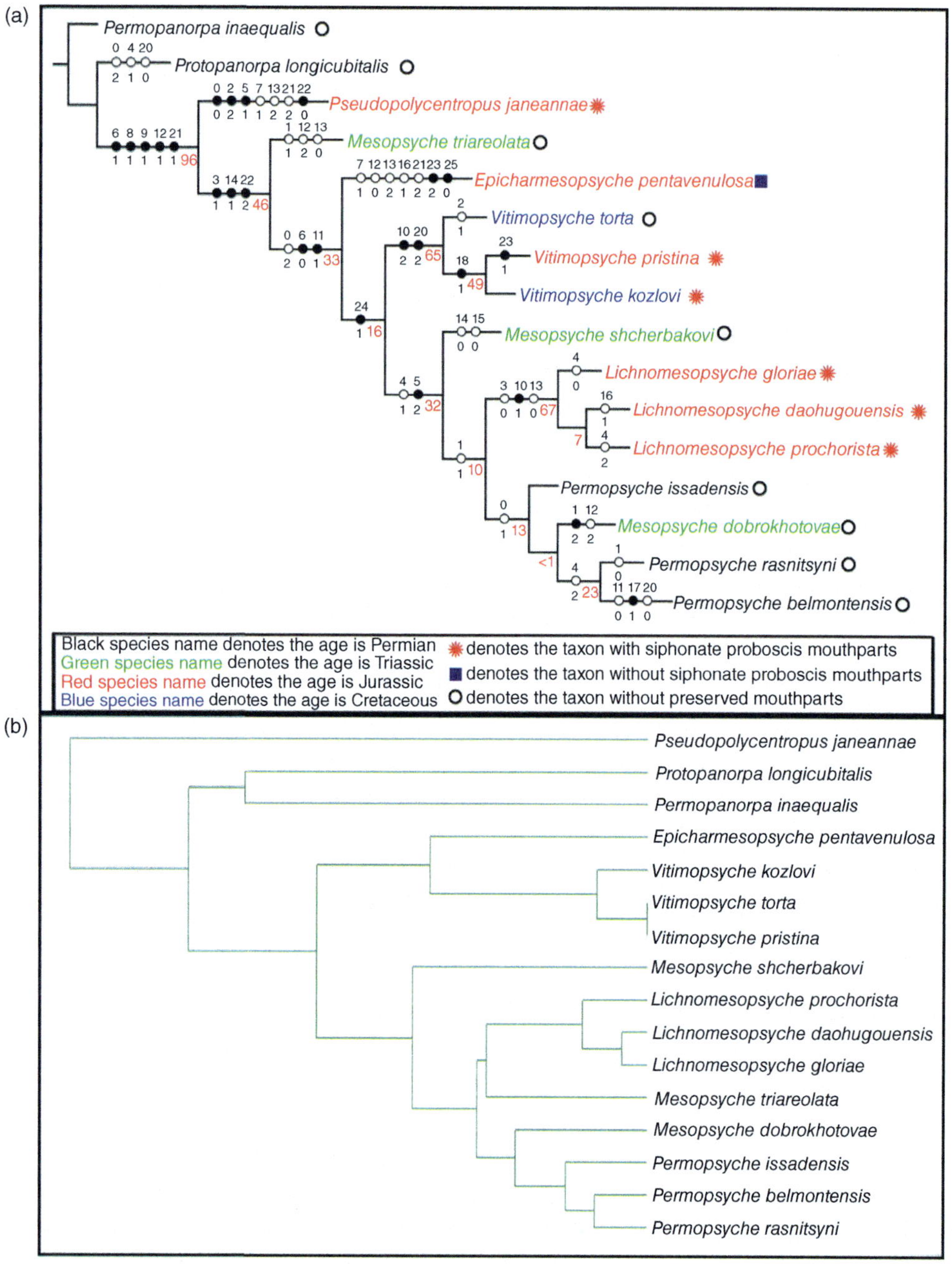

Figure 24.13 Phylogenetic and geometric morphometric results of Mesopsychidae. (a) The most parsimonious tree with bootstrap support by NONA. (b) Resulting tree of the geometric morphometric analyses for 38 landmarks (without showing the tree length). White circles indicate homoplasious characters, and the black circles indicate non-homoplasious characters. The numbers above branches are character numbers, below branches are character states, and the red numbers below the branches are bootstrap values in (a). Source: Modified from [30].

Parapolycentropus may be classified into a separated family from Pseudopolycentropodidae.

Genera included from the Jurassic and Cretaceous of Northern China: *Pseudopolycentropus* Handlirsh, 1906 and *Sinopolycentropus* Shih, Yang, Labandeira & Ren, 2011.

***Pseudopolycentropus* Handlirsh, 1906**

Pseudopolycentropus Handlirsh, 1906-1908, *Verlagvon Wilhelm Engelmann, Leipzig*, 482–483 [59] (original designation).

Type species: *Phryganidium perlaeformis* Geinitz, 1884.

Forewing broad and triangular, Sc short, R_1 single, Rs with 4 branches, M with 5 branches, a discal cell at the center present, CuA and CuP single, A short or absent, and few cross-veins. Hind wing smaller than forewing or some highly reduced.

Distribution and age: Inner Mongolia; Middle Jurassic.

Three species included from the Jurassic of China (see Table 24.1).

Pseudopolycentropus janeannae Ren, Shih & Labandeira, 2010 (Figure 24.14)

Pseudopolycentropus janeannae Ren, Shih & Labandeira, 2010: *Acta Geol. Sin.-Engl.*, 84, 23–24.

Locality and horizon: Daohugou, Ningcheng, Inner Mongolia, China; Middle Jurassic, Jiulongshan Formation.

The specific epithet is dedicated to Ms. Jane Ann Shih, daughter of Dr. Chungkun Shih, for demonstrating initiative, caring and leadership in her study and work, and for providing inspiration and support to paleontology. Head with a long and curved proboscis, distinctly composed of three parts and antenna multi-articulated and filiform, much longer than mouthparts (Figure 24.14a–c).

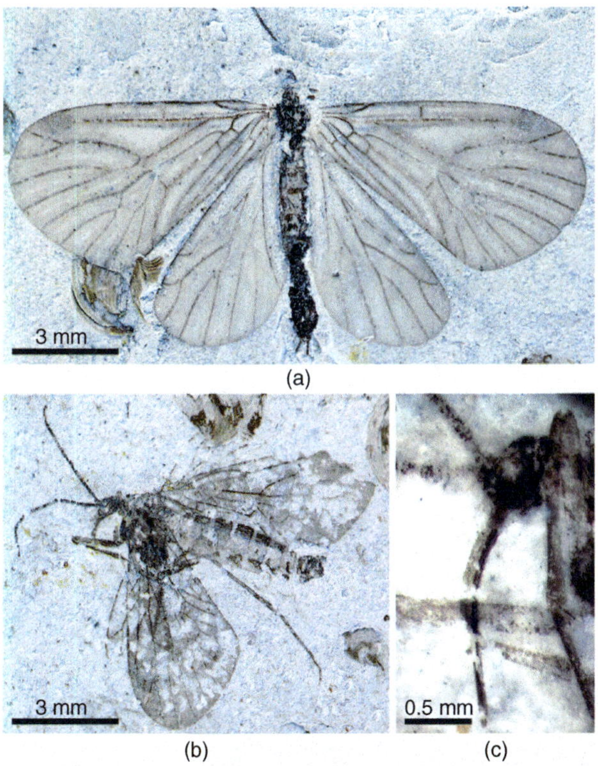

Figure 24.14 *Pseudopolycentropus janeannae* Ren, Shih & Labandeira, 2010 (Holotype, CNU-MEC-NN-2005001p, female; Paratype, CNU-MEC-NN-2005004p, male) (a) Holotype; (b) Paratype; (c) Details of proboscis in (b). Source: Donated by Dr. Chungkun Shih.

The forewing broad, Sc reaching the origin of Rs, c-r cross-vein perpendicular to R_1 and C; the stem of R slightly curved, Rs stem straight, R_{2+3} stem arched with two short branches; M_{4+5} forking proximal to M_{1+3}; m-cua closer to the base of dc cell, a1-a2 proximal to cup-a1. Hind wing distinctly smaller than forewing in size but similar in shape, venation simple, Sc single and short, M with two long branches, CuA single and arched. Hind legs of male covered with annulated pubescence. Abdomen of female has 10 segments, cerci at least two articles (Figure 24.14a). Male abdomen with 9 visible segments, 8–11 forming an enlarged genital bulb (Figure 24.14b) [80].

Sinopolycentropus Shih, Yang, Labandeira & Ren, 2011

Sinopolycentropus Shih, Yang, Labandeira & Ren, 2011, *ZooKeys*, 130, 289 [81] (original designation).

Type species: *Sinopolycentropus rasnitsyni* Shih, Yang, Labandeira & Ren, 2011.

The specific epithet is dedicated to Dr. Alexandr Rasnitsyn for his contribution to paleoentomology and his recognition, with M. V. Kozlov, of the first fossil scorpionfly (*Aneuretopsyche rostrata*) with a documented long proboscis in 1990 [67]. Sc merged with R in the base; R_{2+3} forking proximal to R_{4+5}. Antenna moniliform, robust, covered with annulated setae. Labial palps multi-articulated and covered with long occipital bristles.

Distribution and age: Inner Mongolia; Middle Jurassic.

Only one species included from the Jurassic of China (see Table 24.1).

Sinopolycentropus rasnitsyni Shih, Yang, Labandeira & Ren, 2011 (Figure 24.15)

Sinopolycentropus rasnitsyni Shih, Yang, Labandeira & Ren, 2011: *ZooKeys*, 130, 289–292. **Locality and horizon:** Daohugou, Ningcheng, Inner Mongolia, China; Middle Jurassic, Jiulongshan Formation.

The body length excluding antennae and proboscis is 5.5 mm. Head nearly spheroidal and bulbous compound eyes. Antenna is 2.0 mm long, moniliform and with annulated setae. Siphonate proboscis is 2.0 mm long and slightly decurved, terminus lacks pseudolabellae, labial palps short with three articles (Figures 24.15 and 24.16). Forewing broad, 6.1 mm long and 2.4 mm wide, Sc single, reaching C proximal to the Rs origin, Rs stem straight; M forking slightly proximal to Rs, M with five branches; M + CuA forking proximal to R. Hind wing similar shape to forewing but much smaller in size. Abdomen elongate with nine visible segments, two to five segments distinctly broad, rectangular subgenital is present [81].

Figure 24.15 *Sinopolycentropus rasnitsyni* Shih, Yang, Labandeira & Ren, 2011 (Holotype, CNU-MEC-NN-2010044p).

Significance of Antennal and Mouthpart Modifications

Structural features of the antennae and mouthparts (Figure 24.16) may be relevant for seeking conspecifics and host plants. The unique characteristics of antennal shape, structure and length in *Sinopolycentropus rasnitsyni* might be associated with the sensory detection of mates or food sources. The presence of prominent setae encircling the annular area of each antennomeres, only found in *S. rasnitsyni*, similar to the mouthparts of Aneuretopsychidae having the similar well-developed annulated microtrichia [68], may be involved in detection of conspecific pheromonal cues or specific chemicals from particular plant hosts. Of more importance are the structure, size, and shape of the proboscis, which indicate that it would have been used for access and imbibition of pollination drops or similar ovulate fluids from a variety of smaller gymnospermous fructifications of potential plants such as Bennettites, Coniferae and ancient ginkgo [2, 23, 81].

Suborder Eumecoptera Tillyard, 1919

Family Holcorpidae Willmann, 1989

Holcorpidae, a mysterious and controversial extinct family, have very close relationship with Orthophlebiidae. Fossils of this family are very rare. There are only two genera firmly assigned to holcorpids hitherto, the

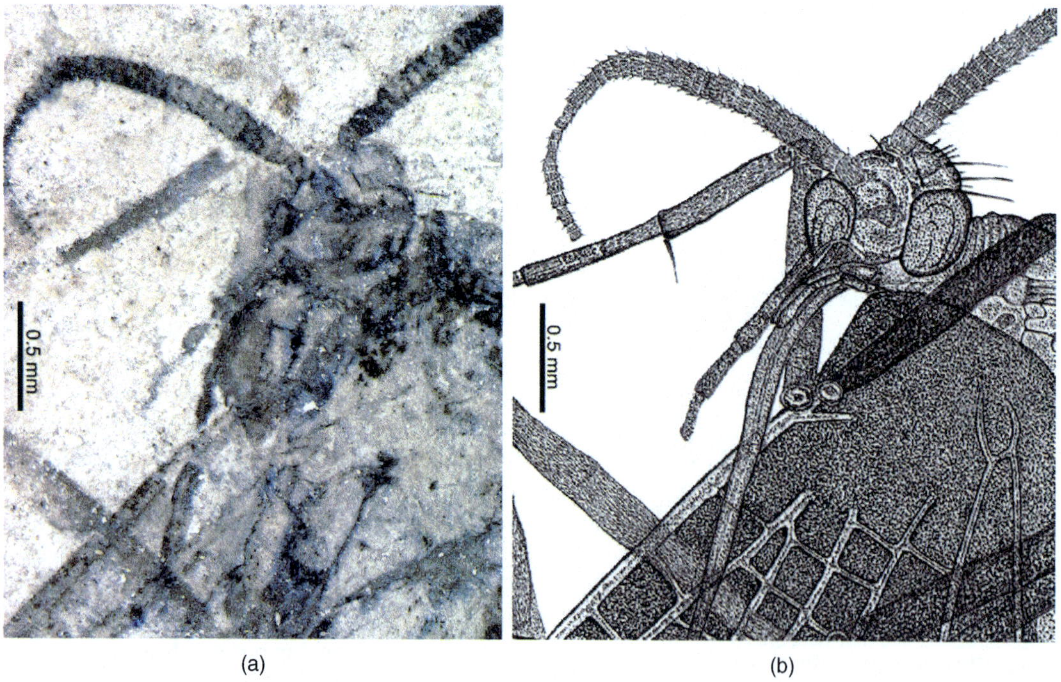

Figure 24.16 Proboscis and antennae of *Sinopolycentropus rasnitsyni* Shih, Yang, Labandeira & Ren, 2011 (Holotype, CNU-MEC-NN-2010044p). (a) Photograph of head and proboscis; (b) Line drawing of head and proboscis. Modified from [81].

type genus *Holcorpa* Scudder, 1878 and *Conicholcorpa* Li, Shih, Wang and Ren, 2017. Holcorpidae have been reported from Colorado, USA [82], British Columbia, Canada [83] and Inner Mongolia, China [84] from the Middle Jurassic to Late Eocene. The diagnostic characters of this family are M with five branches in both fore- and hind wings; A6–A8 elongated in males (female unknown); dististyli extended and slender, without tooth in the base [83].

Although two other genera, *Fortiholcorpa* and *Miriholcorpa,* have been reported to have affinities with Holcorpidae, Wang *et al.* deemed both as Family *incertae sedis* mainly due to significant differences in branching pattern of Media (M) veins and relative length of A8 for *F. paradoxa,* and indiscernible preservation of five-branched M veins in hind wing for *M. forcipata* [85].

Only one genus included from the Jurassic of Northern China: *Conicholcorpa* Li, Shih, Wang & Ren, 2017.

Conicholcorpa Li, Shih, Wang & Ren, 2017

Conicholcorpa Li, Shih, Wang & Ren, 2017, *Acta Geol. Sin.-Engl.*, 91 (3), 798 [84] (original designation).

Type species: *Conicholcorpa stigmosa* Li, Shih, Wang & Ren, 2017.

In forewing, R_1 with three branches; the stem of Rs_{1+2} 1.3 times as long as the Rs_{3+4}, Rs_{1+2} forking more distal to Rs_{3+4}; M and Rs forking almost at the same level. In hind wing, Sc short, at the same level of the middle of C; the stem of Rs_{1+2} twice as long as RS_{3+4}; Rs_{1+2} forking more distal to RS_{3+4}. Spurs are absent at the end of A6.

Distribution and age: Inner Mongolia; Middle Jurassic.

Only one species included from the Jurassic of China (see Table 24.1).

Conicholcorpa stigmosa Li, Shih, Wang & Ren, 2017 (Figure 24.17)

Conicholcorpa stigmosa Li, Shih, Wang & Ren, 2017: *Acta Geol. Sin.-Engl.*, 91 (3), 799–801.

Locality and horizon: Daohugou, Ningcheng, Inner Mongolia, China; Middle Jurassic, Jiulongshan Formation.

The body length from head to the end of genital is 47.4 mm. Head with short chewing mouthparts, compound eyes are large and oval, antenna filiform. Prothorax 0.3 times as long as mesothorax. The total length of the three elongated abdomen segments (A6–A8) is 28.7 mm, and A8 is the longest (about 14.5 mm). Forewing is oval, the length is 21.5 mm and width 5.7 mm; Sc long, nearly reaching the pterostigma area; R single; Rs_{1+2} with four branches, Rs_{3+4} with two branches; M forking distal to Rs; M_{1+2} forking at the same level of Rs; Cu_1 curved in the base; three anal veins long. Hind wing is similar to forewing but much smaller, length is 19.6 mm and width 5.4 mm; Sc short, reaching the middle of the anterior margin. Genital bulb enlarged distinctly, two gonocoxites well-preserved, gonostyli straight and long [84].

Family *Incertae sedis*

Fortiholcorpa Wang, Shih & Ren, 2013

Fortiholcorpa Wang, Shih & Ren, 2013, *PLoS ONE*, 8 (8), e71378 [85] (original designation).

Type species: *Fortiholcorpa paradoxa* Wang, Shih & Ren, 2013.

M with five branches in both fore- and hind wing. Forewing long and narrow, Cu_1 distinctly beyond the midpoint of wing. Hind wing long, apical margin relatively round, Rs_{1+2} forking at same level of Rs_{3+4}, M_5 single. Abdominal segments 6–8 elongate exaggeratedly, A7 and A8 nearly same length, about three times as long as A6. Male genital bulb enlarged and gonostyli extended. The genus is different from the type genus (*Holcorpa* Scudder, 1878) of Holcorpidae in branching pattern of hind wing M veins, length of A8 slightly longer than that of A7 (vs. A8 distinctively longest); and lacking spurs at the terminal part of A6 (vs. two spurs present). In addition, it is not discernible whether dististyli basal tooth is lacking. Therefore, this genus was deemed as Family *Incertae sedis.*

Distribution and age: Inner Mongolia; Middle Jurassic.

Only one species included from the Jurassic of China (see Table 24.1).

Fortiholcorpa paradoxa Wang, Shih & Ren, 2013 (Figure 24.18)

Fortiholcorpa paradoxa Wang, Shih & Ren, 2013: *PLoS ONE*, 8 (8), e71378, 5–8.

Locality and horizon: Daohugou, Ningcheng, Inner Mongolia, China; Middle Jurassic, Jiulongshan Formation.

The body length from head to genital end about 73.5 mm, antenna filiform, compound eyes big and oval. Forewing length about 20.5 mm, M with five branches; m_3-m_4, m_4-m_5 and m_5-cua present; A_1 and A_2 parallel and both straight, A_3 curved in the base. Hind wing slightly shorter than forewing, length about 20.1 mm; Sc slightly distal to mid-wing, with two anterior branches; the base of R, Cu_1 and A_1 quite close to each other; A_1 and A_2 straight and parallel, A_3 curved in the base. Abdomen with eight visible segments, A6 length 7.0 mm, tergal spurs absent; A7 length 21.1 mm, slightly curved in the base, expanded as cone-like; A8 length 21.8 mm, slightly curved in the middle. Genital bulb is large, with a pair of hypovalvae; gonostyli length 3.8 mm, elongated as pincer-like (Figure 24.18 b and d) [85].

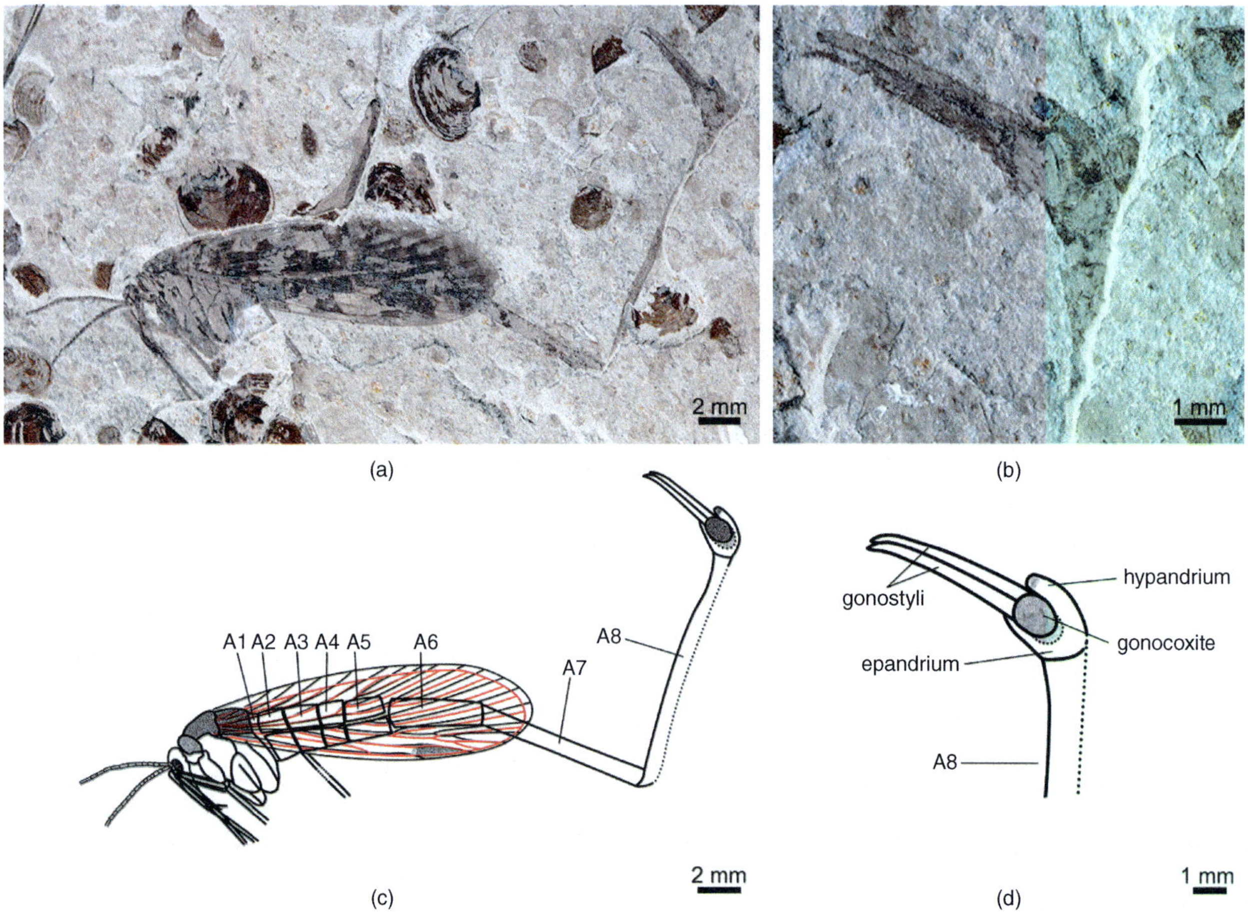

Figure 24.17 *Conicholcorpa stigmosa* Li, Shih, Wang & Ren, 2017 (Holotype, CNU-MEC-NN-2005023, male). (a) Photograph of habitus; (b) Photograph of genitalia; (c) Line drawing; (d) Line drawing of genitalia. Modified from [84].

Miriholcorpa Wang, Shih & Ren, 2013

Miriholcorpa Wang, Shih & Ren, 2013, *PLoS ONE*, 8 (8), e71378, 2 [85] (original designation).

Type species: *Miriholcorpa forcipata* Wang, Shih & Ren, 2013.

Forewing Sc almost at the middle of pterostigma; Rs forking at the same level as M; Rs_{3+4} stem short; R_1 curving around pterostigma; M with five branches; the stem of Rs_{1+2} 1.5–2.0 times as long as the stem of Rs_{3+4}; M_1 single, M_{2+3} forking to M_2 and M_3, M_{4+5} forking to M_4 and M_5; only one cross-vein between Cu_1 and Cu_2. Hind wing M with five branches. Abdomen 6–8 segments elongated, and A8 distinctly longest; male genitalia with slender and extended gonostyli. Due to uncertainty whether hind wing of *Miriholcorpa forcipata* has five-branched M veins, this genus was deemed as Family *Incertae sedis*.

Distribution and age: Inner Mongolia; Middle Jurassic.

Only one species included from the Jurassic of China (see Table 24.1).

Miriholcorpa forcipata Wang, Shih & Ren, 2013 (Figure 24.19)

Miriholcorpa forcipata Wang, Shih & Ren, 2013: *PLoS ONE*, 8 (8), e71378, 2–5.

Locality and horizon: Daohugou, Ningcheng, Inner Mongolia, China; Middle Jurassic, Jiulongshan Formation.

The body length including head to genital end is about 32.7 mm. Antenna filiform, and compound eyes big and oval. Forewing broad, length about 16.6 mm and width about 5.5 mm, with light spots and fascia; Sc extremely long, nearly reach the midpoint of pterostigma; several cross-veins between Sc and R_1; Rs_{1+2} with five branches, Rs_{3+4} with two branches; the stem of Rs_{1+2} 1.65 times as long as the stem of Rs_{3+4}; M and R relatively parallel; M forking at the same level as Rs; Cu_1 curved in the base, cu_1-m and cu_1-cu_2 present; A_1, A_2 and A_3 present. Right hind wing length about 15.2 mm; the stem of Rs_{1+2} 1.36 times as long as the stem of Rs_{3+4}; Rs forking slightly proximal to M, M_{1+2} distal to Rs_{3+4} branching. Genital bulb enlarged,

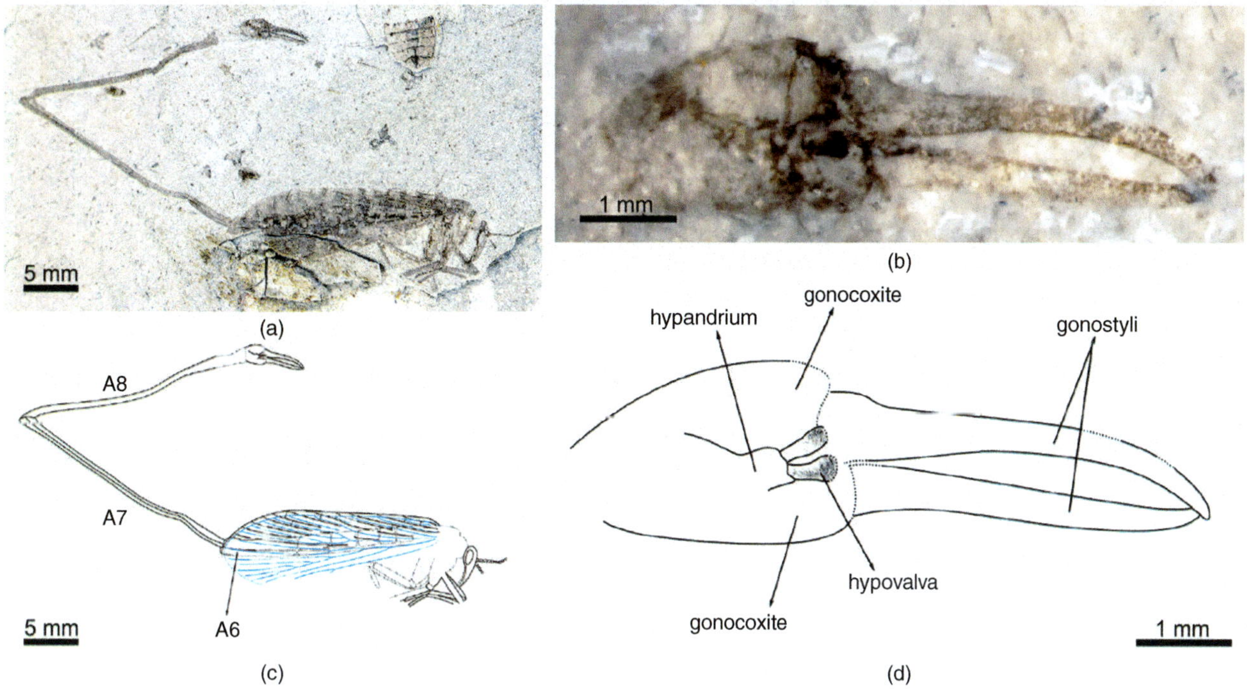

Figure 24.18 *Fortiholcorpa paradoxa* Wang, Shih & Ren, 2013 (Holotype, CNU-MEC-NN-2012002p). (a) Photograph of habitus; (b) Photograph of genitalia; (c) Line drawing; (d) Line drawing of genitalia. Source: Modified from [85]. Donated by Dr. Chungkun Shih.

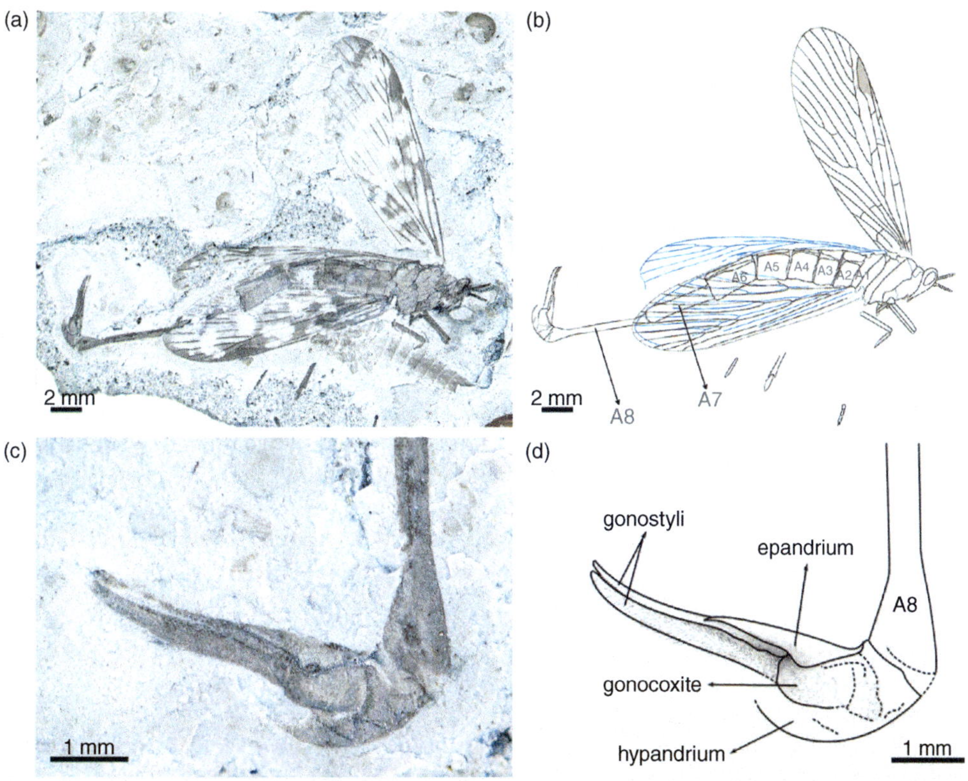

Figure 24.19 *Miriholcorpa forcipata* Wang, Shih & Ren, 2013 (Holotype, CNU-MEC-NN-2012001, male). (a) Photograph of habitus; (b) Line drawing; (c) Photograph of genitalia, under alcohol; (d) Line drawing of genitalia. Source: Modified from [85]. Donated by Dr. Chungkun Shih.

with slender and pincer-like gonostyli, length about 2.8 mm (Figure 24.19) [85].

Sexual Display with Extremely Elongated Male Organs

Conicholcorpa stigmosa Li, Shih, Wang & Ren, 2017, from the Middle Jurassic Jiulongshan Formation, Inner Mongolia, China, is the earliest fossil record of the family Holcorpidae hitherto [84]. Holcorpids have two unique features which are different from all other families of Mecoptera, and even from the other insect groups. Besides the special wing venation, the other important feature is the extremely elongated abdomen segments (A6–A8) and enlarged genitalia in male. Wang *et al.* [85] described two scorpionflies, *Fortiholcorpa paradoxa* and *Miriholcorpa forcipata* Wang, Shih & Ren, 2013, which also have extremely elongated A6–A8 segments, but they were deemed as family *Incertae sedis* as aforementioned. The phenomena of elongated terminal segments of abdomen also appear in two other families, Orthophlebiidae [86, 87] and Panorpidae [88]. However, the length of segments and degree of elongation are quite different. The last two segments of *C. stigmosa* have a length of 24 mm, just similar to other species (21.8–26.3 mm) in this family. The last two segments of *Fortiholcorpa paradoxa* have a length of 42.9 mm, and those of *Miriholcorpa forcipata*, 11.6 mm. All these lengths are much longer than the lengths of those for Orthophlebiidae. These extremely extended abdominal segments from A6 to A8 and enlarged genitalia might have been used for sexual display and, to less extent, for fighting with other males in the competition for mates (Figure 24.20). These three taxa have extended the records of exaggerated male body parts of mecopterans for sexual display and/or selection from the Early Eocene to the latest Middle Jurassic. The similar character present in some extant *Leptopanorpa* of Panorpidae suggests that the sexual display and/or sexual selection due to extremely elongated male abdominal and sexual organs outweigh the negative impact of bulky body and poor mobility in the evolutionary process [84, 85].

Family Parachoristidae Tillyard, 1937

Parachoristidae, a small and extinct family, have close relationship with Permochoristidae and Orthophlebiidae based on previous researches of parachoristids [27, 57, 89, 90]. So far, this family comprise two subfamilies and 9 genera. The subfamilies are Parachoristinae Handlirsch, 1937 and Neoparachoristinae Willmann, 1978. Fossil records are distributed in Australia [91, 92], Kyrgyzstan [27, 57, 93] and China [25, 94], from the Middle Triassic to the Early Cretaceous. Parachoristids have

Figure 24.20 Reconstruction drawing of *Conicholcorpa stigmosa* Li, Shih, Wang & Ren, 2017. Source: Artwork by Dr. Chen Wang.

a combination of the characters of Permochoristidae and Orthophlebiidae, and thus are considered as a transition taxon with several special features, such as MP with at least six branches and Rs_1 with two or three anterior branches in both fore- and hind wings [25].

Only one genus included from the Jurassic of Northern China: *Junggarochorista* Hong, 2009.

Junggarochorista Hong, 2009

Junggarochorista Hong, 2009, *Geological Bulletin of China*, 28 (10), 1382–1389 [25].

Mesopanorpa Zhang, 1996, *Acta Palaeontol. Sin.*, 35 (4), 442–456 [89] (original designation) (assigned to Orthophlebiidae).

Type species: *Junggarochorista tuzigouensis* Hong, 2009.

Sc long straight, single and ending in costal margin; R_1 single; both Rs_1 and MA with two branches; MP with six branches, the stem of MP short and fused with CuA.

Distribution and age: Xinjiang; Early Jurassic.

Only one species included from the Jurassic of China (see Table 24.1).

Suborder Nannomecoptera Hinton, 1981

Family Nannochoristidae Tillyard, 1917

Nannochoristidae, a relict family of Mecoptera, comprise only one extant genus and six extinct genera. The total eight extant species are distributed in the Southern Hemisphere, reported from Australia, South America [56, 95–97] and New Zealand [98]. The adult morphology of Nannochoristidae is already well-known, but the structures of larvae still need further study [99]. The adults often occur on broad-leaved herbs and grasses in moist areas [55, 100], and larvae mostly predacious and aquatic [101, 102]. The fossil records of this family comprise 18 species hitherto, from China [103–105], Mongolia, Siberia, Kazakhstan and Australia [1, 27, 106–110]. The oldest fossil records are known from the Upper Permian of Australia [42], consisting of three genera *Nannochoristella* Riek, 1953, *Neochoristella* Riek, 1953 and *Robinjohnia* Martynova, 1948. *Itaphlebia* Sukatsheva, 1985, the most common genus in the Mesozoic, has several unique characters which are different from those of all extant genera, for instance, Sc with three anterior branches and Rs with four branches.

Only one genus included from the Jurassic of Northern China: *Itaphlebia* Sukatsheva, 1985.

Itaphlebia Sukatsheva, 1985

***Itaphlebia* Sukatsheva, 1985.** In The Jurassic Insects of Siberia and Mongolia. *Rasnitsyn, A.P., ed., Nauka Press, Moscow*, 96–97 [106].

Chrysopanorpa Ren, Lu, Ji & Guo, 1995, *Seismic Publishing House, Beijing*, 91 [103].

Stylopanorpodes Sun, Ren & Shih, 2007, *Acta Zootaxonomica Sinica*, 32, 865–866 [111].

Netropanorpodes Sun, Ren & Shih, 2007, *Acta Zootaxonomica Sinica*, 32, 867 [111].

Protochoristella Sun, Ren & Shih, 2007, *Acta Zootaxonomica Sinica*, 32, 406 [104].

Itaphlebia: Cao, Shih, Bashkuev & Ren, 2015, *Alcheringa*, 40 (1), 2 [29].

Type species: *Itaphlebia completa* Sukatsheva, 1985.

Forewing length is 2.1–15.4 mm, Sc with two or three anterior branches; Rs with four branches, originating distal to M; M with four or five branches. Hind wing Sc short and single, proximal to pterostigma; Rs with four branches; M with four or five branches.

Distribution and age: Inner Mongolia; Middle Jurassic.

Six species included from the Jurassic of China (see Table 24.1).

Itaphlebia amoena Cao, Shih, Bashkuev & Ren, 2015 (Figure 24.21)

Itaphlebia amoena Cao, Shih, Bashkuev & Ren, 2015: *Alcheringa*, 40 (1), 4–7.

Locality and horizon: Daohugou, Ningcheng, Inner Mongolia, China; Middle Jurassic, Jiulongshan Formation.

The body length excluding antenna and cercus is 7.3 mm; the left forewing length is 8.6 mm and width at most 2.9 mm. Compound eyes small and widely spaced; antenna filiform, with 10 and 13 antennomeres

Figure 24.21 *Itaphlebia amoena* Cao, Shih, Bashkuev & Ren, 2015 (Holotype, CNU-MEC-NN-2008209p).

respectively. Maxillary palpi are narrow and covered with short setae. Legs slender, tibia with two apical spurs; all legs are covered with short setae. Abdomen has 10 visible segments. Forewing costal margin slightly convex in the base, Sc with three branches, Sc_1 and Sc_2 ending proximal to the middle of the wing, Sc_3 ending near the pterostigma area; R forking proximal to Sc_2; both Rs and M with four branches; m–cu_1 is somewhat S-shaped; Cu_1 merged with M for a long distance; A_1 and A_2 nearly parallel. Hind wing shorter than forewing, Sc single, c-r_1 almost at the middle of the wing; R_1 curved near the pterostigma area; one cross-vein c–r_1 present [29].

Suborder Pistillifera Willmann, 1987

Family Bittacidae Handlirsch, 1906

Bittacidae are the second most speciose family of Mecoptera, containing about 270 species [112]. They are commonly called "hangingflies" because they usually hang their bodies on some branches of low vegetation by their forelegs (or both fore- and middle legs) and use hind legs to catch preys, such as small flies. Extant bittacids are almost distributed all over the world, they can live in both temperate and tropical areas. The oldest bittacid fossil is *Archebittacus exilis* Riek, 1955 from the Upper Triassic of Australia. The family reached a peak of diversity in the Jurassic with about 24 genera [113–115]. During the Cretaceous, only seven genera with nine species have been reported so far [39, 109, 113, 115–118].

Genera included from the Jurassic and Cretaceous of Northern China: *Orthobittacus* Willmann, 1989, *Sibirobittacus* Sukatsheva, 1990, *Liaobittacus* Ren, 1993, *Preanabittacus* Novokshonov, 1993, *Jichoristella* Ren, Lu, Ji & Guo, 1995; *Megabittacus* Ren, 1997, *Mongolobittacus* Petrulevičius, Huang & Ren, 2007, *Formosibittacus*

Li, Ren & Shih, 2008, *Jurahylobittacus* Li, Ren & Shih, 2008, *Decoribittacus* Li & Ren, 2009, *Karattacus* Li & Ren, 2009, *Exilibittacus* Yang, Ren & Shih, 2012 and *Composibittacus* Liu, Shih & Ren, 2016.

Orthobittacus Willmann, 1989

Orthobittacus Willmann, 1989, *Abhandlungen Sencken.*, 544, 123–124 [119] (original designation).

Type species: *Orthobittacus abshirica* Martynova, 1951.

Sc reaching the anterior margin distal to the forking of R_{4+5}; one or two cross-veins between C and Sc, and one or two Pcv's (Pterostigmal cross-veins) in pterostigmal area; Rs with four branches and M with six or seven branches.

Distribution and age: Inner Mongolia, Middle Jurassic.

Only one species included from the Jurassic of China (see Table 24.1).

Sibirobittacus Sukatsheva, 1990

Sibirobittacus Sukatsheva, 1990, Transactions of the Paleontological Institute of the USSR Academy of Sciences, 93 [107] (original designation).

Type species: *Sibirobittacus undus* Sukatsheva, 1990.

Rs longer than R_{2+3} and R_2 combined, Sc reaching the wing margin distal to the forking of Rs. Forking of M opposite to Rs.

Distribution and age: Liaoning; Early Cretaceous.

Only one species included from the Cretaceous of China (see Table 24.1).

Liaobittacus Ren, 1993

Liaobittacus Ren, 1993, *Acta Geol. Sin.-Engl.*, 67, 378 [120] (original designation).

Type species: *Liaobittacus longantennatus* Ren, 1993.

Antenna longer than the wing; R_1 crossed pterostigma, no pattern in the wing; the forking of Rs is basal to the forking of M; two Pcv's and only one Scv.

Distribution and age: Liaoning; Early Cretaceous.

Only one species included from the Cretaceous of China (see Table 24.1).

Preanabittacus Novokshonov, 1993

Preanabittacus Novokshonov, 1993, *Russian Entomol. J.*, 2, 81 [116] (original designation).

Type species: *Preanabittacus karatavensis* Novokshonov, 1993.

Sc is distal to the forking of R_{4+5} in forewing but distal to Rs in hind wing; both Rs and M with four branches; Rs is basal to the forking of M; A_1 short, reaching margin of the wing basal to the forking of Rs and M.

Distribution and age: Inner Mongolia, Middle Jurassic.

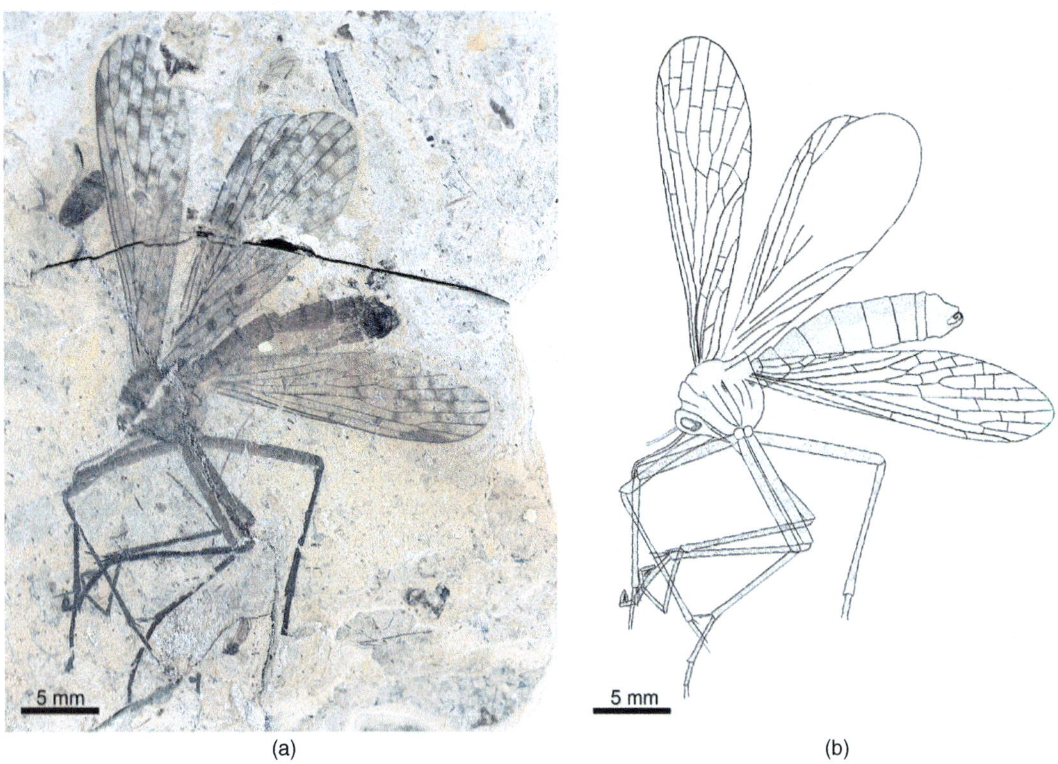

Figure 24.22 *Preanabittacus validus* Yang, Shih & Ren, 2012 (Holotype, CNU-MEC-NN-2010005p). (a) Photograph; (b) Line drawing. Source: Modified from [39].

Only one species included from the Jurassic of China (see Table 24.1).

Preanabittacus validus **Yang, Shih & Ren, 2012 (Figure 24.22)**

Preanabittacus validus Yang, Shih & Ren, 2012: *Alcheringa*, 36 (2), 197–200.

Locality and horizon: Daohugou, Ningcheng, Inner Mongolia, China; Middle Jurassic, Jiulongshan Formation.

The body is 18.5 mm long, the rostrum is long and slender, compound eyes large, the thorax is 4.1 mm long, and the basitergum is fused to the metathorax. The wing shape nearly overlap, basal part of the wing is narrow, lacking a thyridium and gradually broadening from the base toward the apex. Sc with one subapical cross-vein to R_1, R_1 is branched and sharply sagged; Cu_1 and Cu_2 running close to each other and almost parallel; A_1 bent to the posterior basal to the forking of Rs. Hind wing is smaller than forewing, Sc in hind wing is shorter than that of forewing [39].

Jichoristella **Ren, Lu, Ji & Guo, 1995**

Jichoristella Ren, Lu, Ji & Guo, 1995, Seismic Publishing House, Beijing, p. 222 [103].

Type species: *Jichoristella rara* Ren, Lu, Ji & Guo, 1995.

Wing slender, Sc not beyond the middle of the wing; R_1 straight; Rs with three branches, and forking proximal to M; Rs_1 single, Rs_{2+3} forking distal to the forking of M; M with four branches, M_{3+4} forking proximal to M_{1+2}; CuA and M fused for a short distance in the base; CuP single; two anal veins present; many cross-veins on wings.

Distribution and age: Beijing; Early Cretaceous.

Only one species included from the Cretaceous of China (see Table 24.1).

Megabittacus **Ren, 1997**

Megabittacus Ren, 1997, *Acta Zootax. Sin.*, 22, 76 [117] (original designation).

Type species: *Megabittacus colosseus* Ren, 1997.

The wings are narrow, antenna slightly less than one-fourth as long as the forewing; three or four Scv's between Sc and R_1; only one Pcv below pterostigma; Sc ending well beyond the first forking of Rs, both Rs and M with four branches; A_1 long, terminating close to the apex of CuP.

Distribution and age: Liaoning; Early Cretaceous.

Three species included from the Cretaceous of China (see Table 24.1).

Megabittacus spatiosus **Yang, Ren & Shih, 2012 (Figure 24.23)**

Megabittacus spatiosus Yang, Ren & Shih, 2012: *Geodiversitas*, 34 (4), 795–796.

Figure 24.23 *Megabittacus spatiosus* Yang, Ren & Shih, 2012 (Holotype, CNU-MEC-LB-2010003).

Locality and horizon: Huangbanjigou, Beipiao, Liaoning, China; Lower Cretaceous, Yixian Formation.

The wings are much narrower and longer at the basal part than other previously known bittacids, and the apices of wing are much rounded; Sc very long, reaching the wing margin basal to the pterostigma area; R_1 running through the pterostigma area and distal part of the vein present a spoon-like. Genital segment well-preserved, epiandrium long and prong-like, tapering apically with apex curved toward the middle [39].

Mongolobittacus **Petrulevičius, Huang & Ren, 2007**

Mongolobittacus Petrulevičius, Huang & Ren, 2007, *Afr. Invertebr.*, 48 (1), 146 [37] (original designation).

Type species: *Mongolobittacus daohugouensis* Petrulevičius, Huang & Ren, 2007.

Sc reaching the wing margin opposite the forking of Rs; both Rs and M with four branches; distal part of the R_1 is spoon-like and not bifurcated; A_1 short, far basal to the forking of M.

Distribution and age: Inner Mongolia; Middle Jurassic.

Two species included from the Jurassic of China (see Table 24.1).

Formosibittacus **Li, Ren & Shih, 2008**

Formosibittacus Li, Ren & Shih, 2008, *Zootaxa*, 1929, 39 [38] (original designation).

Type species: *Formosibittacus macularis* Li, Ren & Shih, 2008.

Antenna is less than half as long as the body. Sc is distal to the forking of R_{4+5}; R_1 is running through the pterostigma, with only one Pcv; both Rs and M with four branches and forking of Ra is basal to the forking of M; A_1 reaching the margin of wing opposite the first forking of Rs.

Figure 24.24 *Formosibittacus macularis* Li, Ren & Shih, 2008 (Holotype, CNU-M-NN-2007001p).

Distribution and age: Inner Mongolia; Middle Jurassic.

Only one species included from the Jurassic of China (see Table 24.1).

***Formosibittacus macularis* Li, Ren & Shih, 2008 (Figure 24.24)**

Formosibittacus macularis Li, Ren & Shih, 2008: *Zootaxa*, 1929, 42.

Locality and horizon: Daohugou, Ningcheng, Inner Mongolia, China; Middle Jurassic, Jiulongshan Formation.

The length of antenna is less than half as long as the body length. Body 21 mm in length, abdomen at least 10 segments visible. Legs are long and slender, covered with densely short setae, with a single predatory tarsal claw in the distal part. Forewing 23 mm long, hind wing 20 mm long, the basal part of the wing is narrow; a conspicuous thyridium at the first forking of M; A_1 joining A_2 by two cross-veins, A_2 reaching the wing margin distal to the origin of Rs. Hind wing just the same as forewing, except for the apparent fusion of veins A_1 and CuP, and A_2 terminating basal to the origin of Rs [38].

***Jurahylobittacus* Li, Ren & Shih, 2008**

Jurahylobittacus Li, Ren & Shih, 2008, *Zootaxa*, 1929, 42 [38] (original designation).

Type species: *Jurahylobittacus astictus* Li, Ren & Shih, 2008.

Sc is ending in the wing margin basal to the forking of Rs; R_1 is spoon-shaped and running though the pterostigma; both Rs and M with four branches, the forking of Rs is basal to the forking of M.

Distribution and age: Inner Mongolia; Middle Jurassic.

Only one species included from the Jurassic of China (see Table 24.1).

***Decoribittacus* Li & Ren, 2009**

Decoribittacus Li & Ren, 2009, *Acta Zootax. Sin.*, 34 (3), 560 [121] (original designation).

Type species: *Decoribittacus euneurus* Li & Ren, 2009.

Wing slender, Sc very long, reaching the wing margin close to pterostigma area; R_1 running though the pterostigma and turned weakly and smoothly arched before the apex, lacking sharp bend; Rs with four branches and dividing basal to the forking of M.

Distribution and age: Inner Mongolia; Middle Jurassic.

Only one species included from the Jurassic of China (see Table 24.1).

***Karattacus* Li & Ren, 2009**

Karattacus Li & Ren, 2009, *Acta Zootax. Sin.*, 34 (3), 563 [121] (original designation).

Type species: *Karattacus persibus* Novokshonov, 1997.

R_1 running though the pterostigmal and bifurcated at distal part of the wing; Rs four branches, M five branches; A_2 is long, three cross-veins between A_2 and posterior of the wing.

Distribution and age: Inner Mongolia; Middle Jurassic.

Only one species included from the Jurassic of China (see Table 24.1).

***Exilibittacus* Yang, Ren & Shih, 2012**

Exilibittacus Yang, Ren & Shih, 2012, *Geodiversitas*, 34 (4), 788 [39] (original designation).

Type species: *Exilibittacus lii* Yang, Ren & Shih, 2012.

There is one cross-vein in subcostal area, the forking of Rs is basal to the forking of M; A_1 terminating at the posterior margin at the same level or distal to the forking of Rs; Rs with three or four branches in hind wing.

Distribution and age: Inner Mongolia; Middle Jurassic.

Three species included from the Jurassic of China (see Table 24.1).

***Exilibittacus foliaceus* Liu, Shih & Ren, 2014 (Figure 24.25)**

Exilibittacus foliaceus Liu, Shih & Ren, 2014: *ZooKeys*, 466, 87–89.

Locality and horizon: Daohugou, Ningcheng, Inner Mongolia, China; Middle Jurassic, Jiulongshan Formation.

Legs long and slender covered with short setae; the fifth tarsus folded against the fourth and a single claw at the tip of the pretarsus. The basal area of the wing narrow, the color of pterostigma area is slightly dark; Sc reaching the wing margin distal to the forking of Rs, vein R_1 running through the pterostigma without sagging; A_1 reaching the wing margin opposite the Rs forking. Supraanale and cercus covered with small and short setae [122].

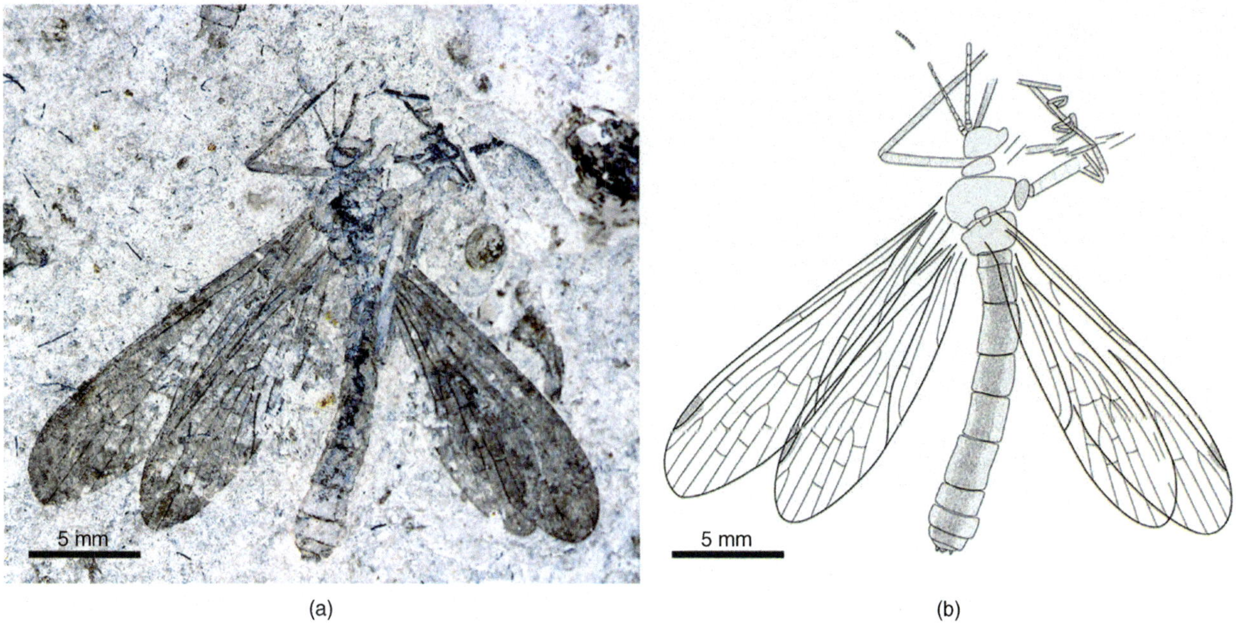

Figure 24.25 *Exilibittacus foliaceus* Liu, Shih & Ren, 2014 (Holotype, CNU-MEC-NN-2013010). (a) Photograph; (b) Line drawing. Modified from [122].

***Composibittacus* Liu, Shih & Ren, 2016**

Composibittacus Liu, Shih & Ren, 2016, *Zootaxa*, 4067 (1), 66 [114] (original designation).

Type species: *Composibittacus bipunctatus* Liu, Shih & Ren, 2016.

Vein Sc long, reaching the wing margin distal to the bifurcation of R_{4+5}; five pterostigmal cross-veins between R_1 and R_2 and R_1 and R_{2+3}; A_1 long, reaching the posterior margin proximal to the Rs forking.

Distribution and age: Inner Mongolia; Middle Jurassic.

Two species included from the Jurassic of China (see Table 24.1).

Hangingflies from Myanmar (Burmese) Amber

Myanmar amber from Kachin (Hukawng Valley) of northern Myanmar (also known as Burmite) has been used as carvings and jewelry in China for at least two thousand years. However, until a century ago, insect inclusions in Myanmar amber were studied scientifically for the first time [123–128]. To date, there are only four specimens of hangingflies reported from the Myanmar ambers [20, 115, 129, 130]. Based on palynology and ammonoid zonation within the surrounding rock units, and insect taxa preserved within the ambers, an age of the earliest Cenomanian (98.79 ± 0.62 Ma) has recently been established for the Myanmar amber [131, 132].

During the Middle Jurassic, the diversity of Bittacidae reached its peak. However, in the Late Cretaceous, it reduced significantly with only eight genera and 10 species documented [113–115, 130]. In 2017, Zhao *et al.* have reported the first hangingfly, *Burmobittacus jarzembowskii*, in Myanmar amber [115]. In 2018, Li *et al.* have reported the earliest fossil record of *Bittacus*, *B. lepiduscretaceus*, which enriched the diversity in genus and species level and also advanced *Bittacus* record to the mid-Cretaceous. Usually, amber has a strong preservation bias toward tiny insects because large insects are not easily trapped by resin [133, 134]. Due to the relative large size of hangingflies (e.g. wings more than 10 mm in length), there are rarely well-preserved specimens in Myanmar amber [115] (see Figure 24.26).

Family Cimbrophlebiidae Willmann, 1977

Cimbrophlebiidae, considered as the sister group of Bittacidae [119, 135], are a little-known extinct family of hangingflies ranging from the Lower Jurassic to the Lower Eocene [136]. Cimbrophlebiidae are similar to Bittacidae by their wing venation and body characters, but Cimbrophlebiidae can be easily distinguished from Bittacidae by their long and branched anal veins [137]. With the high similarity of legs between these two families, Cimbrophlebiidae are also assumed to be predators. Fossil records of Cimbrophlebiidae are fairly rare, with only seven genera described from China hitherto.

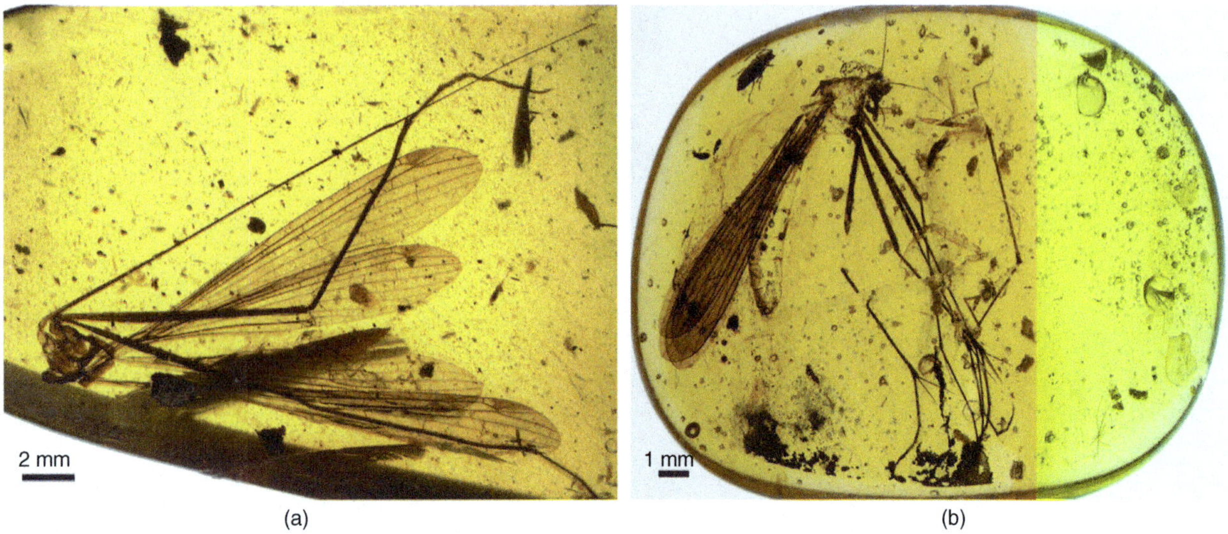

Figure 24.26 Two bittacids in amber. (a) General habitus of *Burmobittacus jarzembowskii* Zhao, Bashkuev, Chen & Wang, 2017 (NIGP164041). (b) General habitus of *Bittacus lepiduscretaceus* Li, Zhang, Shih & Ren, 2018 (BU-001263). Modified from [115, 130].

Genera included from the Jurassic of Northern China: *Cimbrophlebia* Willmann, 1977; *Telobittacus* Zhang, 1993; *Juracimbrophlebia* Wang, Labandeira, Shih & Ren, 2012; *Perfecticimbrophlebia* Yang, Shih & Ren, 2012; *Bellicimbrophlebia* Yang, Shih & Ren, 2013 and *Mirorcimbrophlebia* Yang, Shih & Ren, 2013.

Cimbrophlebia Willmann, 1977

Cimbrophlebia Willmann, 1977, *Neues Jahrbuch für Geologie und Paläontologie, Monatshefte*, 12, 736 [135] (original designation).

Type species: *Cimbrophlebia bittaciformis* Willmann, 1977.

Vein Sc reaching the wing margin basal to the pterostigmal area, Vein M with four branches, Cu_1 and Cu_2 almost parallel, distal part of the Cu_2 bent toward the wing margin.

Distribution and age: Inner Mongolia; Middle Jurassic.

Three species included from the Jurassic of China (see Table 24.1).

Cimbrophlebia amoena Zhang, Shih & Ren, 2015 (Figure 24.27)

Cimbrophlebia amoena Zhang, Shih & Ren, 2015: *Acta Geol. Sin.-Engl.*, 89 (5), 1487–1489.

Locality and horizon: Daohugou, Ningcheng, Inner Mongolia, China; Middle Jurassic, Jiulongshan Formation.

Head is oviform with a slender chewing mouthpart, compound eyes large and oval-shaped. Legs long and slender, covered with short setae. Basal part of the wing narrow, gradually broadening from base toward rounded apex. Sc reaching the wing margin at about 2/3 of the wing. R_1 curved around pterostigmal and join to C. Fork of M distal to the fork of Rs, both Rs and M have five branches, "Bittacid cross" are not aligned. A_1 reaching the posterior wing margin basal to the M_{1+2} forking, A_3 not detected [136].

Figure 24.27 *Cimbrophlebia amoena* Zhang, Shih & Ren, 2015 (Holotype, CNU-MEC-NN-2014072p).

Telobittacus Zhang, 1993

Telobittacus Zhang, 1993, *Palaeoworld*, 2, 56 [138] (original designation).

Type species: *Telobittacus fragosus* Zhang, 1993.

Vein M with five branches; Cu_1 and Cu_2 almost parallel, distal segment of Cu_2 extending toward the wing margin; A_2 with six or more branches.

Distribution and age: Inner Mongolia, Middle Jurassic; Shanxi, Early Cretaceous.

Three species included from the Jurassic of China (see Table 24.1).

***Juracimbrophlebia* Wang, Labandeira, Shih & Ren, 2012**

Juracimbrophlebia Wang, Labandeira, Shih & Ren, 2012, *PNAS*, 20516 [31] (original designation).

Type species: *Juracimbrophlebia ginkgofolia* Wang, Labandeira, Shih & Ren, 2012.

Vein A_2 with six or more pectinate primary branches; Sc reaching wing margin at about 2/3 of wing length; R_1 close to the pterostigma; Rs with five branches and M with four branches, Rs and M fused with Cu toward the base.

Distribution and age: Inner Mongolia; Middle Jurassic.

Only one species included from the Jurassic of China (see Table 24.1).

***Juracimbrophlebia ginkgofolia* Wang, Labandeira, Shih & Ren, 2012 (Figure 24.28)**

Juracimbrophlebia ginkgofolia Wang, Labandeira, Shih & Ren, 2012: *PNAS*, 20516.

Locality and horizon: Daohugou, Ningcheng, Inner Mongolia, China; Middle Jurassic, Jiulongshan Formation.

The body length is 38.5 mm. Forewing, 32.4 mm in length and 8.9 mm in width, has light coloration and a field of transparent spots. Membrane area is distinctly rugose, and the similar condition is in gingko *Yimaia* leaves. Venation is typically cimbrophlebiid-like, vein Sc reaching the wing margin at 2/3 of the wing length; fork of R_1 is close to the pterostigma; vein Rs and M are fused with vein Cu toward the base; vein Cu_2 bent sharply toward posterior wing margin close to the termination; vein A_1 has a single branch but vein A_2 has six or more primary branches and has a short cross-vein occurring close to the wing base. Hind wing is similar to the forewing in venation, but smaller than forewing, with 33.8 mm in length and 8.6 mm in width [31].

Foliaceous Mimicry and Mutualism

The 165 million-year-old fossils of *Juracimbrophlebia ginkgofolia* Wang, Labandeira, Shih & Ren, 2012 represent an interesting case of leaf mimicry and mutualism between an insect and ginkgo leaf. GMA was used to provide direct evidence that these ancient hangingflies possessed significantly spot-on similarity in structure for mimicking ginkgoalean leaves. Although the extant *Ginkgo* is considered to be minimally herbivorized [31, 139–141], insect herbivore damage on Ginkgoites leaves of *Y. capituliformis* from the Middle Jurassic seem to be quite common [31]. Most interesting, their study results suggest that *J. ginkgofolia* developed leaf mimesis either as an anti-predator avoidance device to fool predators such as larger predatory insects, insectivorous dinosaurs and mammals, or alternatively could be a possible predatory strategy to protect its plant host from herbivores, thus gaining a mutual benefit for both the hangingfly and the ginkgo plant. The *Y. capituliformis* host (or model) became extinct around the Jurassic–Cretaceous boundary, as possibly did *J. ginkgofolia* (the mimic), significantly before the initial

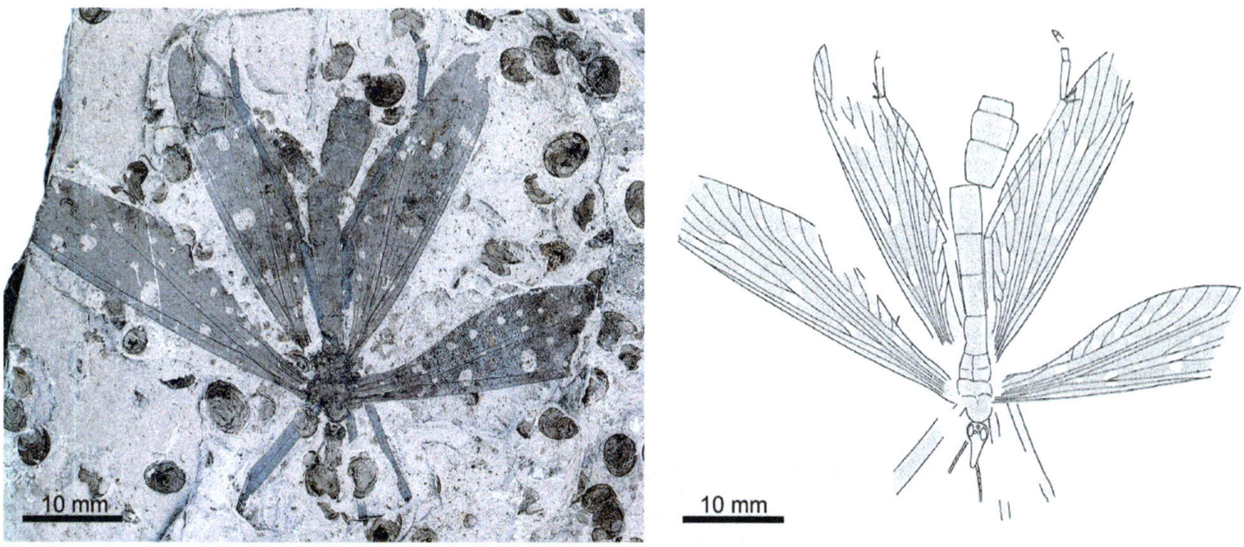

(a) (b)

Figure 24.28 *Juracimbrophlebia ginkgofolia* Wang, Labandeira, Shih & Ren, 2012 (Holotype, CNU-MEC-NN-2010050p). (a) Photograph; (b) Line drawing. Source: Modified from [31].

appearance of angiosperms during the later Early Cretaceous (see Section 29.3.2 and Figures 29.19 and 29.20).

Phylogenetic and Geometric Morphometric Research on Cimbrophlebiidae

In order to clarify the genus-level relationship of Cimbrophlebiidae, a matrix including nine taxa with seven morphological characters were analyzed by PAUP, among them two species of Bittacidae were chosen as outgroups. Although the phylogeny of Cimbrophlebiidae is not completely resolved, this result also provided a perspective on the evolution of Cimbrophlebiidae and showed a possibility direction of the key character transformation. In addition, GMA included 10 specimens of cimbrophlebiids, bittacids and *Yimaia* leaf lobes were completed, and 100 landmarks were selected on the outline of wings and leaves. The shape differences were better revealed by the procrustes distances. Moreover, the final result provided a straightforward way to interpret the relationships among Cimbrophlebiidae, Bittacidae, and ginkgoalean leaves. Bittacids are well-separated from cimbrophlebiids and ginkgoaleans. Meanwhile, the average outline shape of cimbrophlebiids showed more similarity to ginkgoaleans than to bittacids [31] (see Figure 24.29).

Perfecticimbrophlebia Yang, Shih & Ren, 2012

Perfecticimbrophlebia Yang, Shih & Ren, 2012, *Alcheringa*, 36 (2), 196 [40] (original designation).

Type species: *Perfecticimbrophlebia laetus* Yang, Shih & Ren, 2012.

Forewing Sc is long, distal to the forking of R_{5+6}, but hind wing vein Sc is short, basal to the forking of Rs. In both wings, R_1 branched and sagged, with a cross-vein between R_1 and R_{2+3}. M with six and five branches in forewing and hind wing, respectively; Both fore- and hind wings Cu_1 and Cu_2 almost parallel, and A_2 long with two branches.

Distribution and age: Inner Mongolia; Middle Jurassic.

Only one species included from the Jurassic of China (see Table 24.1).

Perfecticimbrophlebia laetus Yang, Shih & Ren, 2012 (Figure 24.30)

Perfecticimbrophlebia laetus Yang, Shih & Ren, 2012: *Alcheringa*, 36 (2), 198–199.

Locality and horizon: Daohugou, Ningcheng, Inner Mongolia, China; Middle Jurassic, Jiulongshan Formation.

The body is 24.5 mm long, the vertex of the head is raised, and the rostrum is long and slender, compound eyes large. The abdomen is 17.5 mm long with nine visible segments. All legs are densely covered with short setae. Forewing Sc long, reaching the wing margin basal to the R_1 and apical to the rise of Rs; only one cross-vein between C and Sc, two cross-veins between Sc and R_1; Cu_1 fused with M for a short distance; A_2 with two branches and connected to A_1 by a short cross-vein [40].

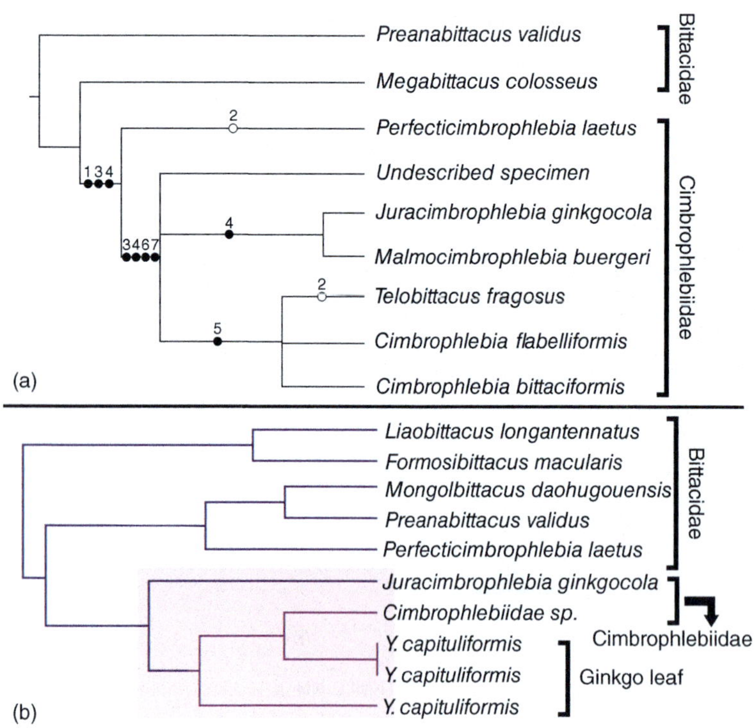

Figure 24.29 Phylogenetic and geometric morphometric results of Cimbrophlebiidae. (a) The best supported tree by PAUP; (b) Phenetic tree of outlined landmarks compiled using NTSYSpc and UPGMA. The black circles represent apomorphic characters, and white circles represent parallelisms and reversals in (a). The pink rectangle denotes the clade of Cimbrophlebiidae forewings and ginkgo leaves in (b). Source: Modified from [31].

Figure 24.30 *Perfecticimbrophlebia laetus* Yang, Shih & Ren, 2012 (Holotype, CNU-M-NN2010004p). Source: Modified from [40].

Bellicimbrophlebia Yang, Shih & Ren, 2013

Bellicimbrophlebia Yang, Shih & Ren, 2013, *Palaeontology*, 56 (4), 718 [142] (original designation).

Type species: *Bellicimbrophlebia cruciata* Yang, Shih & Ren, 2013.

Sc reaching the wing margin far distal to the fork of Rs; both Rs and M have five branches; Forking of Rs is basal to the forking of M; Cu_1 and Cu_2 almost parallel and there are three cross-veins between them.

Distribution and age: Inner Mongolia; Middle Jurassic.

Four species included from the Jurassic of China (see Table 24.1).

Mirorcimbrophlebia Yang, Shih & Ren, 2013

Mirorcimbrophlebia Yang, Shih & Ren, 2013, *Palaeontology*, 56 (4), 718 [142] (original designation).

Type species: *Mirorcimbrophlebia daohugouensis* Yang, Shih & Ren, 2013.

Wing length is about 4.5 times as long as width, M with six branches and veins Cu_1 and Cu_2 almost parallel. The wing has overall light coloration and M_{1+2+3} bifurcated before posterior of "Bittacid cross."

Distribution and age: Inner Mongolia; Middle Jurassic.

Only one species included from the Jurassic of China (see Table 24.1).

Family Eomeropidae Cockerell, 1909

Eomeropidae is a relict family with only one extant species *Notiothauma reedi* M'Lachlan, 1877. These rare insects, known as "living fossils", have been found in the beech forest of southern Chile [96], usually nocturnal and ovipositing under the leaf litter and bark [14]. The fossil records of eomeropids are quite rare. Only six genera and nine species have been reported hitherto from the Early Jurassic to the Paleogene. The oldest fossils were collected from the Lower Jurassic (Sinemurian) of Charmouth Mudstone Formation in Dorset of southern England [143]. There are three genera and four species documented from the Mesozoic of China [144–146] and four Cenozoic species from Canada [147], Colorado [148] and Siberia [149]. The representative characters of this family are: wing venation complicated with the parallel and secondary veins present in costal area; long setae on wings; insect body flat with strong and particularly dispersed spines and chewing mouthparts hypognathous [144, 145].

Genera included from the Jurassic and Cretaceous of Northern China: *Tsuchingothauma* Ren & Shih, 2005; *Typhothauma* Ren & Shih, 2005 and *Jurathauma* Zhang, Shih, Petrulevičius & Ren, 2011.

Tsuchingothauma Ren & Shih, 2005

Tsuchingothauma Ren & Shih, 2005, *Acta Zootax. Sin.*, 30(2), 276 [144] (original designation).

Type species: *Tsuchingothauma shihi* Ren & Shih, 2005.

The generic and specific names are dedicated to Mr. Tsuching Shih, father of Chungkun Shih, for being an excellent role model and providing guidance, motivation and inspiration. Forewing apical margin slightly sharp; costal area narrow at the base, with four veinlets and five rows of cells, all parallel to the costal margin; Sc with one or two anterior branches; R_1 single; Rs and M with nine branches; Cu_1 forking at the same level as Rs from R, and the first branch fused with M for a short distance; Cu_2 with only one anterior branch.

Distribution and age: Inner Mongolia; Middle Jurassic.

Only one species included from the Jurassic of China (see Table 24.1).

Tsuchingothauma shihi Ren & Shih, 2005 (Figure 24.31)

Tsuchingothauma shihi Ren & Shih, 2005: *Acta Zootax. Sin.*, 30 (2), 277–279.

Locality and horizon: Daohugou, Ningcheng, Inner Mongolia, China; Middle Jurassic, Jiulongshan Formation.

The body length excluding antennae about 22 mm, forewing length 28 mm and width 10.5 mm, antenna length at least 5 mm. Head slightly under the pronotum, only part is visible. Mouthparts are hypognathous. Antenna filiform and multi-articulated. Prothorax with large pronotum, meso- and metanota almost similar, scutum and scutellum present. Legs covered with rows of short setae, tibiae with two apical spurs, tarsi with five tarsomeres. Abdomen with 10 visible segments; 7–10 segments distinctly slender than 2–6; 10 and 11 segments fused, cerci at least with two articles [144].

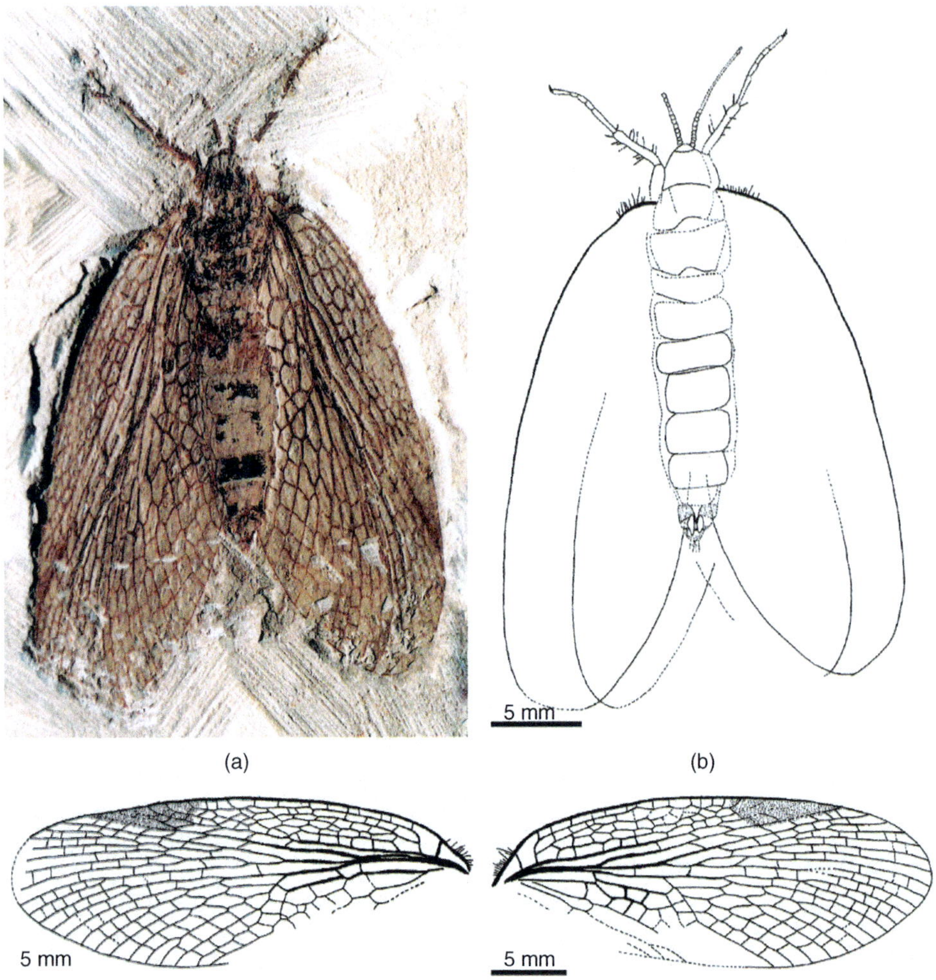

Figure 24.31 *Tsuchingothauma shihi* Ren & Shih, 2005 (Holotype, CNU-MEC-NN-2004001). (a) Photograph; (b) Line drawing; (c) Line drawing of left forewing; (d) Line drawing of right forewing. Source: Modified from [144], Donated by Dr. Chungkun Shih.

***Typhothauma* Ren & Shih, 2005**

Typhothauma Ren & Shih, 2005, *Acta Zootax. Sin.*, 30 (2), 279 [144] (original designation).

Type species: *Typhothauma yixianensis* Ren & Shih, 2005.

The costal area not narrow at the base, with only two veinlets and three rows of cells; humeral vein absent; R_1 single; Rs and M with nine branches; Cu_1 fused with M for a short distance; Cu_2 single; three anal veins single. All four wings have few cross-veins and cells.

Distribution and age: Liaoning; Early Cretaceous.

Two species included from the Cretaceous of China (see Table 24.1).

***Jurathauma* Zhang, Shih, Petrulevičius & Ren, 2011**

Jurathauma Zhang, Shih, Petrulevičius & Ren, 2011, *Zoosystema*, 33(4), 445 [145] (original designation).

Type species: *Jurathauma simplex* Zhang, Shih, Petrulevičius & Ren, 2011.

Relatively simple venation compared with other genera, costal area only with one veinlet and two rows of cells; both Rs and M with five branches; Cu_1 with two branches; Cu_2 single; three anal veins present.

Distribution and age: Inner Mongolia; Middle Jurassic.

Only one species included from the Jurassic of China (see Table 24.1).

***Typhothauma excelsa* Zhang, Shih & Ren, 2012 (Figure 24.32)**

Typhothauma excelsa Zhang, Shih & Ren, 2012: *Acta Zootax. Sin.*, 37 (1), 68–71.

Locality and horizon: Shanyang, Beipiao, Liaoning, China; Lower Cretaceous, Yixian Formation.

Forewing length 13.3 mm and width 5.5 mm; the base of costal margin with long setae (about 0.5 mm long); costal area narrow at the base, with three rows of cells; humeral vein are; Sc with one or two apical branches,

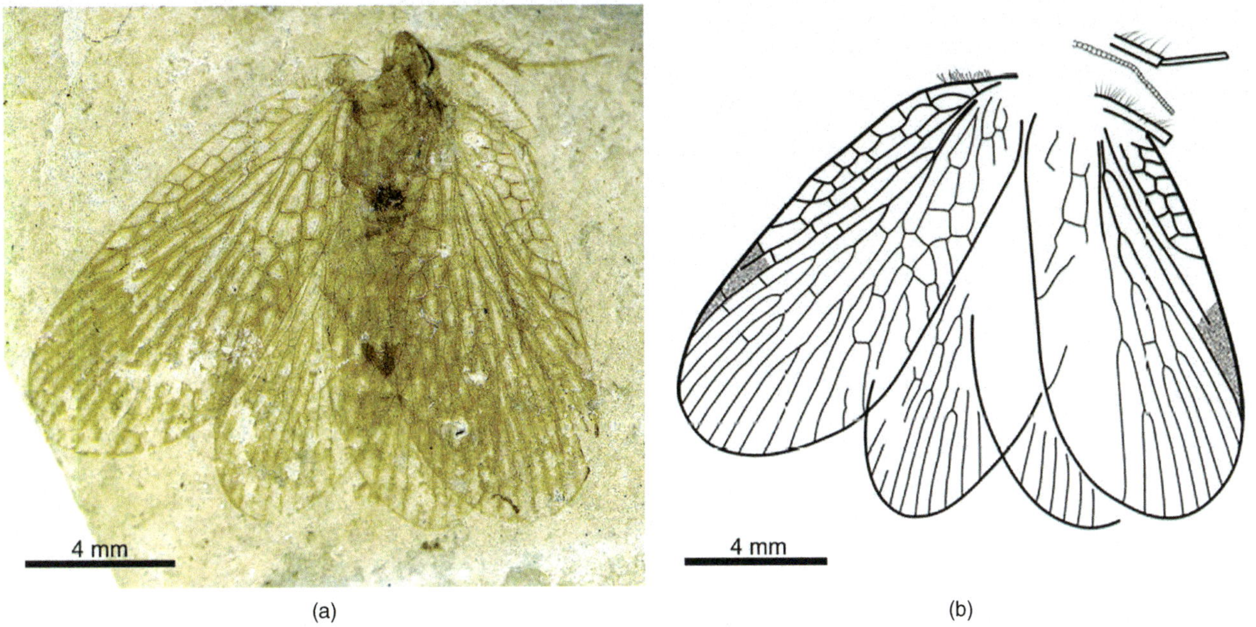

Figure 24.32 *Typhothauma excelsa* Zhang, Shih & Ren, 2012 (Holotype, CNU-MEC-LB-2009004). (a) Photograph; (b) Line drawing. Source: Modified from [146].

four or five cross-veins between the two longitudinal veins; R_1 single; Rs with six branches; M with seven branches; Cu_1 forking at the same level as Rs_{1+2+3} from R, only two apical branches, and fused with M_{3+4} for a short distance; Cu_2 single; pterostigma present. Legs covered with short setae in rows and tibiae slender [146].

Superfamily Panorpidea Willmann, 1987

Family Orthophlebiidae Handlirsch, 1906

Orthophlebiidae, the largest extinct family of Mecoptera, comprise more than 190 species hitherto reported from the Late Triassic to the Early Cretaceous, among which about 30 species are from China [86, 89, 90, 94, 103, 117, 150–153]. The classification of this family has always been controversial. At first, Handlirsch established this family in 1906 with seven genera, and then eight more genera have been assigned to Orthophlebiidae. In 1992, Carpenter revised and combined several genera. Nowadays the generally accepted view is that there are three genera in this family: *Orthophlebia* Westwood, 1845, *Protorthophlebia* Tillyard, 1933, and *Mesopanorpa* Handlirsch, 1906. The morphological characters of Orthophlebiidae, similar to Panorpidae but with many differences: mouthparts short and chewing type; forewing slightly wide, R_1 single, Rs with 3–7 branches and arranged as fan-shaped, MP with 5 branches, CuA not fused to MP but connected by a short cross-vein; hind wing Sc slightly short and CuA fused to MP for a short distance at the base [86].

Genera included from the Jurassic and Cretaceous of Northern China: *Orthophlebia* Westwood, 1845, *Mesopanorpa* Handlirsch, 1906, *Protorthophlebia* Tillyard, 1933 and *Parachorista* Lin, 1976.

Orthophlebia Westwood, 1845

Orthophlebia Westwood, 1845, BRODIE, 399–402 [154] (original designation).

Callopanorpa Handlirsch, 1906, W. Engelmann, 615 [59].

Orthophlebioides Handlirsch, 1906, W. Engelmann, 481 [59].

Orthophlebiites Handlirsch, 1939, *Ann. Natrrhist. Mus. Wien*, 83–84 [77].

Synorthophlebia Handlirsch, 1939, *Ann. Natrrhist. Mus. Wien*, 78–81 [77].

Type species: *Orthophlebia liassica* (Mantell, 1844) Tillyard, 1933.

Forewing Rs_{1+2} and Rs_{3+4} forking at the same level; Rs and MA forking almost at the same level; Rs_1 with at least three branches; hind wing without distal cross-veins between MP and CuA.

Distribution and age: Liaoning and Hebei, Early Cretaceous; Xinjiang, Early Jurassic; Guangxi, Hebei, Inner Mongolia and Shanxi, Middle Jurassic.

Nine species included from the Jurassic and Cretaceous of China (see Table 24.1).

Orthophlebia nervulosa Qiao, Shih & Ren, 2012 (Figure 24.33)

Orthophlebia nervulosa Qiao, Shih & Ren, 2012: *Alcheringa*, 36 (4), 3–4.

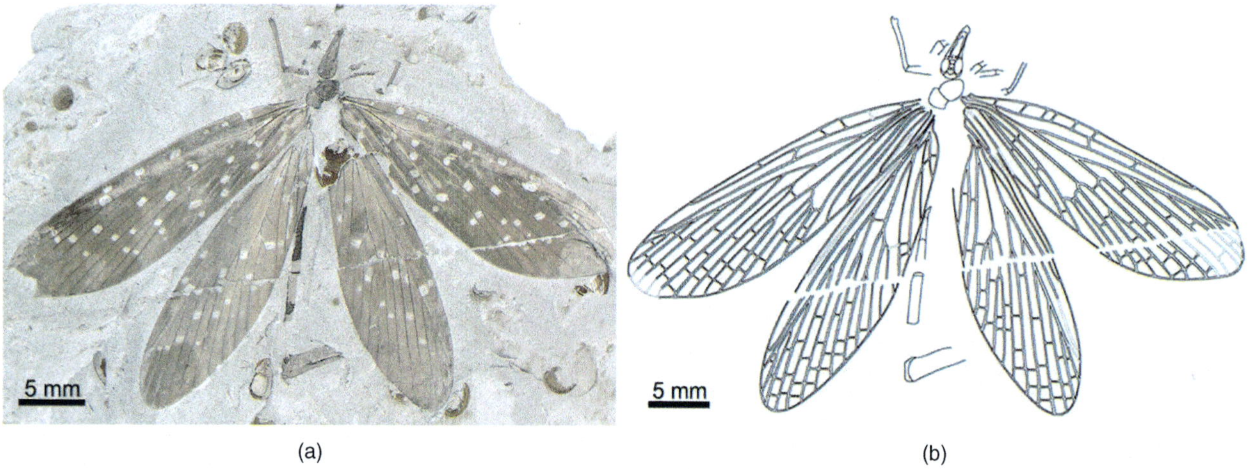

Figure 24.33 *Orthophlebia nervulosa* Qiao, Shih & Ren, 2012 (Holotype, CNU-MEC-NN-2011060). (a) Photograph; (b) Line drawing. Source: Modified from [86].

Locality and horizon: Daohugou, Ningcheng, Inner Mongolia, China; Middle Jurassic, Jiulongshan Formation.

The head is nearly circular with short chewing mouthparts, compound eyes large and oval. Legs covered with dense setae. Forewing length 27.5 mm and width 6.0 mm, Sc long and reaching costal margin; R long and slightly curved; Rs with six branches; the stem of Rs + MA twice as long as the stem of Rs; 5–6 cross-veins between Sc and C; Rs and MA almost forking at the same level; MP with five branches, CuA not fused with MP at the base, but fused with CuP at the base; three anal veins present. Hind wing length 26.7 mm, Sc short, reaching the middle of the wing; Rs with six branches; R single and straight; Rs forking proximal to MA; MP with four branches; CuA fused with MP for a short distance [86].

Mesopanorpa Handlirsch, 1906

Mesopanorpa Handlirsch, 1906, *W. Engelmann*, 615 [59] (original designation).

Type species: *Mesopanorpa hartungi* Brauer, Redtenbacher & Ganglbauer, 1889.

In forewing, the stem of Rs_{1+2} about twice as long as the stem of Rs_{3+4}; Rs with three or more branches, Rs_1 with one or more branches.

Distribution and age: Xinjiang, Early Cretaceous; Beijing, Early Cretaceous; Xinjiang, Early Jurassic; Zhejiang, Early Jurassic; Guangxi, Middle Jurassic; Hebei, Middle Jurassic.

Six species included from the Jurassic and Cretaceous of China (see Table 24.1).

Protorthophlebia Tillyard, 1933

Protorthophlebia Tillyard, 1933, *Fossil Insects, British Museum Natural History*, 3, 28–31 [63] (original designation).

Type species: *Protorthophlebia latipennis* Tliiyard, 1933.

Forewing Rs_1 with two or three branches; Rs forking proximal or distal to MA, or forking at the same level; the stem of Rs shorter or longer than MA; Rs_{1+2} and Rs_{3+4} forking almost at the same level.

Distribution and age: Liaoning, Early Cretaceous; Xinjiang, Early Jurassic.

Four species included from the Jurassic and Cretaceous of China (see Table 24.1).

Parachorista Lin, 1976

Parachorista Lin, 1976, *Acta Pal. Sin.*, 15 (1), 97–116 [65] (original designation).

Type species: *Parachorista miris* Lin, 1976.

Forewing Sc long and single, reaching the costal margin; R straight with one anterior branch; Rs with four branches, M with five branches; CuA single, fused with M_4 in the middle of the wing; two anal veins present.

Distribution and age: Liaoning; Early Cretaceous.

Only one species included from the Cretaceous of China (see Table 24.1).

Family Panorpidae Linnaeus, 1758

Panorpidae are the largest family in extant Mecoptera, with about 400 species and six genera documented hitherto [36, 155]. However, the fossil records are quite rare. There have been only four genera described so far, and three of them only include extinct species, respectively from the Early Cretaceous of China [156], Eocene Baltic amber [3] and the earliest fossil record from the Middle Jurassic Jiulongshan Formation of China [157]. This family are typified by the following characters: mouthparts prolong and hypognathous; wings slender, forewing slightly larger than hind wing, both have similar venation; forewing R_1 with two branches, Rs with five branches, M with four branches; Cu_1 not fused with M at the base; one cross-vein between M_4 and Cu_1; hind wing Rs with five branches and M with four branches [157].

Only one genus included from the Jurassic of Northern China: *Jurassipanorpa* Ding, Shih & Ren, 2014.

Jurassipanorpa Ding, Shih & Ren, 2014

Jurassipanorpa Ding, Shih & Ren, 2014, *ZooKeys*, 431, 81 [157] (original designation).

Type species: *Jurassipanorpa impunctata* Ding, Shih & Ren, 2014.

Forewing Sc almost reaching C at the same level or distal to the middle of the wing; both R_1 and Rs_1 with two branches; A_1 forking distal to the origination of Rs from R_1; three anal veins present; only one cu_1-cu_2 and a_1-a_2.

Distribution and age: Inner Mongolia; Middle Jurassic.

Two species included from the Jurassic of China (see Table 24.1).

Jurassipanorpa impunctata Ding, Shih & Ren, 2014 (Figure 24.34)

Jurassipanorpa impunctata Ding, Shih & Ren, 2014: *ZooKeys*, 431, 81–84.

Locality and horizon: Daohugou, Ningcheng, Inner Mongolia, China; Middle Jurassic, Jiulongshan Formation.

The body length is 12 mm. Head with extended proboscis, compound eyes large and oval. Thorax length 2.9 mm, meso- and metanotum about the same size, and larger than pronotum. Abdomen length 9 mm, with 11 visible segments. Legs covered with short setae, mid legs with two long tibial spurs, and only one spurs in fore- and hind legs. Forewing length 14 mm, and width about 4 mm; Sc reaching C at about the middle of the wing; one cross-vein between Sc and R_1; R_1 long, with one r_1-rs; Rs_1 and Rs_{3+4} forking almost at the same level, Rs_{1+2} forking proximal to Rs_{3+4}; Cu_1 not fused with M at the base; cu_1-cu_2 present; A_1, A_2 and A_3 long, A_1 distal to Rs from R_1. Hind wing length 12.3 mm, and width about 3.9 mm; similar to forewing in shape and venation, but distinctly smaller in size; Sc short, proximal to the middle of the wing; Rs and M

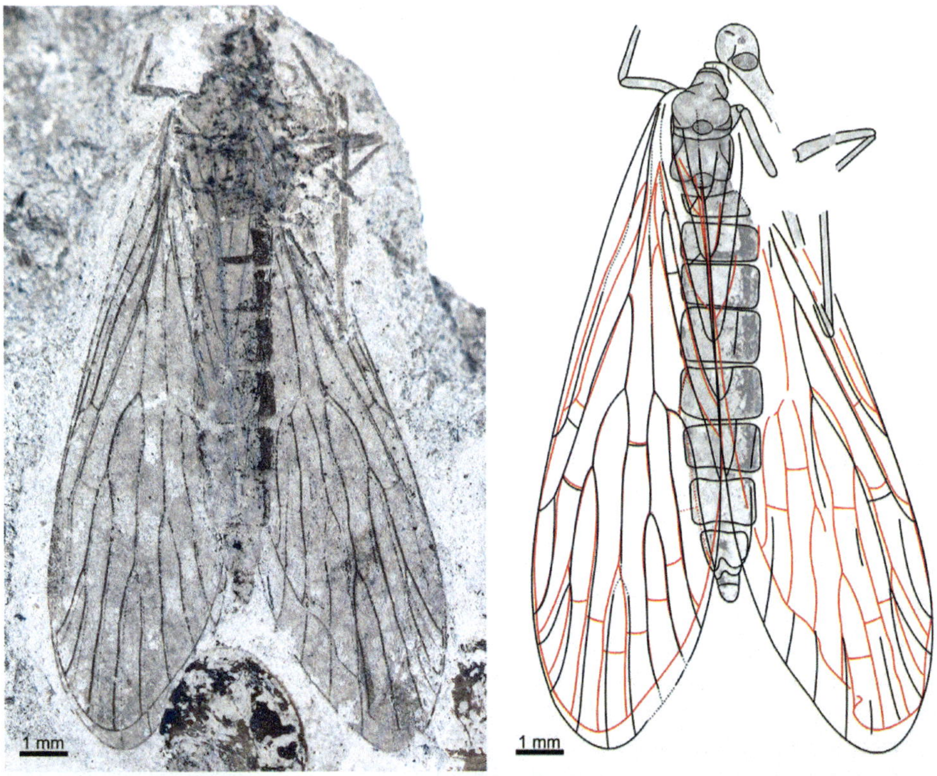

(a) (b)

Figure 24.34 *Jurassipanorpa impunctata* Ding, Shih & Ren, 2014 (Holotype, CNU-MEC-NN-2013006). (a) Photograph; (b) Line drawing, red lines denote hind wing venation. Source: Modified from [157].

forking almost at the same level; M_{1+2} longer than M_{3+4}; Cu_1 fused with M at the base, one m_4-cu_1 present; no cross-veins between anal veins [157].

Suborder *Incertae sedis*

Family Choristopsychidae Martynov, 1937

Choristopsychidae, a little known extinct family of Mecoptera, were erected by Martynov in 1937 based on only a single forewing. Since then, this family were once combined with Agetoparnopidae Carpenter, 1930 [119] and Permochoristidae Tillyard, 1917 [43]. However, many characters, such as wing shape and venation, should have supported that these three families are totally different. Up to date, two genera with four species have been classified into the Choristopsychidae. Except for *Choristopsyche tenuinervis* Martynov, 1937 from the Lower Jurassic of Tajikistan, all other species have been reported from the Middle Jurassic of Jiulongshan Formation in Inner Mongolia, China. In 2013, Qiao *et al.* emended the diagnosis of Choristopsychidae as: forewing relatively broad, oval or subtriangular; C and ScP area relatively broad, ScP with three long branches; RA single, scp-ra and ra-rp cross-veins present; both RP and MA with two branches, MP with five branches, and the MP_{4+5} forking proximal to the MP_{2+3}; MP and CuA merged at the base, CuA strongly bent in the middle; CuP, A_1 and A_2 almost parallel; Hind wing similar to the forewing in shape but slightly smaller, the stem of MP_{4+5} forking proximal to that of forewing and CuA almost straight. Antenna long and filiform and small chewing mouthparts present [158]. For three species, *Choristopsyche tenuinervis* Martynov, 1937, *Choristopsyche perfecta* Qiao, Shih, Petrulevičius & Ren, 2013, and *Paristopsyche angelineae* Qiao, Shih, Petrulevičius & Ren, 2013, their wings have many patterns and spots which are almost symmetrical in left and right wings. These wing patterns, also present in many other mecopteran families, might have been used for protection from predators or for attraction of potential mates.

Genera included from the Jurassic of Northern China: *Choristopsyche* Martynov, 1937 and *Paristopsyche* Qiao, Shih, Petrulevičius & Ren, 2013.

Choristopsyche Martynov, 1937

Choristopsyche Martynov, 1937, *Trudy Paleont. Inst. Akad. Nauk SSSR*, 7, 26–29 [159] (original designation).

Type species: *Choristopsyche tenuinervis* Martynov, 1937.

Forewing ScP long with two anterior branches; both RP & MA with two branches; MP with five branches; the separation of RP + MA from RA distal to MP from CuA; CuA merged with MP at the base, and strongly curved in the middle.

Distribution and age: Inner Mongolia of China, Middle Jurassic; Tajikistan, Early Jurassic.

Three species included from the Jurassic of China (see Table 24.1).

Paristopsyche Qiao, Shih, Petrulevičius & Ren, 2013

Paristopsyche Qiao, Shih, Petrulevičius & Ren, 2013, *ZooKeys*, 318, 101 [155] (original designation).

Type species: *Paristopsyche angelineae* Qiao, Shih, Petrulevičius & Ren, 2013.

Forewing is broadly oval, the separation of RP + MA from RA at about the same level as MP from CuA, RP + MA forking distal to MP, the stem of MP_3 about twice to four times as long as the stem of MP_{2+3}.

Distribution and age: Inner Mongolia; Middle Jurassic.

Only one species included from the Jurassic of China (see Table 24.1).

Paristopsyche angelineae Qiao, Shih, Petrulevičius & Ren, 2013 (Figure 24.35)

Paristopsyche angelineae Qiao, Shih, Petrulevičius & Ren, 2013 *ZooKeys*, 318, 101.

Locality and horizon: Daohugou, Ningcheng, Inner Mongolia, China; Middle Jurassic, Jiulongshan Formation.

The specific epithet is dedicated to Janet Angeline for her professionalism, dedication and accomplishments in her field and providing inspiration and support to Chungkun Shih's paleontology studies. Right forewing length 8.4 mm and width 5.5 mm; the stem of MP_3 about three times as long as the that of MP_{2+3}; only one ma-mp_{1+2+3} and mp_{2+3}-mp_4; CuP, A_1 and A_2 single. Right hind wing length about 7.5 mm and width 4.6 mm; similar to forewing in vernation, but slightly smaller in size. There are many dark spots on wings, which are nearly symmetrical in left and right wings.

Figure 24.35 *Paristopsyche angelineae* Qiao, Shih, Petrulevičius & Ren, 2013 (Holotype, CNU-MEC-NN-2011076p).

In contrast to long and narrow wings of most mecopterans with ratios of forewing length/width ranging from 3.0 to 5.2, choristopsychids have the relatively broadest wings in Mecoptera, either oval or subtriangular-shaped, with length/width ratio from 1.5 to 2.0. The other two families having broad forewings are Pseudopolycentropodidae with forewing length/width ratios ranging from 2.0 to 2.5, while Eomeropidae having ratios from 2.3 to 2.7 [158].

Table 24.1 A list of fossil Mecoptera from the Jurassic and Cretaceous of China.

Family	Species	Locality	Horizon/Age	Citation
Suborder Aneuretopsychina Rasnitsyn & Kozlov, 1990				
Aneuretopsychidae	*Jeholopsyche bella* Qiao, Shih & Ren, 2012	Beipiao, Liaoning	Yixian Fm., K_1	Qiao *et al.* [69]
	Jeholopsyche completa Qiao, Shih & Ren, 2012	Beipiao, Liaoning	Yixian Fm., K_1	Qiao *et al.* [69]
	Jeholopsyche liaoningensis Ren, Shih & Labandeira, 2011	Beipiao, Liaoning	Yixian Fm., K_1	Ren *et al.* [68]
	Jeholopsyche maxima Qiao, Shih & Ren, 2012	Beipiao, Liaoning	Yixian Fm., K_1	Qiao *et al.* [69]
Mesopsychidae	*Epicharmesopsyche pentavenulosa* Shih, Qiao, Labandeira & Ren, 2013	Ningcheng, Inner Mongolia	Jiulongshan Fm., J_2	Shih *et al.* [75]
	Lichnomesopsyche daohugouensis Ren, Labandeira & Shih, 2010	Ningcheng, Inner Mongolia	Jiulongshan Fm., J_2	Ren *et al.* [21]
	Lichnomesopsyche gloriae Ren, Labandeira & Shih, 2010	Ningcheng, Inner Mongolia	Jiulongshan Fm., J_2	Ren *et al.* [21]
	Lichnomesopsyche prochorista Lin, Shih, Labandeira & Ren, 2016	Ningcheng, Inner Mongolia	Jiulongshan Fm., J_2	Lin *et al.* [30]
	Vitimopsyche kozlovi Ren, Labandeira & Shih, 2010	Beipiao, Liaoning	Yixian Fm., K_1	Ren *et al.* [21]
	Vitimopsyche pectinella Gao, Shih, Labandeira, Santiago-Blay, Yao & Ren, 2016	Beipiao, Liaoning	Yixian Fm., K_1	Gao *et al.* [19]
	Vitimopsyche pristina Lin, Shih, Labandeira & Ren, 2016	Ningcheng, Inner Mongolia	Jiulongshan Fm., J_2	Lin *et al.* [30]
Pseudopolycentropodidae	*Pseudopolycentropus daohugouensis* Zhang, 2005	Ningcheng, Inner Mongolia	Jiulongshan Fm., J_2	Grimaldi *et al.* [24]
	Pseudopolycentropus janeannae Ren, Shih & Labanderia, 2010	Ningcheng, Inner Mongolia	Jiulongshan Fm., J_2	Ren *et al.* [80]
	Pseudopolycentropus novokshonovi Ren, Shih & Labanderia, 2010	Ningcheng, Inner Mongolia	Jiulongshan Fm., J_2	Ren *et al.* [80]
	Sinopolycentropus rasnitsyni Shih, Yang, Labandeira & Ren, 2011	Ningcheng, Inner Mongolia	Jiulongshan Fm., J_2	Shih *et al.* [81]
Suborder Eumecoptera Tillyard, 1919				
Holcorpidae	*Conicholcorpa stigmosa* Li, Shih, Wang & Ren, 2017	Ningcheng, Inner Mongolia	Jiulongshan Fm., J_2	Li *et al.* [84]
Mesorthophlebiidae	[a)]*Mesorthophlebia sinica* Hong, 1983	Xiaofanzhangzi, Hebei	Jiulongshan Fm., J_2	Hong [94]
Parachoristidae	[a)]*Jibeiorthophlebia internata* Hong, 1983	Xiaofanzhangzi, Hebei	Jiulongshan Fm., J_2	Hong [94]

(Continued)

Table 24.1 (Continued)

Family	Species	Locality	Horizon/Age	Citation
	[a] *Jibeiorthophlebia xiaofanzhangziensis* Hong, 1983	Xiaofanzhangzi, Hebei	Jiulongshan Fm., J_2	Hong [94]
	Junggarochorista tuzigouensis Hong, 2009	Junggar basin, Xinjiang	Jiulongshan Fm., J_1	Hong [25]
	Suborder Nannomecoptera Hinton, 1981			
Nannochoristidae	*Itaphlebia amoena* Cao, Shih, Bashkuev & Ren, 2015	Ningcheng, Inner Mongolia	Jiulongshan Fm., J_2	Cao *et al.* [29]
	Itaphlebia decorosa Sun, Ren & Shih, 2007a	Ningcheng, Inner Mongolia	Jiulongshan Fm., J_2	Sun *et al.* [111]
	Itaphlebia exquisita Liu, Zhao & Ren, 2010	Ningcheng, Inner Mongolia	Jiulongshan Fm., J_2	Liu *et al.* [105]
	Itaphlebia laeta Liu, Zhao & Ren, 2010	Ningcheng, Inner Mongolia	Jiulongshan Fm., J_2	Liu *et al.* [105]
	Itaphlebia longiovata Cao, Shih, Bashkuev & Ren, 2015	Ningcheng, Inner Mongolia	Jiulongshan Fm., J_2	Cao *et al.* [29]
	Itaphlebia ruderalis Ren, Lu, Ji & Guo, 1995	Beipiao, Liaoning	Haifanggou Fm., J_2	Ren *et al.* [103]
	Suborder Pistillifera Willmann, 1987			
Bittacidae	*Composibittacus bipunctatus* Liu, Shih & Ren, 2016	Ningcheng, Inner Mongolia	Jiulongshan Fm., J_2	Liu *et al.* [114]
	Composibittacus reticulatus Liu, Shih & Ren, 2016	Ningcheng, Inner Mongolia	Jiulongshan Fm., J_2	Liu *et al.* [114]
	Decoribittacus euneurus Li & Ren, 2009	Ningcheng, Inner Mongolia	Jiulongshan Fm., J_2	Li & Ren [121]
	Exilibittacus foliaceus Liu, Shih & Ren, 2014	Ningcheng, Inner Mongolia	Jiulongshan Fm., J_2	Liu *et al.* [122]
	Exilibittacus lii Yang, Ren & Shih, 2012	Ningcheng, Inner Mongolia	Jiulongshan Fm., J_2	Yang *et al.* [39]
	Exilibittacus plagioneurus Liu, Shih & Ren, 2014	Ningcheng, Inner Mongolia	Jiulongshan Fm., J_2	Liu *et al.* [122]
	Formosibittacus macularis Li, Ren & Shih, 2008	Ningcheng, Inner Mongolia	Jiulongshan Fm., J_2	Li *et al.* [38]
	Jichoristella rara Ren, Lu, Ji & Guo, 1995	Xishan, Beijing	Lushangfen Fm., K_1	Ren *et al.* [103]
	Jurahylobittacus astictus Li, Ren & Shih, 2008	Ningcheng, Inner Mongolia	Jiulongshan Fm., J_2	Li *et al.* [38]
	Karattacus longialatus Li & Ren, 2009	Ningcheng, Inner Mongolia	Jiulongshan Fm., J_2	Li & Ren [121]
	Liaobittacus longantennatus Ren, 1993	Beipiao, Liaoning	Haifanggou Fm., J_2	Ren [120]
	Megabittacus beipiaoensis Ren, 1997	Beipiao, Liaoning	Yixian Fm., K_1	Ren [117]
	Megabittacus colosseus Ren, 1997	Beipiao, Liaoning	Yixian Fm., K_1	Ren [117]
	Megabittacus spatiosus Yang, Ren & Shih, 2012	Beipiao, Liaoning	Yixian Fm., K_1	Yang *et al.* [39]
	Mongolbittacus daohugoensis Petrulevičius, Huang & Ren, 2007	Ningcheng, Inner Mongolia	Jiulongshan Fm., J_2	Petrulevičius *et al.* [37]
	Mongolbittacus oligophlebius Liu, Shih & Ren, 2014	Ningcheng, Inner Mongolia	Jiulongshan Fm., J_2	Liu *et al.* [122]
	Orthobittacus maculosus Liu, Shih & Ren, 2016	Ningcheng, Inner Mongolia	Jiulongshan Fm., J_2	Liu *et al.* [114]

(Continued)

Table 24.1 (Continued)

Family	Species	Locality	Horizon/Age	Citation
	Preanabittacus validus Yang, Shih, Ren & Petrulevičius, 2012	Ningcheng, Inner Mongolia	Jiulongshan Fm., J_2	Yang *et al.* [40]
	Sibirobittacus undus Sukatsheva, 1990	Beipiao, Liaoning	Yixian Fm., K_1	Sukatsheva [107]
Cimbrophlebiidae	*Bellicimbrophlebia angusta* Yang, Shih & Ren, 2013	Ningcheng, Inner Mongolia	Jiulongshan Fm., J_2	Yang *et al.* [142]
	Bellicimbrophlebia cruciata Yang, Shih & Ren, 2013	Ningcheng, Inner Mongolia	Jiulongshan Fm., J_2	Yang *et al.* [142]
	Bellicimbrophlebia disvena Yang, Shih & Ren, 2013	Ningcheng, Inner Mongolia	Jiulongshan Fm., J_2	Yang *et al.* [142]
	Bellicimbrophlebia eumorpha Yang, Shih & Ren, 2013	Ningcheng, Inner Mongolia	Jiulongshan Fm., J_2	Yang *et al.* [142]
	Bellicimbrophlebia heteroneura Zhang, Shih, Zhao & Ren, 2015	Ningcheng, Inner Mongolia	Jiulongshan Fm., J_2	Zhang *et al.* [136]
	Cimbrophlebia amoena Zhang, Shih, Zhao & Ren, 2015	Ningcheng, Inner Mongolia	Jiulongshan Fm., J_2	Zhang *et al.* [136]
	Cimbrophlebia gracilenta Zhang, Shih, Zhao & Ren, 2015	Ningcheng, Inner Mongolia	Jiulongshan Fm., J_2	Zhang *et al.* [136]
	Cimbrophlebia rara Wang, Shih & Ren, 2014	Beipiao, Liaoning	Yixian Fm., K_1	Wang *et al.* [160]
	Juracimbrophlebia ginkgofolia Wang, Labandeira, Shih & Ren, 2012	Ningcheng, Inner Mongolia	Jiulongshan Fm., J_2	Wang *et al.* [31]
	Mirorcimbrophlebia daohugouensis Yang, Shih & Ren, 2013	Ningcheng, Inner Mongolia	Jiulongshan Fm., J_2	Yang *et al.* [142]
	Perfecticimbrophlebia laetus Yang, Shih, Ren & Petrulevičius, 2012	Ningcheng, Inner Mongolia	Jiulongshan Fm., J_2	Yang *et al.* [40]
	Telobittacus bellus Yang, Shih & Ren, 2013	Ningcheng, Inner Mongolia	Jiulongshan Fm., J_2	Yang *et al.* [142]
	Telobittacus decorus Zhang, Shih, Zhao & Ren, 2015	Ningcheng, Inner Mongolia	Jiulongshan Fm., J_2	Zhang *et al.* [136]
	Telobittacus fragosus Zhang, 1993	Shangxian, Shanxi	Fengjiashan Fm, K_1	Zhang [138]
Eomeropidae	*Jurathauma simplex* Zhang, Shih, Petrulevičius & Ren, 2011	Ningcheng, Inner Mongolia	Jiulongshan Fm., J_2	Zhang *et al.* [145]
	Tsuchingothauma shihi Ren & Shih, 2005	Ningcheng, Inner Mongolia	Jiulongshan Fm., J_2	Ren & Shih [144]
	Typhothauma excelsa Zhang, Shih & Ren, 2012	Beipiao, Liaoning	Yixian Fm., K_1	Zhang *et al.* [146]
	Typhothauma yixianensis Ren & Shih, 2005	Beipiao, Liaoning	Yixian Fm., K_1	Ren & Shih [144]
Orthophlebiidae	*Mesopanorpa densa* Zhang, 1996	Tuzigou of Karamayi, Xinjiang	Badaowan Fm., J_1	Zhang [89]
	Mesopanorpa enormis Lin, 1986	Zhongshan, Guangxi	Shiti Fm., J_2	Lin [151]
	Mesopanorpa fanshanensis Ren *et al.*, 1995	Xishan, Beijing	Lushangfen Fm., K_1	Ren *et al.* [103]
	Mesopanorpa gambra Lin, 1980	Shouchang, Zhejiang	Laocun Fm., K_1	Lin [156]
	Mesopanorpa monstrosa Zhang, 1996	Tuzigou of Karamayi, Xinjiang	Badaowan Fm., J_1	Zhang [89]
	[a]*Mesopanorpa yaojiashanensis* Lin, 1980	Shouchang, Zhejiang	Laocun Fm., K_1	Lin [156]

(Continued)

Table 24.1 (Continued)

Family	Species	Locality	Horizon/Age	Citation
	Orthophlebia colorata Zhang, 1996	Tuzigou of Karamayi, Xinjiang	Badaowan Fm., J_1	Zhang [89]
	[a]*Orthophlebia deformis* Lin, 1986	Zhongshan, Guangxi	Shiti Fm., J_2	Lin [151]
	Orthophlebia liaoningensis Ren, 1997	Beipiao, Liaoning	Yixian Fm., K_1	Ren [117]
	[a]*Orthophlebia luanpingensis* Hong, 1983	Zhouyingzi, Hebei	Jiulongshan Fm., J_2	Hong [94]
	Orthophlebia nervulosa Qiao, Shih & Ren, 2012	Ningcheng, Inner Mongolia	Jiulongshan Fm., J_2	Qiao *et al.* [86]
	[a]*Orthophlebia quadrimacula* Lin, 1982	Zichang, Shanxi	Zhiluo Fm., J_2	Lin [150]
	Orthophlebia stigmosa Qiao, Shih & Ren, 2012 (=*Longiphlebia stigmosa* Soszyńska-Maj and Krzemiński, 2018)	Ningcheng, Inner Mongolia	Jiulongshan Fm., J_2	Qiao *et al.* [86]; Soszyńska-Maj et al. [161]
	[a]*Orthophlebia yangjuanxiangensis* Hong, 1985	Yuxian, Hebei	[b]Xiahuayuan Fm., K_1	Hong [162]
	[a]*Orthophlebia yaogouensis* Hong, 1983	Zhouyingzi, Hebei	Jiulongshan Fm., J_2	Hong [94]
	Parachorista miris Lin, 1976	Beipiao, Liaoning	Yixian Fm., K_1	Lin [65]
	Protorthophlebia badaowanica Hong & Guo, 2010	Tuzigou of Karamayi, Xinjiang	Badaowan Fm., J_1	Hong & Guo [153]
	Protorthophlebia karamayiensis Hong & Guo, 2010	Tuzigou of Karamayi, Xinjiang	Badaowan Fm., J_1	Hong & Guo [153]
	Protorthophlebia strigata Zhang, 1996	Harsala of Wusu, Xinjiang	[b]Xishanyao Fm., J_2	Zhang [89]
	Protorthophlebia yanqingensis Hong & Xiao, 1997	Yanqing, Beijing	Houcheng Fm., J_3	Hong & Xiao [163]
Panorpidae	*Jurassipanorpa impunctata* Ding, Shih & Ren, 2014	Ningcheng, Inner Mongolia	Jiulongshan Fm., J_2	Ding *et al.* [157]
	Jurassipanorpa sticta Ding, Shih & Ren, 2014	Ningcheng, Inner Mongolia	Jiulongshan Fm., J_2	Ding *et al.* [157]
	Suborder *Incertae sedis*			
Choristopsychidae	*Choristopsyche asticta* Qiao, Shih, Petrulevičius & Ren, 2013	Ningcheng, Inner Mongolia	Jiulongshan Fm., J_2	Qiao *et al.* [158]
	Choristopsyche perfecta Qiao, Shih, Petrulevičius & Ren, 2013	Ningcheng, Inner Mongolia	Jiulongshan Fm., J_2	Qiao *et al.* [158]
	Choristopsyche tenuinervis Martynov, 1937	Ningcheng, Inner Mongolia	Jiulongshan Fm., J_2	Qiao *et al.* [158]
	Paristopsyche angelineae Qiao, Shih, Petrulevičius & Ren, 2013	Ningcheng, Inner Mongolia	Jiulongshan Fm., J_2	Qiao *et al.* [158]
Volitorididae	[a]*Volitoridia fulvis* Hong, 1982	Fengning, Hebei	[b]Yixian Fm., K_1	Hong [66]
Family *Incertae sedis*	*Fortiholcorpa paradoxa* Wang, Shih & Ren, 2013	Ningcheng, Inner Mongolia	Jiulongshan Fm., J_2	Wang *et al.* [85]
	Miriholcorpa forcipata Wang, Shih & Ren, 2013	Ningcheng, Inner Mongolia	Jiulongshan Fm., J_2	Wang *et al.* [85]

a) The species is not presented in the main text because the original description, photo, and line-drawings are not precise and the holotype cannot be rechecked.
b) Horizon/Age revised from the original paper based on updated information and data.

References

1 Grimaldi, D.A. and Engel, M.S. (2005). *The Evolution of the Insects*, 469. New York: Cambridge University Press.

2 Ren, D., Labandeira, C.C., Santiago-Blay, J.A. *et al.* (2009). A probable pollination mode before angiosperms: Eurasian, long-proboscid scorpionflies. *Science* **326** (5954): 840–847. https://doi.org/10.1126/science.1178338.

3 Krzemiński, W. and Soszyńska-Maj, A. (2012). A new genus and species of scorpionfly (Mecoptera) from Baltic amber, with an unusually developed postnotal organ. *Systematic Entomology* **37** (1): 223–228. https://doi.org/10.1111/j.1365-3113.2011.00602.x.

4 Kristensen, N.P. (1999). Phylogeny of endopterygote insects, the most successful lineage of living organisms. *European Journal of Entomology* **96**: 237–253.

5 Byers, G.W. and Thornhill, R. (1983). Biology of the Mecoptera. *Annual Review of Entomology* **28**: 203–228.

6 Chen, Q.X. and Hua, B.Z. (2016). Ultrastructure and morphology of compound eyes of the scorpionfly *Panorpa dubia* (Insecta: Mecoptera: Panorpidae). *PLoS One* **11** (6): e0156970. https://doi.org/10.1371/journal.pone.0156970.

7 Misof, B., Liu, S., Meusemann, K. *et al.* (2014). Phylogenomics resolves the timing and pattern of insect evolution. *Science* **346** (6210): 763–767. https://doi.org/10.1126/science.1257570.

8 Dunford, J.C. and Somma, L.A. (2008). *Scorpionflies (Mecoptera). Ency. entomology (Encyclopedia of entomology)*, 3304–3310. Dordrecht: Springer.

9 Jiang, L. (2016). *Comparative Morphology and Phylogenetic Analysis of the Larvae of Panorpidae (insect: Mecoptera)*. Doctor Dissertation, 13–31. Northwest A & F University.

10 Palmer, C.M. (2010). Diversity of feeding strategies in adult Mecoptera. *Terrestrial Arthropod Reviews* **3** (2): 111–128. https://doi.org/10.1163/187498310X519716.

11 Bockwinkel, G. and Sauer, K.P. (1993). Panorpa scorpionflies foraging in spider webs-kleptoparasitism at low risk. *Bulletin of the British Arachnological Society* **9** (4): 110–112.

12 Tan, J.L. and Hua, B.Z. (2008). Morphology of immature stages of Bittacus choui (Mecoptera: Bittacidae) with notes on its biology. *Journal of Natural History* **42** (31–32): 2127–2142. https://doi.org/10.1080/00222930802209775.

13 Ma, N., Huang, J., and Hua, B.Z. (2013). Functional morphology and sexual dimorphism of mouthparts of the short-faced scorpionfly *Panorpodes kuandianensis* (Mecoptera: Panorpodidae). *PLoS One* **8** (3): e60351. https://doi.org/10.1371/journal.pone.0060351.

14 Peña, L.E. (1968). *Natural History Notes on Notiothauma. Discovery (Yale University)* **4**: 42–44.

15 Wang, J. and Hua, B.Z. (2017). An annotated checklist of the Chinese Mecoptera with description of male *Panorpa guttata* Navás, 1908. *Entomotaxonomia* **39** (1): 24–42. https://doi.org/10.11680/entomotax.2017003.

16 Chapman, R.F. (1998). Wings form. In: *The Insects: Structure and Function*, 4e (ed. S.J. Simpson and A.E. Douglas), 190–192. Cambridge: Cambridge University Press.

17 Gullan, P.J. and Cranston, P.S. (2014). Taxobox 25 Mecoptera (hangingflies, scorpionflies and snowfleas). In: *The Insects: An Outline of Entomology*, 5e, 520. Chichester: Wiley.

18 Daly, H.V., Doyen, J.T., and Ehrlich, P.R. (1978). *Introduction to Insect Biology and Diversity*, vol. **680**. McGraw-Hill Book Company.

19 Gao, T.P., Shih, C.K., Labandeira, C.C. *et al.* (2016). Convergent evolution of ramified antennae in insect lineages from the Early Cretaceous of Northeastern China. *Proceedings of the Royal Society B* **283** (1839): 20161448. https://doi.org/10.1098/rspb.2016.1448.

20 Grimaldi, D.A. and Johnston, M.A. (2014). The long-tongued Cretaceous scorpionfly Parapolycentropus Grimaldi and Rasnitsyn (Mecoptera: Pseudopolycentropodidae): new data and interpretations. *American Museum Novitates* **3793**: 1–24. https://doi.org/10.1206/3793.1.

21 Ren, D., Labandeira, C.C., and Shih, C.K. (2010). New Mesozoic Mesopsychidae (Mecoptera) from Northeastern China. *Acta Geologica Sinica-English Edition* **84** (4): 720–731. https://doi.org/10.1111/j.1755-6724.2010.00244.x.

22 Labandeira, C.C., Kvaček, J., and Mostovski, M.B. (2007). Pollination drops, pollen, and insect pollination of Mesozoic gymnosperms. *Taxon* **56** (3): 663–695.

23 Labandeira, C.C. (2010). The pollination of mid Mesozoic seed plants and the early history of long-proboscid insects 1, 2, 3. *Annals of the Missouri Botanical Garden* **97** (4): 469–513.

24 Grimaldi, D.A., Zhang, J.F., Fraser, N.C., and Rasnitsyn, A. (2005). Revision of the bizarre Mesozoic scorpionflies in the Pseudopolycentropodidae (Mecopteroidea). *Insect Systematic and Evolution* **36** (4): 443–458.

25 Hong, Y.C. (2009). First discovery of fossil Parachoristidae (Insecta: Mecoptera) in China. *Geological Bulletin of China* **28** (10): 1382–1289.

26 Riek, E.F. (1968). *Robinjohnia tillyardi* Martynova, a mecopteron from the Upper Permian of Belmont, New South Wales. *Records of the Australian Museum* **27** (14): 299–302.

27 Novokshonov, V.G. (1997). *Early Evolution of Scorpionflies (Insecta: Panorpida)*, 140. Moscow: Nauka Press [in Russian].

28 Novokshonov, V.G. (1994). Permian scorpionflies (Insecta, Panorpida) of the families Kaltanidae, Permochoristidae and Robinjohniidae. *Paleontologicheskii Zhurnal* **1**: 65–76.

29 Cao, Y.Z., Shih, C.K., Bashkuev, A.S., and Ren, D. (2015). Revision and two new species of *Itaphlebia* (Nannochoristidae: Mecoptera) from the Middle Jurassic of Inner Mongolia, China. *Alcheringa: the Australasian Journal of Palaeontology* **40** (1): 1–10. https://doi.org/10.1080/03115518.2015.1079692.

30 Lin, X.D., Shih, M.J.H., Labandeira, C.C., and Ren, D. (2016). New data from the Middle Jurassic of China shed light on the phylogeny and origin of the proboscis in the Mesopsychidae (Insecta: Mecoptera). *BMC Evolutionary Biology* **16** (1): 1. https://doi.org/10.1186/s12862-015-0575-y.

31 Wang, Y.J., Labandeira, C.C., Shih, C.K. *et al.* (2012). Jurassic mimicry between a hangingfly and a ginkgo from China. *Proceedings of the National Academy of Sciences of the USA* **109** (50): 20514–20519. https://doi.org/10.1073/pnas.1205517109.

32 Vahed, K. (1998). The function of nuptial feeding in insects: A review of empirical studies. *Biological Reviews* **73** (1): 43–78.

33 Xiang, L.B., Xie, G.L., and Wang, W.K. (2016). Application of insect courtship behaviors in classification and identification. *Journal of Environmental Entomology* **38** (5): 883–887.

34 Missoweit, M. and Sauer, K.P. (2007). Not all *Panorpa* (Mecoptera: Panorpidae) scorpionfly mating systems are characterized by resource defence polygyny. *Animal Behaviour* **74** (5): 1207–1213. https://doi.org/10.1016/j.anbehav.2006.10.028.

35 Ma, N., Zhong, W., and Hua, B.Z. (2010). Genitalic morphology and copulatory mechanism of the scorpionfly *Panorpa jilinensis* (Mecoptera: Panorpidae). *Micron* **41** (8): 931–938. https://doi.org/10.1016/j.micron.2010.07.008.

36 Ma, N., Liu, S.Y., and Hua, B.Z. (2011). Morphological diversity of male salivary glands in Panorpidae (Mecoptera). *European Journal of Entomology* **108** (3): 493–499. https://doi.org/10.14411/eje.2011.064.

37 Petrulevičius, J.F., Huang, D.-Y., and Ren, D. (2007). A new hangingfly (Insecta: Mecoptera: Bittacidae) from the Middle Jurassic of Inner Mongolia, China. *African Invertebrates* **48** (1): 145–152.

38 Li, Y.L., Ren, D., and Shih, C.K. (2008). Two Middle Jurassic hanging-flies (Insecta: Mecoptera: Bittacidae) from Northeast China. *Zootaxa* **1929**: 38–46.

39 Yang, X.G., Ren, D., and Shih, C.K. (2012). New fossil hangingflies (Mecoptera, Raptipeda, Bittacidae) from the Middle Jurassic to Early Cretaceous of Northeastern China. *Geodiversitas* **34**: 785–799. https://doi.org/10.5252/g2012n4a4.

40 Yang, X.G., Shih, C.K., Ren, D., and Petrulevičius, J.F. (2012). New Middle Jurassic hangingflies (Insecta: Mecoptera) from Inner Mongolia, China. *Alcheringa: An Australasian Journal of Palaeontology* **36** (2): 195–201. https://doi.org/10.1080/03115518.2012.622143.

41 Ren, D., Shih, C.K., Gao, T.P. *et al.* (2010). *Silent Stories–Insect Fossil Treasures from Dinosaur Era of the Northeastern China*, 322. Beijing: Science Press.

42 Carpenter, F.M. (1992). *Superclass Hexapoda.* In: *Treatise on Invertebrate Paleontology. Part R, Arthropoda 4, vol. 4* (ed. R.L. Kaesler), 279–655. Boulder, Colorado: Geological Society of America.

43 Novokshonov, V.G. (2002). Order Panorpida Latreille, 1802. The scorpionflies. In: *History of Insects* (ed. A.P. Rasnitsyn and Q. DLJ), 194–198. Dordrecht: Kluwer Academic Publisher.

44 Willmann, R. (1987). The phylogenetic system of Mecoptera. *Systematic Entomology* **12**: 519–524.

45 Tillyard, R.J. (1926). Kansas Permian insects; Part 7, The order Mecoptera. *American Journal of Science* (62): 133–164.

46 Tillyard, R.J. (1937). Kansas Permian insects; Part 17, The order Megasecoptera and additions to the Palaeodictyoptera, O donata, Protoperlaria, Copeognatha, and Neuroptera. *American Journal of Science* **194**: 81–110.

47 Carpenter, F.M. (1930). The Lower Permian insects of Kansas. Part 1. Introduction and the order Mecoptera. *Bulletin of the Museum of Comparative Zoology* **52**: 69–101.

48 Martynov, A.V. (1940). Permian Fossil Insects of Tshekarda. *Trudy Paleontologicheskogo Instituta Akademii nauk SSSR* **11** (1): 5–62. [In Russian].

49 Willmann, R. (1978). Mecoptera (Insecta, Holometabola). In: *Fossilium Catalogus, Animalia*, vol. **124**, 1–139.

50 Novokshonov, V.G. (1998). Fossil insects of Chekarda. In: *Chekarda - mestonakhozhdenie permskikh iskopaemykh rastenii i nasekomykh (Chekarda – The Locality of Permian Fossil Plants and Insects)* (ed. V.G. Novokshonov), 25–54. Perm: Perm University Press.

51 Novokshonov, V.G. (1999). New fossil insects (Insecta: Hypoperlida, Panorpida, ordinis incertis) from the Chekarda locality. *Paleontological Journal* **33**: 52–56.

52 Rasnitsyn, A.P., Aristov, D.S., Gorochov, A.V. *et al.* (2004). Important new insect fossils from Carrizo Arroyo and the Permo-Carboniferous faunal boundary. *Bulletin of the New Mexico Museum of Natural History and Science* **25**: 215–246.

53 Bashkuev, A.S. (2008). The first record of Kaltanidae (Insecta: Mecoptera: Kaltanidae) from the Permian of European Russia. *Paleontological Journal* **42** (4): 401–405.

54 Bashkuev, A.S. (2011). Nedubroviidae, a new family of Mecoptera: the first Paleozoic long-proboscid scorpionflies. *Zootaxa* **2895** (1): 47–57.

55 Riek, E.F. (1953). Fossil mecopteroid insects from the upper Permian of New South Wales. *Records of the Australian Museum* **23** (2): 55–87.

56 Riek, E.F. (1954). The Australian Mecoptera or scorpionflies. *Australian Journal of Zoology* **2** (1): 143–168.

57 Novokshonov, V.G. (2001). New Triassic scorpionflies (Insecta, Mecoptera) from Kyrgyzstan. *Paleontological Journal* **35** (3): 281–288.

58 Pinto, I.D. and Ornellas, L.P. (1978). New fossil insects from the White Band Formation (Permian), South Africa. *Pesquisas, Zoologia, Porto Alegre* **10**: 96–104.

59 Handlirsch, A. (1906–1908). *Die fossilen Insekten und die phylogenie der Rezenten Formen: ein Handbuch für paläontologen und Zoologen*, 1430. Leipzig: Verlagvon Wilhelm Engelmann.

60 Handlirsch, A. (1925). Geschichte, literatur, technik, palaeontologie, phylogenie, systematik. In: *Handbuch der Entomologie, Band 3* (ed. C. Schrder), 117–306. Verlag von Gustav Fischer: Jena.

61 Tillyard, R.J. (1917). Mesozoic insects of Queensland. In: *Proceedings of the Linnean Society of NSW*, vol. **42**, 172–200.

62 Tillyard, R.J. (1919). The panorpoid complex: a study of the phylogeny of the holometabolous insects, with special reference to the subclasses panorpoidea and neuropteroidea. Part 3: the wing venation. In: *Proceedings of the Linnean Society of NSW*, vol. **44**, 533–718.

63 Tillyard, R.J. (1933). *The Panorpoid Complex in the British Rhaetic and Lias*, vol. **3**, 7–79. British Museum (Natural History), Fossil Insects.

64 Martynov, A.V. (1927). Jurassic fossil Mecoptera and Paratrichoptera from Turkestan and Ust-Balei (Siberia). *Bulletin of the Academy of Science URSS* **21**: 651–666.

65 Lin, Q.B. (1976). The Jurassic fossil insects from western Liaoning. *Acta Palaeontologica Sinica* **15** (1): 97–121.

66 Hong, Y.C. (1982). *Mesozoic Fossil Insects of Jiuquan Basin in Gansu Province*, 187. Beijing: Geological Publish House.

67 Rasnitsyn, A.P. and Kozlov, M.V. (1990). A new group of fossil insects: scorpionfly with cicada and butterfly adaptations. *Doklady Akademii Nauk SSSR* **310**: 973–976. [in Russian].

68 Ren, D., Shih, C.K., and Labandeira, C.C. (2011). A well-preserved aneuretopsychid from the Jehol Biota of China (Insecta, Mecoptera, Aneuretopsychidae). *ZooKeys* **129**: 17–28. https://doi.org/10.3897/zookeys.129.1282.

69 Qiao, X., Shih, C.K., and Ren, D. (2012). Three new species of aneuretopsychids (Insecta: Mecoptera) from the Jehol Biota, China. *Cretaceous Research* **36**: 146–150. https://doi.org/10.1016/j.cretres.2012.03.004.

70 Zalessky, G. (1936). Sur deux restes d'insectes fossiles provenant du bassin de Kousnetzk et surl'age geologique des depots qui les renferment. *Bulletin de la Société Géologique de France* **5** (5): 687–695.

71 Bashkuev, A.S. (2011). The earliest Mesopsychidae and revision of the family Mesopanorpodidae (Mecoptera). *ZooKeys* **130**: 263–179. https://doi.org/10.3897/zookeys.130.1611.

72 Tscharntke, T., Steffan-Dewenter, I., Kruess, A. *et al.* (2002). Characteristics of insect populations on habitat fragments: a mini review. *Ecological Research* **17** (2): 229–239.

73 Casper, B.B. and Niesenbaum, R.A. (1993). Pollen versus resource limitation of seed production: a reconsideration. *Current Science* 210–214.

74 Novokshonov, V.G. and Sukacheva, I.D. (2001). Fossil scorpionflies of the "Suborder" Paratrichoptera (Insecta: Mecoptera). *Paleontological Journal* **35** (2): 173–182.

75 Shih, C.K., Qiao, X., Labandeira, C.C., and Ren, D. (2013). A new mesopsychid (Mecoptera) from the Middle Jurassic of Northeastern China. *Acta Geologica Sinica-English Edition* **87** (5): 1235–1241. https://doi.org/10.1111/1755-6724.12124.

76 Novokshonov, V.G. (1997). Some Mesozoic scorpionflies (Insecta: Panorpida: Mecoptera) of the families Mesopsychidae, Pseudopolycentrodidae, Bittacidae, and Permochoristidae. *Paleontologicheskii Zhurnal* **1**: 65–71. [In Russian].

77 Handlirsch, A. (1939). Neue Untersuchungen über die fossilen Insekten mit Ergänzungen und Nachträgen sowie Ausblicken auf phylogenetische, palaeogeographische und allgemein biologische

Probleme. *II. Annalen des naturhistorischen Museums in Wien* 78–86.

78 Whalley, P.E.S. (1985). The systematics and palaeogeography of the Lower Jurassic insects of Dorset, England. *Bulletin of the British Museum (Natural History), Geology Series* **39**: 107–189.

79 Papier, F., Nel, A., and Grauvogel-Stamm, L. (1996). Deux nouveaux insectes Mecopteroidea du Buntsandstein supérieur (Trias) des Vosges (France). *Paleontologia Lombarda* **5**: 37–45.

80 Ren, D., Shih, C.K., and Labandeira, C.C. (2010). New Jurassic pseudopolycentropodids from China (Insecta: Mecoptera). *Acta Geologica Sinica-English Edition* **84** (1): 22–30. https://doi.org/10.1111/j.1755-6724.2010.00166.x.

81 Shih, C.K., Yang, X.G., Labandeira, C.C., and Ren, D. (2011). A new long-proboscid genus of Pseudopolycentropodidae (Mecoptera) from the Middle Jurassic of China and its plant-host specializations. *ZooKeys* **130**: 281–297. https://doi.org/10.3897/zookeys.130.1641.

82 Scudder, S.H. (1878). An account of some insects of unusual interest from the Tertiary rocks of Colorado and Wyoming. *Bulletin of the United States Geological and Geographical Survey of the Territories* **4** (2): 519–543.

83 Archibald, S.B. (2010). Revision of the scorpionfly family Holcorpidae (Mecoptera), with description of a new species from Early Eocene McAbee, British Columbia, Canada. *Annales la Societe Entomologique de France* **46** (1–2): 173–182. https://doi.org/10.1080/00379271.2010.10697654.

84 Li, L., Shih, C.K., Wang, C., and Ren, D. (2017). A new fossil scorpionfly (Insecta: Mecoptera: Holcorpidae) with extremely elongate male genitalia from Northeastern China. *Acta Geologica Sinica-English Edition* **91** (3): 797–805. https://doi.org/10.1111/1755-6724.13310.

85 Wang, Q., Shih, C.K., and Ren, D. (2013). The earliest case of extreme sexual display with exaggerated male organs by two Middle Jurassic mecopterans. *PLoS One* **8** (8): e71378. https://doi.org/10.1371/journal.pone.0071378.

86 Qiao, X., Shih, C.K., and Ren, D. (2012). Two new Middle Jurassic species of orthophlebiids (Insecta: Mecoptera) from Inner Mongolia, China. *Alcheringa: the Australasian Journal of Palaeontology* **36** (4): 1–7. https://doi.org/10.1080/03115518.2012.671689.

87 Petrulevičius, J.F. and Ren, D. (2012). A new species of "Orthophlebiidae" (Insecta: Mecoptera) from the Middle Jurassic of Inner Mongolia, China. *Revue de Paléobiologie, Genève* **11**: 311–315.

88 Latreille, P.A. (1802). *Histoire Naturelle, Générale et Particulière des Crustacés et des Insectes*, 467. Paris: Tome Troisième.

89 Zhang, H.C. (1996). The Mesozoic Orthophlebiidae (Insecta, Mecoptera) Junggar basin of Xinjiang. *Acta Palaeontologica Sinica* **35** (4): 442–454.

90 Hong, Y.C. and Zhang, Z.J. (2007). Reclassification of fossil Orthophlebiidae Insecta Mecoptera. *Entomotaxonomia* **29** (1): 26–36.

91 Jell, P.A. and Roberts, J. (1986). *Plants and Invertebrates from the Lower Cretaceous Koonwarra Fossil Bed, South Gippsland, Victoria*, vol. **3**, 196–197. Association of Australasian Palaeontologists.

92 Riek, E.F. (1955). Fossil insects from the Triassic beds at Mt. Crosby, Queensland. *Australian Journal of Zoology* **3** (4): 654–691.

93 Novokshonov, V.G. (1996). Mysterious organs of Jurassic Orthophlebiidae males (Insecta, Mecoptera) from Karatau. *Paleontologicheskii Zhurnal* **75**: 1491–1495.

94 Hong, Y.C. (1983). *Middle Jurassic fossil insects in north China*, 223. Beijing: Geological Publishing House.

95 Byers, G.W. (1989). The Nannochoristidae of South America (Mecoptera). *The University of Kansas Science Bulletin (USA)* **54**: 25–34.

96 Penny, N.D. (1975). Evolution of the extant Mecoptera. *Journal of the Kansas Entomological Society* **48**: 331–350.

97 Kristensen, N.P. (1989). The New Zealand scorpionfly (*Nannochorista philpotti* comb. n.): wing morphology and its phylogenetic significance. *Journal of Zoological Systematics and Evolutionary Research* **27** (2): 106–114.

98 Byers, G.W. (1974). New generic names for Mecoptera of Australia and New Zealand. *Austral Entomology* **13** (2): 165–167.

99 Fraulob, M., Wipfler, B., Hünefeld, F. *et al.* (2012). The larval abdomen of the enigmatic Nannochoristidae (Mecoptera, Insecta). *Arthropod Structure and Development* **41** (2): 187–198. https://doi.org/10.1016/j.asd.2011.11.001.

100 Riek, E.F. (1973). Fossil insects from the Upper Permian of Natal, South Africa. *Annals of the Natal Museum* **21** (3): 513–532.

101 Riek, E.F. (1970). Lower Cretaceous fleas. *Nature* **227** (5259): 746–747.

102 Pilgrim, R.L.C. (1972). Aquatic larva and the pupa of *Choristella philpotti* Tillyard, 1917 (Mecoptera: Nannochoristidae). *Pacific Insects* **14**: 151–168.

103 Ren, D., Lu, L.W., Ji, S.A., and Guo, Z.G. (1995). *Faunae and Stratigraphy of Jurassic-Cretaceous in Beijing and the Adjacent Areas*, 222. Beijing: Seismic Publishing House.

104 Sun, J.H., Ren, D., and Shih, C.K. (2007). Middle Jurassic Nannochoristidae fossils from Daohugou, Inner Mongolia in China (Insecta, Mecoptera). *Acta Zootaxonomica Sinica* **32**: 405–411.

105 Liu, N., Zhao, Y.Y., and Ren, D. (2010). Two new fossil species of *Itaphlebia* (Mecoptera: Nannochoristidae) from Jiulongshan Formation, Inner Mongolia, China. *Zootaxa* **2420**: 37–45.

106 Sukatsheva, I.D. (1985). *Jurassic scorpionflies of South Siberia and West Mongolia*. In: *The Jurassic Insects of Siberia and Mongolia* (ed. A.P. Rasnitsyn), 96–114. Moscow: Nauka Press [in Russian].

107 Sukatsheva, I.D. (1990). Description of fossil insects. Scorpionflies (Panorpida). In: *Late Mesozoic Insects of Eastern Transbaikalia*. Transactions of the Paleontological Institute of the USSR Academy of Sciences, 88–94. Moscow: Nauka Press [in Russian].

108 Sukatsheva, I.D. (1993). New Mesozoic scorpion flies (Nannochoristidae, Mecoptera) from Yakutia. *Paleontological Journal* **27**: 169–171.

109 Novokshonov, V.G. (1997). New and little known Mesozoic Nannochoristidae (Insecta: Mecoptera). *Vestnik Permskogo Universiteta (Biologiya)* **4**: 126–136. [in Russian].

110 Novokshonov, V.G. and Sukatsheva, I.D. (2003). New Jurassic scorpionflies of the families Orthophlebiidae and Nannochoristidae (Insecta: Mecoptera) from Mongolia and Transbaikalia. *Paleontological Journal* **37**: 501–506.

111 Sun, J.H., Ren, D., and Shih, C.K. (2007). Middle Jurassic Mesopanorpodidae from Daohugou, Inner Mongolia, China (Insecta, Mecoptera). *Acta Zootaxonomica Sinica* **32**: 865–874.

112 Krzemiński, W. (2007). A revision of Eocene Bittacidae (Mecoptera) from Baltic amber with the description of a new species. *African Invertebrates* **48**: 153–162.

113 Kopeć, K., Soszynska-Maj, A., Krzemiński, W., and Coram, R.A. (2016). A new hangingfly (Insecta, Mecoptera, Bittacidae) from the Purbeck Limestone Group (Lower Cretaceous) of Southern England and a review of Cretaceous Bittacidae. *Cretaceous Research* **57**: 122–130. https://doi.org/10.1016/j.cretres.2015.08.007.

114 Liu, S.L., Shih, C.K., and Ren, D. (2016). New Jurassic hangingflies (Insecta: Mecoptera: Bittacidae) from Inner Mongolia, China. *Zootaxa* **4067** (1): 065–078. https://doi.org/10.11646/zootaxa.4067.1.5.

115 Zhao, X.D., Bashkuev, A., Chen, L., and Wang, B. (2016). The first hangingfly from mid-Cretaceous Burmese amber (Mecoptera: Bittacidae). *Cretaceous Research* **70**: 147–151. https://doi.org/10.1016/j.cretres.2016.10.008.

116 Novokshonov, V.G. (1993). Mückenhafte (Mecoptera Bittacidae) aus dem Jura, Kreide und Paläogen von Eurasien und ihre phylogenetischen Beziehungen. *Russian Entomological Journal* **2**: 75–86.

117 Ren, D. (1997). Studies Late Jurassic scorpion-flies from Northeast China. *Acta Zootaxonomica Sinica* **22** (1): 75–85.

118 Petrulevičius, J.F. and Jarzembowski, E.A. (2004). The first hangingfly (Insecta: Mecoptera: Bittacus) from the Cretaceous of Europe. *Journal of Paleontology* **78**: 1198–1201.

119 Willmann, R. (1989). Evolution und Phylogenetisches System der Mecoptera (Insecta: Holometabola). *Abhandlungen Sencken* **544**: 1–153.

120 Ren, D. (1993). First discovery of fossil bittacids from China. *Acta Geologica Sinica-English Edition* **67**: 376–381.

121 Li, Y.L. and Ren, D. (2009). Middle Jurassic Bittacidae (Insecta: Mecoptera) from Daohugou, Inner Mongolia, China. *Acta Zootaxonomica Sinica* **3**: 560–567.

122 Liu, S.L., Shih, C.K., and Ren, D. (2014). Four new species of hangingflies (Insecta, Mecoptera, Bittacidae) from the Middle Jurassic of Northeastern China. *ZooKeys* **466**: 77–94. https://doi.org/10.3897/zookeys.466.8047.

123 Cockerell, T.D.A. (1916). Insects in Burmese amber. *American Journal of Science* **42**: 135–138.

124 Cockerell, T.D.A. (1917). Arthropods in Burmese amber. *Psyche* **24**: 40–45.

125 Cockerell, T.D.A. (1917). Insects in Burmese amber. *Annals of the Entomological Society of America* **10**: 323–329.

126 Cockerell, T.D.A. (1920). XXV.–Fossil Arthropods in the British Museum. – IV. *Annals and Magazine of Natural History* **6**: 211–214.

127 Cockerell, T.D.A. (1920). A therevid fly in Burmese amber. *Entomologist* **53**: 69–70.

128 Burn, F.N. (1918). Insects in Burmese amber. *Entomologist* **51**: 102–103.

129 Grimaldi, D.A. and Engel, M.S. (2013). The relict scorpionfly family Meropeidae (Mecoptera) in Cretaceous amber. *Journal of the Kansas Entomological Society* **86**: 253–263. https://doi.org/10.2317/JKES130219.1.

130 Li, S., Zhang, W.W., Shih, C.K., and Ren, D. (2018). A new species of hangingfly (Insecta: Mecoptera: Bittacidae) from the mid-Cretaceous Myanmar amber. *Cretaceous Research* **89**: 92–97. https://doi.org/10.1016/j.cretres.2018.03.005.

131 Cruickshank, R.D. and Ko, K. (2003). Geology of an amber locality in the Hukawng Valley, Northern Myanmar. *Journal of Asian Earth Sciences* **21**: 441–455.

132 Shi, G.H., Grimaldi, D.A., Harlow, G.E. *et al.* (2012). Age constraint on Burmese amber based on U-Pb dating of zircons. *Cretaceous Research* **37**: 155–163. https://doi.org/10.1016/j.cretres.2012.03.014.

133 Martínez-Delclós, X., Briggs, D.E., and Peñalver, E. (2004). Taphonomy of insects in carbonates and amber. *Palaeogeography, Palaeoclimatology, Palaeoecology* **203**: 19–64.

134 Solórzano Kraemer, M.M., Kraemer, A.S., Stebner, F. *et al.* (2015). Entrapment bias of arthropods in Miocene amber revealed by trapping experiments in a tropical forest in Chiapas, Mexico. *PLoS One* **10**: e0118820. https://doi.org/10.1371/journal.pone.0118820.

135 Willmann, R. (1977). Mecoptera aus dem untereozänen Moler des Limfjordes (Denmärk). *Neues Jahrbuch für Geologie und Paläontologie, Monatschefte* **12**: 735–744.

136 Zhang, X., Shih, C.K., Zhao, Y.Y., and Ren, D. (2015). New species of Cimbrophlebiidae (Insecta: Mecoptera) from the Middle Jurassic of Northeastern China. *Acta Geologica Sinica-English Edition* 1428–1496. https://doi.org/10.1111/1755-6724.12559.

137 Archibald, S.B. (2009). New Cimbrophlebiidae (Insecta: Mecoptera) from the early Eocene at McAbee, British Columbia, Canada and Republic, Washington, USA. *Zootaxa* **2249**: 51–62.

138 Zhang, J.F. (1993). A contribution to the knowledge of insects from the Late Mesozoic in Southern Shaanxi and Henan Provinces, China. *Palaeoworld* **2**: 49–56.

139 Major, R.T. (1967). The ginkgo, the most ancient living tree. The resistance of Ginkgo biloba L. to pests accounts in part for the longevity of this species. *Science* **157** (3794): 1270–1273.

140 Wheeler, A.G. Jr., (1975). Insect associates of Ginkgo biloba. *Entomological News* **86**: 37–44.

141 Honda, H. (1997). *Ginkgo Biloba: A Global Treasure* (ed. T. Hori *et al.*), 243–250. Tokyo: Springer.

142 Yang, X.G., Shih, C.K., and Ren, D. (2013). New fossil hangingflies (Insecta: Mecoptera: Raptioedia: Cimbrophlebiidae) from the Middle Jurassic of Inner Mongolia, China. *Palaeontology* **56** (4): 711–726. https://doi.org/10.1111/pala.12009.

143 Soszyńska-Maj, A., Krzemiński, W., Kopeć, K., and Coram, R.A. (2016). Phylogenetic relationships within the relict family Eomeropidae (Insecta, Mecoptera) based on the oldest fossil from the Early Jurassic (Sinemurian) of Dorset, Southern England. *Journal of Systematic Palaeontology* **14** (12): 1025–1031. https://doi.org/10.1080/14772019.2016.1139007.

144 Ren, D. and Shih, C.K. (2005). The first discovery of fossil eomeropids from China (Insecta, Mecoptera). *Acta Zootaxonomica Sinica* **30** (2): 275–280.

145 Zhang, J.X., Shih, C.K., Petrulevičius, J.F., and Ren, D. (2011). A new fossil eomeropid (Insecta, Mecoptera) from the Jiulongshan Formation, Inner Mongolia, China. *Zoosystema* **33** (4): 443–450. https://doi.org/10.5252/z2011n4a2.

146 Zhang, J.X., Shih, C.K., and Ren, D. (2012). A new fossil eomeropids (Insecta, Mecoptera) from the Yixian Formation, Liaoning, China. *Acta Zootaxonomica Sinica* **37**: 68–71. [in Chinese with English abstract].

147 Archibald, S.B., Rasnitsyn, A.P., and Akhmetiev, M.A. (2005). Ecology and distribution of Cenozoic Eomeropidae (Mecoptera), and a new species of Eomerope Cockerell from the early Eocene McAbee locality, British Columbia, Canada. *Annals of the Entomological Society of America* **98** (4): 503–514.

148 Cockerell, T.D.A. (1909). Description of Tertiary insects VI. *American Journal of Science* **27**: 381–387.

149 Ponomarenko, A.G. and Rasnitsyn, A.P. (1974). New Mesozoic and Cenozoic Protomecoptera. *Paleontological Journal* **8** (4): 493–507.

150 Lin, Q.B. (1982). Mesozoic and Cenozoic insects. In: *Atlas of Palaeontology from Northwest Region, Shaanganning Division*, 70–83. Beijing: Geological Publishing House [in Chinese].

151 Lin, Q.B. (1986). Mesozoic fossil insects from South China. *Acta Palaeontologia Sinica* **170**: 82–84.

152 Hong, Y.C. and Zhang, Z.J. (2004). New taxonomy of Orthophlebiidae. *Geological Bulletin of China* **23** (8): 802–808.

153 Hong, Y.C. and Guo, X.R. (2010). Emendation for 3 homonym and 1 species of Orthophlebiidae (Insecta, Mecoptera). *Geological Bulletin of China* **29** (2–3): 188–194.

154 Westwood, J.O. (1845) In a History of the Fossil Insects in the Secondary Rocks of England. BRODIE, P.B., Van Voorst, London, 8vo, xviii, p. 130.

155 Zhong, W. and Hua, B.Z. (2013). Mating behaviour and copulatory mechanism in the scorpionfly *Neopanorpa longiprocessa* (Mecoptera: Panorpidae). *PLoS One* **8** (9): e74781. https://doi.org/10.1371/journal.pone.0074781.

156 Lin, Q.B. (1980). Mesozoic insects from Zhejiang and Anhui Provinces. In Nanjing Institute of Geology and Palaeontology, Chinese Academy of Sciences (eds.),. In: *Division and correlation of the Mesozoic volcano-sedimentary strata in Zhejiang and Anhui Provinces*, 211–234. Beijing: Science Press.

157 Ding, H., Shih, C.K., Bashkuev, A.S. *et al.* (2014). The earliest fossil record of Panorpidae (Mecoptera)

from the Middle Jurassic of China. *ZooKeys* **431**: 79–92. https://doi.org/10.3897/zookeys.431.7561.

158 Qiao, X., Shih, C.K., Petrulevičius, J.F., and Ren, D. (2013). Fossils from the Middle Jurassic of China shed light on morphology of Choristopsychidae (Insecta, Mecoptera). *ZooKeys* **318**: 91–111. https://doi.org/10.3897/zookeys.318.5226.

159 Martynov, A.V. (1937). Liassic insects of Shurab and Kizilkya Mongolia. *Trudy Paleontologicheskogo Instituta Akademii nauk SSSR* **7**: 1–178.

160 Wang, C., Shih, C.K., and Ren, D. (2014). A new fossil hangingfly (Mecoptera: Cimbrophlebiidae) from the Early Cretaceous of China. *Acta Geologica Sinica-English Edition* **88** (1): 29–34. https://doi.org/10.1111/1755-6724.12180.

161 Soszyńska-Maj, A., Krzemiński, W., Kopeć, K. *et al.* (2018). Large Jurassic Scorpionflies Belonging to a New Subfamily of the Family Orthophlebiidae (Mecoptera). *Annales Zoologici* **68** (1): 85–92.

162 Hong, Y.C. (1985). New fossil insects of Xiahuanyuan Formation in Yuxian County, Hebei Province. *Bulletin Tianjin Institute Geological Mineral Research* **13**: 131–138.

163 Hong, Y.C. and Xiao, Z.Z. (1997). *New Fossil Blattodea, Coleoptera and Mecoptera (Insecta) from Houcheng Formation of Yanqing County*, vol. **3**, 1–10. Beijing: Beijing Geology Publishing House.

25

Siphonaptera – Fleas

Taiping Gao[1], *Chungkun Shih*[1,2], *and Dong Ren*[1]

[1] Capital Normal University, Haidian District, Beijing, China
[2] National Museum of Natural History, Smithsonian Institution, Washington, DC, USA

25.1 Introduction to Siphonaptera

Siphonaptera, commonly called "fleas", are a relatively small order of wingless holometabolic insects, comprising more than 2500 species within 15 families [1–3]. Fleas are notorious for their ectoparasitic life, mostly in mammalians and some in Aves [3, 4].

Adult fleas are usual small, with body length under 8 mm, mainly 1–5 mm, adept in crawling and jumping. *Pulex irritans*, a flea living on humans, can jump to the height of 18 cm, and over 30 cm in horizontal distance [5]. It was recognized as the insect jumping champion until it was dethroned by a 6 mm long spittlebug, *Philaenus spumarius*, in 2003. (see box: The High Jump Champion in Section 16.3). Fleas have a laterally flattened body covered by bristles in different lengths, piercing blood-sucking mouthparts with serrate lacinia [6], legs with five tarsal segments and a pair of sharp claws, antennae with compacted flagellomeres freely contracted to antennal fossa, eyes under-development, and ocelli lost. Such specialized morphological features are adaptations to life-long ectoparasitism during evolution [7–9]. To avoid being dislodged and thrown off by movements of hosts, fleas have many rows of widened and hardened bristles (ctenidia) distributed nearly all over their bodies, which are termed frontal bristles, oral bristles, etc. depending on their locations. In addition, a flea has pygidium, a special plate structure with short setae and cuplike pits, located on the abdominal 9th and 10th terga. The pygidium is considered to be an important sensory receptor for a flea [3, 10].

The flea undergoes four different stages in its lifetime: egg, larva, pupa and adult. The adult female flea needs nutrients from blood feeding before laying eggs. A female flea can oviposit 3–18 eggs every time, but it must rest for a period and intake more blood before the next oviposition. The spawning cycle of a female flea can last several months, and a female can lay over 1000 eggs in her life. Fleas' eggs are white, subovate, and often dispersed in the nests of hosts. Eggs will develop to larvae after 2–12 days in proper temperature and humidity. Larvae are maggot-like, some have rows of bristles, eat scraps of adult or dander of hosts rather than ectoparasitism. Larvae will make cocoons after 2–3 moltings about 10 days later, but they can keep the form of larvae for a month, even over 200 days under harsh environments. The pupal stage generally continues for 15–20 days, may be advanced to 3 days or delayed to 2 years depending on living surroundings.

Different groups of fleas have various habits in choosing hosts. Some fleas only live on specific hosts by selecting a specific genus or even species, but others are not strict in choice of host, and able to easily switch to a different host, maybe even in a different family, when the current host dies [3].

Phylogenetic analyses have been carried out based on both morphological and molecular data about the relationship of Siphonaptera and other groups of insect [2, 11], and combined with fossil material [12]. Molecular studies and the result of tissue anatomy suggested that fleas would be the sister-group to Boreidae within scorpionflies [9, 13, 14]. However, a mecopteran family Aneuretopsychidae with long siphonate mouthparts, reported from the Mesozoic of China, indicates a closer relationship with flea than boreids [15, 16].

Rhythms of Insect Evolution: Evidence from the Jurassic and Cretaceous in Northern China, First Edition.
Edited by Dong Ren, Chungkun Shih, Taiping Gao, Yongjie Wang, and Yunzhi Yao.

"Black Death" or Plague

Much worse than their biting and blood-sucking, fleas can spread diseases to people as transmission media for viruses or bacteria. The flea-associated devastating pandemics, "Black Death" or plague, which occurred in Europe during the fourteenth century [17, 18], resulted in deaths of millions of people, and the similar disasters have happened many times in human history.

Though there is still no definite explanation about "Black Death", the mainstream science generally agree that "Black Death" is caused by the bacterium, *Yersinia pestis*, a species within Yersiniaceae of Bacteria, first discovered by Alexandre Yersin in 1894. The bacterium is harmless and common in rodents, but is devastating to human beings. Several groups of fleas (especially *Xenopsylla cheopsis*) live on brown rats. After infecting the fleas, the bacteria will fill up the stomach of fleas quickly. Fleas carrying *Y. pestis* always feel hungry and keep biting the hosts. When the hosts die of the bacteria, fleas quickly transfer to new hosts. During the mid-1300s, lots of people fled to the cities due to poor harvests, resulting in crowding and rubbish-strewn streets, leading to large numbers of rats. *Y. pestis* can be spread quickly through the air (cough) or via fleas' host switching [19, 20]. A person, after being bitten by fleas carrying *Y. pestis*, would develop blackish swellings, leading to severe sickness or ultimately even death.

Nowadays, more and more countries pay attention to sanitation and potential pandemics. In particular, the World Health Organization (WHO), an agency of the United Nations, has done a lot of work in staving off major outbreaks of communicable diseases. Therefore, it is less likely that plague will occur in a large area as it happened in the Middle Ages. However, the plague should not be taken lightly; a recent plague outbreak in Madagascar was reported in November 2014.

25.2 Progress in the Studies of Fossil Fleas

Though fleas are supposed to originate during the Cretaceous based on the estimate of phylogenetic analysis [2], fossil materials of ectoparasites with certain taxonomic placements from the Mesozoic are sparse.

In fact, the only true flea fossils are discovered from Eocene Baltic amber [21, 22] and Miocene Dominican amber [23], and these specimens are placed in the extant genera of *Palaeopsylla* and *Pulex*. Perrichot *et al.* reported an extinct group within Pulicidae of flea, *Eospilopsyllus kobberti*, from Miocene amber [24]. Fleas of the Cenozoic are generally considered to be the modern lineages of fleas, having the similar morphological features as those of the extant fleas [23, 25].

Riek reported the first probable Mesozoic fossil flea, *Tarwinia australis*, from the Lower Cretaceous of Australia, and *T. australis* possessed a wingless thorax, and ctenidia on the tibiae and pygidium [26, 27]. Since Riek did not describe the details of its mouthparts, the systematic position of *T. australis* was widely questioned. Huang re-examined the holotype of *T. australis* and found elongate siphonate mouthparts with the siphon pointing apically and extending beyond the procoxae [28]. *Saurophthirus longipes*, another probable ectoparasite with short sectorial mouthparts, was described from the Lower Cretaceous Zaza Formation of Baissa by Ponomarenko in 1976 [29] and later assigned to Saurophthiridae Ponomarenko, 1986 [30]. Rasnitsyn and Strelnikova recently analyzed the tracheal system of *S. longipes*, and suggested that it has a complex gonotrophic cycle [31].

In 1992, Rasnitsyn described a strange insect, *Strashila incredibilis*, from the Lower or Middle Jurassic of Transbaikalia, which had short antennae and elongated legs, especially the swollen hind coxae and femora. This incompletely preserved specimen was considered to be an ectoparasite at first [32, 33]. Huang *et al.* reported 13 specimens of strashilids from the Middle Jurassic of China, and transferred strashilids to Diptera based on the characters of consolidated thorax, large compound eyes and a pair of lateral ocelli. The males of strashilids have hind leg pincers for grasping the females, and their mouthparts do not have sucking structures [34].

In 2012, Huang *et al.* and Gao *et al.* nearly simultaneously reported several relatively huge "basal fleas", assigned to a new family Pseudopulicidae Gao, Shih & Ren, 2012, from the Middle Jurassic of Jiulongshan Formation and the Lower Cretaceous of Yixian Formation in China [16, 35, 36]. These specimens have stout and elongate sucking stylet mouthparts for piercing, reduced wingless thorax, a row of ctenidia on the tibiae, pygidium, etc. All these characters suggest their close relationships with the modern lineages of fleas. These fossil insects are different from extant fleas by the absence of modified jumping hind legs and the body not laterally compressed. Large body sizes (up to 22.8 mm) and long serrated stylets (up to 5.15 mm) for piercing a tough and thick skin suggest that these ancient groups of basal fleas might have lived on relatively large hosts, such as contemporaneous medium to large-sized vertebrates [16, 35, 37–40].

Huang *et al.* assigned *Pseudopulex wangi*, *Hadropsylla sinica* and *Tyrannopsylla beipiaoensis* to Pseudopulicidae, and transferred *Tarwinia* to a new family Tarwiniidae in 2013 [41]. Gao *et al.* in 2013 placed *Saurophthirus exquisitus* in Saurophthiridae Ponomarenko, 1986, and suggested this group as extinct "transitional fleas" of Siphonaptera based on three well-preserved fossils from the Jehol Biota, Lower Cretaceous of China [31, 42]. Saurophthirids show many morphological transitions from pseudopulicids to extant fleas, such as the small body size, short and slender piercing-sucking stylet mouthparts, rows of short and stiff bristles on the thorax, highly elongated legs, partially invaginated genitalia, etc. In 2014, Gao *et al.* described a female transitional flea, *Pseudopulex tanlan* of Pseudopulicidae from the Early Cretaceous of China, which has an extremely distended abdomen suggesting that it might have consumed its last meal before death [43].

Up to now, all the Mesozoic fleas have been placed within three families Saurophthiridae Ponomarenko, 1986, Pseudopulicidae Gao, Shih et Ren, 2012 and Tarwiniidae Huang, Engel, Cai & Nel, 2013 [31], comprising nine species within five genera (see Table 25.1).

The Mesozoic lineages of fleas are proposed to be a sister group to crown fleas [16, 43], while considering Aneuretopsychidae [15] rather than Boreidae as a sister group of fleas (including Mesozoic fleas) (Figure 25.1). This proposed phylogeny is not necessarily contradictory to the evidence of affinity between fleas and Boreidae, which is based on a different set of characters and does not consider fossils [2, 44]. There are still many questions about the evolutionary process of fleas [12, 45], and future studies and new fossil materials are expected to shed more light to enhance our knowledge.

Are There Basal Fleas?

Dittmar *et al.* have a different view about these Mesozoic fleas. In 2016, they commented on these findings, and concluded that "the current interpretation of the evolutionary trajectory and ecology of these putative dinosaur fleas is based on appeal to probability, rather than evidence. Hence, their taxonomic positioning as fleas, or stem fleas, as well as their ecological classification as ectoparasites and blood feeders is not supported by currently available data." [45, 46].

However, these Mesozoic fossil "fleas", as documented in [16, 35, 41–43], possess the following features to support their flea affinity: the piercing-and-sucking beaks with thick serrate stylets imply a liquid feeding through rather thick and tough body coverings; the soft and distensible abdomen of the female might suggest their adaptation to intake a large amount of food at once; relatively large and wingless body with long but thin legs and long and sickle-shaped claws imply that they adapt to live on a large surface; and body and legs covered with stiff spines and setae in well-organized fashion, all directed backward, are suggestive of adaptation to fix and move on a surface covered with hairs or feathers. Each of the above features separately can be probably found in some insects. However, for insects having all these characters in combination, in our opinion, they should be adapted to blood sucking of vertebrate host with outgrowths like hairs or feathers [47]. These features

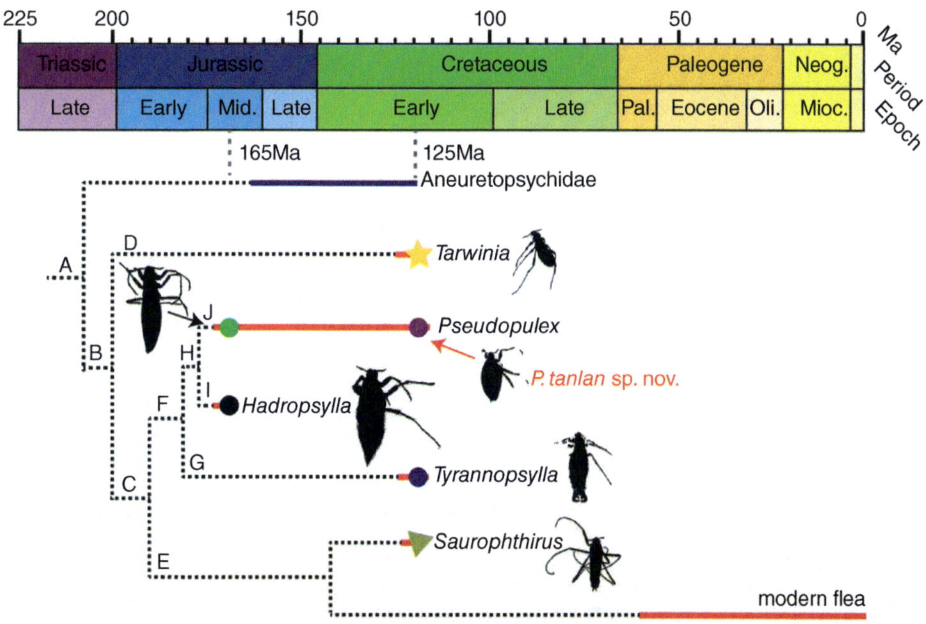

Figure 25.1 Proposed preliminary phylogeny of the Mesozoic fleas. Representative groups: Circle, pseudopulicids; star, tarwiniid; triangle, saurophthirids, Source: modified from [43].

in combination are not consistent with all the known herbivorous insects. Ponomarenko (1976) considered this group as pterosaur parasitism first. We believe any sufficiently big mammals, birds, feathered dinosaurs and pterosaurs with partially naked, elastic skin with hairs, feathers and the like, are possible hosts of parasitism by saurophthirids, pseudopulicids and tarwiniids [16, 35, 41–43, 47].

25.3 Representative Fossils of Siphonaptera from Northern China

Family Pseudopulicidae Gao, Shih & Ren, 2012

Pseudopulicidae only comprise six fossil species within three genera from the Middle Jurassic and the Early Cretaceous of Northeastern China [16, 35, 41, 43]. Pseudopulicidae can be diagnosed by the robust and elongate piercing-sucking mouthparts with serrated stylets, antenna with flagellomeres compact, at least 14 antennomeres, wingless thorax relatively small, elongate legs with distinct ctenidia on the tibiae, body covered with stiff, posteriorly directed setae and bristles. Female is always larger than male, and male genitalia is exposed and large, with elongate gonocoxa and one-segmented gonostylus [41].

Genera included from the Jurassic and Cretaceous of Northern China: *Pseudopulex* Gao, Shih & Ren, 2012, *Hadropsylla* Huang, Engel, Cai & Nel, 2013 and *Tyrannopsylla* Huang, Engel, Cai & Nel, 2013.

Pseudopulex Gao, Shih & Ren, 2012

Pseudopulex Gao, Shih & Ren, 2012, *Curr. Biol.*, 22 (8), 732 [35] (original designation).

Type species: *Pseudopulex jurassicus* Gao, Shih & Ren, 2012.

Body covered with dense setae and bristles. Stylet mouthparts extending to metacoxae, tibiae possessing apical ctenidia, tarsi very long, longer than the femora and tibia combined.

Distribution and age: Inner Mongolia; Middle Jurassic. Liaoning; Early Cretaceous.

Four species included from the Jurassic and Cretaceous of Northern China (see Table 25.1).

Pseudopulex jurassicus Gao, Shih & Ren, 2012 (Figure 25.2)

Pseudopulex jurassicus Gao, Shih & Ren, 2012: *Curr. Biol.*, 22 (8), 732 [35].

Locality and horizon: Daohugou, Ningcheng, Inner Mongolia, China; Middle Jurassic, Jiulongshan Formation.

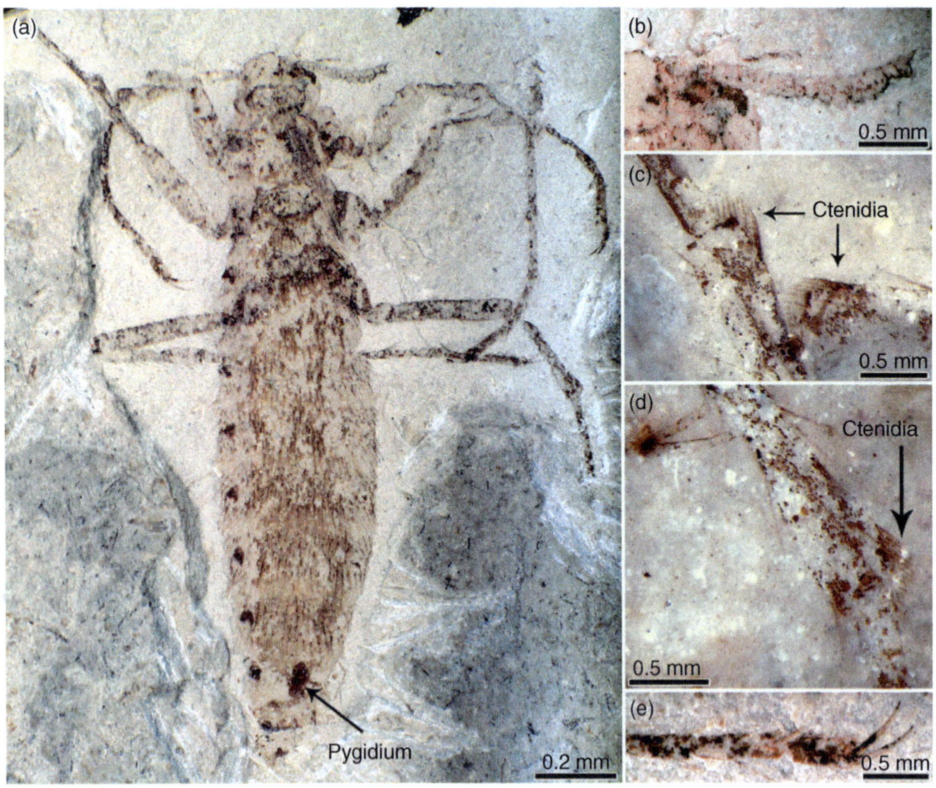

Figure 25.2 *Pesudopulex jurassicus* Gao, Shih & Ren, 2012 (Holotype, CNU-NN2010001). (a) The habitus. (b) Right antenna. (c), (d) Ctenidia of the tibiae. (e) Claw of the left foreleg. Source: Donated by Dr. Chungkun Shih, modified from [35].

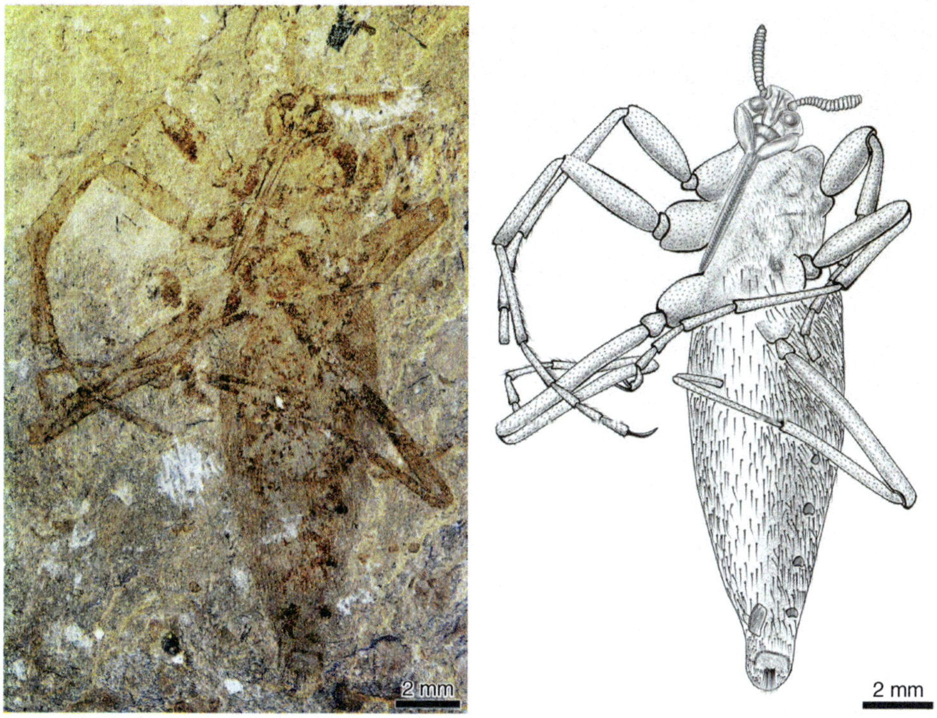

(a) (b)

Figure 25.3 *Pseudopulex magnus* Gao, Shih & Ren, 2012 (Holotype, CNU-NN2010002). (a) Photograph; (b) line drawing Source: modified from [35].

Ocelli absent, antennal short and stout. Mouthparts longer than twice the length of head, labial palps with four segments preserved, maxillary laciniae forming some deeply notched serrations, extending from base to tip. Legs with coxae slightly enlarged, tibiae bearing pseudocombs with short stout bristles, tarsus with a pair of large scythe-like claws on the last tarsomeres. Abdomen more or less dorsoventrally flattened, pygidium ovate [35, 41].

Pseudopulex magnus **Gao, Shih & Ren, 2012 (Figure 25.3)**
Pseudopulex magnus Gao, Shih & Ren, 2012: *Curr. Biol.*, 22 (8), 732.

Locality and horizon: Duolun, Inner Mongolia, China; Lower Cretaceous, Yixian Formation.

Body large, nearly 22.82 mm long excluding antennae; compressed antenna with >16 antennomeres; elongate mouthparts robust, 5.15 mm in length, maxillary laciniae bearing rows of teeth at the edges; abdomen with dense setae and bristles; legs with numerous spurs of various sizes on the inner sides of tibia and tarsus [35].

Pseudopulex tanlan **Gao, Shih, Rasnitsyn & Ren, 2014 (Figure 25.4)**
Pseudopulex tanlan Gao, Shih, Rasnitsyn & Ren, 2014: *BMC Evol. Biol.*, 14 (168), 2.

Locality and horizon: Lingyuan, Liaoning, China; Lower Cretaceous, Yixian Formation.

Medium body size (about 10 mm long), head and thorax relatively small. Body covered with stiff, short bristles and setae. Tibia without ctenidia. Distal abdominal segments not sclerotized. Female with fore tibia about half as long as femur, cerci distinct, reaching end of abdomen. Male with relatively small and short genitalia.

Hadropsylla **Huang, Engel, Cai & Nel, 2013**
Hadropsylla Huang, Engel, Cai & Nel, 2013, *Chinese Sci. Bull.*, 58 (14), 1685 [41] (original designation).

Type species: *Hadropsylla sinica* Huang, Engel, Cai & Nel, 2013.

Female antenna with 19 antennomeres, stylet serrated mouthparts at least extending to the second sternite; scattered ctenidia present apically on all tibiae; tarsi nearly equal to femora and tibiae combined, the fourth tarsomere very small; body covered with posteriorly-directed long setae [41].

Distribution and age: Inner Mongolia; Middle Jurassic.

Only one species included from the Jurassic of Northern China (see Table 25.1).

Tyrannopsylla **Huang, Engel, Cai & Nel, 2013**
Tyrannopsylla Huang, Engel, Cai & Nel, 2013, *Chinese Sci. Bull.*, 58 (14), 1687 [41] (original designation).

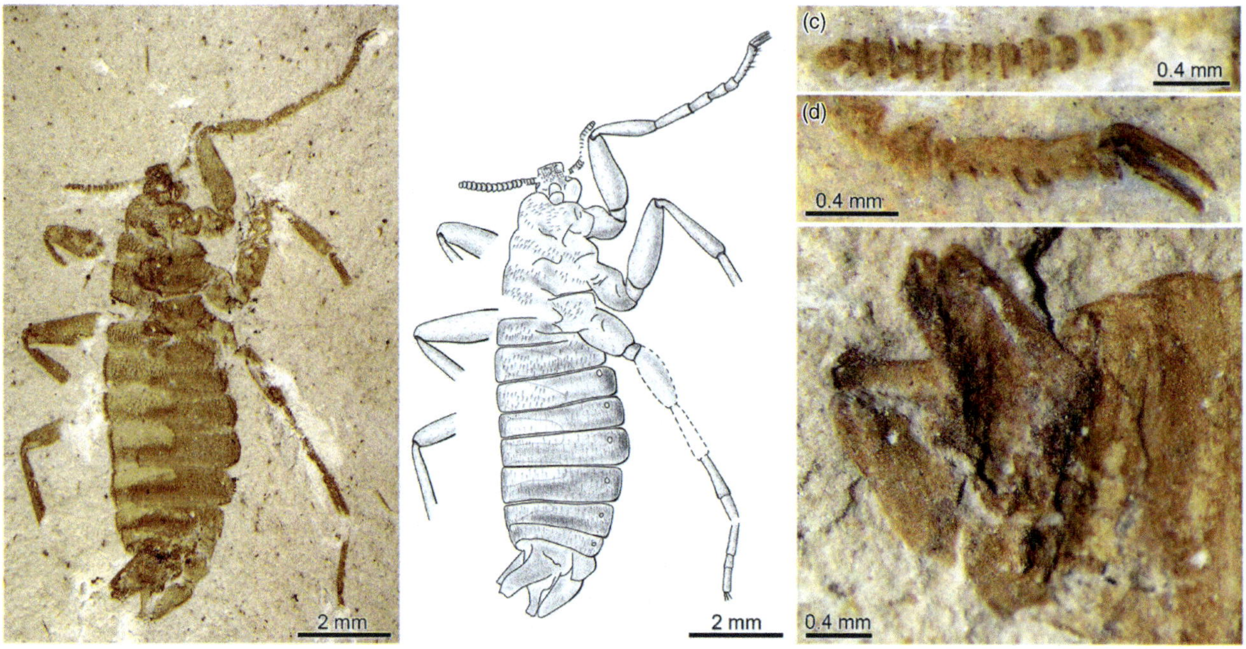

Figure 25.4 *Pseudopulex tanlan* Gao, Shih, Rasnitsyn & Ren, 2014 (Paratype, CNU-SIP-LL2013003). (a, b) Photograph and line drawing. (c) Antenna. (d) Part of the right fore leg. (e) Male genitalia, Source: modified from [43].

Type species: *Tyrannopsylla beipiaoensis* Huang, Engel, Cai & Nel, 2013.

Mouthparts extending to mesocoxae, scattered ctenidia present on apex of tibia; tarsus clearly longer than femur and tibia together, abdominal sternite IV small but sclerotized, sternites V–IX strongly sclerotized. Female slightly larger than male; antenna with 17 antennomeres; sternite VII and IX strongly sclerotized.

Distribution and age: Liaoning; Early Cretaceous.

Only one species included from the Cretaceous of Northern China (see Table 25.1).

***Tyrannopsylla beipiaoensis* Huang, Engel, Cai & Nel, 2013 (Figure 25.5)**

Tyrannopsylla beipiaoensis Huang, Engel, Cai & Nel, 2013: *Chinese Sci. Bull.*, 58 (14), 1687.

Locality and horizon: Beipiao, Liaoning, China; Lower Cretaceous, Yixian Formation.

Male, 14.7 mm long, posterior of head overlapped by pronotum; mouthparts extending median position of meso- and metacoxae; pronotum slightly narrower than mesonotum; middle legs slightly longer than fore legs, hind legs much longer than middle legs; Abdomen covered with posteriorly-directed setae, short setae arranged anteriorly, long setae arranged

Figure 25.5 *Tyrannopsylla beipiaoensis* Huang, Engel, Cai & Nel, 2013, male, (Holotype, NIGP154249a). Source: Photo provided by Dr. Di-ying Huang.

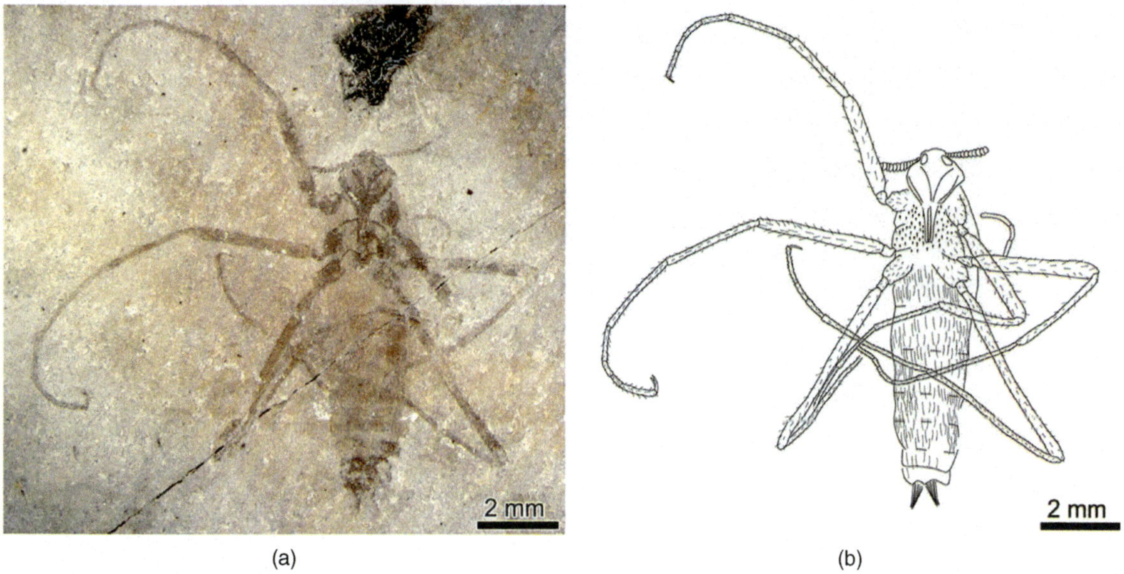

Figure 25.6 Holotype of *Saurophthirus exquisitus* Gao, Shih, Rasnitsyn & Ren, 2013, female (Holotype, CNU-LB2010016c). (a, b) Habitus and line drawing, Source: Donated by Dr. Chungkun Shih, modified from [42].

posteriorly on each segments; male genitalia exposed and large, with elongate gonocoxa bearing a cylindrical, one-segmented broad gonostylus. Female exhibits relatively limited features [41].

Family Saurophthiridae Ponomarenko, 1986

Saurophthiridae Gao, Shih, Rasnitsyn & Ren, 2013, Syn. *Curr. Biol.*, 23 (13), 1261 [42].

Saurophthiridae have been established by Ponomarenko in 1986 [30] for the unique species *S. longipes* Ponomarenko, 1976, which was first reported from the Early Cretaceous of Transbaikalia in 1976 [29]. In 2013, Gao *et al.* assigned another species, *S. exquisitus*, to *Saurophthirus*. Saurophthiridae can be diagnosed based on these characters: head length and width subequal; eye small; antenna with at least 12 antennomeres, fusiform in female and subclavate with palmate penultimate segment in male; hind basitarsus longer than the following tarsal segments combined; abdomen clearly darkened apically [29, 42].

Only one genus included from the Cretaceous of Northern China: *Saurophthirus* Ponomarenko, 1976.

Saurophthirus Ponomarenko, 1976

Saurophthirus Ponomarenko, 1976, *Paleontologicheskii Zhurnal*, 3, 103 [29] (original designation)

Type species: *Saurophthirus longipes* Ponomarenko, 1976.

Medium-sized apterous insects with female slightly bigger than male, legs long and slender, longer than the body length; Head with eyes small, ocelli absent, short piercing-sucking stylet mouthparts including a pair of thin laciniae and wider impair style; Thorax covered with rows of short and stout bristles.

Distribution and age: Liaoning; Early Cretaceous.

Only one species included from the Cretaceous of Northern China (see Table 25.1).

Saurophthirus exquisitus Gao, Shih, Rasnitsyn & Ren, 2013 (Figures 25.6 and 25.7)

Saurophthirus exquisitus Gao, Shih, Rasnitsyn & Ren, 2013: *Curr. Biol.*, 23 (13), 1261.

Locality and horizon: Lingyuan, Liaoning, China; Lower Cretaceous, Yixian Formation.

Female 8–12 mm and male about 7 mm; Head with compact antenna fixed in a deep socket; Coxae round and swollen, fore femora thicker than middle femora, the latter thicker than the hind femora, femur longer than tibia, the latter slightly longer than basitarsus, pretarsus with a pair of long, narrow claws. In the male, abdomen with clear segmental boundary, but the abdomen with integuments not sclerotized; Male genitalia long and narrow, invaginated along with the 9th segment up to base of the 7th segment of the abdomen in repose, and the aedeagus extending beyond the abdomen apex, female possessing a pair of cerci [42].

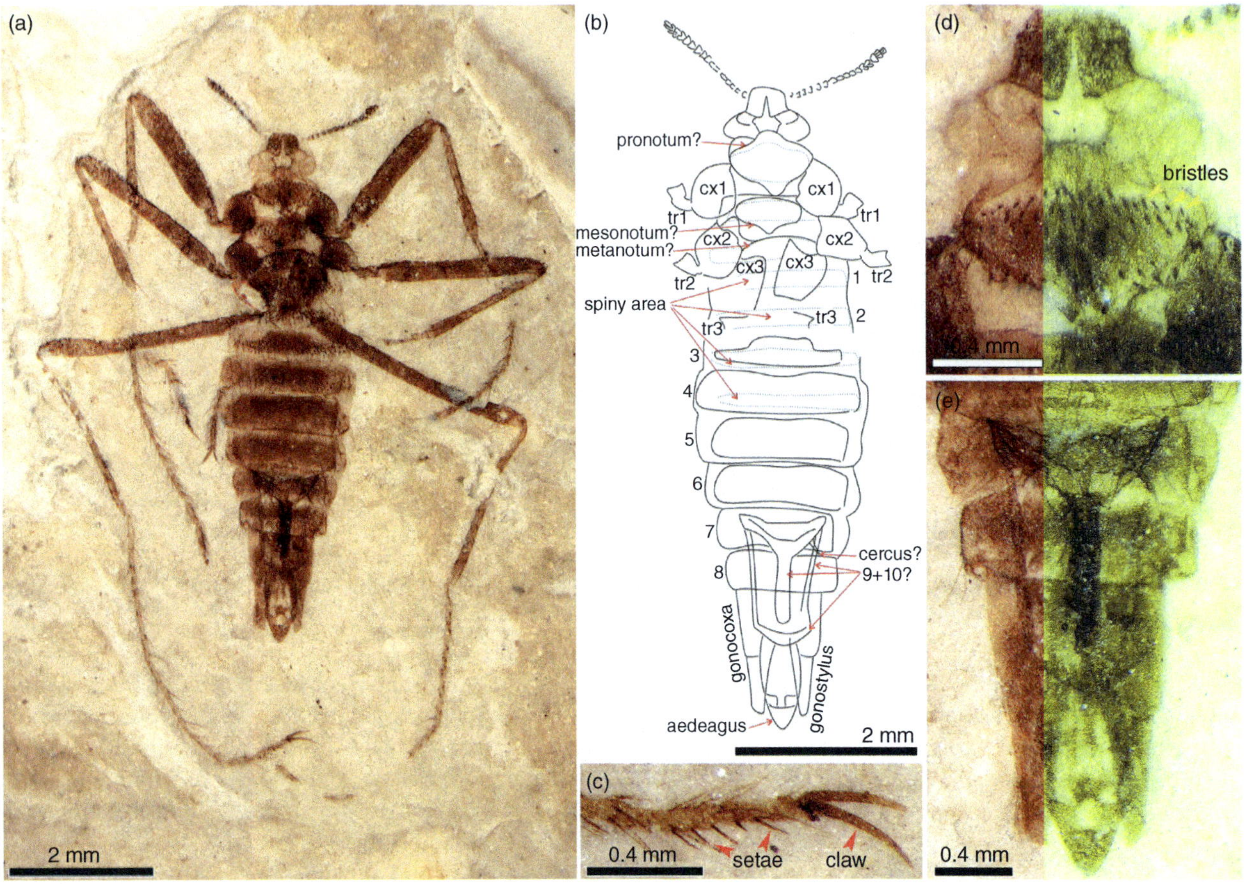

Figure 25.7 Paratype of *Saurophthirus exquisitus* Gao, Shih, Rasnitsyn & Ren, 2013, male (Paratype, CNU-LB2010018). (a) Habitus. (b) Line drawing of the body marking the characters. (c) Claws on the hind leg. (d) Head and pronotum. (e) Terminal of the abdomen under alcohol, Source: modified from [42].

Table 25.1 A list of fossil Siphonapterans from the Jurassic and Cretaceous of China.

Family	Species	Locality	Horizon/age	Citation
Pseudopulicidae	*Pseudopulex Jurassicus* Gao, Shih & Ren, 2012	Ningcheng, Inner Mongolia	Jiulongshan Fm., J_2	Gao *et al.* [35]
	Pseudopulex magnus Gao, Shih & Ren, 2012	Duolun, Inner Mongolia	Jiulongshan Fm., J_2	Gao *et al.* [35]
	Pseudopulex wangi Huang, Engel, Cai & Nel, 2013	Ningcheng, Inner Mongolia	Jiulongshan Fm., J_2	Huang *et al.* [41]
	Pseudopulex tanlan Gao, Shih & Ren, 2014	Beipiao, Liaoning	Yixian Fm., K_1	Gao *et al.* [43]
	Hadropsylla sinica Huang, Engel, Cai & Nel, 2013	Ningcheng, Inner Mongolia	Jiulongshan Fm., J_2	Huang *et al.* [41]
	Tyrannopsylla beipiaoensis Huang, Engel, Cai & Nel, 2013	Ningcheng, Inner Mongolia	Jiulongshan Fm., J_2	Huang *et al.* [41]
Saurophthiridae	*Saurophthirus exquisitus* Gao, Shih, Rasnitsyn & Ren, 2013	Beipiao, Liaoning	Yixian Fm., K_1	Gao *et al.* [42]

References

1 Lewis, R.E. (1998). Resume of the Siphonaptera (Insecta) of the world. *Journal of Medical Entomology* **35** (4): 377–389.
2 Whiting, M.F., Whiting, A.S., Hastriter, M.W., and Dittmar, K. (2008). A molecular phylogeny of fleas (Insecta: Siphonaptera): origins and host associations. *Cladistics* **24**: 677–707. https://doi.org/10.1111/j.1096-0031.2008.00211.x.
3 Krasnov, B.R. (2008). *Functional and Evolutionary Ecology of Fleas: A Model for Ecological Parasitology*. New York: Cambridge University Press.
4 Holland, G.P. (1964). Evolution, classification, and host relationships of Siphonaptera. *Annual Review of Entomology* **9**: 123–146. https://doi.org/10.1146/annurev.en.09.010164.001011.
5 Buckland, P.C. and Sadler, J.P. (1989). A biogeography of the human flea, *Pulex irritans* L. (Siphonaptera: Pulicidae). *Journal of Biogeography* **16**: 115–120. https://doi.org/10.2307/2845085.
6 Michelsen, V. (1997). A revised interpretation of the mouthparts in adult fleas (Insecta, Siphonaptera). *Zoologischer Anzeiger* **235**: 217–223.
7 Snodgrass, R.E. (1946). The skeletal anatomy of fleas (Siphonaptera). *Smithsonian Miscellaneous Collections* **104**: 1–89.
8 Rasnitsyn, A.P. and Quicke, D.L.J. (2002). *History of Insects*. Dordrecht: Kluwer.
9 Grimaldi, D.A. and Engel, M.S. (2005). *Evolution of the Insects*. New York: Cambridge University Press.
10 Medvedev, S.G. (1994). Morphological basis of the classification of fleas. *Entomologicheskoe Obozrenie* **73**: 22–43.
11 Whiting, M.F. (2002). Mecoptera is paraphyletic: multiple genes and phylogeny of Mecoptera and Siphonaptera. *Zoologica Scripta* **31** (1): 93–104. https://doi.org/10.1046/j.0300-3256.2001.00095.x.
12 Zhu, Q., Hastriter, M.W., Whiting, M.F., and Dittmar, K. (2015). Fleas (Siphonaptera) are Cretaceous, and evolved with Theria. *Molecular Phylogenetics and Evolution* **90**: 129–139. https://doi.org/10.1016/j.ympev.2015.04.027.
13 Whiting, M.F., Carpenter, J.C., Wheeler, Q.D., and Wheeler, W.C. (1997). The strepsiptera problem: phylogeny of the holometabolous insect orders inferred from 18S and 28S ribosomal DNA sequences and morphology. *Systematic Biology* **46** (1): 1–68. https://doi.org/10.1093/sysbio/46.1.1.
14 Misof, B., Liu, S., Meusemann, K. *et al.* (2014). Phylogenomics resolves the timing and pattern of insect evolution. *Science* **346**: 763–767. https://doi.org/10.1126/science.1257570.
15 Ren, D., Labandeira, C.C., Santiago-Blay, J.A. *et al.* (2009). A probable pollination mode before angiosperms: eurasian, long-proboscid scorpionflies. *Science* **326** (5954): 840–847. https://doi.org/10.1126/science.1178338.
16 Huang, D.-Y., Engel, M.S., Cai, C.Y. *et al.* (2012). Diverse transitional giant fleas from the Mesozoic era of China. *Nature* **483**: 201–204. https://doi.org/10.1038/nature10839.
17 Haensch, S., Bianucci, R., Signoli, M. *et al.* (2010). Distinct clones of *Yersinia pestis* caused the black death. *PLoS Pathogens* **6** (10): e1001134. https://doi.org/10.1371/journal.ppat.1001134.
18 Welford, M. and Bossak, B.H. (2010). Revisiting the medieval black death of 1347–1351: spatiotemporal dynamics suggestive of an alternate causation. *Geography Compass* **4**: 561–575.
19 Hinnebusch, B.J., Perry, R.D., and Schwan, T.G. (1996). Role of the *Yersinia pestis* Hemin storage (*hms*) locus in the transmission of plague by fleas. *Science* **273** (5273): 367–370. https://doi.org/10.1126/science.273.5273.367.
20 Schmid, B.V., Büntgen, U., Easterday, W.R. *et al.* (2015). Climate-driven introduction of the Black Death and successive plague reintroductions into Europe. *Proceedings of the National Academy of Sciences of the United States of America* **112** (10): 3020–3025. https://doi.org/10.1073/pnas.1412887112.
21 Peus, F. (1968). Über die beiden Bernstein-Flöhe (Insecta, Siphonaptera). *Paläontologische Zeitschrift* **42**: 62–72.
22 Beaucournu, J. and Wunderlich, J. (2001). A third species of *Palaeopshylla* Wagner, 1903, from Baltic amber (Siphonaptera: Ctenophthalmidae). *Entomologische Zeitschrift* **111**: 296–298.
23 Lewis, R.E. and Grimaldi, D.A. (1997). A pulicid flea in Miocene amber from the Dominican Republic (Insecta: Siphonaptera: Pulicidae). *American Museum Novitaties* **3205**: 1–9.
24 Perrichot, V., Beaucournu, J.C., and Velten, J. (2012). First extinct genus of a flea (Siphonaptera: Pulicidae) in Miocene amber from the Dominican Republic. *Zootaxa* **3438**: 54–61.
25 Poinar, G.O. (1995). Fleas (Insecta, Siphonaptera) in Dominican Amber. *Medical Science Research* **23**: 789.
26 Riek, E.F. (1970). Lower Cretaceous fleas. *Nature* **227**: 746–747. https://doi.org/10.1038/227746a0.
27 Jell, P.A. and Duncan, P.M. (1986). Invertebrates, mainly insects, from the freshwater, Lower Cretaceous, Koonwarra fossil bed (Korumburra group),

South Gippsland, Victoria. *Memoirs of the Association of Australasian Palaeontologists* **3**: 111–205.

28 Huang, D.-Y. (2015). *Tarwinia australis* (Siphonaptera: Tarwiniidae) from the Lower Cretaceous Koonwarra fossil bed: morphological revision and analysis of its evolutionary relationship. *Cretaceous Research* **52**: 507–515. https://doi.org/10.1016/j.cretres.2014.03.018.

29 Ponomarenko, A.G. (1976). A new insect from the Cretaceous of Transbaikalia Ussr a possible parasite of Pterosaurians. *Paleontologicheskii Zhurnal* **3**: 102–106.

30 Ponomarenko, A.G. (1986). Insects in the Early Cretaceous ecosystems of Western Mongolia. *Joint Soviet–Mongolian Geological and Paleontological Expedition* **28**: 110–112.

31 Rasnitsyn, A.P. and Strelnikova, O.D. (2017). Tracheal system and biology of the Early Cretaceous *Saurophthirus longipes* Ponomarenko, 1976 (Insecta, ?Aphaniptera, Saurophthiroidea stat. nov.). *Paleontological Journal* **51**: 171–182.

32 Rasnitsyn, A.P. (1992). *Strashila incredibilis*, a new enigmatic mecopteroid insect with possible siphonapteran affinities from the Upper Jurassic of Siberia. *Psyche* **99**: 323–333.

33 Vršanský, P., Ren, D., and Shih, C.K. (2010). Nakridletia ord. n - enigmatic insect parasites support sociality and endothermy of pterosaurs. *AMBA Projekty* **8**: 1–16.

34 Huang, D.-Y., Nel, A., Cai, C.Y. *et al.* (2013). Amphibious flies and paedomorphism in the Jurassic period. *Nature* **495**: 94–97. https://doi.org/10.1038/nature11898.

35 Gao, T.P., Shih, C.K., Xu, X. *et al.* (2012). Mid-Mesozoic flea-like ectoparasites of feathered or haired vertebrates. *Current Biology* **22** (8): 732–735. https://doi.org/10.1016/j.cub.2012.03.012.

36 Poinar, G.O. (2012). Palaeontology: the 165-million-year itch. *Current Biology* **22** (8): 278–280. https://doi.org/10.1016/j.cub.2012.03.014.

37 Zhou, Z.H., Barrett, P.M., and Hilton, J. (2003). An exceptionally preserved Lower Cretaceous ecosystem. *Nature* **421**: 807–814. https://doi.org/10.1038/nature01420.

38 Xu, X. and Zhang, F.C. (2005). A new maniraptoran dinosaur from China with long feathers on the metatarsus. *Naturwissenschaften* **92**: 173–177.

39 Meng, J., Hu, Y.M., Wang, Y.Q. *et al.* (2006). A Mesozoic gliding mammal from Northeastern China. *Nature* **444**: 889–893. https://doi.org/10.1038/nature05234.

40 Huang, D.-Y. (2014). Trace back the origin of recent insect orders - evidence from the Middle Jurassic Daohugou Biota. *Science Foundation in China* **22**: 34–42.

41 Huang, D.-Y., Engel, M.S., Cai, C.Y., and Nel, A. (2013). Mesozoic giant fleas from Northeastern China (Siphonaptera): taxonomy and implications for palaeodiversity. *Chinese Science Bulletin* **58** (14): 1682–1690.

42 Gao, T.P., Shih, C.K., Rasnitsyn, A.P. *et al.* (2013). New transitional fleas from China highlighting diversity of Early Cretaceous ectoparasitic insects. *Current Biology* **23** (13): 1261–1266. https://doi.org/10.1016/j.cub.2013.05.040.

43 Gao, T.P., Shih, C.K., Rasnitsyn, A.P. *et al.* (2014). The first flea with fully distended abdomen from the Early Cretaceous of China. *BMC Evolutionary Biology* **14**: 168. https://doi.org/10.1186/s12862-014-0168-1.

44 Beutel, R.G., Friedrich, F., Hörnschemeyer, T. *et al.* (2011). Morphological and molecular evidence converge upon a robust phylogeny of the megadiverse Holometabola. *Cladistics* **27**: 341–355. https://doi.org/10.1111/j.1096-0031.2010.00338.x.

45 Dittmar, K., Zhu, Q., Hastriter, M.W., and Whiting, M.F. (2016). On the probability of dinosaur fleas. *BMC Evolutionary Biology* **16**: 9. https://doi.org/10.1186/s12862-015-0568-x.

46 De Baets, K. and Littlewood, D.T.J. (2015). The importance of fossils in understanding the evolution of parasites and their ventors. *Advances in Parasitology* **90**: 1–51. https://doi.org/10.1016/bs.apar.2015.07.001.

47 Gao, T.P., Shih, C.K. Rasnitsyn, A.P. *et al.* (2016) Reply to "On the probability of dinosaur fleas" *BMC Evolutionary Biology*, **16**, 9. https://doi.org/10.1186/s12862-015-0568-x

26

Trichoptera – Caddisflies

Mei Wang[1,2], *Weiting Zhang*[2,3], *Chungkun Shih*[2,4], *and Dong Ren*[2]

[1] *Research Institute of Forest Ecology, Environment and Protection, Chinese Academy of Forestry, Haidian District, Beijing, China*
[2] *Capital Normal University, Haidian District, Beijing, China*
[3] *Hebei GEO University, Shijiazhuang, China*
[4] *National Museum of Natural History, Smithsonian Institution, Washington, DC, USA*

26.1 Introduction to Trichoptera

Trichoptera, commonly called "caddisflies", comprise approximately 15 000 described extant species found on all continents except for Antarctica [1]. The body sizes of caddisflies range from 2 to 40 mm in body length. They have a large head with two prominent compound eyes and setal warts on the dorsal part of the thorax. Their body, legs and wings are covered with a lot of setae (hairs). When at rest, the wings are folded roof-like above the body [2].

Martynov, in 1924, classified Trichoptera into two suborders, Annulipalpia and Integripalpia, on the basis of the adult maxillary palps and larval nesting habits [3]. This classification system was widely accepted. Wiggins and Wichard separated the species that have "free living" larvae from Annulipalpia in 1989, and established "Spicipalpia" [4]. But the monophyly of "Spicipalpia" is contentious [2]. In addition, the Permian suborder Protomeropina were initially proposed as a suborder of the Trichoptera [5]. However, Minet *et al.*, in 2010, pointed out this group is definitely polyphyletic and none of the four families, Cladochoristidae, Protomeropidae, Prosepididontidae and Microptysmatidae, could be ascribed to the Trichoptera [6].

Most female caddisflies deposit eggs on aquatic plants, rocks submerged in water, or plants or other objects overhanging the water. Others deposit eggs directly into water. Caddisflies have one generation in one or two years. During most of their lifetime, they stay in water as aquatic larvae in small mobile houses built by themselves. Caddisfly larvae look like caterpillars, but use gills on the abdominal segments for breathing in water. Some larvae feed on aquatic plant matter, small animals or fish eggs, causing damage to fish farming or rice planting. On the other hand, they are food sources for many fish, especially trout. Therefore, besides mayfly, the adult caddisfly is also a favorite artificial fly for fly fishermen. Most adults are active in the evening or at night, but some are active in daytime. They fly around the bush near water, with only one aim in mind – mating and breeding. Most of adult caddisflies don't feed in their short life – a few days only. As such, their mouthparts are reduced. However, some species, living longer than one month, may feed on nectar or plants' sugary fluids.

Indicator for Water Quality

Aquatic caddisfly larvae exhibit a broad range of responses to environmental stressors and are sensitive to water pollution, thus, they are among the most important indicators of water quality [7]. They, together with larvae of mayflies and stoneflies, have been used as the indicator insects for water quality in some countries. Since some species of caddisfly larvae can tolerate a certain degree of pollution and stress in altered environments, the survey of species types and abundance for a stream will provide a good indication of its level of environmental conservation status of preserved, altered or impacted [8].

26.2 Progress in the Studies of Fossil Trichoptera

The study of fossil Trichoptera started in the early nineteenth century. Bosc acted as the pioneer who reported Tertiary caddisfly pupal cases, *Indusia tubulosa*, found in France in 1805 [9]. Carpenter compiled the trichopteran fossils described in the early literatures in 1992 [10]. Sukacheva has described more than 200 fossil species and made significant contributions to the study of fossil

Rhythms of Insect Evolution: Evidence from the Jurassic and Cretaceous in Northern China, First Edition.
Edited by Dong Ren, Chungkun Shih, Taiping Gao, Yongjie Wang, and Yunzhi Yao.

Trichoptera. To date, 43 families with more than 100 genera and 600 species have been reported based on fossils from Russia, USA, Canada, Mongolia, Mediterranean area and China, etc.

Due to the dispute about the position of species within Protomeropina, the origination age of Trichoptera is uncertain based on fossil evidence. Recently, according to molecular data, the origin of Trichoptera is estimated to be around 234 Mya, i.e. the Middle to the Late Triassic [11]. The fossil records of Necrotauliidae and Philopotamidae were found in the Triassic [12]. Although Necrotauliidae has been treated as a component of the Amphiesmenoptera stem-group by some paleontologists [13, 14], Liu *et al.* provided fossil evidence and stated that Necrotauliidae belong to Trichoptera, and represent the stem-group of Integripalpia [15]. Some modern caddisfly taxa with more species and broader diversity did not appear until the Jurassic and the Cretaceous.

Studies of fossil Trichoptera in China were apparently initiated by Hong, who reported an Early Cretaceous Vitimotauliidae species, *Macropteryx xiaoshetaiensis* Hong 1976 discovered in Urad Front Banner, Inner Mongolia [16]. Since then, M.X. Wang, Y. Gao, Y.J. Liu, W.T. Zhang, T.P. Gao, Y.Z. Yao, C.K. Shih and D. Ren have made many important contributions. Up to now, 16 species of nine genera belonging to six families and two species unassigned to families have been described by Chinese paleoentomologists. Most of fossil caddisflies have been found in Inner Mongolia, Liaoning and Hebei. Davis *et al.* described a fossil caddisfly pupa, putatively of the family Vitimotauliidae, from the Lower Cretaceous Yixian Formation in Northeastern China [17]. This is the first fossil trichopteran pupa described from China.

Mobile Home Builders

The Trichoptera are well known for their aquatic larvae, many of which are master builders of mobile homes to disguise and to protect themselves. They use salivary silk to glue together twigs, leaves, sand, small pebbles or debris to construct mobile homes. If necessary, their powerful mandibles are used to chew and tear large material into pieces. There are a wide variety of designs and constructions for these homes, such as straight tubes, spiral tubes, or curved tubes. They carry their mobile homes everywhere, like snails or hermit crabs, while they are feeding on plants or algae. Larvae go over the winter and pupate in the spring, emerging as adults in early summer.

Starting in the early 1980s, an innovative French jewelry artist, Hubert Duprat, came up with an interesting way of using this master builder's unique skills to make custom jewelry. Small beads or flakes of gold, opal, pearl, turquoise, and other semi-precious stones are put at the bottom of an aquarium tank. Caddisfly larvae, collected from a pond, are introduced into the aquarium tank. The larvae start to pick up these beads and pebbles and construct mobile homes of various colors, designs, and shapes for their own hiding. After the larvae pupate inside these shelters and emerge as adults, the caddisfly adults are released back to nature. The casings, abandoned and cast-away by the caddisflies, become strikingly beautiful and colorful tubes for making custom jewelry.

There are other examples of mobile home building by terrestrial insects. For example, caterpillars of bagworm moths (Family Psychidae) build cases from silk, dry leaves, grass, twigs, etc. They carry their cases everywhere for protection and camouflage during feeding. They also pupate inside these mobile homes. Caterpillars of clothes moths (Family Tineidae) use fibers or debris of clothes to build tubular cases and hide inside. They take these mobile homes around while feeding on wool, fur or other clothing fibers. Larvae of green lacewings (Family Chrysopidae) and brown lacewings (Family Hemerobiidae) camouflage themselves with the dry carcass of prey (e.g. aphids), dry leaves, twigs, or debris on their back (see Figure 20.6). They look like dead insects or useless debris to their predators or prey.

The earliest caddis case is known from the Early Jurassic of Transbaikalia [13]. In China, a few small caddis cases have been reported from the Middle Jurassic Daohugou fauna [18].

26.3 Representative Fossils of Trichoptera from Northern China

Suborder Integripalpia Martynov, 1924

Family Vitimotauliidae Sukatcheva, 1968

Vitimotauliidae, an extinct family, have been documented frequently in the Early Cretaceous all over the Eurasia, but scarce or absent in the Jurassic [19]. To date, Vitimotauliidae comprise four genera: *Multimodus* Sukatcheva, 1968 with 14 species [20–24]; *Vitimotaulius* Sukatcheva, 1968 with three species [20, 22]; *Purbimodus* Sukatcheva and Jarzembowski, 2001 with four species [25] and *Sinomodus* Wang & Ren, 2009 with three species [26]. Species of *Multimodus* and *Vitimotaulius* have been described from the Early Cretaceous in Southeastern Siberia, Mongolia, and China, species of *Purbimodus* from the Early Cretaceous in South England, and species of *Sinomodus* from the Early Cretaceous in China. However, vitimotauliids have not been reported in the Southern Hemisphere up to now.

Genera included from the Cretaceous of Northern China: *Multimodus* Sukatcheva, 1968 and *Sinomodus* Wang & Ren, 2009.

Multimodus Sukatcheva, 1968

Multimodus Sukatcheva, 1968, *Palaentol. J.*, 2, 63 [20] (original designation).

Type species: *Multimodus martynovae* Sukatcheva, 1968.

Forewing with Rs short and two-branched; apices of apical forks of F1, F2, F3, F4 located 1/3 of wing length; F5 forking more basal than F1; cells DC (discoidal cell), MC (medial cell) and TC (thyridial cell) closed; cell MC short, almost 2/3 of cell DC; stem of M longer than MC.

Distribution and age: Hebei; Early Cretaceous.

Three species included from the Cretaceous of Northern China (see Table 26.1).

Sinomodus Wang & Ren, 2009

Sinomodus Wang & Ren, 2009, *Cretac. Res.*, 30, 593–594 [26] (original designation).

Type species: *Sinomodus spatiosus* Wang & Ren, 2009.

Forewing with cells DC, MC and TC closed; cell DC equals to MC; apices of F1 and F2 located between 1/2 and 1/3 of wing length; apices of F3 and F4 located between 1/3 and 1/4 of wing length; Cu_{1a} branched at the same level of F1; Cu_2 and 1A reaching posterior margin of forewing beyond mid-length.

Distribution and age: Liaoning; Early Cretaceous.

Three species included from the Cretaceous of Northern China (see Table 26.1).

Sinomodus spatiosus Wang & Ren, 2009 (Figure 26.1)

Sinomodus spatiosus Wang & Ren, 2009: *Cretac. Res.*, 30, 594.

Locality and horizon: Huangbanjigou, Beipiao, Liaoning, China; Lower Cretaceous, Yixian Formation

The body length is 21.0 mm, and forewing, 19.0 mm. The scape of antenna stout, flagellum slender and long, but antenna shorter than forewing. Maxillary palp 4 or 5 segmented, covered with dense setae, sg1–sg4 similar with equal length. On scutellum, two setal warts fused into one. Forewing with Sc straight, reaching costal margin beyond 1/2 of wing length; Rs furcates at 1/3 of wing length; F1 parallel to F2; M branches beyond cross-vein m-cu_1; M_{1+2} and M_{3+4} fork distal to Rs_2 [26].

Figure 26.1 *Sinomodus spatiosus* Wang & Ren, 2009, (Holotype, TNP-42592p) [26].

Locality and horizon: Huangbanjigou, Beipiao, Liaoning, China; Lower Cretaceous, Yixian Formation.

Total body length 29.7 mm (head to apex of anal processes); maximal abdominal width 5.1 mm. Antenna with about 33 antennomeres; scape approximately as long as maxillary palpal segments 2 + 3; pedicel similar in size to basal antennomeres; scape and pedicel slightly thicker than flagellum. Maxillary palpus 5-segmented; segments roughly of equal size and densely setose, except apical segment slightly longer and narrower than preceding segment. Labial palpus with segments indistinct. Antennal setal warts indistinct as preserved; ocellar setal warts faint, subcircular. Pronotal setal warts elongate and elliptical; mesoscutal setal warts with indistinct margins (though setae present and visible); mesoscutellar warts indistinct. Wings densely setose, not fully expanded. Abdomen with distinct lateral fringe apparently along all segments, proceeding slightly mesally on posterior margin of sternite VIII; tergites distinct; anterior of tergite I with transverse setal fringe; paired, elliptical, dorsal hook plates present anteriorly on tergites III–VII; posterior margin of tergite V with elongate, rectangular, paired hook plates with about 6–8 hooks; small clusters of simple, elongate, larval gills present antero-laterally at least on tergites III–VII, apparently

Family Vitimotauliidae? Sukatcheva, 1968

Genus and species *Incertae sedis* Davis, Engel & Ren, 2010 (Figure 26.2)

Genus and species *Incertae sedis* Davis, Engel & Ren, 2010: *Cretac. Res.*, 31, 297.

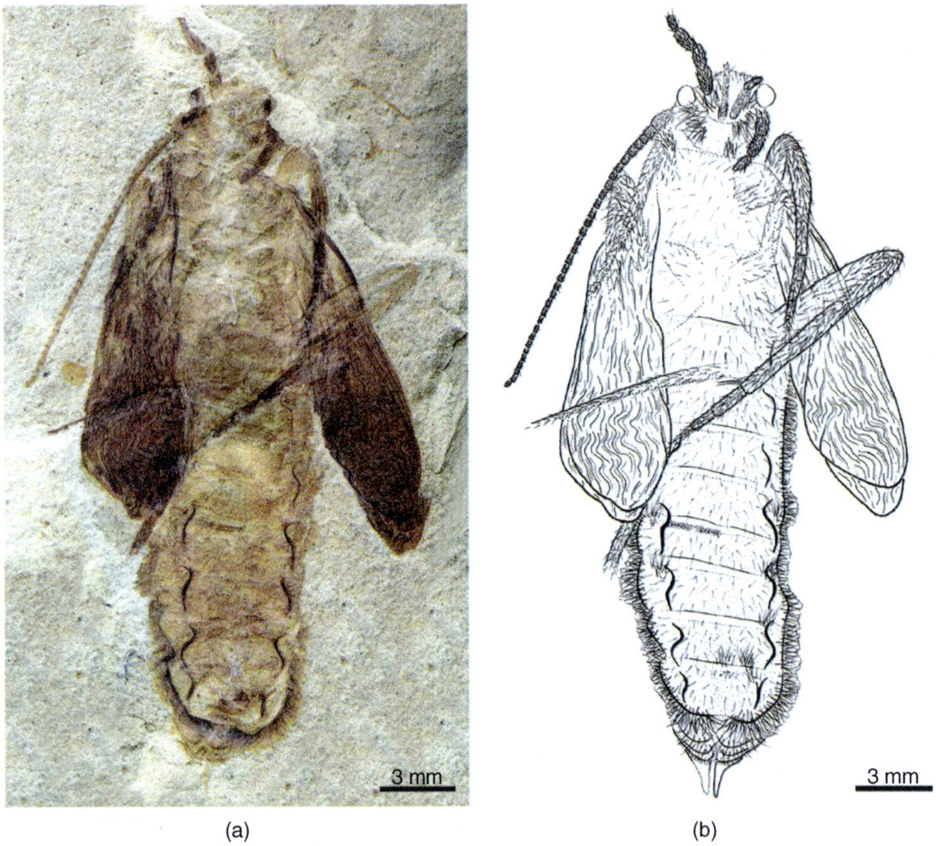

Figure 26.2 Vitimotauliid pupa (CNU-T-LB-2009000). (a). Photograph; (b). Line illustration [17].

also on postero-lateral margins; paired anal processes parallel, approximately as long as tergite VI [17].

Family Ningxiapsychidae Hong & Li, 2004

Ningxiapsychidae, an extinct family, have been erected based on one fossil species collected from the late Early Cretaceous of Ningxia, China. The taxon is characterized by cells DC and MC closed, TC open; M four-branched; F1–F5 straight and present; A_1 and A_2 present.

Only one genus included from the Cretaceous of Northern China: *Ningxiapsyche* Hong & Li, 2004.

Ningxiapsyche Hong & Li, 2004

Ningxiapsyche Hong & Li, 2004, *Acta Zootaxonom. Sin.*, 29 (2), 225 [27] (original designation).

Type species: *Ningxiapsyche fangi* Hong & Li, 2004.

The specific epithet is in honor of Mr. Zhengrong Fang for his collection of this specimen. Forewing with cell DC longer than MC; cell TC almost 2.5 times as long as MC; costal space wide, nearly twice as wide as subcostal area; pterostigma triangle, long and big; three black bands on the wing surface. The forking positions between RS_1 and RS_2, RS_3 and RS_4, M_1 and M_2, M_3 and M_4 form a linear line.

Distribution and age: Ningxia; late Early Cretaceous.

Only one species included from the Cretaceous of Northern China (see Table 26.1).

Family *Incertae sedis*

Cathayamodus Gao, Shih, Labandeira & Ren, 2016

Cathayamodus Gao, Shih, Labandeira & Ren, 2016, *Proc. R. Soc. B*, 283 (1839), 20 161 448 [28].

Type species: *Cathayamodus fournieri* Gao, Shih, Labandeira & Ren, 2016.

The specific epithet is in honor of Dominique Fournier, for his guidance, motivation and encouragement to Dr. Chungkun Shih in his earlier paleoentomological endeavors. Adult moderately large, entire body covered with dense setae, setal warts present on head and thorax. Antennal length slightly shorter than or equal to forewing length; scape stout; flagellum bipectinate, bearing lateral rami on each side. Wings elongate; cells DC and MC closed in forewing. Apices of F1 and F2 located between 1/3 and 1/4 of wing length to the wing apex; apices of F3 and F4 about 1/4 of wing length to the wing apex. Cu_{1a} forking at the level of F3. Cross-veins absent in hind wings.

Distribution and age: Liaoning; Early Cretaceous.

Only one species included from the Cretaceous of Northern China (see Table 26.1).

Figure 26.3 *Cathayamodus fournieri* Gao, Shih, Labandeira & Ren, 2016 (Holotype, CNU-TRI-LB-2009001p, under alcohol). Source: Donated by Dr. Chungkun Shih [28].

Figure 26.4 A 3-D reconstruction of *Cathayamodus fournieri*. Source: Artwork by Dr. Chen Wang [28].

***Cathayamodus fournieri* Gao, Shih, Labandeira & Ren, 2016 (Figures 26.3 and 26.4)**

Cathayamodus fournieri Gao, Shih, Labandeira & Ren, 2016: *Proc. R. Soc. B*, 283 (1839), 20 161 448.

Locality and horizon: Huangbanjigou, Beipiao, Liaoning, China; Lower Cretaceous, Yixian Formation.

Forewing length is 16 mm, hind wing length ca. 12 mm. Adult moderately large. Head covered with setae; antenna slightly shorter than forewing; scape, pedicel, and basal portion of flagellum with rigid, spine-like bristles or less-rigid, non-tapering trichobothria; flagellum slender, bipectinate, with lateral rami on each side of flagellomeres (some branches not preserved); flagellar rami covered by trichobothria. Legs slender; tibiae with fringe of long setae; all tibiae with two apical spurs, metatibia with two preapical spurs. Hind wing F4 absent; cross-veins lacking; Rs and Cu_1 bifurcate at the same level; R_{2+3} and R_{4+5} bifurcate at the same level [28].

Suborder Annulipalta

Family Philopotamidae Stephens, 1829

Philopotamidae, an extant family, are known commonly as the "finger-net caddisflies". The aquatic larvae of these caddisflies spin mesh nets of silk in flowing water to catch food. A larva can spin over a kilometer of extremely thin silk to create its intricate net. It is characterized by five-segmented maxillary palps in both sexes, with the first segment being the shortest, the second one being provided with a mesodistal brush of setae, and the fifth segment long, annulated and flexible, usually at least twice as long as preceding segments [29, 30]. The earliest fossil philopotamid was described from the Kyrgyzstan in the Early Triassic [12]. Philopotamidae comprise three subfamilies containing 17 extant genera and 12 extinct genera. Up to date, 31 fossil species have been described from the Mesozoic, Oligocene and Miocene either on compression fossils of Kyrgyzstan, Mongolia, China, Siberia, Kazakhstan and Russia or in amber from New Jersey, Myanmar, Baltic, Saxonia and Dominica.

Only one genus included from the Jurassic of Northern China: *Liadotaulius* Handlirsch, 1939.

***Liadotaulius* Handlirsch, 1939**

Liadotaulius Handlirsch, 1939, *Ann. Naturhist. Mus. Wien.*, 49, 97 [31] (original designation).

Type species: *Liadotaulius maior* (Handlirsch, 1906).

Forewing with four-branched Rs and M; closed cells of DC and MC; the terminus of Cu_2 being abruptly bent toward the wing margin and desclerotized. Hind wing with complete anal veins.

Distribution and age: Inner Mongolia; Middle Jurassic.

Two species included from the Jurassic of Northern China (see Table 26.1).

***Liadotaulius limus* Zhang, Shih & Ren, 2017 (Figure 26.5)**

Liadotaulius limus Zhang, Shih & Ren, 2017: *Alcheringa*,

Figure 26.5 *Liadotaulius limus* Zhang, Shih & Ren, 2017, (Holotype, CNU-TRI-NN-2014012) [32].

41 (1), 25–27.

Locality and horizon: Daohugou, Ningcheng, Inner Mongolia, China; Middle Jurassic, Jiulongshan Formation.

Forewing length is 6.3 mm, width 2.1 mm. Forewing with rounded apex; Sc straight, reaching costal margin beyond half of wing length; R_1 not bifurcated; Rs furcated at half of the wing length; cross-veins r, s, m and m-cu present; r closer to the base than cross-vein s; cross-vein m-cu oblique; Cu_2 not reaching the posterior margin of forewing. All apical forks present. F1 forking slightly basal to F2 forking; F3 forking distinctly distal to F4 forking [32].

Family *Incertae sedis*

Qinquania Hong, 1982

***Qinquania* Hong, 1982,** Insect fossils in Jiuquan Basin, 158 [33] (original designation).

Type species: *Qinquania combinata* Hong, 1982.

Forewing with SC and R un-branched; the forking of Rs basal to that of M; Rs four-branched. Cells DC and TC open; MC closed; DC > TC > MC; stem of Cu_1 short, branched from M; part of Cu_{1a} merged into Cu_{1b}; two cross-veins m-m and cu_1-cu_2 present.

Distribution and age: Gansu; Early Cretaceous.

Only one species included from the Cretaceous of Northern China (see Table 26.1).

Suborder Spicipalpia Weaver, 1984

Family Rhyacophilidae Stephens, 1836

Rhyacophilidae, the largest family in the Spicipalpia, are an archaic group of caddisflies. The earliest record has been from the Middle Jurassic of Inner Mongolia, China. Larvae of this family are free living and most species are predatory. Up to date, Rhyacophilidae have six genera comprising over 700 reported extant species in North America, Europe, and Asia, India and the tropical areas of Southeastern Asia [29], and 14 extinct species described from the Mesozoic and Cenozoic fossils of China, Siberia, Germany, Russia and Baltic amber [34–40].

Only one genus included from the Jurassic of Northern China: *Declinimodus* Gao, Yao & Ren, 2013.

Declinimodus Gao, Yao & Ren, 2013

Declinimodus Gao, Yao & Ren, 2013, *Acta Geol. Sin.-Engl.*, 87 (6), 1496 [40] (original designation).

Type species: *Declinimodus setulosus* Gao, Yao & Ren, 2013.

Body relatively large. Maxillary palps five-segmented, the first and second button-shaped, shorter than the third segment, the fifth segment terminating in a narrowly rounded distally. Forewing with R_5 reaching wing apex; R_1 forked distally; cell DC closed in forewing; Rs stem nearly as long as DC; Rs and M four-branched; F1–F5 present in forewing but F5 absent in hind wing. Tibia spurs: 3, 4, 4.

Distribution and age: Inner Mongolia; Middle Jurassic.

Only one species included from the Jurassic of Northern China (see Table 26.1).

Declinimodus setulosus Gao, Yao & Ren, 2013 (Figure 26.6)

Declinimodus setulosus Gao, Yao & Ren, 2013: *Acta Geol. Sin.-Engl.*, 87 (6), 1496–1498.

Locality and horizon: Daohugou, Ningcheng, Inner Mongolia, China; Middle Jurassic, Jiulongshan Formation.

Body length is 11.82 mm. Pronotum narrow with a pair of setal warts; mesoscutum broad, with a pair of round symmetrical scutal setal warts. Tarsi of fore legs 1.05 times as long as fore tibiae; mid tibiae 1.64 times as long as mid femora; mid tarsi 1.29 times as long as tibiae. Fore legs with two claws at apex of tarsi. Forewing with Sc straight, with an oblique cross-vein h, inserting on costal margin beyond 1/2 of wing length; M forked at 1/3 of forewing length; cell MC open; Cu_1 forking nearly at the same level of Rs, with Cu_{1a} curving slightly at middle length [40].

Family Hydrobiosidae Ulmer, 1905

Hydrobiosidae, first erected by Ulmer in 1905 as a subfamily of Rhyacophilidae [41], have been elevated to family rank by Schmid (1989) [42]. This large family

Figure 26.6 *Declinimodus setulosus* Gao, Yao & Ren, 2013, (Holotype, CNU-TRI-NN-2011006p) [40].

comprise approximately 50 genera [2]. So far, five genera and five species of the fossil Hydrobiosidae have been reported.

Genera included from the Jurassic of Northern China: *Juraphilopotamus* Wang, Zhao & Ren, 2009 and *Pulchercylindratus* Gao, Yao & Ren, 2013.

Juraphilopotamus Wang, Zhao & Ren, 2009

Juraphilopotamus Wang, Zhao & Ren, 2009, *Prog. Nat. Sci.*, 19, 1428 [43] (original designation).

Type species: *Juraphilopotamus lubricus* Wang, Zhao & Ren, 2009.

The apex of the forewing located at the terminal of R_5; Sc long with a humeral cross-vein and an oblique cross-vein leading to the costal margin; cells of MC and DC closed; the stem of Rs nearly twice as long as that of DC; Rs and M four-branched; F1–F5 complete; Cu_2 and 1A reach posterior wing margin at the same point. Hind wing with cross-veins r and m-cu present; F4 absent, and DC closed.

Distribution and age: Inner Mongolia; Middle Jurassic.

Only one species included from the Jurassic of Northern China (see Table 26.1).

Figure 26.7 *Pulchercylindratus punctatus* Gao, Yao & Ren, 2013, (Holotype, CNU-TRI-NN-2011003) [39].

Pulchercylindratus Gao, Yao & Ren, 2013

Pulchercylindratus Gao, Yao & Ren, 2013, *Foss. Rec.*, 16 (1), 112 [39] (original designation).

Type species: *Pulchercylindratus punctatus* Gao, Yao & Ren, 2013.

Wing moderately broad and smoothly rounded distally; R_5 reaching wing apex; R_1 forking distally; cells DC and MC closed in forewing, but DC open in hind wing; Rs stem nearly twice as long as DC; Rs and M four-branched, respectively; F1–F5 present in forewing; anal cells long. Tibial spurs: 2, 4, 4.

Distribution and age: Inner Mongolia; Middle Jurassic.

Only one species included from the Jurassic of Northern China (see Table 26.1).

Pulchercylindratus punctatus Gao, Yao & Ren, 2013 (Figure 26.7)

Pulchercylindratus punctatus Gao, Yao & Ren, 2013: *Foss. Rec.*, 16 (1), 114–115.

Locality and horizon: Daohugou, Ningcheng, Inner Mongolia, China; Middle Jurassic, Jiulongshan Formation.

Body with length 10.47 mm and maximal width 8.24 mm; forewing length is 9.06 mm, and width 3.47 mm. Head round, distinctly narrower than pronotum. Antenna shorter than forewing, filiform; scape and pedicel broader than flagellomeres. Maxillary palps five-segmented in both sexes, second segment subcylindrical, all segments of subequal length. Ocelli present. Anterior setal warts and posterolateral setal warts present on the head. Prothorax narrow, a pair of pronotal setal warts visible on pronotum [39].

Suborder *Incertae sedis*

Family Necrotauliidae Handlirsch, 1906

Many paleontologists considered Necrotauliidae to be a representative of the amphiesmenopteran stem-group, and proximal to the common ancestor of trichopterans and lepidopterans [13, 14, 44–47]. Liu *et al.* [15] recently argued that Necrotauliidae belong to the Trichoptera, representing the stem-group of Integripalpia, based on the characters of *Acisarcuatus variradius* with a well-preserved body and wings. *Acisarcuatus variradius* has male genitalia with harpagones, which is a synapomorphy of Trichoptera; the structure of the maxillary palps and the absence of cross-vein m represent apomorphies of Integripalpia. However, future discoveries of material belonging to the type species *Necrotaulius parvulus* with more informative body characters are needed to confirm this placement [32].

Necrotauliidae is documented from the Triassic to the Cretaceous. *Necrotaulius proximus* Sukatcheva, 1973 from the Triassic of the Kyrgyz Republic is the oldest fossil record hitherto reported. To date, seven known genera with 25 species have been reported Germany, Russia, China, and United Kingdom [13, 25, 39, 40, 47, 48]. Various necrotauliid specimens have been reported from the late Early Jurassic of Dobberlin, Germany, but many of these have been deemed as junior synonyms of *Necrotaulius parvulus*, *Mesotrichopteridium intermedium* and *Liadotaulius maior* [25, 46, 47].

Genera included from the Jurassic and Cretaceous of Northern China: *Necrotaulius* Handlirsch, 1906 and *Acisarcuatus* Liu, Zhang, Yao & Ren, 2014.

Necrotaulius Handlirsch, 1906

Necrotaulius Handlirsch, 1906, Die fossilen Insekten und die Phylogenie der rezenten Formen: Ein Handbuch für Paläontologen und Zoologen, 483 [44] (original designation).

Type species: *Necrotaulius furcatus* (Giebel, 1856).

Warts present on the head. Forewing densely covered with setae, in the hind wing long setae on the wing margin. In the forewing, Sc ending at 2/3 of the wing length; stem of F1 is longer than that of F2; cross-vein m-cu and thyridium at the medial fork; M_4 usually reduced in the hind wing.

Distribution and age: Hebei, Inner Mongolia; Early Cretaceous, Middle Jurassic

Two species included from the Jurassic and Cretaceous of Northern China (see Table 26.1).

Acisarcuatus Liu, Zhang, Yao & Ren, 2014

Acisarcuatus Liu, Zhang, Yao & Ren, 2014, *PLoS ONE*, 10, 3–5 [15] (original designation).

Type species: *Acisarcuatus variradius* Liu, Zhang, Yao & Ren, 2014.

Ocelli present. Warts present on head and thorax. Tegula with long setae. Tibial spur formula: 0 : 2 : 4. Forewing with Sc bifurcating; DC closed by cross-veins; MC and TC open. Hind wing with Sc not bifurcating. Claspers in male two-segmented.

Figure 26.8 *Acisarcuatus variradius* Liu, Zhang, Yao & Ren, 2014, (Holotype, CNU-TRI-NN-2013001p). (a). Habitus; (b). Line drawing [15].

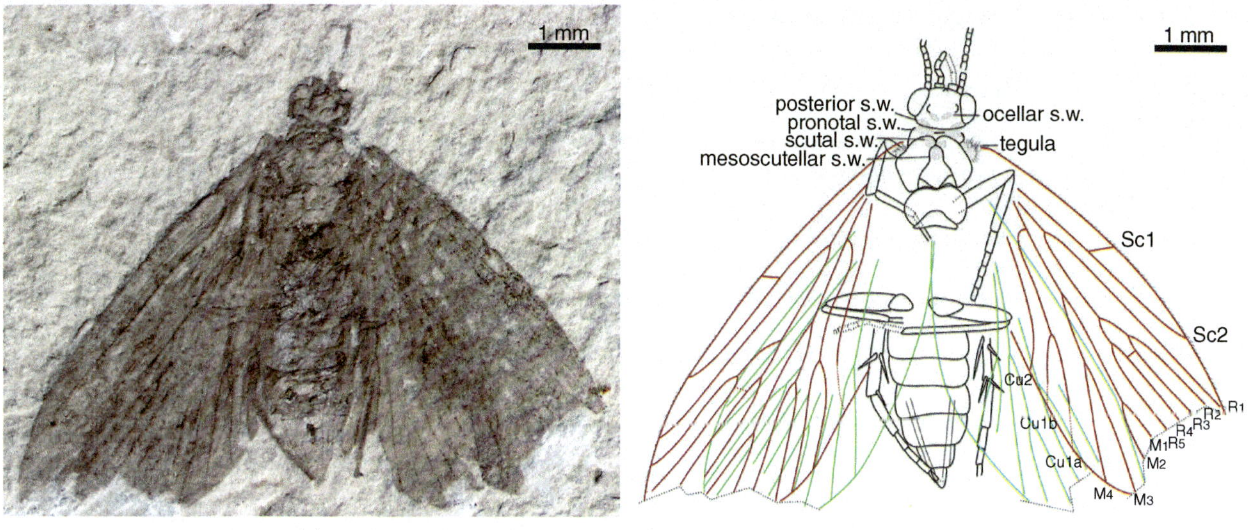

(a) (b)

Figure 26.9 *Acisarcuatus locellatus* Zhang, Shih & Ren, 2017, (Holotype, CNU-TRI-NN-2014011p). (a). Habitus; (b). Line drawing [32].

Distribution and age: Inner Mongolia; Middle Jurassic.

Two species included from the Jurassic of Northern China (see Table 26.1).

Acisarcuatus variradius **Liu, Zhang, Yao & Ren (Figure 26.8)**
Acisarcuatus variradius Liu, Zhang, Yao & Ren, 2014: *PLoS ONE*, 10, 5–8.

Locality and horizon: Daohugou, Ningcheng, Inner Mongolia, China; Middle Jurassic, Jiulongshan Formation.

Body length is 9.92 mm, width 1.74 mm; forewing length is 9.36 mm, width 3.4 mm. Head with saponaceous triangle, compound eyes at both sides of head, oval. Anterior and posterior setal warts present surrounding compound eyes, irregularly oval. Pronotum with one pair of setal warts, symmetrically drop-shaped. Forewing with R_1 unforked distally, straight in basal part and curved in pterostigma area; Rs forked at mid-length of the forewing; DC short and closed by r_3–r_4; F1 forks more basal than F2; Rs_1 slightly bent toward R_1 at terminus; M originating from base of R; M forking before Rs forking; F3 and F4 longer than their stems; F3 forking more distal than F4; cell MC very long and apparently open; Cu_1 bifurcated into Cu_{1a} and Cu_{1b}, and then F5 forking at the same level as Rs forking; cross-vein m-cu_1 present; TC open; Cu_2 straight and simple; anal veins visible; 1A straight, 2A reaching the median of 1A, 3A strongly curved and reaching median of 2A. Hind wing with forks I–V; DC, MC, and TC open; Sc simple; R_1 straight and simple; F1 forking more basal than F2, F3 forking slightly more distal than F4, F5 forking at the most basal position [15].

Acisarcuatus locellatus **Zhang, Shih & Ren, 2017 (Figure 26.9)**
Acisarcuatus locellatus Zhang, Shih & Ren, 2017: *Alcheringa*, 41 (1), 24–25.

Locality and horizon: Daohugou, Ningcheng, Inner Mongolia, China; Middle Jurassic, Jiulongshan Formation.

Body length is 5.5–7.4 mm, width 1.1–1.4 mm; forewing length is 5.4–6.9 mm, width 2.1–2.4 mm. Antenna filiform, nearly 0.6 times as long as forewing. Maxillary palp five-segmented, segment I swollen. Spots present on forewing. Sc bifurcating almost from midlength of stem; R_1 straight and not bifurcated; Rs forking at midlength of the forewing; M forking basal to Rs furcation. All apical forks present. F3 nearly half as long as F4; cell DC closed by cross-vein s; cells MC and TC open; Cu_1 originating from base of M and curving basally; Cu_1 bifurcating into Cu_{1a} and Cu_{1b}; and F5 forking more basal to F4 forking; Cu_2 bending terminally and reaching the posterior margin of forewing. Hind wing with Sc and R_1 not bifurcated; F3 distinctly shorter than F2; F4 absent [32].

Table 26.1 A list of fossil Trichoptera from the Jurassic and Cretaceous of China.

Family	Species	Locality	Horizon/Age	Citation
Vitimotauliidae	*Multimodus dissitus* Ren, 1995	Chengde, Hebei	Yixian Fm., K_1	Ren [24]
	Multimodus stigmaeus Ren, 1995	Chengde, Hebei	Yixian Fm., K_1	Ren [24]
	Multimodus elongatus Ren,1995	Chengde, Hebei	Yixian Fm., K_1	Ren [24]
	Sinomodus spatiosus Wang & Ren, 2009	Beipiao, Liaoning	Yixian Fm., K_1	Wang *et al.* [26]
	Sinomodus peltatus Wang & Ren, 2009	Beipiao, Liaoning	Yixian Fm., K_1	Wang *et al.* [26]
	Sinomodus macilentus Wang & Ren, 2009	Beipiao, Liaoning	Yixian Fm., K_1	Wang *et al.* [26]
Necrotauliidae	*Necrotaulius kritus* Lin, 1986	Zhongshan, Guangxi	Shiti Fm., J_1	Lin [49]
	Necrotaulius qingshilaense Hong, 1984	Shidongzi, Hebei	Yixian Fm., K_1	Hong [50]
	Acisarcuatus variradius Liu, Zhang, Yao & Ren, 2014	Ningcheng, Inner Mongolia	Jiulongshan Fm., J_2	Liu *et al.* [15]
	Acisarcuatus locellatus Zhang, Shih & Ren, 2017	Ningcheng, Inner Mongolia	Jiulongshan Fm., J_2	Zhang *et al.* [32]
Rhyacophilidae	*Declinimodus setulosus* Gao, Yao & Ren, 2013	Ningcheng, Inner Mongolia	Jiulongshan Fm., J_2	Gao *et al.* [40]
Philopotamidae	*Liadotaulius daohugouensis* Wu & Huang, 2012	Ningcheng, Inner Mongolia	Jiulongshan Fm., J_2	Wu and Huang [51]
	Liadotaulius limus Zhang, Shih & Ren, 2017	Ningcheng, Inner Mongolia	Jiulongshan Fm., J_2	Zhang *et al.* [32]
Hydrobiosidae	*Juraphilopotamus lubricus* Wang, Zhao & Ren, 2009	Ningcheng, Inner Mongolia	Jiulongshan Fm., J_2	Wang *et al.* [43]
	Pulchercylindratus punctatus Gao, Yao & Ren, 2013	Ningcheng, Inner Mongolia	Jiulongshan Fm., J_2	Gao *et al.* [39]
Ningxiapsychidae	*Ningxiapsyche fangi* Hong & Li, 2004	Guyuan, Ningxia	Naijiahe Fm., K_1	Hong and Li [27]
Family *Incertae sedis*	*Qinquania combinata* Hong, 1982	Yumen, Gansu	Chijinqiao Fm., K_1	Hong [33]
	Cathayamodus fournieri Gao, Shih, Labandeira & Ren, 2016	Beipiao, Liaoning	Yixian Fm., K_1	Gao *et al.* [28]

References

1 Holzenthal, R.W., Morse, J.C., and Kjer, K.M. (2011). Order Trichoptera Kirby, 1813. In: *Animal Biodiversity: An Outline of Higher-level Classification and Survey of Taxonomic Richness*, Zootaxa, vol. **3148** (ed. Z.Q. Zhang), 209–211.

2 Holzenthal, R.W., Blahnik, R.J., Prather, A.L., and Kjer, K.M. (2007). Order Trichoptera Kirby, 1813 (Insecta), caddisflies. *Zootaxa* **1668**: 639–698.

3 Martynov, A.V. (1924). Rucheiniki (caddisflies [Trichptera]). In: *Prakticheskaya Entomologiya*, vol. **5** (ed. N.N. Bogdanova-Kat'kova). Leningrad, pp. iv + 384 (in Russian).

4 Wiggins, G.B. and Wichard, W. (1989). Phylogeny of pupation in Trichoptera, with proposals on the origin and higher classification of the order. *Journal of the North American Benthological Society* **8**: 260–276.

5 Sukatcheva, I.D. (1980). Evolution of the caddisfly (Trichoptera) larval case construction. *Zhurnal Obshchey Biology* **41**: 457–469.

6 Minet, J., Huang, D.-Y., Wu, H., and Nel, A. (2010). Early Mecopterida and the systematic position of the

Microptysmatidae (Insecta: Endopterygota). *Annales de la Société entomologique de France* **46** (1–2): 262–270.

7 Bowles, D.E., Kleinsasser, L.J., and Jurgensen, T. (2016). Environmental determinates of stream caddisfly (Trichoptera) diversity in eastern Texas, USA. *Transactions of Kansas Academy of Sciences* **119**: 281–298.

8 Pereira, L.R., Cabette, H.S.R., and Juen, L. (2012). Trichoptera as bioindicators of habitat integrity in the Pindaíba river basin, Mato Grosso (Central Brazil). *International Journal of Limnology* **48**: 295–302.

9 Bosc, L. (1805). Note sur un fossile remarquable de la montagne de Saint-Gérand-le-Puy, entre Moulins et Roane, Département de l'Allier, appelé l'Indusie tubuleuse. *Journal des Mines* **17**: 397–400.

10 Carpenter, F.M. (1992). *Treatise on Invertebrate Palaeontology, Part R. Arthropoda 4 Vol 3, Superclass Hexapoda*. Lawrence/Kansas: Geological Society American and University Kansas.

11 Malm, T., Johanson, K.A., and Wahlberg, N. (2013). The evolutionary history of Trichoptera (Insecta): a case of successful adaptation to life in freshwater. *Systematic Entomology* **38**: 459–473.

12 Sukatcheva, I.D. (1973). New caddisflies (Trichoptera) from Mesozoic middle Asia. *Paleontology Journal* **3**: 100–107. (in Russian).

13 Ivanov, V.D. and Sukatcheva, I.D. (2002). Order Trichoptera Kirby, 1813. The caddisflies (=Phryganeida Latrielle, 1810). In: *History of Insects* (ed. A.P. Rasnitsyn and D.L.J. Quicke), 199–220. Dordrecht: Kluwer Academic Publishers.

14 Grimaldi, D.A. and Engel, M.S. (2005). *Evolution of the Insects*. New York, NY: Cambridge University Press.

15 Liu, Y.J., Zhang, W.T., Yao, Y.Y., and Ren, D. (2014). A new fossil of Necrotauliidae (Insecta: Trichoptera) from the Jiulongshan Formation of China and its taxonomic significance. *PLoS One* **9** (12): e114968.

16 Hong, Y.Q. and Wang, W.L. (1976). The volume of Inner Mongolian Autonomous Region. In: *Paleontological Atlas of North China*, 81–87. Beijing: Geological Publishing House (in Chinese).

17 Davis, S.R., Engel, M.S., and Ren, D. (2010). A pupal caddisfly from the Early Cretaceous of China (Trichoptera). *Cretaceous Research* **31** (4): 396–399.

18 Huang, D.-Y., Wu, H., and Dong, F.B. (2009). The discovery and preliminary study of fossil caddis case in China. *Acta Palaeontologica Sinica* **48** (4): 646–653. (in Chinese with English abstract).

19 Sukatcheva, I.D. (1998) The Early Cretaceous Caddisfly Fauna of English. Proceeding of the 9th international symposium on Trichoptera, Thailand, pp. 371–375.

20 Sukatcheva, I.D. (1968). Mesozoic caddisflies (Trichoptera) from the Zabaikalia. *Paleontology Journal* **2**: 59–75. (in Russian).

21 Sukatcheva, I.D. (1982). History development of the Trichoptera. In: *Transactions of Paleontological Institution. Academy of Sciences, USSR*, vol. **197**, 1–111. (in Russian).

22 Sukatcheva, I.D. (1990). Description of fossil insect, Caddisflies, Phryganeida. Late Mesozoic insects of eastern Transbaikalia. In: *Transactions of Paleontological Institution. Academy of Sciences, USSR*, vol. **239**, 94–122. (in Russian).

23 Sukatcheva, I.D. (1992). New fossil Trichoptera from Mongolia. New Species of Fossil Invertebrates, Trans. Joint Soviet-Mongol. *Paleontological Expedition* **4**: 111–116. (in Russian).

24 Ren, D. (1995). Insect: Trichoptera. In: *Faunae and Stratigraphy of the Jurassic-Cretaceous in Beijing and the Adjacent Area* (ed. D. Ren, L.W. Lu, Z.G. Guo and S.A. Ji), 95–97. Beijing: Seismic Publishing House (in Chinese with English abstract).

25 Sukatcheva, I.D. and Jarzembowski, E.A. (2001). Fossil caddisflies (Insecta: Trichoptera) from the Early Cretaceous of South England II. *Cretaceous Research* **22**: 685–694.

26 Wang, M.X., Liang, J.H., Ren, D., and Shih, C.K. (2009). New fossil Vitimotauliidae (Insecta: Trichoptera) from the Jehol Biota of Liaoning Province, China. *Cretaceous Research* **30** (3): 592–3598.

27 Hong, Y.C. and Li, Z. (2004). A new Early Cretaceous family from Liupanshan, Ningxia, China (Insecta, Trichoptera). *Acta Zootaxonomica Sinica* **29** (2): 224–233. (in Chinese with English Abstract).

28 Gao, T.P., Shih, C.K., Labandeira, C.C. *et al.* (2016). Convergent evolution of ramified antennae in insect lineages from the Early Cretaceous of Northeastern China. *Proceedings of the Royal Society of London B: Biological Sciences* **283**: 20161448.

29 Neboiss, A. (1991). Trichoptera (caddis-flies, caddises). In: *The Insects of Australia: A Textbook for Students and Research Workers*, 2e, vol. **2** (ed. I.D. Naumann, P.B. Carne, J.F. Lawrence, *et al.*), 787–816. Carlton, Victoria: Melbourne University Press.

30 Wiggins, G.B. and Currie, D.C. (2008). Trichoptera families. In: *An Introduction to the Aquatic Insects of North America* (ed. R.W. Merritt, K.W. Cummins and M.B. Berg), 439–480. Dubuque, Iowa: Kendall/Hunt Publishing C.

31 Handlirsch, A. (1939). Neue Untersuchungen über die fossilen Insekten mit Ergänzungen und Nachträgen sowie Ausblicken auf phylogenetische, palaeogeographische und allgemein biologische Probleme. II. Teil. *Annalen des naturhistorischen Museums in Wien* **49**: 1–240.

32 Zhang, W.T., Shih, C.K., and Ren, D. (2017). Two new fossil caddisflies (Amphiesmenoptera: Trichoptera) from the Middle Jurassic of Northeastern China. *Alcheringa* **41** (1): 22–29.

33 Hong, Y.C. (1982). *Mesozoic Fossil Insets of Jiuquan Basin in Gansu Province*, 80–91. Beijing: Geological Publishing House (in Chinese).

34 Ulmer, G. (1912). Die Trichopteren des Baltischen Bernsteins. *Beiträge zur Naturkunde Preussens* **10**: 1–380.

35 Botosaneanu, L. and Wichard, W. (1983). Upper-Cretaceous Siberian and Canadian Amber Caddisflies (Insecta: Trichoptera). *Bijdragen tot de Dierkunde* **53**: 187–217.

36 Sukatcheva, I.D. (1985). Jurassic Trichoptera of South Siberia. Jurassic insects of Siberia and Mongolia. *Transactions of Paleontological Institution. Academy of Sciences, USSR* **221**: 115–120. (in Russian).

37 Mey, W. (1988). The caddisflies of the Saxonian Amber (III) (Trichoptera). Deutsch. *Deutsche Entomologische Zeitschrift* **35**: 299–309.

38 Wichard, W. and Neumann, C. (2008). *Rhyacophila quadrata* n. sp., a new caddisfly (Insecta, Trichoptera) from Eocene Baltic amber. *Fossil Record* **11** (1): 19–23.

39 Gao, Y., Yao, Y.Z., and Ren, D. (2013). A new Middle Jurassic caddisfly (Trichoptera, Hydrobiosidae) from China. *Fossil Record* **16** (1): 111–116.

40 Gao, Y., Yao, Y.Z., and Ren, D. (2013). New genus and species of Rhyacophilidae (Insecta: Trichoptera) from the Middle Jurassic of China. *Acta Geologica Sinica (English Edition)* **87** (6): 1495–1500.

41 Ulmer, G. (1905). Zur Kenntniss aussereuropäischer Trichopteren. *Stettiner Entomologische Zeitung* **66**: 3–119.

42 Schmid, F. (1989). Les hydrobiosides (Trichoptera, Annulipalpia). *Bulletin de l'Institute Royal des Sciences Naturelles de Belgique, Entomologie* **59**: 1–154.

43 Wang, M.X., Zhao, Y.Y., and Ren, D. (2009). New fossil caddisfly from Middle Jurassic of Daohugou, Inner Mongolia, China (Trichoptera: Philopotamidae). *Progress in Natural Science* **19** (10): 1427–1431.

44 Handlirsch, A. (1906). Die fossilen insekten und die phylogenie der rezenten formen. In: *Lieferung 4*, 481–640. Leipzig: Wilhelm Engelmann.

45 Willmann, R.h. (1989). Evolution und phylogenetisches system der Mecoptera (Insecta, Holometabola). *Abhandlungen der Senckenbergischen Naturforschenden Gesellschaft*, **544**, 1–153.

46 Ansorge, J. (2002) Revision of the "Trichoptera" described by Geinitz and Handlirsch from the Lower Toarcian of Dobbertin (Germany) based on new material. Proceedings of the 10th International Symposium on Trichoptera. Nova Supplementary Entomology, 15, 55–74.

47 Ansorge, J. (2003). Upper Liassic Amphiesmenopterans (Trichoptera + Lepidoptera) from Germany. *Acta Zoologica Cracoviensia* **46**: 285–290.

48 Ross, H.H. (1967). The evolution and past dispersal of the Trichoptera. *Annual Review Entomology* **12**: 169–206.

49 Lin, Q.B. (1986). *Early Mesozoic Fossil Insect from the South China*, 85. Beijing: Palaeontologica Sinica Science Press (in Chinese).

50 Hong, Y.C. (1984). Class insecta. In: *Paleontological Atlas of North China II. Mesozoic Volume* (ed. Tianjin institute of geology and mineral resources), 128–184. Beijing: Geological Publishing House (in Chinese).

51 Wu, H. and Huang, D.-Y. (2012). A new species of *Liadotaulius* (Insecta: Trichoptera) from the Middle Jurassic of Daohugou, Inner Mongolia. *Acta Geologica Sinica* **86**: 320–324.

27

Lepidoptera – Butterflies and Moths

Weiting Zhang[1,2], Chungkun Shih[2,3], and Dong Ren[2]

[1] Hebei GEO University, Shijiazhuang, China
[2] Capital Normal University, Haidian District, Beijing, China
[3] National Museum of Natural History, Smithsonian Institution, Washington, DC, USA

27.1 Introduction to Lepidoptera

Lepidoptera, commonly known as butterflies or moths, are one of the most speciose lineages of insect herbivores, currently encompassing over 180 000 described species in about 126 families and 46 superfamilies [1], classified into four suborders of Zeugloptera, Aglossata, Heterobathmiina, and Glossata [2]. The Lepidoptera have a sister group relationship with the Trichoptera (caddisflies), together constituting the higher-rank taxon Amphiesmenoptera [3].

Adult lepidopterans range in sizes from very small for some microlepidopterans to very large, with wingspans up to 30 cm for *Thysania agrippina* from Brazil [4, 5]. In all but the most basal forms, feeding by adults is accomplished by siphoning in liquid via a tubular proboscis (haustellum), which usually is elongate and coiled under the head [6]. Adult moths of Micropterigidae have mandibulate mouthparts for feeding on spores and pollen. The dense covering of scales on the wings and parts of the body not only gives the order its scientific name, but also forms the basis for the attractive color patterns present in many species [7]. Wing venation consists predominantly of longitudinal veins with few cross-veins and some large cells, notably the discal [5].

Although Kristensen *et al.* listed 23 autapomorphies for adult Lepidoptera [7], few of these can be observed on compression fossils. Grimaldi [8] specified three critical, external apomorphies that should be present in adults of early lepidopteran fossils: (i) absence of a forewing M_4 vein; (ii) presence of an epiphysis on the foretibia of (most) species; and (iii) the occurrence of wing scales on both fore- and hind wings. A 3-branched medial vein (M) has long been considered a lepidopteran autapomorphy, but exceptions occur in both Lepidoptera and Trichoptera. For example, the M_4 is present in the Aglossata, Mesokristenseniidae, and Ascololepidopterigidae (Lepidoptera), but the M_4 is absent in the extinct family Dysoneuridae (Trichoptera).

All butterflies and moths go through a life cycle consisting of four distinct developmental stages: the egg, the larva or caterpillar, the pupa, and the adult [9]. The ecological and environmental importance of Lepidoptera stems largely from the fact that their larvae consume plants and adults feed on nectar and pollinate for the plants [10]. The majority of lepidopteran larvae live at the expense of living seed plants, including virtually all orders of both gymnosperms and angiosperms, as well as ferns, liverworts, and mosses [3]. Thus, many lepidopteran larvae are treated as pests, like bollworms and stem borers.

Dressed Up by Scales

Because of their beautiful wings, butterflies are one of the most favorite insects mentioned in poems, songs and literature and shown in paintings, photos, and movies. The amazing colors and patterns of butterflies (and some moths) come from a large number of scales that are small in size and arranged in an orderly fashion on wings and parts of the body (Figure 27.1). There are two main sources of the scale colors. One is "chemical color" from the pigment granules produced by the physiological metabolism, while the other is "structural color" produced by light interacting with the scales' microstructure. McNamara *et al.* reported the first example of structurally colored scales in fossil lepidopterans from the 47 million years ago (Mya) Messel oil shale in Germany [11].

Distinctive color patterns of lepidopteran wing play an important role in courtship and intraspecific recognition. In some lepidopterans, scales, and color patterns help the insects protect themselves by camouflage. *Kallima inachus* (Doyère, 1840), in Nymphalidae and commonly

Rhythms of Insect Evolution: Evidence from the Jurassic and Cretaceous in Northern China, First Edition.
Edited by Dong Ren, Chungkun Shih, Taiping Gao, Yongjie Wang, and Yunzhi Yao.

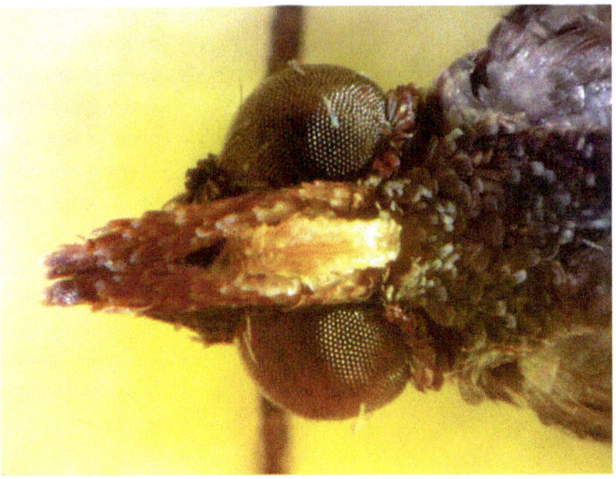

Figure 27.1 A "Flour moth," *Ephestia kuehniella* (Pyralidae), showing scales on the head, proboscis, and body. Source: Photo by Dr. Chungkun Shih.

Figure 27.2 A nymphalid butterfly showing dry-leaf like wings. Source: Photo by Dr. Chungkun Shih.

called "orange oakleaf" or "dead leaf," closely resembles a dry leaf with dark veins (Figure 27.2). When disturbed, an "orange oakleaf" opens up the wings and exposes the dramatic colors on the other sides of wings to surprise the potential predator, then, flies away erratically. If needed, it will drop down into the foliage and take a stationary pose with wings closed to help them blend in with leaves to avoid detection by the predators.

Chinese Butterfly Love Story

Like the Romeo and Juliet love story, the Chinese butterfly love story about Shanbo Liang and Yingtai Zhu has been mentioned and presented in books, poems, songs, movies, Chinese operas, and symphonies. Millions of Chinese have been deeply moved for generations by this story of butterfly lovers (Figure 27.3).

Figure 27.3 "The great Mormon butterflies", *Papilio memnon*, (Papilionidae), with the female at the top of the photo. Source: Photo by Jason Shih.

In ancient China, only boys could attend school. A girl, named Yingtai Zhu, wanted to go to school, so she had to be dressed as a boy. In the school, she had a deep and lasting friendship with her classmate Shanbo Liang. However, Shanbo did not know that Yingtai was a girl. After graduation, Yingtai left school and returned home. Later on, Shanbo visited Yingtai, he found out that she was a pretty girl. Shanbo fell in love with Yingtai. When he decided to ask Yingtai to marry him, Yingtai was already arranged to marry someone else by her father. Shanbo returned home sadly and not long after, Shanbo died of a broken heart. On the day of Yingtai's wedding, she got permission to pass by the tomb of Shanbo during the wedding procession. She insisted on saying goodbye to Shanbo. Suddenly, the sky turned dark, a strong wind blew by and the tomb broke apart. Yingtai jumped into the tomb without hesitation, and the tomb closed. The sky returned to blue and bright again, and two beautiful butterflies flew out of the tomb and danced in the sky.

27.2 Progress in the Studies of Fossil Lepidoptera

In contrast to the documented paleodiversity of other speciose insect orders, such as Coleoptera,

Hymenoptera, and Diptera [12], the Lepidopteran fossil records are poor [13]. Lepidopteran scales make the wing and body resistant to wetting, which may be why lepidopterans are rare as compression fossils [14, 15]. The earliest unequivocal lepidopteran fossil hitherto established is *Archaeolepis mane* Whalley, 1985 from the Early Jurassic, about 190 Mya, from Dorset, England [16]. The next early fossils are eight non-monophyletic genera, dated to approximately 180 Mya, from the uppermost Lias of Dobbertin, Germany [17]. The main grades/clades of the extant moths and butterflies originated during the Cretaceous. Their rapid divergence during the Late Cretaceous was most probably connected with the rapid diversification of the angiosperm plants [18].

The study of fossil Lepidoptera started in the early nineteenth century. Germar and Heer acted as pioneers in reporting Oligocene and Miocene fossil Lepidoptera from Europe [19–21]. To date, described fossil Lepidoptera comprise more than 4000 body fossils and 300 trace fossils [15]. Lepidopteran body fossils have been found worldwide such as North America [14], Asia [22–24], Baltic amber [25–28], Dominican amber [29], and Myanmar amber [30]. Fossils of leaf-mining provided the evidence for angiosperm-Lepidoptera associations [31]. Whalley and Grimaldi reviewed the Mesozoic Lepidoptera [8, 32]. Other reviews include Kozlov, Skalski, and Carpenter [33–35]. Sohn *et al.* compiled all the lepidopteran fossils described or mentioned in the world literature [13].

Studies of fossil Lepidoptera in China were apparently initiated in 1989 by Zhang, who described two Miocene fossil species belonging to Hepialidae and Sphingidae respectively from Shanwang, Shandong [36]. Zhang *et al.* reported one more sphingid from the same locality in 1994 [37]. Huang *et al.* established an extinct family Mesokristenseniidae comprising one genus with three species from the Middle Jurassic Jiulongshan Formation in Inner Mongolia [22]. From the same locality, Zhang *et al.* reported 12 genera and 15 species assigned to Eolepidopterigidae, Mesokristenseniidae, and Ascololepidopterigidae [23, 24].

Figure 27.4 A moth sipping nectar while pollinating the flowers. Source: Photo by Jason Shih.

Figure 27.5 An Eastern tiger swallowtail butterfly, *Papilio glaucus* in Papilionidae, sipping nectar while pollinating the flowers. Source: Photo by Dr. Chungkun Shih.

Pollination Mutualism

Lepidopterans are intimately associated with flowers, and both share a symbiotic and mutualistic relationship. Most of the extant Lepidoptera possess proboscides, coiled under the heads at rest, which consist of a pair of modified mouthparts, the maxillary galeae, joined by legulae. Butterflies extend their proboscides to probe flowers for nectar, or other substrates, e.g. sugary or salty water, fruit juices, or even urine for sugar, salts or other minerals. Owing to the flower's structure of the nectaries, usually hidden in narrow tubes or spurs, butterflies or moths would get in contact with the pollen on their proboscides, head or other body parts, when they forage on nectar (Figures 27.4 and 27.5). The adhered pollen is likely to be transferred when the butterfly flies to another conspecific flower for nectar, resulting in maximizing xenogamy.

The basal family of Lepidoptera do not possess the lepidopteran proboscis, but retain functional mandibles. Nevertheless, many micropterigids visit flowers of angiosperms to feed on pollen [38]. Their maxillary palps are involved with the manipulation of spores and pollen grains during feeding [38]. The micropterigids with functional mandibles and long maxillary palps have been found in the mid-Cretaceous Myanmar (Burmese) amber, which might have played an important role in pollination (Figure 27.6) [30].

Figure 27.6 *Sabatinca cretacea* Zhang, Wang, Shih & Ren, 2017, (Holotype, CNU-LEP-MA2016013) [30].

27.3 Representative Fossils of Lepidoptera from Northern China

Suborder Eolepidopterigina Rasnitsyn, 1983

Family Eolepidopterigidae Rasnitsyn, 1983

The mid Mesozoic Eolepidopterigina with a single family, the Eolepidopterigidae, are an extinct suborder of moths within the Lepidoptera. The moths in Eolepidopterigidae are inferred to have fed on pollen and laid eggs as elongate lesions in plant substrates [39]. The type genus *Eolepidopterix* is represented by *Eolepidopterix jurassica* Rasnitsyn, 1983, restricted to the Upper Jurassic of Transbaikalia, Russia. Eight species in eight genera described from the latest Middle Jurassic of China currently are the oldest known eolepidopterigid records [23].

Genera included from the Jurassic of Northern China: *Akainalepidopteron* Zhang, Shih, Labandeira & Ren, 2013; *Dynamilepidopteron* Zhang, Shih, Labandeira & Ren, 2013; *Grammikolepidopteron* Zhang, Shih, Labandeira & Ren, 2013; *Longcapitalis* Zhang, Shih, Labandeira & Ren, 2013; *Petilicorpus* Zhang, Shih, Labandeira & Ren, 2013; *Quadruplecivena* Zhang, Shih, Labandeira & Ren, 2013; *Seresilepidopteron* Zhang, Shih, Labandeira & Ren, 2013 and *Aclemus* Zhang, Shih, Labandeira & Ren, 2015.

Akainalepidopteron Zhang, Shih, Labandeira & Ren, 2013

Akainalepidopteron Zhang, Shih, Labandeira & Ren, 2013, *PLoS ONE*, 8 (11), 9 [23] (original designation).

Type species: *Akainalepidopteron elachipteron* Zhang, Shih, Labandeira & Ren, 2013.

Mesotibia with few short spines; metatibia with robust spines. Forewing and hind wing bear cilia on the anterior margin; humeral vein present in forewing; Sc and R forked; all furcations of Rs_{1+2}, Rs_{3+4}, and M_{1+2} *ca.* at the same level with each other. Hind wing Sc not forked. R forked; furcations of Rs_{1+2}, Rs_{3+4}, and M_{1+2} almost at the same level. Long piliform scales present on several veins of hind wing. Cross-vein cua-cup present.

Distribution and age: Inner Mongolia; Middle Jurassic.

Only one species included from the Jurassic of Northern China (see Table 27.1).

Akainalepidopteron elachipteron Zhang, Shih, Labandeira & Ren, 2013 (Figures 27.7 and 27.8)

Akainalepidopteron elachipteron Zhang, Shih, Labandeira & Ren, 2013: *PLoS ONE*, 8 (11), 9.

Locality and horizon: Daohugou, Ningcheng, Inner Mongolia, China; Middle Jurassic, Jiulongshan Formation.

The body length 6.3–7.2 mm, and forewing, 6.0–11.1 mm. Head length subequal to width, endowed with dense setae on the anterior margin. Eyes oval, with sparse setae on the outer margin. Forewing with pterostigma; Rs_4 ending slightly below the apex of forewing; Rs_{1+2} stalked; stem Rs_{1+2} subequal to stem Rs_{3+4}; hyaline zones surrounding r-m cross-vein, at Rs_{3+4} and M_{1+2} furcations. Hind wing with a cluster of *ca.* seven frenular bristles (Figure 27.7c), an apparatus for wing coupling, arising near the base of C; Rs_{1+2} forked at the same level with Rs_{3+4}; cross-vein cua-cup slanted obliquely; piliform scales ranging from 0.12 to 0.21 mm in length (Figure 27.7d), present on veins Sc, R_1, R_2, Rs_1, Rs_2, Rs_3, and Rs_4. Female genitalia with short ovipositor [23].

Dynamilepidopteron Zhang, Shih, Labandeira & Ren, 2013

Dynamilepidopteron Zhang, Shih, Labandeira & Ren, 2013, *PLoS ONE*, 8 (11), 12 [23] (original designation).

Type species: *Dynamilepidopteron aspinosus* Zhang, Shih, Labandeira & Ren, 2013.

Body relatively large, *ca.* 7 mm long. Metatibia robust, with strong spines. Forewing and hind wing lack cilia on the anterior margin. Humeral vein present in forewing. Sc and R forked; all furcations of Rs_{1+2} and Rs_{3+4} veins at the same level. Hind wing lacking spines.

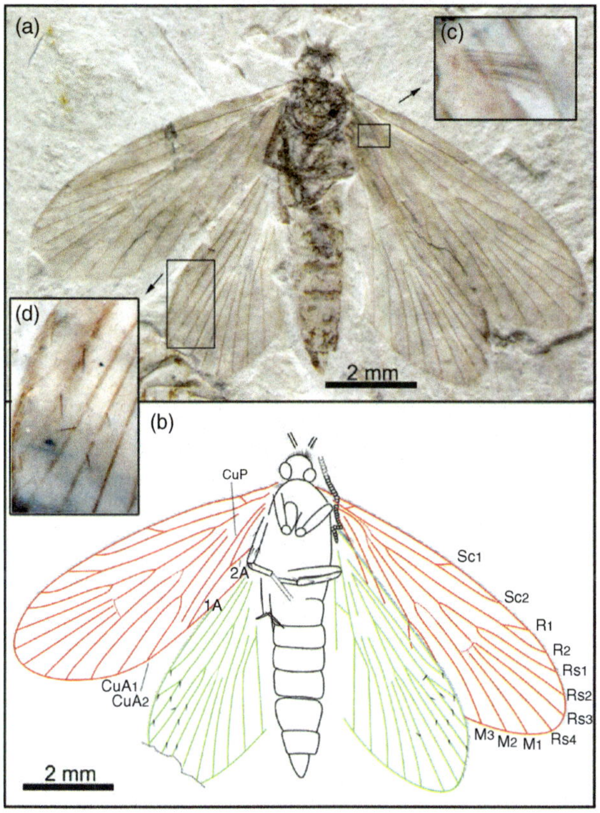

Figure 27.7 *Akainalepidopteron elachipteron* Zhang, Shih, Labandeira & Ren, 2013. Holotype, CNU-LEP-NN-2012024. (a), Fossil specimen; (b), Camera lucida drawing of (a), showing overall habitus; (c), A cluster of frenular bristles (wing coupling apparatus) on hind wing, outlined at the upper-right rectangular template in (a); (d), Piliform scales on hind wing, outlined at the lower-left rectangular template in (a). Source: Donated by Dr. Chungkun Shih.

Figure 27.8 Artwork of 3-D reconstruction by Dr. Chen Wang.

Distribution and age: Inner Mongolia; Middle Jurassic.

Only one species included from the Jurassic of Northern China (see Table 27.1).

***Grammikolepidopteron* Zhang, Shih, Labandeira & Ren, 2013**

Grammikolepidopteron Zhang, Shih, Labandeira & Ren, 2013, *PLoS ONE,* 8 (11), 18 [23] (original designation).

Type species: *Grammikolepidopteron extensus* Zhang, Shih, Labandeira & Ren, 2013.

Body small, less than 4.0 mm. Forewing elongate; humeral vein and sc-r cross-vein absent; Sc and R not forked; Rs_3 terminating only slightly before wing apex; branching points of R and Rs_1 to Rs_3 linearly aligned; M divided into M_3 and M_{1+2}; M_{1+2} subdivided iteratively into veins M_1 and M_2 at an angle of *ca.* 15°; M_1 linear and smooth; CuA bifurcated; CuP simple; three anal veins looping into a double-Y configuration. Apophyses short.

Distribution and age: Inner Mongolia; Middle Jurassic.

Only one species included from the Jurassic of Northern China (see Table 27.1).

***Longcapitalis* Zhang, Shih, Labandeira & Ren, 2013**

Longcapitalis Zhang, Shih, Labandeira & Ren, 2013, *PLoS ONE,* 8 (11), 16 [23] (original designation).

Type species: *Longcapitalis excelsus* Zhang, Shih, Labandeira & Ren, 2013.

Head longer than wide. Metatibiae lacking spines. Anterior margin of wings lacking cilia. Humeral vein in forewing present; Sc forked. R not forked; all furcations of Rs_{1+2}, Rs_{3+4}, and M_{1+2} almost at the same level with each other; anal veins looping into a double-Y configuration. Hind wing with Sc and R not forked. Veins on hind wing lacking spines [23, 24].

Distribution and age: Inner Mongolia; Middle Jurassic.

Only one species included from the Jurassic of Northern China (see Table 27.1).

***Longcapitalis excelsus* Zhang, Shih, Labandeira & Ren, 2013 (Figure 27.9)**

Longcapitalis excelsus Zhang, Shih, Labandeira & Ren, 2013: *PLoS ONE,* 8 (11), 17.

Locality and horizon: Daohugou, Ningcheng, Inner Mongolia, China; Middle Jurassic, Jiulongshan Formation.

Body length 7.9–10.5 mm; forewing length is 8.2–11.8 mm. Head relatively long (*ca.* 0.9 mm); compound eyes elongate oval. Metafemora with a spine at the end of femora. Forewing 2.7 times as long as wide; pterostigma present; Sc forked, at distal 2/5 of the stem; Rs_4 extending to apex of forewing; Rs_{1+2} stalked *ca.* to 0.4 of the total vein length; Rs_3 and Rs_4 free; and Rs_{1+2}

Figure 27.9 *Longcapitalis excelsus* Zhang, Shih, Labandeira & Ren, 2013, (Holotype, CNU-LEP-NN-2012025P).

stem subequal to Rs_{3+4} stem; hyaline zones surround r-m cross-vein at Rs_{1+2} furcation, and at the Rs_{3+4}; M_{1+2} and M_{1+2}–M_3 furcations. Hind wing 2.6 times as long as wide [23].

Petilicorpus Zhang, Shih, Labandeira & Ren, 2013

Petilicorpus Zhang, Shih, Labandeira & Ren, 2013, *PLoS ONE*, 8 (11), 15 [23] (original designation).

Type species: *Petilicorpus cristatus* Zhang, Shih, Labandeira & Ren, 2013.

Body slender. Anterior margins of fore and hind wing with cilia. In forewing, Sc and R forked; Rs branched into Rs_{1+2}, and Rs_{3+4}; Rs_{3+4} forked beyond that of Rs_{1+2}; Rs_{3+4} furcation nearly at the same level as M_{1+2} furcation; cross-vein m present. Hind wing with Sc not forked; R forked. In female, apophyses well-developed.

Distribution and age: Inner Mongolia; Middle Jurassic.

Only one species included from the Jurassic of Northern China (see Table 27.1).

Quadruplecivena Zhang, Shih, Labandeira & Ren, 2013

Quadruplecivena Zhang, Shih, Labandeira & Ren, 2013, *PLoS ONE*, 8 (11), 13 [23] (original designation).

Type species: *Quadruplecivena celsa* Zhang, Shih, Labandeira & Ren, 2013.

Labial palpus 3-segmented. All legs lacking spines. Anterior margins of fore and hind wings with cilia. Forewing with humeral vein; Sc forked; R not forked; Rs_{1+2} forking at same level with Rs_{3+4}, and before M_{1+2} forking. Hind wing Sc not forked; R forked; sc-r_1 cross-vein present.

Distribution and age: Inner Mongolia; Middle Jurassic.

Only one species included from the Jurassic of Northern China (see Table 27.1).

Seresilepidopteron Zhang, Shih, Labandeira & Ren, 2013

Seresilepidopteron Zhang, Shih, Labandeira & Ren, 2013, *PLoS ONE*, 8 (11), 6 [23] (original designation).

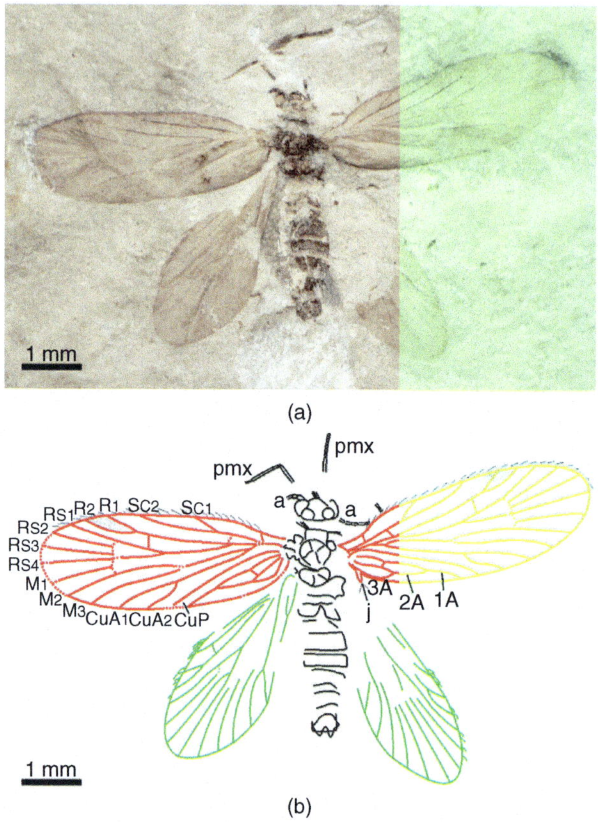

Figure 27.10 *Seresilepidopteron dualis* Zhang, Shih, Labandeira & Ren, 2013. (Male, Holotype, CNU-LEP-NN-2006-001c). (a), Photograph; (b), Camera lucida drawing, showing overall habitus.

Type species: *Seresilepidopteron dualis* Zhang, Shih, Labandeira & Ren, 2013.

Mesothorax slightly longer and wider than metathorax. All legs lacking spines. Forewing with a humeral vein; Sc and R forked; all furcations of Rs_{1+2}, Rs_{3+4}, and M_{1+2} at the same level with each other; humeral vein, sc-r cross-vein and cross-vein between R and Rs present; three anal veins looping into a double-Y configuration; jugum present but short, lacking long setae. Hind wing Sc not forked; R forked; sc-r cross-vein present.

Distribution and age: Inner Mongolia; Middle Jurassic.

Only one species included from the Jurassic of Northern China (see Table 27.1).

Seresilepidopteron dualis Zhang, Shih, Labandeira & Ren, 2013 (Figures 27.10 and 27.11)

Seresilepidopteron dualis Zhang, Shih, Labandeira & Ren, 2013: *PLoS ONE*, 8 (11), 7.

Locality and horizon: Daohugou, Ningcheng, Inner Mongolia, China; Middle Jurassic, Jiulongshan Formation.

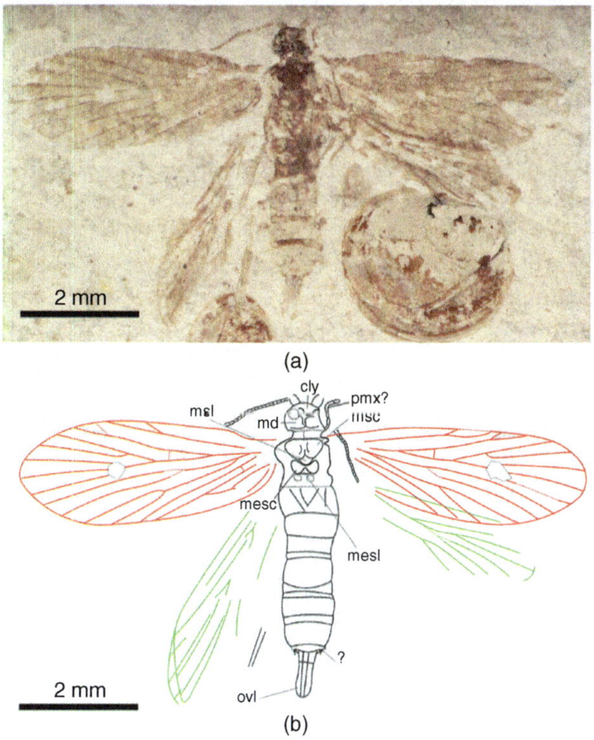

Figure 27.11 *Seresilepidopteron dualis* Zhang, Shih, Labandeira & Ren, 2013. (Female, Paratype, CNU-LEP-NN-2006-002). (a), Photograph, with a conchostracan at lower right. (b), Camera lucida drawing, showing overall habitus. Abbreviations: cly, clypeus; md, mandible; msc, mesoscutum; msl, mesoscutellum; mesc, metascutum; mesl, metascutellum; ovl, ovipositor lobes; pmx, maxillary palpus.

Male (Figure 27.10): Body length is 4.3 mm; forewing length is 4.7 mm. Forewing with Sc gently curved; Rs_4 to the apex of forewing; Rs_{1+2}, Rs_{3+4}, and M_{1+2} furcations arise at the same level in the wing; cross-vein sc-r oblique, proximal to the R–Rs furcation; cross-vein r located midway between R furcation and Rs furcation; the hyaline zones surround the r-m cross-vein at Rs_{1+2}, Rs_{3+4}, and M_{1+2} furcations; CuP slightly curved terminally; cross-vein between 1A and 2A present. A short, digitate jugum present in the basal posterior margin of forewing, lacking long setae. Hind wing with sc-r cross-vein present proximally to R furcation; costal margin bears long bristles [23].

Female (Figure 27.11): Body length is 5.0 mm; forewing length is 4.5 mm. Ovipositor long, well-developed, with a pair of inner apophyses [23]. Wing venation of female similar to male, except for the following differences: (i) Sc_2 extends to anterior wing margin at midlength in male, after midlength in female; (ii) R furcation in female closer to the apex than that in male; (iii) in female, cross-vein between R and Rs located before Rs furcation on right wing, but in male after Rs furcation; (iv) in female, cup-a cross-vein present but no cross-vein in anal area; and (v) in male, cup-a cross-vein absent, but cross-vein in anal area present in the male [23].

Aclemus Zhang, Shih, Labandeira & Ren, 2015

Aclemus Zhang, Shih, Labandeira & Ren, 2015, *J. Paleontol.*, 89(4), 618 [24] (original designation).

Type species: *Aclemus patulus* Zhang, Labandeira & Ren, 2015.

All legs with setae; metatibia with one pair of medial spurs and one pair of apical spurs. Forewings and hind wings lacking cilia on their anterior margins. Forewing veins Sc and R unforked; cross-veins r-m and m_3-cua_1 present. Hind wing with Sc and R unforked; cross-vein m_3-cua_1 present.

Distribution and age: Inner Mongolia; Middle Jurassic.

Only one species included from the Jurassic of Northern China (see Table 27.1).

Aclemus patulus Zhang, Shih, Labandeira & Ren, 2015 (Figure 27.12)

Aclemus patulus Zhang, Shih, Labandeira & Ren, 2015: *J. Paleontol.*, 89 (4), 618.

Locality and horizon: Daohugou, Ningcheng, Inner Mongolia, China; Middle Jurassic, Jiulongshan Formation.

Body length *ca.* 5.2 mm, width 1.3 mm; forewing length 4.8 mm, width 1.8 mm. Eyes oval, with sparse pubescence along the outer ocular margin. Forewing moderately broad, and with a rounded margin distally; wing index of forewing *ca.* 0.25; Rs_4 extending to apex of forewing; Rs_{1+2} and Rs_{3+4} furcations arising at about the same level; CuA furcation beyond M furcation; cross-vein m_3-cua_1 present, originating at 1/3 length of M_3 from M_{2+3} furcation and terminating at 1/3 length of CuA_1 from CuA furcation. Hind wing venation resembles that of forewing; Rs_{1+2} and Rs_{3+4} furcations arising at the same level [24].

Figure 27.12 *Aclemus patulus* Zhang, Shih, Labandeira & Ren, 2015, (Holotype, CNU-LEP-NN-2013001).

Suborder *Incertae sedis*

Family Mesokristenseniidae Huang, Nel & Minet, 2010

Mesokristenseniidae, an extinct family, are considered as the sister group of the Micropterigidae (Huang *et al.* 2010). Currently, all the species are known from China during the Middle Jurassic. The synapomorphies for the Mesokristenseniidae are mesotibia with only one apical spur, wings homoneurous, forewing with R_1 unforked and M vein 4-branched, and ovipositor with anterior apophyses.

Genera included from the Jurassic of Northern China: *Mesokristensenia* Huang, Nel & Minet, 2010 and *Kladolepidopteron* Zhang, Shih, Labandeira & Ren, 2013.

Mesokristensenia Huang, Nel & Minet, 2010

Mesokristensenia Huang, Nel & Minet, 2010, *Acta Geol. Sin.-Engl.*, 84 (4), 875 [22] (original designation).

Type species: *Mesokristensenia latipenna* Huang, Nel & Minet, 2010.

Antenna short, *ca.* 1/3 the length of forewing; scape prominently swollen; pedicel relatively robust. Metatibia with spines and two pairs of spurs. Forewing with Sc forked; cross-vein sc-r absent; M_1, after separation from M_2, subtended by an angle of greater than 60°, sharply angulate at junction with r-m cross-vein; three anal veins looping into a double-Y configuration. Hind wing with Sc and R unforked; Rs_4 to the apex; M 3-branched [22, 23].

Distribution and age: Inner Mongolia; Middle Jurassic.

Four species included from the Jurassic of Northern China (see Table 27.1).

Mesokristensenia trichophora Zhang, Shih, Labandeira & Ren, 2013 (Figure 27.13)

Mesokristensenia trichophora Zhang, Shih, Labandeira & Ren, 2013: *PLoS ONE*, 8 (11), 20.

Figure 27.13 *Mesokristensenia trichophora* Zhang, Shih, Labandeira & Ren, 2013, (Holotype, CNU-LEP-NN-2012032). Source: Donated by Dr. Chungkun Shih.

Locality and horizon: Daohugou, Ningcheng, Inner Mongolia, China; Middle Jurassic, Jiulongshan Formation.

Body length is 5.1 mm; forewing length is 5.0 mm. Antennae scape large and elongate, with setae distally placed. Forewing 2.9 times as long as wide; prominent pterostigma present; humeral vein present; Sc forked at distal 1/3 of the stem; Rs_4 extending to the apex of forewing; Rs_{1+2} stalked *ca.* 1/3 of their total length; M_{1+2} stem longer than M_{3+4} stem; hyaline zones surround r-m cross-vein at Rs_{1+2}, Rs_{3+4}, and M_{1+2} furcations; CuA forked before M_{3+4} furcation; 1A and 2A separate beyond basal half of vein length. Hind wing length *ca.* 2.6 times the width; Rs_{3+4} stem slightly longer than stem Rs_{1+2}. Ovipositor relatively long and robust [23].

Kladolepidopteron Zhang, Shih, Labandeira & Ren, 2013

Kladolepidopteron Zhang, Shih, Labandeira & Ren, 2013, *PLoS ONE*, 8 (11), 21 [23] (original designation).

Type species: *Kladolepidopteron oviformis* Zhang, Shih, Labandeira & Ren, 2013.

Vein M 4-branched in forewing. Prothorax with a pair of oval structures. Metatibia lacking spines. Cross-vein sc-r present at forewing base.

Distribution and age: Inner Mongolia; Middle Jurassic.

Three species included from the Jurassic of Northern China (see Table 27.1).

Kladolepidopteron oviformis Zhang, Shih, Labandeira & Ren, 2013 (Figure 27.14)

Kladolepidopteron oviformis Zhang, Shih, Labandeira & Ren, 2013: *PLoS ONE*, 8 (11), 23.

Locality and horizon: Daohugou, Ningcheng, Inner Mongolia, China; Middle Jurassic, Jiulongshan Formation.

Figure 27.14 *Kladolepidopteron oviformis* Zhang, Shih, Labandeira & Ren, 2013, (Female, Holotype, CNU-LEP-NN-2009007c).

Body length is 6.3 mm; forewing length is 4.8 mm. Body slender. Prothorax relatively large, with a pair of oval, slightly transverse structures that are adjacent medially. Forewings moderately slender, gradually tapering to a subacute apex; length *ca.* 3.0 times the width. Forewing bearing a humeral vein; Sc forked, from distal 1/4 of the stem; Sc_2 extending to costal margin of wing at 3/5 length of wing from base; cross-vein sc-r present; Rs_4 extending to the apex of forewing; stem M_{1+2} longer than stem M_{3+4}; cross-vein m originating near M_{1+2} furcation and terminating near the midpoint of M_3; hyaline zones surrounding r-m cross-vein at Rs_{3+4} and M_{1+2} furcations; 1A and 2A separate at basal half. Hind wing length *ca.* 2.8 times the width; Sc not forked; R not forked; Rs_{3+4} stem *ca.* 2 times as long as stem Rs_{1+2}. Ovipositor long [23].

Family Ascololepidopterigidae Zhang, Shih, Labandeira & Ren, 2013

Ascololepidopterigidae, comprising three genera and three species, are an extinct family, and only known by compression fossils from the Middle Jurassic in China. Ascololepidopterigidae are characterized by their metatibiae lacking medial spurs, forewing R forked, M vein 4-branched and cross-veins m-cua and cua-cup present.

Genera included from the Jurassic of Northern China: *Ascololepidopterix* Zhang, Shih, Labandeira & Ren, 2013, *Pegolepidopteron* Zhang, Shih, Labandeira & Ren, 2013, and *Trionolepidopteron* Zhang, Shih, Labandeira & Ren, 2013.

Ascololepidopterix Zhang, Shih, Labandeira & Ren, 2013

Ascololepidopterix Zhang, Shih, Labandeira & Ren, 2013, *PLoS ONE*, 8 (11), 26 [23] (original designation).

Type species: *Ascololepidopterix multinerve* Zhang, Shih, Labandeira & Ren, 2013.

Body relatively large. Forewing Sc not forked; R forked; cross-veins s and r-m present; stem M_{1+2} shorter than stem M_{3+4}; cross-vein m present. Hind wing R not forked; the distal part of 2A curved upward.

Distribution and age: Inner Mongolia; Middle Jurassic.

Only one species included from the Jurassic of Northern China (see Table 27.1).

Ascololepidopterix multinerve Zhang, Shih, Labandeira & Ren, 2013 (Figure 27.15)

Ascololepidopterix multinerve Zhang, Shih, Labandeira & Ren, 2013: *PLoS ONE*, 8 (11), 27.

Locality and horizon: Daohugou, Ningcheng, Inner Mongolia, China; Middle Jurassic, Jiulongshan Formation.

Body length is 8.1 mm; forewing length is 8.8 mm. Head wider than long. Forewing *ca.* 2.8 times as long

Figure 27.15 *Ascololepidopterix multinerve* Zhang, Shih, Labandeira & Ren, 2013, (Holotype, CNU-LEP-NN-2012028).

as wide; lacking humeral vein; prominent pterostigma present. R_2 curved; Rs_4 to the apex of forewing; stem Rs_{1+2} as long as stem Rs_{3+4}; s cross-vein present. M_1, after separation from M_2, subtends an angle of 63° (right wing) and 53° (left wing), sharply angulated at junction with r-m cross-vein; stem M_{1+2} shorter than stem M_{3+4}; m cross-vein originating near M_{1+2} furcation, ending at M_{3+4}; cross-vein m_{3+4}-cua_1 originating at *ca.* 1/3 length of M_{3+4} from its base and terminating near CuA furcation; cua-cup cross-vein originating at *ca.* 2/3 length of CuA from its base and terminating slightly beyond midpoint of CuP. Hind wing *ca.* 2.6 times as long as wide; stem Rs_{1+2} longer than stem Rs_{3+4}; M_1 and M_2 stalked; furcation of M_{1+2} at the same level with furcation of Rs_{1+2}; cross-vein m_3-cua_1 present [23].

Pegolepidopteron Zhang, Shih, Labandeira & Ren, 2013

Pegolepidopteron Zhang, Shih, Labandeira & Ren, 2013, *PLoS ONE*, 8 (11), 27 [23] (original designation).

Type species: *Pegolepidopteron latiala* Zhang, Shih, Labandeira & Ren, 2013.

Body relatively large. Forewing with Sc forked, R forked; cross-vein r absent; r-m cross-vein weak; stem M_{1+2} longer than stem M_{3+4}; m_{3+4}-cua cross-vein originating near M furcation and terminating at midpoint of CuA; cross-vein cua-cup located at the base of forewing. The distal part of 2A normal, not curved upward. Hind wing R forked.

Distribution and age: Inner Mongolia; Middle Jurassic.

Only one species included from the Jurassic of Northern China (see Table 27.1).

Trionolepidopteron Zhang, Shih, Labandeira & Ren, 2013

Trionolepidopteron Zhang, Shih, Labandeira & Ren, 2013, *PLoS ONE*, 8 (11), 27 [23] (original designation).

Type species: *Trionolepidopteron admarginis* Zhang, Shih, Labandeira & Ren, 2013.

Body small. Forewing with Sc forked; R forked; cross-vein rs_1-rs_2 present; stem M_{1+2} longer than stem M_{3+4}; cross-vein m_{3+4}-cua located at the CuA furcation. Hind wing with both Rs and M three-branched.

Distribution and age: Inner Mongolia; Middle Jurassic.

Only one species included from the Jurassic of Northern China (see Table 27.1).

Table 27.1 A list of fossil Lepidoptera from the Jurassic and Cretaceous of China.

Family	Species	Location	Horizon/Age	Reference
	Suborder Eolepidopterigina Rasnitsyn, 1983			
Eolepidopterigidae	*Akainalepidopteron elachipteron* Zhang, Shih, Labandeira & Ren, 2013	Ningcheng, Inner Mongolia	Jiulongshan Fm., J_2	Zhang *et al.* [23]
	Dynamilepidopteron aspinosus Zhang, Shih, Labandeira & Ren, 2013	Ningcheng, Inner Mongolia	Jiulongshan Fm., J_2	Zhang *et al.* [23]
	Grammikolepidopteron extensus Zhang, Shih, Labandeira & Ren, 2013	Ningcheng, Inner Mongolia	Jiulongshan Fm., J_2	Zhang *et al.* [23]
	Longcapitalis excelsus Zhang, Shih, Labandeira & Ren, 2013	Ningcheng, Inner Mongolia	Jiulongshan Fm., J_2	Zhang *et al.* [23]
	Petilicorpus cristatus Zhang, Shih, Labandeira & Ren, 2013	Ningcheng, Inner Mongolia	Jiulongshan Fm., J_2	Zhang *et al.* [23]
	Quadruplecivena celsa Zhang, Shih, Labandeira & Ren, 2013	Ningcheng, Inner Mongolia	Jiulongshan Fm., J_2	Zhang *et al.* [23]
	Sereslepidopteron dualis Zhang, Shih, Labandeira & Ren, 2013	Ningcheng, Inner Mongolia	Jiulongshan Fm., J_2	Zhang *et al.* [23]
	Aclemus patulus Zhang, Shih, Labandeira & Ren, 2015	Ningcheng, Inner Mongolia	Jiulongshan Fm., J_2	Zhang *et al.* [24]
	Suborder *Incertae sedis*			
Mesokristenseniidae	*Mesokristensenia angustipenna* Huang, Nel & Minet, 2010	Ningcheng, Inner Mongolia	Jiulongshan Fm., J_2	Huang *et al.* [22]
	Mesokristensenia latipenna Huang, Nel & Minet, 2010	Ningcheng, Inner Mongolia	Jiulongshan Fm., J_2	Huang *et al.* [22]
	Mesokristensenia sinica Huang, Nel & Minet, 2010	Ningcheng, Inner Mongolia	Jiulongshan Fm., J_2	Huang *et al.* [22]
	Mesokristensenia trichophora Zhang, Shih, Labandeira & Ren, 2013	Ningcheng, Inner Mongolia	Jiulongshan Fm., J_2	Zhang *et al.* [23]
	Kladolepidopteron oviformis Zhang, Shih, Labandeira & Ren, 2013	Ningcheng, Inner Mongolia	Jiulongshan Fm., J_2	Zhang *et al.* [23]
	Kladolepidopteron parva Zhang, Shih, Labandeira & Ren, 2013	Ningcheng, Inner Mongolia	Jiulongshan Fm., J_2	Zhang *et al.* [23]
	Kladolepidopteron subaequalis Zhang, Shih, Labandeira & Ren, 2013	Ningcheng, Inner Mongolia	Jiulongshan Fm., J_2	Zhang *et al.* [23]
Ascololepidopterigidae	*Ascololepidopterix multinerve* Zhang, Shih, Labandeira & Ren, 2013	Ningcheng, Inner Mongolia	Jiulongshan Fm., J_2	Zhang *et al.* [23]
	Pegolepidopteron latiala Zhang, Shih, Labandeira & Ren, 2013	Ningcheng, Inner Mongolia	Jiulongshan Fm., J_2	Zhang *et al.* [23]
	Trionolepidopteron admarginis Zhang, Shih, Labandeira & Ren, 2013	Ningcheng, Inner Mongolia	Jiulongshan Fm., J_2	Zhang *et al.* [23]

References

1 Mallet, J. (2014). *The Lepidoptera Taxome Project*. London: University College.
2 van Nieukerken, E.J., Kaila, L., Kitching, I.J. *et al.* (2011). Order Lepidoptera Linnaeus, 1758, in animal biodiversity: an outline of higher-level classification and survey of taxonomic richness (ed. Z.Q. Zhang). *Zootaxa* **3148**: 212–221.
3 Kristensen, N.P. (1999). Lepidoptera: moths and butterflies. Vol. 1. (evolution, systematics and biogeography). In: *Handbook of Zoology*, vol. **IV** (Arthropoda: Insecta), Part 35 (ed. M. Fischer), 1–491. Berlin/New York: Walter de Gruyter.
4 Kons, H. Jr., (1998). Largest Lepidopteran wing span, Chapter 32. In: *University of Florida Book of Insect Records* (ed. T.J. Walker), 79–81. Gainesville: University of Florida.
5 Gullan, P.J. and Cranston, P.S. (2004). *The Insects: An Outline of Entomology*, 3e. UK: Blackwell Publishing.
6 Resh, V.H. and Cardé, R.T. (2003). *Encyclopedia of Insects*. San Diego: Academic Press.
7 Kristensen, N.P., Scoble, M., and Karsholt, O. (2007). Lepidoptera phylogeny and systematics: the state of inventorying moth and butterfly diversity. *Zootaxa* **1668**: 699–747.
8 Grimaldi, D.A. (1999). The co-radiations of pollinating insects and angiosperms in the Cretaceous. *Annals of the Missouri Botanical Garden* **86** (2): 373–406.
9 Capinera, J.L. (2008). *Encyclopedia of Entomology*, 2e. Netherlands: Springer Netherlands.
10 Scoble, M.J. (1992). *The Lepidoptera: Form, Function and Diversity*. Oxford: Oxford University Press.
11 McNamara, M.E., Briggs, D.E.G., Orr, P.J. *et al.* (2011). Fossilized biophotonic nanostructures reveal the original colors of 47-million-year-old moths. *PLoS Biology* **9** (11): e1001200.
12 Labandeira, C.C. and Sepkoski, J.J. Jr., (1993). Insect diversity in the fossil record. *Science* **261**: 310–315.
13 Sohn, J.C., Labandeira, C.C., Davis, D., and Mitter, C. (2012). An annotated catalog of fossil and subfossil Lepidoptera (Insecta: Holometabola) of the world. *Zootaxa* **3286**: 1–132.
14 Grimaldi, D.A. and Engel, M.S. (2005). Amphiesmenoptera: the caddisflies and Lepidoptera. In: *Evolution of the Insects* (ed. D.A. Grimaldi and M.S. Engel), 548–555. New York: Cambridge University Press.
15 Sohn, J.C., Labandeira, C.C., and Davis, D. (2015). The fossil record and taphonomy of butterflies and moths (Insecta, Lepidoptera): implications for evolutionary diversity and divergence-time estimates. *BMC Evolutionary Biology* **15**: 12.
16 Whalley, P.E. (1985). The systematics and palaeogeography of the Lower Jurassic insects of Dorset, England. *Bulletin of the British Museum of Natural History (Geology)* **39**: 107–189.
17 Ansorge, J. (2002). Upper Liassic amphiesmenopterans (Trichoptera + Lepidoptera) from Germany–a review. *Acta Zoologica Cracoviensia* **46** (Suppl._Fossil Insects): 285–290.
18 Kozlov, M.V., Ivanov, V.D., and Rasnitsyn, A.P. (2002). Order Lepidoptera Linné, 1758. The butterflies and moths (= Papilionida Laicharting, 1781). In: *History of Insects* (ed. A.P. Rasnitsyn and D.L.J. Quicke), 220–227. Dordrecht, Boston/London: Kluwer Academic Publishers.
19 Germar, E.F. (1837) Fauna insectorum Europae. Fasciculus 19. Insectorum protogaeae specimen sistens insecta carbonum fossilium. Kümmel, Halle.
20 Germar, E.F. (1839). Die versteinerte Insecten Solenhofens. *Nova Acta Leopoldina* **19**: 187–222.
21 Heer, O. (1849). *Die Insektenfauna der Tertiärgebilde von Oeningen und von Radoboj in Croatien*, vol. **2**. Leipzig: Wilhelm Engelmann.
22 Huang, D., Nel, A., and Minet, J. (2010). A new family of moths from the Middle Jurassic (Insecta: Lepidoptera). *Acta Geologica Sinica – English Edition* **84** (4): 874–885.
23 Zhang, W.T., Shih, C.K., Labandeira, C.C. *et al.* (2013). New fossil Lepidoptera (Insecta: Amphiesmenoptera) from the Middle Jurassic Jiulongshan Formation, Northeastern China. *PLoS One* **8** (11): e79500.
24 Zhang, W.T., Shih, C.K., Labandeira, C.C., and Ren, D. (2015). A new taxon of a primitive moth (Insecta: Lepidoptera: Eolepidopterigidae) from the latest Middle Jurassic of Northeastern China. *Journal of Paleontology* **89** (4): 617–621.
25 Skalski, A.W. (1973a). Studies on the Lepidoptera from fossil resins. Part II. *Epiborkhausenites obscurotrimaculatus* gen. et. sp. nov. (Oecophoridae) and a tineid-moth discovered in the Baltic amber. *Acta Palaeontologica Polonica* **28** (1): 153–160.

26 Skalski, A.W. (1973b). Studies on the Lepidoptera from fossil resins. Part VI. *Tortricidrosis inclusa* gen. et spec. nov. from the Baltic amber (Lep., Tortricidae). *Deutsche Entomologische Zeitschrift, N. F.* **20** (4/5): 339–344.
27 Skalski, A.W. (1977). Studies on the Lepidoptera from fossil resins. Part 1. General remarks and descriptions of new genera and species of the families Tineidae and Oecophoridae from the Baltic amber. *Prace Muzeum Ziemi* **26**: 3–24.
28 Mey, W. (2011). On the systematic position of *Baltimartyria* Skalski, 1995 and description of a new species from Baltic amber (Lepidoptera, Micropterigidae). *Zookeys* **130**: 331–342.
29 Peñalver, E. and Grimaldi, D.A. (2006). New data on Miocene butterflies in Dominican amber (Lepidoptera: Riodinidae and Nymphalidae) with the description of a new nymphalid. *American Museum Novitates* **3519**: 1–17.
30 Zhang, W.T., Wang, J.J., Shih, C.K., and Ren, D. (2017). Cretaceous moths (Lepidoptera: Micropterigidae) with preserved scales from Myanmar amber. *Cretaceous Research* **78**: 166–173.
31 Labandeira, C.C., Dilcher, D.L., Davis, D.R., and Wagner, D.L. (1994). Ninety-seven million years of angiosperm-insect association: paleobiological insights into the meaning of coevolution. *Proceedings of the National Academy of Sciences* **91**: 12278–12282.
32 Whalley, P.E.S. (1986). A review of the current fossil evidence of Lepidoptera in the Mesozoic. *Biological Journal of the Linnean Society* **28**: 253–271.
33 Kozlov, M.V. (1988). Paleontology of lepidopterans and problems of the phylogeny of the order Papilionida. In: *The Mesozoic–Cenozoic Crisis in the Evolution of Insects* (ed. A.G. Ponomarenko), 16–69. Moscow: Academy of Sciences (in Russian).
34 Skalski, A.W. (1990). An annotated review of all fossil records of lower Lepidoptera. *Bulletin of the Sugadaira Montane Research Center* **11**: 125–128.
35 Carpenter, F.M. (1992). Superclass Hexapoda. In: *Treatise on Invertebrate Paleontology, Part R (Arthropoda-4), 3 and 4 (Superclass Hexapoda)* (ed. R.L. Kaesler, E. Brosius, J. Keim and J. Priesner). Boulder, Colorado/Lawrence, Kansas: Geological Society of America and the University of Kansas, pp. xxi + 655.
36 Zhang, J.F. (1989). *Fossil Insects from Shanwang, Shandong*. Jinan: Shandong Scientific and Technological Publishing House (in Chinese).
37 Zhang, J.F., Sun, B., and Zhang, X.Y. (1994). *Miocene Insects and Spiders from Shanwang, Shandong*. Beijing: Science Press (in Chinese).
38 Gibbs, G.W. (2014). Micropterigidae (Insecta: Lepidoptera). *Fauna of New Zealand* **72**: 1–127.
39 Rasnitsyn, A.P. (1983). Pervaya nakhodka babochki yurskogo vozrasta. *Doklady Akademiya Nauk SSSR* **269** (2): 467–671.

28

Insect Feeding

Chungkun Shih[1,2], Taiping Gao[1], Yunzi Yao[1], and Dong Ren[1]

[1] Capital Normal University, Haidian District, Beijing, China
[2] National Museum of Natural History, Smithsonian Institution, Washington, DC, USA

28.1 Introduction

Feeding is an essential function for all living organisms so that they can obtain nourishment, chemical compounds, energy and water to maintain body functions, grow to maturity, sustain life, and pass on their genes to the next generations. Insects, adapting to a wide range of environmental factors by having diverse body structures and life styles, have deployed many different feeding strategies. For holometabolous insects with complete metamorphosis, larvae and adult insects even adopt different feeding modes to minimize competition for food resources.

In general, adult extant insects, based on what they feed on, are categorized as herbivores (feeders of leaves, pollen, nectar, seeds, fruits, plant fluids, and/or wood), carnivores, omnivores, scavengers, hematophages (blood-feeders), etc. However, feeding behaviors of insects in deep time are poorly known, due to scarcity of well-preserved fossil records.

With well-preserved morphological characters on compression fossils from Northeastern China and by applying advanced imaging and analytical instruments, it has been demonstrated or inferred that insects at that time were involved in various modes of feeding. Examples presented in this chapter are: pollination drop feeding on gymnosperm plants using elongated siphonate mouthparts by scorpionflies of mesopsychids, pseudopolycentropodids and aneuretopsychids and by lacewings of kalligrammatids; preying other insects using chewing mouthparts and raptorial forelegs by mantispid or dipteromantispid lacewings or using folded raptorial tarsal segments by bittacid and cimbrophlebiid hangingflies; and blood feeding using serrated cutting mouthparts by primitive flea-like pseudopulicids and transitional fleas of saurophthirids and using piercing and sucking mouthparts by true bugs of Torirostratidae. In addition, by examining insect damage types (DTs) on well-preserved plant fossils, it is shown that insects sipped plant fluids with piercing and sucking mouthparts and consumed leaf surface or margin with chewing mandibles. Furthermore, plant insect DTs also show larval feeding by leaf-mining and galling.

28.2 Pollination Mutualism – Feeding on Pollination Drops Before Angiosperms

28.2.1 Scorpionflies with Elongate Siphonate Mouthparts

Insect pollination ensures efficient fertilization for the plants, while plants, in return provide pollen and/or nectar as rewards to pollinating insects [1]. There were limited reports on pollination mutualisms during the mid-Mesozoic [2, 3]. For example, fossil record of insect mouthparts [4], gut contents [5], insect consumption of plant reproductive organs [6], and plant reproductive organs [3, 7] suggest consumption of pollen, pollination drops, and other plant tissues [6]. Modern scorpionflies, a relict clade of approximate 600 species, overwhelmingly are minor predators or consumers of dead organisms [8] and rarely are implicated as floral visitors [9], with the exception of the relict Nannochoristidae having fluid-feeding mouthparts that are inferred to imbibe nectar and other plant exudates [10].

Ren *et al.* [11] reported scorpionflies of three families, Mesopsychidae (Figure 28.1), Aneuretopsychidae (Figure 28.2), and Pseudopolycentropodidae (Figure 24.14) in Aneuretopsychina from the latest Middle Jurassic to the Early Cretaceous having elongate siphonate (tubular) proboscides. Five potential ovulate host-plant taxa co-existed with these insects: a seed fern, conifer, ginkgoopsid, pentoxylalean, and gnetalean. Based on a variety of evidence for these three families of scorpionflies and their inferred pollinated plants, Ren *et al.* [11] demonstrated that Aneuretopsychina,

Rhythms of Insect Evolution: Evidence from the Jurassic and Cretaceous in Northern China, First Edition.
Edited by Dong Ren, Chungkun Shih, Taiping Gao, Yongjie Wang, and Yunzhi Yao.

a major clade of extinct, siphon-bearing, fluid-feeding insects, early members of the order Mecoptera (scorpionflies), were likely feeding on gymnospermous ovulate secretions (Figure 28.3). The evidence includes: (i) Morphology and hairy and obliquely ridged surfaces on the scorpionfly *proboscides* (Figures 28.1 and 28.2). (ii) Co-occurring seed fern and conifer ovulate organs present pollen-receptive areas that are hidden or poorly exposed to wind-dispersed grains and lead into long integumental channels. (iii) The levels of amino acids, sugars, inorganic ions, and other constituents found in modern insect-pollinated gymnosperms suggest that the pollination droplets would sustain high levels of insect activity [15]. (iv) Unisexual pollen-bearing organs typically are separated some distance from the conspecific ovulate organs, either on the same plant or different individual plants, thus favoring outcrossing for more efficient modes of pollination, such as remote insect transport. (v) Pollen extracted from affiliated pollen organs range in length from 17 to 65 μm for the diverse ovulate taxa that are implicated for pollination, approximating the modern 25–50 μm range for entomophilous cycads [16] and within the broader range of 10–80 μm for similarly pollinated gnetaleans [17]. (vi) Cheirolepidiaceous and gnetalean pollen occur as gut contents in several Mesozoic insect lineages [5, 6], serving as a nutritional source and/or as a reward for pollination.

These scorpionfly taxa with siphonate proboscides suggest that they fed on gymnosperm pollination drops or ovular secretions and likely engaged in pollination mutualisms with extinct gymnosperms during the mid-Mesozoic, 65 to 25 million years before the similar and independent coevolution of nectar-feeding flies, moths, bees, and beetles on angiosperms. The fossil records show that all three scorpionfly families became extinct during the later Early Cretaceous, coincident with global gymnosperm-to-angiosperm turnover.

28.2.2 Kalligrammatid Lacewings with Elongate Siphonate Mouthparts

Yang *et al.* [18] described three subfamilies, four genera, twelve species and four unassigned species in Kalligrammatidae (Neuroptera) from the latest Middle Jurassic Jiulongshan Formation and the Lower Cretaceous Yixian

Figure 28.1 Head and siphonate mouthpart features associated with fluid-feeding mesopsychid scorpionflies, *Lichnomesopsyche gloriae* Ren, Labandeira & Shih, 2010 [12]. (a) Specimen CNU-M-NN2005020-1. (b) Enlargement of template in (a) as an overlay drawing of head and mouthparts. Source: Donated by Dr. Chungkun Shih.

Figure 28.2 The head and mouthpart structure of *Jeholopsyche liaoningensis* Ren, Shih & Labandeira, 2011 [13], illustrating ventral body aspect (a) and dorsal view of head and mouthparts in photo (b) and drawing (c) (CNU-M-LB2005002). Source: Donated by Dr. Chungkun Shih.

(a) (b)

Figure 28.3 Pollination of (a). *Pseudopolycentropus janeannae* Ren, Shih & Labandeira, 2010 [14], (b). *Lichnomesopsyche gloriae* for gymnosperms. Source: Artwork by Mary Parrish in Department of Paleobiology, National Museum of Natural History in Smithsonian Institution.

Figure 28.4 Reconstruction of specimen *Oregramma illecebrosa* Yang, Wang, Labandeira, Shih & Ren, 2014 (CNU-NEU-LB-2009-031) above on a bennettitalean host probing a closed strobilus of *Williamsonia* (♀) fructification. Source: Artwork by Vichai Malikul in Department of Entomology, National Museum of Natural History in Smithsonian Institution.

Formations of Northeastern China. These kalligrammatid taxa exhibit diverse morphological characters, such as mandibulate mouthparts in one major clade and elongate siphonate mouthparts in the remaining four major clades. In addition, they show presence or absence of a variety of distinctive wing markings such as stripes, wing spots, and eyespots, as well as multiple major wing shapes.

Mid-Mesozoic kalligrammatid lacewings entered the fossil record 165 million years ago (Mya) and last occurred 125 Mya. By contrast, the extant Cenozoic butterfly clade (Lepidoptera; Papilionoidea) is believed to have originated 80–70 Mya, long after kalligrammatids became extinct.

Labandeira *et al.* [19] used polarized light, epifluorescence photography, environmental scanning electron microscope (ESEM) imaging, and energy dispersive spectrometry (EDS) to examine fossil kalligrammatids and their adjacent fossil matrix. Cladistic analysis on the morphology of specimens was used to build a phylogeny, upon which the evolution of specific traits were mapped. It was suggested that Kalligrammatidae and Papilionoidea convergently evolved wing eyespots, scales on their wings, elongate tubular proboscides, similar feeding style, and probable seed-plant associations (Figure 28.4).

Wing spots and eyespots in both clades appear to share a similar evolutionary trajectory from spot/eyespot absence, to simple circular spots, to simple eyespots, to more complex eyespots with multiple concentric rings of contrasting colors (Figure 28.4).

Mouthparts of kalligrammatids and papilionoids offer another remarkable example of convergent evolution. Kalligrammatid mouthparts evolved from an ancestral mandibulate (chewing) state to a derived long-proboscid (siphoning) state in which maxillary elements were conjoined to form a tube. This parallels the evolution of the proboscis in glossate Lepidoptera, which also originated from mandible-bearing ancestors [20]. The proboscides of more derived kalligrammatids bear a special resemblance to those of Lepidoptera [20]. The kalligrammatid proboscis was long (8–20 mm), flexible, lacked stylets or other piercing structures, smooth or covered with surface hairs, bracketed by multi-segmented maxillary palps, and its terminus typically rounded or truncate, resembling the end of a thick straw The butterfly-like mouthparts indicate long-proboscid kalligrammatid lacewings likely suited to probe larger, sturdier reproductive structures of Bennettitales, cycad-like plants contemporaneous with the Kalligrammatidae [19].

Pollination

On a bright sunny mid-April day in the Pinggu District, on the eastern outskirt of Beijing, the annual Peach Blossom Festival is in full swing. Thousands of visitors enjoy the scenery of a sea of pinkish peach blossoms covering many orchards totaling about 220 thousand acres. Among these blossoms on rows of peach trees, kids are laughing and running, people milling around and snapping phone photos, while others are enjoying their picnic food and drinks. Unbeknown to most of these people, buzzing honey bees are busy sipping nectar and collecting pollen, then, flying to neighboring flowers. Honey bees and flowers, perfect together. Except for the hard-working orchard farmers, very few of these people are contemplating the renowned Pinggu peaches, juicy, sweet, and delicious with an annual production of about 120 million kg (120 000 tons), to be harvested in October. Even fewer appreciate the importance of pollination and significant economic benefits contributed by the hard-working honey bees.

Honey bees, in the genus of *Apis of* the family Apidae, belong to the superfamily Apoidea (Figure 28.5). Other apoideans include stingless bees (in the tribe Meliponini of Apidae), halictid bees (in the Halictidae), leafcutting bees (in the Megachilidae), digger bees (in the tribe Anthophorini of Apidae), carpenter bees (in the genus *Xylocopa of* Apidae), bumble bees (in the genus *Bombus* of Apidae, Figure 28.6), etc. The mouthparts of bees are modified for dipping and sipping nectar. Only some bees, e.g. honey bees, and stingless bees produce and store honey, made from regurgitated nectar and saliva, to provision their colony including larvae. Unlike honey-making by honey bees, bumble bees only store a small amount of nectar for a few days to feed their larvae. Halictid bees, commonly called sweat bees, provide a mixture of nectar and pollen in cells to feed the larvae after their hatching.

Besides body being covered with very fine setae, the worker honey bees have hind legs specially equipped with long setae forming a pollen basket to collect pollen. Pollen is a rich source of proteins, fats, vitamins, and other trace elements needed for larvae to develop and to grow. Synergistically, they pollinate the angiosperm flowers so that fruits and seeds can be formed and dispersed in order to propagate and diversify the genes in the offspring plants.

Hymenopterans have contributed significant commercial values due to their pollination and biological control of insect pests for food crops and other plants. The largest managed pollination by honey bees in the world is associated with the California almonds, with a crop value of $3.6 billion in 2011 [21], where about 1.6 million hives of honey bees are trucked to the almond orchards each spring. It is estimated that in the United States, pollination by honey bees and other insects produced more than $15 billion worth of products in 2000 [22]! In addition, honey bees provide honey for human consumption as an ingredient for food and baked goods, medicine, and beauty care. Beeswax is a component used for wood polishes, candles, adhesives, and cosmetics. In the USA, it is estimated that annual honey production is more than 90 000 tons and beeswax production more than 1800 tons.

In the winter of 2006–2007, beekeepers in the USA noticed a significant percentage of honey bee hive losses, hence they coined the term "Colony Collapse Disorder" (CCD). CCD is characterized by the disappearance of the majority of worker bees in a colony, with the presence of a queen, plenty of food and a few nurse bees to care for the remaining larvae and the queen. In a recent US Department of Agriculture (USDA) report [23]: "As measured by annual surveys conducted by the Bee Informed Partnership and the Apiary Inspectors of America (AIA) and funded by the USDA, winter losses of managed honey bee colonies have been fluctuating in the range of about 22–33% annually between 2007 and 2014, which far exceeds the historical rate of about 10–15% prior to the mid-1980s. This increase in winter colony losses represents a serious threat to both beekeeping and to the production of agricultural crops that depend on honey bee pollination." To address the serious issue, there are many theories on the root causes. Now, most researchers focus on invasive Varroa mites, new or emerging diseases, chemical pesticide poisoning, stress which bees have experienced (e.g. from transportation), disturbance to the habitat of foraging, and/or potential immune-suppressing factors. There is some progress and improvement, but much more need to be done.

Figure 28.5 Honey bees with collected pollen. Source: Photos by Dr. Chungkun Shih.

Figure 28.6 A bumble bee sipping nectar. Source: Photo by Dr. Chungkun Shih.

Figure 28.7 A mantidfly with raptorial forelegs. Source: Photos by Jason Shih.

Figure 28.8 *Clavifemora rotundata* Jepson, Heads, Makarkin & Ren 2013 [29]. (Holotype, CNU-NEU-NN2011001).

28.3 Predation – Preying on Other Insects

28.3.1 Mantispid Lacewings with Raptorial Forelegs

Mantispidae, or "mantidflies" (also known as mantisflies), are a highly specialized group of hemerobiiform Neuroptera characterized by their raptorial forelegs, large eyes, and elongation of the pronotum behind the forelegs [24–26]. Their specialization extends beyond morphology to their behavior and life history, with some species being parasites of Araneae or aculeate Hymenoptera; the larvae developing within the egg cases of spiders (Figure 28.7) or inside the nests of wasps [27, 28].

Jepson *et al.* [29] described three genera and four species of the extinct mantidfly subfamily Mesomantispinae (Insecta: Neuroptera: Mantispidae) from the Lower Cretaceous Yixian Formation of Liaoning and the Middle Jurassic Jiulongshan Formation of Inner Mongolia: *Archaeodrepanicus nuddsi* Jepson, Heads, Makarkin & Ren 2013 (Figure 28.8), *A. acutus* Jepson, Heads, Makarkin & Ren 2013, *Sinomesomantispa microdentata* Jepson, Heads, Makarkin & Ren 2013, (Yixian Formation) and *Clavifemora rotundata* Jepson, Heads, Makarkin & Ren 2013, (Jiulongshan Formation).

These specimens represent the first Mesozoic mantidfly compression fossils to have body parts preserved, including the specialized raptorial forelegs articulated to the prothorax anteriorly, an autapomorphy of the family, which would have served the function of preying other insects. These new taxa further confirm the placement of the subfamily Mesomantispinae within the family Mantispidae; however, the monophyly of Mesomantispinae has not been confirmed, and it is likely that it will prove to be paraphyletic.

28.3.2 Dipteromantispidae with Raptorial Forelegs

Lacewings (Neuroptera) normally bear four well-developed wings. There are a few rachypterous, micropterous, or apterous species, found in several extant families; this wing reduction is usually associated with flightlessness. The only documented fossil neuropteran with reduced hind wings (modified to small haltere-like structures) is the enigmatic minute genus *Mantispidiptera* Grimaldi, 2000 from the Late Cretaceous amber of New Jersey [30].

Makarkin *et al.* [31] reported a taxa from the Lower Cretaceous Yixian Formation of China, *Dipteromantispa brevisubcosta* Makarkin, Yang & Ren, 2013 (Figure 28.9), resembling *Mantispidiptera*. These two genera are placed in the family Dipteromantispidae Makarkin, Yang & Ren, 2013. They bear well-developed forewings with reduced venation, and hind wings that are extremely modified as small structures resembling the halteres of Diptera. They also have specialized raptorial forelegs articulated to the prothorax anteriorly which would have served the function of preying other insects.

Dipteromantispidae might be specialized descendants of some early Berothidae or of stem group Mantispidae + Berothidae. We presume that dipteromantispids were active fliers. This is a remarkable example of parallel evolution of wing structures in this neuropteran family and Diptera.

Figure 28.9 *Dipteromantispa brevisubcosta* Makarkin, Yang & Ren, 2013 [31] having well-developed forewings with reduced venation, while hind wings extremely modified as small halters.

28.3.3 Hangingflies – Bittacidae and Cimbrophlebiidae

Bittacidae, a large extant family of Mecoptera, comprise 16 extant genera and about 270 species [32]. Extant bittacids commonly hang their bodies by their fore legs from branches on low vegetation, while catching soft-bodied insects with folded raptorial tarsal segments [33]. Male hangingflies are noted by presenting nuptial gifts, e.g. captured insects, to females during courtship or copulation.

Cimbrophlebiidae [34, 35] (Figure 28.10), a little-known, extinct family of Mecoptera, are considered to be the sister group of Bittacidae [36] (Figure 28.11), with which they constitute the infraorder Raptipeda [37, 38]. Both families have long, slender legs, each bearing a single, large, raptorial tarsal claw, and their wing venations are similar. But Cimbrophebiidae can be easily distinguished from Bittacidae by their long and branched 2A vein, in contrast to short and unbranched in Bittacidae [39]. Hangingflies have chewing mouthparts [35] (Figure 28.10) to consume their preys.

28.4 Blood Feeding by Ectoparasite Fleas

Ectoparasite insects mainly comprise two clades, lice and fleas. The fleas (Siphonaptera) are notorious for their roles in causing the black plague and killing a huge number of people in human history. Fleas are a group of highly specialized blood-feeding ectoparasitic insects characterized by small size, laterally compressed body, highly modified mouthparts specializing in blood feeding, and long and robust legs specialized in jumping. Six definite flea fossils have been described from Eocene and Miocene ambers: *Palaeopsylla klebsiana* Dampf, 1911 [40]; *Palaeopsylla dissimilis* Peus, 1968

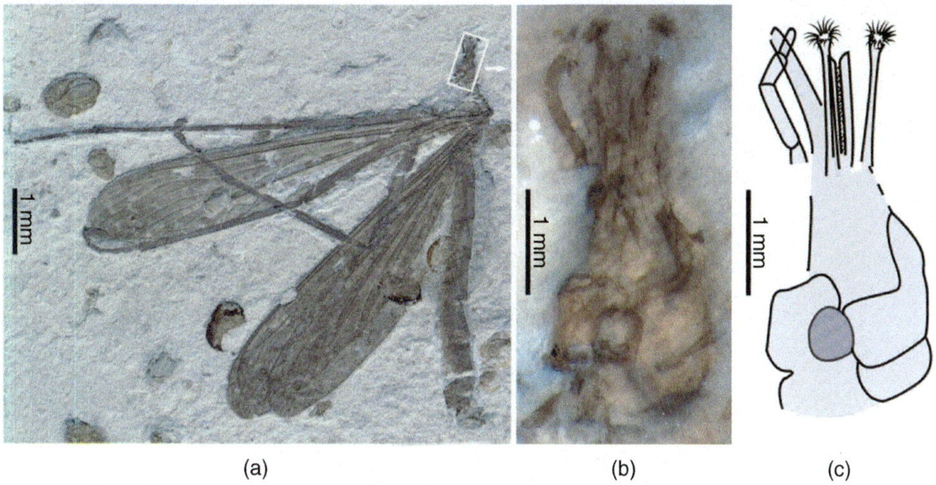

Figure 28.10 *Cimbrophlebia gracilenta* Zhang, Shih, Zhao & Ren, 2015 [35]. Photographs and line drawing. (a) Habitus of Holotype, CNU-MEC-NN2014077p; (b) Mouthparts, under alcohol; and (c) Line drawing of mouthparts.

Figure 28.11 *Exilibittacus lii* Yang, Ren & Shih, 2012 [36] in Bittacidae. (Holotype, CNU-M-NN2010001p).

[41]; *Peusianapsylla baltica* Beaucournu & Wunderlich, 2001 [42]; and *Peusianapsylla groehni* Beaucournu, 2003 [43] from Baltic amber [44]; and *Pulex larimerius* Lewis & Grimaldi, 1997 [45] and *Eospilopsyllus kobberti* Beaucournu & Perrichot, 2012 [46] from the Dominican Republic, all of which are very similar to extant fleas. An Eocene bird louse compression fossil [47, 48] has been reported.

Parasite–host associations among insects and mammals or birds are well attended by neontological studies [49]. But the origin, morphology, and early evolution of parasites and their associations with hosts are poorly known [26, 50] due to sparse records of putative ectoparasites with uncertain classification in the Mesozoic, most lacking mouthpart information and other critical details of the head morphology [51–54].

Huang *et al.* reported three to-be-named taxa of "giant fleas" and provided important information and insights to the morphology and early evolution of ectoparasites while suggesting their earliest hosts were hairy or feathered "reptilians" [55]. Gao *et al.* [56] described two primitive flea-like insects assigned to Pseudopulicidae, *Pseudopulex jurassicus* Gao, Shih & Ren, 2012 (Figure 28.12) from the latest Middle Jurassic (165 Mya) and *Paragomphus magnus* Gao, Shih & Ren, 2012 (Figure 28.13) from the Early Cretaceous (125 Mya) in Northeastern China. These specimens exhibit many features of insect ectoparasites, especially, the long serrated stylets (3.44 mm long for *P. jurassicus* and 5.15 mm for *P. magnus* vs. less than 1.0 mm for extant fleas) for piecing tough and thick skin or hides of hosts. These ectoparasites might have lived on and sucked blood of coexisting feathered dinosaurs, feathered pterosaurs or medium-sized mammals [56]. Huang *et al.* [57] reported three taxa, *Pseudopulex wangi*, *Hadropsylla sinica*, and *Tyrannopsylla beipiaoensis*, from the same stratum of China and classified them to Pseudopulicidae in Siphonaptera.

Gao *et al.* [56] demonstrates similarities among the new genus *Pseudopulex* and extant fleas, including robust and elongate piercing-sucking mouthparts with serrated stylets, small thorax without wings, strong scythe-shaped claws, and body covered with stiff, posteriorly directed setae. In addition, *Pseudopulex* possesses a pygidium with sensilia, which is found only in the extant fleas. However, *Pseudopulex* is different from extant fleas by the absence of ctenidia on the body but presence of ctenidia on the tibiae, lack of the modified jumping hind legs, and the laterally compressed

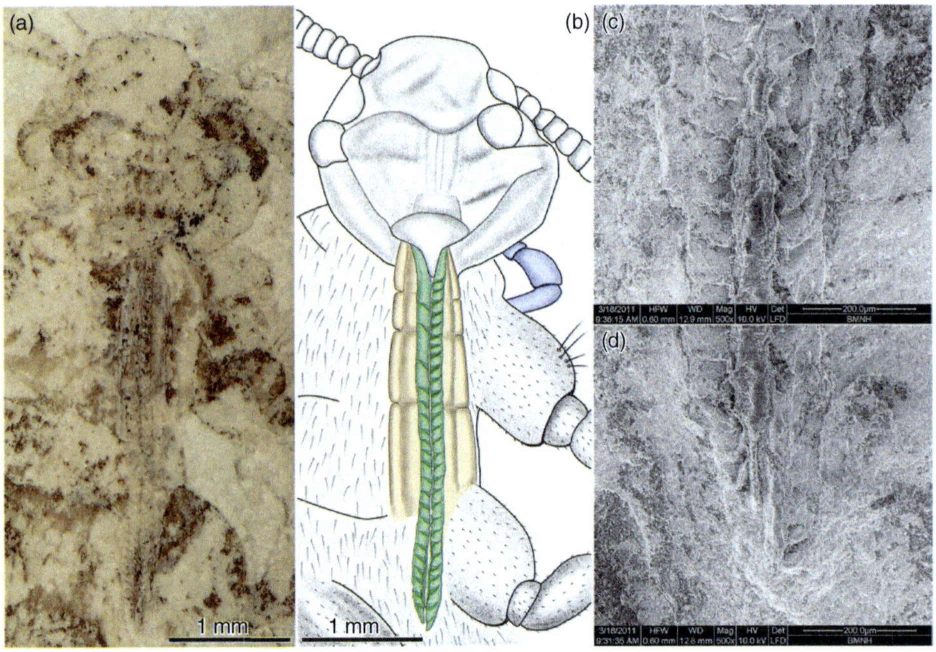

Figure 28.12 Details of *Pesudopulex jurassicus* Gao, Shih & Ren, 2012 (Holotype, CNU-NN2010001). (a) and (b) Photograph and line drawing of the mouthpart. (c) and (d) SEM of mouthpart, middle and tip parts. Source: Donated by Dr. Chungkun Shih. Modified from [56].

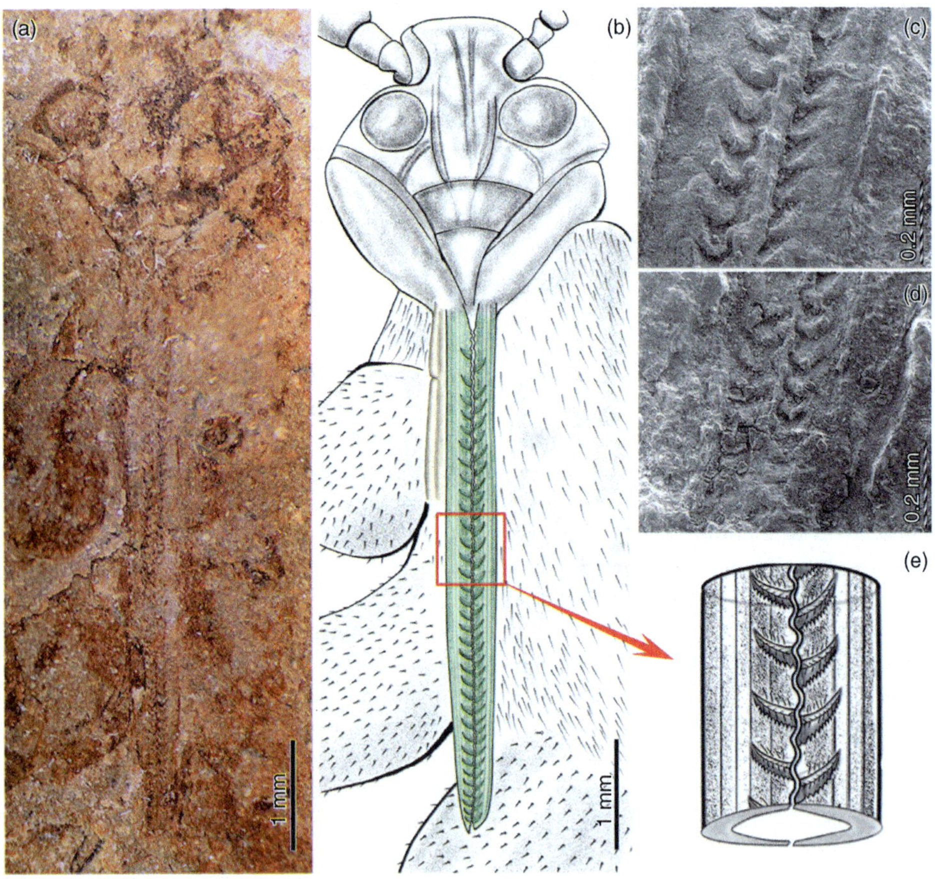

Figure 28.13 *Pseudopulex magnus* Gao, Shih & Ren, 2012 (Holotype, CNU-ND2010002). (a) and (b) Mouthpart photograph and line drawing. (c) and (d) SEM of mouthpart, middle, and tip parts. (e) Enlarged drawing of laciniae. Source: Modified from [56].

body. Furthermore, *Pseudopulex* has more-developed eyes, longer antennae with more segments, and larger body size than extant fleas (17 mm for *P. jurassicus* and 22.8 mm for *P. magnus* versus about 2 mm for extant fleas). They also document that Pseudopulicidae has some striking similarities with the most basal flea family Tungidae. They conclude that Pseudopulicidae is closely related to Siphonaptera and may represent a stem taxon of the flea lineage, but given that it lacks some diagnostic characters of Siphonaptera, it is deemed as Order *Incertae sedis* (of uncertain placement).

To elucidate ectoparasite–host associations, Gao *et al.* under the guidance of Dr. Xu, surveyed coexisting feathered dinosaurs, feathered pterosaurs, mammals, and birds in the contemporaneous ecosystems of the mid-Mesozoic. Their findings suggest that *Pseudopulex*, having very large body sizes and long serrated stylets, would not have found life on small rodent-like mammals with soft dermal tissues suitable owing to their relatively easy detection and removal. Based on the morphology of *Pseudopulex* and the survey of potential hosts, they proposed that these large insects might have fed on feathered dinosaurs (Figure 28.14) feathered pterosaurs, or medium-sized mammals [56].

In 1976, Ponomarenko described *Saurophthirus longipes* as a pre-flea having long legs, haired body, and other important ectoparasitic characters, but lacking some detailed structures of the mouthparts, from the Early Cretaceous of Baissa in Siberia [52]. Gao *et al.* reported a new transitional flea, *Saurophthirus exquisitus* Gao, Shih & Ren, 2013, assigned to Saurophthiridae in Siphonaptera, from the Lower Cretaceous Yixian Formation of Northeastern China [58]. Saurophthirids are more similar to crown fleas than other stem fleas in having a relatively small body size, relatively short and slender piercing-sucking stylet mouthparts, comparably short and compact antennae, rows of short and stiff bristles on the thorax, and highly elongated legs. The new finding greatly improves our understanding of the morphological transition to the highly specialized body plan of extant fleas (Figure 28.15).

Figure 28.14 Three-D reconstruction drawing of a flea-like ectoparasite, *Pseudopulex jurassicus* Gao, Shih & Ren, 2012 [56], parasitizing a coexisting dinosaur, *Pedopenna daohugouensis*, with downy and long pennaceous feathers during the Middle Jurassic in Northeastern China. Source: Artwork by Dr. Chen Wang.

However, saurophthirids also display several features unknown in other fleas, and some of these features are suggestive of a possible ectoparasitic relationship to contemporaneous pterosaurs, though other possibilities exist. Based on well-preserved male and female specimens, we are able to document a striking dimorphic difference of the species (Figure 28.16), in conjunction with previous discoveries, highlight a broad diversity of ectoparasitic insects in the mid-Mesozoic.

Gao *et al.* [59] reported a new flea with transitional characters, *Pseudopulex tanlan* Gao, Shih, Rasnitsyn & Ren, 2014 assigned to Pseudopulicidae, from the Lower Cretaceous Yixian Formation of Liaoning Province, China (Figure 28.17). Different from the previously described pseudopulicids, *P. tanlan* has relatively smaller body size but lacking any ctenidia on the tibiae or body, while the male with comparatively smaller and shorter genitalia. On the other hand, *P. tanlan* has some characters similar to the transitional fleas of saurophthirids, such as, a small head, short compacted antennae, small pygidium and many stiff setae covering the body. Even though other possibilities can not be ruled out, the female specimen with extremely distended abdomen suggests that it might have consumed its last meal before its demise. Compared with other reported female flea fossils, we calculate and estimate that *P. tanlan* might have consumed 0.02 milliliter (ml) of blood, which is about 15 times of the intake volume by extant fleas (Figure 28.18). These new findings further support that fleas had evolved a broad diversity by the Early Cretaceous.

28.5 Blood Feeding by True Bugs

Blood-feeding insects, as vectors of disease for humans and livestock alike, have garnered significant interest [49, 60], but our understanding of their early evolution is hindered by the scarcity of available material and the difficulty in distinguishing early hematophages from nonblood-feeding relatives. Yao *et al.* [61] reported two true bugs in Torirostratidae Yao, Cai, Shih & Engel,

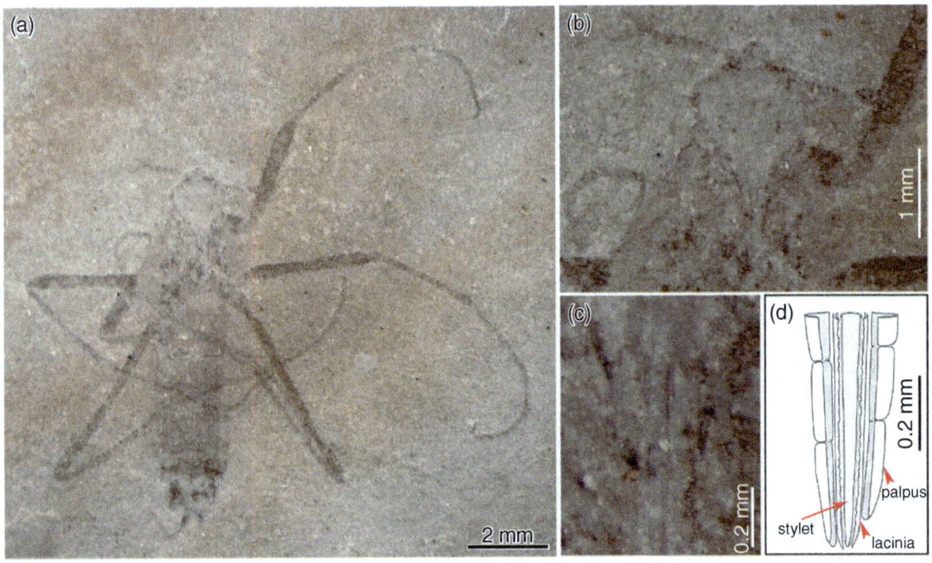

Figure 28.15 *Saurophthirus exquisitus* Gao, Shih, Rasnitsyn & Ren, 2013 (Holotype, CNU-LL2010016P). (a) Habitus; (b) Head; (c) Basal part of the mouthparts; and (d) Line drawing of the terminal of the mouthparts. Source: Donated by Dr. Chungkun Shih. Modified from [58].

Figure 28.16 *Saurophthirus exquisitus* Gao, Shih, Rasnitsyn & Ren, 2013, (Female, Holotype, CNU-LB2010017). (a) Habitus of female; (b) Left eye and antenna under alcohol; (c) Mouthparts under alcohol. Source: Modified from [58].

2014, *Torirostratus pilosus* Yao, Shih & Engel, 2014 (Figure 28.19) and *Flexicorpus acutirostratus* Yao, Cai & Engel, 2014, from the Lower Cretaceous Yixian Formation in Northeastern China.

Blood-feeding insects receive a high iron load from their blood meals, and portions of the iron are incorporated into the tissues of the blood feeders. Utilizing energy-dispersive X-ray spectroscopy (EDS) to analyze the geochemical composition of iron (Fe), the Fe contents of the insect bodies are significantly higher than those in the surrounding matrix. The Fe contents of the fossils are consistent with those of modern Heteroptera.

Figure 28.17 *Pseudopulex tanlan* Gao, Shih, Rasnitsyn & Ren, 2014, (Female, Holotype, CNU-SIP-LL2013002). (a) Photograph; (b) Line drawing; (c) The segmental boundary of the abdomen; (d) Terminalia of the abdomen, ((c) and (d), under alcohol). Source: Modified from [59].

The geochemical analysis suggests that the two species of true bugs had distinctly higher Fe concentrations than those of other lineages, suggesting a concentrated source such as from a blood meal.

Furthermore, morphological and taphonomic data, i.e. 93.0% of vetanthocorids (predaceous bugs), 318 out of 342, are preserved with rostra in outstretched positions [62]. But for fossils of phytophagous bugs, 97.7% of specimens (1768 out of 1809) have rostra appressed to their bodies. Mouthparts of 11 (38%) of the 29 specimens of *Torirostratus* are preserved with the rostra in a laterally stretched orientation and one (3.4%) in an anteriorly extended position. For the 368 specimens of *Flexicorpus*, 115 (31.3%) are preserved with the rostra laterally oriented, while 23 (6.3%) are extended anteriorly.

Generally, modern phytophagous bugs have a long and relatively slender rostrum, usually appressed to the body at rest, which is straight, with more or less parallel sides and/or uniform thickness, while predaceous or hematophagous bugs usually have stout, curving rostra, tapering from base to apex [63]. The rostra of the two fossil taxa are stout with the basal segment swollen, the second segment longest and swollen at its base, and the fourth segment acute distally. The structure of the rostra and their preserved orientations suggest that these new species were predaceous or hematophagous.

Combining all the morphological and taphonomic data and higher Fe content, Yao *et al.* [61] concluded these new species represent the earliest evidence of blood feeding among true bugs, extending the geological record of such lineages by approximately 30 million years. Remarkably, one of the fossil bugs appears to have perished immediately following a blood meal, which may have been from coexisting mammals, birds, or avian-related dinosaurs (Figure 28.20). These records expand the phylogenetic and ecological diversity of

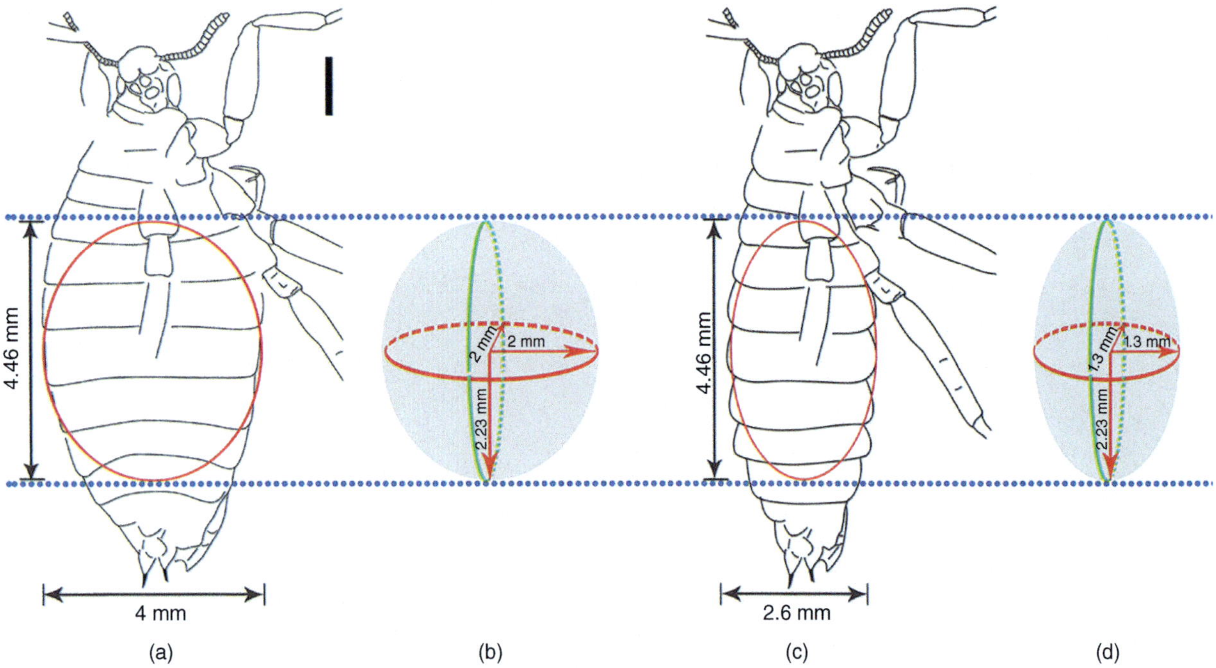

Figure 28.18 Illustration of calculating the intake volume by holotype *Pseudopulex tanlan* Gao, Shih, Rasnitsyn & Ren, 2014. (a) and (b) Selected parts of the fully distended abdomen; (c) and (d) Selected parts of the pseudomorph of non-swelling condition. In our ellipsoids, b = c, the volume change should be: $V1 = 4/3 \times 3.14 \times 2.23 \times 2 \times 2 - 4/3 \times 3.14 \times 2.23 \times 1.3 \times 1.3 \approx 22\ mm^3$. Therefore, we estimate the volume intake by *Yixianpulex tanlan* is nearly 0.02 cc (ml) blood. Source: Modified from [59].

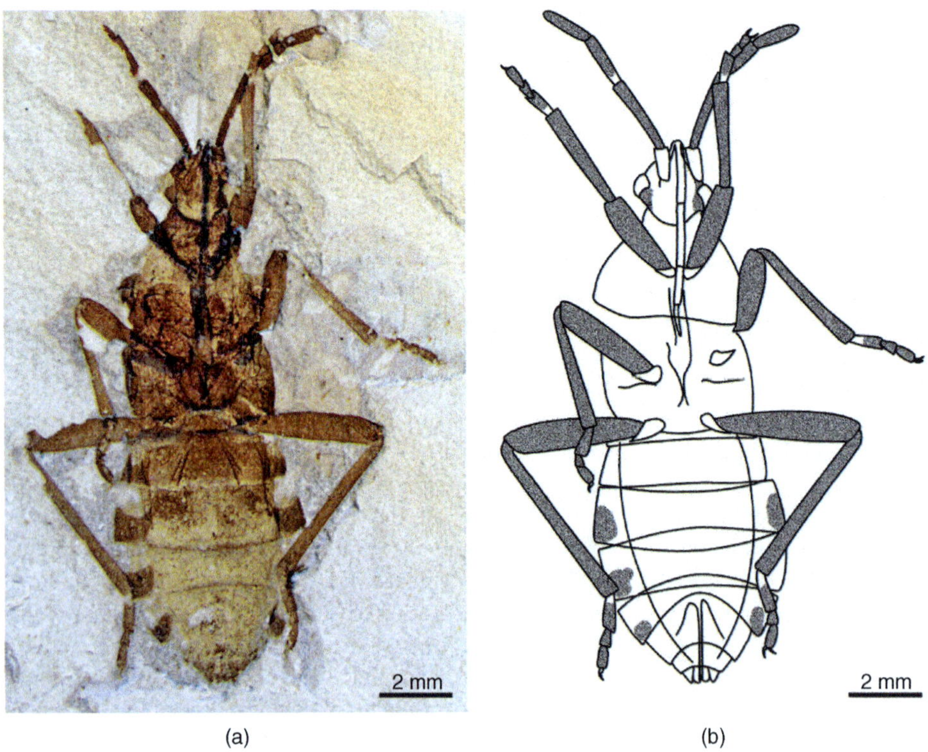

Figure 28.19 *Torirostratus pilosus* Yao, Shih & Engel, 2014, (Paratype, female, CNU-Het-LB2010134p) [61].

Figure 28.20 Ecological reconstruction of blood-feeding true bugs. Several torirostratid true bugs depicted feeding on blood from a sleeping feathered dinosaur in the Early Cretaceous of Northeastern China. Source: Artwork by Dr. Chen Wang.

blood-feeding insects in the Early Cretaceous, enriching our knowledge of paleoecological associations in these ancient environments.

28.6 Herbivores: Feeding on Plant Matter as Evidenced by Insect Damage Types

There are many extant insect herbivores based on observing insect feeding behaviors or examining insect gut contents. For fossil insects, especially those extinct taxa, it is a rather difficult task to obtain direct fossil evidence of herbivorous insects' feeding behaviors, albeit a few gut content inside fossilized insects.

By observing the insect plant DTs as preserved on well-preserved fossil plants, herbivorous insect behaviors can be inferred. Wang *et al.* reported piercing and sucking damage (DT48) and margin feeding (DT12) on Ginkgoite leaves of *Yimaia capituliformis* from the Middle Jurassic Jiulongshan Formation of Northeastern China (Figure 28.21) [64].

Ding *et al.* [65] documented that insect plant DTs were categorized for 343 specimens of the host broad-leaved conifer, *Liaoningocladus boii* Sun, Zheng & Mei 2000 from the Lower Cretaceous Yixian Formation. One of these DTs represented linear patches of intercostal

Figure 28.21 Insect herbivore damage on Ginkgoites leaves of *Y. capituliformis* from the Middle Jurassic Jiulongshan Formation of Northeastern China. Insets (b) and (d) are enlargements of damaged areas circled from respective leaves and using the damage type (DT) system. (a) and (b) Piercing-and-sucking damage (DT48; CNU-PLA-NN-2010-044). (c) and (d) Margin feeding (DT12; CNU-PLA-NN-2010-521). Source: Modified from [64].

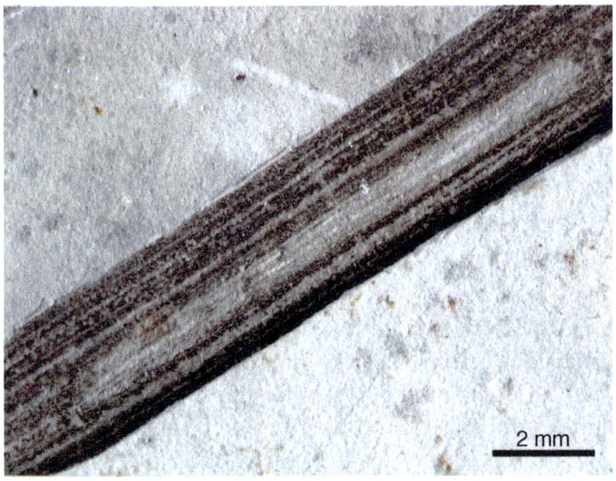

Figure 28.22 Window feeding and surface abrasion (DT103) on *Liaoningocladus boii* Sun, Zheng & Mei, 2000 [66], CNU-PLA-LL-2010-230-1 from the Yixian Formation of Northeastern China.

window feeding, assigned to DT103 likely produced by adults of a leaf miner (Figure 28.22).

Ding *et al.* [67] reported that insect-mediated damage was examined on 756 specimens of three broadleaved conifers *Podozamites, Lindleycladus,* and *Liaoningocladus,* originating from five mid-Mesozoic localities in Northeastern China. These localities are the Late Triassic Yangcaogou Fm. (T3, *ca.* 205 Mya), the latest Middle Jurassic Jiulongshan Fm. (J2, 165 Mya), and the Lower Cretaceous Yixian Fm. (K1, 125 Mya). Plant hosts from these three time intervals harbor five functional feeding groups (FFGs) of herbivores and 23 distinctive DTs, categorized using the widely applied DT system. The DTs were classified into the five FFGs of external foliage feeding (DT01, DT12, DT13, DT30, DT103) and piercing and sucking (DT46, DT48, DT77, DT128, DT138) (Figures 28.23 and 28.24). Potential

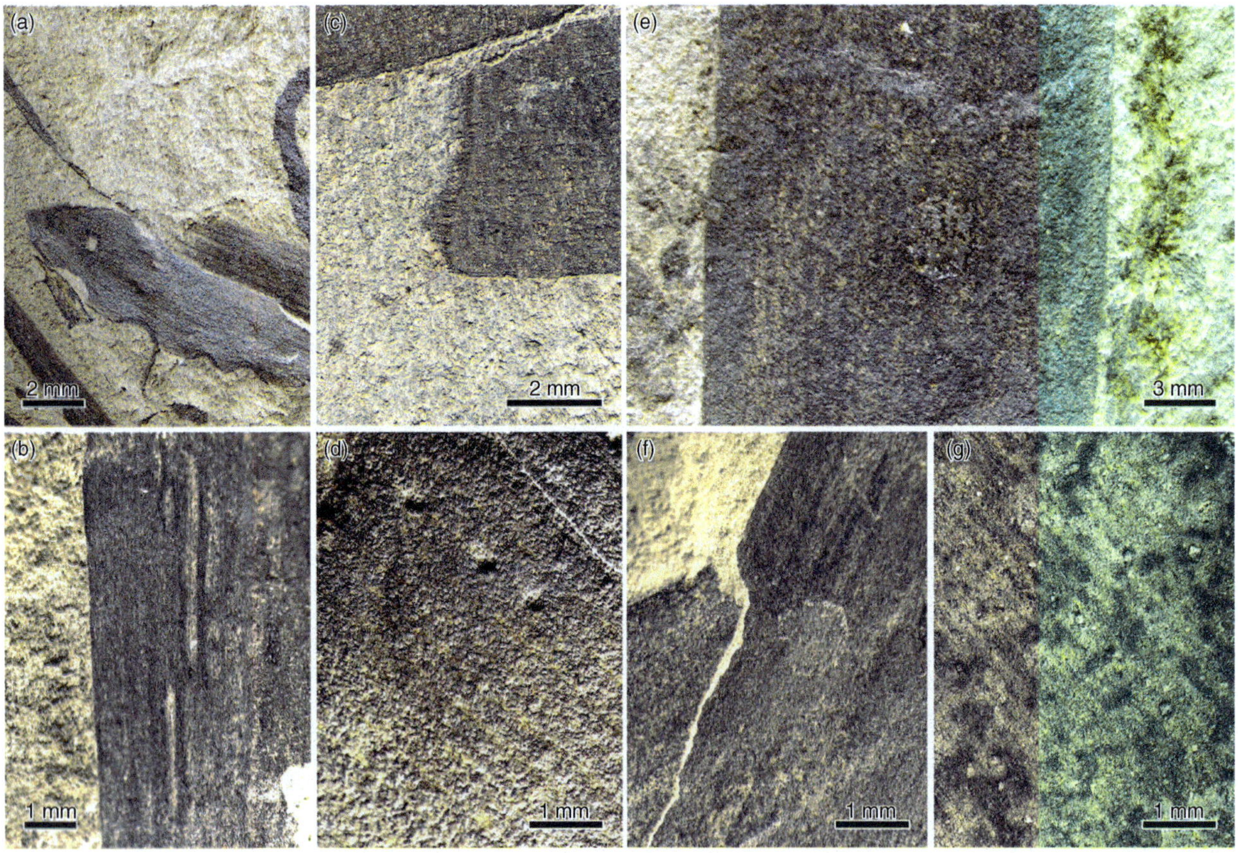

Figure 28.23 Damage types on the broadleaved conifer *Podozamites lanceolatus–Lindleycladus lanceolatus* species complex from the Late Triassic (T3) Yangcaogou Formation. All specimens are from the Yangcaogou locality, Liaoning Province, in Northeastern China. (a) External foliage feeding, consisting of whole feeding (DT01) and margin feeding (DT12) on CNU-CON-LB-2010-103-22. Note reaction rims on DT12. (b) External foliage feeding of linear, slot-like, surface abrasion (DT103) on CNU-CON-LB-2010-042-6. (c) External foliage feeding of margin feeding (DT13) on CNU-CON-LB-2010-106-1. Note veinal stringers at cut leaf margin. (d) Piercing and sucking (DT138) targeting vein vascular tissue on CNU-CON-LB-2010-041-1. Note linear trajectory of punctures that parallel venation. (e) Piercing and sucking indicated by a circular scalemark (DT77) covering the leaf surface of CNU-CON-LB-2010-043P-2. (f) External foliage feeding of surface abrasion (DT30) on CNU-CON-LB-2010-057-1. (g) Piercing and sucking of isolated, randomly positioned punctures (DT46) on surface of CNU-CON-LB-2010-065C-1 [67].

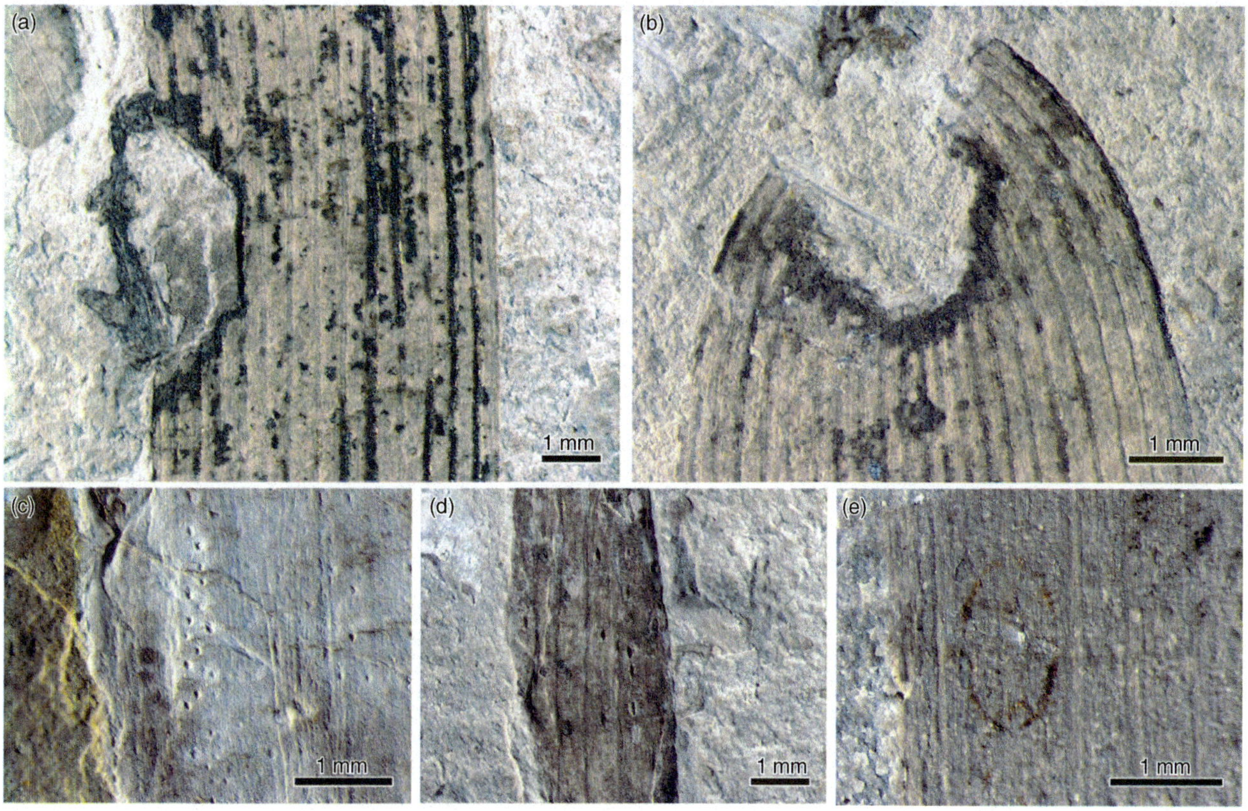

Figure 28.24 Damage on the broadleaved conifer *Podozamites lanceolatus*–*Lindleycladus lanceolatus* species complex from the latest Middle Jurassic (J2) Jiulongshan Formation. All specimens are from the Daohugou 1 and 2 localities from the Inner Mongolia Autonomous Region. (a) External foliage feeding of DT12, a type of margin feeding, on CNU-CON-NN-2011-421P-1. (b) External foliage feeding of DT13, another type of margin feeding, on CNU-CON-NN-2009-646P-1. Note well-developed reaction callus. (c) Piercing and sucking consisting of random punctures (DT46) and linear rows of probes (DT138) on CNU-CON-NN-2011-314P-1. (d) Piercing and sucking (DT48) on CNU-CON-NN-2011-028-1. (e) Piercing and sucking represented by a scale impression mark (DT128) on CNU-CON-NN-2011-633-1 [67]. Note the two concentric lineations on this structure.

Figure 28.25 A cicada might have caused the damages of piercing and sucking.

insects making the damages of piercing and sucking were cicadas and hoppers (Figure 28.25), while foliage feeding likely by grasshoppers with chewing mandibles (Figure 28.26).

Damage-type richness and abundance was established for each FFG, encompassing from 10 to 16 DTs for each of the three time intervals examined. For this 80 million-year-long interval, foliar herbivory on broadleaved conifers was transformed from early predominance of external foliage feeding (T3), later replaced by an emphasis on piercing and sucking (J2), followed by bimodal expansion of endophytic interactions from oviposition and leaf mining (K1).

This trajectory of herbivore succession indicates that, from T3 to K1, plant–insect associations were transformed from earlier reliance on a greater number of exophytic modes of herbivory to a later, increased variety in endophytic consumption. The transformation also was demonstrated by finer-grained partitioning of food resources and specialization on particular host-plant tissue types. This subdivision of tissue types likely promoted greater dietary saturation of tissue space by FFGs.

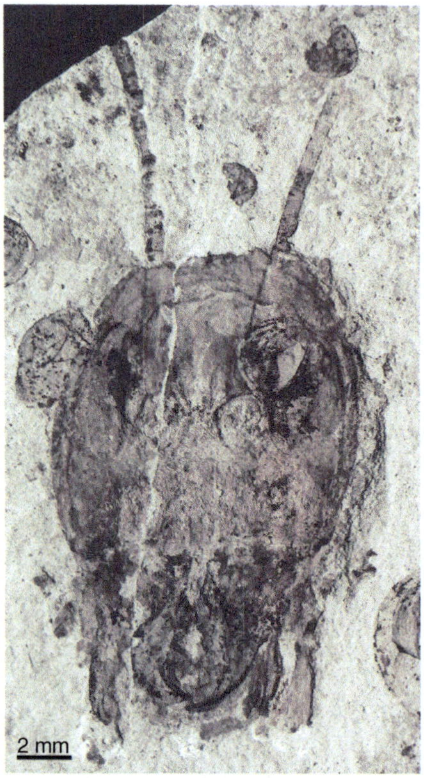

Figure 28.26 Head of *Allaboilus gigantus* Ren & Meng, 2006. Source: Photo by Dr. Chungkun Shih.

Possible explanations for these shifts in herbivory include ecological causes, long-term environmental changes, or both. Ecological factors, such as (i) evolution of a more differentiated plant-host spectrum available for consumption; (ii) long-term changes in plant physiognomy and deployment of anti-herbivore defenses; (iii) change in herbivore partitioning of plant-host tissues; and (iv) emergence of the parasitoid guild for efficient regulation of insect herbivores. Long-term environmental variables may be linked to these shifts in insect herbivory style.

References

1 Proctor, M., Yeo, P., and Lack, A. (1996). *The Natural History of Pollination*. Portland, OR: Timber Press.

2 Grimaldi, D.A. (1999). The co-radiations of pollinating insects and angiosperms in the Cretaceous. *Annals of the Missouri Botanical Garden* **86**: 373–406.

3 Gorelick, R. (2001). Did insect pollination cause increased seed plant diversity? *Biological Journal of the Linnean Society* **74**: 407–427.

4 Labandeira, C.C. (1997). Insect mouthparts: ascertaining the paleobiology of insect feeding strategies. *Annual Review of Ecology and Systematics* **28**: 153–193.

5 Krassilov, V.A., Rasnitsyn, A.P., and Afonin, S.A. (2007). Pollen eaters and pollen morphology: coevolution through the Permian and Mesozoic. *African Invertebrates* **48** (1): 3–11.

6 Labandeira, C.C., Kvaček, J., and Mostovski, M.B. (2007). Pollination drops, pollen, and insect pollination of Mesozoic gymnosperms. *Taxon* **56** (3): 663–695.

7 Axsmith, B.J., Krings, M., and Waselkov, K. (2004). Conifer pollen cones from the Cretaceous of Arkansas: implications for diversity and reproduction in the Cheirolepidiaceae. *Journal of Paleontology* **78** (2): 402–409.

8 Byers, G.W. and Thornhill, R. (1983). Biology of the Mecoptera. *Annual Review of Entomology* **28**: 203–228.

9 Porsch, O. (1957). Alte Insektentypen als Blumenausbeuter. *Plant Systematics and Evolution* **104**: 115–164.

10 Beutel, R.G. and Baum, E. (2008). A longstanding entomological problem finally solved? Head morphology of Nannochorista (Mecoptera, Insecta) and possible phylogenetic implications. *Journal of Zoological Systematics and Evolutionary Research* **46**: 346.

11 Ren, D., Labandeira, C.C., Santiago-Blay, J.A. *et al.* (2009). A probable pollination mode before angiosperms: Eurasian, long-proboscid scorpionflies. *Science* **326**: 840–847.

12 Ren, D., Labandeira, C.C., and Shih, C.K. (2010). New Mesozoic Mesopsychidae (Mecoptera) from

Northeastern China. *Acta Geologica Sinica – English Edition* **84** (4): 720–731.

13 Ren, D., Shih, C.K., and Labandeira, C.C. (2011). A well-preserved aneuretopsychid from the Jehol Biota in China (Insecta: Mecopteroidea: Aneuretopsychidae). *Zookeys* **129**: 17–28.

14 Ren, D., Shih, C.K., and Labandeira, C.C. (2010). New Jurassic Pseudopolycentropodids from China (Insecta: Mecoptera). *Acta Geologica Sinica – English Edition* **84** (1): 22–30.

15 Nepi, M., von Aderkas, P., Wagner, R. *et al.* (2009). Nectar and pollination drops: how different are they? *Annals of Botany* **104**: 205–219.

16 Dehgan, B. and Dehgan, N.B. (1988). Comparative pollen morphology and taxonomic affinities in Cycadales. *American Journal of Botany* **75**: 1501–1516.

17 Tekleva, M.V. and Krassilov, V.A. (2009). Comparative pollen morphology and ultrastructure of modern and fossil gnetophytes. *Review of Palaeobotany and Palynology* **156**: 130–138.

18 Yang, Q., Wang, Y.J., Labandeira, C.C. *et al.* (2014). Mesozoic lacewings from China provide phylogenetic insight into evolution of the Kalligrammatidae (Neuroptera). *BMC Evolutionary Biology* **14**: 126. https://doi.org/10.1186/1471-2148-14-126.

19 Labandeira, C.C., Yang, Q., Santiago-Blay, J.A. *et al.* (2016). The evolutionary convergence of mid-Mesozoic lacewings and Cenozoic butterflies. *Proceedings of the Royal Society B: Biological Sciences* **283**: 20152893. https://doi.org/10.1098/rspb.2015.2893.

20 Krenn, H.W. and Kristensen, N.P. (2000). Early evolution of the proboscis of Lepidoptera: external morphology of the galea in basal glossatan moths, with remarks on the origin of the pilifers. *Zoologischer Anzeiger* **239**: 179–196.

21 Artz, D.R. (2013). Nesting site density and distribution affect *Osmia lignaria* (Hymenoptera: Megachilidae) reproductive success and almond yield in a commercial orchard. *Insect Conservation and Diversity* **6** (6): 715–724. https://doi.org/10.1111/icad.12026.

22 Morse, R.A. and Calderone, N.W. (2000). *The Value of Honeybees as Pollinators of US Crops in 2000.* Cornell University.

23 USDA Report (2015) Colony collapse disorder and honey bee health action plan by CCD and Honey Bee Health Steering Committee on May 21, 2015. United States Department of Agriculture. https://www.ars.usda.gov/is/br/ccd/ccd_actionplan.pdf (Retrieved 23 September 2018).

24 Lambkin, K.J. (1986). A revision of the Australian Mantispidae (Insecta: Neuroptera) with a contribution to the classification of the family. I. General and Drepanicinae. *Australian Journal of Zoology, Supplementary Series* (116): 1–142.

25 New, T.R. (1989). Planipennia: Lacewings. 1–132. In: *Handbuch der Zoologie: Ein Naturgeschichte der Stämme des Tierreiches, Band IV Arthropoda: Insecta, Teilband 30* (ed. M. Fischer). Berlin: Walter de Gruyter.

26 Grimaldi, D.A. and Engel, M.S. (2005). *Evolution of the Insects.* New York: Cambridge University Press, 755 pp.

27 Redborg, K.E. (1998). Biology of the Mantispidae. *Annual Review of Entomology* **43**: 175–194.

28 Wedmann, S. and Makarkin, V.N. (2007). A new genus of Mantispidae (Insecta: Neuroptera) from the Eocene of Germany, with a review of the fossil record and palaeobiogeography of the family. *Zoological Journal of the Linnean Society* **149**: 701–716.

29 Jepson, J.E., Heads, S.W., Makarkin, V.N., and Ren, D. (2013). New Fossil Mantidflies (Insecta: Neuroptera: Mantispidae) from the Mesozoic of Northeastern China. *Palaeontology* **56** (3): 603–613.

30 Grimaldi, D.A. (2000). A diverse fauna of Neuropterodea in amber from the Cretaceous of New Jersey. In: *Studies on Fossil in Amber, with Particular Reference to the Cretaceous of New Jersey* (ed. D.A. Grimaldi), 259–303. Leiden: Backhuys Publishers.

31 Makarkin, V.N., Yang, Q., and Ren, D. (2013). A new Cretaceous family of enigmatic two-winged lacewings (Neuroptera). *Fossil Record* **16** (1): 67–75. https://doi.org/10.1002/mmng.201300002.

32 Krzeminski, W. (2007). A revision of Eocene Bittacidae (Mecotera) from Baltic amber, with the description of a new species. *African Invertebrates* **48**: 153–162.

33 Petrulevicius, J.F., Huang, D.-Y., and Ren, D. (2007). A new hangingfly (Insecta: Mecoptera: Bittacidae) from the Middle Jurassic of Inner Mongolia, China. *African Invertebrates* **48**: 145–152.

34 Yang, X.G., Shih, C.K., Ren, D., and Petrulevičius, J.F. (2012). New Middle Jurassic hangingflies (Insecta: Mecoptera) from Inner Mongolia, China. *Alcheringa: An Australasian Journal of Palaeontology* **36** (2): 195–201. https://doi.org/10.1080/03115518.2012.622143.

35 Zhang, X., Shih, C.K., Zhao, Y.Y., and Ren, D. (2015). New species of Cimbrophlebiidae (Insecta: Mecoptera) from the Middle Jurassic of Northeastern China. *Acta Geologica Sinica – English Edition* **89** (5): 1482–1496.

36 Yang, X.G., Ren, D., and Shih, C.K. (2012). New fossil hangingflies (Mecoptera, Raptipeda, Bittacidae) from

the Middle Jurassic to Early Cretaceous of Northeastern China. *Geodiversitas* **34** (4): 785–799. https://doi.org/10.5252/g2012n4a4.

37 Willmann, R., 1977. Mecopteran aus dem untereozänen Moler des Limfjordes (Dänemark). Neues Jahrbuch für Geologie und Paläontologie, Monatschefte, 735–744.

38 Willmann, R. (1989). Evolution und Phylogenetisches system der Mecoptera (Insecta: Holometabola). *Abhandlungen der Senckenbergischen Naturforschenden Gesellschaft* **544**: 1–153.

39 Archibald, S.B. (2009). New Cimbrophlebiidae (Insecta: Mecoptera) from the early Eocene at McAbee, British Columbia, Canada and Republic, Washington, USA. *Zootaxa* **2249**: 51–62.

40 Dampf, A. (1911). *Palaeopsylla klebsiana* n. sp., eine fossiler Floh aus dem baltischen Bernstein. *Schriften der Physikalisch-Okonomischen Gesellschaft zu Königsberg* **51**: 248–259.

41 Peus, F. (1968). Über die beiden Bernstein-Flöhe (Insecta, Siphonaptera). *Paläontologische Zeitschrift* **42**: 62–72.

42 Beaucournu, J.C. and Wunderlich, J. (2001). A third species of Palaeopsylla Wagner, 1903, from Baltic amber (Siphonaptera: Ctenophthalmidae). *Entomologische Zeitschrift* **111**: 296–298.

43 Beaucournu, J.C. (2003). Palaeopsylla groehni n. sp., quatrième espèce de puce connue de l'ambre de la Baltique (Siphonaptera, Ctenophthalmidae). *Bulletin of Society of Entomology France* **108**: 217–220.

44 Poinar, G.O. (1995). Fleas (Insecta, Siphonaptera) in Dominican Amber. *Medical Science Research* **23**: 789.

45 Lewis, R.E. and Grimaldi, D.A. (1997). A pulicid flea in Miocene amber from the Dominican Republic (Insecta: Siphonaptera: Pulicidae). *American Museum Novitates* **3205**: 1–9.

46 Perrichot, V., Beaucournu, J.C., and Velten, J. (2012). First extinct genus of a flea (Siphonaptera: Pulicidae) in Miocene amber from the Dominican Republic. *Zootaxa* **3438**: 54–61.

47 Wappler, T., Smith, V.S., and Dalgleish, R.C. (2004). Scratching an ancient itch: an Eocene bird louse fossil. *Proceedings, Biological Science* **271** (Suppl 5): S255–S258.

48 Dalgleish, R.C., Palma, R.L., Price, R.D., and Smith, V.S. (2006). Fossil lice (Insecta: Phthiraptera) reconsidered. *Systematic Entomology* **31**: 648–651.

49 Lehane, M.J. (2005). *The Biology of Blood-Sucking in Insects*, 2e. New York: Cambridge University Press, 336 pp.

50 Rasnitsyn, A.P. and Quicke, D.L.J. (2002). *History of Insects*, xii, 517. Boston: Kluwer Academic.

51 Riek, E.F. (1970). Lower Cretaceous fleas. *Nature* **227**: 746–747.

52 Ponomarenko, A.G. (1976). A new insect from the Cretaceous of Transbaikalia Ussr a possible parasite of Pterosaurians. *Paleontologicheskii Zhurnal* **3**: 102–106.

53 Jell, P.A. and Duncan, P.M. (1986). Invertebrates, mainly insects, from the freshwater, Lower Cretaceous, Koonwarra Fossil Bed (Korumburra Group), South Gippsland, Victoria. *Memoirs of the Association of Australasian Palaeontologists* **3**: 111–205.

54 Rasnitsyn, A.P. (1992). Strashila incredibilis, a new enigmatic mecopteroid insect with possible siphonapteran affinities from the Upper Jurassic of Siberia. *Psyche (Cambridge)* **99**: 323–333.

55 Huang, D.-Y., Engel, M.S., Cai, C.Y. *et al.* (2012). Diverse transitional giant fleas from the Mesozoic era of China. *Nature* **483** (7388): 201–204. https://doi.org/10.1038/nature10839.

56 Gao, T.P., Shih, C.K., Xu, X. *et al.* (2012). Mid-Mesozoic flea-like ectoparasites of feathered or haired vertebrates. *Current Biology* **22** (8): 732–735. https://doi.org/10.1016/j.cub.2012.03.012.

57 Huang, D.-Y., Engel, M.S., Cai, C.Y., and Nel, A. (2013). Mesozoic giant fleas from Northeastern China (Siphonaptera): taxonomy and implications for palaeodiversity. *Chinese Science Bulletin* 1–9. https://doi.org/10.1007/s11434-013-5769-3.

58 Gao, T.P., Shih, C.K., Rasnitsyn, A.P. *et al.* (2013). New transitional fleas from China highlighting diversity of Early Cretaceous Ectoparasitic insects. *Current Biology* **23** (13): 1261–1266. https://doi.org/10.1016/j.cub.2013.05.040.

59 Gao, T.P., Shih, C.K., Rasnitsyn, A.P. *et al.* (2014). The first flea with fully distended abdomen from the Early Cretaceous of China. *BMC Evolutionary Biology* **14**: 168–174. https://doi.org/10.1186/s12862-014-0168-1.

60 Adams, T.S. (1999). Hematophagy and hormone release. *Annals of the Entomological Society of America* **92**: 1–13.

61 Yao, Y.Z., Cai, W.Z., Xu, X. *et al.* (2014). Blood-feeding true bugs in the Early Cretaceous. *Current Biology* **24** (15): 1786–1792. https://doi.org/10.1016/j.cub.2014.06.045.

62 Yao, Y.Z., Cai, W.Z., and Ren, D. (2006). Fossil flower bugs (Heteroptera: Cimicomorpha: Cimicoidea) from the Late Jurassic of northeast China, including a new family, Vetanthocoridae. *Zootaxa* **1360**: 1–40.

63 Schuh, R.T. and Slater, J.A. (1995). *True Bugs of the World (Hemiptera:Heteroptera): Classification and Natural History*. Ithaca, NY: Cornell University Press.

64 Wang, Y.J., Labandeira, C.C., Shih, C.K. *et al.* (2012). Jurassic mimicry between a hangingfly and a ginkgo from China. *Proceedings of the National Academy of Sciences of the United States of America* https://doi.org/10.1073/pnas.1205517109.

65 Ding, Q.L., Labandeira, C.C., and Ren, D. (2014). Biology of a leaf miner (Coleoptera) on Liaoningocladus boii (Coniferales) from the Early Cretaceous of Northeastern China and the leaf-mining biology of possible insect culprit clades. *Arthropod Systematics & Phylogeny* **72**: 281–308.

66 Sun, G., Zheng, S.L., and Mei, S. (2000). Discovery of Liaoningocladus gen. nov. from the lower part of Yixian Formation (Upper Jurassic) in western Liaoning, China. *Acta Palaeontologica Sinica* **39** (suppl.): 200–208.

67 Ding, Q.L., Labandeira, C.C., Meng, Q.M., and Ren, D. (2015). Insect herbivory, plant-host specialization and tissue partitioning on mid-Mesozoic broadleaved conifers of Northeastern China. *Palaeogeography, Palaeoclimatology, Palaeoecology* **440**: 259–273. https://doi.org/10.1016/j.palaeo.2015.09.007.

29

Camouflage, Mimicry or Eyespot Warning

Chungkun Shih [1,2], Yongjie Wang [1], and Dong Ren [1]

[1] Capital Normal University, Haidian District, Beijing, China
[2] National Museum of Natural History, Smithsonian Institution, Washington, DC, USA

29.1 Introduction

In an active and well-functioned ecosystem, there are many varieties of organisms living, coexisting, competing and propagating. All organisms must feed to get nutrients and energy (Chapter 28), while avoiding being fed by other organisms. For example, during the Middle Jurassic and the Early Cretaceous in Northeastern China, as indicated by food-chain analysis and energy flow diagrams, most insects played the unfortunate roles of primary and secondary consumers, serving as food for tertiary and top consumers in the ecosystems (figures 23.2 and 23.5 in [1]). To grow, to survive and to propagate, these insects must adopt ways to avoid predation in their ecosystems.

A very common and effective defense by insects against predators is crypsis, which is the ability for insects to conceal their shape so that they can easily blend into their surroundings. For prey, crypsis is adopted to avoid detection by predators, but for predators, it is used to avoid detection by prey. Typical methods of crypsis include camouflage, mimicry or having a lifestyle of nocturnal or subterranean living.

Camouflage is the development of shape, pattern, or coloration by an insect to make itself hard to be observed or detected by predators or prey. Many insects have evolved cryptic body parts and/or coloration so that they visually resemble their surroundings with matched color and/or texture, e.g. a green katydid or a green hopper on green leaves (Figures 29.1 and 29.2). On the other hand, the disruptive camouflage is the usage of streaks of boldly contrasting colors or shades, making insects less visible by breaking up their outlines (Figure 29.3) [2].

Mimicry, the other type of crypsis, is defined as insects, to confuse or betray a predator, developing body parts so that they resemble an object that is common in the environment, but is of no inherent interest to a predator [3]. Mimicry is very common for insects, displayed in various stages of eggs (Figure 29.4), nymphs, larvae, pupae or adults. Leaf-mimicry has been adopted by some extant insects imitating angiosperm leaves as a form of crypsis, e.g. "leaf insects" of the family Phyllidae in the Phasmatodea; "leaf katydids" of the family Tettigoniidae in the Orthoptera; praying mantes of genera of *Gongylus* ("violin mantis") and *Deroplatys* ("dead-leaf mantis") in the Mantodea; and leaf-mimic butterflies of the family Nymphalidae (Figure 29.5) and moths of the family Notodontidae in the Lepidoptera.

Since eye-to-eye has a powerful impact on vertebrates, eyespots become an important anti-predation strategy for diurnal insects and other animals. In today's ecosystems, eyespots are still one of important self-defense methods for insects, especially for extant Lepidoptera. The eyespots can surprise and deter a potential predator (Figure 29.6) or to draw a predator's attention away from the most vulnerable body parts (Figure 29.7) [4].

29.2 Camouflage by Fossil Insects

There are many documented examples of camouflage by various fossil insects of Odonata, Orthoptera, Homoptera, Neuroptera and Mecoptera from the Early Cretaceous Jehol Biota and the Middle Jurassic Yanliao Biota in the Northeastern China. For easy recognition, we expanded the classifications of Neuroptera wing markings [5] and categorized camouflage by these five orders of fossil insects into six types of wing markings which helped to make themselves hard to see or be detected by predators or prey.

29.2.1 Irregular Light and Dark Patches Covering the Entire Wing

Some fossil insects have irregular splotches and markings on their wings that might have imitated insects' resting places of tree bark or branches in the eco-systems.

Rhythms of Insect Evolution: Evidence from the Jurassic and Cretaceous in Northern China, First Edition.
Edited by Dong Ren, Chungkun Shih, Taiping Gao, Yongjie Wang, and Yunzhi Yao.

Figure 29.1 A katydid with green color resembling the leaves. Source: Photo by Jason Shih.

Figure 29.2 A green hopper with green body and wing resembling the leaves. Source: Photo by Jason Shih.

Figure 29.3 A scorpionfly with streaks of boldly contrasting wing colors to break up its shape and outline. Source: Photo by Jason Shih.

Figure 29.4 Mimicry of eggs of a lacewing. Source: Photo by Jason Shih.

Figure 29.5 *Kallima inachus* (Doyère, 1840) ("dried-leaf butterfly") in Nymphalidae, with wings mimicking dried leaves. Source: Photo by Jason Shih.

Figure 29.6 *Caligo placidianus* Staudinger, 1887, ("Placid owl butterfly") in Nymphalidae, with giant owl-like eye spots, at Machu Picchu, Peru. Source: Photo by Dr. Chungkun Shih.

Figure 29.7 *Spindasis seliga* Fruhstorfer, 1912 in Lycaenidae, with rear part of the wing resembling the head. Source: Photo by Jason Shih.

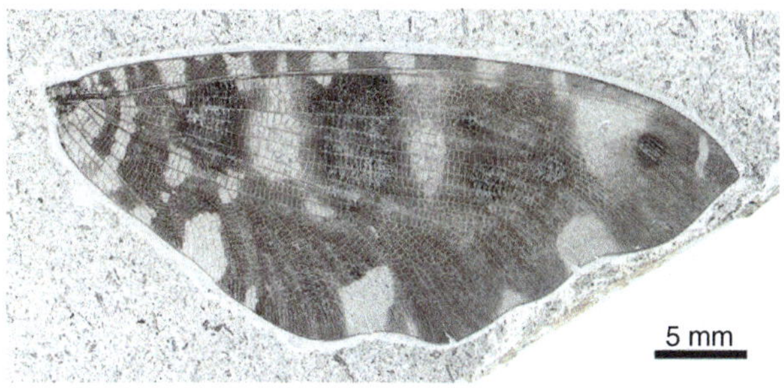

Figure 29.8 *Laccosmylus calophlebius* Ren & Yin, 2003 (CNU-NEU-NN1999013) [6].

Such irregular patches might have provided benefits to the insects to avoid detection by natural enemies or to ambush their prey. Here are some examples:

- *Laccosmylus calophlebius* Ren & Yin, 2003 of Saucrosmylidae, Neuroptera (Figure 29.8) [6].
- *Cretapsychops decipiens* Peng, Makarkin, Yang & Ren, 2010 of Psychopsidae, Neuroptera (Figure 20.34) [7].
- *Daohugoucossus shii* Wang, Ren & Shih, 2007 of Palaeontinidae, Homoptera (Figure 16.16) [8].

29.2.2 Irregular Dark or Light Markings on Part of the Margin and/or Center of Wing

Some fossil insects have irregular spots and markings on parts of their wings that have concealed and broken up their shape. Here are some examples:

- *Eoiocossus conchatus* (Wang, Ren & Shih, 2007) of Palaeontinidae, Homoptera (figure 14.5 in [1]) [9].
- *Archaeodrepanicus nuddsi* Jepson, Heads, Makarkin & Ren, 2013 of Mantispidae, Neuroptera (Figure 20.23) [10].
- *Sialium sinicus* Shi, Winterton & Ren, 2015 of Nymphidae, Neuroptera (Figure 20.28) [11].
- *Spilonymphes minor* Shi, Winterton & Ren, 2015 of Nymphidae, Neuroptera (Figure 29.9) [11].
- *Guithone bethouxi* Zheng, Ren & Wang, 2016 of Ithonidae, Neuroptera (Figure 20.17) [12].
- *Preanabittacus validus* Yang, Shih & Ren, 2012 of Bittacidae, Mecoptera (Figure 24.22) [13].
- *Perfecticimbrophlebia laetus* Yang, Shih & Ren, 2012 of Cimbrophlebiidae, Mecoptera (Figure 24.30) [13].
- *Miriholcorpa forcipata* Wang, Shih & Ren, 2013 of family *Incertae sedis*, Mecoptera (Figure 24.19) [14].

Figure 29.9 *Sialium minor* Shi, Winterton & Ren, 2015 (Paratype, CNU-NEU-LB2014003) [11].

- *Vitimopsyche pectinella* Gao, Shih, Labandeira, Santiago-Blay, Yao & Ren, 2016 of Mesopsychidae, Mecoptera (Figure 24.9) [15].

29.2.3 Dispersed Dark Spots Large or Small

Some fossil insects have dark spots and markings with various sizes irregularly distributed on their wings. Here are some examples:

- *Jurakempynus sinensis* Wang, Liu, Ren & Shih, 2011 of Osmylidae, Neuroptera (Figure 20.30) [16].
- *Protokalligramma bifaciatum* Yang, Makarkin & Ren, 2011 of Kalligrammatidae, Neuroptera (Figure 20.21) [17].
- *Formosibittacus macularis* Li, Ren & Shih, 2008 of Bittacidae, Mecoptera (Figure 24.24) [18].

Figure 29.10 *Choristopsyche perfecta* Qiao, Shih, Petrulevičius & Ren, 2013 (Holotype, CNU-MEC-NN-2011082) [19].

Figure 29.11 *Palaeothyridosmylus septemaculatus* Wang, Liu & Ren, 2009 (Holotype, CNU-NN99032) [20].

- *Choristopsyche perfecta* Qiao, Shih, Petrulevičius & Ren, 2013 of Choristopsychidae, Mecoptera (Figure 29.10) [19].

29.2.4 Dispersed Small Light Spots or Large Light Spots Surrounded by Small Dark Spots

Some fossil insects have light spots and markings with various sizes irregularly distributed on their wings. In some cases, large light spots are surrounded by small dark spots. Here are some examples:

- *Palaeothyridosmylus septemaculatus* Wang, Liu & Ren, 2009 of Osmylidae, Neuroptera (Figure 29.11) [20].
- *Orthophlebia nervulosa* Qiao, Shih & Ren, 2012 of Orthophlebiidae, Mecoptera (Figure 24.33) [21].
- *Orthobittacus maculosus* Liu, Shih & Ren, 2016 of Bittacidae, Mecoptera (figure 4 in [22]).

29.2.5 Regular Transverse (from Anterior to Posterior Margin) Stripes of Light and Dark Bands

There are many examples of this type of disruptive camouflage; the bands on the wings are evenly and regularly

Figure 29.12 *Grammolingia boi* Ren, 2002 (Holotype, NEU99001-1) [23].

distributed showing dark and light stripes. Although they seem to be similar, they differ in details resulting from a long process of natural selection.

- *Grammolingia boi* Ren, 2002 of Grammolingiidae Ren, 2002, Neuroptera (Figure 29.12) [23].
- *Leptolingia jurassica* Ren, 2002 of Grammolingiidae Ren, 2002, Neuroptera (figure 17.7 in [1]) [23].
- *Chorilingia euryptera* Shi, Wang, Yang & Ren, 2012 of Grammolingiidae Ren, 2002, Neuroptera (Figure 20.15) [24].
- *Grammolingia uniserialis* Shi, Wang & Ren, 2013 of Grammolingiidae Ren, 2002, Neuroptera (Figure 20.14) [25].
- *Saucrosmylus sambneurus* Ren & Yin, 2003 of Saucrosmylidae, Neuroptera (figure 17.10 in [1]) [6].
- *Undulopsychopsis alexi* Peng, Makarkin & Ren, 2011 of Psychosidae, Neuroptera (Figure 20.31) [26].
- A female *Parahagla sibirica* Sharov, 1968 of Prophalangopsidae, Orthoptera (figure 11.10 in [1]) [27].
- A male *Parahagla sibirica* Sharov, 1968 of Prophalangopsidae, Orthoptera (figure 11.12 in [1]) [27].
- *Pseudohagla shihi* Li, Ren & Wang, 2007 of Prophalangopsidae, Orthoptera (Figure 29.13) [28].
- *Bacharaboilus jurassicus* Li, Ren & Wang, 2007 of Prophalangopsidae, Orthoptera (figure 11.8 in [1]) [28].
- *Bacharaboilus lii* Gu, Qiao & Ren, 2011 of Prophalangopsidae, Orthoptera (Figure 9.11) [29].
- *Palaeontinodes reshuitangensis* Wang & Zhang, 2007 of Palaeontinidae, Homoptera (Figure 29.14) [30].
- *Cladocossus undulatus* Wang & Ren, 2009 of Palaeontinidae, Homoptera (figure 14.10 in [1]) [31].
- *Sinaeschnidia cancellosa* Ren, 1995 of Aeschnidiidae, Odonata (figure 6.5 in [1]) [32].

29.2.6 Regular Longitudinal (from Base to the Apex) Stripes of Light and Dark Bands

There are a few examples of this type of disruptive camouflage, the longitudinal bands on the wings are evenly

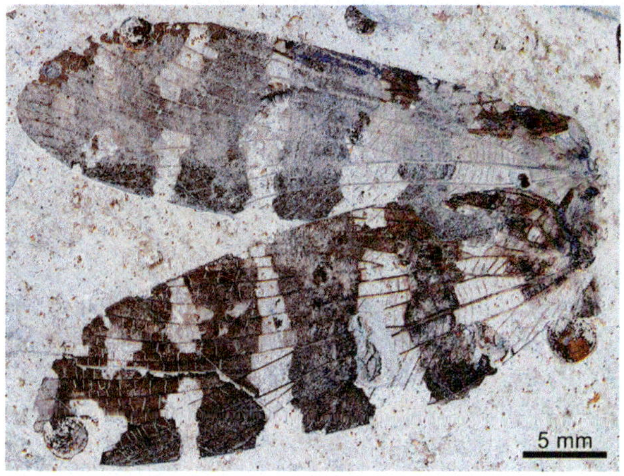

Figure 29.13 *Pseudohagla shihi* Li, Ren & Wang, 2007 of Prophalangopsidae (CNU-O-NN2006011) [28]. Source: Donated by Dr. Chungkun Shih.

Figure 29.14 *Palaeontinodes reshuitangensis* Wang & Zhang, 2007 (CNU-HEM-NN-2006008) [30]. Source: Donated by Dr. Chungkun Shih.

and regularly distributed showing dark and light stripes. Two examples are lacewings from the Middle Jurassic Jiulongshan Formation in Inner Mongolia, China (Figure 29.15).

29.3 Mimicry by Fossil Insects

Mimicry was first proposed in 1862 by an English naturalist, Dr. Bates, to explain the fact that the butterflies of different families have the same appearance in South America. Based on his observation that different butterflies from the same Amazon region had similar

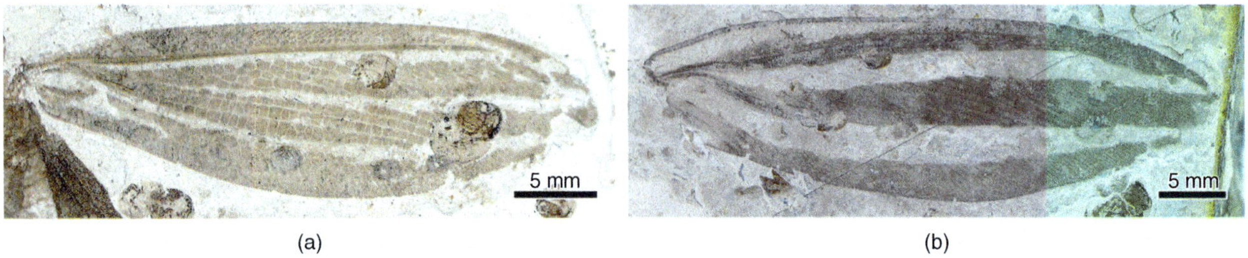

Figure 29.15 (a) *Arbusella magna* Khramov, Liu & Zhang, 2017 with four longitudinal stripes, and one row of small eyespots along the wing margin. (b) An undescribed ithonid, with three longitudinal stripes.

body color and stripe pattern, Bates proposed that this might be caused by and resulting from "natural selection" by predatory birds. In addition, some non-toxic butterflies mimic others with toxins or warning colors [33]. Fisher, in 1958, claimed that this was the most important evidence of "natural selection" since Darwin [34].

A typical mimicry system comprises three parties: mimic, model and dupe. To ensure that mimicry occurs, all three parties should, to some extent, co-exist in the same living ecosystem at the same time. However, some mimicry systems only consist of mimic and model [35]. According to the model for imitation, mimicry is classified into the following types: shape mimicry, color mimicry, acoustic mimicry, optical mimicry, behavioral mimicry, chemical mimicry, etc. In addition, based on scientists' last names, there are Batesian mimicry, Müllerian mimicry, Poultonian mimicry and Wasmannian mimicry.

There is obvious mimicry in Orthoptera, Phasmatodea, Mantodea, Homoptera, Heteroptera, Coleoptera, Lepidoptera, Mecoptera, Diptera, etc., including groups of phytophagy, predation, parasitism and symbiosis. Many studies have indicated that the earliest insect mimicry was found in the Carboniferous. Since then, mimicry co-evolution developed between insects and predators, and between insects and plants.

29.3.1 Ancient Pinnate Leaf Mimicry among Lacewings

A pair of well-preserved lacewing fossils whose wings resemble the feather-like pattern of gymnosperm leaves, *Bellinympha filicifolia* Wang, Ren, Liu & Engel, 2010 and *B. dancei* Wang, Ren, Shih & Engel, 2010, have come to light in Northeastern China (Figure 29.16). Hailing from a bygone era in the Middle Jurassic, the 165 Mya fossil lacewings represent the earliest evidence of leaf mimicry among insects [36]. Many extant insects, including some mantises, treehoppers, and butterflies, mimic the leaves of angiosperm flowering plants to hide from predators. The historical origin of this adaptive mimicry has remained unclear because of a dearth of fossil findings. In 2010, Wang *et al.* [36] documented the remains of two lacewings whose elongated forewings bear undulating margins, coloration resembling leaflets, complex venation, and branches resembling leaf rachides. The

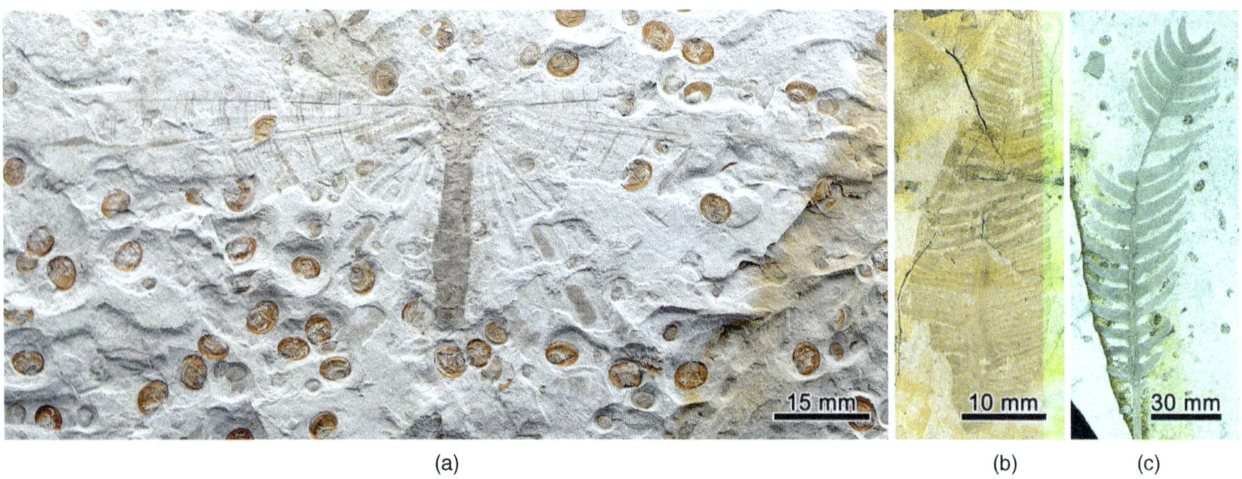

Figure 29.16 (a) *Bellinympha dancei* Wang, Ren, Shih & Engel, 2010 with forewings depicting pinnate leaf markings. (b) Forewing of *B. filicifolia* also depicting distinct pinnate leaf markings. (c) Potential pinnate-leaved model plant of *Nilssonia* (Cycadales) [36]. Source: *B. dancei* fossil donated by Dr. Chungkun Shih.

Figure 29.17 A reconstruction of *Bellinympha* lacewing in the presumed contemporaneous surroundings and the potential predator Source: Artwork by Dr. Chen Wang.

features are strikingly similar to the leaves of certain Mesozoic gymnosperms thought to predate the evolution of angiosperm flowering plants. The findings provide evidence that leaf mimicry evolved before the rise of flowering plants and suggest that the lacewings likely rested and lived on the feather-like leaves, remaining still or swaying in the breeze (Figures 29.16 and 29.17) to fool predators such as insectivorous dinosaurs, primitive birds, and mammals. When these gymnosperms gave way to flowering plants, the lacewings likely became prone to predation, due to the changes of leaf shape, pattern and arrangement. The current fossil record suggests that this enigmatic lineage became extinct during the Early Cretaceous, apparently closely correlated with the decline of Cycadales and Bennettitales at that time, and perhaps owing to the changing floral environment resulting from the rise of angiosperm flowering plants.

29.3.2 Mimicry and Mutualism among Hangingflies and Ginkgo Plants

The 165 Mya fossils of cimbrophlebiid hangingflies, *Juracimbrophlebia ginkgofolia* Wang, Labandeira, Shih & Ren, 2012 (Figure 29.18), represent an interesting case of leaf mimicry and mutualism between an insect and the multilobed leaf of a ginkgo plant, *Yimaia capituliformis*, which occurred in the same deposit [37]. A quantitative analysis, Geometric Morphometric Analysis (GMA), was used to provide direct evidence that particular cimbrophlebiid hangingflies from the Jiulongshan Formation of Northeastern China, possessed significantly spot-on similarity in structure for mimicking ginkgoalean leaves than did other types of hangingflies that were from another related group. The association of *J. ginkgofolia* and the ginkgo leaves considerably extends this phenomenon by adding a more finely tuned example of leaf mimesis wherein the entire insect body participates in the deception (Figure 29.19).

Most interestingly, the results suggest that *J. ginkgofolia* developed leaf mimesis either as an anti-predator avoidance device to fool predators such as larger predatory insects, insectivorous dinosaurs and mammals, or alternatively it could be a possible predatory strategy to protect its plant host from herbivores, thus gaining a mutualism benefit for both the hangingfly and the ginkgo plants [37].

Notably, the *Y. capituliformis* host (the model) became extinct around the Jurassic–Cretaceous boundary, as possibly did *J. ginkgofolia* (the mimic), significantly before the initial appearance of angiosperms during the later Early Cretaceous. These two different examples of leaf mimesis by lacewings [36] and hangingflies [37] during the Middle Jurassic provide unusual insight into a pre-angiospermous world of diverse and close associations between insects and plants.

Mimicry and Defense by Stick and Leaf Insects

Phasmatodea are the masters of camouflage and disguise. Due to elongated and thin cylindrical bodies and green or brownish colors, most stick insects look like stems and twigs, so that they can camouflage and hide themselves for protection. During the day, they stay motionless on the twigs, and only occasionally sway in the wind like twigs [38]. Leaf insects (Phyllidae) have their femora, sometimes tibiae and abdomen strongly foliaceously dilated, imitating angiosperm plant leaves. In some leaf insects, e.g. the giant Malaysian leaf insect, *Phyllium giganteum* [39] and the leaf insect *Phyllium bioculatum* Gray, 1832, the margins of the body even have the appearance of bite marks and discolorations. Some species, e.g. *Pseudodiacantha macklotti* [40] and *Bactrododema centaurum* [41], cover themselves with mossy or lichenous outgrowths to supplement disguise. Some stick insects, like *Bostra scabrinota* [42] and *Timema californica* [43], are able to change color to match their surroundings. By mimicking twigs or leaves accurately, phasmids avoid recognition by their predators, which might have contributed to their survival and propagation since the Jurassic.

Phasmids are good at not only mimicry, but also defense when threatened. One way is to flash their bright colors which are normally hidden, and make a loud noise when the predators encroach them [44]. Some species can release chemical secretions in all directions using a pair of glands at the anterior edge of the prothorax to disturb and deter predators [45]. The stick insects also have a unique skill: if the branches are shaken, they may fall in the grass and retract their thoraxes and legs, feigning death, and then await an

Figure 29.18 Specimens of *Juracimbrophlebia ginkgofolia* Wang, Labandeira, Shih and Ren, 2012, showing an appearance similar to *Y. capituliformis*. (a) A *Ginkgoites* leaf of *Y. capituliformis* (CNU-PLA-NN-2009-085p). (b) A *Ginkgoites* leaf of *Y. capituliformis* (CNU-PLA-NN-2010-044). (c) Holotype of *J. ginkgofolia* (CNU-MEC-NN-2010-050p), with an appearance similar to *Y. capituliformis*. (d) A cimbrophlebiid specimen (CNU-MEC-NN-2010-017p) [37].

opportunity to escape [38]. Some nymphs may sever their legs to run away when attacked. New legs will grow when molting next time.

Extant stick and leaf insects commonly imitate twigs or leaves for protection, with lateral lamellae used to enhance crypsis or achieve mimicry. However, the origin and early evolution of such lateral expansions among Phasmatodea are unknown. Only very few cases of the Mesozoic phasmatodeans' mimicry have been reported due to the scarce fossil evidence. A stem-Phasmatodea *Cretophasmomima melanogramma* Wang, Béthoux and Ren, 2014 (Figure 13.8) has peculiar wing coloration with dark longitudinal veins, which suggest that the leaf-shaped plant organ from the *Membranifolia admirabilis* Sun and Zheng, 2001 in the Lower Cretaceous of Yixian Formation was used as model for crypsis. Wang *et al*. considered the occurrence of leaf mimicry within Phasmatodea should be much earlier than the appearance of twig and bark mimicry [46].

Recently, a new stick insect, *Elasmophasma stictum* Chen, Shih, Gao & Ren, 2018, was described, with well-preserved, thin, lateral lamellae on the thoracic pleura, the terga of abdominal segments I–X, and the ventrolateral margins of all femora [47]. This species, from the mid-Cretaceous amber of northern Myanmar, was assigned to Euphasmatodea. By possessing a clear, stick-like body, *E. stictum* has already started to imitate twigs of trees and bushes during the mid-Cretaceous, around 99 Mya. The abdominal structures of *E. stictum* exhibit traces of multiple expansions of the terga, suggesting that such structure might have been an early development of body expansions used to improve crypsis for stick or leaf insects when they sprawled on twigs or leaves (Figure 29.20).

Figure 29.19 A reconstruction of *Juracimbrophlebia ginkgofolia* mimicking *Ginkgoites* leaves of *Y. capituliformis* while providing mutualism for both insects and plants [37]. Source: Artwork by Dr. Chen Wang.

29.4 Eyespot Warning for Fossil Insects

For extant insects, especially for Lepidoptera, eyespots are one of their important self-defense strategies (Figures 29.6 and 29.7). In the fossil insects, we have reported neuropterans with eyespots (Chapter 20), but not the basal Lepidoptera fossils from the Middle Jurassic (Chapter 27). For example, kalligrammatids of *Kalligramma elegans* Yang, Makarkin & Ren, 2014 [48] (Figure 20.19), *Limnogramma hani* Makarkin, Ren & Yang, 2009 [49], (figure 25.15 in [1]), *Kallihemerobius pleioneurus* Ren & Oswald, 2002 [50] (figure 25.10 in [1]), *Limnogramma mongolicum* Makarkin, Ren & Yang, 2009 [49], (figure 25.16 in [1]), and *Oregramma gloriosa* Ren, 2003 [51], (figure 25.17 in [1]), and many others from the Middle Jurassic and Early Cretaceous possessed various eyespots. These eyespots are delicate, some even with structure simulating the pupil of a vertebrate, which is an efficient defense for vertebrates. Most of the neuropterans were poor flyers, but they were preyed upon by stronger and faster dragonflies, dinosaurs, pterosaurs, mammals or reptiles. Therefore, they evolved various types of eyespots on the wings. However, neuropterans had weak escaping capability; such defensive eyespots were not good enough and they were gradually replaced by more effective cryptic colorations and/or markings [36, 52].

29.4.1 Eyespots and Spots on the Forewings of Kalligrammatids

As shown in Chapter 20, many species of Kalligrammatidae, described from the Middle Jurassic and Early Cretaceous of Northeastern China [53, 54], present eyespots and spots, typically located on the upper surface midway to two-thirds of the proximal-to-distal wing length, to provide warning to potential predators and gain protection (Figure 29.21) [53]. Six distinctive types of forewing eyespots or spots occur on most species of the four derived kalligrammatid clades (Figures 29.22 and 29.23), there are occurrences previously known from some taxa [53, 55, 56], but not others [54]. The basalmost clade has no wing spots or eyespots (Figure 29.24), as seen in almost all modern neuropterans [57]. There are four eyespot types, each consisting of distinctive, differentially pigmented rings surrounding a central pigmented disc with small, whitish, oval-shaped ocules (Types 1–4; Figures 29.22 a–d). In addition, there are taxa with two simple spots, consisting of a round, dark patch lacking concentric rings (Type 5; Figures 29.22 e). Eyespots of Type 6 (Figure 29.23) were not mapped onto the phylogeny, as wing characters of one *Kalligramma* sp. were insufficiently preserved for inclusion in phylogenetic analyses [53]. In Type 1 eyespots, a second ring of dark pigmentation occurs relative to single ringed Types 2–4 and 6 (Figures 29.22 a–d and 29.23) [56].

Forewing eyespot and spot types (Figures 29.22 and 29.23) were mapped onto the phylogenetic tree of Kalligrammatidae (Figure 29.24), revealing major patterns. In the basalmost kalligrammatid clade of Sophogrammatinae, eyespots and spots were absent. The evolution of spots and eyespots likely originated early within the kalligrammatid clade, in the sister lineage to Sophogrammatinae (Figure 29.24). The four derived kalligrammatid clades after this lineage exhibit a variety of spot and eyespot patterns and absences. The most complex eyespot type occurs late in three separate lineages, within Oregrammatinae (Type 1; Figure 29.22 a), Kallihemerobiinae and Kalligrammatinae (Figure 29.24), suggesting that these eyespots derive from simpler ones, a transition that likely happened multiple times. In addition, multiple simple spots were converted to single eyespots in several lineages. These patterns are similar to convergent changes conventionally proposed for nymphalid butterflies in modern Lepidoptera [58–61]. The deployment of a spot of monochromatic pigment between two major veins in basal Kallihemerobiinae, Kalligrammatinae and Oregrammatinae (Figures 29.24 and 29.22 e) has convergently re-evolved in modern, distantly related Psychopsidae and Nemopteridae [57].

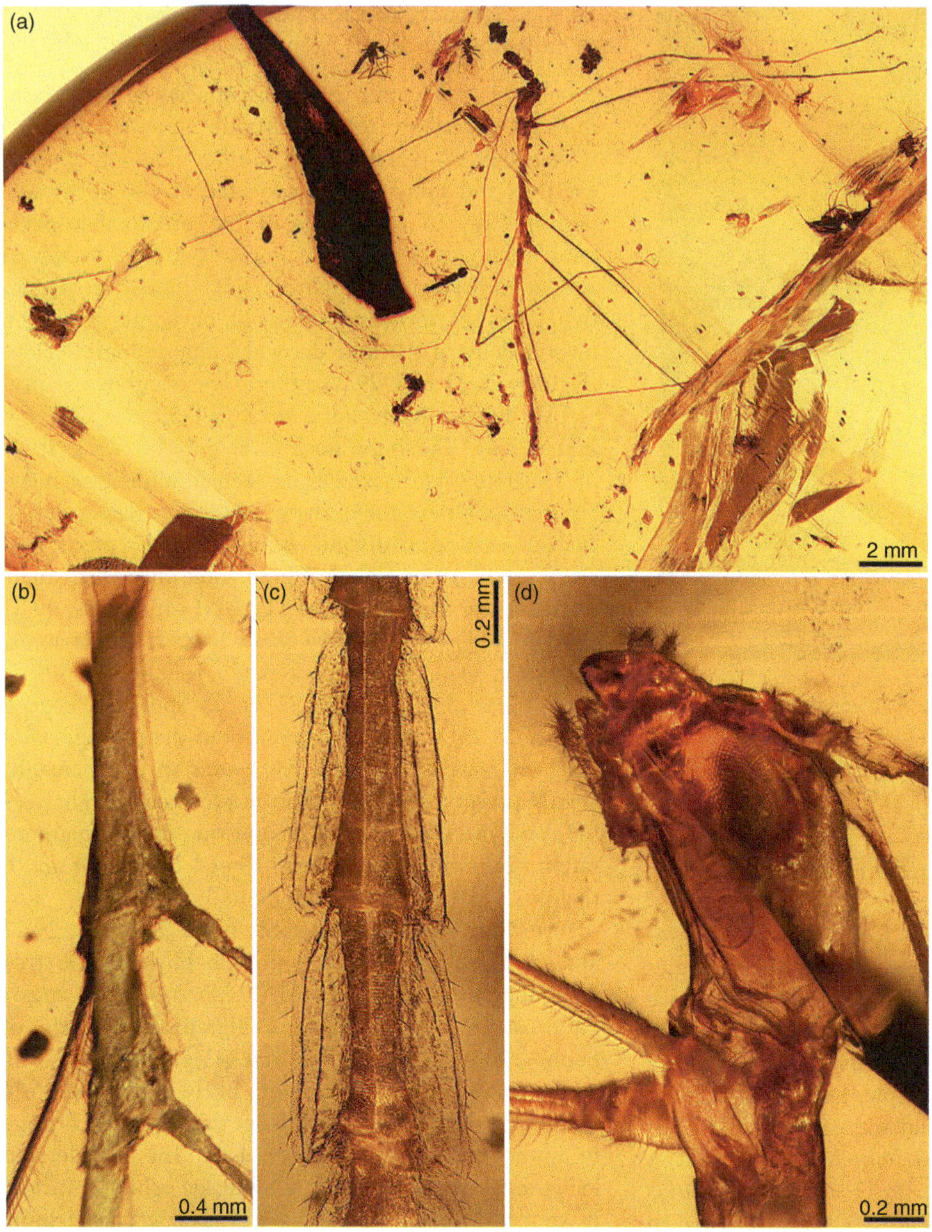

Figure 29.20 *Elasmophasma stictum* Chen, Shih, Gao & Ren, 2018 (Holotype, CNU-PHA-MA2017004). (a) Habitus; (b) An early evolution of lateral lamellae on the thorax, (c) Lateral lamellae on the abdominal segments; (d) Head of the mid-Cretaceous stick insect. Source: modified from [47].

29.5 Summary and Prospects

Well-preserved and highly-diversified fossil insects of Odonata, Orthoptera, Homoptera, Neuroptera and Mecoptera from the Early Cretaceous Jehol Biota and the Middle Jurassic Yanliao Biota in the Northeastern China provide six types of camouflage by their wing markings to make themselves hard to see or be detected by predators or potential prey.

The excellent preservation of the two fossil lacewings, *Bellinympha filicifolia* Wang, Ren, Liu & Engel, 2010 and *B. dancei* Wang, Ren, Shih & Engel, 2010, provided a rare window of opportunity to look into an ancient evolution of insect mimicking feather-like gymnosperm leaves for protection [36]. This enigmatic scenario of interactions between insects and gymnosperms were lost during the course of the evolutionary process. No extant insects have mimicry to imitate the leaves of gymnosperm plants.

Figure 29.21 Eyespots on wings of Neuroptera *Oregramma illecebrosa* Yang, Wang, Labandeira, Shih & Ren, 2014 (CNU-NEU-LB2009-031) [53].

Figure 29.23 *Kalligramma* sp. with wing eyespot of Type 6 (CNU-NEU-NN2010-010p).

Juracimbrophlebia ginkgofolia Wang, Labandeira, Shih and Ren, 2012 developed leaf mimesis as an anti-predator avoidance device to fool predators. Although extant *Ginkgo biloba* is considered to be minimally herbivorized, the distantly related Daohugou ginkgo such as *Yimaia capituliformis* exhibited comparatively high levels of insect consumption, providing potential food to a diverse herbivore fauna. It is possible that the association between *J. ginkgofolia* and *Y. capituliformis* was indeed a mutualism, by which the plant host provided leaves for crypsis while the associated leaf-mimetic hangingfly provided an anti-herbivore function for its plant host [37].

Furthermore, well-preserved kalligrammatid fossils from the Middle Jurassic (165 Mya) and the Early Cretaceous (125 Mya) in Northeastern China unravel a surprising array of morphological and ecological features [53] similar to those in unrelated butterflies. High-tech instrumentation and imaging technology have been applied to analyze kalligrammatid fossils and their environment. The evolution of specific traits mapped onto a kalligrammatid phylogeny indicate that these extinct lacewings convergently evolved wing eyespots that possibly contained melanin, and wing scales, elongate tubular proboscides, similar feeding styles, and seed–plant associations, similar to butterflies, probably originated 80–70 Mya [53, 56].

For future studies of ancient insects' mimicry, based on well-preserved diverse fossils including insects, plants, and other animals with detailed characters, the mimic and the model need to be assessed and the duped predator determined. Different morphologies of the model may result in different outcomes of insects' mimicry. Also, different types of predators in the ecosystems may lead to different defense strategies used by insects. In the future, with discoveries of new fossils and based on the feeding habits of coexisting predators, it is believed that more fossil insects taking rather simple forms of cryptic and/or camouflage markings on the wing or more complicated form of mimicry and/or eyespot warning will broaden and expand our knowledge of this field.

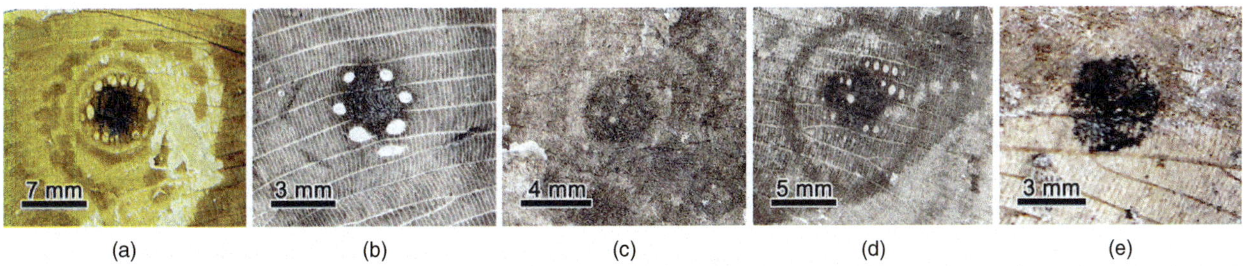

Figure 29.22 Eyespots of Type 1 to Type 5 on wings of kalligrammatids. (a) Type 1 wing eyespot with two outer rings (*O. illecebrosa*, CNU-NEU-LB-2009-031); (b) Type 2 with a single outer ring, lightly hued inter area, and an uninterrupted, pigmented central disc with surrounding, non-contiguous ocules (*Kallihemerobius almacellus*, CNU-NEU-NN-2010-050p), (c) Type 3 with one lightly hued circular area and a few, differently sized ocules within a darkly pigmented central disc (*Ithigramma multinervia*, CNU-NEU-NN-2009-034p); (d) Type 4 with a few ocules embedded within and other ocules surrounding a darkly pigmented central disc, a lightly hued inner area and a surrounding, dark-hued outermost ring (*K. circularia*, CNU-NEU-NN-2010-003); (e) Type 5 of a simple, circular, pigmented, central disc (*Kallihemerobius aciedentatus*, CNU-NEU-NN-2010-008p).

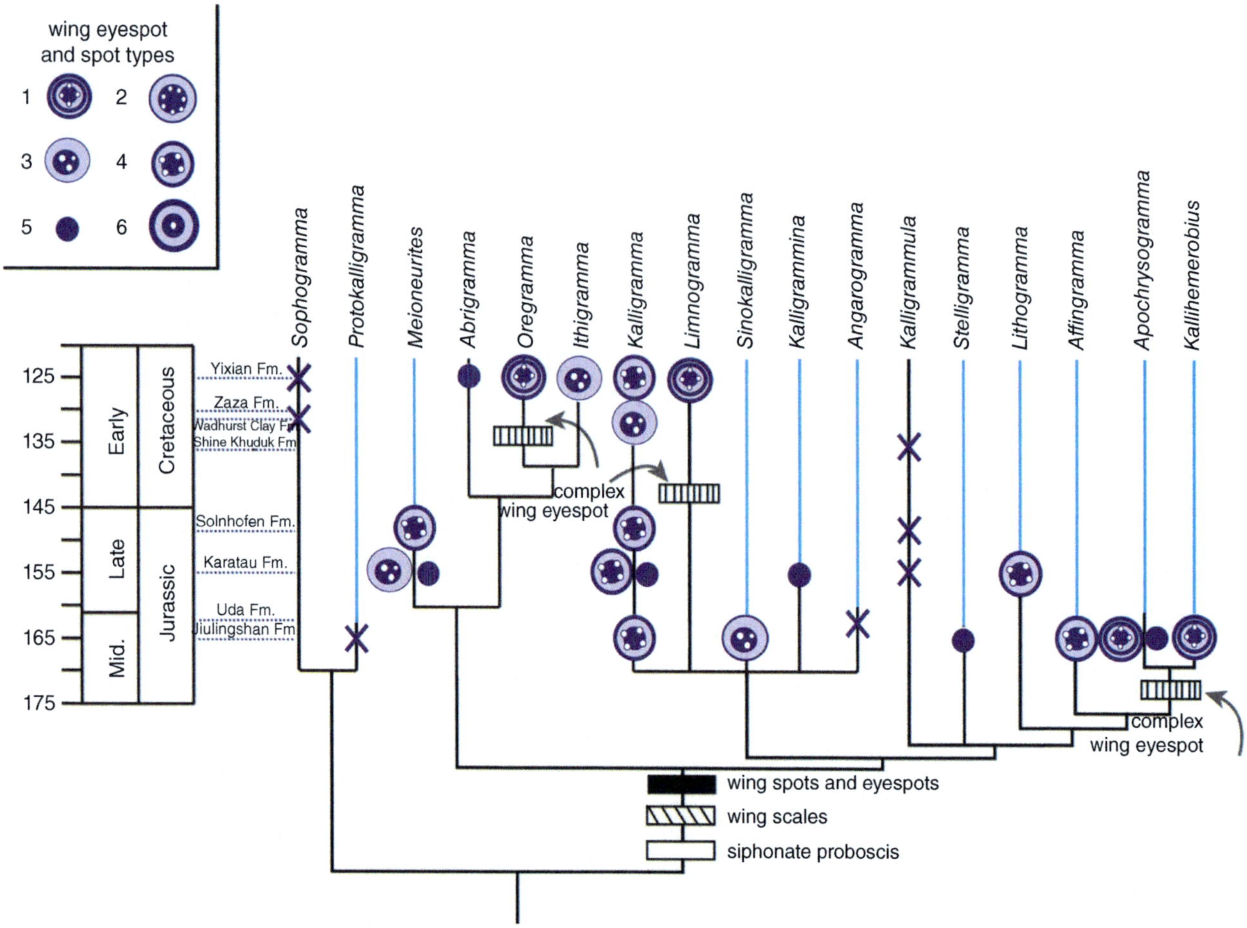

Figure 29.24 Phylogenetic context of wing spots and eyespots in mid-Mesozoic kalligrammatids. Wing eyespot and spot type symbols are upper-left; crosses (X's) are eyespot/spot absences. Source: modified from [53, 56].

During the Early Cretaceous, basal angiosperm plants started to evolve and develop. Later on, they evolved to have more advanced flowers and sweet nectar to attract insects and other animals to pollinate them. This may mark the beginning of significant changes in the biotic community. Myanmar (Burmese) amber dated about 99 Mya, e.g. [47], provides the opportunity to study predation avoidance by insects, plants and animals in their ecosystems, about 26 million years after the Jehol Biota.

References

1 Ren, D., Shih, C.K., Gao, T.P. *et al.* (2010). *Silent Stories–Insect Fossil Treasures from Dinosaur Era of the Northeastern China*, 322. Beijing: Science Press.

2 Cott, H.B. (1940). *Adaptive Coloration in Animals*, 47–67. Methuen: Oxford University Press.

3 Gullan, P.J. and Cranston, P.S. (2005). *The Insects: An Outline of Entomology*, 584. Oxford: Blackwell Publishing.

4 Stevens, M. (2005). The role of eyespots as anti-predator mechanisms, principally demonstrated in the Lepidoptera. *Biological Reviews* **80** (4): 573–588. https://doi.org/10.1017/S1464793105006810.

5 Fu, T., Peng, Y.Y., Wang, Y.J., and Ren, D. (2013). Diversity of wing markings in Neuroptera from Daohugou. *Acta Zootaxonomica Sinica* **38** (4): 741–748.

6 Ren, D. and Yin, J.C. (2003). New 'osmylid-like' fossil Neuroptera from the Middle Jurassic of Inner Mongolia, China. *Journal of the New York Entomological Society* **111** (1): 1–11. https://doi.org/10.1664/0028-7199(2003)111[0001:NOFNFT]2.0.CO;2.

7 Peng, Y.Y., Makarkin, V.N., Yang, Q., and Ren, D. (2010). A new silky lacewing (Neuroptera:

Psychopsidae) from the Middle Jurassic of Inner Mongolia, China. *Zootaxa* **2663**: 59–67.

8 Wang, Y.J., Ren, D., and Shih, C.K. (2007). New discovery of palaeontinid fossils from the Middle Jurassic in Daohugou, Inner Mongolia (Homoptera, Palaeontinidae). *Science in China Series D: Earth Sciences* **50** (4): 481–486.

9 Wang, Y.J., Ren, D., and Shih, C.K. (2007). Discovery of Middle Jurassic palaeontinids from Inner Mongolia, China (Homptera: Palaeontinidae). *Progress in Natural Science* **17** (1): 112–116.

10 Jepson, J.E., Heads, S.W., Makarkin, V.N., and Ren, D. (2013). New fossil mantidflies (Insecta: Neuroptera: Mantispidae) from the Mesozoic of Northeastern China. *Palaeontology* **56** (3): 603–613. https://doi.org/10.1111/pala.12005.

11 Shi, C.F., Winterton, S.L., and Ren, D. (2015). Phylogeny of split-footed lacewings (Neuroptera, Nymphidae), with descriptions of new Cretaceous fossil species from China. *Cladistics* **31** (5): 455–490.

12 Zheng, B.Y., Ren, D., and Wang, Y.J. (2016). Earliest true moth lacewing from the Middle Jurassic of Inner Mongolia, China. *Acta Palaeontologica Polonica* **61** (4): 847–851.

13 Yang, X.G., Shih, C.K., Ren, D., and Petrulevičius, J.F. (2012). New Middle Jurassic hangingflies (Insecta: Mecoptera) from Inner Mongolia, China. *Alcheringa: An Australasian Journal of Palaeontology* **36** (2): 195–201. https://doi.org/10.1080/03115518.2012.622143.

14 Wang, Q., Shih, C.K., and Ren, D. (2013). The earliest case of extreme sexual display with exaggerated male organs by two Middle Jurassic mecopterans. *PloS One* **8** (8): e71378. https://doi.org/10.1371/journal.pone.0071378.

15 Gao, T.P., Shih, C.K., Labandeira, C.C. *et al.* (2016). Convergent evolution of ramified antennae in insect lineages from the Early Cretaceous of Northeastern China. *Proceedings of the Royal Society B* **283** (1839): 20161448. https://doi.org/10.1098/rspb.2016.1448.

16 Wang, Y.J., Liu, Z.Q., Ren, D., and Shih, C.K. (2011). New Middle Jurassic kempynin osmylid lacewings from China. *Acta Palaeontologica Polonica* **56** (4): 865–869.

17 Yang, Q., Makarkin, V.N., and Ren, D. (2011). Two interesting new genera of Kalligrammatidae (Neuroptera) from the Middle Jurassic of Daohugou, China. *Zootaxa* **2873** (6): 60–68.

18 Li, Y.L., Ren, D., and Shih, C.K. (2008). Two Middle Jurassic hanging-flies (Insecta: Mecoptera: Bittacidae) from Northeast China. *Zootaxa* **1929**: 38–46.

19 Qiao, X., Shih, C.K., Petrulevičius, J.F., and Ren, D. (2013). Fossils from the Middle Jurassic of China shed light on morphology of Choristopsychidae (Insecta, Mecoptera). *ZooKeys* **318**: 91–111.

20 Wang, Y.J., Liu, Z.Q., and Ren, D. (2009). A new fossil lacewing genus from the Middle Jurassic of Inner Mongolia, China (Neuroptera: Osmylidae). *Zootaxa* **2034**: 65–68.

21 Qiao, X., Shih, C.K., and Ren, D. (2012). Two new Middle Jurassic species of orthophlebiids (Insecta: Mecoptera) from Inner Mongolia, China. *Alcheringa: the Australasian Journal of Palaeontology* **36**: 467–472. https://doi.org/10.1080/03115518.2012.671689.

22 Liu, S.L., Shih, C.K., and Ren, D. (2016). New Jurassic hangingflies (Insecta: Mecoptera: Bittacidae) from Inner Mongolia, China. *Zootaxa* **4067** (1): 065–078.

23 Ren, D. (2002). A new lacewing family (Neuroptera) from the Middle Jurassic of Inner Mongolia, China. *Insect Science* **9** (4): 53–67.

24 Shi, C.F., Wang, Y.J., Yang, Q., and Ren, D. (2012). Chorilingia (Neuroptera: Grammolingiidae): a new genus of lacewings with four species from the Middle Jurassic of Inner Mongolia, China. *Alcheringa: An Australasian Journal of Palaeontology* **36** (3): 309–318.

25 Shi, C.F., Wang, Y.J., and Ren, D. (2013). New species of Grammolingia Ren, 2002 from the Middle Jurassic of Inner Mongolia, China (Neuroptera: Grammolingiidae). *Fossil Record* **16** (2): 171–178.

26 Peng, Y.Y., Makarkin, V.N., Wang, X.D., and Ren, D. (2011). A new fossil silky lacewing genus (Neuroptera, Psychopsidae) from the Early Cretaceous Yixian Formation of China. *ZooKeys* **130**: 217–228. https://doi.org/10.3897/zookeys.130.1576.

27 Gu, J.J., Qiao, G.X., and Ren, D. (2010). Revision and new taxa of fossil Prophalangopsidae (Orthoptera: Ensifera). *Journal of Orthoptera Research* **19** (1): 41–56.

28 Li, L.M., Ren, D., and Wang, Z.H. (2007). New prophalangopsids from late Mesozoic of China (Orthoptera, Prophalangpsidae, Aboiliane). *Acta Zootaxonomica Sinica* **32** (2): 412–422.

29 Gu, J.J., Qiao, G.X., and Ren, D. (2011). An exceptionally-preserved new species of Bacharaboilus (Orthoptera: Prophalangopsidae) from the Middle Jurassic of Daohugou, China. *Zootaxa* **2909**: 64–68.

30 Wang, B., Zhang, H.C., and Fang, Y. (2007). Palaeontinodes reshuitangensis, a new species of Palaeontinidae (Hemiptera, Cicadomorpha) from the Middle Jurassic of Reshuitang and Daohugou of China. *Zootaxa* **1500**: 61–68.

31 Wang, Y.J. and Ren, D. (2009). New fossil palaeontinids from the Middle Jurassic of Daohugou, Inner Mongolia, China (Insecta, Hemiptera). *Acta Geologica Sinica* **83** (1): 33–38.

32 Ren, D., Lu, L.W., Guo, Z.G., and Ji, S.A. (1995). *Faunae and Stratigraphy of Jurassic-Cretaceous in Beijing and the Adjacent Areas*, 47–49. Beijing: Seismic Publishing House (in Chinese and English).

33 Bates, H.W. (1862). Contributions to an insect fauna of the Amazon Valley. Lepidoptera: Heliconidae. *Transactions of the Linnaean Society* **23**: 495–515.

34 Fisher, R.A. (1958). *The Genetical Theory of Natural Selection*, 291. New York: Dover Publications.

35 Remington, C.L. (1963). Historical backgrounds of mimicry. *International Congress of Zoology* **16** (4): 145–149.

36 Wang, Y.J., Liu, Z.Q., Wang, X. *et al.* (2010). Ancient pinnate leaf mimesis among lacewings. *Proceedings of the National Academy of Sciences* **107** (37): 16212–16215.

37 Wang, Y.J., Labandeira, C.C., Shih, C.K. *et al.* (2012). Jurassic mimicry between a hangingfly and a ginkgo from China. *Proceedings of the National Academy of Sciences of the USA* **109** (50): 20514–20519. https://doi.org/10.1073/pnas.1205517109.

38 Zheng, L.Y. and Gui, H. (1999). *Insect Classification*. Jiangsu: Nanjing Normal University Press.

39 Hausleithner, B. (1984). Eine neue Phyllium-art aus Malaysia (Phasmatodea: Phylliidae). *Entomologische Zeitschrift* **94** (4): 39–42.

40 Haan, W.D. (1842). Bijdragen to de kennis der Orthoptera. In: *Verhandelingen over de Natuurlijke Geschiedenis der Nederlansche Overzeesche Bezittingen*, vol. **2**, 95–138. Leiden: In commissie bij S. en J. Luchtmans, en C.C. van der Hoek.

41 Westwood, J.O. (1859). *Catalogue of Orthopterous Insects in the Collection of the British Museum: Part I. Phasmidæ*. Order of the Trustees.

42 Redtenbacher, J. (1908). *Die Insektenfamilie der Phasmiden. III. Phasmidae Anareolatae (Phibalosomini, Acrophyllini, Necrosciini)*, vol. **3**, 339–589. Leipzig: Verlag Engelmann, pls 16–27.

43 Scudder, S.H. (1895). Summary of the U.S. Phasmids. *The Canadian Entomologist* **27**: 29–30.

44 Robinson, M.H. (1968). The defensive behavior of *Pterinoxylus spinulosus* Redtenbacher, a winged stick insect from Panama (Phasmatodae). *Psyche* **75** (3): 195–207. https://doi.org/10.1155/1968/19150.

45 Dossey, A.T. (2010). Insects and their chemical weaponry: new potential for drug discovery. *Natural Product Reports* **27** (12): 1737–1757. https://doi.org/10.1039/C005319H.

46 Wang, M.M., Béthoux, O., Bradler, S. *et al.* (2014). Under cover at pre-angiosperm times: a cloaked phasmatodean insect from the Early Cretaceous Jehol Biota. *PLoS One* **9** (3): e91290. https://doi.org/10.1371/journal.pone.0091290.

47 Chen, S., Yin, X.C., Lin, X.D. *et al.* (2018). Stick insect in Burmese amber reveals an early evolution of lateral lamellae in the Mesozoic. *Proceedings of the Royal Society B* **285** (1877): 20180425.

48 Yang, Q., Makarkin, V.N., and Ren, D. (2014). Two new species of Kalligramma Walther (Neuroptera: Kalligrammatidae) from the Middle Jurassic of China. *Annals of the Entomological Society of America* **107** (5): 917–925.

49 Makarkin, V.N., Ren, D., and Yang, Q. (2009). Two new species of Kalligrammatidae (Neuroptera) from the Jurassic of China, with comments on venational homologies. *Annals of the Entomological Society of America* **102** (6): 964–969. https://doi.org/10.1603/008.102.0606.

50 Ren, D. and Oswald, J.D. (2002). A new genus of kalligrammatid lacewings from the Middle Jurassic of China (Neuroptera: Kalligrammatidae). *Stuttgarter Beitragezur Naturkunde, Series B* **33**: 1–8.

51 Ren, D. (2003). Two new Late Jurassic genera of kalligrammatids from Beipiao, Liaoning (Neuroptera: Kalligrammatidae). *Acta Zootaxonomica Sinica* **28** (1): 105–109. (in Chinese with English abstract).

52 Wang, S.Z. (2001). Keys to insect success: physiological traits. *Entomological Knowledge* **38** (6): 467–472.

53 Yang, Q., Wang, Y.J., Labandeira, C.C. *et al.* (2014). Mesozoic lacewings from China provides phylogenetic insight into evolution of the Kalligrammatidae (Neuroptera). *BMC Evolution Biology* **14**: 126. https://doi.org/10.1186/1471-2148-14-126.

54 Yang, Q., Zhao, Y.Y., and Ren, D. (2009). An exceptionally well-preserved fossil kalligrammatid from the Jehol Biota. *Chinese Scientific Bulletin* **54**: 1732–1737. https://doi.org/10.1007/s11434-009-0284-2.

55 Panfilov, D.V. (1968). Kalligrammatids (Neuroptera, Kalligrammatidae) in the Jurassic deposits at Karatau. In: *Jurassic Insects of Karatau* (ed. B.B. Rohdendorf), 166–174. Moscow, Russia: Nauka Press.

56 Labandeira, C.C., Yang, Y., Santiago-Blay, J.A. *et al.* (2016). The evolutionary convergence of mid-Mesozoic lacewings and Cenozoic butterflies. *Proceedings Royal Society. B.* **283**: 20152893. https://doi.org/10.1098/rspb.2015.2893.

57 Krenn, H.W., Gereben-Krenn, B., Steinwender, B.M., and Popov, A. (2008). Flower-visiting Neuroptera: mouthparts and feeding behaviour of *Nemoptera sinuata* (Nemopteridae). *European Journal Entomology* **105**: 267–277. https://doi.org/10.14411/eje.2008.037.

58 Wang, X., Li, N., Wang, Y., and Zheng, S. (2009). The discovery of whole-plant fossil cycad from the Upper Triassic in western Liaoning and its significance. *Chinese Scientific Bulletin* **54**: 3116–3119.

59 Monteiro, A. (2008). Alternative models for the evolution of eyespots and serial homology on lepidopteran wings. *BioEssays* **30**: 358–366. https://doi.org/10.1002/bies.20733.

60 Oliver, J.C., Beaulieu, J.M., Gall, L.F. *et al.* (2014). Nymphalid eyespot serial homologues originate as a few individualized modules. *Proceedings Royal Society. B.* **281**: 20133262. https://doi.org/10.1098/rspb.2013.3262.

61 Kodandaramaiah, U. (2009). Eyespot evolution: phylogenetic insights from Junonia and related butterfly genera (Nymphalidae: Junoniini). *Evolutionary Development* **11**: 489–497. https://doi.org/10.1111/j.1525-142X.2009.00357.x.

30

Gene Propagation – Courtship, Mating, and Next Generation

Chungkun Shih[1,2], *Taiping Gao*[1], *and Dong Ren*[1]

[1] *Capital Normal University, Haidian District, Beijing, China*
[2] *National Museum of Natural History, Smithsonian Institution, Washington, DC, USA*

30.1 Introduction

For all living organisms, one of their key functions or "life purposes" is to pass on their genes to the next generation, thus, to sustain the existence of their species. There are many documented studies for extant insects, covering sexual display, sexual competition and sexual selection; acoustic communication and pheromone sensing between males and females; mating behavior and mating strategy; egg laying and brood protection including maternal care; egg hatching, post-embryonic development and larval development. However, due to sparse fossils of well-preserved insects recording and demonstrating these morphologies, functions or behaviors, it is rather difficult to elucidate the early development and long-term trend of the rhythms of insect evolution.

In the past five years, there have been many interesting and important findings of these insect morphologies and functions, based on well-preserved compression fossils of insects and plants from the Middle Jurassic Jiulongshan Formation of Daohugou, Inner Mongolia and the Lower Cretaceous Yixian Formation of Liaoning. For example, three Middle Jurassic male scorpionflies with extremely extended abdominal segments and enlarged genitalia for extreme sexual display [1, 2]; several praesiricid sawflies, from the Lower Cretaceous Yixian Formation, with long and sharp mandibles for sexual display, territory defense and/or mate attraction [3]; one Middle Jurassic katydid with exceptionally well-preserved stridulatory structures producing song and musical singing at low frequencies to attract potential mates [4]; several caddisflies having bipectinate antennae with two rows of comb-like rami along the flagellum, one sawfly bearing plumose (feathery) antennae and one scorpionfly possessing pectinate antennae for locating potential mates or food sources in their Early Cretaceous ecosystems [5]; a pair of Middle Jurassic copulating froghoppers indicating froghoppers' genital symmetry and mating position have remained static for over 165 million years [6]; a few Middle Jurassic parasitoid wasps with very long ovipositors for laying eggs into their hosts [7]; and plant fossils from the Late Triassic to the Early Cretaceous showing various insect damage types (DTs) of oviposition, leaf mines and galls by preserving records of early development of insect larvae in deep time [8, 9].

30.2 Extreme Sexual Display

Many extant male animals exhibit exaggerated body parts for display, defense or fighting in sexual display and selection. Iconic examples are male peacocks, grouse and birds of paradise showing off colorful and elegant plumage, dance movement and song vocalization. Male deer, moose and reindeer bear large structured antlers while engaging in fights during the mating season. For insects, male rhinoceros and stag beetles have huge horn-like structures for fighting and competition and some male *Leptopanorpa* scorpionflies have very long abdominal terminal segments for sexual display and competition.

Fossil records of insects having exaggerated body parts for sexual display are fairly rare. Up to 2010, two male holcorpids with elongate abdominal segments from sixth (A6) to eighth (A8) and enlarged male genitalia have been reported from the Eocene, suggesting evolution of these characters occurred fairly late [10]. Since then, three mecopterans with exaggerated male body parts from the latest Middle Jurassic Jiulongshan Formation in Northeastern China have been reported [1, 2]. All have extremely extended abdominal segments from A6 to A8 and enlarged genitalia, which might have been used for sexual display and, to less extent, for fighting with other males in the competition for mates. Although *Miriholcorpa forcipata* Wang, Shih & Ren,

Rhythms of Insect Evolution: Evidence from the Jurassic and Cretaceous in Northern China, First Edition.
Edited by Dong Ren, Chungkun Shih, Taiping Gao, Yongjie Wang, and Yunzhi Yao.

Figure 30.1 *Miriholcorpa forcipata* Wang, Shih & Ren, 2013, (Holotype, CNU-MEC-NN-2012001). (a) Habitus; (b) Enlarged genitalia [1]. Source: Donated by Dr. Chungkun Shih.

2013 (Figure 30.1) and *Fortiholcorpa paradoxa* Wang, Shih & Ren, 2013 (Figure 30.2) seem to have affinities with Holcorpidae, we deem both as Family *Incertae sedis*, mainly due to significant differences in the branching pattern of hind wing media (M) veins and relative length of A8 for *F. paradoxa*, and indiscernible preservation of five-branched M veins in the hind wing for *M. forcipata* [1]. However, *Conicholcorpa stigmosa* Li, Shih, Wang & Ren 2017 (Figures 30.3 and 30.4) is the earliest confirmed holcorpid hitherto, extending the existence of Holcorpidae from the Early Eocene to the Middle Jurassic by 114 million years [2].

These three taxa have extended the records of exaggerated male body parts of mecopterans for sexual display and/or selection to the latest Middle Jurassic. The similar character present in some extant *Leptopanorpa* of Panorpidae suggests that the advantages of sexual display and/or sexual selection by having extremely elongated male abdominal and sexual organs outweigh the negative impact of a bulky body and poor mobility in the evolutionary process.

Wang *et al.* [3] described *Rudisiricius validus* Wang, Rasnitsyn, Shih & Ren, 2015 (Figure 30.5) in Praesiricidae Rasnitsyn, 1968, collected from Huangbanjigou Village (Yixian Formation, the Lower Cretaceous). It has a long and sharp mandible (about 3.5 mm in length, reaching opposite side of head when closed, with long apical tooth and long, slanting subapical one placed at the mid-length of mandible (Figure 30.5). Extant pamphilioids often have large heads with long and powerful mandibles in both males and females, which are used as weapons for defense [11, 12]. However, their heads and mandibles are clearly smaller than those of the extinct taxa mentioned before. It is likely that these fossil praesiricids with a large head and extremely long and sharp mandibles might have used these mandibles as tools for sexual display, territory defense and/or mate attraction.

30.3 Serenade with Love Songs

Katydids, like many other sound-producing insects, produce species-specific calling songs that form part of the

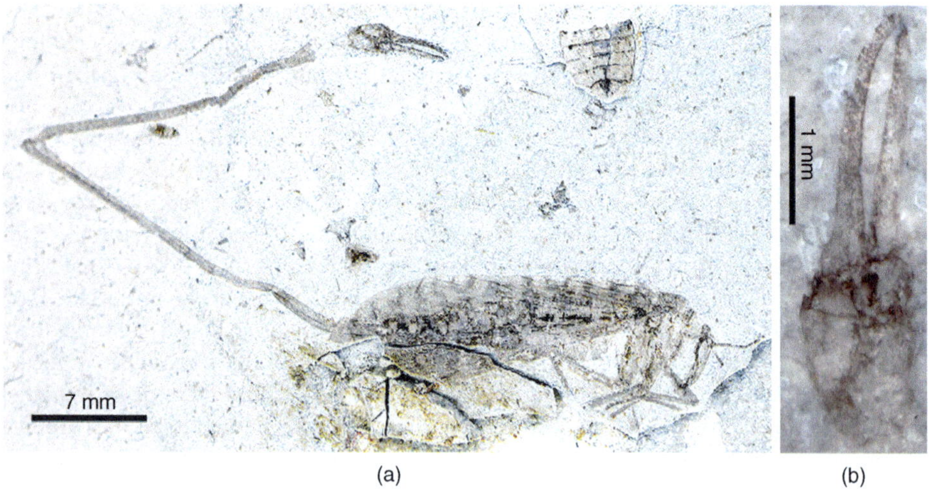

Figure 30.2 *Fortiholcorpa paradoxa* Wang, Shih & Ren, 2013, (Holotype, CNU-NEU-NN-2012002p). (a) Habitus; (b) Enlarged genitalia [1]. Source: Donated by Dr. Chungkun Shih.

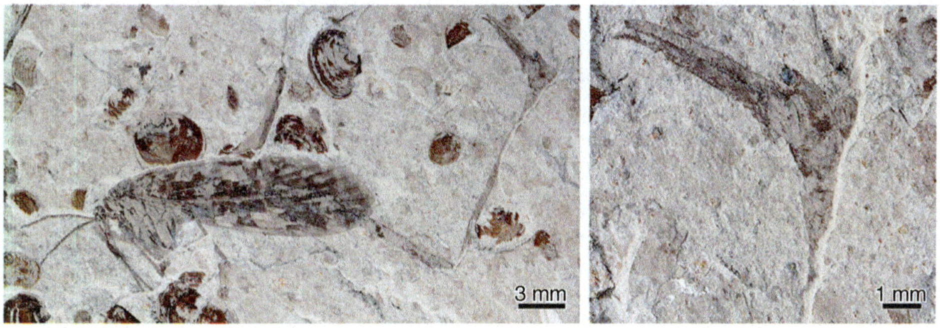

(a) (b)

Figure 30.3 *Conicholcorpa stigmosa* Li, Shih, Wang & Ren, 2017 in the Holcorpidae, (Holotype, CNU-NN-MEC-2015023). (a) Habitus; (b) Genitalia [2].

Figure 30.4 Three-D reconstruction of *Conicholcorpa stigmosa* Li, Shih, Wang & Ren, 2017. Source: Artwork by Dr. Chen Wang.

acoustic ecology of their ecosystems [13, 14]. The males call out to potential mates by rubbing a toothed vein on one wing against a plectrum on the opposite forewing, depending on the species. Based on exceptionally well-preserved stridulatory structures on the wings of a katydid fossil from the latest Middle Jurassic, *Archaboilus musicus* Gu, Engel & Ren, 2012, Gu *et al.* [4] reconstructed the song and musical singing at low frequencies.

From the exceptionally preserved morphology of its stridulatory apparatus in the forewings and phylogenetic comparison with extant species, it was revealed that *A. musicus*, in the extinct family Haglidae (Orthoptera), a group basal to all extant katydids [15, 16], radiated pure-tone (musical) songs using a resonant mechanism tuned at a frequency of 6.4 kHz. This is acoustically compatible with the proposition that the song frequency of *A. musicus* was well-adapted to long-distance communication close to the ground.

Contrary to previous scenarios, musical songs were an early innovation, preceding the broad-bandwidth songs of extant katydids. Providing an accurate insight into paleoacoustic ecology, the low-frequency musical song of *A. musicus* was well-adapted to communication in the lightly cluttered environment of the mid-Jurassic forest populated by coniferous trees and giant ferns [17], suggesting that insectivorous reptiles, amphibians, feathered dinosaurs and mammals could have also heard the song of *A. musicus*. Like many extant katydids

Figure 30.5 *Rudisiricius validus* Wang, Rasnitsyn, Shih & Ren, 2015, (Holotype, Female, CNU-HYM-LB-2012119p). (a) Habitus; (b) Mandible [3]. Source: Donated by Dr. Chungkun Shih.

which stay quiet during the day, it was suggested that *A. musicus* faced the dilemma by serenading love songs in the darkness of the night to potential mates, while avoiding coexisting nocturnal predators (see Figure 30.6).

30.4 Sensing and Locating Potential Mates with Ramified Antennae

Antennae are important sensory organs for insects to communicate with other insects or to detect environmental cues [18]. Therefore, they are principally involved in activities such as locating potential mates [19–21], securing food, and/or targeting biological hosts [22]. Various types of antennal sensilla are located on individual antennomeres and serve as chemoreceptors, mechanoreceptors, thermoreceptors, or hygroreceptors [23–25]. However, the principal sensilla on the antennomeres of most insects are involved in olfaction [22, 26]. Some insects have evolved broadly ramified antennae, ranging from pectinate (a single row of comb-like rami, or branches, along the flagellum), bipectinate (two rows of comb-like rami along the flagellum), to plumose (feathery) or flabellate (fan-shaped) [27]. The consequence of such intricate morphology is an overall expansion in the antennal surface area associated with an increase in the number of sensilla.

Though ramified antennae are common in living insects, occasionally they are present in the Mesozoic fossil records. Gao *et al.* [5] reported the only known caddisfly, *Cathayamodus fournieri* Gao, Shih, Labandeira, Santiago-Blay, Yao & Ren, 2016, having bipectinate antennae (Figure 30.7), the earliest known fossil sawfly, *Jibaissodes bellus* Gao, Shih, Labandeira, Santiago-Blay, Yao & Ren, 2016, bearing plumose (feathery) antennae (Figure 30.8) and a scorpionfly, *Vitimopsyche pectinella* Gao, Shih, Labandeira, Santiago-Blay, Yao & Ren, 2016, possessing pectinate antennae (Figure 30.9), all from the Lower Cretaceous Yixian Formation of Northeastern China, dated at 125 million years ago (Mya). These fossil insects with ramified antennae predate by about 10 Myr the biflabellate antennae borne by *Atefia rasnitsyni* Krogmann, Engel, Bechly & Nel, 2012, a compression fossil from the Lower Cretaceous Crato Formation (115 Mya), which was suggested to indicate the antiquity of insect usage of long-range female attractants [28].

These three insect taxa of unrelated lineages with ramified antennae provide evidence for broad structural convergence that historically has been best demonstrated by features such as convergent mouthparts. In addition, ramified antennae in these mid-Mesozoic lineages likely do not constitute a key innovation, as they are not associated with significantly increased diversification compared with closely related lineages lacking this trait, and nor are they ecologically isolated from numerous, co-occurring insect species with unmodified antennae [5].

30.5 Forever Love – The Hitherto Earliest Record of Copulating Insects

Mating behaviors for extant insects have been studied and documented, for example, for froghoppers [29], scorpionflies [30] and planthoppers [31]. However, fossil records of unequivocal insect mating are fairly sparse.

Figure 30.6 *Archaboilus musicus* Gu, Engel & Ren 2012, (a), (b) Detail of the left wing file; (c), (d) Detail of the right wing file. Line drawings illustrate tooth morphology and spacing along the file. Source: modified from [4].

Figure 30.7 *Cathayamodus fournieri* Gao, Shih, Labandeira, Santiago-Blay, Yao & Ren, 2016, (Holotype, Male, CNU-TRI-LB-2009001p). (a) Habitus under alcohol. (b) The central region of the ramified antenna, showing rami. (c) Scanning electron microscope (SEM) image of flagellomere rami with trichoid sensilla and trichobothria, Source: modified from [5]. Donated by Dr. Chungkun Shih.

(a) (b)

Figure 30.8 *Jibaissodes bellus* Gao, Shih, Labandeira, Santiago-Blay, Yao & Ren, 2016, (Holotype, CNU-HYM-LB-2011009p), (a) Habitus; (b) Right antenna. Source: modified from [5].

(a) (b) (c)

Figure 30.9 *Vitimopsyche pectinella* Gao, Shih, Labandeira, Santiago-Blay, Yao & Ren, 2016, (Holotype, CNU-MEC-LB-2012088p). (a) Habitus; (b) Details of the elongate mouthparts; (c) Left antenna. Source: modified from [5].

Boucot and Poinar [32] listed 33 instances of fossilized mating insects, such as fireflies, mosquitoes, planthoppers, leafhoppers, water striders, bees and ants, 27 of which are preserved in amber, others on compression fossils. They reported that the earliest example of copulation in fossil insects was a pair of chironomids (Diptera) discovered in the Early Cretaceous amber from Lebanon [32, 33]. The limited fossil examples considerably limit our knowledge on insect mating behaviors and genitalia orientation during the Mesozoic, and hinder our understanding of the evolution of insect mating in this major component of modern ecosystems.

Li *et al.* [6] reported a pair of copulating froghoppers, *Anthoscytina perpetua* Li, Shih & Ren, 2013, referable to the Procercopidae, from the Middle Jurassic Jiulongshan Formation in Northeastern China. Procercopidae is an extinct family in the superfamily of froghoppers, Cercopoidea Leach, 1815. Froghoppers get their name because the adults hop around on plants and shrubs like tiny frogs. The nymphs of froghoppers are also called spittlebugs because they cover themselves with foaming spittle, composed of tiny air bubbles trapped in secretions from their Malpighian tubules, which provides protection from predation, parasitism and desiccation [34].

The pair of *A. perpetua* exhibit a belly-to-belly mating position as preserved, with the male's aedeagus inserting into the female's bursa copulatrix (Figures 30.10 and 30.11). Abdominal segments eight to nine of the male are disarticulated suggesting these segments were twisted and flexed during mating. Due to potential taphonomic effect, we cannot rule out that they might have taken a side-by-side position, as in extant froghoppers. The genitalia of male and female, based on paratype specimens, show symmetric structures.

Our findings, consistent with those of extant froghoppers, indicate that their genitalic symmetry and mating position have remained static for over 165 million years. This discovery of the earliest record of copulating insects hitherto sheds light on the evolution of mating behavior in this group of insects [6].

30.6 Long Ovipositors Used for Laying Eggs into Hosts

Li *et al.* [7] described *Proephialtitia acantha* Li, Shih, Rasnitsyn & Ren, 2015 (Figure 30.12), collected from

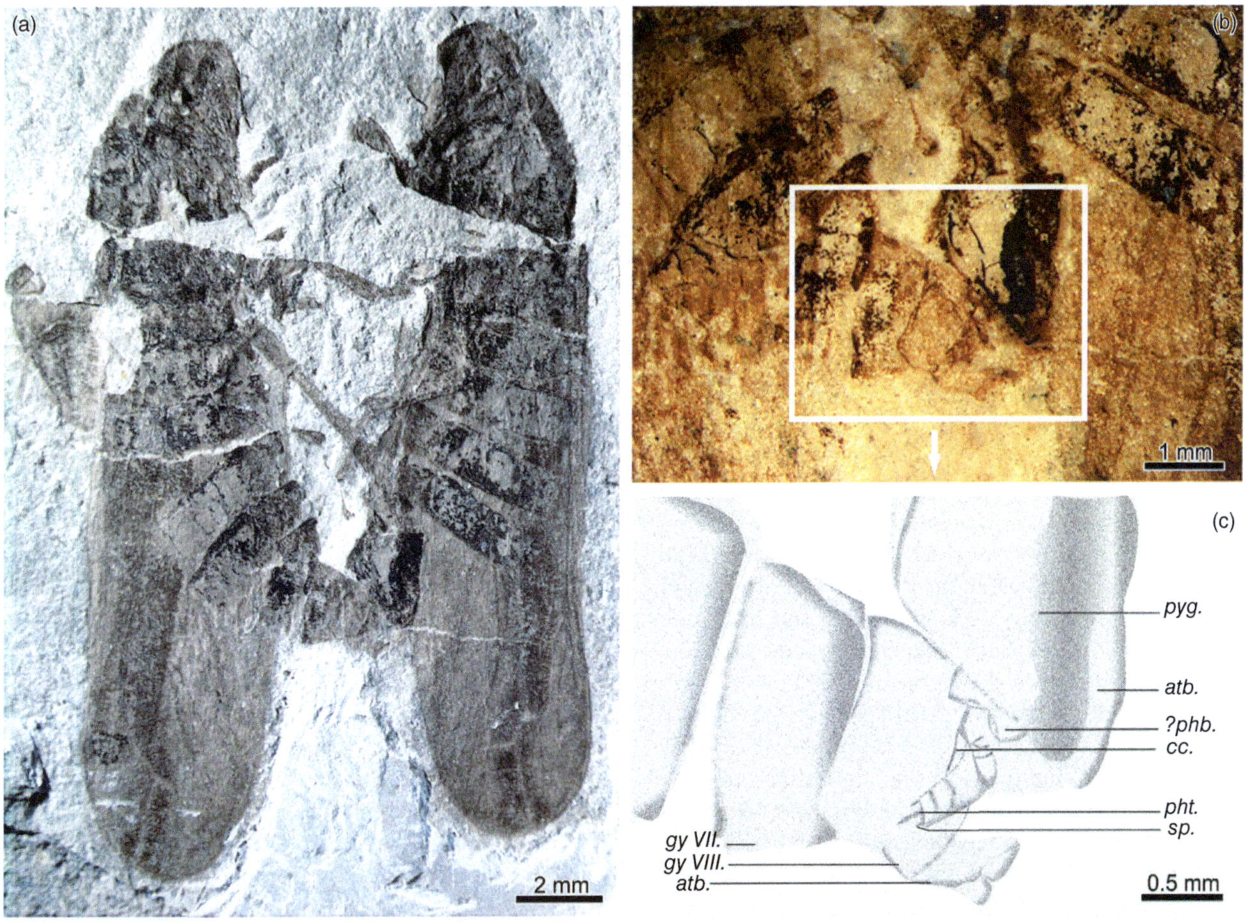

Figure 30.10 *Anthoscytina perpetua* Li, Shih & Ren, 2013, (Holotype, Male, on the right, CNU-HEM-NN2012002p, and Allotype, Female, on the left, CNU-HEM-NN2012003p). (a) Habitus; (b) Male and female genitalia in copulation, under alcohol; (c) Enlargement of template in (b) as interpretative drawing of genitalia in copulation. pyg., pygofer; atb., anal tube; phb., phallobase; cc., corpus connective; pht., phallotrema; sp., sclerotized process; gy., gonapophyses. Source: Modified from [6].

Figure 30.11 Ecological reconstruction of *Anthoscytina perpetua* Li, Shih & Ren, 2013. Source: Artwork by Dr. Chen Wang.

Figure 30.12 *Proephialtitia acantha* Li, Shih, Rasnitsyn & Ren, 2015, (Holotype, CNU-HYM-NN-2014004p) [7]. Source: Donated by Dr. Chungkun Shih.

Daohugou Village (Jiulongshan Formation, the Middle Jurassic). Length of the body excluding antennae and ovipositor is 23.7 mm. Head medium sized, antenna slim with scape swollen, and pedicel distinctly narrower than scape, flagellum very thin. Mesosoma approximately 6.4 mm long and 3.6 mm wide, about 1.8 times as long as wide; pronotum short, nearly as wide as the head; mesonotum slightly wider than pronotum; propodeum long and broad, nearly as long as wide and in contact with mid and hind coxae. Metasoma broadly attaching to propodeum, with eight segments visible, with nearly the same width from the third segment to terminal; the first segment nearly trapezoidal in lateral view, the second segment slightly wider but shorter than the first; the third and fourth segments cylinder-shaped, the third segment nearly as wide and long as the fourth; remaining segments nearly as wide as the fourth segment. Ovipositor slim with a sheath, approximately 38.3 mm long as preserved (1.6 times as long as body) [7].

The first abdominal segment of Apocritan has been incorporated into the thorax as the propodeum. The remaining abdomen, so called metasoma, is connected to this hybrid region via a narrow propodeal–metasomal articulation forming a "wasp waist," which allows for preferable maneuverability in controlling its ovipositor. By observing some ovipositing extant wasps, representatives shown in Figure 30.13, Li *et al.* [7] categorized four different postures of oviposition highlighting that various wasps lay eggs by utilizing their propodeal–metasomal articulation and associated capability to maneuver their ovipositors. These four typical postures of oviposition are as follows: (i) the metasoma is raised vertically above the head and mesosoma forming a "L" shape (Figure 30.13 a), and the long ovipositor is almost vertically inserted into the branches or flowers from the top of the "L" in order to lay eggs flexibly and accurately into hidden host larvae. (ii) The metasoma is bent and parallel to the head and mesosoma forming a "=" shape (Figure 30.13 b), and the ovipositor is used to puncture the host's gut wall, enter the hemocoel and then lay eggs. (iii) The distal part of metasoma is bent downward from the basal part of metasoma forming an inverted "V" shape, the ovipositor and the main body forming an angle from less than 90° to about 90° (Figure 30.13 c), and the ovipositor is inserted into the flowers or plants to lay eggs into hidden host larvae; (iv) The metasoma is not bent from the head and mesosoma, forming a linear "—" shape (Figure 30.13 d), and the ovipositor is bent downward into the flowers

Figure 30.13 Typical postures of oviposition of extant wasps. (a) The metasoma forming an "L" shape; (b) The metasoma forming a "=" shape; (c) The distal part and the basal part of metasomal forming an inverted "V" shape; (d) The metasoma forming a linear "—" shape. Source: modified from [7], Photos by Jason Shih.

or plants to lay eggs. Therefore, we believe that the propodeal–metasomal articulation is a very important factor, which influences the posture and maneuverability of oviposition in Apocrita.

30.7 Breeding – Oviposition, Gall and Leaf Mining

Fossil records of newly emerged insect larvae or nymphs just hatched from eggs are rare, mainly due to their tiny size and delicate body structure. However, well-preserved plant compression fossils from the Northern China have revealed oviposition marks, galls and leaf mines which provide direct evidence of insect oviposition and early development of just emerged offspring.

Ding *et al.* [8] documented that specimens of the broad-leaved conifer, *Liaoningocladus boii* Sun, Zheng & Mei 2000 [35] were evaluated from the Lower Cretaceous Yixian Formation occurring overwhelmingly at Dawangzhangzi, Liaoning Province, Northeastern China. Arthropod-mediated plant DTs were categorized for 343 specimens of this host plant; one of these DTs represented a distinctive leaf-mine, DT280, established as *Fossafolia offae* Ding, Labandeira & Ren 2014, *ichnogen. et ichnosp* [8].

A comparison of DT280 to analogous modern leaf mines was based on: (i) leaf-mine features observed in modern leaf mines; (ii) likelihood of a particular leaf-mine culprit lineage being present, given phylogenetic evidence; and (iii) body fossils of the candidate

(a) (b) (c) (d) (e)

Figure 30.14 Digital photographic images of *Fossafolia offae* Ding, Labandeira & Ren 2014, *ichnogen. et ichnosp.*, a beetle leaf mine (DT280) on leaves of the conifer *Liaoningocladus boii* Sun, Zheng & Mei, 2000 [35]. Late-phase (mature) mines are indicated in a, and c; early-phase (immature) mines are at b, d, and e. (a) A late-phase mine; CNU-PLA-LL-2010-062P-1-2; (b) An early-phase mine; CNU-PLA-LL-2010-031-1-1; (c) a late-phase mine; CNU-PLALL-2010-067C-3-1; (d), (e) An early-phase mine; CNU-PLA-LL-2010-116C-1-1. Source: modified from [8].

culprit occurring in the same or a spatiotemporally proximal deposit. Evidence from these three and other sources of information indicate the most likely miner of *F. offae* was an extinct species of Buprestidae (Coleoptera), perhaps similar to modern leaf-mining tribe Trachyini. Much less likely affiliations were Mordellidae, Chrysomelidae and Curculionoidea.

Fossafolia offae leaf mines were produced by a larva that consisted of four instars, engaged in full-depth tissue feeding, partitioned into a linear, earlier-phase mine with a distinctive frass trail and a more blotch-like, later-phase mine (Figure 30.14). Adults of this leaf miner likely fed on *L. boii*, producing linear patches of intercostal window feeding, assigned to DT103 (Figure 28.22), or less likely, may have been a pollinator [8].

The adult female likely laid eggs singly into leaf tissues between adjacent major veins, resulting in DT101 ovipositional damage (Figure 30.15 b–d). This study enables the use of multiple DTs to document life-history feeding traits for a single herbivore species. Consequently, the *damage-type suite* concept is introduced to refer to two or more different DTs genetically linked to the same culprit herbivore that issue from different developmental stages (larvae, adult), or are produced by different tissue-penetrating insect organs (ovipositors, mouthparts). In addition, this study provides a basis for using three types of general evidence necessary to elucidate identification of the culprit [8].

Ding *et al.* [9] reported that insect-mediated damage was examined on 756 specimens of three broadleaved

Figure 30.15 DT280-DT103-DT101 on *Liaoningocladus boii* Sun, Zheng & Mei, 2000. (a) *Liaoningocladus boii* CNU-PLA-LL-2010-230-1, the sane plant host of the multidamager suite of DT280 leaf mining shown in the Figure 30.14. DT101 oviposition (b–d) at right and lower-left; (b) An oviposition mark of DT101 on *L. boii*, CNU-PLA-LL-2010-282P-1; (c) A second oviposition mark of DT101 on *L. boii*, CNU-PLA-LL-2010-388-1; (d) A third oviposition mark (DT101) on *L. boii*, CNU-PLA-LL-2010-149P-1. Source: modified from [8].

conifers *Podozamites, Lindleycladus* and *Liaoningocladus,* originating from five mid-Mesozoic localities in Northeastern China. These localities are the Late Triassic Yangcaogou Formation (T3, ca. 205 Mya), the latest Middle Jurassic Jiulongshan Formation (J2, 165 Mya), and the Lower Cretaceous Yixian Formation (K1, 125 Mya). Plant hosts from these three time intervals harbor five functional feeding groups (FFGs) of herbivores and 23 distinctive DTs, categorized using the widely applied DT system. The DTs were classified into the five FFGs of external foliage feeding (6 DTs), piercing and sucking (5 DTs) (Figures 28.23 and 28.24), oviposition (3 DTs), galling (8 DTs) and leaf mining (1 DT) (Figures 30.16 and 30.17).

Damage-type richness and abundance was established for each FFG, encompassing from 10 to 16 DTs for each of the three time intervals examined. For this 80 million-year-long interval, foliar herbivory on broadleaved conifers was transformed from early predominance of external foliage feeding (T3), later replaced by an emphasis on piercing and sucking (J2), followed by bimodal expansion of endophytic interactions from oviposition and leaf mining (K1).

This trajectory of herbivore succession indicates that, from T3 to K1, plant–insect associations were transformed from earlier reliance on a greater number of exophytic modes of herbivory to a later, increased variety in endophytic consumption. The transformation also was demonstrated by finer-grained partitioning of food resources and specialization on particular host-plant tissue types. This subdivision of tissue types likely promoted greater dietary saturation of tissue space by FFGs.

Possible explanations for these shifts in herbivory include ecological causes, long-term environmental changes, or both. Ecological factors, such as (i) evolution of a more differentiated plant-host spectrum available for consumption; (ii) long-term changes in plant physiognomy and deployment of antiherbivore defenses; (iii) change in herbivore partitioning of plant-host tissues; and (iv) emergence of the parasitoid guild for efficient regulation of insect herbivores. Long-term

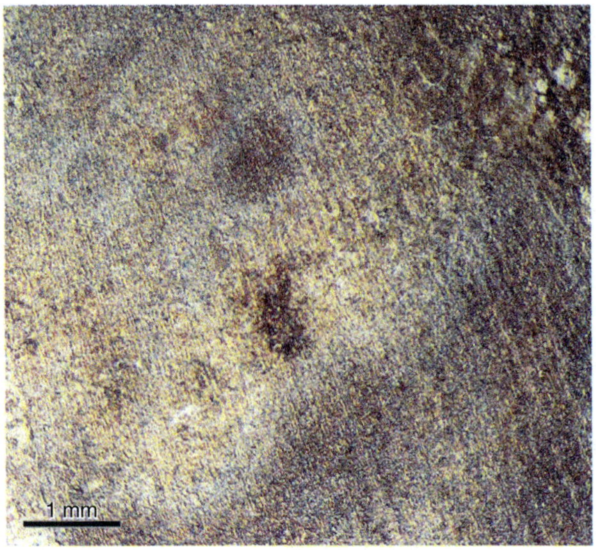

Figure 30.16 Galling on the broadleaved conifer *Podozamites lanceolatus–Lindleycladus lanceolatus* species complex from the Late Triassic (T3) Yangcaogou Formation, the Yangcaogou locality, Liaoning Province, in Northeastern China. Galling of small, circular epidermal galls (DT80) on CNU-CON-LB-2010-104-2. Source: modified from [9].

environmental variables may be linked to these shifts in insect herbivory style [9].

Mutualism in Gene Propagation

Chinese banyan trees, *Ficus microcarpa* L, f., 1782 in Moraceae, are a type of strangler fig trees, distributed abundantly and popularly in China, India, Sri Lanka, Malay Archipelago, Australia etc. Banyan tree may grow fine tendril roots downward from seedlings or branches to the ground, which eventually mature into trunks, while in some cases, smothering and destroying its host tree. As a banyan tree ages, its canopy expands by branches supported by these hardened "air roots". In rural areas of Southern China, the banyan trees with broad canopy coverage provide shade, tranquility and a resting place for local people to congregate. It is not surprising that many banyan trees are associated with parks, temples and schools (Figure 30.18).

Banyan trees are deeply rooted in the Chinese culture. Chinese traditional medicine uses the tendril air roots, leaves, bark and fruit to treat a variety of symptoms. In rural areas, old banyan trees are worshipped by local people as a deity. During the dragon-boat festival, leaves

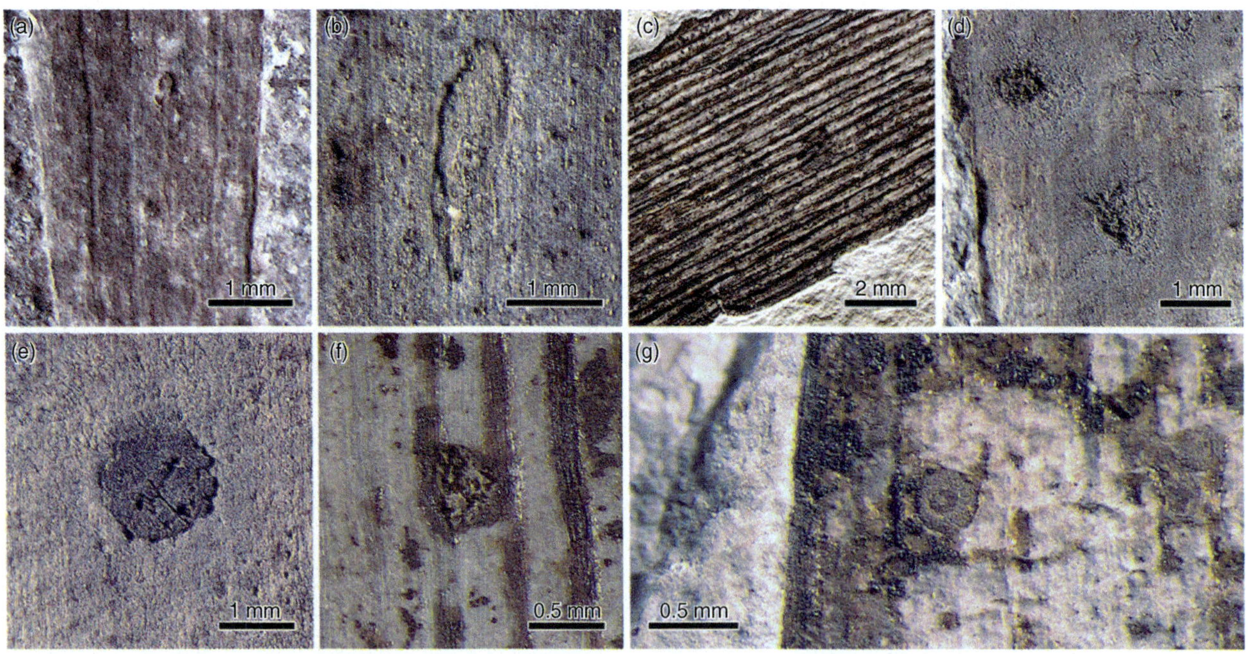

Figure 30.17 Damage on the broadleaved conifer *Podozamites lanceolatus–Lindleycladus lanceolatus* species complex from the latest Middle Jurassic (J_2) Jiulongshan Formation at the Daohugou in the Inner Mongolia. (a) Oviposition (DT76) on a major vein of CNU-CON-NN-2011-634-1; (b) Oviposition (DT101) exhibiting a pronounced outer reaction rim and disrupted inner tissues on CNU-CON-NN-2011-636-1; (c) Galling indicated by a simple gall on major veins (DT33) of CNU-CON-NN-2010-662-2; (d) Another galling, on the minor veins (DT34), on CNU-CON-NN-2009-604-1; (e) Epidermal cover gall (DT161) on CNU-CON-NN-2010-078-1; f and g, The distinctive, columnar, deep-seated gall DT116 on CNU-CON-NN-2011-421P-1 at (f) and a better preserved specimen on CNU-CON-NN-2009-635-1 at (g). Source: modified from [9].

Figure 30.18 A Chinese Banyan tree with expanded canopy supported by hardened "air roots" in the campus of the Tainan First High School. Source: Photo by Dr. Chungkun Shih.

of banyan, together with leaves of "Sweet flag" (*Acorus calamus* L., 1753 in Acoraceae) and "Chinese mugwort" (*Artemisia argyi* in Asteraceae), are hung on the house doors to fend off bad luck or evil spirits. Miniaturized Chinese banyan trees are commonly grown and treasured as Chinese Bonsai.

Banyan trees play host to a variety of birds, reptiles and insects. Birds feed on the banyan fig fruits. Larvae of many taxa of lepidopterans and coleopterans feed on the banyan leaves. In summer, cicadas broadcast their loud "music" while hiding on the branches among the thick leaves.

Hidden in plain view is one of the most interesting mutualism interactions between pollinating banyan fig wasps, *Eupristina verticillata* Waterston, 1921 [36] (Agaonidae in Chalcidoidea) and Chinese banyan trees so that both can propagate their genes. The banyan flowers, as those of figs, lie inside its fruits and are enveloped by a thick layer of fruit flesh. Typically, female pollinating banyan wasps have wings, but the males are wingless. The males emerge from pupae before the females, then the males crawl on the wart flowers to seek the female wasps for mating. After mating, the female banyan wasps crawl to the male flowers inside the fig to gather pollen, then the females exit the banyan fruit by crawling through a tunnel gnawed by a male. The males die after chewing the tunnels. Out of confinement, the pollen-laden winged female wasps fly over long distances before finding another banyan fig to lay their eggs, while pollinating the fig's female flowers. The cycle of mutualism between the fig wasp and the banyan plant starts all over again.

However, there many non-pollinating fig wasps which parasitize fig fruits of Chinese banyan trees. The most speciose wasps are in Pteromalidae, followed by those in Eurytomidae, and only one in the Ormyridae. The non-pollinating wasps have also developed impressive morphological adaptations, such as very long ovipositors, to detect and then deposit eggs inside the banyan fig fruits.

References

1 Wang, Q., Shih, C.K., and Ren, D. (2013). The earliest case of extreme sexual display with exaggerated male organs by two Middle Jurassic Mecopterans. *PLoS ONE* **8** (8): e71378. https://doi.org/10.1371/journal.pone.0071378.

2 Li, L., Shih, C.K., Wang, C., and Ren, D. (2017). A new fossil scorpionfly (Insecta: Mecoptera) with extremely elongate male genitalia from Northeastern China. *Acta Geologica Sinica (English Edition)* **91** (3): 797–805.

3 Wang, M., Rasnitsyn, A.P., Shih, C.K., and Ren, D. (2015). Revision of the Genus *Rudisiricius* (Hymenoptera, Praesiricidae) with six new species from Jehol Biota, China. *Cretaceous Research* **52**: 570–578. https://doi.org/10.1016/j.cretres.2014.02.013.

4 Gu, J.J., Fernando, M.-Z., Robert, D. *et al.* (2012). Wing stridulation in a Jurassic katydid (Insecta, Orthoptera) produced low-pitched musical calls to attract females. *Proceedings of the National Academy of Sciences* **109** (10): 3868–3873. https://doi.org/10.1073/pnas.118372109.

5 Gao, T.P., Shih, C.K., Labandeira, C.C. *et al.* (2016). Convergent evolution of ramified antennae in insect lineages from the Early Cretaceous of Northeastern China. *Proceedings of Royal Society B* **283**: 20161448. https://doi.org/10.1098/rspb.2016.1448.

6 Li, S., Shih, C.K., Wang, C. *et al.* (2013). Forever love: the hitherto earliest record of copulating insects from the Middle Jurassic of China. *PLoS ONE* **8** (11): e78188. https://doi.org/10.1371/journal.pone.0078188.

7 Li, L.F., Shih, C.K., Rasnitsyn, A.P., and Ren, D. (2015). New fossil ephialtitids elucidating the origin and transformation of the propodeal-metasomal articulation in Apocrita (Hymenoptera). *BMC Evolutionary Biology* 15–45. https://doi.org/10.1186/s12862-015-0317-1.

8 Ding, Q.L., Labandeira, C.C., and Ren, D. (2014). Biology of a leaf miner (Coleoptera) on *Liaoningocladus boii* (Coniferales) from the Early Cretaceous of Northeastern China and the leaf-mining biology of possible insect culprit clades. *Arthropod Systematics & Phylogeny* **72**: 281–308.

9 Ding, Q.L., Labandeira, C.C., Meng, Q.M., and Ren, D. (2015). Insect herbivory, plant-host specialization and tissue partitioning on mid-Mesozoic broadleaved conifers of Northeastern China. *Palaeogeography, Palaeoclimatology, Palaeoecology* **440**: 259–273. https://doi.org/10.1016/j.palaeo.2015.09.007.

10 Archibald, S.B. (2010). Revision of the scorpionfly family Holcorpidae (Mecoptera), with description of a new species from Early Eocene McAbee, British Columbia, Canada. *Annales de la Société entomologique de France* **46** (1–2): 173–182.

11 Rasnitsyn, A.P., Zhang, H.C., and Wang, B. (2006). Bizarre fossil insects: web-spinning sawflies of the genus *Ferganolyda* (Vespida, Pamphilioidea) from the Middle Jurassic of Daohugou, Inner Mongolia, China. *Palaeontology Journal* **49**: 907–916.

12 Gao, T.P., Shih, C.K., Rasnitsyn, A.P., and Ren, D. (2013). *Hoplitolyda duolunica* gen. et sp. nov. (Insecta, Hymenoptera, Praesiricidae), the hitherto largest sawfly from the Mesozoic of China. *PLoS ONE* **8** (5): e62420. https://doi.org/10.1371/journal.pone.0062420.

13 Lang, A.B., Kalko, E.K.V., Römer, H. *et al.* (2006). Activity levels of bats and katydids in relation to the lunar cycle. *Oecologia* **146**: 659–666.

14 Diwakar, S. and Balakrishnan, R. (2007). Vertical stratification in an acoustically communicating ensiferan assemblage of a tropical evergreen forest in southern India. *Journal of Tropical Ecology* **23**: 479–486.

15 Gorochov, A.V. (1995). System and evolution of the suborder Ensifera. *Proceedings of the Zoological Institute of the Russian Academy of Sciences* **260**: 3–224.

16 Heads, S.W. and Leuzinger, L. (2011). On the placement of the Cretaceous orthopteran *Brauckmannia groeningae* from Brazil, with notes on the relationships of Schizodactylidae (Orthoptera, Ensifera). *Zookeys* **77**: 17–30.

17 Senter, P. (2008). Voices of the past: a review of Paleozoic and Mesozoic animal sounds. *Historical Biology* **20** (4): 255–287.

18 Cardé, R.T. and Minks, A.K. (1997). *Insect Pheromone Research: New Directions*. New York: Chapman & Hall.

19 Eberhard, W.G. (1985). *Sexual Selection and Animal Genitalia*. Cambridge: Harvard University Press.

20 Tegoni, M., Campanacci, V., and Cambillau, C. (2004). Structural aspects of sexual attraction and chemical communication in insects. *Trends in Biochemical Sciences* **29** (5): 257–264. https://doi.org/10.1016/j.tibs.2004.03.003.

21 Shuker, D.M. and Simmons, L.W. (2014). *The Evolution of Insect Mating Systems*. New York: Oxford University Press.

22 Ventura, M.U. and Panizzi, A.R. (2005). Morphology of olfactory sensilla and its role in host plant recognition by *Neomegalotomus parvus* (Westwood) (Heteroptera: Alydidae). *Brazilian Archives of Biology and Technology* **48** (4): 589–597.

23 Crook, D.J., Kerr, L.M., and Mastro, V.C. (2008). Distribution and fine structure of antennal sensilla in emerald ash borer (Coleoptera: Buprestidae). *Annals of the Entomological Society of America* **101** (6): 1103–1111.

24 Barsagade, D.D., Tembhare, B., and Kadu, S.G. (2013). Microscopic stucture of antennal sensilla in the carpenter ant *Camponotus compressus* (Fabricius) (Formicidae: Hymenoptera). *Asian Myrmecology* **5**: 113–120.

25 Zheng, H.X., Liu, H.X., Guo, S.Y. *et al.* (2014). Scanning electron microscopy study of the antennal sensilla of *Catocala remissa*. *Bull Insectology* **67** (1): 63–71.

26 Benton, R. (2006). On the origin of smell: odorant receptors in insects. *Cellular and Molecular Life Sciences* **63**: 1579–1585. https://doi.org/10.1007/s00018-006-6130-7.

27 Schneider, D. (1964). Insect Antennae. *Annual Review of Entomology* **9** (1): 103–122. https://doi.org/10.1146/annurev.en.09.010164.000535.

28 Krogmann, L., Engel, M.S., Bechly, G., and Nel, A. (2012). Lower Cretaceous origin of long-distance mate finding behaviour in Hymenoptera (Insecta). *Journal of Systematic Palaeontology* **11** (1): 83–89. https://doi.org/10.1080/14772019.2012.693954.

29 Biedermann, R. (2002). Mating success in the spittlebug *Cercopis sanguinolenta (*Scopoli, 1763) (Homoptera, Cercopidae): the role of body size and mobility. *Journal of Ethology* **20**: 13–18.

30 Ma, N., Zhong, W., and Hua, B.Z. (2010). Genitalic morphology and copulatory mechanism of the scorpionfly *Panorpa jilinensis* (Mecoptera: Panorpidae). *Micron* **41**: 931–938.

31 Wang, R.R., Liang, A.P., and Webb, M.D. (2009). A new tropiduchid planthopper genus and species from China with descriptions of in copula genitalic structures (Hemiptera: Fulgoromorpha). *Systematic Entomology* **34**: 434–442.

32 Boucot, A.J. and Poinar, G.O. Jr., (2010). *Fossil Behavior Compendium.* Boca Raton: CRC Press.

33 Boucot, A.J. and Babcock, L.E. (1990). *Evolutionary Paleobiology of Behavior and Coevolution.* Amsterdam/Oxford/New York/Tokyo: Elsevier.

34 Cryan, J.R. and Svenson, G.J. (2010). Family-level relationships of the spittlebugs and froghoppers (Hemiptera: Cicadomorpha: Cercopoidea). *Systematic Entomology* **35**: 393–415.

35 Sun, G., Zheng, S.L., and Mei, S.W. (2000). Discovery of *Liaoningocladus* gen. nov. from the lower part of Yixian Formation (Upper Jurassic) in western Liaoning, China. *Acta Palaeontologica Sinica* **39** (Suppl): 200–208.

36 Waterston, J. (1921). On some Bornean fig-insects (Agaonidae - Hymenoptera Chalcidoidea). *Bulletin of Entomological Research* **12** (1): 35–40.

Index

Rhythms of Insect Evolution: Evidence from the Jurassic and Cretaceous in Northern China, First Edition.
Edited by Dong Ren, Chungkun Shih, Taiping Gao, Yongjie Wang, and Yunzhi Yao.

b

C

d

h

i

j

n

q

r

t

u

Wiley The manufacturer's authorized representative according to the EU General Product Safety Regulation is Wiley-VCH GmbH, Boschstr. 12, 69469 Weinheim, Germany, e-mail: Product_Safety@wiley.com.

Printed and bound by CPI Group (UK) Ltd, Croydon, CR0 4YY
15/09/2026
02217280-0005